World Scientific Proceedings Series on
Computer Engineering and Information Science 7

Uncertainty Modeling in Knowledge Engineering and Decision Making

Proceedings of the 10th International FLINS Conference

World Scientific Proceedings Series on Computer Engineering and Information Science

Series Founding Editor: Da Ruan

Series Editor: Jie Lu, University of Technology Sydney

Vol. 1: Computational Intelligence in Decision and Control
edited by Da Ruan, Javier Montero, Jie Lu, Luis Martínez, Pierre D'hondt and Etienne E. Kerre

Vol. 2: Intelligent Decision Making Systems
edited by Koen Vanhoof, Da Ruan, Tianrui Li and Geert Wets

Vol. 3 Computational Intelligence in Business and Economics
edited by Anna M. Gil-Lafuente and José M. Merigó

Vol. 4 Computational Intelligence: Foundations and Applications
edited by Da Ruan, Tianrui Li, Yang Xu, Guoqing Chen and Etienne E. Kerre

Vol. 5 Quantitative Logic and Soft Computing
edited by Guojun Wang, Bin Zhao and Yongming Li

Vol. 6 Methods for Decision Making in an Uncertain Environment
edited by Jaime Gil-Aluja and Antonio Terceño

Vol. 7 Uncertainty Modeling in Knowledge Engineering and Decision Making
edited by Cengiz Kahraman, Etienne E Kerre and Faik Tunc Bozbura

World Scientific Proceedings Series on
Computer Engineering and Information Science 7

Uncertainty Modeling in Knowledge Engineering and Decision Making

Proceedings of the 10th International FLINS Conference

Istanbul, Turkey 26 – 29 August 2012

editors

Cengiz Kahraman
Istanbul Technical University, Turkey

Etienne E. Kerre
Ghent University, Belgium

Faik Tunc Bozbura
Bahcesehir University, Turkey

World Scientific

NEW JERSEY • LONDON • SINGAPORE • BEIJING • SHANGHAI • HONG KONG • TAIPEI • CHENNAI

Published by

World Scientific Publishing Co. Pte. Ltd.
5 Toh Tuck Link, Singapore 596224
USA office: 27 Warren Street, Suite 401-402, Hackensack, NJ 07601
UK office: 57 Shelton Street, Covent Garden, London WC2H 9HE

British Library Cataloguing-in-Publication Data
A catalogue record for this book is available from the British Library.

World Scientific Proceedings Series on Computer Engineering and Information Science — Vol. 7
UNCERTAINTY MODELING IN KNOWLEDGE ENGINEERING AND DECISION MAKING
Proceedings of the 10th International FLINS Conference

ISBN 978-981-4417-73-0

Printed in Singapore by B & Jo Enterprise Pte Ltd

FOREWORD

FLINS, an acronym originally for Fuzzy Logic and Intelligent Technologies in Nuclear Science (1994), has been launched in line with the Belgian Nuclear Research Center (SCK•CEN)'s objective to give PhD and Postdoc researchers the opportunity to carry out future-orientated research. FLINS was initially built within one of the above-mentioned postdoc positions. Later, FLINS has been extended to include the theoretical and practical aspects of computational intelligent systems.

FLINS 2012 follows a successful conferences series: FLINS 1994 in Mol, FLINS 1996 in Mol, FLINS 1998 in Antwerp, FLINS 2000 in Bruges, FLINS 2002 in Gent, FLINS 2004 in Blankenberge, FLINS 2006 in Genova, FLINS 2008 in Madrid, and FLINS 2010 in Chengdu. FLINS 2012 is co-organized by Istanbul Technical University and Bahcesehir University, and cosponsored by the Belgian Nuclear Research Centre (SCK•CEN) and Ghent University in Belgium, and supported by European Society for Fuzzy Logic and Technology (EUSFLAT). It offers a unique international forum to present and discuss recently developed techniques for uncertainty modeling in Knowledge Engineering and Decision Making.

Prof. Da Ruan, the founder of FLINS conferences, was the general chair of all FLINS conferences. Unfortunately, we unexpectedly lost Prof. Da Ruan in July 31, 2011.We are very sad about this irreparable loss of such a great organizer and extremely friendly human being. He was an excellent researcher who encouraged many researchers from different universities to present their works in the international arena and to make collaborations among them. We sincerely hope that this 10th FLINS conference will take place in the spirit of Prof. Da Ruan and that his legacy will be continued for years to come. We will miss Da forever.

The 10th International FLINS Conference (FLINS 2012) is on Uncertainty Modeling in Knowledge Engineering and Decision Making. FLINS 2012 aims at providing an international forum that brings together those actively involved in areas of interest to the uncertainty modeling in Knowledge Engineering and Decision Making, to report on up-to-the-minute innovations and developments, to summarize the state-of-the-art, and to exchange ideas and advances in all aspects of Uncertainty Modeling in Knowledge Engineering and Decision Making.

The FLINS 2012 proceedings consist of a series of invited lectures by distinguished professors and individual presentations. Out of 310 regular submissions from 35 countries, 212 papers were selected after a peer review process. The four invited speeches are entitled as (1) Fuzzy Relational Calculus and Its Applications by Prof. Etienne Kerre, (2) Can Computing with Words Bring Complex Nuclear Risk Management Issues into Focus? by Prof. Ashok W. Deshpande, (3) Decision Support and Warning Systems for Business Intelligence by Prof. Jie Lu, (4) Historical Development of Fuzzy System Models by Prof. Burhan Turksen.

The submitted papers have been divided into the following seven main titles: Part 1. Decision Making and Decision Support Systems, Part 2. Uncertainty Modeling, Part 3. Foundations of Computational Intelligence, Part 4. Statistics, Data Analysis and Data Mining, Part 5. Intelligent Information Processing, Part 6. Productivity and Reliability, and Part 7. Applied Research.

We are pleased to announce that seven different international journals devoted their special issues to FLINS 2012. These journals are International Journal of Computational Intelligence Systems (SCI-E indexed, Atlantis Press and Taylor & Francis), Journal of Multiple Valued Logic and Soft Computing (SCI-E indexed, Oldcity Publishing), Knowledge-Based Systems (SCI indexed, Elsevier), International Journal of Nuclear Knowledge Management (Inderscience Publication), Journal of Enterprise Information Management (Emerald), International Journal of Information and Communication Technology (Inderscience Publication), and International Journal of Computational Vision and Robotics (Inderscience Publication).

We thank all contributors, referees, regular and invited sessions' chairs, and program committee members of FLINS 2012 for their kind co-operation and enthusiasm for FLINS 2012, especially to Dr. Basar Oztaysi, Dr. Ayberk Soyer, Dr. Seyda Serdar Asan, Dr. Umut Asan, Dr. Sezi Cevik Onar, Dr. Seda Ugurlu, Dr. Ozgur Kabak, Dr. Irem Ucal Sari, Dr. Murat Engin Unal, and Res. Assist. Pinar Dursun for their key roles that made the proceedings of FLINS 2012 possible in a timely manner. We also thank Chelsea Chin (Editor, World Scientific) for her kind advice and help to publish this volume.

Editors
Prof. Cengiz Kahraman
Prof. Etienne E. Kerre
Assoc. Prof. Tunc Bozbura

CONTENTS

INVITED LECTURES

FUZZY RELATIONAL CALCULUS AND ITS WIDE RANGE OF APPLICATIONS

ETIENNE E. KERRE*

Fuzziness and Uncertainty Modelling Research Unit
Department of applied Mathematics and Computer Science
Ghent university
Krijgslaan 281, S9
B9000 Gent, Belgium

From different points of view the concept of a relation is a very fundamental one. Sometimes science has been described as the discovery of relations between objects in a very broad way. Mathematically spoken relations and in particular functional relations or shortly functions are undeniably basic concepts: how could we imagine the development of a mathematical framework without fundamental relations such as : inclusion, membership, equality, smaller than or equal to, strictly greater than? From a practical point of view relations are even more important as suitable tools to describe facts as well as correspondences between facts. In medical diagnosis for example we can consider on one hand a set X of patients and on the other a set of symptoms Y. These two sets can be linked by the relational description: patient x shows symptom y with x belonging to X and y belonging to set Y.

In the first part of the talk we will start with a review of the basic concepts and their properties of the classical binary relations: domain, codomain, afterset, foreset, set-theoretic operations, direct and inverse image, round composition. We will also describe the recent introduced new images and compositions by Bandler and Kohout: sub-image, super-image and square-image and similarly for the compositions. An example from medical diagnosis will end this first part.

In the second part we will tackle the generalization of crisp relational calculus by means of fuzzy relations. Step by step we will follow the same scheme as in the first part, i.e. introduce the basic notions, the fuzzy-set-theoretic operations between fuzzy relations, all kind of fuzzy images of a fuzzy

* etienne.kerre@ugent.be

set under a fuzzy relation, all kind of fuzzy compositions between fuzzy relations.

After the introduction of the theoretical concepts we will illustrate the power of fuzzy relational calculus in different domains: criminal investigation, medical diagnosis, information retrieval, approximate reasoning, ordering of fuzzy quantities, decision making, rough set theory, fuzzy temporal relations and fuzzy spatial relations.

DECISION SUPPORT AND WARNING SYSTEMS FOR BUSINESS INTELLIGENCE

JIE LU*

Decision Systems & E-Service Intelligence Laboratory, Centre for Quantum Computation and Intelligent Systems

School of Software, Faculty of Engineering and Information Technology
University of Technology Sydney, Sydney, Australia

This presentation presents our recent developments in intelligent decision support systems and early warning systems as well as their applications for business intelligence. It reports our research results from two projects in particular: a bi-level decision making technique for competitive strategic bidding optimization in electricity markets, and an intelligent financial warning system using transferable adaptive inference-based fuzzy neural network.

1. Introduction

Decision making methods and decision support systems (DSS) play an increasingly important role in the decision making processes of organizations in every discipline, including health, business, engineering, education and finance. In the 21st century's complex environments, organizational decision makers have increased requirements for advanced knowledge, previous successful experiences, and intelligent technical conditions to support and enable better decisions. Knowledge-based methods and intelligent models have become a necessary component in current advanced DSS and its applications.

Organizational decision making has also sparked an urgent need to build early warning systems (EWS) which can effectively support decision makers to recognize underlying vulnerabilities and implement relevant strategies, enabling them to reduce, or eliminate, future problems and the risk of experiencing a crisis. Powerful predictive ability will ensure that an EWS generates maximal reliable warnings at the onset of crises. Unlike some natural (e.g. earthquake) and technological (e.g. mining disaster) phenomena, prediction of an organizational financial attack is almost impossible to make by traditional mathematical forecasting models or physical monitoring, nor is it possible to have a set of

* jie.lu@uts.edu.au

threshold values for warnings. Such predictions and related decision making are more strongly dependent on domain experts' knowledge and experience of past similar crises, many of which contain vague, uncertain or dynamic elements and can be only described in complex, ill-structured forms. For the kind of non-model driven, high uncertainty crisis prediction problems, case-based reasoning (CBR), as the process of solving new problems based on solutions to similar past problems, neural network, machine learning, fuzzy logic and other computation intelligence approaches provide a suitable framework.

2. Competitive Strategic Bidding Optimization in Electricity Markets using Bi-level Decision Making Technique

Competitive strategic bidding optimization is a key issue in electricity markets. Digital eco-systems provide a powerful technological foundation and support for the implementation of optimization. This project develops a new strategic bidding optimization technique which applies a bi-level decision making model and swarm intelligence. In this project, we first propose a general multi-leader-one-follower non-linear bi-level (MLNB) optimization concept and related definitions based on the generalized Nash equilibrium. By analyzing the strategic bidding behavior of generating companies, we create a specific MLNB decision model for day-ahead electricity markets. The MLNB decision model allows each generating company to choose its biddings to maximize its individual profit, and a market operator can find its minimized purchase electricity fare, which is determined by the output power of each unit and uniform marginal prices. We then develop a particle swarm optimization (PSO)-based algorithm to solve the problem defined in the MLNB decision model. Experiment results on a strategic bidding problem for a day-ahead electricity market have demonstrated the validity of the proposed decision model and algorithm.

3. Intelligent Financial Early Warning System using Transferable Adaptive Inference-based Fuzzy Neural Network

Since the advent of various financial crises in the 1990s and 2000s, and particularly the recent recession in mid-2008, various Financial Early Warning Systems (FEWS) have been developed. Since a FEWS is more than a classical prediction model and should provide an explanatory analysis to describe the reasons behind the failure, the explanatory ability of a system is as important as its predictive accuracy. In addition, a FEWS needs to go beyond prediction and provide a prediction-based decision making component to decide, suggest or recommend solutions for decision makers who have the authority to impede imminent failure. Furthermore, a flexible and transferable FEWS which can be automatically be modified for different domains is highly practical in application, easy to install and cheap to set up. To achieve the aforementioned properties in FEWS, we have developed an Intelligent FEWS which first includes a novel FNN to predict financial status as a prediction component of the system. Second, a Fuzzy CBR model is developed and integrated with Fuzzy Multi Criteria Decision Making techniques to support decision makers in measuring appropriate solutions as the decision component of the system. Third, an innovative Transfer Learning method is developed for the proposed FNN to enable the system to be utilized in different business domains and convey valuable knowledge between them. A set of experiments has been conducted based on two populations of United States banks to test and validate the proposed approaches. This talk will present these new developments in FEWS and the application of intelligent techniques, including a novel FNN to supply a descriptive prediction analysis in the form of fuzzy rules, a new fuzzy transfer learning framework, and fuzzy transfer learning-based prediction techniques that effectively use knowledge learned from different time periods/domains with a variety of features to support prediction and decision making in FEWS. With these developments, a FEWS can make maximum use of previous knowledge to generate interpretable alerts and support decision-makers in their search for solutions, thus reducing the risk of future problems.

A REVIEW OF DEVELOPMENTS FROM FUZZY RULE BASES TO FUZZY FUNCTIONS

I. BURHAN TÜRKŞEN[1]
Department of Industrial Engineering
TOBB-Ekonomics and Teknology University
Sögütözü Cad. No:43, Sögütözü 06560 Ankara/Turkey

We first touch upon a Philosophical Grounding of fuzzy theory expressed by Pierce and Zadeh. Then we first review briefly basic and well known fuzzy rule base models and their variations as well as our fuzzy functions with LSE and their enhanced version. We propose a potential future investigation for the basic structure of fuzzy function models generated with a multiplicative affect of membership values.

1. Philosophical Grounding

It is important to understand an in-depth association of the essential concepts that were treated by Charles S. Peirce and Lotfi A. Zadeh

Peirce's thesis may be represented as"indeterminacy and determinacy" of "symbols".

This view can now be interpreted and expressed with the degree assignment to information granules.

Thus in general a symbol S is indeterminate

$$\text{iff } (\exists P)\sim(S \text{ is } P \text{ OR } S \text{ is } \sim P).$$

Hence, Locke's famous idea of the triangle is stated as:

"It is not the case that a triangle in general is scalene or that is it not scalene".

Peirce's discussion pre-supposes that every symbol is capable of determining an interpretant symbol and that symbols are at least potentially general.

$$(S)\ (\exists P)(S \text{ is } P \text{ OR } S \text{ is } \sim P) \text{ AND } (S)\ (\exists P)\sim(S \text{ is } P \text{ OR } S \text{ is } \sim P).$$

From Zadeh s "meaning representation" of "words" in Computing With Words, CWW, Peirce's representation is re-expressed as:

$$(S)\ (\exists P)(S \text{ is } P \text{ OR } S \text{ is } \sim P)\ (\mu P(S) \in [0,1])$$

---

[1] bturksen@etu.edu.tr

and (S) (∃P)~(S is P OR S is ~P) (1 - μP(S)

Beyond these essential starting points, it is important to recall that fuzzy system developments were enhanced by t-norm and t-conorms introduced by B. Schweitzer, A. Sklar [7a].

2. Introduction

Most commonly known and applied fuzzy system models are "fuzzy rule bases". Such fuzzy rule bases are described by membership functions of the input fuzzy sets that form the left hand sides and the output fuzzy sets that form the right hand sides. This approach was initially proposed by Zadeh [21,22]. There are two well known and basic variations of this approach with various improvements and enhencements:

a) Sugeno-Yasukawa [6] approach where fuzzy sets of both the right and left hand sides are determined either by experts or by fuzzy clustering algorithms such as FCM (Bezdek,[2]).

b) Tagaki-Sugeno [7] approach where fuzzy sets of the left hand sides of a fuzzy rule base are determined either by experts or by fuzzy clustering algorithms such as FCM (Bezdek,[2]) and the right hand sides are functions determined either by function estimation methods.

But there have been new approaches that propose fuzzy regression function developments in place of fuzzy rule bases. These are :

(1) Hathaway and Bezdek [3a] approach where the determination of a classical regression is enhanced by introduction of a diagonal membership matrix in determination of coefficients of a fuzzy regression model where the fuzzy clustering algorithm, FCM (Bezdek,[2]), is used to deteremine the number of such fuzzy regressions required for an affective solution.

(2) Turksen [18] and Celikyılmaz-Turksen [2a] approach where a classical regression is enhanced by introduction membership values and their transformations to improve the regression constant and hence the introduction of fuzzy functions in place of fuzzy rule bases where a fuzzy clustering algorithm such as FCM (Bezdek,[2]) or IFC (Celikyilmaz and Turksen [2a]) is used to deteremine the number of such fuzzy regressions required for an affective solution.

Next we review these approaches and their essential elements with emphasis on our "Fuzzy Functions", fuzzy regression, models generated by membership values and their transformations.

3. Fuzzy Rule Base Models

Let us first review the fuzzy rule base models in order to identify their unique structures and to point out how they differ from each other.

The most commonly applied fuzzy system models are fuzzy rule bases. Here, we only deal with Multi-Input Single Output (MISO) systems. Generally fuzzy system models represent relationships between the input and output variables which are expressed as a collection of IF-THEN rules that utilize linguistic labels, which are represented with fuzzy sets. The general fuzzy rule base structure which is known as Zadeh- Fuzzy Rule Base, Z-FRB, can be written as follows:

$$R : \underset{i=1}{\overset{c^*}{\text{ALSO}}}\left(\text{IF } antecedent_i \text{THEN} consequent_i\right),$$

where c^* is the number of rules in a rule base either given by experts or it is determined by a fuzzy clustering algorithm such as FCM, Fuzzy C Means. The fuzzy rule base structures determined by alternatives (a) and (b) stated above mainly differ in the representation of the consequents in its structure . If the consequent is represented with fuzzy sets then the fuzzy rule base can be categorized as alternative (a). This is the one initially proposed by Zadeh [22] originally applied by Mamdani, et al., [4], and a modified version is proposed by Sugeno and Yasukawa [6]. Whereas, if the consequents are represented with linear equations of input variables, then the rule base structure is the alternative (b) which we call Takagi-Sugeno Fuzzy Rule Base [7]. These models can be formalized as follows.

In general, let nv be the number of selected input variables in the system. Then, the multidimensional antecedent, x, can be defined as $x=(x_1,x_2,\ldots,x_{nv})$, where x_j is the jth input variable of the antecedent and the domain of x in X, can be defined as $X=X_1\times X_1\times\ldots \times X_{nv}$, where $X_j \subseteq\Re$ is the domain of variable x_j. Similarly, the domain of the output variable, y, will be denoted as $Y\subseteq\Re$. Then, the ith rule, R_i, and rulebase, R, in the structure of Sugeno and Yasukawa [6] can be defined as:

$$R_i : \text{ IF } \underset{j=1}{\overset{nv}{AND}}\left(x_j \in X_j \text{ isr } A_{ij}\right) \text{ THEN } \mathrm{y} \in Y \text{ isr } B_i, \forall i = 1,\ldots,c^* \quad (1)$$

$$R: \underset{i=1}{\overset{c^*}{ALSO}}\left(\text{IF } \underset{j=1}{\overset{nv}{AND}}\left(x_j \in X_j \text{ isr } A_{ij}\right) \text{ THEN } \mathrm{y} \in Y \text{ isr } B_i \right) \quad (2)$$

where A_{ij} is the linguistic label, i.e., fuzzy subset, associated with j^{th} input variable of the antecedent in the ith rule, R_i, with membership function

$\mu_i(x_j):X_j \rightarrow [0,1]$ and similarly B_i is the consequent linguistic label, i.e., consequent fuzzy subset, of the ith rule with membership function $\mu_i(y):Y \rightarrow [0,1]$, and c^* is the number of rules in the model. In this structure, the challenges for knowledge representation are: (i) to identify the membership functions of fuzzy sets on the left and right hand sides of the rules and (ii) to identify the most suitable t-norm and t-conorm combinations that represent in a one-to-one correspondence the linguistic "AND" and "OR" for the combination of left hand side fuzzy subsets together with the implication operator, "IMP", that will carry the left hand side membership degree, i.e., the degree of firing, to the right hand side consequent fuzzy subset. As well, one needs to know and be able to apply fuzzy logic to carry out approximate reasoning. It should be recalled that Mamdani, et al., [1977] applied *Min* operator for both "AND and "IMP" which is a very special case whereas Sugeno and Yasukawa [6] model is more general. It is known that the linguistic "AND" and "OR" operators can not be represented in a one-to-one correspondence with a t-norm and a t-conorm, respectively, as it is shown by Turksen [8,9,10,11,12,13,14,15,16,17]. As a result, the FDCF and FCCF, Fuzzy Disjunctive and Conjunctive Canonical Forms, are to be used for the representation of rules and for reasoning with them. However, such models fall into Interval-Valued Type 2 fuzzy systems analyses which are not dealt with in this paper. Finally, one has to carry out defuzzification computations in all fuzzy rule base models. Furthermore the above structure assumes non-interactivity between input variables (Zadeh, [22]). In fact, this is the underlying assumption when the fuzzy subsets for the left and right hand sides are obtained from experts by interview techniques. In order to eliminate the non-interactivity assumption, Delgado et. al. [3] , Babuska et. al. [1], and Uncu and Turksen [19] used multi-dimensional Type 1 fuzzy subsets to represent the antecedent part of the rules. In such investigations, generally a multi-dimensional fuzzy clustering technique, e.g., FCM, is implemented to obtain multi-dimensional fuzzy subsets that capture the interactivity (or joint affect) of input variables. Hence, the initial Zadeh`s Fuzzy Rule Base (Z-FRB) structure can be expressed as follows:

$$R:\ \underset{i=1}{\overset{c^*}{ALSO}}\left(\text{IF } x\in X \text{ isr } A_i \text{ THEN } \mathrm{y}\in Y \text{ isr } B_i\right), \quad (3)$$

where the multi-dimensional antecedent fuzzy subset of ith rule is *Ai*. This multi-dimensional antecedent fuzzy subset determination eliminates the search for the appropriate t-norm for the combination of antecedent fuzzy subsets with "AND".

Thus, in such analyses, the well known variations of Zadeh`s (Z-FRB) in terms of Sugeno-Yasukawa (SY-FRB) and Takagi-Sugeno (TS-FRB) Fuzzy Rule Base structures are:

$$R:\ \underset{i=1}{\overset{c^*}{ALSO}}\left(\text{IF } x\in X \text{ isr } A_i \text{ THEN } \mathrm{y}\in Y \text{ isr } B_i\right) \quad \text{(SY-FRB)} \tag{4.1}$$

$$R: \underset{i=1}{\overset{c^*}{ALSO}}\ (IF\ antecedent_i\ THEN\ y_i = a_i x^T + b_i) \quad \text{(TS-FRB)} \tag{4.2}$$

where, antecedent$_i$ = x $\in X$ isr A_i , and $a_i=(a_{i,1},\ldots, a_{i,nv})$ is the regression coefficient vector associated with the ith rule together with b_i which is the scalar associated with the ith rule. For these special cases of Z-FRB, again each degree of firing, d_i, associated with the-i^{th} rule, is determined directly from the corresponding i^{th} multi-dimensional antecedent fuzzy subset A_i and applied to the consequent fuzzy subset for the SY-FRB or to the classical ordinary regression for the case of TS-FRB.

4. Fuzzy Functions

There are a number of variations of Fuzzy Functions that are structured various transformations of membership values. We will discuss only one alternative, namely, Fuzzy Functions with LSE**, FF-LSE** which introduces the basic starting point of developments in this direction. Here we assume that Least Squares Estimation, LSE, is well known by most readers in this context.

4.1. *Fuzzy Functions with LSE*

The generalization of LSE for Fuzzy Function , known as **FF-LSE**, requires that a fuzzy clustering algorithm, such as FCM, or IFC be available to determine the interactive (joint) membership values of input-output variables in each of the fuzzy clusters that can be identified for a given training data set.

Let (X_k,Y_k), $k=1,\ldots, nd$, be the set of observations in a training data set, such that

$$X_k = (x_{jk} \mid j = 1,\ldots, nv,\ k = 1,\ldots, nd) \tag{5}$$

First, one determines the optimal (m^*, c^*) pair for a particular performance measure, i.e., a cluster validity index, with an iterative search and an application of FCM or IFC algorithm, where m is the level of fuzziness (in our experiments we usually take $m = 1.4,\ldots,2.6$) (Ozkan and Turksen,[6a]), and c is the number of clusters (in our experiments we usually take $c = 2,\ldots,10$). The well known FCM algorithm can be stated as follows:

$$\min\ J(U,V)=\sum_{k=1}^{nd}\sum_{i=1}^{c}(u_{ik})^m(\|x_k-v_i\|)_A$$

$$s.t.\quad 0\le u_{ik}\le 1,\forall i,k$$

$$\sum_{i=1}^{c}u_{ik}=1,\forall k$$

$$0\le\sum_{k=1}^{nd}u_{ik}\le nd,\forall i$$

where J is objective function to be minimized, $\|.\|_A$ is a norm that specifies a distance-based similarity between the data vector x_k and a fuzzy cluster center v_i. In particular,

$A = I$ is the Euclidian Norm and $A = C^{-1}$ is the Mahalonobis Norm, etc. Once the optimal pair (m^*, c^*) is determined with the application of FCM algorithm, one next identifies the cluster centers for $m = m^*$ and $c=1,\dots,c^*$ *as:*

$$v_{X|Y,j}^{m^*}=(x_{1,j}^{c},x_{2,j}^{c},\cdots,x_{nv,j}^{c},y_j^{c}) \tag{6}$$

From this, we identify the cluster centers of the "input space" again for $m=m^*$ *and* $c=1,\dots,c^*$ as:

$$v_{X,j}^{m^*}=(x_{1,j}^{c},x_{2,j}^{c},\cdots,x_{nv,j}^{c}) \tag{7}$$

Next, one computes the normalized membership values of each data sample in the training data set with the use of the cluster center values determined in the previous step. There are generally two steps in these calculations:

(a) First we determine the (local) optimum membership values u_{ik}'s and then determine μ_{ik} 's that are above an α-cut in order to eliminate harmonics generated by FCM as:

$$u_{ik}=\left(\sum_{j=1}^{c}\left(\frac{\|x_k-v_{X,i}\|}{\|x_k-v_{X,j}\|}\right)^{\frac{2}{m-1}}\right)^{-1},\qquad \mu_{ik}=\{u_{ik}\ge\alpha\}, \tag{8}$$

where μik denotes the membership value of the kth vector, $k=1,\dots,nd$, in the ith rule, $i=1,\dots,c^*$ and x_k denotes the kth vector and for all the input variables $j = 1,\dots, nv$, in the input space.

(b) Next, we normalize them as:

$$\gamma_{ij}(x_j) = \frac{\mu_{ij}(x_j)}{\sum_{i'=1}^{c} \mu_{i'j}(x_j)} \tag{9}$$

where $\gamma_{ij}(x_j)$'s are the normalized membership values of x_j , *j=1,...nd,* in the ith rule, *i=1,...,c**, which in turn indicate the membership values that will constitute as a new input variable in our proposed scheme of function identification for the representation of ith cluster. Let $\Gamma_i = (\gamma_{ij} \mid i = 1,..., c^*; j = 1,..., nd)$ be the membership values of *X* data sample in the ith cluster, i.e., i^{th} rule.

Next we determine a new augmented input matrix of *X* for each of the clusters, which could take on several forms depending on which *transformation* of membership values we want to or need to include in our system structure identification for our intended system analyses. For the case of

$$Y_i = \beta_{i0} + \beta_{i1}\Gamma_i + \beta_{i2}X_{ij} ,$$

as an examples of the possible augmented input matrix is:

$$X_i' = [1, \Gamma_i, X] .$$

Alternately, one could consider augmented input matrices of $X_i'' = \left[1, \Gamma_i^2, X\right]$, or $X_i''' = \left[1, \Gamma_i^2, \Gamma_i^m, \exp(\Gamma_i), X\right]$, etc, for various transformations of $\Gamma_i = (\gamma_{ij} \mid i = 1,..., c^*; j = 1,..., nd)$.

The choice depends on whether we want to or need to include just the membership values or some of their transformations as new input variables in order to obtain the best representation of a system behavior. A new augmented input matrix having a single input variable in the original input space when only membership value itself is augmented to the dataset may look like this:

$$X_{ij}' = [1, \Gamma_i, X_{ij}] = \begin{bmatrix} 1 & \gamma_{i1} & x_{i1} \\ \vdots & \vdots & \vdots \\ 1 & \gamma_{ind} & x_{ind} \end{bmatrix} \tag{10}$$

Up to this point, in the proposed system modeling approach, we have defined how the augmented input matrix for each cluster could be formed using FCM algorithm. From this point forward, the estimation of fuzzy functions takes place for each cluster, Different approaches are followed in the estimation of fuzzy functions using the augmented matrices.

Thus the function $Y_i = \beta_{i0} + \beta_{i1}\Gamma_i + \beta_{i2}X_{ij}$, that represents the ith rule corresponding to the ith interactive (joint) cluster in (Y_i, Γ_i, X_j) space, would be estimated with FF-LSE approach as follows:

$$\beta_i^* = (X_{ij}'^T X_{ij}')^{-1}(X_{ij}'^T Y_i) \tag{11}$$

where $\beta_i^* = (\beta_{i0}^*, \beta_{i1}^*, \beta_{i2}^*)$ are the estimates and $X_{ij}' = \left[1, \Gamma_i, X_{ij}\right]$, provided the inverse of covariance matrix, $(X_{ij}'^T X_{ij}')^{-1}$, exists. Therefore the estimate of Y_i would be obtained as:

$$Y^*_i = \beta_{i0}^* + \beta_{i1}^*\Gamma_i + \beta_{i2}^* X_{ij} \tag{12}$$

The overall output value is calculated using each output value one from each cluster and weighting them with their corresponding membership values as follows:

$$Y^*_i = \frac{\sum_{i=1}^{c^*} \gamma_i Y^*_i}{\sum_{i=1}^{c^*} \gamma_i} \tag{13}$$

Within this framework, the general form of the shape of a cluster for the case of a single input variable X_j and for the ith cluster can be conceptually captured by a second order (cone) function when one introduces the square of membership values into the augmented input matrix in the space of $[U \times X \times Y]$ which can be illustrated with a prototype shown in Figure 1.

It ought to be noted for the sake of emphasis that the addition of membership values and or their transformation in fact improves the predictive power of the fuzzy regression equation, for example, in the simplest case:

$$Y^*_i = \beta_{i0}^* + \beta_{i1}^*\Gamma_i + \beta_{i2}^* X_{ij}$$

where $\beta_i^* = (\beta_{i0}^*, \beta_{i1}^*, \beta_{i2}^*)$ are the estimates and $X_{ij}' = \left[1, \Gamma_i, X_{ij}\right]$, provided the inverse of covariance matrix, $(X_{ij}'^T X_{ij}')^{-1}$, exists. Therefore the estimate of Y_i would be obtained with the addition of the second coefficient associated with Γ in $\beta_i^* = (\beta_{i0}^*, \beta_{i1}^*, \beta_{i2}^*)$ as a result of the impact of the membership values included in such a regression.

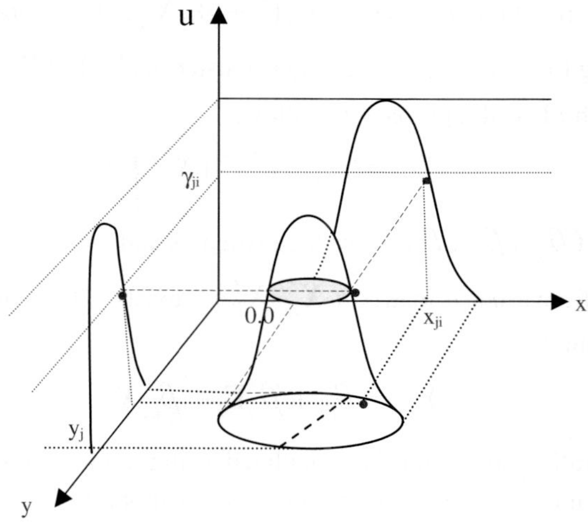

Figure 1. A Fuzzy cluster in [*UxXxY*] space.

In general, in more complex cases, one may require to include what we call **Enhanced Fuzzy Functions** (Celikyilmaz, and Turksen, [2 a]) where membership values represent degrees of **belongingness** with additional **predictors,** which are various transformations of membership values for example:

$$e.g. \quad \boldsymbol{h}_i(\boldsymbol{\tau}_i, \hat{w}_i) = \hat{w}_{0i} + \hat{w}_{1i}\mu_i + \hat{w}_{2i}(1 + \exp(-\mu_{ik}^m))$$

Where β s are now represented by w s where an exp transformation of membership values are included as additional information to enhance constant value of the regression.

5. Conclusions

We have first reviewed the well known variations of fuzzy rule bases. Then we proposed "fuzzy functions" as an alternative to fuzzy rule base approaches to fuzzy system modeling. They can be more easily applied by mathematicians and statisticians without knowing the essential mathematical tools required for building fuzzy rule base system models.

Furthermore in our applied system investigations, we found that "Fuzzy Functions" are better suited for industrial applications and provides better predictions as compared to OLSE, Ordinary Least Squares Estimation, models and to fuzzy rule base models (Turksen, [18]; Turksen and Celikyilmaz, [19]).

Currently we are working on "Multiplicative Fuzzy Functions" structured as:

$$Y = a + (\mu + \mu 2 + \log \mu)\, x$$

In contrast to our previous works based on additive structures:

$$Y = a + b\,\mu + c\,\mu 2 + d \log \mu + x.$$

That is various transformations of μ, membership values, are introduced as a multiplier of input variable x.

References

1. [1] R.Babuska, and H.B. Verbruggen, "Constructing Fuzzy Models by Product Space Clustering", in H. Hellendoorn and D. Driankov(eds.), Fuzzy Model Identification: Selected Approaches, pp.53-90, Springer, Berlin, Germany, 1997.
2. [2] J.C. Bezdek, "Fuzzy Mathematics in Pattern Classification", Ph.D. Thesis, applied Mathematics Centre, Cornell University, Ithaca, 1973.
3. [2a] A. Celikyilmaz, and I. B. Turksen, "Enhanced fuzzy system models with improved fuzzy clustering algorithm", IEEE Trans. Fuzzy Systems, vol.16, pp.779-794, 2008.
4. [3] M. Delgado, A.F. Gomez-Skermata and F. Martin, "Rapid Prototyping of Fuzzy Models", *in H. Hellendoorn and D. Driankov (Eds.), Fuzzy Model Identification: Selected Approaches*, pp. 53-90. Springer, Berlin, Germany, 1997.
5. [3a] Hathaway R.J. and Bezdek J.C., "Switching regression models and fuzzy clustering," *IEEE Transactions on Fuzzy Systems*, vol. 1, no. 3, pp. 195-204, 1993.
6. [4] E. M. Mamdani, "Application of Fuzzy Logic to Approximate Reasoning Using Linguistic Systems" 26,1182-1191, 1977.
7. [5] M. Mizimoto, "Method of Fuzzy Inference Suitable for Fuzzy Control", J.of Soc.Instrument Control Engineering, Vol.58, 1989, pp.959-963.
8. [6] M. Sugeno and T. Yasukawa, *A Fuzzy Logic Based Approach to Qualitative Modelling*. IEEE Trans on Fuzzy Systems, 1993. 1(1): p. 7-31.
9. [6a] I. *Ozkan*, I. B. *Turksen*. Upper and lower values for the level of fuzziness in FCM. Inf. Sci., 2007: 5143-5152
10. [7] T.Takagi and M.Sugeno, "Fuzzy Identification of Systems and Its Applications to Modeling and Control" IEEE Transactions on Systems, Man and Cybernetics, Vol.SMC-15, 1, 116-132, 1985.
11. [7a] B. Schweitzer, A. Sklar, Probabilistic Metric Spaces, North-Holland, New York, 1983
12. [8] I.B. Turksen, "Interval Valued Fuzzy Sets Based on Normal Forms," *Fuzzy Sets and Systems*, vol. 20 (1986) 191- 210.

13. [9] I.B. Turksen, "Four Methods of Approximate Reasoning with Interval-valued Fuzzy Sets", *Int. J. of Approximate Reasoning*, vol. 3 (1989) 121-142.
14. [10] I.B. Turksen, "Interval-valued fuzzy sets and 'compensatory AND," *FuzzSets and Systems*, vol. 51 (1992) 295- 307.
15. [11] I.B. Turksen, "Interval valued fuzzy sets and fuzzy connectives," *Interval Computations*, pp. 125-142, 1993.
16. [12] I.B. Turksen, "Interval valued fuzzy sets and fuzzy measures", *Proc. of First Int. Conf. Of NAFIPS*, pp. 317- 321, 1994.
17. [13] I.B. Turksen, "Fuzzy Normal Forms", *Fuzzy Sets and Systems*, 69, pp. 319-346, 1995.
18. [14] I.B. Turksen, "Type I and "Interval-Valued" Type II Fuzzy Sets and Logics", *In P. Wang (Ed.), Advances in Fuzzy Set Theory and Technology*, 1995.
19. [15] I.B. Turksen, "Non-specificity and interval-valued fuzzy sets," *Fuzzy Sets and Systems*, vol. 80 (1996) 87-100.
20. [16] I.B. Turksen, "Type I and Type II fuzzy system modeling," *Fuzzy Sets and Systems*, vol. 106, pp. 11-34, 1999.
21. [17] I.B. Turksen, "Type 2 representation and reasoning for CWW," *Fuzzy Sets and Systems*, vol. 127 (2002) 17-36.
22. [18] I.B. Turksen, "Fuzzy Functions with LSE," *Applied Soft Computing,* vol.8, 1178-1188, 2008.
23. [19] O. Uncu and I.B. Turksen, "A Novel Fuzzy System Modeling Approach: Multidimensional Structure Identification and Inference", *Proc. of Tenth IEEE International Conference on Fuzzy Systems, Melbourne, Australia,* December, 2001,557-562.
24. [20] N.V.Vapnik, *Statistical Learning Theory,* John Wiley and Sons, New York, 1998.
25. [21] L.A. Zadeh, "Fuzzy sets", *Information and Control*, vol. 8, pp. 338-353, 1965.
26. [22] L.A. Zadeh, "The Concept of a Linguistic Variable and its Application to Approximate Reasoning", *Information Sciences*, 8, 199-249, 1975.

CAN FUZZY LOGIC FORMALISM BRING COMPLEX ENVIRONMENTAL ISSUES INTO FOCUS?

ASHOK DESHPANDE

Founding Chair: Berkeley Initiative in Soft Computing (BISC)
Special Interest Group (SIG) – Environment Management Systems (EMS)
Adjunct Professor: College of Engineering Pune India

In everyday life and in fields such as environmental health, people deal with concepts that involve factors that defy classification into crisp sets—*safe, harmful*, *acceptable, unacceptable,* and so on. A classic example is a regulator carefully explaining the result of a detailed quantitative risk assessment to a community group, only to be asked over and over again, *But are we safe?* In this case, *safe* defies crisp classification because it is a multivariate state with gradations that vary among different individuals and groups.

Multi causality or *"what causes what"* has been one of the quests in scientific research for a long time. One cannot prove/disprove that bathing in polluted waters / inhaling polluted air can cause water borne /airborne diseases. Realizing the complexity in establishing cause–effect relationship in the two important issues: 1. bathing in polluted river and water borne diseases (WBD's), and 2. inhaling polluted air and airborne diseases, an attempt has been made to present a useful method. Some of the important facets of our approach relating to water borne diseases (WBD) include: interviewing student community (bather/ non bather) with a structured questionnaire, collecting information on belief of the resident medical practitioners about bathing in polluted river and WBDs. Furthermore, modelling of epistemic uncertainty in domain expert's belief in supporting their evidence for various WBDs and alike.

Information on the status and changing trends in environmental quality is necessary to formulate sound public policy and efficient implementation of environmental pollution abatement programs. One of the ways of communicating the information to the policy makers and public at large is with indices. In the computation of air/water quality index (AQI/WQI), first a numerical value is computed and then the air/ water quality is described in linguistic terms. There exists aloetary uncertainty in the pollution parametric data and epistemic uncertainty in describing the pollutants by the domain experts in linguistic terms such as *poor, good, very good*, etc. Successes of probability theory have high visibility. But what is unrecognized is that these successes mask a fundamental limitation-the inability to operate on what may be called perception - based information. In this paper an attempt has been made to use fuzzy logic based formalism in modelling these two types of uncertainties, thereby straightway describing air/water quality in linguistic terms with a degree of certainty attached to each term

In order to bring down progressive deterioration in air quality, the Government of India has enacted Air (Prevention and Control of Pollution) Act in 1981 and further stringent guidelines are promulgated in Environment (Protection) Act, 1986. The need for Ambient air quality monitoring programme are needed to determine the existing quality of air, evaluation of the effectiveness of control programme and to develop new programme was recognised As a follow up action, the Central Pollution Control Board (CPCB) initiated National Air Quality Monitoring (NAQM) Programme during 1984-85 at the national level .

A well structured air quality monitoring network involves selection of pollutants, selection of locations, frequency, duration of sampling, sampling techniques, infrastructural facilities, man power and operation, maintenance costs, and alike. The network design also depends upon the type of pollutants in the atmosphere through various common sources, called common urban air pollutants, such as Suspended Particulate Matter (SPM), Respirable Suspended Particulate Matter (RSPM), Sulphur Dioxide (SO_2), Oxides of Nitrogen (NO_x), and Carbon Monoxide (CO) etc. The areas to be chosen primarily are such areas which represent high traffic density, industrial growth, human population and its distribution, emission source, public complaints if any and the land use pattern etc. Generally, most of the times the basis of a network design are the pollution source and the pollution present.

The results of the air quality study are the comparison between the computed AQI and the proposed fuzzy logic based method. It can be revealed that AQI based on the traditional method does not attach any certainty while Type 1 Fuzzy Inference system, we can describe the air quality straightway in linguistic terms with some degree of certainty attached to each term.

Comparison of Conventional AQI, and Fuzzy Description of Air Quality with Degree of Certainty

Monitoring Station	Conventional AQI	AQI with Degree of Certainty
Karve Road	206 Very Poor	0.45 very poor
Bhosari	145.89 Poor	0.43 very poor
Swargate	147.13 Poor	0.54 poor
Nal Stop	120.2 Poor	0.67 fair
PCMC	151.13 Very Poor	0.91 poor

In view of the uncertainties discussed above, the approach outlined in this paper using fuzzy logic could be a better representation of a dynamic system, and thereby providing a new dimension of monitoring air quality. The results of fuzzy modelling for air quality classification in Pune infers that there is an urgent need to carry out air pollution abatement measures on a war footing. Similar is the situation in Indian rivers which are grossly polluted due to domestic waste water and industrial effluents discharged into the river body with no/ inadequate treatment.

PART 1

Decision Making and Decision Support Systems

EVALUATION OF MANUFACTURING TECHNOLOGY OF PHOTOVOLTAIC CELLS

E. CABLES PÉREZ

Depto. Informática. Universidad de Holguín "Oscar Lucero Moya" Holguín. Cuba

M. T. LAMATA JIMÉNEZ[1*], J. L. VERDEGAY[1]

[1]Depto de Ciencias de la Computación e Inteligencia Artificial. Universidad de Granada.18071 Granada. Spain

M. S. GARCIA-CASCALES[2], J. M. SÁNCHEZ-LOZANO[3]

[2]Depto de electrónica, tecnología de computadoras y proyectos.
[3]Depto de Expresión Gráfica.
Universidad Politécnica de Cartagena. Spain

In any decision problem where it is necessary to obtain a ranking to know the decision matrix, the weights of the criteria and the method for evaluating the alternatives must be determined. In the present paper the TOPSIS method is the model employed to achieve this because we have linguistic and numerical values such as the assessment of alternatives for each criterion. The weights of the criteria have been obtained by an OWA operator. This theoretical development has been applied to a decision problem in the field of renewable energy.

1. Introduction

Decision making is a common process in the daily life of human beings, with different alternatives available from which we select the one that seems best or the alternative that satisfies the required needs to the greatest extent [6].

Excessive dependence on non-renewable energy sources in the energy sector, as well as global climate change, provide strong arguments to ensure that sources of energy from renewable sources must not be neglected [2,4,7].

The aim of this paper is to analyze the best technology of photovoltaic cell manufacturing, taking into account both quantitative criteria such as social

* mtl@decsai.ugr.es

and/or technical aspects, as well as qualitative criteria such as social and/or environmental factors. These criteria will be affected by a weight, depending on the importance awarded to each criterion by the decision maker.

In this paper it is proposed to use an operator whose main characteristic is that instead of making pairwise comparisons among all the criteria, only one criterion is compared with the rest. For this, we will consider first the order of importance of the criteria, and then the assessment that the expert gives to C_j on C_{j+1}. Once we know the weights of the criteria and the rating of the alternatives we will apply the TOPSIS decision model.

The paper is organized as follows: the second section details the process to obtain the weights; the third presents the TOPSIS method, and in the fourth section a problem related to renewable energy is developed. Finally, the conclusions are presented.

2. Obtaining weights

In the process of aggregation of the information in multicriteria decision problems, weights play an essential role in indicating the relative importance of criteria.

Many authors have worked with the Analytic Hierarchy Process (AHP) to calculate the weight vector [11], which is based on pairwise comparison between the elements of a set of criteria, with $n(n-1)/2$ being the total number of comparisons needed, where n is the number of criteria.

Instead of applying AHP, we propose to develop a simpler operator with characteristics similar to AHP. This development is based on the specification of a vector that represents the ratio of proportionality between adjacent weights [1,9,10,12]. Therefore, the starting point is the management of the criteria in descending importance and quantification of the importance given by the vector $V=[v_1,v_2,\ldots,v_{n-1}]$ where $v_i \geq 1, \forall i$ and $z_j = 1/v_j \in [0,1]$ represent the proportion of reverse superiority criteria C_j with respect to C_{j+1}. From this, expression (1) allows to obtain a weight vector that verifies the properties of the OWA [10,13,14] and the ratio of proportionality between the criteria.

$$w_i = \begin{cases} \dfrac{1}{\left(\sum\limits_{j=1}^{n-1}\prod\limits_{k=1}^{j} z_k\right)+1} & si \quad i=1 \\ w_1 \prod\limits_{k=1}^{i-1} z_k & si \quad i \neq 1 \end{cases} \tag{1}$$

Example 1: From an organized set of five criteria as follows $C_2 \overset{3}{} C_1 \overset{5/3}{} C_4 \overset{7/5}{} C_5 \overset{9/7}{} C_3$. According to this, the vector indicating the linguistic assessment is $v_j = [3, 5/3, 7/5, 9/7]$ and therefore the inverse vector is given by $z_j = [1/3, 3/5, 5/7, 7/9]$ then, using expression (1) for calculating the weights, we obtain:

$$w_1 = \frac{1}{\left(\sum_{j=1}^{4}\prod_{k=1}^{j} z_k\right)+1} = \frac{1}{\frac{1}{3}+\frac{1}{3}\cdot\frac{3}{5}+\frac{1}{3}\cdot\frac{3}{5}\cdot\frac{5}{7}+\frac{1}{3}\cdot\frac{3}{5}\cdot\frac{5}{7}\cdot\frac{7}{9}+1} = \frac{945}{1689}$$

$$w_2 = \frac{945}{1689}\prod_{k=1}^{1} z_k = \frac{945}{1689}\cdot\frac{1}{3} = \frac{315}{1689}, \text{ etc.}$$

Thus, the resulting weight vector in the order C_1-C_2-C_3-C_4-C_5 is $W = (0.1865,\ 0.5595,\ 0.0621,\ 0.1119,\ 0.0799)$.

3. Decision model

The TOPSIS method is a proposed decision model for sorting preferences by similarity to ideal solution [3,5,8]. The basic principle of the method consists in finding an alternative whose distance from the positive ideal solution (PIS) is less and whose distance from the negative ideal solution (NIS) is greater.

An ideal solution is defined as a collection of scores or values on all attributes considered in the decision-making. However, such a solution may be unattainable.

3.1. *Algorithmic method proposed*

We propose an algorithm to determine the preferred option among all possible alternatives:

Step 1: Identification of evaluation criteria so that the system's capabilities relate to the objectives. Generation of alternatives which identify the finite set of alternatives to achieve the goal. Evaluation of alternatives based on the criteria.

Step 2: Identification of the weight of criteria and construction of the decision matrix.

Step 3: Obtain the associated normalized decision matrix. To achieve this, the values of each criterion are divided by the norm so that the scale is the same for all criteria.

$$\bar{n}_{ij} = \underset{\sim}{x}_{ij} \Big/ \sqrt{\sum_{j=1}^{n}\left(\underset{\sim}{x}_{ij}\right)^2}, \; j = 1,\ldots,n, \; i = 1,\ldots,m \qquad (2)$$

Step 4: Calculate the associated weighted normalized decision matrix.

$$\overline{v}_{ij} = w_j \overline{n}_{ij},\ j = 1,\ldots,n,\ i = 1,\ldots,m, \tag{3}$$

Step 5: Determine the PIS and the NIS, respectively.

$$\overline{A}^{+} = \{\overline{v}_1^{+},\ldots,\overline{v}_n^{+}\} = \{(\max \overline{v}_{ij}, j \in J), (\min \overline{v}_{ij}, j \in J')\} \tag{4}$$

$$\overline{A}^{-} = \{\overline{v}_1^{-},\ldots,\overline{v}_n^{-}\} = \{(\min \overline{v}_{ij}, j \in J), (\max \overline{v}_{ij}, j \in J')\} \tag{5}$$

where J is associated with the criteria that indicate profits or benefits and J' is associated with the criteria that indicate costs or losses.

Step 6: Calculation of the separation of each alternative with respect to the PIS and NIS, respectively:

$$\overline{d}_i^{+} = \left\{ \frac{1}{3} \sum_{j \in J}^{5} \left(\overline{v}_{ij} - \overline{v}_j^{+} \right)^2 \right\}^{\frac{1}{2}},\ \mathrm{i} = 1,\ldots,m \tag{6}$$

$$\overline{d}_i^{-} = \left\{ \frac{1}{3} \sum_{j \in J}^{5} \left(\overline{v}_{ij} - \overline{v}_j^{-} \right)^2 \right\}^{\frac{1}{2}},\ \mathrm{i} = 1,\ldots,m \tag{7}$$

Step 7: Calculation of the relative proximity of each alternative to the PIS and NIS using the index of proximity.

$$\overline{R}_i = \frac{\overline{d}_i^{-}}{\overline{d}_i^{+} + \overline{d}_i^{-}},\ \ \mathrm{i} = 1,\ldots,m \tag{8}$$

4. Decision problem

We wish to carry out a study on the different manufacturing technologies of solar cells (Photovoltaic cells with crystalline silicon A_1, with inorganic thin layer (amorphous silicon) A_2, with inorganic thin layer (CdTe and CIGS) A_3, with advanced III-V thin layer with tracking systems for solar concentration A_4 and advanced cells with low cost, thin layers (Organic and hybrid cells) A_5) whose decision matrix was considered [4], and is reflected in Table 1.

It is important to note that the criteria of efficiency (C_2) and market share (C_3) will be much better with the higher value; this means that we have to maximize them. However, manufacturing cost (C_1), emissions of greenhouse gases (C_4) and energy payback time (C_5) criteria will be better if their values are smaller and thus we will minimize them.

Table 1: Assessment of alternatives for each criterion.

	C_1 (€/Wp)	C_2 (%)	C_3	C_4	C_5
A_1	1.440 ± 0.760	21.0 ± 3.0	VH	m	m
A_2	0.675 ± 0.275	11.5 ± 3.5	L	m	mB
A_3	0.900 ± 0.300	11.5 ± 3.5	L	m	mB
A_4	0.450 ± 0.250	32.5 ± 7.5	L	H	B
A_5	0.600 ± 0.200	9.0 ± 3.0	L	L	B

It should be noted that these ratings are numerical for criteria 1 and 2, whilst C_3, C_4 and C_5 are linguistic.

The linguistic labels to describe the alternatives for each of the criteria and their quantitative assessment and fuzzy numbers are given in Table 2.

Table 2: Linguistic labels for the ratings of the alternatives

Description	Fuzzy numbers
Very Low VL/Very Brief VB	[0, 0, 1]
Low L/Brief B	[0, 1, 3]
Medium low mL/medium brief mB	[1, 3, 5]
Medium m/Medium m	[3, 5, 7]
Medium high mH/Medium long mLg	[5, 7, 9]
High H/Long Lg	[7, 9, 10]
Very High VH/Very Long VLg	[9, 10, 10]

The weights considered are those corresponding to the example detailed in Table 1, $W = (0.1865,\ 0.5595,\ 0.0621,\ 0.1119,\ 0.0799)$.

If we apply the algorithm of the TOPSIS method with our own operations of fuzzy numbers, we obtain the results shown in Table 3.

Table 3. Corresponding values for $d^{\div}$, d^{-} and R.

	$d^{\div}$	d^{-}	R
A_1	0.2532	0.1854	0.4228
A_2	0.3166	0.1559	0.3300
A_3	0.3205	0.1335	0.2941
A_4	0.0880	0.3870	0.8148
A_5	0.3480	0.1879	0.3507

From the results for $d^{\div}$ we can see that the best alternative is A_4, since the distance to the positive ideal alternative is less. With respect to d^{-} and the index

R the best alternative would be the farthest, i.e. the opposite happens, but the best alternative would also be A_4.

5. Conclusions

Weights have been obtained through an operator, and the problems of consistency that could have arisen if the 10 comparisons among the 5 criteria had been gathered, have been eliminated. We also manage to avoid the necessary operations to obtain the values and vectors of the matrix resulting from the comparison.

The problem of multicriteria decision has been solved for obtaining the best technology of photovoltaic cells when the evaluation is done taking into account the manufacturing cost, energy conversion efficiency, market share, emissions of greenhouse gases and energy payback time criteria. With these variables the results obtained show alternative A_4, Photovoltaic cells with advanced III-V thin layer with tracking systems for solar concentration to be the best.

Acknowledgments

This work is partially supported by FEDER founds, the DGICYT and Junta de Andalucía under projects (TIN2008-06872-C04-04, TIN2011-27696-C02-01) and (P07-TIC02970, P11-TIC-8001), respectively.

References

1. E. Cables Pérez and M.T. Lamata, *Caepia*. (2011).
2. F. Cavallaro, *Appl Energ*, **C87**, 496 (2010).
3. S.J. Chen and C.L. Hwang, *Springer-Verlang* (1992).
4. M. S. García-Cascales, M. T. Lamata and J.M. Sánchez-Lozano. *Ann Oper Res*. DOI 10.1007/s10479-011-1009-x .
5. C.L. Hwang, K. Yoon, *Springer- Heidelberg*. (1981).
6. R. Keeney, H. Raiffa: *Wiley*. (1976).
7. C. Kahraman and I. Kaya. Expert Syst Appl, **C37**, 6270 (2010).
8. Y.J. Lai, T.Y, Liu, C.L and Hwang, *Eur J Oper Res* **C76 (3),** 486 (1994).
9. M.T. Lamata and E. Cables Pérez, Mathware & Soft Computing, **C16**, 107 (2009).
10. M.T. Lamata and E. Cables Pérez, Int J Intell Syst, **C27,** 242 (2012).
11. T.L. Saaty, New-York (1980).
12. Z.S. Xu and Q.L. Da, J Southeast Univ. **C33**, 94 (2003).
13. R.R. Yager, Fuzzy Sets and Syst. **C59**, 125 (1993).
14. R.R. Yager, Int J Approx Reason, **C48**, 263 (2008).

A MULTICRITERIA DYNAMIC FLOW MODEL FOR RELIEF OPERATIONS*

G. TIRADO†, B. VITORIANO, M. T. ORTUÑO

Dpt. Statistics and Operational Research, University Complutense of Madrid, 28040 Madrid, Spain

Natural and human-made disasters cause humanitarian crises all around the world. When help is requested from international instances, Humanitarian logistics play an important role. Mathematical models can help to make strategic and tactical decisions in this context, when decisions must be taken in a multicriteria environment. In this work a multicriteria dynamic flow model is developed, allowing a more detailed planning of aid distribution operations than other existing static models.

1. Introduction and Problem Description

Natural and human-made disasters cause humanitarian crises all around the world. When help is requested from international instances, due to the intensity of the phenomenon or to the vulnerability of the country, relief organizations respond in different ways, and Humanitarian logistics play an important role. In Ref. 1 an up-to-date review of Optimization models in emergency logistics can give an idea of the last developments in the field.

The problem addressed in this work concerns last-mile distribution in disaster relief operations. In particular, the problem addressed consists of designing routes for vehicles among nodes that have an available quantity of goods or have a demand of those goods, choosing the types of vehicles more adequate and determining the flow of the aid. More specifically, the problem addressed is described through the following elements:

1. Transportation Network. Nodes representing the places of pick-up, delivery or connection, and main links characterized by distance and average velocity.
2. Goods. Information about quantity available at each node, and quantity required at each node is provided.

---

* This work has been partially supported by project TIN2009-07901 from the Government of Spain.

† : gregoriotd@mat.ucm.es.

3. Vehicles. Several types are considered, characterized by capacity, average velocity, variable and fixed costs and availability in each node of the network.
4. Operation elements. They include the global quantity to be distributed in the operation and the budget available.

The problem consists of designing the vehicles routes so that a fixed quantity of humanitarian aid is distributed.

The optimization process can be guided by different criteria. It has been recognized in the last times that traditional logistic objectives, such as minimizing operation cost, are not the most relevant ones in humanitarian operations. Other criteria, such as time of operation or equity in the distribution, must take a prominent role. In Ref. 2 a goal programming static flow model to deal with this problem is presented, providing an approximation to the best distribution plan using up to six conflicting criteria. In this work the model is extended and a dynamic flow model is proposed, which allows a more detailed planning, as it explicitly considers the movement of vehicles in the different periods of time.

2. Structural Model: Flow and Operation Constraints

In the core of the dynamic distribution model proposed in this work lay two Dynamic Network Flow models, one for load and another one for vehicles, as well as the relations between them. Together with some constraints related to the resources assigned to perform the operation they constitute the basic set of restrictions of the proposed model. A survey on Dynamic Network Flow models can be found in Ref. 3.

2.1. *Main Data*

(N,A): transport network, where N is the set of nodes and A is the set of arcs	i,i': indices to refer nodes $i,i' \in N$
T: planning horizon; the mission can be developed through periods $\{1,\ldots,T\}$	t,t': indices to refer time periods $t,t' \in \{1,\ldots,T\}$

A) Data related to the network.

$dist_{ii'}$: length of arc $(i,i') \in A$

dem_i : demand at node $i \in N$ in units of load

qav_i : supply at node $i \in N$ in units of load

B) Data related to vehicles. (Several types are considered).

V : set of vehicle types, defined by their characteristics	vav_{ji} : availability of vehicle type $j \in V$ at node $i \in N$
j, j' : indices to refer vehicle types, $j, j' \in V$	$cf_{ii'j}$: fixed cost of a vehicle type $j \in V$ through $(i,i') \in A$, per unit of length
cap_j : capacity of vehicle type $j \in V$, in units of load	$cv_{ii'j}$: cost of a vehicle type $j \in V$ through $(i,i') \in A$, per unit of length and load
$tr_{ii'j}$: periods of time needed by vehicle type j to travel from i to i'	

C) Data related to the operation.
$qglobal$: load to be distributed in the operation
b : budget available to perform the operation

2.2. *Variables*

$QLOR_{ii'jt}$: load carried from i to i', $(i,i') \in A$, using vehicle type j, starting in period t	$NL_{ii'jt}$: number of vehicles type j which start travelling from i to i', $(i,i') \in A$ in period t
Q_{it} : load staying (stored or received) at node $i \in N$ in period t	NLF_{ij} : number of vehicles type j whose final destination is node $i \in N$
BQ_t : 1 if all load has been delivered in period t, 0 otherwise	$BLCT_{ii't}$: 1 if arc (i,i') has been used in period t, 0 otherwise

For simplicity, in the following it will be assumed that, unless otherwise stated, indices i, i', j, j', t, t' take values in the sets where they are defined, so that only existing arcs and vehicle types compatible with them are considered in the definition of the constraints.

2.3. *Structural Model*

$$\sum_{i'}\sum_{j}\sum_{t'\le t-tr_{i'ij}} QLOR_{i'ijt'} + qav_i = \sum_{i'}\sum_{j}\sum_{t'\le t} QLOR_{ii'jt'} + Q_{it} \quad \forall i,t \tag{1}$$

$$\sum_{i} Q_{iT} = \sum_{i} qav_i \tag{2}$$

$$\sum_{i'}\sum_{t'\le t-tr_{i'ij}} NL_{i'ijt'} + vav_{ji} \ge \sum_{i'}\sum_{t'\le t} NL_{ii'jt'} \quad \forall i,j,t \tag{3}$$

$$\sum_{i'}\sum_{t\le T-tr_{i'ij}} NL_{i'ijt} + vav_{ji} = \sum_{i'}\sum_{t} NL_{ii'jt} + NLF_{ij} \quad \forall i,j \tag{4}$$

$$\sum_{i} vav_{ji} = \sum_{i} NLF_{ij} \quad \forall j \tag{5}$$

$$QLOR_{ii'jt} \leq cap_j NL_{ii'jt} \quad \forall i, i', j, t \tag{6}$$

$$\sum_{i,i',j,t} \left(2cf_{ii'j} dist_{ii'} NL_{ii'jt} + cv_{ii'j} dist_{ii'} QLOR_{ii'jt} \right) \leq b \tag{7}$$

$$\sum_{i/dem_i > 0} Q_{iT} = qglobal \tag{8}$$

$$QLOR_{ii'jt}, Q_{it} \geq 0 \quad \forall i, i', j, t \tag{9}$$

$$NLF_{ij}, NL_{ii'jt} \geq 0 \text{ and integer} \quad \forall i, i', j, t \tag{10}$$

Condition (1) is the dynamic load flow balance equation at the nodes. Constraint (2) ensures that total load staying at the nodes at the end of the operation is the available load. Conditions (3)-(4) are the dynamic flow balance equations for the vehicles and condition (5) ensures that only the vehicles that are available are used for transportation.

Constraints (6) ensure the capacity of the vehicles is not exceeded and condition (7) states the budget available, while the amount of humanitarian aid to be distributed in the operation is given by equation (8) and the domain of the variables is given in (9)-(10).

3. Attributes and Criteria Aggregation

In a first approximation, criteria concerning the time of operation (measured in two different ways), the equity of the distribution and the cost of operation will be considered. The attribute variables are the following:

COST : total operation cost

T_i : time measure *i*

DX : equity measure

and they are defined by the following constraints:

$$COST = \sum_{i,i',j,t} \left(2cf_{ii'j} dist_{ii'} NL_{ii'jt} + cv_{ii'j} dist_{ii'} QLOR_{ii'jt} \right) \tag{11}$$

$$T_1 = \sum_{t} t^2 BQ_t \tag{12}$$

$$T_2 = \sum_{ii't} t^2 BLCT_{ii't} \tag{13}$$

$$DX \geq 1 - \frac{Q_{iT}}{dem_i} \quad \forall i / dem_i > 0 \tag{14}$$

In order to deal jointly with these attributes, the following goal constraints are included.

$$COST - DV_C \leq tg_C \tag{15}$$

$$T_1 - DV_T \leq tg_T \tag{16}$$

$$DX - DV_D \leq tg_D \tag{17}$$

$$COST, T_1, DX, DV_C, DV_T, DV_V \geq 0 \tag{18}$$

Coefficients tg_C, tg_T, tg_D represent the targets defined by the decision maker for the attributes related to $COST$, T_1 and DX, respectively. Non-negative variables DV_C, DV_T, DV_D are the unwanted deviations from the given targets in the solution. The following relationships must be included in the model to ensure the correct definition of the binary variables used to measure the attributes:

$$BQ_t \geq BQ_{t+1} \; \forall i \tag{19}$$

$$Q_{it} - Q_{iT} \leq \sum_{i'} qav_{i'} BQ_t \; \forall i,t \tag{20}$$

$$Q_{iT} - Q_{it} \leq \sum_{i'} qav_{i'} BQ_t \; \forall i,t \tag{21}$$

$$BQ_t \in \{0,1\} \; \forall t \tag{22}$$

$$\sum_j NL_{ii'jt} \leq \sum_j bd_j BLCT_{ii't} \;\; \forall i,i',t \tag{23}$$

$$BLCT_{ii't} \leq \sum_j NL_{ii'jt} \; \forall i,i',t \tag{24}$$

$$0 \leq BLCT_{ii't} \leq BQ_t \text{ and binary } \forall i,i',t \tag{25}$$

The problem is solved in two phases. Equation (26) is the objective function for the phase 1 model (multi-criteria), in which the sum of weighted normalized deviations from the targets is minimized. Equation (27) is the objective function for the phase 2 model in which, provided that the demand nodes to be served and the vehicles to be used are fixed by phase 1, the time of operation is minimized and the formation of convoys is encouraged.

$$\min\ z_1 = \sum_v \frac{\alpha_v}{t_v} DV_v \tag{26}$$

$$\min\ z_2 = T_2 \tag{27}$$

Phase 1: (1)-(12),(14)-(22),(26).
Phase 2: (1)-(25),(27), fixing variables DV , Q_{iT} and NLF_{ij} .

4. Final Comments

The dynamic model presented in this paper is significantly larger and more complex than the static one introduced in Ref. 2, but also much closer to the real operation. The information given by the static model concerns the flow of vehicles and load along the network, providing a useful approximation to the distribution plan but lacking details about the timing of the operations to be performed. So, we proceed to compare the performance of both models, to evaluate the added value given by the dynamic flow and the capability of the static model to approximate the dynamic operation. As a first study, we have made a comparison based on the case study included in Ref. 2. According to the solution attributes and the amount of humanitarian aid distributed, both the static and the dynamic models build quite similar solutions, delivering the same amount of goods but obtaining some improvements in the itineraries and a much more realistic operations plan when using the dynamic model. These results show that in this case study the static model is able to provide good approximations to the detailed plan given by the dynamic model, being the latter most useful to obtain information about the itineraries timing. In particular, the solution given by the dynamic model provides a detailed distribution plan for each time period, showing the decisions that are taken at the beginning of each time period and allowing an easier replanning in case something unexpected happens at any particular point of time.

References

1. A.M. Caunhye, X. Nie, S. Pokharel. Optimization models in emergency logistics: A literature review. *Socio-Economic Planning Sciences* **46**, 4-13, (2012).
2. B. Vitoriano, M.T. Ortuño, G. Tirado, J. Montero. A multi-criteria optimization model for humanitarian aid distribution. *Journal of Global Optimization* **51**, 189-208 (2011).
3. X. Cai, D. Sha and C. Wong. Time-Varying Network Optimization. Springer, New York (2007).

FUTURE ORIENTED POSITIONING ANALYSIS WITH BAYESIAN NETWORKS

BURAK DERELİ, UMUT ASAN, ÇİĞDEM KADAİFÇİ

Industrial Engineering, Istanbul Technical University, Macka Istanbul,Turkey

This study suggests a future oriented probabilistic approach to the positioning analysis of the videogame console industry. In comparison to the common multivariate positioning techniques, the proposed approach, using Bayesian networks, is less restrictive and attempts to incorporate the potential influences of external factors as well as marketing strategies into the positioning analysis. This provides a better understanding of how the market and its complex environment may evolve in the future and how they can be influenced in favor of one's own product/brand.

1. Introduction

Technology increasingly becomes a more central part of our lives not only in the areas of health, education, and transportation but also in the area of entertainment. Recent developments in game console technology and changing lifestyles have fostered competition even more in the technology-driven entertainment industry and created an evolving market. This intense competition, known as "Console Wars" is in its seventh generation and none of the initial rivals of the war exists today. Winning the Console Wars may depend on product innovations, financial situations, market penetration or customer loyalty and number of fans of the brands [1]. Therefore, console manufacturers need to understand the current and potential developments in their environment and develop marketing strategies that will keep them competitive in the future. To do this, they must figure out how possible marketing activities are perceived by customers. One effective marketing deliberation to reveal and analyze customer perceptions is positioning, which can defined as "the deliberate, proactive, iterative process of defining, measuring, modifying and monitoring consumer perceptions of a marketable object" [2]. Various sophisticated multivariate techniques, such as factor analysis, multidimensional scaling and unfolding, discriminant analysis and conjoint analysis, have been used for positioning analysis [3, 4]. However, these techniques impose rigorous distributional assumptions and are limited in analyzing the impact of complex and uncertain environmental pressures and marketing activities on customer perceptions. This study suggests a less restrictive Bayesian networks (BNs) based approach which incorporates the potential influences of external factors as well as marketing strategies into the positioning analysis. BNs are able to address the uncertainty issue by using a probabilistic approach where the probability of an event is described as a person's degree of belief in that event [5].

2. Bayesian Networks

Based on the Bayes' theorem of conditional probabilities, a BN is a type of probabilistic decision support system specialized in representing and reasoning with uncertain knowledge [6]. As demonstrated in Jones *et al.* [7] several advantages of BNs include i) dealing effectively with uncertainty, ii) modeling complex systems, iii) providing a visual representation of relationships, iv) combining historical data and expert views and v) allowing updating the model as new information becomes available.

BNs are represented with directed acyclic graphs consisting of nodes standing for the random variables and arcs standing for the probabilistic relationships among the variables. The nodes from which the directed arrows originate are defined as parent nodes whereas the nodes where the directed arrows point at are defined as child nodes. The data of conditional relations between nodes is stored in conditional probability tables (CPTs), see Figure 1, which specifies the chance distribution across the states of a child node for each possible combination of states of its parent nodes [8].

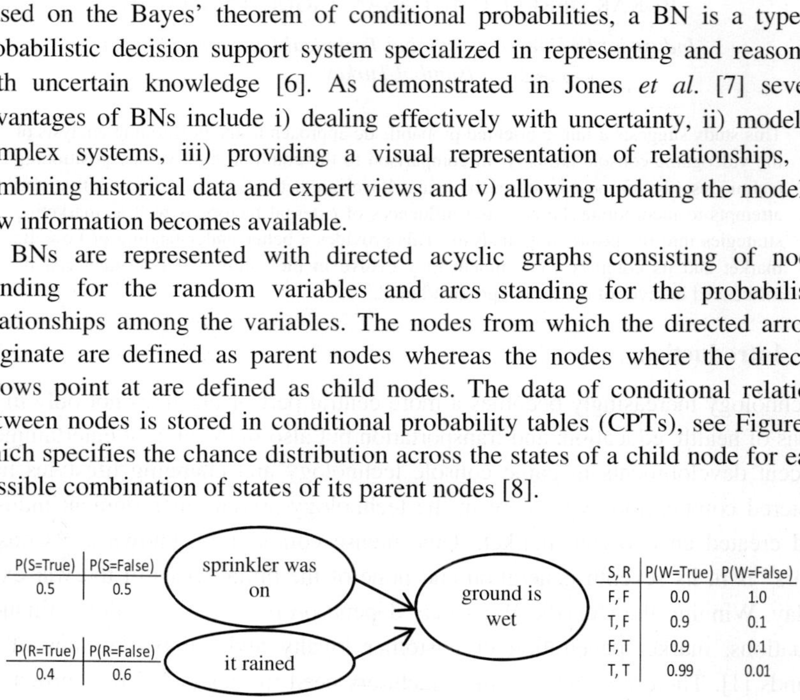

Figure 1. Simple BN Example.

3. Analysis

3.1. *Proposed Model and Data Collection*

The proposed BN model is constructed in two stages, the qualitative stage which involves the development of the network structure and the quantitative stage which involves the elicitation of conditional probabilities and joint probability distributions [see 6]. Figure 2 illustrates a summary view of the conceptual network model that was developed based on a literature review and practical considerations from the videogame console industry. This initial version consisting of variables and their conditional dependencies was reviewed by experts for accuracy and completeness, and was refined accordingly.

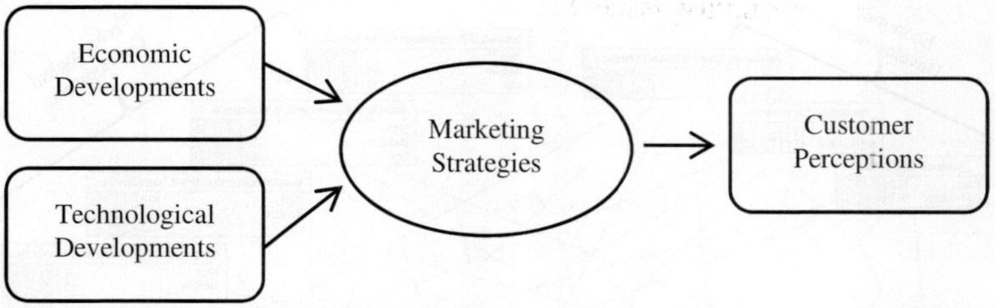

Figure 2. Conceptual Model.

The component "Economic Developments" contains the following variables: economic depression, fluctuation of currencies and purchasing power of target customers. All three variables are specifically analyzed for Turkey. Variables belonging to the component "Technological Developments" describe the intensity of developments in cost reducing technologies, innovative technologies, alternative technologies and supportive technologies in the game console industry. Decisions involving pricing policies, product development, distribution channels, common advertising campaigns and sales promotions are the variables representing the component "Marketing Strategies". Finally, "Customer Perceptions" are modeled based on brand image, economic and physical accessibility, product quality and loyalty. Each of these four variables is represented by five states ranging from "very low" to "very high". All the listed variables, their possible states and the relationships among them can be seen in Figure 3.

In the quantitative stage, in order to elicit conditional probability values, two different questionnaires were designed. One of them includes questions regarding the relationships between economic and technological developments and possible marketing strategies of the console manufacturers. Experts were asked to allocate 100 points across the set of a child's states, reflecting their conditional probabilities given a particular state of a parent node. The second questionnaire was designed to explore console users' perceptions given alternative marketing strategies.

During data collection, changes in average probabilities for each new entry (response) were dynamically observed in order to determine the sufficient sample sizes. Data collection continues until the average value (for each question) converges toward a steady state. Figure 4 and 5 illustrates the convergence of averages for the conditional relations between innovative technological developments and distribution channel development, as well as common advertising campaigns and brand image, respectively.

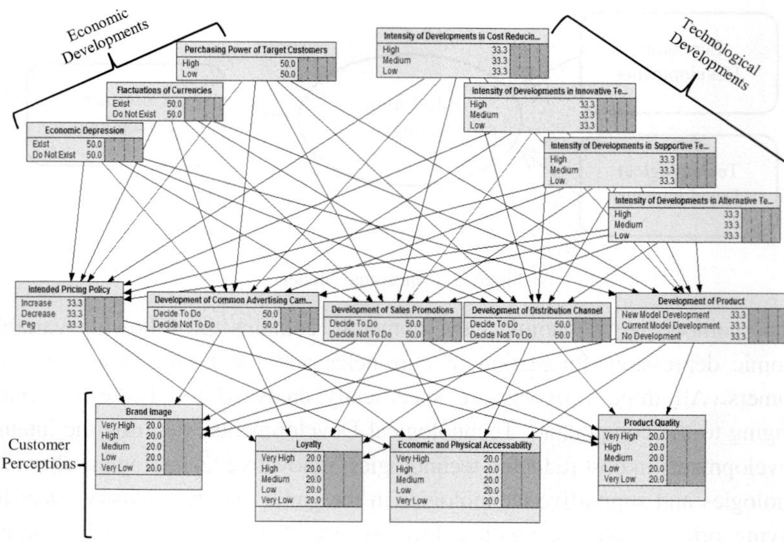

Figure 3. Detailed BN for Positioning Analysis of the Videogame Console Industry.

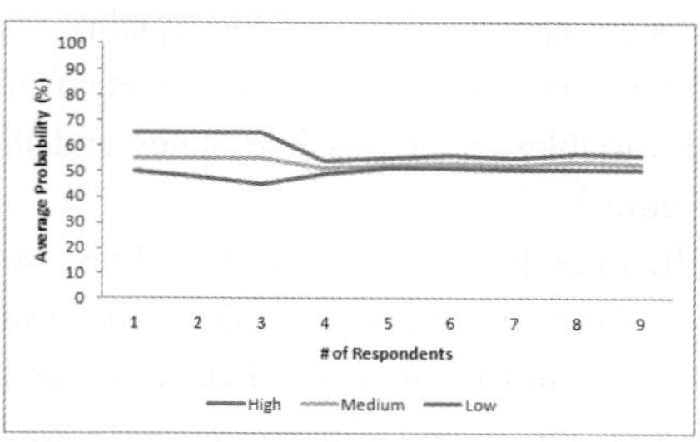

Figure 4. Convergence of averages for innovative technological development-distribution channel development.

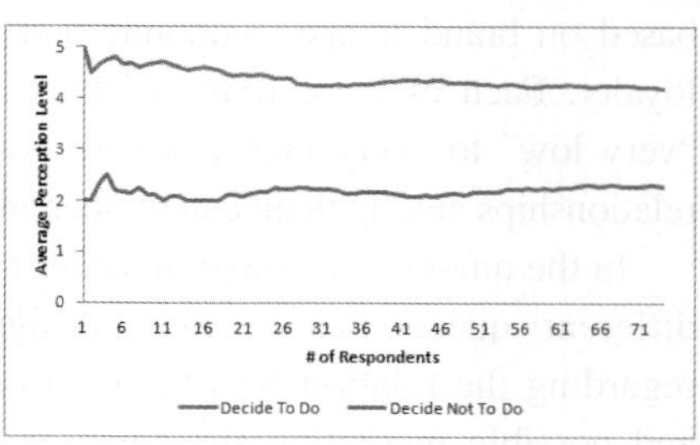

Figure 5. Convergence of averages for common advertising campaigns-image perception.

While obtaining conditional probabilities for nodes with multi-parents, theoretically all possible combinations of parent node states should be evaluated for any child node state. However, just for each marketing strategy variable in the model, the number of combinations to be evaluated becomes 648. Thus, a systematic method for approximating conditional probabilities was used as suggested by Chin et al. [6]. According to this method, a node's probability conditional on each of its parent is assumed to be independent, and all separately obtained conditional probabilities of this node are multiplied and normalized to obtain the node's conditional probability values for each combination.

3.2. *Results*

The network was built up and analyzed using the Bayesian network software Netica (www.norsys.com). The current situation was described and possible future scenarios were built for the game console market in Turkey based on the economic and technological indicators and expert opinions. As a result, the initial probabilities for economic and technological development variables were determined as shown in Table 1. The model was verified by comparing the probability values computed for the current scenario (i.e. theoretical values) with the elicited customer perceptions (i.e. observed values).

Table 1. Initial Probabilities for the Current Situation and Future Scenarios, (%).

	Economic Depression		Fluctuation of Currencies		Target Customer Purchasing Power		Cost Reducing Techs.			Innovative Techs.			Alternative Techs.			Supportive Techs.		
Scenario	Exist	Not Exist	Exist	Not Exist	High	Low	High	Medium	Low	High	Medium	Low	High	Medium	Low	High	Medium	Low
Current	40	60	70	30	40	60	30	40	30	60	30	10	50	40	10	20	30	50
Future	60	40	75	25	50	50	30	30	40	70	20	10	60	30	10	50	30	20
Worst	100	0	100	0	0	100	0	0	100	0	0	100	0	0	100	0	0	100
Best	0	100	0	100	100	0	100	0	0	100	0	0	100	0	0	100	0	0

Table 2 presents the calculated probabilities for the current situation and alternative future scenarios. Surprisingly, the strategy not to change the price dominates the other options across all scenarios. This option was predicted by experts as the most credible one for each state of economic and technological developments. Developing and expanding the distribution channel seems to be another good strategy to improve the position of the brand, since this option has the second highest probability values across the scenarios. Another finding of the scenario analysis is that, running common advertisement campaigns is more preferable than sales promotions under the current and best scenario. However, for the worst case scenario the reverse is true. Furthermore, a change from the best to the worst case scenario as well as from the current situation to the most probable scenario indicates an apparent worsening of the results in terms of customer perceptions. In addition to these findings, in all scenarios, the strategy not to develop a new product or not to improve current products was predicted to be less probable, which indicates a technology based competition also in the future. Finally, brand image gets the highest expected value signaling the need for a clear and unique positioning in the market.

Table 2. Calculated Probabilities for the Current Situation and Future Scenarios, (%).

	Pricing Policies			Advertisement Campaign		Sales Promotion		Distribution Channel Development		Product Development			Expected Values*			
Scenario	Increase	Decrease	Peg	Do	Do Not	Do	Do Not	Do	Do Not	New	Improve	No Devlp.	Image	Loyalty	Accessibility	Quality
Current	0,9	0,2	98,9	89,4	10,6	84,4	15,6	70,7	29,3	52,2	46,9	0,9	4,06	3,74	3,39	3,44
Future	0,5	0,3	99,2	91,1	8,9	89,4	10,6	71,4	28,6	45,4	53,8	0,8	4,06	3,76	3,41	3,45
Worst	0	2,4	97,6	54,9	45,1	87,4	12,6	41,9	58,1	3,3	85,3	11,4	3,5	3,39	3,17	3,17
Best	1,1	0	98,9	98,4	1,6	86,1	13,9	88,5	11,5	94,9	5,1	0	4,38	3,94	3,53	3,61

* based on a five point rating scale

4. Conclusion

This study proposes a probabilistic model for supporting positioning decisions in the videogame console industry. The results presented above suggest that a BN approach offers significant value in addressing the concerns arising from uncertainty and complex relationships among the external environment, console manufacturers and customers. The analysis of various scenarios provides an important tool for exploring the implications of different assumptions about future trends in terms of positioning strategies of the brands. For further research, especially for different industries, the model can be extended and verified to include other external factors, marketing strategies and perceptual dimensions.

References

1. "A Brief History of Game Console Warfare: Nintendo 64". BusinessWeek. http://images.businessweek.com/ss/06/10/game_consoles/source/12.htm, Accessed on 20.12.2011.
2. Arnott, D.C., 1992: Bases of financial services, Unpublished PhD thesis, Manchester Business School, Manchester, UK.

3. Trommsdorff, V., Asan, U. and Becker J., 2004: Marken- und Produktpositionierung, in M. Bruhn (ed.), Handbuch Markenführung, Vol. 1, 2nd edn., Gabler Verlag, Wiesbaden, pp. 541-570.
4. Chiranjeev, K. and Leuthesser, L., 1993: Product positioning: A comparison of perceptual mapping techniques, *The Journal of Product and Brand Management*, Vol. 2, pp. 10-20.
5. Pearl, J., 1988: Probabilistic Reasoning in Intelligent Systems: Networks of Plausible Inference, Morgan Kaufmann, San Mateo, CA.
6. Chin, K. S., Tang, D. W., Jang, J. B., Wong, S. Y. and Wang, H., 2009: Assessing New Product Development Project Risk by Bayesian Network with a Systematic Probability Generation Methodology, Expert Systems with Applications, Vol. 36, pp. 9879-9890.
7. Jones, B., Jenkinson, I., Yang, Z. and Wang, J., 2010: The Use of Bayesian Network Modelling for Maintenance Planning in a Manufacturing Industry, *Reliability Engineering and System Safety*, Vol. 95, pp. 267-277.
8. Verhoeven, M., Arentze, T.A., Timmermans, H.J.P. & Waerden, P.J.H.J. van der, 2006: Modelling consumer choice behaviour with Bayesian belief networks. In Proceedings of the RARCS Conference, Budapest, Hungary, pp. 18.

AGGREGATION PROCEDURE BASED ON CHOQUET INTEGRAL FOR EVALUATING COMPANY'S ENVIRONMENTAL PRACTICES

R. DE ANDRÉS CALLE
Department of Economics, University of Salamanca, Spain

T. GONZÁLEZ-ARTEAGA[A]
Statistics and Operational Research Department, University of Valladolid, Spain

The need of organizations to evaluate their environmental practices is increasing, but there are many possibilities to do it. In this contribution an aggregation procedure is proposed to evaluate company's environmental practices. This proposal integrates quantitative and qualitative information. Since the assumption of independence among criteria is rarely verified in an environmental context, the Choquet integral is proposed as aggregation operator. The final aim is to compute partial and global indicators that can be used by the management team to make their decisions regarding the environmental issues.

1. Introduction

The ability of organizations to manage their environmental performance is emerging as a strategic issue. Environmentally conscious practices refer to programs to improve processes and products environmental performance in diverse forms like design for environment, recycling, waste management, etc.

It is recognized the importance to include multiple viewpoints of the concerned parties to increase internal efficiency and external legitimacy [1, 6]. However, many companies use traditional methods based just on the opinion of a few reviewers. Then, the results of the evaluation process could not capture the company environmental situation properly. Moreover, there are different types of criteria, quantitative and qualitative but most of the current methods provide only a quantitative crisp values modeling for their assessments [7]. This fact can produce a lack of precision in the assessments provided by the reviewers for qualitative criteria due to the difficulty of expressing vague and uncertain knowledge in a precise way. A usual characteristic of the environmental context

[A] E-mail: teresag@eio.uva.es

is the existence of dependence relationships of criteria [5], although this characteristic does not take into account in the traditional evaluation process.

In order to overcome the shortcomings associated with the previous established models to evaluate environmental practices, in this contribution we propose a method where different reviewers have to evaluate company's environmental practices attending to different criteria (quantitative and qualitative), we then consider a flexible heterogeneous framework in which appraisers may express their assessments by means of numerical or linguistic information according to the criteria nature. As well, we propose the discrete Choquet integral as aggregation operator to address the interdependence among criteria in a flexible way.

2. An evaluating model for companies environmental practices

In order to show how a company could carry out an environmental practices evaluation process, in this section, we present our proposal for evaluating company environmental practices based on a classic decision analysis scheme whose accommodation to our problem is showed in Figure 1.

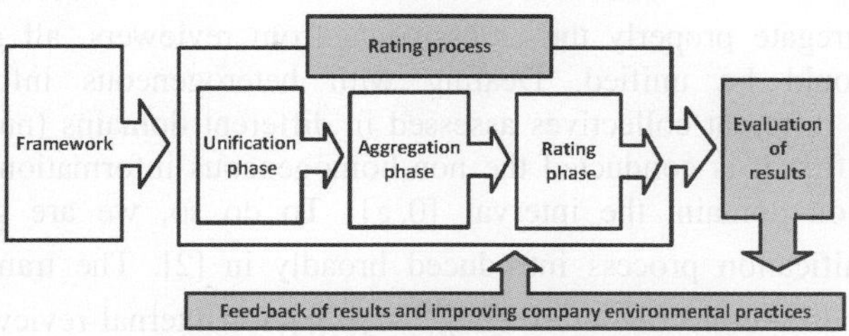

Figure 1. Decision scheme for evaluating company environmental practices

2.1. *Framework*

In this step of the evaluation process, the criteria and the reviewers of the company's environmental practices are identified.

Let us suppose a company which wants to evaluate the environmental practices of its facility sites. It is supposed the company has a set of facility sites $X = \{x_1, \ldots, x_n\}$ to be evaluated by two different collectives: a set of internal reviewers, $A = \{a_1, \ldots, a_r\}$ and a set of external reviewers, $B = \{b_1, \ldots, b_s\}$ attending to two different classes of criteria: management performance criteria, $C^M = \{c_1^M, \ldots, c_p^M\}$ and operational performance criteria, $C^O = \{c_1^O, \ldots, c_q^O\}$. As we have mentioned in introduction, both sets of criteria can have a qualitative or/and quantitative nature. Notice that the same criterion can not be evaluated using different expression domains.

Let $a_{i,j}^{k}$ and $b_{i,j}^{k}$ be the internal and external reviewers' evaluations, respectively, on the facility site x_j in reference to the criterion[a], *k*. Each of these evaluations can belong either to $\Re^{+}$ or to a particular linguistic terms set depending on the specific collective and criterion[b], S_{-}^{k}, that is characterized by its cardinality or *granularity*, $|S_{-}^{k}|$.

2.2. *Ranting process*

To obtain a global environmental performance indicator of each facility site, all information provided by the appraisers must be aggregated. Due to the fact that the framework is heterogeneous this process is carried out in three phases, normalization phase, aggregation phase and rating phase. In the coming subsections we shall present in detail each phase of our model and in order to show how a company or organization could carry out an environmental evaluation process with the proposed model.

2.2.1. *The normalization phase*

In order to aggregate properly the assessments from reviewers, all gathered information should be unified. Dealing with heterogeneous information provided by the different collectives assessed in different domains (numerical, and linguistic), first it is conducted the non-homogeneous information into an unique expression domain, the interval $[0, g]$. To do so, we are going to consider the unification process introduced broadly in [2]. The transformed assessments for each collective are then $\tilde{a}_{i,j,}^{k} \in [0, g]$ for internal reviewers and $\tilde{b}_{i,j,}^{k} \in [0, g]$ for external reviewers.

2.2.2. *Aggregation phase*

Once we have all the gathered information in a common domain, the individual assessments will be aggregated by using aggregation operators. We choose them according to the step of the aggregation phase and considering their properties. As we assume that the reviewers gives their evaluations individually, we propose to use the OWA operator in the first and the second step of our aggregation process, when we need to aggregate the evaluations belonging to several reviewers concerning to the same criterion, because it is anonymous.

---

[a] Abusing notation, on occasions we refer to criterion *k* as c_k^{-}, where superscript denotes the criterion nature.

[b] S_{-}^{k} subscript denotes the reviewers collective.

In such cases the weighting vectors can be computed using the well-known non-decreasing quantifiers proposed by Yager [8]. It is important to note that each aggregation procedure with OWA operators can use a different quantifier. Due to the fact that the Choquet integral is a helpful tool to manage the interdependence among criteria in aggregation procedures, we use Choquet integral as aggregation operator in the rest of the steps of the aggregation process since environmental criteria are interdependent [5, 7]. We now briefly recall some definitions and important needed results for this actual application.

Definition 2.1. Let $C=\{c_1,\dots,c_n\}$ be a finite universe. $P(C)$ is the set of all subsets of C. A *fuzzy measure* is a set function $\mu: P(C)\rightarrow[0,1]$ which satisfies:
1. $\mu(\varnothing)=0$ and $\mu(C)=1$.
2. $A\subseteq B \Rightarrow \mu(A)\le\mu(B)$.

Definition 2.2. Let $f: C\rightarrow\Re^+$ be a function with $y_k=f(c_k)$ for $k\in\{1,\dots,n\}$ and $y_0=0$. The *discrete Choquet integral* of *f* with respect to a fuzzy measure μ is given by

$$F_\mu\left(y_1,\dots,y_n\right)=\sum_{k=1}^{n}\left(y_{(k)}-y_{(k-1)}\right)\mu\left(A_{(k)}\right),$$

where $\left(y_{(1)},\dots,y_{(n)}\right)$ is a non-decreasing permutation of $\left(y_1,\dots,y_n\right)$ and $A_{(k)}=\left\{y_{(k)},\dots,y_{(n)}\right\}$.

Furthermore, it is relevant to point out that the fuzzy measure plays a key role in the Choquet integral comparable with the choice of weights in the weighted mean. The identification of a suitable fuzzy measure to a particular MCDM problem is of the great importance since all the knowledge concerning the criteria is embedded into the fuzzy measure. In order to address this issue different methods have been studied by many authors [3]. We are now in the position to present in further detail each stage of the aggregation process.

1. *Computing collective EPIs for each criterion.* The reviewers' assessments are aggregated for each criterion and each collective by means of a OWA operator, G_-^-. Then, for each collective and for every criterion c_k^-, the process is conducted in the following manner.
   - For internal reviewers: $I_A^k(x_j)=G_A^k(\tilde{a}_{1,j}^k,\dots,\tilde{a}_{r,j}^k)$
   - For external reviewers: $I_B^k(x_j)=G_B^k(\tilde{b}_{1,j}^k,\dots,\tilde{b}_{s,j}^k)$.
2. *Computing EPIs for each criterion.* The previous environmental performance indicators for the x_j facility site are aggregated for each criterion by means of an OWA operator. Let $G^k:[0,g]^2\rightarrow[0,g]$

$$I^k(x_j) = G^k\left(I_A^k(x_j), I_B^k(x_j)\right).$$

3. *Computing a management and operational EPI.* They are obtained by means of aggregating the previous EPIs taking into account the two types of criteria, management and operational, by means of a Choquet integral operator. We propose to determine a fuzzy measure over each set of criteria, management C^M and operational C^O, and then, these fuzzy measures will be used to calculate the associated Choquet integrals, one for the p values coming from the management EPIs and another for the q values coming from the operational EPIs.
   - *Management EPI.* In order to calculate this indicator we aggregate for management criteria by means of $F^M : [0, g]^p \to [0, g]$

     $$I^M(x_j) = F^M(I^1(x_j), \ldots, I^p(x_j)).$$

   - *Operational EPI.* Analogously to previous performance indicator, an operational performance indicator is computed: $F^O : [0, g]^q \to [0, g]$

     $$I^O(x_j) = F^O(I^1(x_j), \ldots, I^q(x_j)).$$

4. *Computing a global EPI.* At this point we have for each facility site two partial EPIs, one from management criteria and another from operational criteria. Since there is interdependence between them, we propose to use another Choquet integral operator. It implies that another fuzzy measure is necessary. $F : [0, g]^2 \to [0, g]$

   $$I(x_j) = F(I^O(x_j), I^M(x_j)).$$

2.2.3. *Rating phase. Evaluation of the results*

In the rating phase, the management team shall classify and order facility sites according to the environmental performance indicators obtained in the previous phase. One of the most important facets of an environmental practices evaluation process is providing performance feedback to facility sites. In this way, manager team knows comprehensive results in the process evaluation and they can improve company's environmental practices.

3. Concluding remarks

In this contribution we have presented an aggregation procedure based on Choquet integral in a MCDM model for evaluating the company's environmental practices. The considered model not only obtains a global environmental performance indicator for each facility site, but also it obtains intermediate environmental performance indicators according to the opinions of each set of reviewers and criterion, and a management and also operational performance indicators. Besides, we deem the Choquet integral as an aggregation operator so that the interdependence among environmental criteria are taking into account in the presented model when several criteria have to be aggregated.

References

1. N. Darnall, I. Seol, J. Sarkis. Perceived stakeholder influences and organiza-tions' use of environmental audits. *Account., Organ. Soc.*,34:170–187,2009.
2. R. de Andrés, T. González-Arteaga and L. Martínez. A MCDM approach for evaluating company's environmental practices. *EUSFLAT. Advances in Intelligent Systems Research*, 1:912–919, 2011.
3. M. Grabisch, I. Kojadinovic, and P. Meyer. A review of methods for capacity identification in Choquet integral based multi-attribute utility theory: Applications of the kappalab R package. *European Journal of Operational Research*, 186(2):766–785, April 2008.
4. F. Herrera, L. Martínez, and P.J. Sánchez. Managing non-homogeneous information in group decision making. *European Journal of Operational Research,* 166:115–132, 2005.
5. M.L.Tseng, A.S.F. Chiu. Evaluating firm's green supply chain management in linguistic preferences. *Journal of Cleaner Production*, in press:1–10, 2010.
6. D. Nawrocka, T. Parker. Finding the connection: environmental management systems and environmental performance. *Journal of Cleaner Production,* 17:601–607, 2009.
7. J. Sarkis. A strategic decision framework for green supply chain management. *Journal of Cleaner Production*, 11:397–409, 2003.
8. R.Yager. On ordered weighted averaging operators in multicriteria decision making. *IEEE Transactions on Systems, Man, Cybernetics*, 18:183–190, 1988.

A FUZZY LINGUISTIC EVALUATION MODEL OF DIGITAL LIBRARIES BASED ON THE LIBQUAL+ METHODOLOGY

F.J. CABRERIZO*

Dept. of Software Engineering and Computer Systems
Distance Learning University of Spain (UNED), Madrid, 28040, Spain

M.A. MARTÍNEZ, J. LÓPEZ-GIJÓN, E. HERRERA-VIEDMA‡

Dept. of Library Science
‡Dept. of Computer Science and Artificial Intelligence
University of Granada, 18071, Granada, Spain

According to the gap model, service quality is the gap between user's expectations and perceptions. When experiences exceed expectations, the quality of the service is high, and vice versa. Following that idea, we present a fuzzy linguistic quality evaluation model based on the LibQUAL+ methodology to evaluate the quality of digital libraries according to user's opinions. To derive the gaps essential for measuring perceptions of service quality, users are asked to establish their judgments across three scales for each digital service of the digital library. It allows the identification of digital services in which service levels should be improved and the identification of digital services satisfied outstandingly by the digital library.

1. Introduction

The World Wide Web is nowadays the main information repository. Its explosive growth has stimulated the development of fast and effective automated systems that support an easy and effective access to the information relevant to specific users' needs.[1] Digital libraries (DLs) are one of these automated systems.

DLs may be defined as information collections that have associated services delivered to user communities using a variety of technologies. Library users are now offered a variety of resources with different forms of interactivity and with different levels of media richness. They can obtain research data and publications as needed without the massive investment of capital and infrastructure to house vast physical collections. As a DL is destined to serve users, if unused, it falls into oblivion and terminates its operation. Hence, one of

* cabrerizo@issi.uned.es.

the main aspects to be considered in DL evaluation is user's perspective, determining the extent to which the DL addresses the real needs of its users.[2]

Cabrerizo et al.[3] presented a model based on fuzzy linguistic information to evaluate the quality of DLs. They defined the quality evaluation of DLs using users' perceptions on the quality of digital services provided through their Web sites. This model provides both the quality assessment of the users on each subjective criterion and the global quality assessment of the users on the DL. However, it does not allow the identification of areas in which service levels should be improved according to the user's expectations. To overcome this drawback, the LibQUAL+ model can be used.[4] LibQUAL+ is based on the well-known, attribute-based gap model SERVQUAL (for SERVice QUALity) which was developed for the for-profit sector in the 1980s by the marketing research group of Parasuraman, Zeithaml, and Berry.[5] Grounded in the Gap Theory of Service Quality, the singular percept of SERVQUAL is that "only customers judge quality; all other judgments are essentially irrelevant".[6] According to the gap model, service quality is the gap between user's expectations and perceptions. When experiences exceed expectations, the quality of the service is high, and vice versa. Following that idea, LibQUAL+ is a survey administered by the Association of Research Libraries to measure library user's perception of library service quality and to help traditional libraries identify service areas needing improvement. The survey data allow the identification of areas in which service levels should be improved, and they have also been used to identify best practices and reallocate resources accordingly.

The aim of this paper is to present a quality evaluation model of DLs, based on the LibQUAL+ methodology, which allows obtaining the digital services which should be improved and the digital services satisfied outstandingly by the DL. To do so, users are asked to indicate their minimum acceptable service level, their desired service level, and the perception of the actual service provided by each digital service of the DL. We assume a fuzzy linguistic modeling to represent the user's opinions and apply tools of fuzzy computing with words based on the LOWA operator[7] to compute the global assessments.

The rest of the paper is set out as follows. Section 2 presents the ordinal fuzzy linguistic modeling for computing with words. Section 3 describes the evaluation model of DLs and Section 4 draws our conclusions.

2. A Fuzzy Linguistic Approach for Computing with Words

The ordinal fuzzy linguistic approach[7] is a very useful kind of fuzzy linguistic approach used for modeling the computing with words process as well as linguistic aspects of the problems. It is defined by considering a finite and totally ordered label set, $S = \{s_i\}, i \in \{0, \ldots, T\}$, in the usual sense, i.e., $s_i \geq s_j$ if $i \geq j$, and with odd cardinality (7 or 9 labels). The mid-term represents an assessment of "approximately 0.5", and the rest of the terms being placed symmetrically around it. The semantics of the label set is established from the ordered structure of the label set by considering that each label for the pair (s_i, s_{T-i}) is equally informative. For example, we can use the following set of nine labels to provide the user's opinions: {T = Total, EH = Extremely High, VH = Very High, H = High, M = Medium, L = Low, VL = Very Low, EL = Extremely Low, N = None}.

In any linguistic approach we need management operators of linguistic information.[7] Usually, the ordinal fuzzy linguistic model for computing with words is defined by establishing (i) a negation operator, $Neg(s_i) = s_j \mid j = T - i$, (ii) comparison operators based on the ordered structure of linguistic terms: maximization operator, $MAX(s_i, s_j) = s_i$, if $s_i \geq s_j$; and minimization operator, $MIN(s_i, s_j) = s_i$, if $s_i \leq s_j$, and (iii) adequate aggregation operators of ordinal fuzzy linguistic information.

The Linguistic Ordered Weighted Averaging (LOWA) is an operator used to aggregate non-weighted ordinal linguistic information, i.e., linguistic information values with equal importance. It is defined as follows:[7]

Definition 2.1. Let $A = \{a_1, \ldots, a_m\}$ be a set of labels to be aggregated, then the LOWA operator, ϕ, is defined as $\phi(a_i, \ldots, a_m) = W \cdot B^T = C^m\{w_k, b_k,\ k = 1, \ldots, m\} = w_1 \odot b_1 \oplus (1 - w_1) \odot C^{m-1}\{\beta_k, b_k,\ k = 2, \ldots, m\}$, where $W = [w_1, \ldots, w_m]$ is a weighting vector, such that, $w_i \in [0,1]$ and $\sum_i w_i = 1$. $\beta_h = w_h / \sum_2^m w_k, h = 2, \ldots, m$, and $B = \{b_1, \ldots, b_m\}$ is a vector associated to A, such that, $B = \sigma(A) = \{a_{\sigma(1)}, \ldots, a_{\sigma(m)}\}$, where, $a_{\sigma(j)} \leq a_{\sigma(i)}\ \forall\, i \leq j$, with σ a permutation over the set of labels A. C^m is the convex combination operator of m labels and if $m = 2$, then it is defined as $C^2\{w_i, b_i,\ i = 1,2\} = w_1 \odot s_j \oplus (1 - w_1) \odot s_i = s_k$, such that, $k = min\{T, i + round(w_1 \cdot (j - i))\}, s_j, s_i \in S, (j \geq i)$, where "round" is the usual round operation, and $b_1 = s_j, b_2 = s_i$. If $w_j = 1$ and $w_i = 0$ with $i \neq j\ \forall i$, then the convex combination is defined as: $C^m\{w_i, b_i,\ i = 1, \ldots, m\} = b_j$.

The LOWA operator is an "or-and"' operator and its behavior can be controlled by means of W. In order to classify OWA operators with regards to their localization between "or" and "and", Yager[8] introduced a measure of *orness*, associated with any vector $W : orness = \frac{1}{m-1}\sum_{i=1}^{m}(m - i)w_i$. This measure characterizes the degree to which the aggregation is like an "or" (MAX) operation. Note that an OWA operator with $orness(W) \geq 0.5$ will be an *orlike*, and with $orness(W) < 0.5$ will be an *andlike* operator.

An important question of the LOWA operator is the determination of the weighting vector W. In [8], it was defined an expression to obtain W that allows to represent the concept of fuzzy majority[9] by means of a fuzzy linguistic nondecreasing quantifier Q:[10]

$$w_i = Q(i/n) - Q((i-1)/n), i = 1, \ldots, n. \quad (1)$$

When a fuzzy linguistic quantifier Q is used to compute the weights of the LOWA operator, ϕ, it is symbolized by ϕ_Q.

3. A Fuzzy Linguistic Evaluation Model of DLs Based On the LibQUAL + Methodology

In order to improve the quality evaluation model of DLs presented in [3], the LibQUAL+ methodology is used. Cabrerizo et al.[3] defined an evaluation scheme of DLs that contemplated four quality dimensions together with their digital quality criteria. In particular, we use the following ten subjective criteria proposed in [3] to measure the quality of the digital services provided by a DL:

- sc_1: You find what you are looking for.
- sc_2: Coverage of the DL about search topics.
- sc_3: Information electronic services about new inputs.
- sc_4: Variety of search tools.
- sc_5: Navigability of the DL Web site.
- sc_6: Understandability of the DL Web site.
- sc_7: Added value information profits.
- sc_8: Satisfaction degree with the computing infrastructure.
- sc_9: Satisfaction degree with the response time.
- sc_{10}: Training received

It is assumed that to measure the users' perceptions on the quality of digital services provided by the DLs through their Web site, users are invited to fill a questionnaire which has 10 questions, $\{q_1, \ldots, q_{10}\}$, where each question corresponds with a subjective criterion. To adapt the quality evaluation model of DLs presented in [3] according to the LibQUAL+ methodology, users are asked for impressions about the above ten subjective criteria according to (i) the minimum service level that they are willing to accept, (ii) the desired service level that they would like to receive, and (iii) the perceived performance level, that is, the actual level of service that they perceive to have been rendered.

The minimum service level, the desired service level, and the perceived performance level behind each question are rated on a linguistic term set S. Once the group of users, $\{e_1, \ldots, e_L\}$, have filled all the questionnaires for a given

DL, A_m, the system calculates for each subjective criterion, sc_i, its global quality assessment of the minimum service level, MSL_i^m, its global quality assessment of the desired service level, DSL_i^m, and its global quality assessment of the perceived performance level, PPL_i^m, by aggregating the opinions provided by the group of users on the subjective criterion, sc_i, by means of the LOWA operator ϕ:

$$MSL_i^m = \phi_Q\left(e_1(q_i^{MSL}), \ldots, e_L(q_i^{MSL})\right), \tag{2}$$

$$DSL_i^m = \phi_Q\left(e_1(q_i^{DSL}), \ldots, e_L(q_i^{DSL})\right), \tag{3}$$

$$PPL_i^m = \phi_Q\left(e_1(q_i^{PPL}), \ldots, e_L(q_i^{PPL})\right), \tag{4}$$

where $e_l(q_i^{MSL}) \in S$ is the minimum service level provided by the user e_l on subjective criterion, sc_i, represented by the question q_i, $e_l(q_i^{DSL}) \in S$ is the desired service level provided by the user e_l on subjective criterion, sc_i, represented by the question q_i, and $e_l(q_i^{PPL}) \in S$ is the perceived performance level provided by the user e_l on subjective criterion, sc_i, represented by the question q_i. MSL_i^m, DSL_i^m and PPL_i^m, are the linguistic measures that represents the minimum service level, the desired service level and the perceived performance level, respectively, of the DL, A_m, with respect to subjective criterion, sc_i, according to the majority (represented by the fuzzy linguistic quantifier Q) of linguistic evaluation judgments provided by the group of users. According to the users' answers, it is defined two scores which can obtain the strengths and weaknesses of a DL, A_m:

$$SA_i^m = \mathrm{I}(PPL_i^m) - I(MSL_i^m), \tag{5}$$

$$SS_i^m = \mathrm{I}(PPL_i^m) - I(DSL_i^m), \tag{6}$$

where SA_i^m is the service adequacy score on the subjective criterion, sc_i, of the DL, A_m, SS_i^m is the service superiority score on the subjective criterion, sc_i, of the DL, A_m, and $I : S \rightarrow \{0, \ldots, T\} \mid I(s_i) = i, \forall i \in S$.

The service adequacy score is calculated by subtracting the minimum score from the perceived score on any given question. It is an indicator of the extent to which DLs are meeting the minimum expectations of their users. A negative service adequacy gap score indicates that the users' perceived level of service quality is below their minimum level of service and can be used by DLs to identify areas needing improvement. The service superiority score is calculated by subtracting the desired score from the perceived score on any given question and is an indicator of the extent to which DLs are exceeding the desired expectations of their users.

4. Concluding Remarks

We have presented a fuzzy linguistic model based on the gap theory of service quality to evaluate the quality of the DLs. In the future, we extend the concept of quality to other new services of the DLs, such as Library 2.0.

Acknowledgments

This paper has been developed with the financing of FEDER funds in FUZZYLING-II Project TIN2010-17876, Andalusian Excellence Projects TIC-05299 and TIC-5991, and "Proyecto de Investigación del Plan de Promoción de la Investigación UNED 2011 (2011V/PUNED/0003)".

References

1. S. Lawrence and C. Giles, Searching the World Wide Web, *Science* **280**, 98 (1998).
2. A. Shiri, Metadata-enhanced visual interfaces to digital libraries, *Journal of Information Science* **34**, 763 (2008).
3. F.J. Cabrerizo, J. López-Gijón, A.A. Ruiz and E. Herrera-Viedma, A model based on fuzzy linguistic information to evaluate the quality of digital libraries, *Int. Journal of Inf. Technology & Decision Making* **9**, 445 (2010).
4. C. Cook, F. Heath, M. Kyrillidou, Y, Lincoln, B. Thompson and D. Webster, *NDSL Evaluation Workshop* (2003).
5. A. Parasuraman, V.A. Zeithaml and L.L. Berry, SERVQUAL: A multiple-item scale for measuring consumer perceptions of service quality, *Journal of Retailing* **64**, 12 (1988).
6. V.A. Zeithaml, L.L. Berry and A. Parasuraman, *Delivering Quality Services - Balancing Customer Perceptions* (The Free Press, New York, 1990).
7. F. Herrera, E. Herrera-Viedma and J.L. Verdegay, Direct approach processes in group decision making using linguistic OWA operators, *Fuzzy Sets and Systems* **79**, 175 (1996).
8. R.R. Yager, On ordered weighted averaging aggregation operators in multicriteria decision making, *IEEE Transactions On Systems Man and Cybernetics* **18**, 183 (1988).
9. J. Kacprzyk, Group decision making with a fuzzy linguistic majority, *Fuzzy Sets and Systems* **18**, 105 (1986).
10. L.A. Zadeh, A computational approach to fuzzy quantifiers in natural languages, *Computers & Mathematics with Applications* **9**, 149 (1983).

A LOCATION-AWARE TOURISM RECOMMENDER SYSTEM BASED ON MOBILE DEVICES

jOSÉ M. NOGUERA*, MANUEL J. BARRANCO, RAFAEL J. SEGURA, LUIS MARTÍNEZ

Department of Computer Sciences, University of Jaén,
Campus Las Lagunillas s/n. Jaén, 23071, Spain

Recommender systems (RS) have been successfully used in different personalization issues. One of the most interesting applications is tourism (hotels, restaurants, monuments, etc.) where users can obtain personalized recommendations according to their profiles. Recently, the increasing use of mobile devices drives RSs to a new trend based on context-awareness. In this contribution we take advantage of mobile devices in RSs and propose a location-aware RS for tourism that provides recommendations to users not only based on their profile but also taking into account their current location.

1. Introduction

Personalization [6] has recently been recognized by experts as a critical factor of e-tourism companies to be successful. Up to date, recommender systems (RS) are the most successful personalization technique [1]. These systems lead users to suitable products by using different information filtering techniques, hence limiting the effects of information overload.

Recently, the fast development and adoption of ubiquitous computing technologies has fostered a considerable interest in the field of recommender systems for mobile devices. This kind of devices (e.g. smartphones, tablets, etc.) provide some unique features that are very appealing for e-tourism, namely, their ubiquity and the possibility of introducing context-awareness capabilities in their recommendations [11].

In this contribution we aim at exploiting these features by presenting a new mobile location-aware RS. This proposal emerges from a complete revision of our previous classic recommender system REJA [12] for restaurants that will be adapted to an ubiquitous environment. First, we extend our system to include user's context information such as distance from the user's current location to

* {jnoguera, barranco, rsegura, martin}@ujaen.es

the points of interest (POIs) considered in the recommendation process. Second, we employ a mobile-friendly user interface for providing georeferenced recommendations on a 2D map.

The rest of the paper is organized as follows. Section 2 provides a necessary background about recommender systems. Section 3 presents our proposal of a location-aware hybrid recommender system designed for mobile tourism. Finally, Section 4 concludes the paper.

2. Preliminaries

Recommender Systems assist people to find out suitable products in e-shops with large databases of items according to their necessities and tastes. There exist different types of RSs depending on the information source and the technique used to rank items. According to recent research, collaborative filtering [5] is the most successful recommendation technique up to date [8, 10]. These systems use customer's ratings to categorize users in groups according to their affinity. Recommendations are then inferred by taking into account ratings of users' in the same group [1].

Despite the success of such techniques, they still present some weakness, mainly the cold start problem [9]. This problem refers to the difficulty to either make recommendations to new users or to recommend new items. Different proposals have been developed to smooth such problem [3, 4, 7, 12].

The advent of ubiquitous computing has motivated a new trend of research in this field. Context-aware recommender systems (CARSs) take full advantage of the unique attributes provided by mobile computing to provide users with relevant information according to their physical context and preferences. This is a very active research field, and we refer the reader to the recent surveys published in [2, 11]. Also, the nature of handheld devices arises unprecedented usability issues that require new technical solutions to tackle the small display sizes and limited input technologies of these devices [11].

3. A collaborative Location-aware Recommender System

In this work, we extend the REJA schema [12] from a web-based system to a map-based mobile location-aware RS. Originally, REJA was formed by a commutation hybridizing system with collaborative and knowledge filtering. Our aim is to integrate context information based on user's location following two different approaches [2], as shown in Figure 1:

1. *Location-aware recommender system:* First, a contextual pre-filter is used to reduce the number of items considered for the recommendation according

to the user's location. These items are used by our RS to generate a top-N list of items potentially interesting for the user.

2. *Distance based re-ranking:* A contextual post-filter re-ranks the previous top-N list according to the physical distance from the user to each item.
3. Following sections present in further detail these two phases, and also describe our client-server implementation and the proposed user interface.

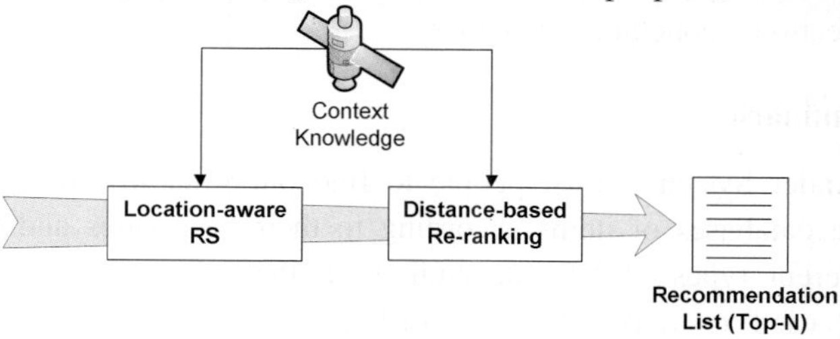

Figure 1. Location-aware Recommender System.

3.1. *Location-aware Recommender System*

We propose a contextual pre-filter that reduces the number of items considered by the RS to provide recommendations. When tourists are visiting a scenic place, they usually move around a limited area that will depend on their transportation (walking, car, bus, etc.). Our system fixes a circular area of radius R_{out} centered on the user's current location. This radius establishes the limits where the items might be recommended. This parameter that can be either user-defined or automatically inferred according to the user's form of transportation.

We assume that our RS deals with a set of recommendable items, $A=\{a_1,...,a_n\}$, geographically located. Therefore, the parameter R_{out} will be used to compute a subset, $A' \subseteq A$, that includes those items that could be suitable to be recommended to the user according to her current location, and ignoring the remaining ones because they are too far away from the user.

This set *A'* of nearby items is provided to a collaborative-based RS. This system groups all registered users according to their preferences and then computes a prediction for every item $a_i \in A'$ that has not been yet experienced by the user. Following, these predictions are used to generate a top-N list of items located close to the user's current location that can be potentially interesting for her. However, if the resulting list is empty or too short (e.g., the user is new and therefore could not be grouped), we are under a cold-start situation. In this case, our system discards this list and provides a new one composed of all the items of A'. In either case, the resulting list is sent to the following phase of our system, see Section 3.2.

3.2. *Distance-based Re-ranking*

In this phase, we apply a contextual post-filtering process that re-ranks the list of top-N recommended items provided by the previous phase using contextual information, in our case, the distance from the items to the user.

Specifically, this filter works as follows. Let $RL=\{(a_i,v_i)\ ,\ i=1,...,N\}$ be a recommendation list of N items where v_i is an estimation of the interest rate of the item a_i for the target user. The distance-based re-ranking filter receives this list as an input and modifies the interest rates v_i according to the distance from each item a_i to the user. The new interest rate after the distance-based re-ranking, w_i, is computed as: $w_i=v_iDD(u,a_i)$, where $DD(u,a_i)$ is the *distance decay function* for the item a_i and the user u. This decay function depends on the parameter R_{out} and may have different forms. We propose a simple to compute lineal decay function as follows:

$$DD(u,a_i) = \begin{cases} 0\ if distance(u,a_i) \geq R_{out} \\ 1 - \dfrac{x}{R_{out}} if distance(u,a_i) < R_{out} \end{cases} \tag{1}$$

Therefore, the recommendation list will be re-ranked according to the new values w_i, in descending order, giving a new list $RL'=\{(a_i,w_i)\ ,\ i=1,...,N\}$. This final list will be shown to the user.

3.3. *Software Design and User Interface*

User interface is an important issue when targeting mobile devices because of the limited screen size and user input. Here, we describe a mobile application that implements the recommendation model proposed in previous sections. Figure 2 shows some screenshots of our prototype under iOS, whose goal is to recommend restaurants in the province of Jaén (Spain).

After start-up, our mobile application shows the user a 2D map centered around its current physical location. The *Google Maps* API is used for this purpose. The location is obtained from the built-in GPS receiver currently included in most mobile devices. A network connection is then established with a remote web-server that implements the proposed hybrid RS described in Section 3.1. Both the user's *id* and current *location* are provided to the server, which in response sends a selection of nearby restaurants depending on the user's profile and location. The initial list of items provided by the server is then re-ranked by the mobile device as explained in Section 3.2. Finally, these items are portrayed on the map, see Figure 2a. The recommendation list is updated in *real-time* as the user moves around the environment.

a b c

Figure 2.Screenshots of the implemented mobile recommender system.

Touching a POI displays its recommendation value w_i according to a 5-stars scale. A descriptive screen with further information about the restaurant is also provided, which allows the user to emit votes, see Figure 2b. The user is also allowed to select a R_{out} value according to her desires, as shown in Figure 2c.

4. Concluding Remarks

Mobile devices are becoming a primary tool for accessing to information services and e-tourism. In this paper we have described the design of a ubiquitous location-aware recommender system tailored to mobile touristic purposes. Our proposal allows on-the-move tourists to download on-demand recommendations on their mobile devices according to their current location and personal tastes. Information is provided by means of a user-friendly interface based on a map.

Acknowledgments

This work has been partially supported by the *Ministerio de Ciencia e Innovación* (Spain) and the European Union (via ERDF funds) through the research project TIN2009-08286; and by the *Junta de Andalucía* (Spain) and the European Union (via ERDF funds) through the research project P07-TIC-02773.

References

1. G. Adomavicius and A. Tuzhilin. Toward the next generation of recommender systems: a survey of the state-of-the-art and possible extensions. *IEEE Transactions on Knowledge and Data Engineering*, 17(6):734–749, 2005.

2. G. Adomavicius and A. Tuzhilin.Context-aware recommender systems. In F. Ricci, L. Rokach, B. Shapira, and P. B. Kantor, editors, *Recommender Systems Handbook*, pages 217–253. Springer US, 2011.
3. H. J. Ahn. A new similarity measure for collaborative filtering to alleviate the new user cold-starting problem. *Information Sciences*, 178:37–51, 2008.
4. M. Díez and P. Villegas. Automatic recommendations for machine-assisted multimedia annotation: A knowledge-mining approach. *In 2nd Int. Conf. on Semantic and Digital Media Technologies*, pp. 95–98, Genova, Italy, 2007.
5. D. Goldberg, D. Nichols, B. M. Oki, and D. Terry. Using collaborative filtering to weave an information tapestry. *Communications of the ACM*, 35(12):61–70, 1992.
6. A. Goy, L. Ardissono, and G. Petrone. The adaptive WEB, chapter Personalization in E-Commerce Applications, page 485520. *Lecture Notes in Computer Science* 4321 Springer, 2007.
7. 7. H. H.-N. Kim, A.-T.Ji, and G. Jo. Error-based collaborative filtering algorithm for top-n recommendation. In *9th Asia-Pacific Web Conf./ 8th Int. Conf. on Web-Age Information Management*, pages 594–605, Huang Shan, Peoples R China, 2007.
8. N. Manouselis, R. Vuorikari and F.V. Assche, Collaborative recommendation of e-learning resources: an experimental investigation. *Journal of Computer Assisted Learning*, 26:227-242, 2010.
9. P. Massa and B. Bhattacharjee. Using trust in recommender systems: an experimental analysis. In *2nd Int. Conf. on Trust Management*, pages 221–235, Oxford, England, 2004.
10. J. Redpath, D. H. Glass, S. McClean, and L. Chen. Collaborative filtering: The aim of recommender systems and the significance of user ratings. In *ECIR 2010*, LNCS 5993, pages 394–406, 2010.
11. F. Ricci. Mobile recommender systems. *International Journal of Information Technology and Tourism*, 12(3):205–231, 2011.
12. R. Rodríguez, M. Espinilla, P. Sánchez, and L. Martínez. Using linguistic incomplete preference relations to cold start recommendations. *Internet Research*, 20:296–315, 2010.

WEB BASED SUPPORT SYSTEM FOR INTEGRAL PERFORMANCE APPRAISAL*

M. ESPINILLA*†, F.J. MARTÍNEZ, L. MARTÍNEZ

Department of Computer Science, University of Jaén, Campus Las Lagunillas s/n
Jaén, 23071, Spain

R. DE ANDRÉS CALLE

University of Salamanca, Edificio FES
Salamanca, 37007, Spain

Integral performance appraisal process is used as a key tool for enterprise competitiveness. It considers different indicators to provide evaluation assessments about employees' performance based on the judgment of different groups of reviewers, who socialize with employees. This contribution presents a Web based Support System for Integral Performance Appraisal, so-called WSSIPA, which implements an integral performance appraisal model that offers a heterogeneous framework in which reviewers can provide their judgments within different domains (numerical, interval-valued and linguistic), according to the nature of criteria and the background of each reviewer. In this contribution, we show the functionality of WSIIPA, conducting a simple case study for an integral performance appraisal process.

1. Introduction

Performance appraisal is a common activity in companies whose aim is to evaluate and analyze employees' capacity to accomplish effectively their work1,2. Currently, integral performance appraisal processes2 are based on the opinion of different groups of reviewers who socialize with employees, since they can truly respond to how an employee develops his/her job. Moreover, this process includes the opinion of employee about her/himself. *Integral evaluation* or *360-degree performance appraisal* [2] overcomes some disadvantages[2] from traditional evaluation such as lack of objectivity, prejudice or halo errors, etc.

The criteria involved in integral performance appraisal processes may have different nature, quantitative or qualitative, and the provided judgments can be

---

* This paper has been partially supported by Projects TIN2009-08286, P08-TIC-3548, ECO2009-17332

† mestevez@ujaen.es.

vague or uncertain. Therefore, it seems necessary and appropriate to provide a heterogeneous evaluation framework in which the involved criteria can be assessed in different expression domains and the final results expressed by linguistic terms to facilitate their understanding.

In this contribution, we present WSSIPA, a Web based system that supports integral performance appraisal processes able to manage a heterogeneous evaluation framework, such that reviewers can provide their judgments within different expression domains (numerical, interval-valued and linguistic), according to the nature of the criteria and the background of each reviewer. Furthermore, in order to offer understandable and interpretable results, WSSIPA presents them by linguistic information. To do so, WSSIPA implements the integral performance appraisal model proposed in [3] that offers these desirable features. The functionality of WSSIPA is shown, conducting a simple case study for an integral performance appraisal process in order to illustrate the usefulness and effectiveness of the system.

This contribution is structured as follows: Section 2 reviews the general scheme of the integral performance appraisal model that we develop in WSSIPA. Section 3 presents the WSSIPA functionality, carrying out a simple case study for an integral performance appraisal process with a heterogeneous evaluation framework. Finally, in Section 4, conclusions are drawn.

2. A Heterogeneous Integral Performance Appraisal Model

Our proposal in this contribution is to develop in WSSIPA an integral performance appraisal model to deal with heterogeneous information that provides linguistic results. Such a model was introduced [3] whose general scheme is shown in Figure 1. To manage heterogeneous information, this model utilizes the heterogeneous computing approach presented in [4].

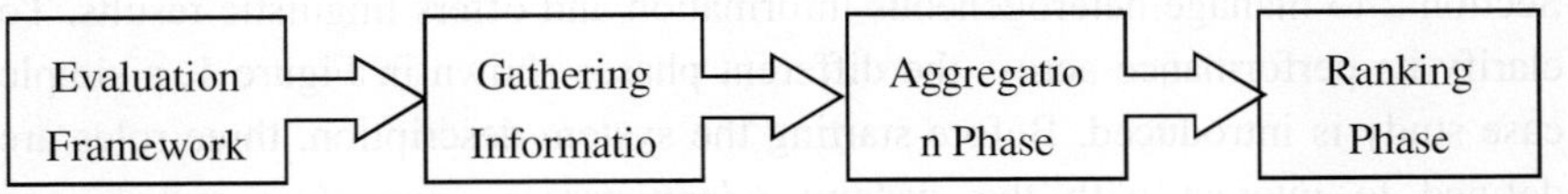

Figure 1: General scheme of the integral performance appraisal model.

a) Evaluation Framework. It fixes the structure of the integral performance appraisal process by defining the main features and terminology of this process in which personnel are evaluated from different points of view. In this case, it defines a heterogeneous context in which reviewers can use different expression domains to assess each criterion.

b) Gathering Information: In this phase, reviewers of the different collectives provide their judgments about the employees, regarding the set of evaluated criteria.

c) Aggregation Phase: The aim of this phase is to compute global values about performance of the set of evaluated employees, regarding to the different criteria and reviewers' collectives. In order to manage the heterogeneous framework, this model follows the approach presented in [4]:

- Choosing a linguistic domain, S_T, to unify the heterogeneous information. This linguistic domain is selected with the aim of keeping as much knowledge as possible. This linguistic domain is a key because evaluation results will be expressed in it.
- Transformation of the information into fuzzy sets in S_T. Each numerical, interval-valued and linguistic judgment is transformed into a fuzzy set on the S_T, by using different transformation functions.
- Transformation of fuzzy sets into linguistic 2-tuples in S_T. In order to simplify the computations and improve the understanding of the results, the fuzzy sets in S_T are transformed into linguistic 2-tuples[5] in S_T.
- Aggregation step. Here, the linguistic 2-tuple computational model is used to aggregate the unified information to obtain global values about the performance of each evaluated employee.

d) Ranking Phase: Finally, the global values expressed in linguistic 2-tuples are sorted to obtain a ranking of employees.

3. Web Based Support System for Integral Performance Appraisal

In this section, we present WSSIPA[‡] that implements the model reviewed in Section 2 to manage heterogeneous information and offers linguistic results. To clarify its performance across the different phases shown in Figure 1, a simple case study is introduced. Before starting the system description, three roles are defined to interact with the system: *administrator*, *appraisers* and *human resources*.

a) Evaluation Framework. *Administrators* define the evaluation framework. In our case study, it includes two employees that will be evaluated by one supervisor, two collaborators, three customers and themselves, according to four criteria. Two criteria are evaluated in a numerical domain and the other two

---

[‡] The system is hosted at the following URL: http://serezade.ujaen.es:8080/mspev2

criteria are evaluated in a linguistic domain with five labels. Obviously, all reviewers do not evaluate all the criteria because some of them do not have sufficient knowledge. Therefore, each evaluator assesses the corresponding criteria with his/her knowledge.

First, the set of criteria and their expression domains are defined in a survey (see Figure 2 and 3), which is associated with an integral performance appraisal process for a set of evaluated employees. The system automatically selects the set of reviewers, who socialize with each employee (see Figure 4). Finally, the system allows manually add or remove any reviewer of the set.

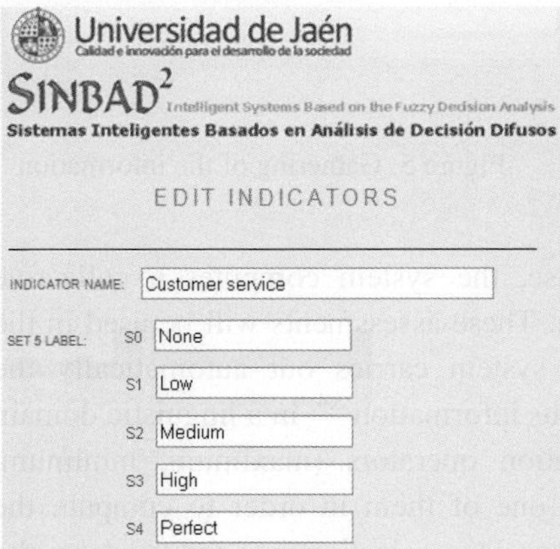

Figure 2: Definition of an indicator

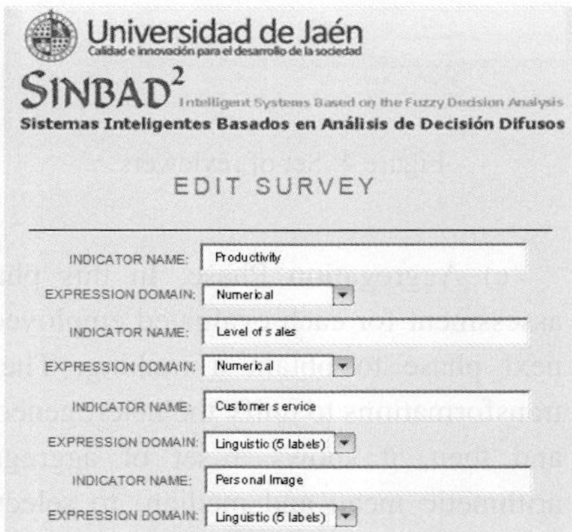

Figure 3: Definition of a survey

b) Gathering Information. The judgments about the employees are provided by *appraisers*. When a user with this role is logged in the system, it shows surveys that he/she should complete. Each survey gathers information from an evaluated employee, according to the criteria defined in the evaluation framework (see Figure 5).

In order to ensure an organization, an *administrator* may impose a deadline for the gathering of the information; when the deadline ends; the reviewers cannot complete or modify their associated surveys.

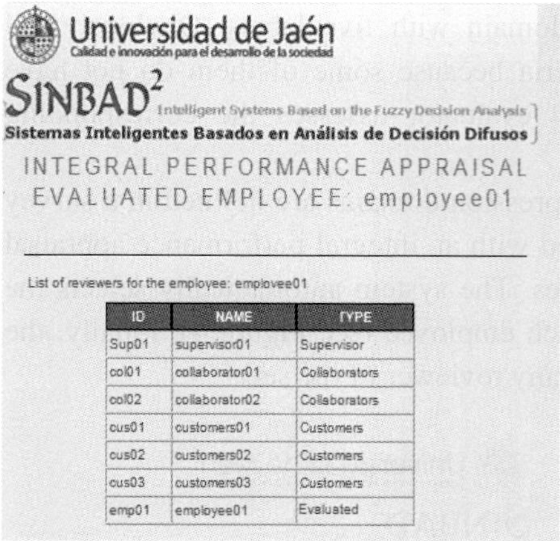

Figure 4: Set of reviewers

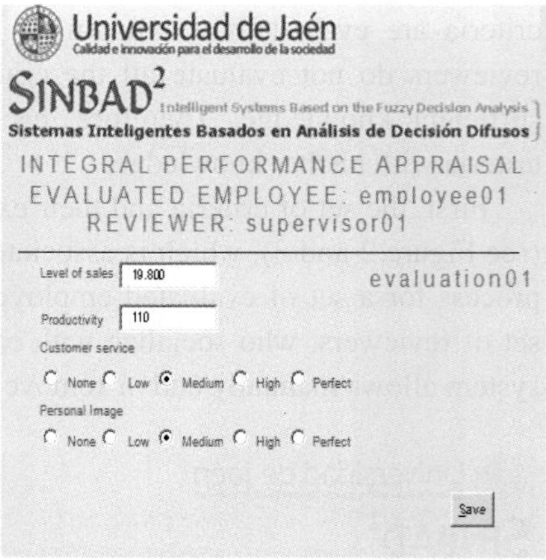

Figure 5: Gathering of the information

c) Aggregation Phase. In this phase, the system computes a collective assessment for each evaluated employee. These assessments will be used in the next phase to obtain a ranking. The system carries out automatically the transformations to unify the heterogeneous information [3,4] in a linguistic domain and then, it shows a set of aggregation operators (maximum, minimum, arithmetic mean and median) to select one of them in order to compute the collective assessment for each employee. This selection is conducted by the *human resources* user. WSSIPA allows multiple resolutions, using different aggregation operators. Despite different options, in our case study, the arithmetic mean operator is selected to aggregate the unified information in a linguistic domain (see Figure 6). The obtained results expressed in linguistic 2-tuples are the following: *(High,-0.2)* for the first employee and *(Medium,0.2)* for the second employee. We can check that these results are close to human natural language in order to be understandable and interpretable by multiple members of the company.

d) Ranking Phase. When the global values of the set of evaluated employees are computed, WSSIPA sorts these values in order to show a ranking of employees to the *human resources* users. In our case study, the employee with the first ranking position corresponds to the first employee with a global values of *(High,-0.2)* (see Figure 7).

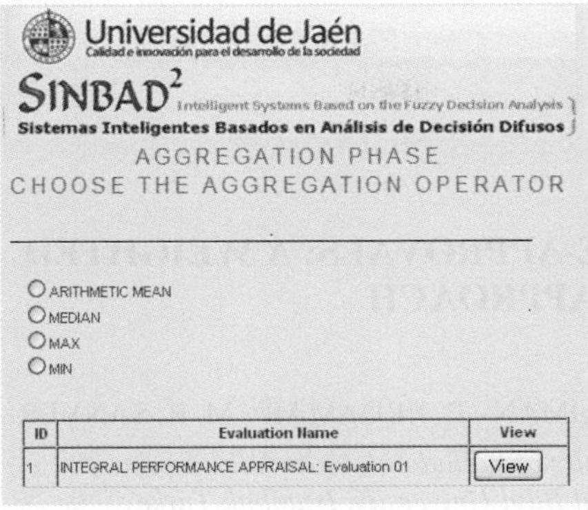

Figure 6: Selection of the aggregation operator

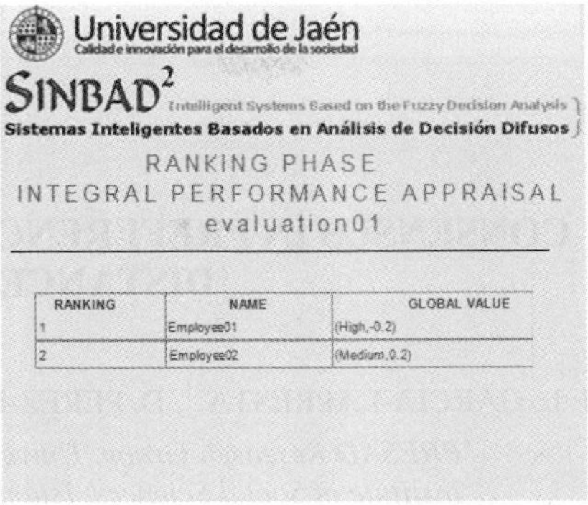

Figure 7: Ranking phase

4. Concluding Remarks

Performance appraisal is a process to determine efficiency and effectiveness of employees. In this contribution, we have presented WSSIPA, a Web based system that supports integral performance appraisal processes with a heterogeneous evaluation framework that provides linguistic results. In this framework, reviewers can provide their judgments within different domains (numerical, interval-valued and linguistic). The linguistic results are close to human natural language in order to be understandable and interpretable by multiple members of the company. The developed system offers companies a software tool for integral performance appraisal processes in an automatic, easy, fast and distributed way.

References

1. J. N. Baron and D. M. Kreps, Strategic Human Resources, *Frameworks for General Managers. Frameworks for General Managers.* (Wiley & Sons, NY, 1999).
2. C. Fisher, L. F. Schoenfeldt and J. B. Shaw, *Human Resources Management* (Houghton Mifflin Company, Boston, 2006).
3. De Andrés, R, García-Lapresta, J.L., Liu, J., Martínez L., Performance Appraisal with heterogenous information. International. Conference on Intelligent System and Knowledge Engineering (2008).
4. F. Herrera and L. Martínez and P.J. Sánchez. Managing non-homogeneous information in group decision making. *European Journal of Operational Research* **166**, 115 (2005).
5. F. Herrera and L. Martínez. A 2-tuple Fuzzy Linguistic Representation Model for Computing with Words. *IEEE Transactions on Fuzzy Systems* **8**, 746 (2000).

CONSENSUS IN PREFERENCE-APPROVALS: A WEIGHTED DISTANCE APPROACH

J. L. GARCÍA-LAPRESTA[*1], D. PÉREZ-ROMÁN[1], B. ERDAMAR[2], M. R. SANVER[2]
[1]*PRESAD Research Group, University of Valladolid, Valladolid, Spain*
[2]*Institute of Social Sciences, Istanbul Bilgi University, Istanbul, Turkey*

Measuring consensus level for a set of preferences requires a proper distance defined on the considered domain. We focus on preference-approvals which are extensions of ordinal preferences by the approval information. For any given set of alternatives, a preference-approval is defined by a weak ordering of the alternative set, subsets of approved and disapproved alternatives and a consistency condition. We propose a method of using weighted distances for ordering and approval components. A consensus measure based on this distance is provided which is sensitive to the position of the disagreements on ordering and approval.

1. Preliminaries

Consider a set of agents $V=\{v_1,\ldots,v_m\}$ with $m\geq 2$ confronting a finite set of alternatives $X=\{x_1,\ldots,x_n\}$ where $n\geq 2$. We assume that each agent ranks the alternatives in X by means of a weak order and additionally, evaluates each alternative as either acceptable or unacceptable by partitioning the alternative set into approved (good) and disapproved (bad) alternatives. These two types of information exhibit the following consistency: Given two alternatives x and y, if x is approved and y is disapproved, then x is ranked above y.

We write $W(X)$ for the set of weak orders on X and $L(X)$ for the set of linear orders on X. Given $R\in W(X)$, we let $\succ$ and $\sim$ stand for the asymmetric and the symmetric parts of R, respectively, i.e., $x_i \succ x_j \Leftrightarrow \text{not } (x_j R x_i)$ and $x_i \sim x_j \Leftrightarrow \big(x_i R x_j$ and $x_j R x_i\big)$.

By $\mathcal{P}(V)$ we denote the power set of V, i.e., $I\in\mathcal{P}(V)\Leftrightarrow I\subseteq V$; and by $\mathcal{P}_2(V)$ we mean the collection of subsets of V with at least two elements that is, $\mathcal{P}_2(V)=\{I\in\mathcal{P}(V)$ Û$\#I\geq 2\}$ where $\#I$ is the cardinality of I. Analogously, we write $\mathcal{P}(X)$ for the power set of X.

* lapresta@eco.uva.es

1.1. *Preference-approval structures*

For any given set of alternatives X, we define preference-approvals by partitioning X into A the set of *acceptable* alternatives and $U = X \setminus A$ the set of *unacceptable* alternatives, where A and U can be empty sets.

Definition 1.1. A *preference-approval* on X is a pair $(R, A) \in W(X) \times \mathcal{P}(X)$ satisfying the following condition $\forall x_i, x_j \in X \ \big((x_i R x_j \text{ and } x_j \in A) \Rightarrow x_i \in A\big)$.

Note that if $x_i R x_j$ and $x_i \in U$, then we have $x_j \in U$. We denote $\mathcal{R}(X)$ for the set of preference-approvals on X. Given $R \in W(X)$ we let R^{-1} be the *inverse* of R, i.e., $x_i R^{-1} x_j \Leftrightarrow x_j R x_i$, for all $x_i, x_j \in X$. Similarly, given a preference-approval $(R, A) \in \mathcal{R}(X)$ we write $(R, A)^{-1} = (R^{-1}, X \setminus A)$ for the preference-approval which is the *inverse* of (R, A)

Definition 1.2. A *profile* is a vector $\mathbf{R} = \big((R_1, A_1), \ldots, (R_m, A_m)\big) \in \mathcal{R}(X)^m$ of preference-approvals, where (R_i, A_i) contains the preference-approval of the agent v_i, with $i = 1, \ldots, m$.

(1) The inverse of $\mathbf{R}$ is $\mathbf{R}^{-1} = \big((R_1, A_1)^{-1}, \ldots, (R_m, A_m)^{-1}\big)$

(2) Given a permutation π on $\{1, \ldots, m\}$ and $\varnothing \neq I \subseteq V$, we denote $\mathbf{R}_\pi = \big((R_{\pi(1)}, A_{\pi(1)}), \ldots, (R_{\pi(m)}, A_{\pi(m)})\big)$ and $I_\pi = \{v_{\pi^{-1}(i)} \ \hat{\mathrm{U}} v_i \in I\}$ i.e., $v_j \in I_\pi \Leftrightarrow v_{\pi(j)} \in I$.

(3) Given a permutation σ on $\{1, \ldots, n\}$ we denote by $\mathbf{R}^\sigma = \big((R_1^\sigma, A_1^\sigma), \ldots, (R_m^\sigma, A_m^\sigma)\big)$ the profile obtained from $\mathbf{R}$ by relabeling the alternatives according to σ, i.e., $x_i R_k x_j \Leftrightarrow x_{\sigma(i)} R_k^\sigma x_{\sigma(j)}$ and $x_i \in A_k^\sigma \Leftrightarrow x_{\sigma(i)} \in A_k$, for all $i, j \in \{1, \ldots, n\}$ and $k \in \{1, \ldots, m\}$

We now introduce a system for codifying each preference-approval structure $(R, A) \in \mathcal{R}(X)$ by means of two vectors: $\mathbf{p}_R \in \mathbb{R}^n$ that represents the position of the alternatives, and $\mathbf{i}_A \in \{0,1\}^n$ that represents acceptable alternatives.

Given $A \subseteq X$, we define $I_A : X™ \{0,1\}$ the *indicator function* (or *characteristic function*) of A:

$$I_A(x_j) = \begin{cases} 1, \text{if } x_j \in A, \\ 0, \text{if } x_j \in X \setminus A. \end{cases}$$

By $\mathbf{i}_A = \big(I_A(x_1), \ldots, I_A(x_n)\big)$ we denote the *indicator vector* of $A \subseteq X$.

Following García-Lapresta and Pérez Román [2], if $R \in W(X)$ we assign the *position* of each alternative x_j in R through the mapping $P_R : X$™ $\mathbb{R}$ defined as $P_R(x_j) = n - \#\{x_i \in X \mid x_j \succ x_i\} - \frac{1}{2}\#\{x_i \in X \setminus \{x_j\} \mid x_i \sim x_j\}$.

By $\mathbf{p}_R = (P_R(x_1), \ldots, P_R(x_n))$ we denote the *position vector* of $R \in W(X)$.

1.2. *Distances and metrics*

Usually, *distance* and *metric* are considered as synonymous. However, we follow the approach given by Deza and Deza [3], where distances and metrics are different concepts.

Definition 1.3. A *distance* on a set $D \neq \varnothing$ is a mapping $d : D \times D$™ $\mathbb{R}$ satisfying the following conditions for all $a, b \in A$:

(1) $d(a,b) \geq 0$ (non-negativity),

(2) $d(a,b) = d(b,a)$ (symmetry),

(3) $d(a,a) = 0$ (reflexivity).If d satisfies the following additional conditions for all $a, b, c \in A$:

(4) $d(a,b) = 0 \Leftrightarrow a = b$ (identity of indiscernibles),

(5) $d(a,b) \leq d(a,c) + d(c,b)$ (triangle inequality),

then we say that d is a *metric*.

Definition 1.4. A set $D \subseteq \mathbb{R}^n$ is *stable under permutations* if for every permutation σ on $\{1, \ldots, n\}$ it holds $(a_1^\sigma, \ldots, a_n^\sigma) \in D$ for every $(a_1, \ldots, a_n) \in D$.

Definition1.5. Given a set $D \subseteq \mathbb{R}^n$ stable under permutations, a distance (or metric) $d : D \times D$™ $\mathbb{R}$ is *neutral* if for every permutation σ on $\{1, \ldots, n\}$ it holds $d\big((a_1^\sigma, \ldots, a_n^\sigma), (b_1^\sigma, \ldots, b_n^\sigma)\big) = d\big((a_1, \ldots, a_n), (b_1, \ldots, b_n)\big)$, for all $(a_1, \ldots, a_n)$, $(b_1, \ldots, b_n) \in D$.

1.3. *Weighted Kemeny distances*

García-Lapresta and Pérez Román [4] was introduced a weighted Kemeny distance on weak orders. We now consider this distance on $W(X)$ normalized to $[0,1]$ and denoted by $d_K^{\mathbf{w}}$ where $\mathbf{w} = (w_1, \ldots, w_{n-1}) \in [0,1]^{n-1}$ is a weighting vector such that $w_1 \geq \cdots \geq w_{n-1}$ and $\sum_{i=1}^{i=n-1} w_i = 1$

Proposition 1.1. For every weighting vector $\mathbf{w} \in [0,1]^{n-1}$ such that $w_1 \geq \cdots \geq w_{n-1}$ and $\sum_{i=1}^{i=n-1} w_i = 1$ $d_K^{\mathbf{w}}$ is a neutral distance on $W(X)$, and it verifies the identity of indiscernibles if and only if $w_{n-1} > 0$ However, it does not always verify the triangle inequality.

1.4. *Approval distances*

Definition 1.6. Let $\vartriangle$ be the symmetric difference.

(1) The mapping $d_a^+ : \mathcal{P}(X)\times\mathcal{P}(X)$™ $[0,1]$ is defined as

$$d_a^+(A,B) = \left(1 - \frac{\#(A\cap B)}{\#X}\right)\cdot\frac{\#(A\vartriangle B)}{\#X}.$$

(2) The mapping $d_a^- : \mathcal{P}(X)\times\mathcal{P}(X)$™ $[0,1]$ is defined as

$$d_a^-(A,B) = \left(1 - \frac{\#\big((X\setminus A)\cap(X\setminus B)\big)}{\#X}\right)\cdot\frac{\#\big((X\setminus A)\vartriangle(X\setminus B)\big)}{\#X}.$$

(3) Given $\lambda\in[0,1]$ the mapping $d_a^\lambda : \mathcal{P}(X)\times\mathcal{P}(X)$™ $[0,1]$ is defined as

$$d_a^\lambda(A,B) = \lambda d_a^+(A,B) + (1-\lambda)d_a^-(A,B).$$

Proposition 1.2. For every $\lambda\in[0,1]$ d_a^λ is a neutral distance on $\mathcal{P}(X)$ and it verifies the identity of indiscernibles. However, it does not verify triangle inequality.

1.5. *Preference-approval distances*

Definition 1.7. Given a weighting vector $\mathbf{w} = (w_1,\ldots,w_{n-1})\in[0,1]^{n-1}$ such that $w_1 \geq \cdots \geq w_{n-1}$ and $\sum_{i=1}^{i=n-1} w_i = 1$ and $\alpha,\lambda\in[0,1]$ the mapping $d_\alpha^{\mathbf{w},\lambda} : \mathcal{R}(X)\times \mathcal{R}(X)$™ $[0,1]$ is defined as

$$d_\alpha^{\mathbf{w},\lambda}\big((R_1,A_1),(R_2,A_2)\big) = \alpha d_K^{\mathbf{w}}(R_1,R_2) + (1-\alpha)d_a^\lambda(A_1,A_2).$$

Proposition 1.3. For every weighting vector $\mathbf{w}\in[0,1]^{n-1}$ such that $w_1 \geq \cdots \geq w_{n-1}$ and $\sum_{i=1}^{i=n-1} w_i = 1$ and all $\alpha,\lambda\in(0,1)$ $d_\alpha^{\mathbf{w},\lambda}$ is a neutral distance on $\mathcal{R}(X)$ bounded by 1 and it verifies the identity of indiscernibles if and only if $w_{n-1} > 0$ However, it does not verify triangle inequality.

2. Consensus measures

Herrera-Viedma, Herrera and Chiclana [1] presented a consensus model for multi-person decision making problems for some preference structures. In the context of linear orders, consensus measures have been introduced and analyzed by Bosch [5]. Subsequently, García-Lapresta and Pérez-Román [2,6] extended this notion to the context of weak orders.

Definition 2.1. A *consensus measure* on $\mathcal{R}(X)^m$ is a mapping $\mathcal{M}: \mathcal{R}(X)^m \times \mathcal{P}_2(V)$™ $[0,1]$ that satisfies the following conditions:

(1) *Unanimity*. For all $\mathbf{R} \in \mathcal{R}(X)^m$ and $I \in \mathcal{P}_2(V)$ it holds

$$\mathcal{M}(\mathbf{R}, I) = 1 \Leftrightarrow (R_i = R_j \text{ and } A_i = A_j, \text{ for all } v_i, v_j \in I).$$

(2) Anonymity. For all permutation π on $\{1,\ldots,m\}$ $\mathbf{R} \in \mathcal{R}(X)^m$ and $I \in \mathcal{P}_2(V)$ it holds $\mathcal{M}(\mathbf{R}_\pi, I_\pi) = \mathcal{M}(\mathbf{R}, I)$

(3) *Neutrality*. For all permutation σ on $\{1,\ldots,n\}$ $\mathbf{R} \in \mathcal{R}(X)^m$ and $I \in \mathcal{P}_2(V)$ it holds $\mathcal{M}(\mathbf{R}^\sigma, I) = \mathcal{M}(\mathbf{R}, I)$

Definition 2.2. Let $\mathcal{M}: \mathcal{R}(X)^m \times \mathcal{P}_2(V)$™ $[0,1]$ be a consensus measure.

(1) $\mathcal{M}$ satisfies *maximum dissension* if for all $\mathbf{R} \in \mathcal{R}(X)^m$ and $v_i, v_j \in V$ such that $i \neq j$, it holds

$$\mathcal{M}(\mathbf{R}, \{v_i, v_j\}) = 0 \Leftrightarrow (R_i, R_j \in L(X),\ R_j = R_i^{-1} \text{ and } A_j = X \setminus A_i).$$

(2) $\mathcal{M}$ is *reciprocal* if for all $\mathbf{R} \in \mathcal{R}(X)^m$ and $I \in \mathcal{P}_2(V)$ it holds

$$\mathcal{M}(\mathbf{R}^{-1}, I) = \mathcal{M}(\mathbf{R}, I).$$

We now introduce our proposal for measuring consensus.

Definition 2.3. Given a distance $d: \mathcal{R}(X) \times \mathcal{R}(X)$™ $\mathbb{R}$ the mapping $\mathcal{M}_d: \mathcal{R}(X)^m \times \mathcal{P}_2(V)$™ $[0,1]$ is defined by

$$\mathcal{M}_d(\mathbf{R}, I) = 1 - \frac{\sum\limits_{\substack{v_i, v_j \in I \\ i<j}} d\big((R_i, A_i), (R_j, A_j)\big)}{\binom{|I|}{2} \cdot \Delta_n},$$

where $\Delta_n = \max\big\{ d\big((R_i, A_i), (R_j, A_j)\big)$ û$(R_i, A_i), (R_j, A_j) \in \mathcal{R}(X) \big\}$

Proposition 2.1. If $d:\mathcal{R}(X)\times\mathcal{R}(X)$™ $\mathbb{R}$ is a neutral distance that satisfies the identity of indiscernibles, then $\mathcal{M}_d$ is a consensus measure.

Proposition 2.2. Let $\mathbf{w}=(w_1,\ldots,w_{n-1})\in[0,1]^{n-1}$ be a weighting vector such that $w_1\geq\cdots\geq w_{n-1}$ and $\sum_{i=1}^{i=n-1} w_i=1$ Let $\lambda,\alpha\in(0,1)$

(1) $\mathcal{M}_{d_\alpha^\lambda}$ is a consensus measure for every $\lambda\in(0,1)$

(2) $\mathcal{M}_{d_\alpha^\lambda}$ is reciprocal if and only if $\lambda=0.5$

(3) $\mathcal{M}_{d_\alpha^\lambda}$ satisfies maximum dissension.

(4) $\mathcal{M}_{d_K^{\mathbf{w}}}$ satisfies unanimity if and only if $w_{n-1}>0$

(5) $\mathcal{M}_{d_K^{\mathbf{w}}}$ is reciprocal if and only if $w_1=\cdots=w_{n-1}=\dfrac{1}{n-1}$

(6) $\mathcal{M}_{d_K^{\mathbf{w}}}$ satisfies maximum dissension if and only if $w_{\text{d}\frac{n+1}{2}\text{t}}>0$

(7) $\mathcal{M}_{d_\alpha^{\mathbf{w},\lambda}}$ satisfies unanimity if and only if $w_{n-1}>0$

(8) $\mathcal{M}_{d_\alpha^{\mathbf{w},\lambda}}$ is reciprocal if and only if $w_1=\cdots=w_{n-1}=\dfrac{1}{n-1}$ and $\lambda=0.5$

(9) $\mathcal{M}_{d_\alpha^{\mathbf{w},\lambda}}$ satisfies maximum dissension if and only if $w_{\text{d}\frac{n+1}{2}\text{t}}>0$

References

1. E. Herrera-Viedma, F. Herrera and F. Chiclana, A consensus model for multiperson decision making with different preference structures, in IEEE Transactions on Systems, Man and Cybernetics-Part A: Systems and Humans, Vol. 32 (pp. 394-402, 2002).
2. J.L. García-Lapresta and D. Pérez-Román, Measuring consensus in weak orders, in Consensual Processes, eds. E. Herrera Viedma et al., Studies in Fuzziness and Soft Computing, Vol. 267 (Springer-Verlag, Berlin, 2011).
3. M. Deza and E. Deza, Encyclopedia of Distances (Springer-Verlag, 2009).
4. J.L. García-Lapresta and D. Pérez-Román, Consensus measures generated by weighted kemeny distances on weak orders, in 10th International Conference on Intelligent Systems Design and Applications (ISDA'10), (Cairo, Egypt, 2010).
5. R. Bosch, Characterizations of voting rules and consensus measures, PhD thesis, (Tilburg University, 2005).
6. J.L. García-Lapresta and D. Pérez-Román, Some consensus measures and their applications in group decision making, in Computational Intelligence in Decision and Control , eds. D. Ruan et al., Computer Engineering and Information Science, Vol. 1 (World Scientific, 2008).

A FUZZY ENVELOPE FOR HESITANT FUZZY LINGUISTIC TERM SETS BASED ON CHOQUET INTEGRAL

R. DE ANDRÉS CALLE
Department of Economics, University of Salamanca, Spain

T. GONZÁLEZ-ARTEAGA
Statistics and Operational Research Department, University of Valladolid, Spain

R.M. RODRÍGUEZ*
Computer Science Department, University of Jaén, Campus las lagunillas s/n, Jaén, 23071, Spain

L. MARTÍNEZ
Computer Science Department, University of Jaén, Campus las lagunillas s/n, Jaén, 23071, Spain

Recently, it has been presented the concept of Hesitant Fuzzy Linguistic Term Sets (HFLTS) to manage hesitant situations under qualitative settings in which experts hesitate among different linguistic terms to express their preferences or assessments. It was also defined the concept of envelope for an HFLTS to carry out the computational processes with them. However, this envelope does not keep the fuzzy representation because it is represented by a symbolic linguistic interval. Therefore, in this contribution we propose a new envelope for HFLTS based on Choquet integral that keeps the fuzzy representation in the computational processes with HFLTS.

1. Introduction

Decision problems defined in context with uncertainty are quite common in real world. Our interest in this contribution is focused on decision situations dealing with qualitative aspects in which the uncertainty is because of the vagueness of meanings of the qualitative values provided by the decision makers. The fuzzy linguistic approach [7] has obtained successful results in the modeling and management of this type of uncertainty. The use of linguistic information in

* rmrodrig@ujaen.es

decision making implies processes of computing with words (CWW) [3, 4] that can be accomplished by different linguistic computing models [3]. Nevertheless, such models present some limitations in qualitative settings when experts prefer using multiple linguistic terms to express their preferences or assessments. In order to manage these hesitant situations, Rodríguez et al. proposed the concept of hesitant fuzzy linguistic term set (HFLTS) [6] to facilitate the elicitation of linguistic information by using comparative linguistic terms. To carry out computational processes in decision making with HFLTS was introduced the concept of envelope of HFLTS that consists of a linguistic interval. This envelope manages in a symbolic way the HFLTS, therefore it does not keep the fuzzy representation. Hence, in this contribution we propose a new HFLTS envelope based on the Choquet integral [5] that computes an aggregated fuzzy number (fuzzy envelope) that represents the hesitant information of an HFLTS. This fuzzy envelope can be strongly used in fuzzy decision models, such as, fuzzy AHP [1], fuzzy TOPSIS [2], etc.

The remaining of the paper is structured as follows: Section 2 revises some basic concepts. Section 3 proposes a fuzzy HFLTS envelope by using the fuzzy Choquet integral and Section 4 points out with some conclusions.

2. Preliminaries

In this section, we review the elicitation of comparative linguistic terms represented by HFLTS and introduce some basic concepts about the fuzzy Choquet integral.

2.1. *Elicitation of Comparative Linguistic Terms*

Rodríguez et al. introduced the concept of HFLTS [6] to manage decision situations in qualitative contexts when experts hesitate among different linguistic terms to express their preferences or assessments.

Definition 1. *[6]* An HFLTS, H_S, is an ordered finite subset of consecutive linguistic terms of S, where $S=\{s_0,\dots,s_g\}$ is a linguistic term set.

Example 1. Let S be a linguistic term set such as, S={s_0:null satisf., s_1:very low satisf., s_2:low satisf., s_3:medium satisf., s_4:high satisf., s_5:very high satisf., s_6:perfect satisf.} and X a linguistic variable, therefore an HFLTS might be:

$$H_S(X)=\{\text{very low satisf., low satisf., medium satisf.}\}.$$

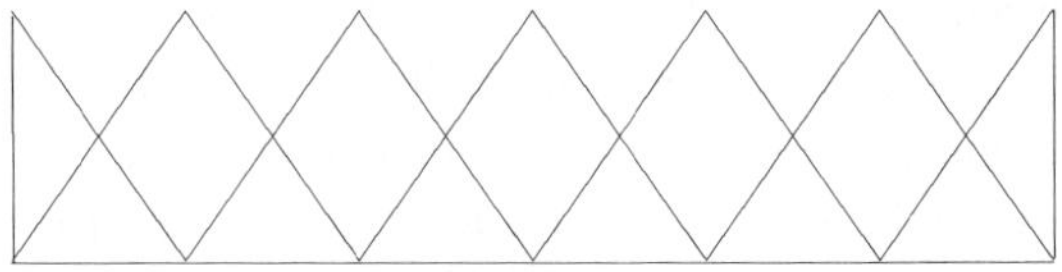

Figure 1: 7-linguistic labels term set

Nevertheless, experts usually do not express their preferences by using multiple linguistic terms, but rather expressions in natural language. Thus, Rodríguez et al. defined a context-free grammar G_H [6], that uses a linguistic term set S, based on fuzzy linguistic approach [7], as is shown in Figure 1, to generate expressions with comparative linguistic terms ll, such as, *greater than* s_i, *lower than* s_i, *between* s_i *and* s_j ($s_i, s_j \in S$).

To obtain an HFLTS from comparative linguistic terms ll, generated by the context-free grammar G_H, was defined the transformation function E_{G_H}.

Definition 2. *[6]* Let E_{G_H} be a function that transforms comparative linguistic terms, ll, obtained by G_H, into HFLTS, , where is the linguistic term set used by G_H.

$$E_{G_H} : S_{ll} \rightarrow H_S.$$

To facilitate the computations with HFLTS was introduced the concept of *envelope* of a HFLTS.

Definition 3. *[6]* The *envelope* of a HFLTS, $env(H_S)$, is a linguistic interval whose limits are obtained by means of upper bound (*max*) and lower bound (*min*):

$$env(H_S) = [H_{S^-}, H_{S^+}], \quad H_{S^-} \leq H_{S^+} \tag{1}$$

where the upper bound is defined as $H_{S^+} = max(s_i)$ and the lower bound is defined as $H_{S^-} = min(s_i)$, $s_i \in H_S$ and $i \in \{1,\dots,g\}$.

Example 2. Following the previous example the envelope of H_S is:

$$env(H_S) = [\text{very low satisfactory, medium satisfactory}].$$

As we can observe in the Example 2, the envelope obtained is represented by a linguistic interval that for computations is managed in a symbolic way losing the fuzzy representation. Therefore, in this contribution we propose a fuzzy HFLTS envelope based on the fuzzy Choquet integral for computing with HFLTS. But first, some basic concepts about the Choquet integral are reviewed.

2.2. *Choquet Integral*

Definition 4. *[5]* Let $C=\{c_1,\ldots,c_n\}$ be a finite universe. A *fuzzy measure* or *capacity* is a set function $\nu : P(C) \to [0,1]$ which satisfies:

1. $\nu(\emptyset)=0$ and $\nu(C)=1$.
2. $A \subseteq B \Rightarrow \nu(A) \le \nu(B)$,

where $P(C)$ is the set of all subsets of C.

Definition 5. [5] Let ν be a fuzzy measure on C, the *Möbius* representation of ν is a set function $m_\nu : P(C) \to R$ given by

$$m_\nu(S) = \sum_{T\subseteq S} (-1)^{|S|-|T|}\ \nu(T)\,, \quad \forall S \subseteq C. \tag{2}$$

Remark 1. Once known the *Möbius* representation is easy to recover the fuzzy measure from which was derived

$$\nu(T) = \sum\nolimits_{S\subseteq T} m_\nu(S), \quad \forall T \subseteq C.$$

This property allows to rewrite the Choquet integral by using only the Möbius representation via

$$F_\nu\left(y_1,\ldots,y_n\right) = \sum_{T\subseteq C} m_\nu(T) \bigwedge_{i\in T} y_i. \tag{3}$$

3. A Fuzzy Envelope for HFLTS

Here, we propose a fuzzy HFLTS envelope based on the fuzzy Choquet integral that increases the flexibility and richness of the computations with HFLTS due to the fact that the use of the fuzzy Choquet integral allows capture and keeps all the subjectivity of HFLTS and the interaction among the linguistic terms in it. To do so, we use fuzzy arithmetic based on extension principle of Zadeh [7] and a fuzzy extension of the Choquet integral defined in [5]:

Definition 6. Let ν be a fuzzy measure on C and m_ν the Möbius representation of ν, the fuzzy Choquet integral is defined by

$$F_\nu\left(y_1,\ldots,y_n\right) = \sum_{T\subseteq C}\left(m_\nu(T) \,\tilde{\cdot} \bigwedge_{i\in T} y_i \right) \tag{4}$$

where it is necessary to use the following operations.

Let y_1, y_2 two fuzzy numbers and μ_{y_1} and μ_{y_2} their membership functions, respectively.

- $y_1 \tilde{+} y_2$ is a fuzzy number with membership function

$$\mu_{y_1 \tilde{+} y_2}(x) = \sup_{a+b=x}\left(\min\{\mu_{y_1}(a) + \mu_{y_2}(b)\}\right)$$

- $\wedge(y_1, y_2)$ is a fuzzy number with membership function

$$\mu_{\wedge(y_1, y_2)}(x) = \sup_{a+b=x}\left(\min\{\mu_{y_1}(a), \mu_{y_2}(b)\}\right)$$

- $\tilde{x} \cdot \tilde{y}$ denote the product of a crisp number x by a trapezoidal fuzzy number $\tilde{y}$. The last is represented in the usual way by $(a_{\tilde{y}}, b_{\tilde{y}}, c_{\tilde{y}}, d_{\tilde{y}})$. $\tilde{x} \cdot \tilde{y}$ is a trapezoidal fuzzy number given by $(xa_{\tilde{y}}, xb_{\tilde{y}}, xc_{\tilde{y}}, xd_{\tilde{y}})$

Therefore, the definition of the fuzzy envelope for HFLTS is as follows,

Definition 7. Let v be a fuzzy measure on H_S and m_v the *Möbius* representation of v. The *fuzzy Choquet integral envelope* of a HFLTS, $env_v(H_S)$, is a fuzzy number given by:

$$env_v(H_S) = \sum_{T \subseteq H_S} \left(m_v(T) \tilde{\cdot} \wedge_{i \in T} s_i \right) \tag{5}$$

where s_i is the corresponding fuzzy number that represents the semantics of the linguistic term s_i, for all $i \in H_S$.

Example 3. Let see how is the new HFLTS envelope of the H_S for the example 1 with two particular fuzzy measures v_1 and v_2. The Figure 2 shows a new HFLTS envelope for v_1 with $v_1(s_1) = v_1(s_2) = v_1(s_3) = 0.1$, $v_1(s_1, s_3) = 0.3$, $v_1(s_1, s_2) = v_1(s_2, s_3) = 0.5$, and the Figure 3 shows a new HFLTS envelope for v_2 with $v_2(s_1) = 0.095$, $v_2(s_2) = 0.189$, $v_2(s_3) = 0.568$, $v_2(s_1, s_2) = 0.298$, $v_2(s_2, s_3) = 0.842$, $v_2(s_1, s_3) = 0.705$.

We can note that the use of this new envelope based on Choquet integral improves the previous one because it not only represents the limits of the HFLTS but also it considers all the linguistic terms included in it and their relationship.

4. Conclusions

This contribution has introduced a fuzzy envelope based on the fuzzy Choquet integral to facilitate the computing processes in those decision problems managing HFLTS. This new fuzzy envelope provides a fuzzy representation of the multiple linguistic terms belong to a HFLTS including their relationship.

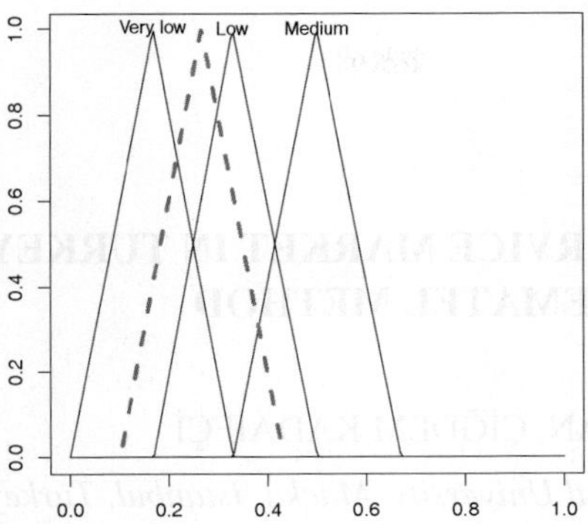

Figure 2: Fuzzy Choquet integral envelope by using the fuzzy measure V_1

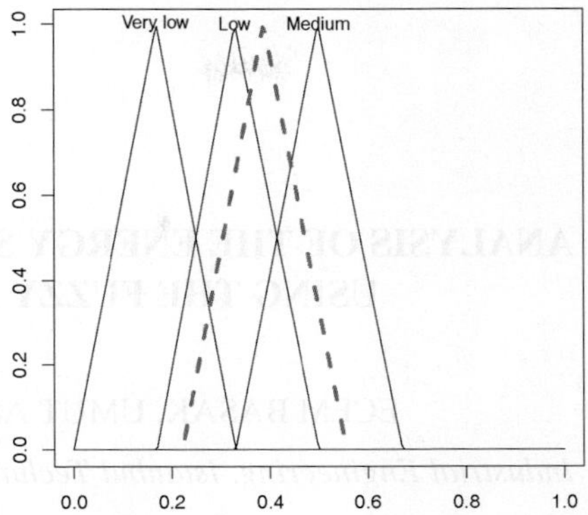

Figure 3: Fuzzy Choquet integral envelope by using the fuzzy measure V_2

Acknowledgements

This work is partially supported by the Research Projects TIN-2009-08286, ECO2009-07332 and FEDER funds.

References

1. S. Aydin and C. Kahraman. Multiattribute supplier selection using fuzzy analytic hierarchy process. *International Journal of Computational Intelligence Systems*, 3(5), 553-565, (2010).
2. F. E. Boran, S. Genc, M. Kurt, and D. Akay. A multi-criteria intuitionistic fuzzy group decision making for supplier selection with topsis method. *Expert Systems with Applications*, 36,11363-11368, (2009).
3. L. Martínez, D. Ruan, and F. Herrera. Computing with words in decision support systems: An overview on models and applications. *International Journal of Computational Intelligence Systems*, 3(4), 382-395, (2010).
4. J.M. Mendel and D. Wu. *Perceptual Computing: Aiding People in Making Subjective Judgments*. IEEE-Wiley, (2010).
5. P. Meyer and M. Roubens. On the use of the choquet integral with fuzzy numbers in multiple criteria decision support. *Fuzzy Sets and Systems*, 157,927-938, (2006).
6. R.M.Rodríguez, L.Martínez, and F.Herrera. Hesitant fuzzy linguistic term sets for decision making. *IEEETransactions on Fuzzy Systems*,20-1,109-119, (2012).
7. L. Zadeh. The concept of a linguistic variable and its applications to approximate reasoning. *Information Sciences, Part I, II, III*, (8,9), 199-249, 301-357, 43-80, (1975).

ANALYSIS OF THE ENERGY SERVICE MARKET IN TURKEY USING THE FUZZY DEMATEL METHOD

ECEM BASAK, UMUT ASAN, ÇİĞDEM KADAİFÇİ

Industrial Engineering, Istanbul Technical University, Macka, Istanbul, Turkey

In this paper we suggest using the fuzzy DEMATEL method to model the energy service contracting market in Turkey and identify factors playing a significant role in its development. The proposed model provides a causative representation of the complex market structure in an uncertain environment. Results are compared to an earlier study applying a fuzzy cognitive mapping approach to explore the critical factors influencing the energy service market in Turkey.

1. Introduction

Energy efficiency, which may increase productivity, avoid waste, reduce costs and thereby increase competitiveness, has recently become one of the prime concerns of many industries. In this study, energy efficiency is defined as reducing the energy consumption without causing any decline in production quality and quantity in industrial establishments (see [1]). Energy efficiency activities can be either outsourced or they can be handled inside the firm by investing to the necessary assets such as workforce, equipment, technology, etc. Experience shows that energy service contracting, a form of outsourcing, plays a significant role in promoting energy efficiency [2]. Such contracts can reduce many of the difficulties that firms often experience in planning, implementing and monitoring energy efficiency projects because they offer a comprehensive package of services. These services are typically provided by energy service companies (ESCOs) that guarantee energy savings to its clients. Their payment is subject to the energy savings achieved [3].

In this paper, we explore the critical factors shaping the energy service market in Turkey. In the past few years the energy industry in Turkey has undergone a change in terms of restructuring and deregulation. However, there are still some barriers hindering the development of this industry such as economic instability, capital inadequacy of local firms, unclear financing-mechanisms, uncompleted structural reforms, and insufficient knowledge of end-users, firms, and financing institutions [3, 4]. To analyze the energy service contracting market and identify factors playing a significant role in its development, we suggest using the fuzzy DEMATEL method, which allows modeling and visualizing a complex network of cause-effect relationships in fuzzy environments. The complexity and uncertainty of the given problem, along with the subjective nature of expert opinions makes it unrealistic and infeasible to acquire exact judgments. Thus, the involvement of a fuzzy set theoretic approach can adequately address this type of uncertainties. The results are compared to an earlier study of Asan *et al.* [5] applying a fuzzy cognitive mapping (FCM) approach to explore the critical factors influencing the energy service contracting market in Turkey. Considering both results provides the basis for developing recommendations for the energy service market in Turkey.

2. Fuzzy DEMATEL

The Decision-Making Trial and Evaluation Laboratory (DEMATEL) method, developed by the Geneva Research Center of the Battelle Institute [6], is used to improve understanding of complex structures of causal relationships among factors/criteria by means of matrices and diagraphs. The analysis of interdependence among factors helps not only dividing the relevant factors into cause and effect groups – that provides valuable insight into the structure of the problem - but also evaluating the importance of these factors [7, 8]. Since we suggest using the fuzzy DEMATEL method, which extends the classical method using crisp values to an uncertain linguistic environment, its main steps are summarized as follows [for more detail see 9, 10];

Step 1: Defining the decision goal and forming a committee of experts.

Step2: Determining the relevant factors and designing the fuzzy linguistic scale indicating the degree of direct influence. The linguistic terms used to represent the six different levels of influence are "No Influence" (No), "Very Low Influence" (VL), "Low Influence" (L), "Moderate Influence" (M), "High Influence" (H), and "Very High Influence" (VH).

Step 3: Generating the initial direct-relation fuzzy matrix ($\tilde{A}$) which consists of triangular fuzzy numbers in form of $\tilde{a}_{ij} = (l_{ij}, m_{ij}, u_{ij})$. These numbers represent the (average) assessments of experts in terms of influence and direction between factors.

Step 4: Obtaining the normalized direct-relation fuzzy matrix $\tilde{X}$ through Equation 1 where s is the maximum raw total of $\tilde{A}$,

$$\tilde{x}_{ij} = \frac{\tilde{a}_{ij}}{s} = \left(\frac{l_{ij}}{s}, \frac{m_{ij}}{s}, \frac{u_{ij}}{s}\right), \quad s = max_{1\leq i\leq n}\left(\sum_{j=1}^{n} u_{ij}\right). \tag{1}$$

Step 5: Deriving the total-relation fuzzy matrix $\tilde{T}$, which reflects the total relationship between each pair of factors, through Equation 2,

$$\widetilde{T} = \lim_{w\to\infty}\left(\widetilde{X} + \widetilde{X}^2 + \cdots + \widetilde{X}^w\right). \quad (2)$$

Suppose that $\tilde{t}_{ij} = (l'_{ij}, m'_{ij}, u'_{ij})$ then,

$$[l'_{ij}] = X_l(I - X_l)^{-1},\ [m'_{ij}] = X_m(I - X_m)^{-1},\ [u'_{ij}] = X_u(I - X_u)^{-1} \quad (3)$$

where I is the identity matrix, and X_l, X_m, X_u are three crisp matrices, whose elements are extracted from $\tilde{X}$.

Step 6: Calculating the sum of rows $\widetilde{D}_i$ and the sum of columns $\tilde{R}_i$ of the total-relation fuzzy matrix $\tilde{T}$. $\widetilde{D}_i$ represents the sum of influence exerted from factor i to the other factors while $\tilde{R}_i$ represents the sum of influence that factor i receives from the other factors.

Step 7: Drawing the causal diagram by using the calculated $\widetilde{D}_i + \tilde{R}_i$ and $\widetilde{D}_i - \tilde{R}_i$ values, representing the horizontal axis and vertical axis respectively. The calculated $\widetilde{D}_i + \tilde{R}_i$ and $\widetilde{D}_i - \tilde{R}_i$ values are turned into crisp values (denoted by def) by using the CFCS (Converting Fuzzy data into Crisp Scores) method [11]. If the value of $(\widetilde{D}_i - \tilde{R}_i)^{def}$ is positive, then the factor belongs to the cause group. If the value of $(\widetilde{D}_i - \tilde{R}_i)^{def}$ is negative, then the factor belongs to the effect group. $(\widetilde{D}_i + \tilde{R}_i)^{def}$ shows how important the factor is.

3. Analysis

3.1. *Proposed Model and Data Collection*

As explained above, in this study, we aim to explore the factors that have the greatest impact on the energy service contracting market in Turkey and on the outsourcing decision of industrial energy consumers. To do this, experts who have knowledge and experience on energy service contracting were involved in the determination of the factors, their causal relationships and in the assessment of the strength of influences by means of fuzzy linguistic terms parameterized with triangular fuzzy numbers. The details of this Delphi based three-stage data collection process are given in Asan *et al.* (2011) and, therefore, will not be discussed here. All factors that have been identified as indispensable and the direct causal relationships among them along with their strengths are summarized in Table 1. The overall influence and dependence values of the factors are determined by applying the equations (1)-(3) and the CFCS method.

3.2. *Results and Comparison*

There is no single, official, scientific interpretation of the fuzzy DEMATEL results. The group of experts/researchers has to ask the right questions and propose explanations. The results presented in Table 2 provide valuable evidence about the importance and classification of the factors, and thereby about the structure of the energy service market in Turkey. The factors are classified according to their influential and dependent capacity. The causal diagram shown in Figure 1 facilitates an easier comparison of the factors. Accordingly, the factors F3, F4, F5, F7, F9, F10, F11, F12, F17, F20, F21 and F22 belongs to the cause group while the rest belongs to the effect group.

Table 1. Factors and their pairwise relations in terms of aggregate linguistic assessment data

	F1. Third Party Financing	F2. Energy Policy of Government	F3. Other Costs	F4. Outsourcing	F5. Economic Developments	F6. Scope of Energy Serv. Contracts	F7. Energy Shortage	F8. Energy Consumption	F9. Cost of Energy	F10. Energy Saving Intention	F11.Energy Efficiency Activities	F12.Opportunity for Energy Perf. Serv.	F13. Financing Difficulties	F14. Local Energy Problems	F15. Insourcing (proportion)	F16. Cost of Quality	F17. Firm's Energy Saving Problem	F18. Economies of Scale	F19. Incentives	F20. Firm Strategy (Energy Policy)	F21. Technical Experience	F22. Technological Developments	F23. Investment Cost
F1	-	No	No	No	No	No	No	No	No	No	No	No	H	No	No	No	No	No	No	No	No	No	No
F2	No	-	No	No	No	No	No	No	No	No	No	No	No	No	No	No	No	No	M	No	No	No	No
F3	No	No	-	L	No	No	No	No	No	No	No	No	No	No	No	No	No	No	No	No	No	No	No
F4	M	No	No	-	No	H	No	No	No	No	No	No	No	No	VH	No	No	H	No	No	No	No	VL
F5	L	M	No	No	-	No	No	H	No	No	No	No	L	No	No	No	No	No	M	L	No	M	VL
F6	M	No	No	No	No	-	No	No	No	No	No	VL	No	No	No	No	No	No	No	No	No	No	No
F7	No	No	No	No	No	No	-	No	No	No	No	No	No	No	No	H	No	No	No	No	No	No	No
F8	No	No	No	No	No	No	No	-	No	No	No	No	No	M	No	No	VL	No	No	No	VL	No	No
F9	No	No	No	M	No	No	No	No	-	No	No	No	No	No	M	No	No	No	No	No	No	No	H
F10	No	No	No	No	No	No	No	No	VH	-	No	No	No	No	No	No	H	No	No	No	No	No	No
F11	No	No	No	No	No	No	No	No	No	H	-	No	No	No	No	No	No	No	No	No	No	No	No
F12	VH	No	No	No	No	No	No	No	No	No	No	-	No	No	No	No	No	No	No	No	No	No	No
F13	H	No	No	VL	VL	No	No	No	No	No	No	No	-	No	No	No	No	No	No	No	No	No	No
F14	No	L	No	No	No	No	No	No	No	No	No	No	No	-	No	No	No	No	No	No	No	No	No
F15	No	No	No	VH	No	No	No	No	No	No	No	No	No	No	-	No	No	No	No	No	No	No	No
F16	No	No	VL	No	No	No	No	No	No	No	No	No	No	No	No	-	No	No	No	No	No	No	No
F17	No	No	No	No	No	No	L	No	No	No	H	No	No	L	No	No	-	No	No	No	No	No	No
F18	No	No	No	No	No	No	No	No	No	No	No	No	No	No	No	No	No	-	No	No	No	No	M
F19	No	No	No	No	No	No	No	No	No	No	No	No	No	No	No	No	No	No	-	No	No	No	L
F20	No	No	No	M	No	No	No	No	No	No	No	No	No	No	M	No	No	No	No	-	No	No	No
F21	No	No	No	VL	No	No	No	No	No	No	No	M	No	No	No	No	No	H	No	No	-	No	No
F22	No	No	No	No	H	H	No	VH	No	No	No	No	No	No	No	No	No	No	No	No	H	-	No
F23	No	No	No	No	No	No	No	No	M	No	No	No	H	No	No	No	No	No	No	No	No	No	-

The factors playing a key role in terms of both influence and dependence (with values over 1.3) are "outsourcing", "economic developments", "third party financing" and "technological developments". As expected, the outsourcing decision was found to be the most strongly connected factor in the system, which was critical in achieving reliable results. Changes in the external environment such as investments on technology, economic stability or macroeconomic equilibrium have also significant impact on the market and the behavior of the firms that wish to implement energy efficiency projects for decreasing their energy costs. The results also indicate that, in the long run, the success of the energy service market may largely depend on the availability of third party financing from a variety of potential institutions (see [12]). Since ESCOs project-financing mechanisms are not clear yet, this highly dependent factor should be influenced and monitored for desirable changes.

Notice, however, that the results of the analysis only partially support the findings reported in the study about the ESCO market in Turkey by Asan et al. (2011). The results of the previous study using fuzzy cognitive mapping are shown in the last column of Table 2. The main reason of the notable differences between the results can be explained by the difference in assumptions the methods rely on. Although both FCM and Fuzzy DEMATEL are able to provide a causative representation of complex systems in an uncertain environment, FCM allows modeling and simulating the dynamic behavior of systems while Fuzzy DEMATEL allows analyzing direct and indirect relationships from a rather static perspective. This implies that the discussed methods are complementary and not substitutes. Therefore, we suggest combining the results of both methods in a joint graph and designating those factors as critical (F4, F5, F9, F22, and F23) that have the highest ranking according to both methods; see Figure 2. These factors deserve more attention and further investigation.

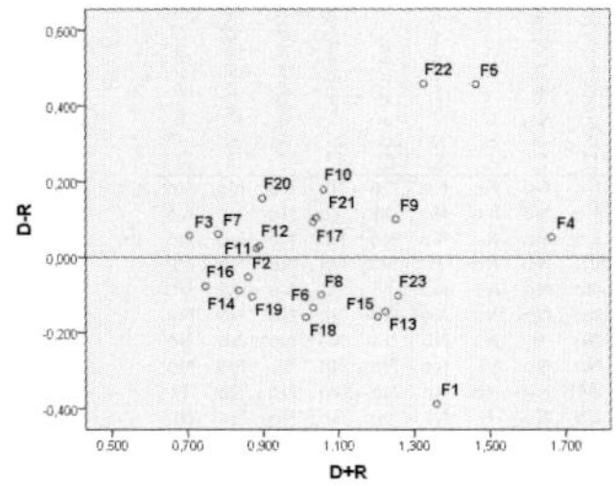

Figure 1. The causal diagram of factors

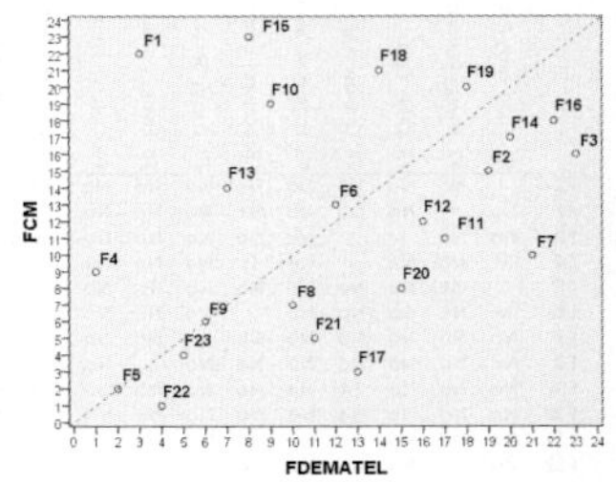

Figure 2. Rank order of the factors

Table 2. The results of Fuzzy DEMATEL and Fuzzy Cognitive Mapping

	Fuzzy DEMATEL								FCM
#	$\widetilde{D}_i + \widetilde{R}_i$			$\widetilde{D}_i - \widetilde{R}_i$			$(\widetilde{D}_i + \widetilde{R}_i)^{def}$	$(\widetilde{D}_i - \widetilde{R}_i)^{def}$	Steady State Value
F1	0.393	0.597	4.923	-2.859	-0.379	1.671	1.358	-0.388	0.467
F2	0.123	0.206	3.944	-2.003	-0.058	1.819	0.859	-0.053	0.746
F3	0,030	0.093	3.637	-1.734	0.042	1.872	0.705	0.058	0.746
F4	0.539	0.861	5.532	-2,420	0.028	2.573	1.663	0.054	0.824
F5	0.381	0.678	5.211	-1,740	0.416	3,090	1.463	0.457	0.954
F6	0,210	0.336	4.285	-2.262	-0.129	1.813	1.031	-0.135	0.769
F7	0.095	0.148	3.789	-1.767	0.043	1.927	0.781	0.062	0.795
F8	0,220	0,360	4.293	-2.202	-0.101	1.871	1.054	-0.100	0.837
F9	0.345	0.514	4.692	-2.026	0.074	2.321	1.251	0.102	0.861
F10	0.264	0.365	4.292	-1.783	0,160	2.245	1.060	0.179	0.699
F11	0.158	0.223	3.983	-1.896	0.012	1.929	0.882	0.024	0.795
F12	0,150	0.236	3.958	-1.884	0.022	1.923	0.889	0.030	0.776
F13	0.273	0.486	4,690	-2.456	-0,140	1.961	1.224	-0.145	0.750
F14	0.104	0.188	3.893	-2.039	-0.089	1,750	0.835	-0.089	0.738
F15	0.338	0.491	4.514	-2.343	-0.155	1.833	1.204	-0.159	0.394
F16	0.072	0.123	3.721	-1.959	-0.073	1.691	0.747	-0.078	0.701
F17	0.199	0.335	4.291	-1.923	0.073	2.169	1.031	0.094	0.895
F18	0.214	0.321	4.245	-2.272	-0.153	1.759	1.012	-0.160	0.491
F19	0.126	0.214	3.966	-2.094	-0.102	1.746	0.870	-0.106	0.549
F20	0.138	0.235	4.011	-1.724	0,130	2.149	0.897	0.156	0.826
F21	0.198	0.342	4.304	-1.917	0.084	2,190	1.039	0.105	0.869
F22	0,390	0,570	4.854	-1.567	0.412	2.896	1.325	0.458	0.980
F23	0.291	0.512	4.761	-2.399	-0.114	2.071	1.257	-0.103	0.881

4. Conclusion

This study reveals the critical factors to focus on that may contribute to the improvement of the energy service market in Turkey. It is demonstrated that

Fuzzy DEMATEL in combination with FCM provides a more effective tool for modeling and visualizing a complex system. The results provide basically soft evidence, which should further analyzed to make more detailed inferences about outsourcing decisions on energy efficiency projects.

References

1. EIE, (2007), Energy Efficiency Law, http://www.eie.gov.tr, (Access: 25.07.2011).
2. Sorrell S., (2007), The Economics of Energy Service Contracts, Energy Policy, Vol. 35, 507-521.
3. Okay E., Okay N., Konukman A.E.Ş. and Akman U., (2008), Views on Turkey's Impending ESCO Market: Is it Promising?, Energy Policy, Vol. 36, 1821-1825.
4. Onaygil S. and Meylani E.A., (2007), Energy Efficiency Consulting Companies: An Overview of the Current Situation in the World (in Turkish), Energy Efficiency Congress, 2007, 41–54.
5. Asan, U., Kutlu, A. C. and Kadaifci, C., (2011), Analysis of Critical Factors in Energy Service Contracting Using Fuzzy Cognitive Mapping, *In the 41st International Conference on Computers and Industrial Engineering, October 23-26, 2011*, Los Angeles, USA.
6. Fontela E. and Gabus A., (1974), DEMATEL report no. 2: structural analysis of the world problematique innovative methods, batelle Geneva research center, Geneva.
7. Li, C.-W. and Tzeng, G.-H., (2009), Identification of a threshold value for the DEMATEL method using the maximum mean de-entropy algorithm to find critical services provided by a semiconductor intellectual property mall, *Expert Systems with Applications*, Vol. 36, pp. 9891–9898.
8. Wu, H. H. and Tsai, Y. N., (2011), A DEMATEL Method to Evaluate the Causal Relations Among the Criteria in Auto Spare Parts Industry, *Applied Mathematics and Computation*, Vol. 218, pp. 2334-2342.
9. Lin, C. J. and Wu, W. W., (2008), A Causal Analytical Method for Group Decision-Making Under Fuzzy Environment, *Expert Systems with Applications*, Vol. 34, pp. 205-213.
10. Jassbi, J., Mohamadnejad, F. and Nasrollahzadeh, H., (2011), A Fuzzy DEMATEL Framework for Modeling Cause and Effect Relationships of Strategy Map, *Expert Systems with Applications*, Vol. 38, pp. 5967-5973.
11. Opricovic, S., and Tzeng, G. H., (2003), Defuzzification within a multicriteria decision model, *International Journal of Uncertainty Fuzziness and Knowledge-Based Systems*, Vol. 11, pp. 635–652.
12. Zhang, Y., Han, Q.M., Liu, C.B. and Sun, J.Y., (2008), Analysis for Critical Success Factors of Energy Performance Contracting (EPC) Projects in China, *Proceedings of the 2008 IEEE IEEM*, Singapore, pp. 675-679.

REPLACEMENT DECISIONS FOR RELIABILITY CENTERED MAINTENANCE

D. ÖZGÜR-ÜNLÜAKIN[†]

Industrial Engineering Dept., Bahçeşehir University, Istanbul, Turkey

T. BİLGİÇ

Industrial Engineering Dept., Boğaziçi University, Istanbul, Turkey

We study reliability centered proactive maintenance of a complex dynamic system consisting of aging components. System reliability can be estimated from the interactions of the components. Maintenance of the system is realized by replacing components in any period. Our aim is to minimize total replacement cost in a discrete horizon such that system reliability never falls below a predetermined threshold value. We represent the problem with dynamic Bayesian networks (DBNs) and develop an algorithm for determining when and what to replace within the DBN framework. We prove that under the existence of a predetermined threshold, this algorithm assures optimum replacement times. Four approaches are proposed to choose the component to replace and are tested on a complex dynamic problem.

1. Introduction

Proactive maintenance can either be realized as a fixed schedule or it can be adaptively determined. We consider the latter case, where the reliability centered maintenance activities of a complex dynamic system consisting of aging components are planned. The aim is to optimize component replacement schedules in a given horizon. We propose a methodology to effectively prepare a predictive maintenance plan using DBNs. We use DBNs to represent the problem and to do fast inference. The failure of the system and its associated costs can also be modeled using influence diagrams. However, we presume that the reliability of the system should be kept over a predetermined threshold value in all periods.

Previous studies related to reliability analysis using DBNs are generally descriptive[1, 2]. Dynamic problem is represented with DBNs and the outcome of

[†] demet.unluakin@bahcesehir.edu.tr

the analysis is how system reliability behaves in time. The impact of maintenance of an element at a specific time on this behavior is also reported in some of them. However optimization of maintenance activities using DBNs is rarely considered which is the main motivation of this study. The proposed proactive maintenance methodology described in this paper is an extension of the methodology presented in [3].

2. Problem Definition

There is a system consisting of several components that are subject to failure. It is impossible or impractical to observe the states of the components and the system due to the following reasons: The system may be far away or strictly hidden to the decision maker, i.e. it may be in space or inside a nuclear tank. Observing the system may be very expensive or time costly which may be the case in machines having long setup times. The system itself has a very low probability of failure. However system reliability can be estimated from the interactions of its components. We presume that the estimated reliability of the system should be kept over a predetermined threshold value in all periods. This is reasonable in mission critical systems where the failure of the system is a very low probability event due to built in redundancy and other structural properties. Components age with a constant transition probability of failure. It is possible to replace components in any period even if they are far away from the decision maker, i.e. by changing software configuration of components. Our aim is to minimize total maintenance cost in a discrete time planning horizon. Furthermore the following assumptions are made: (i) All conditional probability distributions are discrete. (ii) All components and the system have two states ("w": working, "nw": failure). (iii) Components can only fail at the beginning of a time period. Once they fail they will be in state "nw" unless they are repaired. (iv) Components can only be replaced at the beginning of each period. Once they are replaced, their working state probability (i.e., their reliability) becomes 1 in that period. (v) If all components are replaced in a period, system reliability becomes 1 in that period.

3. Proposed Solution

A complex dynamic problem is generated from the well known static auto diagnosis problem[4] from decision theoretic troubleshooting environment by making the components aging in time. We call this problem "dynamic auto problem" (DAP) and propose DBNs to represent it. The resulting DBN is given in Figure 1 for t and $t + 1$. System node is represented with s.

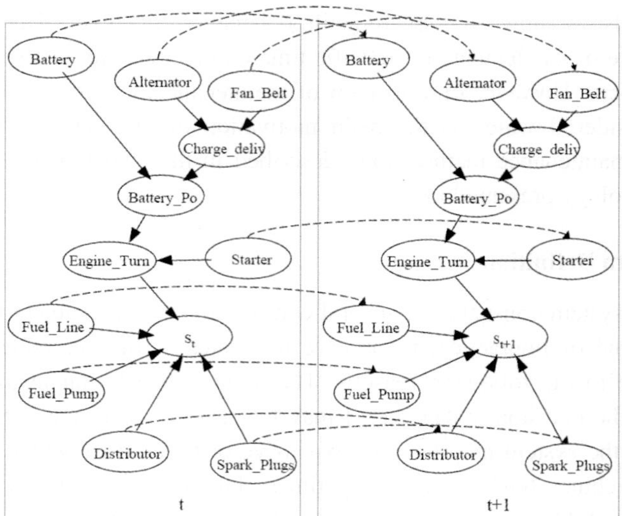

Figure 1. DBN representation of DAP for two successive time slices t and $t + 1$.

3.1. *DBN Algorithm*

In DBNs there are no formally defined action nodes like in influence diagrams. Nevertheless, replacing or not replacing decisions of each component can be handled with DBNs by modeling action nodes with uncertainties. For this purpose, action nodes for each component and decision epoch are also introduced in the DBN representation. Replacements are performed by entering evidence to the actions. Let ε denote the evidence list which is initially empty. When a component is chosen to be replaced in period t, this information is added to the evidence list, i.e., $\varepsilon \leftarrow \varepsilon \cup \{a_{it} \leftarrow rc\}$.

Two important decisions are to be made: when and what to replace. Our methodology decouples these under the assumption of stationary replacement costs. The decision of the replacement time is made by monitoring the first period when the estimated system reliability just falls below a given threshold. This method leads to optimum replacement times given that the estimated system reliability is an increasing function of all components.

Proposition 3.1. *For a given finite horizon T, assume there exists an optimal replacement schedule RS such that it makes at least one replacement when the estimated system reliability is above the threshold. Let RS' be another replacement schedule which is constructed from RS such that all replacements are performed when the estimated system reliability just falls below the threshold and it has the same replacement sequence of components with RS.*

Both schedules are feasible with respect to the threshold constraint. Under the above conditions, RS' is at least as good as RS in minimizing total number of replacements or total replacement cost when costs are stationary.

Proof. It is required to show that in the worst case RS' will have the same number of replacements as with RS in a given planning horizon. In this way total cost can also be improved. □

This result motivates the following DBN algorithm:

(i) Initialize t = 1

(ii) Infer system reliability $P(s_{t'} = w \mid \varepsilon) \quad t \le t' \le T$

(iii) If $P(s_{t'} = w \mid \varepsilon) \ge L \quad \forall t'$, then stop.

(iv) Else find the replacement period, $k = \min_{t \le t' \le T} \{t' : P(s_{t'} = w \mid \varepsilon) < L\}$

(v) Do replacement(s) for period k

  (a) Initialize remaining components list, $I' \leftarrow I$

  (b) Calculate $ef_{ik} \quad \forall i \in I'$

  (c) Select the component to replace, $i^* = \arg\max\{ef_{ik}, i : i \in I'\}$

  (d) Update evidence list, $\varepsilon \leftarrow \varepsilon \cup \{a_{i^*k} \leftarrow rc\}$

  (e) Update remaining components list, $I' \leftarrow I' \setminus \{i^*\}$

  (f) If $P(s_k = w \mid \varepsilon) < L$ then continue with (v.b)

  (g) Else update $t = k + 1$

(vi) If t > T, then stop.

(vii) Else continue with (ii)

In step (v.b) of the DBN algorithm, ef_{ik} is an efficiency measure of component i in period t when a replacement is planned. One of the component selection approaches to be described is performed.

3.2. *Component Selection Methods*

Four approaches are developed to select the component to replace. These are Fault Effect Myopic approach (FEM), Fault Effect Look-ahead approach (FEL), Replacement Effect Myopic approach (REM) and Replacement Effect Look-ahead approach (REL). The efficiency measure of each method is calculated as in Eq. (1-4) respectively where $s_{t+1:T}$ denotes $s_{t+1}, \ldots, s_T$; c_{it} denotes the state of component i in period t, and let t' be the earliest time where the system reliability becomes less than the threshold.

$$ef_{it}^{FEM} = P\left(c_{it} = nw \mid \varepsilon \cup \{s_t \leftarrow nw\}\right). \tag{1}$$

$$ef_{it}^{FEL} = P\left(c_{it} = nw \mid \varepsilon \cup \{s_{t+1:T} \leftarrow nw\}\right). \tag{2}$$

$$ef_{it}^{REM} = P\left(s_t = w \mid \varepsilon \cup \{a_{it} \leftarrow rc\}\right) - P\left(s_t = w \mid \varepsilon\right). \quad (3)$$

$$ef_{it}^{REL} = \min_{t \le t' \le T}\left\{t' : P\left(s_{t'} = w \mid \varepsilon \cup \{a_{it} \leftarrow rc\}\right) < L\right\}. \quad (4)$$

Let FEMc, FELc, REMc and RELc be the cost included versions of the methods which are obtained by dividing the respective efficiency by the replacement cost of the component. RELc uses the same efficiency measure as with REL. However, in case of tie, the component with the smallest replacement cost is chosen. By this slight modification, we improve the REL method to account for also the replacement costs of components while preserving the effort to minimize the total number of replacements.

4. Computational Study

The DBN algorithm is coded in Matlab and uses the Bayes Net Toolbox (BNT)[5] to represent the causal and temporal relations, and to perform inference. The four approaches are compared on DAP. Threshold and planning horizon are taken as 0.50 and 100 respectively. DAP is solved for two cases: to minimize total number of replacements (DAP-PM') and to minimize total cost (DAP-PM).

Table 1. Results of DAP with different component selection approaches

Method	TCost	TRep	St	FB	Al	Ba	Di	SP	FP	FL	Tsec
FEM	346	56	2	8	8	9	7	14	7	1	670
FEL	380	56	5	7	7	13	6	11	6	1	677
REM	345	55	2	8	8	9	7	13	7	1	2098
REL	354	54	3	8	7	9	7	12	7	1	2264
FEMc	386	119	1	11	6	21	4	63	11	2	1194
FELc	348	106	1	10	5	26	3	50	9	2	1079
REMc	322	90	1	10	5	16	4	44	8	2	3301
RELc	320	58	2	8	6	11	6	17	7	1	2518
Rand	723	92	11	12	11	12	12	12	12	10	583
(std)	(83)	(7.5)									

Table 1 shows the results of DAP-PM' with FEM, FEL, REM and REL; and DAP-PM with FEMc, FELc, REMc and RELc respectively. "TCost" and "TRep" stand for total cost and total number of replacements of the solution. "Tsec" shows the solution time in seconds. The remaining columns belong to the components of DAP and show their replacement frequencies. The problem is also solved with a random component selection method where the component to replace is selected randomly at a replacement time. 50 replications are performed and the average solution with its standard deviation is reported.

In DAP-PM', REL finds the best solution although its solution time is the highest. REM and REL perform more inferences than FEM and FEL to determine the component to replace. FEM and REM results are nearly the same except the number of replacements of Spark Plug. Although the total number of replacements found by FEL is similar to the ones found by FEM, REM and REL, its distribution among components is significantly different. In DAP-PM, FELc, REMc and RELc improve total cost while FEMc does not when compared with the results of DAP-PM'. Although FELc and REMc improve total cost, they cause a significant increase in the total number of replacements. RELc finds the best result with respect to total cost and in addition this result has the least total number of replacements. In fact, RELc is a slight modification of REL improving the component selection to account for also component costs. That is why the total number of replacements it finds is still reasonable. The solution of Random method is the worst and its total number of replacements is very far from the solution of others.

5. Conclusion

We develop an algorithm for determining replacement decisions of a complex dynamic system within the framework of DBN representation. We prove that under the existence of a predetermined threshold value, this algorithm assures optimum replacement times. Four alternative component selection approaches with their modified cost versions are proposed among which RELc is the best with respect to both cost and replacement frequencies. Further analysis of the approaches will be performed as a future study.

References

1. R. L. Welch and T. V. Thelen, Dynamic reliability analysis in an operational context: the Bayesian network perspective, in *Proceedings of Dynamic Reliability: Future Directions*, 2000.
2. P.Weber and L. Jouffe, Reliability modeling with dynamic Bayesian networks, in *Proceedings of 5th IFAC Symposium*, 2003.
3. D. Özgür-Ünlüakın and T. Bilgiç, Predictive maintenance using dynamic probabilistic networks, in *Proceedings of 3rd PGM Workshop,* 2006.
4. D. Heckerman, J. S. Breese and K. Rommelse, *Communications of the ACM* **38**, 49 (1995).
5. K. P. Murphy, The Bayes Net Toolbox for Matlab, in *Computing Science and Statistics: Proceedings of the Interface*, 2001.

A NEW PROJECT PLANNING MODEL FOR BETTER HANDLING VAGUENESS AND UNCERTAINTY OF REAL WORLD TASKS WITH AHP AND FUZZY NUMBERS

SIbar Kaan Manga*

ISE Settlement and Custody Bank Inc.
Istanbul, Sisli 34381, Turkey

One of the most important problems of project management is the decision of task durations in the projects. The duration of each task and cumulative durations of the successive tasks generally results in the success or fail of that project. While the common practice is to use natural numbers for task duration, they generally fail to match the real world results. In order to obtain more realistic results, researchers used fuzzy numbers which resulted in an important improvement of estimations but with the vagueness and uncertainty of the business world, they still require improvement. In our study, we used fuzzy numbers with AHP to determine task durations and adapt them to real world, first we separately find the levels of vagueness and uncertainty of tasks using AHP and then we use these categories to manipulate our fuzzy membership function for better prediction of tasks' durations and certainty levels so we can provide a more accurate project plan with more realistic task durations.

1. Introduction

Project Management (PM) can be defined as the orchestration of resources, processes, budget, timelines and constraints to achieve a certain unique business objective [1]. And the nature of Project management is achieving something new so the knowledge base is largely not accurate and in some industries totally unknown. In this environment, task durations have a high level of uncertainty. The natural number task durations lack to represent the actual task durations. So Researches work to find alternative options. One of the most popular ways is the fuzzy numbers which are being used in project planning for a long time as we will see in literature review. In order to utilize the use of fuzzy numbers our model uses Analytic Hierarchy Process (AHP) which is a criteria based hierarchical decision ranking system [2] and one of the most popular Multi Criteria Decision Making (MCDM) methods, to modify task durations in uncertainty and vagueness

* E-mail: smanga@takasbank.com.tr

perspectives, we build two fuzzy membership functions that use the results of AHP processes to represent this perspectives and then we apply this functions to our trapezoidal numbers to best represent the uncertain and vague environment of projects and obtain more accurate task durations.

2. Literature Review

There is a large literature on usage of fuzzy numbers and MCDM tools in project management in the academic world.

An important paper working on the uncertainty in the project environment is [3] the researchers proposed three different project management strategies to handle the real world problems, these strategies are instructionism, learning, and selectionism.

An important papers about fuzzy PM is [4], by fuzzfying the task duration researcher tried to analyze the critical path in a more accurate manner.

Other works on fuzzy numbers, PERT and PM are [5,6]. The researchers used PERT method with fuzzy activity times to measure criticality of project tasks in [5] and try to calculate the degree of criticality of activities in [6].

Another important research is [7]. Researchers used a MCDM tool, TOPSIS; they used fuzzy logic in the process of evaluating their criteria's.

In [8] researchers used a linear programming approach to handle the fuzzy environment of projects.

Another research combines fuzzy project management with neural networks, [9]. The researchers especially used their techniques on software projects as they stated the nature of the software projects is not technique so a large uncertainty is always remain in the process.

There are also researches on project networks with fuzzy times [10]. In this research authors used triangular fuzzy numbers to represent the uncertain activity durations and used probabilistic models to estimate completion times.

In [11] researcher used AHP to evaluate the PM criteria and requirements.

3. Main Concepts

3.1. *Project management*

Project management is the set of actions and initiatives like planning, organizing, securing and handling and directing resources to achieve a specific unique goal [12].

Main element of a project plan is the activities and the activity times, our model, as described in following sections is trying to generate the most accurate activity times for projects success. Another important PM element is Critical Path Critical path is the longest path of activities (the succeeding activities that took longer time) so it also indicates the projects natural earliest finishing time [13]. In our research we will evaluate our results and compare our calculated critical paths.

3.2. *Fuzzy numbers*

Project fuzzy number is an extension of a regular number in the sense that it does not refer to one single value but rather to a connected set of possible values, where each possible value has its own weight between 0 and 1[14].

In our research we used trapezoidal fuzzy numbers to illustrate our activity times. Trapezoidal fuzzy number may be defined by four values, it is a simple representation of fuzzy numbers and fuzzy membership functions.

A Fuzzy number ã=(y1,y2,y3,y4) as shown in Figure 1. [15]

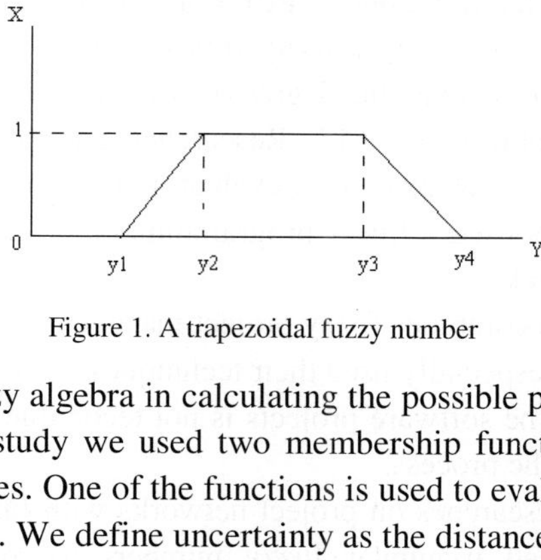

Figure 1. A trapezoidal fuzzy number

We also used fuzzy algebra in calculating the possible paths of project fuzzy algebra [16]. In our study we used two membership functions to evaluate the estimated activity times. One of the functions is used to evaluate uncertainty and other is for vagueness. We define uncertainty as the distance between y1, y2 and y3, y4 while we define vagueness as the distance between y2 and y3. Our model changes this distances like shown in fig's 2, 3 to find a more suitable fuzzy number to represent the vagueness and uncertainty.

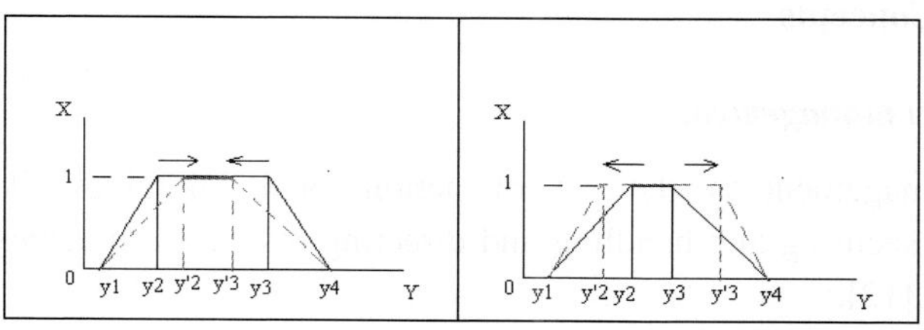

Figure 2. Decrease and Increase of Uncertainty

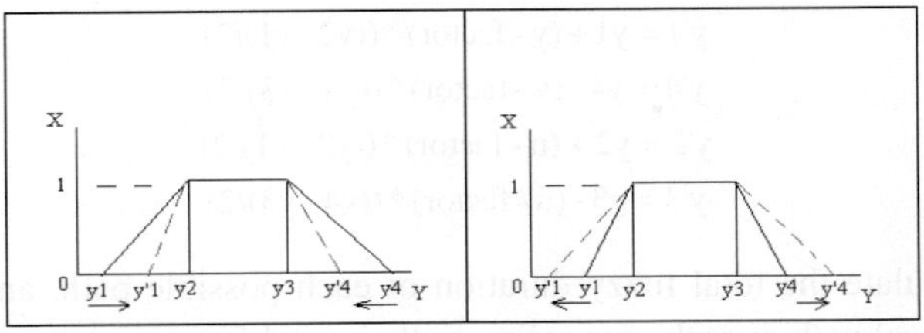

Figure 3. Decrease and Increase of Vagueness

4. Framework and Example

In this research we used a 5 step process to calculate the fuzzy activity times, project timeline and handle uncertainty and vagueness.

We give trapezoidal fuzzy number activity times to each task; this process is like the standard processing of task timing but with fuzzy activity times, fig 4.

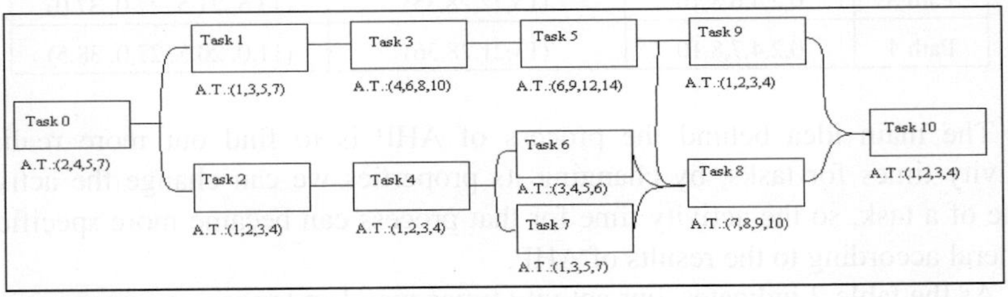

Figure 4. Example Project

Then we calculate the uncertainty and vagueness factor for all tasks with AHP, In our model, there are three alternatives "Pessimistic", "Rationalistic" and "Optimistic" and there are also three criteria which are "Complexity", "Capability" and "Experience" for each task we find out a label, pessimistic, rationalistic or optimistic for both uncertainty and vagueness.

We apply the results of AHP to the fuzzy activity times using their u-factor and v-factors.

In our calculation we use u-factor for uncertainty measure of an activity and v-factor for vagueness, if an activity is pessimistic like we think it will take longer than stated for uncertainty or vagueness, the u-factor or v-factor of that task is set as "-1". If rationalistic means we think the activity time is "OK", they are set as "0" and if they are optimistic, we believe that the task can be completed in less time than stated, u-factor or v-factor are set as "1".

Our general formula can be seen from eq. (1), The fuzzy number M(a)= (y1,y2,y3,y4) changes into a new fuzzy number M(a') = (y'1,y'2,y'3,y'4);

$$
\begin{aligned}
y'1 &= y1 + (v\text{ - factor}) * ((y2 - y1)/2) \\
y'4 &= y4 - (v\text{ - factor}) * ((y4 - y3)/2) \\
y'2 &= y2 + (u\text{ - factor}) * ((y2 - y1)/2) \\
y'3 &= y3 - (u\text{ - factor}) * ((y4 - y3)/2)
\end{aligned} \quad (1)
$$

We calculate the total fuzzy duration of each possible path. and finally we find the critical path or paths according to their total fuzzy activity times, table 1 (* means critical path).

Table 1. Results of our example

	Activity Path	Un-processed Activity Times	Processed Activity Times
Path 1	0,1,3,5,9,10	(15,26,36,46)	(11.5, 24.5, 35.5, 49.5)
Path 2*	0,1,3,5,8,10	(21,32,42,52)	(16.5, 30, 42, 56.5)
Path 3	0,2,4,6,8,10	(15,22,28,35)	(13.5, 21.5, 27.0, 37.0)
Path 4	0,2,4,7,8,10	(13,21,28,36)	(11.0, 20.5, 27.0, 38.5)

The main idea behind the process of AHP is to find out more realistic activity times for tasks, by changing its properties we can change the activity time of a task, so the activity time for that process can became more specific or general according to the results of AHP.

As the table 2 indicates, our activity times may last longer, so our worst case is expanded as our best case scenario. On the other hand the most probable scenarios are also changed. Path 2 is still the critical path, it cannot be completed before 16.5 days and it will most likely to be completed between 30 and 42 days and our worst case is 56.5 days. In a real world project such difference in case scenarios are unacceptable but in our example the activity times are set for easier understanding of the work of our model.

5. Conclusion

In this paper a new framework for deciding and generating activity times in a project is proposed. This research differs from other researches in two main properties; first we used a method to best represent the most realistic activity times both in the perspective of uncertainty and vagueness and second are AHP is used to decide how to change the fuzzy activity times to be more accurate.

So using this framework in projects will supply the project managers and the project stakeholders a better view of project timelines and critical path than standard processes and the success or failure of a project will be evaluated more accurately.

The method represented to handle uncertainty and vagueness in this research can be extended using different MCDM methods like TOPSIS or data mining and regression models and the representation of the fuzzy membership function can be altered into different functions to get more sensitive results in the future studies.

References

1. PMBOK Guide, Project Management Institue (PMI). (2008).
2. Decision Support System Glossary, Definition of AHP, http://www.dssresources.com/glossary/53.php. Ret: 7 Jan. (2012).
3. M. T. Pich, C. H. Loch, Arnoud de Meyer, *Management Science*. **Vol. 48** no. 8 1008-1023 (2002).
4. S.-P. Chen, *European Journal of Operational Research*. **Vol. 183**, Issue 1, Pages 442–459 (2007).
5. C.-T. Chen and S.-F. Huang, *Journal Information Sciences: an International Journal*. **Vol. 177** Issue 12. (2007).
6. S. Chanas, P. Zielinski, *Fuzzy Sets and Systems* **Vol. 122**, 195–204 (2001).
7. F. A. Zammori, M. Braglia and M. Frosolini, *International Journal of Project Management*. **Vol. 27**, Issue 3, Pages 278–291 (2009).
8. T.-F. Liang. *Int. Journal of Sys. Sci.* **Vol. 37** I: 15 (2006).
9. S. Kumar, B. A. Krishna and P. S. Satsangi, *Journal of Applied Intelligence*, **Vol. 4**, Number 1, 31-52 (1994).
10. C.S. McCahona and E.S. Leeb, *Computers & Mathematics with Applications*, **Vol. 15**, Issue 10, Pages 829-838 (1988).
11. K.M.Al-Subhi Al-Harbi, *International Journal of Project Management*, **Vol. 19**, Issue 1, Pages 19-27 (2001).
12. S. Nokes. *The Definitive Guide to Project Management 2nd edition*. London Financial Times, Prentice Hall (2007).
13. J. T. Marchewka, *Information Technology Project Management 1st edition*, Wilet-IT press, 152-53. (2002).
14. O. Wolkenhauer, *Data Eng.: Fuzzy Mathematic in System Theory and Data Analysis*, John Wiley & Sons., 197-212. (2001).
15. W. Kosinski, *International Journal of Applied Mathematics and Computer Science*, **Vol. 16**, No. 1, 51–57 (2006).

EMPLOYING CARDINAL RANK ORDERING OF CRITERIA IN MULTI-CRITERIA DECISION ANALYSIS

MONA RIABACKE

Dept. of Computer and Systems Sciences, Stockholm University, Forum 100, SE-164 40 Kista, Sweden

MATS DANIELSON

Dept. of Computer and Systems Sciences, Stockholm University Forum 100, SE-164 40 Kista, Sweden

ARON LARSSON

Dept. of Information Technology and Media, Mid Sweden University SE-851 70 Sundsvall, Sweden

LOVE EKENBERG

Dept. of Computer and Systems Sciences, Stockholm University Forum 100, SE-164 40 Kista, Sweden

The elicitation of preference information in multi-criteria decision analysis (MCDA) processes and the lack of practical means supporting it is a significant problem in real-life applications of MCDA. The issues at hand are problematical in a multitude of ways, but some of these issues may be remedied by accepting weaker input statements from decision-makers than what is most commonly needed, yet being able to utilize these statements for decision evaluation. In this paper, we propose a fast and practically useful weight elicitation method, which builds on the ideas of rank-order methods, but in addition take imprecise cardinal information into account.

1. Introduction

Multi-Attribute Value Theory, MAVT, and Multi-Attribute Utility Theory, MAUT [1, 2], are the oldest and most widely used MCDA methods in practical applications. The relative importance of each criterion is assessed, as well as, value functions characterizing the satisfaction of the alternatives (according to the decision-maker) under each criterion, and thereafter the overall score of each alternative is calculated. The most common form of value function used is the additive model: $V(a) = \sum_{i=1}^{m} w_i\, v_i(a)$, where $V(a)$ is the overall value of alternative a, $v_i(a)$ is the value score reflecting the alternative's performance on criterion i, and w_i is the weight assigned to criterion i.

2. Weight Elicitation within MCDA

Many papers have discussed problematic elements of the elicitation process, such as cognitive demand, cf., e.g., [3, 4], difficulties in judging and expressing precise input, cf., e.g., [5, 6], biases, and inconsistencies that arise, cf., e.g., [7, 8]. A systematic mistake in practical applications is the absence of adequate (or the employment of inadequate) weight elicitation techniques [9]. Moreover, providing the required input has been found to be demanding for users of decision analysis tools, and a need for less challenging elicitation methods, which are favoured by users, have been seen as essential for the practical applicability of such tools, cf., e.g., [10,11].

2.1 *Methods for Eliciting Weights in MCDA*

There are many different methods for weight elicitation. Most frequently a precise numerical weight is assigned to each criterion to represent the information extracted from the user. Methods adopting this approach range from simple rating procedures, like the frequently used direct rating (DR) and point allocation (PA) methods, to somewhat more advanced procedures, such as the often used SMART [12], SWING [2], SMARTS [13], and trade-off methods.

Ratio weight procedures maintain ratio scale properties of the decision-maker's judgments during the elicitation. Accurate determinations of attribute weights by using ratio weight procedures are tricky to acquire in practice as assessed weights are always subject to response error [14]. Consequently, suggestions on how to use imprecise weights have been proposed instead.

The use of surrogate weights, which are derived from ordinal importance information [3, 11, 15], is one such approach. There are several proposals on how to convert the rankings provided by the decision-maker into numerical weights, e.g., rank sum (RS) weights, rank reciprocal (RR) weights [16], and centroid (ROC) weights [17].

In some decision analysis applications, preferential uncertainties and incomplete information is handled by using intervals, cf., e.g., [18-20], where a range of possible values is represented by an interval. This approach puts less demand on the decision-maker as individual differences in importances and judgments can be represented by value intervals [20].

2.2. *Cardinal Rank Ordering of Criteria*

There is a lot of uncertainty involved in elicitation, and the relaxation of precise importance judgments has been found to be advantageous in order to reduce the gap between theory and practice, cf. [18, 21]. If more precision is needed, the input that needs to be further specified can be identified, which can be a great advantage in real settings.

As mentioned above, to merely provide an ordinal ranking of the criteria is allegedly less demanding on decision-makers and, in a sense, effort-saving. Jia et al. [14] state that due to the relative robustness of linear decision models regarding weight changes, the use of approximate weights often yields satisfactory decision quality, but that the assumption of knowing the ranking with certainty is strong. Instead, they believe that there can be uncertainty regarding both the magnitudes and ordering of weights and that people can be quite confident that some differences in importance are greater than others. Thus, although some weak form of cardinality may exist, cardinal importance relation information is not taken into account in the transformation of rank-orders into weights. As an alternative, if ordinal information as well as imprecise cardinality is taken into account, the resulting input is more in line with a reasonable representation of significance.

An elicitation procedure consists of three parts: i) extraction (extracting information through user input); ii) representation (capturing the information in a formal structure); and iii) interpretation (assigning meaning to the captured information). In order to address the issues above, we propose an approach to more pragmatic weight elicitation allowing for ease of use, imprecise weights, and mixing of ordinal and cardinal information as appropriate. The suggested elicitation method, Cardinal Rank Ordering of Criteria (CROC), takes both ordinal information as well as imprecise cardinal relation information of the importance of the attribute ranges into account during extraction, and interprets the criteria input as regions of significance. In its representational and interpretational aspects, CROC extends rank-order methods into handling imprecise cardinal information. If the cardinal information is omitted and only a ranking of criteria is provided, the results coincide with ROC weights [17].

2.2.1. *Extraction*

The method consists of two stages. In the CROC extraction (user interaction) stage, information on the decision-maker's ranking of the criteria in the criteria set is elicited. After the extraction, the information is represented and interpreted in the second stage. In the following, we denote the set of N criteria **G** =

$\{G_1, ..., G_N\}$ where each criterion $G_i \in \mathbf{G}$ is, during the elicitation, associated with a scale position (value) x_i and a criterion weight variable w_i. The first stage consists of three steps that can be iterated until convergence.

In the first step, only ordinal information is collected. Without loss of generality, we assume that G_1 is more important than G_2 which is more important than G_3 and so forth.

In the second step, the decision-maker is asked to assess the difference in importance between the most and the least important criteria. Denote the highest user score x_H and the lowest x_L. x_H is set to 100 at the outset. The decision-maker is thus asked to express strength in his/her opinion on the difference between the most and least important criteria by setting x_L within the interval (0, 100).

In the third step, the criteria are initially distributed equally along a cardinal slider according to the ordering obtained in the first step and using the end scale points determined in the second step. The scale position difference between criterion G_1 and criterion G_2 is equal to the scale position difference between G_2 and G_3, and so on.

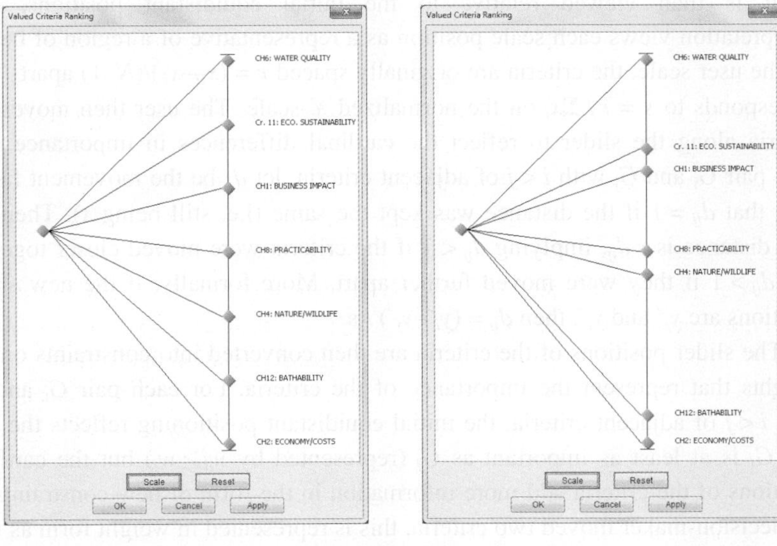

Figure 1. Initial distances (left) and adjusted distances (right) between criteria.

Thereafter, the decision-maker will be able to adjust the distances between the criteria in order to express his or her cardinal importance information between them, e.g., he or she may feel that two criteria are closer together or more apart importance-wise, which is indicated by decreasing or increasing the distance between the criteria (see Fig. 1). If the second and third steps are omitted, the procedure coincides with ROC weights.

2.2.2. *Interpretation*

The interpretation stage is concerned with translating the user input from the extraction stage into sets of statements. Underlying the interpretation is the scaling of the user statements from the slider. Assuming normalized criteria weights, i.e. $\Sigma w_i = 1$ and $0 \leq w_i \leq 1$ for all $i \leq$ N, the user information ($x_i \leq x_H$ for all $i \leq$ N) from the first stage is mapped onto a normalized Y-scale such that $\Sigma x_i \rightarrow 1$; $x_H \rightarrow y_H = x_H / \Sigma x_i$; $x_L \rightarrow y_L = x_L / \Sigma x_i$; and $0 \rightarrow 0$. The middle two are the endpoints of the visible part of the scale; see Fig. 1 for an implementation. Each criterion position x_j in between thus maps onto $y_j = x_j / \Sigma x_i$.

In step one of the extraction stage, an ordinal ranking is obtained. Using comparative statements, this is represented as $w_i \geq w_j$ for $i < j$. In the last step of the extraction, the decision-maker can express cardinality by sliding the different criteria until he/she is satisfied with their relative positions. The end result is then viewed relative to the initial equidistant positions. The interpretation views each scale position as a representative of a region of belief. On the user scale, the criteria are originally spaced $r = (x_H - x_L)/(N-1)$ apart. This corresponds to $s = r / \Sigma x_i$ on the normalized Y-scale. The user then moves the criteria along the slider to reflect the cardinal differences in importance. For each pair G_i and G_j with $i < j$ of adjacent criteria, let d_{ij} be the movement factor such that $d_{ij} = 1$ if the distance was kept the same (i.e. still being s). Then the new distance is $s \cdot d_{ij}$, implying $d_{ij} < 1$ if the criteria were moved closer together and $d_{ij} > 1$ if they were moved further apart. More formally, if the new slider positions are y_i' and y_j', then $d_{ij} = (y_i' - y_j') / \text{s}$.

The slider positions of the criteria are then converted into constraints on the weights that represent the importance of the criteria. For each pair G_i and G_j with $i < j$ of adjacent criteria, the initial equidistant positioning reflects the fact that G_i is at least as important as G_j (represented by $w_i \geq w_j$) but the cardinal positions of the criteria add more information in the form of new constraints. If the decision-maker moved two criteria, this is represented in weight form as $w_i - w_j \geq (d_{ij} - 1) \cdot s$ for each pair of adjacent criteria. The set of inequalities forms an inequality system that defines a polytope. The centroid of the polytope is the weight distribution and can be found by a decision analysis tool such as DecideIT [22]. If the extraction yielded only equidistant criteria, i.e. no cardinal information was supplied, then the CROC centroid coincides with that of ROC given that $x_L = 0$. See Danielson [19] for procedures on expanding centroids to intervals during sensitivity analysis.

3. Results and Discussion

The CROC method was employed in a real-life case study (case 1), set up to be comparable to another case study (case 2), where Point Allocation was employed for the elicitation. The decisions faced in cases 1 and 2 were of the same magnitude, influencing tens of thousands of inhabitants in a city council setting. The decisions were "hot issues" that had been debated for a long time (postponed for several years) and were in the hands of the governing politicians. In both cases, the weight elicitation methods (CROC in case 1, and PA in case 2) were employed twice with some time in between elicitation occasions. Case 1 is described in more detail in Danielson et al. [23] and case 2 in Danielson et al. [24].

The settings in the cases 1 and 2 were similar in terms of type of decision-makers, domain knowledge and decision process, yet these two cases differed a lot. Since the two elicitation procedures employed varied regarding all elicitation components (extraction, representation, and interpretation), it is useful to compare these aspects. The CROC method's extraction part is graphical and its design encourages imprecision, which makes it virtually impossible in practice (other than by chance) to provide the exactly same visual view of criteria preferences on two different occasions. In case 1, statements from both elicitation occasions are represented in the constraint set generated by the interpretation (described in 2.2.2 Interpretation) when changes are minor (which could be attributed to unintended change). In case 2, employing PA, the number of changes were exceedingly more than what would be expected when relating them to the changes in case 1. This can partly be attributed to the fact that minor (unintended) changes in statements on importance cannot be handled realistically in the interpretation of the method. The vagueness even in the ordinal weight information provided by the decision-makers suggests that there is a need for methods that can handle such input. The outcome of the comparison between the cases indicates that the CROC interpretation is more realistic, and the method more robust and persistent to noisy input.

References

1. R. Keeney and H. Raiffa, Decisions with Multiple Objectives: Preferences and Value Trade-offs, (John Wiley, NY, 1976).
2. D. von Winterfeldt and W. Edwards, Decision Analysis and Behavioural Research, (Cambridge University Press, NY, 1986).
3. F. Barron and B. Barrett, Acta Psych. **93**, 23-36 (1996).
4. V. Belton and T. Stewart, Multiple Criteria Decision Analysis: An Integrated Approach, (Kluwer Academic Publishers, UK, 2002).
5. Z. Shapira, Risk Taking: A Managerial Perspective, (Russel Sage Foundation, NY, 1995).
6. J. Corner and P Corner, J. Oper. Res. Soc. **46**, 304-314 (1995).
7. A. Tversky and D. Kahneman, J. Bus. **59**, 252-278 (1986).
8. Lichtenstein, S. and Slovic P. (Eds.), The Construction of Preference, (Cambridge University Press, NY, 2006).
9. J. Pictet and D. Bollinger, Eu. J. Oper. Res. **185**, 1300-1307 (2008).
10. P. Bottomley and J. Doyle, Omega **29**, 553-560 (2001).
11. K. Katsikopoulos and B. Fasolo, IEEE Trans. Syst. Man Cyber. Part A: Syst. Hum. **36**, 960-967 (2006).
12. W. Edwards, IEEE Trans. Syst. Man Cyber. **7**, 326-340 (1977).
13. W. Edwards and F. Barron, Org. Beh. Hum. Dec. Proc. **60**, 306-325 (1994).
14. J. Jia, G.W. Fischer and J Dyer, J. Beh. Dec. Mak. **11**, 85-105 (1998).
15. F. Barron and B. Barrett, Management Science **42**, 1515-1523 (1996).
16. W. Stillwell, D. Seaver and W. Edwards, Org. Beh. Hum. Perf. **28**, 62-77 (1981).
17. F. Barron, Acta Psych. **80**, 91-103 (1992).
18. K.S. Park, IEEE Trans. Syst. Man Cyber. Part A: Syst. Hum. **34**, 601-614 (2004).
19. M. Danielson, Int. J. Inf. Tech. Dec. Mak. **3**, 513–534 (2004).
20. A. Jiménez, S. Rios-Insua and A. Mateos, A., Comp. Oper. Res. **33**, 1081-1101 (2006).
21. A. Larsson, J. Johansson, L. Ekenberg, and M. Danielson, Int. J. Unc. Fuzz. Know.-Based Syst. **13**, 495–509 (2005).
22. M. Danielson, L. Ekenberg, J. Johansson and A. Larsson, Proc. of ISIPTA'03, Bernard, J-M., Seidenfeld, T., and Zaffalon, M. (eds.), 204–217, (Carleton Scientific, 2003).
23. M. Danielson, L. Ekenberg, A. Larsson and M. Riabacke, E-Democracy: A Group Decision and Negotiation Perspective, (Rios Insua, D. and French, S., eds., Springer, 2010).
24. M. Danielson, L. Ekenberg, A. Ekengren, T. Hökby and J. Lidén, Journal of Multi-Criteria Decision Analysis **15**, 15-30 (2008).

A FUZZY INFERENCE SYSTEM FOR NET PRESENT VALUE ANALYSIS OF CONSTRUCTION PROJECTS

ALP USTUNDAG, EMRE CEVIKCAN

Industrial Engineering Department , Istanbul Technical University
Istanbul, Macka 34367, Turkey

Sales forecasting has a great impact on facilitating effective and efficient management of resources for construction projects. Since, having appropriate sales prediction provides appropriate financial activities, especially for large scaled construction projects. This study proposes a numerical model for construction projects which integrates Fuzzy Rule Based System (FRBS) with Net Present Value (NPV) analysis. In the proposed hybrid model, FRBS is used for sales forecasting and provides input for NPV formulation. The proposed model is applied to a construction project in Istanbul so as to demonstrate its performance and practicality for real life problems.

1. Introduction

A major responsibility of marketing is the preparation of sales forecasts. First, market opportunities are identified through marketing research. Then, the size, growth and profitability of each market opportunity are measured and/or forecasted. Accurate sales forecasts are important for financial planning to raise the needed cash for investment and operations. Over-estimates of demand lead to several problems such as increased production and inventory carrying costs.. On the other hand, under-estimates of demand lead to a different set of problems of stock-outs, lost sales. Realizing the detrimental impact of inaccurate forecasts, companies have used a variety of sales forecasting techniques in order to forecast sales accurately. Financial assessment of construction projects has been a critical problem, since construction sector is an industry with high cost, high profit and high risk. In the field of construction projects, the lack of precise information is associated with property investments because: (i) data are usually not readily available in consistent form, (ii) the diversity among properties often resists analysis and interpretation of general trends, and (iii) properties are place-bound, producing even greater uncertainty about the future prospects of a property than for other economic goods that can be moved [1]. Therefore, using linguistic variables and respective fuzzy numbers to evaluate the factors, fuzzy set theory is highly appropriate to be utilized in the stage of sales forecasting so as to determine net present value of a construction project.

Istanbul is the largest city in Turkey with more than 12 million official residents. With its area of 5,712 km2, Istanbul covers only 0.7% of Turkey; however it holds more than 15% of the population within its geographic boundaries. Istanbul is declared as the number one among 27 Europan metropolities for investment potential in real estate by PricewaterhouseCoopers (PwC) and Urban Land Institute (ULI). Istanbul, commercial, financial & cultural capital of Turkey, is located on the crossroads of Europe and Asia. Istanbul accounts for 10% of the housing gap in Turkey. The total market value of the 22 real estate investment companies publicly traded on the Istanbul Stock Exchange reached 11.7 billion TL. According to the report prepared by EVA Real Estate Appraisal Consultancy (2011), 288,000 housings in Istanbul will be supplied for sale in 2012. The number of housings in Istanbul is expected to be 23 million in 2023. It is estimated that only population increase sourced housing requirement is 900,000 in 2017 [2].

In this context, this paper describes a financial decision support systematic for construction projects by integrating Fuzzy Rule Based System (FRBS) with Net Present Value (NPV) analysis. In other words, this paper provides the contribution of integrating fuzzy inference and net present value analysis under fuzzy environment to the relevant literature. The inspiration of this study is the satisfy assessment requirement of construction projects under changing market as well as economic conditions. The proposed approach is twofold. First, a sales forecasting is performed with respect to home loan interest rate, economic stability and housing supply by a FRBS which consists of IF-THEN rules in which input and output variables are fuzzy. Then, FRBS provides the input for NPV formulation. The rest of this paper is organized as follows. Section 2 provides a literature review about sales forecasting in construction projects. In Section 3, the proposed model is presented. In Section 4, the application of the model is included and finally in Section 5, conclusions are provided.

2. Literature Review

Since sales forecasting has recently received an emerging attention, some attempts have been made to predict number of sales in construction projects. Huang et al [3] established a real estate demand-analyzing model founded on the base of the gray theory to demonstrate the factors influencing the real estate market of China. Brooks and Tsolacos [4] analyzed the predictability of real estate asset returns using a number of time series techniques. Magnus et al. [5] showed that the weighted average least squares estimator possesses major advantages over standard Bayesian model averaging estimators. Chen et al. [6] addressed the hedonic modelling of demand elasticity in housing market. They tested their method by using sales of new high rise condominiums in two districts within Shenzhen, China. When considering the relevant literature, it is revealed that there has not been any published study which considers not only factors affecting sales, but also vagueness of data for net present value analysis of construction projects. In this context, this paper has the originality of presenting a net present value systematic which includes a FRB model that predicts real estate sales to address this research gap.

3. Proposed Model

The proposed model for financial assessment of construction projects consists of two phases. In the first phase, the sales forecast of housing units is performed by using the factors of home loan interest rate, consumer confidence index and housing supply by a FRBS. Then in the second phase, NPV analysis is conducted. The related parameters as well as variables are given in Figure 1. In the proposed model, price of each housing unit is equal and constant during planning horizon.

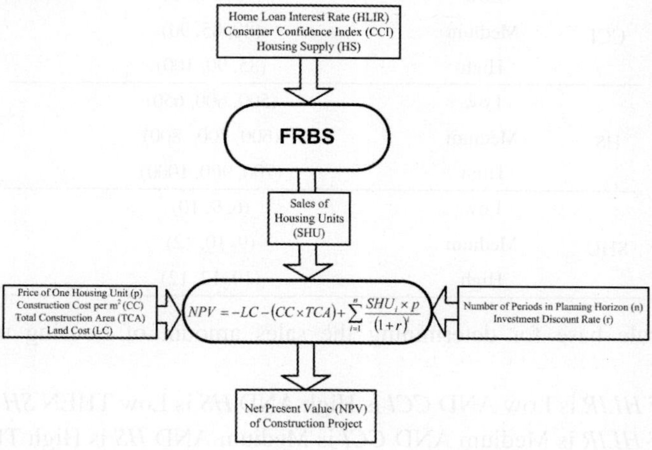

Figure 1. Framework of Proposed Model

4. Application

In our study, the land area is 2,000 m^2 and its cost (LC) is \$6,000,000. Furthermore, the total construction area (TCA) of the project is 5,000 m^2 and the construction cost per m^2 (CC) is \$600. This project consists of 32 housing units, each being 120 m2. The price of one housing unit is determined as \$400,000 which is equal and constant during the planning horizon. It is assumed that the construction project will be finished at the end of the first year and all the housing units are sold in 2 years planning horizon. The length of the period and the number of periods in the planning horizon are 6 months and 4, respectively. Additionally, the investment discount rate is determined as 10%. The sales amount of housing units (SHU) for each period are attempted to be estimated depending on home loan interest rate (HLIR), and consumer confidence index (CCI), and housing supply (HS) in a Mamdani-type FRBS. Table 1 presents the linguistic variables and the corresponding fuzzy numbers used in FRBS for period 2. The membership functions for HLIR, CCI and HS remain same for each period, however the output membership function, SHU, is expected to be different.

Table 1. Linguistic variables for Period 2

	Scale	Corresponding Fuzzy Numbers
HLIR	Low	(0.8, 0.9, 1)
	Medium	(0.9, 1, 1.3)
	High	(1.2, 1.4, 1.5)
CCI	Low	(75, 80, 85)
	Medium	(80, 85, 90)
	High	(85, 90, 100)
HS	Low	(500, 600, 650)
	Medium	(600, 700, 800)
	High	(700, 900, 1000)
SHU	Low	(6, 9, 10)
	Medium	(9, 10, 12)
	High	(10, 12, 12)

The rule base for determining the sales amount of housing units is as follows:

- IF *HLIR* is Low AND *CCI* is High AND *HS* is Low THEN *SHU* is High.
- IF *HLIR* is Medium AND *CCI* is Medium AND *HS* is High THEN *SHU* is Low.
- IF *HLIR* is High AND *CCI* is Medium AND *HS* is Medium THEN *SHU* is Low.
- IF *HLIR* is High AND *CCI* is Medium AND *HS* is Low THEN *SHU* is Medium.
- IF *HLIR* is Medium AND *CCI* is Low AND *HS* is Medium THEN *SHU* is Low.

The sales amount of housing units are determined depending to the specified input values in Table 2 and the NPV is calculated as $1,077,454 in 2-years planning horizon.

Table 2. Sales forecast of housing units

Period	HLIR	CCI	HS	Sales Amount
1	1.15	95	600	6
2	0.88	95	650	10
3	1.3	98	700	8
4	1.35	98	750	8

5. Conclusion

The financial assessment of construction projects can be regarded as a complex problem in free market economies. Even in well-organized, relatively efficient markets, sales amount of housing units is difficult to estimate because of the fact that several factors, since several economic factors have influence on sales amount. In this study, a framework for financial assessment of construction projects is proposed. The framework includes FRBS which has the inputs of home loan interest rate, consumer confidence index and housing supply. As an output of FRBS, estimated sales amount of housing units contributes to the computation of NPV. This study has potential to provide a convenience financial control for construction project owner. For the future study, a decision support system can be developed based on the proposed evaluation framework.

References

1. C. Bagnoli, H. C. Smith, The Theory of Fuzzy Logic and its Application to Real Estate Valuation, J. Real Estate Res., 16 (2), pp. 169-199 (1999).
2. www.evagyd.com
3. Y. Huang,Y. Wang,S. Gai, The Application and Research of a New Combinatorial Analysis and Forecasting Method in Real Estate Area based on Grey System Theory and Multivariate Linear Regression, Procedia Engineering, 15, 4532-4537 (2011).
4. C. Brooks, S. Tsolacos, Forecasting real estate returns using financial spreads, J. Prop. Res., 18(3), 235–248, (2001)
5. J R. Magnus, A. T.K. Wan, X Zhang, Weighted average least squares estimation with nonspherical disturbances and an application to the Hong Kong housing market, Comput. Stat. Data An., 55, 1331–1341, (2011)
6. Y. Chen, J. M. Clapp, D. Tirtiroglu, Hedonic estimation of housing demand elasticity with a markup over marginal costs, J. Hous. Econ., 20, 233–248, (2011)

COMBINING DIFFERENT INFERENCE METHODS FOR MEDICAL DECISION SUPPORT SYSTEMS*

GUVEN KOSE†1, HAYRI SEVER[2], MERT BAL[3], ALP USTUNDAG[4]

[1]*Department of Information Management, Hacettepe University, Ankara, Turkey*
[2]*Department of Computer Engineering, Hacettepe University, Ankara, Turkey*
[3]*Department of Mathematical Engineering, Yildiz Technical University, Istanbul, Turkey*
[4]*Department of Industrial Engineering, Istanbul Technical University, Istanbul, Turkey*

In this study, synthetic data with 100, 1000 and 2000 records have been produced to reflect the probabilities on the ALARM network. In this study, a medical diagnosis system called as DRCAD is presented. DRCAD system is more innovative and interesting than the classical diagnosis support systems. In other words, DRCAD collects possible diagnosis of the patients from two sub modules. Each of these two sub modules gives the possible diagnosis from symptoms in a specific confidence degree. The proposal of two sub modules are combined linearly and diagnosis decisions are presented as a list. Each of the sub modules consist of Bayesian inference and rule-based inference models respectively. As a result, the methods in which the conclusions are combined in a linear manner are 5% more successful than the "Rule Based Method" when applied individually and 30% more successful than the cases where the "Bayesian Network Based Method" is utilized.

1. Introduction

A Decision Support Systems (DSSs) is a computer-based information system that supports organizational and business decision-making activities. Medical Decision Support Systems, which are variants of decision support systems, are intelligent software systems that are designed to improve clinical diagnosis system and to support the healthcare personnel in their decision. Intelligent decision support systems use artificial intelligence system techniques to support the healthcare personnel for selecting the best method for both diagnosis and also for treatment especially when the information about the treatment is incomplete or uncertain. The importance of the decision support systems is increasingly supporting the decision making process in cases of uncertainty and the lack of information and they are widely used in various fields like

* This work is supported TUBITAK, Drcad Medical Decision Support System, No: 3050042
† gkose@hacettepe.edu.tr

engineering, finance, medicine, etc. Bayesian Networks (BNs) also known as causal networks or probabilistic belief networks are widely used knowledge representation and probabilistic reasoning tools for various domains under uncertainty [1-2]. BNs have widely and successfully been used in medical decision support systems. In this study, a new inference mechanism is proposed for Medical Decision Support System.

2. Proposed Inference Method

2.1. *ALARM Network Structure and Datasets*

The network structure, which is used in scientific studies and known as ALARM (A Logical Alarm Reduction Mechanism) Network [3] in literature is used in this study. ALARM network is a network structure that is prepared by using real patient information for many variables and shows the probabilities derived from the real life circumstances. ALARM network calculates the probabilities for different diagnosis based on the current evidences and recently it has been used for many researchers. Totally there are 37 nodes in ALARM network and the relationships and conditional probabilities among these have been defined. The medical information has been coded in a graphical structure with 46 arches, 16 findings and 13 intermediate variables that relates the examination results to the diagnosis problems that represent 8 diagnosis problems. There are three variables named diagnosis, measurements and intermediate variables in the ALARM network. In order to use in the study, 100, 1000 and 2000 records have been produced to reflect the possibilities on the ALARM network. For these operations, based on ALARM network structure, NETICA 3.18 software has been used.

2.2. *Rule Based Method Testing Scenario*

Two approaches have been utilized here in order to determine the rules to be used by the related method on the ALARM network. First of them is to determine the relationships among the defined variables by using the topological structure of the network and constitute the rules and the second one is to derive some meaningful rules from the synthetic data in hand in the network by means of the practices in data mining methods. In the first method, except for the 12 defined triggered effects (input variables) the rules have been composed for all variables by utilizing the network structure. The rules in the first group are the auxiliary ones which are utilized in order to reach the target results. Those rules in the second group are the sort of rules which are utilized in order to reach the target conclusions. 188 rules have been added into the pool of rules which are convenient to examine by implementing machine learning techniques and also capable of generating intermediate conclusions and supplementary or additional diagnosis discovery or generation. These rules are created by utilizing ALARM.net network structure and by setting some possible rules for each variable. The rules which are prepared are, if they are single text based ones, each of them are transferred so as to become and serve as a valid RuleML [4] document and added into the pool of rules. See5 [5] application has been benefitted from in order to determine the meaningful rules on the synthetic data set we have in hand by utilizing the data mining methods. The relationships within the data sets and the definition of the patterns and related rules are derived by evaluating the See5 data of the cases. Thus, the rules intended to the target diagnosis are generated through this application by making use of decision trees. There are 85 rules intended for target diagnosis. Herein, the system as it is seen in Figure 1 is composed of three main architectural and functional parts so that it will be possible to apply and practice the "Rule Based Method".

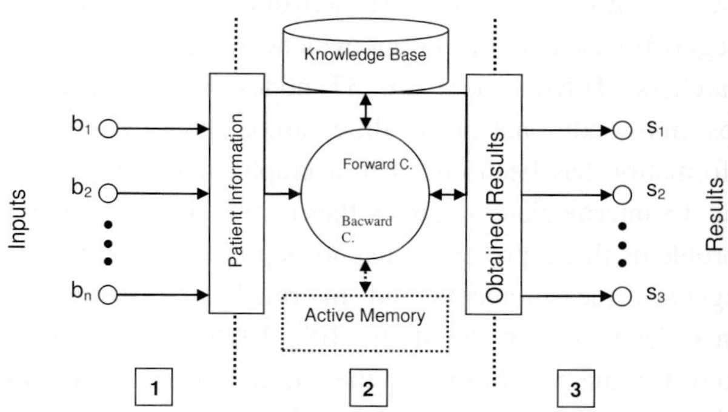

Fig.1. Functional Architecture of the Rule Based Method

In the functional architecture, the first part is the interface which enables us to obtain the information about the patient in question. Obtaining information about a patient means to introduce the facts to the system by means of which the machine will make inferences about the patient. Inferences are made by utilizing these facts and some new facts are obtained. The information in the system must be in the form which will enable us to process all of these pieces of information regarding the patient even if these pieces of information come from other systems which use different software. It has been decided that the bridge to be built between other decision support systems should be XML (EXtensible Markup Language) which has been developed in order to make it possible to operate such various systems together in cooperation with each other compatibly. The function of the bridge is to send the questioning information taken from the users and transfer them to bottom layers, that is, it transfers the information into forms which are possible for the bottom decision systems to process and send them into other layers by modifying them depending on their kinds and origins according to the purpose of the usage. RuleML rule mark up language is used in order to convey meanings of the rules and facts utilized within the developed medical decision support system. In the functional architecture, where the inference operation is actualized can both process the facts which will come from the intermediate layer and it will also undertake the tasks of the intermediate layer. The task of the inference mechanism is to generate results according to questioning performed by utilizing those facts to be sent to the machine and according to the rules. The realized inference machine has been inferring conclusions by utilizing those facts and rules chosen in order to be benefitted from, in accordance with the questioning. In order to carry out the inference operation, it is necessary to provide the system with the facts and they must be given on the basis of rules as well. After defining the field where the rules exist to the system, the inference machine starts making inference operations by utilizing those facts and rules. The other modifications applied to the inference machines are to generate results with the confidence degree which is presented in the structure as a solution to the problem examined. The confidence degree shows the possibility of the correctness of a result which is generated. According to this assumption, their degree of confidence will exist in those results obtained as a result of the inference operations carried out in the inference machine if all those rules and/or the facts are defined in it. Thanks to this, in accordance with the facts utilized within the system and the results obtained in line with the rules on which some comments can also be made saying that they should also be correct and valid. The confidence degrees are the values which may change between 0 and 1. After the completion of the chaining operation and obtaining the conclusions, if there are any, they are arranged according to the degree of their confidence rate starting from the biggest to the smallest ones. Each of the sets within the obtained conclusion sets are recorded as valid RuleML documents. After transferring this phase into practice, the generated solution to the defined problem has been actualized. Thus, conclusions are generated with the facts taken as in the forms of Standard RuleML documents of facts and rules.

2.3. *Bayesian Network Based Method Testing Scenario*

During the development of DRCAD Medical Diagnosis System [6], open source code JavaBayes [7] has been used to implement Bayesian Network based inference method. It is possible to say that JavaBayes software consists of two main modules. First of these modules is the Bayes Editor which has been developed for creating Bayesian network, forming variables and links, making the essential corrections on the network, observing the values of network variables. The second module is the inference algorithm that has been developed for making the essential inference and implementing the query on the network. JavaBayes software retrieves the possible values of variables that are determined according to the given inputs about the network by the help of the related algorithm module. After this phase, it is easy to determine a diagnosis list by ordering the related values. JavaBayes software uses the generalized interpretation of "*Variable Elimination*" [8] method as inference algorithm.

2.4. *The Testing Results of DRCAD Medical Diagnosis Software*

The Bayesian Network Based Inference and the methods of "Rule Based Inferences" which are actualized within the DRCAD Medical Diagnosis Software are observed first of all, independently. These conclusions are obtained from them linearly and all combined together so that it will be possible to reflect the possibilities on the ALARM network structure and then throughout this method records in the numbers of 100, 1000 and 2000 synthetic test data are generated. Each record composed on these data shows the values belonging to 37 variables registered on the network. There are 12 input and 11 conclusion variable on each record and the records of all the intermediate values are also saved and kept for usage. Those tests which are actualized send the input variables for each record to the inference engine and it saves the files of the lists created following each of the inference operation. The correctness of the inference conclusions are decided by comparing the records of the related variables of the recorded test data. 11 probable conclusions are reached for each record. The tests are composed of the linear combination phases of the conclusions obtained throughout both of those methods respectively the application of the "Rule Based Method" on the synthetic data set and by utilizing the "Bayesian Network Based Method". When the conclusions are examined, it can be seen that the Rule Based Method is more successful in the rate of 25 % than the "Bayesian Network Based" method in all dimensions of the data sets. Besides, when both of these methods are combined and utilized together the success rate rises to 80% much higher rates are acquired in comparison to the values obtained by applying these methods individually. In other words, the method in which the conclusions are combined in a linear manner are 5 % more successful than the "Rule Based Method when applied individually and 30% more successful than the cases where the Bayesian Network Based Method is utilized.

3. Conclusion

In this study, for the generated synthetic data, ALARM network structure which is widely used in scientific studies has been used. This network structure is a structure that has been prepared using real patient information for many variables and shows the possibilities derived from the real life circumstances. When the conclusions are examined, it can be seen that the Rule Based Method is more successful in the rate of 25 % than the "Bayesian Network Based" method in all dimensions of the data sets. Besides, when both of these methods are combined and utilized together the success rate rises to 80% much higher rates are acquired in comparison to the values obtained by applying these methods individually. In other words, the method in which the conclusions are combined in a linear manner are 5 % more successful than the "Rule Based Method when applied individually and 30% more successful than the cases where the "Bayesian Network Based Method is utilized. As conclusion, DRCAD system is more innovative than the classical diagnosis support systems and it has more successful results. Various methods can be combined as hybrid methods in order to evaluate the performance of decision support systems in future researches.

References

1. E. Castillo, J.M. Gutierrez, A.S. Hadi, "Expert Systems and Probabilistic Network Models", Springer Verlag, New York (1997)
2. J. Pearl, "Probabilistic Reasoning in Intelligent Systems: Networks of Plausible Inference", Morgan Kaufmann Publishers, California (1988).
3. Beinlich, H. J. Suermondt, R.M. Chavez, G.F. Cooper, "The ALARM Monitoring System: A Case Study with Two Probabilistic Inference Techniques for Belief Networks", In Proc. of the 2nd Eur. Conf. on Art. Int. in Med., 38, 247-256, (1989).
4. http://ruleml.org
5. http://www.rulequest.com/see5-info.html
6. http://netcad.com
7. http://www.cs.cmu.edu/~javabayes
8. F.G. Cozman, "Generalizing Variable Elimination in Bayesian Network", In Proc. of the 7 th Ibero-American Conference on Artificial Intelligence, IBERAMIA/SBIA'00, Sao Paulo, 27-32, 2000.

DECISION MAKING FOR CONCRETE EVALUATION USING FUSION OF NDT TECHNIQUES

Z.M SBARTAÏ

I2M-GCE, University of Bordeaux, CNRS, INRA
351 cours de la libération, Talence, 33405, France

V. GARNIER, M.A. PLOIX

LMA-LCND, University of Mediterranean, Aix-en-Provence, France,

Non-Destructive Testing (NDT) techniques are affected by concrete properties such as porosity, water content, strength, etc. Extracting one concrete property from one NDT measurement may lead to considerable uncertainties. This highlights the benefit of NDT data fusion to evaluate accurately concrete properties.
In this paper, NDT measurements from several NDT techniques were combined to predict more accurately concrete properties such as porosity and water saturation. Two techniques of data fusion were used namely neuro-fuzzy network and theory of possibility. The results obtained show the effectiveness of the statistical modelling to predict the properties of concretes by fusion of NDT measurements.

1. Introduction

For a decade, research investigations have been implemented for testing the potential of NDT techniques for concrete condition evaluation. However, the measurement of NDT physical features such as velocity of ultrasonic waves, electrical resistivity or radar wave are affected by noise "error" either from the accuracy of the measurement and/or the variability of the material. It is important to notice the limited ability of a single NDT technique to directly evaluate concrete property. This highlights the benefit of NDT data fusion to improve the interpretation of measurements and to reduce the uncertainty of predictions due to surface condition of the concrete element, the concrete properties, mix proportions, etc [1].

The results presented in this research work focus on the implementation of a general strategy for enhancing the assessment of concrete structures by NDT techniques. A large laboratory and on site experimental programme was implemented as a part of the SENSO national French project [2]. This paper deals with the implementation of two approaches of data fusion using statistical modelling with an aim of evaluating concrete properties from the measured NDT features. The two tested methods are the Fuzzy Neural Network technique (FNN) and data fusion using the theory of possibility.

2. Experimental database

81 slabs (50 cm x 25 cm x 12 cm), made from eight different compositions of concrete, were used in this study. They were conditioned at various water saturation degrees. The mixes covered a range of water-to-cement ratio sufficiently extended to be representative of what is usually found in structures (0.3 to 0.8). A large number of different NDT measurements were performed (ultrasonic, impact echo, capacitive method, radar, etc).

For the validation of the developed approach, concretes properties were measured using destructive tests. These properties were measured on cylindrical cores (modulus of elasticity, compressive strength, porosity), or by monitoring the slab's weight (degree of saturation). For each slab, the different techniques provided 52 NDT measured physical features (e.g. Ultrasonic Pulse Velocity). About 12,700 numerical values were collected and stored in a database. The first purpose of the data analysis was the reduction of the set of values. The second one was to select the most relevant NDT features for the prediction of each concrete property [3].

3. Data fusion

Several techniques of data fusion are available for combining NDT measurements in order to improve results and reduce uncertainties, in particular for the on site assessment. For instance, the probabilities were already often explored. The Dempster Shaffer theory [4] is well adapted to the classification of information to detect defects [5]. The theory of possibilities is able to take into account the information given by both experts and indicators analysis [6,7]. Artificial Neural Networks for modelling and prediction [8]. In this study, methods based on training laws (fuzzy neural network) and the theory of possibility are implemented and tested.

3.1. *Fuzzy neural network*

Neuro-fuzzy network is the combination of artificial neural network (ANN) and fuzzy logic. This combination allow to a hybrid intelligent system from the human-like reasoning style of fuzzy systems with the learning process of the neural network. FNN have been widely described by several authors [6]. The architecture of the FNN is more advantageous in comparison with ANN because in the case of ANN it is not easy to obtain the optimal structure.

The architectural of FNN is composed of the fuzzification of the imputs (NDT_1 ... NDT_n) and outputs (Y_1....Y_n), the construction of the fuzzy rules (R_1....R_n) and finally the defuzzification for outputs evaluation. The used FNN in this study is an Adaptive Neuro-Fuzzy Inference System (ANFIS) with Takagi–Sugeno type as shown in Figure 1a and Figure 1b.

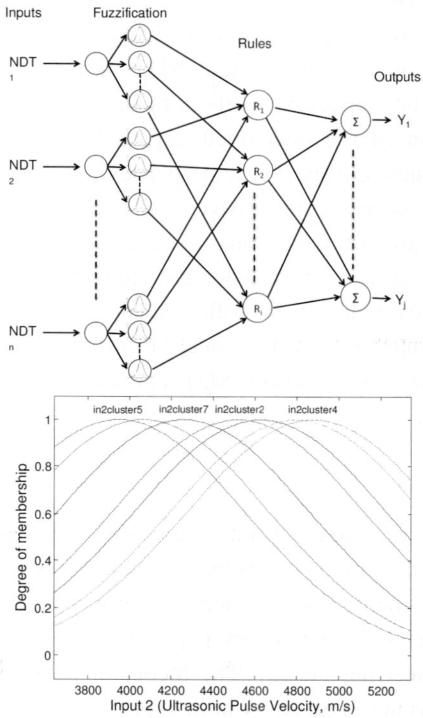

Figure 1. (a) Architecture of the used Adaptative Neuro-Fuzzy Inference System, (b) Fuzzification of the input_2 (Ultrasonic Pulse Velocity m/s).

Training this type of FNN consists in optimizing the model parameters in order to minimize an error function (mean square error) computed from predicted and target data. In the case of this study, back-propagation (BP) algorithm was used. The experimental database was divided into two parts: the training database (involved in the mathematical training process) and the testing database for the validation process.

Figure 2a present typical results of the FNN models trained for concrete porosity evaluation. Good correlation can be seen between actual data and those calculated by the models of FNN. The validation shows lower but acceptable performance. As chowed on Figure 2b, the probability distribution of the error of prediction is lower than 10%. For 95% of probability, the relative error is lower than 5%.

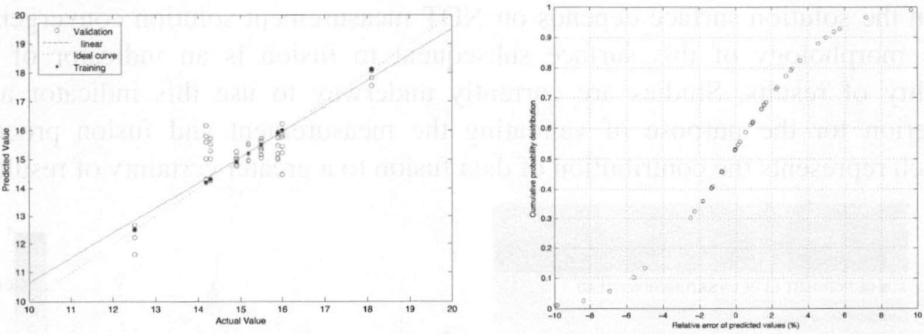

Figure 2. (a) Actual versus predicted concrete porosity for training and validation data. (b) Cumulative probability distribution of the relative error of prediction (validation data)

3.2. *Theory of possibility*

NDT data fusion would sometimes be contradictory. In this case, the decision made to propose a diagnosis would not be straightforward. We must make a choice or offer an "appropriate" response. It is possible to take into account the differences between these results by using fuzzy logic and theory of possibility. The multiplication and combination of measurements, which should lead to additional information, generate conflicts, in particular relative to on-site diagnoses. The data fusion step offers a solution to these conflicts due to the possibility of ambiguous configurations.

During this study, a data fusion model has been developed [7]. We have opted to implement the theory of possibilities, which is able to take into account the information provided by experts as well as an analysis from continuously varying indicators. The principle herein has been founded up on three stages [7]:

1. Knowledge modelling, to create the possibility distributions for indicators starting with non destructive measurements;
2. Data fusion by an operator. We have chosen a Delmotte operator [9] for its capacity to take into account the context and agreement between indicators stemming from each NDT technique;
3. Decision according to the maximum value of the possibility distribution after fusion.

This research has been based on the same experimental campaign and same cross-correlation model relating concrete properties to NDT features [3]. For the two parameters studied in pairs, i.e. (water saturation, porosity), figures 3a show the 2D projections of the possibility distributions for several NDT results before applying the fusion process.

The result in perspective (3D) after the application of fusion operator is provided in Figure 3b. As observed on this perspective, the emergence of a peak from the solution surface depends on NDT measurement solution convergence. The morphology of this surface subsequent to fusion is an indicator of the quality of results. Studies are currently underway to use this indicator as a criterion for the purpose of validating the measurement and fusion process, which represents the contribution of data fusion to a greater certainty of results.

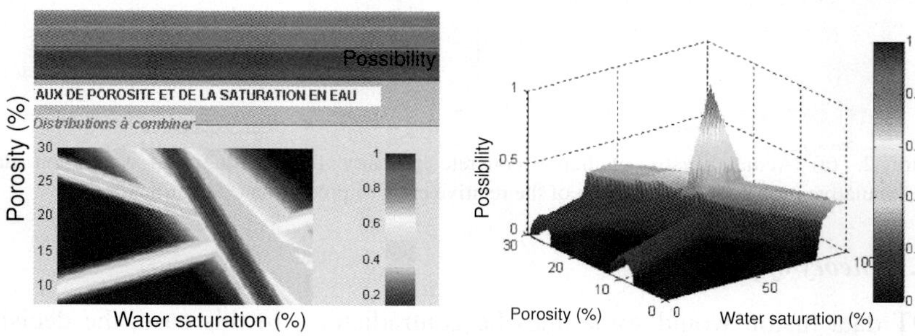

Figure 3. Projection of NDT measurements for porosity and water saturation evaluation, a-overview, b-3D perspective.

4. Conclusions

Measurements from a single non-destructive technique (NDT) may lead to considerable uncertainties. Developing methodologies for the assessment of concrete structures based on fusion of NDT techniques are needed to evaluate concrete properties on site. In this paper, two methodologies were implemented, the theory of possibility and fuzzy neural network (FNN).

The results of this study show that both techniques have satisfactory ability to predict fused NDT measurements for concrete properties. Regarding FNN technique, their prediction ability on new data is more efficient than that of the fusion using theory of possibility. This is because the theory of possibility needs the existing models, which is multi-linear regression in the case of this study. However, it offers a solution conflicts between NDT techniques due to the possibility of ambiguous configurations.

Further works is being implemented which consist of:

1. Testing these approaches on real structures (bridge of Marly at Valenciennes, France);

2. Combining data fusion with the conditional probability in order to evaluate the uncertainty in the support decision-making process.

Acknowledgments

The authors thank the participants involved in the SENSO and ACDC projects for their contributions to this paper. The SENSO and ACDC team is acknowledged for its contributions. The French National Agency for Research (ANR) and the Ministry (MEDDTL) are acknowledged for its financial support.

References

1. D. Breysse, M. Larget, Z.M Sbartaï, J-F. Lataste, J-P. Balayssac. Quality of NDT measurements and accuracy of concrete physical properties quantitative assessment. *European Journal of Environmental and Civil Engineering*. **15(4)**, (2011).
2. JP. Balayssac, SENSO: a French project for the evaluation of concrete structures by combining non-destructive methods. In *SACoMaTiS 2008*, Binda L, di Prisco M, and Felicetti R, editors, Varenna, Italy, 1-2 september, (2008).
3. Z.M Sbartaï, D. Breysse, M. Larget, J-P. Balayssac, Combining NDT Techniques for Improved Evaluation of Concrete Properties. *Cement and Concrete Composites*. **V34(6)**, 725-733, (2012).
4. G.A. Shafer, mathematical theory of evidence, *Princeton University Press*: 297, (1976).
5. X.E. Gros, NDT data fusion, *John Wiley & Sons New York*, 205, (1997).
6. Zhang Shao-Fei Jiang, Chun-Ming Zhang, Shuai. Two-stage structural damage detection using fuzzy neural networks and data fusion techniques. *Expert Systems with Applications*, **V38(1)**, 511-519, (2011).
7. M-A. Ploix, V. Garnier, D. Breysse, J. Moysan, NDE data fusion to improve the evaluation of concrete structures, *NDT & E International*, **44(5)**, 442-448, (2011).
8. Z.M. Sbartaï, S. Laurens, K. Viriyametanont, J_P. Balayssac, G. Arliguie, Non-destructive evaluation of concrete physical condition using radar and artificial neural networks, *Construction and Building Materials*, **23(2)**, 837-845, (2009).
9. Delmotte F. Un nouvel opérateur de fusion adaptatif. *Traitement du Signal*, **17**, 299-311, (2000).

MULTI-PERIOD DECISION MAKING WITH ONE-SHOT DECISION THEORY

PEIJUN GUO[†], YONGGANG LI

Faculty of Business Administration, Yokohama National University, Yokohama, 240-8501 Japan

In this paper, a multi-period decision problem with partially known information is considered. In each stage, a decision maker has one and only one chance to make a decision. The optimal decision in each stage is obtained based on the one-shot decision theory. That is, the decision maker chooses one of states of nature (scenario) of each alternative in every stage with considering the satisfaction of the outcome and its possibility. The selected state of nature is called the focus point. Based on the focus points, the decision maker determines the optimal alternative in each stage by dynamic programming problems.

1. Introduction

It is well known that dynamic programming is a powerful vehicle for deal with multistage decision-making problem. The stochastic dynamic programming problems and fuzzy dynamic programming problems are used for handing the uncertainty existing in the multistage decision problems [1-2,6-7]. The stochastic dynamic programming problems are formulated as a maximization problem of expected value with the assumption that the system under control is a Markov chain while fuzzy dynamic programming problems introduce a fuzzy constraint and a fuzzy goal in each stage following Bellman and Zadeh's framework. Different from the existing approaches, the proposed approaches consider a possibilistic dynamic system under control and in each stage only one chance exists for a decision maker to make a decision. A decision maker makes a decision based on one-shot decision theory, which has been proposed and applied to a duopoly market of a new product with a short life cycle and private real estate investment problem [3-5]. In the one-shot decision theory, the one-shot decision process is separated into two steps. The first step is to identify which state of nature should be taken into account for each alternative with considering possibility of a state of nature and satisfaction of an outcome. The

---

† guo@ynu.ac.jp

states of nature focused are called focus points. The second step is to evaluate the alternatives to obtain the optimal alternative.

The remainder of the paper is organized as follows. In Section 2, the one-shot decision theory is brief introduced. In Section 3, a general approach to multistage decision making based on one-shot decision theory is proposed.

2. Introduction to One-Shot Decision Theory

The first step in decision analysis is the problem formulation. The set of an alternative a is A. The set of a state of nature x is S. The degree to which a state of nature is to occur in the future is characterized by a possibility distribution $\pi(x)$ defined below.

Definition 1. Given a function $\pi: S \to [0,1]$ if $\max_{x\in S} \pi(x) = 1$, then $\pi(x)$ is called a possibility distribution where S is the sample space. $\pi(x)$ is the possibility degree of x. $\pi(x) = 1$ means that it is normal that x occurs and $\pi(x) = 0$ means that it is abnormal that x occurs. The smaller the possibility degree of x, the more surprising is the occurrence of x.

The consequence resulting from the combination of an alternative a and a state of nature x is referred to as a payoff, denoted as $v(x,a)$. The satisfaction level of a decision maker for a payoff can be expressed by a satisfaction function, as defined below.

Definition 2. Denote the set of a payoff $v(x,a)$ as V. The function $u: V \to [0,1]$ is called a satisfaction function if it satisfies $u(v_1) > u(v_2)$ for $v_1 > v_2$.

Because the payoff is the function of x and a, we write the satisfaction function as $u(x,a)$. Since one and only state of nature will come up for a one-shot decision problem, a decision maker needs to decide which state of nature ought to be considered for making a one-shot decision. Each state of nature is equipped with a pair of possibility and satisfaction so that how to determine focus points (focused states of natures) depends on his/her attitudes about possibility and satisfaction. Twelve types of focus points are provided to help a decision maker in finding out his/her own appropriate one.

Type I: $x_\alpha^{1*}(a) = \arg\max_{x\in X^{\geq\alpha}} u(x,a)$ where $X^{\geq\alpha} = \{x \mid \pi(x) \geq \alpha\}$. $x_\alpha^{1*}(a)$ is Type I focus point.

Type II: $x_\alpha^{2*}(a) = \arg\min_{x\in X^{\geq\alpha}} u(x,a)$ where $X^{\geq\alpha} = \{x \mid \pi(x) \geq \alpha\}$ is called as Type II focus point.

Type III: $x_{\alpha}^{3*}(a) = \arg\max_{x\in X^{\leq\alpha}} u(x,a)$ where $X^{\leq\alpha} = \{x \mid \pi(x) \leq \alpha\}$ is called as Type III focus point.

Type IV: $x_{\alpha}^{4*}(a) = \arg\min_{x\in X^{\leq\alpha}} u(x,a)$ where $X^{\leq\alpha} = \{x \mid \pi(x) \leq \alpha\}$ is called as Type IV focus point.

Type V: $x_{\beta}^{5*}(a) = \arg\max_{x\in X^{\geq\beta}(a)} \pi(x)$ where $X^{\geq\beta}(a) = \{x \mid u(x,a) \geq \beta\}$ is called as Type V focus point.

Type VI: $x_{\beta}^{6*}(a) = \arg\min_{x\in X^{\geq\beta}(a)} \pi(x)$ where $X^{\geq\beta}(a) = \{x \mid u(x,a) \geq \beta\}$ is called as Type VI focus point.

Type VII: $x_{\beta}^{7*}(a) = \arg\max_{x\in X^{\leq\beta}(a)} \pi(x)$ where $X^{\leq\beta}(a) = \{x \mid u(x,a) \leq \beta\}$ is called as Type VII focus point.

Type VIII: $x_{\beta}^{8*}(a) = \arg\min_{x\in X^{\leq\beta}(a)} \pi(x)$ where $X^{\leq\beta}(a) = \{x \mid u(x,a) \leq \beta\}$ is called as Type VIII focus point.

Type IX: $x^{9*}(a) = \arg\max_{x\in S} \min[\pi(x), u(x,a)]$. (1)

$x^{9*}(a)$ is Type IX focus point.

Type X: $x^{10*}(a) = \arg\min_{x\in S} \max[\pi(x), u(x,a)]$. (2)

$x^{10*}(a)$ is Type X focus point.

Type XI: $x^{11*}(a) = \arg\min_{x\in S} \max[1-\pi(x), u(x,a)]$. (3)

$x^{11*}(a)$ is Type XI focus point.

Type XII: $x^{12*}(a) = \arg\min_{x\in S} \max[\pi(x), 1-u(x,a)]$. (4)

$x^{12*}(a)$ is Type XII focus point.

For one alternative, more than one state of nature might exist as one type of focus point. We denote the sets of twelve types of focus points of the alternative a as $X_{\alpha}^{1}(a)$, $X_{\alpha}^{2}(a)$, $X_{\alpha}^{3}(a)$, $X_{\alpha}^{4}(a)$, $X_{\beta}^{5}(a)$, $X_{\beta}^{6}(a)$, $X_{\beta}^{7}(a)$, $X_{\beta}^{8}(a)$, $X^{9}(a)$, $X^{10}(a)$, $X^{11}(a)$, and $X^{12}(a)$, respectively.

In one-shot decision, the decision maker contemplates that the focus points are the most appropriate states of nature (scenarios) for him/her and the decision maker chooses one alternative which can bring about the best consequence (highest satisfaction level) once the focus point (scenario) comes true. There are twelve types of optimal alternatives, as shown below.

$$a^{1*}(\alpha) = \arg\max_{a\in A} u(x_{\alpha}^{1*}(a),a),\ a^{2*}(\alpha) = \arg\max_{a\in A} u(x_{\alpha}^{2*}(a),a),$$

$$a^{3*}(\alpha) = \arg\max_{a\in A} u(x_{\alpha}^{3*}(a),a),\ a^{4*}(\alpha) = \arg\max_{a\in A} u(x_{\alpha}^{4*}(a),a)$$

$$a^{5*}(\beta) = \arg\max_{a\in A} \min_{x_{\beta}^{5*}(a)\in X_{\beta}^{5}(a)} u(x_{\beta}^{5*}(a),a),\ a^{6*}(\beta) = \arg\max_{a\in A} \min_{x_{\beta}^{6*}(a)\in X_{\beta}^{6}(a)} u(x_{\beta}^{6*}(a),a),$$

$a^{7*}(\beta) = \arg\max_{a \in A} \min_{x_\beta^{7*}(a) \in X_\beta^7(a)} u(x_\beta^{7*}(a), a)$, $a^{8*}(\beta) = \arg\max_{a \in A} \min_{x_\beta^{8*}(a) \in X_\beta^8(a)} u(x_\beta^{8*}(a), a)$,

$a^{9*} = \arg\max_{a \in A} \max_{x^{9*}(a) \in X^9(a)} u(x^{9*}(a), a)$, $a^{10*} = \arg\max_{a \in A} \min_{x^{10*}(a) \in X^{10}(a)} u(x^{10*}(a), a)$.

$a^{11*} = \arg\max_{a \in A} \min_{x^{11*}(a) \in X^{11}(a)} u(x^{11*}(a), a)$. $a^{12*} = \arg\max_{a \in A} \max_{x^{12*}(a) \in X^{12}(a)} u(x^{12*}(a), a)$.

3. Multistage One-Shot Decision Approaches

Let us consider a multistage decision making problem. If in each stage there is one and only one chance for making a decision, we call such a multistage decision making problem as multistage one-shot decision problem. Such kinds of decision problems are extensively encountered in business, economics and social system. In the multistage one-shot decision problem, the length of decision period is T. $A_t = \{a_{t,1}, a_{t,2}, \cdots, a_{t,m_t}\}$ is the set of the alternatives available in the stage *t*, the cardinality of A_t is m_t. That is, in the stage t, there is one and only one chance to choose one alternative of A_t. $X_{t+1}(x_{t,k}, a_{t,i})$ is the set of a state in the next stage $t+1$ when the state is $x_{t,k}$ and the alternative $a_{t,i}$ is selected in the stage t. $X_{t+1}(x_{t,k}, a_{t,i})$ is as follows:

$$X_{t+1}(x_{t,k}, a_{t,i}) = \{x_{t+1,1}(x_{t,k}, a_{t,i}), x_{t+1,2}(x_{t,k}, a_{t,i}), \cdots, x_{t+1,k_{t+1}(x_{t,k}, a_{t,i})}(x_{t,k}, a_{t,i})\}. \quad (5)$$

Each $x_{t+1,j}(x_{t,k}, a_{t,i}), 1 \le j \le k_{t+1}(x_{t,k}, a_{t,i})$ is one of states in the stage $t+1$ resulted by the alternative $a_{t,i}$ when the state is $x_{t,k}$ in the stage t. $X_{t+1}(x_{t,k}, a_{t,i})$ is governed by a possibility distribution defined below.

Definition 3: A possibility distribution is a function $\pi_{t+1} : X_{t+1}(x_{t,k}, a_{t,i}) \to [0,1]$ satisfying $\max_{x \in X_{t+1}(x_{t,k}, a_{t,i})} \pi_{t+1}(x) = 1$. $\pi_{t+1}(x)$ is the possibility degree of x in the stage $t+1$.

Let $c_{t+1,j}(x_{t,k}, a_{t,i}, x_{t+1,j}) = K(x_{t,k}, a_{t,i}, x_{t+1,j})$ express the transition cost generated from the process $x_{t,k} \to x_{t+1,j}(x_{t,k}, a_{t,i})$ when the state is $x_{t,k}$ and the action we choose is $a_{t,i}$ at time t. The set of the cost corresponding to $X_{t+1}(x_{t,k}, a_{t,i})$ is expressed as

$$C_{t+1}(x_{t,k}, a_{t,i}) = \{c_{t+1,j}(x_{t,k}, a_{t,i}, x_{t+1,j}) \mid 1 \le j \le k_{t+1}(x_{t,k}, a_{t,i})\}. \quad (6)$$

Given the information A_t, $X_{t+1}(x_{t,k}, a_{t,i})$, $\pi_{t+1}(x)$ and $C_{t+1}(x_{t,k}, a_{t,i})$, multi-stage one-shot decision is to choose one alternative in each stage when there is one and only one chance for the decision maker making a decision in each stage. Let us firstly consider the last decision in the stage $T-1$ which is typical one-shot decision. We give the satisfaction function in the stage T as follows:

Definition 4. The function $u_T : C_T(x_{T-1,k}, a_{T-1,i}) \to [0,1]$ is called a satisfaction function of $a_{T-1,i}$ in the stage T if it satisfies $u_T(w_1) > u_T(w_2)$, for any $w_1 < w_2$, $w_1, w_2 \in C_T(x_{T-1,k}, a_{T-1,i})$.

The procedure of one-shot decision in the last stage, $T-1$ is as follows:

Step 1: We seek the focus points for each $a_{T-1,i} \in A_{T-1}$ with the condition that the state in the current stage is $x_{T-1,k}$. Denote the obtained focus point of $a_{T-1,i} \in A_{T-1}$ as $x_T^*(x_{T-1,k}, a_{T-1,i})$, which can be obtained as follows:

$$x_T^*(x_{T-1,k}, a_{T-1,i}) = \arg F(\pi_T(x_{T,j}(x_{T-1,k}, a_{T-1,i})), u_T(c_{T,j}(x_{T-1,k}, a_{T-1,i}, x_{T,j}(x_{T-1,k}, a_{T-1,i})))) \quad (7)$$

where $F(\cdot)$ is a general form representing the approaches for seeking one of twelve types' focus point. For example, $F(\cdot) = \max_x \min[\pi_T, u_T]$ corresponds to the Type IX focus point of $a_{T-1,i}$.

Step 2: The optimal alternative $a_{T-1}^*(x_{T-1,k})$ is determined as follows:

$$a_{T-1}^*(x_{T-1,k}) = \arg f_1(x_{T-1,k}), \quad (8)$$

where $f_1(x_{T-1,k}) = \min_{a \in A_{T-1}} K(x_{T-1,k}, a, x_T^*(x_{T-1,k}, a))$.

Next, we consider the decision in the stage $T-n$, $n \geq 2$. Set

$$\begin{aligned} & G_{T-n+1}(x_{T-n,k}, a_{T-n,i}) \\ & = \{g_{T-n+1,j}(x_{T-n,k}, a_{T-n,i}, x_{T-n+1,j}) \mid 1 \leq j \leq k_{T-n+1}(x_{T-n,k}, a_{T-n,i})\} \\ & = \{c_{T-n+1,j}(x_{T-n,k}, a_{T-n,i}, x_{T-n+1,j}) + f_{n-1}(x_{T-n+1,j}) \mid 1 \leq j \leq k_{T-n+1}(x_{T-n,k}, a_{T-n,i})\} \end{aligned} \quad (9)$$

The satisfaction function in the stage $T-n+1$, given as $u_{T-n+1}(x_{T-n,k}, a_{T-n,i})$: $G_{T-n+1}(x_{T-n,k}, a_{T-n,i}) \to [0,1]$ satisfies $u_{T-n+1}(w_1) > u_{T-n+1}(w_2)$, for any $w_1 < w_2$, $w_1, w_2 \in G_{T-n+1}(x_{T-n,k}, a_{T-n,i})$.

The one-shot decision in the stage $T-n$ is obtained as follows:

Step 1: Finding out the focus points for $a_{T-n,i}$:

Denote the focus point of $a_{T-n,i} \in A_{T-n}$ as $x_{T-n+1}^*(x_{T-n,k}, a_{T-n,i})$, which can be obtained as follows:

$$\begin{aligned} & x_{T-n+1}^*(x_{T-n,k}, a_{T-n,i}) \\ & = \arg F(\pi_{T-n+1}(x_{T-n+1,j}(x_{T-n,k}, a_{T-n,i})), u_{T-n+1}(g_{T-n+1,j}(x_{T-n,k}, a_{T-n,i}, x_{T-n+1,j}(x_{T-n,k}, a_{T-n,i})))) \end{aligned} \quad (10)$$

where the same type of focus point as the one obtained in the previous stage should be sought due to the logic consistency.

Step 2: Obtaining the optimal action in the stage $T-n$:

$$f_n(x_{T-n,k}) = \min_{a \in A_{T-n}} [K(x_{T-n,k}, a, x^*_{T-n+1}(x_{T-n,k}, a)) + f_{n-1}(x^*_{T-n+1}(x_{T-n,k}, a))], \tag{11}$$

where $a^*_{T-n}(x_{T-n,k}) = \arg f_n(x_{T-n,k})$.

Repeating (9), (10) and (11), we can obtain the optimal decision sequence $a^*_0, a^*_1, \cdots, a^*_{T-1}$ for the initial state x_0. We can construct the optimal path under the decision sequence $a^*_0, a^*_1, \cdots, a^*_{T-1}$, that is, x_0, $x^*_1(x_0, a^*_0)$, $x^*_2(x^*_1, a^*_1)$, …. , $x^*_T(x^*_{T-1}, a^*_{T-1})$.

4. Conclusion

In this paper, a new approach for multi-period decision making with one-shot decision theory is proposed. The proposed method provides an efficient vehicle to deal with multistage one-shot decision problems under uncertainty which are extensively encountered in business, economics, and social systems.

References

1. M. P. Cristobal, L. F. Escudero and J. F. Monge, On stochastic dynamic programming for solving large-scale planning problems under uncertainty, *Comput Oper Res*, 36, 2418-2428 (2009).
2. A. O. Esogbue and R. E. Bellman: Fuzzy dynamic programming and its extensions. *TIMS/Studies in the Management Sciences* 20, 147-167 (1984).
3. P. Guo, One-Shot Decision Approach and Its Application to Duopoly Market, *Int J Inf Decis*, 2, 213–232 (2010).
4. P. Guo, Private Real Estate Investment Analysis within One-Shot Decision Framework, *International Real Estate Review*, 13, 238-260 (2010).
5. P. Guo, One-shot decision Theory, *IEEE T Syst Man Cy A*, 41, 917-926 (2011).
6. J. Kacprzyk and A. O. Esogbue, Fuzzy dynamic programming: Main developments and applications, *Fuzzy Sets and Systems* , 81, 31-45 (1996).
7. N. Topaloglou, H. Vladimirou and S. A. Zenios, A dynamic stochastic programming model for international portfolio management, *Eur J Oper Res*, 185, 1501-1524 (2008).

PREFERENCE INTENSITY IN MCDM WHEN AN ADDITIVE UTILITY FUNCTION REPRESENTS DM PREFERENCES

A. MATEOS, A. JIMENEZ, E.A. AGUAYO, P. SABIO

Departamento de Inteligencia Artificial, Universidad Politécnica de Madrid (UPM), 28660 Boadilla del Monte, Madrid, Spain

We propose a new method for ranking alternatives in multicriteria decision-making problems when there is imprecision concerning the alternative performances, component utility functions and weights. We assume decision maker's preferences are represented by an additive multiattribute utility function, in which weights can be modeled by independent normal variables, fuzzy numbers, value intervals or by an ordinal relation. The approaches are based on dominance measures or exploring the weight space in order to describe which ratings would make each alternative the preferred one. On the one hand, the approaches based on dominance measures compute the minimum utility difference among pairs of alternatives. Then, they compute a measure by which to rank the alternatives. On the other hand, the approaches based on exploring the weight space compute confidence factors describing the reliability of the analysis. These methods are compared using Monte Carlo simulation.

1. Introduction

In multicriteria decision-making, the classical additive multiattribute utility model is considered to be a valid approach in many practical situations[2]. Incorporating imprecision concerning weights and/or component utilities is one way of extending the model to closely describe real situations: less information is usually available than is needed to determine the best alternative[9]. Sarabando and Dias[7] give a brief overview of approaches within the multiattribute utility theory framework to deal with incomplete information.

A recent approach is to use information about each alternative's intensity dominance, known as *dominance measuring methods*. Ahn and Park[1] were the first to propose a dominance measuring method. This method computes both dominating and dominated measures and then derives a *net dominance*. Net dominance is used as a measure of the strength of preference in the sense that a greater net value is better. Mateos et al.[5,6] propose and compare two alternative methods aimed at improving Ahn and Park's methods. Mateos et al.[5] consider uniformly distributed intervals to model imprecision concerning weights, whereas Mateos et al.[6] consider ordinal relations among attribute weights.

A second approach to deal with imprecision is the Stochastic Multicriteria Acceptability Analysis (*SMAA*)[3]. *SMAA* computes *acceptability indices*, which measure the variety of different preferences that rank each alternative as best. This information can be used to classify the alternatives as more or less acceptable and as unacceptable. However, *SMAA* ignores information about the other ranks. This problem is solved in Lahdelma and Salminen[4] using the SMAA-2 method, which extends the analysis to the sets of weight vectors for any rank from best to worst for each decision alternative.

In this paper, we extend and improve the methods proposed in Mateos et al.[5,6]. Instead of weight intervals for each attribute or ordinal relations among attribute weights, we assume weights can follow independent normal distributions or fuzzy numbers (triangular or trapezoidal). In Section 2 we introduce the extension of dominance measuring methods proposed in Mateos et al.[5,6]. Section 3 reports a simulation study of the methods outlined in previous sections. Finally, some conclusions are considered.

2. A New Preference Intensity Method (*NPIM*)

We consider a set of alternatives $A_1, \ldots, A_m$ evaluated on attributes $X_1, \ldots, X_n$, a utility function $u_j(x_{ij})$ on the performance x_{ij} of alternative A_i under attribute X_j, a set of weights w_j and the well-known functional form $u(A_i) = \sum_{j=1}^{n} w_j u_j(x_{ij})$, $\forall i$. This represents the utility of each alternative A_i, $\forall i$. The incomplete information about input parameters has been incorporated into the decision-making process: (1) alternative performances under uncertainty and imprecision concerning utility function assessment, $u_j(x_{ij}) \in \boldsymbol{U}_{ij}$; and (2) uncertainty about weights, which can be represented by: (a) ordinal relations, $\boldsymbol{W} = \{w: w_1 \geq w_2 \geq \cdots \geq w_n\}$; (b) value intervals, $\boldsymbol{W} = \{w: w_j \in [w_j^L, w_j^U], \forall j\}$; (c) intervals for weight ratios, $\boldsymbol{W} = \{w: w_j/w_k \in [w_{jk}^L, w_{jk}^U]\}$; (d) linear inequality constraints for weights, $\boldsymbol{W} = \{Aw \leq c\}$; (e) nonlinear inequality constraints for weights, $\boldsymbol{W} = \{g(w) \leq 0\}$; (f) independent normal distributions, $\boldsymbol{W} = \{w: w_j \sim N(\mu_j, \sigma_j^2), \forall j\}$; (f) fuzzy numbers, $\boldsymbol{W} = \{\widetilde{w}: \widetilde{w}_j, \forall j\}$, where $\widetilde{w}_j$ are triangular or trapezoidal fuzzy numbers.

In this section we propose a method based on the concept of dominance. Given two alternatives A_k and A_l, alternative A_k dominates A_l if $D_{kl} \geq 0$, D_{kl} being the optimum value of the optimization problem:

$$D_{kl} = \min_{\substack{w \in \boldsymbol{W} \\ u_j(x_{kj}) \in \boldsymbol{U}_{kj}, \forall j \\ u_j(x_{lj}) \in \boldsymbol{U}_{lj}, \forall j}} \left\{ u(A_k) - u(A_l) = \sum_{j=1}^{n} w_j \left(u_j(x_{kj}) - u_j(x_{lj}) \right) \right\} \quad (2)$$

The method is implemented in the following 4 steps.

1. Compute D_{kl} for alternatives A_k and A_l ($\forall k,l$) following (2) and the intervals $\boldsymbol{I_{kl}} = [D_{kl}, -D_{lk}]$, if uncertainty about weights can be represented by style a), b), c), d) or e), see Section 2.
2. Compute proportions P_{kl} as follows:

a. If uncertainty about weights is represented by style a), b), c), d) or e), see Section 2, then

$$P_{kl} = \begin{cases} d(\boldsymbol{I_{kl}}, 0), & if\ D_{kl} \geq 0 \\ \frac{-D_{lk}+D_{kl}}{-D_{lk}-D_{kl}} d(\boldsymbol{I_{kl}}, 0), & if\ D_{kl} < 0\ and -D_{lk} > 0; \\ -d(\boldsymbol{I_{kl}}, 0), & if\ -D_{lk} \leq 0 \end{cases}$$

b. If uncertainty about weights is represented by style f), then

$$P_{kl} = \left(\int_0^{\infty} f_{kl}(x)dx\right) d(\boldsymbol{I_{kl}}, 0) - \left(1 - \int_0^{\infty} f_{kl}(x)dx\right) d(\boldsymbol{I_{kl}}, 0)$$

where $f_{kl}(x)$ is the density function of the variable Dkl and $I_{kl} = [\mu - 2\sigma, \mu + 2\sigma]$;

a. If uncertainty about weights is represented by style g), then

$$P_{kl} = \frac{\int_0^{\infty} f_{kl}(x)dx}{\int_{-\infty}^{\infty} f_{kl}(x)dx} d(D_{kl}, 0) - \frac{\int_{-\infty}^{0} f_{kl}(x)dx}{\int_{-\infty}^{\infty} f_{kl}(x)dx} d(D_{kl}, 0)$$

where $f_{kl}(x)$ is the membership function of the fuzzy number D_{kl}. Here $d(\cdot,\cdot)$ is a distance measure[8].

3. Compute a preference intensity measure for each alternative A_k: $P_k = \sum_{l=1, l\neq k}^{m} P_{kl}$.
4. Rank alternatives according to the P_k values, where the best (rank 1) is the alternative with greatest P_k and the worst is the alternative with the least P_k.

3. Simulation Study

Let us compare the proposed method (*NPIM*) with the Mateos et al's[5,6] (*PIM*), Ahn and Park's[1] (*AP*), *SMAA*[3] and *SMAA-2*[4] methods.

We propose to carry out a simulation study of the above methods to analyze their performance. The process would be as follows: (1) Randomly generate component utilities for each alternative in each attribute from a uniform distribution in (0,1), leading to an $m \times n$ matrix. Remove dominated alternatives; (2) Generate attribute weights. Note that these weights are the TRUE weights, and the derived ranking of alternatives will be denoted as the TRUE ranking. The resulting weights will sum 1 and be uniformly distributed in the weight space; (3) To derive the corresponding weight intervals, add and subtract the same quantity to precise values, leading to the lower and upper endpoints of the weight intervals. We used the quantities, q, of 0.05, 0.1, 0.15, 0.2 and 0.25 that represent 10%, 20%, 30%, 40% and 50% imprecision, respectively. In other words, $[w_i^L, w_i^U] = [w_i^T - q, w_i^T + q]$. If $w_i^T - q < 0$, then $w_i^T - q = 0$, and if $w_i^T + q > 1$, then $w_i^T + q = 1$ is considered. Throughout the simulation process weights will be randomly generated from these weight intervals, $[w_i^T - q, w_i^T + q]$; (4) Compute the ranking of alternatives for each method according to their procedures and compare with the TRUE ranking, computed in step 2. We used two measures of efficacy, *hit ratio* and *rank-order correlation*[1]. The hit ratio is the proportion of all cases in which the method selects the same best alternative as in the TRUE ranking. Rank-order correlation represents how similar the overall structures ranking alternatives are in the TRUE ranking and in the ranking derived from the method. It is calculated using Kendall's τ.

Four different levels of alternatives (m = 3, 5, 7, 10) and five different levels of attributes (n = 3, 5, 7, 10, 15) were considered in order to validate the results output. Also, 20,000 trials were performed for each of the 20 design elements (alternatives × attributes).

Table 1. Average hit ratios.

Alt.	*Atrib.*	*NPIM*	*PIM*	*AP*	*SMAA*	*SMAA-2*
3	3	0.9937	0.9923	0.9813	0.9962	0.9962
	5	0.9878	0.9856	0.9744	0.9900	0.9901
	7	0.9778	0.9765	0.9610	0.9820	0.9824
	10	0.9652	0.9596	0.9459	0.9697	0.9698
	15	0.9300	0.9248	0.9043	0.9469	0.9470
5	3	0.9902	0.9880	0.9697	0.9947	0.9951
	5	0.9815	0.9773	0.9540	0.9856	0.9857
	7	0.9709	0.9645	0.9404	0.9743	0.9752
	10	0.9504	0.9404	0.9163	0.9600	0.9606
	15	0.9110	0.90085	0.8701	0.9271	0.9273
7	3	0.9906	0.9867	0.9667	0.9935	0.9935
	5	0.9768	0.9711	0.9446	0.9814	0.9818
	7	0.9657	0.9549	0.9249	0.9727	0.9732
	10	0.9438	0.9311	0.89985	0.9509	0.9512
	15	0.9025	0.8853	0.8455	0.9081	0.9087
10	3	0.9895	0.9826	0.9603	0.9924	0.9926
	5	0.9753	0.9658	0.9317	0.9790	0.9798
	7	0.9602	0.9431	0.9079	0.9620	0.9631
	10	0.9379	0.9191	0.8807	0.9397	0.9406
	15	0.8944	0.8742	0.8326	0.8921	0.8925

Table 1 exhibits the average hit ratio for each of the 20 design elements when the interval length is 0.1, i.e., the average values of 20,000 trials.

If we consider the average rank-order correlation or other levels of imprecision, the results are very similar.

We can conclude that: (1) The new method and *SMAA-2* are similar, and the difference between them is negligible; (2) The new method is easier to apply than *SMAA-2* because the multidimensional integrals generally have to be computed using numerical techniques in *SMAA-2*; (3) The new method can be applied when uncertainty about weights is represented by fuzzy numbers, whereas *SMAA-2* is not applicable in these cases; (4) All methods return a worse hit ratio when imprecision increases, as is to be expected.

4. Conclusions

In this paper we consider a new method to ranking alternatives when there is imprecision in the DM's preferences. This method is based on dominance measures. This method has the following advantages. First, it is applicable when the uncertainty about the weights of the additive multiattribute utility function can be represented by ordinal relations, value intervals, intervals for weight ratios, linear inequality constraints for weights, independent normal distributions or fuzzy numbers. Second, the method is one of the best considering the proportion of all cases in which the method selects the same best alternative or the rank-order correlation, as two measures of the goodness of the methods. Third, the method operations are very straightforward. The most complicated operation is the optimization problem, whereas integrals have to be solved to compute volumes to apply the *SMAA* or *SMAA-2* methods. Numerical techniques are generally required to compute these integrals. Monte Carlo simulation was used to apply *SMAA* and *SMAA-2*.

A future research line is to analyze new preference intensity measures that consider the centroid values for the imprecision.

Acknowledgments

The paper was supported by Madrid Regional Government project S2009/ESP-1685, the Spanish Ministry of Education and Science project TIN 2008-06796-C04-02 and the Spanish Ministry of Science and Innovation project MTM2011-28983-C03-03.

References

1. B.S. Ahn and K.S. Park, Comparing Methods for Multiattribute Decision Making with Ordinal Weights, *Comp. Oper. Res.*, 35, 1660-1670 (2008).
2. R.L. Keeney and H. Raiffa, *Decision with Multiple Objectives: Preferences and Value Tradeoffs*, Wiley, New York (1976).
3. R. Lahdelma, J. Hokkanen and P. Salminen, SMAA – Stochastic Multiobjective Acceptability Analysis, *Eur. Oper. Res.* 106, 1, 137-143, (1998).
4. R. Lahedelma and P. Salminen, SMAA-2: Stochastic Multicriteria Acceptability Analysis for Group Decision Making, *Oper. Res.*, 49, 444-454 (2001).
5. A. Mateos, A. Jiménez and J.F. Blanco, Ranking Methods Based on Dominance Measures Accounting for Imprecision, In: F. Rossi and A. Tsoukis (Eds.), *ADT 2009*, LNAI 5783, Springer-Verlag, 328-339 (2009).
6. A. Mateos A, A. Jiménez and J.F. Blanco, Two MCDM Ranking Methods Based on Preference Intensity Measures: Computational Study, *Group Dec. and Negot.*, (in revision), 2012.
7. P. Sarabando and L.C. Dias, Simple Procedures of Choice in Multicriteria Problems without Precise Information about the Alternatives Values, *Comp. Oper. Res.*, 37,12, 2239-2247 (2010).
8. L. Tran and L. Duckstein, Comparison of Fuzzy Numbers Using a Fuzzy Distance Measure, *Fuzzy Sets and Systems* 130, 331-341 (2002).
9. M. Weber, Decision Making with Incomplete Information, *Eur. J. of Oper. Res.*, 28, 44-57 (1987).

FUZZY TOPSIS WITH LINGUISTIC EVALUATIONS USING α–CUTS

SEZA ÖZGE GÖNEN
Department of Chemical Engineering, Istanbul Technical University, Istanbul, 34469, Turkey

SERDAR BAYSAN
Department of Industrial Engineering, Istanbul Technical University, Istanbul, 34367, Turkey

The aim of this paper is to develop a fuzzy TOPSIS method with linguistic evaluations by using α–cuts. In this paper, the rating of each alternative and the weight of each criterion are described by linguistic terms which can be expressed in α–cuts. A numerical example is presented to highlight the procedure of the proposed method.

1. Introduction

TOPSIS was developed by Hwang and Yoon [1]. It is one of the major techniques in dealing with MCDM (Multiple Criteria Decision Making) problems. In the traditional formulation of TOPSIS, the personal judgments are represented by numerical values. However, in real world applications, it is difficult to assign numerical values to the judgments.

Bellman and Zadeh [2] were the first to introduce the theory of fuzzy sets in MCDM problems as an effective approach to treat vagueness, lack of knowledge and ambiguity inherent in the human decision making process, which are known as fuzzy MCDM.

Chen [3] was one of the researchers, who extended TOPSIS to the fuzzy environment where the rating of each alternative and the weight of each criterion are described by linguistic variables and expressed in triangular fuzzy numbers. After that, many extended TOPSIS methods; i.e., Fuzzy TOPSIS [3, 4, 5, 6], Fuzzy AHP and TOPSIS [7, 8], TOPSIS with interval data [9, 10], TOPSIS with grey relation analysis [11], and group TOPSIS [10, 12, 13] have been proposed in order to meet the specific requirements of decision making problems of real world. However, the details of Fuzzy Set Theory and Fuzzy TOPSIS are not included in this text due to space limit.

In this study, a fuzzy TOPSIS method that also uses linguistic variables is proposed. Nevertheless, linguistic variables are not expressed in triangular fuzzy numbers but in the form of α–cuts. Therefore, the proposed method allows linguistic variables that are not defined by triangular fuzzy numbers to be also used.

2. Proposed Method

The steps of proposed fuzzy TOPSIS method using α–cuts are as follows:

1. Decision makers choose the linguistic variables to rate the alternatives and the importance weights of the criteria. The rating $\tilde{x}_{ij}$ of alternative A_i with respect to criterion C_j and the importance weight $\tilde{w}_j$ of the criterion C_j are determined by using Eqs. (1) and (2), in which $\tilde{x}_{ij}^k$ and $\tilde{w}_j^k$ are the rating and the importance weight.

$$\tilde{x}_{ij} = \sum_{k=1}^{N} \frac{\tilde{x}_{ij}^k}{N} = \left[a_{ij}, b_{ij}\right] \tag{1}$$

$$\tilde{w}_j = \sum_{k=1}^{N} \frac{\tilde{w}_j^k}{N} = \left[c_j, d_j\right] \tag{2}$$

2. Problem can be expressed in matrix format as shown in Eq. (3). Also, the importance weights of the criteria can be written in matrix format as given in Eq. (4).

$$D = \begin{matrix} & C_1 & C_2 & \cdots & C_n \\ A_1 \\ A_2 \\ \vdots \\ A_m \end{matrix} \begin{bmatrix} \tilde{x}_{11} & \tilde{x}_{12} & \cdots & \tilde{x}_{1n} \\ \tilde{x}_{21} & \tilde{x}_{22} & \cdots & \tilde{x}_{2n} \\ \vdots & \vdots & \cdots & \vdots \\ \tilde{x}_{m1} & \tilde{x}_{m2} & \cdots & \tilde{x}_{mn} \end{bmatrix} = \left[\tilde{x}_{ij}\right]_{m\times n} \tag{3}$$

$$W = [\tilde{w}_1 \quad \tilde{w}_2 \quad \cdots \quad \tilde{w}_n] = [\tilde{w}_j]_{1\times n} \tag{4}$$

3. The decision matrix must first be normalized as shown in Eq. (5).

$$R = \left[\tilde{r}_{ij}\right]_{m\times n} = \left[\frac{a_{ij}}{b_j^*}, \frac{b_{ij}}{b_j^*}\right], \qquad \text{where } b_j^* = \max_i\left(b_{ij}\right) \text{ for } \alpha = 0 \tag{5}$$

4. The weighted normalized matrix (V) is determined by using Eq. (6).

$$V = [\, \tilde{v}_{ij} \,]_{m\times n} = [\, \tilde{r}_{ij} (\times) \tilde{w}_j \,]_{m\times n} = [e_{ij}, f_{ij}]_{m\times n} \tag{6}$$

5. In the case of j being a benefit criterion, the fuzzy positive ideal solution (A^+) and the fuzzy negative ideal solution (A^-) are taken as in Eqs. (7) and (8), respectively. However, vice versa for j being a cost criterion.

$$A^+_{j\alpha} = [\max_i (e_{ij}), \max_i (f_{ij})] = [A^+_{1j\alpha}, A^+_{2j\alpha}] \quad (7)$$

$$A^-_{j\alpha} = [\min_i (e_{ij}), \min_i (f_{ij})] = [A^-_{1j\alpha}, A^-_{2j\alpha}] \quad (8)$$

6. The distances from the fuzzy positive ideal solution (FPIS) and the fuzzy negative ideal solution (FNIS) are calculated by using Eqs. (9) and (10), respectively.

$$d_i^+ = \frac{\sum_{\alpha=0}^{1}\left(\sum_{j=1}^{n}\left[\frac{\left(e_{ij\alpha} - A^+_{1j\alpha}\right)^2 + \left(f_{ij\alpha} - A^+_{2j\alpha}\right)^2}{2}\right]^{0.5}\right)}{number\ of\ \alpha - cuts} \quad (9)$$

$$d_i^- = \frac{\sum_{\alpha=0}^{1}\left(\sum_{j=1}^{n}\left[\frac{\left(e_{ij\alpha} - A^-_{1j\alpha}\right)^2 + \left(f_{ij\alpha} - A^-_{2j\alpha}\right)^2}{2}\right]^{0.5}\right)}{number\ of\ \alpha - cuts} \quad (10)$$

7. The closeness coefficient (CC_i) of the alternative A_i with respect to positive ideal solution is expressed by Eq. (11) and the alternatives are ranked according to the closeness coefficient values.

$$CC_i = \frac{d_i^-}{d_i^+ + d_i^-} \quad (11)$$

3. Numerical Example

A numerical example from Chu's paper [5] is used and the results obtained by the proposed method are compared with the ones from Chu's method.

Step 1: The decision-makers use the linguistic variables to assess the importance weights of the criteria (Table 1) and the rating of alternatives with respect to the criteria (Table 2).

Step 2: The importance weights and the ratings are chosen by the decision makers (D_1, D_2 and D_3) as presented in Tables 3 and 4, respectively.

Table 1: Linguistic variables for assessing the ratings of alternatives with respect to criteria.

Linguistic variable	Abbreviation	Fuzzy number	α–cut expression
Very Poor	VP	(0, 0, 3)	[0, 3-3α]
Poor	P	(0, 3, 5)	[3α, 5-2α]
Fair	F	(2, 5, 8)	[2+3α, 8-3α]
Good	G	(5, 7, 10)	[5+2α, 10-3α]
Very Good	VG	(7, 10, 10)	[7+3α, 10]

Table 2: Linguistic variables for assessing the importance weights of the criteria.

Linguistic variable	Abbreviation	Fuzzy number	α–cut expression
Very Low	VL	(0, 0.1,0.3)	[0.1α, 0.3-0.2α]
Low	L	(0.1, 0.3, 0.5)	[0.1+0.2α, 0.5-0.2α]
Medium	M	(0.3, 0.5, 0.7)	[0.3+0.2α, 0.7-0.2α]
High	H	(0.5, 0.7, 0.9)	[0.5+0.2α, 0.9-0.2α]
Very High	VH	(0.7, 0.9, 1)	[0.7+0.2α, 1-0.1α]

Table 3: The importance weights of the criteria and the averaged weights.

	D_1	D_2	D_3	Averaged weights $\widetilde{w}_j$
C_1	VH	H	H	[0.57+0.20α, 0.93-0.17α]
C_2	H	M	H	[0.43+0.20α, 0.83-0.20α]
C_3	L	M	M	[0.23+0.20α, 0.63-0.20α]
C_4	VH	VH	VH	[0.70+0.20α, 1.00-0.10α]

Table 4: The ratings of the three candidates under all criteria and the averaged ratings.

		D_1	D_2	D_3	Averaged ratings $\widetilde{x}_{ij}$
C_1	A_1	G	G	VG	[5.67+2.33α, 10.00-2.00α]
	A_2	VG	G	G	[5.67+2.33α, 10.00-2.00α]
	A_3	VG	VG	G	[6.33+2.67α, 10.00-1.00α]
C_2	A_1	F	G	F	[3.00+2.67α, 8.67-3.00α]
	A_2	G	F	G	[4.00+2.33α, 9.33-3.00α]
	A_3	G	G	F	[4.00+2.33α, 9.33-3.00α]
C_3	A_1	P	G	G	[3.33+2.33α, 8.33-2.67α]
	A_2	G	F	G	[4.00+2.33α, 9.33-3.00α]
	A_3	G	G	F	[4.00+2.33α, 9.33-3.00α]

Among the four criteria, the fourth criterion ("investment cost" – C_4) is a cost criterion and assessed by decision makers for each alternative as follows: A_1 = [77+5α, 87-5α], A_2 = [78+4α, 89-7α], and A_3 = [82+5α, 92-5α].

Step 3: The normalized and the weighted fuzzy decision matrices are constructed as in Table 5.

Step 4: The distances of each candidate from FPIS and FNIS are calculated, respectively, as shown in Table 6.

Step 5: The closeness coefficients of each candidate are obtained as follows: CC_1 = 0.256, CC_2 = 0.689, and CC_3 = 0.744. According to the closeness coefficients, the ranking order of the three candidates is A_3, A_2, and A_1. Obviously, the best selection is location A_3. These results have the exact same ranking order assessed by Chu [5].

Table 5: The normalized and the weighted fuzzy decision matrices.

		$\tilde{r}_{ij}$	$\tilde{v}_{ij}$
C_1	A_1	[0.57+0.23α, 1.00-0.20α]	[0.32+0.25α+0.05α^2, 0.93-0.35α+0.03α^2]
	A_2	[0.57+0.23α, 1.00-0.20α]	[0.32+0.25α+0.05α^2, 0.93-0.35α+0.03α^2]
	A_3	[0.63+0.27α, 1.00-0.10α]	[0.36+0.28α+0.05α^2, 0.93-0.26α+0.02α^2]
C_2	A_1	[0.32+0.29α, 0.93-0.32α]	[0.14+0.19α+0.06α^2, 0.77-0.45α+0.06α^2]
	A_2	[0.43+0.25α, 1.00-0.32α]	[0.19+0.19α+0.05α^2, 0.83-0.47α+0.06α^2]
	A_3	[0.43+0.25α, 1.00-0.32α]	[0.19+0.19α+0.05α^2, 0.83-0.47α+0.06α^2]
C_3	A_1	[0.36+0.25α, 0.89-0.29α]	[0.08+0.13α+0.05α^2, 0.57-0.36α+0.06α^2]
	A_2	[0.43+0.25α, 1.00-0.32α]	[0.10+0.14α+0.05α^2, 0.63-0.40α+0.06α^2]
	A_3	[0.43+0.25α, 1.00-0.32α]	[0.10+0.14α+0.05α^2, 0.63-0.40α+0.06α^2]
C_4	A_1	[0.84+0.05α, 0.95-0.05α]	[0.59+0.21α+0.01α^2, 0.95-0.15α+0.01α^2]
	A_2	[0.85+0.04α, 0.97-0.08α]	[0.59+0.20α+0.01α^2, 0.97-0.17α+0.01α^2]
	A_3	[0.89+0.05α, 1.00-0.05α]	[0.62+0.22α+0.01α^2, 1.00-0.15α+0.01α^2]

Table 6: The distances of each candidate from FPIS and FNIS.

	d_i^-	d_i^+	CC_i	Rank
A_1	0.048	0.139	0.256	3
A_2	0.129	0.058	0.689	2
A_3	0.139	0.048	0.744	1

4. Conclusion

In multi–criteria decision making problems, the alternatives with respect to criteria and the importance weight of criteria are often suitable to be expressed by linguistic variables instead of numerical values, which leads to a requirement for solving these type of problems under fuzzy environment.

In this study, a new method using α–cuts to convert linguistic variables into fuzzy numbers is proposed to extend fuzzy TOPSIS procedure into field of non–linear membership functions. To test the ability of the proposed method for solving decision making problems, a numerical example taken from Chu's paper [5] is solved by the proposed method and then, compared in terms of results with each other. Further research can apply the proposed method to other decision making problems in which linguistic terms can be expressed in α–cuts.

References

1. C.L. Hwang, and K.P. Yoon, *Multiple attributes decision making methods and applications*, Springer-Verlag, Berlin (1981).
2. R.E. Bellman, and L.A. Zadeh, *Management Science*, **17** (4), 141–164 (1970).
3. C.T. Chen, *Fuzzy Sets and Systems*, **114** (1), 1–9 (2000).
4. C.T. Chen, C.T. Lin, and S.F. Huang, *International Journal of Production Economics*, **102** (2), 289–301 (2006).
5. T.C. Chu, *International Journal of Advanced Manufacturing Technology*, **20** (11), 859-864 (2002).
6. T.C. Chu, and Y.C. Lin, *International Journal of Advanced Manufacturing Technology*, **21** (4), 284–290 (2003).
7. S. Balli, and S. Korukoglu, *Mathematical and Computational Applications*, **14** (2), 119–130 (2009).
8. I. Ertugrul, and N. Karakasoglu, *Expert Systems with Applications*, **36** (1), 702–715 (2009).
9. G.R. Jahanshahloo, F.H. Lotfi, and M. Izadikhah, *Applied Mathematics and Computation*, **175** (2), 1375–1384 (2006).
10. F. Ye, and Y.N. Li, *Expert Systems with Applications*, **36** (5), 9350–9357 (2009).
11. M.F. Chen, and G.H. Tzeng, *Mathematical and Computer Modelling*, **40** (13), 1473–1490 (2004).
12. H.S. Shih, H.J. Shyur, and E.S. Lee, *Mathematical and Computer Modelling*, **45** (7-8), 801–813 (2007).
13. Y.J. Wang, and H.S. Lee, *Computers and Mathematics with Applications*, **53**, 1762–1772 (2007).

AN EVOLUTIONARY ALGORITHM FOR INTEGER MULTICRITERIA OPTIMIZATION (EVALIMCO)*

VASSIL GULIASHKI, LEONEED KIRILOV, KRASIMIRA GENOVA

Section "Optimization and Decision Making", Institute of Information and Communication Technologies, Sofia 1113, "Acad. G. Bonchev" str., Bl. 2, Bulgaria

An interactive population-based algorithm called EVALIMCO is presented in the paper. It is designed to solve multi-criteria convex integer optimization problems. A heuristic procedure is used to accelerate the search process. In this way the algorithm performs faster than the usual population-based algorithms. The performance of the algorithm is demonstrated on an illustrative example.

1. Introduction

The evolutionary multi-objective optimization (EMOO) is a popular and useful field for research and development of algorithms which solve many real-life multi-objective problems (for example, see [2, 4]). EMOO has been evaluated as a very fast growing field of research and application (see [2, 4]).

The Evolutionary Optimization (EO) algorithms use a population-based approach, in which the iterations are performed on a set of solutions (called population) and more than one solution is generated at each iteration. The main reasons for the popularity of EO algorithms are the following:

i. They do not require any derivative information;
ii. EO algorithms are relatively simple to implement;
iii. EO algorithms are flexible and robust, i.e., they perform very well on a wide spectrum of problems (see [3]).

The shortcoming of using a population of solutions is the computational cost and the memory needed for the execution of one iteration.

2. Problem formulation

The integer multi-objective convex optimization problem can be stated as follows:

---

* This work is supported by the Bulgarian National Science Fund, Grant No DTK02/71 "Web-Based Interactive System, Supporting the Building Models and Solving Optimization and Decision Making Problems".

$$\text{Min } f(x) = [\, f_1(x), f_2(x), \ldots, f_k(x)]^{\mathrm{T}} \tag{1}$$

subject to $\qquad g_j(x) \le 0, \qquad j = 1, 2, ..., m, \qquad (2)$

$$x_i^{(L)} \le x_i \le x_i^{(U)}, \; i = 1, 2, \ldots, n, \tag{3}$$

$$x \in Z^n, \tag{4}$$

where $\boldsymbol{g_j(x)}$, $j = 1, 2, \ldots, m$ are convex functions;

$\boldsymbol{x_i}^{(L)}$ and $\boldsymbol{x_i}^{(U)}$, $i = 1, 2, \ldots, n$ are the known lower and upper bound of the variable x_i respectively;

$f_i(\boldsymbol{x})$, $i = 1, 2, \ldots, k$ are convex nonlinear functions.

Further on the solution $\boldsymbol{x} \in \boldsymbol{Z}^n$ denotes a vector of n decision variables: $\boldsymbol{x} = (x_1, x_2, \ldots, x_n)^{\mathrm{T}}$. The constraints (2)-(4) constitute a feasible decision domain $\boldsymbol{V} \subset \boldsymbol{Z}^n$.

$\boldsymbol{S} = \boldsymbol{f}(\boldsymbol{V}) = \{\, \boldsymbol{s} = \boldsymbol{f}(\boldsymbol{x}), \boldsymbol{x} \in \boldsymbol{V} \}$ is a k-dimensional objectives' region, $\boldsymbol{S} \subset \boldsymbol{R}^k$.

We shall use the term "solution" as a vector of variables in the decision space and the term "point" as a vector of the criteria values in the objectives' space.

Definition: A solution $x^{(1)}$ is said to dominate the solution $x^{(2)}$, if the following two conditions are true:

1. The solution $x^{(1)}$ is not worse than $x^{(2)}$ in all the objectives. Thus, the solutions are compared based on their objective function values.

2. The solution $x^{(1)}$ is strictly better than $x^{(2)}$ in at least one objective.

All the points which are not dominated by any other point $\boldsymbol{s} \in \boldsymbol{S}$, are called Pareto-optimal points. They constitute together the Pareto-optimal front (see [1, 2]) in the objectives' space.

The algorithm EVALIMCO uses an accelerating technique, designed to improve the efficiency of the developed multiple objective population based algorithm. The search process does not explore the entire Pareto-optimal front, but it is organized to explore only a small part of it in details, according to DM's preferences. The performance of the algorithm is demonstrated on an illustrative example and compared to the Strength Pareto Evolutionary Algorithm (SPEA) (see [9]).

There are some shortcomings of SPEA (see [2]):

1) The non-dominated sorting of the whole population is not used for fitness assigning and for this reason the fitness values do not favour all non-dominated solutions in the same way equally. This feature is dependent on the exact population and densities of the solutions in the search space.

2) In SPEA fitness assignment, an external solution, which dominates more solutions, gets worse fitness. This assignment is justified when all dominated

solutions are concentrated near the dominating solution. Since this is not true in most of the cases, the crowding effect should come only from the clustering procedure used. Otherwise, this type of a fitness function may provide wrong selection pressure for the non-dominated solutions.

3) The number of objectives is considered as a convergence factor in [7]. The results showed that the performance of SPEA2 deteriorates substantially as the number of the objectives increases.

To overcome these shortcomings we propose the following improvements:

- We use a heuristic procedure to accelerate the moving the whole population towards the Pareto front. It is similar to those, described in [5]. In this way we avoid the slow speed performance of the evolutionary algorithms.
- We include an interaction step, where the Decision Maker (DM) sets a reference point f^r in the objectives' space (see [6, 8]. The DM has the possibility to change his/her preferences periodically and to replace the former reference point by a new one. This step ensures the convergence of the proposed algorithm to a desired non-dominated solution.

3. The proposed evolutionary algorithm

We use an internal population P of solutions and an explicitly defined external population Pe. Pe contains the best non-dominated solutions found during the search.

We propose a *heuristic procedure* to move quickly the internal population to the Pareto-optimal front. For this purpose we calculate the direction $y = Ce - Ci$, where Ce is the weight center of Pe and Ci is the weight center of 10% of solutions in P, relevant to points with worst objectives values if the points in P are ordered in an ascending order according to the weighted sum of the objectives' values. The y vector is directed to the Pareto-optimal front, because the members of Pe dominate all members of P. Then we move the population as close as possible to the Pareto-optimal front (reaching eventually the boundaries of the system (2)-(3)). We perform consecutive steps calculating solution $x' = x + \alpha.y$, where α is the step length. In case x' violates any constraint in the system (2) or in case the current step in y – direction leads to deteriorating the criteria values, the corresponding feasible solution is calculated using the Golden section method for line search along the segment xx' and by rounding it to an integer solution.

3.1. *Scheme of the algorithm*

Step 1. Set the iteration counter h = 0. Generate uniform distributed solutions' vectors around the Tchebycheff center *Ch* of the feasible domain $N + k$ by using a deviation of $\pm\delta$, where δ is a % of the corresponding component variation (for example, δmax = ±5%). Use them to create the initial population P_h and the external population *Pe*.

Step 2. Perform the *heuristic procedure* to move P_h towards the Pareto-optimal front.

Step 3. Arrange the points in P_h in an ascending order according to the weighted sum of the objectives' values. Compute the weight centers C^{best} and C^{worst} correspondingly of first 10% and of last 10% of solutions, relevant to the points in reordered P_h. Form the moving direction $d = C^{best} - C^{worst}$. Compute a series of solutions $t^j = C^{best} + j*d$, $j = 1,\ldots$; and present the corresponding points $f(t^i)$ to the DM as possible reference points. The DM chooses one of them.

Step 4. Form a direction with an origin Tchebycheff center *Ch* and an end – the reference solution x^r, corresponding to f^r. Then move this reference solution as close as possible to the Pareto-optimal front (reaching the boundaries of the system (2)-(3)) along this direction by using the Golden section method for line search as in the heuristic procedure. The solution x^{r*} is obtained. Let us denote the corresponding reference point by f^{r*}.

Step 5. Compute the weight center C^i of population P_h and the vector d^r with an origin C^i and an end – the solution, corresponding to f^{r*}. Move the population P_h with a step size 1 along this direction: $\{Pnew\} = \{P_h\} + 1.d^r$.

Step 6. Some solutions in the *Pnew* may be infeasible. For each infeasible solution x^i in *Pnew* perform the procedure in *Step 4* along $x^i Ch$ to move it to the feasible region.

Step 7. Arrange all the points in *Pnew* according to their Euclidean distance to the reference point f^{r*}. The DM evaluates visually the first ten points in *Pnew* and if he/she is satisfied by one of them, go to *Step 8*, otherwise set $h=h+1$, $Ph=Pnew$, and go to *Step 3*.

Step 8. End.

4. Illustrative example

We consider the following problem:

$$\text{Min } f_1 = 1/(x_1+1), \text{ Min } f_2 = 1/(x_2+1),$$

subject to: $$x_1^2 + 100x_2^2 \le 10^6; \quad - x_1 + 15x_2 - 1200 \le 0;$$
$$0 \le x_1 \le 1000; \quad 0 \le x_2 \le 100;$$
$$x_1, x_2 \in Z.$$

The search process of one iteration is presented on Figure 1.

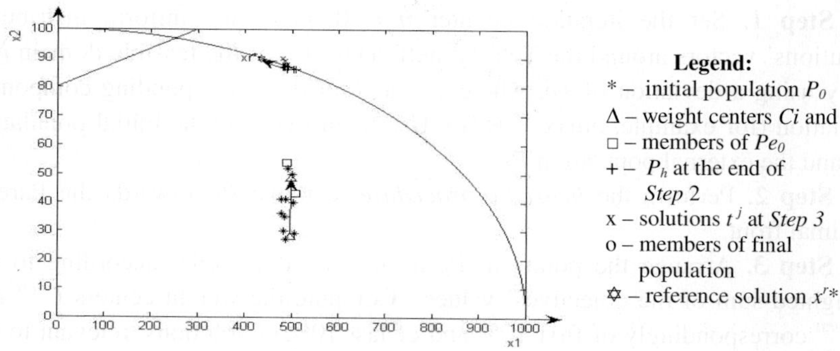

Figure 1. One algorithmic iteration

We denote values $s^i_r = s^i.10^3$. The initial internal population at *Step* 1 is: $x^1 = (500; 50)$, $s^1_r = (1.996; 19.608)$, $x^2 = (506; 29)$, $s^2_r = (1.972; 33.333)$, $x^3 = (482; 30)$, $s^3_r = (2.070; 32.258)$, $x^4 = (493; 52)$, $s^4_r = (2.024; 18.868)$, $x^5 = (477; 41)$, $s^5_r = (2.092; 23.810)$, $x^6 = (485; 35)$, $s^6_r = (2.058; 27.778)$, $x^7 = (504; 40)$, $s^7_r = (1.980; 24.390)$, $x^8 = (487; 41)$, $s^8_r = (2.049; 23.810)$, $x^9 = (488; 27)$, $s^9_r = (2.045; 35.714)$, $x^{10} =(479; 36)$, $s^{10}_r =(2.083; 27.027)$;. Its weight center is $Ci = (497; 28)$. The initial external population is $x^1=(508; 43)$, $s^1_r=(1.9646; 22.7272)$, $x^2=(489; 54)$, $s^2_r=(2.0408; 18.1818)$;. Its weight center is $Ce = (498.5; 48.5)$. Worst so0lutions are x^2 and x^9. $Ci = (497; 28)$. The direction y at *Step* 2 is $y = [1.5; 20.5]$. The obtained weight centers at *Step* 3 are: $C^{best} = (492.5; 87)$ and – $C^{worst} = (496; 86)$. Direction $d = (–3.5; 1)$. The following solutions: $t^1=(489; 88)$, $t^2=(485; 89)$, $t^3=(481; 90)$ are computed at *Step* 3. The corrected reference solution and the reference point calculated at *Step* 4 are $x^r= (435; 90)$, $f^r = (2.2936; 10.989)$. The final population after the first iteration is as follows: $x^1 = x^2 = x^7 = (455; 89)$, $s^1_r = s^2_r = s^7_r = (2.1930; 11.1111)$, $x^3 = x^4 = x^8 = x^9 = x^{10} = (435; 90)$, $s^3_r = s^4_r = s^8_r = s^9_r = s^{10}_r = (2.2936; 10.989)$, $x^5 = (433; 90)$, $s^5_r = (2.3041; 10.989)$, $x^6 = (442; 89)$, $s^6_r = (2.2573; 11.1111)$;

The DM is satisfied with the last solution found x^3 and the search procedure terminates.

For comparison at the same starting conditions the obtained population after the first SPEA iteration is: $x^1 = (500; 50)$, $s^1_r = (1.996; 19.608)$, $x^2 = (506; 29)$, $s^2_r = (1.972; 33.333)$, $x^3 = (482; 30)$, $s^3_r = (2.070; 32.258)$, $x^4 = (493; 52)$, $s^4_r = (2.024; 18.868)$, $x^5 = (477; 41)$, $s^5_r = (2.092; 23.810)$, $x^6 = (508; 54)$, $s^6_r = (1.964; 22.727)$, $x^7 = (504; 40)$, $s^7_r = (1.980; 24.390)$, $x^8 = (489; 43)$, $s^8_r = (2.041; 22.727)$, $x^9=(493; 40)$, $s^9_r =(2.024; 24.390)$, $x^{10} =(504; 52)$, $s^{10}_r =(1.980; 18.868)$;

5. Conclusion

The basic characteristics of the proposed interactive evolutionary algorithm solving convex integer multi-criteria problems may be summarized as follows:

- it is an interactive algorithm in contrast to SPEA algorithm;
- the algorithm proposed explores only a desired part of the Pareto optimal front in contrast to SPEA, where the purpose is to obtain a representative sample of the whole Pareto-optimal front;
- an accelerated approach to the Pareto front is achieved, leading to better efficiency;
- The increasing the number of objectives in the optimization problem does not have essential influence on the algorithm performance.
- The Decision Maker is supported in the choice of a suitable reference point, so that he/she can easily direct the search process to the desired region.

The algorithm EVALIMCO is a module in a web-based interactive system for multiple objective optimization.

References

1. J. Branke J., K. Deb, K. Miettinen, R. Słowiński. Multiobjective Optimization. Interactive and Evolutionary Approaches. Berlin Heidelberg, Springer-Verlag, (2008).
2. K. Deb, Multi-Objective Optimization Using Evolutionary Algorithms. Wiley-Interscience Series in Systems and Optimization. Chichester, John Wiley & Sons, (2001).
3. D. E. Goldberg, Genetic Algorithms in Search, Optimization and Machine Learning, Addison Wesley Reading, Mass, (1989).
4. V. Guliashki, H. Toshev, C. Korsemov, Problems of Engineering Cybernetics and Robotics, Vol. 60, pp. 42-54, (2009).
5. V. Guliashki, L. Kirilov, Cybernetics and Information Technologies – Journal of Bulgarian Academy of Sciences, vol.9, No4, pp.93-101, (2009).
6. K. Miettinen, L. Kirilov, Journal of Multi-Criteria Decision Analysis, vol.13, pp. 115-123, (2005).
7. K. Praditwong, Xin Yao, In: *Proc. of the 2007 IEEE Congress on Evolutionary Computation (CEC'07)*, Singapore, 25-28 September (2007), 3959-3966.
8. V. Vassilev, Genova K., Vassileva M., Narula S., International Journal on Information Theories and Applications, vol.10, No 3, 266-270, (2003).
9. E. Zitzler, L. Thiele, In: IEEE Transactions on Evolutionary Computation. Vol. 3, November (1999), No 4, 257-271.

THE APPLICATIONS OF ENERGY ALTERNATIVES IN TURKEY USING MULTICRITERIA DECISION MAKING PROCESSES

ABİT BALIN, PELİN ALCAN, HÜSEYİN BAŞLIGİL
Yildiz Technical University, The Departmant of Industrial Engineering, Besiktas, Turkey

As we search to achieve sustainable development and substance, energy security and efficiency should be viewed not only through the perspective of addressing short-term challenges, but also a necessity for long-term growth of the economy. In this paper, PROMETHEE and Electre approaches are suggested for the selection among energy policies. Also, the methodology is based on these two methods via comparison. In the application of the proposed methodologies, the best energy policy is launched into determining for Turkey.

1. Introduction

Decision making is the study of identifying and choosing alternatives based on the values and Preferences of the decision maker. Making a decision implies that there are alternative choices to be considered, and in such a case we want not only to identify as many of these alternatives as possible but to choose the one that best fits with our goals, objectives, desires, values, and so on [1]. Since the early 1980s, Turkish energy policy has concentrated on market liberalization in an effort to stimulate investment in response to increasing internal energy demand. [2]. In the literature, multi-criteria decision analysis is applied in the form of the Electre III model in order to evaluate alternative scenarios of RES exploitation in the paper of Papadopoulos and Karagiannidis [3]. The aim of the paper of Ghafghazi et al. [4] is to evaluate and rank energy sources available for a case of district heating system in Vancouver, Canada, based on multiple criteria (Promethee) and the viewpoints of different stakeholders, and to show how communication would affect the ranking of alternatives. The article of Tsoutsos et al. [5] exploits the multi-criteria methodology for the sustainable energy planning on the island of Crete in Greece. This paper is organized as follows: In Section 2, electre and promethee methods are listed and briefly explained. The applications are presented in Section 3. Conclusion and future research directions are summarized in Section 4.

2. Methods

2.1. *ELECTRE Method*

ELECTRE methods comprise two main procedures: construction of one or several outranking relation(s) followed by an exploitation procedure. The construction of one or several outranking relation(s) aims at comparing in a comprehensive way each pair of actions. The exploitation procedure is used to elaborate recommendations from the results obtained in the first phase. The nature of the recommendations depends on the problematic (choosing, ranking or sorting). Hence, each method is characterized by its construction and its exploitation procedures [6].

Step 1: Firstly, all criteria convert to the benefit criteria. Then, vector normalization method is applied to the process.

$$r_{ij} = \frac{x_{ij}}{\sqrt{\sum_{i=1}^{m} x_{ij}^2}} \tag{1}$$

Step 2: Weighted normalization is the second part of the electre method. The weights are by order of a1 -0.2, a2 -0.5, a3 -0.1, a4 -0.2.

$$v_{ij} = w_{ij} * r_{ij} \tag{2}$$

Step 3: In the third step, accordance and dissonance sets are composed. Criteria turn into two different sets. These sets are; A_p and A_q (1,2,........,m *and* A_{pq}). In the accordance set, alternative A_p is preferred to the alternative A_q.

$$C(p,q)=\{ jIv_{pj} \geq v_{qj} \} \tag{3}$$

If A_p is worser than A_q $A_p$alternative is generated as discordance set.

$$D(p,q)=\{ jIv_{pj} < v_{qj} \} \tag{4}$$

Step 4: Concordance and discordance index created. Each set of concordance is measured by the relative strength index of concordance. Concordance index Cpq, shows us how sure from the result of a binary comparison. Concordance index C(p,q) is defined as follows.

$$C_{pq} = \sum W_j^* \tag{5}$$

Discordance index measures the power of D(p,q) and is defined as follows.

$$D_{pq} = \frac{(\sum_{j^0} Iv_{pj^0} - v_{qj^0} I)}{(\sum_j Iv_{pj^0} - v_{qj} I)} \tag{6}$$

Step 5: To compare alternative for defining how A_p dominant to alternativeA_q, is about concordance index C_{pq} how bigger and how discordance index D_{pq} smaller. If $C_{pq} \geq C$ and $D_{pq} < D$, A_p alternative preferred A_q alternative. The value C and D will be average of C_{pq} and D_{pq}

2.2. *PROMETHEE Method*

Preference Ranking Organization Method for Enrichment Evaluation (PROMETHEE), a multi-criteria decision making method of outranking nature is used to rank the alternatives. This technique has all the advantages of the outranking methods, combined with ease of use and decreased complexity. It performs a pair-wise comparison of alternatives in order to rank them with respect to a number of criteria [7]. The PROMETHEE II method is used to rank the alternatives against six criteria of cost, GHG emissions, PM emissions, maturity of technology, traffic load, and local source. In this paper, PROMETHEE is determined with 7 steps as follows;

Step 1: First of all, $w = (w1, w2, \ldots, wk)$ with weights of criteria $c = (f1, f2, \ldots, fk)$ are assessed by the alternatives $A = (a, b, c, \ldots)$ on the Weighting Matrix. Table 1 can be constructed as given;

Step 2: In step 2, criteria is defined for the preference function. Also, method uses 6 different preference functions which are shown in Table 1.

Step 3: Common preference functions are determined for alternative pairs based on the preference functions. Common preference functions for alternatives a and b are assessed by Eq (7).

$$P(a,b) = \begin{cases} 0, & f(a) \leq f(b) \\ p[f(a) - f(b)], & f(a) > f(b) \end{cases} \tag{7}$$

Step 4: Preference indices are defined using common preference functions for each pair of alternatives. $wi\ (fi = 1,2, \ldots k)$ with weights judged by the criteria of k alternatives a and b preference indexes are calculated with Eq. (8).

$$\pi(a,b) = \frac{\sum_{i=1}^{k} w_i P_i(a,b)}{\sum_{i=i}^{k} w_i} \tag{8}$$

Step 5: The ranking of alternatives is conducted with the assistance of the positive and negative flows using the following:

$$\phi^{+}(a) = \sum \pi(a,x) \qquad x = (b, c, d, \ldots) \tag{9a}$$

$$\phi^{-}(a) = \sum \pi(x,a) \qquad x = (b, c, d, \ldots) \tag{9b}$$

Step 6: The classification rules are given in this step:

Any of the following conditions are met; the alternative "a" is preferable to alternative "b".

$$\phi^+(a) > \phi^+(b).and\, \phi^-(a) < \phi^-(b) \quad (10a)$$

$$\phi^+(a) > \phi^+(b).and\, \phi^-(a) = \phi^-(b) \quad (10b)$$

$$\phi^+(a) = \phi^+(b).and\, \phi^-(a) < \phi^-(b) \quad (10c)$$

Any of the following conditions are met; the alternative "a" is incomparable to alternative "b".

$$\phi^+(a) > \phi^+(b).and\, \phi^-(a) > \phi^-(b) \quad (11a)$$

$$\phi^+(a) < \phi^+(b).and\, \phi^-(a) < \phi^-(b) \quad (11b)$$

Step 7: Notice that, with the use of PROMETHEE I, some actions cannot be classified, since they are not comparable. On the other hand the PROMETHEE II model can produce a total classification by assuming that [5] :

If $\phi(a) > \phi(b)$, a is superior alternative (12a)

If $\phi(a) = \phi(b)$, a and b are no different from the alternatives (12b)

$$\phi(a) = \phi^+(a) - \phi^-(a) \quad (12c)$$

3. Application

In this paper, ELECTRE and POMETHEE methods were used to select the energy sources of Turkey. The main and sub-criteria which are obtained by taking into account the above works are used to evaluate alternative energy policies in Table 1. Pairwise comparisons of criteria were optained form paper of Kaya and Kahraman [8] which is shown below in Table 2.

Table 1.Criteria taken into account to select the best energy policy [8]

Main Criteria	Sub-Criteria
C1:Technological	C11: Feasibility
	C12: Risk
	C13: Reliability
	C14: The Duration Of Preparation Phase
	C15: The Duration Of Implementation Phase
	C16: Continuity And Predictability Of Performance
	C17: Local Technical Know How
C2:Environmental	C21: Pollutant Emission
	C22: Land Requirements
	C23: Need Of Waste Disposal
C3: Socio-Political	C31: Compatibility With The National Energy Policy
	C32: Political Acceptance
	C33: Social Acceptance
	C34: Labor Impact
C4: Economic	C41: Implementation Cost
	C42: Availability Of Funds
	C43: Economic Value (PW, IRR, B/C)

Table 2: Fuzzy AHP Weighting Matrix [8]

	a1	a2	a3	a4
b1	6,85	7,71	7,87	8,48
b2	3,88	4,55	4,93	5,51
b3	6,89	7,46	7,71	8,01
b4	6,52	7,08	7,36	7,90
b5	6,06	6,57	6,77	7,14
b6	3,70	4,21	4,35	5,18
b7	3,55	4,12	4,44	5,09
b8	7,18	7,80	8,10	8,28
b9	7,68	8,30	8,59	8,78

3.1. *Application of the ELECTRE Method*

Following, all criteria are converted to the criteria of utility, vector normalization is applied. By the time, the other steps of ELECTRE method are exerted, solution values are achieved using the value of the C and D as shown in the Table 3. When table 3 is analyzed, it is seen that ELECTRE method does not allow any evaluation of this research. Therefore ELECTRE method is not suitable for such a question.

Table 3: ELECTRE Comparison Table

1,2	**1**	**YES**	**1,2**	**0**	**NO**	**NO**
1,3	0,8	YES	1,3	0,0409	NO	NO
1,4	**1**	YES	1,4	0	NO	NO
1,5	1	YES	1,5	0	NO	NO
1,6	1	YES	1,6	0	NO	NO
1,7	1	YES	1,7	0	NO	NO
1,8	0,2	NO	1,8	0,8207	YES	NO
1,9	0	NO	1,9	1	YES	NO
2			2			
2,1	0	NO	2,1	1	YES	NO
2,3	0	NO	2,3	1	YES	NO
2,4	0	NO	2,4	1	YES	NO
2,5	0	NO	2,5	1	YES	NO

3.2. *Application of PROMETHEE Method*

PROMETHEE is created by referring to a function for each criterion after the initial matrix. Then, the positive and negative values are calculated for each option. The rest of the other steps are completed by applying the method. By calculations made by means of PROMETHEE II, and evaluating all the alternatives and the precise priority values the full ranking is determined under stable circumstances. As shown in Fig. 1. The results of the calculations acquired by the methodology of PROMETHEE are shown in Table 4. One of the alternative energy sources the wind power "b9" is pondered as primarily preferable one. Consequently, other sources of energy are ranged as b9-(Wind), b8-(Solar), b1-(Biomass), b3-(geothermal), b4-(Hydroelectric), b5-(Natural Gas), b2-(Coal and Lignite), b6-(Nuclear) and b7-(Oil) as featured in the Table. The wind power is concluded to be more efficient source compared to other ones according to PROMETHEE method. On the other hand ELECTRE method is not recommended for a appropriate solution for this type of question as shown in Table 4.

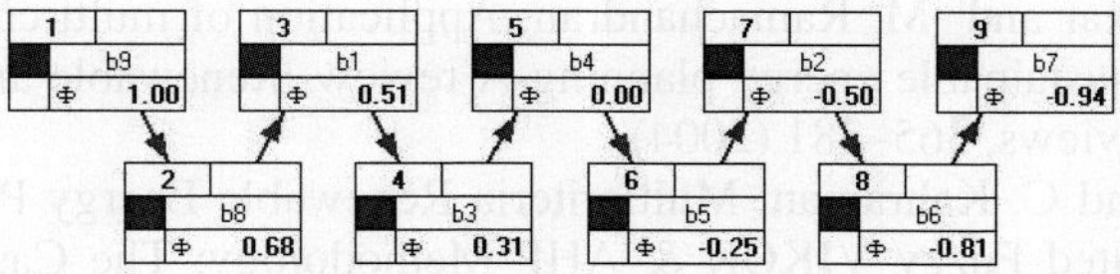

Figure 1. PROMETHEE II methodology final table for energy policy making

Table 4. Comparison of the results obtained with ELECTRE and PROMETHEE Methods

Energy Alternatives	PROMETHEE	ELECTRE
Biomass	3	…..
Coal and Lignite	7	…..
Geothermal	4	…..
Hydropower	5	…..
Natural Gas	6	…..
Nuclear	8	…..
Oil	9	…..
Solar	2	…..
Wind	1	…..

4. Conclusion

Wind energy is determined as the most attractive energy source to use in Turkey. Also the solar and biomass energies are determined second, and third, respectively. The Government of Turkey should invest in and evaluate the projects which are related these alternatives.

References

1. R. Harris, Introduction to Decision Making, Virtual Salt. http://www.virtualsalt.com/crebook5.htm (1998).
2. O. Ozyurt, Energy issues and renewables for sustainable development in Turkey, Renewable and Sustainable Energy Reviews, 14, 2976–2985 (2010).
3. A. Papadopoulos and A. Karagiannidis, Application of the multi-criteria analysis method Electre III for the optimisation of decentralised energy systems, Omega, 36, 766–776 (2008).

4. S. Ghafghazi, T. Sowlati, S. Sokhansanj, S. Melin, A multicriteria approach to evaluate district heating system maxims, Applied Energy, 87, 1134–1140 (2010)
5. T. Tsoutsos, M. Drandaki, N. Frantzeskaki, E. Iosifidis, I. Kiosses, Sustainable energy planning by using multi-criteria analysis application in the island of Crete, Energy Policy, 37, 1587–1600 (2009).
6. [http://www.lamsade.dauphine.fr/dea103/ens/bouyssou/Outranking_Moussea u.pdf]
7. S.D. Pohekar and M. Ramachandran, Application of multi-criteria decision making tosustainable energy planning-A review, Renewable and Sustainable Energy Reviews, 365–381 (2004).
8. T. Kaya and C. Kahraman, Multicriteria Renewable Energy Planning Using An İntegrated Fuzzy VIKOR & AHP Methodology: The Case of Istanbul, Energy 35, 2517–2527(2010).

A MCDM MODEL FOR ENERGY POLICY EVALUATION

MEHMET KABAK*
Department of Industrial and System Engineering, Turkish Military Academy, Bakanlıklar, Ankara, 06654, Turkey

METİN DAĞDEVİREN
Department of Industrial Engineering, Gazi University Maltepe, Ankara, 06570, Turkey

SERHAT BURMAOĞLU
Department of Business, Turkish Military Academy Bakanlıklar, Ankara, 06654, Turkey

Energy has a vital role in the development of countries. The situation of global increasing demand for energy, lack of domestic resources, energy prices, etc. force the policy makers to plan energy policies in detail. In this paper, the strategic energy situation of Turkey is analyzed. This paper presents a hybrid model that combines strength, weakness, opportunity and threat (SWOT) analysis with the fuzzy analytic network process (FANP) method. In the evaluation process, it is occasionally impossible to take precise information from decision makers (DMs) who can say their assessments relatively and approximately. It is also needed to use fuzzy logic to alleviate this problem. Since the evaluation and selection of alternative energy policies is a complex multi-criteria decision making (MCDM) problem, one of the most popular MCDM methods, ANP, is used under fuzziness in this paper. For this aim, four factors, twenty one sub-factors are determined and seven alternative policies are evaluated. It is computed that the most important three sub-factors are "major companies' interest for offshore drilling in Mediterranean and Black Sea", "developing economy of the country" and "rapidly increasing energy demand", respectively. As a result, Turkish energy policy makers are advised to focus on the privatization of electric transmission and distribution facilities, integrating the national electric system with European transmission systems and turning the country into an energy hub and terminal to be able to get the rapid economic growth and mitigate future energy problems.

1. Introduction

The rapid economic growth has forced countries to work on energy policies in detail. All countries try to provide alternative energy policies which will

* mkabak@kho.edu.tr

mitigate future problems by making energy services available without excessive costs, reducing dependence on oil and gas and developing competitive sustainable energy technologies. The alternative energy technologies of wind, biofuels, solar thermal and photovoltaics are showing maturity and the ultimate promise of cost competitiveness [1, 2]. Policy-makers and investors around the world try to determine energy policies of their countries [3, 4, 5, 6].

The energy demand of Turkey was doubled between the years 2000–2010 and will be fivefold between 2000 and 2025. It was 129 and 211 billion kilowatt hour (kWh) in 2002 and 2010, respectively and the amount of 450 billion kWh is proposed for energy demand in Turkey. The major energy demand is currently fulfilled from the conventional fossil energy resources like coal, petroleum and natural gas [7]. With a fast growing population, rapid urbanization, booming industry Turkey also needs to develop new energy policies. The detailed policies about coming energy application might be determined as follows:

Policy 1 (P1): The privatization of electric transmission and distribution facilities

Policy 2 (P2): Integrating the national electric system with European transmission systems

Policy 3 (P3): Turning the country into an energy hub and terminal

Policy 4 (P4): Spreading the natural gas consumption and establishing stock agency

Policy 5 (P5): Integration of the nuclear energy into the electricity production

Policy 6 (P6): Completion of the ongoing projects

Policy 7 (P7): Increasing the contribution of native resources

In this study, a new model is proposed for strategic planning in energy sector with the cooperation of related institutions. The aim of the model is to see where Turkey's policies in the global energy sector, what policies the country can develop to strengthen its position and to determine the priorities of energy policies on a scientific basis.

2. Methods

There are many approaches and techniques to analyze strategic cases in the strategic management process. Hence Strengths, Weaknesses, Opportunities and Threats (SWOT) analysis enables the managers to discover and collect the facts; it is the most important tool for decision makers (DM) in this process [8, 9].

If SWOT analysis is done correctly, it can provide a good basis for strategy formulation. On the other hand, there are weaknesses in the measurement and evaluation steps of the analysis. Therefore, researchers developed hybrid methods to eliminate the weaknesses in the measurement and evaluation steps of the SWOT analysis [10]. In this study, we propose an integrated SWOT FANP methodology. The hybrid model would identify energy policies by considering the effects of interdependence among the factors, as well as using triangular fuzzy numbers (TFN) in order to enter the vagueness and uncertainty on judgments of the DMs.

3. The application of methodology in the energy sector

In the application, firstly the experts, who will take a part in strategy selection process, are determined to analyze the internal and external factors. Two of the six experts are academic personnel who have published energy papers, the other one is an expert on energy affairs in Ministry of Energy and Natural Resources of Turkey and the last ones are the authors of this paper. Based on these analyses, the strategically important sub-factors, i.e. the sub-factors which have very significant effects on the success of the country, were determined. Using the SWOT sub-factors, the SWOT matrix and alternative policies based on these sub-factors were developed as in Table 1. It can be seen that the country has seven alternative policies. The P1, P2 and P3 policies identified involves making good use of opportunities by using the existing strengths of the country. The P4 policy seeks to gain benefit from the opportunities by taking into account the weaknesses of the country. P5 and P6 are the policies associated with using the country's strengths to remove or reduce the effects of threats. The last one is P7, in which the country tries to reduce the effects of its threats by taking its weaknesses into account.

After construction of the SWOT matrix, the problem is converted into a hierarchical structure in order to transform the sub-factors and alternative policies into a state in which they can be measured by the FANP technique. The schematic structure established is shown in Figure 1. The overall fuzzy priorities of the alternative policies, reflecting the interrelationships within the SWOT factors, are calculated as shown below. All calculated scores of the policies are defuzzified. Crisp values and ranks are determined as shown in Table 2. The FANP analysis results indicate that P1, P2 and P3 are the best three strategies, respectively.

According to the FANP analysis, the most important strategic factor is "strengths factor" (0.326, 0.427, 0.550). Other strategic factors are ranked as opportunities, threats and weaknesses factor, respectively. The most important sub-factors are "major companies' interest for off-shore drilling in Mediterranean and Black Sea" (O4), "developing economy of the country" (S2) and "rapidly increasing energy demand" (S3) in this order. According to the

Table 6, alternatives policies are ordered as P1, P2, P3, P6, P5, P4 and P7, respectively.

Table 1. SWOT Matrix for Turkish Energy Sector

External factors	Internal factors	
	Strengths (S)	**Weaknesses (W)**
	Partnerships between public and private organizations (S1)	Tax and price policies (W1)
	Developing economy of the country (S2)	Insufficient participation by the private sector (W2)
	Rapidly increasing energy demand (S3)	Inadequate domestic reserves of oil or natural gas (W3)
	Policies and incentives in favor of renewable energy (S4)	Increasing the dependency of importation (W4)
	Strategic location in the transmission of energy (S5)	Social resistance issue with nuclear power (W5)
Opportunities (O)	**SO Strategies**	**WO Strategies**
Regional privatization tendency in energy sector (O1)	The privatization of transmission and distribution facilities (P1)	Spreading the natural gas consumption and establishing stock agency (P4)
High attractiveness for investment in energy sector (O2)	Integrating the national electric system with European systems (P2)	
Potential for high and stable energy demand in the region (O3)	Turning the country into an energy hub and terminal (P3)	
Major companies' interest for off-shore drilling in Mediterranean and Black Sea (O4)		
Increasing share of renewable energy resources in electricity production (O5)		
Threats (T)	**ST Strategies**	**WT Strategies**
Russia and Iran's search for alternative routes excluding Turkey (T1)	Integration of the nuclear energy into the electric production (P5)	Increasing the contribution of native resources (P7)
High cost of building and maintaining pipelines in the region (T2)	Completion of the ongoing projects (P6)	
Global economic downturn (T3)		
Environmental damages (T4)		
Regional countries' unstable political situations (T5)		
Political problems with some neighbor countries (T6)		

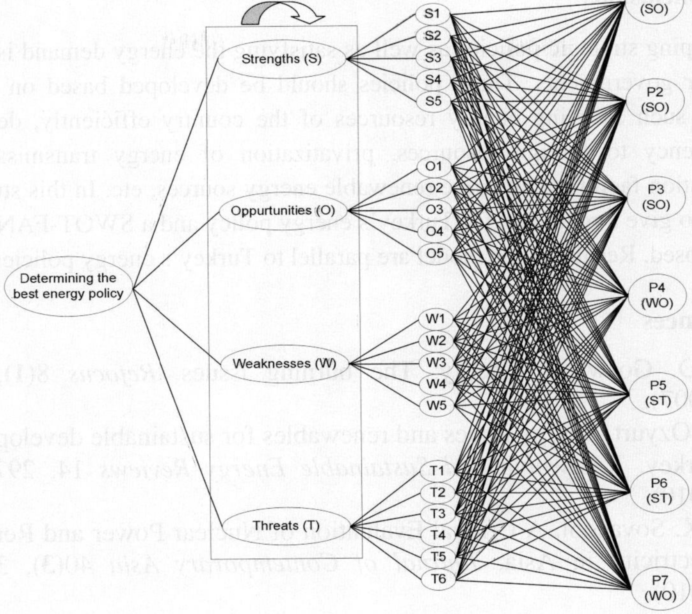

Figure 3. The hierarchical structure for the proposed model

$$\tilde{W}_{alternatives} = \begin{bmatrix} P1 \\ P2 \\ P3 \\ P4 \\ P5 \\ P6 \\ P7 \end{bmatrix} = \tilde{W}_4 x \tilde{W}_{sub-factors(global)} = \begin{bmatrix} 2.564 & 5.900 & 10.450 \\ 2.565 & 5.711 & 10.120 \\ 2.600 & 5.639 & 9.671 \\ 2.466 & 5.364 & 9.165 \\ 2.162 & 5.174 & 9.179 \\ 2.531 & 5.545 & 9.530 \\ 2.062 & 4.939 & 9.128 \end{bmatrix}$$

Table 2. Crisp values and ranks

Alternatives	ANP values	ANP ranks	AHP values	AHP ranks
P1	4.597	1	3.871	1
P2	4.421	2	3.724	2
P3	4.236	3	3.307	7
P4	4.021	6	3.578	4
P5	4.064	5	3.623	3
P6	4.181	4	3.553	5
P7	4.002	7	3.434	6

4. Conclusions

Developing strategic policies as well as satisfying the energy demand is a major task for governments. These policies should be developed based on strategic factors such as using energy resources of the country efficiently, decreasing dependency to foreign resources, privatization of energy transmission and distribution facilities, integrate renewable energy sources, etc. In this study, it is aimed to give a road map for Turkey's energy policy and a SWOT-FANP model is proposed. Results of the model are parallel to Turkey's energy policies.

References

1. Y.D. Goswami, Energy: The burning issues. *Refocus* 8(**1**), 22–25 (2007).
2. O. Ozyurt, Energy issues and renewables for sustainable development in Turkey. *Renewable and Sustainable Energy Reviews* 14, 2976–2985 (2010).
3. B.K. Sovacool, A Critical Evaluation of Nuclear Power and Renewable Electricity in Asia. *Journal of Contemporary Asia* 40(**3**), 369–400 (2010).
4. M.E. Ertem, S. Şen, G. Akar, C. Pamukcu and S. Gurgen, Energy balance analysis and energy saving opportunities for Erdemir slab furnace #3. *Energy Sources* Part A, 32, 979–994 (2010).
5. M. Tavana, M. Pirdashti, D.T. Kennedy, J.P Belaud and M. Behzadian, A hybrid Delphi-SWOT paradigm for oil and gas pipeline strategic planning in Caspian Sea basin. *Energy Policy* 40, 345–360 (2012).
6. A. Florini and B.K. Sovacool, Bridging the gaps in global energy governance. *Global Governance* 17, 57–74 (2011).
7. B. Yüksel, Energy production and sustainable energy policies in Turkey. *Renewable Energy* 35, 1469–1476 (2010).
8. M. Babaesmailli, B. Arbabshirani and V. Golmah, Integrating analytical network process and fuzzy logic to prioritize the strategies –A case study for tile manufacturing firm. *Expert Systems with Applications* 39, 925–935 (2012).
9. M. Sevkli, A. Oztekin, O. Uysal, G. Torlak, A. Turkyilmaz and D. Delen, Development of a fuzzy ANP based SWOT analysis for the airline industry in Turkey. *Expert Systems with Applications* 39, 14–24 (2012).
10. C. Kahraman, N.C. Demirel and T. Demirel, Prioritization of e-government strategies using a SWOT-AHP analysis: the case of Turkey. *European Journal of Information Systems* 16, 284–298 (2007).

PROJECT SELECTION BY USING FUZZY TOPSIS METHOD: A REAL APPLICATION IN CONSTRUCTION SECTOR

BERSAM BOLAT
Department of Management Engineering, Istanbul Technical University, Macka, Istanbul, 34367, Turkey

GÜL TEKİN TEMUR*
Department of Management Engineering, Istanbul Technical University, Macka, Istanbul, 34367, Turkey

PINAR DURSUN
Department of I Industrial Engineering, Istanbul Technical University, Macka, Istanbul, 34367, Turkey

BURÇ ONURSAL
Department of Management Engineering, Istanbul Technical University, Macka, Istanbul, 34367, Turkey

Many companies operating in construction sector have the projects resulted with failure because of making unsuccessful decisions on project selection. The goal of this study is to propose a decision making approach for project selection by taking strategic goals and organizational expectations into consideration for the firms operating in construction sector. In this study, project selection criteria have been determined by discussing with the firms producing similar type of houses and by considering the related literature. Then a real application has been achieved in a large scale construction company. In the proposed model, a fuzzy technique for order preference by similarity to ideal solution (TOPSIS) method is used for evaluating and prioritizing the candidate construction projects. Consequently, it is determined that proposed model gives successful results and may help decision makers choose the best alternative projects.

1. Introduction

In the competitive business environment, difficulties in competing with conditions caused by fast developments experienced in economical and social field give no chance to organizations for making mistakes. In order to be adapted to the dynamic business environment, organizations should utilize from effective

* temurg@itu.edu.tr.

decision making tools that are improved for overcoming difficulties resulted from changing conditions. For instance, construction companies usually face with difficulties in project selection. The decision makers can usually confuse at project selection processes because it is required to control many factors that have impact on the process: (1) project condition (profitability, difficulty, relationship with order), (2) firm (need for work, resources and capabilities), (3) competition (keenness of competitors, competitiveness of competitors) and (4) risk (project execution risk, financial risk) [1]. This multi dimensional environment increases uncertainty and makes development of an effective project selection process complicated. Hence, it becomes very important to determine factors have impact on the process and evaluate the projects by considering their influences. In this study, a model is proposed for selecting construction projects in compliance with strategic goals and organizational expectations in the direction of resource in portfolio management. Fuzzy TOPSIS is a powerful tool to deal with complex decisions including multi dimensional factors. Therefore, alternative projects are evaluated by utilizing from fuzzy TOPSIS method. In the view of these goals, the rest of paper proceeds as follows: Section 2 describes the literature related with project selection problem in construction sector. Section 3 addresses the factors affecting project selection. Section 4 describes the methodology including fuzzy TOPSIS approach development and a real application. The study is concluded with the suggestions for future research.

2. Construction Project Selection Problem

A successful project selection makes companies utilize resources at optimal level and provides increase in advantages obtained from selected projects [2]. Therefore the managers and researchers attach too much importance to project selection problem area. Cheng and Li [2] use the analytic network process (ANP) as a decision making tool to prioritize alternative projects. As an alternative method to traditional benefit/cost (B/C) analysis, Dikmen et al. [3] develop an ANP model that considers both quantitative and qualitative factors. There are four attributes in the study: benefits, costs, opportunities and risks. Ravanshadnia et al. [4] suggest a construction project selection model which takes the impact of existing projects into account. The study improves a software program called as fuzzy multi attribute decision making project selection to help contractors in evaluating alternative projects. Tan et al. [1] propose a fuzzy TOPSIS approach to give support to contractors on project selection decisions. Although there are a few studies that propose decision making approach by using fuzzy TOPSIS approach in many industries, there is lack of study focusing on construction project selection [1]. Major drawback of fuzzy TOPSIS based construction project selection studies is; they do not generate a real case based methodology, they commonly have illustrative examples with numeric values. This paper aims to use fuzzy TOPSIS method by considering many factors that have impact on construction project selection, and apply the method in a real decision process at a large scale construction company.

3. Factors Affecting Project Selection

In construction sector, complexity of project management increase owing to technical, environmental, commercial, operational, organizational and/or human based factors [5]. Similarly, Cheng and Li [2] who follow the categorization of Mohanty [6] describe the factors that influence project selection process by grouping them into six main categories: operational, managerial, financial, technological, legal and environmental. Puthamont and Charoenngam [7] consider unique factors for Thailand, such as factors related to nation, national security, project rationale, project feasibility, investment analysis, readiness for implementation, project benefits and evaluation, project impact and socio economic and political environment. Ravanshadnia et al. [4] categorize factors into five groups: organizational considerations in bidding, project characteristics, risk, financial considerations and project synergy, correlation and portfolio effects. All categorizations include subcriteria and due to the limited space, subcriteria could not be clarified at this section. It should be noticed that although similar criteria are considered, various approaches on categorization can be developed. In practical, it is not possible to use all criteria, because considerable criteria differ from market to market and product to product. The criteria list considered in this study is explained in the next section.

4. Fuzzy TOPSIS Methodology

In this study real alternative construction projects are evaluated with regard to judgments of the managers. Therefore, Chen's [8] fuzzy TOPSIS method which takes into account linguistic variables is utilized in this study. The first step of the algorithm is identifying the group of experts and criteria which will be used for evaluation of the alternatives. Then the linguistic variables and corresponding fuzzy scores for the weights of criteria and ratings are formed as Table 1 and Table 2. Before constructing and normalizing the fuzzy decision matrix, the aggregated fuzzy weights of criterion is calculated and each alternative is evaluated using these criteria by each expert.

Table 1: Linguistic variables and fuzzy scores for the weights of criteria

Very low (VL)	(0, 0, 0.1)
Low (L)	(0, 0.1, 0.3)
Medium low (ML)	(0.1, 0.3, 0.5)
Medium (M)	(0.3, 0.5, 0.7)
Medium high (MH)	(0.5, 0.7, 0.9)
High (H)	(0.7, 0.9, 1.0)
Very high (VH)	(0.9, 1.0, 1.0)

Table 2: Linguistic variables and fuzzy scores for the ratings of alternatives

Very Poor (VP)	(0, 0, 1)
Poor (P)	(0, 1, 3)
Medium Poor (MP)	(1, 3, 5)
Fair (F)	(3, 5, 7)
Medium Good (MG)	(5, 7, 9)
Good (G)	(7, 9, 10)
Very Good (VG)	(9, 10, 10)

After calculating the distance from fuzzy positive ideal and fuzzy negative ideal solution, the closeness coefficient which will be used to determine the ranking order is computed for each alternative. Interested reader may refer to excellent work for detailed explanation on calculation steps and discussion on formulations by Chen [8].

5. Application

Firstly the criteria used to evaluate alternative projects identified by interviewing with managers of four construction companies. First company has operated in the sector since1996 managing big housing and shopping center projects. Second company has taken place since 1969 in the sector of paper, packaging, cement, tourism and retailing. Third company develops landed estate projects for national and international market. The last one manages luxury land escape projects for customers with high income. By considering the results of interviews and searching related literature twenty one criteria are filtered and used in the model. These criteria are as follows: *expected profit, return on investment, financing cost, ability to attract foreign investor, development cost, quality and easiness of construction, adaptiveness to different concepts, constraints, project completion time, equipment and technical knowledge, customer satisfaction, compliance with the target market, type of investment, location of the project, sales ratio, market share, competitive advantage, compliance with new developments, prestige, economic contribution to the company, compliance with strategic goals.* The managers evaluated the criteria and the aggregated importance weights of the criteria are obtained. Table 3 gives the results of the weights of evaluation criteria and Figure 1 figures out the hierarchical structure for the selection of construction projects. Location of projects, sale ratio, market share and compliance of the project with the target market are considered as the most important factors by experts.

Table 3: Weights of criteria

C_1	(0.633, 0.800, 0.900)	C_8	(0.233, 0.433, 0.633)	C_{15}	(0.900, 1.000, 1.000)
C_2	(0.433, 0.633, 0.800)	C_9	(0.366, 0.566, 0.766)	C_{16}	(0.700, 0.900, 1.000)
C_3	(0.633, 0.833, 0.933)	C_{10}	(0.233, 0.433, 0.633)	C_{17}	(0.366, 0.566, 0.766)
C_4	(0.233, 0.433, 0.633)	C_{11}	(0.500, 0.700, 0.866)	C_{18}	(0.233, 0.433, 0.633)
C_5	(0.233, 0.433, 0.633)	C_{12}	(0.766, 0.933, 1.000)	C_{19}	(0.300, 0.500, 0.700)
C_6	(0.500, 0.700, 0.866)	C_{13}	(0.700, 0.900, 1.000)	C_{20}	(0.233, 0.433, 0.633)
C_7	(0.300, 0,500, 0.700)	C_{14}	(0.900, 1.000, 1.000)	C_{21}	(0.566, 0.766, 0.900)

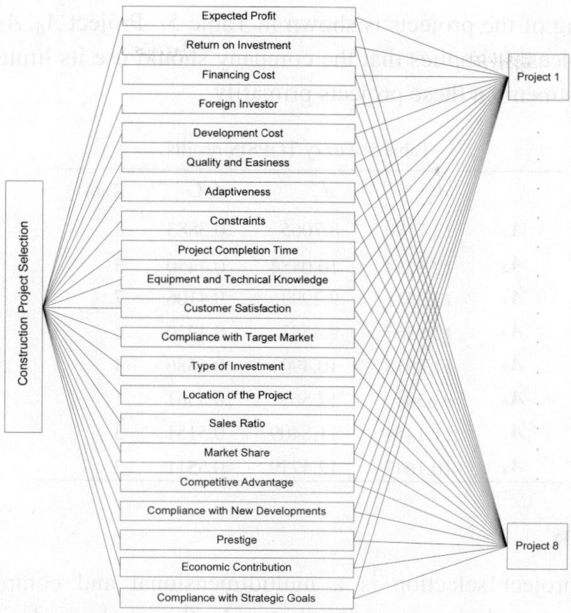

Figure 1. The hierarchical structure for the selection of construction projects

The company where the case study carried out faced with decision of project selection. Eight potential projects are assessed by three experts working in the company: project manager, general manager and construction supervisor. Also the evaluation of construction supervisor is given in Table 4 as an example.

Table 4: Evaluation scores of one decision maker

	C_1	C_2	C_3	C_4	C_5	C_6	C_7	C_8	C_9	C_{10}	C_{11}
A_1	VG	MG	MP	VP	P	MG	MP	F	P	MP	G
A_2	F	G	P	MP	MG	P	F	MP	MG	F	G
A_3	G	MP	MG	VP	MP	MG	F	F	MG	MP	G
A_4	MG	F	G	P	F	MG	G	F	G	F	F
A_5	F	G	P	F	F	MP	MP	P	P	F	G
A_6	VG	VG	G	MG	G	F	G	MG	MG	MG	VG
A_7	G	MG	G	F	MG	MG	VG	MG	MG	MG	G
A_8	MG	G	G	F	F	MG	G	G	G	MG	VG

	C_{12}	C_{13}	C_{14}	C_{15}	C_{16}	C_{17}	C_{18}	C_{19}	C_{20}	C_{21}
A_1	F	VG	P	MG	MG	F	MP	P	MP	MP
A_2	G	MG	G	G	G	MG	MG	G	MG	F
A_3	G	MG	MP	MP	MP	MP	F	MP	MP	MP
A_4	MP	VG	G	P	MP	P	P	MP	MP	MP
A_5	F	G	VG	VG	MG	F	MG	G	F	F
A_6	VG	VG	VG	VG	VG	G	G	VG	G	G
A_7	G	G	G	MG	MG	F	G	G	MG	MG
A_8	G	G	VG	VG	VG	G	G	VG	MG	MG

The ranking of the projects is shown in Table 5. Project A_6, A_7 and A_8 will be the best choices. It implies that the company should use its limited resources and make investment on these projects primarily.

Table 5: Fuzzy TOPSIS results

	d_i^-	d_i^*	CC_i	Rank
A_1	13.7180	8.7065	0.3883	8
A_2	12.5264	10.0454	0.4450	5
A_3	13.3311	9.2993	0.4106	7
A_4	12.9567	9.4582	0.4220	6
A_5	11.8978	10.4901	0.4686	4
A_6	9.6304	12.5637	0.5561	1
A_7	11.1459	11.8409	0.5151	3
A_8	10.1647	12.4779	0.5511	2

6. Conclusions

Construction project selection is a multidimensional and complex decision which requires considering many attributes. In this study, a decision making methodology which uses fuzzy TOPSIS method is proposed. The developed methodology is successfully applied to a real decision process in a large construction company. It is noticeable to say that the contribution of the study results from the use of real data in the uncertain nature of construction sector. As a future research, studies based on other fuzzy multi criteria decision making techniques such as fuzzy ELECTRE/ PROMETHEE can be conducted and can be compared with fuzzy TOPSIS results.

References

1. Y.T. Tan, L.Y. Shen, C. Langston and Y. Liu, *J. Model. Manage*. 5, 3, 302–315 (2010)
2. E.W.L. Cheng and H. Li, *J. Constr. Eng. Manage*, **131**, 459–466 (2005).
3. I. Dikmen, M.T. Birgonul and B. Ozorhon, *J. Civ. Eng*., **34**, 7, 786–792(2007).
4. M. Ravanshadnia, H. Rajaie and H.R., *J. Civ. Eng*., **37**, 1082–1093 (2010).
5. N.J. Smith, Managing Risk in Construction Projects, (1999).
6. Mohanty, R. P, *Int. J. Proj.Manage*., **10**, 1 , 31–38 (1992).
7. S. Puthamont and C. Charoenngam, *Int. J. of Project Manage*., **25**, 178–188(2007).
8. C.T. Chen, *Fuzzy Sets and Systems*, **114**, 1-9,(2000).

LOCATION SELECTION OF EMERGENCY LOGISTICS CENTERS USING AN INTEGRATED DEMATEL-ANP APPROACH

UMUT RIFAT TUZKAYA*, KADRIYE BUSRA YILMAZER
Department of Industrial Engineering, Yildiz Technical University, Besiktas, Istanbul, 34349, Turkey

GULFEM TUZKAYA
Department of Industrial Engineering, Marmara University, Göztepe, Istanbul, 34722, Turkey

In this study, a location selection methodology using an integrated multi-criteria decision making approach for a logistics network model design in the presence of health emergency situations for Turkey is proposed. It is an important issue to urgent response to such unpredictable disasters. The proposed methodology tries to select the locations of temporary emergency centers for affected areas while considering eleven decision making criteria. Decision Making Trial and Evaluation Laboratory (DEMATEL) and Analytic Network Process (ANP) methods are integrated to determine the convenient locations considering the relations between criteria and alternative locations.

1. Introduction

When a public health emergency occurs, the main purpose is mitigation the severe of disasters such as epidemic, earthquake, terrorist attacks, etc. During an emergency, there comes up important problems related to dispatching right commodities (medical materials, shelters, etc.) needed for the humanitarian operations to the best possible locations as quick as possible and determining the amount of these commodities. Therefore, government and disaster relief organizations have to use local and global resources effectively, coordinate the process with qualified staff and apply an efficient logistic network design (LND) after a disaster.

With the aim of quick response to emergency situations like catastrophic events; a rapid, flexible and efficient LND for emergency is very important to

* tuzkaya@yildiz.edu.tr.

help government and disaster relief organizations. In order to being efficient, some of the materials like disaster kits, medications, nutritional supplements and resources have to be prepared before the emergency situations and have to be dispatched from the conveniently located facilities. In this study, location selection methodology using an integrated multi-criteria decision making approach for a LND model in the presence of health emergency situations for Turkey is proposed. In Turkey, various disastrous incidents can be seen but the most widespread one is earthquake. Therefore, it is an important issue to urgent response to such unpredictable disasters. The proposed methodology tries to select the locations of temporary emergency centers for affected areas while considering eleven criteria. DEMATEL and ANP methods are integrated to determine the relations between criteria and alternative locations. By the end of applying these methods, weights for alternative emergency centers are obtained to use in a mathematical model for LND.

Due to the importance of emergency response to disaster areas and LND, various researches have been carried out. For instance, there are studies about key success factors of emergency logistics systems, suggestions on transportation of first-aid commodity problems and selection of temporary emergency center locations. Some of these studies are mentioned below.

Hong and Xiaohua [1], investigate a location selection problem. With the intention of choosing an appropriate place for emergency logistics centers, a multi-objective location selection model is proposed and solved by using MATLAB simulation results based on AHP theory. Another location selection study is conducted by Balcık and Beamon [2]. They show a mixed integer programming model that determines the number and locations of the distribution centers for disaster relief operations. Also, they try to determine the amount of needed materials at each distribution centers. Thus, this model considers location and inventory decisions, budgetary constraints, time requirements and capacity restrictions.

Besides, there is a study conducted by Canbolat and Massow [3] which they intend to minimize the expected maximum risk through the minimization of expected maximum distance to emergencies while locating an emergency response facility. In this study, they suggest a simulation model to illustrate the model.

Jia et al. [4] propose a model about facility location of medical supplies for large-scale emergencies where exist demand uncertainty and limited medical supply. In order to solve the model, a genetic algorithm heuristic, a locate–allocate heuristic, and a Lagrangean relaxation heuristics are developed.

Murali et al. [5] also study on locating capacitated facilities under demand uncertainty such as in the presence of large scale bio-terror attack. They formulate a maximal covering location problem and solve it with a locate-allocate heuristic.

Moreover, Wei Yi et al. [6], discuss an integrated location and distribution problem which is showed on an earthquake scenario for Istanbul. In this study, an effective coordination of dispatching medical materials to distribution centers in disaster areas is intended.

In our study, we tried to determine convenient locations for Emergency Logistics Centers (ELCs) in a logistics network. An integrated multi-criteria decision making technique is used considering eleven criteria for ten alternative cities. The rest of the paper includes definitions of DEMATEL and ANP techniques. Application of this method to ELC location selection problem for a large scale problem about Turkey is realized. And the conclusions are examined.

2. The DEMATEL and the ANP Methods

The aim of using an integrated approach is, firstly, to categorize the influencing factors as cause group and effect group by using DEMATEL method. After determining the interdependencies between decision criteria by DEMATEL, ANP method is used to evaluate the weights of alternatives and to choose the most convenient one at the end.

2.1. *DEMATEL*

DEMATEL method is developed by The Battelle Memorial Institute through Geneva Research Centre [7]. It is used for evaluating complicated and intertwined multi-criteria decision problems [8]. In order to apply the DEMATEL method, the steps below should be followed:

Step 1: Calculating the initial average matrix,

Step 2: Calculating the initial direct influence matrix,

Step 3 : Calculating the full direct/indirect influence matrix,

Step 4 : Setting a thresholds value and obtaining impact-digraph diagram.

For detailed information on DEMATEL, [7,8,9] can be seen.

2.2. *ANP*

ANP is a multi-criteria decision making method for complex problems which cannot be structured hierarchically. It is developed by Saaty (1990's) [7]. The steps of applying ANP are in the following:

Step 1: Pairwise comparisons between criteria should be done by giving importance weights. This can be determined by using the valuation scales which range from 1 to 9 for representing equal importance to extreme importance. After pairwise comparison matrix A (where $a_{ij}.a_{ji}=1$) is calculated, the largest eigen value is computed (priority vector). In order to examine the data consistency, Consistency Index (CI) is calculated and divided by Random Inconsistency (RI) to evaluate Consistency Ratio (CR). When CR<0.10, matrix A can be concluded as consistent [7,9,10].

Step2 : The formalization and normalization of this matrix realized by priority vectors (Eigen-vectors). [9]

Step3 : Finally, the supermatrix is raised to the largest power until converging to the solution. [9,10]

3. Application of the proposed method

The application of the model is realized for weighting ELC alternatives considering 11 criteria: route flexibility (C1), transportation mode opportunities (C2), damaging risks (C3), international accessibility (C4), proximity to non-governmental organizations (C5), seasonal and environmental effects (C6), transportation vehicle reachability (C7), presence of the 3PL firms (C8), communication and coordination conditions (C9), shortest starting time (C10), proximity to Red Crescent depots (C11). These eleven criteria determined based on the studies in the literature with taking into account the strategic and tactical decisions of Turkish disaster relief organizations in the presence of health emergency cases. Based on the population densities of the Turkish cities, 10 alternative locations are selected to locate ELCs: Istanbul (A1), Ankara (A2), Şanlıurfa (A3), Antalya (A4), Bursa (A5), Erzurum (A6), İzmir (A7), Kırklareli (A8), Samsun (A9), Tekirdağ (A10). The first step of the integrated approach is determining the initial direct relation matrix which describes the influence of the row elements on column elements. The initial direct relation matrix is given in Table 1.

Afterwards, normalized direct relation matrix and total-relation matrix are generated. A threshold value is chosen as 0.14 and if a value in total-direction matrix is higher than threshold value, it means that row criteria are strongly affecting the column criteria (Table 2).

Table 1. The initial direct relation matrix

	C1	C2	C3	C4	C5	C6	C7	C8	C9	C10	C11
C1	0	1	1	3	2	1	2	3	1	2	1
C2	4	0	1	4	1	0	3	3	2	3	2
C3	3	1	0	3	1	1	2	2	4	3	2
C4	1	1	0	0	1	1	2	3	1	3	1
C5	1	0	0	1	0	0	1	1	1	3	1
C6	3	3	2	2	2	0	2	1	2	2	1
C7	1	1	0	1	1	0	0	2	3	3	2
C8	1	1	0	2	2	1	3	0	3	3	2
C9	1	1	1	2	1	0	3	1	0	3	1
C10	1	1	2	0	1	0	1	1	1	0	1
C11	1	0	1	1	2	0	1	1	2	3	0

Table 2. The total-relation matrix

	C1	C2	C3	C4	C5	C6	C7	C8	C9	C10	C11	Di
C1	0.06	0.07	0.06	**0.17**	0.13	0.05	**0.15**	**0.18**	0.11	**0.18**	0.09	1.26
C2	**0.21**	0.05	0.08	**0.23**	0.11	0.03	**0.21**	**0.20**	**0.17**	**0.25**	**0.15**	1.68
C3	**0.17**	0.09	0.04	**0.19**	0.11	0.06	**0.17**	**0.16**	**0.23**	**0.25**	0.14	1.61
C4	0.08	0.07	0.03	0.05	0.08	0.05	0.13	**0.16**	0.10	**0.19**	0.08	1.03
C5	0.06	0.02	0.02	0.06	0.03	0.01	0.07	0.07	0.07	**0.16**	0.06	0.63
C6	**0.18**	**0.15**	0.11	**0.16**	0.14	0.02	**0.17**	0.13	**0.16**	**0.21**	0.11	1.53
C7	0.08	0.06	0.03	0.09	0.08	0.01	0.07	0.12	**0.16**	**0.19**	0.11	1.00
C8	0.09	0.07	0.04	0.13	0.12	0.05	**0.18**	0.07	**0.18**	**0.22**	0.13	1.27
C9	0.08	0.06	0.06	0.12	0.08	0.01	**0.16**	0.10	0.06	**0.19**	0.08	1.02
C10	0.07	0.06	0.09	0.04	0.07	0.01	0.08	0.08	0.08	0.07	0.07	0.71
C11	0.07	0.02	0.06	0.08	0.10	0.01	0.09	0.08	0.12	**0.18**	0.04	0.84
Ri	1.16	0.73	0.61	1.32	1.05	0.30	1.48	1.35	1.45	2.08	1.06	

One of the most important purpose of using DEMATEL is saving ANP pairwise evaluations from becoming an intractably complex system. Hence, pairwise analysis among the criteria in ANP supermatrix is realized only for the criteria whose influence values are above the threshold value.

Firstly, the ANP initial supermatrix and its general sub-matrices for weighting the ELC location alternatives are prepared. The weights of the criteria with respect to the goal are shown in Table 3. The most important three criteria are shortest starting time, communication and coordination conditions, and route flexibility, respectively. The least important criterion is proximity to Red Crescent depots. Then normalized supermatrix is calculated. After raising the power of normalized supermatrix to 16, it is converged.

Table 3. Weights of the criteria with respect to the goal

C1	C2	C3	C4	C5	C6	C7	C8	C9	C10	C11
0.138	0.122	0.044	0.104	0.044	0.071	0.118	0.051	0.140	0.148	0.020

At the conjunction of goal and alternatives cells of limited supermatrix, the weights of the alternative locations are obtained. According to these results, Istanbul (15.75%) is the most important location to locate an ELC. Next, Bursa (11.85%), İzmir (11.84%), Antalya (11.60%), Ankara (11.15%), Samsun (9.12%), Kırklareli (7.49%), Şanlıurfa (7.45%), Tekirdağ (7.30%) and Erzurum (6.45%) are coming respectively.

4. Conclusions

Vitally important point in an emergency LND is to choose the most convenient locations considering various criteria because the amount of dispatched materials with taking into account the demand points of dispersion and finding the most rapid way of transferring these emergency materials is closely related to the location of the ELC. The proposed integrated method is applied for Turkey's emergency LND to choose best locations considering various criteria. As future studies, this paper can be extended by constituting an emergency LND. In an emergency response LND model, there can be a set of demand points and their corresponding locations as well as a set of candidate sites weighted by DEMATEL-ANP integrated method for siting main and regional ELCs. One of the objective functions would be maximizing total weight for transferred materials over high weighted locations and the other would be minimizing the total cost or transportation time.

References

1. L. Hong and Z. Xiaohua, Procedia Engineering 15, 2128 (2011).
2. B. Balcik and B. M. Beamon, International Journal of Logistics: Research and Applications 11, 101 (2008).
3. M. S. Canbolat, and M. V. Massow, Expert Systems with Applications 38, 10099 (2011).
4. H. Jia, F. Ordonez, and M. M. Dessouky, Computers & Industrial Engineering 52, 257 (2007).
5. P. Murali, F. Ordonez, and M. M. Dessouky, Socio-Economic Planning Sciences 46, 78 (2012).
6. W. Yi and L. Özdamar, European Journal of Operational Research 179, 1177 (2007).
7. G. H. Tzeng and J. J. Huang, CRC Press, 978-1-4398-6157-8, (2011).
8. H. H. Wu, J. I. Shieh, Y. Li and H. K. Chen, Info. Tech. Journal (9) 3, 569 (2010).
9. S. J. Hung, Expert Systems with Applications 38, 9053 (2011).
10. S. Jharkharia, and R. Shankar, Omega 35, 274 (2007).

GARAGE LOCATION SELECTION FOR PUBLIC TRANSPORTATION SYSTEM IN ISTANBUL USING FUZZY AHP AND FUZZY AXIOMATIC DESIGN TECHNIQUES

ÖZGE NALAN ALP, NURGÜL DEMİRTAŞ, HAYRİ BARAÇLI, UMUT RIFAT TUZKAYA*

Department of Industrial Engineering, Yildiz Technical University, Besiktas, Istanbul, 34349, Turkey

In this study, the best location for a bus garage, in which maintenance and repair activities are operated, is tried to determine for public transportation system in Istanbul. An integrated multi-criteria decision making technique is used to obtain reliable results. Firstly, various criteria related with garage location selection are specified and weighted by fuzzy AHP. Then, these weights are used in fuzzy axiomatic design technique to determine the precedencies of the alternative garage locations.

1. Introduction

With the increase of the living standards in Istanbul, the population is asking for a higher service and quality. This requirement is also including the quality of the public transportation system. IETT, which carries a huge part of the passenger density as a public transportation corporation, is the most important actor in Istanbul traffic. Serving in a high quality is very difficult in Istanbul, when the constraints are considered like road conditions, number of vehicles, and high density of the passengers especially in the peak hours. The number of the buses is increasing day by day in the proportion of the population of the city. Besides, this increase brings extra problems caused by the ages of the buses, accidents in the traffic, lack of the road infrastructures, and over using the buses. Especially maintenance activities have a vital importance on journeys to eliminate these problems and to ensure customer satisfaction without failures.

In this study, we try to determine the best locations for bus garages to operate the maintenance and repair activities. There are many important criteria that affect the location decisions like cost, time, infrastructure, environment, accessibility, other macro factors, etc. Istanbul is a growing city and the land use is very important in terms of the city master plans. Therefore, probable places

* tuzkaya@yildiz.edu.tr.

are determined by the İETT authorities and best alternative is selected considering the above mentioned criteria. An integrated multi criteria decision making technique, fuzzy AHP & fuzzy Axiomatic Design (AD), is used for this purpose. Fuzzy AHP is used to weight the criteria according to the precise and imprecise judgments of the decision makers. Then the obtained weights of the criteria are utilized as an input data in fuzzy AD to determine the precedencies of the alternative locations.

The literature includes many multi criteria decision making techniques integrations for various areas [5], [6], [7], [8]. Alp et al. [9] also studied on the bus garage location selection using fuzzy AHP and PROMETHEE techniques. Even though fuzzy AHP and fuzzy AD techniques are combined for solving the decision making problems of the different areas ([3] and [4]), this study is novel due to the applying them for a garage location selection for public transportation system.

2. Methodology

2.1. *Fuzzy AHP*

Buckley's Fuzzy AHP method is used to fuzzify the hierarchical analysis by allowing fuzzy numbers for the pairwise comparisons and find the fuzzy weights [10].

Step 1: Construct pairwise comparison matrices among all the elements/criteria in the dimensions of the hierarchy system. Assign linguistic terms to the pairwise comparisons by asking which is the more important of each two elements/criteria.

Step 2: To use geometric mean technique to define the fuzzy geometric mean and fuzzy weights of each criterion by Buckley as follows:

$$\tilde{r}_i = (\tilde{a}_{i1} \otimes \tilde{a}_{12} \otimes \ldots \otimes \tilde{a}_{in})^{1/n}, \quad \tilde{w}_i = \tilde{r}_i \otimes (\tilde{r}_1 \oplus \ldots \oplus \tilde{r}_n)^{-1}, \tag{1}$$

where $\tilde{a}_{in}$ is fuzzy comparison value of criterion i to criterion n, thus, $\tilde{r}_i$ is geometric mean of fuzzy comparison value of criterion i to each criterion, $\tilde{w}_i$ is the fuzzy weight of the *i*th criterion, can be indicated by a TFN, $\tilde{w}_i = (Lw_i, Mw_i, Uw_i)$. Here Lw_i, Mw_i and Uw_i stand for the lower, middle and upper values of the fuzzy weight of the *i*th criterion.

Step 3: Take $\tilde{E}_{ij}^k$ to indicate the fuzzy performance value of evaluator k towards alternative i under criterion j, and all of the evaluation criteria will be indicated by $\tilde{E}_{ij}^k = (LE_{ij}^k, ME_{ij}^k, UE_{ij}^k)$.

$$\tilde{E}_{ij} = (1/m) \otimes (\tilde{E}^1_{ij} \oplus \tilde{E}^2_{ij} \oplus ... \oplus \tilde{E}^m_{ij}) \quad (2)$$

The end-point values LE_{ij}, ME_{ij} and UE_{ij} can be solved by the method put forward by Buckley, that is,

$$LE_{ij} = (\sum_{k=1}^{m} LE^k_{ij}) / m; \quad ME_{ij} = (\sum_{k=1}^{m} ME^k_{ij}) / m; \quad UE_{ij} = (\sum_{k=1}^{m} UE^k_{ij}) / m. \quad (3)$$

Step 4: The criteria weight vector $\tilde{w} = (\tilde{w}_1, ..., \tilde{w}_j, ..., \tilde{w}_n)^t$, The fuzzy performance matrix of each of the alternatives $\tilde{E} = (\tilde{E}_{ij})$. The final fuzzy synthetic decision matrix $\tilde{R} = \tilde{E} \circ \tilde{w}$.

$\tilde{R}_i = (LR_i, MR_i, UR_i)$, where LR_i, MR_i and UR_i are the lower, middle and upper synthetic performance values of the alternative i, that is:

$$LR_i = \sum_{j=1}^{n} LE_{ij} \times Lw_j; \quad MR_i = \sum_{j=1}^{n} ME_{ij} \times Mw_j; \quad UR_i = \sum_{j=1}^{n} UE_{ij} \times Uw_j. \quad (4)$$

Step 5: The procedure of defuzzification is to locate the Best Non-fuzzy Performance value (BNP). To utilize the COA(Center of area) method to find out the BNP is a simple and practical method, and there is no need to bring in the preferences of any evaluators, so it is used in this study. The BNP value of the fuzzy number $\tilde{R}_i$ can be found by the following equation:

$$BNP_i = [(UR_i - LR_i) + (MR_i - LR_i)]/3 + LR_i \qquad \forall i. \quad (5)$$

2.2. *Axiomatic Design*

Axiomatic Design (AD) is a method developed to design scientific fields for products, system and process by Suh [2]. The most important concept in AD is the existence of the design axioms. The first design axiom is known as the Independence Axiom and the second is known as the Information Axiom [1].Information is defined in terms of the information content, I_i that is related in its simplest form to the probability of satisfying the given Functional Requirements (FRs) [5]. The probability of density function (p_i) identifies the satisfaction level of design characteristics with respect to pre-defined FRs. Eq.(6) gives the I_i [1]:

$$I_i = \log_2 (1/p_i) \quad (6)$$

$$p_i = \left(\frac{\text{common range}}{\text{system range}}\right) \qquad I_i = \log_2 \left(\frac{\text{system range}}{\text{common range}}\right) \quad (7)$$

In the fuzzy case, we have incomplete information about the system and design ranges. The system and design range for a certain criterion will be expressed by using 'over a number', 'around a number' or 'between two numbers. If the decision-maker wants to assign a different weight (w_j) for each criterion, the following weighted multi-attribute AD approach can be used. Eq. (8) is proposed for the weighted multi-attribute AD approach [4]:

$$\left\{\begin{array}{ll} [\log_2(1/p_{ij})]^{1/w_j} & , \quad 0 \le I_{ij} \le 1 \\ [\log_2(1/p_{ij})]^{w_j} & , \quad I_{ij} \ge 1 \\ w_j & , \quad I_{ij} = 1 \end{array}\right\} \tag{8}$$

3. Application

In this study we propose a methodology to select the most appropriate garage location in the European Region of Istanbul. The fuzzy AD and weighted fuzzy AD which is weighted by fuzzy AHP are used for the evaluation and results of both methods are compared.

The alternative areas are Beylikdüzü, Arnavutköy and Silivri. These alternatives are identified according to the experts' opinions in IETT.

The number of main-criteria that we use to solve the problem is five and the number of sub-criteria is fourteen. The criteria are explained below:

Cost: The *"investment cost (C1)"* is the total cost that is necessary for set up of the garage. *"Spare parts transportation cost (C2)"* is the cost to transfer the spare parts from their suppliers to the garages. *"Vehicle transport cost (C3)"* occurs when a bus travels without passengers to go to the garage or to the start of the lines. *"Operation cost (C4)"* is the total maintenance, cleaning and other operations' cost.

Infrastructure: *"Municipality services (C5)"* are electric, water and gas requirements supply. *"Technological infrastructure (C6)"* is important for telecommunication and internet supply.

Social and economic structure: *"The population structure (C7)"* is the density of the population in the regions. *"Urbanization of the region (C8)"* means immigration to the region and the population increase rate. *"Neighborhood response (C9)"* is the response of the neighborhood when a new garage is set up. *"Behaviors of the passengers (C10)"* is the satisfaction degree of the passengers those live around the new garage.

Macro factors: *"Government policy (C11)"* means the taxes and incentives. *"Reconstruction and building plans (C12)"* is the features of the garage building.

Environmental factors: *"Effects on the open land (C13)"* means the effect to the social/urban life. *"Convenience of the land (C14)"* means the grade of the land, stream bed situation and earthquake risk.

Firstly we determine the weights of main and sub-criteria by using the fuzzy AHP. In first stage, the pairwise comparisons of main and sub-criteria are made by experts. The pairwise comparison matrix is shown in the Table 1.

Table 1. The main-criteria pairwise comparison matrix

The comparisons of main criteria		Cost			Infrastructure			Social and economic structure			Macro factors			Environmental factors		
1	Cost	1.00	1.00	1.00	0.25	0.33	0.50	2.00	3.00	4.00	0.25	0.33	0.50	0.33	0.50	1.00
2	Infrastructure	2.00	3.00	4.00	1.00	1.00	1.00	4.00	5.00	6.00	4.00	5.00	6.00	6.00	7.00	8.00
3	Social and economic structure	0.25	0.33	0.50	0.17	0.20	0.25	1.00	1.00	1.00	1.00	1.00	2.00	0.25	0.33	0.50
4	Macro factors	2.00	3.00	4.00	0.17	0.20	0.25	0.50	1.00	1.00	1.00	1.00	1.00	0.25	0.33	0.50
5	Environmental factors	2.00	3.00	4.00	0.13	0.14	0.17	2.00	3.00	4.00	2.00	3.00	4.00	1.00	1.00	1.00

After comparisons, the Buckley's fuzzy AHP methodology is applied and the criteria weights are calculated. The detailed results are shown in Table 2.

Table 2. The criteria weights found by fuzzy AHP

	Cost												Infrastructure					
	C1			C2			C3			C4			C5			C6		
Fuzzy weights of main-criteria	0.06	0.10	0.19	0.06	0.10	0.19	0.06	0.10	0.19	0.06	0.10	0.19	0.35	0.52	0.77	0.35	0.52	0.77
Fuzzy weights of sub-criteria	0.21	0.27	0.33	0.03	0.04	0.05	0.49	0.62	0.78	0.06	0.07	0.10	0.68	0.83	1.02	0.14	0.17	0.21
Fuzzy importance degrees	0.01	0.03	0.06	0.00	0.00	0.01	0.03	0.06	0.15	0.00	0.01	0.02	0.24	0.44	0.78	0.05	0.09	0.16
Weights of main-criteria	0.119			0.119			0.119			0.119			0.546			0.546		
Weights of sub-criteria	0.270			0.040			0.632			0.075			0.843			0.171		

	Social and economic structure												Macro factors						Environmental factors					
	C7			C8			C9			C10			C11			C12			C13			C14		
Fuzzy weights of main-criteria	0.05	0.07	0.12	0.05	0.07	0.12	0.05	0.07	0.12	0.05	0.07	0.12	0.06	0.11	0.16	0.06	0.11	0.16	0.12	0.20	0.30	0.12	0.20	0.30
Fuzzy weights of sub-criteria	0.11	0.17	0.25	0.50	0.66	0.87	0.06	0.09	0.12	0.06	0.08	0.13	0.68	0.83	1.02	0.14	0.17	0.21	0.41	0.50	0.83	0.29	0.50	0.59
Fuzzy importance degrees	0.01	0.01	0.03	0.02	0.05	0.11	0.00	0.01	0.01	0.00	0.01	0.02	0.04	0.09	0.17	0.01	0.02	0.03	0.05	0.10	0.25	0.04	0.10	0.18
Weights of main-criteria	0.081			0.081			0.081			0.081			0.112			0.112			0.206			0.206		
Weights of sub-criteria	0.177			0.680			0.089			0.088			0.843			0.171			0.581			0.460		

After determining the weights of the criteria, we apply the AD manually to find final decision. We use crisp AD for the "Cost" main criterion and fuzzy AD for other main criteria.

The system and design range are identified by experts in IETT to implement the AD methodology (Table 3). In the first stage of AD, we make the calculations without the criteria weights. The triangular fuzzy numbers for intangible factors are Excellent: (16,20,20), Very good: (12,15,18), Good: (8,11,14), Fair: (4,7,10), Poor: (0,0,6). The results are shown in the Table 4.

Table 3. The system and design range

	Alternatives			
	System Range			Design Range
	Beylikdüzü	Arnavutköy	Silivri	
C1	55.000.000-65000000	34.000.000-35000000	40.000.000-45000000	30.000.00-60.000.000
C2	250.000-300000	175.000-200000	375.000-450000	150.000- 400.000
C3	1250.000-1500000	800.000-1000000	2300.000-2500000	900.000-2.400.000
C4	4700.000-4950000	7200.000-7425000	12.000.000-12375000	4.500.000-12.250.000
C5	8.11.14	8.11.14	8.11.14	min 12
C6	4.7.10	8.11.14	4.7.10	min 8
C7	4.7.10	8.11.14	8.11.14	min 8
C8	4.7.10	12.15.18	8.11.14	min 8
C9	12.15.18	8.11.14	8.11.14	min 8
C10	8.11.14	8.11.14	8.11.14	min 4
C11	8.11.14	8.11.14	12.15.18	min 12
C12	4.7.10	8.11.14	12.15.18	min 8
C13	4.7.10	8.11.14	12.15.18	min 8
C14	4.7.10	12.15.18	8.11.14	min 8

Table 4. Unit Index for unweighted information contents

	I1	I2	I3	I4	I5	Total I	
Beylikdüzü	0.250	4.407	3.354	4.407	5.492	17.909	
Arnavutköy	0.250	2.822	1.769	2.822	1.809	9.471	***
Silivri	0.792	4.407	2.025	1.292	1.809	10.326	

The alternative "Arnavutköy" is selected bacause it has the minimum information content.

In the second stage, weighted fuzzy AD calculations are made and the weights which are calculated in fuzzy AHP are taken into account. The results in Table 5-7 are obtained.

Table 5. The weights of criteria and the information contents for alternatives

The weights of main-criteria	**W1**				**W2**		**W3**				**W4**		**W5**	
	0.119				0.546		0.081				0.112		0.206	
The weights of sub-criteria	**W11**	**W12**	**W13**	**W14**	**W21**	**W22**	**W31**	**W32**	**W33**	**W34**	**W41**	**W42**	**W51**	**W52**
	0.270	0.040	0.632	0.075	0.843	0.171	0.177	0.680	0.089	0.088	0.843	0.171	0.581	0.460
Information contents	**I11**	**I12**	**I13**	**I14**	**I21**	**I22**	**I31**	**I32**	**I33**	**I34**	**I41**	**I42**	**I51**	**I52**
Beylikdüzü	1.000	0.000	0.000	0.000	3.322	5.492	5.492	5.492	1.295	1,136	3.322	5.492	5.492	5.492
Arnavutköy	0.000	0.000	1.000	0.000	3.322	2.322	2.322	1.295	2.322	1.136	3.322	2.322	2.322	1.295
Silivri	0.000	1.585	1.000	0.585	3.322	5.492	2.322	2.322	2.322	1.136	1.290	1.295	1.295	2.322

Table 6. Unit Index for weighted Information contents of sub attribute

	W1	W2	W3	W4	W5
	0.119	0.546	0.081	0.112	0.206
	Av. I1	**Av. I2**	**Av. I3**	**Av. I4**	**Av. I5**
Beylikdüzü	0.068	2.044	1.643	2.044	2.439
Arnavutköy	0.158	1.953	1.111	1.953	1.379
Silivri	0.413	2.044	1.256	1.142	1.318

Table 7. Unit Index for weighted Information contents

	I1	I2	I3	I4	I5	Total I
Beylikdüzü	0.000	1.478	1.041	1.083	1.202	4.804
Arnavutköy	0.000	1.442	1.009	1.078	1.068	4.596
Silivri	0.001	1.478	1.019	1.015	1.059	4.571

In addition, to validate the results of this methodology, we have developed software based on AD principles, which is called ADSolver and this solver is used to solve garage location selection problem. The same fuzzy AHP results, design and system range data are used. The screenshots of the software coded in C# are shown in the Figure 1.

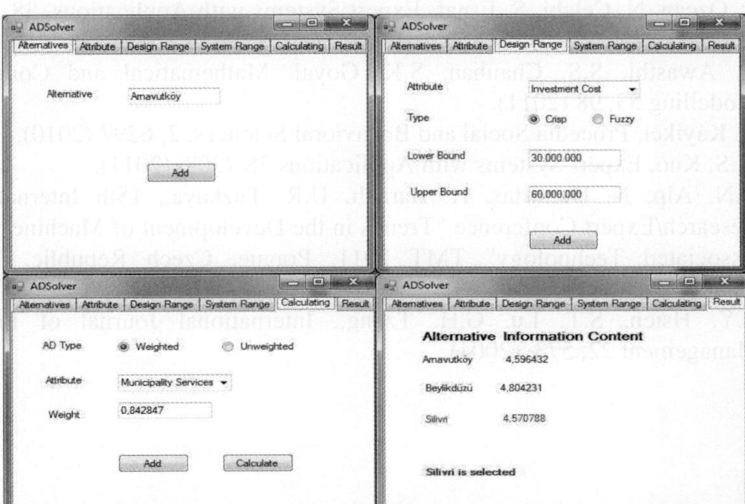

Figure 1. The screenshots of the AD application software

4. Conclusion

Istanbul is a growing city and the land use is very important in terms of the city master plans. Considering this reality, probable places are determined by the İETT authorities in this study. Best alternative is selected by applying fuzzy AD and weighted fuzzy AD in which the weights are calculated using fuzzy AHP manually. The aim of using AD is taking into account the design ranges for all attributes. This helps the decision maker to select the most appropriate alternative that is suitable with his/her design features. In fuzzy AD Arnavutköy is selected and in weighted fuzzy AD Silivri is selected because it has the minimum information content.

After meetings with experts, it is considered that all evaluation criteria don't have the same importance weights. Because of this, in ADSolver we have used the weighted fuzzy AD principles. In conclusion, Silivri is selected like in manual calculations of weighted fuzzy AD.

References

1. N.P. Suh,, Microelectronics Reliability, 37, 538 (March 1997).
2. N.P. Suh, D.S. Cochran, C.L. Paulo, Annals of the CIRP, 47, 627 (1998).
3. C. Kahraman, I. Kaya, S. Cebi, Energy, 34, 1603 (2009).
4. O. Kulak, C. Kahraman, Information Sciences, 170, 191 (2005).
5. T. Özcan, N. Çelebi, Ş. Esnaf, Expert Systems with Applications, 38, 9773 (2011).
6. A. Awasthi, S.S. Chauhan, S.K. Goyal, Mathematical and Computer Modelling 53, 98 (2011).
7. Y. Kayikci, Procedia Social and Behavioral Sciences, 2, 6297 (2010).
8. M.S. Kuo, Expert Systems with Applications 38 7208, (2011).
9. Ö.N. Alp, N. Demirtaş, H. Baraçlı, U.R. Tuzkaya,, 15th International Research/Expert Conference "Trends in the Development of Machinery and Associated Technology" TMT 2011, Prague, Czech Republic, 12-18 September 2011.
10. T.Y. Hsieh, S.T. Lu, G.H. Tzeng,, International Journal of Project Management 22, 573, (2004).

EVALUATION OF CRITERIA TO SELECT APPROPRIATE CANDIDATES FOR SURGICAL SCIENCES

İPEK NUR AKSU, F. TUNÇ BOZBURA, AHMET BEŞKESE

Bahçeşehir University, Faculty of Engineering, Industrial Engineering Department
Çırağan Caddesi No:4/6, Beşiktaş 34353, İSTANBUL

Today's companies, perform a work which aims to enhance individual and corporate efficiency and effectiveness. Companies need to provide establish and provide valid and applicable performance management system in order to become successful on sectoral. In a competitive environment, companies' staff performance evaluation as indicate retrospective performance level, identify prudential potential performance and provide new perspective to studies about enhance performance. The purpose of this study is to specify the recruitment criteria of position that clinical chief of surgery. Research is done on a group including 2 experts that have a career in health sector for many years. In this research, staff recruitment criteria, recruitment process evaluation and the ranking of importance for the corporation are involved for the non-profit organization. This study includes identifying the criteria for selection, some information about criteria are get from the experts which working at hospitals in Istanbul then all in information are combined to build hierarchy of recruitment criteria. By using Analytical Hierarchy Process (AHP) in fuzzy environment criteria are ranked and the importance is identified and this method is used to select appropriate candidate for the position. As a result, according to FAHP methods recruitment criteria and process evaluation for the surgeons are identified.

1. Introduction

Along with globalization, companies must continuously develop their competition skills and keep up with the alteration in order to survive and to be float. One of the most important ways to ensure competitive success is to invest in human resources and use this resource efficiently.

Employees are the most valuable assets of a company in human resources management, which has doing set of activities carried out in order to manage effectively. The aim of human resource management is maximize the contributions of employee with business, to ensure integration of the business and increase satisfaction.

Human resources management; operation will be a new addition to the selection of employees for the performance evaluation of existing employees in

the making, finding the differences between employees and managers and employees play an important role in drawing up of the relationship between the features that affect.
This study is aim to clarify the method of fuzzy with an application to identify selection process of surgeons in health sector. This study consists of two parts.

The first section focuses on fuzzy sets to decide criteria for recruitment process and second section identifies the importance of criteria to ranking selection criteria.

2. Literature Review

AHP is a common method of multi-criteria decision making which is developed by Thomas L. Saaty (1980). AHP paired comparison process may be inadequate in dealing with situations of uncertainty and instability therefore fuzzy AHP method has been developed to solve hierarchical problems. However, for real world environmental management problems that involve many stakeholders and conflicting viewpoints, the traditional AHP method is insufficient. Buckley (1985) applies the fuzzy theory to the AHP method to avoid neglecting extreme values.

Many researchers studies about selection with AHP and fuzzy AHP in terms of different perspectives in literature. Kahraman, Cebeci and Ulukan (2003) for multi-criteria supplier selection, Zouggari and Benyoucef (2011) for multi-criteria group decision supplier selection, Mahmoodzadeh, Shahrabi, Pariazar, and Zaeri (2007) for project selection, Ballı and Korukoğlu (2009) for operating system selection, Güngör, Serhadlıoğlu, Kesen (2009) for personnel selection. Dursun and Karsak (2010) for a Fuzzy MCDM approach for personnel selection, Büyüközkan and Çiftçi (2012) for A combined fuzzy AHP and fuzzy TOPSIS based strategic analysis of electronic service quality in healthcare industry, Gibney and Shang (2007) for Decision making in academia: A case of the dean selection process, Kabak, Burmaoğlu and Kazançoğlu (2012) for A fuzzy hybrid MCDM approach for professional selection. In literature there are many studies about personnel selection problems by using AHP and Fuzzy AHP in firms in all sectors. Today's personnel selection are getting more important for firms and also health sector especially hospitals although there is no adequate studies in literature about staff selection in health sector.

3. Application

The proposed model for Surgical Science Candidate Selection problem is following;

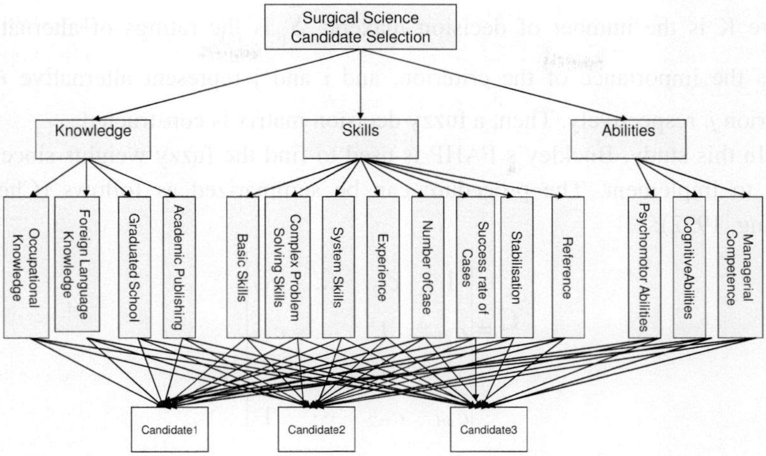

Figure 1. The Hierarchical Structure of Selection Criteria

According to hierarchical model, identified the importance level of recruitment criteria by decision makers with respect to the pair-wise comparisons matrix.

Step 1: Decision makers decide the importance level of criteria by using pair-wise comparison matrix with linguistic scale as shown in Table 1. (Kahraman & Cebi 2009) (Hsieh et al., 2004)

Table 1. Linguistic Scale for Weight Matrix

Linguistic scales	Scale of fuzzy number	
(1, 1, 3)	Equally important	(Eq)
(1, 3, 5)	Weakly important	(Wk)
(3, 5, 7)	Essentially important	(Es)
(5, 7, 9)	Very strongly important	(Vs)
(7, 9, 9)	Absolutely important	(Ab)

The aggregated decision matrix is constructed to satisfy each decision maker in the group because in this study there is a group decision. Therefore, to obtain the aggregation of the importance weight of each criterion and the rating of each alternative (Chen,2000) as follows;

$$\tilde{S}_{ij} = \frac{1}{K}\tilde{S}_{ij}^{1} + \tilde{S}_{ij}^{2} + ... + \tilde{S}_{ij}^{t} + ... + \tilde{S}_{ij}^{K}, \qquad \tilde{S}_{ij}^{t} = (a_{ij}, b_{ij}, c_{ij}),$$

$$\tilde{w}_{j} = \frac{1}{K}\tilde{w}_{j}^{1} + \tilde{w}_{j}^{2} + ... + \tilde{w}_{j}^{t} + ...\tilde{w}_{j}^{K}, \qquad \tilde{w}_{j}^{t} = (w_{jt}, w_{jm}, w_{ju}),$$

where K is the number of decision makers; $\tilde{S}_{ij}$ is the ratings of alternatives; $\tilde{w}_j$ is the importance of the criterion, and i and j represent alternative *i* and criterion *j*, respectively. Then, a fuzzy decision matrix is constructed.

In this study, Buckley's FAHP is used to find the fuzzy weights since it is easy to implement. The procedure can be summarized as follows (Chen & Hwang, 1992):

$$\tilde{C} = \begin{vmatrix} 1 & \tilde{c}_{12} & \cdots & \tilde{c}_{1n} \\ \tilde{c}_{21} & 1 & \cdots & \tilde{c}_{2n} \\ \vdots & \vdots & \cdots & \vdots \\ \tilde{c}_{m1} & \tilde{c}_{m2} & \cdots & 1 \end{vmatrix}$$

Where $\tilde{C}$ pair-wise comparison matrix and

$$\tilde{c}_{ij} = \begin{Bmatrix} i > j, (1,1,3), (1,3,5), (3,5,7), (5,7,9,), (7,9,9), \\ i = j, 1, \\ i < j, (1,1,3)^{-1}, (1,3,5)^{-1}, (3,5,7)^{-1}, (5,7,9)^{-1}, (7,9,9)^{-1} \end{Bmatrix}$$

Aggregated fuzzy matrix for mail goal can be obtained as follows in Table 2.

Table 2. Fuzzy Aggregated Decision Matrix

	Knowledge			**Skills**			**Abilities**		
Knowledge	1,00	1,00	1,00	0,12	0,17	0,26	0,17	0,26	0,57
Skills	3,87	5,92	7,94	1,00	1,00	1,00	1,00	1,00	3,00
Abilities	1,73	3,87	5,92	0,33	1,00	1,00	1,00	1,00	1,00

Step 2: Then, the fuzzy weight matrix is calculated by Buckley's Method as follows and stated in Table 4.

$$\tilde{r}_i = (\tilde{c}_{i1} \otimes \tilde{c}_{i2} \otimes \cdots \otimes \tilde{c}_{in})^{1/n},$$

$$\tilde{w}_i = (\tilde{r}_i \otimes (\tilde{r}_1 + \tilde{r}_2 + \cdots \tilde{r}_n)^{-1},$$

where $\tilde{c}_{in}$ is the fuzzy comparison value of criterion *i* to criterion *n*, $\tilde{r}_i$ is the geometric mean of fuzzy comparison value of criterion *i* to each criterion.

Table 3. The geometric mean of fuzzy comparison values

$\tilde{r}_i$			
0,27	0,35	0,53	$\tilde{r}_1$
1,57	1,81	2,88	$\tilde{r}_2$
0,83	1,57	1,81	$\tilde{r}_3$

Step 3: After the importance of weight matrix is obtained, defuzzification process which converts a fuzzy number into a crisp value is utilized. At first, fuzzy numbers will be defuzzified into crisp values and then normalization procedure will be applied. For the defuzzification process, centroid method, which provides a crisp value based on the center of the gravity, is selected since it is the most commonly used method (Opricovic & Tzeng, 2004). Following equation presents both defuzzification and normalization procedure in one formula as stated in Table 5.

$$w_r = \frac{\tilde{w}_r}{\sum_{j=1}^{n} \tilde{w}_j} = \frac{w_{rl} + w_{rm} + w_{ru}}{\sum_{j=1}^{n} \tilde{w}_j},$$

where the importance of rth criterion, w_r, is a non-fuzzy number and n is the number of the criteria.

Table 4. Fuzzy weight matrix

$\tilde{w}_i$			
0,05	0,09	0,20	$\tilde{w}_1$
0,30	0,54	1,08	$\tilde{w}_2$
0,16	0,42	0,68	$\tilde{w}_3$

Table 5. Fuzzy weights after defuzzification and normalization

$\tilde{w}_r$	
0,10	**Knowledge**
0,54	**Skills**
0,36	**Abilities**

The fuzzy weight vector of the criteria obtained by Buckley formulations is presented in Table 4. After that defuzzification procedure is done and Table 5 is obtained. In Figure 2, the hierarchical structure of importance of the criteria is given after the application of procedure for all criteria.

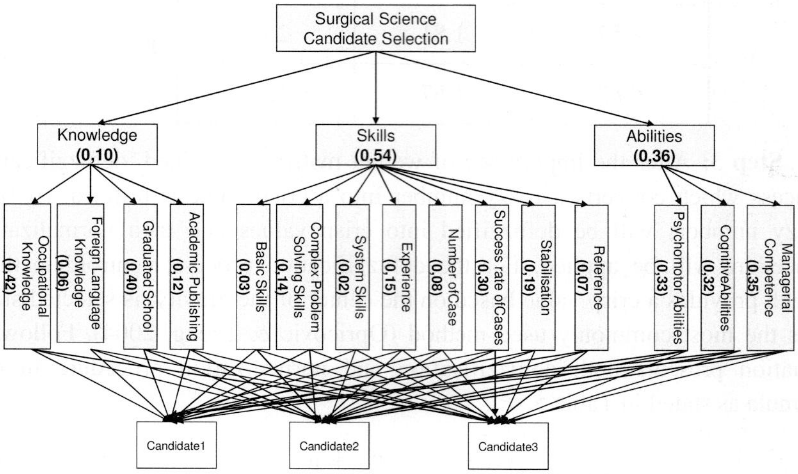

Figure 2. The Hierarchical Structure of criteria for importance level

4. Conclusion

The aim of this study is to evaluate criteria for selecting candidate to the position. First identified the selection criteria for the clinical chief of surgery and then determined the weights of criteria by using pair-wise comparisons matrix by decision makers for all criteria. One decision maker is the chief of clinic and one of is head doctor of hospital.

According to decision makers' evaluations for criteria, the problem solved by using fuzzy AHP which includes Buckley Method. As a result Skills is the most important criteria as 0.54 and Success Rate of Cases is the most important sub-criteria as 0.30. According to these results, the most important criteria is Skills and the most important sub-criteria is Success Rate of Cases should be taken into account for the candidate selection of clinical chief of surgery. Selection of candidates with respect to the weight of criteria getting from FAHP can be performed by using Fuzzy TOPSIS which will be the future work of this study.

References

1. Kahraman, C., Cebeci, U. & Ulukan, Z, 2003. Multi-criteria supplier selection using fuzzy AHP. Logistics Information Management, 16(6), pp. 382-394.
2. Ballı, S., Korukoğlu, S., 2009. Operating System Selection Using Fuzzy AHP and TOPSIS Methods. Mathematical and Computational Aplications, 14(2), pp. 119-130.
3. Zouggari, A., Benyoucef, L., 2011. Multi-criteria Group Decision Supplier Selection Problem Using Fuzzy TOPSIS Based Approach.
4. Mahmoodzadeh, S., Shahrabi, J., Pariazar, M.& Zaeri, M. S. 2007. Project Selection by Using Fuzzy AHP TOPSIS Technique. World Academy Science, Engineering and Technology.
5. Güngör, Z., Serhadlıoğlu, G., Kesen, S. E. 2009. A Fuzzy AHP Approach to Personnel Selection Problem. Applied Soft Computing, pp. 641-646.
6. Dursun, M., Karsak, E. E. 2010. A Fuzzy MCDM Approach for Personnel Selection. Expert Systems with Applications, 37(6), pp. 4324-4330.
7. Büyüközkan, G., Çiftçi, G. 2012 .A combined fuzzy AHP and fuzzy TOPSIS based Strategic Analysis of Electronic Service Quality in Healthcare Industry. Expert Systems with Applications, 39(3), pp. 2341–2354.
8. Gibney, R., Shang, J. 2007 .A Decision making in academia: A case of the dean selection process. Mathematical and Computer Modelling, 46(7-8), pp. 1030-1040.
9. Kabak, M., Burmaoğlu, S. & Kazançoğlu, Y. 2012 . A fuzzy hybrid MCDM approach for professional selection,, 39(3), pp. 3516-3525.
10. Opricovic, S., & Tzeng, G. H. (2004). Compromise solution by MCDM methods: A comparative analysis of VIKOR and TOPSIS. European Journal of Operational Research, 156(2), 445–455
11. Chen, S. J., & Hwang, C. L. (1992). Fuzzy multi attribute decision making: methods and applications. New York: Springer-Verlag.
12. Kahraman, C., Cebı, S., (2009) Expert Systems with Applications 36 pp. 4848–4861
13. Hsieh, T. Y., Lu, S. T., & Tzeng, G. T. (2004). Fuzzy MCDM approach for planning and design tenders selection in public office buildings. International Journal of Project Management, 22, 573–584.
14. Chen, C. T. (2000). Extensions of the TOPSIS for Group Decision-Making under Fuzzy Environment, Fuzzy Sets and Systems, 114, p.1-9.

AN EXTENDED METHOD FOR FUZZY MULTIPLE ATTRIBUTE GROUP DECISION-MAKING BASED ON INTERVAL TYPE-2 FUZZY SETS

ZAFER EREN, C. ERHAN BOZDAG
Industrial Engineering, Istanbul Technical University, Faculty of Management, Maçka, 34367, Istanbul, Turkey

In this paper, a new method is presented to deal with fuzzy multiple-attribute group decision-making problems based on weights vector of decision makers. In this paper all evaluations that are given by decision makers are interval type-2 fuzzy trapezoidal numbers, since type-2 fuzzy sets involve more uncertainties than type-1 fuzzy sets. We also use a numerical example for location selection problem to utilize the proposed method.

1. Introduction

The purpose of multiple attributes decision-making (MADM) is to choose a best candidate from a set of alternatives by means of evaluating multiple attributes of the alternatives [1]. There are many existing methods for solving MADM problems in the literature [1] and they have been extensively applied to various areas such as society, economics, management, military and engineering [2]. In real life the judgment of the decision makers may not be correct as it depends upon the uncertainty of human judgment. In such cases fuzzy set theory is used to model the human judgment [3]. If we can deal with fuzzy multiple attributes group decision-making problems based on interval type-2 fuzzy sets, then there is room for more flexibility [1]. Ashtiani et al. [4] proposed an interval-valued fuzzy TOPSIS method to provide a more accurate modeling. Chen et al. [1] presented a new method for ranking interval type-2 fuzzy sets and they placed it into fuzzy multiple attributes group decision-making model.

In this paper, a new method based on interval-valued type-2 fuzzy trapezoidal numbers is proposed to find the weights of attributes and group decision matrix. In the proposed method, each decision maker evaluates the others by considering the level of their capability. The contribution of a decision maker to the final weights of attributes is determined by multiplying the level of capability and the weight given by the decision maker. The group decision matrix is also determined by multiplying the level of decision maker's capability with individual decision. The method is integrated into Liu's ranking method [2] and Chen & Lee's TOPSIS method [1].

2. The basic concept of the interval type-2 fuzzy numbers

2.1. *Interval type-2 fuzzy numbers*

(i) Chen and Lee [1] represented the interval type-2 fuzzy numbers

$\tilde{\tilde{A}}_i = [\tilde{A}_i^L, \tilde{A}_i^U] = \left[\left(a_{i1}^L, a_{i2}^L, a_{i3}^L, a_{i4}^L; H_1(\tilde{A}_i^L), H_2(\tilde{A}_i^L)\right), \left(a_{i1}^U, a_{i2}^U, a_{i3}^U, a_{i4}^U, H_1(\tilde{A}_i^U), H_2(\tilde{A}_i^U)\right)\right]$ as shown in Figure 1, where $0 \le a_{i1}^L \le a_{i2}^L \le a_{i3}^L \le a_{i4}^L \le 1, 0 \le a_{i1}^U \le a_{i2}^U \le a_{i3}^U \le a_{i4}^U \le 1, 0 \le H_1(\tilde{A}_i^L) \le H_1(\tilde{A}_i^U) \le 1, 0 \le H_2(\tilde{A}_i^L) \le H_2(\tilde{A}_i^U) \le 1$ and $\tilde{A}_i^L \subset \tilde{A}_i^U$.

(ii) The operations on interval type-2 trapezoidal fuzzy numbers will not be given here because of space constraint. People who are interested can look up [6].

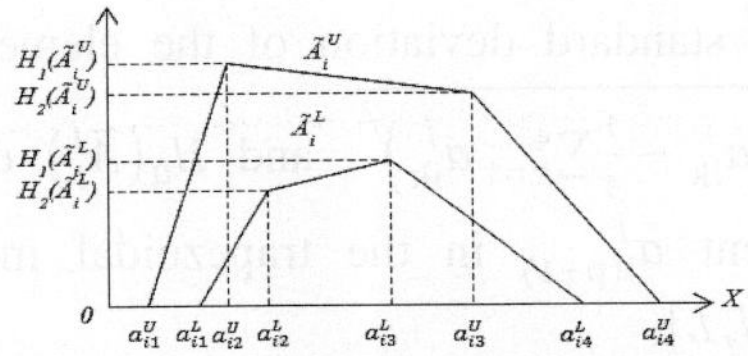

Figure 1. Interval type-2 trapezoidal fuzzy numbers.

2.2. *The distance of interval type-2 trapezoidal fuzzy numbers*

Wei and Chen [2] represented the coordinate of COG (the center of gravity) point is used to the comparison of two interval type-2 trapezoidal fuzzy numbers.

(i) Calculate the coordinate of COG point $(x_{\tilde{A}^L}, y_{\tilde{A}^L}), (x_{\tilde{A}^U}, y_{\tilde{A}^U})$ which belongs to the type-1 trapezoidal fuzzy numbers $\tilde{\tilde{A}}^L, \tilde{\tilde{A}}^U$ respectively.

(ii) The center of gravity of interval type-2 trapezoidal fuzzy numbers $\tilde{\tilde{A}}$ are $(x_{\tilde{\tilde{A}}}, y_{\tilde{\tilde{A}}})$, then:

$$\begin{cases} x_{\tilde{\tilde{A}}} = \dfrac{x_{\tilde{\tilde{A}}^L} + x_{\tilde{\tilde{A}}^U}}{2} \\ y_{\tilde{\tilde{A}}} = \dfrac{y_{\tilde{\tilde{A}}^L} + y_{\tilde{\tilde{A}}^U}}{2} \end{cases} \tag{1}$$

(iii) Find the distance between origin point and $\tilde{\tilde{A}}$

$$d_{\tilde{\tilde{A}}} = \sqrt{x_{\tilde{\tilde{A}}}^2 + y_{\tilde{\tilde{A}}}^2} \quad (2)$$

(iv) The comparison of $\tilde{\tilde{A}}$ and $\tilde{\tilde{B}}$:

If $d_{\tilde{\tilde{A}}} > d_{\tilde{\tilde{B}}}$, then $\tilde{\tilde{A}} > \tilde{\tilde{B}}$ and $\tilde{\tilde{A}}$ is larger or better than $\tilde{\tilde{B}}$.

2.3. *The ranking value of the trapezoidal interval type-2 fuzzy number*

The ranking value $Rank\left(\tilde{\tilde{A}}_i\right)$ of the trapezoidal interval type-2 fuzzy number $\tilde{\tilde{A}}_i$ is calculated as follows [1]:

$$Rank\left(\tilde{\tilde{A}}_i\right) = M_1(\tilde{A}_i^U) + M_1(\tilde{A}_i^L) + M_2(\tilde{A}_i^U) + M_2(\tilde{A}_i^L) + M_3(\tilde{A}_i^U) + M_3(\tilde{A}_i^L) - \frac{1}{4}\left(S_1(\tilde{A}_i^U) + S_1(\tilde{A}_i^L) + S_2(\tilde{A}_i^U) + S_2(\tilde{A}_i^L) + S_3(\tilde{A}_i^U) + S_3(\tilde{A}_i^L) + S_4(\tilde{A}_i^U) + S_4(\tilde{A}_i^L)\right) + H_1(\tilde{A}_i^U) + H_1(\tilde{A}_i^L) + H_2(\tilde{A}_i^U) + H_2(\tilde{A}_i^L) \quad (3)$$

where $M_{\mathrm{p}}(\tilde{A}_i^j)$ denotes the average of the elements a_{ip}^j and $a_{\mathrm{i(p+1)}}^j$, $M_{\mathrm{p}}(\tilde{A}_i^j) = \left(a_{\mathrm{ip}}^j + a_{\mathrm{i(p+1)}}^j\right)/2$, $1 \le p \le 3$, $S_{\mathrm{q}}(\tilde{A}_i^j)$ denotes the standard deviation of the elements a_{iq}^j and $a_{\mathrm{i(q+1)}}^j$, and $S_{\mathrm{q}}(\tilde{A}_i^j) = \frac{1}{2}\sum_{k=q}^{q+1}\left(a_{\mathrm{ik}}^j - \frac{1}{2}\sum_{k=q}^{q+1} a_{\mathrm{ik}}^j\right)^2$, $1 \le q \le 3$, $S_4(\tilde{A}_i^j)$ denotes the standard deviation of the elements $a_{\mathrm{i1}}^j, a_{\mathrm{i2}}^j, a_{\mathrm{i3}}^j, a_{\mathrm{i4}}^j$, and $S_4(\tilde{A}_i^j) = \sqrt{\frac{1}{2}\sum_{k=q}^{q+1}\left(a_{\mathrm{ik}}^j - \frac{1}{4}\sum_{k=1}^{4} a_{\mathrm{ik}}^j\right)^2}$, and $H_{\mathrm{q}}(\tilde{A}_i^j)$ denotes the membership value of the element $a_{\mathrm{i(p+1)}}^j$ in the trapezoidal membership function $\tilde{A}_i^j$, $1 \le p \le 2, j \in \{U, L\}$.

3. Group decision making method

3.1. *Decision making problems*

Let $D = \{d_1, d_2, \dots, d_q\}$ be the set of decision makers, $A = \{A_1, A_2, \dots, A_m\}$ be set of alternatives, and $C = \{C_1, C_2, \dots, C_n\}$ be set of attributes. Suppose that $\tilde{\tilde{a}}_{ijk} = [(a_{ijk1}^L, a_{ijk2}^L, a_{ijk3}^L, a_{ijk4}^L; h_{ijk}^L,), (a_{ijk1}^U, a_{ijk2}^U, a_{ijk3}^U, a_{ijk4}^U; h_{ijk}^U,)]$ be the attribute value given by decision maker d_k, where $\tilde{\tilde{a}}_{ijk}$ is a interval type-2 trapezoidal fuzzy number for alternative A_i with respect to the attribute C_j, $\tilde{\tilde{w}}_{kj} = [(w_{kj1}^L, w_{kj2}^L, w_{kj3}^L, w_{kj4}^L; h_{kj}^L,), (w_{kj1}^U, w_{kj2}^U, w_{kj3}^U, w_{kj4}^U; h_{kj}^U,)]$ be the weight of attribute C_j given by the decision maker d_k, and $\tilde{\tilde{w}}_j$ is weight of attribute C_j,. $\tilde{\tilde{\lambda}}_{kl} = [(\lambda_{kl1}^L, \lambda_{kl2}^L, \lambda_{kl3}^L, \lambda_{kl4}^L; h_{kl}^L,), (\lambda_{kl1}^U, \lambda_{kl2}^U, \lambda_{kl3}^U, \lambda_{kl4}^U; \lambda_{kl}^U,)]$ is the decision maker weight for d_k given by the d_l decision makers, where $\tilde{\tilde{\lambda}}_{kl}$ is a interval type-2 trapezoidal fuzzy number, and $k \ne l$, and $\tilde{\tilde{\lambda}}_k$ is weight of decision maker k.

3.2. *Evaluation the alternatives*

In most methods, weights of decision makers are accepted equal or at least real number. In this paper, weights of decision makers are interval type-2 fuzzy numbers and it is integrated with Liu ranking method [2] in Method 1 and Chen & Lee Topsis method [1] in Method 2.

Method 1:

Step 1: Calculate the weight of decision maker. Using average value of $\tilde{\tilde{\lambda}}_{ij}$ to calculate the weight of decision maker *i* is with respect to Eq. (4).

$$\tilde{\tilde{\lambda}}_k = (\textstyle\sum_{l=1}^{q} \tilde{\tilde{\lambda}}_{kl})/(q-1) \tag{4}$$

Step 2: Calculate the weight of attribute.

$$\tilde{\tilde{w}}_j = (\textstyle\sum_{k=1}^{q} \tilde{\tilde{\lambda}}_k \times \tilde{\tilde{w}}_{kj})/q \tag{5}$$

Step 3: Calculate the value of alternative.

$$\tilde{\tilde{a}}_{ij} = (\textstyle\sum_{k=1}^{q} \tilde{\tilde{\lambda}}_k \times \tilde{\tilde{a}}_{ijk})/q\ (\sum_{k=1}^{q} \tilde{\tilde{\lambda}}_k \times \tilde{\tilde{a}}_{ijk})/(q) \tag{6}$$

Step 4: Calculate the weighted value of alternative.

$$\tilde{\tilde{a}}_i = (\textstyle\sum_{j=1}^{n} \tilde{\tilde{w}}_j \times \tilde{\tilde{a}}_{ij})/n \tag{7}$$

Step 5: Calculate the center of gravity (COG) coordinates of interval type-2 trapezoidal fuzzy numbers of alternatives.

Step 6: Calculate the distance from the starting point.

Step 7: Rank the alternatives.

Method 2:

Step 1, *Step 2*, and *Step 3* is same as in Method 1

Step 4: Calculate the weighted value of alternative.

Step 5: Based on Eq. (3), calculate the $Rank(\tilde{\tilde{a}}_{ij})$ of alternative *i*.

Step 6: Determine the positive ideal solution $a^+ = (a_1^+, a_1^+, \dots, a_n^+,)$ and the negative ideal solution $a^- = (a_1^-, a_1^-, \dots, a_n^-,)$ where

$$a_j^+ = \begin{cases} max_{1\le i\le m}\{Rank(\tilde{\tilde{a}}_{ij})\}, & if\ c_j\ is\ benefit\ attribute \\ min_{1\le i\le m}\{Rank(\tilde{\tilde{a}}_{ij})\}, & if c_j\ is\ cost\ attribute \end{cases} \tag{8}$$

$$a_j^- = \begin{cases} min_{1\le i\le m}\{Rank(\tilde{\tilde{a}}_{ij})\},, & if\ c_j\ is\ benefit\ attribute \\ max_{1\le i\le m}\{Rank(\tilde{\tilde{a}}_{ij})\} & if c_j\ is\ cost\ attribute \end{cases} \tag{9}$$

Step 7: Calculate the distance $d^+(a_i)$ between each alternative a_i and the positive ideal solution a^+, shown as follows:

$$d^+(a_i)=\sqrt{\sum_{j=1}^{n}\left(Rank\left(\tilde{\tilde{a}}_{ij}\right)-a_j^+\right)^2},\ 1\le i\le m \tag{10}$$

Calculate the distance $d^-(a_i)$ between each alternative a_i and the negative ideal solution a^-, shown as follows:

$$d^-(a_i)=\sqrt{\sum_{j=1}^{n}\left(Rank\left(\tilde{\tilde{a}}_{ij}\right)-a_j^-\right)^2},\ 1\le i\le m \tag{11}$$

Step 8: Calculate the relative degree of closeness $C(a_i)$ of a_i with respect to the positive ideal solution , shown as follows:

$$C(a_i)=\left(d^-(a_i)\right)/\left(d^+(a_i)+d^-(a_i)\right),\ 1\le i\le m \tag{12}$$

Step 9: Rank the alternatives with respect to the values of $C(a_i)$ in a descending sequence

4. Numerical example

Assume that there are three decision maker D1, D2 and D3 who are board member of a medical group to open a new medical center and assume that there are six candidate district Bahçelievler (a_1), Başakşehir (a_2), Beylikdüzü (a_3), Kağıthane (a_4), Levent (a_5), and Maslak (a_6) and five criterion "population density" (c_1), "average income" (c_2), "number of medical centers of district" ($1/c_3$), "Location" (c_4), and "Land Price" ($1/c_5$). "number of medical centers of district" and "Land Price" are cost attribute, then c_3 and c_5 are evaluated "1/attribute" by decision makers. Suppose that three decision maker use the linguistic terms shown in Table 1 to represent the weights of each decision maker, criteria as shown in Table 2 and Table 3, and to represent the evaluating values of the alternatives with respect to criterion, as shown in Table 4.

Table 1. Seven-member interval type-2 linguistic term set.

Linguistic Terms (for decision makers and attribute values)	Linguistic Terms (for weights)	Interval type-2 trapezoidal fuzzy numbers
Very-poor (VP)	Very-low (VL)	[(0,0,0,0.05,0.8),(0,0,0,0.1,1.0)]
Poor (P)	Low (L)	[(0.05,0.1,0.1,0.2,0.8),(0,0.1,0.1,0.3,1.0)]
Medium-poor (MP)	Medium-low (ML)	[(0.2,0.3,0.3,0.4,0.8),(0.1,0.3,0.3,0.5,1.0)]
Medium (F)	Medium (M)	[(0.4,0.5,0.5,0.6,0.8),(0.3,0.5,0.5,0.7,1.0)]
Medium-good (MG)	Medium-high (MH)	[(0.6,0.7,0.7,0.8,0.8),(0.5,0.7,0.7,0.9,1.0)]
Good (G)	High (H)	[(0.8,0.9,0.9,0.95,0.8),(0.7,0.9,0.9,1.0,1.0)]
Very-good (VG)	Very-high (VH)	[(0.95,1.0,1.0,1.0,0.8),(0.9,1.0,1.0,1.0,1.0)]

Table 2.The evaluation information of each decision maker given by decision makers.

	DM1	DM2	DM3
DM1		G	VG
DM2	G		G
DM3	MG	VG	

Table 3. Evaluating values of criterion given by decision makers.

	c1	c2	c3	c4	c5
DM1	H	H	VH	H	MH
DM2	VH	H	MH	VH	VH
DM3	MH	VH	VH	H	M

Table 4. Evaluating values of the alternatives given by the decision makers with respect to criterion.

DM1	Criterion					DM2	Criterion					DM3	Criterion				
Alt.	c1	c2	c3	c4	c5	Alt.	c1	c2	c3	c4	c5	Alt.	c1	c2	c3	c4	c5
a1	VG	G	ML	G	MG	a1	VG	G	M	MG	M	a1	VG	MG	ML	G	M
a2	G	MG	VG	MG	G	a2	M	MG	G	M	G	a2	G	G	G	M	G
a3	G	G	G	M	VG	a3	G	G	VG	G	VG	a3	G	G	G	MG	VG
a4	VG	MG	G	G	G	a4	VG	M	VG	MG	VG	a4	VG	M	G	MG	G
a5	G	VG	MG	VG	M	a5	G	G	M	G	M	a5	MG	VG	M	G	ML
a6	MG	VG	VG	VG	M	a6	ML	VG	VG	VG	ML	a6	M	VG	VG	VG	ML

Method 1:

Step 1, 2, 3 and 4:

$$\tilde{\tilde{\lambda}}_1 = [(0,700;0,800;0,800;0,875;0,80);(0,600;0,800;0,800;0,950;1,00)]$$

$$\tilde{\tilde{w}}_1 = [(0,639;0,778;0,778;0,862;0,80);(0,513;0,778;0,778;0,950;1,00)]$$

$$\tilde{\tilde{a}}_{11} = [(0,776;0,900;0,900;0,942;0,80);(0,660;0,900;0,900;0,983;1,00)]$$

$$\tilde{\tilde{a}}_1 = [(0,335;0,512;0,512;0,642;0,80);(0,208;0,512;0,512;0,795;1,00)]$$

Step 5,6 and 7:

Table 5: Evaluation result of Method 1

	COG	Distance	Rank
a1	[(0,335;0,512;0,512;0,642;0,800);(0,208;0,512;0,512;0,795;1,000)]	0,584	6
a2	[(0,367;0,560;0,560;0,655;0,800);(0,228;0,560;0,560;0,863;1,000)]	0,617	3
a3	[(0,421;0,625;0,625;0,754;0,800);(0,272;0,625;0,625;0,902;1,000)]	0,671	1
a4	[(0,394;0,587;0,587;0,718;0,800);(0,254;0,587;0,587;0,869;1,000)]	0,643	2
a5	[(0,353;0,536;0,536;0,665;0,800);(0,222;0,536;0,536;0,816;1,000)]	0,601	5
a6	[(0,384;0,559;0,559;0,670;0,800);(0,253;0,559;0,559;0,798;1,000)]	0,615	4

Rank of alternatives is $a_3 > a_4 > a_2 > a_6 > a_5 > a_1$. Beylikdüzü ($a_3$) is an ideal candidate to open new medical center.

Method 2:

Step 1, Step 2, and Step 3 is same as in Method 1

Step 4: Calculate the weighted value of alternative.

$$\tilde{\tilde{a}}_{11} = [(0,496;0,700;0,700;0,812;0,80);(0,339;0,700;0,700;0,934;1,00)]$$

Step 5,6,7,8 and 9:

Table 6: Evaluation result of Method 2

	c1	c2	c3	c4	c5	a+	a-	C	Rank
a1	7,593	7,196	5,166	7,196	5,555	3,113	2,654	0,460	6
a2	6,837	6,954	7,444	6,057	6,680	2,252	3,288	0,593	3
a3	7,203	7,480	7,469	6,696	7,011	1,363	3,860	0,739	1
a4	7,593	6,057	7,469	6,907	6,797	2,125	3,704	0,635	2
a5	6,940	7,753	5,955	7,607	5,110	2,703	2,785	0,507	5
a6	5,599	7,903	7,730	7,903	4,867	2,928	3,660	0,556	4
+ ideal	7,593	7,903	7,730	7,903	7,011				
- ideal	5,599	6,057	5,166	6,057	4,867				

Rank of alternatives is $a_3 > a_4 > a_2 > a_6 > a_5 > a_1$. Beylikdüzü ($a_3$) is an ideal candidate to open new medical center.

5. Conclusion

In this paper, a new method based on interval type-2 fuzzy trapezoidal numbers has been presented to find the weights of attributes and group decision matrix in group decision making problems. In the proposed method, each decision maker evaluates the others to attain a weight vector of decision makers. The weights of attribute and decision matrix which are given by a decision maker is affected from the weight of the decision maker. The proposed method provides more reliable results and a useful way to deal with MADM problems.

References

1. S. M. Chen and L. W. Lee, *Expert Systems with Application*, **Vol: 37**, 2790–2798 (2010).
2. P. Liu, *Expert Systems with Application*, **Vol: 38**, 1053–1060 (2011).
3. R. K. Singh, L. Benyoucef, *Engineering Applications of Artificial Intelligence*, **Vol: 24**, 437–448 (2011).
4. B. Ashtiani, F. Haghighirad, A. Makui, G. A. Montazer, *Applied Soft Computing* , **Vol: 9**, 457–461 (2009).
5. S. J. Chen and S. M. Chen (2003), *Cybernetics and Systems*, **Vol: 34-2**, 109–137 (2003).
6. S. H. Wei and S. M. Chen (2009), *Expert Systems with Application,* **Vol: 36**, 2285–2299 (2009).

GROUP MCDM METHOD WITH ALTERNATIVE VOTING SYSTEM BASED ON MULTI-INFORMATION ENVIRONMENT*

CHEN-TUNG CHEN†, CHUN-CHE HUANG, PING-FENG PAI
Department of Information Management, National United University, MiaoLi, Taiwan,

WEI-ZHAN HUNG

Department of International Business Studies, National ChiNan University, NanTou, Taiwan

The goal of this research is to present a novel group multi-criteria decision-making (GMCDM) method with alternative vote system (AVS). In the process of solving a GMCDM problem, it is one of most important issue to aggregate the ranking order of alternatives by each decision maker. The best alternative can be selected by minimizing the total unsatisfied points of members in decision group. By considering the importance of each criterion in GMCDM problem, the AVS is applied to minimize the total unsatisfied point of members in decision group and to determine the final ranking order of alternatives. A numerical example is implemented to illustrate the advantage of the proposed method. Finally, conclusions are discussed at the end of paper.

1. Introduction

In group decision making situation, voting is a usually rational and effective method for dealing with conflict problem although it neglects a small number of opponents' opinions. In reality, a common consensus of group is usually a compromise alternative because all of member usually choose the alternative with some unsatisfied points for themselves. The best alternative can be usually selected by minimizing the total unsatisfied point for the group. From this point of view, a few of multi-criteria decision-making methods used this concept to select suitable alternative. TOPSIS is choosing the best alternative according to the relative position, which is the shortest distance from the positive ideal solution (PIS) and the farthest from the negative ideal solution (NIS), in all of the alternatives [1]. ELECTRE is based on the axiom of partial comparability which is suitable for alternative selection. PROMETHEE can cope with problems where a finite number of alternative actions are to be ranked by

* This work was supported by National Science Council, Taiwan. The project numbers are "99-2410-11-239-009-MY2" and "99-2221-E-260-006".

† E-mail: ctchen@nuu.edu.tw

considering several, sometimes conflicting, criteria [2, 3]. VIKOR determines a compromise solution that provides the maximum group utility for the majority and a minimum of individual regret for the opponent [4]. In past literature, only VIKOR is considered the unsatisfied point of member in the decision group. However, VIKOR is just considered one most unsatisfied criterion for influencing alternative choice.

For minimizing total unsatisfied point for decision group, this research presents a novel GMCDM method with alternative voting system (AVS). AVS is already used in election process such as Australia election [5]. The concept of AVS is choosing the absolute better candidate based on voters. It is that the selected candidate must get the support votes more than 50% of all voters. In this paper, the concept of AVS is applied to develop a group MCDM method.

2. Evaluation Information

There are many types of information will be considered in dealing with MCDM problem. This information can be divided into crisp values, interval values and linguistic variables. The interval is a bounded subset of real numbers as $IV=[a,b]$, where a is lower bound and b is upper bound of interval value [6].

Definition 2.1. Supposed that $A=[a_{low},a_{upp}]$ and $B=[b_{low},b_{upp}]$ are two interval values. The minimum function of interval A and B is

$$\min(A,B)=[\min(a_{low},b_{low}),\min(a_{upp},b_{upp})] \tag{1}$$

The maximum function of interval A and B is

$$\max(A,B)=[\max(a_{low},b_{low}),\max(a_{upp},b_{upp})] \tag{2}$$

If $\max(A,B)=A$, then $A>B$. If $\min(A,B)=A$, then $A<B$.

Definition 2.2. A positive triangular fuzzy number (PTFN) $\tilde{T}$ can be defined as $\tilde{T}=(l,m,u)$, where $l\le m\le u$ and $l>0$. The membership function $\mu_{\tilde{T}}(x)$ is defined as [7]

$$\mu_{\tilde{T}}(x)=\begin{cases}\dfrac{x-l}{m-l}, & l<x<m\\ \dfrac{u-x}{u-m}, & m<x<u\\ 0 & , otherwise\end{cases} \tag{3}$$

Let $\tilde{P}_i=(p_{il},p_{im},p_{iu}),\ i=1,2,...,n$ be n positive triangular fuzzy numbers. The average of them can be calculated as [8]

$$\frac{1}{n}(\tilde{P}_1+\tilde{P}_2+...+\tilde{P}_n)=(\frac{1}{n}\sum_{i=1}^{n}p_{il},\frac{1}{n}\sum_{i=1}^{n}p_{im},\frac{1}{n}\sum_{i=1}^{n}p_{iu}) \tag{4}$$

Let $\tilde{P}=(p_l,p_m,p_u)$ be a positive triangular fuzzy number. Then transfer $\tilde{P}$ into crisp value as $\Delta(\tilde{P})=(p_l+p_m+p_u)/3$, where $\Delta()$ is the transfer function [9]. If $\Delta(\tilde{P}_1)>\Delta(\tilde{P}_2)$, then $\tilde{p}_1>\tilde{p}_2$.

3. The Proposed Method

In fact, a group MCDM problem can be described by the following sets.

(i) A set of experts (decision-makers) is called $\mathrm{E}=\{\mathrm{E}_1,\mathrm{E}_2,\cdots,\mathrm{E}_k\}$;

(ii) A set of alternatives is called $\mathrm{A}=\{\mathrm{A}_1,\mathrm{A}_2,\cdots,\mathrm{A}_m\}$;

(iii) A set of criteria $\mathrm{C}=\{\mathrm{C}_1,\mathrm{C}_2,\cdots,\mathrm{C}_n\}$ where the crisp values are expressed in criterion $C_j,\ j=1,2,...,p$, the interval values are expressed in criterion $C_j,\ j=p+1,p+2,...,q$, the fuzzy numbers are expressed in criterion $C_j,\ j=q+1,q+2,...,n$.

(iv) A decision matrix can be expressed as $D=[x_{ij}]_{m\times n}$, where x_{ij} are crisp values, $\bar{x}_{ij}$ are interval values, $\tilde{x}_{ij}$ are fuzzy numbers.

(v) A set of criterion weight is called $\mathrm{W}=\{\mathrm{W}_1,\mathrm{W}_2,\cdots,\mathrm{W}_n\}$.

If the performance of i-th candidate with respect to the j-th qualitative criterion decided by the k-th expert ($\tilde{x}_{ij}^k$) can be represented as $(x^k_{lij},x^k_{mij},x^k_{uij})$. The performance of i-th candidate with respect to the j-th criterion can be computed as $\tilde{x}_{ij}=\sum_{y=1}^{k}\left(\tilde{x}_{ij}^y\right)/k$. The importance of the j-th criterion can be computed as $\tilde{w}_j=\sum_{y=1}^{k}\left(\tilde{w}_j^y\right)/k$, where $\tilde{w}_j^y$ is linguistic weight of j-th criterion by y-th expert. $\tilde{w}_j$ is the linguistic weight of j-th criterion. The normalized crisp weight of j-th criterion (w_j) can be calculated as

$$w_j=\Delta(\tilde{w}_j)/\sum_{j=1}^{n}(\Delta(\tilde{w}_j)) \tag{5}$$

If we consider alternatives as candidates and criteria as voters, then AVS can be applied to select the best alternative. The process of group MCDM method with AVS is illustrated as follows (Refer to Figure 1).

Step1. Collect the multiple information of each alternative.

Step2. Experts use linguistic variable to express their opinions about the important of each criterion and the performance of each alternative with respect to qualitative criterion.

Step3. Aggregate the opinions of experts of each alternative with respect to qualitative criterion.

Step4. Construct the decision matrix.

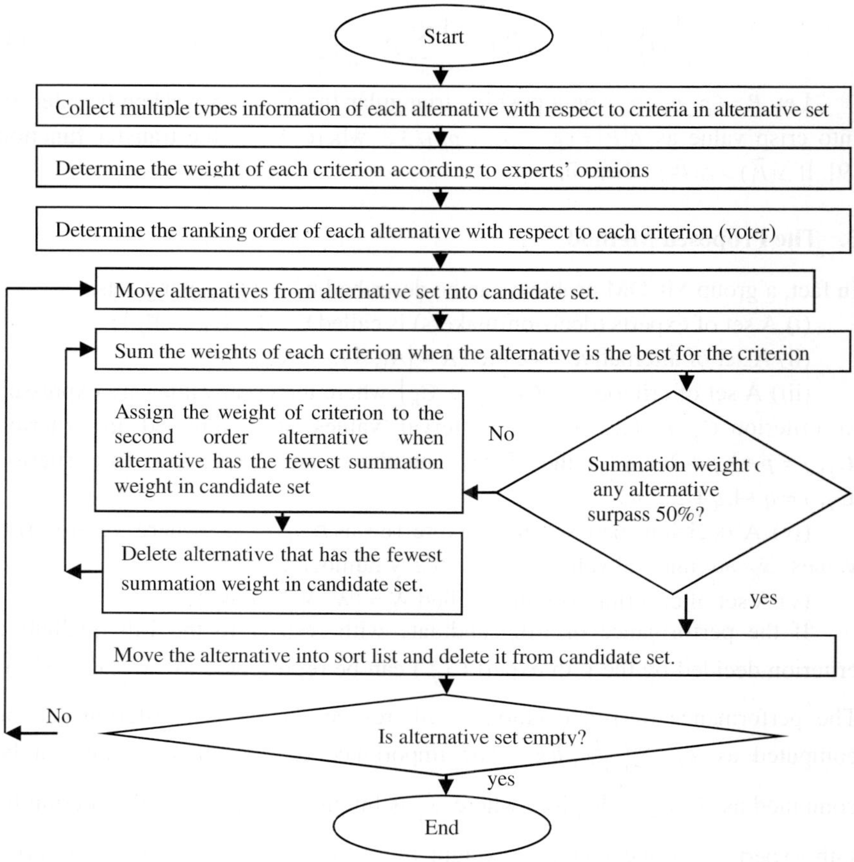

Figure 1. The conceptual process of the proposed method.

Step5. Aggregate experts' opinions about the important of each criterion which is expressed as fuzzy number and compute the weight of each criterion.

Step6. Compare the multiple information of each alternative with respect to each criterion. And then, the criterion preference matrix (CP) can be determined as $CP = \lfloor p_{ij} \rfloor_{m \times n}$, where p_{ij} means the order of A_i with respect to C_j.

Step7. Select the best alternative according to the process of AVS.

Step8. Repeat the process of AVS until the alternative set is empty.

4. Numerical Example

Suppose that a mobile phone manufacturer wants to develop a new product. Factory director assigns three experts to choose the best product plan from four product plans according to seven criteria. The crisp value criterion is unit product profit. The interval value criteria are product developed cost, market volume and product life time. The qualitative criteria are product appearance, product function and product quality. According to the proposed method, the computational procedures of the problem are summarized as follows.

Step1. Collect crisp and interval information of each alternative as Table 1.

Table 1. Crisp value and interval value information of each alternative

	C_1	C_2	C_3	C_4
A_1	50	[70,100]	[100,200]	[1.5,3]
A_2	60	[80,120]	[120,300]	[2,4.5]
A_3	55	[100,180]	[140,180]	[1,2]
A_4	40	[90,150]	[220,350]	[1,4]

Step2. Aggregate the linguistic performances as Table 2.

Table 2. The aggregated performance of each alternative with respect to qualitative criteria

	A_1	A_2	A_3	A_4
C_5	(0.5,0.625 , 0.75)	(0.208,0.333 ,0.458)	(0.333,0.458, 0.583)	(0.458, 0.583, 0.708)
C_6	(0.417,0.542, 0.667)	(0.5,0.625 ,0.75)	(0.575, 0.708, 0.792)	(0.583, 0.708,0.833)
C_7	(0.25, 0.375,0.5)	(0.542,0.667,0.792)	(0.375, 0.5, 0.625)	(0.417,0.542, 0.667)

Step3. According to the importance evaluations of experts, the weight of each criterion is computed as W_1 =0.1557, W_2 =0.0969, W_3 =0.1522, W_4 =0.1142, W_5 =0.2007, W_6 =0.1349, W_7 =0.1453.

Step4. Construct the criterion preference matrix (CP) as Table 3.

Table 3. Criterion preference matrix

	C_1	C_2	C_3	C_4	C_5	C_6	C_7
A_1	3	1	3	3	1	4	4
A_2	1	2	2	1	4	3	2
A_3	2	4	4	4	3	2	3
A_4	4	3	1	2	2	1	1

Step5. Execute AVS step by step to select the best alternative as Table 4.

Table 4. The steps of alternative voting system

		A_1	A_2	A_3	A_4
Step 1	Round 1	C_2 C_5 0.2976	C_1 C_4 0.2699	0	C_3 C_6 C_7 0.4324
	Round 2			Delete A_3	
	Round 3	C_1 C_2 C_5 0.4634	Delete A_2		C_3 C_4 C_6 C_7 0.5466
Step 2		A_1	A_2	A_3	
	Round 1	C_2 C_5 0.2976	C_1 C_3 C_4 C_7 0.5674	C_6 0.1349	
Step 3		A_1		A_3	
	Round 1	C_2 C_3 C_4 C_5 0.5641		C_1 C_6 C_7 0.4359	

Step6. According to the Table 5, the final ranking order is $A_4 > A_2 > A_1 > A_3$.

5. Conclusions

In this paper, a group MCDM method with alternative voting system is developed to deal with the MCDM problem. The strong point of proposed method is that it can help decision maker copy with decision making problem based on multiple types information environment effectively such as crisp value, interval value, fuzzy number, type 2 fuzzy numbers, linguistic variable etc. In the proposed method, it doesn't need to transfer the different kinds of information into the same type of information for aggregating the data of each alternative with respect to criteria. In additional, the proposed method provides an objective and systematic way to determine the ranking order of all alternatives based on the steps of AVS. In the future, proposed method can be compared with same traditional aggregation method such as Borda method and Copeland method by computer simulation.

References

1. C. T. Chen, and W. Z. Hung, A new decision-making method for stock portfolio selection based on computing with linguistic assessment, *J. Appl. Math. Decis. Sci.*, ***Vol. 2009***, Article ID 897024, 20 pages (2009).
2. M. Goumas, and V. Lygerou, An extension of the PROMETHEE method for decision making in fuzzy environment: ranking of alternative energy exploitation projects, *Eur J of Oper Res*, ***123***, 606-613 (2000).
3. J. P. Brans, and Ph. Vincke, A preference ranking organization method (the PROMETHEE method for MCDM), *Manage Sci*, ***31***, 647-656 (1985).
4. C. T. Chen, P. F. Pai, and W. Z. Hung, Applying linguistic VIKOR and knowledge map in personnel selection, *Asia Pacific Management Review*, ***16***(4), 15-26(2011).
5. G. A. Almond, G. Bingham Powell and Jr. eds, Comparative Politics Today: A World View, 7th Edition, (New York: Harper Collins, 2003).
6. V. Behnam, H. K. J. Amir, R. Vahid and Z. Mostafa, Extension of the ELECTRE method for decision-making problems with interval weights and data, *Int J Adv Manuf Tech*, ***50***, 793–800 (2010).
7. C. T. Chen, Extensions of the TOPSIS for group decision making under fuzzy environment, *Fuzzy Set Syst*, ***114***, 1-9 (2000).
8. A. Kaufmann, and M. M. Gupta, Introduction to fuzzy arithmetic: theory and applications, International Thomson Computer Press (1991).
9. E. S. Lee and R. J. Li, Comparison of fuzzy numbers based on the probability measure of fuzzy events, *Comput Math Appl*, ***15***, 887–896 (1988).

A RANKING METHODOLOGY USING HİERARCHICAL GREY RELATIONAL ANALYSIS: THE CASE OF TURKISH BANKING SYSTEM *

MEHMET EMİN BAYSAL†, BÜŞRA AKSU

Department of Industrial Engineering, Selcuk University
Selçuklu, Konya, 42250, Turkey

Financial indicators which have been monitoring and announcing to public periodically are indispensable tools particularly in the banking sector. In this study, it has been chosen 22 different Banks in Turkey. The 39 performance related criteria for those banks were evaluated by an integrated AHP approach and their weights were computed. Then, the grey relational degrees of 22 banks were computed by Grey Relational Analysis Method. On the other hand banks were ranked by using Financial Ratio Analysis. The obtained results from both methods were presented comparatively.

1. Introduction

Since the Turkey has a rapid growth rate recently, financial sector which has a vital role in an economy is very dynamic in nature. When it is considered the last decade of the banking system in Turkey, it has been witnessed lots of reforms and new practices along with the whole development. Even, Turkey has been affected by the economic crisis or bankruptcy particularly in banking there has been a good progress related to stabilization in governance. In these circumstances the bank performance evaluation in the banking system can be done more precisely and this topic is very popular nowadays.

This study was organized as follows: after a short introduction, in the second section it was described the problem as the performance evaluation problem. It explains and defines the set of decision making units and the criteria. Section 3 gives the Hierarchy Grey Relational Analysis (HGRA) used in the study. Section 4, emphasizes the implementation and results. The last section reveals the conclusion and the further research.

---

* This work is supported by the Selcuk University Scientific Research Coordination Centre (BAP).
† mebaysal@selcuk.edu.tr.

2. Problem Statement

In this study, activities of 22 banks which operated in Turkish banking sector, were considered. Activities were performed with respect to evaluation of 39 criteria. The determination of these 39 criteria is based on criteria which were used by Ho [1]. These 39 criteria were used as sub-criteria of the 5 main criteria. It was shown the names of the banks and their representation in the Table 1.

Table 1. The names of the banks under evaluations

Representation	Banks	Representation	Banks
B1	Akbank	B12	Turkish Ekonomy Bank
B2	Alternatifbank	B13	Turkish Garanti Bank
B3	Anadolubank	B14	Turkish Halk Bank
B4	Citibank	B15	Turkish Is Bank
B5	Denizbank	B16	Turkish Vakif Bank
B6	Finansbank	B17	Yapikredi Bank
B7	Fortis Bank	B18	Ziraat Bank
B8	Hsbc Bank	B19	Albaraka Turk Participation Bank
B9	Ing Bank	B20	Asya Participation Bank
B10	Sekerbank	B21	Kuveyt Turk Participation Bank
B11	Tekstil Bank	B22	Turkiye Finans Participation Bank

The data for evaluating the banks were provided from official website of Banking Regulation and Supervision Agency (BRSA), and the solo data reports from September 2010 were used. The main criteria used in the study are: safety, profitability, liquidity, efficiency and growth. Main and Sub-criteria were presented in the Table 2.

3. A Ranking Methodology by Using HGRA

Grey system theory has been set up by Deng (1982). Grey system is defined as a system having fractional information. Its nature has uncertainty so it is used to solve the problems consisted of discrete data and fractional information. Grey system theory's distinctive feature is that it can handle smaller data easily and can be achieved in good results. It carries out this by putting the data in its regular place with proper treatment [2,3,5,6].

Table 2. Main and sub criteria (ratios)

Code	Main Criteria	Code	Sub-Criteria
C1	Safety	S1	Deposits/Stockholders' Equity
		S2	Stockholders' Equity /Total Assets
		S3	Liabilities/Stockholders' Equity
		S4	Stockholders' Equity/Investment and Loans
		S5	Total Assets/ Stockholders' Equity
		S6	Loans and Lease Financing/ Stockholders' Equity
		S7	Stockholders' Equity/Deposits
		S8	Loans/Stockholders' Equity
		S9	Reserve For Credit Losses To Loan/Loans and Lease Financing
C2	Profitability	P1	Operatin Income/Operating Revenues
		P2	Earning Before Taxes/ Operating Revenues
		P3	Earning Before Taxes/Average Stockhoders' Equity
		P4	Earning Before Taxes/Average Total Assets
		P5	Operating Profit/Total Assets
		P6	Operating İncome/Average Stockhoders' Equity
		P7	Net Earning/Total Assets
		P8	Net Earning/Average Stockhoders' Equity
C3	Liquidity	C1	(Cash+Deposits İn Other Banks+Government Bonds)/Deposits
		C2	First and Second Reserve/Deposits
		C3	Cash+Deposits İn Other Banks /Deposits
		C4	Current Reserve/Deposits
		C5	(Cash+T-Bills)/Total Assets
		C6	Loans/Total Assets
C4	Efficiency	E1	Operating Revenues/Fixed Assests
		E2	Operating Revenues/Total Assets
		E3	Noninterest İnterest Expenses/Operating Revenues
		E4	Noninterest İnterest Expenses/ Total Assets
		E5	Reserve For Credit Losses To Loan/Loan and Lease Financing
		E6	Reserve For Credit Losses To Loan / Total Assets
		E7	(Deposits+İnterest Expenses)/Deposits
		E8	Interest Income on Loans/Avarega Total Loans
		E9	Operating Expenses/Operating Revenues
		E10	Deposits and Borrowed Fund/ Operating Revenues
C5	Growth	G1	(Operating Revenues of Current and Last Period)/Operating Revenues of Last Period
		G2	(Loans and Lease Financing of Current and Period)/ Loans and Lease Financing of Last Period
		G3	(Total Assets of Current Period-Total Asstes of Last Period)/Total Assets of Last Period
		G4	Operating Revenues of Current Period/Operating Revenues of Last Period
		G5	Total Assets of Current Period/Total Assets of Last Period
		G6	(Deposits of Current Period-Deposits of Last Period)/Deposits of Last Period

Modification was done by using GRA as follows, [4,7]:

1) Let the reference sequence be

$$x_0 = (x_0(1), x_0(2), \ldots , x_0(n)) \tag{1}$$

2) Represented the m sequences to be compared as

$$x_i = (x_i(1), x_i(2), \ldots , x_i(n)), \quad i = 1, 2, \ldots , m \tag{2}$$

3) Normalize the sequences to guarantee that all of them are in the same order, and the normalized sequences can be denoted as

$$x_i^* = (x_i^*(1), x_i^*(2), \dots, x_i^*(n)) \quad i = 1, 2, \dots, m \tag{3}$$

4) The grey relational coefficient between the compared sequence, x_i, and the reference sequence, x_0, for the jth factor (j=1, 2, . . . , n), was defined as:

$$\xi_i(j) = \frac{\min_i(\min_j |x_0(j) - x_i^*(j)|) + \rho \max_i(\max_j |x_0(j) - x_i^*(j)|)}{|x_0(j) - x_i^*(j)| + \rho \max_i(\max_j |x_0(j) - x_i^*(j)|} \tag{4}$$

where $x_i^*(j)$ is the grey relational coefficient between the ith index of the jth bank to be evaluated and the ith element of the reference (or optimal) sequence; where $\xi_i(j) \in [0, 1]$, $x_i^*(j)$ is the value of factor j of grid i, ρ is the distinguishing coefficient, $\rho \in [0, 1]$, and typically $\rho = 0{,}5$.

5) The aggregated evaluation model could thus be written as follows:

$$S_i = \sum_{j=1}^{n} w_j x \xi_i(j) \tag{5}$$

where S_i is the integrated evaluation value of grid i, w_j is the weight for factor *j* of grid *i*, and $S_i \in [0,1]$.

4. Implementation and Results

In this study, it has been chosen 22 different Banks in Turkey. The 39 performance related criteria for those banks were evaluated by an integrated AHP approach and their weights were computed. Then, the grey relational degrees (GRD) of 22 banks were computed by Grey Relational Analysis Method. On the other hand banks were ranked by using Financial Ratio Analysis (FRA). The obtained results from both methods were presented comparatively in Table 3.

Table 3. Summary results for the evaluation by two methods

	BANKS	GRD.	FRA	Ratings		BANKS	GRD	FRA	Ratings
1	B18	0.689	B18	75.10	12	B3	0.444	B5	55.53
2	B14	0.621	B19	69.01	13	B5	0.440	B21	54.67
3	B17	0.598	B14	65.69	14	B20	0.427	B9	53.60
4	B4	0.561	B2	65.18	15	B19	0.416	B20	52.82
5	B6	0.535	B6	60.91	16	B12	0.378	B15	51.91
6	B10	0.535	B12	59.59	17	B16	0.373	B13	51.46
7	B21	0.502	B4	59.32	18	B2	0.369	B17	49.48
8	B13	0.462	B22	59.30	19	B9	0.354	B1	48.67
9	A1	0.453	B16	59.30	20	B11	0.347	B7	42.35
10	B8	0.452	B10	56.72	21	B7	0.346	B8	41.85
11	B15	0.446	B3	55.76	22	B22	0.321	B11	39.47

5. Conclusion

Based on the ranking by the HGRA, it was shown that the Ziraat Bank is the first with 0.689 grd; Halkbank is the second with 0.621grd; and the Yapı Kredi Bank is the third with 0.598 grd. The last one is Türkiye Finans Participation Bank with 0.321 grd. which is a profit sharing bank. There are two public banks in the first two places with respect to evaluation by HGRA whereas the third one is a private bank. So it can be inferred that public banking is better than private banking in operations. On the other hand, based on the evaluation by FRA Halk Bank is the third, Albaraka Turk Participation Bank which is a profit sharing bank is in second place instead. This deviation can be seen as a remarkable point when it is focused on the classification of private and public banking. Because it should be mentioned also to the different operations between profit sharing banks and private banks. Besides these differences, there is a consistency between two evaluations. Furthermore, based on the different selection of the criteria it can be obtained different results. The important thing is the selection of the criteria. It may be classified the domains such as public, private and profit sharing banks in the future study.

References

1. C. T. Ho, "Measuring bank operations performance: an approach based on grey relation analysis", *Journal of the Operational Research Society,* **57**, 337-349 (2006).
2. 2A. Sarucan, M. E. Baysal, C. Kahraman and O. Engin, "A Hierarchy grey relational analysis for selecting the renewable electricity generation

technologies", *World Congress on Engineering 2011,* London, U.K., 1149-1154 (2011).

3. 3G. Sasikumar, V. Selladurai and A. Sivasangari, "Development of a supplier selection system by combining analytic hierarchy process and grey relational analysis: An automobile industry case study," *The Icfai University Journal of Supply Chain Management*, **5**, 82-94, (2008).
4. 4F. Yang, G. M. Zeng, C. Du, L. Tang, J. Zhou and Z. Li, "Integrated geographic information systems–based suitability evaluation of urban land expansion: A Combination of analytic hierarchy process and grey relational analysis," *Environmental Engineering Science*, **26**, 1025-1032, (2009).
5. J. L. Deng, Essential Topics on Grey System: Theory and Applications, Wuhan: Huazhong University of Science and Technology Press, (1987).
6. P. Li, T. C. Tan and J. Y. Lee, "Grey relational analysis of amine inhibition of mild steel corrosion in acids", *Corrosion*, **53**, 186-194, (1997).
7. 7S. Cheng and R. Wang, "Analyzing hazard potential of typhoon damage by applying grey analytic hierarchy process," *Natural Hazards,* **33**, 77–103, (2004).

AN ARTIFICIAL BEE COLONY ALGORITHM FOR SOLVING NURSE SCHEDULING PROBLEMS*

AHMET SARUCAN†

Department of Industrial Engineering, Selcuk University, Konya, Turkey

KADIR BÜYÜKÖZKAN

Department of Industrial Engineering, Karadeniz Technical University, Trabzon, Turkey

The nurse scheduling problems (NSP) deal with assignment of the nurse to the specific shifts in specific days is taking into account both the hospital's need and nurse preference. In the hospitals, nurses are serving under difficult circumstances. To improve the quality of service, work schedules should be prepared in a fair manner and has to provide the desired conditions. In this study, it is aimed at developing a model which can be used in real applications and can be rapidly reached to the solution for the nurse scheduling problem. For this purpose, the formulated problem was solved by using artificial bee colony (ABC) algorithm, which is one of the new meta-heuristic methods. It has been coded a program with a user interface using the Visual Studio programming language for running the model. The designed interface is able to take into account the different level of the demand and the number of staff. The model was tested on a real problem and found better solution than the current situation.

1. Introduction

Companies aim to plan their personnel in the best way that they need to sustain the activities within the organization using the Personnel Scheduling techniques. One of the most important personnel scheduling problem encountered in hospitals is the Nurse Scheduling Problem (NSP).

There are different classifications for NSP and different solution approaches have proposed in the literature [1-3]. Some of these approaches are meta-heuristics such as Genetic Algorithm [4, 5], Tabu Search [6], Neighborhood Search [7, 8], Simulated Annealing [9], Ant Colony [10] and Evolutionary [11] algorithms.

NSPs strive to meet multiple constraints and objective simultaneously. In the hospital that was conducted the study, the schedule is created by the chief nurse.

* This work is supported by the Selcuk University Scientific Research Coordination Centre (BAP).

† sarucan@selcuk.edu.tr

So it was not easy to create a high quality solution manually. In this study, it was developed a program with the ABC algorithm for the creation of a higher quality schedule. After having solution with the proposed algorithm it was shown that the study is the first study in order to show the efficiency of ABC algorithm to the nurse scheduling problem. The paper organized as follows: after a short introductory part, ABC was explained. The third chapter devoted to implementation in a hospital. The fourth chapter is the last chapter with a conclusion.

2. Artificial Bee Colony Algorithm

ABC algorithm is a novice heuristic approach developed recently. As well as other colony algorithms, the basic idea of ABC is mimicking the species collective action in the nature. ABC algorithm is inspired by the behavior of the bees' food searches. It works by following steps to ensure the optimum solution [12];

1. Create an initial population of virtual bees,
2. Encode the optimization function into virtual food,
3. Define the criterion for communicating food location with others,
4. March all the virtual bees randomly to new positions for virtual food searching, find food and communicate with neighboring bees by virtual waggle dance,
5. Evaluate the encoded intensity /locations of bees,
6. Decode the results to obtain the solution to the problem.

The applications of ABC algorithm in a wide range of fields applied to obtain optimum solution for many problems have increased rapidly in recent years [12-14]. In this study it was investigated a proposed ABC algorithm to solving nurse scheduling problems.

3. An Application of ABC Algorithm to NSP

The NSP considered in this study is a problem of assigning work to nurses meeting the demand in a certain period. The data required for the application has provided from the Medical Faculty Hospital of Black Sea Technical University, Department of Nephrology and Endocrinology.

In the existed system it is chief nurse's responsibility to determine which nurse will work on what days and shift. Chief Nurse creates this schedule manually and then print schedule and transmits to the service. The constraints discussed in this study were determined by taking into consideration literature and existing applications.

Weekly demand is constant and was given in table 1 for a month planning horizon. The number of nurses is 13 and working shifts are given as 8-16, 16-24 and 24-8 respectively. The post-natal nurses shouldn't work at night and weekend shifts. The nurse-8 was on annual leave between 8th and 16th days of the schedule.

Table 1. Weekly nurse demand

	Monday	Tuesday	Wednesday	Thursday	Friday	Saturday	Sunday
8-16	4	4	4	4	4	2	2
16-24	3	3	3	3	3	2	2
24-8	2	2	2	2	2	2	2

The hard and soft constraints used in the study were presented in the table 2. The steps of ABC algorithm for NSP are as follows:

Step 1. Create a random weekly chart for the initial solution.

Step 2. Find the penalty values for each schedule.

Step 3. Sort the schedules according to penalty value and select the schedule to the number of scout bees,

Step 4. Create followers for scout bees with neighborhood search algorithm,

Step 5. Find the penalty values for each followers,

Step 6. Choose new scout bees according to the penalty values,

Step 7. Repeat 4, 5 and 6th steps until stopping criterion is reached,

Step 8. Stop and show results.

3.1. *Developing a Program for Solving NSP*

To solve the NSP it was coded a program that has a user interface by using the Visual Studio C# programming language. It is shown the main interface of the program in the figure 1. The data for the schedule and the existed schedule between 3 and 31 October was secured from hospital.

The number of bees, scout bees, follower bees and iterations were determined 250, 15, 15, and 1000 respectively. The schedule has been created by running the program in order to four weeks and then imported to Ms-Excel file.

Table 2. Comparison of the results.

Const. No	Hard Constraints	ABC - Week 1	2	3	4	Existing -Week 1	2	3	4
1	Each nurse have to work minimum 40 hours in a week	0	0	0	0	0	0	0	0
2	Maximum consecutive work is 16 hours.	0	0	0	0	0	0	0	0
3	At least once a week consecutive 48 hours rest is required,	0	0	0	0	0	1	0	2
4	Head nurse must work the day shift on weekdays only,	0	0	0	0	0	0	0	0
5	Consecutive two night shift is not allowed,	0	0	0	0	0	1	0	0
6	After 8 hour working at least 16 hour rest is required,	0	0	0	0	0	0	0	0
7	After 16 hour working at least 24 hour rest is required,	0	0	0	0	0	3	0	0
8	Nurse's permission preference must be fulfillment.	0	0	0	0	0	0	0	0
9	A nurse may start only one shift per day.	0	0	0	0	0	0	0	0
10	it should be taken into account that nurse not wish to work on some days	0	0	0	0	0	0	0	0
	Soft Constraints								
11	For every nurse, number of night shift works should be equal	1	2	3	3	2	4	5	6
12	After 16 hour working at least 48 hour rest is better,	0	0	5	0	8	14	17	18
13	For every nurse, number of overtime works should be equal	0	0	0	0	0	2	0	0
14	Nurse shouldn't work both two days on weekend	1	2	2	3	2	2	1	2

3.2. *Outputs and Assessments*

The consistency of the resulting schedule obtained from the program and the actual schedule were comparatively shown on the Table 2. While the ABC algorithm has yielded all the hard constraints, in the existed schedule there has been some cases which do not satisfy the third, fifth and seventh hard constraints.

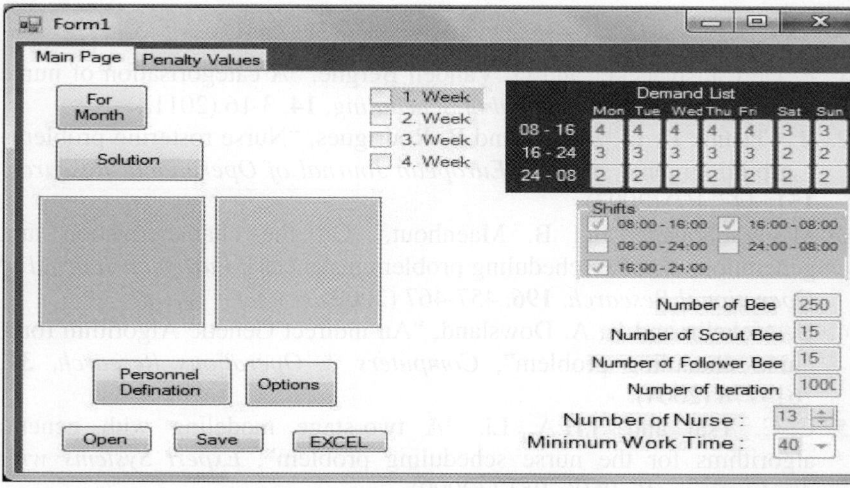

Figure 1. Main interface of the program

And also it was achieved better results with the program than existed schedule in terms of the soft constraints. For instance, the minimum night shift work (24:00-08:00) is one while the maximum night shift work is two based on the solution given by the program for the first week. It is resulted to the value of the first soft constraint is 1. Its value is 2 in the actual schedule.

4. Conclusion

In this study, the nurse scheduling problem is solved by using the ABC algorithm. The proposed solution method has given better values regarding to satisfactions of the hard and soft constrains. Overall performance of the ABC algorithm can easily picked up if it is reviewed the table 2. The ABC solution satisfies all the hard constraints including 3rd, 5th and 7th constraints while they were not satisfying in the actual schedule. The better results were obtained for the soft constraints as well with the high quality and fair schedule.

This study shows that the ABC algorithm can be applied successfully to the nurse scheduling problems. The conventional method and the ABC algorithm based method were compared in the study but the comparison could be done with other heuristic applications in the future. The quality of the solution by the ABC algorithm depends on the parameters and the coefficients of penalty values of both soft and hard constraints. By conducting comprehensive parameter optimization and sensitivity analysis under different demand and nurse constraints more accurate and faster solutions can be obtained and tested.

References

1. P. De Causmaecker and G. Vanden Berghe, "A categorisation of nurse rostering problems", *Journal of Scheduling*, **14**, 3-16 (2011).
2. B. Cheang, H. Li, A. Lim and B. Rodrigues, "Nurse rostering problems – a bibliographic survey", *European Journal of Operational Research,* **151**, 447-460 (2003).
3. M. Vanhoucke and B. Maenhout, "On the characterization and generation of nurse scheduling problem instances", *European Journal of Operational Research,* **196**, 457-467 (2009).
4. U. Aickelin and K. A. Dowsland, "An indirect Genetic Algorithm for a nurse-scheduling problem", *Computers & Operations Research,* **31**, 761–778 (2004).
5. C.-C. Tsai and S.H.A. Li, "A two-stage modeling with genetic algorithms for the nurse scheduling problem", *Expert Systems with Applications*, **36**, 9506–9512 (2009).
6. A. Oughalime, W.R. Ismail and L.C. Yeun, "A tabu search approach to the nurse scheduling problem", IEEE, (2008).
7. E.K. Burke, J. Li and R. Qu, "A hybrid model of integer programming and variable neighbourhood search for highly-constrained nurse rostering problems", *European Journal of Operational Research*, **203**, 484-493 (2010).
8. F. Bellanti, G. Carello, F. Della Croce and R. Tadei, "A greedy-based neighborhood search approach to a nurse rostering problem", *European Journal of Operational Research*, **153**, 28-40 (2004).
9. M. Cheng, H. I. Ozaku, N. Kuwahara, K. Kogure and J. Ota, "Simulated Annealing Algorithm for Daily Nursing Care Scheduling Problem", *IEEE Conference on Automation Science and Engineering*, 507-512 (2007).
10. W.J. Gutjahra and M.S. Raunerb, "An ACO algorithm for a dynamic regional nurse-scheduling problem in Austria", *Computers & Operations Research*, **34**, 642-666 (2007).
11. B. Maenhout and M. Vanhoucke, "An evolutionary approach for the nurse rerostering problem", *Computers & Operations Research*, **38**, 1400–1411 (2011).
12. X-S. Yang, "Engineering optimizations via nature-inspired virtual bee algorithms", *IWINAC 2005,* 317–323 (2005).
13. B. Suri, and Snehlata, "Review of artificial bee colony algorithm to software testing", *International Journal of Research and Reviews in Computer Science,* **2**, 706-711 (2011).
14. P. Tapkan, L. Özbakır and A. Baykasoğlu, "Bees algorithm and generalized assignment problem: comparison of different neighborhood structures", *Journal of Industrial Engineering*, **21**, 2-13 (2008).

A GREEDY ALGORITHM FOR MULTIOBJECTIVE FUZZY FLOW-SHOP SCHEDULING PROBLEM*

ORHAN ENGIN†
Selcuk University Department of Industrial Engineering, Konya, TURKEY

M. KERIM YILMAZ
Kocaeli University Department of Industrial Engineering, Kocaeli, TURKEY

M. CABIR AKKOYUNLU, M. EMIN BAYSAL, AHMET SARUCAN
Selcuk University Department of Industrial Engineering, Konya, TURKEY

In this paper, a new greedy algorithm is proposed to solve the fuzzy multiobjective flow shop scheduling problem. We use the two approaches; the possibility measure and the area of intersection for multi objective fuzzy flow shop scheduling problem. The proposed new greedy algorithm is tested on the benchmark problems in the literature. The performance parameters of proposed greedy algorithm are determined by full factorial design of experiments (DOE). The performance of the proposed greedy algorithm is compared with the scatter search method.

1. Introduction

A flow shop problem consists in scheduling n jobs on m machines, each job consists in m operations and the operating sequences of the jobs are the same on every machine. Over the past six decades, extensive research has been done on the flow-shop problem. The flow-shop scheduling problems can be classified into the following three categories: deterministic flow-shop problem, stochastic flow-shop problem, and fuzzy flow-shop problem [1]. In the recent years, fuzzy flow shop scheduling problems have been done extensively in the literature. Ishibuchi et al. [2] formulate fuzzy flow-shop scheduling problems with fuzzy processing time. They give the processing time of each job at each machine a fuzzy number. They select the makespan, maximum delay and total delay as performance criteria. They suggest a multi-objective genetic algorithm for large

*This work is supported by Selcuk University Scientific Research Coordination Centre (BAP)
†orhanengin@yahoo.com

size fuzzy flow-shop scheduling problems. Kilic and Kahraman [3] deal with a flowshop scheduling problem with fuzzy processing times and flexible due dates. They proposed an ant colony optimization metaheuristic approach for solve that problem. Engin et al. [4] propose a scatter search method for multiobjective fuzzy permutation flow shop scheduling problem. The objectives the minimizing average tardiness and the number of tardy jobs are selected. Kahraman et al. [5] propose an artificial immune system algorithm for multiobjective fuzzy flow shop problems. The objectives are minimize the average tardiness and the number of tardy jobs.

Huang et al. [6] propose a modified genetic algorithm to search optimal solutions to the flow shop scheduling problems with fuzzy processing times and fuzzy due dates. Lai and Wu [7] propose a modified virus-evolutionary Genetic Algorithms to minimize the fuzzy makespan and total weighted fuzzy completion time using the possibility and necessity measures to propose the ranking concept among fuzzy numbers.

In this paper, a new greedy algorithm is developed to solve the fuzzy multiobjective flow shop scheduling problem. The effectiveness of the developed method is tested with the Kahraman et al. [5] artificial immune system algorithm. To the best of our knowledge, there is no greedy algorithm applied to multi objective fuzzy flow shop scheduling problem in the literature.

The remainder of the paper is organized as follows. The fuzzy flow shop scheduling is explained in Section 2. In section 3, the proposed greedy algorithm is devoted. The computational results are given in Section 4, and finally the conclusions are made in Section 5.

2. Fuzzy Flow shop Scheduling

In a static permutation flow shop, the processing time for each job and due dates are usually assumed to be known exactly, but in many real world applications, processing times and due dates vary dynamically due to human factors or operating faults [4,5]. In this study, fuzzy sets are used to model the uncertain processing times and due dates for flow shop scheduling problems. The notation of multiobjective fuzzy flow shop scheduling problem is given as follow; n number of jobs, $\tilde{t}_{ij}$ fuzzy processing times of job i on machine j, $\tilde{T}_j$ fuzzy tardiness of job j, m number of machines, $\tilde{d}_j$ fuzzy due date of job j, C_j fuzzy completion time of job j. Fuzzy processing times $\tilde{t}_{ij}$ are modeled by triangular membership functions represent by a triplet $(t_{ij}^1, t_{ij}^2, t_{ij}^3)$.The membership function of fuzzy processing time is shown in Figure 1. Also the due date $\tilde{d}_j$ of each job is modeled by a trapezoidal fuzzy set and represented by a doublet $(d_j^1,\ d_j^2)$. In this study the, the following two objectives are considered to minimize [4,5,8].

1. *The average tardiness C_{AT}:*

$$C_{AT} = \frac{1}{n}\sum_{j=1}^{n} T_j; \qquad T_j = \max\{0, \tilde{C}_j - \tilde{d}_j\} \qquad j = 1,.......,n \qquad (1)$$

2. *The number or tardy jobs C_{NT}:*

$$C_{NT} = \sum_{j=1}^{n} u_j \qquad u_j = 1 \quad if \quad T_j > 0, \quad otherwise \quad u_j = 0 \qquad (2)$$

3. Greedy Algorithms

The Greedy Algorithm (GA) is a well-known simple approach to solve the optimization problems. The GA works in phases. At each phase, a decision is made that appears the best at the moment, without regard for future consequences. It means that choosing a local optimum at each step, when the algorithm terminates, we hope that the local optimum is equal to the global optimum. Greedy Randomized Adaptive Search Procedure (GRASP) was first proposed by Binato et al. [9].

We proposed a greedy algorithm for solving the multi objective fuzzy flow shop scheduling problem with Cmax criteria. The steps of the GA approach are Start, Initialization, Subgroup size, Created sub populations, Rearranged population, Destruction, Construction, Local search, Elitist Strategy, Stopping criteria and End. GA is applied two phases: destruction and construction[10].

In the recent years, greedy algorithm has been used in the literature for scheduling problems. Some of these studies are given below. Ruiz and Stützle [11] propose an iterated greedy algorithm for the permutation flowshop scheduling problem. Also, Ruiz and Stützle [12] propose two new Iterated Greedy algorithms for a complex flowshop problem. Li and Zhao [13] propose a greedy algorithm solving flexible flow shop scheduling problem. Akgöbek et al. [14] propose a parallel greedy metaheuristic algorithm for solving open shop scheduling problem. Sheibani [15] propose a fuzzy greedy heuristic for solving permutation flow-shop scheduling problems. Kahraman et al. [10] propose a parallel greedy algorithm to solve Hybrid Flow Shop Scheduling with Multiprocessor Task problem. They applied the parallel greedy algorithm by two phases iteratively, called destruction and construction. Ribas et al. [16] propose an iterated greedy algorithm for solving the blocking flowshop scheduling problem for makespan minimization. Minella et al. [17] propose a Iterated Greedy technique for solving the multi-objective permutation flowshop problem.

4. Computational Results

The proposed Greedy Algorithm is compared with an Artificial Immune System (AIS) algorithm [5]. In this study, full factorial design of experiment has been used. The application involves five parameters with different possible values each. The best parameters sets for proposed GA are shown in Table 1. In this study we proposed two constructive heuristic methods. These are initially positions (IP) constructive methods and end positions (EP) constructive methods.

Table 1. The Parameters Proposed by GA for Multiobjective Fuzzy Flow Shop Scheduling Problem

Parameter	**Range**	**Best**
Population size	15, 30	30
Subgroups number	2,3,4,5,6,7,8,9	7
Greedy ratio	0.1, 0.2, ..., 0.9	0.6
Construction method	IP, EP	EP
Iteration number	50,100,150,200,250	200

The algorithm was implemented in Borland Delphi and the computational experiments were performed on an Intel i7-720QM with 1.6 (max 2.8) GHz Quad Core processor and 4 GB memory. A problem is designed according to literature [18]. The problem is solved by the proposed GA and AIS. For AIS algorithm, Kahraman et al. [5]'s parameter settings are adopted. The parameters of AIS algorithm are given in Table 2.

Table 2. The parameter sets for the AIS algorithm

Parameter	**Value**
Number of iteration to determine the initial immune strings	1000
Initial parameter of the antibodies significance level	0.6
The ratio of increasing the significance level of the antibody	0.1
Number of iteration	100

Multi objective fuzzy flow shop scheduling problems are formulated by two objectives. These are to minimize the average tardiness and to minimize the number of tardy jobs [5]. To compare the fuzzy completion time of a job with its fuzzy due date, two approaches are used. These are PM introduced by Dubois et al. [19] and AIM introduced by Sakawa et al. [20].The improvement rate is calculated as;

$$\text{Improvement rate} = \frac{GA - AIS}{AIS} \tag{3}$$

The computational results are recorded in Table 3. Three aggregation operators' averages, standard deviations, and maximum values are calculated in Table 3. As it seen in which *average* indicates the average value of the best solutions found in the problem, *best* denote the best solution and *worst* denote the worst solution.

Table 3. Computational Results of GA and AIS Algorithms

Fitness value		AIS algorithm		GREEDY algorithm		Improvement Rate	
		AIM	PM	AIM	PM	AIM	PM
Average of the satisfaction grades	Average	0.78181	0.76928	0.96700	0.97717	0.24	0.27
	Std. Dev.	0.06932	0.05607	0.01208	0.01322	4.74	3.24
	Max	0.94390	0.92017	0.99329	0.99500	0.05	0.08
Minimum of the satisfaction grades	Average	0.95153	0.78844	0.99550	1.00000	0.05	0.27
	Std. Dev.	0.08200	0.16280	0.01190	0.00000	5.89	Grdy bet.
	Max	1.00000	1.00000	1.00000	1.00000	0.00	0.00
Average weighted sum of the satisfaction grades	Average	0.98225	0.89146	0.99798	0.96835	0.02	0.09
	Std. Dev.	0.01192	0.09654	0.00541	0.07152	1.20	0.35
	Max	1.00000	1.00000	1.00000	1.00000	0.00	0.00

5. Conclusion

In this study, a new greedy algorithm is proposed to solve the multiobjective fuzzy flow shop scheduling problem. Multiobjective fuzzy flow shop scheduling problems are formulated by to minimize the average tardiness and to minimize the number of tardy jobs. The proposed GA is compared with AIS. For the benchmark problem the proposed GA is obtained the best value for averages and maximum values. The proposed GA is a good problem solving technique for the fuzzy multiobjective flow shop scheduling problem. For further research, the proposed GA can be tested on real world data collected at industrial firm.

References

1. J. Peng, K. Song, Fuzzy Flow-Shop Scheduling Models Based on Credibility Measure, *The IEEE International Conference on Fuzzy Systems*, (2003).
2. H. Ishibuchi, T. Murata and K. H. Lee, Formulation of Fuzzy Flowshop Scheduling Problems with Fuzzy Processing Time, *Proceedings of the Fifth IEEE International Conference on Fuzzy Systems*, (1996).
3. S. Kilic, C. Kahraman, Solution of a fuzzy flowshop scheduling problem using a necessity measure, *Journal of Multiple Valued Logic and Soft Computing*,.15, 1, 51-64, (2008).
4. O. Engin, C. Kahraman, M. K. Yılmaz, A scatter search method for multiobjective fuzzy permutation flow shop scheduling problem: A Real world application, in *Computational Intelligence in Flow Shop and Job Shop Scheduling* (Uday K. Chakraborty (Editor)) 230, 169-189 Verlag (Heidelberg), (2009).
5. C. Kahraman, O. Engin, M. K. Yılmaz, A new artificial immune system Algorithm for multiobjective fuzzy flowshop problems, *International Journal of Computational Intelligence Systems*, 2, 3, 237- 248, (2009).
6. C .S. Huang, Y. C. Huang, P. J. Lai, Modified genetic algorithms for solving fuzzy flow shop scheduling problems and their implementation with CUDA, *Expert Systems with Applications*, Article in Press, (2011).

7. P. J. Lai, H. C. Wu, Evaluate the fuzzy completion times in the fuzzy flow shop scheduling problems using the virus-evolutionary genetic algorithms, *Applied Soft Computing* 11, 4540–4550, (2011).
8. C. Fayad, S. Petrovic, A fuzzy genetic algorithm for real-world job shop scheduling. University of Nottingham, http://www.cs.nott.ac.uk/~cxf~sxp, (2003).
9. S. Binato, W.Hery, D.Loewenstern,, M. Resende, , A GRASP for job shop scheduling in: C. Riberio, P. Hansen (Eds), *Essays and Surveys on Metaheuristic*, Kluwer Academic Publishers, 59-79, (2001).
10. C. Kahraman, O. Engin, I. Kaya, R. E. Ozturk, Multiprocessor task scheduling in multistage hybrid flow-shops:A parallel greedy algorithm approach, *Applied Soft Computing* 10, 1293–1300, (2010).
11. R. Ruiz, T. Stützle, A simple and effective iterated greedy algorithm for the permutation flowshop scheduling problem, *Eur J Oper Res,* 177,2033–2049, (2007).
12. R. Ruiz, T. Stützle, An Iterated Greedy heuristic for the sequence dependent setup times flowshop problem with makespan and weighted tardiness objectives, *Eur J Oper Res,* 187, 1143–1159, (2008).
13. X. L. H. Zhao, Greedy Algorithm Solution of Flexible Flow Shop Scheduling Problem, IJCSNS *International Journal of Computer Science and Network Security*, 9, 11, (2009).
14. Ö. Akgöbek, S. Kaya, Ü. Değirmenci, O. Engin, Parallel Greedy Metaheuristic Algorithm for Solving Open Shop Scheduling Problem, *NWSA e-Journal of New World Science Academy, Engineering Sciences*, 6, 1, 421- 427, (2011).
15. K. Sheibani, A fuzzy greedy heuristic for permutation flow-shop scheduling, *Journal of the Operational Research Society* 61, 813–818, (2010).
16. I.Ribas, R. Companys, X. T. Martorell, An iterated greedy algorithm for the flowshop scheduling problem with blocking, *Omega* 39, 293–301, (2011).
17. G. Minella, R. Ruiz, M. Ciavotta, Restarted Iterated Pareto Greedy algorithm for multi-objective flowshop scheduling problems, *Computers & Operations Research* 38, 1521–1533, (2011).
18. E. Taillard, Benchmarks for basic scheduling problems, *Eur J Oper Res*, 64, 2, 278–285, (1993).
19. D. Dubois, H. Prade, Possibility theory: an approach to computerized processing of uncertainty, New York. (1988).
20. M. Sakawa, R. Kubota, Fuzzy programming for multiobjective job shop scheduling with fuzzy processing time and fuzzy duedate through genetic algorithms, *Eur J Oper Res,* 120, 393- 407, (2000).

A GROUP DECISION MAKING MODEL WITH INTUITIONISTIC FUZZY PREFERENCE RELATIONS

HÜLYA BEHRET*

Erciyes University, Department of Industrial Engineering, 38039, Kayseri, Turkey

In this study a priority weight generation method on the intuitionistic fuzzy preference relations is proposed. The method consists of two linear programming models which are generalized to the case of group decision making with the weight information defined by each DM. In the models, the collective preference relations are obtained by using the intuitionistic fuzzy weighted geometric averaging operator and additive consistency is considered. Finally, a numerical example is give to verify the validity and applicability of the models.

1. Introduction

In the decision making process, a decision maker is usually asked to give his/her preferences over alternatives. In some real life decision making problems, DM may provide his/her preferences for alternatives to a certain degree, but it is possible that he/she is not so sure about it [1]. The reason of this situation maybe the lack of precise or sufficient level of knowledge related to the problem or the difficulty in explaining explicitly the degree to which one alternative are better than others. In such situations, it is very suitable to express the DM's preference values with the use of intuitionistic fuzzy values rather than exact numerical values or linguistic variables [2]. Atanassov (1986) introduced the concept of intuitionistic fuzzy sets (IFS) characterized by a membership function and a nonmembership function [3]. Gau and Buehrer (1993) presented the concept of vague sets and Bustince and Burillo (1996) showed that vague sets are intuitionistic fuzzy sets [4][5]. Xu and Yager (2006) developed some new geometric aggregation operators with intuitionistic fuzzy information [6]. Recently, some researchers have shown great interest in the IFS theory and applied it to the field of decision making [7][8]. Xu (2007b) defined the concept of consistent intuitionistic fuzzy preference relation (IFPR) and established a method for estimating criteria weights from intuitionistic fuzzy preference

* E-mail:hbehret@erciyes.edu.tr

relations [9]. Qian and Feng (2008) studied the consistency and priority method of intuitionistic preference relation [10].

The aim of this paper is to propose a GDM model with IFPRs. The rest of the paper is organized as follows: In Sec. 2, some basic concepts about IFPR is explained. Sec. 3 introduces the notion of consistent intuitionistic preference relation. In Sec. 4, linear programming models are established to get the priority weight of collective IFPR. In Sec. 5, a numerical example is given to illustrate the validity and practicality of the proposed methods and Sec. 6 concludes this paper.

2. Preliminaries

The notion of intuitionistic fuzzy sets is introduced as [3];

$$A = \{< x_i, \mu_A(x_i), v_A(x_i) > | x_i \in X\} \quad (1)$$

which is characterized by a membership function $\mu_A: X \to [0,1]$ and a non-membership function $v_A: X \to [0,1]$ with the condition:

$$0 \leq \mu_A(x_i) + v_A(x_i) \leq 1, \quad \forall x_i \in X \quad (2)$$

The value, $\pi_A(x_i) = 1 - \mu_A(x_i) - v_A(x_i)$ is called the indeterminacy degree or hesitation degree of x_i to A, $0 \leq \pi_A(x_i) \leq 1$. Especially, if $\pi_A(x_i) = 1 - \mu_A(x_i) - v_A(x_i) = 0$, $\forall x_i \in X$ then, the intuitionistic fuzzy set A is reduced to a common fuzzy set.

Definition 2.1 [2]: An intuitionistic fuzzy preference relation B on X is defined as a matrix $B = (b_{ij})_{n\times n} \subset X \times X$ where b_{ij} is the intuitionistic fuzzy value, composed by the certainty degree μ_{ij} to which x_i is preferred to x_j and the certainty degree v_{ij} to which x_i is non-preferred to x_j, and $\pi_{ij} = 1 - \mu_{ij} - v_{ij}$ is interpreted as the hesitation degree to which x_i is preferred to x_j. Moreover, μ_{ij} and v_{ij} satisfy the following condition,

$$0 \leq \mu_{ij} + v_{ij} \leq 1, \mu_{ji} = v_{ij}, \mu_{ii} = v_{ii} = 0.5, for\ all\ i,j = 1,2,\dots,n \quad (3)$$

Definition 2.2 [6]: Let $b_{ij}^{(k)} = (\mu_{ij}^{(k)}, v_{ij}^{(k)}), k = 1,2,..,m$ be a collection of intuitionistic fuzzy values and let $\lambda^k = (\lambda_1, \lambda_2, \dots, \lambda_m)^T$ be the weight vector of $b_{ij}^{(1)}, b_{ij}^{(2)}, \dots, b_{ij}^{(m)}, \lambda^k > 0, \sum_{k=1}^{m} \lambda^k = 1$, then the aggregated value $\overline{b_{ij}}$ of $b_{ij}^{(k)}$ is also an intuitionistic fuzzy value, where $\overline{b_{ij}}$ is obtained by using the intuitionistic fuzzy weighted geometric (IFWG) averaging operator. $(k = 1,2,\dots,m)$:

$$\overline{b_{ij}} = \prod_{k=1}^{m} (b_{ij}^{(k)})^{\lambda^k} = \left(\prod_{k=1}^{m} (\mu_{ij}^{(k)})^{\lambda^k}, 1 - \prod_{k=1}^{m} (1 - v_{ij}^{(k)})^{\lambda^k}\right), \quad (4)$$

3. Consistency of Intuitionistic Preference Relation

Consider that each intuitionistic fuzzy value $b_{ij} = (\mu_{ij}, v_{ij})$ must satisfy the condition $0 \le \mu_{ij} + v_{ij} \le 1$, i.e., $\mu_{ij} \ge 1 - v_{ij}$. This condition is the same as the condition under which two real numbers form an interval [11]. As a result, an intuitionistic fuzzy preference relation can be transformed into an interval fuzzy preference relation [11].

Definition 3.1 [12]: An interval fuzzy preference relation $R = (r_{ij})_{n\times n}$ where $r_{ij} = (\mathrm{r_{ij}}^{-}, r_{ij}^{+})$ is additive consistent if there exists a normalized priority vector, $w = (w_1, w_2, \dots, w_n)^T = ([w_{1l}, w_{1u}], [w_{2l}, w_{2u}], \dots, w_{nl}, w_{nu})^T$ such that,

$$r_{ij}^{-} \le 0.5(w_i - w_j + 1) \le r_{ij}^{+}, for\ all\ i = 1,2, \dots, n-1; j = i+1, \dots, n, \quad (5)$$

$$w_i \ge 0, i = 1,2, \dots, n, \textstyle\sum_{i=1}^{n} w_i = 1, \quad (6)$$

where R is called an additive consistent interval fuzzy preference relation. Since an intuitionistic fuzzy preference relation can be transformed into an interval fuzzy preference relation where $r_{ij}^{-} = \mu_{ij}$, $r_{ij}^{+} = 1 - v_{ij}$, for all $i, j = 1,2, \dots, n$, Eq. (5) can be rewritten as,

$$\mu_{ij} \le 0.5(w_i - w_j + 1) \le 1 - v_{ij} \quad (7)$$

And also we can transform the interval priority vector w to the intuitionistic fuzzy numbers,

$$w = ([w_{1l}, 1 - w_{1u}], [w_{2l}, 1 - w_{2u}], \dots, w_{nl}, 1 - w_{nu})^T.$$

Definition 3.2: An intuitionistic fuzzy preference relation $B = (b_{ij})_{n\times n}$ is additive consistent if there exists a vector $w = (w_1, w_2, \dots, w_n)^T$ such that Eq.(6) and Eq(7) holds where $\dot{w} = ([w_{1l}, 1 - w_{1u}], [w_{2l}, 1 - w_{2u}], \dots, [w_{nl}, 1 - w_{nu}])^T, i = 1,2, \dots, n$, otherwise we call B an inconsistent intuitionistic fuzzy preference relation.

4. Priority Vector Generation Method for Collective IFPR

If $B = (b_{ij})_{n\times n}$ is an additive consistent intuitionistic fuzzy preference relation, then there exists a priority vector $w = (w_1, w_2, \dots, w_n)^T$ of B which satisfies Eq(7) and Eq.(8). If $b_{ij}^{(k)} = (\mu_{ij}^{(k)}, v_{ij}^{(k)}), k = 1,2, .., m$ is a collection of intuitionistic fuzzy preference relations where $\lambda^k = (\lambda_1, \lambda_2, \dots, \lambda_m)^T$ is the weight vector of $b_{ij}^{(1)}, b_{ij}^{(2)}, \dots, b_{ij}^{(m)}, \lambda^k > 0, \sum_{k=1}^{m} \lambda^k = 1$, then the aggregated value $\overline{b_{ij}}$ of $b_{ij}^{(k)}$ is also an intuitionistic fuzzy preference relation, where $\overline{b_{ij}}$ is obtained by using the intuitionistic fuzzy weighted geometric (IFWG) averaging operator. If the collective intuitionistic fuzzy preference relation $\bar{B} = \overline{b_{ij}}$ is additive consistent then there exists a priority vector $\bar{w} = (\overline{w_1}, \overline{w_2}, \dots, \overline{w_n})^T$ of $\bar{B}$ which satisfies,

$$\bar{w}_i \geq 0, i = 1,2,\dots,n, \sum_{i=1}^{n} \overline{w_i} = 1, \tag{8}$$

$$\prod_{k=1}^{m} \left(\mu_{ij}{}^{(k)}\right)^{\lambda k} \leq 0.5(w_i - w_j + 1) \leq \prod_{k=1}^{m} \left(1 - v_{ij}{}^{(k)}\right)^{\lambda k}, \tag{9}$$

$$for\ all\ i = 1,2,\dots,n-1; j = i+1,\dots,n; k = 1,2,\dots,m$$

Thus, the following linear programming model is developed by using Eq(8) and Eq.(9):

$$w_{il} = Min\ \overline{w_i}\ and\ w_{iu} = Max\ \overline{w_i} \qquad \textbf{[M-1]}$$

$$\begin{aligned} s.t.\ \ & 0.5(\overline{w_i} - \overline{w_j} + 1) \geq \prod_{k=1}^{m} \left(\mu_{ij}{}^{(k)}\right)^{\lambda k} \\ & 0.5(\overline{w_i} - \overline{w_j} + 1) \leq \prod_{k=1}^{m} \left(1 - v_{ij}{}^{(k)}\right)^{\lambda k} \\ & \bar{w}_i \geq 0, i = 1,2,\dots,n, \sum_{i=1}^{n} \overline{w_i} = 1, \\ & i = 1,2,\dots,n-1; j = i+1,\dots,n; k = 1,2,\dots,m \end{aligned}$$

If $\bar{B} = \overline{b_{ij}}$ is not an consistent collective intuitionistic fuzzy preference relation, then we relax Eq.(9) by introducing the non-negative deviation variables $d_{ij}{}^{-}$ and $d_{ij}{}^{+}$, $i = 1,2,\dots,n-1; j = i+1,\dots,n$:

$$\prod_{k=1}^{m} \left(\mu_{ij}{}^{(k)}\right)^{\lambda k} - d_{ij}{}^{-} \leq 0.5(w_i - w_j + 1) \leq \prod_{k=1}^{m} \left(1 - v_{ij}{}^{(k)}\right)^{\lambda k} + d_{ij}{}^{+},$$

$$for\ all\ i = 1,2,\dots,n-1; j = i+1,\dots,n \tag{10}$$

Especially, if both deviation values are equal to zero, then Eq.(10) reduces to Eq.(9). As the deviation variables $d_{ij}{}^{-}$ and $d_{ij}{}^{+}$ gets smaller, $\bar{B}$ gets closer to an additive collective consistent intuitionistic fuzzy preference relation. As a result to find the smallest deviation variables the following model is developed:

$$\delta^* = Min \sum_{i=1}^{n-1} \sum_{j=i+1}^{n} \left(d_{ij}{}^{-} + d_{ij}{}^{+}\right) \qquad \textbf{[M-2]}$$

$$\begin{aligned} s.t.\ \ & 0.5(\overline{w_i} - \overline{w_j} + 1) + d_{ij}{}^{-} \geq \prod_{k=1}^{m} \left(\mu_{ij}{}^{(k)}\right)^{\lambda k} \\ & 0.5(\overline{w_i} - \overline{w_j} + 1) - d_{ij}{}^{+} \leq \prod_{k=1}^{m} \left(1 - v_{ij}{}^{(k)}\right)^{\lambda k} \\ & \bar{w}_i \geq 0, i = 1,2,\dots,n, \sum_{i=1}^{n} \overline{w_i} = 1, d_{ij}{}^{-}, d_{ij}{}^{+} \geq 0 \\ & i = 1,2,\dots,n-1; j = i+1,\dots,n; k = 1,2,\dots,m. \end{aligned}$$

Based on the optimal deviation values $d_{ij}{}^{-}$ and $d_{ij}{}^{+}$, we establish the following linear model:

$$w_{il} = Min\ \overline{w_i}\ and\ w_{iu} = Max\ \overline{w_i} \qquad \textbf{[M-3]}$$

$$\begin{aligned} s.t.\ \ & 0.5(\overline{w_i} - \overline{w_j} + 1) + d_{ij}{}^{-} \geq \prod_{k=1}^{m} \left(\mu_{ij}{}^{(k)}\right)^{\lambda k} \\ & 0.5(\overline{w_i} - \overline{w_j} + 1) - d_{ij}{}^{+} \leq \prod_{k=1}^{m} \left(1 - v_{ij}{}^{(k)}\right)^{\lambda k} \end{aligned}$$

$$\overline{w}_i \geq 0, i = 1,2,\dots,n, \sum_{i=1}^{n} \overline{w_i} = 1,$$
$$i = 1,2,\dots,n-1; j = i+1,\dots,n; k = 1,2,\dots,m$$

Solving the model (M-3), we get the collective interval fuzzy priority vector $\overline{w} = ([\overline{w_{1l}}, \overline{w_{1u}}], [\overline{w_{2l}}, \overline{w_{2u}}], \dots, [\overline{w_{nl}}, \overline{w_{nu}}])^T$. We can transform this solution to the collective intuitionistic fuzzy priority vector as $\overline{w} = ([\overline{w_{1l}}, 1 - \overline{w}_{1u}], [\overline{w_{2l}}, 1 - \overline{w}_{2u}], \dots, [\overline{w_{nl}}, 1 - \overline{w}_{nu}])^T$.

5. Numerical Example

This section presents an example to show the validity and practicality of the proposed methods.

Example: Consider the example which was discussed by Gong et al. [13]. A DM invites three experts to help him buy a house, and the weights for each DM are 0.3, 0.4 and 0.3, respectively. There are three alternatives (houses) $X = \{x_1, x_2, x_3\}$ to be chosen. The IFPR $B^{(k)} = \left(b_{ij}\right)^{(k)}$ presented by the kth, k=1,2,3 expert are as follows:

$$B^{(1)} = \begin{pmatrix} (0.5,0.5) & (0.1,0.6) & (0.6,0.3) \\ (0.6,0.1) & (0.5,0.5) & (0.8,0.2) \\ (0.3,0.6) & (0.2,0.8) & (0.5,0.5) \end{pmatrix} B^{(2)} = \begin{pmatrix} (0.5,0.5) & (0.3,0.7) & (0.7,0.2) \\ (0.7,0.3) & (0.5,0.5) & (0.6,0.2) \\ (0.2,0.7) & (0.2,0.6) & (0.5,0.5) \end{pmatrix}$$

$$B^{(3)} = \begin{pmatrix} (0.5,0.5) & (0.3,0.6) & (0.9,0.1) \\ (0.6,0.3) & (0.5,0.5) & (0.7,0.2) \\ (0.1,0.9) & (0.2,0.7) & (0.5,0.5) \end{pmatrix}$$

When we apply the model (M-2), we get δ^* =0,070. Thus, by Definition 3.2, collective intuitionistic fuzzy preference relation $\bar{B}$ is inconsistent and the deviation values are, $d_{12}{}^{-} = 0; d_{12}{}^{+} = 0{,}040;\ d_{13}{}^{-} = 0{,}030; d_{13}{}^{+} = 0; d_{23}{}^{-} = 0;\ d_{23}{}^{+} = 0$. Based on the optimal deviation values, we solve the model (M-3) and the following collective interval fuzzy priority vector $\overline{w} = ([0{,}40; 0{,}40], [0{,}60; 0{,}60], [0; 0])^T$ and the collective intuitionistic fuzzy priority vector as $\overline{w} = ([0{,}40; 0{,}60], [0{,}60; 0{,}40], [0; 0])^T$. The optimal ranking of the alternatives is given by $x_2 \succ x_1 \succ x_3$ which is accorded with the result of Gong et al. [13].

6. Conclusion

In this study, linear programming models are established to get the priority weight of a collective intuitionistic fuzzy preference relation where the collective preference relation is obtained by using the intuitionistic fuzzy weighted geometric averaging operator. The method can be applied to multi-criteria group decision making problems in many fields such as, supply chain management, project evaluation, risk management and etc. This study considers additive consistency; the models can be extended for multiplicative consistency. Additionally, estimating the priority weight vector of incomplete IFPRs can be examined as a future study.

References

1. G. Deschrijver, E.E. Kerre, On the composition of intuitionistic fuzzyrelations, *Fuzzy Sets and Systems,* **136**, 333–361 (2003).
2. Z.S. Xu, Intuitionistic preference relations and their application in groupdecision making. *Information Sciences*, **177(11)**, 2363-2379 (2007).
3. K. Atanassov, Intuitionistic fuzzy sets, *Fuzzy Sets and Systems* **20**, 87-96 (1986).
4. W.L. Gau, D.J. Buehrer, Vague sets, *IEEE Transactions on Systems, Man,and Cybernetics*, **23**, 610–614 (1993).
5. H. Bustince, P. Burillo, Vague sets are intuitionistic fuzzy sets, *Fuzzy Sets and Systems,* **79**, 403–405 (1996).
6. Z. S. Xu, R. Yager, Some geometric aggregation operators based on intuitionistic fuzzy sets, *International Journal of General Systems*, **35**, 417–433 (2006).
7. K. Atanassov, G. Pasi, R. R., Yager, Intuitionistic fuzzy interpretations of multi-criteria multi-person and multi-measurement tool decision making, *International Journal of Systems Science*, **36**, 859–868 (2005).
8. D. F. Li, Multiattribute decision making models and methods using intuitionistic fuzzy sets, *Journal of Computer and System Sciences*, **70**, 73-85 (2005).
9. Z. S. Xu, A Method for Estimating Criteria Weights from Intuitionistic Preference Relations, *Fuzzy Information and Engineering (ICFIE*), ASC, 503-512 (2007).
10. Qian, G. and X.Q. Feng, Intuitionistic weight generation approach from intuitionistic preference relations. *Proceedings of 2008 International Conference on Machine Learning and Cybernetics*. July, 12-15. Kunming, 536-541 (2008).
11. F. Herrera, L. Martinez and P.J. Sanchez, Managing non-homogeneous information in group decision making, *European Journal of Operational Research,* **166**, 115–13(2005).
12. Z.S. Xu, J. Chen, Some models for deriving the priority weights from interval fuzzy preference relations, *European Journal of Operational Research*, **184**, 266–280 (2008).
13. Zai-Wu Gong , Lian-Shui Li , Fei-Xue Zhou , Tian-Xiang Yao, Goal programming approaches to obtain the priority vectors from the intuitionistic fuzzy preference relations, *Computers and Industrial Engineering*, **57(4)**, 1187-1193 (2009).

FUZZY ANALYTIC HIERARCHY PROCESS WITH TYPE-2 FUZZY SETS

CENGIZ KAHRAMAN[†]

Industrial Engineering Department, Istanbul Technical University, Maçka, İstanbul, 34367, Turkey

İREM UÇAL SARI

Industrial Engineering Department, Istanbul Technical University, Maçka, İstanbul, 34367, Turkey

EBRU TURANOĞLU

Industrial Engineering Department, Selçuk University, Konya, Turkey

Some fuzzy multicriteria methods have recently been developed by using type-2 fuzzy sets. Fuzzy TOPSIS is one of these methods. While these need excessive arithmetic operations with respect to type-1's, they can handle the uncertainty in defining membership functions. We modified Buckley's fuzzy AHP method by using type-2 fuzzy sets.

1. Introduction

The concept of a type-2 fuzzy set was introduced by Zadeh [9] as an extension of the concept of an ordinary fuzzy set called a type-1 fuzzy set. Such sets are fuzzy sets whose membership grades themselves are type-1 fuzzy sets; they are very useful in circumstances where it is difficult to determine an exact membership function for a fuzzy set; hence, they are useful for incorporating linguistic uncertainties, e.g., the words that are used in linguistic knowledge can mean different things to different people [5]. The fuzzy AHP technique can be viewed as an advanced analytical method developed from the traditional AHP. Despite the convenience of AHP in handling both quantitative and qualitative criteria of multi-criteria decision making problems based on decision makers' judgments, fuzziness and vagueness existing in many decision-making problems may contribute to the imprecise judgments of decision makers in conventional AHP approaches [3]. In this paper, fuzzy AHP approaches with type-1 fuzzy

[†] kahramanc@itu.edu.tr

sets are examined by using type-2 fuzzy sets. This will help for us to take care of the uncertainty of membership functions.

2. Type-2 Fuzzy Sets

In this section, it is briefly explained some definitions of type-2 fuzzy sets and interval type-2 fuzzy sets from [6].

A type-2 fuzzy sets $\tilde{\tilde{A}}$ in the universe of discourse X can be represented by a type-2 membership function $\mu_{\tilde{\tilde{A}}}$, shown as follows [9]:

$$\tilde{\tilde{A}} = \left\{\left((x,u), \mu_{\tilde{\tilde{A}}}(x,u)\right) \mid \forall x \in X, \forall u \in J_x \subseteq [0,1], 0 \le \mu_{\tilde{\tilde{A}}}(x,u) \le 1\right\}, \quad (1)$$

where $J_x$denotes an interval [0,1]. Moreover, the type-2 fuzzy set $\tilde{\tilde{A}}$ also can be represented as follows [6]:

$$\tilde{\tilde{A}} = \int_{x \in X} \int_{u \in J_x} \mu_{\tilde{\tilde{A}}}(x,u)/(x,u), \quad (2)$$

where $J_x \subseteq [0,1]$ and $\int\int$ denote union over all admissible x and u.

Let $\tilde{\tilde{A}}$ be a type-2 fuzzy set in the universe of discourse X represented by the type-2 membership function $\mu_{\tilde{\tilde{A}}}$. If all $\mu_{\tilde{\tilde{A}}}(x,u) = 1$, then $\tilde{\tilde{A}}$ is called an interval type-2 fuzzy set [1]. An interval type-2 fuzzy set $\tilde{\tilde{A}}$ can be regarded as a special case of a type-2 fuzzy set, represented as follows [6]:

$$\tilde{\tilde{A}} = \int_{x \in X} \int_{u \in J_x} 1/(x,u), \quad (3)$$

where $J_x \subseteq [0,1]$.

Definition 2.1 The upper membership function and the lower membership function of an interval type-2 fuzzy set are type-1 membership functions, respectively.

It is presented a method to use interval type-2 fuzzy sets for handling fuzzy multiple attributes group decision making problems, where the reference points and the heights of the upper and the lower membership functions of interval type-2 fuzzy sets are used to characterize interval type-2 fuzzy sets. A trapezoidal interval type-2 fuzzy set is illustrated as

$$\tilde{\tilde{A}}_i = \left(\tilde{A}_i^U, \tilde{A}_i^L\right) = \left(\left(a_{i1}^U, a_{i2}^U, a_{i3}^U, a_{i4}^U; H_1\left(\tilde{A}_i^U\right), H_2\left(\tilde{A}_i^U\right)\right), \left(a_{i1}^L, a_{i2}^L, a_{i3}^L, a_{i4}^L; H_1\left(\tilde{A}_i^L\right), H_2\left(\tilde{A}_i^L\right)\right)\right)$$

where $\tilde{A}_i^U$ and $\tilde{A}_i^L$ are type-1 fuzzy sets, $a_{i1}^U, a_{i2}^U, a_{i3}^U, a_{i4}^U, a_{i1}^L, a_{i2}^L, a_{i3}^L$, and a_{i4}^L are the references points of the interval type-2 fuzzy set $\tilde{\tilde{A}}_i$, $H_j(\tilde{A}_i^U)$ denotes the membership value of the element $a_{j(j+1)}^U$ in the upper trapezoidal membership function $\tilde{A}_i^U$, $1 \le j \le 2$, $H_j(\tilde{A}_i^U)$ denotes the membership value of the element $a_{j(j+1)}^L$

in the lower trapezoidal membership function $\widetilde{A}_i^L$, 1≤j≤2, $H_1(\widetilde{A}_i^U) \in [0,1]$, $H_2(\widetilde{A}_i^U) \in [0,1]$, $H_1(\widetilde{A}_i^L) \in [0,1]$, $H_2(\widetilde{A}_i^L) \in [0,1]$, and 1≤i≤n [4].

Definition 2.2 The addition operation between the trapezoidal interval type-2 fuzzy sets $\tilde{\tilde{A}}_1 = (\tilde{A}_1^U, \tilde{A}_1^L) = (a_{11}^U, a_{12}^U, a_{13}^U, a_{14}^U; H_1(\tilde{A}_1^U), H_2(\tilde{A}_1^U)), (a_{12}^L, a_{13}^L, a_{14}^L, a_{12}^L;$ $H_1(\tilde{A}_1^L), H_2(\tilde{A}_2^L)))$ and $\tilde{\tilde{A}}_2 = (\tilde{A}_2^U, \tilde{A}_2^L) = (a_{21}^U, a_{22}^U, a_{23}^U, a_{24}^U; H_1(\tilde{A}_2^U), H_2(\tilde{A}_2^U)),$ $(a_{21}^L, a_{22}^L, a_{23}^L, a_{24}^L; H_1(\tilde{A}_2^L), H_2(\tilde{A}_2^L))$ is defined as follows[4]:

$$\tilde{A}_1 \oplus \tilde{A}_2 = (\tilde{A}_1^U, \tilde{A}_1^L) \oplus (\tilde{A}_2^U, \tilde{A}_2^L) = ((a_{11}^U + a_{21}^U, a_{12}^U + a_{22}^U, a_{13}^U + a_{23}^U, a_{14}^U + a_{24;}^U \min(H_1(\tilde{A}_1^U), H_1(\tilde{A}_2^U)), \min(H_2(\tilde{A}_1^U), H_2(\tilde{A}_2^U))), (a_{11}^L + a_{21}^L, a_{12}^L + a_{22}^L, a_{13}^L + a_{23}^L, a_{14}^L + a_{24;}^L \min(H_1(\tilde{A}_1^L), H_1(\tilde{A}_2^L)), \min(H_2(\tilde{A}_1^L), H_2(\tilde{A}_2^L)))) \quad (4)$$

Definition 2.3 The subtraction operation between the trapezoidal interval type-2 fuzzy sets $\tilde{\tilde{A}}_1 = (\tilde{A}_1^U, \tilde{A}_1^L) = (a_{11}^U, a_{12}^U, a_{13}^U, a_{14}^U; H_1(\tilde{A}_1^U), H_2(\tilde{A}_1^U)), (a_{12}^L, a_{13}^L, a_{14}^L, a_{12}^L;$ $H_1(\widetilde{A}_1^L), H_2(\widetilde{A}_2^L)))$ and $\tilde{\tilde{A}}_2 = (\widetilde{A}_2^U, \widetilde{A}_2^L) = (a_{21}^U, a_{22}^U, a_{23}^U, a_{24}^U; H_1(\widetilde{A}_2^U), H_2(\widetilde{A}_2^U)),$ $(a_{21}^L, a_{22}^L, a_{23}^L, a_{24}^L; H_1(\widetilde{A}_2^L), H_2(\widetilde{A}_2^L))$ is defined as follows[4]:

$$\tilde{A}_1 \ominus \tilde{A}_2 = (\tilde{A}_1^U, \tilde{A}_1^L) \ominus (\tilde{A}_2^U, \tilde{A}_2^L) = ((a_{11}^U - a_{24}^U, a_{12}^U - a_{23}^U, a_{13}^U - a_{22}^U, a_{14}^U - a_{21;}^U \min(H_1(\tilde{A}_1^U), H_1(\tilde{A}_2^U)), \min(H_2(\tilde{A}_1^U), H_2(\tilde{A}_2^U))), (a_{11}^L - a_{24}^L, a_{12}^L - a_{23}^L, a_{13}^L - a_{22}^L, a_{14}^L - a_{21;}^L \min(H_1(\tilde{A}_1^L), H_1(\tilde{A}_2^L)), \min(H_2(\tilde{A}_1^L), H_2(\tilde{A}_2^L)))) \quad (5)$$

Definition 2.4 The multiplication operation between the trapezoidal interval type-2 fuzzy sets $\tilde{\tilde{A}}_1 = (\widetilde{A}_1^U, \widetilde{A}_1^L) = (a_{11}^U, a_{12}^U, a_{13}^U, a_{14}^U; H_1(\widetilde{A}_1^U), H_2(\widetilde{A}_1^U)),$ $(a_{12}^L, a_{13}^L, a_{14}^L, a_{12}^L); H_1(\widetilde{A}_1^L), H_2(\widetilde{A}_2^L)))$ and $\tilde{\tilde{A}}_2 = (\widetilde{A}_2^U, \widetilde{A}_2^L) =$ $(a_{21}^U, a_{22}^U, a_{23}^U, a_{24}^U; H_1(\widetilde{A}_2^U), H_2(\widetilde{A}_2^U)), (a_{21}^L, a_{22}^L, a_{23}^L, a_{24}^L; H_1(\widetilde{A}_2^L), H_2(\widetilde{A}_2^L))$ is defined as follows [4]:

$$\tilde{A}_1 \otimes \tilde{A}_2 = (\tilde{A}_1^U, \tilde{A}_1^L) \otimes (\tilde{A}_2^U, \tilde{A}_2^L) = ((a_{11}^U \times a_{21}^U, a_{12}^U \times a_{22}^U, a_{13}^U \times a_{23}^U, a_{14}^U \times a_{24;}^U \min(H_1(\tilde{A}_1^U), H_1(\tilde{A}_2^U)), \min(H_2(\tilde{A}_1^U), H_2(\tilde{A}_2^U))), (a_{11}^L \times a_{21}^L, a_{12}^L \times a_{22}^L, a_{13}^L \times a_{23}^L, a_{14}^L \times a_{24;}^L \min(H_1(\tilde{A}_1^L), H_1(\tilde{A}_2^L)), \min(H_2(\tilde{A}_1^L), H_2(\tilde{A}_2^L)))) \quad (6)$$

Definition 2.5 The arithmetic operations between the trapezoidal interval type-2 fuzzy sets $\tilde{\tilde{A}}_1 = (\widetilde{A}_1^U, \widetilde{A}_1^L) = (a_{11}^U, a_{12}^U, a_{13}^U, a_{14}^U; H_1(\widetilde{A}_1^U), H_2(\widetilde{A}_1^U)),$ $(a_{12}^L, a_{13}^L, a_{14}^L, a_{12}^L);$ $H_1(\widetilde{A}_1^L), H_2(\widetilde{A}_2^L)))$ and the crisp value k is defined as follows [4]:

$$k\tilde{\tilde{A}}_1 = \left(\left(k \times a_{11}^U, k \times a_{12}^U, k \times a_{13}^U, k \times a_{14}^U; H_1(\tilde{A}_1^U), H_2(\tilde{A}_1^U)\right), \left(k \times a_{11}^L, k \times a_{12}^L, k \times a_{13}^L, k \times a_{14}^L; H_1(\tilde{A}_1^L), H_2(\tilde{A}_1^L)\right)\right), \quad (7)$$

$$\frac{\tilde{\tilde{A}}_1}{k} = \left(\left(\frac{1}{k} \times a_{11}^U, \frac{1}{k} \times a_{12}^U, \frac{1}{k} \times a_{13}^U, \frac{1}{k} \times a_{14}^U; H_1(\tilde{A}_1^U), H_2(\tilde{A}_1^U)\right), \left(\frac{1}{k} \times a_{11}^L, \frac{1}{k} \times a_{12}^L, \frac{1}{k} \times a_{13}^L, \frac{1}{k} \times a_{14}^L; H_1(\tilde{A}_1^L), H_2(\tilde{A}_1^L)\right)\right), \quad (8)$$

where $k > 0$.

3. Type-1 Fuzzy AHP

There are several fuzzy AHP applications which use fuzzy set theory to identify the linguistic variables. Laarhoven and Pedrycz [8] are proposed the first algorithm in fuzzy AHP by describing compared fuzzy ratios with triangular fuzzy membership functions. The computation steps in fuzzy AHP are the same as the steps in crisp AHP which is proposed by Saaty [7]. Logarithmic least square method is used to derive fuzzy weights and fuzzy performance scores.

Buckley [1] presented fuzzy priorities of comparison ratios whose membership functions are trapezoidal. He also extended Saaty's AHP method to incorporate fuzzy comparison ratios. He mentioned those linear equations which do not always have a unique solution and using triangular fuzzy weights which are not always obtained as triangular fuzzy numbers are the main two problems of van Laarhoven and Pedrycz's method. To overcome these difficulties, Buckey used the geometric mean method to derive fuzzy weights and performance scores [3].

Chang [2] proposed a methodology with the use of triangular fuzzy numbers for pairwise comparison scale of fuzzy AHP, and the use of the extent analysis method for the synthetic extent values of the pairwise comparisons. The extent analysis method is used to consider the extent of an object to be satisfied for the goal. On the basis of the fuzzy values for the extent analysis of each object, a fuzzy synthetic degree value is obtained.

4. Type-2 Fuzzy AHP

The procedure of FAHP for determining the evaluation weights are explained as follows:

Step 1: Construct fuzzy pairwise comparison matrices among all the criteria in the dimensions of the hierarchy system. The result of the comparisons is constructed as fuzzy pairwise comparison matrices $(\tilde{\tilde{A}})$ as following;

$$\tilde{\tilde{A}} = \begin{bmatrix} 1 & \tilde{\tilde{a}}_{12} & \cdots & \tilde{\tilde{a}}_{1n} \\ \tilde{\tilde{a}}_{21} & 1 & \cdots & \tilde{\tilde{a}}_{2n} \\ \vdots & \vdots & \ddots & \vdots \\ \tilde{\tilde{a}}_{n1} & \tilde{\tilde{a}}_{n2} & \cdots & 1 \end{bmatrix} = \begin{bmatrix} 1 & \tilde{\tilde{a}}_{12} & \cdots & \tilde{\tilde{a}}_{1n} \\ 1/\tilde{\tilde{a}}_{12} & 1 & \cdots & \tilde{\tilde{a}}_{2n} \\ \vdots & \vdots & \ddots & \vdots \\ 1/\tilde{\tilde{a}}_{1n} & 1/\tilde{\tilde{a}}_{2n} & \cdots & 1 \end{bmatrix} \tag{9}$$

where

$$1/\tilde{\tilde{a}}_{ij} = \left(\left(\frac{1}{a_{ij4}^U}, \frac{1}{a_{ij3}^U}, \frac{1}{a_{ij2}^U}, \frac{1}{a_{ij1}^U}; H_1\left(\tilde{a}_{ij}^U\right), H_2\left(\tilde{a}_{ij}^U\right) \right), \left(\frac{1}{a_{ij4}^L}, \frac{1}{a_{ij3}^L}, \frac{1}{a_{ij2}^L}, \frac{1}{a_{ij1}^L}; H_1\left(\tilde{a}_{ij}^L\right), H_2\left(\tilde{a}_{ij}^L\right) \right) \right)$$

Step 2: Examine the consistency of the fuzzy pairwise comparison matrices. Assume $A = \lfloor a_{ij} \rfloor$ is a positive reciprocal matrix and $\tilde{\tilde{A}} = \lfloor \tilde{\tilde{a}}_{ij} \rfloor$ is a fuzzy positive reciprocal matrix. If the result of the comparisons of $A = \lfloor a_{ij} \rfloor$ is consistent, then it can imply that the result of the comparisons of $\tilde{\tilde{A}} = \lfloor \tilde{\tilde{a}}_{ij} \rfloor$ is also consistent.

Step 3: Compute the fuzzy geometric mean for each criterion. If $\tilde{\tilde{r}}_i$ is geometric mean of each row of $\tilde{\tilde{A}} = \lfloor \tilde{\tilde{a}}_{ij} \rfloor$ then it is calculated as below;

$$\tilde{\tilde{r}}_i = \left[\tilde{\tilde{a}}_{i1} \otimes \ldots \otimes \tilde{\tilde{a}}_{in}\right]^{1/n} \tag{10}$$

where

$$\sqrt[n]{\tilde{\tilde{a}}_{ij}} = \left(\left(\sqrt[n]{a_{ij1}^U}, \sqrt[n]{a_{ij2}^U}, \sqrt[n]{a_{ij3}^U}, \sqrt[n]{a_{ij4}^U}; H_1\left(\tilde{a}_{ij}^U\right), H_2\left(\tilde{a}_{ij}^U\right) \right), \left(\sqrt[n]{a_{ij1}^L}, \sqrt[n]{a_{ij2}^L}, \sqrt[n]{a_{ij3}^L}, \sqrt[n]{a_{ij4}^L}; H_1\left(\tilde{a}_{ij}^L\right), H_2\left(\tilde{a}_{ij}^L\right) \right) \right)$$

Step 4: Compute the fuzzy weights by normalization. The fuzzy weight $\tilde{\tilde{w}}_i$ of the ith criterion is calculated as below;

$$\tilde{\tilde{w}}_i = \tilde{\tilde{r}}_i \otimes \left[\tilde{\tilde{r}}_1 \oplus \ldots \oplus \tilde{\tilde{r}}_i \oplus \ldots \oplus \tilde{\tilde{r}}_n\right]^{-1} \tag{11}$$

where

$$\frac{\tilde{\tilde{a}}_{ij}}{\tilde{\tilde{b}}_{ij}} = \left(\frac{a_{ij1}^{u}}{b_{ij4}^{u}}, \frac{a_{ij2}^{u}}{b_{ij3}^{u}}, \frac{a_{ij3}^{u}}{b_{ij2}^{u}}, \frac{a_{ij4}^{u}}{b_{ij1}^{u}}, \min\left(H_1\ (\tilde{a}_{ij}^{U}), H_1\ (\tilde{b}_{ij}^{U})\right) \min\left(H_2\ (\tilde{a}_{ij}^{U}) H_2\ (\tilde{b}_{ij}^{U})\right)\right),$$
$$\left(\frac{a_1^l}{b_4^l}, \frac{a_2^l}{b_3^l}, \frac{a_3^l}{b_2^l}, \frac{a_4^l}{b_1^l}, \min\left(H_1\ (\tilde{a}_{ij}^{L}), H_1\ (\tilde{b}_{ij}^{L})\right) \min\left(H_2\ (\tilde{a}_{ij}^{L}) H_2\ (\tilde{b}_{ij}^{L})\right)\right) \quad (12)$$

Step 5: Defuzzify fuzzy numbers in order to determine the importance ranking of the criteria.

5. Conclusion

In recent years, TOPSIS has been developed by using type-2 fuzzy sets. In this paper we modified Buckley's fuzzy AHP approach using type-2 fuzzy sets. Type-2 fuzzy AHP can consider the footprints of uncertainty. Nonsymmetrical type-2 membership functions may create significant differences in the results with respect to the ones of type-1 AHP methods. For further research, other fuzzy AHP methods like Chang's [2] or Laarhoven and Pedrycz's [8] may be examined by type-2 fuzzy sets.

References

1. Buckley, J. J., (1985), Fuzzy hierarchical analysis, *Fuzzy Sets and Systems*, 17, 233–247.
2. Chang, D.Y, (1996), Applications of the extent analysis method on fuzzy AHP, *European Journal of Operational Research*, 95, 649–655.
3. Chen, S. J., Hwang, C. L., Hwang, F. P., (1992), Fuzzy multiple attribute decision making: Methods and applications, Springer-Verlag, Berlin.
4. Chen, S-M., & Lee, L-W., (2010), Fuzzy multiple attributes group decision-making based on the interval type-2 TOPSIS method, *Expert System with Applications*, 37, 2790–2798.
5. Karnik, N. N. and Mendel, J. M., (2001), Operations on type-2 fuzzy sets, Fuzzy Sets and Systems, 122, 327–348.
6. Mendel, J.M., John, R.I., & Liu, F.L., (2006), Interval type-2 fuzzy logical systems made simple, *IEEE Transactions on Fuzzy Systems*, 14(6), 808-821
7. Saaty, T. L., (1980), The analytic hierarchy process: planning, priority setting, resource allocation, New York: McGraw-Hill.
8. van Laarhoven, P. J. M. and Pedrycz, W. (1983), A fuzzy extension of Saaty's priority theory, *Fuzzy Sets and Systems*, 11, 229–241.
9. Zadeh, L.A. , (1975), The concept of a linguistic variable and its application to approximate reasoning – 1, Inform. Sci., 8 199–249.

FUZZY DEA APPROACH TO SOLVE THE LOCATION SELECTION OF WIND ENERGY

NIHAL ERGINEL, MERVE YALÇINKAYA

Industrial Engineering Department, Anadolu University, Iki Eylul Campus, 26550 Eskişehir,TURKEY

SEVIL ŞENTURK

Statistics Department, Anadolu University, Yunus Emre Campus, 26470 Eskişehir, TURKEY

Wind energy is an important energy source among the renewable energies, in nowadays. Turkey has many wind energy locations due to the appropriate weather conditions. The problem of location selection of wind energy depends on several criteria: wind speed, air density, surface roughness coefficient, and altitude. Therefore, the location selection of wind energy is a difficult multi-criteria decision making problem. In this study, this multi-criteria problem is solved by fuzzy data envelopment analysis (FDEA) with triangular fuzzy numbers (TFN).

1. Introduction

Renewable energy like wind energy, solar energy, geothermal energy, and hydropower is an alternative for the fossil fuels. Wind energy is getting a popular renewable energy source, especially for Turkey, because Turkey is surrounded by seas on three sides. And also, wind is aroused from the circulation between land and seas.

The literature has several studies about wind energy. Sahin [1] mentioned the progress and recent trends of wind energy. Saidur et al. [2] paper includes a comparative study of wind energy, problems, solutions and suggestion as a result of the implementation of wind turbine. Furkan [3] emphasized the presented study, which comprehensively reviews wind energy in terms of three aspects, namely status, potential and policies analyses and assessments, for the first time to the best of the author’s knowledge. Ebubekir et al. [4] collected wind data in Balıkesir, and showed the difference between wind energy production using extrapolated wind data and energy production. Tuğrul [5] presented the prevailing energy types and the expected energy situation and energy demand. Wind energy potential in Turkey has been also discussed.

The location selection of wind energy is one of the curricle problems on the renewable energy, because the wind energy has close relationship with the weather conditions of the location. There are several techniques of location selection in literature. Data envelopment analysis (DEA) is a well-known technique for measuring and evaluating the relative efficiencies for crisp inputs and outputs. Also, FDEA is evaluated the fuzzy efficiency of decision-making units with the given fuzzy input and output data. To deal with imprecision, Zadeh [6] introduced the notion of fuzziness. Kao and Liu [7] developed a method to find the membership functions of the fuzzy efficiency measures when some observations are fuzzy numbers. Guo and Tanaka [8] proposed fuzzy DEA models and Charnes, Cooper, and Rhodes extended (CCR) model to more general forms where crisp, fuzzy and hybrid data can be handled easily. Leon et al. [9] used fuzzy linear programming to provide a new approach to the problem of assessing efficiency with DEA models. Saati and Memariani [10] suggested a procedure to find a common set of weights (CSW) in fuzzy DEA. Saati and Memariani's model upper levels of the weights are determined by solving some linear programming problems and then a CSW is determined by solving a linear programming problem. Also, authors have developed and used many fuzzy DEA models [11], [12], [13].

Due to the fact that location selection of wind energy has multi criteria, the multi-criteria decision making model is required. In this study, FDEA is used to evaluate the efficiency of many locations in TURKEY by considering the inputs: wind speed, air density, surface roughness coefficient, altitude; and the potential of wind power is defined as an output. The rest of this paper is organized as follows. Section 2 presents fuzzy data envelopment analysis. Section 3 is devoted to an illustrative problem on the location selection of wind energy in Turkey by using fuzzy data envelopment analysis. Section 4 presents a conclusion.

2. Fuzzy Data Envelopment Analysis

Data envelopment analysis with crisp data was initially proposed by Charnes, Cooper, and Rhodes in 1978. Inputs and outputs data for DEA model are assumed to be measured with precision. However, uncertainty in a situation is indispensable part of real life problems. In these cases, the fuzzy approach is particularly suitable for the theory of DEA models. Also, fuzzy numbers and fuzzy membership functions are evaluated in the FDEA models. There are several Fuzzy DEA models in literature. In this paper, we can use Saati-

Memariani [10] method for DEA analysis. Steps of Saati-Memariani models are given in the following:

Step 1: To determine the upper bounds on input and output weights, consider the following problems:

Upper bounds of output (p)

$$\max u_p$$

$$\text{st} \sum_{i=1}^{m} v_i \tilde{X}_{ij} \le 1 \qquad j=1,2,\dots,n \qquad (1)$$

$$\sum_{r=1}^{s} u_r \tilde{Y}_{rj} - \sum_{i=1}^{m} v_i \tilde{X}_{ij} \le 0 \quad j=1,2,\dots,n$$

$$v_i, u_r \ge \varepsilon \qquad r=1,2,\dots,s \quad i=1,2,\dots,m$$

Upper bounds of input (t)

$$\max v_t$$

$$\text{st.} \sum_{i=1}^{m} v_i \tilde{X}_{ij} \le 1 \qquad j=1,2,\dots,n \qquad (2)$$

$$\sum_{r=1}^{s} u_r \tilde{Y}_{rj} - \sum_{i=1}^{m} v_i \tilde{X}_{ij} \le 0 \quad j=1,2,\dots,n$$

$$v_i, u_r \ge \varepsilon \quad r=1,2,\dots,s \quad i=1,2,\dots,m$$

Among the various types of fuzzy numbers, triangular fuzzy numbers are more important. The inputs and outputs of the triangular fuzzy numbers:

Let: $\tilde{X}_{ij} = \left(x_{ij}^L, x_{ij}^M, x_{ij}^U\right)$, $\tilde{Y}_{rj} = \left(y_{rj}^L, y_{rj}^M, y_{rj}^U\right)$ and α-cut sets for upper and lower bounds are determined as follows for $\mu \ge \alpha$:

$$\tilde{X}_{ij} = \left|\alpha x_{ij}^M + (1-\alpha) x_{ij}^L, \alpha x_{ij}^M + (1-\alpha) x_{ij}^U\right| \; \tilde{Y}_{rj} = \left|\alpha y_{rj}^M + (1-\alpha) y_{rj}^L, \alpha y_{rj}^M + (1-\alpha) y_{rj}^U\right|$$

Step 2: Common sets of weights are determined with the following models

$$\max \varphi$$

$$\text{Subject to: } \sum_{r=1}^{s} \bar{y}_{rj} - \sum_{i=1}^{m} \bar{x}_{ij} \le 0 \qquad j=1,2,\dots,n$$

$$v_i \left|\alpha x_{ij}^M + (1-\alpha) x_{ij}^L\right| \le \bar{x}_{ij} \le v_i \left|\alpha x_{ij}^M + (1-\alpha) x_{ij}^U\right| \qquad r=1,2,\dots,s \quad i=1,2,\dots,m \quad j=1,2,\dots,n \qquad (3)$$

$$u_r \left|\alpha y_{rj}^M + (1-\alpha) y_{rj}^L\right| \le \bar{y}_{rj} \le u_r \left|\alpha y_{rj}^M + (1-\alpha) y_{rj}^U\right| \qquad r=1,2,\dots,s \quad i=1,2,\dots,m \quad j=1,2,\dots,n$$

$$\varepsilon + \varphi(U_r - \varepsilon) \le u_r \le (1-\varphi)(U_r - \varepsilon) \qquad r=1,2,\dots,s \qquad (4)$$

$$\varepsilon + \varphi(V_i - \varepsilon) \le v_i \le (1-\varphi)(V_i - \varepsilon) \qquad i=1,2,\dots,m$$

U_r : Upper bounds of output r from previous models

V_i : Upper bounds of inputs i from previous models

φ: Deviation in weights

Step 3: u_r^*, v_i^* are the optimum values of the second step: The efficiency of decision making units is calculated as follows:

$$\left(\tilde{E}_o\right)_\alpha = \frac{\sum_{r=1}^{s} u_r^* \tilde{Y}_{ro}}{\sum_{i=1}^{m} v_i^* \tilde{X}_{io}} \tag{5}$$

Applying the α-midrange transformation techniques by using Eq. 6, the fuzzy efficiencies are converted to the crisp efficiency measures:

$$S_{mr,j}^{\alpha} = \frac{(a_j + c_j) + \alpha[(b_j - a_j) - (c_j - b_j)]}{2} \tag{6}$$

3. The Location Selection of Wind Energy in TURKEY with Fuzzy Data Envelopment Analysis

In Turkey, the potential of wind energy is determined as output and wind speed, air density, surface roughness coefficient, and altitude are determined as the inputs of FDEA models. The locations are defined from previous studies on wind energy. Also, inputs are obtained from literature and the several web addresses [14], [15]. The inputs and output include long term data. So, data are measured on interval scale given in Table 1. Therefore, data are defined as triangular fuzzy numbers and fuzzy DEA is used to find the efficiency of the determined locations in Turkey.

Table 1. Fuzzy inputs and fuzzy output on wind energy

Inputs						Output
City	District	Wind speed (m/s)	Air density (kgm^3)	Surface roughness coefficient	Altitude (m)	Wind energy potential (MW)
İzmir	Çeşme	5.6-7.6	1.103-1,303	8-10	4-6	146.55-148.55
Balıkesir	Bandırma	4.9-6.9	1.122-1,322	6-8	62-64	12900-131.00
Manisa	Akhisar	5.3-7.3	1.110-1,310	6-8	93-95	42.75-44.75
Çanakkale	Center	6.0-8.0	1.115-1,315	2-4	49-51	182.90-184.90
Hatay	Samandağ	5.9-7.9	1.114-1,314	6-8	639-641	139.00-141.00
Muğla	Datça	4.9-6.9	1.116-1,316	0-2	69-71	110.60-112.60
Aydın	Didim	3.7-5.7	1.108-1,308	6-8	159-161	230.00-232.00
İzmir	Karaburun	5.5-7.5	1.107-1,307	4-6	30-32	230.55-232.55
Tekirdağ	Muratlı	3.1-5.1	1.113-1,313	6-8	179-181	106.80-108.80
Kırklareli	Center	4.0-6.0	1.137-1,337	4-6	202-204	171.00-173.00

Saati- Memariani [10] model for solving FDEA is used to evaluate the efficiency of location selection of wind energy. The theoretical structure of Saati- Memariani model is given in Section 2. In the first step of model, the upper bounds of inputs and output weights are determined as given in Table 2 while $\alpha = 0.75$. In the second step, common sets of weights are determined as following Table 3.

Table 2.The upper bounds of inputs and output weights

	U_1	V_1	V_2	V_3	V_4
$\alpha = 0.75$	0.00421	0.14064	0.78611	0.11415	0.00156

Table 3. Common sets of weights

	U_1^*	V_1^*	V_2^*	V_3^*	V_4^*
$\alpha = 0.75$	0.00201	0.06556	0.36620	0.05322	0.00078

In the final step, the efficiency values of each location are calculated by using Eq.5. The lower, middle and the upper efficiency values are given as Table 4. Applying α-midrange transformation techniques by using Eq.6, the fuzzy efficiencies are converted to the crisp efficiency measures as Table 5.

Table 4. Efficiencies of locations by FDEA model

City	District	$\left(E_o^L\right)$	$\left(E_o^M\right)$	$\left(E_o^U\right)$
İzmir	Çeşme	0.1890	0.2189	0.2488
Balıkesir	Bandırma	0.1772	0.2082	0.2392
Manisa	Akhisar	0.0545	0.0676	0.0806
Çanakkale	Center	0.2823	0.3356	0.3890
Hatay	Samandağ	0.1415	0.1592	0.1769
Muğla	Datça	0.1924	0.2389	0.2854
Aydın	Didim	0.3207	0.3725	0.4242
İzmir	Karaburun	0.3424	0.4022	0.4620
Tekirdağ	Muratlı	0.1491	0.1769	0.2047
Kırklareli	Center	0.2446	0.2871	0.3297

Table 5.Crisp efficiency values

City	District	$S_{mr,j}^{\alpha}$	Rank order
İzmir	Karaburun	0.4022	1
Aydın	Didim	0.3725	2
Çanakkale	Merkez	0.3357	3
Kırklareli	Merkez	0.2872	4
Muğla	Datça	0.2389	5
İzmir	Çeşme	0.2190	6
Balıkesir	Bandırma	0.2083	7
Tekirdağ	Muratlı	0.1769	8
Hatay	Samandağ	0.1593	9
Manisa	Akhisar	0.0676	10

The best efficient location in Turkey is İzmir-Karaburun according to the FDEA model. The second and the third efficiency locations are Aydın-Didim and Çanakkale-Center, respectively.

4. Conclusion

The wind energy is one of the most popular renewable energies in the world. In this study, the efficiencies of the location of wind energy in Turkey are found according to the several inputs and one output by using FDEA model. FDEA model is preferred due to the data that include long term and interval data. Other fuzzy multi-criteria models can be used for selecting the locations of wind energy for the future study.

References

1. A.D. Sahin, Progress and recent trends in wind energy, Prog in En and Combus Sc, 30, 501–543, (2004).
2. R. Saidur , N.A. Rahim, M.R. Islam, K.H. Solangi, Environmental impact of wind energy, Renew and Sust En Rev, 15, 2423–2430, (2011).
3. F. Dinçer, The analysis on wind energy electricity generation status, potential and policies in the world, Renew and Sust En Rev 15, 5135–5142,(2011)
4. E. Fırtın, O. Guler , S. A. Akdag, Investigation of wind shear coefficients and their effect on electrical energy generation, App En 88, 4097–4105,(2011).
5. R. T. Oğulata, Energy sector and wind energy potentialin Turkey, Renew and Sust Ener Rev, 7, 469–484,(2003).
6. L.A Zadeh, Fuzzy Sets, Information and Control, 8, 338–353 (1965).
7. K. Chiang and L, Shiang Fuzzy efficiency measures in data envelopment analysis, Fuz Sets and Sys, 113, 427–437, (2000).
8. G, Peijun and T, Hideo,. Fuzzy DEA: a perceptual evaluation method, Fuz Sets and Sys, 119, 149–160. (2001).
9. T, Leon, V Liern, J.L Ruiz and I Sirvent, A fuzzy mathematical programming approach tothe assessment of efficiency with dea models. Fuz Sets and Sys, 139, 407–419, (2003).
10. S Saati and A.Memariani, Reducing weight flexibility in fuzzy DEA. App Math and Comp, 161, 811–822, (2005).
11. S. Lertworasirikul, S. C. Fang, J. A. Joines and H. L. W. Nuttle, Fuzzy data envelopment analysis (DEA): a possibility approach, Fuz Sets and Sys, 139, 2, 379–394, (2003).
12. H. T. Lin, Personnel selection using analytic network process and fuzzy data envelopment analysis approaches, Com Ind Eng, 59, 937–944, (2010).
13. P. Guo, Fuzzy data envelopment analysis and its application to location problems, Inf Sci, 179, 820–829, (2009).
14. www.eie.gov.tr
15. www.mfa.gov.tr

AN INTEGRATED METHODOLOGY COMBINING FUZZY GOAL PROGRAMMING AND FUZZY AHP: CASE OF ELT IN TURKEY[*]

YESIM KOP[†]

Galatasaray University Industrial Engineering Department, Ciragan Cad. No: 36 Ortakoy, Istanbul, 34357, Turkey

MUJDE EROL GENEVOIS

Galatasaray University Industrial Engineering Department, Ciragan Cad. No: 36 Ortakoy, Istanbul, 34357, Turkey

HAKKI ZIYA ULUKAN

Galatasaray University Industrial Engineering Department, Ciragan Cad. No: 36 Ortakoy, Istanbul, 34357, Turkey

This study presents an application of a fuzzy goal programming approach to the problem of end-of-life tyres (ELT). The proposed model attempts to find out the percentage of ELT that must be used in different recovery techniques. The model is solved by using LINDO computer package and yields an efficient compromise solution with the multiple fuzzy goal values.

1. Introduction

This work introduced a fuzzy goal programming (FGP) approach for solving the end-of-life tyres (ELT) decision problems with multiple goals. ELT is evaluated currently by three different methods: Mechanical pulverization, Waste to energy process in cement factories and Fuel/Raw material extraction by pyrolysis. Expert opinions have been evaluated through Fuzzy extended Analytic Hierarchy Process (FAHP) and the results indicate that Waste to Energy process in cement factories is the best-suited ELT recovery method in Turkey. The other alternatives have approximately the same weight.

---

[*] This research has been financially supported under the project 11.402.006 by Galatasaray University Research Fund.

[†] ykop@gsu.edu.tr

The aim of this study is to evaluate these two alternatives according to quantitative and qualitative data through FGP method and find out the percentages of ELT that must be used in each of them. The remainder of this work is organized as follows. Section 2 describes the FGP weighted with FAHP. Section 3 describes the problem, details the assumptions and formulates the model. Finally, conclusions are drawn in Section 4.

2. FGP Weighted with FAHP

In 1965 Zadeh invented the fuzzy set notion to represent the real world imprecise data [1]. The integrated use GP and fuzzy sets theory has already been reported in the literature, Hannan, (1981) [2], Leberling (1981) [3], Luhandjula (1982) [4], Rubin and Narasimhan (1984) [5], Tiwari et al. (1987) [6], Wang and Fu (1997) [7], Chen and Tsai (2001) [8], Yaghoobi and Tamiz (2007) [9] further integrated several fuzzy linear and multiobjective programming techniques.

To solve our FGP model we use the algorithm proposed by Huey-Kuo Chen, which is a modified version of the method developed by Tiwari, Dharmar and Rao in 1994 [10].

We consider the special type of fuzzy values, namely, triangular fuzzy numbers with piecewise linear membership functions. A set of goals G=T is a set of triangular fuzzy numbers

$$\tilde{T} = \{\tilde{g}; \tilde{g} = (\underline{g}, g, \overline{g}), \underline{g} \geq 0, \overline{g} \geq 0\} \tag{1}$$

Our solving algorithm uses symmetrically triangular membership functions of fuzzy goals [10]. This means that

$$(\overline{g_i} - g_i) = (g_i - \underline{g_i}) = \Delta_i \tag{2}$$

We can represent m imprecise fuzzy goals and membership functions as follows:

$$\begin{aligned} &G_1 : a_{11}x_1 + a_{12}x_2 + \ldots + a_{1n}x_n \approx g_1 \\ &G_2 : a_{21}x_1 + a_{22}x_2 + \ldots + a_{2n}x_n \approx g_2 \\ &\vdots \\ &G_m : a_{m1}x_1 + a_{m2}x_2 + \ldots + a_{mn}x_n \approx g_m \\ &x_j \geq 0, \quad j = 1, \ldots, n \end{aligned} \tag{3}$$

The symbol $\approx$ refers to the fuzzification of the aspiration level. The triangular membership function of this FGP is given in Equation 4 [10]. The aim of the method proposed by Tiwari et al. [6] is to find the maximum membership value by enumerating all possible combinations, which cover the entire feasible region. Different combinations of membership functions of fuzzy goals can be constructed as in Equations 5-6. This algorithm is solved by LINDO.

$$\mu_i(AX)=\begin{cases} 1 & if\,(AX)_i = g_i \\ 0 & if\,(AX)_i \le \underline{g}_i \\ \dfrac{(AX)_i-\underline{g}_i}{\Delta_i} & if\,\underline{g}_i \le (AX)_i \le g_i \\ \dfrac{\overline{g}_i-(AX)_i}{\Delta_i} & if g_i \le (AX)_i \le \overline{g}_i \\ 0 & if\,\overline{g}_i \le (AX)_i \end{cases} \tag{4}$$

$$\max_{x\ge 0}\lambda=\left[\min_i\left\{\frac{(AX)_i-\underline{g}_i}{\Delta_i}\right\}\right] \tag{5}$$

$$g_i \le (AX)_i \le g_i \qquad i=1,\ldots,m$$

and

$$\max_{x\ge 0}\lambda=\left[\min_i\left\{\frac{\overline{g}_i-(AX)_i}{\Delta_i}\right\}\right] \tag{6}$$

$$g_i \le (AX)_i \le \overline{g}_i \qquad i=1,\ldots,m$$

3. Integrated Methodology and Application

3.1 *Tyre Recovery*

According to Sharma et al., incinerators with energy recovery systems are technically and economically more advantageous compared to the tyre pulverization or the pyrolysis process [11]. Amari et al. examine the technologies used to recover the ELT for energy or as rubber in USA and they affirmed that reusing tyres offers the best strategy [12]. In their study Corti and Lombardi compare different processes for the end life treatment of exhausted tyres, using life cycle assessment [13]. Mergias et al. used PROMETHEE to select the best compromise scheme for End-of-Life Vehicles management [14].

3.1 *Goal Weighting with Fuzzy AHP*

In a previous study, ELT recovery method selection has been made considering different conflicting quantitative and qualitative perspectives as environmental protection, energy saving, user friendliness, application risks and costs (Figure 1). The selection has been made using the FAHP. The three ELT recovery methods existing in Turkey are Mechanical Pulverization (A1), Waste to Energy Process in a cement factory (A2) and Pyrolysis (A3). Expert opinions have been evaluated through FAHP and the results indicate that Waste to Energy process in cement factories is the best-suited ELT recovery method in Turkey (A2: 0.4403). The other alternatives have approximately the same weight (A1: 0.2658 and A3: 0.2939).

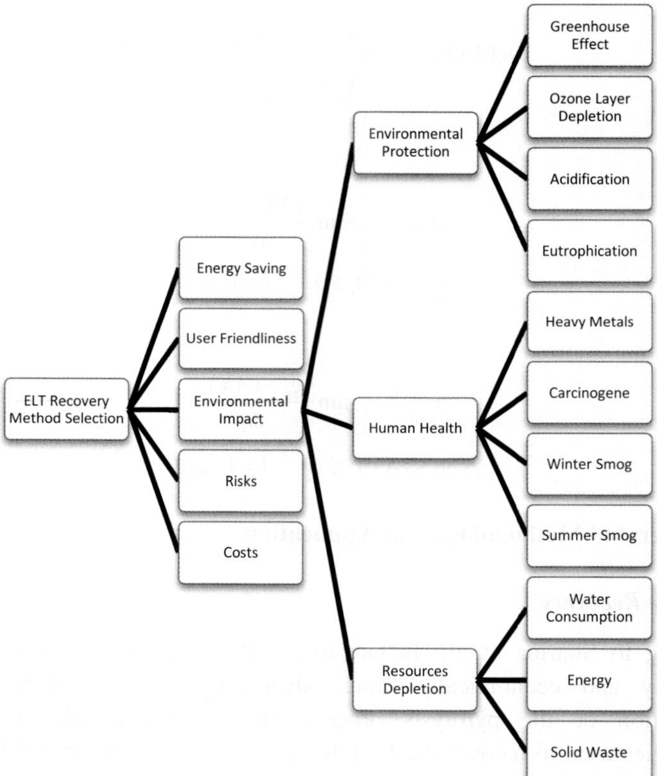

Figure 1. AHP Hierarchy.

The second step of this work consists of making a distinction between these two alternatives. Therefore, we evaluate them according to quantitative and qualitative data through fuzzy goal programming method and find out the percentages of ELT that must be used for each of them. Thereby the decision variables are:

X1 : Percentage of ELT that will be recovered by the pyrolysis method.

X2 : Percentage of ELT that will be recovered by the mechanical pulverization method.

13 criteria from AHP hierarchy are selected as goals because of their relatively important weights. Each goal has a different weight specified by using fuzzy AHP and the table below includes weights of each goal in the same priority level (Table 1).

Table 1. Weights of the Goals.

Goals	Weights
G1 Greenhouse effect	0.0096
G2 Ozone layer depletion	0.0096
G3 Acidification	0.0096
G4 Eutrophication	0.0096
G5 Heavy metals	0.0689
G6 Carcinogens	0.1366
G7 Water depletion	0.0011
G8 Energy using	0.0025
G9 Solid waste	0.0003
G10 Energy saving	0.2479
G11 User friendliness	0.1051
G12 Risks	0.1512
G13 Costs	0.2479

There are 2^{13} sub-problems to solve. In order to decrease the number of sub-problems, the goals having weights less than 5% are eliminated from the model. The resulting model is solved by LINDO.

4. Conclusion

In this study we proposed an application of a fuzzy goal programming approach with different importance and priorities. The proposed model attempts to find out the percentage of ELT that must be used in different recovery techniques. The model is solved by using LINDO computer package and yields an efficient compromise solution with the multiple fuzzy goal values. Computational methodology developed in this work can easily be extended to any other situations. Future researchers can adopt other techniques for exploring the relative weights among multiple objective functions to make it better suited to practical applications.

References

1. L. A. Zadeh, Information and Control, **8**, 338, (1965)
2. E.L. Hannan, Decision Sciences, **12**, 522, (1981)
3. H. Leberling, Fuzzy Sets and Systems, **6**, 105, (1981)
4. M.K. Luhandjula, Fuzzy Sets and Systems, **8**, 245, (1982)

5. P.A. Rubin, and R. Narasimhan, Fuzzy Sets and Systems, **14**, 115, (1984)
6. R.N. Tiwari, S. Dharmar, and J.R. Rao, Fuzzy Sets and Systems, **24**, 27, (1987)
7. H.F. Wang, and C.C. Fu, Computers and Operations Research, **24**, 819, (1997)
8. L.H. Chen, and F.C. Tsai, European Journal of Operational Research, **133**, 548, (2001)
9. M. A. Yaghoobi and, M. Tamiz, European Journal of Operational Research, 1**77**, 1580, (2007)
10. H.K. Chen, Fuzzy Sets and Systems, **62**, 287, (1994)
11. V.K. Sharma, M. Mincarini, F. Fortuna, F. Cognini and G. Cornacchia, Energy Convers., **39**, 511, (1997)
12. T. Amari, N.J. Themelis, and I.K. Wernick, Resources Policy, **25**, 179, (1999)
13. A. Corti and A., L. Lombardi, Energy, **29**, 2089, (2004)
14. I. Mergias, K. Moustakas, K., A. Papadopoulos and M. Loizidou, Journal of Hazardous Materials, **147**, 706, (2007)

PRIORITIZING THE STAKEHOLDER REQUIREMENTS IN QFDE WITH FAHP*

ILKE BEREKETLI†

Industrial Engineering Department, GalatasarayUniversity, Ciragan Cad., No:36, Besiktas, Istanbul, Turkey

MUJDE EROL GENEVOIS

Industrial Engineering Department, Galatasaray University, Ciragan Cad., No:36, Besiktas, Istanbul, Turkey

Concerning the environmental parameters in product development is crucial. However the studies which focus on environmental product developmentput mostly aside cost and quality parameters. In this study, a multi aspect QFD for Environment (QFDE) methodis proposed. In order to define the priorities of the stakeholders' requirements, Fuzzy Analytic Hierarchy Process (FAHP) Extent Analysis technique is used. The method is applied for the product "hand blender".

1. Introduction

Increasing the quality and reduction of costs were the main drivers of production management since the last century.In the last decades, due to significant environmental problems, an interest on environmental parameters in product development started growing very fast. However most of the studies in sustainable production area focus only on the environment and put aside the cost and quality parameters.

In this study, a QFDE method will be proposedin order to reflect not only the end users', but also the other stakeholders' (government, recyclers, suppliers, environment, etc.) requirements for an environmental product. To find the weights of the stakeholder requirements (SR) in QFDE, FAHP technique will be used as a prioritization technique.

---

* This work is financially supported by Galatasaray University Research Fund.

† ibereketli@gsu.edu.tr

2. Methodology

QFD for Environment (QFDE) is a method developed by Masui *et al.*[1] by incorporating environmental aspectsinto QFD to handle the environmental and traditional product quality requirements together. Unlike traditional QFD, not only the end users but also B2B customers, recyclers, the government and the environment itself are considered to be customers.In this multi aspect method, SR weights play an important role, hence it's crucial to give a realistic approach.

There are some other studies which combines decision making methods with QFD in order to weight SR or Engineering Metrics (EM) [2-6]. With some given criteria, SR and/or EM are prioritized as alternatives. According to Zadeh [7], it is very difficult for conventional quantification to define the complex situations, so the notion of a linguistic variable, whose values are words or sentences, is necessary.

In this paper, to assess the relative importance of the criteria and to evaluate the alternatives,the FAHP approach is introduced, with the use of Triangular Fuzzy Numbers for pairwise comparison scale of FAHP according to the method of Chang's [8] fuzzy extent analysis by applying correct normalization formula given later by Wang *et al.* [9]. The steps of Chang's extent analysis can be given as follows[9].First, by fuzzy arithmetic operations, take the sum of each row of the fuzzycomparison matrix.

$$RS_i = \sum_{j=1}^{n} \tilde{a}_{ij} = \left(\sum_{j=1}^{n} l_{ij},\ \sum_{j=1}^{n} m_{ij},\ \sum_{j=1}^{n} u_{ij} \right),\quad i = 1,\ldots,n \tag{1}$$

Using fuzzy synthetic extent analysis, the value of fuzzy synthetic extent with respect to the ith object x_i, $i = 1,2,\ldots,n$ that represents the overall performance of the object across all goals can be determined by the following normalization formula given by Wang *et al.* [9]:

$$\tilde{S}_i = \frac{RS_i}{\sum_{j=1}^{n} RS_j} = \left(\frac{\sum_{j=1}^{n} l_{ij}}{\sum_{j=1}^{n} l_{ij} + \sum_{\substack{k=1 \\ k\neq i}}^{n} \sum_{j=1}^{n} u_{kj}},\ \frac{\sum_{j=1}^{n} m_{ij}}{\sum_{k=1}^{n} \sum_{j=1}^{n} m_{ij}},\ \frac{\sum_{j=1}^{n} u_{ij}}{\sum_{j=1}^{n} u_{ij} + \sum_{\substack{k=1 \\ k\neq i}}^{n} \sum_{j=1}^{n} l_{kj}} \right),\quad i = 1,\ldots,n \tag{2}$$

Compute the degree of possibility of $\tilde{S}_i \geq \tilde{S}_j$ by the following equation where $\tilde{S}_i = (l_i, m_i, u_i)$ and $\tilde{S}_j = (l_j, m_j, u_j)$:

$$V(\tilde{S}_i \geq \tilde{S}_j) = \begin{cases} 1, & \text{if} \quad m_i \geq m_j, \\ \dfrac{u_i - l_j}{(u_i - m_i) + (m_j - l_j)}, & \text{if} \quad l_j \leq u_i, \quad i,j = 1,\ldots,n; i \neq j \\ 0, & \text{otherwise}, \end{cases} \tag{3}$$

To compare $\tilde{S}_i = (l_i, m_i, u_i)$ and $\tilde{S}_j = (l_j, m_j, u_j)$, we need both the values of $V(\tilde{S}_i \geq \tilde{S}_j)$ and $V(\tilde{S}_j \geq \tilde{S}_i)$. The degree of possibility for a convex fuzzy number $\tilde{S}$ to be greater than k convex fuzzy numbers $\tilde{S}_i$, i=1,…,k can be defined by:

$$V(M \geq M_1, \ldots, M_k) = V\left[(M \geq M_1) \wedge \ldots \wedge (M \geq M_k)\right] = \min V(M \geq M_i) \text{ for } i = 1,2,\ldots,k \quad (4)$$

Assume that for the alternative A_i,

$$d'(A_i) = \min V(S_i \geq S_j), \text{ for } j = 1,2,\ldots,n;\ j \neq i \quad (5)$$

Then the weight vector is given by,

$$W' = \left(d'(A_1), d'(A_2), \ldots, d'(A_n)\right)^T \quad (6)$$

Via normalization, the normalized weight vectors are,

$$W = \left(d(A_1), d(A_2), \ldots, d(A_n)\right)^T \quad (7)$$

3. Case Study: Measuring the environmental priorities of a Hand blender with Fuzzy AHP

A hand blender is a kitchen appliance to blend ingredients or puree food in the container in which they are being prepared. It will be used as a case study subject in this paper.The list of stakeholder requirements and product and environmental engineering metrics are as follows:

Environmental stakeholder requirements: Cheap (SR1), Easy to use (SR2), Energy saving (SR3), Durable (SR4), Lightweight (SR5), Easy to maintain (SR6), Easy to repair (SR7), Quiet (SR8), Ergonomic hold (SR9), Reliable (SR10), Aesthetics appearance (SR11), Easy to clean (SR12), Safe to use (SR13), Easy to reuse (SR14), Easy to recycle (SR15), Easy to disassemble (SR16), Free of hazardous substances (SR17), Less material type usage (SR18),Environmentally safe (SR19), andLess transportation (SR20).

Product and Environmental Engineering Metrics (EM): Weight, Volume, Lifetime, Functionality, Materials used, Production technology, Production waste, Packaging, Usability, Reparability, Maintenance, Transportation, Fasteners and joints, Time for disassembly, Rate of reusability, Rate of recyclability, Problematic materials, Energy Consumption, Emissions in use, Supply parts environmental performance, Waste.

Cost, quality and environmental concerns are accepted as criteria to weight the SR, which are the alternatives (A_i) of the prioritization. The hierarchy of this decision problem is given in Figure 1.

The triangle fuzzy conversion scale of the linguistic values in the weighting set used by decision makers is shown in Table 1.

Table 1: The triangular fuzzy conversion scale

Equally important	(1,1,2)	Equally to moderately	(1,2,3)
Moderately more important	(2,3,4)	Moderately to strongly	(3,4,5)
Strongly more important	(4,5,6)	Strongly to very strongly	(5,6,7)
Very strongly more important	(6,7,8)	Very strongly to extremely	(7,8,9)
Extremely more important	(8,9,9)		

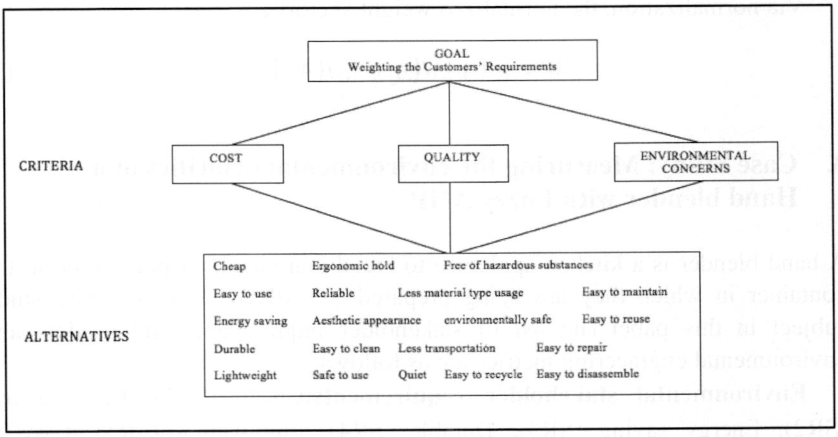

Figure 1Hierarchy of the weighting problem

The pairwise comparison matrices are formed by three decision makers. From these comparison matrices, using FAHP method, the following results presented in Table 2 and Table 3 are obtained. Considering cost, quality, and environmental concerns, a "*Cheap*" (SR1) hand blender is still the most important feature.

Table 2 Criteria weights

Criteria	**C1**	**C2**	**C3**
Weights	0.338	0.126	0.536

House of Quality is prepared with SR weightsand the relationships between SR and EM given by decision makers. The results are given on Figure 2.

Table 3 Stakeholder requirements weights as alternatives

Alternatives	A1	A2	A3	A4	A5	A6	A7	A8	A9	A10
Weights	0.126	0.100	0.110	0.071	0.026	0.027	0.029	0.029	0.044	0.018
Alternatives	**A11**	**A12**	**A13**	**A14**	**A15**	**A16**	**A17**	**A18**	**A19**	**A20**
Weights	0.008	0.040	0.027	0.100	0.089	0.034	0.018	0.040	0.017	0.108

			Weight	Volume	Supply parts env. Performance	Lifetime	Functionality	Materials used	Production technology	Production waste	Packaging	Transportation	Usability	Energy Consumption	Waste (generated)	noise and vibrations	Emissions in use	Maintenance	Reparability	Fasteners and joints	Time for dissasembly	Rate of reusability	Rate of recyclability
		unit	kg	m³		h					kg	km		kWh		dB					h	%	%
		direction of improvement	↓	↓	↑	↑	↑	↓		↓	↓	↓	↑	↓	↓	↓	↓	↓	↑		↓	↑	↑
	SR	w	1	2	3	4	5	6	7	8	9	10	11	12	13	14	15	16	17	18	19	20	21
Cheap	1	12,61	3	1				3	9		1	3		1								1	1
Easy to use	2	3,879518	9	9			9	1					9				1						
Energy saving	3	10,96					3		3					9		1							
Durable	4	7,08	1		3	9							9			3		3	3				
Lightweight	5	2,60	9					9			3		9										
Easy to maintain	6	2,72				3												9		1			
Easy to repair	7	2,91			3	3													9	3	3	1	
Quiet	8	2,91														9							
Ergonomic hold	9	4,36	3	9				1					9										
Reliable	10	1,84				3												9	9				
Visually attractive-Aesth.	11	0,75		3				9			1												
Easy to clean	12	4,01						3										3		1			
Safe to use	13	2,73					1						9				9						
Easy to reuse	14	10,04			3	3									9						3	9	
Easy to recycle	15	8,94						9							3						3		9
Easy to dissassemble	16	3,45			3										9					9	9	9	
Free of hazardous substances	17	1,78	1	1	3			9	1	3	9	1			9								9
Less material type usage	18	3,96	1					9			3				1								1
Environmentally safe	19	1,69			3				9				1		9		9						
Less transportation	20	10,78						1	1			9											
		w-importance in %	0,0556592	0,0414294	0,0368767	0,0530234	0,0321639	0,1054315	0,0794319	0,0024405	0,0223849	0,0623230	0,0855356	0,0507390	0,0836477	0,0266454	0,0199025	0,0338733	0,0291899	0,0211946	0,0440968	0,0624281	0,0515827
		R w	122,04	90,84	80,86	116,26	70,52	231,17	174,16	5,35	49,08	136,65	187,55	111,25	183,41	58,42	43,64	74,27	64,00	46,47	96,69	136,88	113,10

Figure 2: House of quality (HoQ)

As it is seen on HoQ, "*Materials used*" is found as the most important EM to fulfill the multi aspect stakeholder requirements.

4. Conclusion

In this study, a QFDE approach for the environmental product development of a hand blender is conducted. Weighting the requirements of stakeholders having different perspectives for a product is crucial. In order to realize this important step in QFDE, FAHP Extent Analysis technique is used by determining SR as alternatives and by setting cost, quality, and environmental concerns as criteria of the decision problem. *Cheapness* is found as the SR with highest weight. Via the relationship levels given by decision makers, house of quality is prepared and "*Materials used*" is found as the most important EM. Hence it can be recommended to the producers that they should first focus on "*Materials used*" in order to improve their product to fulfill the multi aspect stakeholder requirements. Considering future work, a full 4-phased QFDE can be completed and environmental improvement strategies can be proposed according to the results.

References

1. Masui K., T. Sakao, M. Kobayashi, A. Inaba, *International Journal of Quality & Reliability Management*, **20(1)**, 90 (2003).
2. D.K. Pal, B. Ravı, L.S. Bhargava. *International Journal of Computer Integrated Manufacturing*, **20(4),** 338 (2007).
3. A. Andronikidis, A.C. Georgiou, K. Gotzamani, K. Kamvysi. *Total Quality Management Journal*, **21(4),** 319 (2009).
4. M. Çelik, S. Cebi, C. Kahraman, I.D. Er, *Expert Systems with Applications*, **36(3)**, 6227 (2009).
5. A.H.I. Lee, H. Kang, C. Yang, C. Lin. *International Journal of Production Research*, **47**, 1 (2009).
6. Y.H. Lin, H.P. Cheng, M.L. Tseng, J.C.C. Tsai. *Expert Systems with Applications*, **37(3)**, 2186 (2010).
7. Lotfi A. Zadeh. *Information and Control*, **8**, 338 (1965).
8. D. Y. Chang, *European Journal of Operational Research,* **95**, 649 (1996).
9. Y.-M. Wang, Y. Luo, Z. Hua, *European Journal of Operational Research*, **186**, 735 (2008).

FUZZY MULTI ATTRIBUTE VALUATION OF SOFTWARE PROJECTS IN BANKING VIA FUZZY REAL OPTIONS USING BALANCED SCORECARD*

OZLEM AFACAN†

Industrial Engineering, IBTECH, 34357 Ortaköy, İstanbul, Turkey

A.CAGRI TOLGA

Industrial Engineering, Galatasaray University, 34357 Ortaköy, İstanbul, Turkey

This paper presents fuzzy multi attribute valuation of new software projects in banking industry by balanced score card based on the fuzzy TOPSIS model which includes fuzzy real option. Balanced scorecard method is used to determine the decision criteria by applying survey to managers. The economic side of the evaluation is dynamic and risky in nature in software development projects. Fuzzy real options method is used for surmounting these risky situations. Hence the evaluation of the software development projects has many attributes to investigate; fuzzy TOPSIS method based fuzzy real options method is offered in this paper. There are three software projects to be evaluated and investment ranking will be made according to the result of the method offered in this paper.

1. Introduction

Information technology in banking sector is getting bigger and bigger day by day and it becomes more competitive. Trade rivalry drives the companies to extreme measures. However in daily life many projects are generated in enterprises. Choosing the right project may preserve the companies from extreme measures decision. In this paper the company's work area is based on software and technology development for banking industry. New software and technology projects are very important in being powerful and to have great position in banking sector. One of greatest banks in Turkey requests to develop and implement one of the three new software projects by real options integrated multi criteria decision making method. New offered technique is used to decide decision criteria and make selection among these projects. Hence financial data

* This work is financially supported by Galatasaray University scientific research project funds.

† ozlemafacan@gmail.com

has strong importance in banking; the authorities of the company do not let the data out the corporation. In addition while the experts' ideas used in balanced scorecard and TOPSIS methods are vague in nature fuzzy logic is preferred in this paper. Balanced scorecard method is used for determining the criteria according to the top management's vision and then we apply these criteria to fuzzy TOPSIS with fuzzy real options method.

First project is "Our Bank" project which aims to gain new or old customers whose ages are between twenty and forty. It is new special internet banking system for only special customers including almost every operation which can be done without going to bank branch. Second project is called "New Faced Internet Banking" everybody who is customer of bank or not can use this site. Old one is changing with new one which is using new technology and getting user friendly. The last project is called "Information Technology Qualified User Creating and Canceling Software" which aims to make information technology user creation and deletion process easy and reportable. It will be useful to give exact reports to audit department from one and main point.

Multi-criteria decision making mentions to select the best opinion from all of the probable options in the existence of multiple, conflicting criteria. TOPSIS is one of the well known outranking methods for multiple-criteria decision-making and can be easily used for ranking alternatives [1]. In fuzzy TOPSIS model the subjective and objective criteria are simultaneously considered in the real life [2]. Using the fuzzy set theory, decision makers have more descriptive power to describe uncertain and flexible project information [3]. Also the real options methodology helps to use real options to structure investments in different ways that increase the value of projects [4]. Strong evidence is found that practitioners indeed follow the logic of real options in managing the risk of their IT investments, though largely based on intuition [5]. A strategic planning study such as balanced scorecard is very useful from vision to action [6]. The balanced scorecard allows companies to evaluate whether they are meeting their objectives, based on both financial and non-financial measures using tangible and intangible assets [7].

The rest of the study is arranged as follows: The second section includes the balanced scorecard method for software project selection. In Section 3, selected criteria are represented. The preceding section contains fuzzy TOPSIS model. The fuzzy real option value is explained in Section 5. Steps of offered model are represented in Section 6. The last part of study is the conclusion section.

2. Balanced Scorecard in Software Selection

The balanced scorecard method is a technique companies use to translate their strategies into objectives and measures [8]. Scorecard allows companies to evaluate whether they are meeting their objectives, based on both financial and non-financial measures using tangible and intangible assets. The balanced scorecard aim measures organizational performance according to four balanced perspective: financial perspective, customer perspective, internal business perspective and learning and growth perspective. Each of the four scorecards are interconnected and related to the company's strategic plan which is represented in Figure 1.

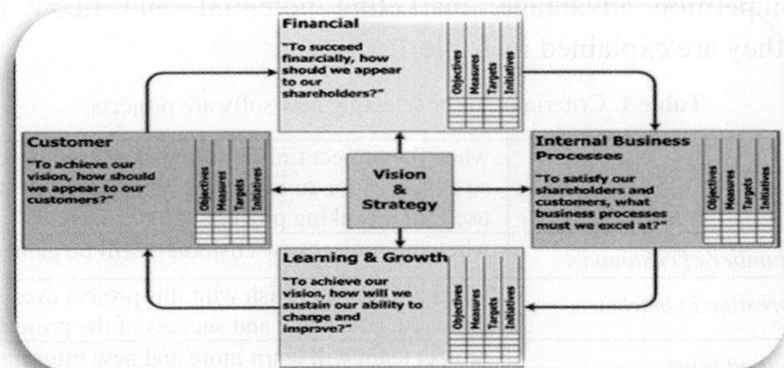

Figure 1. The structure of balanced scorecard developed by Kaplan and Norton [8].

Financial Perspective: financial objectives serve as the focus for the objectives and measures of the other three perspectives. Every measure should be part of a cause-and-effect relationship culminating in long-term sustainable financial performance.

Customer Perspective: financial success is closely linked to customer satisfaction. Satisfied customers mean repeat business, referrals and new business and thereby contribute to the financial results of the company.

Internal Operations Perspective: customer satisfaction is directly achieved through the operational activities of the company. The objectives and measures for this perspective thus enable a company to focus maintaining and improving the performance of processes that deliver the established objectives those are keys to satisfying customers, which in turn satisfy shareholders.

Learning and Growth Perspective: the ability, flexibility and motivation of staff support all of the financial results, customer satisfaction and operational activities measured in the other three quadrants of balanced scorecard. The balanced scorecard shows how the overall strategic objectives are translated into the performance drivers that the company has identified as critical success factors.

3. Criteria Used in Selection of Software Projects

Balanced scorecard method is used to determine the criteria. For that reason some questions are asked to managers, employees, customers like below: **How can customers be satisfied? *How can we operate more efficiently? *How we develop the abilities we need to; execute our strategies? *How can we satisfy our stakeholder*?

After responds to these questions nine criteria are specified to make selection among new software projects. These criteria are as follows: decreasing expenses, increasing number of customers, increasing prestige in market, technological additive, operation time income, compliance with the current system, competition advantage, marketing potential, and fuzzy real option values and they are explained in Table 1.

Table 1. Criteria used for selecting new software projects.

Decreasing expenses	when the project finishes number of employee decrease so expenses for an employee will gain or electric that is used for a banking process will decrease
Increasing number of customers	when the project new customers will be gained
Increasing prestige in market	when the project finish whit this project everybody will know the advantage and success of the project
Technological additive	project team will learn more and new information about the new technology and knowledge will increase
Operation time income	time that is used to consume while doing operation in bank will decrease with new project
Compliance with the current system	new project must be adaptable and work with the current system without any problem any extra development
Competitive advantage	being more competitive in market in other competitors who are using new technology
Marketing potential	can new project sell to other banks
Economic utility	Fuzzy real options values

4. Fuzzy TOPSIS

A fuzzy number is characterized by a given interval of real numbers, each with a grade of membership between 0 and 1 as in Eq. 1. It is possible to use different fuzzy numbers according to the situation. Generally in practice triangular and trapezoidal fuzzy numbers are being used. It is often simple to work with triangular fuzzy numbers (TFNs). The Fuzzy TOPSIS formulae used in this study are taken from Tolga's [3] work (for further investigation see).

$$\mu(x/\tilde{A}) = \begin{cases} 0, & x < a_1, \\ (x - a_1)/(a_2 - a_1), & a_1 \leq x \leq a_2, \\ (a_3 - x)/(a_3 - a_2), & a_2 \leq x \leq a_3, \\ 0, & x > a_3 \end{cases} \tag{1}$$

Let the fuzzy numbers $\tilde{A} = (a_1, a_2, a_3)$ and $\tilde{B} = (b_1, b_2, b_3)$ be triangular, then the basic operations are given in Eq.s (2-4) [1].

$$\tilde{A} \otimes \tilde{B} = (a_1 \times b_1, a_2 \times b_2, a_3 \times b_3) \quad (2)$$

$$\tilde{A} \oplus \tilde{B} = (a_1 + b_1, a_2 + b_2, a_3 + b_3) \quad (3)$$

$$(\tilde{A})^{-1} \approx (1/a_3,\ 1/a_2,\ 1/a_1) \quad (4)$$

5. Fuzzy Real Options Value

Projects usually have private risks, which usually cannot be predicted without extra knowledge. Using soft-computing methods, experts can use their experience to make subjective estimation of private risk.

Fuzzy systems theory is suitable for presentment of expert ability. Fuzzy real options are purposed to contend with risks that are hard to predict with an open mind. The fuzzy also think of uncertainty and require less data so it is preferred to provide pleasurable outcomes.

$$E(\tilde{A}) = \frac{a_2+a_3}{2} + \frac{a_1+a_4-a_2-a_3}{6} \quad (5)$$

$$\sigma^2(\tilde{A}) = \frac{(a_3-a_2)^2}{4} + \frac{(a_3-a_2)(a_2+a_4-a_1-a_3)}{6} \frac{(a_2+a_4-a_1-a_3)^2}{24} \quad (6)$$

With these formulae given above, which are used in Tolga et. al.'s study [9], fuzzy real option value (FROV) can be calculated as below:

$$\text{FROV} = \tilde{S}_0 e^{-T} N(d_1) - \tilde{X} e^{-rT} N(d_2) \quad (7)$$

$$d_1 = \frac{\ln(E(\tilde{S}_0)/E(\tilde{X})) + (r+\sigma^2/2)T}{\sigma\sqrt{T}} \quad (8)$$

here $d_2 = d_1 - \sigma\sqrt{T}$, and where $\tilde{S}_0$ denotes the possible values of the present value (PV) of expected cash flows (CF), in a similar manner $\tilde{X}$ quantifies the possible values of investment cost, $E(\tilde{S}_0)$ is the possibilistic mean value of the PV of expected CF, $E(\tilde{X})$ stands for the possibilistic mean value of expected costs, $\sigma(\tilde{S}_0)$ denotes the possibilistic variance of the PV of the expected CF.

6. Steps of the Offered Model

Step 1: Produce criteria by balanced score card; use it creating initial TOPSIS matrix.

Step 2: Construct $\tilde{I}_{MA}$ matrix according to the data received from respondents.

Step 3: Construct $\tilde{I}_{SA}$ matrix and if there are sub-sub-attributes, then also construct $\tilde{I}_{SSA}$ matrix according to the data received from the respondents.
Step 4: For fuzzy real options value, calculate $E(\tilde{S}_0), E(\tilde{X})$ and $\sigma(\tilde{S}_0)$ values for each alternative separately via Eqs. (5) and (6).
Step 5: Calculate d_1 *and* d_2 values respectively by Eq. (8).
Step 6: Find $FROV$ for each alternative by Eq. (7) which is used by [3].
Step 7: Construct $\tilde{I}_A$ matrix according to the data received from respondents and values calculated at Step5.
Step 8: Include $\tilde{I}_A$ matrix to fuzzy TOPSIS algorithm described in Tolga's work [3]. Then apply the algorithm of fuzzy TOPSIS.
Step 9: Rank the alternatives in descending order according to the C_i index found from fuzzy TOPSIS and implement the best alternative.

7. Conclusion

In today's competitive banking sector it is important to have new project before nobody else has done or to do the right project on right time. Using the fuzzy set theory, decision makers have more descriptive power to describe uncertain project information. In this paper a fuzzy ROV integrated TOPSIS model by balanced scorecard is used to select new projects in banking sector. FROV can improve valuation of risky projects. The recommended model can help decision makers to decide about new project which using four perspectives to improve company's strategy. For further research new criteria can be used like innovativeness to obtain more detail.

References

1. S. Ballı and S. Korukoğlu, MCA. **Vol. 14, No. 2,** pp. 119-130, (2009).
2. J. Ding , JMTS. **Vol. 19, No. 4,** pp. 341-352, (2011).
3. A. Ç. Tolga, *JIFS.* **Vol. 19,** pp. 359–371, (2008).
4. T. Petravicius, TIBE. **Vol. 8, No. 3, (18),** (2011).
5. M. Benaroch, Y. Lichtenstein, K. Robinson, MIS **Vol. 30 No. 4,** pp. 827-864, (2006).
6. U. Cebeci, ESA, **36,** pp. 8900–8909, (2009).
7. M. J. Schniederjans, J. L. Hamaker, A. M. Schniederjans, Inf. Tech. Inv. (Dec.-Making Meth.), WSPC. (2004).
8. R. S. Kaplan, D. P. Norton, Using Balanced Scorecard as a Str. Mngt. Sys., HRWBS Press (1996).
9. A. Ç. Tolga, C. Kahraman, M. L. Demircan, JMLVSC. **Vol. I6.,** pp. 135-154, (2009).

A COGNITIVE GROUP DECISION SUPPORT SYSTEM FOR PROJECTS EVALUATION

FAHIMEH RAMEZANI*, JIE LU
Decision Systems & e-Service Intelligence Lab
Centre for Quantum Computation & Intelligent Systems
School of Software, Faculty of Engineering and Information Technology
University of Technology, Sydney, P.O. Box 123, Broadway, NSW 2007, Australia

In any organization there are some main goals and lots of projects for achieving these goals. For any organization, it is important to determine how much these projects affect on achieving the main goals. This paper proposes a new fuzzy multiple attribute-based decision support system (DSS) for evaluating projects in promoting the goals as such a selection may involve both quantitative and qualitative assessment attributes. In addition the proposed DSS has ability to choose the most appropriate fuzzy ranking method for solving given MADM problem. Also it contains sensitivity analysis system which provides opportunity for analyzing the impacts of attributes' weights and project s' performance on achieving organizations' goals, and assess the reliability of the decision making process. The proposed DSS can be applied for solving every FMADM problem which needs to rank some alternatives according to some attributes.

1. Introduction

Project ranking before investment decision is a complex problem that is often faced by the decision makers involved in the planning process. The necessity to take into account several decision parameters apart from purely economic ones, such as socio-political, technical, institutional and environmental, lead to the use of multi-criteria methods. Moreover, most of the times such decisions are taken in a group environment [1].

Sanna *et al*. [2] proposed the creation of a hierarchy among the different projects based on ratings expressed as fuzzy numbers. Henriksen and Palocsay [3] proposed an excel-base, user-friendly decision support system for evaluating and ranking R&D projects. Kumar Dey [4] proposed a decision support system, which analyses projects with respect to market, technicalities, and social and environmental impact in an integrated framework using analytic hierarchy process (AHP), a multiple-attribute decision-making (MADM) technique.

* Fahimeh.Ramezani@student.uts.edu.au.

Chiang and Che [5] also applied the fuzzy AHP and fuzzy data envelopment analysis (DEA) to develop an evaluation and ranking methodology to select high added value NPD projects. Marques *et al.* [6] proposed a new multi-dimensional Project Performance Measurement System that would enable managers to deal with the volume of data. Baykasoglu *et al.* [7] proposed a practical fuzzy rating and ranking approach to prioritize project activities with fuzzy attributes.

However there are many researches in this field, none of them applied the fuzzy ranking methods which are mentioned in this paper, and in previous works, the proposed methodology is applicable for a particular MADM problems and they did not assess the reliability of the final decision. This research proposes a new DSS which is applicable to solve every FMADM problem which needs to rank some alternatives according to some attributes, and have ability to select the most appropriate fuzzy ranking algorithm for solving a given FMADM problem. In addition, it contains sensitivity analysis system which provides opportunity to assesses the reliability of the decision making process.

2. The Main Fuzzy Ranking Algorithms

Two fuzzy ranking algorithms: (1) SAW and (2) Chen and Hwang fuzzy ranking algorithms [8], are chosen to solve FMADM problem for valuating projects. These algorithms can be applied for solving both MADM and FMADM problems with both crisp and fuzzy numbers. Moreover, despite some algorithms like AHP, these two algorithms give DMs an opportunity to assess attributes individually.

3. Conditions for Choosing Appropriate Fuzzy Ranking Algorithm

In our previous work [9], we proposed a new method for choosing the most appropriate fuzzy ranking algorithm for solving a given MADM problem, based on the type and the number of attributes and the number of alternatives, considering the least computation and time consumption for ranking alternatives. Assume *m* be the number of alternatives and *n* the number of attributes for a MADM problem.

Let $J = \{j | n_j$ is a benefit attribute$\}$ and $J' = \{j | n_j$ is a cost attribute$\}$
If (m>1 and n>1) then
  If ($J=\emptyset$ or $J'=\emptyset$) then
    SAW algorithm is appropriate
  Else
    Chen and Hwang algorithm is appropriate

4. Sensitivity Analysis

In this section (1) sensitivity analysis on the attributes' weight and (2) sensitivity analysis on the projects' performance, will be described.

4.1. *Sensitivity Analysis on the Attributes' Weight*

In this paper for sensitivity analysis on the attributes' weight, the method which is proposed in [10] is applied. Assume in the MADM problem the weights of attributes are normalized, that is:

$$\sum_{j=1}^{n} \mathrm{w_j} = 1 \tag{1}$$

With this assumption, if the weight of one attribute changes, then the weight of other attributes will change accordingly. If in the MADM problem the weight of attribute pth changes as Δ_p , then the weight of other attributes change as:

$$\Delta_j = \frac{\Delta_k * w_j}{w_k - 1} \quad j = 1, 2, \ldots, n, j \neq k \tag{2}$$

In the MADM model of SAW, if the weight of attribute pth changes as Δ_p, then the final score of alternative ith would change as δ_i that is:

$$\delta_i = \Delta_k * n_{ik} + \sum_{\substack{j=1 \\ j\neq k}}^{n} \frac{\Delta_k * w_j}{w_k - 1} * n_{ij} \quad i = 1, 2, \ldots, m \tag{3}$$

In addition in the MADM model of TOPSIS if the weights of attributes change, the new alternatives' score can be calculated. As the results, in the MADM model, if the weight of attribute pth changes as Δ_p, then the final score of alternative ith would change as:

$$cl'^{+}_{i} = \frac{d'^{+}_{i}}{d'^{+}_{i} + d'^{-}_{i}} \tag{4}$$

where: $\gamma = \frac{1 - w_k - \Delta_k}{1 - w_k}$,

$$d'^{+}_{i} = \{\gamma^2 * {d^{+}_{i}}^2 + (1 - \gamma^2)(v_{ik} - v^{+}_{k})^2 + \Delta_k^2 (n_{ik} - n_{lk})^2 + 2\Delta_k (v_{ik} - v^{+}_{k})(n_{ik} - n_{lk})\}^{1/2},$$

$$d'^{-}_{i} = \{\gamma^2 * {d^{-}_{i}}^2 + (1 - \gamma^2)(v_{ik} - v^{-}_{k})^2 + \Delta_k^2 (n_{ik} - n_{l'k})^2 + 2\Delta_k (v_{ik} - v^{-}_{k})(n_{ik} - n_{l'k})\}^{1/2},$$

$$l = \begin{cases} arg_{\max_i} v_{ik} \ \textit{if (k is a benefit attribute)} \\ arg_{\min_i} v_{ik} \ \textit{if (k is a Cost attribute)} \end{cases}, \quad l' = \begin{cases} arg_{\min_i} v_{ik} \ \textit{if (k is a benefit attribute)} \\ arg_{\max_i} v_{ik} \ \textit{if (k is a Cost attribute)} \end{cases}$$

4.2. *Sensitivity Analysis on the Projects' Performance*

In this paper for sensitivity analysis on the projects' performance following equation is proposed:

$$A_k = \sum_{j=1}^{n} W_j * \left(\sum_{i=1}^{m} \gamma_{ij} * Y_i\right) \quad (5)$$

where γ_{ij} is the score of *i*th project in *j*th attribute. Y_i is the project's performance percentage and W_j is the weight of *j*th attribute. According to equation (5), if the *i*th project's performance percentage changes, we can determine how projects' performance affects on achieving *k*th goal.

5. Decision Making System Simulation

To simulate decision making process for assessing projects, all decision making features are considered and a DSS conceptual model is designd as a use case diagram (Figure 1).

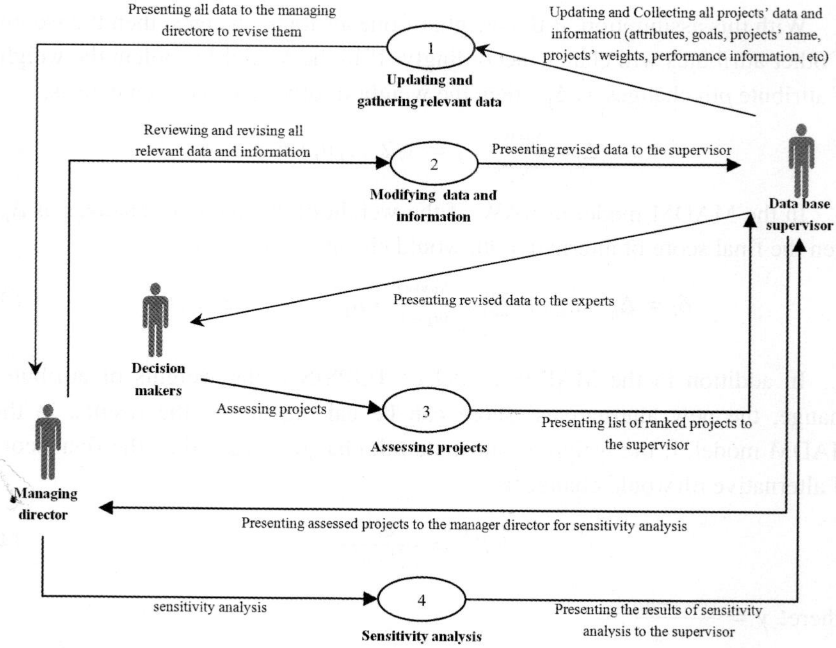

Figure 1. The conceptual model (use case diagram)

There are three types of users for this system: Managing director, Experts (DMs) and Data base supervisor. They follow four fundamental actions:

1. Updating and gathering relevant data: The initial procedure for gathering projects' data and information includes the following steps:
   - Determining organizations' goals
   - Set of criteria (attributes): definition criteria related to every goal

- Set of projects: classifying projects considering the goals
- Gadering projects' performance data via contorol project system

2. Modifying data and information
3. Assessing projects for each criterion
4. Sensitivity analysis

6. A New Project Evaluation Algorithm

In this paper we developed a new project evaluation algorithm on the basis of proposed DSS model for project evaluation, which is describes as follow:

Step 1: Collecting all data includes: goals, attributes, alternatives (projects' name), attributes' weights, projects' performance information, etc.

Step 2: Modifying all projects' data and information.

Step 3: Assessing all projects considering relevant attributes to determine the score of projects' efficiency in achieving the organization goals.

Step 4: Choosing the most appropriate fuzzy ranking algorithm based on the conditions which are explained in Section 3.

Step 5: Ranking all projects by chosen algorithm using group's final decision results: in this step calculated weight and alternatives scores in every attribute, are applied by selected fuzzy ranking algorithm as *inputs* and produce the list of ranked projects as *outputs*.

Step 6: Final assessment: evaluation the projects' importance and their efficiency by (1) sensitivity analysis on the attributes' weight and (2) sensitivity analysis on projects' performance.

Step 7: Choosing the list of the most efficient and important projects for achieving the organization goals. Considering this list, manager director can reschedule projects' time table, focus on important projects and manage investments to enhance these projects' performance.

This algorithm facilitates decision making process and helps managers to choose the most efficient projects in achieving organization goals. Using this algorithm, managers will be able to steer organization resources towards the most inportant projects.

7. Conclusion

In regard to the importance of evaluating and choose the most effective projects in achieving the organizations' goals, which is a FMADM problem with both qualitative and quantitative attributes, this paper has proposed a new DSS for evaluating projects which has ability to choose the most appropriate fuzzy ranking method for solving the given problem. In addition, concerning the importance of determining the weight of all attributes and the score of alternatives under every attribute, which are obtained through the decision makers' judgement, the sensitivity analysis system is developed. Via this sensitivity analysis system, the reliability of the decision making process and the impacts of attributes' weights and projects' performance on achieving organizations' goals could be assessed. The proposed model simplifies and accelerates the evaluating process. Applying this model not only helps organizations to choose the optimal projects for sustainable development, but also helps them to assess the reliability of the decision making process.

References

1. Y. Goletsis, J. Psarras and J.E. Samouilidis, *Project Ranking in the Armenian Energy Sector Using a Multicriteria Method for Groups.* Annals of Operations Research, 120 (2003) 135-157
2. U. Sanna, C. Atzeni and N. Spanu, *A fuzzy number ranking in project selection for cultural heritage sites.* Journal of Cultural Heritage, 9 (2008) 311-316
3. A.D. Henriksen and S.W. Palocsay, *An excel-based decision support system for scoring and ranking proposed R&D projects.* International Journal of Information Technology & Decision Making, 7(3) (2008) 529-546
4. P. Kumar Dey, *Integrated project evaluation and selection using multiple-attribute decision-making technique.* International Journal of Production Economics, 103 (2006) 90-103
5. T.A. Chiang and Z.H. Che, *A fuzzy robust evaluation model for selecting and ranking NPD projects using Bayesian belief network and weight-restricted DEA.* Expert Systems with Applications, 37 (2010) 7408–7418
6. G. Marques, D. Gourc and M. Lauras, *Multi-criteria performance analysis for decision making in project management.* International Journal of Project Management, 29 (2010) 1057-1069
7. A. Baykasoglu, T. Goc Ken and V. Kaplanoglu, *A Practical Approach to Prioritize Project Activities Through Fuzzy Ranking.* Cybernetics and Systems. An International Journal, 42 (2011)165-179
8. S. J. Chen and C. L. Hwang, *Fuzzy multiple attribute decision making.* Springer-Verlag Berlin (1992)
9. F. Ramezani and J. Lu, *A new approach for choosing the most appropriate fuzzy ranking algorithm for solving MADM problems.* Proceedings of the PhD Seminar Autonomous Systems (PSAS), Springer-Verlag, Spain, Mallorca (2011)
10. A. Memariania, A. Aminib and A. Alinezhadc, *Sensitivity Analysis of Simple Additive Weighting Method (SAW): The Results of Change in the Weight of One Attribute on the Final Ranking of Alternatives.* Journal of Industrial Engineering, 4 (2009) 13-18

FUZZY FORECASTING OF INTEREST RATES IN INVESTMENT DECISIONS

İREM UÇAL SARI[†] AND CENGIZ KAHRAMAN

Industrial Engineering Department ,Istanbul Technical University, Maçka, İstanbul, 34367, Turkey

Interest rate is one of the most used tools to determine the time value of money in investment projects. In this paper fuzzy regression analysis method is used to determine the relation among interest rate, inflation and the increases on export and import. The interest rate is formulated by using past data of interest rates, inflation, export and import sizes in Turkey. An illustrative example is given.

1. Introduction

Forecasting investment parameters is very important since investment decisions are directly connected with the success of a company. In a competitive market new investments could make the company distinctive and as a result of that market share of the company is increased. Especially when a large-scale investment is considered the importance of forecasting parameters rises.

Another important issue in forecasting is to define relations among parameters due to their effects on predictions. In many problems, two or more variables are related to each other and it is of interest to model and explore this relationship. Regression is a solution for this problem. The relationship between these variables is characterized by a mathematical model called a regression model [4].

Humans are insufficient in making predictions quantitatively. Mostly predictions include linguistic variables such as; approximately, probably, likely…etc. Fuzzy logic which is proposed by Zadeh (1965) enables us to determine linguistic variables in terms of numerical values by using membership functions [7].

In forecasting, if insufficient data are available, fuzzy regression should be preferred. There are lots of examples of using fuzzy regression to determine the relation between variables in the literature. One of the most used fuzzy

[†] ucal@itu.edu.tr

regression method is proposed by Tanaka in 1987. He proposed a fuzzy regression analysis method for non-fuzzy inputs with symmetric fuzzy coefficients [5]. Ishibuchi and Nii (2001) expanded their model with asymmetric fuzzy coefficients and they proposed a linear programming for fuzzy inputs [3].

The aim of our study is to determine the relationship among interest rate, increase on export size, increase on import size and inflation. To achieve this aim, a fuzzy linear model of interest rate is formulated. Fuzzy regression analysis using triangular asymmetric coefficients is used to determine the relation.

The rest of the paper is organized as follows: In Section 2, fuzzy regression methodology is determined. In Section 3, forecasting fuzzy interest rate by fuzzy regression is developed. In Section 4, forecasting of interest rate with the proposed linear fuzzy model is applied on a project's evaluation. The paper is concluded in the last section.

2. Fuzzy Regression

The linear fuzzy model is given in Eq.1 where fuzzy coefficients are triangular fuzzy numbers (TFN) $\tilde{A}_i = (a_i^L, a_i^c, a_i^U)$.

$$\tilde{Y} = \tilde{F}(x) = \tilde{A}_0 + \tilde{A}_1 x_1 + \cdots + \tilde{A}_n x_n \quad (1)$$

The center and spreads of linear fuzzy model are determined in Eq.s 2-4 where $f^c(x)$ denotes its center, $f^L(x)$ denotes its lower bound and $f^U(x)$ denotes its upper bound [3]:

$$f^L(x) = a_0^L + \sum_{\substack{i=1 \\ x_i \geq 0}}^{n} a_i^L . x_i + \sum_{\substack{i=1 \\ x_i < 0}}^{n} a_i^U . x_i \quad (2)$$

$$f^C(x) = a_0^C + \sum_{i=1}^{n} a_i^C . x_i \quad (3)$$

$$f^U(x) = a_0^U + \sum_{\substack{i=1 \\ x_i \geq 0}}^{n} a_i^U . x_i + \sum_{\substack{i=1 \\ x_i < 0}}^{n} a_i^L . x_i \quad (4)$$

An α-cut of $F(x) = (f^L(x), f^C(x), f^U(x)$ is calculated as the interval which is given in Eq.5.

$$[F(x)]^\alpha = [\alpha f^C(x) + (1-\alpha) f^L(x), \alpha f^C(x) - (1-\alpha) f^U(x)] \quad (5)$$

Ishibuchi and Nii (2001) proposed a methodology to determine triangular fuzzy coefficients of the linear programming problem. There are 2 steps in the methodology [3]:

Step 1. Determine the center of the linear fuzzy regression model $F(x)$ by the least squares regression. In this step, the center of each triangular fuzzy coefficient a_i^c is determined.

Step 2. Determine the lower limit a_i^L and upper limit a_i^U of $F(x)$ by solving the following linear programming problem:

$$\text{Minimize } z = \sum_{p=1}^{m}\{f^U(x_p) - f^L(x_p)\} \tag{6}$$

Subject to $\alpha f^C(x_p) + (1-\alpha)\cdot f^L(x_p) \le y_p \le \alpha f^C(x_p) + (1-\alpha)\cdot f^U(x_p),$

$$p = 1,2,\dots,m \tag{7}$$

$$a_i^L \le a_i^C \le a_i^U,\ \ i = 0,1,\dots,n \tag{8}$$

3. Forecasting Investment Parameters

Discounted cash flows (DCF) analysis methods are the most preferred methods to evaluate investments. The main parameters of DCF methods are cash flows, interest rate and life of the investment project. There are several methods to forecast cash flows of an investment in the literature. In this section forecasting interest rate using the fuzzy set theory is examined. A fuzzy linear model for interest rate is formulated.

Time value of money is an important concept in financial management. It is one of the important tools used in project evaluations to compare various investment alternatives.

Exports and imports also affect the financial instruments. Past data of interest rates, inflation, export and import sizes in Turkey are examined and it is observed that there are synchronous movements of interest rate and the incremental ratio of export and import sizes. The first variable of fuzzy linear model is chosen as the ratio of increase on export and increase on import.

It is well known that there is a relation between interest rate and inflation. Interest rates are used by central banks to control inflation, so the second variable of fuzzy linear model is chosen as inflation in this paper.

Scatter plots are used to determine the relation among export, import, inflation and interest rate. There is linearity among interest rate, inflation and the ratio of increases on export and import sizes. This linearity is shown in Figure 1.

Past data of interest rate, inflation and increase on export over increase on import ratio in Turkey are given in Table 1 [1,6].

Table 1. Past Data

Year	Annual average interest rate	Increase on export/Increase on import	Inflation
2011	14.19	3.81	6.45
2010	14.99	2.63	8.58
2009	17.2	4.90	6.28
2008	22.92	7.76	10.43
2007	22.26	7.13	8.78
2006	21.47	5.29	9.59
2005	19.88	5.36	8.19
2004	23.62	5.64	10.66

The visual representation of the relation between variables is shown in Figure 1.

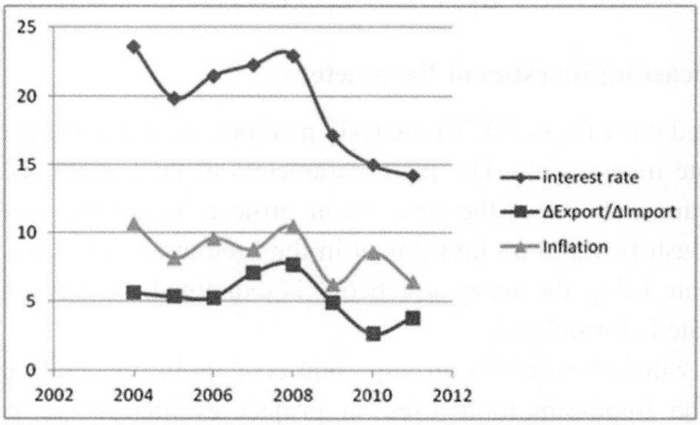

Figure 1. Relation between interest rate, increase on export/increase on import ratio and inflation.

Interest rate is denoted as $\tilde{\mathrm{i}}$ and formulated in Eq. 9 where x_1 denotes increase on export over increase on import ratio and x_2 is inflation.

$$\tilde{\mathrm{i}} = \tilde{\mathrm{F}}(\mathrm{x}) = \tilde{\mathrm{A}}_0 + \tilde{\mathrm{A}}_1\mathrm{x}_1 + \tilde{\mathrm{A}}_2\mathrm{x}_2 \qquad (9)$$

Step 1. The center of the linear fuzzy regression model $F(x)$ is determined by the least squares regression. Centers of the coefficients are determined as $a_0^c = 0.087$, $a_1^c = 1.345$ and $a_2^c = 1.430$.

$$\mathrm{f}^{\mathrm{C}}(\mathrm{x}) = 0.087 + 1.345\mathrm{x}_1 + 1.430\mathrm{x}_2$$

Step 2. α is determined as 0.5 and the lower limit a_i^L and upper limit a_i^U of $F(x)$ are found by solving the linear program using excel-solver.

$a_0^L = a_0^U = 0.087$, $a_1^L = 1.335$ $a_1^U = 1.345$, $a_2^L = 0.221$ and $a_2^U = 1.430$.

$$\tilde{\mathrm{Y}} = \tilde{\mathrm{F}}(\mathrm{x}) = (0.087, 0.087, 0.087) + (1.335, 1.345, 1.345)\mathrm{x}_1 + (0.221, 1.430, 1.430)\mathrm{x}_2$$

Table 2. Forecasted interest rates

Year	Forecasted interest rate		
	Y^l	Y^c	Y^u
2011	6.60	14.44	14.44
2010	5.50	15.90	15.90
2009	8.01	15.66	15.66
2008	12.75	25.44	25.44
2007	11.55	22.24	22.24
2006	9.27	20.92	20.92
2005	9.05	19.01	19.01
2004	9.98	22.92	22.92

Chiu and Park used the formula which is given in Eq. 9 to defuzzify fuzzy number $\widetilde{A}_i = (A_i^L, A_i^C, A_i^U)$. A_i^P denotes the preference value of $\widetilde{A}_i$, where A_i^L is the lowest possible value of $\widetilde{A}_i$, A_i^C is the most promising value (the mean) of $\widetilde{A}_i$, A_i^U is the highest possible value of $\widetilde{A}_i$ and w denotes preference weight determined by the magnitude of the most promising value. If the magnitude of the most promising value is important, a larger preference weight such as w = 0.3 is recommended otherwise smaller preference weight such as w = 0.1 is recommended [2].

$$A_i^P = \left(\frac{A_i^L + A_i^C + A_i^{uU}}{3}\right) + \mathrm{w}A_i^c \qquad (10)$$

Forecasted interest rates are defuzzified by using weighted average method with 0.3 preference weight. Error and error rates are also calculated and given in Table 3.

Table 3. Error calculations

Year	Annual average interest rate	Defuzzified forecasted interest rate	Error	Error/annual average interest rate
2011	14.19	14.26	0.07	0.005
2010	14.99	17.2	2.21	0.147
2009	17.2	17.81	0.61	0.035
2008	22.92	28.84	5.92	0.258
2007	22.26	25.35	3.09	0.139
2006	21.47	23.31	1.84	0.086
2005	19.88	21.39	1.51	0.076
2004	23.62	25.48	1.86	0.078
				Average 0.103

It is found that the formula found by fuzzy regression has 0.10 error rate.

4. An Illustration

In this section fuzzy net present value of an investment project which has annual payments of $\tilde{A} = (10000{,}14000{,}17000)\$$ for 5 years is calculated. If increase on export/increase on import ratio is assumed as 3.5% and inflation rate is assumed as 6.5% during this 5 years period, fuzzy interest rate is calculated as $\tilde{\imath} = (6.2,14.1,14.1)\%$ by the proposed fuzzy linear model. Fuzzy present value of a project which has annual payments is calculated by Eq. 11.

$$\widetilde{PV} = \tilde{A} * \frac{(1+\tilde{\imath})^n - 1}{(1+\tilde{\imath})^n \tilde{\imath}} \tag{11}$$

$$\widetilde{PV} = (10000,14000,17000) * \frac{(1.062,1.141,1.141)^5 - 1}{(1.062,1.141,1.141)^5 (0.062,0.141,0.141)}$$

$$\widetilde{PV} = (21811,81266,241113)\$$$

Fuzzy present value of a project is calculated as $(21811,81266,241113)\$$. This means that in the worst case the project has a present value of 21811$, in the best case it has a present value of 241113$ and the most promising present value of the project is 81266$.

5. Conclusion

Interest rate is an important parameter on investment evaluation. In this paper, we proposed an interest rate formula which depends on inflation, increase on exports and increase on imports. A fuzzy regression model with fuzzy asymmetric coefficients is used to determine the relation between parameters.

As a further research, it is suggested to use fuzzy regression analysis in forecasting fuzzy cash flows of a project. Grey theory is another alternative to forecast investment parameters.

References

1. Central Bank Of The Republic Of Turkey, www.tcmb.gov.tr
2. Chiu, C. Y. & Park, C. S., 1994, Fuzzy cash flow analysis using present worth criterion, *The Engineering Economist*, Volume 39, Issue 2, Pages 113–138.
3. Ishibuchi, H. and Nii, M., 2001, Fuzzy regression using asymmetric fuzzy coefficients and fuzzified neural networks, *Fuzzy Sets and Systems*, Volume 119, Issue 2, Pages 273-290.
4. Montgomery, D. C., 2009, Statistical Quality Control –A modern Introduction, 6th Edition, John Wiley&Sons (Asia) Pte. Ltd. , Asia, Page 150.
5. Tanaka H. ,1987, Fuzzy data analysis by possibilistic linear models, *Fuzzy Sets and Systems,* Volume 24, Issue 3, Pages 363-375.
6. Turkish Statistical Institute, www.tuik.gov.tr
7. Zadeh, L .A., 1965, Fuzzy sets, *Information and Control*,Volume 8, Pages 338-353.

PERFORMANCE EVALUATION OF TURKISH RETAIL FIRMS USING THE FUZZY AHP, PROMETHEE, ELECTRE AND VIKOR METHODS

UFUK BÖLÜKBAŞ[1]†, ERKAN ÇELİK[2], ALİ FUAT GÜNERİ[1]

[1]*Department of Industrial Engineering, Yildiz Technical University, Beşiktaş, Istanbul, 34349, Turkey*

[2]*Department of Industrial Engineering, Tunceli University, Tunceli, 62000, Turkey*

The performance evaluation of firms determines capabilities of firms to compete in the sector and has a critical importance for the development of the sector. In this study, a fuzzy multi-criteria decision model is proposed to evaluate the performances of retail firms. The eight Turkish retail firms are analyzed and firms are evaluated in terms of ten different financial ratios in six separate groups. Fuzzy Analytic Hierarchy Process (FAHP), Preference Ranking Organization Method for Enrichment Evaluation (PROMETHEE), Elimination Et Choix Traduisant la Realité (ELECTRE) and Multicriteria Optimization and Compromise Solution (VIKOR) methods are integrated in the proposed model. A hierarchical model of financial performance evaluation is structured by the decision makers. FAHP is used to determine the weights of the criteria. These criteria weights and financial ratio values are input to the PROMETHEE, ELECTRE and VIKOR method to rank the firms. The rankings of methods are compared to each other's in the conclusion.

1. Introduction

Retail sector is one of the leading sectors in Turkey. This is dynamic trade area as an energy, education and health sector. Especially, Istanbul, Izmir, Ankara and Bursa are growing rapidly so then big retail firms are widespread in these cities in Turkey. Eight retail firms are evaluated based on multi criteria decision making (MCDM) methods.

In this study, the importance weight of the criteria is determined with FAHP and ranking of the firms is calculated with PROMETHEE, ELECTRE and VIKOR methods. Financial ratios of firms are computed from Istanbul Stock Exchange (ISE) database for three years. Financial measures of firms have basically been used to evaluate the company's financial situation and performance [1].

†e-mail: bolukbas@yildiz.edu.tr

According to the three expert' advice, ten financial ratios are selected, representing mainly liquidity, leverage, solvency and profitability ratios as a measure of ranking the firms. This study is aimed to introduce an approach for evaluating the financial performance of retail firms.

In the literature, there are many model applications about these three sorting methods. The PROMETHEE method is used for evaluating financial performance [2]. Financial performances of Turkish manufacturing firms and banking sector are evaluated by Yalcin et al.[3,4].

2. Financial Performance Indicators

In this study, decision makers determined to use ten different financial indicators in six separate groups representing liquidity, operation activity, capital intensity, leverage, solvency, profitability and equity productivity. These performance measures are important for evaluation in the retail sector. The criteria are explained in Table 1.

Table 1.The criteria set for the evaluation of firms

Criterion	Definition	Direction	Perspective
Current Ratio	Current assets / Current liabilities [6]	Max	Liquidity
Currency Ratio	Cash and cash equivalents / Current	Max	Liquidity
Quick Ratio	Curent assets-Inventory / Current	Max	Liquidity
Accounts	Net sales / Average receivables	Max	Operation Activity
Inventory Turnover	Net sales income /Average inventory	Max	Operation Activity
Assets Turnover	Net sales / Current liabilities	Max	Capital intensity
Debt Ratio	Total liabilities / Equity capital	Min	Financial leverage
Earnings per Share	Net income / Paid in capital	Max	Profitability
Return on Equity	Net income / Equity	Max	Profitability
Equity Capital	Net sales / Equity capital	Max	Equity productivity

Values of ten ratios explained briefly below are derived from the balance sheets and detailed income statements of the firms from consecutive three years, 2006, 2007 and 2008.

2.1. *Determining the Evaluation Criteria Weights*

In this study the Buckley's method is used for FAHP to fuzzify hierarchical analysis by allowing fuzzy numbers for the pairwise comparisons and find the fuzzy weights (Table 2). We summarize concepts and calculation of fuzzy hierarchical evaluation. Buckley extended Saaty's AHP to the case where the decision makers are allowed to employ fuzzy ratios in place of exact ratios to handle the difficulty for people to assign exact ratios when comparing two criteria and derive the fuzzy weights of criteria by geometric mean method [7].

Table.2 Membership function of linguistic scale

Fuzzy Numbers	Linguistic scales	Scale of Fuzzy Number
1	Equally important	(1, 1, 3)
3	Weakly important	(1, 3, 5)
5	Essentially important	(3, 5, 7)
7	Very strongly important	(5, 7, 9)
9	Absolutely important	(7, 9, 9)
1/*x*	Between two adjacent judgments	(1/*u*, 1/*m*, 1/*l*)

The procedure for determining the evaluation criteria weights by FAHP can be summarized below. Pairwise comparison matrices among all the criteria are constructed. Linguistic terms to the pairwise comparisons are assigned by asking which is the more important of each two criteria. Fuzzy numbers are applied in the accordance comparison matrix of criteria in terms of linguistic values of Table 2. In this study, geometric mean method is not used since decision makers used accordance solution for criteria comparisons. Synthetic pairwise comparison matrix is constructed. Fuzzy weights of criteria are obtained for the consistency of the fuzzy pair-wise comparison matrices. In this study, we utilized the center of area method owing to its calculation easiness [8, 9]. The weights of criteria will be obtained as follows. Fuzzy weights of criteria used in the methods for sorting firms are obtained in Table.3

Table.3 Determining the weights of criteria in Fuzzy AHP approach

Criteria	BNP w(i)	Fuzzy Weights	Normalized weights
Current Ratio	$BNPw_1$	0.033	% 2,36 (9)
Currency Ratio	$BNPw_2$	0,032	% 2,27 (10)
Quick Ratio	$BNPw_3$	0,054	% 3,83 (8)
Accounts Receivable Turnover	$BNPw_4$	0,299	% 21,14 (2)
Inventory Turnover	$BNPw_5$	0,324	% 22,92 (1)
Assets Turnover	$BNPw_6$	0,071	% 5,02 (7)
Debt Ratio	$BNPw_7$	0,105	% 7,44 (6)
Earnings per Share	$BNPw_8$	0,212	% 15 (3)
Return on Equity Capital	$BNPw_9$	0,173	% 12,24 (4)
Equity Capital Turnover	$BNPw_{10}$	0,110	% 7,78 (5)

3. Application Methods

This study aims to evaluate Turkish retail firms by using MCDM methods. The weight of the criteria is obtained by using FAHP, firstly. PROMETHEE, ELECTRE and VIKOR methods determined the ranking of the firms.

3.1. *PROMETHEE Method*

PROMETHEE is an outranking method which initial references are prepared by [10,11,12]. The evaluation of alternatives with PROMETHEE is required information on relative importance and decision makers' preference function

[13,14]. In the first step a generalized preference function for each criterion is specified. The preference index is a measure for the intensity of preference of the decision maker for an alternative in comparison with an alternative for the simultaneous consideration of all criteria. In this paper, the Gaussian criterion is determined as preference function for each criterion. Then the leaving flow and the entering flow are calculated. A partial preorder is determined in PROMETHEE I [10]. The complete ranking can be obtained via PROMETHE II. For the complete ranking calculations, net flow values of alternatives are calculated. The bigger net flow values show the more suitable alternative.

3.2. *ELECTRE Method*

The ELECTRE method was first introduced [15]. The basic concept of the ELECTRE method is to deal with "outranking relations" by using pair wise comparisons among alternatives under each one of the criteria separately. A decision matrix A_{ij} containing the alternatives and criteria is defined. Normalized decision matrix is calculated and normalized X_{ij} matrix is obtained. And then weighted decision matrix Y_{ij} is calculated which weighted of criteria is defined. Set of concordance and discordance index is defined which based on comparing the alternatives. C_{kl} is sum of weighted which is contained concordance matrix. Average value of concordance and discordance is calculated. Alternatives are ranked with respect to net concordance index.

3.3. *VIKOR Method*

The VIKOR method is introduced by Yu [16] and Zeleny [17]. This method is based on an aggregating function representing closeness to the reference point(s). The VIKOR method is introduced an aggregating function, representing the distance from the ideal solution by [3]. The worst and best value of all each criterion is determined in the first step of the VIKOR. S_j, R_j and Q_j are computed. The acceptable stability in decision making is analyzed.

4. Ranking the Alternatives

The assessment procedure utilizes FAHP method in combination with multi-criteria analyses methods, PROMETHEE, ELECTRE and VIKOR method. The results of three methods are demonstrated in table 4. Ranking of the first firm, BIMAS is same for three methods; best performance is related to BIMAS in the sector. MIPAZ is the last firm in the performance rating, nearly in 2007 and 2008 years.

Table 4. Rankings of the retail firms throughout the three years

	ELECTRE			PROMETHEE			VIKOR (v=0,5)		
	2006	2007	2008	2006	2007	2008	2006	2007	2008
BIMAS	1	1	1	1	1	1	1	1	1
SELEC	4	7	6	5	7	5	6	6	5
KIPA	2	5	4	2	4	4	2	5	4
CARFA	8	2	2	8	5	2	4	3	3
BOYNR	6	6	7	7	6	7	8	7	7
MIPAZ	7	8	8	6	8	8	7	8	8
VAKKO	5	3	5	4	3	6	5	4	6
SANKO	3	4	3	3	2	3	3	2	2

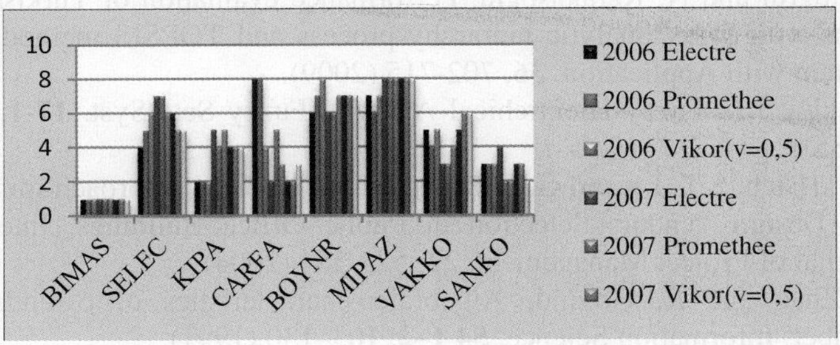

Fig.1. Comparison of the retail firms by the methods throughout the three year

5. Conclusion

A model composed of FAHP and there different MCDM techniques are proposed for performance evaluation of retail firms which compared there years data. FAHP method is utilized to determine the weights of the criteria of the performance evaluation. In the second part of the study, proposed methods are used in determining the ranking of the retail firms by comparing the firms according to the criteria determined. In the last part of the study, ranking of the firms are evaluated according to the three years.

In future studies other MCDM methods can be used to evaluate the performance of retail firms. Also various different kinds of MCDM methods under fuzzy environment can be used regarding performance evaluation which can consider other financial criteria.

References

1. E. Kızılkaya Aydoğan, Performance measurement model for Turkish aviation firms using the rough-AHP and TOPSIS methods under fuzzy environment Expert Systems with Applications. **38**, 3992–399 (2011).
2. N. Kalogeras, G. Baourakis, C. Zopounidis and G.V. Dijk, Evaluating the financial performance of agri-food firms: a multicriteria decision-aid approach. Journal of Food Engineering **70,** 365–37 (2005).

3. N. Yalcin A. Bayrakdaroglu and C. Kahraman, Application of fuzzy multi-criteria decision making methods for financial performance evaluation of Turkish manufacturing industries. Expert System with Application, **39**, 350-364 (2012).
4. N. Yalcin A. Bayrakdaroglu and C. Kahraman, Fuzzy performance evaluation in Turkish Banking Sector using Analytic Hierarchy Process and TOPSIS. Expert System with Application. **36**, 11699-11709 (2009).
5. H. Youn, Z .Gu, Predicting Korean lodging firm failures: An artificial neural network model along with a logistic regression model. International Journal of Hospitality Management, **29**, 120–127 (2010).
6. İ.Ertuğrul and N. Karakaşoğlu, Performance evaluation of Turkish cement firms with fuzzy analytic hierarchy process and TOPSIS methods. Expert System with Application. **36**, 702-715 (2009).
7. Buckley J.J., Fuzzy Hierarchical Analysis Fuzzy Sets Syst. **17-1**, 233–47, (1985).
8. T.Y. Hsieh, S.T. Lu and G.H. Tzeng, Fuzzy MCDM Approach for Planning and Design Tenders Selection in Public Office Buildings. International Journal of Project Management, **22**, 573-584 (2004).
9. R. Zhao and R. Govind, Algebraic characteristics of extended fuzzy number. Information Science, **54 1–2**, 103–130 (1991).
10. J.P. Brans, B. Mareschal and P.P. Vincke, PROMETHEE: A new family of outranking methods in MCDM. Operational Research'84. Amsterdam: North-Holland. 477–490 (1984).
11. J.P. Brans, , P. Vincle and B. Mareschal, How to select and how to rank projects: The PROMETHEE method. European Journal of Operational Research, **24**, 228–238 (1986).
12. J.P. Brans and P.Vincle, A preference ranking organization method. Management Science, **31**, 647–656 (1985).
13. G. Tuzkaya, B. Gulsun, C. Kahraman and D. Ozgen, An integrated fuzzy multi-criteria decision making methodology for material handling equipment selection problem and an application. Expert Systems with Applications. **37** 2853–2863 (2010).
14. M. Dağdeviren, Decision making in equipment selection: an integrated approach with AHP and PROMETHEE. Journal of Intelligent Manufacturing. **19**, 397–406 (2008).
15. R., Benayoun, B., Roy and B. Sussmann, ELECTRE:Une Methode Pour Guider le Choix en Presence de Points de Vue Multiples. Direction Scientifique, Paris,**49** (1966).
16. P.L. Yu, A class of solutions for group decision problems. Management Science **19-8** 936–946 (1973).
17. M. Zeleny, Multiple criteria decision making. McGraw-Hill, New York, (1982).

FUZZY AXIOMATIC DESIGN SIMULATION FOR EVALUATING THE CONCEPTUAL AIRCRAFT DESIGNS*

AHMET KANDAKOGLU†, CENGIZ KAHRAMAN, Y.ILKER TOPCU

Department of Industrial Engineering, Management Faculty, Istanbul Technical University, Macka, Istanbul, 34367, Turkey

Selecting the best conceptual aircraft design is a crucial problem during an aircraft development process. To find out a solution for this complex problem, there is only few information available with uncertainty as obtaining more precise measurements and accurate preference information is difficult or sometimes impossible. Therefore, this study proposes a realistic simulation approach based on Fuzzy Axiomatic Design (FAD) method and a novel Random Weight Generation (RWG) algorithm to select an aircraft design that meets the requirements the most. This approach gives an opportunity to the decision-makers to evaluate alternative designs under multiple qualitative and quantitative criteria by considering the operational requirements. Finally, an empirical study is presented to illustrate the applicability of the proposed approach.

1. Motivation

Aircraft design is a complex long-term process initiated by the operational requirements, followed by the three central phases as conceptual, preliminary and detailed design, and finally ended with the fabrication of the aircraft (Figure 1) [1, 2]. First, the customer provides a specific set of design requirements including parameters, constraints and targets such as range, payload, endurance, speed, acceleration, climb rate, turn rate for the design [1]. Then, the conceptual design phase addresses the basic questions of design arrangements, size, weight and performance about the proposed aircraft. In particular, the main requirements and desired functions are considered, and normally a number of potential designs are presented to the customer. The result of the conceptual phase is a design which best matches the requirements. In other words, to start the preliminary design phase, a decision has to be made to freeze the design configuration. At this point, selecting one candidate design for development becomes a critical decision making problem [3].

---

*The views expressed in this paper are those of the authors and do not reflect the official policy or position of the Turkish Air Force, Turkish General Staff or Department of Defense.

†kandakoglu@itu.edu.tr.

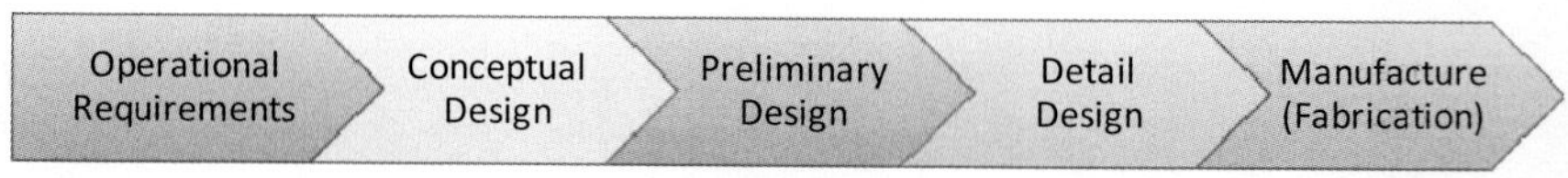

Figure 1. Simple schematic of the aircraft design process

This study asserts that selecting an aircraft design from a set of proposed designs at the end of the conceptual phase is a MCDM (Multi Criteria Decision Making) problem. However, both the requirements and the proposed values are uncertain, and preference information is imperfect. Hence, this study proposes a realistic simulation approach based on Fuzzy Axiomatic Design (FAD) method and a novel Random Weight Generation (RWG) algorithm to select an aircraft design under the constraints of operational requirements.

The rest of this paper is organized as follows: Section 2 overviews the fundamentals of the FAD and the RWG algorithm. In section 3, the developed FAD simulation is introduced. An illustrative case study is represented in section 4 and the final section addresses the conclusion and further studies.

2. Methodological Background

2.1. *Fuzzy Axiomatic Design (FAD)*

FAD is a recent decision making method developed by Kulak and Kahraman for handling multiple criteria problems including uncertain information [4]. Actually, it is a fuzzy extension of axiomatic design (AD) methodology which was first proposed by Suh in 1990 [5]. The main aim of AD is to establish a theoretical foundation based on logical and rational thought process and tools to improve the design activities. The AD method consists of independence and information axioms. The independence axiom states that the independence of functional requirements (*FRs*) must always be maintained, where *FRs* are defined as the minimum set of independent requirements that characterizes the design goals [5]. On the other hand, the information axiom states that among those designs that satisfy the independence axiom, the design that has the smallest information content is the best design [6].

In FAD method, the design and the system range for a certain criterion is mostly represented by a triangular fuzzy number (TFN). The common area is the intersection area of the design range and the system range TFNs (Figure 2). Finally, the information content is defined as follows:

$$I = \log_2\left(\frac{TFN\ of\ System\ Range}{Common\ Area}\right) \quad (1)$$

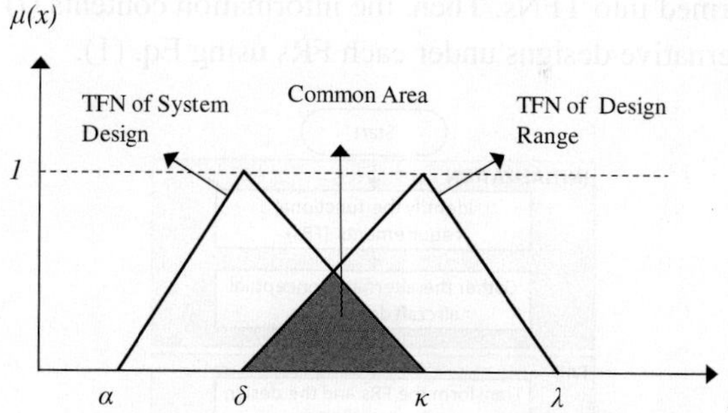

Figure 2. The common area of system and design ranges

2.2. *Random Weight Generation (RWG) Algorithm*

Determining the actual weights of different criteria is a subjective process in decision models as it is driven by the decision makers' (DMs) experiences [7]. Different weight elicitation methods yield different results [8]. Furthermore, it is extremely difficult to obtain preference information from the DMs. In some situations, the DMs may provide partial preference information: exact numerical values for some weights, ranking for a set of weights, and interval constraints for some weights. Several weight generation ways are proposed to handle these different types of information [7, 9]. However, they are not perfect due to their high computation complexity and/or acceptance-rejection filter. In this study, a novel random weight generation algorithm is developed to overcome this issue. It has a low complexity and produces uniform criteria weights as an output. The main steps of the algorithm are:

Step 1. Randomly select criterion C_i from the set$\{C_1, C_2,\ldots,C_n\}$.
Step 2. Determine the new lower (w_i^{min}) and upper (w_i^{max}) bounds of C_i.
Step 3. Randomly generate a number r from the uniform (0, 1) distribution.
Step 4. Set $w_i = w_i^{min} + r \cdot (w_i^{max} - w_i^{min})$.
Step 5. Repeat Step 1 through 4 n-1 times.
Step 6. The remaining weight is determined so that sum of all weights is one.

3. Proposed FAD Simulation

The proposed simulation based on FAD method and RWG algorithm is shown in Figure 3. In the initialization phase, the customer provides an Operational

Requirement Document (ORD) to the aviation company including a set of FRs for the aircraft. Then, the company develops different designs by considering these FRs. In order to apply the FAD method, the FRs and the proposed design data are transformed into TFNs. Then, the information contents (*I*) are calculated for different alternative designs under each FRs using Eq. (1).

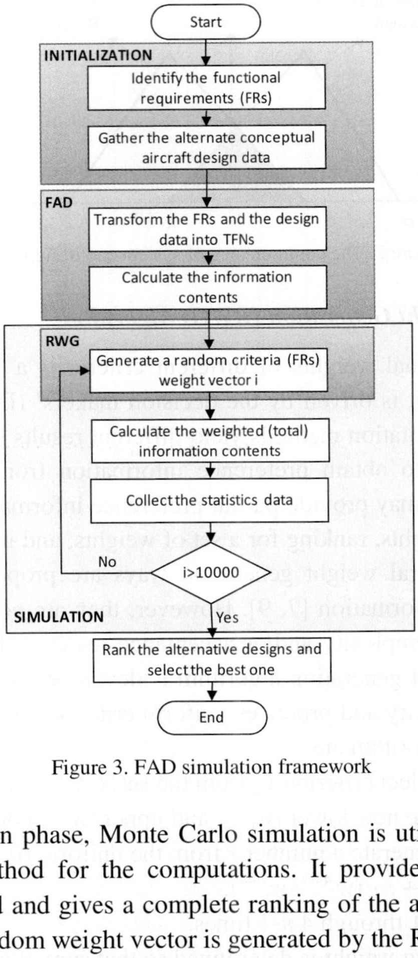

Figure 3. FAD simulation framework

In the simulation phase, Monte Carlo simulation is utilized as an easy and well-established method for the computations. It provides result that can be statistically analyzed and gives a complete ranking of the alternatives after each iteration. First, a random weight vector is generated by the RWG algorithm using the preference information for the FRs. Then, the total information content of the design *i* (TIC_i) can be expressed as

$$TIC_i = \sum_{j=1}^{n} I_{ij} w_j \qquad i = 1,2,..,n \tag{2}$$

where I_{ij} is the information content of design i with respect to FR_j , w_j is the importance weight of FR_i. The alternative which has the minimum TIC_i value is the best alternative or vice-versa. This process is conducted for 10000 iterations to produce accurate results. At each step the best, worst and weighted average ranks are collected as statistics. Finally, the best alternative design is selected according to these results.

4. Case Study: Fighter Aircraft Design Selection

A fighter aircraft is ordered to be designed to fulfill a military mission and meet the following requirements: FR_{CR}=Combat radius: at least 400 nm; FR_{MLC}= Mission load capacity: in the range of 17500 to 20000 lbs; FR_E=Endurance: over 2 hours at 30,000 ft; FR_{MS}=Maximum speed: at least 1.8 mach at 30,000 ft; FR_{AC}=Absolute ceiling: higher than 30,000 ft; FR_{TOR}=Take-off run: at most 2000 ft; FR_A=Acceleration: in the range of 11 to 15 sec from 0.3 mach to 0.9 mach; FR_{ROC}=Rate of climb: in the range of 1000 to 1200 ft/sec; FR_{SL}= Structural limit: in the range of +7 to +9 g; FR_{PC}=Production cost: at most 100 million TL; FR_S=Stealth: at least very good; FR_M=Maintainability: at least good

The aviation company proposed seven alternative designs at the end of the conceptual design phase (Table 1). As preference information, the decision makers only provided a complete ranking for the weights of the FRs as follows: $FR_{MS}>FR_{ROC}>FR_A>FR_S>FR_{CR}>FR_{PC}>FR_M>FR_E>FR_{SL}>FR_{MLC}>FR_{AC}>FR_{TOR}$

The FRs and the system data in Table 1 are transformed into the TFNs. Then, the information contents for all alternative designs under each FRs are calculated using the FAD method. The random weights are generated ensuring the above ranking of the FRs using the RWG algorithm and total information content is calculated using Eq. (2). The simulation results (best, worst and weighted average ranks of designs) for 10000 iterations are illustrated in Figure 4. According to these results, Design C with its 1.080 weighted average rank is selected as the best design that satisfies the operational requirements.

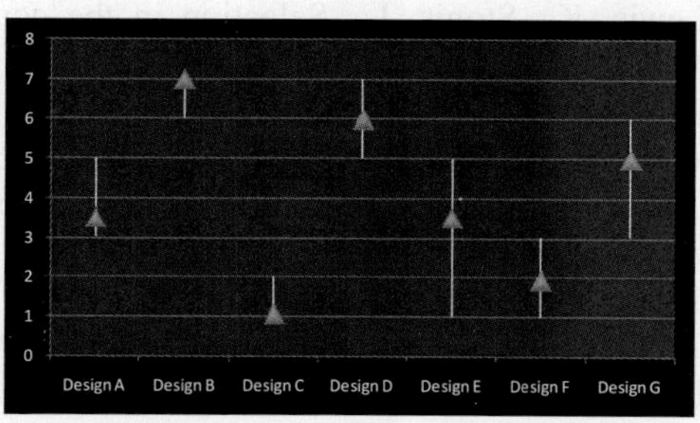

Figure 4. Simulation results (ranking results for the designs)

Table 1.Conceptual aircraft design information

	Conceptual aircraft design alternatives						
	Design A	Design B	Design C	Design D	Design E	Design F	Design G
FR_{CR}	500	450	650	600	450	600	600
FR_{MLC}	18000	18500	19000	20000	19000	18750	20000
FR_{E}	4	2,5	2,5	3	2,5	3	4
FR_{MS}	2,10	1,90	2,30	2,10	2,40	2,25	2,00
FR_{AC}	40000	45000	55000	48000	60000	50000	55000
FR_{TOR}	1500	1650	2000	1850	1800	1300	1900
FR_{A}	12	14,5	13	15	14	12	14
FR_{ROC}	1150	1050	1175	1030	1075	1100	1150
FR_{SL}	9	8	9	8	9	9	9
FR_{PC}	95	75	100	85	80	85	60
FR_{S}	very good	good	very good	good	excellent	very good	good
FR_{M}	good	very good	good	fair	good	very good	very good

5. Conclusion

In this study, a simulation approach based on FAD method and RWG algorithm is proposed for conceptual aircraft design selection problem. This is a well suited tool for this kind of decision making problems that includes incomplete and uncertain information. Furthermore, RWG algorithm can help the decision makers to perform high-dimensional robustness and sensitivity analysis, which has the flexibility to change all weights at the same time.

References

1. Raymer, D.P., Aircraft Design: A Conceptual Approach, AIAA Education Series, CA, 1992.
2. Price, M., Raghunathan, S., Curran R., (2006), *Progress in Aerospace Sciences*, 42, 331–376.
3. Mistree, F., Lewis, K., Stonis, L., Selection in the conceptual design of aircraft, AIAA-94-4382-CP.
4. Kulak, O., Kahraman, C., (2005), *IJPE*, 95, 415-424.
5. Suh, N.P., The Principles Of Design. Oxford University Press, 1990.
6. Suh, N.P., Axiomatic Design: Advances and Applications. Oxford University Press, New York, 2001.
7. Butler, J., Jia, J.,Dyer, J., (1997), *EJOR*, 103 531-546.
8. Borcherding, K., Eppel, T., von Winterfeldt, D., (1991), *Management Science*, 37 (12), 1603-1619.
9. Tervonen, T., Lahdelma, R., (2007), *EJOR*, 178, 500-513.

CRITERIA WEIGHING AND 4P'S PLANNING IN MARKETING USING A FUZZY METRIC DISTANCE AND AHP HYBRID METHOD*

TUNCAY GÜRBÜZ†, Y. ESRA ALBAYRAK, ELIF ALAYBEYOĞLU
Industrial Engineering Dept., Galatasaray University, Çırağan cad. No.36 34357 Ortaköy, İstanbul, Turkey

Production and consumption relationship shows that marketing plays an important role in enterprises. In the competitive market, it is very important to be able to sell rather than produce. Nowadays, marketing is customer-oriented and aims to meet the needs and expectations of customers to increase their satisfaction. While creating a marketing strategy, an enterprise must consider many factors. Which is why, the process can and should be considered as a multi-criteria decision making (MCDM) case. In this study, marketing strategies and marketing decisions in the new-product-development process has been analyzed in a macro level. To deal quantitatively with imprecision or uncertainty, fuzzy sets theory has been used throughout the analysis.

1. Introduction

Nowadays the development of technology and the supply which exceeds demand create difficult conditions in the market and competition. Marketing is the process of planning and executing the conception, pricing, promotion, and distribution of ideas, goods, and services to create exchanges that satisfy individual and organizational goals. Marketing Mix (MM) is probably the most famous marketing term. Its elements are the basic, tactical components of a marketing plan. MM does provide a handy framework for marketing analysis. The tactical section of a marketing plan summarizes how the business intends to use each element of the MM. Also known as the 4P's, the MM elements are price, place, product, and promotion.

The first P is the price of the product and decisions surrounding the overall pricing strategies of company. Pricing is the process of determining what a company will receive in exchange for its products. *Place* is also known as channel, distribution, or intermediary. It is the mechanism through which goods

* *This research has been financially supported by Galatasaray University Research Fund.*
† tgurbuz@gsu.edu.tr.

and/or services are moved from the manufacturer/ service provider to the user or consumer. *Product* is simply the tangible, physical entity that they may be buying or selling. This should define the characteristics of the product or service that will meet customers' needs of the target segment. *Promotion* includes all of the tools available to the marketer for marketing communication. This may include advertising, personal selling, sales promotions and public relations.

In this study, the criteria that are effective in marketing strategies and marketing decisions in the new-product-development process will be defined and analyzed at a macro level *i.e.* all the sub-criteria that can be effective in this decision process will not be included in this stage of the study. A novel hybrid MCDM model will be proposed to develop optimal marketing-mix plans weigh the criteria and to define the best strategic marketing scenario in dynamic competitive markets. Analytically, decision-makers are asked to express their opinions on comparative importance of various factors in linguistic terms. Since the crisp values are inadequate to model real-life situations where human perception is a factor, these linguistic variable scales, such as "high", "medium" and "low", are then converted into fuzzy numbers. As opposed to regular applications of fuzzy AHP, in this study there will not be a defuzzification step and final fuzzy weights for the alternatives of the model will be ranked using a fuzzy metric distance method. Section 2 of the study gives an overview of the proposed method.

We illustrated the application of this hybrid methodology by studying a new product of a company producing fast moving consumer goods in the food sector in the launch market.

2. Methodology

2.1. *Analytic Hierarchy Process - AHP*

AHP is a very well known and widely used MCDM method proposed by Saaty [2]. It is a theory of measurement that has been extensively applied in modeling the human judgment process. It decomposes a complex decision into a multi-level hierarchical structure enabling people to effectively combine both measurable and subjective factors in the decision process.

2.2. *Ranking Fuzzy Numbers with Fuzzy Metric Distance Method*

In this section, we present the method of Cheng [3] for ranking fuzzy numbers based on metric distance method. Cheng's method use the inverse functions of the left and the right membership functions to calculate the metric distance and this method can use in negative and positive ranking of fuzzy numbers. The distance method is based on calculating the centroid point, where the distance means the distance from original point to the centroid point $(\overline{x_0}, \overline{y_0})$, which is the geometric center of a fuzzy number $\tilde{A} = (l, m, u)$ defining its membership function as follows:

$$\mu_{\tilde{A}}(x) = \begin{cases} f_{\tilde{A}}^{L}(x) = \dfrac{x-l}{m-l} & l \le x \le m \\ f_{\tilde{A}}^{R}(x) = \dfrac{x-u}{m-u} & m \le x \le u \\ 0 & otherwise \end{cases} \quad (1)$$

With $g_{\tilde{A}}^{L}$ and $g_{\tilde{A}}^{R}$ the inverse functions of $f_{\tilde{A}}^{L}$ and $f_{\tilde{A}}^{L}$ respectively, the centroid point $(\overline{x_0}, \overline{y_0})$ for $\tilde{A}$ is defined as follows:

$$\overline{x}_0(\tilde{M}) = \frac{\int_l^m x f_{\tilde{M}}^{L}(x)dx + \int_m^u x f_{\tilde{M}}^{R}(x)dx}{\int_l^m f_{\tilde{M}}^{L}(x)dx + \int_m^u f_{\tilde{M}}^{R}(x)dx} \qquad \overline{y}_0(\tilde{M}) = \frac{\int_0^1 y g_{\tilde{M}}^{L}(y)dy + \int_0^1 y g_{\tilde{M}}^{R}(y)dy}{\int_0^1 g_{\tilde{M}}^{L}(y)dy + \int_0^1 g_{\tilde{M}}^{R}(y)dy} \quad (2)$$

Finally the ranking depends on the following rule: Greater the distance from the origin, greater the fuzzy number. In cases where the centroid point of two different fuzzy numbers are the same, then the tie breaking will be handled with the coefficient of variation method proposed by Cheng [3].

2.3. *Step by Step Procedure for Proposed Method*

A step by step procedure of the proposed method can be stated as follows:

- Step1: Pairwise comparisons of criteria with respect to goal.
- Step2: Pairwise comparisons of alternatives with respect to criteria.
- Step3: Finding the fuzzy eigenvectors of pairwise comparison matrices using the geometric mean technique [4-5].
- Step4: Calculate the final aggregated fuzzy scores for the alternatives.
- Step5: Rank the alternatives using the fuzzy metric distance method.

3. Proposed Model

When marketing their products firms need to create a successful mix of the right product, sold at the right price, in the right place and using the most suitable promotion. The objective of the model is to determine a marketing strategy for the company to launch the new product to the market and to weight the elements of the MM.

3.1. *The Alternatives of the Model*

Each P represents different strategies for the marketing of the new product and is vital for the success. It is a framework which helps to structure the approach to each market. The mix is a bundle of variables which are offered to the customer. These include the product or service itself (its advantages); its availability (the place where and when it is available, delivered or distributed); its image (the way it is promoted) and the price which should be charged. This model aims to give percentages to these strategies, to show their importance and to help the company to decide how to make investment and how to give importance to these strategies. Consequently, the 4P's are the four alternatives of the model [6].

3.2. *The Criteria of the Model*

In order to adequately evaluate the MM decision, we considered four critical criteria. These four major evaluation factors are company, market, consumer and product life cycle. *Company:* Organizational relationships, organizational resources and organizational culture may have a strong impact on marketing mix factors. Many businesses in markets have realized the importance of relationship building and have made this a priority in their business planning activities. *Consumer buying behavior:* Researchers in marketing have studied most areas of consumer behavior including the impact of everything on how people behave and how they consume products. *Market:* The choice of target market affects the mix. The mix must adapt itself to the market. *Product life cycle:* A new product progresses through a sequence of stages from introduction to growth, maturity and decline. This sequence is known as the product life cycle and is associated with changes in the marketing situation, thus impacting the marketing strategy and the marketing strategy (see Figure 1).

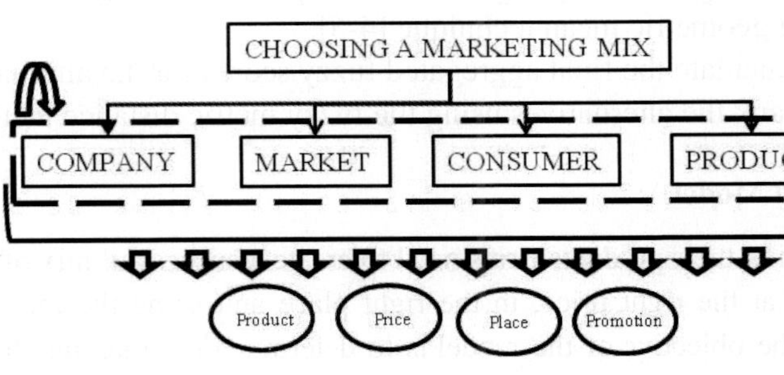

Figure 1. Proposed Model

4. Numerical Application

The application of the proposed model will be carried on the launch market for a new product of a company producing fast moving consumer goods in the food sector where it serves for 40 years and holds second place in the market share. The criteria and the alternatives of the model are weighed with the consensus decision given by a group of experts who are chosen by the company: two engineers from the production department, two managers from the marketing department and two customers of the company.

Performing the steps 1 to 4 explained previously, the fuzzy priority vectors of the criteria and the fuzzy scores of the alternatives with respect to each criterion are obtained. With step 5, the final aggregated fuzzy scores of the alternatives have been calculated. From those scores, percentages have been found. These values and the final aggregated fuzzy scores for the alternatives can be observed in Table 1.

Table 1. Fuzzy Weights and Final Calculations

CRITERIA	*COMPANY*			*CONSUMER*			*MARKET*			*PRODUCT*		
	0,13	0,23	0,38	0,21	0,36	0,58	0,16	0,28	0,54	0,08	0,13	0,2
ALTERNATIVES												
PRICE	0,15	0,23	0,38	0,28	0,41	0,59	0,14	0,22	0,35	0,15	0,29	0,54
PRODUCT	0,28	0,42	0,61	0,15	0,25	0,37	0,17	0,25	0,41	0,1	0,17	0,32
PLACE	0,09	0,12	0,18	0,07	0,1	0,14	0,08	0,11	0,16	0,11	0,2	0,39
PROMOTION	0,14	0,23	0,36	0,16	0,25	0,4	0,28	0,42	0,6	0,19	0,34	0,57
FINAL AGGREGATION	*PRICE*			*PRODUCT*			*PLACE*			*PROMOTION*		
	0,11	0,3	0,79	0,1	0,28	0,74	0,05	0,12	0,31	0,11	0,3	0,81
DISTANCE TO ORIGIN	**0.52**			**0.50**			**0.37**			**0.53**		
PERCENTAGES	*27.1*			*26.1*			*19.3*			*27.4*		

5. Conclusion

According to the result of the model, the company must spend:

- 26.1% of the total investment (financial resources, effort, etc.) to the product itself,
- 27.1% for the price strategy,
- 27.4% for the promotion and
- 19.3% for the distribution.

The reason that the majority is dedicated to the promotion is that the company is making category extension. As the product has new brand, the company must give significant weight on promotion activities. Therefore, it is normal that the share of the promotion will have a much higher percentage than others. It is observed that the price has a higher percentage compared to the product and distribution. Pricing is a really important aspect of a business and is a crucial component of the marketing mix. The right pricing can help to achieve greater sales and greater profitability, whereas poor pricing can leave the products failing to sell or even to sell too cheaply.

As the company must give a new product, the share of the product is higher than this of the distribution. The company will conduct R&D and market research; it will create a new production line, new brand and will waste time and money in the process of decision.

The importance of the distribution is lower than others because of the current presence of a distribution chain and the absence of the need to invest more just for this new product.

These four P's are the parameters that the marketing manager can control, subject to internal and external constraints of the marketing environment. The goal is to make decisions that center the four P's on the customers in the target market in order to create perceived value and generate a positive response. The marketing mix framework was particularly useful in the early days of the marketing concept when physical products represented a larger portion of the economy. Today, with marketing more integrated into organizations and with a wider variety of products and markets, it is useful to extend the study by proposing a fifth P, such as people, packaging, process etc. Today however, the marketing mix most commonly remains based on the 4P's. Despite its limitations and perhaps its simplicity, the use of this framework remains strong and this study has been organized around it.

References

1. P. Kotler, Principles of Marketing. Upper Saddle River, NJ: Prantice Hall, (2001).
2. Saaty, T.,L., The Analytic Hierarchy Process, McGraw Hill, New York, (1980).
3. Cheng, C.H.,(1998) A new approach for ranking fuzzy numbers by distance method, Fuzzy Sets and Systems, 95, 307-318.
4. Buckley, J.J., (1984). Fuzzy Hierarchical Analysis, Fuzzy Sets and Systems, 17, 233-247
5. Mikhailov, L., (2003). Deriving priorities from fuzzy pairwise comparison judgments, Fuzzy Sets and Systems, 134, 365-385.
6. P. Kotler, Le Marketing Selon Kotler. Paris: Village Mondial, (1999).

EVALUATION OF URBAN REGENERATION STRATEGIES USING VIKOR METHOD: THE CASE OF ISTANBUL - GALATA DISTRICT

NİHAN ÇETİN DEMİREL*, ÖMER ONUR KAYA AND TUFAN DEMİREL

Department of Industrial Engineering, Yildiz Technical University, Besiktas, Istanbul, 34349, Turkey

The Galata district, which is one of the first settlements in Istanbul, has lost its unique architecture and urban texture due to radical changes after the 1950s. Therefore, a need of urban regeneration has emerged. In Istanbul, the concept of 'urban regeneration' has gained ground since the early 2000s. In this study, while considering five main criteria, five different urban regeneration strategies for the Galata district are prioritized via the VIKOR method. The weights of criteria are determined according to the judgements of multiple decision-makers.

1. Introduction

Settlements may be posterior to contemporary changes in the especially after the 20th century because of rapid technical improvements and increases in population. Therefore, they may lose social and commercial liveliness each day. For these reasons, some precautions should be taken like urban regeneration in order to interrupt social and commercial deterioration. Urban regeneration is demolition and rebuilding of urban slums and rehabilitation of physical, social, economic and environmental circumstances. One of these places that urban regeneration should be applied is the Galata district which is an historical settlement in Istanbul. Hereby, we aimed to make Galata more popular, central and preferred district.

In this study, we tried to prioritize alternative urban regeneration strategies for Istanbul-Galata district.

2. Urban Regeneration

After the Industrial Revolution, rapid developments in industrial cities result in lack of infrastructure, environmental pollution, and accordingly lower living

* nihan@yildiz.edu.tr

standards. Thereupon, in the late 19th century the concept of 'Urban regeneration' was emerged so as to build better planned cities. The term urban regeneration, which refers to 'demolition and rebuilding of urban slums' in dictionary, actually comprehends various strategies that differ by visions and objectives. Also, it includes a set of interventions to improve the economic, physical, social and environmental conditions of urban areas. There are more than ten urban regeneration strategies in the literature; however only a limited number of these are extensively implemented in Turkey:

Renewal (A_1)*:* Renewal is destruction of dilapidated urban regions and construction of the new ones according to the new social and economic needs of the time. Renewal is a radical intervention and it reacts to the oldness. This kind of intervention is a solution for urban areas which are physically risky and have lower quality of life. However, this process threatens historical urban areas [5].

Regeneration (A_2)*:* Roberts and Sykes (2000) defines urban regeneration as comprehensive and integrated vision and action which leads to the resolution of urban problems and which seeks to bring about a lasting improvement in the economic, physical, social and environmental condition of an area that has been subject to change [4].

Rehabilitation (A_3)*:* Rehabilitation can be defined as renewal of urban spaces, which dilapidated and outmoded but didn't lost their original properties, are realized by repairs and reorganization [5]. Urban rehabilitation projects are especially applied in historical heritage sites, which contain distinctive architectural style of buildings and take an important value in cities' history [2].

Conservation (A_4)*:* Conservation is preventing the disappearance of the physical structure, which represents cultural heritage, social and economic conditions of a society, because of transformation and development of city. The goals of conservation are providing the cultural assets for the benefits of society by integrating contemporary civilization and urban fabric [3]. Conservation is the most frequent applied urban regeneration strategy in historical sites.

Gentrification (A_5)*:* It is the process of restructuring the deteriorated urban neighborhoods in terms of socio-cultural and socio-economic classes by settling middle-class or affluent people to places where low-income people reside. This process leads to an increase in rents and taxes, which frequently pushes out lower-class populations. Gentrification and its consequences have attracted more criticism than other urban regeneration strategies. According to Carmon (1999), displacement is the most hotly debated effect of gentrification. However, gentrification also has important benefits such as revitalization of economy and reduction in crime rates [1].

3. The Criteria for Determining the Urban Regeneration Strategies

As we have described before, five convenient urban regeneration strategies are selected for Galata district. While evaluating alternative strategies, five main criteria are considered (see Fig. 1). These criteria are briefly summarized below:

Economic factors, (C_1) which are perhaps the most important factors, express government/project budget, credit availability, gross national product (GNP), potential unearned income and so forth. *Social and cultural factors* (C_2), connote demographic structures, social facilities, social infrastructures, historical textures etc. *Political factors and laws* (C_3) are relevant to voter profiles, local authorities, municipal plans, zoning regulations, conservation/preservation laws etc. *Ecological and environmental factors* (C_4) comprise natural disasters, environmental pollution, climate and so on. And lastly, *influencing factors* (C_5) include the effects of media and non-governmental organization (NGO), sponsorship of private sector, outcomes of research studies and so forth.

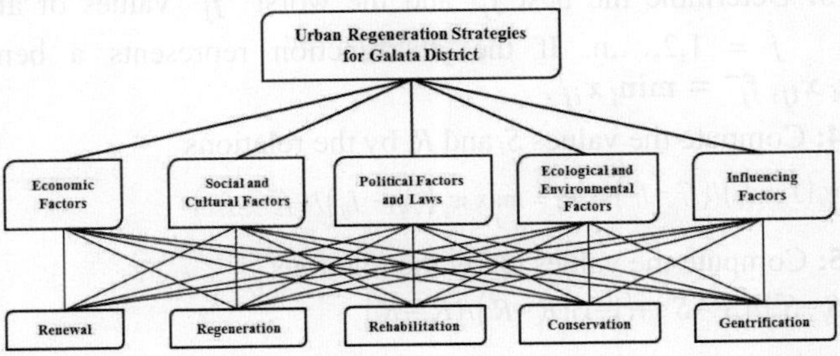

Figure 1. Convenient strategies for urban regeneration and evaluation criteria.

4. Steps of the Extended VIKOR Method

The VIKOR method was developed for multi-criteria optimization of complex systems and this method focuses on ranking and selecting from a set of alternatives, and determines compromise solutions for a problem with conflicting criteria, which can help the decision makers to reach a final decision [6]. Assuming that each alternative is evaluated according to multiple criterion functions, the compromise ranking is performed by comparing the measure of "closeness" to the "ideal" solution [8]. In this paper the problem is evaluated under fuzzy environment with fuzzy sets. The main steps of the algorithm are taken from Sanayei et al.'s (2010) [6] and Fouladgar et al.'s (2011) study [7]:

Step 1: Form decision matrices for all alternatives with respect to each criterion. Pull the decision makers' opinions to get the aggregated fuzzy weight of criteria, and aggregated fuzzy rating of alternatives and construct a fuzzy decision matrix: Let the fuzzy rating and importance weight of the k th decision maker be $\tilde{x}_{ijk} = (x_{ijk1}, x_{ijk2}, x_{ijk3}, x_{ijk4})$ and $\tilde{w}_{jk} = (w_{jk1}, w_{jk2}, w_{jk3}, w_{jk4})$; i = 1, 2,... ,m and j = 1, 2, …, n respectively.

A suitable urban regeneration strategy selection problem can be concisely expressed in matrix format as follows:

$$\tilde{D} = \begin{bmatrix} \tilde{x}_{11} & \tilde{x}_{12} & \cdots & \tilde{x}_{1n} \\ \tilde{x}_{12} & \tilde{x}_{22} & \cdots & \tilde{x}_{2n} \\ \vdots & \vdots & \cdots & \vdots \\ \tilde{x}_{m1} & \tilde{x}_{m2} & \cdots & \tilde{x}_{mn} \end{bmatrix}, \quad \tilde{W} = [\tilde{w}_1, \tilde{w}_2, \ldots, \tilde{w}_n],$$

where $\tilde{x}_{ij}$ the rating of alternative A_i with respect to C_j, $\tilde{w}_{ij}$ the importance weight of the jth criterion.

Step 2: Defuzzify the fuzzy decision matrix and fuzzy weight of each criterion into crisp values: This calculation is done by using center of area defuzzification method.

Step 3: Determine the best f_j^* and the worst f_j^- values of all criterion ratings, $j = 1,2,\ldots,n$. If the *j*th function represents a benefit then: $f_j^* = \max_i x_{ij}$, $f_j^- = \min_i x_{ij}$.

Step 4: Compute the values S_i and R_i by the relations

$$S_i = \sum_{j=1}^{n} w_j \left(f_j^* - f_{ij}\right) / \left(f_i^* - f_i^-\right), \quad R_i = \max_j w_j \left(f_j^* - f_{ij}\right) / \left(f_i^* - f_i^-\right). \tag{1}$$

Step 5: Compute the values Q_i by the relations

$$Q_i = v\left(S_i - S^*\right) / \left(S^- - S^*\right) + (1-v)\left(R_i - R^*\right) / \left(R^- - R^*\right). \tag{2}$$

where $S^* = \min_i S_i$, $S^- = \max_i S_i$, $R^* = \min_i R_i$, $R^- = \max_i R_i$ and v is introduced as a weight for the strategy of maximum group utility, whereas *1-v* is the weight of the individual regret.

Step 6: Rank the alternatives, sorting by the values S, R and Q in ascending order. Propose as a compromise solution the alternative ($A^{(1)}$) which is the best ranked by the measure *Q* (minimum) if the following two conditions are satisfied;

C1. Acceptable advantage: $Q(A^2) - Q(A^1) \geq DQ$, (3)

where $A^{(2)}$ is the alternative with second position in the ranking list by *Q*;

DQ = 1 / (J – 1).

C2. Acceptable stability in decision making:

The best alternative, ranked by *Q*, is the one with the minimum value of *Q*. The main ranking result is the compromise ranking list of alternative, and the compromise ranking list of alternatives, and the compromise solution with the "advantage rate".

5. Application of the Extended VIKOR Method in the Evaluation Urban Regenerations Strategies

As an application, the proposed model has been applied to evaluate urban regeneration strategies in Galata district. A committee of four decision makers, D1, D2, D3 and D4, has been formed to determine the most suitable urban regeneration strategy with the criteria. Application of the rest of the method can be seen in below.

The importance weights of the criteria determined by these four decision makers are shown in Table 1. The crisp values for decision matrix and weight of each criterion are computed as shown in Table 2.

Table 1. Importance weight of criteria from four decision makers.

Criteria	Decision makers			
	D1	D2	D3	D4
Economic Factors	VH	H	VH	VH
Social and Cultural Factors	MH	VH	H	H
Political Factors and Laws	H	MH	M	MH
Ecological and Environmental Factors	L	VL	VL	ML
Influencing Factors	M	M	MH	M

Table 2. Crisp values for decision matrix and weight of each criterion.

	Criteria				
	C1	C2	C3	C4	C5
Weight	0.88	0.77	0.65	0.22	0.58
A1	0.88	0.77	0.70	0.29	0.55
A2	0.92	0.71	0.72	0.28	0.41
A3	0.85	0.88	0.58	0.28	0.50
A4	0.88	0.64	0.58	0.40	0.57
A5	0.77	0.88	0.58	0.14	0.65

Table 3. The values of S, R and Q for all alternative strategies.

	Alternatives (Ranking)				
	A1	A2	A3	A4	A5
By S	1.08 (1)	1.23 (2)	1.55 (3)	1.87 (5)	1.74 (4)
By R	0.355 (1)	0.577 (2)	0.65 (3)	0.768 (4)	0.875 (5)
By Q (v=0)	0.00 (1)	0.428 (2)	0.566 (3)	0.794 (4)	1.00 (5)
By Q (v=0.5)	0.00 (1)	0.306 (2)	0.576 (3)	0.897 (4)	0.917 (5)
By Q (v=1)	0.00 (1)	0.183 (2)	0.585 (3)	1.00 (5)	0.835 (4)

The best and the worst values of all criterion ratings are determined as follows:

$f_1^* = 0.92$ $f_2^* = 0.88$ $f_3^* = 0.72$ $f_4^* = 0.40$ $f_5^* = 0.65$

$f_1^- = 0.77$ $f_2^- = 0.64$ $f_3^- = 0.58$ $f_4^- = 0.14$ $f_5^- = 0.41$

The value of S, R and Q which are calculated for all alternative strategies and the ranking of the alternative strategies by S, R, and Q in decreasing order are shown in Table 3. The ranking [A1-A2-A3-A4-A5] is same for values of R, Q (for v=0) and Q (for v=0.5).

6. Conclusions

Considering the weightings of four decision makers, the extended VIKOR method is applied in order to choose the most convenient urban regeneration strategy for Galata district. According to the results, firstly, implementation of alternative 1 (Renewal) is recommended. Afterwards, alternative 2 (Regeneration), alternative 3 (Rehabilitation), alternative 4 (Conservation) and alternative 5 (Gentrification) are suggested, respectively. Moreover, it should be noted that A1 is the most suggested strategy for each criteria that considered in the method. Finally, consideration of different criteria and alternatives while evaluating different regeneration districts is worthy of future studies.

References

1. N. Carmon, "Three generations of urban renewal policies: Analysis and policy implications", Geoforum 30, 145 (1999).
2. E. Mutlu, "Criteria for a 'good' urban renewal project: The case of Kadifekale Urban Renewal Project (Izmir, Turkey)", Unpublished Graduate Thesis, Izmir Institute of Technology, (2009).
3. S. Polat and N. Dostoğlu, "Kentsel dönüşüm kavramı üzerine: Bursa'da Kükürtlü ve Mudanya örnekleri", Uludağ Üniversitesi Mühendislik-Mimarlık Fakültesi Dergisi 12 (1), 61 (2007).
4. P. W. Roberts and H. Sykes, "Urban Regeneration: A Handbook", SAGE Publications, 0-7619-6716-8, (2007).
5. G. Tarkay, "Evaluation of urban regeneration issues for an early 20th century quarter: Kadıköy-Yeldeğirmeni", Unpublished Graduate Thesis, Middle East Technical University, (2010).
6. Sanayei, A., Mousavi, S.F., Yazdankhah, A.,"Group decision making process for supplier selection with VIKOR under fuzzy environment", Expert Systems with Applications, 37, 24-30, (2010).
7. M.M. Fouladgar, A. Yazdani-Chamzini, S.H. Yakhchali, M.H. Ghasempourabadi, N.Badri, "Project portfolio Selection Using VIKOR Technique under Fuzzy Environment", 2nd International Conference on Construction and Project Management IPEDR vol.15, (2011).
8. S.Opricovic, G.H. Tzeng "Compromise solution by MCDM methods: A comparative analysis of VIKOR and TOPSIS", European Journal of Operational Research, 156, 445-455, (2004).

WEIGHTING THE CRITERIA OF GREEN SUPPLY CHAIN MANAGEMENT FOR SUPPLIER SELECTION IN AUTOMOTIVE INDUSTRY USING FUZZY ANP

SERHAT TÜZÜN, TUFAN DEMIREL*

Department of Industrial Engineering, Yıldız Technical University, Yıldız Campus, Beşiktaş, Istanbul, 34349, Turkey

The aim of the present study is to evaluate the weights of the criteria in green supply chain management to aid the supplier selection process in automotive industry. The problem was considered as a multi-criteria decision making problem. The main and the sub criteria affecting green supply chain management are taken into account in this paper, and their weights are evaluated through fuzzy ANP in order to determine their importance on automotive industry.

1. Introduction

Supply chain management (SCM) is the management of process wherein raw materials are converted into final products, then delivered to the end-consumer [1]. Since this process involves extraction and exploitation of the natural resources [2], it is important to consider environmental sustainability. Beginning in the early 1990's, manufacturers have been forced to adapt their supply chains with sustainability [3]. Adding the 'green' to the SCM adds a new paradigm where the supply chain will have a direct relation to the environment.

As the environmental awareness begins to gain an important role in consumer behavior, manufacturers pay attention to organize their supply chain activities to decrease the harm they gave to the environment with the help of green supply chain management (GrSCM). GrSCM is now an important business issue affecting an organization's supply chain or logistics network and the results of applying GrSCM must be quantified to aid the customers in the selection process.

This study aims to make a research over the factors affecting GrSCM and evaluate their importance levels in automotive industry. To satisfy this necessity, multi-criteria decision making methods are proposed as innovative approach [4].

* tdemirel@yildiz.edu.tr

And to avoid the problem of transforming improportionate data to a numerical scale, MCDM methods have been combined with fuzzy logic [5], [6].

2. Green Supply Chain Management in Automotive Industry

Greening the automotive industry has been disputed in international energy and environmental policy studies. According to major car producers, the vision and operational implementation of sustainable development is part of the overall objective to satisfy the need of the society [7, 8].

Green supply chain in automotive industry has become the main interest in many industrial fields. Its performance evaluation is essential when environmental issues have been addressed all over the world [9]. However, there have been few studies exploring the issue of GrSCM performance evaluation. Hence, applying green concepts into automotive manufacturing is essential to reduce environmental impacts, enhance market competition, and ensure regulation compliance [10].

To explore the overall performance of automobile manufacturers in terms of GrSCM, the effecting factors and their relations should be evaluated. In order to provide this, the factors from Büyüközkan and Vardaoğlu's study [11] have been taken as the base and the network in Fig. 1 has been established.

3. Multi-Criteria Decision Making

Selecting or prioritizing alternatives from a set of available alternatives with respect to multiple criteria, is often refer to Multi-Criteria Decision Making. Academics have developed and used various fuzzy AHP and fuzzy ANP methods for solving different problems [12], [13], [14], [15]. In this study, criteria of green supply chain management for supplier selection plant are evaluated using Promentilla et al.'s [14] fuzzy ANP method.

In this paper, the computational procedure of using multi-criteria decision making methodology is summarized as follows [14]:

Step 1: Decompose the multiple criteria evaluation problem into a strongly connected hierarchical network.

Step 2: Elicit value judgments from DM to compare each pair of elements with respect to an element. Then, construct the fuzzy pairwise comparison matrix such that:

$$\hat{A} = \begin{bmatrix} 1 & \hat{a}_{12} & \cdots & \hat{a}_{1n} \\ \hat{a}_{21} & 1 & \cdots & \hat{a}_{2n} \\ \vdots & \vdots & \ddots & \vdots \\ \hat{a}_{n1} & \hat{a}_{n2} & \cdots & 1 \end{bmatrix} \quad (1)$$

Step 3: Solve the fuzzy eigenvalue problem.

Step 4: Assign an optimism λ index value. The optimism index is a linear convex combination defined as

$$\hat{a}_{ij}^{\alpha} = (1-\lambda)\, a_{ijl}^{\alpha} + \lambda a_{iju}^{\alpha}, \quad \forall \lambda \in [0,1]. \quad (2)$$

Step 5: With α fixed, construct the following matrix after setting the λ parameters:

$$\hat{A} = \begin{bmatrix} 1 & \hat{a}_{12}^{\alpha} & \cdots & \hat{a}_{1n}^{\alpha} \\ \hat{a}_{21}^{\alpha} & 1 & \cdots & \hat{a}_{2n}^{\alpha} \\ \vdots & \vdots & \ddots & \vdots \\ \hat{a}_{n1}^{\alpha} & \hat{a}_{n2}^{\alpha} & \cdots & 1 \end{bmatrix} \quad (3)$$

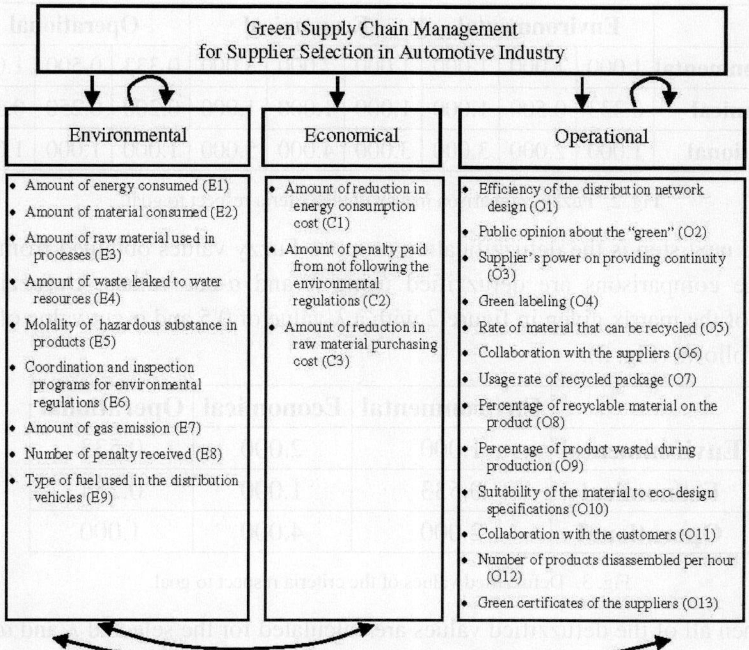

Fig. 1. The network of the problem.

Step 6: From Eq. (3), calculate the local priorities or the normalized principal right eigenvector for all α-cut values [0,1] at a given fixed λ value.

Step 7: Compute the overall priorities from principal column eigenvector of the initial supermatrix of the hierarchical network.

Step 8: Repeat calculations from *Steps 4* to *7* to derive the set of desired priorities for all α-cut at different λ values.

4. The Weighting of the Criteria

The problem discussed here is about evaluating the criteria of green supply chain management for supplier selection in automotive industry. Hierarchy network that is established in section 2 is determined from its main and sub criteria by Büyüközkan and Vardaoğlu's study [11].

After the network has been established, fuzzy comparison matrices are formed. The data used in the matrices are obtained from experts' views, previous studies and reports. Then, these data has been transformed into fuzzy values with the δ value of 1. Fuzzy comparison matrix of the criteria respect to goal can be seen in figure 2. All of the fuzzy comparison matrices are formed like in figure 2.

	Environmental			**Economical**			**Operational**		
Environmental	1.000	1.000	1.000	1.000	2.000	3.000	0.333	0.500	1.000
Economical	0.333	0.500	1.000	1.000	1.000	1.000	0.200	0.250	0.333
Operational	1.000	2.000	3.000	3.000	4.000	5.000	1.000	1.000	1.000

Fig. 2. Fuzzy comparison matrix of the criteria respect to goal.

The next step is the defuzzification process. Fuzzy values obtained from the pairwise comparisons are defuzzified using λ and α-cut values. Defuzzified values of the matrix given in figure 2 with a λ value of 0.5 and α-cut value of 0.5 are as follows (Fig. 3):

	Environmental	**Economical**	**Operational**
Environmental	1.000	2.000	0.533
Economical	0.533	1.000	0.254
Operational	2.000	4.000	1.000

Fig. 3. Defuzzified values of the criteria respect to goal.

When all of the defuzzified values are calculated for the selected λ and α-cut values, these values are brought together to form the unweighted supermatrix. Then, the weights of the criteria are put to use to calculate the weighted supermatrix. Last, the limit matrix is calculated using the formula given in step 6 in the previous section.

The computations that were made for λ value of 0.5 and α-cut value of 0.5 and the results are obtained as follows:

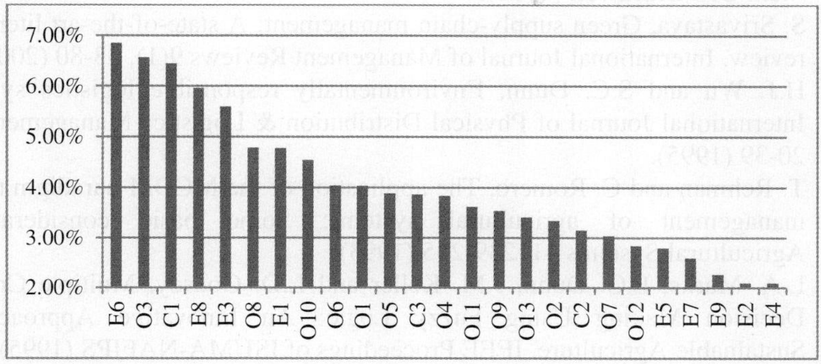

Fig. 4. Weights and ranking of the criteria in the descending order.

We have calculated the overall weights for each criterion by letting α= 0.5 and λ=0.5 (for a moderate DM) and determine its corresponding ratings. The results in fig. 4 show that "Coordination and inspection programs for environmental regulations (E6)" is the most effecting criterion for supplier selection in automotive industry in terms of GrSCM; followed by "Supplier's power on providing continuity (O3)" and "Amount of reduction in energy consumption cost (C1)."

5. Conclusion and Further Suggestions

In the present study, after main and sub-criteria were determined, the criteria are ranked with their effect on GrSCM for supplier selection in automotive industry by using fuzzy ANP method. This methodology allows reflecting the DM's attitude towards the fuzziness of judgment.

The results show us that, not all of the criteria are essential for the selection process in the given conditions. Most of the factors from Büyüközkan and Vardaoğlu's study [11] do not have a direct effect over the process. The criteria with less than 3% can be extracted from the network. As for the criteria stationed between 3% and 5%, mostly the economical and operational factors, it can be said that these factors only have indirect effects over the process.

As future research, the output can be used to rank the suppliers to determine the most suitable one in automotive industry corresponding to GrSCM.

References

1. C.M. Harland, Supply Chain Management, Purchasing and Supply Management, Logistics, Vertical Integration, Materials Management and Supply Chain Dynamics. In: Slack, N (ed.) Encyclopedic Dictionary of Op. Man. UK: Blackwell (1996).
2. S. Srivastava, Green supply-chain management: A state-of-the-art literature review. International Journal of Management Reviews 9(1), 53-80 (2007).
3. H.J. Wu and S.C. Dunn, Environmentally responsible logistics system. International Journal of Physical Distribution & Logistics Management 25, 20-39 (1995).
4. T. Rehman and C. Romero, The application of the MCDM paradigm to the management of agricultural systems: Some basic considerations. Agricultural Systems 41, 239-255 (1993).
5. L.A. Marks, E.G. Dunn, J.M. Keller and L.D. Godsey, Multiple Criteria Decision Making Using Fuzzy Logic: An Innovative Approach to Sustainable Agriculture. IEEE Proceedings of ISUMA-NAFIPS (1995).
6. T. Yakupoğlu, N. Özdemir and I. Ekberli, Fuzzy Logic Applications in Soil Erosion Studies (in Turkish). J. Of Fac. Of Agric. 23(2), 121-130 (2008).
7. Environmental Report. http://www.mobility-and-sustainability.com/.
8. Guiding principles at Toyota. http://www.toyota.co.jp/en/sustainability/index.html.
9. E.U. Olugu, K.Y. Wong and A.M. Shaharoun, Development of key performance measures for the automobile green supply chain. Resources, Conservation and Recycling 55(6), 567-579 (2010).
10. L. Gan, Globalization of the automobile industry in China: dynamics and barriers in greening of the road transportation. Energy policy 31, 537- 551 (2003).
11. G. Büyüközkan and Z. Vardaloğlu, Green Supply Chain Management (in Turkish). Lojistik Dergisi 8, 66-73 (2008).
12. H. Deng, Multicriteria analysis with fuzzy pairwise comparison. International Journal of Approximate Reasoning 21, 215-231 (1999).
13. C. Kahraman, U. Cebeci and D. Ruan, Multi-attribute comparison of catering service companies using fuzzy AHP: The case of Turkey. International Journal of Production Economics 87, 171–184 (2004).
14. M. A. B. Promentilla, T. Furuichi, K. Ishii and N. Tanikawa, A fuzzy analytic network process for multi-criteria evaluation of contaminated site remedial countermeasures. Journal of Environmental Management 88, 479-495 (2008).
15. T. Demirel, H. Muşdal, N. C. Demirel and G. N. Yücenur, Multi-Criteria Evaluation of Land Cover Policies Using Fuzzy AHP and Fuzzy ANP: The Case of Turkey. Journal of Human and Ecological Risk Assessment 15(4), 746-764 (2009).

MODELLING A PAY STRUCTURE BY USING GREY RELATIONS ANALYSIS INTEGRATED WITH FUZZY AHP

AHMET CAN KUTLU†
Department of Industrial Engineering, Istanbul Technical University, Macka, Istanbul, 34367, Turkey

Pay structure is the ordering of pay rates for jobs or groups of jobs within an organization and plays an important role in the employment relationships affecting the efficiency of an organization. Design of a pay structure is a complex process requiring the evaluation many factors including both quantitative and qualitative data. Grey Relations Analysis (GRA) a useful mathematical tool for dealing with uncertainty and system analysis with limited information is introduced for creating a pay structure. Fuzzy Analytical Hierarchy Process (FAHP) is also used to determine the factor weights of payment types by pair-wise comparisons. The hypothetical case study shows the applicability of the model.

1. Introduction

Pay is an important and contentious part in the employment relationship. In any organization, morale and productivity of the workforce are affected by the level and distribution of pay and benefits and this issue influences the efficiency of the organization. Therefore an ideal pay system should be efficient and should reduce inequalities rewarding workers fairly for the work they perform.

Design of a pay structure is a complex process requiring an integration of a group of experts' knowledge and experiences for evaluating many factors that have effects on the payment of an employee. In the study, a grey relationship analysis (GRA) is proposed for structuring an effective pay system. GRA which is part of grey system theory developed by Deng (1982) [1] is an appropriate tool for solving problems with complicated interrelationships between multiple factors and variables [2]. The combinations of both qualitative and quantitative data are associated to GRA to obtain a monetary term. The importance weights of the criteria used in the analysis are calculated by FAHP.

The rest of the paper is organized in four sections. In the following section the proposed model for establishing a pay structure is introduced by explaining the methods GRA and FAHP. Third section involves an illustrative case study

† kutluac@yahoo.com

that expresses the applicability of the model. In the last section, paper is concluded by expressing results, remarks and future recommendations.

2. The Proposed Model

The study proposes a GRA model shown in Figure 1 to determine the payment of an employee including two paying types as constant pay and variable pay. Constant pay is obtained by the job evaluation process which is a technique based on systematically determining a relative internal value of a job by comparing the relative similarities and differences in the content and the value of jobs. The most common job evaluation method is the point-factor method where points are given to the jobs according to their worth by the assessments of experts with respect to the identified compensable factors. Subsequently the jobs are assigned to job classes determined by the overall evaluation.

The variable pays given to an employee are identified by the employee's performance evaluation, seniority and qualifications. Performance evaluation (appraisal) covers to what degree a person performs his/her job well. It is the process of information analysis about the relative worth of an employee to the organization executed by the supervisor. Seniority refers to the length of the working time of an employee in the organization. Seniority-based pay policy is a good motivator in employee retention with pay increase upon employee's length of time spent on the job. Pay increases may be based on qualifications such as job-related skills and knowledge. The purpose of this competency-based pay system is to motivate employees to gain additional skills, competencies and knowledge that will increase the value to the firm [3].

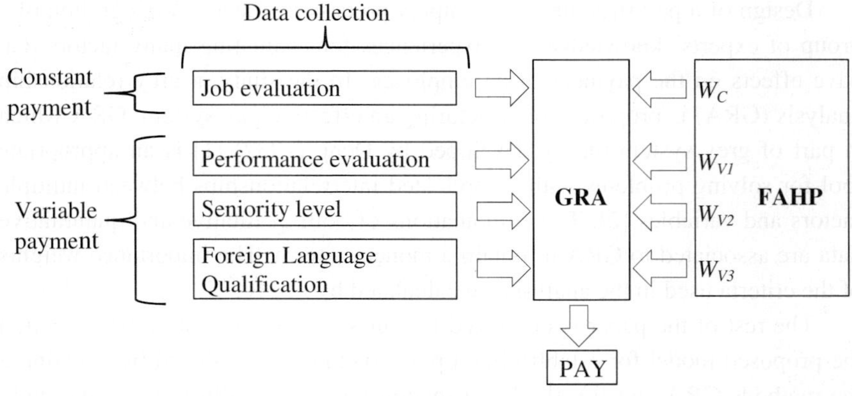

Figure 1. The flowchart of the proposed model.

The outcome of the point-factor job evaluation process is used to calculate constant pay whereas data of performance evaluation score of the employee evaluated by the managers, the seniority level and one foreign language examination score are the inputs for variable pay calculation. Thus since the problem involves both qualitative and quantitative data, employing GRA is appropriate. Finally the importance weights of the constant and variable pay criteria are determined by using FAHP for concluding payments calculation.

2.1. *GRA*

Grey theory was introduced by Deng (1982) [1] in order to study the degree of relationship between various criteria in a multi criteria decision making (MCDM) problem [4]. Deng (1982) defines the grey systems as the systems that lack information, parameters, operation mechanism and system behavior [5]. GRA, is used with the purpose of exploring the qualitative and quantitative relationships among abstract and complex sequences and capturing their dynamic characteristics during the development process with the measuring the relative influence of the compared series on the reference series [6].

The assumption is that alternatives are denoted as $a_1, a_2,..,a_i,..,a_m$ and are assessed by criteria $c_1, c_2,\ldots, c_i,\ldots,c_m$, respectively. $W_i(j)$ symbolizes the weight/importance of criterion j for alternative a_i; $a_i(j)$ is the performance value for alternative a_i at criterion j. $A = \{a_i | i = 0, 1, \ldots, m\}$ is an alternative set for the grey relation, where $a_o \in A$ is the referential sequence, and $a_i \in A$ $(i = 1, 2,\ldots, m)$ is the comparative sequence. Finally the grey relational coefficient in $a_i(j)$ corresponding to $a_o(j)$ is $\gamma(a_o(j), a_i(j))$. Moreover $a^*(j)$ is defined as best/aspired $a(j)$ and $a^-(j)$ is defined as worst/tolerable $a(j)$ [7]. Table 1 describes the variables and parameters.

Table 1. The performance table for the GRA model [7]

Alternatives		Criteria				
		c_1	$\cdots$	c_j	$\cdots$	c_{n_i}
		w_1	$\cdots$	w_j	$\cdots$	w_{n_i}
a_1		$a_1(1)$	$\cdots$	$a_1(j)$	$\cdots$	$a_1(n_1)$
$\vdots$		$\vdots$		$\vdots$		$\vdots$
a_i		$a_i(1)$	$\cdots$	$a_i(j)$	$\cdots$	$a_i(n_i)$
$\vdots$		$\vdots$		$\vdots$		$\vdots$
a_m		$a_m(1)$	$\cdots$	$a_m(j)$	$\cdots$	$a_m(n_m)$
Aspired value	a^*	$a^*(1)$	$\cdots$	$a^*(j)$	$\cdots$	$a^*(n_i)$
The worst value	a^-	$a^-(1)$	$\cdots$	$a^-(j)$	$\cdots$	$a^-(n_i)$

[Note] Data matrix: normalization

Grey relational coefficient in a_i corresponding to aspired level a^* and worst level a^-, $\gamma(a^*(j), a_i(j))$ can be derived from Eq. (1) and Eq. (2) consecutively [7] where $\gamma \in A$ [0,1] is the distinguished coefficient whose value is generally set at $\gamma = 0.5$ [8].

$$\gamma(a^*(j), a_i(j)) = \frac{\min_i \min_j |a^*(j)-a_i(j)| + \zeta \max_i \max_j |a^*(j)-a_i(j)|}{|a^*(j)-a_i(j)| + \zeta \max_i \max_j |a^*(j)-a_i(j)|} \quad (1)$$

$$\gamma(a^-(j), a_i(j)) = \frac{\min_i \min_j |a^-(j)-a_i(j)| + \zeta \max_i \max_j |a^-(j)-a_i(j)|}{|a^-(j)-a_i(j)| + \zeta \max_i \max_j |a^-(j)-a_i(j)|} \quad (2)$$

2.2. *FAHP*

Analytical Hierarchy Process (AHP), developed first by Saaty [9], is a widely-used methodology in MCDM. The fuzzy AHP methodology extends traditional AHP by combining it with the fuzzy set theory making the comparison process more flexible and capable to explain experts' preferences [10]. In this study, Buckley's [11] fuzzy extent analysis method is utilized to determine the weight vector of the payment type criteria. The theoretical formulation of the Buckley's fuzzy extent analysis for AHP can be found in Hsieh *et. al* [12]. The hierarchical structure for the payment criteria adopted in this study is shown in Figure 2.

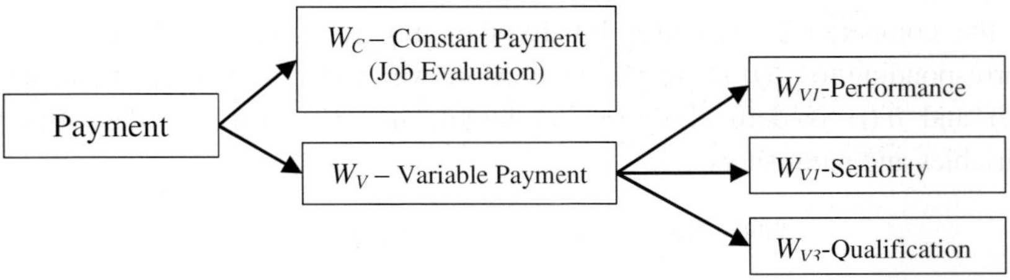

Figure 2. The hierarchical structure for payment types.

3. Case Study

In the case study, the proposed model is applied for calculating the payments of 6 employees. At first, point-factor job evaluation process is applied to obtain the base wages. Experts qualitatively assess the job positions which are later assigned to five classes in relation with their factor points with a minimum value of 300 and maximum value of 800. Table 2 expresses the original and normalized job class factor points.

Table 2. The grey numbers for job classes

Classes	Boundary Points	Normalized Points	Mean Value
Class 1	[300-400]	[37.5-50]	43.75
Class 2	[400-500]	[50-62.5]	56.25
Class 3	[500-600]	[62.5-75]	68.75
Class 4	[600-700]	[75-87.5]	81.25
Class 5	[700-800]	[87.5-100]	93.75

In the second phase, the performances of the employees are evaluated by 3 managers with a scale of linguistic variables shown in Table 3. Grey numbers corresponding to the seniority level are scaled in Table 4. The foreign language examination score is directly used in the analysis.

Table 3. The grey numbers for performance evaluation

Linguistic Variables		Grey Numbers	Linguistic Variables		Grey Numbers
Very Poor	VP	[0, 15]	Medium Good	MG	[55,70]
Poor	P	[15,30]	Good	G	[70,85]
Medium Poor	MP	[30,45]	Very Good	VG	[85,100]
Fair	F	[45,55]			

Table 4. The grey numbers for seniority levels

Seniority Level (years)	Normalized values	Seniority Level (years)	Normalized values
[0-0.99]	20	[5-9.99]	80
[1-2.99]	40	[10-inf.]	100
[3-4.99]	60		

After collecting quantitative and expert opinions' qualitative data, the results are aggregated in Table 5. In the next step, results are normalized and GRA is conducted referring to Eq. (1). The weights of the criteria are also obtained by utilizing FAHP. Final results with the calculated payments are shown in Table 6.

Table 5. The aggregated results for the employees

Employees	Expert Evaluations			Grey number	Mean values	Seniority level	Test Score	Job class
Employee 1	G	MG	G	[65, 80]	**72.50**	**100**	**85**	**81.25**
Employee 2	G	G	G	[70, 85]	**77.50**	**80**	**80**	**81.25**
Employee 3	VG	G	VG	[80, 95]	**87.50**	**60**	**90**	**68.75**
Employee 4	F	F	MG	[48.33, 60]	**54.17**	**80**	**70**	**56.25**
Employee 5	MG	G	MG	[60, 75]	**67.50**	**20**	**60**	**56.25**
Employee 6	F	MP	MP	[35, 48.33]	**41.67**	**40**	**0**	**43.75**

Table 6. Final results for the employees

	W_{V1} 0.18	W_{V2} 0.08	W_{V3} 0.03	W_C 0.71		*Maximum Price 3000 TL/month*
Employees	**Mean values**	**Seniority level**	**Test Score**	**Job class**	**FINAL RESULTS**	**Payments**
Employee 1	0.6	1	0.9	1	**0.93**	**2777.37**
Employee 2	0.7	0.67	0.82	0.67	**0.91**	**2739.59**
Employee 3	1	0.5	1	0.5	**0.68**	**2028.00**
Employee 4	0.41	0.67	0.69	0.4	**0.45**	**1355.16**
Employee 5	0.53	0.33	0.6	0.4	**0.45**	**1335.21**
Employee 6	0.33	0.4	0.33	0.33	**0.34**	**1016.00**

4. Conclusion

The results of the case study show that Employee 1 gets the highest and Employee 6 gets the lowest pay according to the evaluations and scores. The second employee's payment is very close to first one since their job classes are same and this criterion has the greatest share in the payment policy. As a concluding remark, the proposed model provides the flexibility for managers for deciding their payment policy like job, performance or skill oriented by just changing the importance weights of the criteria with a small amount of data. In future studies, the results or this paper can be compared with other MCDM methods that can be applied for complex and uncertain conditions.

References

1. J. Deng. *Systems and Control Letters*. **5**(2) 288 (1982).
2. G.-W. Wei. *Expert Systems with Applications*. 38 4824 (2011).
3. M. Conyon. S. Peck and L. Read. *European Man. Jour.* **19**(1) 73 (2001).
4. J. J. H. Lioua. C.-C. Hsu. W.-C. Yeh and R.-H. Lin. *Tourism Man.* **32**(6) 1381 (2011).
5. R.-H. Liang. *Electrical Power and Energy Systems*. **21** 357 (1999).
6. S. J. Lina. I. J. Lub and C. Lew. *Energy Policy*. **35** 1948 (2007).
7. J. Deng. *The Journal of Grey Systems*. **1**(1) 1(1989).
8. H. Guo, *The Journal of Fuzzy Mathematics*. **3**(2) 55 (1985).
9. T.L. Saaty. The Analytic Hierarchy Process. McGraw-Hill Book Company. New York. (1980).
10. C. Kahraman. U. Cebeci and Z. Ulukan. *Logistics Information Man.* **16**(6) 382 (2003).
11. J. J. Buckley, *Fuzzy Sets Syst*. **17**(1) 233 (1985).
12. T. Hsieh. S. Lu and G. Tzeng. *Inter. Jour. of Project Man.* **22** 573 (2004).

AN INTEGRATED MULTI-CRITERIA DECISION MAKING METHODOLOGY FOR FACILITY LOCATION SELECTION PROBLEM UNDER FUZZY ENVIRONMENT: APPLICATION IN CEMENT SECTOR

İREM OTAY, FERHAN ÇEBİ

Dept. of Management Engineering, Faculty of Management, Istanbul Technical University, Macka-Istanbul, 34367, TURKEY

The study developed an integrated approach for facility location problem based on a real life case from the Cement Sector in Turkey. A fuzzy multi-criteria decision making approach integrating Fuzzy AHP and Fuzzy VIKOR was designed in the study by concerning the vague and complicated nature of the information. The application was consisted of two stages: the region selection and site selection.

1. Introduction

Location selection is one of the critical issues effecting organizations' survival, performance and competitiveness in the long terms. The process for evaluating and selecting where the facility should be located requires a comprehensive and systematic analysis by considering social and economic factors, qualitative and quantitative factors as well as strategic and operational factors.

The factors that are important in decision where to locate a facility and their impact level on making decision are mostly differs according to the type of the facility such as heavy manufacturing sectors, light-industry sector and service facilities [20]. Although site selection in manufacturing sector is usually based on the costs, in recent years environmental and social factors have been growing importance on location selection decision for plants in the sector as indicated by several authors [8,20]. In a study conducted in cement sector, the authors [8] presented technical, environmental, social and national planning factors for site selection by emphasizing the importance of the site selection for supporting cement production. On the other hand, the authors also indicated location selection problems as a multi-stage problem, in which the factors affecting the decision differs from the country selection to site selection [16,20]. This study also focused on site selection problem for a cement factory as a two-stage decision making problem with multi-criteria.

In the literature, there have been huge amount of studies proposing the methods, techniques, models, and approaches with multi-criteria for the selection of facility locations. The studies have introduced various economical, social and environmental criteria, and various objectives such as distance minimization, demand, profit maximization and environmental factors depending on their objectives for location selection problem [1,6,10,22]. On the other hand, the authors have presented different approaches for solving site selection problem by using the methodologies of data envelopment analysis (DEA), analytic hierarchy process (AHP), analytic network process (ANP), ordered weighted average (OWA), heuristic methods, simulation and mathematical approaches [8,11,12,13,14,17,18,19,23,24,25,26,29].

The authors also developed the models to find a solution to facility location selection problem by considering vagueness and imprecise data. The researchers applied fuzzy methodologies such as a fuzzy simple additive weighting system, integrated with the simplex algorithm, fuzzy simulations and a genetic algorithm, a hybrid intelligent algorithm, a hybrid modified particle swarm optimization approach, fuzzy AHP, fuzzy TOPSIS in order to select the facility locations [2,5,9,15,21,27,28].

In this study, it was aimed to develop an integrated fuzzy multi-criteria decision making approach for two-stage facility location evaluation and selection problem. By using a systematic approach, the study was conducted to help the managers to make efficient and effective decision on region & site selection of a company seeking a location for a cement factory.

2. Hierarchical Structure for Facility Location Selection Problem

As stated above, the paper was conducted to address facility location selection problem of a private engineering company which sets up a turnkey cement manufacturing factories in Turkey. The company has been operating over the two decades in the sector. The company's service offering to customers is to design the manufacturing facilities including site selection, building construction, facility layout and material handling. This study was dealt with only their facility location problem in which the company has to select the best appropriate region among three alternatives and then to select the best site in the selected region. So, the evaluation and selection process was consisted of mainly two stages as illustrated in Figure 1.

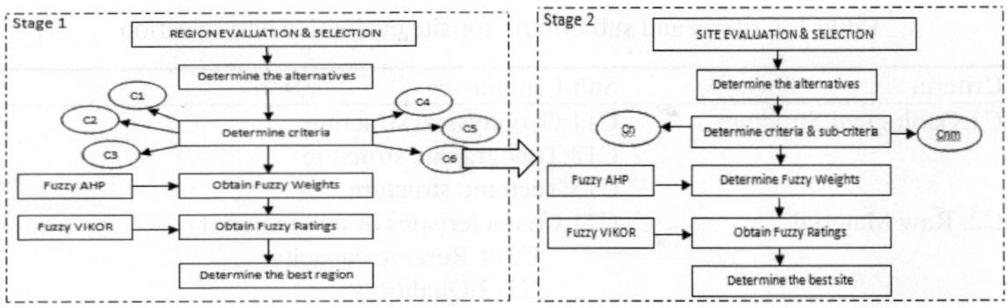

Figure 1. Facility Location Evaluation & Selection Procedure

2.1. *Region Evaluation and Selection*

Initially, face to face meetings with three experts from the project research & management department of the company were conducted. During the meetings, the experts were informed about the methodology of the study, and the factors affecting the region selection decision were discussed.

The model was consisted of three levels. The first level included the main goal of choosing the most appropriate region. The second level contained six main criteria: "Total cost (C1)" such as land cost, raw material cost, energy cost etc.; "proximity to sources (C2)" in terms of mines, labor, energy; "strategic factors (C3)" such as competition in the region, market condition etc.; "distribution systems (C4)" with respect to availability of waterway, highways etc.; "government policies (C5)" such as taxes, incentives and "environmental factors (C6)" including climate of the region. Finally, the last level indicated region alternatives (R1, R2 and R3) determined by the management.

2.2. *Site Evaluation and Selection*

During the meeting the factors affecting the site selection were also discussed with the company management. By using the similar procedures at the first stage, the hierarchical structure which will be used for evaluating and selecting the sites was developed. The hierarchical model was consisted of five levels as illustrated in Table 1.

3. Application of Hybrid Approach

An integrated approach was applied to the hierarchical models to determine the best appropriate region and the best appropriate site. The proposed approach used Fuzzy AHP and Fuzzy VIKOR. Buckley's Fuzzy AHP approach was employed to calculate the weights of the criteria and sub-criteria in the model of each stage [3,4]. Fuzzy VIKOR using the weights obtained from FAHP was implemented to evaluate both types of alternatives, regions and sites.

Table 1. Criteria and sub-criteria for site evaluation and selection

Criteria	Sub-Criteria
C1:Geological Structure	C11:Geophysical structure
	C12:Topographic structure
	C13:Tectonic structure
C2: Raw Material	C21:Characteristics of raw material
	C211:Reserve capacity
	C212:Quality
	C213: Sustainability of raw material
	C214:Proximity to surface
	C22:Distance to mines (coal, limestone)
C3-Infrastructure System	C31:Utilities
	C32:Energy
C4-Social Factors	C41:Population
	C42:Public acceptance
	C43:Contribution to the society
	C44:Distance to the nearest villages
	C45:Distance to the historical sites
C5- Transportation	C51:Distance to distribution systems
	C52:Potential transportation facilities
C6- Labor	C61:Availability of blue-collar employees
	C62:Availability of white-collar employees

3.1. *Selecting the best region*

Considering the hierarchical model, three experts of the company were asked to indicate the pairwise comparison judgments for the main criteria and the decision alternatives by using the linguistic scale. In the evaluation process, the judgments on which the experts arrived a consensus were used. Then, the fuzzy weight matrix was calculated by using Buckley's Method as follows [7].

$$\tilde{r}_i = \left(\tilde{c}_{i1} \otimes \tilde{c}_{i2} \otimes \ldots \otimes \tilde{c}_{in}\right)^{1/n} \quad \tilde{w}_i = \tilde{r}_i \otimes \left(\tilde{r}_1 \otimes \tilde{r}_2 \otimes \ldots \otimes \tilde{r}_n\right)^{-1} \tag{1}$$

In the formulas, $\widetilde{C_{1n}}$ is the fuzzy comparison value of criterion i to criterion n, $\widetilde{r_1}$ is the geometric mean of fuzzy comparison value of criterion i to each criterion. Centroid method was used as a defuzzification process.

The weight vector was obtained as (0.19,0.20,0.23.0.13.0.10,0.21). The result indicated that "strategic factors" was the most important main criteria with the priority of "0.23".This criterion was closely followed by the "environmental factors" and "proximity to sources" respectively. "Government policies" had the lowest priority with the value of "0.10".

The weights obtained from Buckley's FAHP approach were used as weights of criteria ($\widetilde{w_j}$) in the Fuzzy VIKOR method. By using formulas (2)-(5), the fuzzy best value (FBV,$\widetilde{f_j^*}$), the fuzzy worst value (FWV,$\widetilde{f_j^-}$), the separation measure of alternatives from the fuzzy best value and the fuzzy worst value

($\widetilde{S}_i, \widetilde{R}_i$) were calculated [15]. In the Formula (5), S^*, S^-, R^* and R^- represent $\min_i S_i$, $\max_i S_i$, $\min_i R_i$ and $\max_i R_i$ respectively. When S, Q and R values were ranked in the ascending order, Region 1 was found as the best region to be chosen.

$$\widetilde{f_j^*} = \max_i \widetilde{x_{ij}}\,,\ j \in B;\ \widetilde{f_j^-} = \min_i \widetilde{x_{ij}}\,,\ j \in C \quad (2)$$

$$\widetilde{S_i} = \sum_{j=1}^{n} \widetilde{w_j}\ (\widetilde{f_j^*} - \widetilde{x_{ij}})/(\widetilde{f_j^*} - \widetilde{f_j^-}) \quad (3)$$

$$\widetilde{R_i} = \max_j\ [\widetilde{w_j}\ (\widetilde{f_j^*} - \widetilde{x_{ij}})/(\widetilde{f_j^*} - \widetilde{f_j^-})] \quad (4)$$

$$Q_j = v\frac{S_i - S^*}{S^- - S^*} + (1 - v)\frac{R_i - R^*}{R^- - R^*} \quad (5)$$

3.2. *Selecting the best site*

The hybrid approach was also employed to evaluate alternative sites in Stage 1. By using similar evaluation process in finding weights of the region criteria, the weight vector of the main criteria in the site evaluation hierarchical model was obtained as (0.20,0.23,0.18.0.13.0.15,0.10). Then, the fuzzy VIKOR methodology was applied for ranking the sites. The results showed that Site 4 having the lowest Q value was the best appropriate site for establishing the cement factory.

4. Discussion and Conclusion

The study was proposed a systematic and integrated approach using fuzzy multi-criteria decision making methods to find out an effective solution to facility location selection problem on which a company has to select the best appropriate location for constructing a cement factory in Turkey. The results obtained from the application indicated that such an approach can be used for finding the most appropriate facility location.

In the study, the criteria and hierarchical structure of the study was inherently based on the company's requirements and objectives with regard to location selection for the cement factory although most of the criteria in the model were the ones widely used in location selection decision. On the other hand, the location selection differs with regard to the type of business as indicated by the authors [20]. The importance and evaluation criteria can be changed from sector to sector as well as company to company. It is suggested that organizations should identify suitable criteria for location selection by assessing carefully their own requirements and strategies. In the study, the selection for both region and sites was based on the criteria weight without making a sensitivity analysis. Therefore, future studies conducting this type of analysis are suggested.

References

1. A.B.Arabani and R.Z.Farahani, *Comput Ind Eng.* 62,408 (2012).
2. A.Awasthia, S.S.Chauhanb and S.K.Goyal, *Math Comput Model.* 53,98 (2011).
3. J.J. Buckley, Fuzzy Sets and Systems.17,233 (1985).
4. S.J.Chen and C.L.Hwang, Lecture Notes in Economics and Mathematical Systems), Springer (1992).
5. S.Y.Chou,Y.H.Chang and C.Y.Shen, *Eur J Oper Res.* 189,132(2008).
6. J.Current, H.Min and D.Schilling, *Eur J Oper Res.* 49,3,295(1990).
7. S.Cebi and C.Kahraman, *Knowledge Based Systems.*23,3(2010).
8. P.K.Dey and E.K.Ramcharan, *J Environ Manage.* 88, 1384(2008).
9. P.Dheena and G.Mohanraj, *Expert Syst Appl.*38(2011).
10. I.Dogan, *Expert Syst Appl.*39,1092(2012).
11. T.Drezner,Z.Drezner and S.Salhi, *Eur J Oper Res.*142, 138(2002).
12. R.L.Francis, L.F.McGinnis and J.A.White, Eur J Oper Res.12,220(1983).
13. P.V.Gorsevski, K.R.Donevska, C.D.Mitrovski and J.P.Frizado, *Waste Manage,* 32,287(2012).
14. C.Kahraman, D.Rhan and I.Dogan, *Inform Sciences.* 157,135(2003).
15. T.Kaya and C.Kahraman, *Energy.* 35,2517(2010).
16. B.Kobu, Üretim Yönetimi (Production Mngt,11thEd.),Istanbul, 2003.
17. J.Korpela and M.Tuominen, *Int J Prod Econ.* 45,169(1996).
18. F.Y.Partovi, Omega. 34,41(2006).
19. M.Rezaei and M.H.Fazel Zarandi, *Applied Soft Computing.*11,5330(2011).
20. R. Russell and B.W. Taylor. Operations Mngt., 4th Ed.,Prentice Hall, (2003).
21. C.Y.Shen and K.T.Yu, *Expert Syst Appl.*36,3,4709(2009).
22. M.Tabari,A.Kaboli,M.B.Aryanezhad,K.Shahanaghi and A.Siadat, *Appl Math Comput.* 206,598(2008).
23. P.Thomas,Y.Chan, L.Lehmkuhl and W.Nixon, *Eur J Oper Res.*141, 495 (2002).
24. L.Tong and W. Zhong-tuo, *Systems Engineering-Theory & Practice,* 28,12,107(2008).
25. G.Tuzkaya,S.Onut,U.R.Tuzkaya and B.Gulsun, *J Environ Manage.* 88, 970 (2008).
26. V.Verter, *Comput Oper Res.*29,583(2002).
27. S.Wang and J.Watada, *Inform Sciences.* In Press(2010).
28. M.Wen and K.Iwamura, *Eur J Oper Res.*184,627,(2008).
29. J.Yang and H.Lee, *Facilities.*15(9/10),241(1997).

A DECISION SUPPORT SYSTEM FOR TWO-ECHELON SERVICE SUPPLY CHAINS

BARIŞ SELÇUK*

Department of Industrial Engineering,Bahçeşehir University, Yıldız mah. Çırağan cad. Beşiktaş, İstanbul,Turkey

In this paper, we provide a decision support system for a two-echelon spare parts inventory distribution system that consists of a single central warehouse and a local warehouse. These warehouses are organized as separate entities that make their individual spare parts inventory provisioning decisions independently. The goal is to minimize the total backorders seen by the local warehouse such that the total inventory investment does not exceed a given budget. We provide a decision support system that can be used to coordinate these warehouses in a way to reach this goal. We used a real-life case data obtained from an original equipment manufacturer located in Europe and provided numerical results.

1. Introduction

In equipment intensive industries, spare part inventory management has crucial importance for uninterrupted continuation of operations. Therefore, customers require high availability levels of the spare parts in stock. Since these customers are usually scattered over a disperse geographical area, many original equipment manufacturers (OEM) use an extensive distribution network of multi-echelon inventory locations. Optimal spare parts inventory allocation can be achieved by a single monolithic optimization model. However, in some real-life cases the warehouses must be treated as separate independent intities that plan their individual spare parts. In such a case, a coordination mechanism between different warehouses must be established. For this purpose we develop a decision support system that will help the decision makers coordinate the allocation of inventories between different locations.

* baris.selcuk@bahcesehir.edu.tr

1.1. *Case Description*

We analyze the service supply chain of an OEM in Europe. The OEM operates a single central warehouse located in Europe and provides service to its customers through a network of local warehouses all over the world. Base-stock policy is commonly used in spare parts inventory systems, and this company is no exception. Base-stock policy is applied at all local and central warehouses. That is, for each part demanded from the local warehouse a replenishment order from the central warehouse is issued, and for each part demanded from the central warehouse a replenishment order from the supplier is issued. Since, it is difficult to estimate and control separate lead times for each spare part, the company prefers to plan according to a single planned lead time parameter for all spare parts. In our case, the planned lead time from the supplier to the central warehouse is taken as 65 days. Similarly, a fixed common planned lead time is set for all spare parts and is communicated to the local warehouse so that the local warehouse can plan its stocks accordingly. Currently, the company uses a fixed planned lead time of 14 days between the central and the local warehouses. Demand for each spare part is assumed to be Poisson distributed with a fixed mean. In this case, we consider 200 different spare parts. The prices of spare parts change between 100 and 100000 Euros, and their failure rates change between 0.0002 and 2 per year. The company is willing to minimize the total average backorders as seen by its customer under a budget limitation for the spare parts investment.

1.2. *Related Literature*

The literature on multi-echelon inventory systems for service parts covers over 40 years of research. Sherbrooke's Multi-Echelon Technique for Recoverable Item Control model, (METRIC by Sherbrooke (1968)), is the most well-known model in this topic. Since then, variations of METRIC have been proposed by Muckstadt (1973), Sherbrooke (1986) and many others. All these models assume that there exists a single central decision maker who can jointly optimize stock levels of all locations. However, this may not be always possible in real life. Companies, as the OEM that we are analyzing in this paper, may resort to decentralized decision making in different outsourced or proprietary facilities. Muckstadt and Thomas (1980) provided a critical analysis of multi-echelon models and analyzed the added value of a centralized multi-echelon model over de-centralized single-echelon models for the special case of low-demand high-cost items. Hausman and Erkip (1994) extended Muckstadt and Thomas (1980), and looked at the amount of sub-optimization which can occur if multi-echelon systems are managed as independent single-echelon systems. They consider low-demand, high-cost items with a continuous review base-stock inventory policy where backorders are not allowed (in case of stock-out an emergency order is triggered). However, they did not propose guidance on how to align de-centralized decision makers when a centralized model is not applicable from an organizational point of view. In this paper, we target this interesting aspect of the problem, and we aim to provide managerial insights on the way that a de-centralized system can be made efficient by utilizing planning parameters.

2. Problem Description

Let c_j denote the price of item *j*, and $S_{i,j}$ denote the base stock level of item *j* at location *i* for, $j \in J$, *i* = 0, 1. *i* = 0 refers to the central warehouse and *i* = 1 refers to the local warehouse. L_0 is the lead time from the supplier to the central warehouse, and in our case $L_0 = 65$ days. t_0 is the transportation time from the central warehouse to the local warehouse, and in our case $t_0 = 5$ days. The average replenishment lead time from the central warehouse to the local warehouse depends on the stock level of the central warehouse. If the central warehouse always has available parts in stock then the average replenishment lead time equals the transportation time, which is 5 days. This is not always possible, and replenishment orders sometimes have to wait for a part to become available at the central warehouse. For example, if the central warehouse keeps zero stock of a particular spare part then it takes 65 + 5 = 70 days to deliver that part to the local warehouse. The average lead time for a spare part *j* at the local warehouse is then written by $L_j(S_{0,j}) = t_0 + E\lfloor B_{0,j}(S_{0,j})\rfloor / \lambda_j$, where $E\lfloor B_{0,j}(S_{0,j})\rfloor$ is the expected backorder level for item *j* at the central warehouse. λ_j is the failure rate of item *j*. Similarly $E\lfloor B_{1,j}(S_{0,j}, S_{1,j})\rfloor$ is the expected backorder level fors pare part *j* at the local warehouse.

$$E\left[B_{0,j}(S_{0,j})\right] = \sum_{x=S_{0,j}+1}^{\infty} (x - S_{0,j}) e^{\lambda_j L_0} \frac{(\lambda_j L_0)^x}{x!}, \text{ and}$$

$$E\left[B_{1,j}(S_{0,j}, S_{1,j})\right] = \sum_{x=S_{1,j}+1}^{\infty} (x - S_{1,j}) e^{\lambda_j L_j(S_{0,j})} \frac{(\lambda_j L_j(S_{0,j}))^x}{x!}$$

The centralized optimization model is then written as follows.

$$\text{Minimize} \sum_{i \in I} E\left[B_{1,j}(S_{0,j}, S_{1,j})\right] \quad \text{subject to} \quad \sum_{i=0}^{1}\sum_{j \in J} c_i S_{ij} \leq TB$$

where *TB* is the total budget. This is a computationally complex optimization model and finding the exact optimum for realistic problems as the one in this paper is not possible. Therefore, we use heuristics to find the near optimal solutions. The heuristic algorithm for the centralized model and our decision support system are presented below.

3. Centralized Model

The centralized model jointly considers stock levels in the central warehouse and the local warehouse. This model assumes that there exists a central decision maker who has the decision power over all locations in the service supply chain. The following heuristic tries to find near optimum base stock levels both for the central warehouse and the local warehouse.

Inventory Planning Algorithm:

1. For all *i* and *j* set $S_{i,j} \leftarrow 0$.
2. For all $j = 1,2,...|J|$ do
   a. If $c_j > TB$ then, $\Delta_{0,j} \leftarrow 0, \Delta_{1,j} \leftarrow 0$.
   b. Else, $\Delta_{0,j} \leftarrow \dfrac{E\lfloor B_{1,j}(S_{0,j}, S_{1,j})\rfloor - E\lfloor B_{1,j}(S_{0,j}+1, S_{1,j})\rfloor}{c_j}$, and

   $$\Delta_{1,j} \leftarrow \frac{E\lfloor B_{1,j}(S_{0,j}, S_{1,j})\rfloor - E\lfloor B_{1,j}(S_{0,j}, S_{1,j}+1)\rfloor}{c_j}.$$
3. $i^* \leftarrow \arg\max_{j \in J}\{\Delta_{i,j}\}$, and $j^* \leftarrow \arg\max_{i \in 0,1}\{\Delta_{i,j}\}$.
   a. If $\Delta_{i^*,j^*} = 0$ then, the solution is found go to Step 4.
   b. Else, $S_{i^*,j^*} \leftarrow S_{i^*,j^*} + 1$, and $TB \leftarrow TB - c_j$, go back to Step 2.
4. STOP.

4. Decision Support System

We provide a de-centralized framework where base-stock levels of the central and the local warehouses are planned independently. Our decision support system acts as a coordination mechanism between the central warehouse and the local warehouse. We consider two fine-tuning parameters: 1) The allocation of total budget between the central and the local warehouse, and 2) The planned lead time communicated to the local warehouse. We first decide on how to allocate the total budget between the central and the local warehouse. At the same time, we set the planned lead time parameter for the local warehouse. Then we compute the base stock levels of the local warehouse and the central warehouse separately by the *Inventory Planning Algorithm* given above. Then we compute the expected total backorder faced by the local warehouse. For different budget allocation and different planned lead time values, the results may differ as we show in the following section.

5. Numerical Analysis

In our numerical tests, we consider different planned lead time and budget settings and provide managerial insights on how de-centralized systems can be made efficient. The results of our numerical tests can been in Figures 1, 2, and 3 below. In each figure the horizontal axis refers to the budget allocation, and the vertical axis refers to the expected total backorder faced by the local warehouse. Budget allocation 0.0 means all budget is given to the local warehouse, and budget allocation 1.0 means all budget is given to the central warehouse.

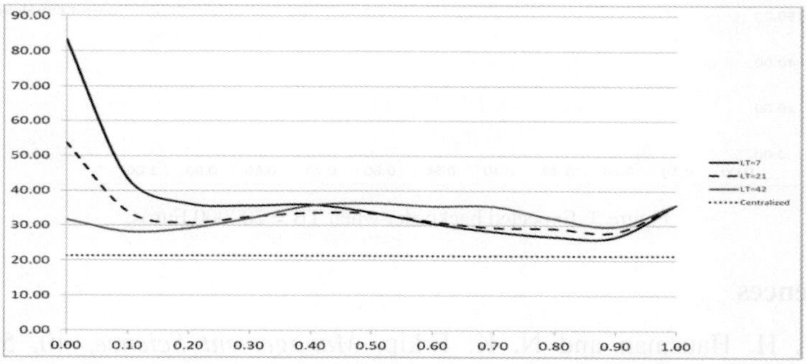

Figure 1. Expected backorder when TB = 1000000 Euros.

Our numerical results suggest that for a robust de-central planning that is insensitve to the budget allocation, a long planned lead time must be used. In this case, budget allocation does not become critical. However, it is critical when a short planned lad time is used. Especially when a limited budget will be allocated to the central warehouse, then the backorders can be much higher than expected. In this case, the central warehouse will not able to satisfy the short planned lead time, and the impact on the backorder levels will also be high. The de-centralized system becomes most efficient when a short planned lead time is used and most of the budget is allocated to the central warehouse.

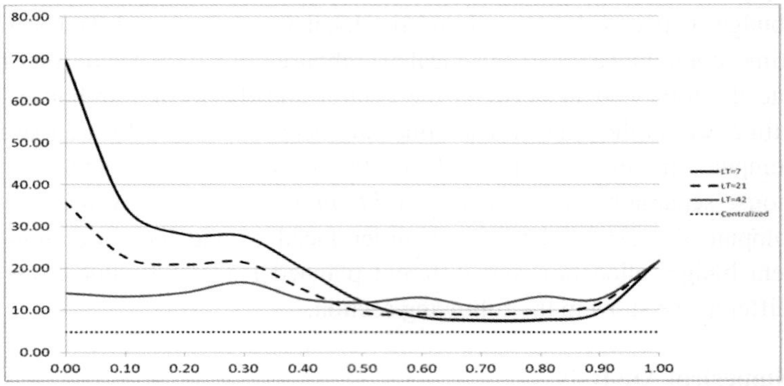

Figure 2. Expected backorder when TB = 1500000 Euros.

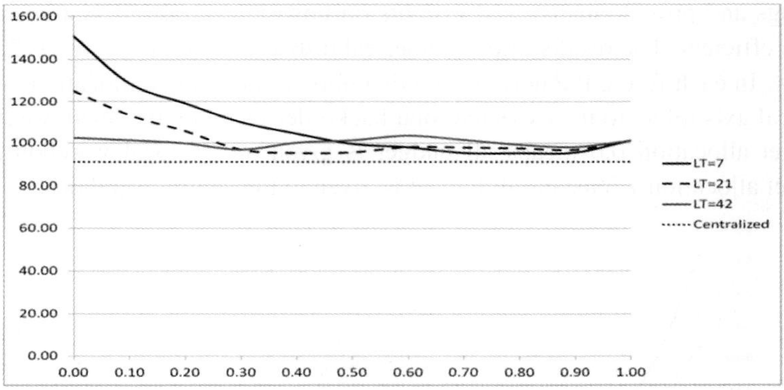

Figure 3. Expected backorder when TB = 200000 Euros.

References

1. W. H. Hausman and N. K. Erkip, *Management Science,* 40, 597–602 (1994).
2. J. A. Muckstadt, *Management Science,* ***20***, 471–481 (1973).
3. J. A. Muckstadt and L. J. Thomas, *Management Science,* 26, 483–494 (1980).
4. C. C. Sherbrooke, *Operations Research,* **16**, 122–141 (1968).
5. C. C. Sherbrooke, *Operations Research,* **34**, 311–319 (1986).

A FUZZY ANALYTIC HIERARCHY PROCESS METHODOLOGY FOR THE SUPPLIER SELECTION PROBLEM

PINAR MIZRAK OZFIRAT

Department of Industrial Engineering, Celal Bayar University, Muradiye Kampusu, Muradiye, Manisa, 45140, Turkey

GOKCECICEK TUNA*

Department of Industrial Engineering, Celal Bayar University, Muradiye Kampusu, Muradiye, Manisa, 45140, Turkey

GONCA TUNCEL

Department of Industrial Engineering, Dokuz Eylul University, Tinaztepe Kampusu, Buca, Izmir, 35160, Turkey

Purchasing activities of a company form a very important part in the operation of the company. The quality of production and services mainly rely on the performance of the suppliers. In addition, a large amount of the product's total cost belongs to the purchasing cost of its materials. Therefore, selecting the most suitable suppliers is very important for the firm. Unfortunately, most of the time, suppliers cannot provide the highest quality parts on-time at the lowest costs in the market. In other words, quality, delivery and cost objectives conflict with each other. In this study a fuzzy analytic hierarchy process approach is employed for the supplier selection problem of a textile company. First the selection criteria according to company's objectives are determined. Then the pair-wise comparisons are made on a fuzzy basis. Similarly, in the next step, the six alternative suppliers are compared by fuzzy means. Finally, the two suppliers which achieved the highest performance are advised to the company.

1. Introduction

Purchasing activities of a company form a very important part in the operation of the company. The quality of production and services mainly rely on the performance of the suppliers. In addition, a large amount of the product's total cost is made up of the purchasing cost of its materials. Therefore effective purchasing is of crucial importance and can bring competitive advantages to the firm. However, most of the time, suppliers cannot provide the highest quality parts on-time at the lowest costs in the market. In other words, quality, delivery and cost objectives conflict with each other.

The question of 'who to buy from and how much to buy' is simply the ***Supplier Selection Problem.*** An effective selection process offers enormous potential for decreased costs and effective control of resources. Researchers state that selecting the sources of supply is the most important function of purchasing department, as an unsuitable selection can significantly affect performance of the company.

Supplier selection process starts with the realization of the need for a new supplier. Then, the selection criteria that will guide the decision making process is identified and formulated. Decision criteria vary according to a number of qualitative and quantitative factors. Dickson identified 23 criteria for supplier selection based on a survey of 273 purchasing managers from United States and Canada [3]. Although there is no common identification of factors guiding the supplier selection process, the majority of the studies showed that quality is the most prominent criteria followed by delivery and performance history [1]. After the elaboration of decision criteria, initial screening and final selection of suppliers from a large set is performed.

Although AHP has been extensively used to solve the multi-criterion decision making problems including supplier selection and resource allocation, its ability is limited in handling the uncertainties and vagueness inherent in real life decision making environment. Therefore, recently, there has been a growing interest in merging AHP and fuzzy set theory to extend the ability to catch decision maker's uncertain preferences. In this paper, a fuzzy analytic hierarchy process (FAHP) is used to solve the supplier selection problem of a textile company. First the selection criteria according to the company's objectives are determined and the alternative suppliers are listed together with the decision makers of the system. Then fuzzy comparison matrices are built both for selection criteria and the suppliers. By use of fuzzy prioritization method, priorities are computed for each selection criteria and similarly local priorities for each supplier are found. In the last step global priorities are computed for each supplier and the best performing suppliers in terms of company objectives are revealed.

The intention of our study is to provide a comprehensive methodology for prioritization of the preference intensities by dealing with subjective and imprecise assessments and ensuring proper alignment between performance improvement initiatives in supplier selection process and operational requirements of the firm.

In the following of the paper, the FAHP methodology employed is explained in detail in section 2. In section 3, the methodology is applied to solve the supplier selection problem of a textile company. Finally, concluding remarks are given in section 4.

2. Fuzzy Analytic Hierarchy Process

In the proposed approach, classical analytic hierarchy process is fuzzified and fuzzy prioritization method is employed in order to compute priorities of both selection criteria and the alternative suppliers.

Analytic hierarchy process is composed of mainly three steps which are;

- problem structuring,
- evaluation of local priorities,
- computing the global priorities of alternatives to come to a decision.

In the first stage, which is problem structuring, the decision maker states the objectives, defines the selection criteria and picks the alternative choices to be selected from. In the second stage, pairwise comparisons are made, local priorities of selection criteria and alternatives are determined. Finally, in the third stage global priorities of each alternative are computed. Among these the best one is selected by the decision maker.

In the fuzzy analytic hierarchy process methodology employed in this paper, the second stage is fuzzy applications stage. In other words, fuzzy techniques are employed when determining the pair-wise comparisons and the priorities. By this way, it would be easier for the decision maker to make comparisons and it would be more accurate.

When making fuzzy comparisons, one would say that "Objective 1 is likely to be strongly important than objective 2, but it is between 4 and 6 compared to objective 2".

By this way the lower bound, most likely situation and the upper bound on the comparison is given. In order to develop the fuzzy comparison matrices, let:

l_{ij}: lower bound on the comparison between objective i and j.

m_{ij}: most likely value on the comparison between objective i and j.

u_{ij}: upper bound on the comparison between objective i and j.

The fuzzy pairwise comparison matrix is A^{Fuzzy} given in (1).

$$A^{Fuzzy} = \begin{bmatrix} \frac{w_1}{w_1} & \frac{w_1}{w_2} \dots & \frac{w_1}{w_n} \\ \frac{w_2}{w_1} & \frac{w_2}{w_2} \dots & \frac{w_2}{w_n} \\ \frac{w_n}{w_1} & \frac{w_n}{w_2} \dots & \frac{w_n}{w_n} \end{bmatrix} =$$

$$\begin{bmatrix} 1 & (l_{12}, m_{12}, u_{12}) \dots & (l_{1n}, m_{1n}, u_{1n}) \\ (l_{21}, m_{21}, u_{21}) & 1 \dots & (l_{2n}, m_{2n}, u_{2n}) \\ (l_{n1}, m_{n1}, u_{n1}) & (l_{n2}, m_{n2}, u_{n2}) \dots & 1 \end{bmatrix} \quad (1)$$

In addition, the reader should note that:

$$l_{ij}=1/\, u_{ji} \;,\; m_{ij}=1/\, m_{ji} \;,\; u_{ij}=1/\, l_{ji} \quad (2)$$

Once the fuzzy comparison matrices (A^{Fuzzy}) are built, local priorities (w_i) should be computed. In the proposed approach the fuzzy prioritization method, developed by Mikhailov & Tsvetinov [4], is employed to compute the priorities of the selection criteria and the alternative suppliers.

The fuzzy membership functions are defined as in Equation (3).

$$\mu_{ij}\left(\frac{w_i}{w_j}\right) = \begin{cases} if \;\; \frac{w_i}{w_j} \le m_{ij} \;\; then & \frac{\left(\frac{w_i}{w_j} - l_{ij}\right)}{m_{ij} - l_{ij}} \\ if \;\; \frac{w_i}{w_j} > m_{ij} \;\; then & \frac{\left(u_{ij} - \frac{w_i}{w_j}\right)}{u_{ij} - m_{ij}} \end{cases} \quad (3)$$

In order to make a consistent analysis in selecting the best supplier, all membership functions should be maximized as much as possible. This situation can be stated mathematically as in formulation (4):

$$\begin{aligned} & \textbf{maximize } \lambda \\ & \textbf{subject to} \\ & \lambda \le \mu_{ij}\left(\frac{w_i}{w_j}\right) \quad \forall i,j \\ & \textstyle\sum_i w_i = 1 \\ & w_i > 0 \end{aligned} \quad (4)$$

In addition, when we replace $\mu_{ij}\left(\frac{w_i}{w_j}\right)$ with (3) , the new formulation turns into (5).

$$\begin{aligned} & \textbf{maximize } \lambda \\ & \textbf{subject to} \\ & \lambda w_j(m_{ij} - l_{ij}) \le w_i - l_{ij} w_j \quad \forall i,j \\ & \lambda w_j(u_{ij} - m_{ij}) \le u_{ij} w_j - w_i \quad \forall i, \\ & \textstyle\sum_i w_i = 1 \\ & w_i > 0 \end{aligned} \quad (5)$$

It can be seen that this formulation is a non-linear programming model and can be solved by a non-linear programming tool. In the optimum solution the value of λ is the consistency level. In literature, it is desired to be more than 0,9 (In other words inconsistency level should be less than 0,1). Otherwise it is determined that the decision maker is not consistent and the pair-wise comparisons should be revised.

In this paper, OPL Studio* optimization software is employed to solve the model and find the priorities.

3. Application of FAHP to a Supplier Selection Problem

The application is carried out at a firm manufacturing sports outer clothing out of knitted fabric. As a nature of textile industry the number of materials is quite low whereas product variety is quite high. Firstly, ABC analysis is carried out. Looking at the total yearly purchasing cost of all materials, one of the class A items is considered in this study. However, the procedure can easily be applied to all class A and class B items. The item considered in this study has six alternative suppliers. Among these one of them is selected using FAHP.

3.1. *Determination of the Selection Criteria and The Pair-wise Comparisons*

At this step, together with the decision makers of the system, four main selection criteria are determined. These are:

Quality (Q): In the incoming quality control, some of the lots may be rejected due low quality which is an undesirable situation. Therefore, quality of the lots delivered by the supplier is an important selection criterion.

Lead-Time (L): Late delivery of lots is also undesirable. Therefore, lead-time is included in the list of supplier selection criteria.

Delivery Performance (DP): Delivery performance of a supplier is the ratio of orders it can fill. In other words, some of the lots delivered by the supplier may be smaller than the order or some of the orders may not be satisfied at all. This type of situation would decrease the reliability of the supplier. Therefore delivery performance is also included in the selection criteria.

Capacity (C): Capacity of the supplier is also important in order to understand the magnitude and the economic status of the supplier.

The fuzzy comparisons between each pair are given in Table 1. The comparison values are given in the form of (l_{ij}, m_{ij}, u_{ij}) being the lower bound,

* Software can be downloaded from www.ibm.com/developerworks/university/academicinitiative/ for free

most likely and upper bound values. It can be seen that only the values above the diagonal are given. The values below the diagonal can be found by Equation 2.

Table 1. Fuzzy comparison matrix of selection criteria.

	Q	L	DP	C
Q	1	(2, 4, 6)	(2, 3, 4)	(3, 4 ,5)
L		1	(1/5, 1/3, 1)	(1,2,3)
DP			1	(2, 3, 4)
C				1

After developing pair-wise comparisons formulation 5 is solved using OPL Studio 3.7 optimization software. The priorities belonging to each selection criterion and the consistency level (optimum value of λ) are found to be as in Table 2. Since the consistency level is greater than 0,9 these priorities are acceptable.

Table 2. Priorities of selection criteria computed by FAHP.

Criteria	Priority	Consistency
Q	**0,525**	**0,96**
L	**0,159**	
DP	**0,211**	
C	**0,105**	

3.2. *Listing Alternative Suppliers and Their Pair-wise Comparisons*

The next step in the proposed approach is listing alternative suppliers and making the pairwise comparisons. In the problem under study, there exists six alternative suppliers. These suppliers are evaluated in terms of their quality, lead-time, delivery and capacity performances. The pair-wise comparisons in between alternative suppliers can be seen in Tables 3, 4, 5 and 6 respectively. Similar to Table 1, the values below the diagonal can be found by Equation 2.

Table 3. Fuzzy comparison matrix of suppliers according to quality.

	S1	S2	S3	S4	S5	S6
S1	1	(1/3,1/2/,1)	(2,3,4)	(1,2,3)	(2,3,4)	(1/4,1/3,1/2)
S2		1	(4,5,6)	(2,3,4)	(4,5,6)	(1/2,1,2)
S3			1	(1/3,1/2/,1)	(1,2,3)	(1/8,1/7,1/6)
S4				1	(1,2,3)	(1/4,1/3,1/2)
S5					1	(1/8,1/7,1/6)
S6						1

Table 4. Fuzzy comparison matrix of suppliers according to lead time.

	S1	S2	S3	S4	S5	S6
S1	1	(1/3, 1/2, 1)	(1/2, 1, 2)	(1, 2, 3)	(1/2, 1, 2)	(1/2, 1, 2)
S2		1	(2, 3, 4)	(2, 3, 4)	(2,3,4)	(1/2, 1, 2)
S3			1	(1/2, 1, 2)	(1/2, 1, 2)	(1/4, 1/3, 1/2)
S4				1	(1/2, 1, 2)	(1/4, 1/3, 1/2)
S5					1	(1/4, 1/3, 1/2)
S6						1

Table 5. Fuzzy comparison matrix of suppliers according to delivery performance

	S1	S2	S3	S4	S5	S6
S1	1	(1/3, 1/2, 1)	(2,3,4)	(1/2, 1, 2)	(1,2,3)	(1/3, 1/2, 1)
S2		1	(6,7,8)	(2,3,4)	(6,7,8)	(1/2, 1, 2)
S3			1	(1/4, 1/3, 1/2)	(1,2,3)	(1/8, 1/7, 1/6)
S4				1	(1,2,3)	(1/4, 1/3, 1/2)
S5					1	(1/8, 1/7, 1/6)
S6						1

Table 6. Fuzzy comparison matrix of suppliers according to capacity.

	S1	S2	S3	S4	S5	S6
S1	1	(1/4, 1/3, 1/2)	(1,2,3)	(1,2,3)	(1,2,3)	(1/4, 1/3, 1/2)
S2		1	(6,7,8)	(4,5,6)	(6,7,8)	(2,3,4)
S3			1	(1/3, 1/2, 1)	(1/3, 1/2, 1)	(1/6, 1/5, 1/4)
S4				1	(1,2,3)	(1/4, 1/3, 1/2)
S5					1	(1/6, 1/5, 1/4)
S6						1

Table 7. Priorities of suppliers computed by FAHP

		Criteria			
		Q	L	DP	C
Suppliers	S1	0,147	0,162	0,123	0,112
	S2	0,297	0,28	0,327	0,478
	S3	0,063	0,113	0,053	0,076
	S4	0,096	0,111	0,123	0,084
	S5	0,053	0,113	0,047	0,069
	S6	0,344	0,221	0,327	0,18
Cosistency (λ)		**0,956**	**0,918**	**0,993**	**0,927**

After settling the pair-wise comparisons, the priorities of each supplier are computed by formulation 5. This model is solved for quality, lead-time, delivery performance and capacity factors separately. In other words, to find the priorities of each supplier according to quality purposes, the fuzzy comparison matrix given in Table 3 is used with formulation 5. Similarly, to compute priorities of suppliers for lead-time, delivery performance and capacity purposes, matrices in Table 4, 5 and 6 are used respectively. The priorities found for suppliers under each category together with the corresponding consistency levels can be seen in Table 7.

From Table 7, it can be seen that all consistency levels are greater than 0,9. Therefore, all priorities are valid and can be used to compute the overall priorities of suppliers.

3.3. *Computing Overall Priorities of Suppliers*

At the last step of the procedure, the overall priorities of each alternative supplier is found which is the weighted sum of all supplier priorities (quality, lead-time, delivery performance, capacity). This is found by multiplying the priorities of each selection criterion (from Table 2) and the corresponding priority of the supplier (from Table 7). Equation (6) gives the computation for supplier S1. All other final scores are computed in the same manner and given in Table 8.

$$\begin{bmatrix} 0,147 & 0,162 & 0,123 & 0,112 \end{bmatrix} \times \begin{bmatrix} 0,525 \\ 0,159 \\ 0,211 \\ 0,105 \end{bmatrix} = 0,141 \tag{6}$$

Table 8. Overall Priorities of suppliers

Supplier	Final Priority
S1	0,141
S2	0,320
S3	0,070
S4	0,103
S5	0,063
S6	0,304

As seen from Table 8, the best performing suppliers are S2 (0,320) and S6 (0,304). These two suppliers can be selected as the strategic partners of the firm. The overall score of S1 is 0,141. This supplier may be employed at certain times. However performance of S3, S4 and S5 are very low. These suppliers should not be selected at all.

4. Conclusion

In this study, supplier selection problem of a textile company is handled and a FAHP procedure is employed for the solution. Four main selection criteria which are quality, lead-time, delivery performance and capacity, are defined. Their fuzzy comparisons are made by experts of the system. Then by fuzzy prioritization method, weight coefficient of each criterion is found. Similarly, among the six alternative suppliers pair-wise comparisons are made on each factor (quality, lead-time, delivery performance, capacity) and priorities are computed for each supplier. Finally, overall performances are computed of all suppliers. Suppliers S2 and S6 turned out to be the most suitable vendors for the company.

References

1. Amin, S.H., Razmi J., 2009, An integrated fuzzy model for supplier management: A case study of ISP selection and evaluation, Expert Systems with Applications, Volume 36, Issue 4, Pages 8639-8648.
2. De Boer L, Van der Wegen L, Telgen J., 1998, Outranking methods in support of supplier selection. European Journal of Purchasing and Supply Management, 4:109–18.
3. Dickson, G. W. (1966). An analysis of vendor selection system and decisions. Journal of Purchasing, 2(1), 28–41.
4. Mikhailov, L., Tsvetinov P., 2004. Evaluation of Services using a Fuzzy Analytic Hierarchy Process. *Applied Soft Computing*. *5*(1), pp. 23-33.
5. Winston, W., L., 1994. Operations Research: Applications and Algorithms, Third Edition, Duxbury Press, USA.

PART 2

Uncertainty Modeling

FUZZY MODELING AND OPTIMIZATION

ANGEL GARRIDO
Faculty of Sciences UNED, Madrid, Spain

PIEDAD YUSTE
Faculty of Philosophy UNED, Madrid, Spain

The word "modeling" comes from the Latin word "modellus". It describes a typical human way of coping with the reality. So, a model will be a simplified version of something that is real. Before to the study of the typical methods of Fuzzy Logic we expose briefly which have been the historical origins of so forceful tool of the current science and technology. Because this search of their origins and evolution not only possess an intrinsic interest of mere cultural character, it can allow us to open new perspectives to know where it can advance in the future. Fuzzy Modeling will be defined as the construction of the Fuzzy Inference process. We analyze here this process, very useful for many interesting applications in modeling and optimization problems.

1. Theoretical Model

The creation of a theoretical model for explaining the workings of natural phenomena and for organizing and interpreting the data gathered from experience is a sign that marks the step from empirical knowledge to the scientific process. Theoretical models are shaped around a set of assumptions or hypotheses that are the culmination of a long process of observation. We can find a first theoretical model in some of the texts of the Chaldean astronomers; it is not expressed as such, but in the form of a rule or algorithm with which the learned men of Babylon calculated the movements and positions of the stars (*Ephemerides*, *Goal Year Texts*) [5, 6]. Some Hippocratic medical schools also conceived a theoretical model when they conjectured that our organism is made up of four humors that coexist in equilibrium and harmony (*On the Nature of Man*) [7]; they thought that the imbalance of these substances gave rise to illness or provoked death. Other doctors of the same period attributed illness to an excess of heat or cold, humidity or dryness. None of them had irrefutable proof that demonstrated the truth of their assertions; however, they designed a model to which they adjusted their observations.

In other areas of research, such as geometry, no one opposed the Greek mathematicians' use of hypotheses in their proofs. Plato, admiring the results, introduced these kinds of formulations into philosophical argumentation (*Meno* 86b-e).

The mere observation of natural phenomena is not a sufficient condition for the appearance of a science; not even the extraction of a general law applicable to all the events of the same kind is, nor is the knowledge of the causes that trigger these phenomena. All these factors are necessary but not sufficient for creating a science. The learned Greeks, contrary to their eastern predecessors, sought the origin of phenomena in nature itself (*physis*), not in the whim or will of gods and demons. But nature showed itself to be unstable and changing. The constant flow made it impossible to establish a permanent and lasting knowledge of the beings that make it up. How is it possible to know something that is so today but ceases to be so tomorrow? The first Greek cosmologists sought the substance that endures as opposed to change; the unity of the being as opposed to the plurality of its forms. They formulated so many hypotheses that, in the end, some philosophers, such Protagoras, concluded that all knowledge is reduced to sensation: "Man is the measure of all things" (Sextus Empiricus, *Adversus Mathematicos* VII, 60) [8]. The difficulty of establishing scientific knowledge led Parmenides of Elea to formulate the *principle of non-contradiction*: "it is impossible for something to be and not to be at the same time" (*On Nature*) [8]. This basic and unprovable formulation, in Aristotle's (*Metaphysics* IV, 4, 1009a-1010a) judgment, together with other axioms with analogous qualities, lays the foundation for western science. This principle rises above the discordances that we notice in nature: it stops time at that instant and captures the immutable being that does not experience change. For Parmenides of Elea, everything in nature is Being, to the point that it is impossible to conceive of Non-Being. From our point of view, this hypothesis is hard to accept because we all see how things transform themselves from what they are not now to what they will be later on. In short, all organisms seem to be and not to be at the same time. Plato gave being to the Non-Being Parmenidean (*Teaethetus* 257c). How can this be explained? A man is an animal, a mammal, etc., but not a bird or a stone or a fountain. Shortly afterward, Aristotle went farther with this idea and proposed an ontological model that got around these contradictions: natural beings are here and now, actualized, but they possess potential qualities that they will develop later on (*Metaphysics* IV, 5 1010a). Thus, a boy is a potential man and a seed will become a tree in time. Aristotle attempted in this way to achieve a universal and necessary knowledge, from its causes, rising above appearances and sensations. In his cognitive model, the *principle of non-contradiction* acquires the role of a basic, irrefutable axiom: "the same attribute cannot at the same time belong and not belong to the same subject" (*Metaphysics* IV, 3 1005a). Aristotle reduces the formulations of science to propositions in which a predicate is affirmed or denied regarding one or several subjects. But between both extremes of the contradiction – possessing and not-possessing a certain quality - there sometimes exists a range of attributes that are more or less close to these extremes: "there is a more and a less in the nature of beings" (*Metaphysics* IV 4 1009a). Aristotle knows that, in cases such as the move from white to black, from large to small, from slow to fast, there is not one but many intermediate terms located more or less close to both extremes. However, in his theoretical model, he only contemplates those qualities that do not admit the existence of an element that simultaneously denies the extremes of the contradiction (*Metaphysics* IV, 7 1011b), such as when we decide between even and odd numbers. Thus, he avoids the vagueness existing in the sensible world and chooses for the science the concepts which designating only the actualized essence. He draws a bivalent universe, without nuance, lopsided towards the ends, too rigid to be a mirror of reality. The Aristotelian paradigm is clearly insufficient to decide on events involved in human discernment and natural language, but was not questioned until well into the 20th century, when thinkers such as Max Black, Bertrand Russell, and Jan Lukasiewicz already had enough theoretical tools available to rescue and interpret the notion of vagueness. And after these the seminal papers of Lofti Asker Zadeh [4], from which derives many of the more innovative and interesting new Mathematics, with very surprising applications [1-3].

2. Fuzzy Modeling

Fuzzy Logic can be conceptualized as a generalization of Classical Logic, dealing mathematically with imprecise information usually employed by humans. As a Multi-Valued Logic, it extends Boolean Logic, usually employed in classical science. *Fuzziness* describes event ambiguity. Hence, it measures the degree to which an event occurs, not whether it occurs, whereas *Randomness* describes the uncertainty of event occurrence. Whether an event occurs is *random;* to what degree it occurs is *fuzzy*. It may be applied to generalize the Classical Set Theory of Cantor, giving the Fuzzy Set Theory. A *Fuzzy Set* is any point in the cube $I^n = [0, 1]^n$. So, given U as the Universe of Discourse, (U, I^n) is the fundamental measurable space of fuzzy theory.

A linear combination like a fuzzy model is clearly understandable. The fuzzy model proposed by Takagi, Sugeno, and Kang (TSK, by acronym) is described through fuzzy IF-THEN rules which represents local input-output relations of nonlinearity. The main feature of a TSK fuzzy model is to express the local dynamics of each fuzzy implication (rule) by a linear system model. The overall fuzzy model is achieved by fuzzy "blending" of the linear system models.

Fuzzy Modeling is many times used to transform the knowledge of an expert into a mathematical model. The emphasis is on constructing a fuzzy expert system that replaces the human expert. A fuzzy model represents the real system in a form that corresponds closely to the way humans perceive it. Thus, the model is easily understandable, and each parameter has a readily perceivable meaning. The model can be easily altered to incorporate new phenomena and if its behavior is different than expected it is usually easy to find which rule should be modified, and how. Furthermore, the mathematical procedures used in fuzzy modeling have been tried and tested many times, and their techniques are relatively well documented.

3. Fuzzy Rule-Based Systems

The Rules shows great advantages over Classical Logics [1,3]. In Classical Logics, Reasoning is Monotonic, with inferences without contradiction with the pre-existing. Whereas in RBS we can delete facts or assertions from the FB, as the new inferences prove them wrong. This makes this kind of reasoning Non-Monotonic, because we can modify the conclusion.

What happens in case that we have more than one suitable rule at the same time? Which one should be applied first? In each step, we call the Conflict Set the set of applicable rules; it is, obviously, a dynamic set. The subjacent decision problem is called Resolution of Conflicts or Control of Reasoning.

The introduction of concepts and methods of *Fuzzy Logic* is very usefulness covering adequately the indetermination or imprecision of the real world. We define the *"world"* as a complete and coherent description of how things are or how they could have been. In the problems related with this "real world", which is only one of the "possible worlds", the Monotonic Logic does not work with frequency. Such type of Logic is the classical in formal worlds, such as in Mathematics. But it is necessary to provide our investigations with a mathematical construct that can express all the "grey tones", not the classical representation of real world as either black or white, all or nothing, but as in the common and natural reasoning, through progressive gradation.

Logic analyzes the notion of consequence. So, it deals with Propositions or Set of Propositions, and their mutual relationships. Formal Logic attempts to represent all this by means of well-defined logical calculi. Some calculi differ in their definitions of sentences and their respective notion of consequence. In Medicine, a fuzzy proposition may be true to some degree, and this degree may be evolving with time. So, the sentence "the patient is old" is true in some degree: the older the age of the patient, the truer the sentence. Standard examples of fuzzy propositions may include linguistic variables, such as the related with age, being possible values: young, medium, old, with intermediate degrees, by fuzzy modifiers, as almost, something, enough, very, etc.

Fuzzy Logic has two different meanings, wide FL, and narrow FL. In this last sense, *narrow Fuzzy Logic, FLn* by acronym, is a logical system that attempts to formalize the approximate reasoning. So, it will be an extension of Multi-Valued Logic. Therefore, this FLn has a much wider range of applications than traditional logical systems, because the apparition of new concepts. Instead of this, if we consider *Fuzzy Logic in its widest sense, FLw* by acronym, it will be a synonymous with Fuzzy Set Theory (FST, by acronym). Indeed, it is the theory of classes with un-sharp boundaries. Therefore, FST is much broader than FLn, including this latter as one of their branches. So, in the broad sense, everything dealing with fuzziness may be called a Fuzzy Logic. But in the narrow sense, the base of Fuzzy Logic will be the formal calculus of Multi-Valued Logic.

4. Fuzzy Optimization

The Mamdani method is the most used in applications, due to its simple structure of "min-max" operations. It proceeds in four steps, evaluate the antecedent of each Rule; obtain each Rule´s conclusion, aggregate conclusions, and defuzzification. If we only take into account the factors that really matter in the problem, it is enough to write a set of rules that model the problem. Another advantage of using the fuzzy approach [2] is that, should we want to add more variables to the problem, all we would have to do is write new rules or edit the existing ones. This means a lesser amount of effort than rewriting an algorithm. So, Fuzzy Logic is adaptable, simple and easily applied.

Mamdani´s method is useful when there are a very small number of variables. Otherwise, we will find certain difficulties, as may be: the number of rules increases exponentially with the number of variables in the antecedent; the more Rules we construct, the harder is to know if they are suitable for our problem; and if is too large the number of variables in the antecedent, it results difficult to understand the causal relationship between them (the antecedent and the consequents); hence, constructing new Rules may be harder.

The second Fuzzy Inference method was introduced by Takagi, Sugeno, and Kang (TSK method, by acronym) in 1985. It is very similar to Mamdani´s method in many respects. The first two steps are the same. The essential difference between them is that in Sugeno method the output membership functions are either constant or linear.

5. Conclusions

Fuzzy Modeling may be used to transform the knowledge of an expert into a mathematical model. The emphasis is on constructing a fuzzy expert system that replaces the human expert. Also as a tool that can assist human observers in the difficult task of transforming their observations into a model. In many fields of science, human observers have provided linguistic descriptions and explanations of various systems. For this, there is a need to construct a suitable mathematical model, a process that usually requires a very subtle mathematical understanding. This new modeling is a many more direct and natural approach for transforming the linguistic description into such model. It represents the real system in a form that corresponds closely to the way humans perceive it. Thus, the model is easily understandable, and each parameter has a readily perceivable meaning. The model can be altered to incorporate new phenomena, and if its behavior is different than expected, it is easy to find which rule should be modified, and how. The mathematical procedures used have been tried and tested many times. So, there are different ways to implement Fuzzy Rules, and by these, to improve tools for Fuzzy Reasoning. We have showed here a general vision for modeling and optimization problems, by searching the historical foundations of these topics.

References

1. A. Garrido, "Searching the Arcane Origins of Fuzzy Logic", *BRAIN* 2(2): 51-57, Bacau (2011).
2. A. Garrido, and V. Postoliça, *Modern Optimization,* Matrix-Rom Editura, Bucharest (2011).
3. A. Garrido, "A Brief History of Fuzzy Logic", *BRAIN* 3(1): 71-77 (2012).
4. L. Zadeh, "Fuzzy Sets", *Information and Control,* 8: 338-55 (1965).
5. O. Neugebauer, *Astronomical Cuneiform Texts,* 3 vols. Springer (1983).
6. A.J. Sachs, "A Classification of the Babylonian Astronomical Tablets of the Seleucid Period", *Journal of Cuneiform Studies* 2: 271-290 (1948).
7. W.H.S. Jones, *Hippocrates Collected Works*, Cambridge: Harvard Univ. Press (1968).
8. H. Diels ; W. Kranz, *Die Fragmente der Vorsokratiker,* Berlin (1903).

ANALYZING MATURITY LEVELS OF TURKISH SMES IN RESPECT TO ERP AND LEAN MANUFACTURING APPLICATIONS: A RELATIONAL MODEL

UFUK CEBECI

Department of Industrial Engineering, Istanbul Technical University, Istanbul, 34367, Turkey

ÇAGATAY IRIS[†]

Department of Industrial Engineering, Istanbul Technical University, Istanbul, 34367, Turkey

Interaction with rapid development of information technology and progress in modern production management science has emerged. In this respect, ERP (Enterprise Resource Planning) and Lean Manufacturing applications are important not only for big companies but also for Small and Medium Size Enterprises (SMEs) to become more competitive. The aim of this study is to analyze the maturity levels of Turkish SMEs on ERP systems and Lean Manufacturing applications. A questionnaire is prepared and applied to White goods enterprises (mostly subcontractors). The derived correlation between these two main function areas is also used to design such new systems. Additionally, ERP and Lean Manufacturing concepts are analyzed together in questionnaire which is applied by regular visits to facilities. Hence, related performance model also gives information about effects of each performance on one another.

1. Introduction

Models that focus on both distinct areas (ERP and lean manufacturing) have not analyzed maturity levels in a combined aspect [1]. The aim of this study is to understand how well SMEs perform ERP usage in module aspect and lean production associated. The study also conducts a relational model that try to formulate correlation between usage of ERP systems and lean production implementation in White Goods Manufacturing facilities of Turkey.

Most of the studies in literature focus on these tool in separate way. Schlichter and Kraemmergaard [3] have published a literature survey on ERP studies and emphasized that affect of ERP usage on other business activities of facilities are not analyzed in detail. Federici [4] has explained factors

[†] Email: ciris@itu.edu.tr

influencing ERP outcomes in SMEs. And, it is found out that one of the most influential factors on ERP performance is attaining valuable business activities. In this respect, it is vital to evaluate maturity level of ERP usage. Tagliavini et. al. [5] have explored usage of ERP by SMEs and tested numerous hypotheses. In this study, performance evaluation in respect to ERP application is done according to modules usage efficiency. Shah and Ward [6] have defined parameters that can be used to test Lean production activities. The aspects those are limited to 7 fundamental requirements are used in this study. Determining impact of ERP usage on lean production is not analyzed in literature. However, Yang and Su [1] have established a relational model that reflects relationship between ERP usage and supply chain management performance. The methodology used in study is based on canonical correlation which aims to build a correlation between multivariate parameters. Riezebos et. al. [2] have focused whether there is a connection or contradiction between ERP usage and lean production. It is emphasized that ERP software lack supporting lean production activities. Organization of paper is as follows; in second section, methodology of study will be detailed, and after that, initial factor analysis results for ERP usage and Lean production will be presented separately, and finally relational model to reflect correlation will be analyzed.

2. Methodology

A questionnaire was constructed based on an extensive review of the literature in the areas of ERP and lean production implementation. Fifty Turkish White-Goods SMEs were determined as the candidates to participate. One manager of each enterprise who was responsible for implementing ERP was visited to rate the level of success in achieving improvements and lean production maturity.

Maturity of ERP usage for each of fourteen modules is determined by using two main parameters. It is a combined version of effective usage of modules (E1: percentage of tasks done by using ERP) and efficient usage of modules (E2: correctly usage of ERP modules). Hence, overall utilization of ERP module is a non-metric parameter to be used in statistical inferences (Overall= E1*E2). As it is mentioned, each ERP module is tested with strategy proposed in [4]. Lean production is also tested by considering six fundamental aspects as; 5S, SMED, Pull-based Production (KANBAN), Total Preemptive Maintenance, Cellular Manufacturing, Quality Management. Each of listed main frames has sub-questions to understand actual level of lean production maturity. This data gathering questions used a five-point Likert scale.

Initially, data is analyzed by means of fourteen ERP modules and basic parameters are obtained for each module (mean usage, standard deviation). After determining overall means, factor analysis is applied to understand actual characteristics of modules. The same procedure is applied for lean production. Finally, by using results of factor analysis a relational model based on MANOVA is established.

3. Survey results

3.1. *ERP module usage*

Initial results obtained from SMEs are presented in Table 1. It may be possible to understand usage related modules by controlling overall usage.

Table 1. Basic statistic from survey (Ref Table 2)

	PR		BOM		PM		IM		SL		CRM		FN	
	E1	E2	E1	E2	E1	E2	E1	E2	E1	E2	E1	E2	E1	
Mean	.81	.79	.95	.90	.82	.76	.85	.83	.71	.72	.46	.41	.81	.8
Std. Dev.	.18	.22	.08	.20	.15	.19	.13	.14	.26	.25	.29	.29	.18	.18
Overall		.64		.86		.63		.71		.51		.19		.65
	AC		PL		QM		HR		BT		DA		MM	
Mean	.96	.96	.73	.72	.58	.59	.53	.53	.86	.85	.28	.30	.38	.39
Std. Dev.	.10	.08	.27	.28	.32	.29	.37	.35	.17	.19	.37	.37	.35	.32
Overall		.91		.52		.34		.28		.73		.08		.15

Bill-of-Material and Accounting modules are most commonly used, while CRM and Automated data gathering systems are not widely used. It is required to consider standard deviation of answers in each topic. It is observed that there is a huge fluctuation in usage of HR and Data automation. The loss of usage in PP and IM is mostly based on lack of ERP packages (e.g. capacity planning, scheduling, inter-depot transfer etc.).

Factor analysis is conducted on explanatory variables, with the objective of determining minimum number of factors representing maximum variance of data obtained in ERP module usage results. For related data set, Kaiser-Meier Olkin test statistics is obtained as 0,61 which corresponds to an applicable set with Bartlett's test significance zero. Overall statistics show that data set is applicable for factor analysis. Varimax rotation is used with principle component analysis. Factors having an eigenvalue greater than 1 are selected to be presented in the factors list. For rotated component matrix, there are 5 different factors which are illustrated in Table 2 are obtained.

Table 2. Rotated component matrix

Factors	Purchasing (PR)	Bill-of-Material (BOM)	Production Management (PM)	Inventory Management (IM)	Sales (SL)	Customer Relations Management (CRM)	Finance (FN)	Accounting (AC)	Planning (PL)	Quality Management (QM)	Human Resources (HR)	Budgeting (BT)	Data Automat (DA)	Maintenance Management (MM)
F1		0.28	0.48		0.29			0.33	0.36	0.82	0.67			0.83
F2			0.44			0.88	0.25		0.52			0.30	0.89	0.27
F3	0.84							0.74		0.36	0.25	0.76		
F4		0.80	0.58	0.81	0.31				0.61					
F5					0.81		0.80				0.42	0.49		

According to rotated component matrix; F1 (including; QM, MM, HR dominantly) represents overall OEE parameters which all together corresponds efficiency of manufacturing systems. F2 (including; CRM, DA dominantly) corresponds to new technologies adapted to ERP systems recently. While, F3 (including; PR, AC, BT) directly represents purchasing based accounting and budgeting activities. F4 (including; BOM, IM, PM, PL) directly reflect overall production planning and control activities. Finally, F5 (including; SL, FN, BT) gives direct information on sales and finance activities. Factor derived from survey is very sensible in respect to usage comments that are made in previous sections. Factors obtained in this phase will be used in further analysis.

3.2. *Lean manufacturing maturity level*

As mentioned before, maturity level of SMEs in respect to Lean Production is analyzed and it is found out that facilities are not properly working according to value-addition principle. Table 3. represents results of regular visits. KANBAN based pull-production strategy is main problem area in most of facilities. It seems that facilities mostly apply lean rules in initial phases. Further steps which require analytical projects are mostly neglected. It is also found out that ERP packages used by facilities do not support lean production activities. It is very well-known that easy-to-implement strategies are very common between facilities. But, whenever issue gets complicated (in which ERP system should have to be integrated), means decrease.

Table 3. Delta determination test (Four different probability distributions)

	5S (7S) Applications								Pull Production					Total Productive Maintenance				
	Continuous 5S	5S Standardization	5S Plans	5S Control Charts	Red Tag Application	Visualization of 5S	Safety factors of 5S	Security factors	Implication of Zero Inventory	KANBAN Usage	KANBAN Calculation	Visual Inventory	One-piece Flow	TPM Policy and Goals	Autonomous Maintenance	Planned Maintenance	TPM Trainings	OEE Calculation
M	2.6	2.5	2.5	2.4	2.0	2.9	2.8	2.6	2.7	1.6	1.4	2.2	2.1	2.8	2.6	3.3	2.1	2.0
ST	1.2	1.5	1.2	1.3	1.2	1.1	1.1	1.1	1.4	1.2	1.0	1.1	1.1	1.2	1.3	1.2	1.1	1.3

	SMED			Cellular Manufacturing					Quality Assurance									
	Setup Clustering	SMED Plans	Effc. Implementation	Part and Product lists	Part-Machine Matrices	Production Cells	One-piece flow in cell	Visual Inventory Man.	Pareto and Fishbone Diag.	ANOVA tests	Value-stream mapping	Kaizen activities	Hoshin Kanri	Heijunka	Shojinka	Poke-Yoke	Education and Training	Work Study&Measurement
M	2.2	2.0	2.1	3.6	3.4	2.8	2.4	2.4	3.3	1.9	1.5	3.1	2.0	2.0	2.2	2.9	2.9	2.7
ST	1.3	1.3	1.4	1.1	1.1	1.2	1.2	1.1	1.4	1.2	1.0	1.1	1.2	1.2	1.1	1.3	1.2	1.3

According to obtained results, there should be six factors which are direct representatives of main frames that we have proposed in this study. Hence, means are calculated for each type of lean production aspect and these values are used in MANOVA analysis.

3.3. *MANOVA analysis for erp usage-lean production*

Multivariate analysis of variance strategy is used to test interaction between these variables. Due to the fact that, ERP usage is non-metric, formulation is updated. Box M's test statistic (sign:0) and Levene's test of equality of error variance results in computational adequacy. Table 4. reflects that factor score means are statistically different not only for Roy's Largest rule (Wilk's Lambda) but also other tests. Hence, MANOVA is applicable for data set.

Table 4. Multivariate test results

Effect		Value	F	Hypothesis df	Error	Sig.
Intercept	Pillai's Trace	.945	48.733	6.000	17.000	.000
	Roy's Largest Root	17.200	48.733	6.000	17.000	.000

Table of Tests of Between-Subjects Effects eventually establish correlation within related data. It is derived that intercept parameter between all of factors is valuable. Table 5. gives information about multiple comparisons and affects of each factor on each variable which reflect lean production outcomes.

Table 5. Relational model between ERP and Lean Production

Dependent	Corrected	Intercept	FAC1_1	FAC2_1	FAC3_1	FAC4_1	FAC5_1
VAR001	0.01	0.00	0.35	0.00	0.06	0.41	0.40
VAR002	0.63	0.00	0.88	0.15	0.48	0.62	0.51
VAR003	0.14	0.00	0.04	0.05	0.69	0.79	0.89
VAR004	0.04	0.00	0.86	0.00	0.37	0.60	0.72
VAR005	0.21	0.00	0.99	0.04	0.98	0.10	0.70
VAR006	0.47	0.00	0.98	0.09	0.62	0.52	0.83

Results show that there is a strong correlation of 5S activities with F1 (which reflect maintenance), F4 (Production activities), F5(Managerial issues). SMED applications seem to be effected by most of modules used in ERP systems. Pull-based production strategies are mostly affected by F4 and F5 which are production dependent factors. It is also detected that there is an obvious correlation between F1(quality tools) and VAR6 (quality assurance) systems. Finally, obtained results VAR4 (TPM) activities are mostly affected by F1 which has HR, MM together.

4. Conclusion

In this study, we have analyzed maturity levels of ERP and lean production of Turkish SMEs. It is found out that there is some strong correlation between specific parameters. Hence, SMEs may concentrate on these topics to increase productivity. Results are presented in different organizations to inform SMEs.

References

1. C. Yang, Y. Su, "The relationship between benefits of ERP systems implementation and its impacts on firm performance of SCM", *JEIM*, Vol. 22, No.6, pp. 722-752, 2009.
2. J. Riezebos, W. Klingenberg, C. Hicks, "Lean production and Information Technology: Connection or Contradiction?", *Computers in Industry*, Vol. 60, pp. 237-247, 2009.
3. B. Schlichter, P. Kraemmergaard, "A comprehensive literature review of the ERP research field over a decade", *JEIM*, Vol.23(4), 486-520, (2010).
4. T. Federici, "Factors influencing ERP outcomes in SMEs: a post introduction assessment", *Journal of Enterprise Information Management,* Vol. 22, No: 1, pp. 81-98, (2009).
5. M. Tagliavini, P. Faverio, A. Ravarini, F. Pigni, G. Buonanno, "Exploring use of ERP systems by SMEs", *6th WMCSCI*, 2002.
6. R. Shah, P. Ward, "Defining and developing measures of lean production", *Journal of Operations Management*, Vol.25, 785-805, 2007.

A POSSIBILISTIC MATHEMATICAL PROGRAMMING APPROACH FOR BACKBONE SELECTION AND DEMAND ALLOCATION PROBLEM IN TELECOM NETWORKS

H. H. TURAN[1], N. SENER

Department of Industrial Engineering, University of Yalova,Yalova, 77100, Turkey

In this study, single objective profit maximization problem of a telecommunication intermediary is analyzed under uncertainty. In telecom market, various uncertainties exist in the related cost and pro t margins, which can affect the optimization processes and the decision schemes generated. Therefore, the fuzzy set theory is employed in order to handle the non-deterministic problem parameters rising due to the presence of vagueness and imprecision. Two different pricing policies namely all-you-can send and pay-per-byte are implemented into novel possibilistic mathematical programming model. As solution approach, two commonly used different solution methodologies are applied to resulting integer programming model with fuzzy objective function coefficients. Finally, performances of suggested methods are tested on several randomly generated scenarios. The results obtained indicate that both approaches provide almost identical solutions.

1. Introduction

Backbone network providers, bandwidth brokers (intermediaries) and customers constitute current telecommunication market. The backbone provider owns the network infrastructure which is necessary for the service provision. The intermediary does not have a network and in order to provide the service he has to lease the necessary network capacity from the backbone provider. [1] Thus, the bandwidth broker has to decide how much capacity to lease from backbones to satisfy end customers bandwidth requests.

As the bandwidth resources required to support the quality of various applications are different, end customers similarly have different bandwidth requirements.[2] Therefore, telecom backbones provide a variety of capacity purchasing options to intermediaries under the condition that each type of purchasing option has different sale price, a unique bandwidth restriction and distinct Quality of Service (QoS) level. QoS is the collective summation of service measures, which determines the degree of user satisfaction of the

[1] email: Hasan.turan@yalova.edu.tr

service. [3] Common measures of QoS are throughput, latency, packet loss, delay and jitter.

A variety of charging schemes is modeled as a function of the amount transmitted, the transmission rate, bandwidth and QoS in the literature. In this study, two different pricing schemes embedded into objective function of telecom intermediary. Firstly, it is assumed that intermediary is charged by all-you-can send pricing policy when he purchases capacity from back-bones. Secondly, pay-per-byte pricing scheme is implemented into objective function when intermediary charges his end-customers for their network capacity acquisitions. In the case of all-you-can send pricing, customer (intermediary) will buy bandwidth at a fixed price (pay per resource) for a fixed amount of time.[4,5] This pricing structure encourages the customer to use as much bandwidth as possible in order to maximize his utility. The cost function is depicted in Fig. 1. According to Fig. 1, if the customer buys a resource with bandwidthβ_i^0, he has to pay $c_i^{\beta^0}$ regardless of how much of the bandwidth he uses. Similarly, one has to pay $c_i^{\beta^1}$ forβ_i^1. The challenge for the customer is to acquire just enough capacity. For Pay-per-byte pricing policy the total charge of a task is a function of the total number of packets successfully carried.[6,7] Bandwidth broker aims for maximizing his profit by making decisions about choosing right backbone providers to lease network capacity and allocating realized customers' demand into acquired network capacity.

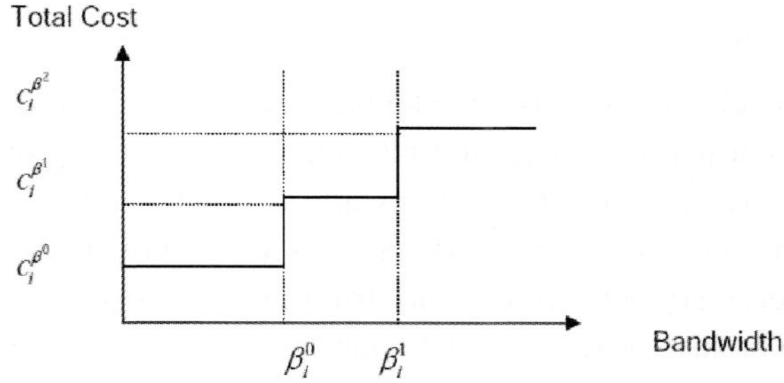

Fig. 1. The *all-you-can* pricing scheme.

However, much of the decisions undertaken in the real world take place in an environment where the objectives, or constraints or parameters are not known precisely. Thus, fuzzy set theory may provide alternative approach for absorbing the imprecise nature of the real world. Fuzzy set theory was introduced by Zadeh [8] and implemented in mathematical programming since 1970 when Bellman and Zadeh [9] introduced the concept of fuzzy goals, fuzzy constraints and decisions.

The rest of the paper is organized as follows. Section 2 presents problem definition, notation and the proposed possibilistic integer programming model with fuzzy objective coefficients. In Section 3, algorithms are summarized to unravel the problem. Results of computational studies are also demonstrated in this section. Finally, Section 4 drowns conclusions and describes future goals respectively.

2. Mathematical Model

The problem is modeled from the intermediarys (bandwidth brokers) point of view and assumes intermediary is rational decision maker. We model single period optimization problem that he faces when acquiring and selling bandwidth at a market in which there exist several backbone providers and end users. Table 1 documents all notations used in mathematical modeling formulation.

FBSDAM :

$$Z = \max \sum_{j\in J} \mu_j \beta_j \tilde{v}_j^{+} - \sum_{i\in I} x_i \tilde{c}_i - \sum_{j\in J} \beta_j \tilde{v}_j^{-} (1-\mu_j) \quad (1)$$

Subject to:

$$\sum_{j\in J} y_{ij}\beta_j \le x_i \beta_i^{max} \quad (2)$$

$$y_{ij} \le x_i \;, \forall i \in I, \forall j \in J \quad (3)$$

$$y_{ij} q_j \le Q_i \;, \forall i \in I, \forall j \in J \quad (4)$$

$$\sum_{i\in I} y_{ij} = \mu_j \;, \forall j \in J \quad (5)$$

$$x_i, y_{ij}, \mu_j \in \{0,1\} \;, \forall i \in I, \forall j \in J \quad (6)$$

The Fuzzy Backbone Selection and Demand Allocation Model's (FBS-DAM) objective function (1) represents pro t maximization goal of band-width broker under all-you-can-send pricing scheme. The objective function includes one fuzzy profit term for gained revenue and two fuzzy cost terms. The first cost term represents total capacity acquisition cost. The second term reflects opportunity or penalty cost due to not meeting the end users' capacity requests.

Furthermore, all of the fuzzy coefficients are modeled as triangular fuzzy number (TFN) such as $\tilde{v}_j^{+} = (\tilde{v}_j^{+L}, \tilde{v}_j^{+M}, \tilde{v}_j^{+U})$.Eq. (2) ensures that it is not possible to allocate total end users' demand that is more than purchased capacity from corresponding backbone provider. Eq. (3) guarantees that demands can be assigned to only purchased capacity. Eq. (4) assures that bandwidth requests of end customers are allocated to the backbones satisfying the quality of service requirements. Eq. (5) together with Eq. (6) guarantee that demands are met from only one back-bone. Moreover, Eq. (6) also represents binary restrictions on all of decision variables.

Table 1. Problem parameters and Decision variables.

Index Sets		
	I:	Set of backbone telecommunication providers
	J:	Set of end users
Parameters		
	β_j:	Bandwidth demand of end user j
	$\tilde{v}_j^+$:	Fuzzy unit revenue earned by meeting demand j
	$\tilde{v}_j^-$:	Fuzzy unit penalty cost for unsatisfied demand j
	$\tilde{c}_i$:	Fuzzy cost of leasing capacity from backbone i
	β_i^{max}:	Maximum amount of bandwidth can be purchased from backbone i
	Q_i:	Quality of service level of backbone i
	q_j:	Quality of service level requested by demand j
Decision variables		
	x_i :	1, if backbone i is selected; 0, otherwise
	μ_j :	1, if demand j is satisfied; 0, otherwise
	y_{ij} :	1, if demand j is allocated into backbone provider ; 0, otherwise

3. Solution Algorithms and Computational Study

In this section, two well-known and widely used solution methodologies are applied to solve FBSDAM. The details of first methodology (ZA) can be found in Ref. 10, and Ref. 11 discusses the second solution algorithm (HA). Flow of both algorithms is summarized below.

(1) HA Algorithm

Set $\tilde{v}_j^+ \leftarrow \frac{4v_j^{+M}+v_j^{+L}+v_j^{+U}}{6}, \tilde{v}_j^- \leftarrow \frac{4v_j^{-M}+v_j^{-L}+v_j^{-U}}{6}$ $and \tilde{c}_i^+ \leftarrow \frac{4c_i^M+c_i^L+c_i^U}{6}$

Solve FBSDAM

(2) ZA Algorithm

Calculate Z^{NIS} by **Setting** $\tilde{v}_j^+ \leftarrow v_j^{+L}, \tilde{v}_j^- \leftarrow v_j^{-U} and\ \tilde{c}_i \leftarrow c_i^U$ and

Solving FBSDAM

Calculate Z^{PIS} by **Setting** $\tilde{v}_j^+ \leftarrow v_j^{+U}, \tilde{v}_j^- \leftarrow v_j^{-L} and\ \tilde{c}_i \leftarrow c_i^L$

Solving FBSDAM

Solve

$$Max\ \lambda \tag{7}$$

Subject to

$$\lambda \leq \frac{Z - Z^{NIS}}{Z^{PIS} - Z^{NIS}} \tag{8}$$

$$\lambda \geq 0 \tag{9}$$

and constraint sets (2) – (6)

Table 2 summarizes the averaged solutions obtained by running 10 different test instances where $|I| = 10$ and $|J| = 20$ under suggested approaches. Both algorithms provide almost identical solutions in terms of profit, revenue and costs. One major difference is the running time where HA algorithm is faster compared to ZA due to the less preprocessing labor needed to start running integer programming model.

4. Conclusion

Selecting the right backbones to lease network capacity plays important role in broker's efficiency and effectiveness. As a result, the main contribution and innovation of proposed work is the development of a mechanism for solving complex backbone selection and demand allocation problems in communication networks.

Table 2. The comparison of solution quality by different approaches.

	ZA	HA
# of Backbone Selected	7.6000	7.8000
# of Satisfied Demand	19.5000	19.2000
	0.4955	*****
Revenue	69.9677	69.9247
Cost of Buying Capacity	29.9204	29.9069
Opportunity Cost	0.1983	0.1963
Net Profit	39.8491	39.8215
Purchased Capacity Amount	9.6269	9.8050
Capacity Utilization	0.9669	0.9350
Demand Fulfillment Ratio	0.9759	0.9620
CPU Time	2.8236	0.2430

Based on the fact that accurate information needed for making these decisions is not available. The computational experiments revealed the fact that both algorithms provide nearly equal performances. As a future research direction, extending the model to multiperiod and taking into account fuzziness of QoS parameters would be worthwhile theoretical and managerial contributions to current literature.

References

1. Audestad, J. A., Gaivoronski, A. A., Werner, A., (2006). Extending the stochastic programming framework for the modeling of several decision makers: pricing and competition in the telecommunication sector, *Annals of Operational Research*, 142, pp.19-39.
2. You, P. S., Hsieh, Y., Ikuta, S. (2011), A heuristic to bandwidth allocation and sales limit setting for Internet service providers, International Journal of Systems Sciece, DOI:10.1080/00207721.2011.565132.
3. Ahsan, S. M., (2006). A framework for QoS computation in web service and technology selection, Computer Standards and Interfaces, 28, pp.714-720.
4. Courcoubetis, C., Kelly, F., Weber, R., (2000). Measurement-based usage charges in communications networks, *Operations Research*, 48(4), pp.535-548.
5. Kelly, F., P., (1994). On tariffs, policing and admission control for multi-service networks, *Operations Research Letters*, 15, pp.1-9.
6. Altmann, J., Chu, K., (2001). How to charge for network services at rate or usage-based? *Computer Networks*, 36, pp.519-531.
7. Gupta, A., D. O. Stahl, A. B. Whinston, (1997). Priority pricing of the integrated services networks. L. W. McKnight, J. P. Bailey, eds. Internet Economics, 3rd ed. The MIT Press, Cambridge, MA, pp.323-352.
8. Zadeh, L.A. (1965). Fuzzy sets, *Information and Control*, 2, pp.338-353.
9. Bellman, R.E., Zadeh, L.A., (1970). Decision making in a fuzzy environment, *Management Science*, 17, pp.141-164.
10. Zimmermann, H.J., (1978), Fuzzy programming and linear programming with several objective functions. *Fuzzy sets and systems*, 1, pp.45-55.
11. Lai, Y.J. and Hwang, C.L., (1992). A new approach to some possibilistic linear programming problem, *Fuzzy Sets and Systems*, 49(2), pp.121-133.

A MODIFIED QFD TECHNIQUE FOR PRODUCT R&D

SELCUK CEBI[1], EMRULLAH DEMIRCI
Department of Industrial Engineering, Karadeniz Technical University
Trabzon, 61080, Turkey

Quality Function Deployment (QFD) is an effective methodology to use product research and development (R&D) and it has lots of applications in the literature. However, none of the presented studies takes into account the dependencies among technical characteristics. Therefore a modified QFD approach by integrating DEMATEL technique into QFD has been proposed to fill in the gap in the literature. To illustrate the proposed algorithm, a numerical example has been presented in the scope of the paper.

1. Introduction

Nowadays, in order to stay in competitive market, the companies try new products. A new product development is a key factor for business success. The successful new products put into application. But, product life cycles are getting shorter in parallel with rapid developments in technology [1]. A new product development is rapidly becoming the focus point in many companies for the fulfillment of customer needs. Therefore, Quality Function Deployment (QFD) has been used for product development. QFD is one of the tools that can be used to identify customer needs and link the needs the product design [2]. Furthermore, QFD is one of the techniques that provide a structured frame work to incorporate the voice of customer into product design [3]. This study proposes a modified QFD approach by integrating decision-making trial and evaluation laboratory (DEMATEL) technique into QFD in order to take into consideration interdependencies among technical characteristics. DEMATEL is an effective technique which presents the interdependencies among the design parameters through a causal diagram [4].

2. Literature Review

In the literature, there are many methodological studies on QFD of which aim is to improve effectiveness of QFD. These studies can be classified into four

[1] E-mail: scebi@ktu.edu.tr

categories; *(i) uncertainty based models, (ii) quality management based models, (iii) mathematical modeling based models,* and *(iv) MCDM based models*

For *uncertainty based models*, there are several papers which integrated fuzzy set theory into QFD structure [5-9]. Furthermore, robustness index by Kim et al. [10] rough numbers by Zhai et al. [11] and Yang et al.[12], grey relational analysis by Chen and Chou [13], fuzzy least-squares regression by Kwong et al [14], and grey theory and the Taguchi method by Chen et al [15] integrated into QFD structure to cope with uncertainties on the identification of relationships between customer requirements and technical characteristics.

Following studies can be given to illustrate *quality management based models*: Sakao [16] proposed an ecodesign methodology which integrates life cycle assessment (LCA) and QFD. Su and Lin [17] combined QFD, TRIZ and Fuzzy set theory. In another study, Utne [18] combined QFD, life-cycle cost, and LCA. Yeh et al [19] integrated TRIZ into QFD structure. Almannai et al [20] combined QFD and the failure mode and effects analysis (FMEA) techniques. Based on QFD and FMEA structure, fuzzy linear programming by Chen and Ko [21], fuzzy objective nonlinear modeling by Erginel [22], and activity based costing by Hassan et al. [23] are integrated.

The main studies dealing with *mathematical modeling based model* are as follows: Min and Kim [24] extended QFD technique by integrating mathematical modeling. Chen and Ngai [25] integrated fuzzy set theory into QFD framework by providing mathematical modeling. Chaudhuri and Bhattacharyya [26] presented an approach combining QFD, integer programming, and conjoint analysis. Chen and Ko [27] used fuzzy linear programming and QFD structure for new product development. Cherif [28] used imprecise goal programming approach and Delice and Gungor [29-30] integrated mixed integer goal programming into QFD structure.

An and Park [31] combined product planning, product design, process planning and process control planning phases by modifying QFD technique. Bottani [32] used similar structure by adding fuzzy sets into QFD. Smilarly, Liu [33] integrated pairwise comparison into same structure. Carmignani [34] integrated pairwise comparisons in to all stages of QFD structure. Celik et al [35] integrated fuzzy analytic hierarchy process and fuzzy axiomatic design algorithms into the QFD frame. Furthermore, Bhattacharya et al. [36] and Ho and Dey [37] adopted analytical hierarchical process (AHP) and Carnevalli et al. [38] inserted axiomatic design into QFD structure. Khademi-Zare et al [39] developed a structure based on TOPSIS, AHP, and QFD techniques. Okur et al. [40] and Pal et al. [41] integrated ordered weighted averaging aggregation and analytic network process (ANP), respectively. In the paper, the relationship between voice of customers and technical characteristics in QFD are determined based ANP procedure [42]. Lee et al [43] proposed a structure combining ANP, QFD, and multi-choice goal programming techniques. IN another study, Lin et al. [44] used fuzzy set theory based on ANP and QFD structure to evaluate critical criteria.

Chakrabort and Dey [45] and Zhai et al. [46] proposed an expert system based on QFD methodology. Ni et al. [47] integrated data-mining techniques. Tontini [48] and Sireli et al. [3] integrated Kano's model into QFD.

Although there are lots of papers dealing with the improving the structure of QFD to cope with complex problems, none of them interests in the interactions among the design parameters quantitatively. Therefore, the main objective of this paper is to take interdependencies among design parameters into consideration quantitatively by using DEMATEL method.

3. Modified QFD

The structure of the proposed algorithm is as follows;

Phase 1. Determine the scope of voice of customer: In this phase, customer requirements (CRs) are determined based on surveys and feedbacks obtained from customers. Then, to determine importance degrees of CRs, pairwise comparisons are conducted. In this study, a modified fuzzy AHP proposed by Zeng et al. [49] is used to calculate importance degrees of CRs. The steps of the method are given below [49];

Step 1. CRs are compared by using any linguistic scale.

Step 2. Preferences on CRs are converted into the standardized triangular fuzzy numbers (STFN).

Step 3. The fuzzy numbers are aggregated by applying the fuzzy weighted averaging operator, which is defined by

$$\tilde{a}_{ij} = \tilde{a}_{ij1} \otimes w_{e1} \oplus \tilde{a}_{ij2} \otimes w_{e2} \oplus .... \oplus \tilde{a}_{ijm} \otimes w_{em} \tag{1}$$

where w_{ej} is the weight of the jth expert and $\tilde{a}_{ij}\,(a_{ij}^{l}, a_{ij}^{m}, a_{ij}^{u})$ is the aggregated fuzzy scale of ith CR to jth CR.

Step 4. STFN scales are defuzzified by

$$a_{ij} = \frac{a_{ij}^{l} + 4a_{ij}^{m} + a_{ij}^{u}}{6} \tag{2}$$

Step 5. The priorities of CRs are obtained by;

$$w_i^{CR} = \frac{1}{n}\sum_{j=1}^{n} \frac{a_{ij}}{\sum_{k=1}^{n} a_{kj}} \qquad i, j = 1,2,3,...,n \tag{3}$$

where w_i^{CR} is the weight of *i*th CR

Phase 2. Technical characteristics (TCs) and the relation among the design parameters are determined. The importance degrees of TCs are calculated based on DEMATEL technique as follows [4]:

Step 6. Each expert evaluates the direct influence between any two TCs by an integer score ranging from 0 to 3, representing ''*no influence*'', ''*low influence*'', ''*medium influence*'', and ''*high influence*'', respectively. Then Average matrix is calculated by

$$A_o^{TC} = \frac{1}{m}\sum\nolimits_{j=1}^{m} A_j \tag{4}$$

where A_j is the preference of *j*th expert for TCs.

Step 7. Normalize initial direct-relation matrix *D* by

$$D^{TC} = \frac{1}{\left[\max_{1<i,j<k} \sum_{i,j=1}^{k} a_{i,j}\right]} A_o^{TC} \tag{5}$$

where *k* is the number of TCs

Step 8. The total relation matrix T^{TC} is calculated by

$$T^{TC} = D(I-D)^{-1} \tag{6}$$

Assume that *r* and *c* be *kx1* and *1xn* vectors representing the sum of each rows and sum of each columns of T^{TC}, respectively. Then r_i summarizes both direct and indirect effects given by a TC *i* to the other TC. If denotes the sum of jth column in matrix T, then c_j shows both direct and indirect effects by TC *j* from the other TCs. When $j = i$, the sum $(r_i + c_j)$ shows the total effects given and received by TC *i*. Hence, this sum presents the importance degree of TC [4].

Step 9. The interdependencies among the TCs are represented by a flow diagram.

Phase 3. The relation matrix (S) between TCs and CRs, are determined by each expert individually. Then, by using Eq.(1) aggregated relation matrix obtained.

Phase 4. The relative importance degree of TCs are calculated by

$$w_j^{TC} = (r_j + c_j) \times \sum_{i=1}^{n} w_{ij}^{CR} \times s_{ij} \tag{7}$$

4. Illustrative Example

A numerical example has been presented in Figure 1a. We cannot give processing steps of the numerical example in detailed for shake of shortening the paper. The fields numbered with 1, 2, and 3 in the figure represent the pairwise comparisons for CRs, dependencies among the TCs, and relations between RCs and TCs, respectively. When the mathematical procedure described above is followed step by step, the highlighted weights in Figure 1b are obtained.

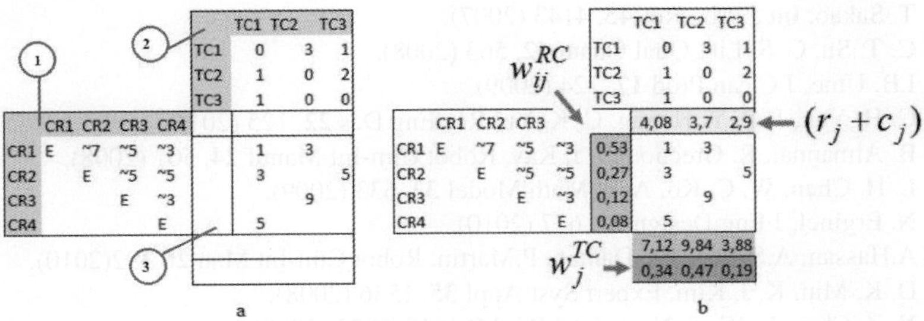

Figure 1. Illustrative example

5. Conclusion

As QFD is the most used method to design a new product/system or to improve current product/system, there are numerous papers dealing with developing mathematical foundation of QFD and its applications in the literature. The common objective of these studies is to extend QFD techniques with different tools to cope with difficulties in understanding customer requirements or technical characteristics. Fuzzy sets, mathematical modeling, quality management tools such as TRIZ, FMEA, and MCDM techniques are the mainly used techniques which are integrated into QFD structures. The originality of this paper is to present a new approach by adopting DEMATEL techniques into QFD structure to take into interdependencies among the design parameters.

For the further research, proposed algorithm can be applied on a real case application and the performance of the proposed algorithm can be compared with the older modified QFD methodologies proposed to literature.

References

1. H. Yüksel, Int.J.of Sustainable Engineering, **3**, 170 (2010).
2. A.Rahman, A.Rahim, M.Baksh, Ind.Man.DataSys, 103, 373-387 (2003).
3. Y. Sireli, P. Kauffmann, E. Ozan, Ieee T Eng Manage **54**, 380 (2007).
4. H. H. Wu, Y. N. Tsai, Appl Math Comput **218**, 2334 (2011).
5. Y.K.Juan, Y.H.Perng, D.C.Lacouture, K. S. Lu, Autm Constr **18**,139(2009).
6. S. Vinodh, S. K. Chintha, Int J Prod Res **49**, 1627 (2011).

7. C. W. Chen, S. T. Huang, Total Qual Manag Bus **22**, 387 (2011).
8. G. Z. Jia, M. Bai, Comput Ind Eng 60, 445 (2011).
9. L.Wang,Y.Juan, J.Wang, K.Li, C.L.Ong, Exprt Syst Appl **39**, 1484 (2012).
10. K. J. Kim, D. H. Kim, D. K. Min, Qual Reliab Eng Int **23**, 31 (2007).
11. L. Y. Zhai, L. P. Khoo, Z. W. Zhong, Adv Eng Inform **23**, 222 (2009).
12. M. Yang, F. I. Khan, R. Sadiq, P. Amyotte, J Clean Prod **19**, 1513 (2011).
13. Y. T. Chen, T. Y. Chou, J Acad Libr **37**, 237 (2011).
14. C. K. Kwong, Y. Chen, K. Y. Chan, X. Luo, J Eng Design **21**, 601 (2010).
15. M. S. Chen, C. C. Lin, Y. Y. Tai, M. C. Lin, Con Eng-Res A **19**, 35 (2011).
16. T. Sakao, Int J Prod Res **45**, 4143 (2007).
17. C. T. Su, C. S. Lin, Qual Quant **42**, 563 (2008).
18. I.B. Utne, J Clean Prod **17**, 724 (2009).
19. C. H. Yeh, J. C. Y. Huang, C. K. Yu, Res Eng Des **22**, 125 (2011).
20. B. Almannai, R. Greenough, J. Kay, Robot Cim-Int Manuf **24**, 501 (2008).
21. L. H. Chen, W. C. Ko, Appl Math Model **33**, 633 (2009).
22. N. Erginel, J Eng Design **21**, 677 (2010).
23. A.Hassan, A.Siadat, J.Y.Dantan, P.Martin, Robot Cim-Int Man **26**,392(2010).
24. D. K. Min, K. J. Kim, Expert Syst Appl **35**, 1546 (2008).
25. Y. Z. Chen, E. W. T. Ngai, Int J Prod Res **46**, 6823 (2008).
26. A. Chaudhuri, M. Bhattacharyya, Int J Prod Res **47**, 6633 (2009).
27. L. H. Chen, W. C. Ko, Eur J Oper Res **201**, 619 (2010).
28. M. S. Cherif, H. Chabchoub, B. Aouni, Int T Oper Res **17**, 85 (2010).
29. E. K. Delice, Z. Gungor, Int J Prod Res **49**, 2941 (2011).
30. E. K. Delice, Z. Gungor, Comput Ind Eng **57**, 906 (2009).
31. Y. An, S. Lee, Y. Park, Int J Serv Ind Manag **19**, 621 (2008).
32. E. Bottani, Int J Prod Econ **119**, 380 (2009).
33. H. T. Liu, Expert Syst Appl **36**, 11131 (2009).
34. G. Carmignani, Qual Reliab Eng Int **25**, 241 (2009).
35. M. Celik, S. Cebi, C. Kahraman, I. D. Er, Exprt Syst Appl **36**, 6227 (2009).
36. A. Bhattacharya, J. Geraghty, P. Young, Appl Soft Com **10**, 1013 (2010).
37. W. Ho, P. K. Dey, M. Lockstrom, Supply Chain Manag **16**, 446 (2011).
38. J. A. Carnevalli, P. A. C. Miguel, F. A. Calarge, Int J Pr Ecn **125**, 1 (2010).
39. H.Khademi-Zare, M.Zarei, A.Sadeghieh, M.Owlia, Telecom Policy **34**,747(2010).
40. Okur, E. N. Nasibov, M. Kilic, M. Yavuz, Qual Quant **43**, 999 (2009).
41. D. K. Pal, B. Ravi, L. S. Bhargava, Int J Comput Integ M **20**, 338 (2007).
42. E. E. Karsak, S. Sozer, S. E. Alptekin, Comput Ind Eng **44**, 171 (2003).
43. A.H. I.Lee, H.Y.Kang, C.Y.Yang, C.Y. Lin, Int J Prod Res **48**, 3977 (2010).
44. Y.H.Lin, H.P.Cheng, M.L.Tseng, J.C.C.Tsai, Expt Syst Ap**37**, 2186 (2010).
45. S. Chakraborty, S. Dey, Exprt Syst Appl **32**, 1208 (2007).
46. L. Y. Zhai, L. P. Khoo, Z. W. Zhong, Expert Syst Appl **37**, 8888 (2010).
47. M. Ni, X. Xu, S. Deng, Int J Comput Integ M **20**, 280 (2007).
48. G. Tontini, Total Qual Manag Bus **18**, 599 (2007).
49. J. Zeng, M. An, N. J. Smith, Int J of Project Man **25**, 589 (2007).

AN INCREMENTAL COLLABORATIVE FILTERING ALGORITHM FOR RECOMMENDER SYSTEMS

MAYTIYANIN KOMKHAO[1], ZHONG LI, WOLFGANG A. HALANG

Chair of Computer Engineering, Fernuniversität in Hagen, 58084 Hagen, Germany.

JIE LU

School of Software, Faculty of Engineering and Information Technology, University of Technology Sydney, P.O. Box 123, Broadway, NSW 2007, Australia.

Recommender systems are effective approaches to implement personalised e-services. In recent years, they have gained widespread applications in e-commerce. Current recommender systems still need, however, further improvements with respect to the accuracy of prediction and to solve the scalability problem. To this end, an incremental collaborative filtering (InCF) algorithm based on the Mahalanobis distance is presented for recommender systems. Furthermore, the Mahalanobis radial basis function with ellipsoidal shape is employed to determine the decision boundaries of clusters. Experimental results show that the algorithm proposed can lead to improved prediction accuracy and that it turns out to be scalable in recommender applications.

1. Introduction

The vastly increasing amount of data available on the World Wide Web makes it very hard to find the information one really needs. Recommender systems aim to solve this problem by recommending products or services that may be of interest to users based on prior ratings of such items, opinions and preferences and, thus, provide a personalised e-service [1]. As an efficient technique, collaborative filtering is often employed in recommender systems [2] to base recommendations of new items on. To make recommendations, initially learning systems were applied to find sets or models (called peers in [3]) of users whose rating profiles are most similar to the requirements of the users seeking recommendations [4]. Then, systems predict whether interested users will like particular items by calculating weighted composites of the similar users' ratings given on particular items. There are two major versions of collaborative filtering [3], [4]. Memory-based collaborative filtering bases recommendations on entire

[1] E-mail: maytiyanin.komkhao@fernuni-hagen.de

matrices of previous user-item ratings. To avoid extensive searching in these rating matrices, model-based collaborative filtering, on the other hand, uses them to build models by grouping similar users into clusters. To address the problem of learning from large databases, incremental learning techniques have been proposed to build such models continuously, leading to incremental improvements of recommendations.

For use in recommender systems here we propose an incrementally working collaborative filtering algorithm (called InCF), which is based on the Mahalanobis distance. In InCF, a user-item rating matrix (a matrix containing the ratings given by users to items) is utilised to build a model, which relies on the similarities between given user ratings in order to predict the ratings an interested user may have on particular items. This paper focuses on two main challenges, namely accuracy of prediction and scalability. To improve prediction accuracy, it is proposed to apply the Mahalanobis distance [5] instead of ordinary ones like the Euclidean distance in model-based collaborative filtering. The Mahalanobis distance has proven useful [6] to determine the decision boundaries of clusters. InCF employs the Mahalanobis membership function to express partial (or fuzzy) membership in overlapping clusters. Addressing the other challenge, the scalability problem, InCF is designed to be able to predict the ratings of particular items to interested users incrementally and, thus, to cope with the tremendous growth of both items and interested users present on the web. Furthermore, redundant attributes are removed in the user-item rating matrix. Correlation-based feature subset selection (CFS) has been used [7] to reduce the datasets in size and to remove irrelevant attributes. Thus, CFS can help learning algorithms to focus on the features important for particular problems, and can enable them to operate faster and more effectively [7].

2. Similarity Measure and Mahalanobis Distance

Similarity measures are used to identify the similarity of users. The relationship between dissimilarity and similarity is given by $S_{ij} = 1 - \delta_{ij}$, where S_{ij} is the similarity between users i and j, and δ_{ij} is the dissimilarity between users i and j. Similar users are grouped into the same cluster, and dissimilar ones are grouped into different clusters.

Distances are used to represent similarity and dissimilarity between users. If the ratings of two users are most similar, then the distance between them is minimised; if the ratings are most dissimilarity, then the distance between the users is maximised. Let d_{ij} be the distance between features i and j, then d_{ij} is always positive or zero and symmetric, i.e. $d_{ij} = d_{ji}$.

There are many ways to measure similarity and dissimilarity, respectively, between users based on feature types or quantitative, ordinal, binary and nominal variables. In this paper, we focus on quantitative variables such as Euclidean, Manhattan and Mahalanobis distance.

For quantitative variables, most existing algorithms use the Euclidean and the Manhattan distances as proximity functions, because they are easy to implement. The Mahalanobis distance [5] is different from the Euclidean one, the most popular similarity measure. It considers the orientation of data in the sample space, and groups them by mean (called centroid in this paper) and covariance matrices [5]. The shapes of clusters are hyper-ellipsoidal rather than hyper-spherical as obtained when working with the Euclidean distance.

Let a user $\mathbf{x}_j, j = 1, \ldots, N$, be characterised by a vector of n>0 attributes $\mathbf{x}_j = \left[x_{1,j}, \ldots, x_{n,j}\right]^{\mathrm{T}}$, and let $\bar{\mathbf{x}}=1/N\sum_{j=1}^{\mathrm{N}}\mathbf{x}_j$. Then, the Euclidean distance (dist_E) of two users $\mathbf{x}_j$ and $\mathbf{x}_s$ is generalised by the Mahalanobis distance [5] as $\mathrm{dist}_M(j,s) = [(\mathbf{x}_j - \mathbf{x}_s)^{\mathrm{T}}\mathbf{K}^{-1}(\mathbf{x}_j - \mathbf{x}_s)]^{1/2}$ with the users' covariance matrix [8] $\mathbf{K} = \frac{1}{N-1}\sum_{j=1}^{N}(\mathbf{x}_j - \bar{\mathbf{x}})^T(\mathbf{x}_j - \bar{\mathbf{x}})$.

3. Incremental Collaborative Filtering with the Mahalanobis Distance

InCF creates models, in form of sets of clusters, from the user-rating matrix representing groups of similar input vectors. Each cluster has its own centroid. InCF uses the Mahalanobis distance and Mahalanobis radial basis functions to identify which cluster is the best match with an input vector, which is called the winning cluster. The best match is defined by the degree of membership based on the similarity distance. The algorithm compares input vectors joining the system with known clusters. In making predictions for an interested user, by incremental learning InCF either associates the user with an existing cluster as target cluster or includes the user into the model as a new cluster. As discussed below, InCF has two main phases, viz. the *learning* and the *prediction phase*.

The *learning phase* starts with correlation-based subset selection utilising the user-item rating matrix to select relevant features of users' ratings, and with assigning initial values to the similarity distance threshold and the covariance matrix. The similarity distance threshold $0 < \mathrm{d}_{th} < 1$ limits the cluster boundaries, and the covariance matrix is initialised non-singularly with $0 < \sigma < 1$ as $\mathbf{K}_{\text{initial}} = \sigma\,\mathbf{I}$. As final initialisation a cluster is defined by an input vector taken from the user-item rating matrix, and from this single cluster a cluster model will develop incrementally by performing the algorithm's main part. The latter consists of carrying out the following operations for any further input vector available. The covariance matrix is updated by adding $\sigma \mathbf{I}$, and using the Mahalanobis distance the distances between the input vector and the centroids of each cluster in the model are calculated. Furthermore, the degrees of membership of the input vector in each cluster are computed as $exp(-1/2\text{dist}_M^2)$. Then, for the input vector the winning cluster is identified as the one, for which the highest degree of membership is assumed. In order to decide whether to merge the input vector with the winning cluster, the input vector's Mahalanobis distance to the winning cluster's centroid is compared with the similarity distance threshold d_{th}. If the distance is lower than the threshold, the input vector is merged with the winning cluster, the parameters of this cluster are updated and the covariance matrix is newly calculated as described in [9], [10]. If, on the other hand, the distance exceeds the threshold, a new cluster is created with the input vector as sole element.

In the subsequent *prediction phase*, interested, but not yet rated users are associated with respective winning clusters. From the latter rating profiles are derived for the former. To an interested user the highest ranking item from its winning cluster will be recommended.

4. Experimental Results

As benchmark the dataset "MovieLens" is selected, which contains real movie ratings gathered by the GroupLens research laboratory at the University of Minnesota, and available online at http://www.grouplens.org/node/73.

To show the proposed algorithm's feasibility for use in recommender systems, we first compare the proposed algorithm with traditional clustering and, then, investigate the effect of correlation-based feature subset selection (CFS) by measuring the average Normalised Mean Absolute Error (*NMAE*).

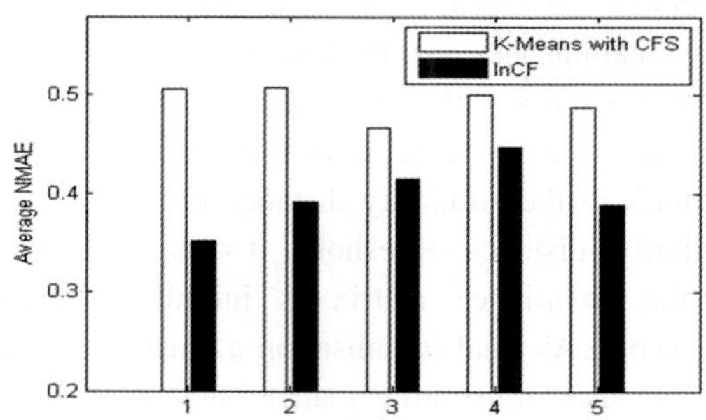

Figure 1: Average *NMAE* values obtained by K-Means with CFS and by InCF for five different datasets and ten clusters each.

For comparison purposes, model-based collaborative filtering was implemented using classical K-Means clustering. CFS was applied as well in order to reduce the dataset's dimensions, and to decrease the average *NMAE*. Figure 1 shows for our experiments with ten clusters and five datasets that the average *NMAE* value of InCF is lower than the one for K-Means with CFS. The lower average value of *NMAE* for InCF (0.399) as compared to the one for K-Means with CFS (0.494) is due to the ellipsoidally shaped fit of the decision boundaries better matching feature distributions than spheres.

Finally, the InCF algorithm was applied to recommend movies to interested users. In particular, five users were selected randomly, and the algorithm suggested five example movies to each user based on the model created.

Table 1: Information on selected interested users from MovieLens dataset.

ID	Age	Gender	Job
463	48	F	Healthcare
474	51	M	Executive
94	26	M	Student
727	25	M	Student
234	60	M	Retired

Table 1 presents information on the selected interested users consisting of age, gender and job. The user "463", for instance, is 48 years old and employed in healthcare. The algorithm suggests five movies to her, viz. Vertigo (1958) with rating 4.556, Rear Window (1954) with rating 4.778, North by Northwest (1959) with rating 4.556, Casablanca (1942) with rating 4.445 and The Bridge on the River Kwai (1957) with rating 4.445. Since the movie Rear Window has the highest rating in the model, it is finally recommended to user "463".

5. Conclusion

An incremental collaborative filtering algorithm based on the Mahalanobis distance was presented to be employed in recommendation systems. The algorithm was developed in order to improve the accuracy of predictions as substantiated by experimental results that average *NMAE* values of InCF are lower than those of ordinary collaborative filtering based on K-Means. To handle the scalability problem, instead of batch learning incremental on-line learning was employed for its ability to detect new clusters and to label them. Thus, it can cope with the tremendous growth of both items and interested users.

According to the current tendency in medicine to reduce hazardous complications by detecting symptoms of a disease before it occurs, we plan to develop a medical recommender system. It is to use the InCF algorithm, with the difference that knowledge-based filtering is to be extended by taking medical data into account as well. The system's objective is to suggest a ranked list of predicted risks and, then, take actions based on the similarity of a considered patient's vector with others in medical history databases.

Acknowledgment

This work was partly supported by the Australian Research Council under discovery project grant DP110103733, a Matching Funds scholarship by the German Academic Exchange Service and a Mercator visiting professorship by the German Research Foundation.

References

1. J. Lu, Q. Shambour, Y. Xu, Q. Lin and G. Zhang, A Hybrid Semantic Recommendation System for Personalized Government-to-Business e-Services. Internet Research, Vol. 20(3): 342–365 (2010).
2. R. D. Burke, Hybrid Recommender Systems: Survey and Experiments. User Model. User-Adapt. Interact, Vol. 12(4): 331–370 (2002).
3. G. Adomavicius and A. Tuzhilin, Towards the Next Generation of Recommender Systems: A Survey of the State-of-the-Art and Possible Extensions. IEEE Trans. Knowl. and Data Eng., Vol. 17(6): 734–749 (2005).
4. M. C. Pham, Y. Cao, R. Klamma and M. Jarke, A Clustering Approach for Collaborative Filtering Recommendation Using Social Network Analysis. J. of Universal Computer Science, Vol. 17(4): 583–604 (2011).
5. P. C. Mahalanobis, On the Generalised Distance in Statistics. Proc. National Institute of Science of India, Vol. (2)1: 49–55 (1936).
6. R. Tanner and D. G. M. Cruickshank, RBF Based Receivers for DS-CDMA with Reduced Complexity. Proc. IEEE ISSSTA, Vol. 1/3 (1998).
7. M. A. Hall, Correlation-based Feature Selection for Machine Learning, PhD Thesis, Dept. Computer Science, University of Waikato, NZ (1999).
8. A. P. Dempster, Covariance Selection, Biometrics, Vol. (28): 157–175 (1972).
9. G. G. Yen and P. Meesad, An Effective Neuro-fuzzy Paradigm for Machinery Condition Health Monitoring. IEEE Trans. Systems, Man, and Cybernetics, Part B: Cybernetics, Vol. 31(4): 523–536 (2001).
10. H. Xu and M. Vuskovic, Mahalanobis Distance-Based ARTMAP Network. Proc. IEEE Intl. Joint Conf. on Neural Networks, Vol. (3): 2353–2359 (2004).

A JUDGEMENT METHOD FOR EARTHQUAKE EARLY WARNING INFORMATION

HONGCHANG AN†[1, 2],JIE LU[2], GUANGQUAN ZHANG[2]

[1] *Department of Safety Engineering, China Institute of Industrial Relations, 45 Zengguang Road, Haidian, Beijing 100048, China*

[2] *Decision Systems & E-Service Intelligence Research Laboratory, Centre for Quantum Computation & Intelligent Systems, Faculty of Engineering & Information Technology, University of Technology, Sydney, PO Box 123, Broadway, NSW 2007, Australia*

Earthquake early warning (EEW) is very important to earthquake mitigation, but EEW information before an earthquake is usually ignored. China Public Safety Early Warning Information Systems (CPSEWIS) including EEW information before, amid and after an earthquake is introduced. More important is that a method is designed to judge credibility of EEW information for the CPSEWIS to obtain useful EEW information and help people mitigate.

1. Introduction

Earthquake early warning (EEW) has been a popular subject since 2007 because Japan Meteorological Agency started to issue real-time earthquake information for particular users from August 1, 2006, and warning information for public users, through TV and radio, from October 1, 2007 [1], but the degree that people understand EEW is distinct in different scholars [1-3]. EEW and its applications in Japan are universal such as [4-6]. EEW is also applied in other countries and areas such as in Istanbul [7], in USA [8], in Bucharest [9], in Italy [10], in Taiwan [11], in Vrancea [12].

However, Wenchuan Earthquake occurred in China in 2008, Haiti Earthquake, Chile Earthquake and Yushu Earthquake occurred respectively in Haiti, Chile and China in 2010, and East Japan Earthquake occurred in Japan in 2011. Therefore, it is very important and necessary to collect EEW information, i.e. omens before earthquakes as well as early warning information after an earthquake occurs and adopt some earthquake mitigation measurements to make successful earthquake prevention besides earthquake early warning system and its wide application like in Japan.

† hongchang.an@uts.edu.au

The current problem in the field of EEW is to pay more attention to EEW information after an earthquake occurs and to ignore EEW information before an earthquake. To solve these problems, we have developed a comprehensive EEW information system named as China Public Safety Early Warning Information Systems (CPSEWIS). We also design a method to judge credibility of EEW information for the CPSEWIS.

In section 2, the CPSEWIS is briefly introduced. In section 3, a method of judging credibility of EEW information is expressed in detail. Conclusions occur in section 4.

2. China Public Safety Early Warning Information Systems

After Wenchuan Earthquake and Haiti Earthquake, the CPSEWIS has been set up. Figure 1 shows architecture of the CPSEWIS.

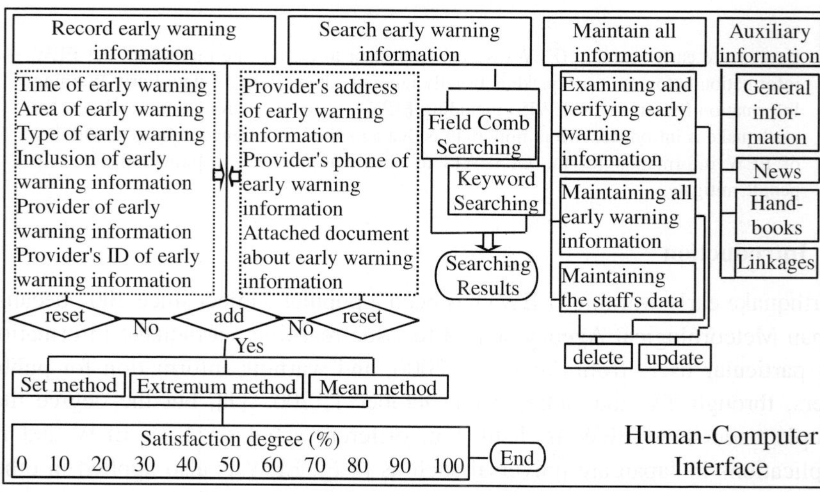

Figure 1. Architecture of the CPSEWIS

3. A Method of Judging Credibility of EEW Information

To judge credibility of EEW information as objective, quantitative, active and automatic as possible, a method of judging EEW information is designed at follows. The feature of this method deals with both capacity and skill and attitude and mentality of a provider of early warning information.

As shown in Figure 1, complete EEW information includes time of early warning (TMoEW), area of early warning (AoEW), type of early warning (TPoEW), inclusion of early warning information (IoEWI), provider of early warning information (PoEWI), provider's ID of early warning information

(PIDoEWI), provider's address of early warning information (PAoEWI), provider's phone of early warning information (PPoEWI), attached document(s) about early warning information (ADaEWI). As well, note that someone issuing and providing fake information is guilty and it is better to believe that it exists than it does not. So, we define credibility of EEW information (C_{EEWI}) as a key index to judge EEW information.

3.1. *Quantification of TMoEW, AoEW, TPoEW*

TMoEW has fixed format. Let the start time of TMoEW as s, the end time of TMoEW as e, current time as c and credibility of TMoEW as C_{TM},

$$C_{TM} = \begin{cases} 0 & e < c \\ 50\% & s < c \le e \vee c + 5\,yrs < s \\ 70\% & c + 2\,yrs < s \le c + 5\,yrs \\ 75\% & c + 1\,yr < s \le c + 2\,yrs \wedge c + 5\,yrs < e \\ 80\% & \begin{array}{l}(c + 1\,yr < s \le c + 2\,yrs \wedge c + 2\,yrs < e \le c + 5\,yrs) \\ \vee (c \le s \le c + 1\,yr \wedge c + 5\,yrs < e)\end{array} \\ 85\% & \begin{array}{l}(c + 1\,yr < s \le c + 2\,yrs \wedge c + 1\,yrs < e \le c + 2\,yrs) \\ \vee (c \le s \le c + 1\,yr \wedge c + 2\,yrs < e \le c + 5\,yrs)\end{array} \\ 90\% & c \le s \le c + 1\,yr \wedge c + 1\,yr < e \le c + 2\,yrs \\ 95\% & c \le s \le c + 1\,yr \wedge c \le e \le c + 1\,yr \end{cases} \tag{1}$$

As of AoEW, let province of AoEW as p, city of AoEW as i, county of AoEW as u and credibility of AoEW as C_A,

$$C_A = \begin{cases} 50\% & p = '\text{All China}' \\ 70\% & p \ne '\text{All China}' \wedge i = '\text{Select}' \wedge u = '\text{Select}' \\ 80\% & p \ne '\text{All China}' \wedge i \ne '\text{Select}' \wedge u = '\text{Select}' \\ 90\% & p \ne '\text{All China}' \wedge i \ne '\text{Select}' \wedge u \ne '\text{Select}' \end{cases} \tag{2}$$

About TPoEW, let type of TPoEW as t, subtype of TPoEW as b and credibility of TPoEW as C_{TP},

$$C_{TP} = \begin{cases} 50\% & t = '\text{Other}' \\ 70\% & t \ne '\text{Other}' \wedge b = '\text{Other}' \\ 90\% & t \ne '\text{Other}' \wedge b \ne '\text{Other}' \end{cases} \tag{3}$$

3.2. *Quantification of IoEWI, PoEWI, PIDoEWI, PAoEWI, PPoEWI, ADaEWI*

With regard to IoEWI, if subtype of type of early warning, any type of number form, month and day among the start time and the end time of time of early warning, any of area of early warning are included, credibility of IoEWI (C_I) will be given 100%. If any three of subtype of type of early warning, any type of number form, month and day among the start time and the end time of time of early warning, any of area of early warning are included, C_I will be given 90%. If any two of subtype of type of early warning, any type of number form, month and day among the start time and the end time of time of early warning, any of area of early warning are included, C_I will be given 80%. If any one of subtype of type of early warning, any type of number form, month and day among the start time and the end time of time of early warning, any of area of early warning is included, C_I will be given 70%. If any one of subtype of type of early warning, any type of number form, month and day among the start time and the end time of time of early warning, any of area of early warning is not included and IoEWI is not empty, C_I will be given 50%. If IoEWI is empty, C_I will be given zero.

Concerning PoEWI, if set name except the item 'Other' is selected, credibility of PoEWI (C_P) will be given zero. If 'Other' is selected and the textbox following 'Other' includes set character, C_P will be given 90%. If 'Other' is selected and the textbox following 'Other' does not include set character, C_P will be given 50%. (If we are authorized by relative government(s), PoEWI will be matched given name list provided by the relative government(s) to judge C_P as either 100%, or 50%, or zero.)

On PIDoEWI, once this information is submitted successfully, credibility of PIDoEWI (C_{PID}) will be given 90%. (If we are authorized by relative government(s), PIDoEWI will be matched given ID list provided by the relative government(s) to judge C_{PID} as either 100%, or 50%, or zero.)

Regarding PAoEWI, if it follows general address rules and is relative to area of early warning, credibility of PAoEWI (C_{PA}) will be given 90%. If it follows general address rules and is not relative to area of early warning, C_{PA} will be given 80%. If it does not follow general address rules, C_{PA} will be given zero. (If we are authorized by relative government(s), PAoEWI will be matched given address list provided by the relative government(s) to judge C_{PA} as either 100%, or 50%, or zero.)

As for PPoEWI, once this information is submitted successfully, credibility of PPoEWI (C_{PP}) will be given 90%. (If we are authorized by relative government(s) or organizations, PPoEWI will be matched given phone list provided by the relative government(s) or organizations to judge C_{PP} as either 100%, or 50%, or zero.)

As to ADaEWI, if there is/are attached document(s) and attached document(s) deal(s) with subtype of type of early warning, credibility of ADaEWI (C_{AD}) will be given 90%. If there is/are attached document(s) and attached document(s) do(es) not deal with subtype of type of early warning, C_{AD} will be given 50%. If there is not attached document, C_{AD} will be given zero.

3.3. *Expression of C_{EEWI}*

Based on quantification of items included in EEW information, C_{EEWI} may be expressed with the following methods.

1. Set method

$$C_{EEWI} = (C_{TM}, C_A, C_{TP}, C_I, C_P, C_{PID}, C_{PA}, C_{PP}, C_{AD}) \tag{4}$$

2. Extremum method

$$C_{EEWI} = \begin{cases} \min(C_{TM}, C_A, C_{TP}, C_I, C_P, C_{PID}, C_{PA}, C_{PP}, C_{AD}) \\ \max(C_{TM}, C_A, C_{TP}, C_I, C_P, C_{PID}, C_{PA}, C_{PP}, C_{AD}) \end{cases} \tag{5}$$

3. Mean method

$$C_{EEWI} = \frac{1}{9}\sum(C_{TM}, C_A, C_{TP}, C_I, C_P, C_{PID}, C_{PA}, C_{PP}, C_{AD}) \tag{6}$$

When a provider of EWI or his/her assistant clicks the button 'add', any one of three kinds of methods may be selected. When set method is selected, C_{TM}, C_A, C_{TP}, C_I, C_p, C_{PID}, C_{PA}, C_{PP} and C_{AD} with their value will be shown, so you can see assessment of all information. If Extremum method is selected, minimum and maximum among C_{TM}, C_A, C_{TP}, C_I, C_P, C_{PID}, C_{PA}, C_{PP} and C_{AD} with their own and their corresponding values will be shown, thus you can see which information is enough and which information is not enough or all information is even. When mean method is selected, a result will appear that includes C_{EEWI} and corresponding suggestion as strongly suggested to provide more detailed and more accurate information and materials if C_{EEWI} is not more than 50%, suggested to further provide more detailed and more accurate information and materials if C_{EEWI} is between 50% and 70%, suggested to further provide more detailed materials about EEW information or more accurate basic information if C_{EEWI} is between 70% and 80%, provide further evidences together with information provider if C_{EEWI} is between 80% and 95%, or information will be revised appropriately and sent as a true early warning if C_{EEWI} is not less than 95%. Then the provider of early warning information or his/her assistant may click the button 'satisfactory' or 'dissatisfactory' (or different degree of satisfaction) to further interact with the CPSEWIS.

4. Conclusions

While EEW information after an earthquake is paid attention to, EEW information before an earthquake should paid more attention to. The CPSEWIS deals with all EEW information before, amid and after an earthquake and adopts a method to judge credibility of EEW information to obtain useful EEW information. Clustering method, case-based grouping, statistics-based method and mixed method will be discussed in future study to improve use of EEW information in the CPSEWIS.

Acknowledgment

The work presented in this paper was partly supported by Australian Research Council (ARC) under Discovery Project DP0880739.

References

1. K. Doi, Soil Dyn. Earthquake Eng. **31**, 119 (2011).
2. S. Fujita, K. Minagawa, G. Tanaka and H. Shimosaka, Soil Dyn. Earthquake Eng. **31**, 223 (2011).
3. C. Satriano, Y. Wu, A. Zollo and H. Kanamori, Soil Dyn. Earthquake Eng. **31**, 106 (2011).
4. Y. Nakamura, J. Saita and T. Sato, Soil Dyn. Earthquake Eng. **31**, 127 (2011).
5. T. Kubo, Y. Hisada, M. Murakami, F. Kosuge and K. Hamano, Soil Dyn. Earthquake Eng. **31**, 231 (2011).
6. M. Erdik, Y. Fahjan, O. Ozel, H. Alcik, A. Mert and M. Gul, Bull. Earthquake Eng. **1**, 157 (2003).
7. H. Alcik, O. Ozel, Y. Wu, N. Ozel and M. Erdik, Soil Dyn. Earthquake Eng. **31**, 181 (2011).
8. V. Grasso, J. Beck and G. Manfredi, Eng. Struct. **29**, 3455 (2007).
9. F. Wenzel and G. Marmuraenu, Pure App. Geo. **164**, 929 (2007).
10. C. Satriano, L. Elia, C. Martini, M. Lancieri, A. Zollo and G. Iannaccone, Soil Dyn. Earthquake Eng. **31**, 137 (2011).
11. N. Hsiao, Y. Wu, L. Zhao, D. Chen, W. Huang, K. Kuo, T. Shin and P. Leu, Soil Dyn. Earthquake Eng. **31**, 201 (2011).
12. A. Mărmureanu, C. Ionescu and C. Cioflan, Soil Dyn. Earthquake Eng. **31**, 163 (2011).

MEDICAL SKILLS ASSESSMENT IN TRAINING BASED ON VIRTUAL REALITY USING A POSSIBILISTIC APPROACH[1]

LILIANE S. MACHADO*

Department of Informatics, Federal University of Paraíba, João Pessoa/PB, Brazil

RONEI M. MORAES

Department of Statistics, Federal University of Paraíba, João Pessoa/PB, Brazil

Virtual reality have been used to provide training systems in several areas, in particular in medicine. In that area, user's interactions in a virtual environment are modeled and they are compared with predefined classes of performance to know how much users are prepared to perform that procedure on human beings. In this paper, it is proposed a new approach for online Single User's Assessment System (SUAS) using Possibility and Necessity measures. Those fuzzy measures provide a interval estimator for the compatibility between an user's interactions and previous stored classes of performance and it is the kernel of a decision support system.

1. Introduction

An important focus of Virtual Reality simulators are risky tasks, that can cause injures, or those related to expensive procedures. In some areas, VR systems for training have been used to provide metrics to a proficiency criterion of learning, as in comercial aviation [5]. The risk present in several medical procedures turns it a potential area for VR simulators.

Assessment systems have been used in VR simulators to allow objectively determine the user performance. In some areas, it is done by an expert, which observes those actions and movements, and afterwards emits an assessment. In others, the assessment is based on videotapes post-analysis [8]. However, this task could be helped by a decision support system (DSS) coupled to the VR system. The main idea of those systems relays on the definition of performance classes from expert knowledge to classify users actions [5]. Thus, an assessment

[1] Work partially supported by grants 312375/2009-3 and 310339/2009-0 of the Brazilian Council for Scientific and Technological Development, CNPq. This work is related to the National Institute of Science and Technology for Medicine Assisted by Scientific Computing also supported by CNPq.

* Corresponding author: *liliane@di.ufpb.br*

system is a knowledge-based system which analyze data from users' interactions in a training system based on VR and provides a decision making support on the users' proficiency in performing real procedures.

Since 90's, several Single User's Assessment System (SUAS) were proposed [5,9], mainly for medical training. With continuous advances on capacity of computers, SUAS evolved too. Nowadays a SUAS must continuously monitor all user interactions on VR environment and compare his performance with pre-defined expert's classes of performance to recognize user's level of training. Basically, there are two types of SUAS: off-line and on-line. Off-line SUAS can be defined as methods coupled or not to VR systems, whose assessment results are provided some time (which can be minutes, hours or days) after the end of the training. On the other hand, on-line SUAS are coupled to the training system and collect user data to provide a result about his/her performance at the end of the simulation [5,6], which is characterized by time of response lower than one second. An on-line SUAS must have low complexity to does not compromise VR simulations performance, but it must have high accuracy to does not compromise the assessment.

This paper presents a new training assessment system which use Possibility and Necessity Measures. This method uses fuzzy measures to provide a weighted interval estimator to provide users' assessment support based on their actions in VR simulator. Next section presents some requirements of VR training simulators. Section 3 brings some theoretical aspects of the assessment methodology proposed on Section 4. Results obtained, as well as their analysis, are presented in Section 5, followed by some considerations in Section 6.

2. Training Simulators Based on VR

Training simulators based on VR are real time computer systems used to provide realistic environments for practice of some specific activity. The specificity of the activity, which is simulated, will determine the features of the system and also the variables necessary to be monitored by the system. VR systems explore human senses, as sight, hearing and touch, to provide interactive and immersive virtual worlds. The quality of the systems can be related to the graphics quality, textures, use of deformable objects in visual feedback, lighting features, stereo sounds and physical properties of the objects [6]. However, the integration and synchronization of tasks for simulation of senses depends on the necessities of the procedure since obtaining high levels of realism can compromise the real time.

Specific devices are used to allow users to interact with the simulation in a natural way, as when manipulating real tools. Examples of specific device are the goggles for stereoscopic view and the haptic devices for touch and force feedback. Thus, interaction devices can be used not only by users, but also by an assessment system (SUAS) coupled to the simulator in order to monitor interactions and movements and verify user performance.

3. Theoretical Aspects

For better understanding of the assessment method proposed, some considerations and definitions are previously provided. First, it is defined the concept of fuzzy sets, followed by definitions of measures of possibility and necessity. Finally it is presented the assessment method.

Definition 1. Let X be a universal set, in which each element is denoted by x. Then, a fuzzy set A in X is given by $A = \{x, \mu_A(x)\}, x \in X$, where $\mu_A(x)$ is called the membership function, which provides a grade of membership of a element x in the fuzzy set A and $\mu_A: X \to [0;1]$. The value 0 denotes that x does not belong to the set A and value 1 indicates that x fully belongs to the set A. This interpretation is analogous to classical sets [10].

Definition 2. Let *F*(U) a set of fuzzy subsets of U. The possibility measure Π is a function of *F*(U) to [0;1], which satisfies the following axioms [1]:

(1) $\Pi(\varnothing) = 0$;

(2) $\Pi(U) = 1$;

(3) For any collection A_i of subsets of U, $\Pi(\cup_i A_i) = \sup_i \Pi(A_i)$.

Definition 3. Let *F*(U) a set of fuzzy subsets of U. The necessity measure N is a function of *F*(U) to [0;1], which satisfies the following axioms [3]:

(1) $N(\varnothing) = 0$;

(2) $N(U) = 1$;

(3) For any collection A_i of subsets of U, $N(\cap_i A_i) = \inf_i N(A_i)$.

Definition 4. Let A be a subset of U, with its membership function μ_A and let X be a variable which assumes values u in U. Then, the possibility distribution function associate to X is defined as [10]:

$$\pi_X(u) = \mu_A(u) \quad (1)$$

It is important to note that possibility distribution is not equivalent to a probability distribution. More discussions about it can be found in [1]. As it can be observed above, possibility and the necessity measures are dual. Other relations between them were provide in [1]:

$$\Pi (A) = 1 - N(\overline{A}) = \sup \{\pi (u) \mid x \in A\}$$
$$N (A) = 1 - \Pi (\overline{A}) = \inf \{1 - \pi (u) \mid x \notin A\}$$
$$\Pi (A) \geq N (A)$$
$$N (A) > 0 \Rightarrow \Pi (A) = 1$$
$$\Pi (A) < 1 \Rightarrow N (A) = 0$$

Definition 5. Let A and B be fuzzy subsets of U, with membership function μ_A and μ_B, respectively. Let X be a variable which assumes values u in U. Using (1), the conditional possibility and necessity measures are given by [1]:

$$\Pi (A \mid B) = \max_{u \in X} \min (\mu_A (u), \pi_B (u)) \text{ and}$$
$$N (A \mid B) = \min_{u \in X} \max (\mu_A (u), 1- \pi_B (u)). \quad (2)$$

For training assessment, during its execution, some features are monitored and measured from user's manipulation. After the end of training, from these features, a training assessment method classify user's performance. Formally, let be the classes of performance in space of decision $\Omega=\{1,...,M\}$ where M is the total number of classes of performance. Let be w_i, $i \in \Omega$ the class of performance for a user. In this case, the assessment system computes conditional possibilities and necessities to predict the class of a vector of training data $\boldsymbol{X}$, according to sample data D, where $\boldsymbol{X}$ is a vector with n features obtained when a training is performed, i.e. $\boldsymbol{X}=\{X_1, X_2, ..., X_n\}$.

From (2) is possible to construct an interval for the real value of the class of performance w_i, given each feature X_j , with $j=\{1,...,n\}$, from the training data $\boldsymbol{X}$ from a user is:

$$\mu_{wi} (X_j) \in [N_j (w_i \mid X_j); \Pi_j (w_i \mid X_j)] \quad (3)$$

The global matching between the class of performance w_i, given the training data $\boldsymbol{X}$ from a user is an interval where the minimum value is the minimum compatibility and the maximum value is the maximum compatibility:

$$\mu_{wi} (\boldsymbol{X}) \in [\text{compat}_{min} ; \text{compat}_{max}]. \quad (4)$$

As the class of performance w_i is expressed by a conjunction of features X_j , then this aggregation is performed by a t-norm. In this case, the "min" operator preserves the semantics of possibility and necessity measures [2]:

$$\text{compat}_{min} = \min_{1 \leq j \leq n} (N_j (w_i \mid X_j))] \text{ and}$$
$$\text{compat}_{max} = \min_{1 \leq j \leq n} (\Pi_j (w_i \mid X_j))], \quad (5)$$

After all intervals were obtained, the problem is provide an answer to training assessment for the user. A possible solution is provide information about each class of performance and its interval respectively. However, the decision maker may not be prepared to interpret this result. Certainly, a crisp answer can be better for interpretation by decision maker. Among several ideas, it is possible use a defuzzification method as the centroid C of interval and take the maximum one. Thus, for this case, the decision rule is:

$$\text{select performance class } w_i \text{ if } \quad w_i = \arg\max_{1 \leq i \leq M} \; C\,[\mu_{wi}\,(\boldsymbol{X})], \tag{6}$$

where $C\,[\mu_{wi}\,(\boldsymbol{X})]$ is the centroid between $compat_{min}$ and $compat_{max}$ for the pertinence function of class w_i , acording to $\boldsymbol{X}$. It is interesting to note that it was not found in literature another SUAS with interval nature as the one presented here [7].

4. Training Assessment

A computational tool for training assessment must supervise user movements during the training and assess the training according to M possible classes of performance. The assessment of trainee intends to monitor the training quality and provide some feedback about the user performance. Users interactions, as spatial movements, applied forces, angles and torque can be collected to assess them in a simulation.

As mentioned before, several assessment methods have been proposed and tested for the bone marrow harvest simulator proposed by Machado et al. [4]. That simulator has as goal to training new doctors to execute the bone marrow harvest, one of the stages of the bone marrow transplant. The simulator uses a haptic device [6] that operates with six degrees of freedom and provides force feedback to give to the user the tactile sensations felt during the penetration in the patient's body. Sight and touch are the two senses identified as fundamental in the procedure. The touch is necessary for the palpation of the virtual body and for the puncture with the needle. In this case, the use of a haptic device is mandatory to simulate the tissues properties and provide touch feedback for users. The perception of touch allows identifying texture of objects and force feedback allows to fell position and movement associated to the interaction with objects. In the system the haptic device simulates the needle used in the real procedure and the virtual body visually represented has the tactile properties of the real tissues. Users can fell those properties when manipulate the haptic device and at same time, data of that manipulation can be stored.

From interaction data, the computational tool for training assessment based on possibility and necessity measures acquire membership functions for each feature X_j , with $j=\{1,...,n\}$. From that, parameters to construct possibility and necessity distributions are provided, according each class of performance previously defined. When a user perform his training, data are collected from that. From data, possibility and necessity measures are calculated according (2) and also the intervals for the real value of the class of performance w_i, given each feature X_j, according (3). After that global matching between the class of performance w_i, given the training data $\boldsymbol{X}$ is calculated using (4) and (5). Finally, performance class is selected using (6).

5. Results

The calibration of the SUAS based on Possibilistic approach was made before any assessment of training. For that, an expert executed the procedure several times, where each execution is labeling according one of class of performance. In this case, M=3 classes of performance were defined by the expert: 1) “correct procedure” 2) “acceptable procedure” and 3) “badly executed procedure”. After, for a controlled and impartial analysis, several users used the system and 300 training procedures were monitored, where 100 were for each class of performance. The data collected from these trainings were manually labeled according to the expert specifications. These same cases were labeled using the SUAS based on Possibilistic approach and it generated the classification matrix showed in Table 1. The diagonal of that matrix shows the correct assessment. In the other cells, can be observed the mistakes of assessments. It was used the Kappa Coefficient to perform the comparison of the assessments agreement. From the classification matrix obtained, the Kappa coefficient for all samples was K=95.50% with variance 0.000218%.

The SUAS made mistakes in only 9 cases. It is important to note that for the class “acceptable procedures”, all classifications were correct. That performance is acceptable and it shows the good adaptation of SUAS based on Possibilistic approach in the solution of this assessment problem. Another important result is the computational performance of the assessment tool: with a Intel i3 (M330, 2.13GHz), with 4GB of RAM, the average time of CPU consumed by the assessment was 0.030 seconds of CPU. Then, it is possible to affirm the SUAS based on Possibilistic approach has low computational complexity and it has properties of an on-line SUAS. It allows the inclusion of other variables in the assessment tool without degradation of the performance of the VR simulation.

Table 1. Classification matrix for the SUAS based on Possibilistic approach.

Class of performance according to experts	Class of performance according to SUAS based on Possibilistic approach		
	1	2	3
1	91	0	9
2	0	100	0
3	0	0	100

6. Final Considerations

This paper presented a new methodology for SUAS of medical skills in VR simulators. That methodology is based on a Possibilistic approach, which use Possibility and Necessity Measures as a kernel of a DSS. Those fuzzy measures provide an interval estimator for the compatibility between an user's interactions and previously defined classes of performance by experts.

It was not found in literature another SUAS with interval nature as the one presented in this paper. This approach is general and can be applied also to other kind of training. Results obtained showed good accuracy of SUAS based on Possibilistic approach, helping to know skills and competences. The computational performance measure show that system proposed here can be characterized as an on-line SUAS.

References

1. D. Dubois, H. Prade, *Possibility Theory*, New York: Plenum Press (1988).
2. D. Dubois, H. Prade, C. Testemale, *Fuzzy Sets & Systems* **28**:313-331(1988).
3. T. Flaminio, L. Godo, E. Marchioni, *J. of Logic and Computation Advance* **21**(3):429-446 (2011).
4. L. Machado et al., *Studies in Health Tech. and Inf.* **81**: 293-297 (2001).
5. L. Machado, R. Moraes, In: *Comp. Int. in Complex Decision Systems* 85-123, chapter 4. Atlantis Press (2010).
6. R. Moraes, L. Machado, *Virtual Reality*, chapter 15. Intech (2011).
7. R. Moraes, L. Machado, *Proc. of VII ICECE*, CD-ROM Portugal (2011).
8. D.J. Scott et al., J. *Laparoendoscopic & Advanced Surgical Tech.* **10**(4): 183-190 (2000).
9. J. Solis et al., *IEEE Int. Conf. Robotics and Automation* 4219-4224 (2009).
10. Zadeh, L. A. *Fuzzy Sets and Systems* **1**(1): 3-28 (1978).

A NOVEL CLASS-ATTRIBUTE INTERDEPENDENCY DISCRETIZATION ALGORITHM

HONGLIANG ZHENG[1], LI ZOU[1,2], DEQIN YAN[1]

[1]*School of Computer and Information, Liaoning Normal Un., Dalian, 116029, China*
[2]*School of Computer Science and Technology, Dalian University of Technology, Dalian 116024, China*

The discretization of continuous value is important to machine learning and data mining. In this paper, based on novel class-attribute interdependency criterion, a discretization algorithm (called NCAIC) is proposed, which considers data distribution and the interdependency between all the classes and attributes, and adopts upper approximations as an part of the discretization criterion. In NCAIC algorithm, the class-attribute mutual information is adopted which can automatically control and adjust the extent of the discretization of continuous value, ensure discretization of real value attributes reasonable.

1. INTRODUCTION

Discretization of real or contiuous value attributes is indispensable in data mining, pattern recognition, machine learning and rough set analysis domain. There are five different axes by which the proposed discretization algorithms can be classified [12,16]. The two representative unsupervised algorithms are equal-width and equal-frequency discretizations [2]. In the literature [1], based on information theory a discretization algorithm (CADD) was proposed, which took class-attribute interdependence redundancy (CAIR) [19] as an important discretization criterion and selected the candidate cut point that could lead to better correlation between class and discrete intervals. But, this algorithm has the following disadvantages [10]: It uses a user-specified number of intervals when initializing the discretization intervals. The significance test used in the algorithm requires training for selection of a confidence interval. With class-attribute interdependency maximization (CAIM) criterion CAIM algorithm [10] avoids the disadvantages of CADD. However, CAIM algorithm has two drawbacks [17] : first, it gives a high factor to the number of generated intervals when it discretizes an attribute, hence, usually generates a simple discretization scheme. Secondly, it considers only the distribution of the major target class. Aiming to overcome the drawback, class-attribute contingency coefficient

(CACC) algorithm [17] gives a new criterion (formula). However, the formula of CACC brings mistakes in some cases.

In this paper, we define a novel class-attribute interdependency criterion, and propose a discretization algorithm (called NCAIC).The rest of the paper is organized as follows. In section 2, the preliminaries about rough set and discretization are presented. NCAIC discretization algorithm is presented in section 3. Finally, the conclusions are presented in section 5.

2. PRELIMINARIES

2.1. *Rough Set*

Let $U = \{x_1, x_2, \cdots, x_n\}$ be a universe and R be an equivalence relation on U. The equivalence relation R partitions the universe U into a collection of equivalence classes $\tilde{R} = \{E_1, E_2, \cdots, E_k\}$. Let X be a subset of finite universe U, according to Pawlak Rough Sets model [14], rough set of X is defined as

$$R\_(X) = \bigcup\{E \in \tilde{R} : E \subseteq X\},$$

$$R^{-}(X) = \bigcup\{E \in \tilde{R} : E \bigcap X \neq \emptyset\}.$$

Where $R\_(X)$ and $R^{-}(X)$ are called the lower and the upper approximation of X respectively.

2.2. *Basic Concept in Discretization*

Discretization is a technique to transform a continuous attribute's values into a finite number of intervals, and then assign a numerical or discrete value to each interval [10], [17]. Assuming that there is a training dataset consisting of M samples, where each sample belongs to only one of S classes. A indicates any of the continuous attributes in the dataset. A discretization scheme D on A would discretize the continuous domain of attribute A into n discrete intervals bounded by the pairs of numbers:

$$D : \{[d_0, d_1], (d_1, d_2], \cdots, (d_{n-1}, d_n]\} \tag{1}$$

Where, d_0 is the minimal value and d_n is the maximal value of attribute A, and the values in Equ.(1) are arranged in ascending order. These values constitute the boundary set $\{d_0, d_1, d_2, \cdots, d_{n-1}, d_n\}$ for discretization scheme D. d_i $(i = 0, 1, \cdots, n)$ are called cut points.

Each value of attribute A can be classified into only one of the n intervals defined in (1). Membership of each value within a certain interval for attribute A may change with the change of the discretization D. The class variable and the discretization variable of attribute A are treated as two random variables defining a two-dimensional frequency matrix (called quanta matrix) that is shown in Table 1. Where, q_{ir} is the total number of continuous values belonging to the *ith* class that are within interval $(d_{r-1}, d_r]$. C_i is the class label of the *ith* class. M_{i+} is the total number of objects belonging to the *ith* class and M_{+r} is the total number of continuous values of attribute A that are within the interval $(d_{r-1}, d_r]$, for $i = 1, 2, \cdots, S$, and $r = 1, 2, \cdots, n$.

The Shannon entropy and class-attribute information [9] of the given matrix are defined, respectively, as:

$$H(C, D \mid A) = \sum_{i=1}^{S} \sum_{r=1}^{n} p_{ir} \cdot \log_2 (1/p_{ir}),$$

$$I(C, D \mid A) = \sum_{i=1}^{S} \sum_{r=1}^{n} p_{ir} \cdot \log_2 (p_{ir}/(p_{i+} p_{+r}))$$

Where, $p_{ir} = q_{ir}/M$ is probability of the occurrence that attribute A values are within the interval $(d_{r-1}, d_r]$ and belong to class C_i. $q_{i+} = M_{i+}/M$ is probability of the occurrence that attribute A values belong to class C_i. $p_{+r} = M_{+r}/M$ is probability of the occurrence that attribute A values are within the interval $(d_{r-1}, d_r]$.

Table 1 Quanta Matrix for Attribute A and Discretization Scheme *D*

Class	Interval					Class Total
	$[d_0, d_1]$	…	$(d_{r-1}, d_r]$	…	$(d_{n-1}, d_n]$	
C_1	q_{11}	…	q_{1r}	…	q_{1n}	M_{1+}
:	:	…	:	…	:	:
C_i	q_{i1}	…	q_{ir}	…	q_{in}	M_{i+}
:	:	…	:	…	:	:
C_S	q_{S1}	…	q_{Sr}	…	q_{Sn}	M_{S+}
Interval Total	M_{+1}	…	M_{+r}	…	M_{+n}	M

3. NCAIC DISCRETIZATION ALGORITHM

3.1. *Related works*

In 2004, L.A. Kurgan etc. [10] proposed the CAIM discretization algorithm, the criterion used in the algorithm is:

$$CAIM(C,D \mid A)=\frac{\sum_{r=1}^{n}\frac{\max_r^2}{M_{+r}}}{n}$$

Where, n is the number of intervals, $\max_r$ is the maximum value among all q_{ir} values (maximum value within the rth column of the quanta matrix).

CACC is a newly proposed discretization algorithm [17], of which the discretization criterion is:

$$CACC(C,D \mid A)=\sqrt{\frac{x}{x+M}}$$

Where, $x=M\left[\left(\sum_{i=1}^{S}\sum_{r=1}^{n}\frac{q_{ir}^2}{M_{i+}M_{+r}}\right)-1\right]\Big/\log(n)$.

Comparing with CAIM, this criterion has considered the data distribution. Moreover, In the formula, the role of $\log(n)$ is: (a) speeds up the discretization process (b) controls continuous attribute values to be discretized excessively.

3.2. *NCAIC Discretization Algorithm*

Despite CACC discretization algorithm makes a better improvement on the previous ones, it suffers following drawbacks:

1. In CACC criterion, it is possible that the value of $\frac{x}{x+M}$ is negative, it happens in the case of $-M<x<0$. For example, for the data set in Table 2, we can see $\frac{x}{x+M}=-0.6143$.

2. $f(x)=\frac{x}{x+M}$ is a monotone increasing function, in the sense of criterion, it plays the role of x. So $f(x)$ makes the calculation complicated.

3. All continuous attribute are treated with same operation by dividing $\log(n)$, brings imprecise and immoderate discretization.

A good discretization algorithm should not only maximize the interdependency the real value attribute and its class labels, consider distribution of samples, but also, which is important, make full use of particular information of each attribute. Based on the idea, we define a discretization criterion called NCAIC (Novel Class-Attribute Interdependency Criterion):

$$NCAIC(C,D \mid A)=\left(\sum_{i=1}^{s}\frac{M_{i+}}{P^{-}(i)}\cdot\sum_{r=1}^{n}\frac{\prod_{i=1}^{s}\sqrt[s]{q_{ir}}}{M_{+r}}\right)\Big/\log_c(n)$$

Where $P^{-}(i)$ denotes upper approximation of the ith class (The term of rough set theory [14]), $c = \frac{H(C,D \mid A)}{I(C,D \mid A)}$.

The NCAIC criterion considers data distribution and the interdependency between all classes and attributes, adopts upper approximations of classes, divides $\log_c(n)$ to limit excessive discretization.

In the formula NCAIC, $\sum_{r=1}^{n}\left(\prod_{i=1}^{s}\sqrt[s]{q_{ir}}\Big/M_{+r}\right)$ represents the information of data distribution, the relationship of all the intervals and classes. $\sum_{i=1}^{S}\left(M_{i+}/P^{-}(i)\right)$ counts the sum of proportions of each class in the interval corresponding to a cut point. $\log_c(n)$ indicates the extent of disretization.

With NCAIC criterion, the proposed discretization algorithm (called NCAIC algorithm) is:

```
Step1: input data: Dataset with N continuous attributes, M examples and S target classes.

Step2: For each continuous attribute A_i
  {
  1. Find the maximum d_n and the minimum d_0 values of A_i
  2. Form values of A_i in ascending order
  3. Initialize all possible intervals and build candidate cut point set B
  4. Calculate the midpoints of all the adjacent pairs in the set B
  5. Set the initial discretization scheme as D :{ [d_0, d_n]} and vncaic = 0
  6. Initialize k = 1
  7. For each candidate cut point in the set B
     {
  8. Add it into D
  9. Calculate the corresponding NCAIC value ncaic
  10. Create a new discretization scheme D* with the highest ncaic value
       If (ncaic > vncaic or k < S )
     {
  Replace D with D*
  vncaic = ncaic
  k = k + 1
  }
  Else
    D* = D
  }
  11. Output the Discretization scheme D* with k intervals for continuous
attribute A_i
  }
```

4. THE EXPERIMENTAL RESULTS AND ANALYSIS

In experiments, we adopt the data sets of UCI machine learning database [13] (see table 2). NCAIC and other five discretization algorithms CADD, CAIM, CACC, Equal-F, MDL [1],[10],[17],[2],[5] are used for comparison. We perform C4.5 [15] on the discreted data, choosing randomly 80 percent of examples for training, and the rest for testing. The average predictive accuracys are shown in table 4.. The time comparison of building tree by C4.5 is shown in Table 5. and also in Figure 2..

Table 2. 13 UCI Datasets

Data Sets	Continuous attributes attributes attributes	Discrete attributes attributes	Number of Classes	Examples
iris	4	0	3	150
auto	5	2	3	392
breast	9	0	2	683
Ionosphere	34	0	2	351
pima	8	0	2	768
glass	9	0	7	214
wine	13	0	3	178
machine	7	0	8	209
heart	5	8	2	296
sonar	60	0	2	208
vehicle	18	0	4	846
vowel	10	3	6	990
bupa	6	0	2	345

Table 3. The comparison of C4.5 performance on 13 UCI real datasets

		Data Sets												
Criterion	Algorithm	iris	auto	breast	ionosphere	pima	glass	wine	machine	heart	sonar	vehicle	vowel	bupa
predictive accuracy (%)	CADD	90.0	64.6	92.0	86.9	63.6	72.1	91.7	83.3	73.3	64.3	47.1	38.9	65.2
	CAIM	90.0	59.5	93.4	85.9	63.6	67.4	91.7	76.2	80.0	69.0	46.5	38.9	63.8
	CACC	93.3	68.4	94.2	83.1	76.0	67.4	86.1	83.3	75.0	69.0	44.1	40.9	65.2
	NCAIC	96.7	72.2	94.9	88.7	76.0	67.4	94.4	85.7	76.7	64.3	51.8	49.0	72.5
	Equal-F	90.0	72.2	92.0	85.9	67.5	58.1	91.7	66.7	78.3	64.3	47.1	47.1	58.0
	MDL	90.0	72.2	94.2	84.5	67.5	58.1	83.3	83.3	71.7	64.3	50.6	40.9	60.9

From Table 2. and Table 3., we can see that, compared with other algorithms, NCAIC can obtain very good accuracy, and the accuracy on some data (such as wine, vowel, bupa) have raised greatly.

5. CONCLUSIONS

In this paper, a novel discretization algorithm (NCAIC) for real value attributes based on rough set theory and mutual information is proposed. In NCAIC algorithm, a new Class-Attribute Interdependency Criterion (called NCAIC criterion) is defined, which considers data distribution and the interdependency between all the classes and attributes, and regards upper approximations conception in rough set theory as an important part of the discretization criterion. Besides, in NCAIC criterion, the class-attribute mutual information is used to automatically control and adjust the extent of the discretization of continuous values, which makes discretization of real value attributes exactly and reasonably. Comparison experiments are performed with NCAIC and other five discretization algorithms for prediction by using C4.5. The experiment results show that the presented algorithm is effective and promising.

Acknowledgements

This work is partly supported by National Nature Science Foundation of China (Grant No. 61105059, 61175055, 61173100), Liaoning Excellent Talents in University(LJQ2011116), ichuan Key Technology Research and Development Program under Grant No.2011FZ0051, Sichuan Key Laboratory of Intelligent Network Information Processing(SGXZD1002-10) and Key Laboratory of Radio Signals Intelligent Processing (Xihua University) (XZD0818-09).

REFERENCES

1. J.Y. Ching, A.K.C. Wong and K.C.C. Chan. Class-Dependent Discretization for Inductive Learning from Continuous and Mixed-Mode Data. IEEE Transactions on Pattern Analysis and Machine Intelligence, 17(7), 641-651,1995
2. D. Chiu, A. Wong, and B. Cheung. Information Discovery through Hierarchical Maximum Entropy Discretization and Synthesis. Knowledge Discovery in Databases, G.Piatesky-Shapiro and W.J. Frowley, ed., MIT Press, 1991
3. Y Ding, LM Zhu, XJ Zhang. A full-discretization method for prediction of milling stability. International Journal of Machine Tools and Manufacture, 5(5), 502-509, 2010.
4. XJ Zhang, CH Xiong. Improved full-discretization method for milling chatter stability prediction with multiple delays. Intelligent Robotics and Applications, Lecture Notes in Computer Science, 6425, 541-552, 2010
5. U.M. Fayyad, K.B. Irani. On the handling of continuous-valued attributes in decision tree generation. Machine Learning, 8: 87-102, 1992
6. SH Ha, L Zhuang, Y Zhou, Z Pei. Treatment Method after Discretization of Continuous Attributes Based on Attributes Importance and Samples Entropy. Proceedings of Intelligent Computation Technology and Automation (ICICTA), 1169 - 1172, 2011
7. Alexis Bondu, Marc Boullé, Vincent Lemaire. A non-parametric semi-supervised discretization method. Knowledge and Information Systems, 24(1), 35-57,2010
8. J. Dougherty, R. Kohavi and M. Sahami. Supervised and Unsupervised Discretization of continuous feature. In: Proceeding of Twelfth International Conference on Machine Learning, 194-202, 1995
9. R. Kerber. ChiMerge: discretization of numeric attributes. Proceedings Ninth National Conference on Artificial Intelligence, AAAI Press: 123-128, 1992
10. L.A. Kurgan and K.J. Cios. CAIM Discretization Algorithm. IEEE Transactions on Knowledge and Data Engineering, 16(2): 145-153, 2004
11. C.I. Lee, C.J. Tsai, Y.R. Yang, W.P. Yang. A top-down and greedy method for discretization of continuous attributes. in: Proceedings of Fourth

International Conferenceon Fuzzy Systems and Knowledge Discovery, Haikou, China, 2007
12. X. Liu, H. Wang. A discretization algorithm based on a heterogeneity criterion. IEEE Transactions on Knowledge and Data Engineering, 17(9): 1166-1173, 2005
13. S. Hettich, S. D. Bay. The UCI KDD Archive [DB/OL]. http://kdd.ics.uci.edu/, 1999
14. Z. Pawlak. Rough set. International Journal of Information and Computer Science, 11(5): 341–356, 1982
15. J.M. Quinlan. C4.5: programs for machine learning. San Mateo, Calif.: Moorgan Kaufmann, 1993
16. Chao-Ton Su, Jyh-Hwa Hsu. An Extended Chi2 Algorithm for Discretization of Real Value Attributes. IEEE Transactions on Knowledge and Data Engineering, 17(3): 437-441, 2005
17. Cheng-Jung Tai, Chien-I. Lee, Wei-Pang Yang. A discretization algorithm based on Class-Attributes Contingency Coefficient. Information Sciences, 178: 714-731, 2008
18. E.H. Tay, L. Shen. A Modified Chi2 Algorithm for Discretization. IEEE Transactions on Knowledge and Data Engineering. 2002, 14(3): 666-670
19. A.K.C. Wong and D.K.Y. Chiu. Synthesizing Statistical Knowledge from Incomplete Mixed-Mode Data. IEEE Transactions on Pattern Analysis and Machine Intelligence, 1987, 9: 796-805

THE TASK OF MINIMUM COST FLOW FINDING IN TRANSPORTATION NETWORKS IN FUZZY CONDITIONS*

ALEXANDR BOZHENYUK[1], EVGENIYA GERASIMENKO[†1], IGOR ROZENBERG[2]

[1]*Southern Federal University, Oktyabrskaya Square, 4, Taganrog, 347922, Russia*[2]*Public corporation "Research and development institute of railway engineers",27/1, Nizhegorodskaya, Moscow, 109029, Russia*

This article describes the task of minimum cost flow finding in the transportation network in fuzzy conditions. Technique for solving the task of minimum cost flow determining in the graph with upper and lower fuzzy flow bounds is examined. This task is observed in fuzzy conditions according to the criteria of real life. The numerical example is presented.

1. Introduction

The minimum cost flow finding is fundamental problem, arising in the analysis of transportation networks. This problem can be formulated as follows: let $G=(X,A)$ be a directed network (graph), where X is the set of nodes, A is the set of arcs. Each arc $(i,j)\in A$ has a capacity q_{ij} determining the maximum amount of flow, that can pass along the arc (i,j), and transmission cost c_{ij} of one flow unit along the arc (i,j). The essence of this task is to send given flow value from the source to the sink in the network by the best in terms of costs way, taking into account bounded arc capacities. The minimum cost flow problem can be stated as follows:

$$\sum_{(x_i,x_j)\in A} c_{ij}\cdot\xi_{ij} \to \min,$$

$$\sum_{x_j\in\Gamma(x_i)} \xi_{ij} - \sum_{x_k\in\Gamma^{-1}(x_i)} \xi_{ki} = \begin{cases} v, x_i = s, \\ -v, x_i = t, \\ 0, x_i \neq s,t, \end{cases} \tag{1}$$

$$0 \le \xi_{ij} \le q_{ij}, \forall (x_i,x_j)\in A,$$

---

* *This work is supported by the Russian Foundation for Basic Research, project 11—01—00011 a*

† *e.rogushina@gmail.com*

where c_{ij} — transmission cost of one flow unit along the arc (i,j); ξ_{ij} — amount of flow, passing along the arc (i,j); v — given flow value; s — initial node (source); t — terminal node (sink); q_{ij} — capacity of the arc (i,j) in Eq. (1). The task has a solution if the given flow value does not exceed maximum flow in the graph.

In practice the arc capacities, the flow values can not be accurately measured according to their nature. Weather conditions, emergencies on the roads, traffic congestions and repairs influence arc capacities. Variations in petrol prices, tolls can either influence transmission costs. Therefore, these parameters should be presented in a fuzzy form, such as fuzzy triangular numbers [1]. Therefore, we obtain a problem statement of minimum cost flow finding in fuzzy conditions.

2. Literature Review

Many researchers have examined the task of minimum cost flow finding in crisp conditions in the literature. Methods of its solution can be divided into graph techniques and the methods of linear programming. In particular, solutions by the graph methods are considered in Ref. [2,3]. The advantages of this approach are great visualization and less cumbersome. In Ref. 3, 4 a task of minimum flow determining is considered as a task of linear programming. This method is more universal, as researchers do not have to be an experts in flow problems. The best possibility of automation is another advantage of this approach.

There are methods of "fuzzy linear programming", which were widely reported in the literature [5,6]. The task of "fully fuzzy linear programming" was considered in Ref. 7. These methods are cumbersome and can not lead to optimal solutions in the minimum cost flow determination. The solution of fuzzy linear programming task by the comparison of fuzzy numbers based on ranking functions is examined in Ref. 8. The algorithm, described in Ref. 9, is implied in order to allow alternative solutions that may arise in such problems. There are approaches that describe the problem solution as a linear parametric model based on the membership function of fuzzy parameter λ, [10] but these approaches are described only for intervals, rather for fuzzy triangular (trapezoidal) numbers.

3. The task of minimum cost flow determining with upper and lower fuzzy flow bounds

Suppose that the considering transportation network consists of highways (roads) and air roads. Air roads use airplanes as means of transportation, which can fly at a specific level of load. Thus, it is necessary to set limits on the lower bounds of the flows. For example, the maximum number of seats on the plane is 100, minimum — 35, thus, the capacity of the airplane varies from 35 to 100.

Consider the problem of minimum cost flow finding in a graph, that has arcs with upper flow bounds and arcs with either upper, or lower bounds. This problem can be presented considering fuzzy (triangular) arc capacities as follows:

$$\sum_{(x_i,x_j)\in A} c_{ij}\cdot\xi_{ij} \to \min,$$

$$\sum_{x_j\in\Gamma(x_i)} \xi_{ij} - \sum_{x_k\in\Gamma^{-1}(x_i)} \xi_{ki} = \begin{cases} v, x_i = s, \\ -v, x_i = t, \\ 0, x_i \neq s,t, \end{cases} \tag{2}$$

$$\tilde{l}_{ij} \le \xi_{ij} \le \tilde{q}_{ij}, \forall (x_i,x_j)\in A,$$

where $\tilde{l}_{ij}$ — fuzzy lower flow bound of the form (l_1,l_2,l_3), $\tilde{q}_{ij}$ — fuzzy upper flow bound (arc capacity) of the form (q_1,q_2,q_3), where l_2,q_2 are the centers, l_1,q_1 — the left borders, l_3,q_3 — the right borders of flow bounds, which have triangular membership functions in Eq. (2).

Let us define the membership function, which reflects satisfying a condition for the capacity constraints $\mu(\tilde{l}_{ij} \le \xi_{ij} \le \tilde{q}_{ij})$:

$$\mu(\tilde{l}_{ij} \le \xi_{ij} \le \tilde{q}_{ij}) = \begin{cases} 0, q_3 < \xi_{ij} < l_1; \\ \dfrac{\xi_{ij}-l_1}{l_3-l_1}, l_1 \le \xi_{ij} \le l_3; \\ 1, l_3 < \xi_{ij} < q_1; \\ 1-\dfrac{\xi_{ij}-q_1}{q_3-q_1}, q_1 \le \xi_{ij} \le q_3. \end{cases} \tag{3}$$

Inequality $\tilde{l}_{ij} \le \xi_{ij} \le \tilde{q}_{ij}$ means that the flow does not exist in the range of $[0,l_1)$ and $(q_3,\infty]$; the flow increases in the range of $[l_1,l_3]$; the flow equals to "1" in the range from l_3 to q_1; the flow decreases in the range of $[q_1,q_3]$, as shown in Fig. 1.

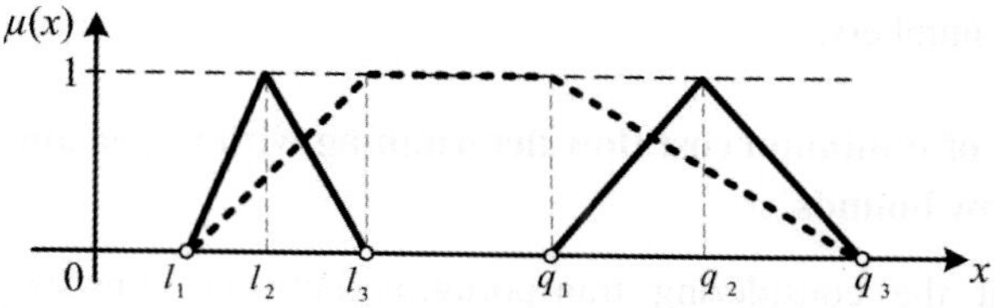

Fig. 1. Transforming triangular lower and upper flow bounds to trapezoidal.

According to Ref. 10 λ –level of membership function is:

$$\xi_{ij}^{\lambda} = \left\{ \xi_{ij} \in R^n : \mu(\tilde{l}_{ij} \le \xi_{ij} \le \tilde{q}_{ij}) \ge \lambda \right\}, \forall (x_i, x_j) \in A. \tag{4}$$

Convert the second and the fourth expressions of Eq. (3), considering Eq. (4), we obtain:

$$\begin{cases} \dfrac{\xi_{ij} - l_1}{l_3 - l_1} \ge \lambda, \\ 1 - \dfrac{\xi_{ij} - q_1}{q_3 - q_1} \ge \lambda. \end{cases} \tag{5}$$

Transforming Eq. (5), we obtain a system Eq. (6), which is true for any arcs of the graph:

$$\begin{cases} \xi_{ij} \ge l_1 + \lambda(l_3 - l_1), \\ \xi_{ij} \le q_3 - \lambda(q_3 - q_1). \end{cases} \tag{6}$$

We turn to special form of fuzzy number of L–R type from triangular membership function in the case when arcs have only upper flow bounds, as described and solved in Ref. 10.

According to membership function of such a number of L–R type and Eq. (4), we obtain the second inequality of Eq. (6).

Finally, we can obtain the model of the minimum cost flow finding task with fuzzy lower and upper flow bounds, combining Eq. (6) and the original problem statement Eq. (2). Thus, presented model has crisp form:

$$\sum_{(x_i, x_j) \in A} c_{ij} \cdot \xi_{ij} \to \min,$$

$$\sum_{x_j \in (x_i)} \xi_{ij} - \sum_{x_k \in^{-1} (x_i)} \xi_{ki} = \begin{cases} v, x_i = s, \\ -v, x_i = t, \\ 0, x_i \ne s, t, \end{cases} \tag{7}$$

$$\begin{cases} \xi_{ij} \ge l_1 + \lambda(l_3 - l_1), \\ \xi_{ij} \le q_3 - \lambda(q_3 - q_1), \end{cases}$$

$$\lambda \in [0,1].$$

Eq. (7) is a linear parametric model, which basic solution method is parametric simplex method of linear programming. It calculates the optimal solution, depending on the parameter λ. Thus, let us consider an example presenting the solution of this problem. Consider the graph in Fig. 2 with fuzzy values of arc flow bounds given above the arcs of the graph and arc transmission costs of one flow unit under the arcs. Our task is calculating minimum transmission cost for given flow value of $v = 13$ flow units. Other words, we must define optimal paths (in terms of costs) of sending of 13 flow units. The task will have an optimal solution if observed flow value (in our case $v = 13$) is less than maximum flow value in the graph. The model has the following form:

$$F = 8 \cdot \xi_{sx_1} + 12 \cdot \xi_{sx_2} + 15 \cdot \xi_{x_1x_3} + 20 \cdot \xi_{x_2x_4} + 6 \cdot \xi_{x_3x_2} + 7 \cdot \xi_{x_3t} +$$
$$+9 \cdot \xi_{x_4t} \to \min, \xi_{sx_1} + \xi_{sx_2} = 13; \xi_{x_1x_3} - \xi_{sx_1} = 0;$$
$$\xi_{x_2x_4} - \xi_{sx_2} - \xi_{x_3x_2} = 0; \xi_{x_3t} + \xi_{x_3x_2} - \xi_{x_1x_3} = 0;$$
$$\xi_{4t} - \xi_{24} = 0; -\xi_{3t} - \xi_{4t} = -13; \xi_{sx_1} \geq 3 + 3 \cdot \lambda;$$
$$\xi_{sx_1} \leq 35 - 11 \cdot \lambda; \xi_{sx_2} \leq 14 - 7\lambda, \xi_{x_1x_3} \leq 16 - 5 \cdot \lambda;$$
$$\xi_{x_2x_4} \geq 5 + 6 \cdot \lambda; \xi_{x_2x_4} \leq 48 - 17 \cdot \lambda; \xi_{x_3x_2} \leq 29 - 9 \cdot \lambda;$$
$$\xi_{x_3t} \leq 9 - 4 \cdot \lambda; \xi_{x_4t} \leq 40 - 12 \cdot \lambda;$$
$$\lambda \in [0;1].$$

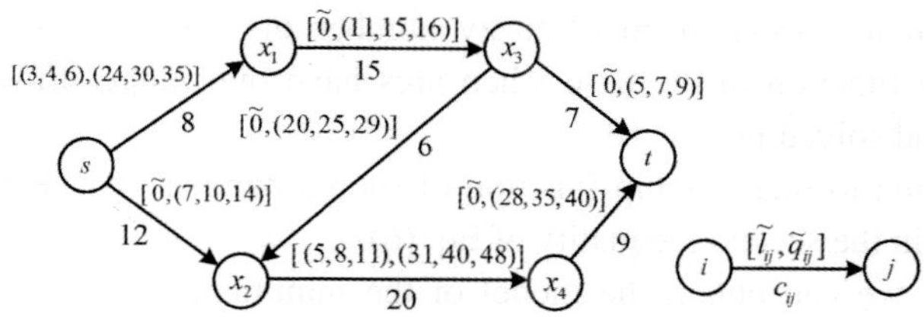

Fig. 2. Graph with specified arc capacities as fuzzy triangular numbers and crisp costs.

$\tilde{0} = (0,0,0)$, because it means the absence of lower bound. Solve the problem by the simplex method of linear programming and obtain the optimal solution, depending on the parameter $\lambda \in [0;1]$. Thus, the optimal solutions are found and represented in the Table 1.

Let us carry out an interpretation of the results. It can be argued that all obtained admissible solutions are optimal, but with different degrees of optimality. Consequently, the λ–level equals to "1" is considered the most optimal and the corresponding solutions will be most feasible.

Table 1. The optimal solutions depending on the parameter λ.

	$\xi^*_{sx_1}$	$\xi^*_{sx_2}$	$\xi^*_{x_1x_3}$	$\xi^*_{x_2x_4}$	$\xi^*_{x_3x_2}$	$\xi^*_{x_3t}$	$\xi^*_{x_4t}$	F^*
$\lambda = 0$	8	5	8	5	0	8	5	445
$\lambda = 0.5$	5	8	5	8	0	5	8	478
$\lambda = 1$	6	7	6	11	4	2	11	579

Therefore the objective function $F^* = 579$ conventional units and flows corresponding to the "solution 3" are considered optimal.

4. Conclusion

The problem of minimum cost flow finding in the transportation network in terms of uncertainty is presented in this article. In particular, the method of minimum cost flow finding in transportation network which arcs have upper and lower fuzzy flow bounds is examined and solved as a linear parametric model based on the membership function of fuzzy parameter λ. This task has been solved according to the criteria of real life, i.e. the fuzzy nature of the arc capacities and costs was taken into account.

References

1. H. J. Zimmermann, *Fuzzy Set Theory and Its Applications*, 5th edn. (Kluwer Academia Publishers, Boston, Dordrecht, London, 1991).
2. N. Christo_des, *Graph Theory: An Algorithmic Approach* (Academic Press, London, 1976).
3. E. Minieka, *Optimization Algorithms for Networks and Graphs* (Marcel Dekker, New York, 1978).
4. T.C. Hu, *Integer Programming and Network Flows* (Addison–Wesley Publishing Company, Ontario, 1970).
5. P. A. Thakre, D. S. Shelar and S. P. Thakre, Solving Fuzzy Linear Programming Problem as Multi Objective Linear Programming Problem in *Proc. of the World Congress on Engineering (WCE 2009)* **II**, (London, U.K., 2009).
6. K. Ganesan and P. Veeramani, Fuzzy Linear Programs with Trapezoidal Fuzzy Numbers, *Ann Oper Res.* **143**, 305—315 (2006).
7. A. Kumar, J. Kaur and P. Singh, Fuzzy Optimal Solution of Fully Fuzzy Linear Programming Problems with Inequality Constraints, *International Journal of Mathematical and Computer Sciences* **6**, 37—41 (2010).
8. K. P. Yoon, A Probabilistic Approach to Rank Complex Fuzzy Numbers, *Fuzzy Sets and Systems* **80**, 167—176 (1996).
9. H. R. Maleki and M. Mashinchi, Fuzzy Number Linear Programming: a Probabilistic Approach, *J. Appl. Math. and Computing* **15**, 333—341 (2004).
10. A. V. Bozhenyuk, I. N. Rozenberg and T. A. Starostina, *Analysis and Investigation of Flows and Vitality in Transportation Networks in Fuzzy Data* (Scientific World, Moscow, 2006).

PATH IDENTIFYING IN POINTS-TO ANALYSIS FOR JAVA WITH ANSWER SET PROGRAMMING*

YANG BO

Department of Physics and Electronics Information Science, Guiyang University
Guiyang, 550005, China

ZHANG YING[2], ZHANG MING-YI[†1,2]

[1] *Chongqing University of Arts and Science, Chongqing, 402160, China*
[2] *Guizhou Academy of Science, Guiyang, 550001, China*

In this paper a set of rules of Answer Set Programming is presented to represent and reason about points-to information in program written in object-oriented language like Java. Especially, path identification problems such as forward and backward path problems in context of aliasing are well solved with these rules. A simple Java program is given to show the effect of our method.

1. Introduction

In the past two decades, many researchers dedicated to find a well tradeoff between precision and computation cost of alias analysis for some programming languages[1,2,3,4]. Recent works showed that aliasing is also a key step in program debugging[5]. For object-oriented language like Java, reference variables are apt to alias to each other. Operation on a variable perhaps leads to an inexpectant change to other variables. Thus, alias in Java program is one of the main causes for program errors.

Alias occurs when two or more variables refer to a same memory location. In some aliasing algorithms for C or C++[1,2], alias information is represented by alias pair in form of (x, y), where x, y are access paths. Points-to relation introduced by Emami et al is more concision and informative than alias pair[3]. Referrence-set alias representation adopted in [4] is essentially same as points-to relation, but less space are occupied. Almost all methods mentioned above computed alias information according to data flow equation of program static

* This work is supported by the Natural Science Foundation of China(No. 60963009), The Open Project of Shanghai Key Laboratory of Trustworthy Computing(No.07dz22304201108), Natural Science Foundation of Guizhou Province(No.[2010]2092 and No.[2012]2021).
† email: zhangmingyi045@yahoo.com.cn

analysis, some imprecision is introduced when dealing with the conditional statement, and all of them need to implement a software tool to compute aliases.

Logic programming has got successful applications in software engineering[6,7]. Especially, works of Lam et al showed that using logic programming on points-to and alias analysis is feasible and promised[6]. Motivated by their ideas, we tried to discuss the alias problem from the view of Answer Set Programming(ASP), and found that propagation of points-to information along execution paths of program accords with the feature of frame problem that arises from reasoning actions and their effects in a dynamic system. ASP provides elegant solution for frame problem. So in [8] we gave a set of rules to describe the propagating of points-to information in a natural and concise way. These rules allow us easily reason about points-to information of variables at each program point. However, if we want to apply these rules on real Java program, a more complicate problem--path identifying--need to be taken into consideration. Now we expand rules presented in [8] to accommodate this problem. So we can determine the propagation path of points-to information more accurately, and hence improve the precision of points-to analysis.

In section 2, path identification problem is described based on an example. Section 3 presents an expansion of our previous set of rules, and an example is given to show the effects of these rules. Finally, section 4 concludes our work.

2. Path Identification Problem

We describe path identification problem with the following example program of Java.

Example 1 An example Java program

```
    public class shape                   public class triangle extends shape
s1: { public void draw( ){ }             { public void draw( )
    }                                 s3: {System.out.println("t.draw( )"; }
                                         }

    public class circle extends shape
    { public void draw( )
s2: {System.out.println("c.draw( )"); }
    }

public class shapes{                  s7:      sh=c;
   public static void main(String[ ] args )  s8:   else  sh=t;
   { shape sh;                        s9:    sh.draw( );
s4:  circle c=new circle( );          s10:   circle c1=new circle( );
s5:  triangle t=new triangle( );      s11:   c1.draw( )
s6:      if(cond)                            } }
```

Calling graph(CG) and control flow graphs(CFGs) for this program are shown in Fig.1. CFGs of methods *main*(Fig.1(b)) and *draw*(Fig.1(c)~(e)) are connected through three pairs of arrows. Call sites s_9 and s_{11} are split into two parts, where the former(s_9 and s_{11}) denote the pre-call nodes and the latter(s_9' and s_{11}') are post-call nodes. Node s_m is introduced as a merging node of conditional statement s_6~s_8.

Path identifying refers to guaranteeing points-to information propagates along valid path. It involves three sub-problems, and they are: (1) forward path problem; (2) backward path problem; (3) type inference.

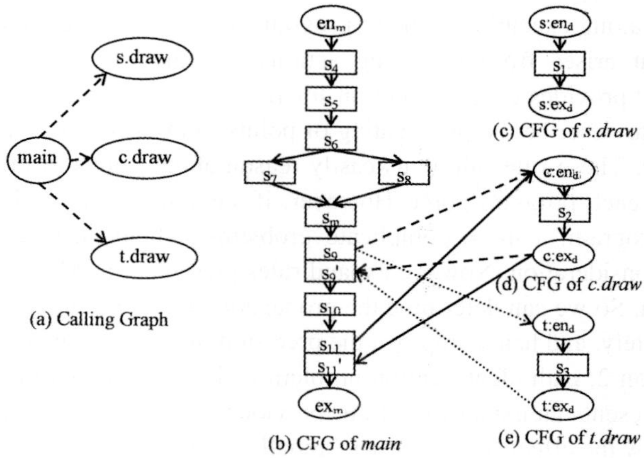

Figure 1 calling graph and control flow graphs for example 1

Forward path problem occurs when points-to information from different paths converge at a merging node, e.g. node s_m and $c{:}en_d$ in Fig.1(b) and (d). When alias pair representation is adopted, invalid alias relation maybe introduced. For instance, if we get alias pair (x,y) from s_7 and (x,z) from s_8, then invalid alias pair (y,z) is generated at node s_m. So, source information of an alias relation should be recorded, such as statement generating the alias relation and call site from which this alias relation propagated into the called procedure.

Backward path problem occurs when aliases are returned back to inappropriate caller(s). In Fig.1 (b) and (d), if there is an alias relation (x,y) is propagated from call site s_9 into procedure *c.draw* and kept at node $c{:}ex_d$, then (x,y) is returned back to call site s_9' and s_{11}', where the latter is an improper accepter. This makes spurious aliases are propagated forward. Thus, when a procedure is invoked by multi-callers, we need to identify the call site where a given alias relation comes from. Type inference refers to statically determine the run-time type of an object. For object-oriented programming language like Java, inheritance and polymorphism allow a reference of a base class to refer to an object of its derived class. When this reference is used to invoke an overridden method that exists in both base class and its derived class(es) at call site C, it is difficult to determine which method is called exactly because of dynamic binding of method call. Which one of these overridden methods is actually called depends on the runtime type of the reference used to invoke it at *C*. In Fig.1, the two pairs of dashed arrows represent *possible* calling relationship between method *main* and *c.draw*, *t.draw*. Which *draw* method is called depends on the type of reference *sh* at node s_9. Many aliasing algorithms included mechanisms to statically determine runtime types of variables.

3. Solution to path identification problem

In [8], some sets and predicates are defined to represent knowledge of a given program. To solve problems presented in Sec.2, we must redefine some of them and introduce some new sets and predicates:

- Set S of nodes: each $s \in$ S represents a node in CFGs of methods;
- Set X of references: each $x \in$ X is a reference variable defined in program;
- Set F of fields: collection of fields such that each $f \in$ F is also a reference;
- Set O of objects: $o_s \in$ O represents an object defined at node *s*;
- Set M of methods: each *m* in M is a method in a program;
- Set C of classes: each $c \in$ C is a class defined in a given program;
- ref(*x*,*o*,*cs*,*s*): reference *x* refers to object *o* after execution of node *s*, where *cs* is the latest node calling method in which *s* is included;
- reff(*o*,*f*,*o'*,*cs*,*s*): field *f* of object *o* refers to object *o'* after execution of *s*;
- in(*s*,*c*,*m*): *s* is one of the node in CFG of method *m* of class *c*;
- asn(*s*,*x*,*y*): node *s* assigns reference *y* to *x*;
- store(*s*,*x*,*f*,*y*): node *s* is of the form *x.f*=*y*;
- load(*s*,*y*,*x*,*f*): node *s* is a load operation *y*=*x.f*;
- new(*s*,o_s,*x*): node *s* performs a *new* operation creating object o_s referred by *x*;
- class(*o*, *c*): object *o* is of class *c*;
- if(*s*): node *s* is the beginning node of a conditional statement;
- branch(*s'*,*s*): node *s'* is one of the immediate successors in branches of the conditional statement beginning from *s*;
- inbranch(*s*): node *s* is in a branch of a conditional statement;
- same(*s*,*s'*): node *s* and *s'* are in the same branch of a conditional statement;
- pred(s,s'): node s is the immediate predecessor of node s';
- pair(s,s'): node s and s' are pre-call node and post-call node respectively of a calling statement;
- aP(ap,i,m): reference ap is the ith actual parameter when calling m;
- fP(fp,i,m,c): fp is the ith formal parameter of method defined in class c;
- callby(x,m,s): reference x is used to call method m at node s, and the method containing s is called by cs;

- call(m,c,s,cs): method m defined in class c is called at node s, and the method containing s is called by node cs ;
- entry(c,m,si): node si is the entry node of method m of class c;
- exit(c,m,so): node so is the exit node of method m of class c;
- local(x,c,m): reference x is a local reference variable of method m of class c.

Rules dealing with intra-procedural analysis are as follows:

$\text{ref}(x,o,cs,s')\leftarrow\text{pred}(s,s'), \text{ref}(x,o,cs,s), \text{not } \neg\text{ref}(x,o,cs,s')$ (R1)
$\text{reff}(o,f,o',cs,s')\leftarrow\text{pred}(s,s'), \text{reff}(o,f,o',cs,s), \text{not } \neg\text{reff}(o,f,o',cs,s')$ (R2)
$\text{ref}(x,o_s,cs,s)\leftarrow\text{new}(s,o_s,x), \text{call}(m,c,cs,cs'), \text{in}(s,c,m), \text{not inbranch}(s)$ (R3)
$\text{ref}(x,o_s,cs,s)\leftarrow\text{new}(s,o_s,x), \text{call}(m,c,cs,cs'), \text{in}(s,c,m), \text{pred}(s_1,s_2), \text{if}(s_1),$
$\text{branch}(s_2,s_1), \text{inbranch}(s), \text{same}(s_2,s)$ (R4)
$\neg\text{ref}(x,o',cs,s')\leftarrow\text{pred}(s,s'), \text{new}(s',o,x), \text{ref}(x,o',cs,s)$ (R5)
$\text{ref}(x,o_y,cs,s')\leftarrow\text{pred}(s,s'), \text{ref}(y,o_y,cs,s), \text{asn}(s',x,y)$ (R6)
$\neg\text{ref}(x,o_x,cs,s')\leftarrow \text{pred}(s,s'), \text{ref}(x,o_x,cs,s), \text{asn}(s',x,y), \text{not sp}(s')$ (R7)
$\text{reff}(o_x,f,o_y,cs,s')\leftarrow\text{pred}(s,s'), \text{ref}(y,o_y,cs,s), \text{ref}(x,o_x,cs,s), \text{store}(s',x,f,y)$ (R8)
$\neg\text{reff}(o_x,f,o,cs,s')\leftarrow\text{pred}(s,s'), \text{ref}(x,o_x,cs,s), \text{reff}(o_x,f,o,cs,s), \text{store}(s',x,f,y),$
$\text{not sp}(s')$ (R9)
$\text{ref}(y,o,cs,s')\leftarrow\text{pred}(s,s'), \text{ref}(x,o_x,cs,s), \text{reff}(o_x,f,o,cs,s), \text{load}(s',y,x,f)$ (R10)
$\neg\text{ref}(y,o_y,cs,s')\leftarrow\text{pred}(s,s'), \text{ref}(y,o_y,cs,s), \text{load}(s',y,x,f), \text{not sp}(s')$ (R11)
$\text{sp}(s')\leftarrow \text{pred}(s,s'), \text{ref}(x,o_x,cs,s), \text{ref}(y,o_y,cs,s), o_x=o_y , \text{asn}(s',x,y)$ (R12)
$\text{sp}(s')\leftarrow\text{pred}(s,s'), \text{ref}(x,o_x,cs,s), \text{reff}(o_x,f,o,cs,s), \text{ref}(y,o_y,cs,s), o=o_y,$
$\text{store}(s',x,f,y)$ (R13)
$\text{sp}(s')\leftarrow\text{pred}(s,s'), \text{ref}(y,o_y,cs,s), \text{ref}(x,o_x,cs,s), \text{reff}(o_x,f,o,cs,s),$
$o_y=o, \text{load}(s',y,x,f)$ (R14)
$1\{\text{pred}(s,s'):\text{branch}(s',s)\}1\leftarrow\text{if}(s)$ (R15)

Rules R1~R15 characterize propagating of points-to information and influence of operation such as *new*, *store*, *load* and *assignment* on points-to relation. Variable *cs* in predicates ref( ) and reff( ) is used to record the nearest call site from which the points-to relation comes. For *main* method in a program, call(*main*,c_m,*null*,*null*) always holds, where *null* is a special element in set S of nodes and c_m is the class containing method *main*. More detailed explanation for rules R1~R15 see [8].

R15 is a "choice rule" with numerical bounds[9], it says that if *s* is a beginning node of a conditional statement, then from all possible successors of *s*, choose at least 1 and at most 1 successor *s'* to make pred(*s*,*s'*) holds. Introducing R15 results in more than one answer sets for logic program Π that contains this rule. This means each branch of a conditional statement is analyzed independently. Analysis result for each branch is represented in different answer set of Π.

As for inter-procedural points-to analysis, the following rules are used:

call(m,c,s,cs) ←callby(x,m,s), ref(x,o,cs,s), class(o,c) (R16)
ref(fp,o,s,s_i)←call(m,c,s,cs), entry(c,m,s_i), aP(ap,i,m), fP(fp,i,m,c), ref(ap,o,cs,s) (R17)
ref(x,o,s,s_i)←call(m,c,s,cs), entry(c,m,s_i), ref(x,o,cs,s) (R18)
reff(o,f,o',s,s_i)←call(m,c,s,cs), entry(c,m,s_i), reff(o,f,o',cs,s) (R19)
ref(x,o,cs,s')←call(m,c,s,cs), exit(c,m,s_o), ref(x,o,s,s_o), pair(s,s'), not local(x,c,m), not fP(x,i,m,c) (R20)
reff(o,f,o',cs,s')←call(m,c,s,cs), exit(c,m,s_o), reff(o,f,o',s,s_o), pair(s,s') (R21)

To determine the runtime type of a reference used to invoke an overridden method at call site *s*, points-to information of this reference is very helpful. R16 decides the correct method to be invoked at node *s*. Once the appropriate method *m* to be called is determined, R17~R19 specify the points-to information of formal parameters of *m* and points-to relations come from *s* at entry node s_i of *m*.

With the source information recorded by variable *cs* we can deal with the backward path problem. Rule R20 and R21 are used to indicate the correct post-call node to which given points-to relations should return back. Because of the last predicates in R20, points-to information of formal parameters and local variables in current method will not be propagated back to its caller.

Example 2 Points-to analysis for Example 1 using R1~R21

We will show the effects of R1~R21 on points-to information for program listed in Example 1. Particularly, we will give our attention to points-to information at pivotal nodes----node creating object, pre-call and post-call nodes of a call site. Now we show the key steps.

Basic facts of the program are represented as the following predicates:

P={pred(en_m,s_4), pred(s_4,s_5), pred(s_5,s_6), pred(s_7,s_m), pred(s_8,s_m), pred(s_m,s_9), pred(s_9',s_{10}), pred(s_{10},s_{11}), pred(s_{11}',ex_m), pred($s{:}en_d,s_1$), pred($s_1,s{:}ex_d$), pred($c{:}en_d$, s_2), pred($s_2,c{:}ex_d$), pred($t{:}en_d,s_3$), pred($s_3,t{:}ex_d$)}

N={in($s_4,c_m,main$), in($s_5,c_m,main$), in($s_{10},c_m,main$), entry($shape,draw,s{:}en_d$), exit($shape$, $draw$, $s{:}ex_d$), entry($circle$, $draw$, $c{:}en_d$), exit($circle$, $draw$, $c{:}ex_d$), entry($triangle$, $draw$, $t{:}en_d$), exit($triangle$, $draw$, $t{:}ex_d$), if(s_6), branch(s_7, s_6), branch(s_8,s_6), pair(s_9,s_9'), pair(s_{11},s_{11}')}

A={new(s_4,o_4,c), new(s_5,o_5,t), new($s_{10},o_{10},c1$), asn(s_7,sh,c), asn(s_8,sh,t), class($o_4,circle$), class($o_5,triangle$), class($o_{10},circle$), callby($sh,draw,s_9$), callby($c1$, $draw,s_{11}$), call($main,shapes,null,null$)}

ASP program Π containing P, I, A and R1~21 have two answer sets, the first one contains P, I, A and the following predicates:

pred(s_6,s_7), ref($c,o_4,null,s_4$),
ref($c,o_4,null,s_5$), ref($t,o_5,null,s_5$),
ref($c,o_4,null,s_6$), ref($t,o_5,null,s_6$),
ref($c,o_4,null,s_7$), ref($sh,o_4,null,s_7$), ref($t,o_5,null,s_7$),
ref($c,o_4,null, s_m$), ref($sh,o_4,null,s_m$), ref($t,o_5,null,s_m$),
ref($c,o_4,null,s_9$), ref($sh,o_4,null,s_9$), ref($t, o_5,null,s_9$),call($draw,circle,s_9,null$),
ref($c,o_4,s_9,c{:}en_d$), ref($sh,o_4,s_9, c{:}en_d$), ref($t,o_5,s_9, c{:}en_d$),
ref(c,o_4,s_9,s_2), ref(sh,o_4,s_9,s_2), ref(t,o_5,s_9,s_2),
ref($c,o_4,s_9,c{:}ex_d$), ref($sh,o_4,s_9, c{:}ex_d$), ref($t,o_5,s_9, c{:}ex_d$),
ref($c,o_4,null,s_9'$), ref($sh,o_4,null,s_9'$), ref($t,o_5,null,s_9'$),
ref($c,o_4,null,s_{10}$), ref($sh,o_4,null,s_{10}$), ref($t,o_5,null,s_{10}$), ref($c1,o_{10},null,s_{10}$)
ref($c,o_4,null,s_{11}$), ref($sh,o_4,null,s_{11}$), ref($t,o_5,null,s_{11}$), ref($c1,o_{10},null,s_{11}$),
call($draw,circle,s_{11},null$)
ref($c,o_4,s_{11},c{:}en_d$), ref($sh,o_4,s_{11},c{:}en_d$), ref($t,o_5,s_{11},c{:}en_d$), ref($c1, o_{10},s_{11},c{:}en_d$)
ref(c,o_4,s_{11},s_2), ref(sh,o_4,s_{11},s_2), ref(t,o_5,s_{11},s_2), ref($c1,o_{10},s_{11},s_2$)
ref($c,o_4,s_{11},c{:}ex_d$), ref($sh,o_4,s_{11},c{:}ex_d$), ref($t,o_5,s_{11},c{:}ex_d$), ref($c1,o_{10}, s_{11},c{:}ex_d$)
ref($c,o_4,null,s_{11}'$), ref($sh,o_4,null,s_{11}'$), ref($t,o_5,null,s_{11}'$), ref($c1,o_{10},null,s_{11}'$)
ref($c,o_4,null,ex_m$), ref($sh,o_4,null,ex_m$), ref($t,o_5,null,ex_m$), ref($c1,o_{10},null, ex_m$)

And the second answer set of Π contains P, I, A and the following predicates:

pred(s_6,s_8), ref($c,o_4,null,s_4$),
ref($c,o_4,null,s_5$), ref($t,o_5,null,s_5$),
ref($c,o_4,null,s_6$), ref($t,o_5,null,s_6$),
ref($c,o_4,null,s_8$), ref($t,o_5,null,s_8$), ref($sh,o_5,null,s_8$),
ref($c,o_4,null,s_m$), ref($t,o_5,null,s_m$), ref($sh,o_5,null,s_m$),
ref($c,o_4,null,s_9$), ref($t,o_5,null,s_9$), ref($sh,o_5,null,s_9$), call($draw,triangle,s_9,null$)
ref($c,o_4,s_9,t{:}en_d$), ref($t,o_5,s_9, t{:}en_d$), ref($sh,o_5,s_9,t{:}en_d$),
ref(c,o_4,s_9,s_3), ref(t,o_5,s_9,s_3), ref(sh,o_5,s_9,s_3),
ref($c,o_4,s_9,t{:}ex_d$), ref($t,o_5,s_9,t{:}ex_d$), ref($sh,o_5,s_9,t{:}ex_d$),
ref($c,o_4,null,s_9'$), ref($t,o_5,null,s_9'$), ref($sh,o_5,null,s_9'$),
ref($c,o_4,null,s_{10}$), ref($t,o_5,null,s_{10}$), ref($sh,o_5,null,s_{10}$), ref($c1,o_{10},null,s_{10}$)
ref($c,o_4,null,s_{11}$), ref($t,o_5,null,s_{11}$), ref($sh,o_5,null,s_{11}$), ref($c1,o_{10},null,s_{11}$),
call($draw,circle,s_{11},null$)
ref($c,o_4,s_{11},c{:}en_d$), ref($t,o_5,s_{11},c{:}en_d$), ref($sh,o_5,s_{11},c{:}en_d$), ref($c1,o_{10}, s_{11},c{:}en_d$)
ref(c,o_4,s_{11},s_2), ref(t,o_5,s_{11},s_2), ref(sh,o_5,s_{11},s_2),ref($c1,o_{11},s_{11},s_2$)
ref($c,o_4,s_{11},c{:}ex_d$), ref($t,o_5,s_{11},c{:}ex_d$), ref($sh,o_5,s_{11},c{:}ex_d$), ref($c1,o_{10},s_{11}, c{:}ex_d$)
ref($c,o_4,null,s_{11}'$), ref($t,o_5,null,s_{11}'$), ref($sh,o_5,null,s_{11}'$),ref($c1,o_{10},null,s_{11}'$)
ref($c,o_4,null,ex_m$), ref($t,o_5,null,ex_m$), ref($sh,o_5,null,ex_m$), ref($c1,o_{10},null, ex_m$)

The result verifies that rules presented above are valid for reasoning about points-to information with accurate identification of propagation path. Predicates in the two answer sets list points-to relationship at every program point more accurately, from them alias information of each realizable path is easily implied.

4. Conclusion

In this paper, we presented a set of rules of Answer Set Program evolved from our previous work to describe the propagation process of points-to information in Java program. Path identification problem is well dealt with when using these rules to reasoning about points-to information of references at every program point. This makes the points-to analysis result more precise, which is desirable in context of diagnosing alias-caused program faults. An example is given to illustrate the effect of our method, computed result of this example shows that problems including forward path, backward path and type inference are solved in a natural and concise way. Next, we will discuss the theoretic proof and analysis for validity and computational efficiency of these rules.

References

1. W. Landi, B.G. Gyder. A Safe Approximation Algorithm for interprocedural Pointer Aliasing. Proc. of the ACM SIGPLAN'92 Conf. on PLDI,235-248(1993).
2. J. Choi, M. Burke, and P. Carini. Efficient Flow-Sensitive Interprocedural Computation of Pointer-Induced Aliases and Side Effects. Twenties Annual ACM SIGACT SIGPLAN Symp. on POPL,232-245(1993).
3. M. Emami, R. Ghiya, L.J. Hendren. Context-Sensitive Interprocedural Points-to Analysis in the Presence of Function Pointers. Proc. of the ACM SIGPLAN'94 Conf. on PLDI,242-256(1994).
4. J. Woo, J-L Gaudiot, A.L. Wendelborn. Alias Analysis in Java with Reference-Set Representation for Hign-Performance Computing. J. of Para. Prog,32:39-76 (2004).
5. B. Meyer. Towards a theory and calculus of aliasing. J. of Object Technology,9(2):37-74(2010).
6. J. Whaley, D. Avots, M. Carbin, M. Lam. Using Datalog with Binary Decision Diagrams for Program Analysis. APLAS 2005. LNCS 3780. Springer, Heidelberg.97-118(2005).
7. W. Song. Using horn clauses and binary decision diagram for program analysis. Ph.D. dissertation. State university of New York, (2010).
8. B. Yang, M. Zhang, Y. Zhang. Applying Answer Set Programming to Points-to Analysis of Object-Oriented Language. LNCS 6838,676-685(2011).
9. V. Lifschitz. What is Answer Set Programming. Proc. of the 23th AAAI Conf. on AI,1594-1597(2008).

A HEURISTIC BASED ON MULTI OBJECTIVE LINEAR PROGRAMMING UNDER FUZZINESS FOR THE VEHICLE ROUTING PROBLEM

GULCIN DINC YALCIN[†], NIHAL ERGINEL
Anadolu University, Department of Industrial Engineering, Eskisehir, Turkey

A heuristic is proposed for the Vehicle Routing Problem in this study. The heuristic has two phases that are named as clustering phase and routing phase, and also it has a local search. Firstly customers are assigned to vehicles with proposed multi objective linear programming and fuzzy approaches are used to solve this model in clustering phase. Secondly clusters are routed with TSP classical integer mathematical model in routing phase. Finally local search is used to improve the solution. Some problems from literature are solved and results are given.

1. Introduction

The Vehicle Routing Problem (VRP), which is introduced by Danzig and Ramser [7] in 1959, is to set routes that start from depot, delivery goods to customers and finish to depot with the constraints of vehicle capacity and customers demand with minimizing total travel distance or total delivery time. Several heuristic approaches have been proposed for VRP in literature. They can be divided into constructive heuristics, two phase heuristics, and improvement heuristics [11]. Constructive heuristics build a solution step by step and do not improve the solution such as saving algorithm [6]. Two phase heuristic is divided into cluster-first route-second methods such as Fisher and Jaikumar Algorithm [9] and Sweep algorithm [10] and route-first cluster-second methods such as Dijkstra [8]'s algorithm. Improvement heuristics are based on Lin's λ-opt approach [12]. Metaheuristic approaches have been developed for VRP. Alfa et. al [2] and Van [17] proposed simulated annealing algorithms; Osman [16] and Barbarosoglu et. al. [4] developed tabu search algorithms; Osman [16] and Lin et. al. [13] proposed algorithms that are a hybrid of simulated annealing and tabu search; Mazzeo et. al. [17], Bin et. al. [5] and Zhang et. al. [18] developed ant based algorithms; Baker et. al. [3] used genetic algorithm; Ai et. al. [1]

† gdinc@anadolu.edu.tr

proposed particle swarm optimization; Marinakis et. al. [14] developed a hybrid of genetic and particle swarm optimization algorithm to solve VRP.

Multi Objective Linear Programming (MOLP) has K objectives under the problem constraints. Fuzzy solution approaches have been proposed to solve MOLP with using membership functions that defines the closeness of objectives to optimal value and linear membership functions are used in this study. The first fuzzy approach for MOLP is proposed by Zimmermann [19] and named as max-min (MO) operator that aims maximizing the minimum membership degree. Then Li, Zhang and Li [20] added a second phase to MO and named as two phase approach (TPA) that aims improving the degrees of memberships from MO. Tiwari, Dharmar and Rao [21] proposed weighted additive model (WAM) whose purpose is to maximize weighted sum of membership degrees. Lin [22] proposed a weighted max-min model (WMM) that finds degree of membership functions as close as the ratio of weights. The weight of objectives should be defined for these approaches except MO. In this study, a method that is proposed by Dinc Yalcin and Erginel [23] is used to find weights of objectives. The method is based on two-person zero-sum game with mixed strategies under fuzziness.

This study presents a heuristic based on MOLP under fuzziness for VRP. The paper is organized as follows: the proposed heuristic is defined in Section 2 while the results of benchmark problems are discussed in Section 3. Some conclusion remarks are made in Section 4.

2. A Heuristic Based on Fuzzy Multi Objective Linear Programming

The customers are assigned to vehicles in the clustering phase with proposed multi objective linear programming that consists of two objectives. First objective is to minimize total distance in clusters and the second objective function is to maximize saving value in clusters which takes place in saving algorithm. The model chose the seed customers that are equal to number of vehicles and assign other customers to seed customers. The clusters can be thought as TSP after clustering phase. TSP classical integer programming is used to solve sub problems in routing phase. Local search is used to improve the solution that is generated after the two phases. Insertion and interchange operations are used for this purpose. A customer from its original vehicle is taken and inserted in another vehicle if the vehicle capacity is enough for demand of the customer in insertion operation. New route is occurred if the total distance is reduced after this operation. Two customers that are from different vehicles are exchanged and positioned the best place under the constraints of vehicle capacity. New route is obtained if there is a decrease in total distance after the interchange operation. The proposed integer programming model for the cluster phase is as:

Sets : i, j customer nodes (i=j=1,…, N)

Constants :

N	number of customers	a_i	demand of customer i
d_{ij}	distance between node i and j	C	capacity of vehicles
s_{ij}	Saving value between node i and j	K	number of vehicles

Variables: $x_{ij} = \begin{cases} 1 \text{ if node } i \text{ is assigned to node } j \\ 0 \text{ otherwise} \end{cases}$

Formulation : $minimize\ z_1 = \sum_i \sum_j d_{ij} x_{ij}$ (1)

$maximize\ z_2 = \sum_i \sum_j s_{ij} x_{ij}$ (2)

Subject to: $\sum_{j=1}^{N} x_{ij} = 1 \quad \forall i = 1, \dots, N$ (3)

$\sum_{i=1}^{N} a_i x_{ij} \leq C x_{jj} \quad \forall j = 1, \dots, N$ (4)

$\sum_{i=1}^{N} \sum_{j=1}^{N} x_{jj} = K$ (5)

$x_{ij} \in \{0,1\}$ (6)

The first objective of the model (1) is to minimize the total distance that is the sum of distance between a customer and the relevant seed customer in each cluster. The second objective of the model (2) is to maximize the total saving value in clusters. Constraint (3) represents that each customer must be assign only one cluster. Constraint (4) says that total demand of customers that is assigned to a cluster cannot be exceeded capacity of vehicle and if a customer is seed customer than other customers can be assigned. Constraint (5) indicates that the number of clusters is equal to the number of vehicle.

3. Computational Experiments

Benchmark problems that are proposed by Christofides et. al. [24] are used to test proposed heuristic. Some problems that do not have maximum tour length limitation are solved. The integer programming models are solved by Gams with Cplex 9.0 and Local search are codded on Excel Visual Basic application that is also used to arrange the data for Gams on 2.20 GHz, 1GB RAM computer. Solutions of benchmark problems are given in Table 1. CPU1 is for clustering phase, CPU2 is for routing phase, and CPU3 is for local search and total CPU is sum all of them. CPU1 includes the time of finding ideal and anti-ideal values of objectives in fuzzy approaches. Also it includes the time of finding weights of objective, except min operator.

The proposed heuristic found better results of case C12 for all solutions due to best known solution. Average percentage gap is 4.12 for the first objective, 1.93 for the second objective, 5.32 for MO, 5.37 for TPA, 3.06 for WSM and

5.76 for WMM. The minimum gap is usually obtained by the second objective. The gaps of fuzzy approaches are usually close each other.

Table 1: Results of the benchmark problems

Name	The name of solution phase	Best Known	Proposed heuristic	% Gap	CPU 1	CPU 2	CPU3	Total CPU
	z_1		537.34	2.43	1.671	0.873	0.033	2.58
	z_2		538.25	2.60	0.5	1.107	0.036	1.64
C1	Min op.	524.61	537.72	2.5	9.186	0.794	0.034	10.01
	Two-phase		537.34	2.43	7.139	0.873	0.031	8.043
	W. sum		537.34	2.43	7.139	0.873	0.031	8.043
	Max-min		537.34	2.43	7.139	0.873	0.031	8.043
	z_1		890.85	6.66	92.22	0.947	0.045	93.21
	z_2		853.11	2.14	395.22	0.745	0.04	396
C2	Min op.	835.26	869.26	4.07	872.03	0.841	0.04	872.91
	Two-phase		869.26	4.07	921.48	0.841	0.04	922.36
	W. sum		869.26	4.07	814.53	0.841	0.04	815.41
	Max-min		890.85	6.66	843.20	0.728	0.05	843.98
	z_1		879.42	6.45	38.66	2.386	2.44	43.48
	z_2		838.54	1.5	52.34	12.575	1.59	6.51
C3	Min op.	826.14	862	4.34	199.25	9.481	1.89	210.62
	Two-phase		854.82	3.47	166.58	3.716	2.25	172.54
	W. sum		854.82	3.47	169.84	3.716	2.25	172.81
	Max-min		859.11	3.99	181.9	1.324	2.95	186.18
	z_1		1101.38	7.09	1000	4.26	5.13	1009.39
	z_2		1073.43	4.38	1000	92.399	4.66	1097.06
C4	Min op.	1028.42	1073.43	4.38	1000	92.399	4.66	1097.06
	Two-phase		1078.66	4.89	4287.8	47.619	4.49	4339.494
	W. sum		1078.66	4.89	4198.06	47.619	4.49	4250.20
	Max-min		1078.66	4.89	4009.06	47.619	4.49	4061.20
	z_1		1354.63	4.89	1000	5.2	10.09	1015.29
	z_2		1335.08	3.38	1000	71.85	12.08	1083.93
C5	Min op.	1291.45	1567.66	21.39	4017.45	3773.42	11.42	7802.3
	Two-phase		1567.66	21.39	4023.77	3773.42	11.42	7808.64
	W. sum		1359.62	5.28	5000	3.27	12.41	5015.71
	Max-min		1567.66	21.39	4017.16	3773.42	11.42	7802.03
	z_1		1061.24	1.84	1000	2071.62	0.1	3071.72
	z_2		1046.02	0.38	72.25	5002.218	0.14	5074.61
C11	Min op.	1042.11	1053.83	1.12	1387.19	2002.81	0.12	3390.12
	Two-phase		1061.24	1.84	2076.45	2085.48	0.06	4161.99
	W. sum		1061.24	1.84	2076.45	2076.94	0.1	4153.49
	Max-min		1061.24	1.84	2076.45	2071.4	0.07	4147.93
	z_1		**815.24**	**-0.53**	0.58	1.106	2.1	3.78
	z_2		**812.23**	**-0.89**	0.70	48.854	1.91	51.47
C12	Min op.	819.56	**815.24**	**-0.53**	31.16	0.87	2.06	34.09
	Two-phase		**815.24**	**-0.53**	29.56	0.869	2.06	32.08
	W. sum		**815.24**	**-0.53**	28.89	0.869	2.06	31.82
	Max-min		**812.23**	**-0.89**	29.51	48.854	1.91	80.28

CPU for clustering phase is increased while the number of customers is increased. Maximum CPU is 4287.8 seconds with Case C4 and TPA for clustering phase; is 5002.218 seconds with Case C11 and the second objective for routing phase; is 12.21 seconds with Case C5 and WSM for local search. Minimum CPU is 0.5 seconds with Case C1 and the second objective for clustering phase; is 0.728 seconds with Case C2 and WMM for routing phase; is 0.031 seconds with Case 1 and TPA, WSM, WMM for local search. Maximum total CPU is 7088.64 seconds with Case C5 and TPA while minimum total CPU is 1.64 with Case C1 and the second objective.

4. Conclusion

A heuristic that is based on multi objective linear programming is proposed for VRP in this study. The heuristic is similar to cluster-first route-second methods. In clustering phase, customers are assigned to vehicles with proposed MOLP. The model determines the seed customers and assigns other customers to seed customers under the constraints of vehicle capacity. Also one customer can be assigned to only one seed customer. The objectives of the model are to minimize total distance and maximize total saving value in clusters. Fuzzy approaches are used to solve the model. In route phase, each cluster is routed as TSP with its classical integer model. The solution is improved by local search. The heuristic generated better results for one benchmark problem. Average gap of the solutions for other problems is 4.26. The heuristic found solutions in acceptable time. CPU of three problems is below 200 seconds and CPU of one problem is 1000 seconds and CPU of three problems is above 4000 seconds.

References

1. T. J. Ai and V. Kachitvichyanukul, "Particle swarm optimization and two solution representations for solving the capacitated vehicle routing problem" Computers & Industrial Engineering, 56, 380-387 (2009).
2. A. S. Alfa, S. S. Heragu and M. Chen, "A 3-opt based simulated annealing algorithm for vehicle routing problems" Computer & Industrial Engineering, 21, 635-639 (1991).
3. B. M. Baker and M. A. Ayechew, "A genetic algorithm for the vehicle routing problem" Computers & Operations Research, 30, 787-800 (2003).
4. G. Barbarosoglu and D. Ozgur, "A tabu search algorithm for the vehicle routing problem" Computers and Operations Research, 26, 255-270 (1999).
5. Y. Bin, Y. Zong-Zhen and Y. Baozhen, "An improved ant colony optimization for vehicle routing problem" European Journal of Operational Research, 196, 171-176 (2009).
6. G. Clarke and J. V. Wright, "Scheduling of vehicles from a central depot to a number of delivery points" Operations Research, 12, 568-581 (1964).
7. G. B. Dantzig and J. H. Ramser, "The Truck dispatching problem" Management Science, 6(1), 80-91 (1959).

8. E. W. Dijkstra, "A note on two problems in connection with graphs" Numerische Mathematik, 1, 269-271 (1959).
9. Fisher, M. L. and Jaikumar, R., 1981, "A generalized assignment heuristic for the vehicle routing problem" Networks, 11, 109-124.
10. B. E. Gillett and L. R. Miller, "A heuristic algorithm for the vehicle dispatch problem" Operations Research, 22, 340-349 (1974).
11. G. Laporte and F. Semet, "Chapter 5: Classical Heuristics for the Capacitated VRP" appears in The Vehicle Routing Problem, Toth, P. and Vigo, D. (eds.), SIAM, Bologna, 109-128 (2001).
12. S. Lin, "Computer solutions of the traveling salesman problem" Bell System Technical Journal, 21, 498-516 (1965).
13. S. Lin, Z. Lee, K. Ying and C. Lee, "Applying hybrid meta-heuristics for capacitated vehicle routing problem" Expert Systems with Applications, 36, 1505-1512 (2009).
14. Y. Marinakis and M. Marinaki, "A hybrid genetic-particle swarm optimization algorithm for the vehicle routing problem" Expert System with Applications, 37, 1446-1455 (2010).
15. S. Mazzeo and I. Loiseau, "An ant colony algorithm for the capacitated vehicle routing" Electronic Notes in Discrete Mathematics, 18, 181-186 (2004).
16. I. H. Osman, "Metastrategy simulated annealing and tabu search algorithms for the vehicle routing problems" Annals of Operations Research, 41, 421-451 (1993).
17. B. Van, "A. Improvement heuristics for the vehicle routing problem based on simulated annealing" European Journal of Operational Research, 86, 480-490 (1995).
18. X. Zhang and L. Tang, "A new hybrid ant colony optimization algorithm for the vehicle routing problem" Pattern Recognition Letters, 30, 848-855 (2009).
19. H. J. Zimmermann, "Fuzzy programming and linear programming with several objective functions", Fuzzy Sets and System, 1, 44-55 (1978).
20. X.-Q. Li, B. Zhang and H. Li,"Computing efficient solutions to fuzzy multiple objective linear programming problems", Fuzzy Sets and System, 157, 1328-1332 (2006).
21. R.N. Tiwari, S. Dharmar and J. R. Rao, "Fuzzy goal programming- An additive mode", Fuzzy Sets and System, 24, 27-54 (1987).
22. C.-C. Lin, "A weighted max-min model for fuzzy goal programming", Fuzzy Sets and System, 142, 407-420 (2004).
23. G. Dinc Yalcin and N Erginel, "Determining weights in multi-objective linear programming under fuzziness", Word Congress in Engineering, London, England (2011).
24. N. Christofides, A. Mingozzi and P. Toth, "The Vehicle Routing Problem" appears in Combinatorial Optimization, N. Christofides, A. Mingozzi, P. Toth and C. Sandi (Eds.) Wiley, Chichester (1979).

RANK REVERSALS AND TESTING OF PAIRWISE COMPARISONS BASED NON-NUMERICAL RANKINGS

RYSZARD JANICKI*

Department of Computing and Software, McMaster University,Hamilton, Ontario L8S 4K1, Canada

YUN ZHAI

The News Group,Burlington, Ontario L7N 3M6, Canada

Rank reversal is one of the major problems in all ranking techniques. The problems of testing and test results for rank reversals, are discussed for pairwise comparison based non-numerical rankings. The results indicate that some algorithms that use non-numerical approach produce almost no rank reversals.

1. Introduction

The *pairwise comparisons* method is based on the observation that it is much easier to rank the importance of *two* objects than it is to rank the importance of *several* objects [2]. This very old idea goes back to Ramon Llull (end of XIII century) and Marquis de Condorcet (1785), but it was Thurstone (1927) and Saaty (1977) who provided its mathematical foundations [2,6,10–12]. The *rank reversal is a flaw of ranking techniques* It occurs when adding some elements to a given sample, and then applying the same ranking technique, the ranking relation changes for some objects† [3]. The ranking is *numerical* if numbers are used to create it.

At present, pairwise comparison *numerical* ranking is often equalized with the Saaty's Analytic Hierarchy Process, or *AHP* first proposed in 1977 [12,13]. Despite many practical applications, many researchers still consider it as a flawed procedure that may produce arbitrary rankings [3,6,11]. The rank reversal occurs quite often [3]. There are also many other problems [3,11].

A pairwise comparison non-numerical ranking was proposed in 1996 by Janicki and Koczkodaj [8] and developed in 2007-2011 by Janicki and Zhai

* janicki@mcmaster.ca

† A ranking procedure results in $a < b$ if c is not involved and in $b < a$ if c is added.

[6,9,10]. It based on the concepts of a *partial order* and an *approximation of an arbitrary binary relation by a partial order* [7].

In this paper we will discuss basis of pairwise comparisons based non-numerical ranking techniques [6,9] as well asproblems of testing and rank reversals for these techniques.

2. Relations, Partial Orders and Approximations

Let X be a finite set and let $\|X\|$ denotes its number of elements. For every relation $R \subseteq X \times X$, let $R^+ = \bigcup_{i=1}^{\infty} R^i$, denote the *transitive closure* of R, $id = \{(x,x) \hat{U} x \in X\}$ denote the identity relation, $R^{\circ} = R \cup id$ denote the *reflexive closure* of R, $R^{sym} = R \cup R^{-1}$ denote the *symmetric closure* and for each $a \in X$ let $Ra = \{x \hat{U} xRa\}$ and $aR = \{x \hat{U} aRx\}$. A relation R is *acyclic* if and only if $\neg(xR^+x)$ for all $x \in X$.

A relation $< \in X \times X$ is a (*sharp partial order* if it is irreflexive and transitive, i.e. if $\neg(a<a)$ and $a<b<c \Longrightarrow a<c$, for all $a,b,c \in X$ [4,5].

We write $a \sim_< b$ if $\neg(a<b) \wedge \neg(b<a)$ that is if a and b are either *distinctly incomparable* (w.r.t. $<$ or *identical* elements, and define $g_<(x)$, a *global score function* $g_<(x) = \|\{z \hat{U} z<x\}\| - \|\{z \hat{U} x<z\}\|$.

A partial order [4,5] is *total* or *linear* if $\forall a,b \in X.\, a=b \vee a<b \vee b<a$, and *weak* or *stratified* if $a \sim_< b \sim_< c \Rightarrow a \sim_< c$.

It is assumed that any ranking is either weak or total order [5]. Numerical rankings are usually totally ordered.

An order $<^w$ defined as $a <^w b$– $g_<(a) < g_<(b)$, is a *proper weak extension* of $<$ [4], and $<^w$ is considered a *weak order approximation* of $<$ [6,9].

For every relation R, define the relations R^{cyc}, % R_{id}^{cyc} $R^{\bullet}$ and $R^{\subset}$ as

- $aR^{cyc}b$– $aR^+b \wedge bR^+a$ and $aR^{\bullet}b$– $aRb \wedge \neg(aR^{cyc}b)$,
- $aR^{\subset}b$– $bR^{\circ} \subset aR^{\circ} \wedge R^{\circ}a \subset R^{\circ}b$.

We will call $R^{\bullet}$ the *acyclic refinement of* R and $R^{\subset}$ an *inclusion property kernel of* R [6,7]. Note that if R is a partial order then $R = R^{\subset} = R^+ = R^{\bullet}$ [7].

Theorem 2.1. ([7]) $(R^{\bullet})^{\subset}$ $(R^{\bullet})^+$ $(R^+)^{\bullet}$ *are partial orders and* partial order approximations‡ *of* R, $(R^{\bullet})^{\subset} \subseteq (R^{\bullet})^+ \subseteq (R^+)^{\bullet}$ *and* $(R^{\bullet})^{\subset} \subseteq R$. □

‡ Due to pages limit no detail definition will be discussed.

3. Pairwise Comparison Based Non-Numerical Ranking

Non-numerical ranking procedure, consists of the following steps: define some relations that represent empirical ranking data, construct appropriate partial order approximations, refine the outcome according to some consistency rules, and eventually build weak order representations [6,9].

Let X be a set of objects and $\approx, \sqsubset, \subset, <, \prec$ be *disjoint* relations on X such that $\approx$ is *symmetric* and *reflexive* $\sqsubset, \subset, <, \prec$ are *asymmetric* and *irreflexive* and $\forall a,b \in X.(x \approx y) \vee (x \sqsubset^{sym} y) \vee (x \sqsubset^{sym} y) \vee (x <^{sym} y) \vee (x \prec^{sym} y)$. These relations are interpreted as follows $a \approx b$: a and b are *indifferent* $a \sqsubset b$: *slightly in favor* of b, $a \subset b$: *in favor* of b, $a < b$: b is *strongly better* $a \prec b$: b is *extremely better*. The list $\sqsubset$, $\subset$, $<$ $\prec$ may be shorter or longer, but not empty and not much longer. [§].

If no other axioms are added, the tuple $R = (X, \approx, \sqsubset, \subset, <, \prec)$ is called *pairwise comparison (empirical) ranking data* [6].

We define the relations $\hat{\sqsubset}$ $\hat{\subset}$ $\hat{<}$ and $\hat{\prec}$ as follows:

$$\hat{\prec} = \prec, \quad \hat{<} = \prec \cup <, \quad \hat{\subset} = \prec \cup < \cup \subset, \quad \hat{\sqsubset} = \prec \cup < \cup \subset \cup \sqsubset .$$

The relations $\hat{\sqsubset}$ $\hat{\subset}$ $\hat{<}$ and $\hat{\prec}$ are interpreted as *combined preferences* i.e. $a \hat{\sqsubset} b$: *at least slightly in favor* of b, $a \hat{\subset} b$: *at least in favor* of b, etc.

Definition 3.1. ([6])

1. A tuple $R=(X, \approx, \sqsubset, \subset, <, \prec)$ is a *pairwise comparison ranking system* if it satisfies the axioms for a pairwise comparison ranking data and the relations $\hat{\sqsubset}$ $\hat{\subset}$ $\hat{<}$ $\hat{\prec}$ are partial orders.
2. R is *derived from* $R'=(X, \approx', \sqsubset', \subset', <, \prec')$ if each $R \in \{\sqsubset, \subset, <, \prec\}$ is a *partial order approximation* of $R' \in \{\sqsubset', \subset', <', \prec'\}$. □

If one is interested in one preference relation only, then the relation $\hat{\sqsubset}$ will do it. The relation $\hat{\sqsubset}$ is called a *principal order* of $\mathcal{R}$.

A ranking $R=(X, \approx, \sqsubset, \subset, <, \prec)$ is *weakly ordered* if $\hat{\sqsubset}$ is a weak order.

Definition 3.2 ([9]). The tuple $R = (X, \approx, \sqsubset, \subset, <, \prec)$ is a *pairwise comparison* **consistent** *ranking system* if it satisfies the axioms for a pairwise comparison ranking data and the following rules (called *consistency rules*)are also satisfied:

1. $(a \approx b \wedge b \approx c) \Rightarrow (a \approx c \vee a \sqsubset c \vee c \sqsubset a)$

2.1. $(a \approx b \wedge b \sqsubset c) \vee (a \sqsubset b \wedge b \approx c) \Rightarrow (a \approx c \vee a \sqsubset c \vee a \subset c)$

2.2. $(a \approx b \wedge b \subset c) \vee (a \subset b \wedge b \approx c) \Rightarrow (a \sqsubset c \vee a \subset c \vee a < c)$

2.3. $(a \approx b \wedge b < c) \vee (a < b \wedge b \approx c) \Rightarrow (a \subset c \vee a < c \vee a \prec c)$

3. $(a \approx b \wedge b \prec c) \vee (a \prec b \wedge b \approx c) \Rightarrow (a < c \vee a \prec c)$

4.1. $(a \sqsubset b \wedge b \sqsubset c) \Rightarrow (a \sqsubset c \vee a \subset c \vee a < c)$

4.2. $(a \subset b \wedge b \sqsubset c) \vee (a \sqsubset b \wedge b \subset c) \Rightarrow (a \subset c \vee a < c \vee a \prec c)$

[§] Due to limitations of the human mind, 'magical numbers: 7 and 4' [1].

4.3. $(a<b \wedge b \sqsubset c) \vee (a \sqsubset b \wedge b<c) \Rightarrow (a<c \vee a \prec c)$
5.1. $(a \sqsubset b \wedge b \prec c) \vee (a \prec b \wedge b \sqsubset c) \Rightarrow (a \prec c)$
5.2. $(a \subset b \wedge b \prec c) \vee (a \prec b \wedge b \subset c) \Rightarrow (a \prec c)$
5.3. $(a<b \wedge b \prec c) \vee (a \prec b \wedge b<c) \Rightarrow (a \prec c)$
5.4. $(a \prec b \wedge b \prec c) \vee (a \prec b \wedge b \prec c) \Rightarrow (a \prec c)$
6.1. $(a \subset b \wedge b<c) \vee (a<b \wedge b \subset c) \Rightarrow (a \prec c)$
6.2. $(a<b \wedge b<c) \Rightarrow (a \prec c)$
7. $(a \subset b \wedge b \subset c) \Rightarrow (a<c \vee a \prec c)$
8.1. $(a \approx b \wedge b \sqsupset c) \vee (a \sqsupset b \wedge b \approx c) \Rightarrow (a \approx c \vee a \sqsupset c \vee a \supset c)$
8.2. $(a \approx b \wedge b \supset c) \vee (a \supset b \wedge b \approx c) \Rightarrow (a \sqsupset c \vee a \supset c \vee a>c)$
8.3. $(a \approx b \wedge b>c) \vee (a>b \wedge b \approx c) \Rightarrow (a \supset c \vee a>c \vee a \succ c)$
8.4. $(a \approx b \wedge b \succ c) \vee (a \succ b \wedge b \approx c) \Rightarrow (a>c \vee a \succ c)$
9.1. $(a \sqsupset b \wedge b \sqsupset c) \Rightarrow (a \sqsupset c \vee a \supset c \vee a>c)$
9.2. $(a \sqsupset b \wedge b \supset c) \vee (a \supset b \wedge b \sqsupset c) \Rightarrow (a \supset c \vee a>c \vee a \succ c)$
9.3. $(a \sqsupset b \wedge b>c) \vee (a>b \wedge b \sqsupset c) \Rightarrow (a>c \vee a \succ c)$
9.4. $(a \sqsupset b \wedge b \succ c) \vee (a \succ b \wedge b \sqsupset c) \Rightarrow (a \succ c)$
10.1. $(a \supset b \wedge b \supset c) \Rightarrow (a>c \vee a \succ c)$
10.2. $(a \supset b \wedge b>c) \Rightarrow (a \succ c)$
10.3. $(a \supset b \wedge b \succ c) \Rightarrow (a \succ c)$
11.1. $(a>b \wedge b>c) \Rightarrow (a \succ c)$
11.2. $(a>b \wedge b \succ c) \vee (a \succ b \wedge b>c) \Rightarrow (a \succ c)$
12. $(a \succ b \wedge b \succ c) \Rightarrow (a \succ c)$
13.1. $(a \sqsupset b \wedge b \sqsubset c) \vee (a \sqsubset b \wedge b \sqsupset c) \Rightarrow (a \approx c \vee a \sqsubset c \vee a \sqsupset c)$
13.2. $(a \sqsupset b \wedge b \subset c) \vee (a \subset b \wedge b \sqsupset c) \Rightarrow (a \approx c \vee a \sqsubset c \vee a \subset c)$
13.3. $(a \sqsupset b \wedge b<c) \vee (a<b \wedge b \sqsupset c) \Rightarrow (a \sqsubset c \vee a \subset c \vee a<c)$
13.4. $(a \sqsupset b \wedge b \prec c) \vee (a \prec b \wedge b \sqsupset c) \Rightarrow (a \subset c \vee a<c \vee a \prec c)$
14.1. $(a \supset b \wedge b \sqsubset c) \vee (a \sqsubset b \wedge b \supset c) \Rightarrow (a \approx c \vee a \sqsupset c \vee a \supset c)$
14.2. $(a \supset b \wedge b \subset c) \vee (a \subset b \wedge b \supset c) \Rightarrow (a \approx c \vee a \sqsubset c \vee a \sqsupset c)$
14.3. $(a \supset b \wedge b<c) \vee (a<b \wedge b \supset c) \Rightarrow (a \approx c \vee a \sqsubset c \vee a \subset c)$
14.4. $(a \supset b \wedge b \prec c) \vee (a \prec b \wedge b \supset c) \Rightarrow (a \sqsubset c \vee a \subset c \vee a<c \vee a \prec c)$
15.1. $(a>b \wedge b \sqsubset c) \vee (a \sqsubset b \wedge b>c) \Rightarrow (a \sqsupset c \vee a \supset c \vee a>c)$
15.2. $(a>b \wedge b \subset c) \vee (a \subset b \wedge b>c) \Rightarrow (a \approx c \vee a \sqsupset c \vee a \supset c)$
15.3. $(a>b \wedge b<c) \vee (a<b \wedge b>c) \Rightarrow (a \approx c \vee a \sqsubset c \vee a \sqsupset c)$
15.4. $(a>b \wedge b \prec c) \vee (a \prec b \wedge b>c) \Rightarrow (a \approx c \vee a \sqsubset c \vee a \subset c \vee a<c \vee a \prec c)$
16.1. $(a \succ b \wedge b \sqsubset c) \vee (a \sqsubset b \wedge b \succ c) \Rightarrow (a \supset c \vee a>c \vee a \succ c)$
16.2. $(a \succ b \wedge b \subset c) \vee (a \subset b \wedge b \succ c) \Rightarrow (a \sqsupset c \vee a \supset c \vee a>c \vee a \succ c)$
16.3. $(a \succ b \wedge b<c) \vee (a<b \wedge b \succ c) \Rightarrow (a \approx c \vee a \sqsupset c \vee a \supset c \vee a>c \vee a \succ c)$
16.4. $(a \succ b \wedge b \prec c) \vee (a \prec b \wedge b \succ c) \Rightarrow$
$(a \approx c \vee a \sqsupset c \vee a \sqsubset c \vee a \supset c \vee a \subset c \vee a>c \vee a<c \vee a \succ c \vee a \prec c)$ □

The rules above implement the idea that composition may not change precedences in a drastic way, only by 'one or two steps', or not at all.

Proposition 3.1 ([9]). *If* $R=(X,\approx,\sqsubset,\subset,<,\prec)$ *is a pairwise comparison* consistent *ranking system, then* $\hat{\sqsubset},\hat{\subset},\hat{<},\hat{\prec}$ *are partial orders.* □

Hence any pairwise comparison consistent ranking system (Def. 3.2) is also a pairwise comparison ranking system (Def. 3.1).

Since the starting pairwise comparison data $R'=(X,\approx',\sqsubset',\subset',<',\prec')$ are seldom even partial orders, we need some credible algorithms that could transform R' into an appropriate pairwise comparison ranking system or pairwise comparison consistent ranking system.

For any relation R, $\mathsf{A}_{\lhd}(R)$ will denote a partial order approximation, and in this paper we assume $\mathsf{A}_{\lhd}(R)$ to be either $(R^{\bullet})^{\subset}$ or $(R^{\bullet})^{+}$ or $(R^{+})^{\bullet}$

Algorithm 3.1 ([6]).

1. Calculate $\hat{\sqsubset}'$ $\hat{\subset}'$ $\hat{<}'$ and $\hat{\prec}'$
2. Calculate $\hat{\prec}=\mathsf{A}_{\lhd}(\hat{\prec}')$ $\hat{<}=\mathsf{A}_{\lhd}(\hat{<}'\cup\hat{\prec})$ $\hat{\subset}=\mathsf{A}_{\lhd}(\hat{\subset}'\cup\hat{<})$ $\hat{\sqsubset}=\mathsf{A}_{\lhd}(\hat{\sqsubset}'\cup\hat{\subset})$
3. Calculate $\approx=(X\times X)\setminus(\hat{\sqsubset}\cup\hat{\sqsubset}^{-1})$
4. Calculate $\prec=\hat{\prec}$ $<=\hat{<}\setminus\hat{\prec}$ $\subset=\hat{\subset}\setminus\hat{<}$ $\sqsubset=\hat{\sqsubset}\setminus\hat{\subset}$ □

The tuple $R=(X,\approx,\sqsubset,\subset,<,\prec)$ produced by Algorithm 3.1. *is a Pairwise Comparisons Ranking System, but not necessarily consistent* The below algorithm starts *either* with raw ranking data *or* with a ranking system and produces a pairwise comparison *consistent* ranking system.

Algorithm 3.2 ([9.14]). *Let $\mathcal{R}'$ be a given ranking data/system.*

1. *Find all pairs that violate consistency rules, if none, stop, otherwise pick the pairs which violate the biggest number of rules.*
2. *Revise the relation between the pairs violating a rule, by appropriately* lowering preferences *for example from $<$ to $\subset$, etc., and go to 1.* □

The tuple $\mathcal{R}$ produced by Algorithm 3.2 is a pairwise comparisons consistent ranking system. The algorithm may end with $\approx=X\times X$ ($\approx$is *indifference*, but always stops. Time complexity of both algorithms is $O(\|X\|^3)$.

If one wants a *weakly ordered* ranking or consistent ranking, it suffices to calculate $\hat{\sqsubset}^{w}$ using the global score function $g_{\hat{\sqsubset}}(x)$. (in $O(\|X\|^2)$ time.

Despite Proposition 3.1 we do need Algorithm 1. Using only Algorithm 2 might result in a huge $\approx$ (i.e. *indifference* so the outcome may be of no practical value . Usually we need to use the Algorithm 1 and Algorithm 2 in this order, especially when the initial data are 'messy' [9].

4. Rank Reversal and Testing

Let X be a finite set of objects, α a given ranking procedure and $\lhd_X^{\alpha}\subset X\times X$ be the ranking relation (weak order) produced by α applied to X.

Definition 4.1 (Rank Reversals).

1. We will say that α may produce a (standard) *rank reversal* on X, if there exists a nonempty set $Y\subset X$ and there are $x,y\in Z=X\setminus Y$, such that: $(x\lhd_X^{\alpha} y\wedge y\lhd_Z^{\alpha} x)$.
2. We will say that α may produce a *weak rank reversal* on X, if there is a nonempty set $Y\subset X$ and there are $x,y\in Z=X\setminus Y$, such that

$(x \lhd_X^\alpha y \wedge y \lhd_Z^\alpha x)$ or $(x \approx_{\lhd_X^\alpha} y \wedge x \lhd_Z^\alpha y)$ or $(x \lhd_X^\alpha y \wedge x \approx_{\lhd_Z^\alpha} y)\ \}$ □

Clearly each reversal is also a weak reversal. If α always produces a total order than weak rank reversal is a rank reversal. However if α may produce weak order, the classical definition of rank reversal may be too crude in same cases, so we introduced the concept of weak rank reversal.

Because of Arrow's Impossibility Theorem [5,8], reversal ranking may occur in every nontrivial pairwise comparisons based ranking technique, the only question is 'how often'. Standard Saaty's AHP [3] is very prone to rank reversal and it may occur quite often and quite easily. *This is often considered as the major and most serious flaw of AHP* [3]. Our tests show that this problem is far less serious for some non-numerical rankings.

Definition 4.2. We will say that a ranking $R = (X, \approx, \sqsubset, \subset, <, \prec)$ obtained from $R' = (X, \approx', \sqsubset', \subset', <', \prec')$ by applying a procedure α produces a rank (weak rank) reversal, if $\lhd_X^\alpha = \hat{\sqsubset}$ produces it. □

• Testing means that there are some data and results that are known to be correct, and then the technique is applied to the same data. Hence testing models such as the one presented above is problematic since *it is not obvious what should be tested against* What are the correct results for given data? The used the following testing method (suggested by W. Koczkodaj).

• *A blindfolded person compared the weights of stones. The person put one stone in his left hand and another in his right hand, and then decided which of the relations $\approx$, $\sqsubset$, $\subset$, $<$ or $\prec$ held.*

Our prehistoric ancestors probably used this technique to decide which stone was better to kill an enemy or an animal.

• *In this experiment the stones can be weighted using a precise scale, so we have the precise results to test against.*

The results of those experiments (60 stones, 50 experiments) and their partial analysis can be found in Zhai's Thesis [14] (see Appendix B in it).

5. Tests Results for Rank Reversals

Table in Figure 1 shows the summary of tests results for rank reversal and weak rank reversal. Two disjoint sets of experiments have been analyzed. Each set consisted of four disjoint subsets according to the two properties of the relation $\hat{\sqsubset}'$ acyclicity and transitivity. The set #1 consisted of 40 samples, 10 samples for each case (i.e. 10 for $\hat{\sqsubset}'$ both acyclic and transitive, etc.). The set #2 has identical structure but the number of samples in each subset was 20 The samples in the set #1 had between 20 and 35 elements, the samples in #2 were shorter, between 4 and 18 elements. We then combine the sets #1 and #2 into one set, the set #3

Algorithms	*Partial Order Approximation*	$\widehat{\sqsubseteq}'$ *of* $\mathcal{R}'$ *is acyclic*	$\widehat{\sqsubseteq}'$ *of* $\mathcal{R}'$ *is transitive*	**% of Rank Reversals of $\widehat{\sqsubseteq}^w$ found**					
				Number of experiments in sets					
				set #1 $4 \times (10)$		set #2 $4 \times (20)$		set #3 $4 \times (30)$	
				R	WR	R	WR	R	WR
All below		Yes	Yes	0%	0%	0%	0%	0%	0%
Alg. 3.2 only	Not applied	No	Yes	20%	60%	25%	80%	23%	73%
		Yes	No	30%	70%	15%	75%	20%	73%
		No	No	30%	100%	20%	100%	23%	100%
Alg. 3.1 only	$(R^\bullet)^\subset$	No	Yes	20%	40%	10%	55%	13%	50%
		Yes	No	20%	40%	10%	60%	13%	53%
		No	No	10%	100%	5%	80%	7%	87%
	$(R^\bullet)^+$	No	Yes	0%	0%	0%	0%	0%	0%
		Yes	No	0%	0%	0%	0%	0%	0%
		No	No	20%	90%	25%	55%	23%	67%
	$(R^+)^\bullet$	No	Yes	0%	0%	0%	0%	0%	0%
		Yes	No	0%	0%	0%	0%	0%	0%
		NO	**NO**	0%	0%	0%	0%	0%	0%
Alg. 3.1 followed by Alg. 3.2	$(R^\bullet)^\subset$	No	Yes	20%	30%	15%	50%	17%	43%
		Yes	No	20%	30%	10%	55%	13%	47%
		No	No	10%	80%	5%	70%	7%	73%
	$(R^\bullet)^+$	No	Yes	0%	0%	0%	0%	0%	0%
		Yes	No	0%	0%	0%	0%	0%	0%
		No	No	20%	70%	30%	45%	27%	53%
	$(R^+)^\bullet$	No	Yes	0%	0%	0%	0%	0%	0%
		Yes	No	0%	0%	0%	0%	0%	0%
		NO	**NO**	0%	0%	0%	0%	0%	0%

Figure 1. Summary of tests results for rank reversal. Most important parts are shaded. The darker the more important. The sets #1and #2are disjoint, the set #3 is the union of #1and #2. 'R' means *reversal* while 'WR' means *weak reversal*

• Note that while weak reversal occur more often than reversal, no *qualitative difference* between them was observed, namely ' $\text{WR} = 0\% \Leftrightarrow \text{R} = 0\%$.

• The results vastly depended on procedures used to transform %ranking data $\mathcal{R}'$ into $\mathcal{R}$ but there is *no much difference between the set* $\#1$ *and the set* $\#2$.

When $\hat{\sqsubseteq}'$ of $\mathcal{R}'$ is both acyclic and transitive, i.e. $\hat{\sqsubseteq}'$ is already a partial order (but not necessarily weak, and $\mathcal{R}'$ may still differ from $\mathcal{R}$, then no weak rank reversal was observed. As expected using Algorithm 3.2 alone produced the worst results, weak rank reversal occurred quite often. Due to the nature of this algorithm, when we have $x \hat{\sqsubseteq}' y$, z is added and $z \approx' y$, the algorithm may easily ends up with $x \approx y$. For (standard) rank reversal the results were much better, but still worse than when Alg. 3.1 was involved.

When $\hat{\sqsubseteq}'$ is either acyclic or transitive - which appears to happen not so seldom, then the partial order approximations $(R^\bullet)^+$ and $(R^+)^\bullet$ *did not produce any weak rank reversal at all* independently of using Alg. 3.1 alone (i.e. consistency is not important), or Alg. 3.1 followed by 3.2.

Using the partial order approximation $(R^\bullet)^\subset$ called *inner* because of the property $(R^\bullet)^\subset \subseteq R$ (c.f. [7]), produced both weak rank reversal and rank reversal when $\hat{\sqsubset}'$ either contained a cycle or was not transitive. This was probably because of the '$^{\subset}$' ('inclusion property kernel') operator. However (standard) rank reversal occurred not so often. The approximation $(R^{\bullet})^{\subset}$ also produced slightly less rank reversals when Algorithm 3.1 followed by Algorithm 3.2, than when Algorithm 3.2 was used alone.

When the partial order approximation $(R^{\bullet})^{+}$ was used, both weak rank reversals and rank reversals were produced only when $\hat{\sqsubset}'$ contained a cycle and was not transitive. It also produced slightly less weak rank reversals and rank reversals when Alg. 3.1 followed by Alg. 3.2, than when Alg. 3.1 was used alone. Surprisingly, when $\hat{\sqsubset}'$ contained a cycle and was not transitive, $(R^{\bullet})^{+}$ performed *worse* than $(R^{\bullet})^{\subset}$ but this could be a statistical fluke.

• The partial order approximation $(R^{+})^{\bullet}$ *never produced any weak rank reversal nor rank reversal* hence from the point of view of not producing rank reversal, its performance was the best.

• The most important case is when $\hat{\sqsubset}'$ of $\mathcal{R}'$ is neither acyclic nor tran-sitive. For this case only when the partial order approximation $(R^{+})^{\bullet}$ was used, *neither weak rank reversal nor rank reversal was discovered*

It appears that all non-numerical ranking considered in this paper methods are much less prone to suffer rank reversals than AHP. When avoiding rank reversal or weak rank reversal is important one should apply Alg. 3.1 followed by Alg. 3.2, preferably with $(R^{+})^{\bullet}$ as an order approximation.

References

1. Cowan N., Behavioural and Brain Sciences **24**, 87 (2001).
2. David H.A., The Method of Paired Comparisons (Oxford U. Press, NY, 1988).
3. Dyer J. S., Management Science **36**, 244 (1990).
4. Fishburn P. C., Interval Orders and Interval Graphs (J. Wiley, NY, 1985).
5. French S., Decision Theory (Ellis Horwood, New York, 1986).
6. Janicki R., Fundamenta Informaticae **94**, 197 (2009).
7. Janicki R., Transactions on Rough Sets **13**, 17 (2011).
8. Janicki R., Koczkodaj W. W., Computers Math. Applic. **32**, 51 (1996).
9. Janicki R., Zhai Y., Logic Journal of IGPL, doi:10.1093/jigpal/jzr018 (2011).
10. Janicki R., Zhai Y.,Lecture Notes in Art. Int. **6954**, 290 (2011).
11. Koczkodaj. W. W., Mathematical and Computer Modelling **18**, 79 (1993).
12. Saaty T. L., Journal of Mathematical Psychology **15**, 234 (1977).
13. Saaty T. L., The Analytic Hierarchy Process (McGraw-Hill, New York, 1980).
14. Zhai Y., Pairwise comparisons based non-numerical ranking, PhD thesis, Mc-Master University, (Hamilton, Ontario Canada, 2010).

MODELING A BOILING PROCESS BY MEANS OF A TAKAGI-SUGENO-FUZZY-MODEL

DANIEL FISS

IPM, Zittau/Goerlitz University of Applied Sciences, Theodor-Koerner-Allee 16 Zittau, D-02763, Germany

MICHAEL WAGENKNECHT

IPM, Zittau/Goerlitz University of Applied Sciences, Theodor-Koerner-Allee 16 Zittau, D-02763, Germany

RAINER HAMPEL

IPM, Zittau/Goerlitz University of Applied Sciences, Theodor-Koerner-Allee 16 Zittau, D-02763, Germany

The heat transfer during boiling depends on a variety of factors (boiling stage, material parameters, geometrical parameters, state variables and flow conditions). All variations are based on empirical relationships gained from experimental data because there is still no comprehensive theory. With increasing wall superheat (Tw-Ts), the evaporation changes from convective boiling to nucleate boiling and then to film boiling. For each type of boiling, separate equations for the calculation of the heat transfer coefficient do exist. This paper presents possibilities to take account of the existing uncertainties in the dynamic simulation of boiling. For this reason a Takagi-Sugeno-fuzzy-model was developed which includes the fuzzy transitions between the boiling stages.

1. Introduction

Steam production is the basic process for the generation of electrical energy in nuclear or coal power plants.

By means of the flow type, the boiling process can be distinguished between pool boiling and flow boiling. Pool boiling occurs in free flow. In a forced flow the evaporation process is called flow boiling. Undercooled boiling is not investigated. The medium has reached its saturation temperature.

The boiling process can be subdivided in three stable stages:

- convective boiling
- nucleate boiling
- and film boiling.

The transitions between the boiling stages are not sharp and they are determined by the wall superheat ΔT. The wall superheat is the difference between the wall temperature T_W and the saturation temperature of the fluid T_S. The saturation temperature remains constant in the boiling process. The heat transfer is determined by the heat transfer coefficient α. The correlations are shown in the following figure. The abscissa is the logarithmic temperature difference (T_W-T_S) and the ordinate shows the heat flux density.

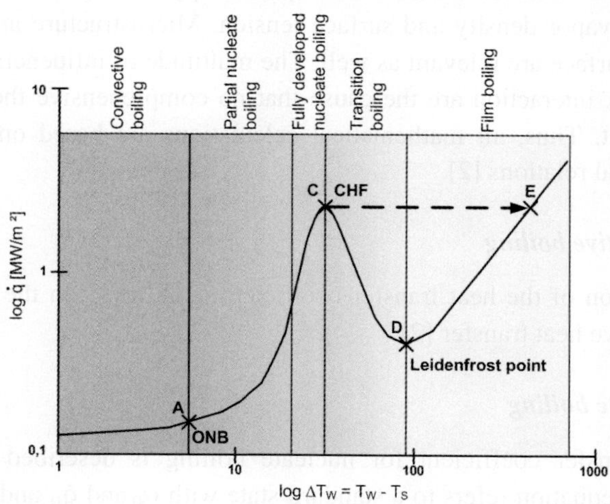

Figure 1: Explanation of the boiling stages based on the boiling curve of Nukiyama

The boiling process begins with convective boiling. The wall temperature is only a few Kelvin above the saturation temperature. No or only very few bubbles are formed. The increase of the wall temperature increases the bubble formation and the convective boiling changes over to nucleate boiling. Point A marks the onset of nucleate boiling, ONB [1]. The bubbles form on cavities or scratches on the surface containing pre-existing gas/vapor nuclei. The rising bubbles mix the fluid and thus improve the heat transfer. This is demonstrated by the growing increase of the heat transfer coefficient. Further increase of the heat supply causes also an increase in bubble formation and the flow of fluid to the wall is lower. At a maximum heat flux density (point C, called the critical heat flux CHF) forms a closed vapor film restraining the heat transfer. This means that the transferred heat flow decreases, then reaches a minimum (Leidenfrost point) and increases. This behavior is achieved by setting the wall temperature. The evaporation process in nuclear or coal power plants is setting the heat flow. This means that due to the suddenly worsened heat transfer, the wall temperature increases drastically (dashed line from C to E) [2].

2. Mathematical description of the boiling process

The describing variable for the modeling of the heat transfer is the heat transfer coefficient α. For each boiling stage many formulas do exist which mostly are based on experimental data. Besides viscosity, thermal conductivity, density, thermal expansion coefficient and geometrical parameters the heat transfer depends on many influencing factors. During heat transfer with phase change, the number of variables is extended by enthalpy of vaporization, saturation temperature, vapor density and surface tension. Microstructure and material of the heating surface are relevant as well. The multitude of influencing factors and their complex interaction are the cause that no comprehensive theory could be developed yet. Thus, all mathematical calculations are based on empirical or semi-empirical relations [2].

2.1. *Convective boiling*

The calculation of the heat transfer coefficient α is based on the equations for free convective heat transfer [2].

2.2. *Nucleate boiling*

The heat transfer coefficient for nucleate boiling is described by empirical models. The equation refers to a standard state with α_0 and $\dot{q}_0$ and considers the relative effects of wall roughness C_W, the boiling pressure by F (p*) and n [2].

2.3. *Film boiling*

The heat transfer coefficient for film boiling is composed of the heat transfer coefficient α_L and α_S. The heat transfer coefficient α_L is determined by the heat conduction process through the vapor layer and α_S through the heat radiation process [2]. The explicit equation of Bromley's proximity for the technical area of interest is defined in [4].

3. Takagi-Sugeno-fuzzy-model

To model the fuzzy transition between the boiling stages, a Takagi-Sugeno fuzzy model is suitable.

The height of the wall superheat is the decision criterion, which boiling stage is present. For this reason it is fuzzified and ΔT is defined as a linguistic variable (Figure 2), which consists of three membership functions "small", "medium" and "large". Convective boiling is definitely given if the wall superheat is located between 0 K and 7 K ("small"), we speak of nucleate boiling , when ΔT is between 20 K and 35 K ("medium") and stable film boiling starts at a high temperature of 100K ("large"). The transition regions are modelled linearly. The data were taken from Figure 1.

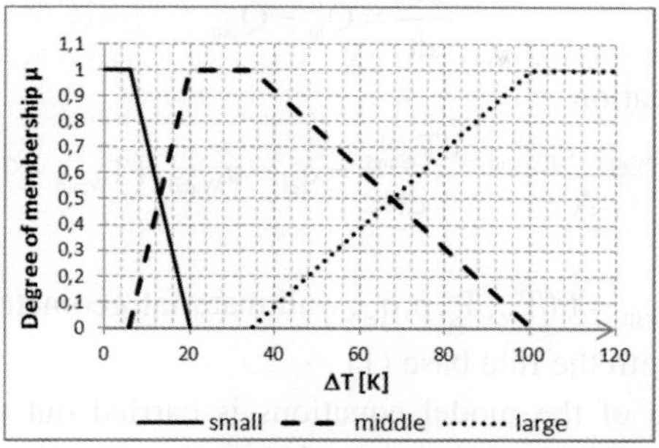

Figure 2: Linguistic variable "wall superheat ΔT"

The following rule base is derived:

If ΔT = "small" then α = convective boiling
If ΔT = "medium" then α = nucleate boiling
If ΔT = "large" then α = film boiling (1)

Subsequently, the accumulation is performed to determine the resulting heat-transfer coefficient.

4. Modeling a boiling process

To illustrate the methods / procedures developed, the following boiling process is applied. Water boils in a container, the heat is supplied electrically. In Figure 3, the model is depicted schematically.

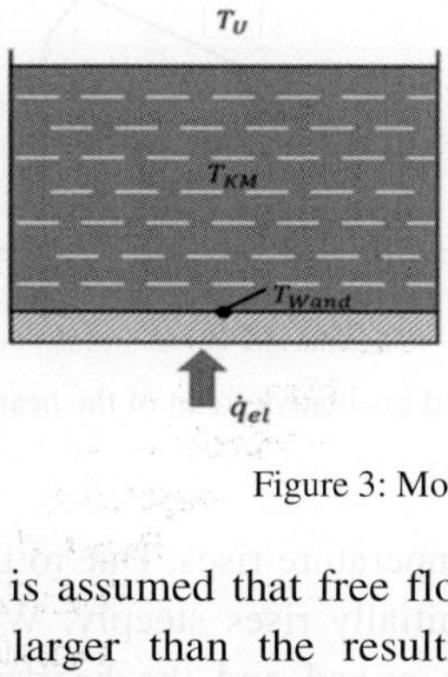

T_U – ambient temperature
T_{KM} – Fluid temperature
T_{Wand} – Wall temperature
$\dot{q}_{el}$ – heat flux

Figure 3: Model for describing a boiling process

It is assumed that free flow does exist (pool boiling), that the container is much larger than the resulting bubbles, the ambient temperature remains constant, the fluid temperature is equal to the saturation temperature, the level remains constant, the container is ideally insulated, the ambient pressure is 1 bar and the temperature distribution is homogeneous (point model).

Starting from the energy balance

$$\frac{dQ}{dt} = \dot{Q}_{in} - \dot{Q}_{out} \tag{2}$$

differential equation

$$\frac{m_{Wand} \cdot c_{PWand}}{A} \cdot \frac{dT_{Wand}}{dt} = \dot{q}_{el} - \alpha_{Wand} \cdot \left(T_{Wand} - T_S\right) \tag{3}$$

follows with

$$\alpha_{Wall} = f\left(T_W, T_S, p, \dot{q}_{el}, \text{wall material, geometry}\right) \tag{4}$$

with the rule base (1)

The realization of the model equations is carried out with the simulation system "DynStar". The simulation software is suitable for modeling and simulating processes which are described by system models based on algebraic equations and differential equations. Thus, both the static and dynamic simulation of the behavior of the process itself and its automation-related components (process measuring and control technology device) are possible.

The heat flux density is given. After 10 seconds the heat flux density increases from 1000 W/m² to 1000000 W/m² at 110 seconds (Figure 4, left). At the beginning of the simulation, the fluid temperature is already at saturation temperature. This value remains constant throughout the boiling process. The simulation starts with the state of convective boiling. The heat transfer coefficient is low with about 3000 $\frac{W}{m^2 \cdot K}$. For the first 10 seconds a steady state appears. The wall temperature is a few Kelvin above the saturation temperature.

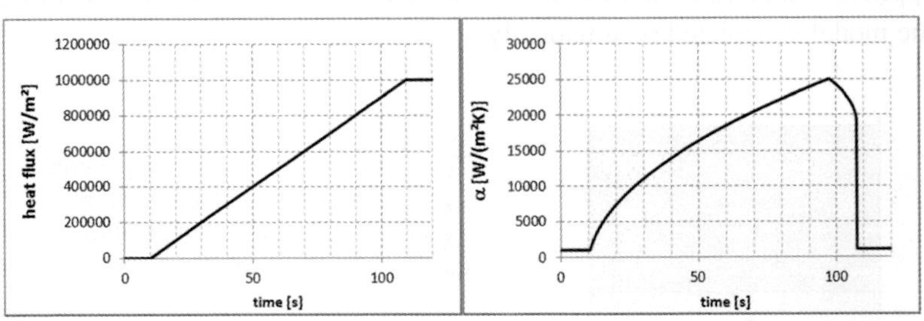

Figure 4: Temporal course of the prescribed heat flux (left) and qualitative trend of the heat transfer coefficient α (right)

With increasing heat flux density the wall temperature rises. Due to the low heat transfer coefficient, the wall temperature initially rises steeply. With the onset of bubble production the water is getting mixed and the heat transfer improves. This means that the heat transfer coefficient increases and the wall temperature rises slower despite constantly increasing heat supply. The heat transfer to the water improves until the critical point is reached, so that the bubbles form a closed vapor film. The steam has considerably poorer heat transfer properties ($\alpha = 1100\frac{\mathrm{W}}{\mathrm{m^2 \cdot K}}$). That means for a given heat flux density, that the wall temperature (Figure 5) changes almost in a jump-like way and the process switches to stable film boiling.

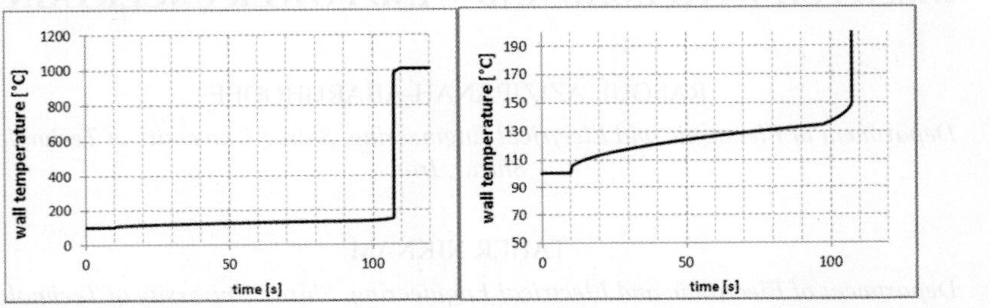

Figure 5: Qualitative temporal course of the wall temperature, left) overview right) zoomed

5. Summary and outlook

The article presents possibilities to take account of various uncertainties. In a first step, the heat transfer coefficient for the dynamic simulation of boiling process is realized with a Takagi-Sugeno-fuzzy-model and integrated into the differential equation. The dynamic simulation of the process example represents the process behavior realistically. To extend the model, the parameter uncertainties for the major variables (eg pressure, temperature) are considered as a fuzzy set and incorporated into the simulation.

Acknowledgments

The research project is funded by the European Social Fund and the Free State of Saxony.

References

1. Kandlikar, S. G..: Handbook of phase change: boiling and condensation, (Taylor & Francis 1999)
2. Stephan, K.: Wärmeübergang beim Kondensieren und beim Sieden, (Springer-Verlag, Berlin Heidelberg 1988)
3. Gorenflo, D.: VDI Heat Atlas, chapter Hab, (Springer-Verlag, Berlin Heidelberg 2006)
4. Bromley, L. A.: Heat Transfer in Stable Film Boiling, (Chem. Eng. Progress, V.46, No 5. 1950)

A NEW IMPROVED BAT ALGORITHM FOR FUZZY INTERACTIVE MULTI-OBJECTIVE ECONOMIC/EMISSION DISPATCH WITH LOAD AND WIND POWER UNCERTAINTY

RASOUL AZIZIPANAH-ABARGHOOEE

Department of Electronic and Electrical Engineering, Shiraz University of Technology, Shiraz, Iran

TAHER NIKNAM

Department of Electronic and Electrical Engineering, Shiraz University of Technology, Shiraz, Iran

This paper focuses on the stochastic 2*m*+1 Point Estimated Method (PEM) for Economic/Emission Load Dispatch (EELD) in a power system considering both thermal and wind turbine units. In solving this problem, the goal is to find optimal generation output of units to serve load demand while both cost and emission are minimized simultaneously. To this end, a novel evolutionary algorithm using Improved Bat Algorithm (IBA) is employed. To enhance the convergence characteristics and avoid being trapped in local optima, a new mutation strategy is suggested. Also, an interactive fuzzy satisfying technique is used to deal with the multi-objective optimization problem. The effectiveness and superiority of the proposed framework is validated with numerical results.

1. Introduction

Recently, due to many advantages of Wind Turbines (WTs), these units are incorporated in conventional power stations. The primary problem associated with the presence of WTs into the EELD model is the intermittent nature of wind speed. Thus to handle the stochastic sources, different methods were provided. In [1], a development of a stochastic model for economic load dispatch was presented. In [2], the weibull Distribution Function (DF) was implemented to characterize the wind power uncertainty. Accordingly, the System Operator (SO) needs to employ an accurate stochastic programming to handle the uncertainty in the input random variables. In this paper, the 2*m*+1 PEM is implemented as a useful tool to model the uncertainty in power systems.

This study is concerned to determine the available optimal generation output among the various depletable units (e.g. Thermal Units) and interminable units (e.g. WTs) to serve load demand in the EELD problem. In solving this complex non-linear, non-smooth, non-convex and multi-constraint problem two objective functions i.e. cost and emission are considered to both minimized simultaneously. An interactive fuzzy satisfying method based on a novel Improved Bat Algorithm (IBA) is presented to find the optimal operating scheduling of the WTs and TUs for the proposed multi-objective problem. The effectiveness of the proposed technique is verified on a 14-unit test system.

2. Deterministic/Stochastic Models for EELD with WTs

2.1. *Deterministic EELD (DEELD)*

Cost minimization [3]:

$$F_1(\mathbf{X}) = \sum_{i=1}^{N_{TU}} \left(a_i P_i^2 + b_i P_i + c_i + \left| e_i \sin(f_i (P_i^{\min} - P_i)) \right| \right) + \sum_{f=1}^{N_f} \sum_{w=1}^{N_{WT}} \left(C_{w,f} P_{w,f} + C_{ues,w,f} (\tilde{P}_{w,f} - P_{w,f}) + C_{oes,w,f} (P_{w,f} - \tilde{P}_{w,f}) \right) (\$) \tag{1}$$

a_i, b_i, c_i, e_i, f_i are the cost coefficients of the ith TU. $C_{w,f}$, $C_{ues,w,f}$, and $C_{oes,w,f}$ are the direct, underestimation, and overestimation cost of the wth WT in wind farm f [3]. P_i is the power output of the ith TU. $P_i^{\min}$ is the lower limit of the ith TU power output. $P_{w,f}$ and $\tilde{P}_{w,f}$ are the schedule and available power output of the wth WT in wind farm f, respectively. N_{TU} and N_{WT} are the number of TUs and WTs, respectively. N_f , $\mathbf{X}$ are the number of wind farm and vector of control variables, respectively.

Emission minimization [3]:

$$F_2(\mathbf{X}) = \sum_{i=1}^{N_{TU}} 10^{-2} \left(\alpha_i + \beta_i P_i + \gamma_i P_i^2 \right) + \xi_i \exp(\lambda_i P_i) \text{ (ton)} \tag{2}$$

$\alpha_i, \beta_i, \gamma_i, \xi_i$, and λ_i are the emission coefficients of the ith TU.

$$\sum_{i=1}^{N_{TU}} P_i + \sum_{f=1}^{N_f} \sum_{w=1}^{N_{WT}} P_{w,f} - P_{loss} - P_D = 0; P_{loss} = \sum_{iw=1}^{NU} \sum_{iw'=1}^{NU} P_{iw} B_{iwiw'} P_{iw'} + \sum_{iw=1}^{NU} B_{0iw} P_{iw} + B_{00} \tag{3}$$

$$P_i^{\min} \le P_i \le P_i^{\max} \; i = 1,\ldots,N_{TU}; 0 \le P_{w,f} \le P_{r,w,f} \; w = 1,\ldots,N_{WT}; f = 1,\ldots,N_f \tag{4}$$

where, P_{loss} and P_D are the transmission network losses and load demand, respectively. $B_{iwiw'}, B_{0iw}, B_{00}$ are the B-loss coefficients. $P_i^{\max}$ is the capacity of the ith TU. NU is the number of all wind-thermal units. $P_{r,w,f}$ is the rated power for the wth WT in wind farm f.

Since the objective functions simultaneously optimized in the problem have the competing nature inherent in human judgments in multi-objective optimization, the preferences of the Decision Maker (DM) can be determined using fuzzy sets. In this regard, a linear membership function is defined for each objective function and scenario as follows:

$$\mu_q(\mathbf{X}) = \begin{cases} 1, & F_q(\mathbf{X}) \le F_{q,\min} \\ \dfrac{F_{q,\max} - F_q(\mathbf{X})}{F_{q,\max} - F_{q,\min}}, & F_{q,\min} \le F_q(\mathbf{X}) \le F_{q,\max} \\ 0, & F_q(\mathbf{X}) \ge F_{q,\max} \end{cases} \tag{5}$$

Also, a reference membership value $\mu_{q,ref}$, is defined for each objective q. To select the best solution, i.e., the closest to the requirements of the DM, the following fitness function is selected for control variables vector $\mathbf{X}$:

$$F(\mathbf{X}) = \max_{q=1,2} \left[\mu_{q,ref} - \mu_q(\mathbf{X}) \right] \tag{6}$$

2.2. *Stochastic procedure*

It should be pointed out that the PEM was developed firstly by Rosenblueth [4] for symmetric input random variables. To cope with the non-symmetric input random variables, Rosenblueth proposed another method in [5]. For a model involving m random parameters, the Rosenblueth method required 2^m model evaluations. The Rosenblueth's algorithm was inefficient due to their huge number of simulations. Later, Harrs presented a new PEM that reduced the number of model computations from 2^m to $2m$ in [6]. Although the Harr's method is computationally more efficient than the prior ones, it is limited to symmetric variables. Finally, Hong proposed an effective and computationally efficient PEM which are proper for both the symmetric and non-symmetric input random variables [7]. This paper uses the Hong's $2m$+1 PEM to characterize the uncertainty in load demand and wind speed [7].

3. Improve Bat Algorithm (IBA)

3.1. *Original BA*

BA is a new population based evolutionary algorithm which inspire from the food foraging of magical animals i.e. bats. In each iterate k, each bat

$X_i^k = [x_{i,iw}^k, ..., x_{i,NU}^k]$ move with velocity V_i^k. The velocity of each bat is dynamically adjusted based on the flying experience of the best bats among all of them (*Gbest*) as follows [8]:

$$X_i^k = X_i^{k-1} + V_i^k; V_i^k = V_i^{k-1} + f_i(Gbest^{k-1} - X_i^{k-1});\ f_i = f_{i,\min} + rand(.)(f_{i,\max} - f_{i,\min}) \quad (7)$$

where, $f_{i,min}$ and $f_{i,max}$ are the minimum and maximum pulse frequency of bat *i*, respectively. After this movement, a new solution for each bat is generated:

If rand(.)> r_i^k $\qquad X_{i,new}^k = X_i^k + \varepsilon A_{mean}^k (Pbest_i^{k-1} - X_i^{k-1}) \quad (8)$

where, ε is a random number in the range [-1,1]. $Pbest_i^{k-1}$ is the best position for bat *i* till to this iterate. A_{mean}^k is the average loudness of all bats in this iterate.

Else $\qquad X_{i,new}^k = X_i^k + \varepsilon A_{mean}^k (X_j^{k-1} - X_i^{k-1});\ j \neq i \quad (9)$

End

If (*rand*< A_i^k or $F(\mathbf{X}_{i,new}^k) < F(\mathbf{X}_{i,new}^{k-1})$)

Accept the new solution for bat *i* as the initial population for the next iteration. It should be noted that the A_i^k and r_i^k are updated based to the following way [8]:

$$A_i^{k+1} = \alpha A_i^k; r_i^{k+1} = r_i^1(1 - \exp(-\gamma(k+1))) \quad (10)$$

where, α and γ are constants [8].

Else select $\mathbf{X}_i^k$ for bat *i* as the initial population for the next iteration.

End If

In this step, the $Pbest_i^k$ and $Gbest^k$ are updated, consequently.

3.2. *Mutation operator*

This paper presents a new method for information exchange between bats. This new approach makes the algorithm much more powerful. This approach prevents premature convergence and trapping in local optima efficiently. In this scheme, four bats m_1, m_2, m_3, and m_4 for each bat *i* are selected randomly such that $m_1 \neq m_2 \neq m_3 \neq m_4 \neq i$. A new bat $\mathbf{X}_{i,novel}^k = [x_{i,1,novel}^k, ..., x_{i,NU,novel}^k]$ is created:

$$\mathbf{X}_{i,novel}^k = \mathbf{X}_{m_1}^k + rand_1(.)(\mathbf{X}_{m_2}^k - \mathbf{X}_{m3}^k) + rand_2(.)(Gbest^k - \mathbf{X}_{m4}^k) \quad (11)$$

If the $\mathbf{X}_{i,novel}^k$ is better than the $\mathbf{X}_i^k$, then it is replaced.

4. Simulation Results

The test system consists of 14 TUs and its data can be found in [9]. In this study, the 2th, 13th, and 14th TUs of this test system are consequently replaced by 75 2-MW, 85 1-MW and 55 1-MW WTs, respectively to analyze the stochastic scheme [3]. For a better illustration of the proposed method performance, two cases are considered as follows: (i) *Case 1:* Deterministic single-objective Economic Load Dispatch (DELD1) and Emission Load Dispatch (DELD2) without WTs. (ii) *Case 2*: SEELD (with WTs).

Case 1: In order to study the effectiveness of the proposed optimization method, the single-objective DELD 1 & 2 is implemented. The complete comparison of the results and performance of IBA with respect to the BA and those of the DE method reported in [9] are given in Table 1. The detailed results of the best solution of IBA for DELD1 & 2 are shown in Table 2.

Table 1. Comparison of the results obtained by different methods.

Evolution Method	IBA	BA	DE [9]
Best cost (DELD1) ($)	9181.12	9,219.36	9,592.40
Best emission (DELD2) (ton)	2887.83	2,961.46	3,232.20

Table 2. Best dispatch found by IBA for different cases.

1. Unit	1. DELD1 (MW)	1. DELD2 (MW)	1. SEELD (Interaction 0)	1. Unit	1. DELD1 (MW)	1. DELD2 (MW)	1. SEELD (Interaction 0)
P_1	329.52	189.52	220.178	P_8	109.87	177.92	205.41
P_2	299.60	150.00	93.95	P_9	124.73	162.00	160.61
P_3	94.80	130.00	129.66	P_{10}	137.20	160.00	159.99
P_4	119.73	130.00	129.73	P_{11}	58.00	80.00	79.92
P_5	249.73	162.51	182.62	P_{12}	80.00	80.00	79.99
P_6	184.87	270.41	326.38	P_{13}	62.40	85.00	34.35
P_7	135.00	199.84	220.31	P_{14}	52.40	55.00	11.01

Case 2: For the multi-objective optimization of SEELD problem, the min-max approach is adopted to obtain a global non-inferior solution in the first interaction for the initial reference membership values $\mu_{q,ref}, q=1,2$. The DM terminates the interactive steps if the solution be satisfactory; otherwise, it selects a new $\mu_{q,ref}$ for the new interactive steps. This procedure continues until the DM is satisfied with the solution. The results of five interactions are shown in Table 3. Also, the best dispatch results of interaction 0 are given in Table 2.

Table 3. Results of the interactive fuzzy satisfying procedure.

Interaction	$\mu_{1,ref}$	$\mu_{2,ref}$	F_1 ($)	F_2 (ton)
0	1	1	27,538.54	3,197.40
1	1	0.8	23,773.71	3,402.66
2	1	0.6	23,272.89	3,508.43
3	0.8	1	32,652.86	2,980.90
4	0.6	1	35,607.26	2,867.38

5. Conclusion

This Paper provides a novel IBA method to cope with a SEELD problem in which the valve point effects and generation limits are modeled. Also using stochastic 2*m*+1 PEM approach to characterize the load and wind speed in the 14-unit test system can lead to a more efficient utilization of energy resources and the emission can be reduced significantly using wind power. Moreover, the proposed method can resolve the tradeoff between the objectives of the SEELD problem through an interactive procedure by which the DM can update the solutions based on its preferences through the optimization process. Numerical results were showed accuracy of presented SEELD.

References

1. D. W. Bunn and S. N. Paschentis, *Eur. J. Oper. Res.* **27**, 179 (1986).
2. J. Hetzer, D. C. Yu, K. Bhattarai, *IEEE Trans. Energy Conv.* **32**, 603 (2008).
3. R. Azizipanah-Abarghooee, T. Niknam, A. Roosta, A. R. Malekpour, M. Zare, *Energy* **37**, 322 (2012).
4. E. Rosenblueth, *P. Natl. Acad. Sci. USA.* **72**, 3812 (1975).
5. E. Rosenblueth, *Appl. Math. Model.* **33**, 329 (1981).
6. M. E. Harr, *Appl. Math. Model.* **25**, 180 (1989).
7. H. P. Hong, *Reliab. Eng. Syst. Saf.* **59**, 261 (1998).
8. X.-S. Yang, *Stud. Comp. Intell.* 284, 65 (2010).
9. K. K. Mandal, N. Chakraborty, *Int. J. Emerg. Electric Power Syst.* **9**, 1 (2008).

HANDLING NONLINEAR LARGE-SCALE PROBLEMS WITH MODIFIED FIREFLY ALGORITHM

BAHMAN BAHMANIFIROUZI

Department of Electrical Engineering, Marvdasht Branch, Islamic Azad University, Marvdasht , Iran

ABDOLLAH KAVOUSI FARD

Electronic and Electrical Department, Shiraz University, Shiraz, Iran

TAHER NIKNAM[†]

Electronic and Electrical Department, Shiraz University of Technology, Shiraz, Iran

In this paper, an efficient optimization method called Modified Firefly Algorithm (MFA) is proposed to handle complex nonlinear optimization problems. In order to enhance the exploratory ability of the algorithm, a new modification process is integrated sufficiently. The proposed method is examined on several benchmark optimization problems with different characteristics and results are compared to some of the other well-known algorithms in the area.

Introduction

Most of the traditional techniques require gradient information and hence it is not possible to solve non-differentiable functions with the help of such techniques [1]. Therefore, in the last few years, a new optimization field referred to as nature-inspired meta-heuristic optimization is proposed which have resulted to satisfying outcomes in different types of optimization problems [2]. The good performance of the evolutionary algorithms in solving difficult nonlinear optimization problems has allowed a wide variety of the algorithms such as Genetic Algorithm (GA), Particle Swarm Optimization (PSO), Ant Colony Optimization (ACO), Artificial Bee Colony (ABC), etc. Nevertheless, there are some limitations in the performance of the above evolutionary algorithms. The dependability of these algorithms on their initial parameters and the possibility of trapping in local optima are some of the present deficiencies. In this regard, the enhancement can be done by the use of two procedures: 1) applying modification and 2) hybridization of the present algorithms. According to the above descriptions, in this paper a new method based on Firefly Algorithm (FA) [4] and a novel modification is proposed to handle different types of complex nonlinear multi-objective optimization problems efficiently. The proposed Modified FA (MFA) is able to improve the overall performance of the FA for both local and global search exploration

2. Firefly Algorithm

2.1. Original Firefly Algorithm Descriptions

Dr Xin-She Yang at the Cambridge University proposed FA by defining three simple rules [5]: 1) all fireflies are supposed to be unisex. 2) The firefly with less brightness is attracted by the brighter one. 3) If no firefly brighter than a given firefly exits, then the given firefly can move randomly in the entire search space. In FA, the attractiveness of each firefly is determined by its brightness (light intensity). When the distance between two fireflies increases, the brightness is decreased. Therefore, the attractiveness of each firefly can be modeled by any monotonically decreasing function as follows [5]:

$$\beta(r) = \beta_0 \times \exp(-\gamma r^m) \quad ;m \geq 1 \tag{1}$$

where r is the distance of two fireflies, β0 is the attractiveness at r=0 and γ is the absorption coefficient to control the light intensity. The distance can be calculated in different frameworks like Cartesian distance, Manhattan or Mahalanobis distance [5]. The distance between ith firefly (Xi) & jth firefly (Xj) in the Cartesian framework is defined as follows:

$$r_{ij} = \left\| X_i - X_j \right\| = \sqrt{\sum_{k=1}^{d} (X_{i,k} - X_{j,k})^2} \tag{2}$$

where $X_{i,k}$ is the kth element of the special vector X_i associated with the ith firefly. Also, d is the dimension of the problem investigated. Suppose two fireflies i & j which the brightness of the ith firefly is more than the jth firefly. Consequently, the jth firefly will move toward the ith firefly which is modeled as follows:

$$X_j = X_j + \beta_0 \times \exp(-\gamma r^m) \times (X_i - X_j) + u_j; \quad u_j = \alpha(rand - \frac{1}{2}) \tag{3}$$

In this equation the third term models the random behavior of a firefly when no brighter firefly is sensed. The constant value α, is called randomization parameter which is in the range of (0,1).

2.2. Modified Firefly Algorithm (MFA)

In this section, a modification phase is proposed such that the ability of the algorithm is enhanced while the possibility of trapping in local optima is reduced. After implementing the steps mentioned in the last sections, for each firefly i in the population, three fireflies z_1, z_2 and z_3 are chosen from the fireflies' population randomly such that $z_1 \neq z_2 \neq z_3 \neq i$. Now, by the use of these three fireflies a new improved individual is generated as follows:

$$X_{improv} = X_{z_1} + \sigma \times (X_{z_2} - X_{z_3});$$
$$X_{improv} = [x_{improv,1}, \ldots, x_{improv,d}]; X_{z_i} = [x_{z_i,1}, x_{z_i,2}, \ldots, x_{z_i,d}]; i = 1,2,3 \quad (4)$$

where σ is a random value in the range of [0,1]. Now by the use of $X_{improved}$, the best firefly (X_{best}) and X_i (the ith firefly), the following two fireflies are generated:

$$X_{best,1} = [x_{best,1}, x_{best,2}, \ldots, x_{best,d}]; \qquad x_{firefly1,j} = \begin{cases} x_{improv,j}, & \text{if } \kappa_1 \leq \kappa_2 \\ x_{best,j}, & \text{otherwise} \end{cases}$$
$$X_{firefly,2} = \psi \times X_{best} + \zeta \times (X_{best} - X_{I_{rand}}) \quad (5)$$

where κ_1, κ_2, ψ and ζ are random values in the range of [0,1]. Also, $X_{I_{rand}}$ is a firefly which is selected from the population randomly. Now the most attractive firefly among $X_{firefly,1}$, $X_{firefly,2}$ is selected and compared to the ith firefly (X_i). If it better than X_i, then it replaces X_i else it remains in its position.

3. Experimental Studies

3.1. Experiment 1

In this section, 5 different benchmarks are utilized as the test functions as shown in Tables 1&2 [6]. Here, two criteria are defined: 1) success percentage (Suc%) which shows the stability of MFA, 2) Average function evaluations (AFE). Also, the algorithm is supposed to be successful if the difference between the achieved optimum and the global optimal solution is less than 0.001 [6] & the number of runs is 100. The population size is 20 (except the Rosenbrock for which it is 50). It can be seen from Table 2 that MFA has similar results for

Sec% criterion while it has better AFE values regarding the other algorithms in the table.

Table 1. Detailed data of the benchmarks considered in experiment 1 [6]

Bench. no	Function	Range	Global optimum
1	De Jong	[-2.048,2.048]	F_{min} = 3905.93, X = (1,1)
2	Goldstein and Price	[-2, 2]	F_{min} = 3, X = (1,1)
3	Martin and Gaddy	[0,10]	F_{min} = 0, X = (5, 5)
4 (a)	Rosenbrock (D = 1)	(a) [-1.2, 1.2]	F_{min} = 0, X = (1, 1)
4 (b)		(b) [-10,10]	F_{min} = 0, X = (1, 1)
5	Rosenbrock (D = 3)	[-1.2, 1.2]	F_{min} = 0, X = (1, 1, 1, 1)
6	Hyper Sphere (D = 6)	[-5.12, 5.12]	F_{min} = 0, X = (0, 0, 0, 0, 0,0)

Table 2. Values of Sec% & MFE for GA, ACO, ABC, Grenade Explosion Method (GEM) & AMFA

Bench. no	GA		ACO		ABC		GEM		(AMFA)	
	Suc%	AFE	Suc%	AFE	Suc%	AFE	Suc%	AFE	Suc%	AFE
1	100	10160	100	6000	100	868	100	746	100	704
2	100	5662	100	5330	100	999	100	701	100	660
3	100	2488	100	1688	100	526	100	258	100	240
4 (a)	100	10212	100	6842	100	631	100	572	100	545
4 (b)	-	-	100	7505	100	2306	100	2289	100	1123
5	-	-	100	8471	100	28529	100	82188	100	2611
6	100	15468	100	22050	100	7113	100	423	100	332

3.2. Experiment 2

In the Experiment 2, six objective functions are utilized which are shown in Table. 3 [7]. In order to see the appropriate performance of the proposed algorithm, similar comparisons with PSO and the hybrid of PSO and Nelder Mead simplex search (called NM-PSO) are done. The number of simulation runs is 200 times. The population size is 20. In Table 4, it can be seen that the proposed method has a very good consistency with Sec=100% for all benchmarks. Also, regarding the AFE and Error criteria, the superiority of the proposed algorithm is evident.

Table 3. Detailed data of the benchmarks considered in experiment 2 [7]

Bench. no	Function	Range	Global optimum
1	Powell badly scaled function	[-50, 50]	$F_{min} = 0$, X = (1.098 e5, 9.106)
2	B2 function	[-50, 50]	$F_{min} = 0$, X = (0,0)
3	Booth function	[-50, 50]	$F_{min} = 0$, X = (1, 3)
4	Griewank (D = 10)	[-50, 50]	$F_{min} = 0$, X = (0,0, …)
5	Rastrigin (D = 10)	[-50, 50]	$F_{min} = 0$, X = (0, 0)
6	Sphere (D = 30)	[-50, 50]	$F_{min} = 0$, X = (0, 0, …)
7	Griewank (D = 50)	[-50, 50]	$F_{min} = 0$, X = (0, 0, …)

Table 4. The values of Sec%, MFE & Error for PSO, NM-PSO and AMFA

Bench. no	PSO			ACO			AMFA		
	Suc%	AFE	Error	Suc%	AFE	Error	Suc%	AFE	Error
1	94	20242	9.89 e-6	100	2971	3.78 e-6	100	2920	6.2312 e-6
2	100	4188	1.40 e-8	100	1124	3.23 e-10	100	1087	8.1221 e-11
3	100	3848	2.60 e-8	100	1065	1.26 e-9	100	901	5.8361 e-11
4	0	504657		82	14076	1.04 e-11	100	1125	5.8492 e-15
5	30	510050	1.08 e-4	60	12353	1.91 e-11	100	2131	0
6	0	4530150		100	87004	2.76 e-11	100	3312	0
7	0	2550250		82	378354	9.96 e-12	100	2123	0

4. Conclusion

In this paper an appropriate modified algorithm based on FA is proposed to handle continuous nonlinear large-scale optimization problems. The proposed modification is consisted of two main ideas 1) using a mutation operation to improve the diversity of the fireflies and 1) treating the population to avoid the promising local solutions. The simulation results show the stability and strong ability of the proposed algorithm in finding the global optimal solutions.

References

1. R.V. Rao, V.J. Savsani and D.P. Vakharia, Teaching–Learning-Based Optimization: An optimization method for continuous non-linear large scale problems, Information Sciences, in press, (2011).

2. C. A. Coello Coello, D. A. Van Veldhuizen, and G. B. Lamont, Evolutionary Algorithms for Solving Multi-Objective Problems, Norwell, MA: Kluwer, ISBN 0-3064-6762-3, (2002).
3. J. Holland, Adaptation in Natural and Artificial Systems, University of Michigan Press, Ann Arbor, (1975)
4. T. Apostolopoulos and A. Vlachos, Application of the Firefly Algorithm for Solving the Economic Emissions Load Dispatch Problem, International Journal of Combinatorics, Article ID 523806, 23 (2011)
5. X. S. Yang, Nature-Inspired Metaheuristic Algorithms, Frome: Luniver Press, (2008).
6. A. Ahrari, A.A. Atai, Grenade explosion method – a novel tool for optimization of multimodal functions, Applied Soft Computing 10, 1132–1140, (2010).
7. K.S.F. Shu, Z. Erwie, A hybrid simplex search and particle swarm optimization for unconstrained optimization, European Journal of Operational Research 181, 527–548, (2007).

AN EFFICIENT FUZZY PROGRAMMING APPROACH TO THE VEHICLE ROUTING PROBLEM WITH SOFT TIME WINDOWS

SEZGIN KILIC*

Industrial Engineering Department, Turkish Air Force Academy, Yesilyurt Istanbul, 34149, Turkey

SEZGIN KAPLAN

Industrial Engineering Department, Turkish Air Force Academy, Yesilyurt Istanbul, 34149, Turkey

This paper considers the vehicle routing problem with soft time windows (VRPSTW) where the customers' preferences and tolerances on the latest delivery times are handled by use of a fuzzy programming model. An indirect enumeration strategy is proposed to generate pseudo-efficient (Pareto) solutions to the VRPSTW. The proposed model and the solution approach make easy to implement distributors' various strategies. A simulated annealing (SA) algorithm is implemented to solve the test problems. The importance and advantages of the proposed approach are illustrated by numerical results.

1. Introduction

The vehicle routing problem with time windows (VRPTW) is a well-known and complex combinatorial optimization problem concerned with finding efficient routes to serve a number of customers within their time window constraints. Due to its importance and frequency in practice, there is a vast literature that addresses modeling and solution aspects of the VRPTW. A special type of the VRPTW where the time windows need not to be satisfied exactly is called as VRPTW with soft time windows (VRPSTW). With regard to previous studies [1-4] and their results, it can be seen that VRPSTW has many practical applications and its solutions provide some key advantages for distributors.

Evidently, soft time windows relax the problem resulting in better solutions and make the delivery firm operate more efficiently by utilizing its vehicles' capacity in any case. However, understanding, modeling and managing this relaxation is needed to exploit these benefits in reality. In this work, we

* email: *s.kilic@hho.edu.tr*

proposed a model that covers vagueness type of uncertainty which stems from customers' preferences and tolerances on due dates.

2. Previous Works

Against the extensive literature of VRPTW, works on the soft windows case is very limited. The research efforts in modeling the imprecision on the boundaries of delivery time intervals can be classified under two groups.

In the first group, authors widen the precise time window intervals by using some coefficients and obtain broader ones but still in a crisp manner. Violating hard time window intervals as much as to that extended ones is permitted with a penalty cost. Figliozzi (2010) is a recent work in this group and gives a brief survey of previous works [2]. But, violation of customers' time windows cannot be measured by a penalty cost. In practice, customer satisfaction levels may decline those results in customer losts in long term [3].

In the second group, researchers handle the uncertainty in due dates using fuzzy set theory. Tang *et al.* (2009) as the one of the former works in this area, applied fuzzy sets to model the customer satisfaction levels. They solved the problem via defuzzification by setting a lowest service level for each customer [3]. Moreover, Xu *et al.* (2011) handled the customers' preferences by ambiguous type of uncertainty and modeled the time windows by fuzzy random variables [1]. They solved a crisp model by transforming fuzzy random variables initially to fuzzy numbers then to deterministic numbers. However, both of these works did not consider the numbers of vehicles as a cost factor and vehicles are allowed to wait on customers or routes without generating any cost.

3. Proposed Model

Depending on the nature of uncertainty, the type of uncertainty should be well defined to clarify the use of proper mathematical tool. Usually ambiguity and vague type of uncertainties are confused [5]. The model proposed in this work covers vagueness type of uncertainty which represents the customers' preferences or tolerations. The uncertainty in the decision makers preferences sourcing from lack of imprecision associated with the use of language is named as vagueness and fuzzy sets are capable of representing this type of uncertainty. Accordingly the proposed model is called as a fuzzy programming model.

The problem is defined on a network as classical VRPTWs. The objective of the model is to satisfy all customers' demands while minimizing the travelling costs as the number of vehicles (NV) and total traveling distance (TD) hierarchically. It is assumed that vehicles start their deliveries as early as possible. Arbitrary waiting at customers or on roads are not permitted. A minimum satisfaction level for each customer, δ (it may be identical or different for each customer), is assumed to be determined by the distributor and these levels are achieved by corresponding constraints in the proposed solution approaches.

In reality, however the customers may declare crisp time windows for deliveries; they usually have preferences or tolerances on the boundaries of these intervals. In addition, customers may prefer to declare their delivery time windows by vague terms which cannot be modeled by probabilistic measures. Fuzzy set theory gave rise to very suitable, capable and practical tools for modeling and computing this type of uncertainty. In view of that, soft time windows are handled by fuzzy sets. An earliest delivery time (EDT), a latest delivery time (LDT) and an endurable latest delivery time (ELDT) is defined for each customer. Deliveries before EDT or after ELDT are not accepted. It is assumed that customer satisfaction or service level decreases as the delivery time passes the LDT. In this concept, soft delivery time window is modeled by fuzzy sets representing the customer's satisfaction level as seen on Fig. 1.

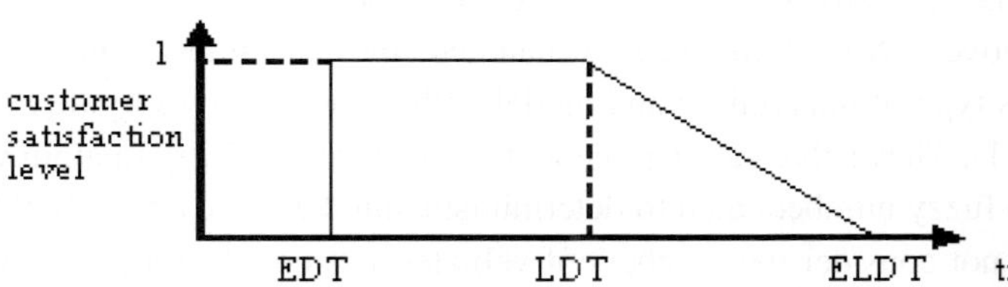

Figure 1. Soft time windows representing the customer's satisfaction level by fuzzy sets.

Let v_i^g represents the arrival time of vehicle g to customer i. We can calculate the satisfaction level of a customer i, (π_i) by Eq. 1 for linearly decreasing satisfaction level.

$$\pi_i = \mu(v_i^g) = \text{maks}\left(0, \min\left(1, \frac{ELDT - v_i^g}{ELDT - LDT}\right)\right) \geq \delta \tag{1}$$

4. Simulated Annealing Approach

A simulated annealing algorithm is implemented to solve the proposed model. A sample feasible solution reflecting the routes of a solution is shown in Fig. 2. The solution is illustrated by a $(2n+1)$ sized string which has enough digits to include all possible solutions where n is the number of customers. A central depot was represented by dummy customer (indexed with 0).The customer sequence of each route starts from and ends at the customer depot. Thus, the additional zeros at this solution give the number of vehicles that are not needed in serving the customers.

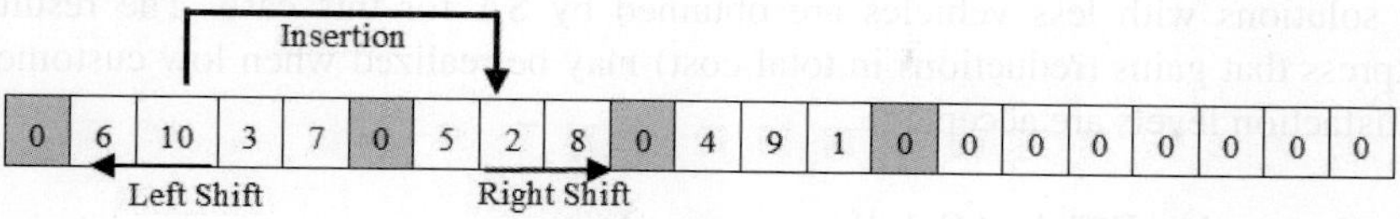

Figure 2. A sample solution

An insertion operation is used for the SA to perturb the current solution locally. As seen on Fig. 2, customer insertion is executed by inserting a randomly selected customer into a position on the same or other route and shifting the customers to the right on that route. Ensuring the hierarchical objective function, we defined total cost as; TC=TDxNV. Multi factorial design of experiments has been applied to set the parameters for better solutions.

5. Computational Results

We used the well-known Solomon VRPTW benchmark problems with 25 customers [6]. We set ELDT=0.30xLDT to generate VRPSTW instances. Table 1 represents the TD and NV (in parenthesis) of generated solutions. Second column gives optimal solutions generated by exact methods for original problems [7]. Columns for $\delta=\{1, 0.75, 0.50, 0.25, 0.01\}$ stand for the best solutions generated by SA algorithm after 10 replications when the minimum customer satisfaction of a solution (Π_{min}) is set to $\Pi_{min} \geq \delta$.

Table 1. Results for R1 type VRPSTW instances.

Problem	optimal	δ=1	δ=0.75	δ=0.50	δ=0.25	δ=0.01
R101	617.1(8)	618.33(8)	601.52(7)	575.57(6)	525.79(5)	505.71(5)
R102	547.1(7)	548.11(7)	513.53(6)	482.45(5)	462.4(5)	434.98(4)
R103	454.6(5)	474.17(4)	433.32(4)	421.45(4)	419.41(4)	421.17(3)
R104	416.9(4)	417.96(4)	403.58(4)	394.48(4)	390.28(4)	374.66(3)
R105	530.5(6)	531.8(6)	506.14(5)	489.95(4)	472.31(4)	449.63(4)
R106	465.4(5)	466.48(5)	447.41(4)	434.98(4)	417.7(3)	401.1(3)
R107	424.3(4)	425.27(4)	410.66(4)	420.76(3)	412.24(3)	375.72(3)
R108	397.3(4)	398.29(4)	390.28(4)	376.09(3)	367.97(3)	361.51(3)
R109	441.3(5)	460.52(4)	453.66(4)	430.64(4)	420.51(4)	433(3)
R110	444.1(4)	445.8(4)	426.06(4)	409.36(4)	400.89(3)	397.69(3)
Average	473.86(5.2)	483.5(4.92)	463.12(4.54)	447.19(4.1)	432.22(3.88)	420.06(3.4)

Since exact methods typically use monolithic total distance as the objective function and use low precision distance and time calculations, our results are not directly comparable with optimal solutions. However, the case $\delta=1$represents the SA solution for original crisp VRPTW. It was seen that either optimal solutions or solutions with less vehicles are obtained by SA for this case. The results express that gains (reductions in total cost) may be realized when low customer satisfaction levels are accepted.

6. Generating Efficient Solutions

In real life decision makers usually deal with a trade-off problem between costs and customer satisfaction levels. In this section, we propose an approach that helps to manage this trade off easily with a less computational effort. Suppose that we have to choose one among two solutions S and S´ for which TC<TC´ and $\Pi_{min}>\Pi'_{min}$ are the case. It is clear that decision maker should prefer route S and S is said to be better than S´ or S dominates S´. A solution is considered efficient with respect to total cost and customer satisfaction, when there does not exist a solution which dominates it. This preliminary rationale directs to search only considering for efficient solutions. Thus, the following meta-algorithm which generates efficient solutions for VRPSTW can be proposed.

Step 1. Set $\delta=0.01$

Step 2. Solve VRPSTW subject to $\Pi_{min}\geq\delta$. If there is no solution STOP.

Step 3. Let the solution found in *Step 2* have $\Pi_{min}=\Omega$. Set $\delta=\Omega+0.01$

Step 4. If $\delta<=1$ continue to *Step 2*. Otherwise, STOP.

As the algorithm used in this work use a heuristic approach, the generated solutions should be called heuristically efficient/dominant ones.

Table 2. Heuristic efficient solutions for R110 instance.

δ	TD	NV	TDxNV	Π_{min}
0.01	397.09	3	1191.3	0.12
0.13	397.86	3	1193.6	0.19
0.20	400.89	3	1202.7	0.27
0.28	408.4	3	1225.2	0.51
0.52	405.85	4	1623.4	0.55
0.56	408.82	4	1635.3	0.69
0.7	426.06	4	1704.2	0.87
0.88	445.8	4	1783.2	1

The computational results for proposed meta-algorithm are represented in Table 2. Fig. 3 presents a decision making tool to choose among heuristic efficient solutions in Table 2.

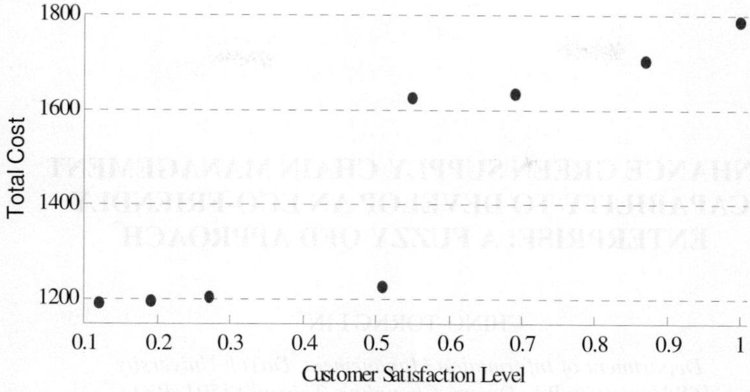

Figure 3. Heuristic efficient solutions for R110 instance

7. Conclusions

The proposed solution approach exposed a practical decision making tool that can be easily used to generate efficient solutions for distributers in terms of multi-objectives, the total cost and the minimum customer satisfaction level. The meta-algorithm may also be used with exact methods. As stated by [2], the proposed approach and the algorithm for the VRPSTW should be simple, flexible, highly interactive, and allow for manual improvements and manipulation of constraints and customer priorities, not too many parameters difficult to be calibrated or understand in order to be useful as commercial software in practice. In this work a promising approach that is more close to targets above is proposed rather than the previous works on related literature.

References

1. J. Xu, F. Yan and S. Li, *Transport Res. E-Log.* **47**, 1075 (2011).
2. M.A. Figliozzi, *Transport Res. C-Emer.* **18**, 668, (2010).
3. J. Tang, Z. Pan, R.Y.K. Fung and H. Lau, *Fuzzy Set Syst.* **160**, 683, (2009).
4. W.C. Chiang and R.A. Russell, *J. Oper. Res. Soc.* **55**, 1298, (2004).
5. W.A. Lodwick, K.D. Jamison, *Fuzzy Set Syst.* **158**, 1861, (2007).
6. M.M. Solomon, *Oper. Res.* **35(2)**, 254, (1987).
7. B. Kallehauge, J. Larsen, O.B.G. Madsen, *Comput. Oper .Res.* **33**, 1464, (2009)

ENHANCE GREEN SUPPLY CHAIN MANAGEMENT CAPABILITY TO DEVELOP AN ECO-FRIENDLY ENTERPRISE: A FUZZY QFD APPROACH*

CHING-TORNG LIN †

Department of Information Management, DaYeh University
168 University Rd., Dacun, Changhua, Taiwan 51591, R.O.C.

Green or environmental concerns have been drawn more and more attention both in academia and industry. Nowadays, few firms have all the skills and resources to produce complex products entirely in-house and supply chain management has become the dominant vehicle. Thus how to measure and enhance capability of green supply chain management (GSCM) plays an important role in developing an eco-friendly enterprise. The purpose of eco-friendly development planning is to unite the resources of an enterprise to create business value. For an enterprise to enhance GSCM, deploying and starting with green drivers of the business environment, as well as green supply chain (GSC) capabilities and providers, and ultimately transforming all these attributes into strategic competitive edges are all critical. However, the published literature on GSCM hitherto has failed to sufficiently address the relevant perspectives in such analyses. The relationship matrix in the quality function deployment (QFD) method provides an excellent tool for deploying important concepts and linking processes. Furthermore, due to the ill-defined and ambiguous relationships among GSC drivers, capability and provides, the conventional evaluation approaches not can handle such decisions suitably and effectively. This study suggests an operational strategy for deploying the GSC drivers, capabilities as well as providers by using the relationship matrix in the QFD method and fuzzy logic. An illustrative example is presented to demonstrate the applicability of the proposed method.

1. Introduction

The environmental impacts have becoming a key factor to constraint the socio-economic development of human in the 21st century. Environmental concerns from consumers, governments and academics have encouraged businesses to introduce and promote business practices that help to ease the negative impacts of their actions on environment, such as green management, green production and innovation, etc. are now being pursued. Nowadays, GSCM is a modern dominant

* This work is partially supported by grant 99-2221-E-212-011 of the National Science Council of the Republic of China, Taiwan.
† charllin@mail.dyu.edu.tw

strategy in consideration of environmental impacts and resource efficiency in the entire supply chain management. The theoretical foundation of GSCM includes the theory of green manufacturing and supply chain management technology. Its aims are to make the impact on the environment the smallest, resources efficient the highest for making products from the acquisition of raw materials, processing, to waste disposal and recycling the entire process.

The embracing green strategy has several benefits for companies [1-3]. GSC has been the subject of increased academic interest in recent years. Previous researches have focused on the concept and connotation of GSC [4-9]. Although each study or technique above contributes to an understanding of the problem, each - functioning alone - is insufficient for handling the problem in its entirety because the capabilities and assessment should be closely linked and coordinated with the drivers and the providers [10]. It is therefore necessary to examine the problem from a broader perspective.

The purpose of eco-friendly development planning is to unite the resources of an enterprise to create business value. According to researchers studied [11, 12], firm value can be maximized and competitive threat minimized only by selecting a strategy based on all facets of the business not merely as isolated organizational strategy islands within the company [13]. Thus, for a successful organizational strategy such as eco-friendly enterprise, an effective integrated procedure within the business to assure that the green providers can satisfy the green capabilities and cope with green drivers, ultimately transforming all of these attributes into strategic competitive edges are critical and must be created [9, 10, 14]. It is difficult for a firm to achieve eco-friendly enterprise because of the lack of an efficient approach for eco-friendly development planning. The focus of current researches [5, 6, 10] is on setting eco-friendly enterprise. However, there is a lack of sufficient details on how these objectives are translated into action plans.

The relationship matrix in the QFD method provides an excellent tool for aligning important concepts and linking processes. Due to an either "imprecise" or "vague" definitions of green attributes and matrix relationships, the aligning and integrating evaluation process is associated with uncertainty and complexity. To compensate for the lack of an efficient approach that can deal directly with the interface and alignment issues among green drivers, capabilities and the choice of appropriate green providers to assist managers in more efficiently achieving eco-friendly enterprise, a new systematic methodology for eco-friendly enterprise development planning, based on fuzzy logic and the relationship matrix in the QFD is devised to align the overall relationship that exists from the green drivers in the business environment down to green providers. Moreover, a fuzzy GSCM capability index (FGSCMCI) for an enterprise composed of GSC attributes ratings and the total relation-weight with GSC drivers was developed for measurement of the GSCM capability index of an enterprise.

2. Eco-Friendly Enterprise Conceptual Model

The goal of an eco-friendly enterprise is incorporating environmental concerns to enrich/satisfy customers and employees. The rise of international environmental regulations and popular environmental consciousness of consumers would bring significant impacts to businesses in the world [2]. Therefore, corporate environmental management will play an important role, nowadays. From a review of the relevant literature [3, 15-18], the author has developed an eco-friendly enterprise conceptual model, as shown in Figure 1.

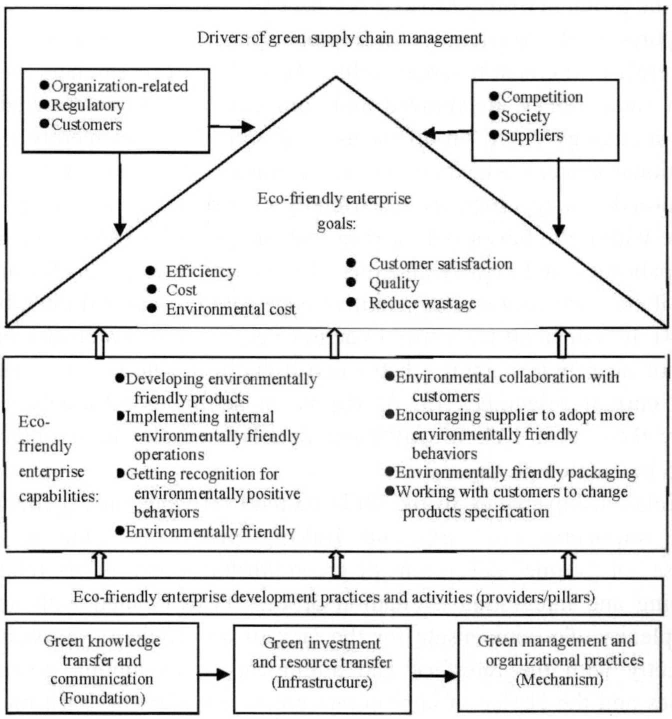

Figure. 1. A conceptual framework of an eco-friendly enterprise

Furthermore, to assist managers in more efficiently achieving eco-friendly, based on the eco-friendly enterprise conceptual model and by using the relationship matrix in the QFD approach, an original systematic model for aligning and integrating to assure that the green providers can satisfy the green capabilities and cope with green drivers can be constructed as shown in Figure 2. This model is described as follows:

- Aligning green strategy: to identify the green capabilities that can provide the required strength for responding to the problems of environmental impacts and searching for competitive advantage using the close linkage and coordination of green drivers and green capabilities.
- Aligning green ingredients: to find green providers constituting the means by which the so-called needs of an enterprise related to green capabilities that can be achieved by close linkage and coordination of green capabilities and green providers.

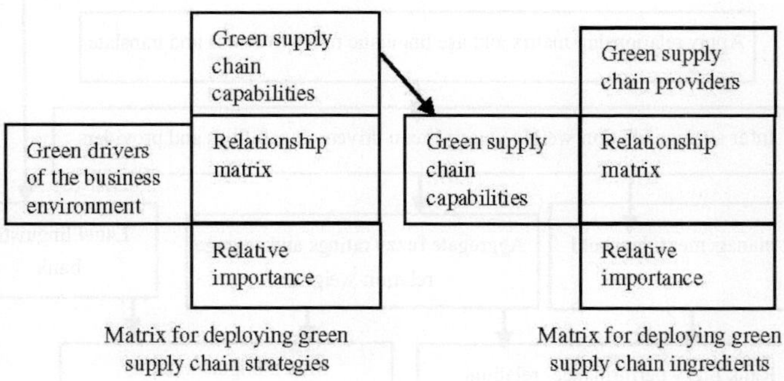

Figure. 2. Framework for deploying green supply chain drivers, capabilities and providers

3. Fuzzy QFD-Based Algorithm for Evaluation of Green Supply Chain Management Capabilities

To assist managers in more efficiently enhancing eco-friendly using the QFD relationship matrix and fuzzy logic, a fuzzy logic-QFD eco-friendly-enhancing model (FLQFDEFEM) composed of four major parts (as shown in Figure 3) was devised for eco-friendly enterprise development and evaluation. The green drivers are identified first based on a survey of the business operation environment. The environmental goals are then determined and the requirements for measuring the capabilities are identified. The required green providers for assessment are then selected. The relationship matrix is then aligned to analyze the fuzzy average relation-weight among the drivers, capabilities and providers. The fuzzy ratings and average relation-weights of the capabilities are synthesized to obtain the fuzzy GSCM capability index (FGSCMCI) of the enterprise and match the FGSCMCI with an appropriate linguistic term to label the green level. The fuzzy ratings and average relation-weights of the providers are then synthesized to obtain the fuzzy importance-performance index (FIPI) for each. They are ranked to identify the major barriers to enabling proactive implementation of the appropriate ameliorating measures

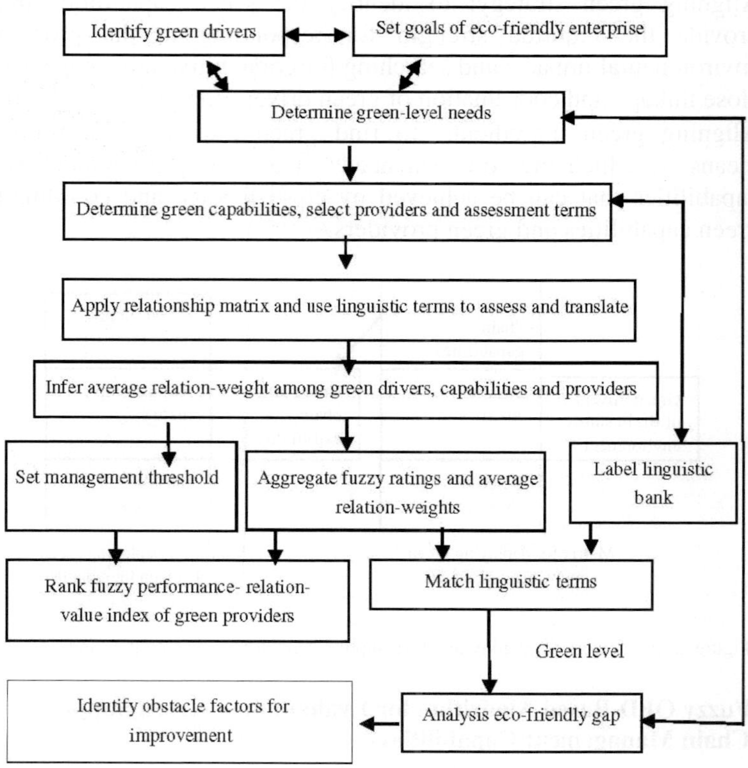

Figure 3. A method for evaluating and achieving eco-friendly enterprise

4. Numerical Example

Electronic AU company continues to face challenges with how they manage the competitive advantages of GSCM practices and the firm has to follow the green aspects to develop their criteria from a competitive market. To achieve a large-scale global and extremely eco-friendly-enterprise, AU company applied the fuzzy logic-QFD eco-friendly-enhancing model to evaluate their GSCM capability:

1. Identify green drivers, determine green capabilities and select green providers for assessment: ten green drivers, seven green capabilities and fourteen green providers were selected for assessment.

2. Determine the preference scale for measurement: seven linguistic terms and their associated membership functions for important levels and merit ratings assessment; five linguistic terms and their associated membership functions for relationship-levels assessment. (Appendix Table 1)
3. The relationship matrix was applied and linguistic terms were used to assess the green attributes and relationship-levels. (Appendix Table 2)
4. Analyze the fuzzy average relation-weight in the relationship matrix.
5. Fuzzy weighted average was used to aggregate the fuzzy ratings and fuzzy average relation-weights into an FGSCMCI. FGSCMCI = (0.37, 0.56, 0.75)
6. Match FGSCMCI with an appropriate linguistic level: as shown in Figure 4.
7. Analysis and suggestions.

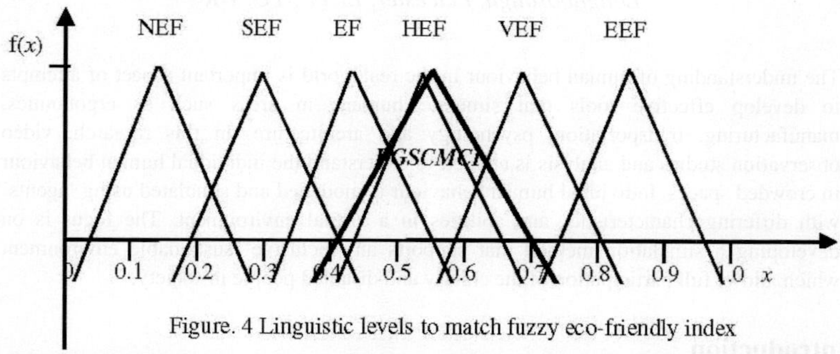

Figure. 4 Linguistic levels to match fuzzy eco-friendly index

{Non-Eco-Friendly [NEF] (0, 0.15, 0.3); Slightly Eco-Friendly [SEF] (0.15, 0.3, 0.45); Eco-Friendly [EF] (0.3, 0.45, 0.6); Highly Eco-Friendly [HEF] (0.4, 0.55, 0.7); Very Eco-Friendly [VEF] (0.55, 0.7, 0.85); Extremely Eco-Friendly [EEF] (0.7, 0.85, 1.0)}

5. Discussion and Conclusions

According to the comments from case study, this approach resolves some of the problems in traditional strategic business planning methods, having several advantages compared to previous methods:

- This method provides a structured procedure for green operational strategy planning, not only green measurement.
- This method gives the analyst more convincing, informative and reliable results.

References

1. A. Paulraj, Environmental motivations: a classification scheme and its impact on environmental strategies and practices. Business Strategy and the Environment, 18 (7), 453–468 (2009).
2. Y. S. Chen, S. B. Lai and C. T. Wen, The influence of green innovation performance on corporate advantage in Taiwan, Journal of Business Ethics, 67(4), 331-339(2006). **(References available upon request from authors)**

AGENT-BASED MODELLING AND SIMULATION THROUGH VIDEO OBSERVATION ANALYSIS

SHAHROL MOHAMADDAN[†], KEITH CASE
Mechanical and Manufacturing Engineering, Loughborough University
Loughborough, Leicester, LE11 3TU, UK

The understanding of human behaviour in the real world is important aspect of attempts to develop effective tools that simulate humans in areas such as ergonomics, manufacturing, transportation, psychology and architecture. In this research, video observation studies and analysis is applied to understand the individual human behaviour in crowded spaces. Individual human behaviour is modelled and simulated using 'agents' with differing characteristics and abilities in a virtual environment. The focus is on developing a simulation method that supports an 'inclusive' sustainable environment which allows full participation of the elderly and disabled people in society.

1. Introduction

Fire hazards, accidents, earthquakes and other emergencies are examples of phenomena that always threaten human live. Such phenomena become major issues when they occur in crowded spaces such as stadia, spiritual places, train stations and other public facilities. Considerable research has been conducted to understand the phenomena and increase crowd safety through methods such as human crowd modelling. Most of this research focuses on human movement and behaviour during the evacuation process, especially in emergency situations. In the evacuation process, the humans in the crowd is approached from a macroscopic level where the movement and behaviour is considered to be similar to the movement of fluid [1], gases [2] and the movement of a school of fish or a flock of birds responding to external stimuli.

Macroscopic approaches have limitations in representing the human in crowd as individuals with different characteristics and abilities. The research work in [1-3] are examples of the application of the macroscopic approach. However, the current research is focused on a microscopic approach where individual human behaviours and interactions within the crowd are considered. For example, different types of activity can be found within the crowd at train

[†] S.Mohamaddan@lboro.ac.uk

station such as queuing for ticket machines, social interaction, moving with luggage and shopping. In order to model and simulate these kinds of behaviour within the microscopic approach, agent-based modelling is applied.

The agent-based modelling is a 'bottom up' computational model that simulates individual humans as 'agents' within the virtual environment. Work in [4] is an example of agent-based modelling. In the current research, heterogeneous characteristics of humans, particularly for older and disabled people, is developed as part of major research project known as AUNT-SUE (**A**ccessibility and **U**ser **N**eeds in **T**ransportation for **S**ustainable **U**rban **E**nvironments) [5].

2. Conceptual Behaviour

In order to acquire relevant information about individual human behaviour through video observation, the conceptual behaviour is developed as a foundation for what is going to be observed and analysed. In this way, better insight into human behaviour can be achieved whilst avoiding any confusion during the observation process due to the large volume of movement and behaviour within the recorded video. Conceptual behaviour is considered as a set of behaviours that are acquired from the major human movements of free movement, opposite direction movement and same direction movement.

Free movement is considered as a basic human movement from one point of interest to another point of interest. In free movement, the human is considered to have enough space to move without facing any obstacle or constraint during the movement. There are two types of behaviours that can occur during free movement; Moving Through Behaviour and Move-Stop-Move behaviour. Moving Through is behaviour where the human just move through the observation area without any specific concern. The only difference that can be observed between humans is the speed of movement. Move-Stop-Move is the behaviour where humans have to stop a few seconds or a few minutes to perform some other related task during the movement such as buying a ticket, using a mobile phone, waiting for someone or looking for information.

Opposite direction refers to the condition where a human meets another human travelling in the opposite direction, for example when moving along a pathway. Upon meeting each other during the movement there are two possible behaviours that can occur; Avoiding Behaviour and the Passing Through behaviour. Avoiding refers to the behaviour where the human gives way to other people or performs an avoidance route. On the other hand, Passing Through refers to the behaviour where humans do not consider others or their surroundings during the movement. The major concern is to reach the targeted area within the planned route. Same direction movement consists of Queuing Behaviour and Competitive Behaviour. The Queuing and Competitive Behaviours can easily be observed when humans enter or exit a door. Queuing Behaviour is the manifestation of self-organization where individual humans create a line to enter or exit the door. On the other hand, in Competitive Behaviour humans have less consideration for each other and each individual tries to enter or exit the door as quickly as possible.

Figure 1. The observation area

3. Observational Studies and Analysis

The observation studies were conducted to understand and analyse the individual human movement and behaviour in crowded spaces. The observation studies were conducted at one of the largest transportation hubs in Kuala Lumpur, Malaysia. For this research, the observation studies are focusing on the exit door as shown highlighted in yellow in Figure 1. The selected exit door is a crowded interchange location that shows multi-directional movement with humans moving to the airport train and other public transportation areas for buses and taxis. In addition, there also a ticket machine, a stall and a shop located in the exit door area.

The observation studies were conducted since the method supplies more information on 'real' movement and behaviour of individual humans when compared with questionnaires and laboratory experiments. In laboratory experiments, the incentive and early instruction given to the participant can influence their psychological condition, and there are many factors that affect the real world traffic so that for example different crowd is observed at train station and shopping malls. Both conditions cannot be observed under laboratory conditions. In this research, a video camera is used to record the 'real' human movement and behaviour. The video camera is situated at the opposite side of the exit door and provides an aerial view of the human activities.

Table 1. Number of subjects by behavior

No.	Behaviour	No. of Older People	No. of Disabled	No. of Adult	No. of Others	Total
	Free Movement					
1.	Moving Through (MT)	97	6	3,444	63	3,610
2.	Move-Stop-Move (MSM)	17	0	260	4	281
	Opposite Direction Movement					
3.	Avoiding (AB)	1	0	109	0	110
4.	Passing Through (PT)	6	0	56	2	64
	Same Direction Movement					
5.	Queuing (QB)	0	0	82	0	82
6.	Competitive (CB)	2	0	23	0	25
	Total	123	6	3,974	69	**4,172**

The observation studies were conducted during the weekend and weekday, one hour each starting from 7.30 am to 8.30 am in the morning. In total 4,172 humans were observed (1,152 during the weekend period and 3,020 during the weekday period). The start and end times of the observations were determined as the peak-time on a weekday when people start using the transportation services to travel to the workplace and leisure time during the weekend. There are four different types of subject that form the focus for the studies, the elderly, disabled, adult and others (children or workers that moving around with trolley and cleaning devices). The number of subjects by behaviours is shown in Table 1. Based on the two hours video observation analysis, 95.3 per cent of the subjects or 3,974 people are adult. Disabled people are the group of subjects that are rarely observed with only 6 people or 0.1 per cent in the two hours recording. In general, most of the human behavior was free movement especially during the weekend when there are small numbers of humans and bigger spaces to move around.

4. Behavioural Model

The observational studies and analysis shows that there are several factors affecting individual human behaviour within the crowded spaces. Based on the limited two hours observation, there are four factors that are considered to affect human behavior; personal objectives, field of view, speed of movement and personal space. The four factors are the basis on which the behavioural model or set of rules for the simulation were constructed.

4.1. *Personal Objective*

Personal objective refers to the goal that human has during the movement. The observation studies and analysis shows that humans have different types of personal objectives and it can be observed based on the physical appearance within the recorded video, that although there is no direct communication between individuals in the crowd. For example, there is the human moving with luggage or backpack, looking for information, buying a ticket at the ticket machine and other activities. The personal objective is the main reason that determines human decision making and behaviour during movement in crowded spaces.

4.2. *Field of View*

Field of view (FOV) refers to the extent of the observable range seen by a human at any given moment, where in the human FOV is around 120 to 180 degrees [6]. FOV is important since humans develop their perception of the environment during the movement primarily from visual data. The importance of FOV is observed during the Move-Stop-Move Behaviour as shown where human movement becomes slower when looking for information during the journey. There is the possibility of performing the Move-Stop-Move Behaviour in order to acquire persistent information so that a personal objective is achieved. In general, older and disabled people having a smaller FOV compared to young and able-bodied which is a contributory cause of the difficulty older and disabled people have when moving within the crowded spaces.

4.3. *Speed of Movement*

Speed of movement refers to the walking speed of humans, classified into slow, average, fast and extra fast [7]. Humans are observed to have slow or average speed of movement when performing the Moving Through and Move-Stop-Move Behaviours. However, when performing the Competitive and Passing Through Behaviour humans are observed to have faster (fast or extra fast) speed of movement.

4.4. *Personal Space*

Personal space refers to the invisible surrounding or territory around human beings. Personal space is considered as a social condition that affects communication or contact between humans. The personal space is divided into four categories known as the Intimate, Personal, Social Consultative and Public Zones which are based on the distance between humans within the public spaces. In general, older and disabled people have a larger personal space compared to the young and able-bodied. The workers with trolleys, families with children in strollers, humans with luggage and humans in group are observed to need a greater personal space.

5. Simulation

In this paper, the simulation conducted is focused on two types of movement which are the opposite direction movement and same direction movement. The simulation is conducted using the DarkBasic Professional software. Individual humans are represented by an agent in circular shape with different colours representing adults, older and disabled people which have differing characteristics and abilities. Each agent is programmed to move to the selected objective with different FOV, speed of movement and personal space similar to the real world movement and behaviour. Work in [7] shows an example of simulation based on London Paddington station.

6. Conclusion

This paper presents the development of agent-based modelling and simulation system through video observational studies and analysis based on a conceptual behaviour model. Video observational studies and analysis provide a better input towards understanding human behaviour in real life situations. The simulation conducted offers valuable understanding of human behaviour and environments which is potentially valuable to the planners and designers. However, the observational studies conducted are limited and there is a need to increase the observational time in order to get better insight of human behaviour. A full-scale case study for the research is also need to be conducted in order to test the parameters.

References

1. P. Thompson and E. Marchant, *Safety Science*, 18, 277 (1995).
2. D. Helbing, M. Isobe, T. Nagatani and K. Takimoto, *Physics Review E*, 67, 067101 (2003).
3. L. Yang, D. Zhao, J. Li and T. Fang, *Building and Environment*, 40(3), 411 (2005).
4. F. Cherif and R. Chighoub , *Journal of Computing*, 2(4), 48 (2010).
5. AUNT-SUE, *http://www.aunt-sue.info/*, (2011).
6. S. Sukhatme, *http://www.artinarch.com/vp05.html/*, (2011).
7. K. Case, M. Freebody, S. Mohamaddan and S.T. Syed Shazali, in *Proceedings of the 9th International Conference on Manufacturing Research*, edited by D.K. Harrison, B.M. Wood and D. Evans (2011), p. 31.

A BAYESIAN FRAMEWORK FOR UNCERTAINTY FORMULATION OF ENGINEERING DESIGN PROCESS

M. RAJABALINEJAD[†], C. SPITAS

Faculty of Industrial Design Engineering, TUDelft, 2628CE Delft, the Netherlands

Uncertainties in the design process are investigated in this paper. A formal Bayesian method is presented for designers to quantify uncertainties in design process. The uncertainties are implemented in a decision support system that plays a key role in design of complex projects where a large and multidisciplinary team of engineers are involved. The proposed method produces the probability distribution function of the model score or uncertainty. Two example applications for design and empirical demonstrate application of the method.

1. Introduction

In this paper, a formal approach is provided to formulate the uncertainties of the design process by integration of the prior information and data. These are important elements in the design process since designers start the design with a program of requirements which include a large amount of uncertainty [1, 2]. Then, data and information are used to reduce uncertainty and to improve the prior information of the designer. As a result, implementation of a technique that can integrate the prior information of the designer and the accumulated data to assess the overall uncertainty of the problem is of great value to design process in system analysis and product development. For this purpose, we use a Bayesian method based on the global uncertainty concept [3, 4]. This robust concept helps us to unify different uncertainties into one. In other words, the uncertainties sourced from experts, designers, users and clients are unified using the concept of the global uncertainty. These results are used for a decision support tool which leads a designer to robust product design [5].

The use of graphs and drawings is preferred to effectively communicate with users. Because designers are usually strong in visual communication, we embrace visual communication with designers to present the uncertainties or confidence visually. Having this said, we used a simple table for this purpose with a scale factor ranging from 0 to 100. This is shown in Figure 1. The values

[†] email: M.Rajabalinejad@tudelft.nl

of 0 and 100 represent two extreme limits respectively, for no chance of acceptance, and absolutely satisfied criterion for a certain concept. Figure 1(b) represents a concept that satisfies a certain criterion between 20% and 80%.

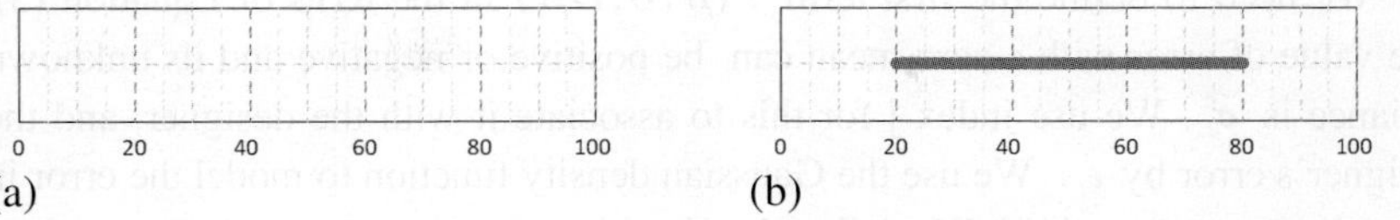

Figure 1. (a) Survey form for an expert to present his opinion within a confidence interval. (b) The confidence interval presented by the expert.

Assume that there are a number of criteria i for a few given concepts. Given these, we wish to make a rational decision by choosing between the competing concepts. The expert opinion for each concept is given by a set of random variables with Gaussian distribution, $s = \{s^{(1)}, s^{(2)}, ..., s^{(i)}\}$. On the basis of the rational frame work, we consider the first moment of these random variables as the scoring function [4, 6]. Assuming different weight factor for each criteria, we use the following equation to rank the candidate concepts and select the concept with the highest score:

$$\mu = \frac{1}{\sum_{k=1}^{i} \lambda_k} \left[\sum_{k=1}^{i} \lambda_k \int \frac{1}{\sqrt{2\pi}\sigma^{(k)}} \exp\left(\frac{\left(x - \mu\left(s^{(k)}\right)\right)^2}{2\left(\sigma^{(k)}\right)^2} \right) x\,dx \right], \tag{1}$$

where μ is the given score (rank) to the j-th concept, i is the total number of criteria, λ_i is the weight factor of criterion i.

2. Mathematical Formulation

Consider a model score function μ that we wish to estimate based on some prior information, I, and data, D. The prior information is the result of the designer's knowledge, while the data comes out of the interview with stakeholders, experts, users, etc. Let $P(\mu \mid D, I)$ be the univariate probability density function (pdf) representing the model score as a random variable. The data D and prior information I are obtained from the designers, experts or interviews (surveys). Let σ be the global uncertainty which relates all local uncertainties. The global uncertainty σ (e.g., global standard deviation) was first used in [7-9] to define a nuisance random variable. With marginalization and the global uncertainty, we can write

$$P(\mu \mid D, I) = \int P(\mu, \sigma \mid D, I)\, d\sigma. \tag{2}$$

By application of Bayes Theorem, we find

$$P(\mu \mid D, I) = \int P(\mu \mid \sigma, D, I)\, P(\sigma \mid D, I) d\sigma. \tag{3}$$

We need to define the first term $P(\mu \mid \sigma, D, I)$ in the RHS of Equation (3). The value of error with a zero mean can be positive or negative and its unknown variance is σ^2. We use index j for this to associate it with the designer, and the designer's error by e_j. We use the Gaussian density function to model the error in a standard error form [10]. We define the Gaussian error as

$$P(e_j \mid \sigma_j) = \frac{1}{\sqrt{2\pi}\sigma_j} \exp\left\{-\frac{1}{2\sigma_j^2} e_j^2\right\}, \tag{4}$$

where e_j is the error and σ_j^2 is the unknown variance. This error represents the difference between the model score μ and the designer's estimate $\hat{\mu}_j$, $e_j = \mu - \hat{\mu}_j$. By making the change of variable from e_j to μ, we obtain the following multivariate pdf for the pixels μ as:

$$P(\mu \mid \sigma_j, D, I) = \frac{1}{\sqrt{2\pi}\sigma_j} \exp\left\{-\frac{1}{2\sigma_j^2} (\mu - \hat{\mu}_j)^2\right\}. \tag{5}$$

Following the approach used in [7], we can associate the global and local uncertainties by defining

$$\sigma_j = \alpha_j \sigma, \tag{6}$$

where $\alpha_j = \sqrt{|\mu - \hat{\mu}_j|}$. We apply the Gaussian distribution in Equation (4) to determine the error of the i-th measured data point e_i with unknown variance σ_i^2 and a mean value $\hat{u}_i$ obtained from the collected data, D. For n data points, the error of the model score for any of the observed data points may be defined as:

$$e_i = \mu - d_i, \tag{7}$$

where $i = 1, \ldots, n$. Assuming the error in Equation (7) has a zero mean (e.g., the error may have positive or negative values) with an unknown variance σ_i^2, it can be concluded that

$$P(\sigma \mid \mu, d_1, \ldots, d_n) = \frac{1}{(2\pi)^{n/2} (B\sigma^n)} \exp\left\{-\frac{1}{2\sigma^2} \sum_{i=1}^{n} \frac{(d_i - \mu)^2}{\alpha_i^2}\right\}, \tag{8}$$

where B is

$$B = \alpha_1 \cdots \alpha_n. \tag{9}$$

Multiplication of Equations (5) and (8) according to Equation (3) and integration over σ gives the posterior $P(\mu \mid D, I)$ as

$$P(\mu \mid D, I) \propto \frac{1}{B'}\left[\frac{\dfrac{(\mu-\hat{\mu}_j)^2}{\alpha_j^2}}{\sum_{i=1}^{n}\left\{\dfrac{(d_i-\mu)^2}{\alpha_i^2}\right\}}+1\right]^{-\frac{1}{2}n}, \tag{10}$$

where $B' = B\alpha_j$. The numerical algorithm best for implementing these formulas is summarized next.

3. Application Example: COLD FACTS Project

The Cold Facts is a program of the Dutch World Wide Fund (WWF Netherlands) established or climate change studies in the Polar Regions to build a reliable and lightweight weather station deployed in North Pole. The design team for this project is composed of 15 designers from different faculties of TUDelft. Decision-making in the design process was a challenge at the managerial level because of the experience level and the diversity of designers' knowledge. This challenge was explicitly addressed through a series of reports and presentations.

To clarify the implementation aspect of our method, we selected three criteria for selecting the final concept. These criteria were: reliability, aesthetics, and light weight. We considered these to be the most important requirements of the model. In the next step, Concept 1 and Concept 2 were presented to an expert, asking for his expert opinion. The designer's prior knowledge for either of the concepts is the same as given in Figure 2. For the demonstration purpose, we present here the results of integration of three different experts with a designer's opinion on either of the concepts. These are shown in Figure 4, representing the probability density function (pdf) of the model score. Having calculated the first moment of these pdfs, the best concept can now be chosen.

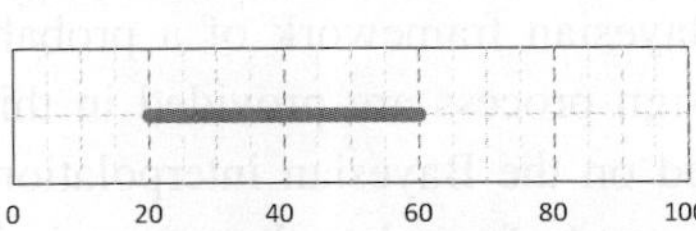

Figure 2. The designer's opinion on the reliability of either of the concepts.

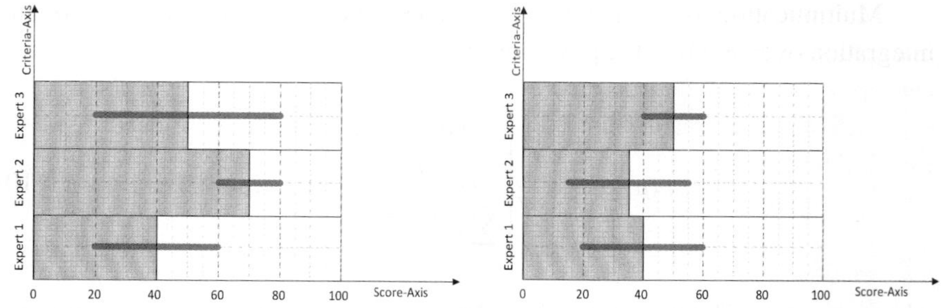

(a) (b)

Figure 3. (a) The reliability of Concept 1 is scored by three different experts. (b) The reliability of Concept 2 is scored by three different experts.

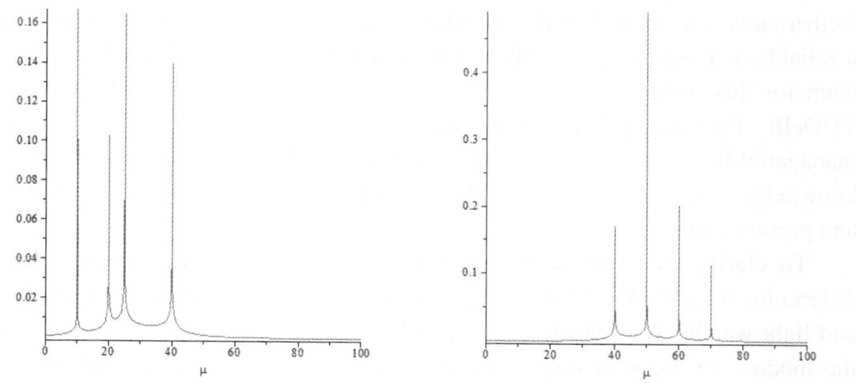

(a) (b)

Figure 4. This figure presents two different probability density functions assigned to the concepts for the Cold Facts project.

4. Conclusions

A detailed description of the theoretical formulation and numerical implementation for a Bayesian framework of a probabilistic approach and its integration with the design process are provided in this paper . The proposed hybrid approach is based on the Bayesian interpolation method. We present a methodology that enables us to formulate the uncertainties of different elements in design process and integrate the prior information with the data. This is important in the design process because at each stage, there is a prior information, which needs to be integrated to the prior information in order to reduce the uncertainty and make more educated decisions.

The application of graphical tool is demonstrated here in a numerical example for the Cold Facts project. This project was chosen because of an interdisciplinary team of designers and experts who had different level of experience and expertise. The proposed and applied method relies on experts' and designers' knowledge, and integrates these uncertainties into the decision-making process.

References

1. Rajabalinejad, M. and C. Spitas, *Incorporating Uncertainty into the Design Management Process.* Design Management Journal, 2011: p. 11.
2. Spitas, C., *Analysis of systematic engineering design paradigms in industrial practice: Scaled experiments.* Journal of Engineering Design, 2011. **22**(7): p. 447-465.
3. Rajabalinejad, M. and T. Mahdi, *The inclusive and simplified forms of Bayesian interpolation for general and monotonic models using Gaussian and Generalized Beta distributions with application to Monte Carlo simulations.* Natural Hazards, 2010. **55**(1): p. 29.
4. Rajabalinejad, M., *Reliability Methods for Finite Element Models*. I ed2009, Amsterdam, the Netherlands: IOS Press.
5. Du, X. and W. Chen, *Efficient uncertainty analysis methods for multidisciplinary robust design.* AIAA journal, 2002. **40**(3): p. 545-581.
6. Cover, T.M. and J.A. Thomas, *Elements of information theory*2004: Wiley Online Library.
7. Rajabalinejad, M., *Reliability Methods for Finite Element Models*. I ed2009, Amsterdam, the Netherlands: IOS Press.
8. Rajabalinejad, M., P. van Gelder, and N. van Erp, *Application of Bayesian Interpolation in Monte Carlo Simulation*, in *Safety, Reliability and Risk Analysis (ESREL)*2008, Taylor and Francis Group, London, UK. p. 705-713.
9. Rajabalinejad, M., *Bayesian Monte Carlo method.* Reliability Engineering and System Safety, 2010. **95**(10): p. 1050-1060.
10. Jaynes, E.T., *Probability theory, the logic of science.* 2003.

QUANTUM GATES AND ACTIVE SET THEORY OF AGENTS

GERMANO RESCONI
Dept. of Mathematics and Physics, Catholic University Brescia, I-25121, Italy

CHRIS HINDE
Department of Computer Science, Loughborough University,
Leicestershire LE11 3TU UK.

An active set of agents is a unifying space being able to act as a "bridge" for transferring information, ideas and results between distinct types of uncertainties and different types of logic. An active set is a set of agents who independently deliver true or false values for a given proposition. Difference between an ordinary or fuzzy set and active set is that the ordinary set has passive elements with values of the attributes defined by an external agent, In the active set any element is an agent that internally defines the value of a given attribute for a passive element. The active set is a set of agents in epistemic modal logic. The difference between active set and modal logic is that all the agents (worlds) in the modal logic are separate but in active set are joined in the evaluation of the given proposition. The agents of the active set evaluate propositions includes the Hilbert multi-dimensional space where is possible to simulate quantum logic gate.

1. Introduction

In the papers [1, 2, 7] and in the paper "Belief merging and voting" [6] we found that in the voting model there are several conflicting demands/preferences and we are looking for a collective compromise. Voting is concerned with the aggregation of individual preferences in order to select a collectively preferred alternative. This problem is extensively studied by social choice theory [1, 2, 7]. Probably the most famous method for the aggregation of preferences is the one proposed in the 18th century by the Marquis de Condorcet [3]. Given a set of individual preferences, we compare each of the alternatives in pairs. For each pair, we determine the winner by majority voting, and the final collective ordering is obtained by a combination of all partial results. Unfortunately, this method led to the first aggregation problem, known as the Condorcet paradox: the pair wise majority rule can lead to cycles in the collective ordering. In other words, this ordering cannot be used to select an overall preferred candidate.

In [4] given a fuzzy set defined in a space X, let P be a representative population of persons. For a given x that belongs to X each member of P is asked to accept or reject x as satisfying the fuzzy set condition. A binary decision must be made. Now for a group of 10 persons, Equation 1, the voting pattern could then be

$$X = \{a, b, c, d, e, f, g\} \tag{1}$$

with a voting pattern such as in Table 2,

$$\begin{bmatrix} 1 & 2 & 3 & 4 & 5 & 6 & 7 & 8 & 9 & 10 \\ a & a & a & a & a & a & a & a & a & a \\ & & & b & b & b & b & b & & b \\ & & & c & & c & & c & & c \\ & & & d & & d & & & & \end{bmatrix} \tag{2}$$

would result in the fuzzy set 3,

$$f = \{1.0 \mid a + 0.7 \mid b + 0.5 \mid c + 0.2 \mid d\} \tag{3}$$

Now in the voting pattern we allow repetitions of the elements in X in a way to agree to the fuzzy set definition, we can order the elements so that the highest count is to the left and the counts decrease towards the right, see Table 4

$$\begin{bmatrix} 1 & 2 & 3 & 4 & 5 & 6 & 7 & 8 & 9 & 10 \\ a & a & a & a & a & a & a & a & a & a \\ b & b & b & b & b & b & b & & & \\ c & c & c & c & & & & & & \\ d & d & & & & & & & & \end{bmatrix} \tag{4}$$

If the elements are ordered and we grant that any voter who votes for b will vote for a then we get a unique representation. The new freedom for which we can have many voting patterns opens the door to introduce not only min, max logic rules but many more logic operations. Now this extension is denoted Active set model of agents where we fuse the classical vote model with all other types of uncertainty and logics.

2. Properties and definition of the Active Set

Any active set is a set of Superpose Agents, any agents give a value for the same proposition p. Active set appear similar to the Epistemic logic evaluation but the difference is connected with the superposition of the worlds or agents which judgment is not related to one agent but to the set of agents.

$$\Omega(p)=\begin{pmatrix} Agent & 1 & 2 & 3 \\ Logic\text{ value} & true & true & true \end{pmatrix},\begin{pmatrix} Agent & 1 & 2 & 3 \\ Logic\text{ value} & true & true & false \end{pmatrix},\ldots.$$

$$\ldots\ldots,\begin{pmatrix} Agent & 1 & 2 & 3 \\ Logic\text{ value} & false & false & false \end{pmatrix}$$

In a more formal way we have

$$SS(p)=\left\{A,\Omega(p)|A=\text{set of agents},\Omega(p)=\text{ power set }2^{A}\text{ of the evaluations}\right\}$$

Given the proposition p, we denote as Criteria C one of the possible evaluation p in the set Ω(p). For example with three agents we have eight criteria to evaluate the proposition itself so we can write

$$\Omega(p,C_1)=\begin{pmatrix} Agent & 1 & 2 & 3 \\ Logic\text{ value} & true & true & true \end{pmatrix},\ldots,$$

$$\ldots\ldots,\Omega(p,C_8)=\begin{pmatrix} Agent & 1 & 2 & 3 \\ Logic\text{ value} & false & false & false \end{pmatrix}$$

We remark that the set of Criteria is a mathematical lattice. The agents set A is an ordinary set with normal intersection union and complementary operator. For the logic evaluation we have two different operations.

1) Operation among criteria for the same proposition. Because we have the same proposition with two different criteria, we cannot compose the logic values that are heterogeneous. So we have the rule

$$C_i\oplus C_j=\begin{pmatrix} Agent & 1 & 2 & \ldots & n \\ C_i & v_{1,1} & v_{1,2} & \cdots & v_{1,n} \\ C_j & v_{2,1} & v_{2,2} & \cdots & v_{2,n} \end{pmatrix}$$

So we increase the dimension of the space of the evaluation. For example given ten agents and two criteria we have

$$\Omega(p,C_i,C_j)=\begin{pmatrix} Agents & 1 & 2 & 3 & 4 & 5 \\ p,C_i & f & f & t & t & f \\ p,C_j & t & t & f & t & f \end{pmatrix}$$

2) For two different propositions p and q we have the composition rule for the active set

$$\Omega(p \wedge q, C) = \begin{pmatrix} Agents & 1 & \ldots & n \\ p,C & v_{1,p} & \ldots & v_{n,p} \end{pmatrix} \wedge \begin{pmatrix} Agents & 1 & \ldots & n \\ q,C & v_{1,p} & \ldots & v_{n,p} \end{pmatrix}$$

$$= \begin{pmatrix} Agents & 1 & 2 & \ldots & n \\ p,q,C & v_{1,p} \wedge v_{1,q} & v_{2,p} \wedge v_{2,q} & \ldots & v_{n,p} \wedge v_{n,q} \end{pmatrix}$$

$$\Omega(p \vee q, C) = \begin{pmatrix} Agents & 1 & \ldots & n \\ p,C & v_{1,p} & \ldots & v_{n,p} \end{pmatrix} \vee \begin{pmatrix} Agents & 1 & \ldots & n \\ q,C & v_{1,p} & \ldots & v_{n,p} \end{pmatrix}$$

$$= \begin{pmatrix} Agents & 1 & 2 & \ldots & n \\ p,q,C & v_{1,p} \vee v_{1,q} & v_{2,p} \vee v_{2,q} & \ldots & v_{n,p} \vee v_{n,q} \end{pmatrix}$$

For the negation operator we have

$$\Omega(\neg p) = \begin{pmatrix} Agents & 1 & \ldots & n \\ value & \alpha_1(\neg v_1)+(1-\alpha_1)(v_1) & \ldots & \alpha_n(\neg v_n)+(1-\alpha_n)(v_n) \end{pmatrix}$$

3) Aggregation rule for active set

Given an active set we associate to any active set evaluation a number by an aggregation function that can be linear or non linear. For the linear case the aggregation can be simple aggregation or can be weighted aggregation. For example for simple linear aggregation rule we have the aggregation rule

$$for\ \Omega(p, C_1) = \begin{pmatrix} Agent & 1 & 2 & 3 \\ Logic\ \text{value} & true & true & true \end{pmatrix},$$

$$\mu(p, C_1) = \frac{1}{3}\left|true\right\rangle + \frac{1}{3}\left|true\right\rangle + \frac{1}{3}\left|true\right\rangle = \frac{1}{3} + \frac{1}{3} + \frac{1}{3} = 1$$

Where Q is the linear superposition of the logic value for the active set.

3. Quantum gate and Active Set of agents

The Mutual Exclusive (ME) principle [5] states that an object cannot be in two different positions at the same time. In quantum mechanics, we have *non-local phenomena* for which the same particle can be in many different positions at the same time. This is a clear violation of the ME principle. The non-locality is essential for quantum phenomena. Given proposition p = "The particle is in position x in the space", agent g_i can say that p is true, but agent g_j can say that the same p is false.

Individual states of the particle include its position, momentum and others. The *complete state* of the particle is a *superposition* of the quantum states of the particle w_i at different positions x_i in the space, This superposition can be a global wave function:

$$\psi = w_1|x_1\rangle + w_2|x_2\rangle + ... + w_n|x_n\rangle$$

where $|x_i\rangle$ denotes the state at the position. If we limit our consideration by an individual quantum state of the particle and ignore other states then we are in Mutual Exclusion situation of the classical logic (the particle is at x_i or not). In the same way in AUT, when we observe only one agent at the time the situation collapses to the classical logic that may not adequately represent many-valued logic properties. Having multiple states associated with the particle, we use Active Set as a way to model the multiplicity of states or agents. This situation is very complex because measuring position x_i of the particle changes the superposition value ψ. However, this situation can be modelled by the first order conflict, because each individual agent has no self-conflict. The modern concept of the quantum computer is based on following statements:

1. Any state as $|n\rangle$ is a field of complex number on the reference space (position and time). In our notation, the quantum state is a column of values for different points (objects).
2. Any combination of states $|nm\rangle = |n\rangle|m\rangle$ is a product of fields.
3. The space H is the space which dimension is the number of elementary fields or states.
4. Any classical bit became a field of uncertainty or a qubit.
5. Any space of 2^m -1 dimensions is represented by H and UH is the unitary transformation of H by which computations are made in quantum physics. Given the unitary transformation

$$U = \begin{bmatrix} 1 & 0 & 0 & 0 \\ 0 & 1 & 0 & 0 \\ 0 & 0 & 0 & 1 \\ 0 & 0 & 1 & 0 \end{bmatrix}$$

With two states $|1\rangle = true, |0\rangle = false$ we have

$$B = \begin{bmatrix} & W_1 & W_2 & W_3 & W_4 \\ p & false & true & false & true \\ q & false & false & true & true \end{bmatrix}$$

and

$$H = B^T = \begin{bmatrix} false & false \\ true & false \\ false & true \\ true & true \end{bmatrix} \qquad (5)$$

with $2^2 = 4$ qubits we have the quantum XOR

$$\mathrm{UH} = \begin{bmatrix} 1 & 0 & 0 & 0 \\ 0 & 1 & 0 & 0 \\ 0 & 0 & 0 & 1 \\ 0 & 0 & 1 & 0 \end{bmatrix} \begin{bmatrix} false & false \\ false & true \\ true & false \\ true & true \end{bmatrix} = \begin{bmatrix} false & false \\ false & true \\ true & true \\ true & false \end{bmatrix} = \begin{bmatrix} p & p\ XOR\ q \end{bmatrix} \qquad (6)$$

By the quantum transformation U.

4. Conclusion

The quantum gate use superposition of quantum states to speed up the computation process. At the moment there is a lack of knowledge of the superposition phenomena and also we have no formal description of the superposition itself. In this paper we suggest a formal description of the superposition by active set theory. So Active set theory includes the superposition as logic conflict that generate fuzziness, inconsistency and many valued logic in physics. In this way we can give a formal description of the many quantum paradoxes as wave and particle paradox, particle entanglement paradox for which two particles are one and so on. In quantum mechanics we can not solve the paradox because we use by classical logic and classical set theory but with the new logic in the Active set we can solve the paradox.

References

1. Arrow, K.: Social choice and individual values: second edition. Cowles Foundation Monograph Series (1963).
2. Arrow, K., Sen, A., Suzumura, K.: Handbook of Social Choice and Welfare. Elsevier (2002).
3. Baker, K.: Condorcet: From Natural Philosophy to Social Mathematics. University of Chicago Press (1975).
4. Baldwin, J., Martin, T., Pilsworth, B.: Fril - Fuzzy and Evidential Reasoning in Artificial Intelligence. Research Studies Press Ltd, Taunton UK (1995).
5. D'Espagnat, B. (1999) Conceptual Foundation of Quantum mechanics, second edition, Perseus Books.
6. Gabbay, D., Rodrigues, O., Russo, A.: Revision by Translation. Kluwer (1999).
7. Sen, A.: Collective Choice and Social Welfare. Holden Day (1990)

MIXTURE AUGMENTED LAGRANGE MULTIPLIER METHOD FOR TENSOR RECOVERY*

HUACHUN TAN[†,1], BIN CHENG[1], JIANSHUAI FENG[1], GUANGDONG FENG[1], WUHONG WANG[1], YU-JIN ZHANG[2]

[1]*Department of Transportation Engineering, Beijing Institute of Technology, Beijing 100081, P.R. China*

[2]*Department of Electronic Engineering, Tsinghua University, Beijing 100084, P.R. China*

The problem of data recovery in multi-way arrays (i.e., tensors) arises in many fields such as image processing and computer vision, etc. In this paper we propose a scalable and fast algorithm for recovering a low-n-rank tensor with an unknown fraction of its entries being arbitrarily corrupted. In the new algorithm, the tensor recovery problem is formulated as a mixture convex multilinear RPCA optimization problem by minimizing a sum of the nuclear norm and the ℓ_1-norm. The problem is well-structured in both the objective function and constraints. We apply augmented Lagrange multiplier method which can make use of the good structure for efficiently solving this problem.

1. Introduction

A tensor is a multidimensional array. It is the higher-order generalization of vector and matrix.1. In the real world, nearly all of the existing high-dimensional real world data either has the natural form of tensor (e.g. multi-channel images and videos) or can be grouped into the form of tensor (e.g. tensor face2.). For some reasons, one wants to capture the underlying low-dimensional structure of the tensor data or seeks to detect the irregular sparse patterns of the tensor data, such as image compression3. 3. , foreground segmentation4. and saliency detection.5. As a consequence, it is desirable to develop algorithms that can capture the low-dimensional structure or the irregular sparse patterns in the high-dimensional tensor data.

In the matrix case, the "rank" and "sparsity" are the most useful tools for matrix-valued data analysis. Wright et al.6 demonstrated that if the irregular sparse matrix S is sufficiently sparse, one can accomplish the sparse and low-rank recovery by solving the following convex optimization problem:

---

* The research in this chapter was supported by NSFC-61171118, NSFC-91120015, NSFC-50878023, and Beijing Natural Science Foundation (4122067).

* E-mail: tanhc@bit.edu.cn

$$min_{L,S} : \|L\|_* + \mu\|S\|_1, \qquad s.t.\ A = L + S, \tag{1}$$

This optimization method is called the Robust Principal Component Analysis (RPCA)**Error! Reference source not found.** or the Principal Component Pursuit (PCP) due to its ability of exactly recovering the underlying low-rank matrix even in the presence of being corrupted by large entries or outliers.

Although the low-rank matrix recovery problem has been well studied, there is not much work on tensors. In this paper, a new algorithm for low-n-tensor recovery, which is termed as Mixture Augmented Lagrange Multiplier Method for Tensor Recovery (MALM-TR), is proposed. In the new algorithm, we convert the tensor recovery problem into a mixture convex optimization problem. It can be treated as a multilinear extension of the RPCA problem and subsumes the matrix RPCA problem as a special case. Then we extend the augmented Lagrange multiplier method (ALM)7. '8. to solve the multilinear RPCA problem.

2. MALM-TR

In this section, the low-n-tensor recovery problem is converted into a multilinear RPCA problem. And we present the details of the proposed algorithm to solve this problem.

2.1. *The multilinear RPCA problem*

N-mode tensors are denoted as $\mathcal{A} \in \mathbb{R}^{I_1 \times I_2 \times \ldots \times I_N}$. Its elements are denoted as $a_{i_1 \ldots i_k \ldots i_n}$. The mode-n unfolding of a tensor $\mathcal{A} \in \mathbb{R}^{I_1 \times I_2 \times \ldots \times I_N}$ is defined as $unfold(\mathcal{A}, n) = A_{(n)}$. Accordingly, its inverse operator fold can be defined as $fold\left(A_{(n)}, n\right) = \mathcal{A}$. The n-rank of a N-dimensional tensor $\mathcal{A} \in \mathbb{R}^{I_1 \times I_2 \times \ldots \times I_N}$, denoted by r_n, is the rank of the mode-n unfolding matrix $A_{(n)}$.The inner product of two same-size tensors $\mathcal{A}, \mathcal{B} \in \mathbb{R}^{I_1 \times I_2 \times \ldots \times I_N}$ is defined as the sum of the products of their entries $\langle \mathcal{A}, \mathcal{B} \rangle = \sum_{i_1} \sum_{i_2} \cdots \sum_{i_N} a_{i_1 \ldots i_k \ldots i_n} b_{i_1 \ldots i_k \ldots i_n}$. Moreover, denote $\|\mathcal{A}\|_F = \sqrt{\langle \mathcal{A}, \mathcal{A} \rangle}$ and $\|\mathcal{A}\|_1 := \sum_{i_1 \ldots i_k \ldots i_n} \left| a_{i_1 \ldots i_k \ldots i_n} \right|$ as the Frobenius norm and ℓ_1 norm of a tensor. It is clear that $\|\mathcal{A}\|_F = \left\|A_{(n)}\right\|_F$ and $\|\mathcal{A}\|_1 = \left\|A_{(n)}\right\|_1$ for any $1 \leq n \leq N$.

As tensors are the generalizations of the matrix concept, we generalize the recovery problem from the matrix (i.e. 2nd-order tensor) case to higher-order tensors by solving the following optimization problem:

$$min_{\mathcal{L},\mathcal{S}} : \|\mathcal{L}\|_* + \mu\|\mathcal{S}\|_1, \qquad s.t.\ \|\mathcal{A} - \mathcal{L} - \mathcal{S}\|_F^2 \leq \varepsilon_1, \tag{2}$$

where $\mathcal{A}, \mathcal{L}, \mathcal{S}$ are n-mode tensors with identical size in each mode. $\mathcal{A}$ is the given tensor to be recovered; $\mathcal{L}$ is the low-n-rank component; $\mathcal{S}$ is the sparse component; Liu et al.9. have proposed the definition of the nuclear norm of an n-mode tensor as $\|\mathcal{X}\|_* = \frac{1}{n}\sum_{i=1}^{n}\|X_{(i)}\|_*$. Based on this definition, the optimization in Eq. (2) can be written as:

$$min_{\mathcal{L},\mathcal{S}} : \|\mathcal{L}\|_* + \eta\|\mathcal{S}\|_1 \equiv \frac{1}{n}\sum_{i=1}^{n}\lambda_i\|L_{(i)}\|_* + \frac{1}{n}\sum_{i=1}^{n}\eta_i\|S_{(i)}\|_1,$$

$$s.t.\ \|A_{(i)} - L_{(i)} - S_{(i)}\|_F^2 \le \varepsilon_1. \tag{3}$$

Problem (3) is still hard to solve due to the interdependent trace norm and ℓ_1 norm constraint. In order to simplify the problem, we introduce additional auxiliary matrix $M_{(i)} = L_{(i)}$, $N_{(i)} = S_{(i)}$. Then, we relax the equality constrains by $\|L_{(i)} - M_{(i)}\|_F^2 \le \varepsilon_2$ and $\|S_{(i)} - N_{(i)}\|_F^2 \le \varepsilon_3$. Finally, we get the relaxed form of Eq. (3) which can be seen as a multilinear RPCA problem:

$$\min_{\mathcal{L},\mathcal{S}} \frac{1}{n}\sum_{i=1}^{n}\lambda_i\|M_{(i)}\|_* + \frac{1}{n}\sum_{i=1}^{n}\eta_i\|N_{(i)}\|_1,$$

$$\text{s.t. } \|A_{(i)} - L_{(i)} - S_{(i)}\|_F^2 \le \varepsilon_1, \|L_{(i)} - M_{(i)}\|_F^2 \le \varepsilon_2, \|S_{(i)} - N_{(i)}\|_F^2 \le \varepsilon_3. \tag{4}$$

2.2. *Optimization process*

It is observed that Eq. (4) is well-structured and the separable structure emerges in both the objective function and constraint conditions. We convert Eq. (4) into its augmented Lagrange form with proper α_i, β_i and γ_i . The augmented Lagrange of Eq. (4) is

$$\min_{\mathcal{L},\mathcal{S},M_{(i)},N_{(i)}}: \sum_{i=1}^{n}\lambda_i\|M_{(i)}\|_* + \sum_{i=1}^{n}\eta_i\|N_{(i)}\|_1 + \sum_{i=1}^{n}\langle Y_i, L_{(i)} - M_{(i)}\rangle + \sum_{i=1}^{n}\frac{\alpha_i}{2}\|L_{(i)} - M_{(i)}\|_F^2 + \sum_{i=1}^{n}\langle Z_i, S_{(i)} - N_{(i)}\rangle + \sum_{i=1}^{n}\frac{\beta_i}{2}\|S_{(i)} - N_{(i)}\|_F^2 + \sum_{i=1}^{n}\langle W_i, A_{(i)} - M_{(i)} - N_{(i)}\rangle + \sum_{i=1}^{n}\frac{\gamma_i}{2}\|A_{(i)} - M_{(i)} - N_{(i)}\|_F^2 \tag{5}$$

Eq. (5) can be simplified into its equivalent form:

$$\min_{\mathcal{L},\mathcal{S},M_{(i)},N_{(i)}}: \sum_{i=1}^{n}\lambda_i\|M_{(i)}\|_* + \sum_{i=1}^{n}\eta_i\|N_{(i)}\|_1 + \sum_{i=1}^{n}\frac{\alpha_i}{2}\|M_{(i)} - L_{(i)} - Y_i/\alpha_i\|_F^2 + \sum_{i=1}^{n}\frac{\beta_i}{2}\|N_{(i)} - S_{(i)} - Z_i/\beta_i\|_F^2 + \sum_{i=1}^{n}\frac{\gamma_i}{2}\|M_{(i)} + N_{(i)} - A_{(i)} - W_i/\gamma_i\|_F^2 \tag{6}$$

The core idea of solving the optimization problem in Eq. (6) is to optimize a group of variables while fixing the other groups.

Computing $M_{(i)}$: The optimal $M_{(i)}$ with all other variables fixed is the solution to the following sub-problem:

$$\min_{M_{(i)}}: \lambda_i\|M_{(i)}\|_* + \frac{\alpha_i}{2}\|M_{(i)} - L_{(i)} - Y_i/\alpha_i\|_F^2 + \frac{\gamma_i}{2}\|M_{(i)} + N_{(i)} - A_{(i)} - W_i/\gamma_i\|_F^2 \tag{7}$$

As shown in [10], the optimal solution of Eq. (7) is given by

$$M_{(i)} = U_i D_{\lambda_i/\alpha_i+\gamma_i}(\Lambda) V_i^T \tag{8}$$

where $U_i \Lambda V_i^T$ is the singular value decomposition given by

$$U_i \Lambda V_i^T = \left(\alpha_i\left(L_{(i)} + Y_i/\alpha_i\right) + \gamma_i\left(A_{(i)} - N_{(i)} + W_i/\gamma_i\right)\right)/(\alpha_i + \gamma_i), \tag{9}$$

and $\mathcal{D}_\tau$ is the "shrinkage" operation. The "shrinkage" operator $\mathcal{D}_\tau(x)$ with $\tau > 0$ is defined as $\mathcal{D}_\tau(x) = x - \tau, if\ x > \tau; \mathcal{D}_\tau(x) = x + \tau, if\ x < -\tau;\ \mathcal{D}_\tau(x) = 0, otherwise$. The operator can be extended to the matrix or tensor case by performing the shrinkage operator towards each element.

Computing $N_{(i)}$: The optimal $N_{(i)}$ with all other variables fixed is the solution to the following sub-problem

$$\min_{N_{(i)}} : \eta_i \left\|N_{(i)}\right\|_1 + \frac{\beta_i}{2}\left\|N_{(i)} - S_{(i)} - Z_i/\beta_i\right\|_F^2 + \frac{\gamma_i}{2}\left\|M_{(i)} + N_{(i)} - A_{(i)} - W_i/\gamma_i\right\|_F^2. \tag{10}$$

By the well known ℓ_1 norm minimization[11], the optimal solution of Eq. (10) is

$$N_{(i)} = \mathcal{D}_{\eta_i/\beta_i+\gamma_i}\left(\left(\beta_i\left(S_{(i)} + Z_i/\beta_i\right) + \gamma_i\left(A_{(i)} - M_{(i)} + W_i/\gamma_i\right)\right)/(\beta_i + \gamma_i)\right). \tag{11}$$

Computing $S_{(i)}, L_{(i)}$ respectively: The optimal $\mathcal{S}, \mathcal{L}$ with all other variables fixed are the solutions to the following two sub-problems:

$$\min_{\mathcal{S}} : \sum_{i=1}^n \frac{\beta_i}{2}\left\|N_{(i)} - S_{(i)} - Z_i/\beta_i\right\|_F^2, \quad \min_{\mathcal{L}} : \sum_{i=1}^n \frac{\alpha_i}{2}\left\|M_{(i)} - L_{(i)} - Y_i/\alpha_i\right\|_F^2 \tag{12}$$

It is easy to show that the solutions to the two sub-problems are given by

$$\mathcal{S} = \sum_{i=1}^n \beta_i fold\left(N_{(i)} - \frac{Z_i}{\beta_i}, i\right)/\sum_{i=1}^n \beta_i \quad \mathcal{L} = \sum_{i=1}^n \alpha_i fold\left(M_{(i)} - \frac{Y_i}{\alpha_i}, i\right)/\sum_{i=1}^n \alpha_i \tag{13}$$

3. Experiments

In this section, using both the numerical simulations and the real world data, we evaluate the performance of our proposed algorithm and then compare the results with the rank sparsity tensor decomposition (RSTD) algorithm[12] on the low-n-rank tensor recovery problem. For comparing with RSTD, we also use the difference of $\mathcal{L}$ and $\mathcal{S}$ in successive iterations against a certain tolerance as the stopping criterion. All the experiments are conducted and timed on the same desktop with an Pentium (R) Dual-Core 2.50GHz CPU that has 4GB memory, running on Windows 7 and Matlab.

In numerical simulations experiments, the N-way Tensor Toolbox is used to generate the low-n-rank tensor $\mathcal{L}$ and the sparse tensor $\mathcal{S}$. The simulated tensor used in the experiments is size of $40 \times 40 \times 40$, varying the n-rank r and the sparse ratio *spr*. The entries of sparse tensor $\mathcal{S}$ are independently distributed, each taking on value 0 with probability *1-spr*, and each taking on impulsive value with probability *spr*. The input tensor $\mathcal{A}$ is generated as $\mathcal{A} = \mathcal{L} + \mathcal{S}$. The quality of recovery is measured by the relative square error (*RSE*) to $\mathcal{L}$, which is defined as $RSE = \|\hat{\mathcal{L}} - \mathcal{L}\|_F / \|\mathcal{L}\|_F$. Table 1 presents the average results (across 10 instances) for different sparse ratio. The results demonstrate that our proposed algorithm MALM-TR outperforms RSTD on either efficiency or accuracy.

Table 1: Correct recovery for random problems of size 40×40×40. Left: the n-rank of the data is set as [5, 5, 5] and the spr varies from 5% to 35%. Right: the n-rank of the data is set as [10, 10, 10] and the spr varies from 5% to 20%.

spr	MALM-TR RSE (e-5)	MALM-TR Time (s)	RSTD RSE (e-5)	RSTD Time (s)
0.05	0.01	17.7	450	19.5
0.15	0.06	19.5	910	27.4
0.25	1.1	29.1	1510	44.4
0.35	2010	41.9	5620	44.1

spr	MALM-TR RSE (e-5)	MALM-TR Time (s)	RSTD RSE (e-5)	RSTD Time (s)
0.05	0.2	29.4	400	33.8
0.10	10.1	43.5	520	51.3
0.15	120	62.5	650	68.5
0.20	1250	88.3	2650	88.1

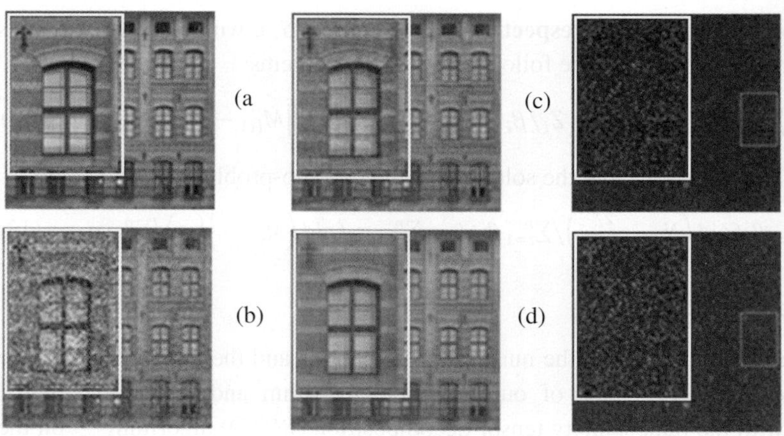

Figure 1. Comparisons in terms of visual effects. (a) the original image. (b) image corrupted by 25% sparse impulsive noise. (c) recovered image by our algorithm. (d) recovered image by RSTD. (e) recovered impulsive noise by our algorithm. (f) recovered impulsive noise by RSTD .

One straightforward application of our algorithm is the image restoration. In the real world data experiments, we apply the algorithm on image restoration of the façade image, which is also used in9. [, 12.] . We add 25 percent of random impulsive noise to the image and compare MALM-TR with RSTD. The results produced by both of the algorithms are shown in Figure 1.

4. Conclusion

In this paper, we extend the matrix recovery problem to low-n-rank tensor recovery and propose an efficient algorithm based on mixture augmented Lagrange multiplier method. In the future, we would like to investigate how to automatically choose the parameters in our algorithm and develop more efficient method for tensor recovery problem. Also we will explore additional applications of our method in computer vision and graphics.

References

1. T.G. Kolda and B.W. Bader, Tensor decompositions and applications, *SIAM Review*, 51(3): 455-500, (2009).
2. M.A.O. Vasilescu and D. Terzopoulos, Multilinear analysis of image ensembles: Tensorfaces, *European Conference on Computer Vision (ECCV)*, 447-460, (2002).
3. A. Lewis and G. Knowles, Image compression using the 2-d wavelet transform, *IEEE Transactions on Image Processing,* 1: 244 –250, (1992).
4. Y. Sheikh and M. Shah, Bayesian modeling of dynamic scenes for object detection, *IEEE Transactions on Pattern Analysis and Machine Intelligence,* 27: 1778-1792, (2005).
5. L. Itti, C. Koch and E. Niebur, A model of saliency-based visual attention for rapid scene analysis, *IEEE Transactions on Pattern Analysis and Machine Intelligence,* 20: 1254-1259, (1998).
6. E. J. Candès, X. Li, Y. Ma and J. Wright, "Robust principal component analysis?" *preprint*, (2009).
7. Z. Lin, M. Chen, L. Wu and Y. Ma, The augmented Lagrange multiplier method for exact recovery of a corrupted low-rank matrices, *Mathematical Programming*, (2009).
8. D. Bertsekas, Constrained Optimization and Lagrange Multiplier Method, *Academic Press,* (1982).
9. J. Liu, P. Musialski, P. Wonka and J. Ye. Tensor completion for estimating missing values in visual data. *International Conference on Computer Vision (ICCV),* (2009).
10. J. F. Cai, E. J. Candes and Z. Shen, A singular value thresholding algorithm for matrix completion. *SIAM J. on Optimization*, 20(4): 1956–1982, (2008).
11. E.T. Hale, W. Yin and Y. Zhang, Fixed-point continuation for ℓ_1-minimization: Methodology and convergence. *SIAM Journal on Optim*, 19: 1107–1130, (2008).
12. Y. Li, J. Yan, Y. Zhou and J. Yang, Optimum Subspace Learning and Error Correction for Tensors. *The 11th European Conference on Computer Vision (ECCV)*, Crete, Greece, (2010).

ASYMPTOTIC RESULTS FOR THE ERGODIC DISTRIBUTION OF AN INVENTORY MODEL WITH FUZZY DEMANDS*

TAHIR KHANIYEV†

Department of Industrial Engineering, TOBB University of Economics and Technology, Ankara, Turkey,
Institute of Cybernetics, Azerbaijan National Academy of Science, Baku, Azerbaijan

CİHAN AKSOP

Science and Society Department, The Scientific and Technological Research Council of Turkey, Ankara, Turkey,
Department of Statistics, Gazi University, Ankara, Turkey

BAŞAK GEVER

Department of Industrial Engineering, TOBB University of Economics and Technology, Ankara, Turkey

In this paper, a renewal-reward process which represents a fuzzy inventory model of type (s, S) with a discrete interference of chance is investigated. By utilizing the expression of the ergodic distribution of this process in terms of a fuzzy renewal function, asymptotic expansions for the α-cuts of this ergodic distribution is obtained when the interferences have a Weibull distribution with a fuzzy parameter.

1. Introduction

Renewal–reward processes and their modifications are very important stochastic processes with various applications in finance, inventory control and many other fields. On this topic, there exist very important studies in the literature when the whole system is modeled with a crisp distribution (see [1, 2, 6]). From a practical viewpoint, the exact values of the parameters of the underlying distribution are rarely known. Although these values can be estimated by using various parameter estimation methods, there can be fuzziness in the dataset (see [4, 11]). Therefore we need to combine randomness and fuzziness to obtain a better description of the data.

* This study was partially supported by TUBITAK (110T559 coded project).
† tahirkhaniyev@etu.edu.tr.

Recently, renewal-reward processes with fuzzy parameters are studied under various assumptions (see [3, 10, 12] and the references therein). But inventory model of type (s, S) with discrete interference of chance are not studied enough. These kinds of processes (for the crisp distributions) have been studied with various additional assumptions (see, [8, 9]). Recently, in Gever [7], these processes are studied under the assumption that some parameters of the process are fuzzy numbers. In this study, we obtain asymptotic expansion for the ergodic distribution of these kind processes when the demands have fuzzy characteristics. Following section gives a mathematical construction of studied process. In section 3, asymptotic expansion for the ergodic distribution of this process is presented, and in the last section some discussions are given.

2. Mathematical Construction of the Process X_t

Let ξ_n and η_n, $n \geq 1$ be independent random variables where ξ_n is interpreted as the inter-arrival time between consecutive demands, and η_n is interpreted as the amount of nth demand. Let denote the times with τ_n, when the demands cannot be met. Define ζ_n as the unmet part of the demand which occurs at time τ_n. Moreover, let $\Phi(t) \equiv P\{\xi_n \leq t\}$, $F(x) \equiv P\{\eta_n \leq x\}$, $n = 1,2,\ldots$ and η_n has a Weibull distribution with parameters (r, β); that is, $F(x) = W_{r,\beta}(x) = 1 - \exp\{-(\beta x)^r\}, x, \beta, r > 0$.

Denote the time of the occurrence of the nth demand by T_n, and total amount of demands till the time T_n with S_n; that is, $T_0 = S_0 = 0, T_n = \sum_{i=1}^{n} \xi_i, S_n = \sum_{i=1}^{n} \eta_i$, $n = 1,2,\ldots$ Moreover, let λ be an arbitrary positive constant, $\zeta_0 = z \geq 0$, $N_0 = 0$, and

$$N_n \equiv N_n(\lambda\zeta_{n-1}) = inf\{k \geq N_{n-1} + 1: \lambda\zeta_{n-1} - (S_k - S_{N_{n-1}}) < 0\}, \quad (1)$$

$$\zeta_n \equiv \zeta_n(\lambda\zeta_{n-1}) = |\lambda\zeta_{n-1} - (S_{N_n} - S_{N_{n-1}})|. \quad (2)$$

Now, the desired stochastic process X_t can be defined as follows:

$$X_t = \sum_{n=0}^{\infty}[\lambda\zeta_n - (S_{\nu(t)} - S_{N_n})]I_{[\tau_n,\tau_{n+1})}(t), \quad (3)$$

where, $I_{(A)}(t)$ denotes the indicator function of the set A. This process is called "renewal – reward process with generalized reflecting barrier". A realization of this process is given in Figure 1.

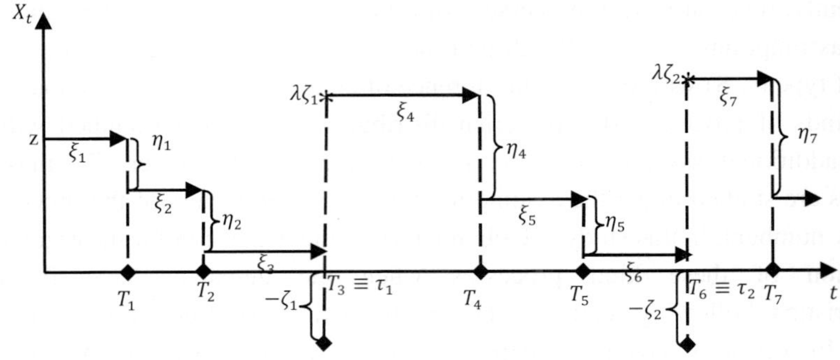

Figure 1. A realization of the process X_t.

3. Main Results

In this paper, one of the parameters of the Weibull distribution is assumed to be a fuzzy number. Before giving the main results, let define the membership function $\mu_{\tilde{\beta}}(x)$ of the fuzzy number $\tilde{\beta}$ as follows:

$$\mu_{\tilde{\beta}}(x) = \begin{cases} \mu_{1\beta}(x), & \beta_1 \le x < \beta_2 \\ \mu_{2\beta}(x), & \beta_2 \le x < \beta_3 \\ 0, & otherwise. \end{cases} \tag{4}$$

Here $0 < \beta_1 < \beta_2 < \beta_3 < \infty$, $\mu_{1\beta}(x)$ is a monotone increasing function on $[\beta_1, \beta_2]$, $\mu_{2\beta}(x)$ is a monotone decreasing function on $[\beta_2, \beta_3]$ and $\mu_{1\beta}(\beta_2) = \mu_{2\beta}(\beta_2) = 1$, $\mu_{1\beta}(\beta_1) = \mu_{2\beta}(\beta_3) = 0$. To obtain the asymptotic expansions for the α – cuts of ergodic distribution of the process X_t, we need the following theorem from Gever [7].

Theorem 3.1. (Gever [7]) Let $\xi_1, \eta_1 > 0$ have finite first moment, and η_1 be a non-arithmetic random variable. Moreover, assume that η_1 has a Weibull distribution with parameters $(r, \tilde{\beta})$, where $r > 0$ and $\tilde{\beta}$ is a fuzzy number with a membership function defined as in (4). Then, α-cuts of the ergodic distribution $\tilde{Q}(x)$ of the process $\tilde{X}_t$ can be expressed as follows, when $\lambda \to \infty$:

$$\left[\tilde{Q}(x)\right]^{\alpha} \approx [Q_1(\alpha, x); Q_2(\alpha, x)], \quad \alpha \in [0, 1). \tag{5}$$

Here

$$Q_1(\alpha, x) = \left[1 - \frac{\int_x^{\infty} U_{r,1}\left(\lambda(z-x)\mu_{2\beta}^{-1}(\alpha)\right)\bar{W}_{r,1}\left(z\mu_{1\beta}^{-1}(\alpha)\right)dz}{\int_0^{\infty} U_{r,1}\left(\lambda z\mu_{1\beta}^{-1}(\alpha)\right)\bar{W}_{r,1}\left(z\mu_{2\beta}^{-1}(\alpha)\right)dz}\right]^{+} \tag{6}$$

$$Q_2(\alpha, x) = 1 - \frac{\int_x^\infty U_{r,1}\left(\lambda(z-x)\mu_{1\beta}^{-1}(\alpha)\right)\overline{W}_{r,1}\left(z\mu_{2\beta}^{-1}(\alpha)\right)dz}{\int_0^\infty U_{r,1}\left(\lambda z\mu_{2\beta}^{-1}(\alpha)\right)\overline{W}_{r,1}\left(z\mu_{1\beta}^{-1}(\alpha)\right)dz}, \tag{7}$$

and $U_{r,1}(x)=\sum_{n=0}^{\infty} W_{r,1}^{*n}(\beta x)$, $\overline{W}_{r,1}(x) \equiv 1 - W_{r,1}(x)$.

Lemma 3.1. (Feller [5, page 366]) Assume that η_1 is a non-arithmetic random variable with finite first two moments $(m_k \equiv E[\eta_1^k] < \infty, k = 1,2)$. Then, the renewal function $U_{r,1}(x)$ can be written as follows, when $x \to \infty$:

$$U_{r,1}(x) = \frac{x}{m_1} + \frac{m_2}{2m_1^2} + o(1). \tag{8}$$

Lemma 3.2. Let $g: \mathbb{R} \to \mathbb{R}$ be a bounded function with $g(x) \to 0$ as $x \to \infty$. Then, the following limit holds for all $x \geq 0$,

$$lim_{\lambda\to\infty} \int_x^\infty g\left(\lambda c_1(z-x)\right) exp\{-c_2 z^r\}\, dz = 0, \quad c_1, c_2 > 0. \tag{9}$$

Proof. Let $M = \sup_{x\geq 0} g(x)$, $y = c_1(z-x)$ and $\varepsilon > 0$. Then,

$$\left|\int_x^\infty g\left(\lambda c_1(z-x)\right) exp\{-c_2 z^r\}\, dz\right| \leq \int_0^{\delta(\varepsilon)} |g(\lambda y)|\, exp\left\{-c_2\left(\frac{y}{c_1}+x\right)^r\right\}\frac{dy}{c_1} + \int_{\delta(\varepsilon)}^\infty |g(\lambda y)|\, exp\left\{-c_2\left(\frac{y}{c_1}+x\right)^r\right\}\frac{dy}{c_1} \tag{10}$$

Here, $\delta(\varepsilon) = \sup\left\{\delta > 0: \int_0^\delta \exp\{-c_2(y/c_1 + x)^r\}\frac{dy}{c_1} \leq \varepsilon/K\right\}$, and K is a fixed integer which will be defined later. Let $z(\varepsilon) = \inf\left\{z > 0: \sup_{u\geq z}|g(u)| \leq \frac{\varepsilon}{K}\right\}$. Then, we can choose λ so large such that $\lambda\delta(\varepsilon) \geq z(\varepsilon)$ holds. Then, we get;

$$\int_0^{\delta(\varepsilon)} |g(\lambda y)|\, exp\{-c_2(y/c_1 + x)^r\}\frac{dy}{c_1} \leq M\frac{\varepsilon}{K} \tag{11}$$

and $\int_{\delta(\varepsilon)}^\infty |g(\lambda y)| \exp\left\{-c_2\left(\frac{y}{c_1}+x\right)^r\right\}\frac{dy}{c_1} \leq \frac{\varepsilon}{K}\int_0^\infty \exp\left\{-c_2\left(\frac{y}{c_1}+x\right)^r\right\}\frac{dy}{c_1} = \frac{\varepsilon}{K}$.

Thus, we have $\left|\int_x^\infty g\left(\lambda c_1(z-x)\right)\exp\{-c_2 z^r\}\, dz\right| \leq \frac{\varepsilon}{K}(M+1)$. Choosing $K = [M+1]$ yields to $\left|\int_x^\infty g\left(\lambda c_1(z-x)\right)\exp\{-c_2 z^r\}\, dz\right| < \varepsilon$. Since, this inequality is true for all $\varepsilon > 0$, we have the desired limit. □

Lemma 3.3. For all $x \geq 0$,the following asymptotic expansion holds, as $\lambda \to \infty$,

$$J(x; c_1, c_2) \equiv \int_x^\infty U_{r,1}(\lambda(z-x)c_1)\overline{W}_{r,1}(zc_2)dz \tag{12}$$

$$= \frac{c_1}{m_1}\left(\frac{c_2^{-2/r}}{r}\Gamma\left(\frac{2}{r}, c_2 x^r\right) - x\overline{W}_{r,c_2}(x)\right)\lambda + \frac{m_2}{2m_1^2}\overline{W}_{r,c_2}(x) + o(1). \tag{13}$$

Proof. By using the asymptotic expansion for $U_{r,1}(x)$ in Lemma 3.1 and then applying Lemma 3.2 yields to desired result. □

Theorem 3.2. In addition to the conditions of Theorem 3.1, let η_1 has a finite second moment. Then the following asymptotic expansion can be written for the α-cuts of the ergodic distribution $[\tilde{Q}(x)]^{\alpha}$ of the process X_t, as $\lambda \to \infty$,

$$[\tilde{Q}(x)]^{\alpha} = [Q_1(\alpha, x), Q_2(\alpha, x)], \tag{14}$$

$$Q_1(\alpha, x) = 1 - \frac{d(k_2(\alpha), k_1(\alpha); x)}{d(k_1(\alpha), k_2(\alpha); 0)} + R_{k_1(\alpha), k_2(\alpha)}(x)\frac{1}{\lambda} + o\left(\frac{1}{\lambda}\right), \tag{15}$$

$$Q_2(\alpha, x) = 1 - \frac{d(k_1(\alpha), k_2(\alpha); x)}{d(k_2(\alpha), k_1(\alpha); 0)} + R_{k_2(\alpha), k_1(\alpha)}(x)\frac{1}{\lambda} + o\left(\frac{1}{\lambda}\right), \tag{16}$$

Here $d(a, b; x) = \frac{a}{m_1}\left(\frac{b^{-2/r}}{r}\Gamma\left(\frac{2}{r}, bx^r\right) - x\overline{W}_{r,b}(x)\right)$; $k_i(\alpha) \equiv \mu_{i\beta}^{-1}(\alpha)$; $i = 1,2$ and $R_{a,b}(x) = \frac{m_2}{2m_1^2}\frac{a}{d(a,b;0)}\left(\overline{W}_{r,b}(x) - \frac{d(b,a;x)}{d(a,b;0)}\right)$.

Proof. Observe that

$$Q_1(\alpha, x) = 1 - \frac{J(x; k_2, k_1)}{J(0; k_1, k_2)}; \quad Q_2(\alpha, x) = 1 - \frac{J(x; k_1, k_2)}{J(0; k_2, k_1)} \tag{17}$$

Substituting the values of the integrals $J(x; c_1, c_2)$ from Lemma 3.3 into Eq. (17) yields to desired result.

4. Conclusion

The lack of knowledge on the parameters drives us to model the systems with fuzzy numbers. Therefore, in this paper an inventory control model of type (s, S) is investigated when the demands have a Weibull distribution with a fuzzy parameter. By using the results of Gever [7] and Feller [5], we obtain an asymptotic expansion for the ergodic distribution of the process $\tilde{X}_t$, as $\lambda \to \infty$, in Theorem 3.2. The applied methodology serves as an important tool to obtain considerable results on the fuzzy inventory theory.

References

1. Borovkov, A. A. (1976) *Stochastic Processes in Queuing Theory*, Springer-Verlag, New York.
2. Brown, M., Solomon, H. A. (1975) Second-order Approximation for the Variance of a Renewal-Reward Process, *Stochastic Processes and Their Applications*, 3, 301-314.
3. Buckley, J. J., Feuring, T., Hayashi, Y. (2002) Solving Fuzzy Problems in Operations Research: Inventory Control. *Soft Computing*, 7: 121-129.

4. Celikyilmaz, A., Turksen, B. (2009) *Modeling Uncertainty with Fuzzy Logic*. Springer-Verlag Berlin.
5. Feller, W. (1971) *Introduction to Probability and Its Applications II*, John Wiley.
6. Gihman, I. I., Skorohod, A. V. (1975) *Theory of Stochastic Processes II*, Springer, Berlin.
7. Gever, B. (2011) Investigation of the Renewal Reward Process with Generalized Reflecting Barrier by Asymptotic Methods, M.Sc. thesis, TOBB Economics and Technology University Institute of Natural and Applied Sciences.
8. Khaniyev, T., Aksop, C. (2011) Asymptotic Results for an Inventory Model of Type (s, S) with a Generalized Beta Interference of Chance, *TWMS Journal of Applied Engineering and Mathematics*, 1(2): 223-236.
9. Khaniyev, T., Atalay, K. D. (2010) On the Weak Convergence of the Ergodic Distribution of an Inventory Model of Type (s, S), *Hacettepe Journal of Mathematics and Statistics*, 39(4):599-611.
10. Khaniyev, T., Turksen, I. B., Gokpinar, F., Gever, B., Ergodic Distribution for a Fuzzy Inventory Model of Type (s,S) with Gamma Distributed Demands, *Expert Systems with Applications*, 2012 (is accepted).
11. Lee, K. H. (2005) *First Course on Fuzzy Theory and Applications*. Springer-Verlag Berlin.
12. Wang, X. (2011) Continuous Review Inventory Model with Variable Lead Time in a Fuzzy Environment. *Expert Systems with Applications*, 38, 11715-11721.

PROCESS CHAIN MODEL FOR CORPORATE SOCIAL RESPONSIBILITY PROJECTS

FUNDA ÇINAR[1], ESRA BAS

Istanbul Technical University, Department of Industrial Engineering, Macka 34367 Istanbul

Corporate Social Responsibility (CSR) has taken an increasingly important role in business. It's observed that the literature for CSR projects has more focus on some parts of the topic such as the effects of the CSR project to the financial performance of the company. In this study, the process chain model originally developed for business logistics is proposed to be adapted to CSR project life cycle for an overall success. This model is aimed to be a backbone model for determining all possible project risks throughout CSR project life cycle. The final objective of the model is to obtain an integrated view to a CSR project for maintaining its sustainability.

1. Introduction and Literature Review

Corporate social responsibility (CSR) is commitment of business to contribute to sustainable economic development, working with employees, their families, the local community and society at large to improve their quality of life (Tsai et al. (2010) [1], World Business Council for Sustainable Development (2000) [2]). Although some studies noted that the beginning of the academic interest in CSR could ascend to the 1850s, it is only in the last years that it has acquired its final position within research institutions, non-governmental organizations (NGOs), corporations, governments and agencies (Garriga and Mele (2004)) [3].

Although the concept of CSR is widely discussed in theory and practice, there is no general agreement about its definition. According to the definition of Commission of the European Communities (2001) [4], CSR is ''a concept whereby companies decide voluntarily to contribute to a better society and a cleaner environment'' by integrating "social and environmental concerns in their business operations and in their interaction with their stakeholders". Regarding CSR often described concept is its "contribution to sustainable development" (OECD (2001)) [5]. With this regard, a growing number of worldwide firms have undertaken serious efforts to integrate CSR into various aspects of their

[1] : fundacinar@itu.edu.tr

businesses (Jo and Harjoto (2011) [6], El Ghoul et al. (2011) [7]). Although the literature in CSR has been relatively rich, most papers have a narrow focus on the issue, and to the best of our knowledge, no paper provides a holistic view to CSR project life cycle. Henderson (2007) [8] indicated that CRS has a valuable contribution to the sustainability of tourism industries. Rodriguez and del Mar Armas Cruz (2007) [9] found that hotel firms' CRS projects have positive impact on their financial performance. Tsai and Hsu (2008) [10] proposed an integrated approach to solve the decision problem of CSR project selection in the airline industries. Tsai et al. (2010) [1] also developed scientific methods for CSR project selection decision and cost evaluation to overcome the deficiencies of methods developed for hospitality industry. Weber (2008) [11] explored the question how to measure the company specific value of CSR activities.

In this paper, a holistic and integrated view to CSR project life cycle is proposed, and process chain model of Baumgarten and Darkow (2002) [12] developed for business logistics is adapted to CSR project life cycle. The structure of the paper is as follows: In Section 2, process chain model of Baumgarten and Darkow (2002) [12] is briefly summarized, and its adaptation to CSR project life cycle is illustrated, and detailed. In Section 3, conclusions and discussions for future research are given.

2. Process Chain Model Adapted to CSR Project Life-Cycle

Process chain model has originally been developed for designing integrated processes of material, information, and financial flows from product development to returns processes in a company (Baumgarten and Darkow (2002)) [12]. Its adaptation to CSR project life cycle is illustrated in Figure 1. The original process chain model constitutes four main processes; namely product development, supply, order management, and returns (Baumgarten and Darkow (2002) [12], Atac (2004) [13]). Supply process includes requirements planning, purchasing, and procurement logistics; order management process comprises order entry, production planning and scheduling, production and distribution; and finally returns process covers return of product by the customer, its treatment by the company, and material flows and information flows sent back to the process cycle (Baumgarten and Darkow (2002) [12], Atac (2004) [13]). The process chain approach gives the opportunity to monitor the effects of actions in a sub-process to the other sub-process for an overall success of the project.

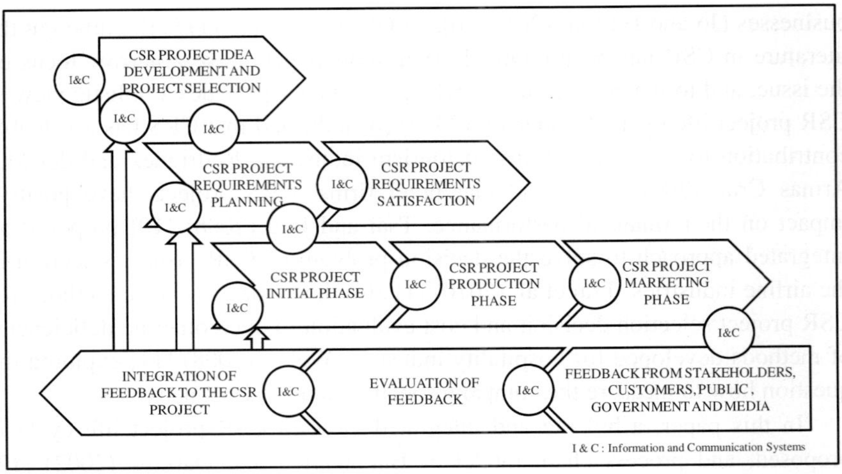

Figure 1. Process chain model adapted CSR project life cycle (adapted from Baumgarten and Darkow, 2002)

In this study, a process chain model approach is proposed for the integration of the processes throughout the CSR project life cycle. As a special project of a company, a CRS project is taken into effect not only for an overall value to the stakeholders and customers, but also to all society in the short-term and long-term horizon. Thus, the processes in a CRS project should be planned, executed, and controlled for maintaining its sustainability with the objective of maximizing overall economic effects to the company, and social effects to all society. In this model, the main processes throughout CSR project life cycle are proposed to be CSR project idea development and project selection, CSR project requirements planning and satisfaction, CSR project initial phase, production phase and marketing phase, and finally CSR feedback phase.

2.1. *CSR project idea development and project selection*

The original process chain model by Baumgarten and Darkow (2002) [12] incorporates research and development, and prototype production as the sub-processes of product development (Atac (2004)) [13]. With regards to CSR project life cycle, this phase could include understanding the expectations of the stakeholders and society needs, CSR project idea development by using the data provided from company's internal processes and analyzing CSR projects of competitors by national or global benchmarking, economic value analysis in accordance with the firm's overall goals and its corporate culture, social value analysis with the possible long-term effects to the society, and optimum selection of projects by considering limited resources of the company. Key performance indicators (KPIs) can be determined at this phase to evaluate the performance of the CSR project throughout its life cycle.

2.2. *CSR project requirements planning and satisfaction*

In the original process chain model, supply process includes all activities from material requirements planning to the availability of the purchased materials for the production (Baumgarten and Darkow (2002) [12], Atac (2004) [13]). With regards to CSR projects, it could be more relevant to view this phase as a make or buy decision, i.e. decision whether to be plan, execute, and control the CSR project by the internal resources including human resources, or outsourcing the project. In case of in sourcing, the selection of the team members, the training of the employees for overall commitment to the project should be planned, and new design of the daily processes of the firm could also be expected. Project planning including technical and financial planning is also proposed to take place in this phase.

2.3. *CRS project initial phase, production phase and marketing phase*

The original process chain model is designed for order management at this phase that incorporates all activities from the order entry to the delivery of the product to the customer (Baumgarten and Darkow (2002) [12], Atac (2004) [13]). With regards to CSR project life cycle, a CSR project is proposed to have three sub-processes, i.e. initial phase, production phase, and marketing phase. The initial phase is crucial, since some feedback could already be obtained from stakeholders, customers, public, government and media. The participation of the employees should be fostered by increasing their motivation, and external network should be built with public organizations, media, and governmental organizations for their support. The production phase could merely include material flows including primarily technical processes or mostly information flows including information and communication processes with the participants. The success of this phase also depends on the effective design of integrated marketing processes such as promotion and distribution of values from CSR project to the selected segments of the society.

2.4. *CSR project feedback phase*

Unlike a physical return of the products to the firm for a treatment as in the original model, this phase basically comprises processes of feedback collection from the stakeholders, customers, public and media, operational results such as waste reduction and reduction of resource consumption, and self-evaluation of CSR projects by KPIs. Possible negative feedback should also be considered as part of risk management. The final goal in this phase is not only the adaptation of CSR project as a result of feedbacks, but also to create new CSR projects for the sustainability.

3. Conclusions and Discussions

In this study, a process chain model for CSR project lifecycle is proposed for a holistic view to the processes. A CSR project may have considerable contributions to all society including the stakeholders, customers, public, and government. However, its success depends primarily on effective design and management of all processes from CSR project idea development to the feedback of the results from the society to the company. This model is also proposed as a backbone for CSR project risk management. A CSR project risk management is crucial, since any unexpected results from CSR project may have detrimental effect to the company's public relations and reputation. As a next research, the sub-processes of the main processes proposed as a part of process chain model of a CSR project life cycle should be detailed, and the sub-processes should be mapped to possible risks.

References

1. W.-H. Tsai., J.-L. Hsu, C.-H. Chen, W.-R. Lin, S.-P.Chen, An Integrated Approach for Selecting Corporate Social Responsibility Programs and Cost Evaluation in the International Tourist Hotel, *International Journal of Hospitality Management* 29, 385-396 (2010).
2. World Business Council for Sustainable Development, Corporate Social Responsibility: Making Good Business Sense (2000). Available at http://www.wbcsd.org/web/publications/csr2000.pdf
3. E. Garriga and D. Mele, Corporate Social Responsibility Theories: Mapping the Territory, *Journal of Business Ethics* 53,51-71 (2004).
4. Commission of the European Communities (COM), Green paper: Promoting a European framework for Corporate Social Responsibility, 366 Final, Brussels (2001). Available at: http://eur-lex.europa.eu/LexUriServ/site/en/com/2001/com2001_0366en01.pdf
5. OECD: Corporate social responsibility: Partners for progress, Paris (2001). Available at: http://www.oecd.org/officialdocuments/publicdisplaydocumentpdf/?cote=PAC/COM/PUB(2001)36&docLanguage=En
6. H. Jo and M.A. Harjoto, Corporate Governance and Firm Value: The Impact of Corporate Social Responsibility, *Journal of Business Ethics* 103, 351-383 (2011).

7. S. El Ghoul, O. Guedhami, C.C.Y. Kwok, D.R. Mishra, Does Corporate Social Responsibility Affect the Cost of Capital?, *Journal of Banking & Finance* 35, 2388-2406 (2011).
8. J.C. Henderson, Corporate Social Responsibility and Tourism: Hotel Companies in Phuket, Thailand, After the Indian Ocean Tsunami, *International Journal of Hospitality Management* 26, 228-239 (2007).
9. F.J.G. Rodriguez and Y. del Mar Armas Cruz, Relation Between Social-Environmental Responsibility and Performance in Hotel Firms, *International Journal of Hospitality Management* 26, 824-839 (2007).
10. W.-H. Tsai and J.-L. Hsu, Corporate Social Responsibility Programs Choice and Cost Assessment in the Airline Industry- A Hybrid Model, *Journal of Air Transport Management* 14, 188-196 (2008).
11. M. Weber, The Business Case for Corporate Social Responsibility: A Company-Level Measurement Approach for CSR, *European Management Journal* 26, 247– 261 (2008).
12. H. Baumgarten and I.-L. Darkow, Logistikprozesse: Management von Logistikprozessen (2002). In: H. Baumgarten, H.-P Wiendahl and J. Zentes, Logistik-Management: Strategien-Konzepte-Praxisbeispiele, Part 2.03.01, Springer Verlag, Berlin et al. (2002), pp. 1-18
13. E. Atac, Implementations of Advanced Planning and Scheduling Systems and Potentials of Logistics Service Providers, Unpublished M.Sc. Thesis, Technical University of Berlin (2004).

MULTI MODE RESOURCE CONSTRAINED PROJECT SCHEDULING PROBLEMS WITH SOLVING TABOO SEARCH

OMER ATLI[*1], CENGIZ KAHRAMAN[2]

[1]*Aeronautics and Space Technologies Institute, Turkish Air Force Academy Yesilyurt, Bakirkoy, Istanbul 34149, Turkey*

[2]*Department of Industrial Engineering, Istanbul Technical University, Istanbul Turkey*

In this paper we consider the resource-constrained project scheduling problem with multiple execution modes for each activity. The objective function is the minimization of the project completion time. Heuristics based on priority rule are considered as initial solution procedures for this problem. The proposed Taboo search algorithm (TSA) is computationally compared, the results are analyzed and discussed, and some conclusions are given. Results obtained on six classes of test problems and comparison with other algorithms from the literature show that our algorithm gives better solutions.

1. Introduction

Project scheduling problems (PSP) as the core content of project management has been extensively studied in recent decades, but its complexity for the multi modes, dynamic randomness, fuzziness and multi-objective attributes have not yet solved by a systematic method or theory. There is a big gap between the theoretical research and practical application of PSP. This paper discusses the Multi Mode Resource Constrained Project Scheduling (MM-RCPS) problems which exist widely in construction, software development, aircraft and ship manufacturing, designing a computer system and planning a military invasion, introducing a new product, and other single or small batch production mode of the enterprise. The problem of scheduling activities under resource and precedence restrictions with the objective of minimizing the project duration is referred to in the literature as the RCPSP [1]–[3]. The resource constraints refer to limited renewable resources such as manpower, materials and machines which are necessary for carrying out the project activities. It can be clearly seen that the RCPSP is a generalization of the job shop, flow shop, assembly line balancing, related scheduling problem and hence belongs to the class of NP hard problems. [4]–[6].

* atliomer@gmail.com

In the following, we briefly summarize the related literature: Drexl and Grunewald [2] suggest regret based biased random sampling approach, Slowinski et al. [3] describe a single-pass approach, a multi-pass approach, and a simulated annealing algorithm and Kolisch and Drexl [4] present a local search procedure. Özdamar [5] proposed a genetic algorithm based on a priority rule encoding. Bouleimen and Lecocq [6] suggested a simulated annealing heuristic. Sprecher and Drexl [1] developed a branch-and-bound procedure which is, according to the results obtained by Hartmann and Drexl [7], the currently most powerful algorithm for solving exactly the MM-RCPSP. Sprecher and Drexl [1] suggested using it as a heuristic by imposing a time limit. Jarboui et al. [8] investigated the particle swarm optimization to solve this challenging problem. Boctor [9, 10] presented heuristics for multi-mode problems without non-renewable resources. The differential evolution (DE) is a direct search method based on population generation. It was proposed by Storn and Pricein [11]. DE was applied to problems involving multiple criteria, as a spread sheet solver application [12]. Lorenzoni et al. [13] designed an algorithm based on an extension of the DE algorithm to solve a MM-RCPSP. The remainder of the paper is organized as follows: Section 2 describes the MM-RCPSP. Then we propose TS algorithm to the MM-RCPSP. Finally, Section 3 computational experiments result concludes the paper and future remarks for the researcher.

2. Problem Description: RCPS Model with Multiple Modes

When there are several different modes that may be selected for an activity, the problem is known as a MM-RCPSP. In this problem, activities of the project may be executed in more than one operating mode and renewable/non-renewable resource constraints exist. Each activity operation mode has a different duration and requires different amount of renewable resources. Each mode corresponds to a different time resource trade-off option for the activity under consideration. A feasible schedule specifies the implementation mode, as well as the start and finish times for each activity. The general features of the MM-RCPS model relies on some of the basic assumptions of RCPS model:

A project represented using an activity-on-node (AON) representation.
A project consists of N activities
Activity *i* may be performed in any one of the modes $j=1,\ldots,\ M_i$.
Scheduling activity *i* in mode *j* take d_{ijm} time units (duration)
Activity *i* cannot start unless all of its predecessors have been completed.
Activity preemption is not allowed.

There are K kinds of the renewable/non-renewable resources, where resource k is available in quantity $k \in K^{\nu}$ and $k \in K^{\rho}$ per period.

Each activity must be performed in a mode where each activity-mode combination has a fixed duration and requires a constant amount of one or more types of renewable resources when activity is being executed.

Mode switching is not allowed when activities are resumed after splitting.

2.1. *Heuristics for The Multi-Mode Case*

In order to solve the multi-mode RCPSP heuristically, often a procedure with two stages is applied. While in one stage the mode assignment is determined, in the other stage a corresponding schedule is calculated. If this is done in a hierarchical way (first fixing the modes), in the second stage a classical single mode RCPSP has to be solved. The taboo algorithm based on activity lists has been extended to the multimode RCPSP by introducing an additional mode list which assigns a mode to each activity.

2.2. *Starting Solution of The Resource Constraint Scheduling Based on The TSM*

In the only two kinds of methods presented in here. A simple heuristic method is generally used as the starting solution of the TS method, such as Minslack, ROT and etc. Priority given to the activity with the maximum ROT value. The ROT value is calculated by dividing the sum of each activity's resource requirements by the duration of the activity and then finding the maximum ROT value that an activity controls through the network on any one path. Schedule an activity with the minimum value of minslack where t now denotes the current clock time.

2.3. *Mode Priority Rules*

Once activity *ij* is chosen, assign mode *m* by one of the following rules for m=1,…,. M_{ij}

1. Minimum duration mode (Min-m) rule: $\min_m \; d_{ijm}$.
2. Maximum Duration Mode (Max-m) rule: $\max_m \; d_{ijm}$.
3. Minimum Resource Demand Mode (Min-rdm): $\min_m \; \sum_{r=1}^{R} r_{ijm} d_{ijm}$.
4. Maximum Resource Demand Mode (Max-rdm): $\max_m \; \sum_{r=1}^{R} r_{ijm} d_{ijm}$.
5. Choose Any Mode Number: Perform activity ij in mode m by using the constant any mode.
6. Randomly choose mode (Rand-m) rule: Randomly perform activity ij in mode m by derived from a uniform distribution from 1 to M_{ij} .

At first some superfluous modes and resources may be eliminated in a preprocessing step. A mode $m \in M_j$ of an activity j is called non-executable if its resource requirements cannot be fulfilled in any feasible schedule, i.e. if $r_{jkm}^{\rho} > R_k^{\rho}\, k$ for a renewable resource $k \in K^{\rho}$ or if $r_{jkm}^{\nu} + \sum_{j=1}^{n} \min_{\mu \in M_j} \left\{ r_{jk\mu}^{\nu} \right\} > R_k^{\nu}$ for a non-renewable resource $k \in K^{\nu}$ holds.

2.4. *Taboo Search Method and Proposed Taboo Search Algorithm*

Taboo search algorithms (TSA) was first developed by F.Glover in 1975, based on Artificial intelligence systems. Each search iteration stands for a possible solution to a problem. After several iterations, the algorithm that hopefully represents the optimal or near optimal solution to a problem can be found. The TS method, a tool to solve the optimization problem, could be expected to be applied in this research to solve the MM-RCPSP and obtain the optimal solution of this problem. A balance between exploitation and exploration in the search space is one of the important factors when using TSA. To provide this balance, determination of the design strategy for TSA parameters like maximum iteration is one of the critical issues.

2.5. *Parameters*

Memory Structure: The memory structure of Taboo list is mainly used to memorize the several precedent moves, that is, to memorize the several precedent moving paths, which is divided into longer term memory and short term memory. In this study utilizes the memory list with the short term memory.

Neighborhood: Neighbors of the current solution are generated by swapping activities. Changes the execution mode of each activity to every other executable mode for an activity; as a result, a set of neighboring solutions is created, each of them differing from the current solution.

Termination criterion: The stopping rule of the TS method is flexible. If satisfying the condition that searching objective is a criterion, the algorithm will set the stopping rule as the following two conditions, either one of the following these two conditions is satisfied and then stop calculating. Maximum iteration and the maximum number of computation on the condition of not finding better objective value.

Taboo list: The taboo list (TL) is a queue of a given length. Whenever a move is performed, its reverse is added to TL in order to avoid going back to a solution already visited. The oldest existing move is removed from the front of the list (according to the FIFO policy). All moves existing on TL are taboo.

The Taboo list size (TL): TL size is a parameter that is preset or determined experimentally. Our experiments have demonstrained that the better size of TL size should be around activity size by $\sqrt{N-2}$.

2.6. *The Procedure of Taboo Search Algorithm*

In this paper, we present a new approach based on the improved TSA for solving MM-RCPSP. In the following, we give the fundamentals of TSA. The procedure of TSA is summarized as follows:

Step 1. Identify activities in project, establish precedence relationships of all activities and construct project network.

Step 2. Read data file.

Step 3. Resource k is sufficiently available independent of the assigned modes.

Step 4. Mode priority rules.

- **Once activity *ij* is chosen, assign mode *m* by one of the following rules.**

Step 5. Use CPM and calculate project parameters from the CPM algorithm.

Step 6. Choose and use initial solution method (Minslack or ROT priority rules)

- **Define a starting feasible sequence, denoted as S with priority rules.**

Step 7. Use Taboo search algorithm (The basic steps of the proposed approach).

Step 8. Initialize variables used in Taboo search.

- **Create initial schedule for Taboo Search previous step 6.**
- **At the beginning the Taboo list is empty.**
- **Select values for Taboo list size ($\sqrt{N-2}$).**
- **Select values for stopping criteria maximum iterations.**

Step 9. Create a candidate list which consists of number of swap moves.

- **Two activities are randomly selected and the positions of these two activities are interchanged. Then proceed swap rules for the neighborhoods.**

Step 10. Choose the best admissible candidate move.

Step 11. Make the best admissible candidate move.

Step 12. Check stopping criteria.

- **Max_Iter = N*2 assigned for the standardization.**
- **If iter < Max_Iter then schedule is reported as the solution, else go to Step 13.**

Step 13. Update Taboo restrictions and aspiration criteria and go to Step 6.

3. Computational Results

In this section, we present the results of a computational experiment concerning the implementations of the three search methods for the considered MM-RCPSP. The implementations were coded and compiled in C # and ran on a standard laptop. We used the standard benchmark sets for MM-RCPSP problems with 10, 12, 14, 16, 18, and 20 non-dummy activities. The files are available in the project scheduling problem library [16].

Table 1. Solutions for the MM-RCPSP benchmark instances

		Our Method (TS)		Simulating Annealing		Partical Swarm Optinization		Differential Evolution	
Benchmark Problem	Problem instances	Average deviation	Percentage of optimal found	Average deviation	Percentage of optimal found	Average deviation	Percentage of optimal found	Average deviation	Percentage of optimal found
10	536	0,046891	95,3	0.21	96.30	0.06	99.25	0.09	99.30
12	547	0,034031	96,6	0.19	91.20	0.14	98.47	0.29	99.30
14	551	0,034199	96,58	0.92	82.60	0.44	91.11	0.63	97.60
16	550	0,033299	96,67	1.43	72.80	0.59	85.91	0.83	96.38
18	552	0,032611	96,74	1.85	69.40	0.99	79.89	1.58	94.43
20	554	0,033367	96,66	2.10	66.90	1.21	74.19	1.91	92.75

We compared our results with those obtained through the simulated annealing algorithm of [6] and the particle swarm optimization of [8] and differential evolution algorithm of [15] in Table 1. The solution determined using the TS also depends on the starting solution. In order to explore diversified paths so that a better solution may be found, one can repeat the procedure described above several times with different starting solutions. Table shows that TS strictly dominates all other alternatives. The results of the computational experiment show that TS seems to be an efficient algorithm for solving the MM-RCPSP.

References

1. A. Sprecher and A. Drexl, *EJOR*, 107:431–50 (1998).
2. A, Drexl and J. Grunewald, *IIETransactions*, 25:74–81 (1993).
3. R. Slowinski, B. Soniewicki and J. Weglarz, EJOR, 79:220–9 (1994).
4. R. Kolisch and A. Drexl, *IIEE Transactions*, 29:987–99 (1997).
5. L. Ozdamar, *IIEE Transactionson*, 29:44–59 (1999).
6. K. Bouleimen and H. Lecocq, *EJOR*, 149:268–81 (2003).
7. S. Hartmann and A. Drexl, Networks, 32:283–97 (1998).
8. B.Jarboui, N.Damak, P.Siarry and A.Rebai, *App.Math.&Comp*, 299-308 (2008).
9. F F. Boctor, *Int.l Journal of Production Research*, 31:2547–58 (1993).
10. F F. Boctor, EJOR, 90:349–61 (1996).
11. R. Storn and K. Price, *IEEE Trans. on Evolutionary Comp*. 3(1): 22–34 (1999).
12. P K. Bergey, AISAMCIS Milwaukee, USA, p.966–8 (1999).
13. L. Lorenzonil, H. Ahonen and A.G. Alvarenga, *Comp.&Ind. Eng*. 50:55–6 (2006).
14. R. Kolisch, A. Sprecher and A. Drexl, Mang. Science 41:1 693–703 (1995).
15. M. Mika, G. Waligora and J. Weglarz, *EJOR* 187 1238–1250 (2008).
16. PSPLIB from the web service (http://129.187.106.231/psplib/)

A POWER MODEL OF ENTERPRISE INTERESTS' ALLOCATION*

XIAOHONG LIU

College of Management, Southwest University for Nationalities, Chengdu, Sichuan Prov., 610041, China

It mainly includes the enterprise profit distribution, payment of wages and occupation opportunities and so on of enterprise interests' allocation. Because it relates to many types of stakeholders, such as the shareholders, managers, employees, government and banks, the allocation of business interests is a complex management. In order to achieve the purpose of the enterprise, a power model of business interests' allocation is presented in this paper. The goal function of this model is the power to maximize and the constraint is the total interests limited.

1. Introductions

It is well known that the enterprise isn't only an activity main body and cell of national economy, but also is main creators of social wealth. At the same time, it's very important for the enterprise to meet the different needs of the stakeholders, such as the distribution of profits, the payment of wages and occupation opportunity etc. Enterprise stakeholders are different types of individuals or groups who can affect the achievement of enterprise goals or be the realization of business objectives influenced, which include internal stakeholders (such as contributive person, managers, employees and so on), external stakeholders (such as suppliers, distributors, consumers and other stakeholders) and indirect (such as government, non-governmental organizations, media, community, natural environment etc.). Therefore, the enterprise is facing opportunities and challenges in the case of resources limited, especially in the course of allocation of its interests. The core of allocation of enterprise interests is how to meet the different needs of the stakeholders. The allocation of enterprise interests as a kind of inventive means plays an important role in the

* This work is supported by grant 09SZYZJ05 of the Fundamental Research Funds for the Central Universities, Southwest University for Nationalities, and supported by grant 2010GXS5D253 of National Soft Science of China, and supported by grant 2011XWD-S1202 of Southwest University for Nationalities.

management of human resources. Because of enterprise's internal and external conditions, especially its own economic constraints, it can not satisfy all stakeholders requirement of the allocation. Therefore, enterprise should achieve its main objective. Firstly, the enterprise should stimulate and mobilize the motivation and enthusiasm of its key stakeholders; secondly it should pay more attention to the oriented and dynamic action of the allocation, make the enterprise get the reasonable allocation of its interests. A power model of the allocation of enterprise interests is presented in this paper, and the basic idea of this model is that transfer the enterprise interests' allocation into the power of interest allocation, by definition the concept related to this model, the objective function of this model is established.

2. The Concept of Interest Group and Allocation Power

In order to build the model of allocation power of enterprise interests, it's necessary that the basic concepts are explained in this paper. The basis concepts of interest group and allocation power are defined.

2.1. *Interest Group*

The interest group is a crowd that has the same or similarity interest. There exist two fuzzy concepts, one is the big or small of the interest, and another is the scale of the crowd.

Set A_i is one interest group, we have $A_i = f(\tilde{P}_A, \tilde{n}_A)$, here $\tilde{P}_A$ and $\tilde{n}_A$ are fuzzy sets that denote the big or small of interest and the scale of the crowd.

In general, we use the average interest $\overline{P}_A$ and the average number of human being (n) to describe. For example, $A_i(\overline{P}_A, n_{Ai})$ denotes the interest group A_i and its average interest are $\overline{P}_{Ai}$ and the number of human being is n_{Ai}.

For the same interest group, we have $\forall x \in A_i$, $|P_{Ax} - \overline{P}_A| \le \delta$, there δ is a coefficient that is given to judge the group of interest.

In the scope of discussing enterprise S, if preset δ, there exist m interest groups, and have:

1) if $i \neq j$, then $A_i \cap A_j = \phi$

2) $A_1 \cup A_2 \cup ... \cup A_m = S$

We call $A_1, A_2, ..., A_m$ is in the condition of δ to divide S.

2.2. *Allocation Power*

The allocation power is generated by the interest groups that get interest in the current or the future gain interests. The former known as the dominant allocation power; the latter is called recessive allocation power. The allocation power is decided by the increased effectiveness and number of the crowd of the interest group.

An allocation of enterprise interests will make the current benefit of a number of interest group (Set it is m_1) increased, the future of benefit of a number of interest group (Set it is m_2) will be increased. Therefore they will support the allocation of enterprise interests by different power. The allocation power of enterprise interests is composed of dominant and recessive allocation power. We use F_P、 F_{P1} and F_{P2} to denote the general allocation power, the dominant allocation power and the recessive allocation power, therefore,

$$F_P = F_{P1} + F_{P2} \tag{1}$$

The effect of aggregation will be produced in the course of allocation of enterprise interests. We use $h_1(\sum_{j=1}^{m_1} F_{P1j}, m_1)$ and $h_2(\sum_{k=1}^{m2} F_{P2k}, m2)$ denote the result of the effect of aggregation of the dominant allocation power and the recessive allocation power respectively, set,

$$F_{P1} = \sum_{j=1}^{m_1} F_{P1j} + h_1(\sum_{j=1}^{m_1} F_{P1j}, m_1) \tag{2}$$

$$F_{P2} = \sum_{k=1}^{m_2} F_{P2k} + h_2(\sum_{k=1}^{m_2} F_{P2k}, m_2) \tag{3}$$

In (2), the factors of deciding F_{P1j} mainly are the principle of adjust encourage of enterprise reform w (it reflects the factor of increasing relatively), the utility reduced of group of interest v_P (it reflects the factor of increasing absolutely) and the number of human being of group of interest n, then $F_{P1j} = f_{P1}(w, v_{Pj}, n_j)$

In (3), the factors of deciding the F_{P2k} mainly are the principle of adjust encourage of enterprise reform w and the number of human being of group of interest n, then $F_{P2k} = f_{P2}(w, n_k)$

3. Analysis on Allocation Power of Enterprise Interests

3.1. *A Model of Allocation Power of Enterprise Interests*

The model of allocation power of enterprise interest is divided into two basic especially model. One is in the case of not newly increasing interests, then to increase some interest groups' befits and reducing another interest groups' befits, i.e. the model of power based on the present interest adjustment (it is called model 1). Another is in the case of having newly increasing interest but the distribution is not average, i.e. the model of power based on the distribution not average (called model 2). The allocation power relate to the adjustment interest about three interest groups are shown in figure 1 and figure 2 respectively.

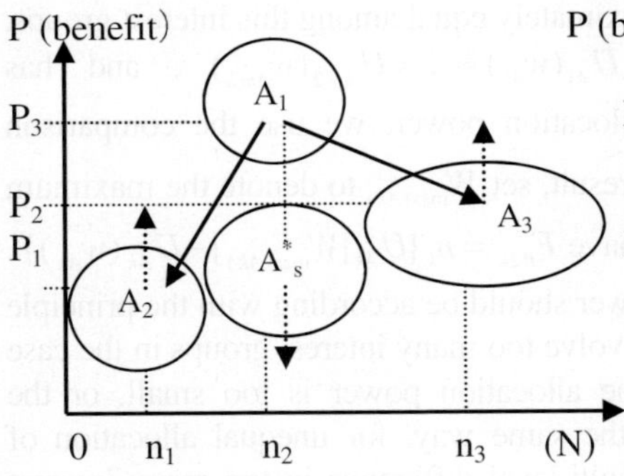

Figure 1 Model 1

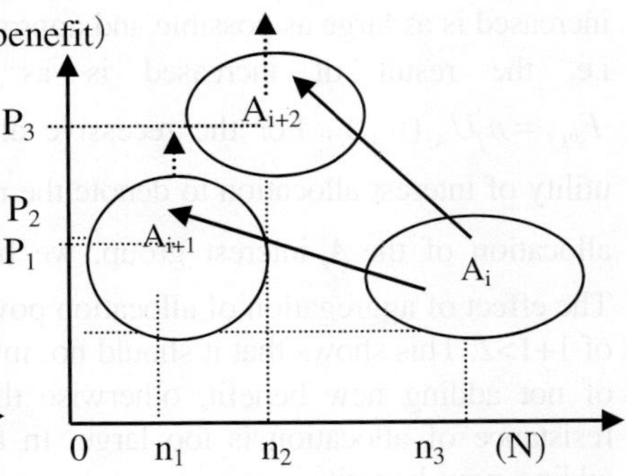

Figure 2 Model 2

In the model 1, we have

the state of reducing benefit :

$A_1(P_3, n_2) \rightarrow A_s^{\ *}(P_s, n_2)$,

the state of increasing benefit :

$A_2(P_1, n_1) \rightarrow A_i^{\ *}(P_i, n_1)$, $A_3(P_2, n_3) \rightarrow A_j^{\ *}(P_j, n_3)$,

The sources of allocation power of enterprise interests are produced by $A_2(P_1, n_1)$ and $A_3(P_2, n_3)$, and the condition is the total of benefit increased is not more than the reduced.

In the mode 2, we have

the state of increasing benefit :

$A_i(P_1, n_3) \rightarrow A_{i+1}^{\ \ *}(P_{i+1}, n_1) + A_{i+2}^{\ \ *}(P_{i+2}, n_2)$, and $P_{i+2} > P_{i+1}$

The sources of allocation power of enterprise interests are produced by $A_{i+1}(P_{i+1}, n_1)$ and $A_{i+2}(P_{i+2}, n_2)$, and the condition is the total of benefit increased is not more than the newly increased.

3.2. *Discuss on Allocation Power of Enterprise Interests*

The allocation power will be produced in two cases, one is that the benefit of the vested interest groups is increased in current, which is called absolutely allocation power, and it belongs to the dominant allocation power. Another is that the benefit of interest groups will be increased in the foreseeable future, which is called respectively allocation power, and it belongs to the recessive allocation power. In order to discuss the dominant allocation power of enterprise interests, firstly, we present a relatively sorted of the interest group vested interest in increasing in enterprise inside, set the sorted results are $A_1, A_2, ..., A_{m2}$ by the benefit increased is decreasing, and then accordingly set their utility function are $U_{A1}, U_{A2}, .., U_{Am2}$. Secondly, according to one certain principle, namely, to increase their benefit of vested interest groups priority who are sensitive to interests, and the result of utility increased is as large as possible and approximately equal among this interest groups, i.e. the result of increased is as $U_{A1}(w_{A1}) \approx ... \approx U_{Am2}(w_{Am2})$, and has $F_{P1j} = n_j U_{Aj}(w_{Aj})$. For the recessive allocation power, we use the comparison utility of interest allocation to denote the result, set $W_{\max(Ak)}$ to denote the maximum allocation of the A_k interest group, we have $F_{P2k} = n_k \{U_{Ak}[W_{\max(Ak)}] - U_{Ak}(w_{Ak})\}$. The effect of aggregation of allocation power should be according with the principle of 1+1>2. This shows that it should not involve too many interest groups in the case of not adding new benefit, otherwise the allocation power is too small, or the resistance of allocation is too large. In the same way, for unequal allocation of adding new benefit, it cannot generate multi-level difference in too many interest groups, otherwise the allocation power is too small, or the resistance of allocation is too large also.

Set, $h_1(\sum_{j=1}^{m_1} F_{P1j}, m_1) = (\sum_{j=1}^{m_1} F_{P1j})^{\lambda_1 m_1}$ and $h_2(\sum_{k=1}^{m_2} F_{P2k}, m_2) = (\sum_{k=1}^{m_2} F_{P2k})^{\lambda_2 m_2}$

Here λ_1 and λ_2 is the relatively importance of the vested interest groups while increasing benefit absolutely and increasing benefit relatively absolutely, and $\lambda_1, \lambda_2 \in [0,1]$ 。

The （2） is modified as

$$F_{P1} = \sum_{j=1}^{m_2} n_{PAj} U_{PAj}(w_{Ai}) + (\sum_{j=1}^{m_2} n_{PAj} U_{PAj}(w_{Aj}))^{\lambda_2 m_2} = (\sum_{j=1}^{m_2} n_{PAj} U_{PAj}(w_{Aj}))^{1+\lambda_2 m_2} \quad (4)$$

The （3） is modified as

$$F_{P2} = \{\sum_{k=1}^{m_2} n_{Oj} \{U_{PAk}[W_{\max(Ak)}] - U_{PAk}(w_{Ai})\}\}^{1+\lambda_2 m_2} \quad (5)$$

According the（4）and （5）, has

$$F_P=(\sum_{j=1}^{m_1} n_{PAj}U_{PAj}(w_{Aj}))^{1+\lambda_1 m_1}+\{\sum_{k=1}^{m_2} n_{Pj}\{U_{PAk}[W_{\max(Ak)}]-U_{PAk}(w_{Ai})\}\}^{1+\lambda_2 m_2} \quad (6)$$

Integrated the basic concepts and assumptions, the model of allocation power enterprise interests is put forward as following.

$$\max \ F_P \quad (7)$$

$$s.t: 1) \ \sum_{i}^{m_1+m_3} w_{aAi}\, n_{aAi} \le P_1 + \sum_{j=1}^{m_2} w_{dAj}\, n_{dAj}$$

$$2) \ m_1 + m_2 + m_3 = m$$

$$3) \ w_{aAi}, n_{aAi}, w_{dAj}, n_{dA\,j} \ge 0; m_1, m_2, m_3, m \in Z^+$$

The explanation of the model: the goal function denotes the allocation power to maximum, i.e. the comparison of the power and the resistance to maximum. The condition is the total of resource interest groups obtained is not more than the total allocation.

4. Conclusion

This paper presents an analysis model of allocation power of enterprise interests, and this model mainly is used to explain the actions of enterprise how to make the allocation power to maximize in the condition of resources limited. The key problem of enterprise interest allocation is to come on the goal of enterprise as soon and as low cost possible. This paper built the concept of interest group and described its effectiveness change and the effect of aggregation to support reaching the goal of enterprises. The optimization results of allocation of enterprise interests should be meet the needs of the key stakeholders.

References

1. J.C.Rode, Job satisfaction and life satisfaction revisited: longitudinal test of an integrated model, Human relation,45(2004).
2. W.Grant, Business Interests, Organizational Development and Private Interest Government: An International Comparative Study of the Food Processing Industry Walter de Gruyter(1987).
3. X.H.Liu, An optimization model of resistance of enterprise reform in E-business, Proceeding of 2010 International Conference on Management Science and Engineering, Orient Academic Forum, Sydney-Australia(2010).

EVALUATION OF AIRPORT PASSENGER TERMINALS USING FUZZY COGNITIVE MAPPING

ÇİĞDEM KADAİFÇİ

Industrial Engineering, Istanbul Technical University, Macka Istanbul,Turkey

ILKER TOPÇU

Industrial Engineering, Istanbul Technical University, Macka Istanbul,Turkey

This study aims to provide a decision support tool to improve the quality of the service at an airport terminal. For this purpose, a fuzzy cognitive mapping based model is utilized. The fuzzy cognitive mapping method allows dynamic modeling of problem by considering the complex network structure of model, i.e. the effects of factors on the quality of terminal services. Simulations are performed to analyze the changes in the system.

1. Introduction

Airports are important infrastructures of modern life and play a major role in the transportation of passengers for tourism, business, and other purposes. As the air transportation passengers want rapidity, comfort, and security; airports should be well managed to meet these demands. Especially, services in airport terminals, where passengers spend a significant amount of time, should be well planned and organized to provide satisfaction of air transportation passengers. Level of service in airport passenger terminals has been research topics since 1990s. More recently, studies have been initiated to identify the passenger terminal problem in general and service measures in particular.

This study refers to fuzzy cognitive maps (FCMs) as causal representations of the relations and interaction between factors related to airport terminal services. The paper is organized into four sections. In the following section, the literature review about FCM is expressed. Third section gives the details and the steps of the methodology and the paper is concluded in the fourth section.

2. Fuzzy Cognitive Mapping

A Cognitive Map (CM), which is an image of cognitive processes [1], is generated from relationships between the ideas and events that are both measurable and immeasurable [2]. The origin of Fuzzy Cognitive Mapping Method (FCM) is Cognitive Mapping Method [3] is an illustrative causative representation of complex systems and can be used to model and manipulate the dynamic behavior of systems. FCM, created in 1986 by Kosko as a combination of fuzzy logic and neural networks, is utilized for modeling complex systems, especially for situations where fuzziness and uncertainty exist [4].

A FCM can be represented either as a signed, weighted graph consisting nodes and arcs symbolize relationships between concepts [5], or as an adjacency matrix, which has entries w_{ij}'s indicating the direct relationship between concept i and concept j. A simple FCM consisting of five concepts C_i (i=1, …,5) where w_{ij} represents the influence degree from cause C_i to effect C_j is shown in Figure 1. FCM does not allow any direct connections between a concept and itself, thus all w_{ii} elements equal to zero. All other w_{ij} elements take values in [-1, 1] and the meaning of these values as [3];

- wij > 0 indicates a causal increase
- wij < 0 indicates causal decrease
- wij = 0 indicates no causality

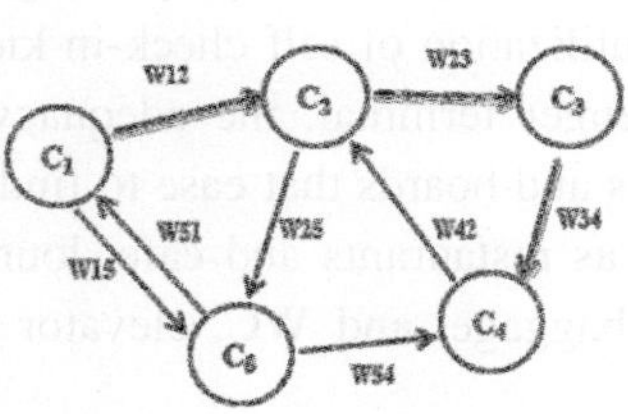

$$E = \begin{pmatrix} 0 & w_{12} & 0 & 0 & w_{15} \\ 0 & 0 & w_{23} & 0 & w_{25} \\ 0 & 0 & 0 & w_{34} & 0 \\ 0 & w_{42} & 0 & 0 & 0 \\ w_{51} & 0 & 0 & w_{54} & 0 \end{pmatrix} \quad -1 < w_{ij} < 1$$

Figure 1. A hypothetical FCM model and the adjacency matrix.

The mathematical formula of the graphical or matrix representation is given by;

$$A_i^{(k+1)} = f(A_i^{(k)} + \sum_{\substack{j=1 \\ j \neq i}}^{n} A_j^{(k)} w_{ji} \tag{1}$$

In Equation 1, $A_i^{(k+1)}$ is the value of concept at step (*k*+1), $A_i^{(k)}$ is the value of concept at step (*k*), W_{ji} is the weight of interconnection between C_j and C_i [7]. *f* is the threshold function (activation function) used to reduce the concept value into a normalized range [3]. In this study, hyperbolic tangent function is chosen to show not only the increase or decrease but also in degree of the change in concept.

3. Methodology

3.1. *Model*

FCM is chosen to express the complex network structure and dynamic behavior of airport terminals through the causal relationship between factors. First, factors affecting the efficiency of time passengers spend in airport terminal were determined among alternatives by reviewing literature on airport terminal services and interviewing experts who were traveling by plane over 50 per year. After eliminating alternative factors according to the expert views, the model was built with selected 26 factors especially focused on the delays occurring during terminal services. These delays occur in the following consecutive processes: the security controls, check-in counters, and passport controls, security controls before lounge, boarding pass controls, and transferring the passengers to the aircraft by bus or boarding bridge. Factors including the take-off or landing delays of previous flights, scheduling the flights incorrectly, and waiting the late comer passengers were considered as the other reasons of delays. The number of X-rays, check-in counters, terminal employees, gates, baggage claims, flights (landing/take off), the utilization of self check-in kiosk, appropriateness of layout of the airport passenger terminal, the adequacy of investment plans, the capacity of terminal, signs and boards that ease to find the right direction, the idle time alternatives such as restaurants and cafe, lounges and shopping stores, the duration of waiting baggage, and WC, elevator and escalator alternatives are other factors in model.

A questionnaire was designed and used to collect data from experts for determining the degree of effects of each factor on others. A cognitive map was designed by using the data collected from 31 experts and was converted to a FCM shown in Figure 2. According to the combined adjacency matrix, the most affecting factors were the capacity and appropriateness of layout of the airport passenger terminal, the most affected factors was delays caused by waiting the late comer passengers and the central factor was the capacity of the airport passenger terminal.

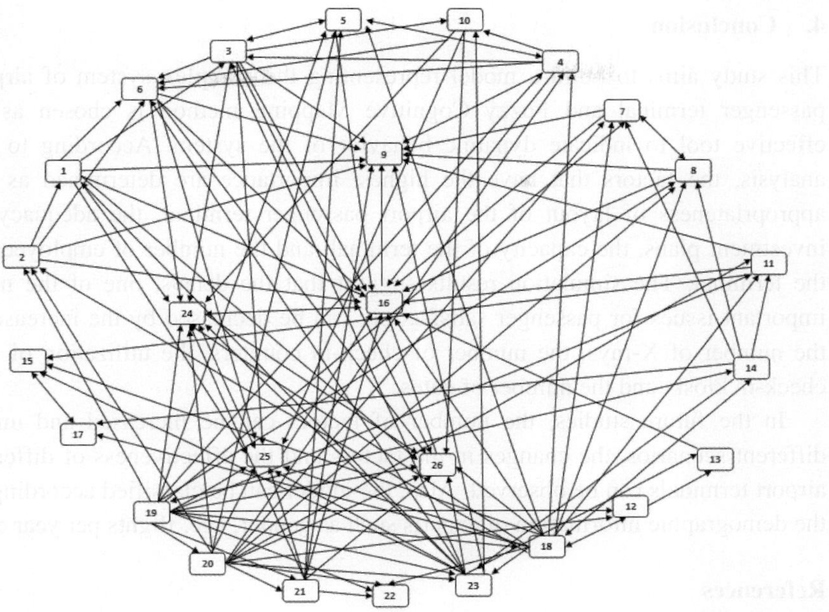

Figure 2. Cognitive Map of The Factors Affecting the Quality of Airport Passenger Terminal Services.

3.2. *Results*

The constructed fuzzy cognitive map (FCM) was used to perform simulations in order to predict possible changes and to observe whether the system converges toward a steady state. Scenarios were performed for each factor represented as concepts. In each scenario, the FCM was initialized with a value between -1 and +1 for the current state based on the expert views. For instance, $A^{(0)}$ = [0,1,0,0,0,0,0,0,0,0,0,0,0,0,0,0,0,0,0,0,0,0,0,0,0,0] was an initial state for the number of X-rays was activated. After 19 iterations, the simulation reached an equilibrium state and was stopped. According to this scenario, the delays caused by the security controls before lounge had the highest value as $A_6^{(19)}$=-99%, so by increasing the number of x-rays, the delays caused by the security controls before lounge will decrease. Also, the utilization of self check-in kiosk is the second factor decreasing under the determined scenario with the value of -92%. This means that, if the number of x-rays is increased, passengers will not prefer to use the self check-in kiosk instead of check-in counters and they prefer to spend their time in restaurant, cafe, lounge or shopping stores which values are 96%, 95% and 96% respectively.

4. Conclusion

This study aims to build a model representing the complex system of airport passenger terminal and Fuzzy Cognitive Mapping method is chosen as an effective tool to indicate dynamic behavior of the system. According to the analysis, the factors that have the highest importance are determined as the appropriateness of layout of the airport passenger terminal, the adequacy of investment plans, the capacity of the terminal, and the number of employees in the terminal. The simulation results suggest that the delays, one of the most important issues for passenger satisfaction, can be decreased by the increase in the number of X-rays, the number of check-in counters, the utilization of self check-in kiosk, and the number of gates.

In the future studies, the number of factors can be increased and under different scenarios, the changes in the factors and the effectiveness of different airport terminals can be observed. Also, the model can be classified according to the demographic information of experts such as gender, age, flights per year etc.

References

1. Ülengin, F., ve Topçu, Y. İ., 1997: Cognitive Map: KBDSS Integration in Transportation Planning, *Journal of the Operational Research Society*, Vol. 48, pp. 1065-1075.
2. Hasıloğlu, S. B., 2009: Evaluation of the Product with Cognitive Mapping Method, One of the Elements of Internet Marketing Mix, *Journal of Applied Sciences*, Vol. 9, no. 14, pp. 2606-2612.
3. Bueno, S., ve Salmeron, J. L., 2009: Benchmarking Main Activation Functions in Fuzzy Cognitive Maps2, *Expert Systems with Applications*, Vol. 36, pp. 5221-5229.
4. Papakostas G.A., Boutalis Y.S., Koulouriotis D.E. and Mertzios B.G., 2008: Fuzzy Cognitive Maps for Pattern Recognition Applications, *International Journal of Pattern Recognition And Artificial Intelligence*, 22, 1461-1486.
5. Georgopoulos, V. C., Malandraki, G. A. ve Chrysostomos, D. S., 2003: A Fuzzy Cognitive Map Approach to Differential Diagnosis of Specific Language Impairment, *Artificial Intelligence In Medicine*, Vol. 29, pp. 261-278.
6. Papageorgiou, E. I., 2010: A New Methodology for Decisions in Medical Informatics Using Fuzzy Cognitive Maps Based on Fuzzy Rule-Extraction Techniques, *Applied Soft Computing*, doi:10.1016/j.asoc.2009.12.010.
7. Tsadiras, A. K., 2008: Comparing The Inference Capabilities Of Binary, Trivalent and Sigmoid Fuzzy Cognitive Maps, *Information Sciences*, Vol. 178, pp. 3880-3894

IMPLEMENTATION OF MAXIMIN AND MAXIMAL SOLUTIONS FOR LINEAR PROGRAMMING PROBLEMS UNDER UNCERTAINTY*

NATHAN HUNTLEY[1], ROLANDO QUIÑONES[1], KEIVAN SHARIATMADAR[1], ERIK QUAEGHEBEUR[1], GERT DE COOMAN[1], ETIENNE KERRE[2]

[1]*SYSTeMS Research Group, Faculty of Engineering and Architecture, Ghent University*

[2]*Fuzziness and Uncertainty Modeling, Faculty of Sciences, Ghent University*

We present a software implementation of the methods for solving linear programming problems under uncertainty from previous work. Uncertainties about constraint parameters can be expressed as intervals or trapezoidal possibility distributions. The software computes the solutions for the optimality criteria maximin and maximality. For maximality with possibility distributions, only an approximate solution is obtained.

1. Introduction

Linear programming has been widely developed and used in optimization problems [1]. We consider linear programming problems with uncertainty of the form

$$\text{maximize } c^T x$$

$$\text{subject to } Ax \leq B\,,\; x \geq 0\,,$$

where x in $\mathbb{R}^n$ is an optimization vector, c in $\mathbb{R}^n$ is a vector of objective function coefficients, A is a $m \times n$ matrix of uncertain constraint coefficients A_{ij}, and B is an m-vector of uncertain constraint variables B_i.

In earlier work [2] we reformulated this problem into a decision problem. Uncertainty about A_{ij} and B_i is modeled by crisp intervals $[\underline{a}_{ij}, \overline{a}_{ij}]$ and $[\underline{b}_i, \overline{b}_i]$ or by possibility distributions [3]. Independence is assumed between all constraint parameters. These models can be seen as special cases of coherent wer and upper previsions $\underline{P}$ and $\overline{P}$ [4], the framework of which we use to lyze the problem. Solution methods for the *maximin* and *maximality* lity criteria [5] were given elsewhere [6].

---

by the IWT SBO project 60043, "Fuzzy Finite Element Method". We thank the heir helpful comments.

In this paper we discuss a Matlab implementation of these solution methods. It allows specification of the uncertainties in the A_{ij} and B_i as single values (no uncertainty), intervals, triangular possibility distributions, and trapezoidal possibility distributions. These specifications are made in a text file, which also includes the values of m, n, and the components of c. This file is loaded into a graphical user interface, which also allows editing of the parameters and displays graphs of the possibility distributions chosen. The user then selects the type of uncertainty model to use: intervals or possibility distributions. Finally, the user chooses whether to use maximin or maximality.

The interface then displays a plot of the corresponding solution and exports the solution to a file for further investigation. In Section 3 we explain the implementation and interface, and in Section 4 we give an illustration. A version of the software is available upon request.

2. Theoretical Solution

The following decision problem was introduced in earlier work [2,6]. The gain $G_x(a,b)$ for a particular optimization vector x and a particular realization (a,b) of the constraint parameters (A,B) is $c^T x$ if $ax \leq b$ and L if $ax \nleq b$. L is a penalty that must be smaller than $c^T x$ for any x in the *outer feasibility space*: those x for which there is a realization (a,b) of (A,B) such that $ax \leq b$.

We consider two optimality criteria for upper and lower previsions. *Maximin* selects all x that maximize the lower prevision $\underline{P}(G_x)$ *Maximality* selects all x such that, for all other potential optimization vectors z in the outer feasibility space, $\overline{P}(G_x - G_z) \geq 0$ holds.

Maximality is often strengthened by requiring x to be *undominated* by any other z: either $G_x(a,b) > G_z(a,b)$ for some (a,b), or $G_x = G_z$. We cannot yet deal with this in our software, so we do not consider dominance in this paper.

The solutions in this section were derived in an earlier paper [6], which is based on a more general treatment [2].

Interval model. In the interval case the maximin problem (on the left) becomes a standard linear programming problem (on the right):

$$\begin{array}{rl} \text{maximize} & c^T x \\ \text{subject to} & Ax \leq B,\, x \geq 0 \\ \text{with} & \underline{a} \leq A \leq \overline{a},\, \underline{b} \leq B \leq \overline{b} \end{array} \quad \rightarrow \quad \begin{array}{l} \text{maximize } c^T x \\ \text{subject t-} \end{array}$$

where $\overline{a}$ is the matrix of the $\overline{a}_{ij}$ and $\underline{b}$ is the vect

The maximality problem becomes a vertex en

$$\begin{array}{ll} \text{maximize } c^T x & \\ \text{subject to } Ax \le B, x \ge 0 & \rightarrow \\ \text{with } \underline{a} \le A \le \overline{a}, \underline{b} \le B \le \overline{b} & \end{array} \quad \begin{array}{l} \text{find all } x \\ \text{subject to } \overline{a}x \le \overline{b}, x \ge 0 \\ c^T x \ge \max_{\overline{a}z \le \underline{b}} c^T z \end{array}$$

Possibility distribution model. Let the A_{ij} or B_i be modeled by trapezoidal possibility distributions π_{ij} or π_i. Special cases of trapezoidal functions are admitted: triangular functions, crisp intervals, and crisp numbers. Independence implies $\pi_A(a) := \min_{1 \le i \le m, 1 \le j \le n} \pi_{ij}(a_{ij})$ $\pi_B(b) := \min_{1 \le i \le m} \pi_i(b_i)$ and $\pi(a,b) := \min\{\pi_A(a), \pi_B(b)\}$ The maximin problem becomes a nested optimization problem

$$\begin{array}{ll} \text{maximize } c^T x & \\ \text{subject to } Ax \le B, & \\ x \ge 0 & \\ \text{with given } \pi & \end{array} \rightarrow \begin{array}{l} \text{maximize } L + (1-t) \\ \text{subject to } \quad 0 \le t < 1 \end{array} \begin{pmatrix} \text{maximize } c^T x - L \\ \text{subject to } \overline{a}^t x \le \underline{b}^t, \\ x \ge 0 \end{pmatrix}$$

where $\overline{a}_{ij}^t = \max\{a_{ij} : \pi_{ij}(a_{ij}) \ge t\}$ $\underline{b}_i^t = \min\{b_i : \pi_i(b_i) \ge t\}$ $\overline{a}^t$ is the matrix with components $\overline{a}_{ij}^t$ and $\underline{b}^t$ is the vector with components $\underline{b}_i^t$.

For maximality, $\overline{P}(G_x - G_z)$ is difficult to calculate, and is of the form

$$\overline{P}(G_x - G_z) = \alpha(x,z,\pi) + \beta(x,z)\gamma(x,z,\pi), \tag{1}$$

where α is an easily-calculated function, β is an easily-calculated positive function, and γ is a function taking values in $[0,1]$ that we have found no efficient way to calculate. In the next section we explain how the software deals with this problem.

3. Software Implementation

Methods. For the interval case, we solve the linear programming and vertex enumeration problems using standard toolboxes. For maximin with possibility distributions, the inner maximization is a linear program. The outer maximization is unimodal in t , so we use the golden section method [7, § 10.2]

For maximality, the upper prevision in Eq. [1] is difficult to calculate efficiently. However, we only need to know whether it is nonnegative, that is, whether

$$\gamma(x,z,\pi) \ge -\alpha(x,z,\pi) / \beta(x,z). \tag{2}$$

Since $\gamma(x,z,\pi)\in[0,1]$ if the right-hand side is 0 or less, then the left-hand side is not smaller. If the right-hand side is greater than 1 then the left hand side is smaller. If the right-hand side is in $(0,1]$ it turns out we can check Eq. (2) by solving at most m linear programs.

To find an approximate set of maximal points, we discretize the outer feasible region into a grid of candidate vectors, and test these against one another. As soon as a candidate is found to be non-maximal it can be removed from all further comparisons, because the ordering corresponding to maximality is transitive. Any candidate removed by this method is non-optimal, but a candidate x not removed is not necessarily maximal: it could be that none of the z such that $\overline{P}(G_x - G_z) < 0$ were in the grid. This can often happen when the grid resolution is too coarse or when the candidate vector is close to a maximal vector.

```
Input File
m=1
n=3
Objective
function
-5 -3.75 -2.5
A
4.8 5 5.3
8 9 10
8.3 9.3 10.4
B
8
resolution=40
L=-10
```

Interface and Input. On the right, we show an example input file, used for the illustration of Section 4. `n` sets the dimension of x. `m` sets the number of constraint rows. The numbers under `Objective function` give the values of the c_i. Each row of numbers under `A` gives the uncertainty model for a particular element A_{ij}: in this case (in order) A_{11}, A_{12} A_{13} all of which here are triangular possibility distributions. Each row of numbers under `B` gives the uncertainty model for a particular B_i in this case B_1 only, which here is a crisp value 8 As an example, adding another constraint row would mean there are four new uncertain values: A_{21} A_{22} A_{23} and B_2 To add these to the problem, we would add three more rows of numbers underneath `8.3 9.3 10.4` (corresponding to A_{21} A_{22} A_{23}, and one row of numbers underneath `8` (corresponding to B_2. `L` is the penalty L, and `resolution` is the grid resolution for the method for possibility distributions and maximality. These last two are optional.

In Figures 1–3 we show the three panels of the graphical user interface.

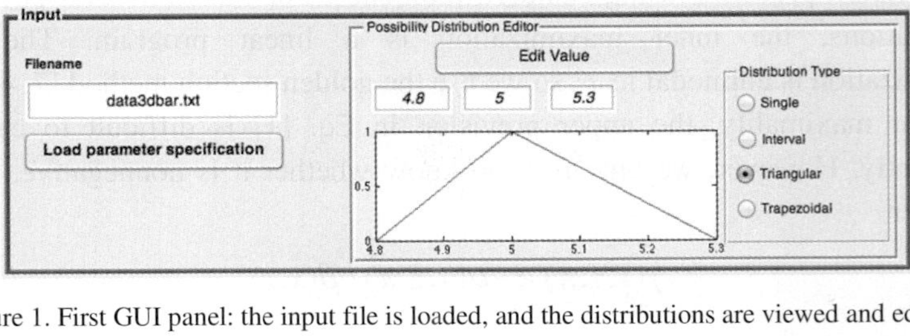

Figure 1. First GUI panel: the input file is loaded, and the distributions are viewed and edited.

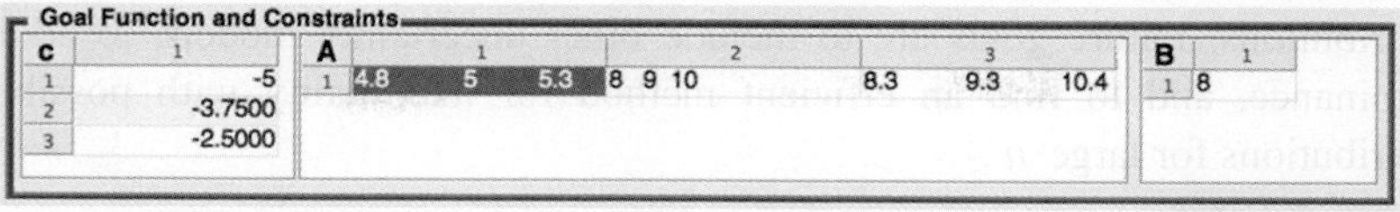

Figure 2. Second GUI panel: the values for c and the parameters for the possibility distributions of A and B are all shown. Clicking on a cell displays the possibility distribution in the possibility distribution editor (see Fig. 1).

4. Illustration: minimizing the cost of a beam

We apply the method to a loaded beam problem. We want to minimize the total cost of a beam of length ℓ and cross section area a that may consist of n different segments i of different material, each with a relative cost per meter of c_i. Under a load f, the beam should not stretch more than a prescribed length d. So we must find relative segment lengths x_i that minimize the cost. The constraint can be approximated by a linear one determined by a finite element analysis [8] in which we take beam segments as elements. There is uncertainty about the elastic compliances S_i (inverses of elastic or Young moduli) that determine how much each material stretches. The problem can be formulated as the linear program with one uncertain constraint on the right.

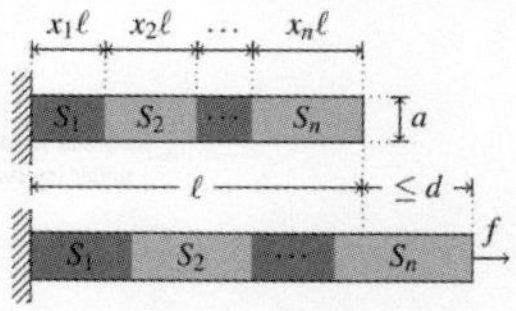

maximize $\sum_{i=1}^{n} -c_i \ell x_i$
subject to $\sum_{i=1}^{n} S_i x_i \leq \frac{da}{f\ell}$,
$\sum_{i=1}^{n} x_i = 1, x \geq 0.$

We imagine a beam with $n = 3$ segments with the following materials, using triangular possibility distributions for the elastic compliances (units are mm^2/MN): wrought iron with $S_1 \sim (4.8;5.0;5.3)$; brass with $S_2 \sim (8.0;9.0;10.0)$; bronze with $S_3 \sim (8.3;9.3;10.4)$. The quantities ℓ, a, f, and d are chosen such that $\frac{da}{f\ell} = 8.0\text{mm}^2/\text{MN}$ The relative costs are $3c_1 = 4c_2 = 6c_3$ and the penalty is $L = -10$ The results of the analysis of this problem with our program are given in Fig. 4.

5. Conclusion

We implemented the approaches in previous work \cite{Quaeghebeur-Huntley-Shariatmadar-DeCooman-2012-IPMU} for linear programming problems under uncertainty. An intuitive user interface allows visualization and specification of the uncertain parameters, and offers a choice of two solution methods (maximin and maximality) for two uncertainty models (intervals and possibility distributions). The current version allows uncertainties to be given as any combination of crisp values, intervals, and triangular or trapezoidal possibility distributions. Future goals are to include other uncertainty models, to address dominance, and to find an efficient method for maximality with possibility distributions for large n.

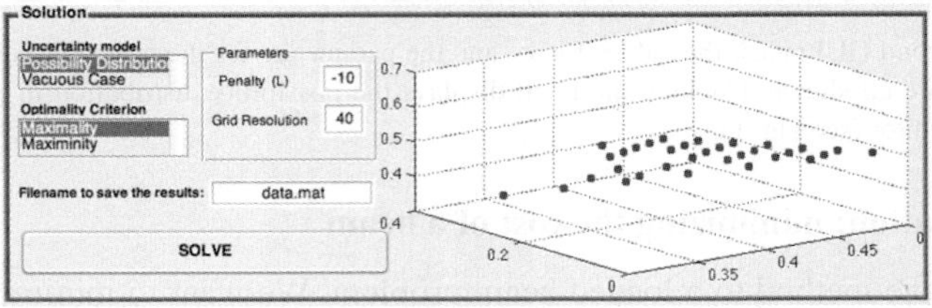

Figure 3 Third GUI panel: the choice of uncertainty model to use can be made (if 'Vacuous Case' is chosen, then all triangular and trapezoidal possibility distributions are reduced to crisp intervals). When 'Possibility Distribution' is chosen, the optimality criterion to use must be selected. When 'Vacuous Case' is chosen, solutions for both criteria are calculated. When n is 2 or 3 a plot of the solution is displayed in the right panel. Finally, the solution is exported to a specified file, along with c, L, and the distributions for the A_{ij} and B_i.

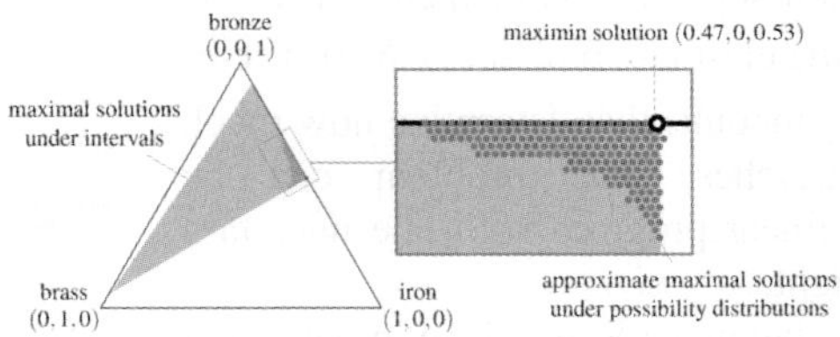

Figure 4. Solution of beam problem for interval and possibility distribution cases. Each point in the simplex corresponds to the optimization vector x with that ratio of materials. The light grey region is the maximal region using intervals. The dark grey points show the approximate maximal region using possibility distributions. The maximin solution is the same for both uncertainty models.

References

1. G. Dantzig, Linear Programming and Extensions (1998).
2. E. Quaeghebeur, K. Shariatmadar and G. De Cooman, Fuzzy Sets and Systems (in press).
3. D. Dubois and H. Prade, Possibility Theory (1988).
4. P. Walley, Statistical Reasoning with Imprecise Probabilities (1991).
5. M. C. M. Troffaes, International Journal of Approximate Reasoning **45**, 17 (2007).
6. E. Quaeghebeur, N. Huntley, K. Shariatmadar and G. De Cooman, Maximin and maximal solutions for linear programming problems with possibilistic uncertainty, accepted by IPMU-2012.
7. W. Press, S. Teukolsky, W. Vetterling and B. Flannery, Numerical Recipes (2007).
8. O. Zienkiewicz, R. L. Taylor and J. Z. Zhu (eds.), The Finite Element Method: Its Basis and Fundamentals (2005).

A NEW $C \times K$ - NEAREST NEIGHBOR LINKAGE APPROACH TO THE CLASSIFICATION PROBLEM*

GÖZDE ULUTAGAY†
Department of Industrial Engineering, Izmir University, Gursel Aksel Blv. No.14 Izmir, 35350, Turkey

EFENDI NASIBOV
Department of Computer Science, Dokuz Eylul University, Tinaztepe Campus, Buca Izmir, 35160, Turkey

A new $C \times K$ nearest neighbor algorithm with a new point of view is proposed. In this algorithm K neighbors from each of the classes are taken into account instead of the well-known K neighbor algorithm in which only the total number of neighbors are considered. After experiments with well-known classification datasets, we conclude that K-NN, weighted K-NN, and average linkage neighbors results are between the single-linkage and complete-linkage algorithms. After the evaluation of the average accuracy results, we realized that the best results are obtained for the values of K between 1 and 10. On the other hand, it is determined that using single-linkage strategy provides high values of the results at most of the times.

1. Introduction

K-nearest neighbor (K-NN) algorithm is a well-known non-parametric approach in which a new unknown data is assigned to the closest class in the learning set [1,2]. The similarity is determined by using distance measures. K-NN algorithm gives equal importance to each of the objects in assigning class label to the input vector. This is one of the challenges of the K-NN algorithm since such an assignment could reduce the accuracy of the algorithm if there is a strong overlapping degree amongst the data vectors. There are some other approaches that increase the accuracy of the algorithm by weighting the nearest neighbors in various ways.

In this study, a new $C \times K$ nearest neighbor algorithm reflecting a different kind of viewpoint is proposed. By using K-NN algorithm, a new point is assigned to the class in which most of the neighbors exist. In the proposed

* This work is supported by TUBITAK Grant No.111T273.
† gozde.ulutagay@izmir.edu.tr.

approach, K neighbors from each of the classes are taken into account instead of only the total number of K neighbors. Such a point of view results in a better classification since it considers totally $C \times K$ number of points. The aggregated distance of the classified point between its K-nearest neighbors in each class could be calculated by using different strategies and such a property increases the flexibility of the proposed algorithm [3,6]. In the study, single-linkage, complete-linkage, and average-linkage approaches which are widely used in hierarchical clustering are considered. After conducting some experiments with well-known classification datasets, we conclude that K-nearest neighbors, weighted k-nearest neighbors, and average linkage neighbors results are between the single-linkage and complete-linkage algorithms. The experiments are conducted on the well-known datasets such as Wine, Glass, and Iris for $K=1$ through $K=50$. For these values of K, the experiments are repeated for 10 times on randomly selected learning and test sets. After the evaluation of the average accuracy results, we realized that the best results are obtained for the values of K between 1 and 10. On the other hand, it is determined that using single-linkage strategy provides high values of the results at most of the times.

2. Single-Linkage Approach

The inputs for the single linkage algorithm could be distances or similarities between pairs of objects. Classes are constructed from the individual entities by joining nearest neighbors, where the term nearest neighbor connotes the largest similarity or smallest distance [1,2]. Let $Dist(A,B)$ be the distance between clusters A and B,and y_i and z_j be the elements of clusters A and B, respectively. Then the single-linkage method defines the inter-cluster distance as the distance between the elements in each of the two clusters that are nearest:

$$Dist(A,B) = \min_{y_i \in A,\ z_j \in B} [d_{ij}]. \quad (1)$$

Since single-linkage joins clusters by the shortest link between them, the technique cannot discern poorly separated clusters. On the other hand, single-linkage is one of the few clustering methods that can delineate non-ellipsoidal clusters. The tendency of single-linkage to pick out long string-like clusters is known as chaining. Chaining can be misleading if items at opposite ends of the chain are, in fact, quite similar [7].

The clusters formed by the single-linkage method will be unchanged by any assignment of distance or similarity that gives the same relative orderings as the initial distances.

3. Complete-Linkage Approach

Complete-linkage works similar to single-linkage clustering, except that at each stage of clustering process, the distance or similarity between clusters is determined by the distance between the two elements from different clusters that are most distant [1,2,4]. Similar to the single-linkage, complete-linkage method defines the inter-cluster distance as the distance between the elements in each of the two clusters that are most distant [8]:

$$Dist(A,B) = \max_{y_i \in A,\ z_j \in B} [d_{ij}] \tag{2}$$

Thus, complete-linkage ensures that all items in a cluster are within some maximum distance or minimum similarity of each other.

4. Average-Linkage Approach

Average-linkage treats the distance between two clusters as the average distance between all pairs of items where one member of a pair belongs to each cluster:

$$Dist(A,B) = \frac{1}{n_a}\frac{1}{n_b}\sum_{i \in A}\sum_{j \in B} d_{ij}\ . \tag{3}$$

For average-linkage clustering, changes in the assignment of distances or similarities can affect the arrangement of the final configuration of clusters, even though the changes preserve relative orderings.

5. $C \times K$ –Nearest Neighbor Algorithm

In K -NN algorithm, a new unknown data is put in to the closest class in the learning set. The similarity is determined by using distance measures. Let $X = \{x_1, x_2, ..., x_n\}$ be a set of n labeled samples. K -NN algorithm gives equal importance to each of the objects in assigning class label to the input vector which is one of the challenges of the K -NN algorithm since such an assignment could reduce the accuracy of the algorithm if there is a strong overlapping degree amongst the data vectors.

In the proposed approach totally $C \times K$ numbers of neighbors are being considered for the new data with unknown class label. The difference of the $C \times K$ -nearest neighbor algorithm is that a point is assigned to a class, which is closest to the point among its other nearest neighbors. The distance between the point and its nearest neighbor could be handled as single-linkage, average-linkage, complete-linkage or any other aggregated distance.

Let $X = \{x_1, x_2, ..., x_n\}$ be a set of n labeled samples, $C_j = \{x_1^j, x_2^j, ..., x_{n_j}^j\}$ $j = 1, 2, ..., C$ be the a priori known classes, n_j be the number of elements in class C_j where $n_1 + n_2 + ... + n_c = n$. The pseudo code of the $C \times K$ nearest neighbors algorithm is given below:

BEGIN

An unclassified input sample x

Set K, $1 \le K \le n$

FOR EACH CLASS C_j

DO UNTIL (K -nearest neighbor found in class C_j)

Set $i = 1$.

Calculate the distance between x and x_i^C

IF ($i \le K$ and $x_i^{C_j}$ is closer to x than any other previous neighbor in class C_j)

THEN

Delete the farthest sample in the set of K -nearest neighbors in class C_j .

Assign $x_i^{C_j}$ to the set of K -nearest neighbors for class C_j .

END IF

$i = i + 1$

END DO UNTIL

Calculate the distance of x from K -nearest neighbors of class C_j

END FOR

Mark the class with minimum distance as j^*

Assign x to C_{j^*}

END

6. Experimental Results

In our study, we compare single-linkage, complete-linkage, and average-linkage based $C \times K$ nearest neighbor classification results for known datasets such as Glass, Iris, and Wine from UCI machine learning repository. Note that approximately 2/3 of data for each data set is used as learning, and the rest is as test dataset. First, a model is constructed on learning set, then this model is used on test data set and classification accuracy of the results is measured. We handle the following formula as the classification accuracy rate [6]:

$$CA = \frac{correctly\ classified\ number\ of\ data}{total\ number\ of\ data} . \quad (4)$$

In Fig. 1, the classification accuracy results of single-linkage, complete-linkage, and average-linkage approaches are shown. It is obvious that $C \times K$

nearest neighbor approach with single-linkage provides better results than those of other methods.

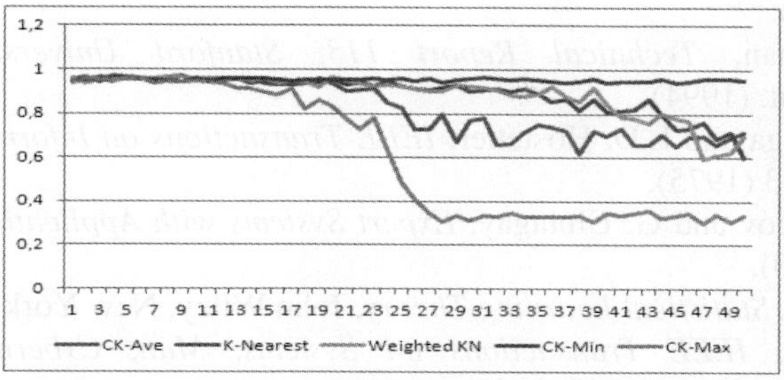

(a)

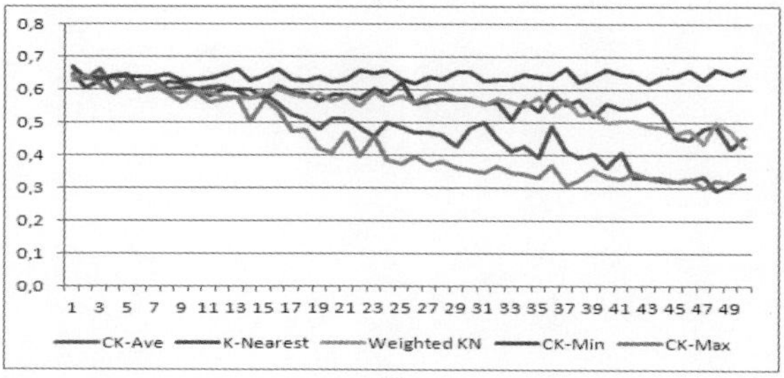

(b)

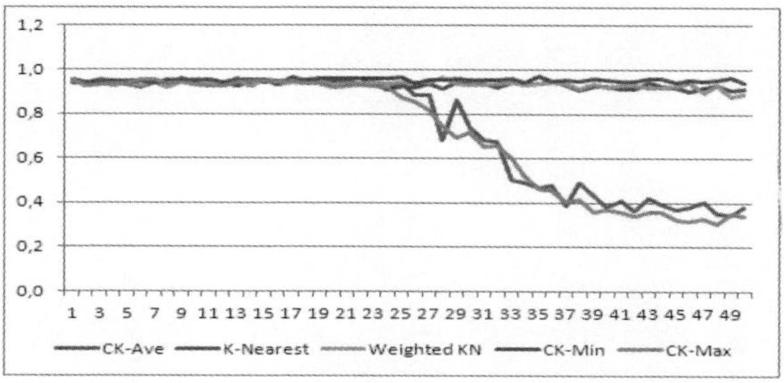

(c)

Figure 1. Average accuracy versus K for (a) Iris, (b) Wine, (c) Glass data set

Acknowledgments

This work has been supported by Scientific and Technological Research Council of Turkey (TUBITAK) with grant number 111T273.

References

1. T.M. Cover and P.E. Hart, *IEEE Transactions on Information Theory.* **13,** 21-27 (1967).

2. R.O. Duda and P.E. Hart. *Pattern Classification and Scene Analysis*, New York: Wiley (1973).
3. S.A. Dudani, *IEEE Transactions on Systems, Man and Cybernetics*, **18**, 325-327 (1975).
4. J. Friedman, *Technical Report 113, Stanford University*, Statistics Department. (1994).
5. K. Fukunaga and L.D. Hostetler, *IEEE Transactions on Information Theory*, **21**, 285-293 (1975).
6. E.N. Nasibov and G. Ulutagay, *Expert Systems with Applications,* **37**, 2495-2504 (2010).
7. V. Vapnik, *Statistical Learning Theory*, John Wiley, New York (1998).
8. R.R.Yager, *IEEE Transactions on Systems, Man, Cybernetics-Part B: Cybernetics,* **30,** 835-845 (2000).

ANALYZING COMPETITIVENESS OF AUTOMOTIVE INDUSTRY THROUGH CUMULATIVE BELIEF DEGREES*

ÖZGÜR KABAK

Industrial Engineering Department, Istanbul Technical University, Macka Istanbul 34367, Turkey

FÜSUN ÜLENGIN, ŞULE ÖNSEL, ÖZAY ÖZAYDIN

Industrial Engineering Department, Dogus University, Acibadem Zeamet Sok. Istanbul 34722, Turkey

EMEL AKTAŞ

Business School, Brunel University, Kingston Lane Uxbridge, UB8 3PH, UK

This study aims to analyze the automotive industry from competitiveness perspective using a novel cumulative belief degrees (CBD) approach. For this purpose, a mathematical model based on CBD is proposed to quantify the relations among the variables in a system. This model is used to analyze the Turkish Automotive Industry through scenario analysis.

1. Introduction

Competitiveness is generally defined as the set of institutions, policies and relevant factors that determine the level of productivity of a country [1]. Each year, selected organizations, such as the World Economic Forum (WEF) and the Institute for Management Development (IMD), apply several hundred objective and subjective indicators to assess the wealth created by the world's nations, and subsequently publish rankings of national competitiveness. As traditional competition becomes global, businesses fail to take the required measures on their own to become more competitive. Hence, in a globally competitive environment, national improvement has also become vital. Businesses need to utilize and be supported by countries' international competitiveness. The purpose of this study is to analyse the system of automotive industry that enable

* This research is supported by SEDEFED (Federation of Industrial Associations), REF (TÜSİAD Sabanci University Competitiveness Forum), and OSD (Automotive Manufacturers Association)

national competitive advantage through a novel approach. In this way, it is aimed to facilitate the selection and prioritization of policies to be followed in order to improve Turkish Automotive Industry.

This study is a part of a large-scale project [2] that aims to analyse the competitiveness of the automotive industry. In this particular paper, we focus on the quantification of the relations in the system of automotive industry. The second section summarises the project. The third section gives the details of the proposed model. The fourth section presents the application of the model to Turkish Automotive Industry. Finally the fifth section concludes the paper.

2. Background of the study

The methodology used in the project consists of three- stages. In the first stage, we collect information from various stakeholders to define the problem situation of the automotive industry. Indicators of the competitiveness are listed depending on the WEF indicators [3] and expert knowledge using a survey that is conducted to the stakeholders of the automotive industry. Furthermore; the relations among the indicators are determined through a workshop, participants of which are also the stakeholders. The results of the workshop are integrated to the mathematically found relations. As a result the indicators and the causal relations given in Table 1 are specified. The third column of the table shows the relations between indicators. For instance *Foreign market size index* (ID#2) is affected from *Domestic market size index* (ID#1), and *Automotive foreign market effectiveness* (ID#17).

Table 1. Indicators and causal relations determined in Stage 1.

ID #	Indicator	Affecting indicator
1	Domestic Market Size Index	12, 13, 14, 16
2	Foreign market size index	1, 17
3	Quality of Innovation	4
4	Quality of scientific research institutions	5, 6
5	Company spending on Research and Development (R&D)	13, 14
6	Availability of scientists and engineers	14
7	University-industry collaboration in R&D	3, 4, 5, 14
8	Local Supplier Quality	3, 4, 11
9	Production process sophistication	5, 7, 8
10	Firm-level technology absorption	3, 5, 7
11	Availability of latest technologies	7, 10
12	Ease of access to loans	
13	Extent and effect of taxation	
14	Total tax rate	
15	Degree of customer orientation	9
16	Domestic automotive market size	8, 12, 13, 14
17	Automotive foreign market effectiveness	15
18	Automotive production process sophistication	10, 18, 17

In the light of the results of first stage, a mathematical model based on Bayesian Causal Networks (BCNs) was set up. It is aimed to formulate the relationships that will enable to calculate the results of the various policies under different scenarios. During the second stage we realised that some of the observed evidence required for building BCNs are missing or not adequate. The stakeholders prefer to use linguistic terms instead of numerical expressions that improves the understanding of the system. Therefore in this study we develop a novel model based on cumulative belief degrees (CBDs) to reveal the relationships among the indicators using fuzzy linguistic terms.

3. CBDs approach for quantification of causal relations

The CBDs approach was developed for nuclear safeguards evaluation based on fuzzy linguistic terms and belief structure [4]. Basically, CBDs enable mathematical operations on the belief structures. In this research we use CBDs to quantify the causal relations among the variables in a system. For instance, suppose that we intent to analyze the relations in the system given in Figure 1, where variables A and C are inputs and variable D is the output. Our aim here is to quantify the given relations (i.e., A→B; B→D; C→D) so that we can make interpretations about the output when the inputs are changed.

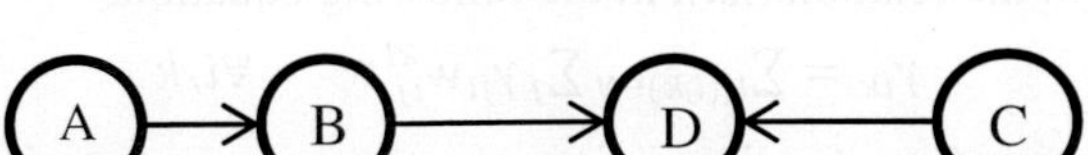

Figure 1. An example of a system with four variables

Quantification of the relations in such a system can be achieved by the use of past data and/or expert opinions. In this study we assume that all the data and/or expert opinions are represented by belief structures and linguistic terms. Given the fact that most of the different type of information represented by different scales can be converted to belief structures [4] it is a valid assumption.

3.1. *CBDs Defined*

We use fuzzy linguistic terms [5] to represent the information by the belief structure. Let S = $\{s_i\}$, $i \in \{0,...,m\}$ be a finite and totally ordered term set. Any label, s_i, represents a possible value for a linguistic variable. Linguistic term sets can be defined according to the nature of the problem. For the current study we use a five-term set, S = $\{s_i\}$, $i \in \{0,...,4\}$, in which the following meanings to the terms are assigned. s_0: very low, s_1: low, s_2: medium, s_3: high, s_4: very high.

The belief structure represents general belief of the level of an indicator as a result of past data or expert evaluations. Such that, if the past data of an indicator is available, the fuzzy linguistic sets are defined and the membership degrees of evidence to the fuzzy sets are specified as belief degrees.

In general, the belief structure can be defined as follows:

$$B(I_k) = \{(\beta_{ik}, s_i), i = 0, \dots, m)\}, \forall k, \qquad \sum_{i=0}^{m} \beta_{ik} \leq 1, \forall k \tag{2}$$

where k and i are indices for indicators and linguistic terms, respectively, and β_{ik} is the belief degree for the level of indicator k at s_i level.

The CBD at certain linguistic term level can be defined as the aggregated belief degrees of greater or equal terms of the related linguistic term. The cumulative belief structure can be defined as follows:

$$C(I_k) = \{(\gamma_{ik}, s_i), i = 0, \dots, m)\}, \forall k, \qquad \gamma_{ik} = \sum_{j=i}^{m} \beta_{jk} \tag{3}$$

where γ_{ik} is CBD related to indicator k at threshold level i.

3.2. *Proposed model*

We assume a system of variables (or indicators) with relations among them. The aim is to quantify the relations given past data or expert opinions in CBDs. Suppose that N, the set of relations between indicators such that if indicator l affects indicator k, then $(l,k) \in$ N, is known and we try to find w_{ij}^{kl}, importance weight related to the relation (i,k), in the following equation:

$$\gamma_{ik} = \sum_{l|(l,k)\in N} \sum_j \gamma_{jl} w_{ij}^{kl} \qquad \forall i, k \tag{4}$$

where i and j are indices for linguistic terms (i corresponds to indicator k, and j corresponds to indicator l). Here we assume that CBD of any indicator at each linguistic term set level is affected from CBDs of affecting indicators at all linguistic term set levels. The importance weights can be derived from the given data or expert opinion as a regression-based model as follows.

For k, which is an affected indicator; l, which are affecting indicator of k.

Find w_{ij}^{kl}, $\forall i$; By minimizing $\sum_i \sum_n (e_{ik}^n)^2$ (5)

Subject to $e_{ik}^n = \gamma_{ik}^n - \sum_{l|(l,k)\in N} \sum_j \gamma_{jl}^n w_{ik}^{jl}$; $\sum_l \sum_j w_{ij}^{kl} = 1 \quad \forall i$

where n is an index for the evident data, e_{ik}^n is the error related to dataset n, for calculating γ_{ik}. This model can be solved like a classical regression model or by any non-linear optimization method.

4. Analyzing the Turkish Automotive Industry using CBDs approach

The proposed CBDs approach is used for quantification of the relations in Turkish Automotive Industry, which are given in Table 1. We used the data of 28 countries provided from WEF report [3]. The data first normalized to 0-1 interval (the best score is 1 and the worst score is 0). Then the membership values of the normalized scores to the fuzzy sets defined in Figure 2 are determined as the belief degrees. For instance if the normalized score is 0.55 then the related belief structure is $B(I) = \{(s_2, .8), (s_3, .2)\}$ (See Figure 2)

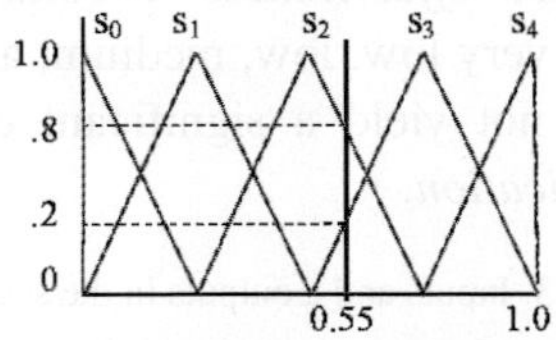

Figure 2. Fuzzy sets for transforming the data to belief structure

The proposed model is built for 15 indicators (all indicators except #12, #13, and #14) that are affected from other indicators. For instance for indicator #9, the affecting indicators are #5, #7, and #8, and the weights calculated as given in Table 2. Results show that #8 has the highest impact on #9.

Table 2. Weight related to indicator #9

	w_{i0}^{95}	w_{i1}^{95}	w_{i2}^{95}	w_{i3}^{95}	w_{i4}^{95}	w_{i0}^{97}	w_{i1}^{97}	w_{i2}^{97}	w_{i3}^{97}	w_{i4}^{97}	w_{i0}^{98}	w_{i1}^{98}	w_{i2}^{98}	w_{i3}^{98}	w_{i4}^{98}
w_{0j}^{9l}	.333	0	0	0	0	.333	0	0	0	0	.333	0	0	0	0
w_{1j}^{9l}	0	.092	0	0	0	0	.099	0	0	0	0	.604	.205	0	0
w_{2j}^{9l}	0	.066	.061	.001	0	0	0	0	.154	0	0	0	.471	.247	0
w_{3j}^{9l}	0	0	.106	.097	0	0	0	0	.269	0	0	0	0	.518	.011
w_{4j}^{9l}	0	0	0	.014	.598	0	0	0	0	0	0	0	0	0	.389

After finding all the weights they are used for analyzing different scenarios for Turkish Automotive Industry. Optimistic and pessimistic scenarios are designed for this purpose [2]. In the scenario analysis the input indicators, namely #3, #5, #7, #8, #11, and #13, are selected based on previous analysis of the industry [2]. The output indicators are performance indicators of the automotive industry, namely indicators #16, #17, and #18. In the optimistic scenario each input indicator set as a next upper linguistic term level. in the pessimistic scenario, each input indicator drops to next lower level. The levels of input indicators in the scenarios are presented in Table 3.

The resulting levels of the output indicators are given in the last three rows of Table 3. The results are consistent with the expectations. In the optimistic scenario, the possibility of outputs' level increases while the possibility of outputs level decreases in the pessimistic scenario. According to the results, if Turkey shows an improvement defined in the optimistic scenario, the *Domestic automotive market size* will be *medium* with .43 and *high* with .32 possibilities, the *Automotive foreign market effectiveness* will be *medium* with .43 and *high* with .26 possibilities; and *Automotive production process sophistication* will be *medium* with .45 and *high* with .29 possibilities. However, when the situation worsens (i.e., the pessimistic scenario), the *Domestic automotive market size* will decrease to *low* with .41 and *medium* with .43 possibilities. The level of the *Automotive foreign market effectiveness* become very uncertain, as the possibility is distributed to very low, low, medium, and high almost evenly. The pessimistic scenario does not yield a significant change for the *Automotive production process sophistication*.

Table 3. Inputs and a outputs in the scenarios

ID #	Type	Current Level	Level in Optimistic Scenario	Level in Pessimistic Scenario
3	Input	$\{(s_0, .5), (s_1, .5)\}$	$\{(s_2, 1.0)\}$	$\{(s_0, 1.0)\}$
5	Input	$\{(s_0, .64), (s_1, .36)\}$	$\{(s_2, 1.0)\}$	$\{(s_0, 1.0)\}$
7	Input	$\{(s_0, .43), (s_1, .57)\}$	$\{(s_2, 1.0)\}$	$\{(s_0, 1.0)\}$
8	Input	$\{(s_1, .72), (s_2, .28)\}$	$\{(s_3, 1.0)\}$	$\{(s_1, 1.0)\}$
11	Input	$\{(s_2, 1.0)\}$	$\{(s_3, 1.0)\}$	$\{(s_1, 1.0)\}$
12	Input	$\{(s_0, .05), (s_1, .95)\}$	$\{(s_2, 1.0)\}$	$\{(s_0, 1.0)\}$
13	Input	$\{(s_1, .6), (s_2, .4)\}$	$\{(s_3, 1.0)\}$	$\{(s_1, 1.0)\}$
16	Output	$\{(s_2, 1.0)\}$	$\{(s_0,.05),(s_1, .20),(s_2, .43),(s_3, .32)\}$	$\{(s_0,.09),(s_1, .41),(s_2, .43),(s_3, .06)\}$
17	Output	$\{(s_2, .57), (s_3, .43)\}$	$\{(s_0, .09), (s_1, .17), (s_2, .43), (s_3, .26), (s_4, .05)\}$	$\{(s_0, .34), (s_1, .16), (s_2, .23), (s_3, .23), (s_4, .04)\}$
18	Output	$\{(s_1, .37), (s_2, .63)\}$	$\{(s_0, .07), (s_1, .16), (s_2, .45), (s_3, .29), (s_4, .03)\}$	$\{(s_0, .18), (s_1, .30), (s_2, .44), (s_3, .07), (s_4, .01)\}$

5. Conclusions

One of the novel contributions of this study is the use of the CBD approach to quantify the relations among the variables in a system. The method can be applied to any data as long as they are transformed to belief structures. This property can be useful when different types of data available, such as expert opinions, numerical values, linguistic expressions etc., in a single problem. The applicability of the problem is justified by the automotive industry application.

Further studies may be done to investigate to application of the proposed model to analyze other sectors.

References

1. E.W. Artto, Management International Review; 27, 47-58 (1987)
2. F. Ulengin, Ş. Önsel, E. Aktaş, Ö. Kabak, www.sedefed.org/documents/RK_2010_Raporlar_-_Otomotiv_Sektoru_Raporu_2010.pdf (2010)
3. WEF, Global Competitiveness Report 2010-2011 (2010)
4. Ö. Kabak and D. Ruan, IEEE Transactions on Knowledge and Data Engineering; 23, 1441-1454 (2011)
5. L.A. Zadeh, Information Sciences, 8(2) 301-357 (1975)

USING SURROGATE MODELS FOR PROCESS DESIGN AND OPTIMIZATION*

A.SHOKRY, A.D. BOJARSKI, A. ESPUÑA†
Department of Chemical Engineering, Universitat Politecnica de Catalunya
Av. Diagonal, 020280, Barcelona, Spain.

In this paper we present a methodology for rigorous optimization of nonlinear programming problems in which the objective function can be represented using black box functions. The specific application is process design and operation in which the process is modeled using modular process simulators. These models consume large CPU time to converge, derivatives are not available, in some cases generate noise, and they are seen as black box models. Different techniques are available for replacing those models by simpler and computational inexpensive models. In this case a kriging metamodel is used to replace them. Kriging has high prediction accuracy and capabilities to estimate the prediction variance. Coupled to kriging the "Expected improvement" technique (EI) [3] is used to find the global solution for the kriging metamodel. The methodology is tested with two mathematical functions, and finally, applied to the optimization of the operation parameters of a gas turbine case study, which is simulated using AspenPlus.

1. Introduction

In chemical engineering, many decisions are taken based on the analysis of a process model results. Detailed process models are complex to be coded, so specialized simulation software packages are used, most of them have modular architectures (Aspen Plus), and they appear to the user as black boxes. But they consume large CPU time to converge, moreover, in case of optimization major difficulties appear: gradient based optimizers, which try to estimate derivatives, end up using noisy estimates because of errors introduced by these simulators (e.g. by termination criteria); consequently, their accuracy is affected badly [2].

To tackle this problem the original model is used as "physical experiment" for generating data points [5]. These data points are used to construct a simpler but accurate model which is called "metamodel" or "surrogate model". These

* Financial support received from the "Generalitat de Catalunya" with the European Social Fund (FI grants), and the projects TOLERANT (DPI2006-05673) and EHMAN (DPI2009-09386) financed by the Spanish Ministry of Education and Science are fully appreciated.
† antonio.espuna@upc.edu; ahmed.shokry@upc.edu.

models are accurate empirical approximation describing the relation between input variable(s) and response value(s) of a process. Jones [3] analyzed the most types of metamodels and concluded that non-interpolating (regression) metamodels are unreliable in optimization, because they do not appropriately capture the function shape; it is usually better to use surfaces that interpolate the data with linear combinations of basis functions. Jones [3] showed that even when using an interpolating metamodel, exploring the metamodel with an arbitrary optimizer can fail even to find local optima. Consequently, there is a need for optimization techniques that not only consider the metamodel prediction, but also consider the uncertainty about this prediction [10]. One of such techniques is the "maximum expected improvement" which has been presented by Schonlau, [9] and Sasena et al. [8], and tested by Jones [3, 4]. In this work, an optimization methodology of process operation, in which a complex process is modeled by process simulators, is presented. The methodology is based on fitting a kriging metamodel for the whole process flowsheet, and then using the EI improvement technique for optimization. The rest of the paper provides an overview of kriging and maximum expected improvement technique. The methodology is tested on mathematical functions and, finally on a gas turbine case study modeled using AspenPlus.

2. Methodology

2.1. *Sampling*

The selection of sample points used to construct the metamodel is a key point that significantly affects the metamodel prediction accuracy. There are several techniques are available, but because sampling is not the main goal of the paper we just mention that the Hammersley has been used here because its simplicity.

2.2. *Kriging*

The kriging assumes a stochastic process, in which the error in the predicted value is also a function of the input variables (x). The kriging predictor $\hat{Y}(x)$ is then composed of two parts, a polynomial term $f(x)$ (in most cases $f(x) = \mu$) and a departure term $Z(x)$ from that polynomial. So $\hat{Y}(x) = f(x) + Z(x)$.where $Z(x)$ is a stochastic Gaussian process that represents the uncertainty about the mean of Y(x) with expected value zero $E\,(Z(x)) = 0$, and a covariance for two points x_i , x_j calculated as: $cov(Z(x_i),Z(x_j)) = \sigma^2\ R(x_i,x_j)$, where σ^2 is the process variance, and $R\ (x_i\ ,x_j)$ is the spatial correlation function (SCF) which is usually selected exponential [3,4,7]. The final predictor of the kriging method is given by the following equation (1).

$$\hat{Y}(x_{new}) = \hat{\mu} + r^T R^{-1}(Y - 1\hat{\mu}) \quad (1)$$

Where: r is the $n\times 1$ vector of correlations $R(x_{new}, x_i)$ between the point to be predicted x_{new} and the sample design points. The mean squared error of the predictor is *given* by:

$$\hat{S}^2(x_{new}) = \sigma^2\left(1 - r^T R^{-1} r + \frac{(1 - r^T R^{-1} r)^2}{r^T R^{-1} r}\right) \quad (2)$$

A detailed mathematical derivation can be founded in [3]. Figure 1 shows the accuracy of the kriging prediction when tested with Peaks [3] and Branin [2] functions and compared with thin plate spline radial basis function (RBF) and artificial neural network (ANN) metamodels.

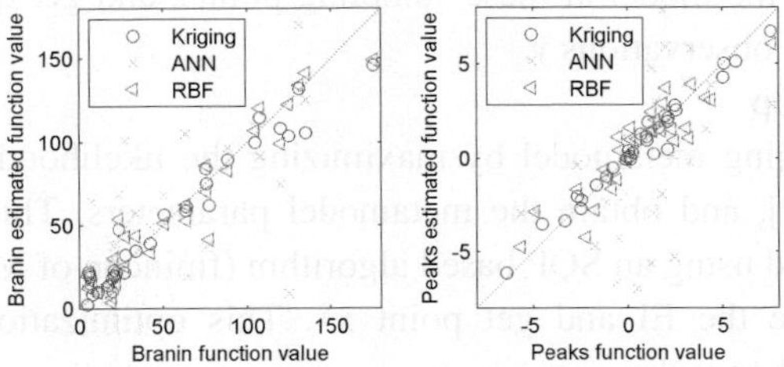

Figure 1. Leave one out cross validation (LOOCV) of kriging, thin plate spline RBF, and ANN metamodels of Branin and Peaks functions. Dotted line shows best possible prediction.

2.3. *Optimization using the method of Maximum Expected Improvement*

The expected improvement (EI) approach uses the expected value of the function at a certain untrained data point x, represented by the random variable $Y(x)$ which is normally distributed, with the mean equal to the kriging prediction at this point $\hat{Y}(x)$, and variance equal to the kriging estimated variance $S^2(x)$. It is assumed that the current best minimum value of the objective is F_{min}, hence, the best current solution is expected to get improvement with by an amount I. The expected improvement is given by equation (5), [3].

$$E[I(x)] = (f_{\min} - \hat{y})\phi\left(\frac{f_{\min} - \hat{y}}{s}\right) + s\varphi\left(\frac{f_{\min} - \hat{y}}{s}\right) \quad (3)$$

Where Φ and φ are the normal cumulative distribution function, and density function respectively. The EI value at a point x in the metamodel domain, is a function of the predicted value of the metamodel $\hat{Y}(x)$, and the estimated kriging variance $S(x)$ at this point. Thus the point that maximizes the expected improvement is the point in the metamodel domain that has minimum prediction and maximum variance. The method has been tested by [6] and [3]; both found that it converges to a global optimum.

2.4. *Algorithm*

The main steps in the proposed algorithm can be summarized as follows:

Initialization (sampling design and computer experiment)

1. The process flowsheet is explored to identify the independent variables affecting the objective function that will be optimized.
2. Specify the independent variable bounds (metamodel domain).
3. Over the metamodel domain, perform a sampling plan x with a specific design and certain number of sample points.
4. Evaluate the model at these sampling points, and get the corresponding matrix of observations y.

Optimization loop

5. Fit a kriging metamodel by maximizing the likelihood of the observed data [x y], and obtain the metamodel parameters. This optimization is performed using an SQP based algorithm (fmincon of Matlab).
6. Maximize the EI and get point x*. This optimization is done using simulated annealing.
7. Evaluate the real model at x* and get y*.
8. Add the new point [x* y*] to the original matrix of observations [X Y].
9. Return to step (5) and continue iteration.
   a) Stop if no further improvement in the objective function value y^* (the difference between objective value y^* in two successive iterations is less than specific tolerance) is observed.

3. Applications

3.1. *Mathematical Examples*

The methodology has been applied to the Peaks and Branin functions, which are good test functions due to their nonlinearity and multiple peaks and valleys. A Hammersley design sample is used to generate 31 points used to fit the kriging metamodel. The results are summarized in Table 1, which shows how the proposed algorithm provides accurate results with a few number of function evaluations.

Table 1 . Results of the optimization algorithm using the Peaks and Branin functions.

	Global optimum	*Optimum (proposed algorithm)*	*Function eval. (kriging model)*	*Function eval., (optimization stage)*
Peaks fun.	8.106 (0.009, 1.58)	8.099 (0.004, 1.59)	31	3
Branin fun.	0.397 (-3.14, 12.27)	0.408 (-3.09, 12.15)	31	11

3.2. *Gas turbine*

The methodology has been also applied to the optimization of the operation parameters of a gas turbine (GT). The GT cycle is composed by the system of compressor-combustion chamber-turbine that uses natural gas. A saturation column, before the clean gas combustion, saturates this stream with vapor and nitrogen. GT air cooling has been modeled by taking into consideration four stages GT, thus the air compressor has been modeled as a four steps process as well. The combustion chamber is considered to operate at 15 bars, while air is feed to compressor at atmospheric pressure. All turbines and compressors model units are taken from the AspenPlus model library and are considered to be isentropic. The combustion chamber is modeled as a Gibbs reactor [1]. Each corresponding stage could have the same pressure loss or gain ratio, but in this case the intermediate pressures have been left unspecified and are subject of optimization. The objective function pursued is the maximization of net power (*NP*) obtained, by modifying the cooling air splits fractions and the intermediate stage pressures (of compressors and turbines).

Table 2. Results for the Gas Turbine case study optimization.

	Case 1 (kriging+EI)	**Case 2 (kriging)**	**Case 3 (A+SQP)**
SF_1	0	0	0.0797
SF_2	0	0.1249	0.1011
SF_3	0	0.1507	0.1000
SF_4	0.11329	0	0.0010
PC_2	8.116	9	4
PC_3	14.921	11.0555	10
PT_2	2.0000	2.8225	3
PT_4	1.21334	0.8764	1.5
NP	**239571**	**239490**	**239610**
# metamodel	60	60	0
# optimization	7	21	293

The algorithm has been tested under three different conditions: 1) optimization using kriging and the EI, 2) optimization using kriging, 3) optimization using the original AspenPlus model and Matlab's SQP algorithm. In the 1st and 2nd cases, the algorithm uses a Hammersley design of 60 points. The results show a significant number of function evaluations in the 3rd case. When using the kriging metamodel only (2nd column), the number of function evaluations significantly decreased but the optimizer goes to a local optimum, finally when using the EI coupled with the kriging metamodel the number of function evaluations is further decreased and also the objective is improved. In case of EI, a better solution can be obtained by decreasing the tolerance in the stopping criteria in the algorithm (step 9), but the number of evaluations will increase, so a trade off should be considered between the optimum accuracy and the computational effort.

4. Conclusions

One important goal of using metamodels to replace first principle models is to reduce the original model function evaluations (experiments). The EI technique shows a great behavior in meeting this goal, in front of other methods currently in use. The results clearly show how the use of the proposed methodology avoids local optima, provides better solutions and reduces the number of required function evaluations.

References

1. Bojarski, A.D; Pérez-Fortes, M.; Velo, E.; Puigjaner, L. (2010 (ECOS2010). Book of Abstracts 1493-1501.
2. Caballero, J.A., Grossmann, I. E. (2008). AICHE J. 54, 2633.
3. Jones, D.R., (2001. J. Glob. Optim. 21(4), 345-383.
4. Jones, D.R.; Schonlau, M.; Welch, W. J. (1998). J. Glob Optim. 13(4), 455 492.
5. Fang, K-T, Li R, Sudjianto A. (2006). Chapman & Hall / CRC.
6. Locatelli M. (1997). Journal of Global Optimization 10, 57–76.
7. Sacks J, Welch WJ, Mitchell WJ. Design and Analysis of Computer Experiments. Stat Sci. 1989; 4:409–435.
8. Sasena, M.J., Papalambros, P.Y. and Goovaerts, P. Proceedings of the 8th AIAA / USAF / NASA / ISMMO Symposium of Multidisciplinary Analysis & Optimization, September 2000. AIAA Paper 2000-4921.
9. Schonlau, M., Welch, W., Jones, D. (1998. New Developments and Applications in Experimental Design, N. Flournoy, W.F. Rosenberger, and W.K. Wong (editors), Institute of Mathematical Statistics, Hayward, California, Vol. 34, 11-25
10. Shinkyu Jeong, M. and Yamamoto, K. (2005). Journal of Aircraft 42, 412-420.

A CHANCE CONSTRAINED PROGRAMMING MODEL FOR SHORTEST PATH PROBLEM WITH FUZZY CONSTRAINTS

PINAR DURSUN*, ERHAN BOZDAG

Department of Industrial Engineering, Istanbul Technical University, Macka, Istanbul, 34367, Turkey

There are several methods in the literature for finding the shortest path under uncertain constraints. In this study a fuzzy chance constrained model based on genetic algorithm is proposed for shortest path problem with fuzzy constraints. The fuzzy simulation algorithm which checks whether fuzzy total travel time is in time window is presented in detail. A numerical example is given, and solved by using proposed method.

1. Introduction

To find the shortest path from a resource node to a destination node is a fundamental problem in graph theory. The cost of a path is the sum of the weights of the nodes on the path and weights may be distance, time, cost and so forth. Although in classical graph theory, the weights of the nodes in an SPP are supposed as real numbers; most practical applications have parameters that are not naturally precise [1]. Since these quantities are difficult or impossible to be estimated precisely in real life, uncertainty cannot be ignored [2]. There are two kinds of uncertainty-probability and possibility-the former may be measured by repetitions of experiments and the latter is employed to measure subjective belief in a fuzzy set [3]. Hui et al. [4] proposed chance constrained programming (CCP) model based on particle swarm algorithm embedded with stochastic simulation. Usually the probability distribution of arc lengths is difficult to estimate due to lack of historical data. However, these quantities may be given as membership functions by experienced expert. In this case, fuzzy set theory is a powerful tool to deal with this problem and the fuzzy shortest path problem appears in a natural way [2], [1]. Liu et al. [5] used CCP and dependent-chance programming based on credibility theory for the matching problems with fuzzy weights. Li et al. [6] constructed fuzzy chance constrained programming model based on the credibility theory for emergency resource scheduling problem with

* dursunpi@itu.edu.tr

fuzzy times. In this study, a fuzzy CCP model based on genetic algorithm is proposed to solve shortest path problem with fuzzy travel time constraint.

2. Chance Constrained Model

The mathematical formulation of the shortest path problem with fuzzy travel time constraint is represented below. Let ξ be a trapezoidal fuzzy number determined by quadruplet (u_1, u_2, u_3, u_4) and membership function $\mu(u)$. Then possibility measure of the fuzzy event $\{\xi \geq c\}$ can be represented by $\text{Pos}\{\xi \geq c\} = \sup_{u \geq c} \mu(u)$. The binary decision variable x_{ij} is 1 if the link from *i*th node to *j*th node is on the path, 0 otherwise. The discrete parameter c_{ij} denotes the travel cost and fuzzy parameter $\tilde{t}_{ij}$ denotes the travel time on link (*i*, *j*). The vehicle starts trip at time *T*. The arrival time to the destination must be between *a* and *b* namely time window of the destination.

$$\min \ \sum_{i=1}^{N} \sum_{j=1}^{N} x_{ij} \cdot c_{ij} \tag{1}$$

$$\text{s.t.} \quad \text{Pos}\left\{\left(\sum_{i=1}^{N} \sum_{j=1}^{N} x_{ij} \cdot \tilde{t}_{ij} + T\right) \in [a, b]\right\} \geq \alpha \tag{2}$$

$$\sum_{j=1, j \neq i}^{N} x_{ij} - \sum_{j=1, j \neq i}^{N} x_{ji} = 1 \qquad \text{for } i = 1 \tag{3}$$

$$\sum_{j=1, j \neq i}^{N} x_{ij} - \sum_{j=1, j \neq i}^{N} x_{ji} = -1 \qquad \text{for } i = N \tag{4}$$

$$\sum_{j=1, j \neq i}^{N} x_{ij} - \sum_{j=1, j \neq i}^{N} x_{ji} = 0 \qquad \text{for } i \neq 1 \text{ and } i \neq N \tag{5}$$

$$\sum_{j=1, j \neq i}^{N} x_{ij} \leq 1 \qquad \text{if } i \neq N \tag{6}$$

$$\sum_{j=1, j \neq i}^{N} x_{ij} = 0 \qquad \text{if } i = N \tag{7}$$

$$x_{ij} \in \{0, 1\} \qquad \text{for } \forall i, \forall j \tag{8}$$

3. Fuzzy Simulation based Genetic Algorithm

Genetic algorithms are population-based heuristic algorithms which draw inspiration from natural selection. In many complex optimization problems, significant success in providing good solutions has been obtained by using genetic algorithm. Fuzzy simulation is used to check whether a solution is feasible or not under fuzzy constraints. The steps of fuzzy simulation which are explained in detail in [3] are given in Figure 1. The proposed fuzzy simulation based genetic algorithm is explained in the rest of this section.

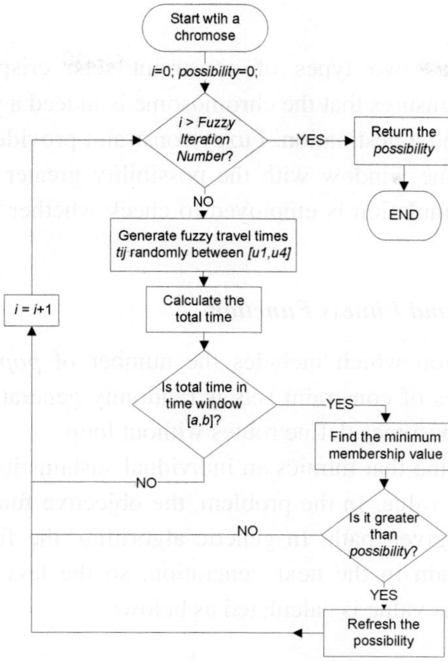

Figure 1. Fuzzy Simulation

3.1. *Database Matrix*

Firstly we constructed a symmetric Nx(N+1) database matrix keeps the information about the network. For any row of the matrix, the first N columns show the cost of the link which connects row node to column node. The last column keeps the number of available connections for each row node, namely the number of positive entries in a given row. This matrix is used to obtain the path from source to destination and to calculate the cost of the path.

3.2. *Representation*

In the proposed algorithm a chromosome is formed as a sequence of positive integers representing the nodes. The length of a chromosome is variable and the first locus is reserved for the source node as in [7]. The number of nodes that a route includes can be at most N which is the total number of nodes in the network and the last locus keeps the destination node. A node cannot be shown more than one time in a chromosome. For instance, $V = (s, n_2, n_6, \text{d})$ is a chromosome with four nodes.

3.3. *Feasibility*

Proposed model has two types of constraint sets: crisp and fuzzy. Crisp constraint set (3-8) ensures that the chromosome is indeed a path (without loops) between a source and a destination. Fuzzy constraint provides that the total time of the path is in time window with the possibility greater than a given target value. The fuzzy simulation is employed to check whether the fuzzy constraint holds or not.

3.4. *Initialization and Fitness Function*

The initial population which includes the number of *pop_size* chromosomes providing both types of constraint sets is randomly generated. It is certain that the generated chromosomes define routes without loop.

Each chromosome that mimics an individual sustains its own presence with regard to its fitness value. In the problem, the objective function is minimizing the total cost of a given path. In genetic algorithm, the fitter chromosome is more likely to remain in the next generation, so the less total cost is better. Therefore, the fitness value is calculated as below:

$$f = \frac{1}{\sum_{i=1}^{N}\sum_{j=1}^{N} x_{ij} \cdot c_{ij}} \tag{9}$$

3.5. *Selection*

In the proposed method the roulette wheel is used to select the best parents before crossover operation. Roulette wheel method assigns a probability of reproduction to each chromosome, so the probability of being selected is proportional to its fitness value relative to the other chromosomes in the population.

3.6. *Crossover*

A *crossover_proportion* that represents the expected proportion of the population which will undergo crossover operation is defined. Parents for crossover operation are chosen randomly. Except source and destination node, at least one of the nodes of parents must be common. If more than one node is common then one of them is selected randomly. The partial routes, from source node to selected node and from selected node to destination node are swapped and we get two offsprings. Suppose that $V_1 = (s, n_2, n_5, n_3, n_6, \text{d})$ and $V_2 = (s, n_5, n_4, \text{d})$ are two chromosomes which will be crossed over. The only common node is fifth node n_5 and it is located at third locus in the first parent and second locus in the second parent. Then after crossover operation, new chromosomes will be $V_1' = (s, n_2, n_5, n_4, d)$ and $V_2' = (s, n_5, n_3, n_6, d)$. Rarely, in a new offspring there may be a loop. In this case, a repair function finds and eliminates the defective loci forming loop [7]. The fuzzy simulation is applied to new chromosomes after crossover operation. Only if both of them are feasible under fuzzy constraint new chromosomes are changed with the original ones.

3.7. *Mutation*

Mutation makes possible to diversify the population and prevent premature convergence to local optimal solutions. The parameter m*utation_proportion* is defined to choose a chromosome which will be undergo mutation operation. For the chosen chromosome one of the loci except last one (destination node) is chosen randomly, and then the remaining partial route from this locus is refreshed by considering both types of constraint sets.

4. Numerical Example

We adapted the proposed method to the numerical experiment in [4]. There are 14 nodes in the network, node 1 is source and node 14 is destination. The network is shown in Figure 2. The distance (as cost) of the links are crisp and displayed in the figure.

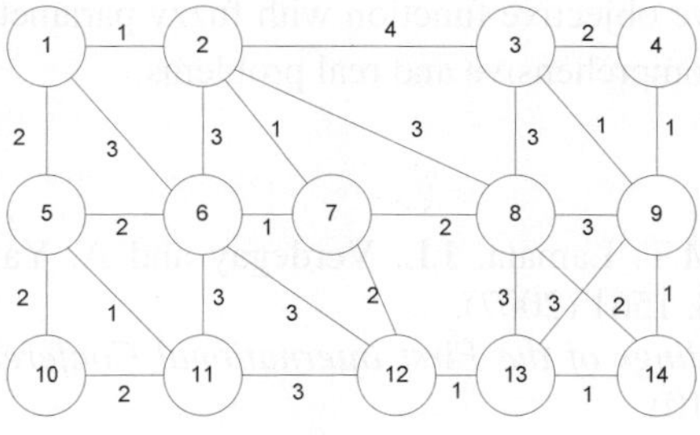

Figure 2. Network

The travel time of the links is trapezoidal fuzzy number (u_1, u_2, u_3, u_4) as given in Table 1. We assumed that there is only one vehicle starts travelling from node 1 and try to reach node 14 with minimum cost. But there is a time window that the vehicle must arrive to destination. The starting time of the vehicle is zero and the arriving time should be less than 9. Target possibility of fuzzy constraint is 0.90. So the fuzzy constraint will be $\text{Pos}\{\sum_{i=1}^{N}\sum_{j=1}^{N} x_{ij} \cdot \tilde{t}_{ij} \leq 9\} \geq 0.90$. If there is no time window, the optimum solution may be found by using deterministic methods and the optimal route will be 1-2-7-12-13-14 with the distance 6. A computer program was developed for the proposed method in C programming language and the example was solved. We run the program with the following parameters; *crossover_proportion* is 0.4, *mutation_proportion* is 0.1 and *fuzzy iteration number* is 1000. The algorithm found the solution which is stated in [4] in average five iterations with quite small population sizes as 6. According to the results, the best path is 1-6-12-13-14 with the distance 8 and possibility of being in time window is 0.99.

Table 1. Fuzzy Travel Times

Links	u_1	u_2	u_3	u_4
(1,2)	8.7	8.85	9.15	9.3
(1,5)	1.7	1.85	2.15	2.3
(1,6)	2.7	2.85	3.15	3.3
Others	0.7	0.85	1.15	1.3

5. Conclusion

In this paper, a chance constrained programming model was proposed to solve SPP with fuzzy parameters. According to the results, the model is efficient to solve this kind of problems. The model may be extended to other problems such as vehicle routing problem with time windows in uncertain environment and problems which have objective function with fuzzy parameters. Also, the model may be applied to comprehensive and real problems.

References

1. F. Hernandes, M.T. Lamata, J.L. Verdegay and A. Yamakami, *Fuzzy Sets and Systems* **158**, 1561 (2007).
2. W. Liu, *Proceedings of the First International Conference on Uncertainty Theory* **148**, (2010).
3. B. Liu and K. Iwamura, *Fuzzy Sets and Systems* **94**, 227 (1998).
4. F. Hui, Z. Zi and H. Gang, *International Conference on Logistics Engineering and Intelligent Transportation Systems (LEITS),* 1, (2010).
5. L. Liu, Y. Li and L. Yang, *Applied Mathematics and Computation* **181**, 662 (2006).
6. D. Li, G. Liu and Y. Gao, *Second International Symposium on Knowledge Acquisition and Modeling* (2009).
7. C. W. Ahn, *IEEE Transactions on Evolutionary Computation* **6** (2002).

PROACTIVE AND REACTIVE SCHEDULING WITH FUZZY ACTIVITY TIMES

MAREK JANCZURA

Institute of Organisation and Management, Faculty of Computer Science and Management, Wrocław University of Technology, Wrocław, Poland

DOROTA KUCHTA

Institute of Organisation and Management, Faculty of Computer Science and Management, Wrocław University of Technology, Wrocław, Poland

In this paper we develop a method of proactive and reactive scheduling, proposed earlier by Janczura et al. [3] in which fuzzy task durations are used. We construct a tool to create a robust schedule that might also be useful in repairing a schedule during the project realization. One of the benefits of this approach is supporting a Project Manager in project monitoring in the planning and realization phase. In this paper we propose a modification of an algorithm used to defuzzyficate the fuzzy task durations. This approach is based on defining a project security level and allows to control the project risk on the whole project level and not only separate task levels. Moreover, it provides a tool to a build secure schedule during the planning phase. For determining the crisp representations of the activity times the non-linear programming algorithm is used with the criteria of minimization the project duration given a project security level. In this paper we propose a modification of a defuzzyfication algorithm to consider the whole information included in a given fuzzy number.

1. Fuzziness as a measure of uncertainty

Planning activity durations is a challenge. First of all, very often we have to estimate the length of an activity with innovative character. In such situations we are not able to use information from the preceding projects and we have to cope with uncertainty.

Among many methods dealing with uncertainty we have selected fuzzy numbers as this approach allows to rewrite human subjective judgment into math formulas in a quite easy way.

We choose the following notation for fuzzy number $\tilde{T}_{ij}$ that determines the duration time of an activity (i, j):

$$\tilde{T}_{ij}, = (\underline{t}_{ij}, \overline{t}_{ij}, l_{ij}, u_{ij}) \tag{1}$$

Where $\underline{t}_{ij}$ and $\bar{t}_{ij}$ are the lower and upper bounds of the time period with the highest chance of realization, respectively. The two other parameters l_{ij} and u_{ij} define the boundaries of the intervals with diversified chance of realization, thus they define the risk of wrong time estimation. The fuzzy number is completely described by the membership function. Here we use the well-known concept of fuzzy numbers of L-R type [1] in which the membership function is defined as

$$m_{ij}(t) = \begin{cases} \dfrac{l_{ij} - \underline{t}_{ij} + t}{l_{ij}}, & t < \underline{t}_{ij} \\ 1, & \underline{t}_{ij} \leq t \leq \bar{t}_{ij} \\ \dfrac{u_{ij} - t + \bar{t}_{ij}}{u_{ij}}, & t > \bar{t}_{ij}. \end{cases} \tag{2}$$

In our research we use a linear function defining diversified boundaries of the activity time. Then fuzzy number is represented as the trapezoidal figure (see Fig.1).

2. Proactive and reactive scheduling

For the planning and realization phase of a project we propose to keep both schedules - the fuzzy and the crisp one. One of the main reasons is to monitor the variability of uncertainty during the project realization. The risk and uncertainty of the project realization are important not only during the planning phase, but usually remain significant till the end of the project. Hence, it is important to be able to reschedule the project during it's realization. The crisp schedule actualization is based on a fuzzy activity times update and their recalculation into the crisp form.

In our previous paper [4] we proposed a method of a fuzzy schedule actualization. In the first step we update the fuzzy duration estimates $\tilde{T}_{ij} = (\underline{t}_{ij}, \bar{t}_{ij}, l_{ij}, u_{ij})$ of unfinished activities on the basis of the information of the activity realization so far. Then we re-estimate the project total duration time, also in the form of a fuzzy number. In the real project defuzzyfication is required to enable to see the real cost estimations or even to allocate resources. For the conversion of numbers from the fuzzy to the crisp form we proposed [3] three strategies model (optimistic, average and pessimistic). This approach requires making decision at the given task level, and this is associated with increase of the project manager effort in a very huge projects. Therefore, we proposed a method that is based on defining a strategy (so called project security level) at the whole project level and algorithm that computes the corresponding crisp times for all activities.

2.1. *Project security level approach*

In this paper we propose a new method of robust schedule construction. It is based on the fuzzy task durations and a project security level (PSL) defined by a project manager. The main idea is to transform fuzzy schedule into the crisp one, using PSL for the whole project. The algorithm for planning and realization phase is as follows:

0. The task durations are given by fuzzy numbers.
1. A project manager sets a desired Project Security Level (PSL) for the whole project (e.g. 90%).
2. The project manager may define some constraints for a specific task or for a group of tasks (e.g. critical path).
3. The algorithm computes (using nonlinear programming) the best, at the given constraints, crisp times KSL_{ij} for all activities (i, j) in the project.

The project security level allows to determine in what extent we want to take into account the uncertainty of the task duration times estimation, which is included in fuzzy numbers. The knowledge about uncertainty can be used in variety of ways, but the settlement of the approved risk on the whole project level allows to control the entire project. It will be the project manager task to choose the desired PSL values, which can be interpreted as:

- PSL = 100% – the complete uncertainty is considered (pessimistic approach)
- PSL < 100% - the partial uncertainty is considered at PSL

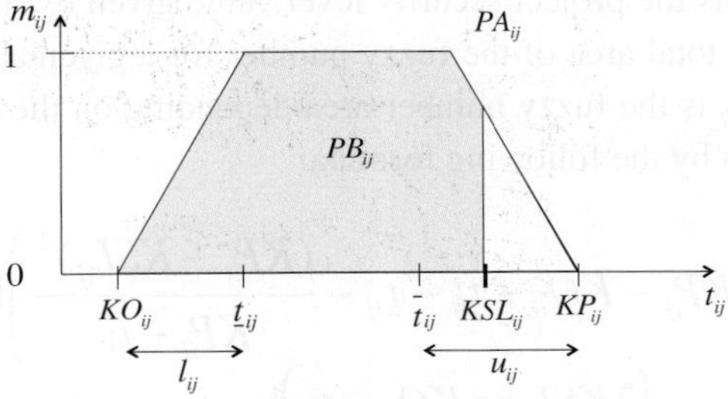

Figure 1. Fuzzy duration time for an activity (*i*, *j*).

We propose the measure of the considered uncertainty for a given task (*i*, *j*) to be PB_{ij}/PA_{ij} - the ratio of an area corresponding to the estimated duration (KSL_{ij} and PB_{ij} – see Fig 1) for the given task to the entire fuzzy number area. (KP_{ij} and PA_{ij} – see Fig 1). Therefore, the closer to 1 is the ratio, the more pessimistic estimation is. Such an approach is a modification and improvement of the method proposed by Janczura *et al.* [4]. The current approach allows us to take into account the whole information about the uncertainty and risk for each task included in their fuzzy duration estimates.

The level of the considered uncertainty for all project tasks can be controlled by the ratio $\frac{\sum_{ij} PB_{ij}}{\sum_{ij} PA_{ij}}$.

2.2. *Non-linear problem formulation*

A problem of finding the proper crisp time KSL_{ij} for each task (i, j) with respect to the given project security level (PSL) can be solved by the nonlinear programming methods based on minimization of the following objective function

$$\underset{(i,j)\in C}{Min} \sum_{ij} KSL_{ij},$$

where KSL_{ij} are the decision variables and C is the set of activities, for which we want to minimize their duration times. This can be all activities from the project or only these located on the critical path.

The above objective function is minimalized with the following constraints:

1) For the entire project we have:

$$PSL \le \frac{\sum_{ij} PB_{ij}}{\sum_{ij} PA_{ij}}, \tag{3}$$

where PSL is the project security level value given by the decision maker. PA_{ij} indicates the total area of the fuzzy number for a given activity $(i,\ j)$ and it is a constant. PB_{ij} is the fuzzy number area depending on the decision variables KSL_{ij} and is given by the following formula:

$$PB_{ij} = \frac{1}{2}\Bigg[\left(KP_{ij} - KO_{ij} + \bar{t}_{ij} - \underline{t}_{ij} - \frac{(KP_{ij} - KSL_{ij})^2}{KP_{ij} - \bar{t}_{ij}}\right)1_{\{\bar{t}_{ij} < KSL_{ij} \le KP_{ij}\}} + \left(2KSL_{ij} - KO_{ij} - \underline{t}_{ij}\right)1_{\{\underline{t}_{ij} < KSL_{ij} \le \bar{t}_{ij}\}} + \frac{\left(KSL_{ij} - KO_{ij}\right)^2}{\underline{t}_{ij} - KO_{ij}}1_{\{KO_{ij} \le KSL_{ij} \le \underline{t}_{ij}\}}\Bigg]. \tag{4}$$

In the above definition for PB_{ij} we have binary variables, as the area of the corresponding PB_{ij} depends on which range KSL_{ij} belongs to. The symbol $1_{\{\}}$ denotes indicator function.

2) Lower and upper constraints for the KSL_{ij} value can be also given. One of the possible applications might be securing all activities on critical path with a proper constraint.

$$KSL_{ij} \in \left[KO_{ij} * \phi_{ij}, KP_{ij} * \delta_{ij}\right]: \quad \phi \in \left[1, \frac{KP_{ij}}{KO_{ij}}\right], \delta_{ij} \in \left[\frac{KO_{ij}}{KP_{ij}}, 1\right].$$

The ϕ and δ_{ij} parameters are used to narrow the interval of the possible KSL_{ij} values.

3. Summary

In this paper we propose new approach for proactive and reactive scheduling with fuzzy activity times. In the planning phase we combine the fuzzy representations of the activities duration together with a approved risk (*PSL*) on the whole project to create a robust schedule. Moreover, during realization phase of a project we propose to keep both schedules - the fuzzy and the crisp one. The proposed tool of schedule actualization and reparation allows project manager to control the variability of uncertainty during the project realization. To illustrate the proposed approach, the algorithm will be tested on the data from a real project.

A verification of this algorithm on several different projects with big number of activities will be a topic for future research. We also plan to compare it with other commonly used methods. It may help to calibrate and deeply verify the proposed approach.

References

1. S. Chanas, P. Zieliński, *Critical path analysis in the network with fuzzy activity times*, "Fuzzy Sets and Systems" 122, 195-204 (2001).
2. S.J. Chen, C.L. Hwang, Fuzzy Multiple Attribute Decision Making, Springer Verlag, Berlin – Heidelberg (1997).
3. M. Janczura, D. Kuchta, *Reactive scheduling and risk in the project*, "Zeszyty naukowe Uniwersytetu Ekonomicznego w Katowicach". Modelowanie Preferencji a Ryzyko. Red. T. Trzaskalik (2011). (in polish)

4. M. Janczura, D. Kuchta, *Reactive scheduling with fuzzy activity times* – 31th International Conference ISAT2010 (Information System, Architecture, and Technology), Szklarska Poręba (2010).
5. D. Kuchta, *Time and cost in the project.* "Oficyna Wydawnicza Politechniki Wrocławskiej", Wrocław (2011). (in polish)
6. D. Kuchta, *Soft mathematic in management*, „Oficyna Wydawnicza Politechniki Wrocławskiej", Wrocław (2001).(in polish)
7. O. Lambrechts, E. Demeulemeester, W. Herroelen, *Proactive and reactive strategies for resource-constrained project scheduling with uncertain resource availabilities*, Springer (2007).
8. Vonder de S.V, F. Ballestín, E. Demeulemeester, W. Herroelen, *Heuristic procedures for reactive project scheduling* , "Computers & Industrial Engineering", 52(1), p.11-28 (2007).
9. Vonder de S.V., E. Demeulemeester, R. Leus, W. Herroelen, *Proactive/reactive project scheduling – trade-offs and procedures*, "Perspectives in modern project scheduling", Józefowska J., Węglarz J. (red.), Springer, Berlin-Heidelberg-Nowy Jork, p. 25-51 (2006b).

USING FUZZY SET THEORY FOR WATER QUALITY EVALUATION OF WULIHU LAKE, CHINA*

YUANKUN WANG[1], DONG SHENG[2], DONG WANG[1†], JICHUN WU[1]

[1]*Department of Hydrosciences, School of Earth Sciences and Engineering, State Key Laboratory of Pollution Control and Resource Reuse, Nanjing University, Nanjing, 210093, P. R. China*

[2]*Hunan Water Resources and Hydropower Research Institute, Changsha, 410007, P. R. China*

This paper reports the application of fuzzy set theory for water quality assessment of Wulihu Lake, China. Methodology with ranking feature value (RFV) which transforms fuzzy combination vector into assessment level is used to express the quality of water in the environment of monitored data. The results show that the proposed method gives the certainty water quality levels. The water quality level of Wulihu Lake grew better because of ecological projects. This study could provide a scientific basis for analyzing and evaluating the water quality for environment management.

1. Introduction

Many countries have introduced a scheme for river and lake water quality assessment, examining separate stretches of freshwater in terms of their chemical, biological and nutrient constituents and overall aesthetic condition [1, 2, and 3]. Assessment of lake environmental quality can provide an overall understanding of environmental conditions and assist in establishing priorities for management purposes in terms of environmental protection.

Environmental quality assessment, a quantitative description of environmental quality, is a challenge within the study of the environment [4]. Many methods have been proposed for water quality assessment in the literature, including expert assessment [5], cluster analysis [2], and neural networks [6]. However, in all environmental quality assessments, because of the inconsistency and peculiarity of each pollutant, there is a vagueness or fuzziness in relation to environmental quality [3]. Thus, the classification criteria used and the

* This work is supported by the National Natural Science Fund of China (No. 41071018, 41030746), China Doctoral Program of Higher Education (20100091120059), the Skeleton Young Teachers Program and Excellent Disciplines Leaders in Midlife-Youth Program of Nanjing University.

† E-mail: wangdong@nju.edu.cn

boundaries between different classes should be somewhat fuzzy [4, 7]. Fuzzy theory introduced by Zadeh (1965) [8] was designed to interpret the fuzziness and uncertainties of real situations. It became popular in the 1980s and was used mainly by researchers in electronics and computer engineering. Nowadays, an immense number of studies are carried out using this method. The fuzzy set theory has been proved effectively in solving problems of fuzzy boundaries and controlling the effect of monitoring errors on assessment results [4, 9]. These researches find that this model is fit to describe fuzzy character of classified bounds for water quality and can reflect the actual water quality on objectiveness.

The objectives of this paper were: (a) to investigate the status of pollution of Wulihu Lake of Taihu Lake Basin and (b) to assess the current situation of overall environmental quality by applying proposed assessment model. The findings could provide a scientific basis for improved water resource management and protection of the Taihu Lake Basin.

2. Materials and Methods

2.1. *Study site and data*

Wulihu Lake, located in north of Taihu Lake Basin, with an area of 5.15km^2, is a tourist attraction for Wuxi City (Fig. 1). Since the 1980s, the water of Wulihu Lake has become increasingly polluted, with its water quality being degraded over the passing years, to the degree that it became hypertrophic. This hampers the sustainable social and economical development of the whole basin.

Water samples were collected from the Wulihu Bay from 1998-2006. Standard procedures were followed for the collection of water samples [10]. The water samples were transported to the laboratory and analyzed for total nitrogen (TN), total phosphorus (TP), ammonia nitrogen (NH3-N), dissolved oxygen (DO), biological oxygen demand (BOD), permanganate index (COD_{Mn}), volatile phenol (VP), and petroleum (Pe) by standard methods.

2.2. *Assessment methods and procedure*

The following procedure describes ranking feature value (RFA) method based on fuzzy set theory [3, 4, and 11].

Step1: Select assessment parameters and establish assessment criteria

It is crucial to select assessment parameters which are representative, rational and accurate to form an assessment factor set *U*. The *U* is based on the actual local situation, and can be expressed as:

$$U=\{u_1, u_2, \ldots, u_n\} \tag{1}$$

where n is the number of selected assessment parameters. Eight main parameters, TN, TP, NH3-N, DO, BOD, COD_{Mn}, VP, and Pe, are selected for assessment parameters of the water quality in Wulihu Lake.

Then the assessment criteria set ***V*** is expressed as:

$$V=\{v_1, v_2, \ldots, v_m\} \tag{2}$$

where m is the number of assessment criterion categories. For Wulihu Lake, the water quality is classified into five levels (class I–V). And the assessment criterion set V (Table 1) is established based on National Surface Water Environmental Quality Standards of China (GB3838-2002).

Step2: Establish membership functions of each assessment parameter to assessment criteria at each level

The value of fuzzy membership function of each assessment parameter to assessment criteria at each level can be calculated by a set of formulae. In the present paper, the degree of affecting factors belong to certain class is determined through fuzzy theory. The triangle function is selected as membership function.

Table 1. National quality standards of surface waters (GB3838-2002) of China (units of mg/l).

Index	*I*	*II*	*III*	*IV*	*V*
DO	12-7.5	7.5-6	6-5	5-3	3-2
NH3-N	0-0.15	0.15-0.5	0.5-1.0	1.0-1.5	1.5-2.0
TN	0-0.2	0.2-0.5	0.5-1.0	1.0-1.5	1.5-2.0
BOD_5	0-3	3	3-4	4-6	6-10
COD_{Mn}	0-2	2-4	4-6	6-10	10-15
TP	0-0.02	0.02-0.1	0.1-0.2	0.2-0.3	0.3-0.4
VP	0-0.002	0.002	0.002-0.005	0.005-0.01	0.01-0.1
Pe	0-0.05	0.05	0.05	0.05-0.5	0.5-1

Step3: Calculate the membership function

The fuzzy matrix R is characterized by a membership function. Every element of matrix is the membership value of one factor to each evaluation class, which is calculated based on the membership function $\mu(x)$. The fuzzy matrix $\tilde{R}$ can be described as

$$\tilde{R} = \begin{pmatrix} r_{11} & r_{12} & \cdots & r_{1m} \\ r_{21} & r_{22} & \cdots & r_{2m} \\ \cdots & \cdots & \cdots & \cdots \\ r_{n1} & r_{n2} & \cdots & r_{nm} \end{pmatrix}_{n\times m} \tag{3}$$

where n is the number of assessment factors, m is the number of assessment class. r_{ij} is the membership degree of the ith assessment parameter at the jth level.

Step4: Calculate the weights matrix

Allocate weight of each assessment parameter to get matrix $\tilde{W}$ with the Analytical Hierarchy Process method. $\tilde{W} = \{w_1, w_2, \cdots, w_n\}$, $\sum_{i=1}^{n} w_i = 1$, n is the is the number of assessment factors.

Step5: Carry out the fuzzy algorithm

Determination of the fuzzy algorithm of $\tilde{W} \circ \tilde{R}$ can be computed by matrix multiplication:

$$\text{C=W} \times R = [C_1 \ C_2 \ \ldots \ C_m] \tag{4}$$

Step6: Obtain the assessment result

Finally, by using ranking feature value method proposed by Chen (1998) [12], water quality level can be obtained by

$$S\text{=S(u)=hC}^{\text{T}} \tag{5}$$

where h is the vector of ranking feature value, $\text{h}=\{1, 2, \cdots, \text{k}\}$. Here, k is the quantity of assessment class, and k=5 in this paper.

3. Results and Discussion

Using the values of assessment indices of the Wulihu Lake during 1998–2006, step by step application of the model based fuzzy theory is implemented. Take the data in 1998 as example, the assessment process is shown as follows.

In the present study, the AHP method is used to get the weight of every factor. The weight of every factor is shown as follows

$$\text{W} = [0.114\ 0.116\ 0.125\ 0.143\ 0.151\ 0.175\ 0.091\ 0.085]$$

Fuzzy combination expressions:

$$\text{C=W} \times R = [0.0437\ 0.0855\ 0.2336\ 0.329\ 0.3083]$$

S value:

$$S\text{=S(u)=}[1\ 2\ 3\ 4\ 5][0.0437\ 0.0855\ 0.2336\ 0.329\ 0.3083]^{\text{T}} = 3.773$$

Similarly, assessment results of all data during 1998–2006 are shown in Table 2. From Table 2, with the original fuzzy comprehensive assessment (FCA) method, we got the assessment results that the water quality level fall in class V during 1999-2006, except class V in 1998. In the original fuzzy comprehensive assessment method, the maximum membership principle is widely used [3, 11]. The maximum value of b_j, determines risk class of the study case (Class = max (b_j), j _ 1, 2, . . . , 5). If we judge assessment class with the maximum membership principle, then contradictions may arise when there is more than one the maximum values of b_j. For example, membership values in 2006 is b_1 =0.1877, b_2 = 0.1730, b_3 = 0.2324, b_4 = 0.1659, and b_5 = 0.2410. The results should be class *V* by the maximum membership principle. However, b_3 is very close to b_5, the assessment results may be not exact.

Table 2. Assessment results of Wulihu Lake (1998 - 2006)*.

Year	*I* (c_1)	*II* (c_2)	*III* (c_3)	*IV* (c_4)	*V* (c_5)	c_{max}	*Class*	*RFV*
1998	0.0437	0.0855	0.2336	0.3290	0.3083	**0.3290**	*IV*	3.7730
1999	0.0914	0.0378	0.2022	0.2846	0.3840	**0.3840**	*V*	3.8320
2000	0.0259	0.1033	0.2022	0.2846	0.3840	**0.3840**	*V*	3.8975
2001	0.1169	0.0881	0.1870	0.2240	0.3840	**0.3840**	*V*	3.6701
2002	0.0262	0.1485	0.1356	0.2816	0.4082	**0.4082**	*V*	3.8974
2003	0.0241	0.1405	0.3326	0.1803	0.3225	**0.3225**	*V*	3.6366
2004	0.0321	0.0970	0.2837	0.2353	0.3518	**0.3518**	*V*	3.7774
2005	0.0951	0.2092	0.2258	0.1355	0.3259	**0.3259**	*V*	3.3624
2006	0.1877	0.1730	0.2324	0.1659	0.2410	**0.2410**	*V*	3.0995

*Ranking feature value: RFV.

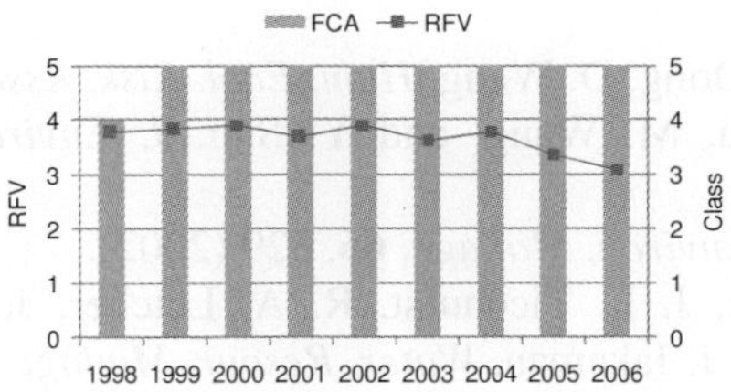

Figure 2. Changes of water quality in Wulihu Lake during 1998–2006.

With ranking feature value (RFV) based on fuzzy comprehensive assessment, the assessment results are also shown in Table 2. One can see from Table 2, the ranking feature values vary from 3.0995 to 3.8975. Water quality values of the Wulihu Lake belong to grade of 3-4. The water quality changes in Wulihu Lake from 1998 to 2006 are plotted in Figure 2. It had become better since 2004 with RFV while always in class *V* since 1999 with the maximum membership principle. In fact, great efforts had been devoted to improve the water quality of Taihu Lake Basin. Many important water quality improvement ecological projects of Wulihu Lake have been carried out since 2004 [13]. The ecological engineering did prevent the water quality further deterioration. Through fuzzy comprehensive assessment results, one can know that the water quality level fell into class *V* from 2004 to 2006. The assessment results did not reflect the real status of Wulihu Lake water quality. However, by RFV the water quality reduced from 3.774 to 3.009, close to class III. Thus, the assessment results of the proposed model here was more exact. Recently, blue-algae blooming occurred in Wulihu Lake, which seriously impacts water's function. Therefore, more efforts should be paid to protect the water quality of Wulihu Lake further.

4. Conclusions

In the present paper, the fuzzy set theory is applied for water quality assessment of Wulihu Lake, China. The method showed its great advantages in the case study of China. The ranking feature value (RFV) based on fuzzy set theory provides a scientific basis for analyzing and evaluating the environmental quality of lake. According to the Chinese Environmental Quality Standard for water, by the method proposed here the water quality in the Wulihu Lake region belongs to class IV, and gradually close to class III.

References

1. J. Dojlido, J. Raniszewski, and J. Woyciechowska, *Water. Sci. Technol.* **30**, 57 (1994).
2. Y. K. Wang, S. Dong, D. Wang. *Hum. Ecol. Risk. Assess.* **17**, 212 (2011).
3. G. Shen, Y. Lu, M. Wang, and Y. Sun, *J. Environ. Manage.* **76**, 355 (2005).
4. H. Y. Wang, *J. Environ. Manage*, **66**, 329 (2002).
5. B. E. W. Croke, J. L. Ticehurst, R. A. Letcher, J. P. Norton, L. T. H. Newham, and A. J. Jakeman, *Water. Resour. Manage.* **21**, 351 (2007).
6. K. P. Sudheer, I. Chaubey, and V. Garg, *J. Am. Water. Resour. As.* **42**, 1683 (2006).
7. T. Li, S. Cai, H. Yang, X. Wang, S. J. Wu, and X. Ren, *Environ. Eng. Sci.* **26,** 451 (2009).
8. L. A. Zadeh, *Inf. Control.* **8**, 338 (1965).
9. O. E. Guleda, D. Ibrahim, and H. Halil, *Atmos. Environ.* **38**, 3809 (2004).
10. Chinese Environmental Protection Agency. China Environment Press, Beijing (in Chinese), (2002).
11. Y. K. Wang, D. Wang, and J. C. Wu, *J. Mult-valued. Log. S.* **Accepted**. (2012)
12. S. Chen, Engineering Fuzzy Set Theory and Application, Natl Def Ind Press, Beijing (in Chinese), (1998).
13. K. N. Chen, C. H. Bao, and W. P. Zhou, *Ecol. Eng.* **35**, 1646 (2009).

NEWSVENDOR PROBLEM WITH ONE-SHOT DECISION THEORY

PEIJUN GUO[†]

Faculty of Business Administration, Yokohama National University, Yokohama, 240-8501 Japan

XIUYAN MA

Faculty of Business Administration, Yokohama National University, Yokohama, 240-8501 Japan

In this paper, a single-period inventory problem with partially known demand information characterized by possibility distributions is analyzed. Decision models are proposed based on one-shot decision theory. Two choices are used for determining which state of nature (demand) should be considered for each order quantity. Based on the selected states of nature (focus point), the optimal order quantity is obtained. Different kinds of focus points (demands) lead to different results (optimal order quantities) which reflecting a decision maker's attitudes about possibility of demand and satisfaction of decision.

1. Introduction

The newsvendor problem, also known as newsboy or single-period problem is a common inventory problem. In general, the newsvendor problem has the following characteristics. Prior to the season, the buyer must decide the quantity of the goods to purchase/produce. The procurement lead-time tends to be quite long relative to the selling season so that often there is not enough opportunity to replenish inventory once the season has begun. Excess stock can only be salvaged at a loss once the season is over. Many extensions of the problem have been studied, such as problems with different objective and utility functions, supplier pricing policies and newsvendor pricing policies [7-9, 11]. Most of the extensions have been made in the probabilistic framework; i.e. the uncertainty of demand and supply is characterized by the probability distribution, and the objective function is expressed as maximizing the expected profit or probability measure of achieving a target profit.

[†]E-mail address: guo@ynu.ac.jp (P. Guo)

In this research, we consider a new product with a short life cycle. Because the product is new, there is no data available for forecasting the upcoming demand via statistical analysis. Since the life cycle of the product is short, how to determine the optimal order amount is a typical one-shot decision problem where a decision is experienced only once.

Decision theories under uncertainty are theories of choice under uncertainty where the objects of choice are probability distributions (for Expected Utility Theory, Subjective Expected Utility and their varieties), or prospects framed in terms of gains and losses (for Prospect Theory), or possibility distributions (regarded as possibilistic lotteries [2]). In fact, for one-shot decision problems, there is one and only one chance for only one state of nature occurring. In the paper [3, 4], the one-shot decision framework is proposed for dealing such one-shot decision problems. The one-shot decision process is separated into two steps. The first step is to identify which state of nature should be taken into account for each alternative with considering possibility of a state of nature and satisfaction of an outcome. The states of nature focused are called focus points. The second step is to evaluate the alternatives to obtain the optimal alternative. As an application, a duopoly market of a new product with a short life cycle is analyzed. In the paper [5], private real estate investment problem is analyzed within one-shot decision framework. In the paper [6], one-shot decision theory is proposed.

The paper is organized as follows: In section 2, one-shot decision theory is introduced. In section 3, the newsvendor problem is considered based on the one-shot decision theory. A summary of our results as well as concluding results and directions for future study are provided in section 4.

2. Introduction to one-shot decision theory

The first step in decision analysis is the problem formulation. The set of an alternative a is A. The set of a state of nature x is S. The degree to which a state of nature is to occur in the future is characterized by a possibility distribution $\pi(x)$ defined below.

Definition 1. Given a function $\pi : S \rightarrow [0,1]$ if $\max_{x \in S} \pi(x) = 1$, then $\pi(x)$ is called a possibility distribution where S is the sample space. $\pi(x)$ is the possibility degree of x. $\pi(x) = 1$ means that it is normal that x occurs and $\pi(x) = 0$ means that it is abnormal that x occurs. The smaller the possibility degree of x, the more surprising is the occurrence of x.

The consequence resulting from the combination of an alternative a and a state of nature x is referred to as a payoff, denoted as $v(x,a)$. The satisfaction level of a decision maker for a payoff can be expressed by a satisfaction function, as defined below.

Definition 2. Denote the set of a payoff $v(x,a)$ as V. The function $u:V \to [0,1]$ is called a satisfaction function if it satisfies $u(v_1) > u(v_2)$ for $v_1 > v_2$.

Because the payoff is the function of x and a, we write the satisfaction function as $u(x,a)$. Since one and only state of nature will come up for a one-shot decision problem, a decision maker needs to decide which state of nature ought to be considered for making a one-shot decision. Each state of nature is equipped with a pair of possibility and satisfaction so that how to determine focus points (focused states of natures) depends on his/her attitudes about possibility and satisfaction. Twelve types of focus points are proposed to help a decision maker in finding out his/her own appropriate one [6]. We considered two of them in this paper, which are shown as follows:

$$x^*(a) = \arg\max_{x \in S} \min[\pi(x), u(x,a)] . \tag{1}$$

$\arg\max_{x \in S} \min[\pi(x), u(x,a)]$ is for seeking the state of nature with the higher possibility degree and satisfaction level. The selected state of nature $x^*(a)$ is called an active focus point of an alternative a.

$$x_*(a) = \arg\min_{x \in S} \max[1 - \pi(x), u(x,a)] \tag{2}$$

$\arg\min_{x \in S} \max[1 - \pi(x), u(x,a)]$ is for seeking the state of nature with the higher possibility degree and lower satisfaction level. The selected state of nature $x_*(a)$ is called a passive focus point of an alternative a.

For one alternative, more than one state of nature might exist as one type of focus point. We denote the sets of two types of focus points of the alternative a as $X^*(a)$, $X_*(a)$, respectively.

In one-shot decision, the decision maker contemplates that the focus points are the most appropriate states of nature (scenarios) for him/her and the decision maker chooses one alternative which can bring about the best consequence (highest satisfaction level) once the focus point (scenario) comes true. Two types of optimal alternatives are shown below.

$$a^* = \arg\max_{a \in A} \max_{x^*(a) \in X^*(a)} u(x^*(a), a),$$

$$a_* = \arg\max_{a \in A} \min_{x_*(a) \in X_*(a)} u(x_*(a), a).$$

3. Newsvendor model based on one-shot decision theory

The plausible information of demand x is represented by a possibility distribution. The possibility distribution of x, denoted as $\pi_D(x)$, is defined by the following continuous function

$$\pi_D : [d_l, d_u] \rightarrow [0,1], \tag{3}$$

$\exists d_c \in [d_l, d_u]$, $\pi_D(d_c) = 1$, $\pi_D(d_l) = 0$ and $\pi_D(d_u) = 0$. π_D increases within $x \in [d_l, d_c]$ and decreases within $x \in [d_c, d_u]$. d_l and d_u are the lower and upper bounds of demand, respectively, d_c is the most possible amount of demand.

Because demand is inside the interval $[d_l, d_u]$, a reasonable supply also should lie inside this region. The highest profit of retailer is $r_u = (R - W)d_u$. That is, the retailer orders the most d_u and the demand is the largest d_u. The lowest profit is $r_l = (d_l R + (d_u - d_l)S_o - d_u W) \wedge (d_l R - (d_u - d_l)S_u - d_l W)$, which is determined by the minimum of two cases, one is that the retailer order the most but the demand is the lowest, the other is that the retailer order the lowest but the demand is the highest. Without loss of generality, the assumption $W \geq S_o + S_u$ is made, which leads to $r_l = d_l R + (d_u - d_l)S_o - d_u W$.

The satisfaction function of retailer is defined by the following continuous strictly increasing function of profit r,

$$U : [r_l, r_u] \rightarrow [0,1], \tag{4}$$

where $U(r_l) = 0, U(r_u) = 1$. (4) gives a general form of satisfaction function of decision-maker where the satisfaction of the lowest profit is 0 and the satisfaction of the highest profit is 1. Considering the active focus point as

$$x_o^*(q) = \arg\max_x \min[\pi_D(x), U(r(x,q))], \tag{5}$$

the optimal active order is

$$q_o^* = \arg\max_q \max_{x_o^*(q) \in X_o(q)} U(r(x_o^*(q), q)), \tag{6}$$

where $X_o(q)$ is the set of active focus points when the quantity of order is q. $U(r(q,q))$ is the maximum of $U(r(x,q))$ and also a strictly increasing function of q. $U(r(q,q))$ and $\pi_D(q)$ has an intersection within (d_l, d_c) and another one within (d_c, d_u), their horizontal coordinates are d_{ol}^* and d_{ou}^*, respectively.

Theorem 1. The optimal active order q_o^* is as follows:

$$q_o^* = \arg \max_{q \in [d_{ol}^*, d_{ou}^*]} \max_{x_o^*(q) \in X_o(q)} U(r(x_o^*(q), q))$$
$$= \arg \max_{q \in [d_{ol}^*, d_{ou}^*]} U(r(q,q)) = d_{ou}^* \quad (7)$$

which is the solution of $U(r(q,q)) = \pi_D(q)$ with $q \in (d_c, d_u)$.

Let us consider the passive focus point as follows:

$$x_p^*(q) = \arg \min_x \max[1 - \pi_D(x), U(r(x,q))]$$

The optimal passive order is

$$q_p^* = \arg\max_q \min_{x_p^*(q) \in X_p(q)} U(r(x_p^*(q), q)) \quad (9)$$

where $X_p(q)$ is the set of passive focus points when the quantity of order is q.

Theorem 2. The optimal passive order quantity q_p^* is as follows:

$$q_p^* = \frac{(R - S_o) d_{pl}^{*3}(q) + S_u d_{pu}^{*3}(q)}{(R - S_o) + S_u}, \quad (10)$$

where $d_{pl}^{*3}(q)$ and $d_{pu}^{*3}(q)$ are the horizontal coordinates of the intersections of $U(r(x,q))$ and $1 - \pi_D(x)$ within $[d_l, \min(q, d_c)]$ and $[\max(q, d_c), d_u]$, respectively.

4. Conclusions

In this paper, the newsvendor problem is analyzed with one-shot decision theory. The analysis results provide insights into different kinds of retailers. The procedures for calculating optimal active order quantity and optimal passive order quantity have been concluded.

The analysis of newsvendor model is just the beginning of our research, in future, the supply chain coordination with contracts between a single supplier and a single retailer or a single supplier and several retailers will be studied.

References

1. D. Dubois, H. Prade, Possibility theory as a basis for qualitative decision theory, *Proceedings of the 14th International Joint Conference on Artificial Intelligence*, 1924-1930 (1995).

2. D. Dubois, H. Prade and R. Sabbadin, Decision-theoretic foundations of possibilty theory, *European Journal of Operational Research*, 128, 459-478 (2001).
3. P. Guo, Decision analysis based on active focus point and passive focus point, *Proceedings of the International Workshop of Fuzzy Systems and Innovational Computing*, 39-44 (2004).
4. P. Guo, One-shot decision approach and its application to duopoly market, *International Journal of Information and Decision Sciences*, 2, 213–232 (2010).
5. P. Guo, Private Real estate investment analysis within one-shot decision framework, *International Real Estate Review*, 13, 238-260 (2010).
6. P. Guo, One-shot decision theory, *IEEE Transactions on SMC: Part A*, 41, 917-926 (2011).
7. M. Khouja, The single-period (news-vendor) problem: Literature review and suggestion for future research, *Omega*, 27, 537-553 (1999).
8. S.A. Lippman, K.F. McCardle, The competitive newsboy, *Operations Research*, 45, 54-65 (1997).
9. N.C. Petruzzi, M. Dada, Pricing and the newsvendor problem: A review with extensions, *Operations Research,* 47, 183-194 (1999).
10. H. Prade, Modal semantics and fuzzy set theory. *In Recent Developments in Fuzzy Set and Possibility Theory*, R. R. Yager, Ed. 232-246. Pergamon Press, New York (1982).
11. G. Raz, E.L. Porteus, A fractiles perspective to the joint price/quantity newsvendor model, *Management Science*, 52, 1764-1777 (2006).
12. R.R. Yager, A foundation for a theory of possibility, *J. Cybernetics*, 10, 177-204 (1980).
13. L.A Zadeh, Fuzzy sets as a basis for a theory of possibility, *Fuzzy Sets and Systems* 1,3-28 (1978).

A TWO-PHASED ADDITIVE APPROACH FOR MULTIPLE OBJECTIVE SUPPLIER SELECTION WITH FUZZY DEMAND LEVEL

FEYZAN ARIKAN*
Industrial Engineering Department, Gazi University,
Ankara, 06570-Maltepe, Turkey.

In today's severe competitive environment the selection of appropriate suppliers is a significantly important decision for effective supply chain management. In this study a multiple sourcing supplier selection problem is considered as a multiple objective linear programming problem with fuzzy demand level. Three objective functions are minimization of costs, maximization of quality and maximization on-time delivery respectively. In order to solve the problem, a two-phased additive approach is proposed. In the first phase, ideal solutions are utilized for finding aspiration levels for each objective. In the second phase Chen&Tsai's fuzzy model is adopted for the problem. The adopted model can be efficiently used to obtain a non-dominated solution. The approach is illustrated by a numerical example.

1. Introduction

The selection of appropriate suppliers is a multi-criteria decision making problem which includes both qualitative and quantitative factors some of which may conflict. Moreover, informational vagueness because of the tangible and intangible factors of the problem must be taken into account to reach effective configuration and coordination of supply chains. Supplier selection problem has been a focus area of research since 1960. Basically there are two kinds of supplier selection problem: *(1) Single sourcing* and *(2) Multiple sourcing* in which buyers need to make two decisions: which suppliers are the best, and how much should be purchased from each selected supplier?

In this study multiple sourcing supplier selection is considered as a multiple objective decision making problem with informational vagueness. The solution approaches of fuzzy multiple objective decision making problems can be found in [1]. In literature a few of the studies have been addressed both multiple objectives and fuzziness in supplier selection problem. Details and classification

* E-mail: farikan@gazi.edu.tr

of these studies can be found in [2] and [3]. They have been employed five solution approaches which are Zimmermann's max-min approach; Tiwari *et al.*'s additive model, Fuzzy goal programming with weights, Fuzzy programming with modified fuzzy or operator, and sequential quadratic programming. Zimmermann's max-min approach [4] uses min operator which corresponds to the set-theoretic intersection in fuzzy mathematical modeling. Tiwari *et al.* [5] proposed an additive model in which membership functions are combined using the add operator. The model maximizes achievement levels in total and the solution may include zero or low level achievement(s) [6]. In such a case, Chen&Tsai's model will be appropriate for the solution. The model guarantees non-dominated solution whereas Zimmermann's max-min does not [6]. In this study Chen&Tsai's fuzzy model [7] is utilized in a two phased solution approach for a typical multiple objective supplier selection problem

2. Multiple Objective Supplier Selection Model

A typical linear model for supplier selection problem [2]-[3] is as follows:

$$\text{Minimize } z_1 = \sum_{i=1}^{n} P_i(x_i) \qquad (1)$$

$$\text{Maximize } z_2 = \sum_{i=1}^{n} F_i(x_i) \qquad (2)$$

$$\text{Maximize } z_3 = \sum_{i=1}^{n} S_i(x_i) \qquad (3)$$

$$\text{s.t. } \sum_{i=1}^{n} x_i = D; \quad x_i \le C_i, \forall i; \quad x_i \ge 0 \text{ and integer} \qquad (4), (5), (6)$$

Where i is index for suppliers, for all $i=1,2,\dots,n$; x_i *is* the number of units purchased from the $i^{th.}$ supplier; D is aggregate demand of the item over a fixed planning period, n is the number of suppliers competing for selection, P_i is the per unit net purchase cost from supplier i, F_i is the percentage of the accepted units delivered by the supplier i, S_i is the percentage of the on-time deliveries by the supplier i,, C_i capacity of ith supplier. Objective function (1) minimizes total monetary cost. Objective function (2) maximizes total quality of purchased items. Objective function (3) maximizes service level of purchased items. Constraint (4) ensures that the overall demand satisfied. Constraint (5) means that order quantity of each supplier should be equal or less than its capacity. Constraint (6) prohibits negative orders.

3. Preliminary Definitions of Fuzzy Mathematical Programming

Consider the fuzzy multiple objective programming problem (7) with k fuzzy objective functions and s fuzzy constraints:

$$
\begin{aligned}
& Find \quad x \\
& s.t. \\
& c_l x \gtrsim z_l \qquad && l \in I_1 \\
& c_l x \lesssim z_l \qquad && l \in I_2 \\
& a_r x \cong b_r \qquad && r \in T \\
& x \in X
\end{aligned}
\tag{7}
$$

Where $c_k x = \sum_{i=1}^{n} c_{ki} x_i$, $k = 1,\ldots,l$; $a_r x = \sum_{i=1}^{n} a_{ri} x_i$, $r=1,\ldots,s$; $I_1 \cup I_2 = \{1,\ldots,l\}$, $I_1 \cap I_2 = \emptyset$, and X is a set of deterministic linear constraints and sign restrictions. For $l \in I_{1,2}$, z_l is the imprecise aspiration level for the $l^{th.}$ objective function.. $z_l \in [z_l^L, z_l^U]$ denotes the imprecise lower and upper, respectively, bounds for the lth. fuzzy objective function.. $b_r \in [b_r^L, b_r^U]$ denotes the imprecise lower and upper, respectively, bounds for the rth. fuzzy constraints. According to fuzzy mathematical programming, each fuzzy objective and constraint are defined in terms of fuzzy subsets with the appropriate membership functions denoted by $\mu_l(c_l x)$ for $i \in I_{1,2}$ and $\mu_r(a_r x)$ *for* $r \in T$. Assuming that membership functions are linear, mathematical definitions (8) to (10) are as follows:

$$
\mu_l(c_l x) = \begin{cases} 1 & if \quad c_l x \ge z_l^U \\ \dfrac{(c_l x) - z_l^L}{z_l^U - z_l^L} & if \quad z_l^L \le c_l x \le z_l^U, \ \forall l \in I_1 \\ 0 & if \quad c_l x \le z_l^L \end{cases} \tag{8}
$$

$$
\mu_l(c_l x) = \begin{cases} 1 & if \quad c_l x \le z_l^L \\ \dfrac{z_l^U - (c_l x)}{z_l^U - z_l^L} & if \quad z_l^L \le c_l x \le z_l^U, \quad \forall l \in I_2 \\ 0 & if \quad c_l x \ge z_l^U \end{cases} \tag{9}
$$

$$
\mu_r(a_r x) = \begin{cases} 0 & if \quad a_r x \le b_r^L \\ \dfrac{a_r x - b_r^L}{b_r - b_r^L} & if \quad b_r^L \le a_r x \le b_r \\ \dfrac{b_r^U - (a_r x)}{b_r^U - b_r} & if \quad b_r \le a_r x \le b_r^U \\ 0 & if \quad a_r x \ge b_r^U \end{cases}, \quad \forall r \in T \tag{10}
$$

Chen&Tsai's fuzzy model [7] is adopted for model (7) in (11). Variables denoted by λ_k, λ_r and λ_p represent achievement levels of fuzzy objective functions and fuzzy constraints respectively. Parameters α_k, α_r and α_p represent the minimum acceptable achievement levels for the $k^{th.}$ objective, $r^{th.}$ constraint $p^{th.}$ constraint respectively determined by the DM. This model allows the DM to assign explicitly a desirable achievement degree for each fuzzy aspiration [7].

$$
\begin{aligned}
&\max.\ \left(\sum_{k=1}^{l}\lambda_k+\sum_{r=1}^{s}\lambda_r+\sum_{p=1}^{q}\lambda_p\right)/(l+s+q) && \\
&s.t. && \\
&\lambda_k \le \mu_k(c_k x) && k\in I_l \cup I_2 \\
&\lambda_r \le \mu_r(a_r x) && r\in T \\
&\lambda_p \le \mu_p(a_p x) && p\in P \\
&\lambda_k \ge \alpha_k, && k=1,\dots,l \\
&\lambda_r \ge \alpha_r && r=1,\dots,s \\
&\lambda_p \ge \alpha_p && p=1,\dots,q \\
&x\in X && \\
&\alpha_k,\alpha_r,\alpha_p,\lambda_i,\lambda_r,\lambda_p \in [0,1] && k=1,\dots,l;\ r=1,\dots,s;\ p=1,\dots,q
\end{aligned}
\tag{11}
$$

4. Two Phased Additive Approach

Phase 1. *Step 1.* Construct model (7). *Step 2.* Determine the type of the membership function for fuzzy demand level. *Step 3.* According to the bounds of fuzzy demand level solve model (7) construct corresponding payoff tables and present them to the DM. Determine suitable upper and lower bound as aspiration level for each objective. Construct linear memberships for objectives.
Phase 2. *Step 4.* Construct additive model and solve it. Present the solution to the DM as references for the minimum acceptable achievements for each objective and demand level. *Step 5.* Construct model (11) and solve it. If the DM satisfied with the solution, then stop. Otherwise go to step 3

5. Illustrative Example

The following example problem is adopted from Yücel and Güneri's study [8] for multiple objective supplier selection problem with fuzzy demand level. The suppliers' quantitative information is given in Table 1.

Step 1-2) Multiple objective supplier selection model (7) is constructed according to the quantitative information in Table 1 with fuzzy demand which is assumed as a fuzzy triangular number (10) and it is defined with [750, 800, 875].

Table 1. Suppliers' Source Data for the Illustrative Example

Supplier number i	P_i Net price	F_i Quality (%)	S_i Delivery (%)	C_i Capacity
1	5	80	90	400
2	7	90	80	450
3	4	85	85	450

Step 3) Considering 750 and 875 demand levels, ones at a time, the ideal solutions for each objective in model (7) are found. Constructed payoff tables are Table 2 and Table 3. The lower and upper bounds for each objective according to payoff tables are determined as (z_1^L, z_1^U)=(3300, 4850), (z_2^L, z_2^U)=(617.5, 766.25), (z_3^L, z_3^U)=(615, 762.5). Bounds are assumed as acceptable for the decision maker. The related membership functions are constructed by using (8-9) equations with the mentioned bounds of aspirations.

Table 2. Payoff table (demand level chosen as 750)

$b_l-\Delta_{lL}=750$	$z_1(x_1^*)$	$z_2(x_2^*)$	$z_3(x_3^*)$
Min z_1	3300	4350	3400
Max z_2	622.50	660	617.5
Max z_3	652.5	615	657.5

Table 3. Payoff table (demand level chosen as 875)

$b_l+\Delta_{lL}=875$	$z_1(x_1^*)$	$z_2(x_2^*)$	$z_3(x_3^*)$
Min z_1	3975	4850	3975
Max z_2	725	766.25	725
Max z_3	762.5	721.25	762.5

Table 4 Solution Summaries for the Illustrative Example

Approach	Fuzzy Additive Model	Chen &Tsai's Fuzzy Model
x_1,x_2,x_3	350,0,450	369,17,450
Min z_1 [3550,4900]	3975.0	3764.0
Max z_2 [660, 702.5]	725.0	693.0
Max z_3 [657.5, 700]	762.5	728.2
Demand [750, 800,875]	875.0	836.0
$\mu_1(c_1x)$	0.839	0.701
$\mu_2(c_2x)$	0.303	0.508
$\mu_3(c_3x)$	0.559	0.767
$\mu_4(a_1x)$	1.000	0.520

Step 4). Model (11) is utilized for the illustrative example and the fuzzy optimal solution is obtained by GAMS computer programming package. The additive model solution is given in Table 4.

Step 5) Minimum acceptable aspiration levels attained by the DM assumed as (0.7, 0.5, 0.7) for α_k, k=1,2,3 and (0.5) for α_r r=1. Then model (11) is constructed and solved (Table 4). Chen&Tsai's fuzzy model (11) reflects the DM's preferred achievement levels. The algorithm is terminated with the assumption that the DM is satisfied with the current fuzzy optimal solution.

6. Conclusions

In this study a novel two phased fuzzy additive solution approach is proposed for a typical multiple objective supplier selection model which considers three crisp objective functions with fuzzy demand level. The procedure gives an opportunity to the DM to obtain her/his own preferred achievement levels for the objectives and for the demand level. The proposed solution approach can be employed not only for supplier selection problem but for any kind of multiple objective programming problems.

References

1. Arikan, F., Güngör, Z., A two-phased solution approach for multi-objective programming problems with fuzzy coefficients, *Information Sciences* 177, 5191-5202 (2007).
2. Arikan F., A Fuzzy Solution Approach for Multiple Objective Supplier Selection, *Expert Systems with Applications, forthcoming in* the special issue of The 2nd. International Fuzzy Systems Symposium (FUZZYSS'11), Hacettepe University Beytepe Campus, Ankara-Turkey, 17-18th. November (2011).
3. Arikan F., An Augmented Max-Min Model For Multiple Objective Supplier Selection, 9th International Congress on Logistics and Supply Chain Management, Yaşar University, İzmir-Turkey, 27-29th. October (2011).
4. Zimmermann H.J, Fuzzy programming and linear programming with several objective functions, *Fuzzy Sets and Systems* 1(1), 45–55 (1978).
5. Tiwari R.N., Dharmar S., Rao J.R., Fuzzy Goal Programming an Additive Model, *Fuzzy Sets and Systems* 24, 27-34 (1987).
6. Lee L., and Li R.J., Fuzzy multiple objective programming and compromise programming with pareto optimum, *Fuzzy Sets and Systems* 53, 275-288 (1993).
7. Chen L.H., Tsai F.C., Fuzzy goal programming with different importance and priorities, *European journal of Operational Research* 133, 548-556 (2001)
8. Yücel A., Güneri A.F., A weighted additive fuzzy programming approach for multi criteria supplier selection, *Expert Systems with Applications* 38, 6281-6286 (2011).

SELECTION OF REVERSE LOGISTICS PROVIDER FOR A PAPER FIRM

JIHENE JLASSI[1]
LOGIQ, University Of Sfax Tunisia

ABEDERRAHMAN EL MHAMEDI
MGSI, IUT de Montreuil, France

HABIB CHABCHOUB

Reverse Logistics focuses on the backward flow of materials from customer to supplier (or alternate disposition) with the goals of maximizing value from the returned item or minimizing the total reverse logistics cost. The selection of best reverse logistics provider becomes the most important issue in the whole supply chain system. In this study, a multi-criteria decision making method in fuzzy environment is developed to guide the selection process of best Reverse Logistics Provider. The proposed decision-making methodology has been applied to a case of paper recycling industry in Tunisia where this material is reused by manufacturing companies.

1. Introduction

There are many motivations for planning the reverse logistics channel. Some are economic and others are environmental. The area of reverse logistics (RL) management is currently attracting more interest both commercially and academically. RL is defined by REVLOG (Markus *et al*, 2007) as "The process of planning, implementing and controlling flows of raw materials, in process inventory, and finished goods, from the point of use back to a point of recovery or point of proper disposal".

Some industries have outsourced their reverse logistics services. Reverse logistics may take place either through original network or through specialized reverse logistics providers. In fact, by outsourcing reverse logistics, companies can reduce costs and increasing customer satisfaction. The evaluation of reverse logistics provider may involve several and different types of criteria. The selection process should start with valid criteria. In order to select the Reverse Logistics Provider in this paper, we focus on the application of a fuzzy PROMETHEE II method.

---

[1] jihene2000tn@yahoo.fr

2. Literature Review

A number of RL network issues have been explored in the literature. For example Klavdij *et al* (2006) presented that the general model, based on the principles of reverse logistics, was developed and applied with the aim of reducing the extent of internal aluminium scrap transportation required between certain production units of an aluminium manufacturing plant.

Serrato *et al* (2007) also suggest outsourcing RL activities when the returns are more variable.

Akram El korchi and Dominique Millet (2011) presented that remanufacturing end-of-life products requires setting up an economically and environmentally viable reverse logistics channel for supplying reusable used modules to the production chain. The paper introduced a framework to allow generating and assessing different reverse logistics channel structures.

3. Proposed Model

The proposed decision-making methodology has been applied to a case of paper recycling industry in Tunisia where this materiel is reused by manufacturing companies.

After the use of different types of notebooks (notebooks full bindings, school book,…). A collection of papers to be made to make recycling and then the paper will be manufacturing for the realization of other quality specifications below.

The company to choose the most reverse logistics provider among four providers that have as activities collection, recycling and transport of the papers.

The selection of evaluation of reverse logistics provider is based on some criteria which determined by literature review. In fact Gunasekaran *et al* (2001) the criteria are

Table1. Different criteria (Gunasekaran et al, 2001)

Criteria	Definition
Quality (Q)	Covers product performance, accuracy, etc.
Delivery (D)	It refers the ability to meet delivery schedules. It covers flexibility and reliability of delivery, lead time, etc.
Rejection rate (R)	It refers to inability of the returned product to meet the quality specifications consistently for the recycling process.

For Meade and Sarkis (2002) the criterion is

Table2. Criteria for selection of reverse logistics provider (Meade and Sarkis, 2002)

Criteria	Definition
Reverse logistics cost (C)	It includes inspection cost, inventory cost, transportation cost, packaging cost, material handling cost, warehousing cost, etc.

3.1. *Proposed Method*

The PROMETHEE method (Preference Ranking Organization METhod for Enrichment Evaluation) is a multicriteria decision-making method developed by Brans *et al* (1985). It is a quite simple ranking method in conception and application compared with other methods for multicriteria analysis.

The weights can be determined according to various methods (Nijkamp *et al*, 1990). The preference function (Pj) translates the difference between the evaluations obtained by two alternatives (a and b) in terms of a particular criterion, into a preference degree ranging from 0 to 1.

The PROMETHEE approach is based on an extension of the criterion notion through the introduction of a function θ, expressing the decision maker's preference for an alternative in relation to another. For a criterion j to be maximized and two alternatives, a and b, the evaluation of the alternative a according to the criterion j can be defined as Cj (a), and the difference of the evaluations as dj (Pomerol *et al*, 1980)

In other words,

$$d_j = C_j(a) - C_j(b) \tag{1}$$

The evaluation function θ will vary in the interval [0, 1]. Also, two threshold values, q as the indifference threshold and p as the strict preference one, can be introduced. Thus, θ can be presented as follows:

$$\theta_j(a, \quad b) = \theta_j(d_j) = \begin{Bmatrix} 0 & if\ d_j < q \\ 1 & if\ d_j > p \\ H_j(d_j)\ if\ q \le d_j \le p \end{Bmatrix} \tag{2}$$

Where Hj can be one of the six proposed functions. These functions and their detailed definitions can be found. Once the type of Hj criterion function to be used for each criterion is defined, the preference index C(a, b) can be calculated as follows:

$$C(a, \quad b) = \sum_j w_j \theta(a, \quad b)_j \tag{3}$$

Where wj represents the weight of criterion j. For a proper application of the PROMETHEE method, weights must be strictly positive and normalized.

The next step is to calculate the flows for an alternative a. There are two types of flows. The leaving flow at a indicates a preference of the alternative a over all other actions. It shows how "good" the alternative a is. The leaving flow is calculated as follows:

$$\varphi^{+}(a)=\sum_{x\in A}C(a,b) \tag{4}$$

The entering flow at a, on the other hand, indicates a preference of all other alternatives, compared to a. It shows how "weak" the alternative a is. The entering flow is calculated as follows:

$$\varphi^{-}(a)=\sum_{x\in A}C(x,a) \tag{5}$$

According to the PROMETHEE I, the action a is superior to the action b if the leaving flow of a is greater than that of b and the entering flow of a is smaller than that of b. Using our notation, a outranks b if $\varphi^{+}(a)\geq\varphi^{+}(b)$ and $\varphi^{-}(a)\leq\varphi^{-}(b)$. a and b cannot be compared when $\varphi^{+}(a)\geq\varphi^{+}(b)$ and $\varphi^{-}(a)\geq\varphi^{-}(b)$ or $\varphi^{+}(a)\leq\varphi^{+}(b)$ and $\varphi^{-}(a)\leq\varphi^{-}(b)$. To overcome the issue of incomparability and to obtain a complete ranking, the PROMETHEE II method can be applied. The PROMETHEE II takes into consideration the net flows, φ the difference of leaving minus entering flows. The alternative with a higher net flow is better than the ones with lower net flows.

Because of the underlying structure for multicriteria decision problems, the consideration of fuzzy logic within MCDM seems to be almost self-evident. The fuzzy PROMETHEE method is a combination of the PROMETHEE method and fuzzy numbers. Integration of fuzzy set theory and the PROMETHEE method was first proposed by Le Téno and Mareschal (1989). This suggestion was further developed by Geldermann et al (2000) and Goumas and Lygerou.(2000). Geldermann et al (2000) proposed the application of triangular fuzzy numbers for the interpretation of linguistic variables, and Goumas and Lygerou (2000) developed an easy-to-use methodology to apply the fuzzy PROMETHEE to ranking problems.

In this work, we adopt the fuzzy PROMETHEE method as it is described by Goumas and Lygerou (2000). In this case, calculations for evaluation of alternatives, described in the precedent Section, will be executed using fuzzy numbers, whereas preferences of the decision maker, that is, the weights, will remain as crisp numbers. Setting alternative evaluations as fuzzy numbers will help to translate qualitative information and vagueness in the decision maker's opinions to a solid mathematical expression. Preference threshold values q and p will be crisp numbers. If they were taken fuzzy, evaluation might be unclear due to the stretched form of a fuzzy number (Goumas and Lygerou, 2000). The

PROMETHEE method interprets a difference of the evaluations, expressed as dj, using evaluation functions called Hj. In this case, for certain criteria, we choose a linear preference function with indifference and strict preference threshold values q and p, respectively. Thus, the evaluation function θ will become

$$\theta_j(a,b) = \theta_j(d_j) = \begin{cases} 0 & if \quad d_j < q \\ \dfrac{d_j - q}{p-q} & if \quad q \le d_j \le p \\ 1 & if \quad d_j > p \end{cases} \tag{6}$$

When using the fuzzy PROMETHEE, dj will be introduced as a fuzzy number (n, c, d) and Equation 6 will be as follows:

$$\theta_j(a,b) = \theta_j(d_j) = \begin{cases} 0 & if \quad (n-c) < q \\ \dfrac{(n,c,d)-q}{p-q} & if \quad q \le (n-c) and (n+d \le p) \\ 1 & if \quad (n+d) > p \end{cases} \tag{7}$$

At the end of the calculations, fuzzy numbers will be obtained. To derive a solution, these numbers must be ranked, which means that fuzzy numbers have to be compared. Goumas and Lygerou suggested the use of the index proposed by Yager to overcome this comparison problem.

The fuzzy PROMETHEE method has been applied as described in the precedent and the evaluation phase outputs are shown in this table.

Table3. Application of fuzzy PROMETHEE II method

	RLP1	RLP2	RLP3	RLP4	φ^+	φ
RLP1	0	0.35	0.25	0.24	0.84	0.06
RLP2	0.78	0	0	0.45	1.23	0.53
RLP3	0	0	0	0.95	0.95	0.45
RLP4	0	0.35	0.25	0	0.6	-1.04
φ^-	0.78	0.70	0.5	1.64		

The results indicate that the Reverse Logistics providers are classified as follows: RLP2, RLP3, RLP1 and RLP4.

The results of the application of the fuzzy PROMETHEE method to help the company that has taken in our case study to make a decision and choose the supplier for the outsourcing and recycling of the papers to end manufacture other notebooks.

4. Conclusion

Reverse logistics involves many problems in the supply chain whose importance has been heightened by increasing concerns about the environment, customer service, and cost reduction. In order to ensure efficiency, many organizations outsource their reverse logistics activities by engaging third-party logistics providers that implement reverse logistics programs designed to gain value from returned products.

Therefore, the selection of best reverse logistics provider becomes the most important issue in the whole supply chain system. Today, when evaluating supplier performance, companies usually consider conventional factors.

This paper aims at contributing to the literature by multiple criteria decision method solution for a reverse logistics service provider selection problem in an imprecise environment. We use the fuzzy PROMETHEE II.

References

1. Akram El korchi, Dominique Millet. Designing a sustainable reverse logistics channel: the 18 generic structures framework. Journal of Cleaner Production 19 588-597 (2011).
2. Brans, J.P. et Vincke. P.H. PROMETHEE, a new family of outranking methods in MCDM. Manage. Sci, Vol.31, N°6, pp. 647–656 (1985).
3. Geldermann, J., Spengler, T. et Rentz, O. Fuzzy outranking for environmental assessment. Case study: Iron and steel making industry. Fuzzy Set Syst, Vol. 115, pp. 45– 65 (2000).
4. Goumas, M. et Lygerou, V. An extension of the PROMETHEE method for decision making in fuzzy environment: Ranking of alternative energy exploitation projects. Eur J Oper Res, Vol.123, pp. 606– 613 (2000).
5. Gunasekaran A, Patel C, Tirtiroglu E. Performance measures and metrics in a supply chain environment. International Journal of Production and Operations Management 21(1/2):71–87 (2001).
6. Klavdij Logôzar , Gregor Radonji^c, Majda Basti^c. Incorporation of reverse logistics model into in-plant recycling process: A case of aluminium industry. Resources, Conservation and Recycling 49 49–67 (2006).
7. Markus Biehl, Edmund Prater, Matthew J. Realff. Assessing performance and uncertainty in developing carpet reverse logistics systems. Computers & Operations Research 34 443–463(2007).
8. Meade L, Sarkis J. A conceptual model for selecting and evaluating third-party reverse logistics providers. Supply Chain Management: An International Journal;7(5):283–95 (2002).
9. Serrato M, Ryan S, Gaytan J. A Markov decision model to evaluate outsourcing in reverse logistics. International Journal of Production Research ;45(18–19):4289–315(2007).

A FUZZY COGNITIVE MAP APPLICATION FOR CAUSES OF BULLWHIP EFFECT IN SUPPLY CHAINS

SECIL ERCAN, MINE ISIK, SEYDA S. ASAN AND GULGUN KAYAKUTLU

Industrial Engineering Department, Istanbul Technical University, Istanbul, Macka, 34376 Turkey

A growing problem of the supply chain, the bullwhip effect is studied in this paper. In order to develop effective strategies to mitigate the bullwhip effect, we need to understand the factors that cause it and the interactions among them. This paper uses Fuzzy Cognitive Map to analyze the interaction among the causes of the bullwhip effect and determines the leading factors that are the root causes of this problem.

1. Introduction

The importance of practicing supply chain management (SCM) has long been recognized as a way of achieving competitive advantage and creating value for the companies as well as for the whole chain. As supply chains grow, management becomes more complex. One of the problems is the bullwhip effect, which is the amplification and demand distortion that occurs within a supply chain. The bullwhip effect is a widely studied subject and the main focus of research is to analyze the potential causes and solutions for this problem. However, precautions can only be taken for the true root causes, which are rather difficult to discover.

This study is attempt influents to determine the grounds of bullwhip effect in supply chains and the interactions between them, so that the bullwhip effect can be avoided in advance rather than planning for the reactions. Fuzzy Cognitive Maps (FCM), a method based on the expert intelligence, is used as a tool to identify the interactions.

The paper is so organized that, a review of the literature about the bullwhip effect and the FCM is presented in the next section. Section 3 will be devoted to clarify the proposed model and present the application. The concluding remarks will be given in the last section.

2. Literature Review

The bullwhip effect is the increase in variability due to the distorted information that flows up and down the supply chain. Both practitioners and academicians studied the subject with different approaches. The latter defend the idea that the main factors that contribute to the bullwhip effect can be traced back to the complex interactions in a supply chain. Forrester [1], Towill [2] and Sterman [3] mentioned the destructive effect of wrong feedback mechanisms in Supply Chain (SC). Another destructive driver is found to be the irrationality throughout the SC mechanism due to the wrong perception of the decision makers, so the problems have been adapted into the mathematical models. On the contrary, practitioners prefer to handle each factor individually.

Since the expert judgments have uncertain characteristics, fuzzy cognitive mapping is preferred in this study. Moreover, some of the concepts included in the proposed model such as miscommunication among distribution channels, personalized managerial decision etc. do not have an exact structure, so they are well-defined with fuzzy numbers rather than crisp values.

2.1. *Causes of Bullwhip Effect*

As long as the factors that contribute to the bullwhip effect can be analyzed and clarified, the control of the bullwhip effect would become easier. The main factors which have been determined to cause the bullwhip effect in the related literature are listed in Table 1.

Table 1. The literature on the causes of bullwhip effect

Factor	Literature
Miscommunication Among Distribution Channels	Metters [4] , Ouyang [5], Kelepouris [6]
Order Batching	Miragliotta [7], Lee et al. [8], Potter & Disney [9]
Price Fluctuation	Miragliotta [7], Lummus et al. [10], Lee et al. [8], Paksoy & Keskin [11]
Time Delays	Sterman [3], Blackburn [12], Makui & Madadi [13]
Shortage Gaming	Lee et al. [8], Wu et al. [14]
Demand Forecast Updating	Naish [15], Lee et al. [8], Zhang [16], Jaksic&Russian[17]
Lead Time Variability	Lee et al.[8], Makui & Madadi [13], Agrawal et al. [18]
Demand Fluctuation	Lee et al. [8], Hong & Wang [19], Wang et al [20]
Product Features	Nepal et al. [21]
Personalized Managerial Decision	Senge [22]

2.2. *Fuzzy Cognitive Maps*

Cognitive maps (CMs) are tools for analysis of interactions among the elements of a system. Cognitive maps are composed of concepts represented by nodes and relations represented by arrows [23]. A pairwise comparison matrix is constructed by the experts to determine if the relation is positive, negative or neutral [24]. It draws the causal perspective of a system [25].

Causal factors are defined by their outdegree, indegree and centrality. Outdegree is the row sum of absolute values of a variable in the comparison matrix. It shows the cumulative strength of connections triggering the concept. Indegree is the column sum of absolute values of a concept. It shows the cumulative strength of criteria entering the concept. The centrality of a concept is the sum of its outdegree and indegree [26].

Fuzzy Cognitive Map (FCM) is a qualitative alternative approach to system dynamics [27], early stated as the fuzzy version of CM by Kosko [28], and was developed by Zhang and his colleagues [29]. The system is based on the negative-positive-neutral (NPN) interval [-1, 1] [30]. Hence, the strength of impact is calculated as a weight in the interval [-1, 1].

FCM has two significant characteristics that differ from CM [31, 32]:

1. Causal relationships between concepts are evaluated by not only using signs but also with numbers to express the degree of causality.

2. What makes the system be dynamic is that, the effect of change in one concept influences also the other concepts.

3. Proposed Model and Sample Application

This paper uses Fuzzy Cognitive Maps to study the relations among the causes of the bullwhip effect and determines the leading factors that are the root causes of the bullwhip effect.

Five membership functions are considered to represent the strength of the interaction between two factors in each evaluation. Using five linguistic variables ("negative strong", "negative weak", "neutral", "positive weak", "positive strong"), an expert can describe the influence of one factor on another and can distinguish different degrees of influence. The assessments of the experts are composed as an overall linguistic weight. This overall weight is produced through aggregation. Defuzzification process is utilized on aggregated results.

The Center of Area (CoA) defuzzification method is used for the transformation of the linguistic weight to a crisp number that belongs to [-1, 1] interval [32]. Considered relations are defined as triangular fuzzy numbers in FCM. The linguistic scales are shown in Table 2. In addition, the membership functions of linguistic terms are depicted in Figure 1.

Table 2. Linguistic scales

Linguistic scales			Scale of fuzzy number	
-1.0	-1.0	-0.6	Negatively Strong	(NS)
-0.8	-0.4	-0.1	Negatively Weak	(NW)
-0.2	0.0	0.2	Neutral	(N)
0.1	0.4	0.8	Positively Weak	(PW)
0.6	1.0	1.0	Positively Strong	(PS)

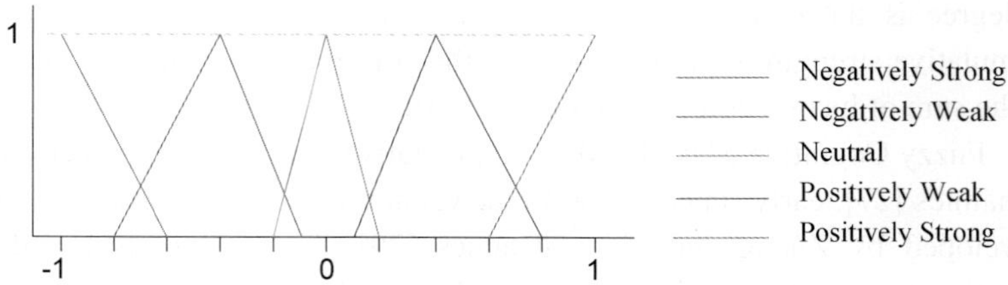

Figure 1. Membership functions of linguistic terms

The weight matrix that belongs to the considered factors after defuzzification process is given in Table 3.

Table 3. Weight matrix of the effects of the factors

	C1	C2	C3	C4	C5	C6	C7	C8	C9	C10
C1	0	0.12	0.16	0.48	-0.12	0.52	0.4	0.2	0.2	0.32
C2	-0.08	0	0.28	0.24	0.2	0.2	0.48	0.4	-0.12	0.28
C3	0.08	0.08	0	0.08	0.36	0.28	0.12	0.76	0	0
C4	0.08	0.4	0	0	0.36	0.4	0.6	0.16	-0.08	0.4
C5	0.2	0.08	0.16	0.08	0	0.4	0.2	0.2	0.2	0.48
C6	-0.08	-0.08	-0.04	0	0.16	0	-0.08	-0.08	0	-0.08
C7	0.32	0.52	0.28	0.6	0.24	0.56	0	0.48	0	0.6
C8	0.32	0.4	0.52	0.12	0.48	0.48	0.32	0	0.4	0.6
C9	0.12	0.36	0.36	0.28	0.48	0.28	0.28	0.56	0	0.44
C10	0.48	0.28	0.28	0.4	0.04	0.08	0.4	0	0	0

Apparently, the strongest effect obtained from C3 to C8 is 0.76, on the contrary C3 has neutral effect on C9. Remaining effects among factors can be depicted from the matrix.

Table 4. Results of the FCM

Concepts		Indegree	Outdegree	Centrality
C7	Lead Time Variability	2.88	3.60	6.48
C8	Demand Fluctuation	2.84	3.64	6.48
C10	Personalized Managerial Decision	3.20	1.96	5.16
C4	Time Delays	2.28	2.48	4.76
C2	Order Batching	2.32	2.28	4.60
C5	Shortage Gaming	2.44	2.00	4.44
C1	Miscommunication Among Distribution Channels	1.76	2.52	4.28
C9	Product Features	1.00	3.16	4.16
C3	Price Fluctuation	2.08	1.76	3.84
C6	Demand Forecast Updating	3.20	0.60	3.80

Fuzzy cognitive matrix is constructed with pair wise comparisons. Indegree, outdegree and centrality values are listed in Table 4. C7 and C8 have the same centrality value with 6.48 that is the highest centrality. When outdegree is taken into account, C8 has the strongest impact on the other concepts with an outdegree value of 3.64.

4. Conclusion

Uncertainty in business environment causes the analysis of priority of the business dynamics. From the supply chain management point of view, bullwhip effect is considered as a crucial issue to cope with. This research has been conducted to analyze the causes of bullwhip effect and the relations among them. The Fuzzy Cognitive Map that relies on human expert intelligence is used in analysis.

Ten factors causing the bullwhip effect have been identified through literature review. After applying FCM, according to the centrality results, lead-time variability and demand fluctuation are observed as the leading factors followed by personalized managerial decision factor. On the contrary, price fluctuation and demand forecast update can be considered the quasi neutral.

The results obtained will be beneficial for the Supply chain managers when dealing with the bullwhip effect.

References

1. J.W. Forrester, TIMS Studies in the Management Sciences, **C14**, 7-21 (1980).
2. D.R. Towill, IJOR, **C20**, 671-687 (1982).
3. J.D. Sterman, Man. Sci., **C35**, 321-339 (1989).
4. R. Metters, JOM, **C15**, 89-100 (1997).
5. Y. Ouyang, EJOR, **C182**, 1107-1121 (2007).
6. T. Kelepouris, P. Miliotis, and K. Pramatari, COR, **C35,** 3657-3670 (2008).
7. G. Miragliotta, IJPE, **C104**, 365-381 (2006).
8. P.H. Lee, V. Padmanabhan and S. Whan, SMR, **C38**, 93-102 (1997).
9. A. Potter and S.M. Disney. IJPE, **C104**, 408-418 (2006).
10. R.R. Lummus, L.K. Duclos, R.J. Vokurka, Supply Chain Management, **C8,** 317-323 (2003).
11. T. Paksoy and E. Keskin, Selçuk Üni. Sosyal Bilimler Enstitüsü Dergisi (2006).
12. J.D. Blackburn, Time based competition, business one Irvin Homewood (Chapter 11) (1991).
13. A. Makui and A. Madadi, AM&C, **C189,** 35-40 (2007).
14. S. Wu, W. Gan and F. Wei, Procedia Eng., **C15,** 4276-4281 (2011).

15. H.F. Naish, QJE, **C104,** 864-875 (1994).
16. X. Zhang, IJPE, **C88**, 15-27 (2004).
17. M. Jaksic and B. Russian, EJOR, **C184,** 946-961 (2008).
18. S. Aggraval, R.N. Sengupta and K. Shanker, EJOR, **C192,** 576-593 (2009).
19. L. Hong and P. Wang, Systems Engineering Theory Practice, **C27(7),** 26-33 (2007).
20. J.L. Wang, J.H. Kuo, S. Chou and S. Wang, Expert Syst. Appl., **C37(7)**, 4726-4736 (2010).
21. B. Nepal, A. Murat and R.B. Chinnam, IJPE, **C136(2)**, 318-331 (2011).
22. P.M. Senge, The Fifth Discipline, Doubleday, New York (1990).
23. A. Altay and G. Kayakutlu, Procedia Comput. Sci., **C3**, 1111–1119 (2011).
24. S.B. Hasiloglu, J. Appl. Sci., **C9**(14), 2606-2612 (2009).
25. A. Pena, H. Sossa, A. Gutierrez, Expert Syst. Appl., **C35**, 2-18 (2008).
26. U. Özesmi and S.L. Özesmi, Ecol. Model., **C176**, 43-64 (2004).
27. J.P. Carvalho and J.A.B. Tome, Fuzzy Systems, The 10th IEEE International Conference on, 280-283 (2001).
28. B. Kosko, Int. J. Man-Machine Studies, **C24,** 65-75 (1986).
29. W.R. Zhang, S.S. Chen, J.C. Bezdek, IEEE Transactions on Systems, Man and Cybernetics, **C19,** 31-39 (1989).
30. E.I. Papageorgiou, C. Stylios, P.P. Groumpos, Int. J. Human-Computer Studies, **C64,** 727-743 (2006).
31. K.E. Parsopoulos, E.I. Papageorgiou, P.P. Groumpos and M.N. Vrahatis, LNCS, **C3102**, 402-413 (2004).
32. E.I. Papageorgiou and P.P. Groumpos, ASOC, **C5,** 409-431 (2005).

UNCERTAINTIES AND HANDLING METHODS IN SUPPLY CHAIN: A LITERATURE REVIEW

ERHAN YAZICI, MURAT BASKAK
Industrial Engineering Department, Istanbul Technical University, Istanbul, Turkey

GÜLÇİN BÜYÜKÖZKAN
Industrial Engineering Department, Galatasaray University, Istanbul, Turkey

In this paper, we make an attempt to review literature in supply chain (SC) by uncertainty aspect. Using resource based approach [65], main uncertainty sources which are demand, process and supply of supply chain are taken into account for analysis. Uncertainties are given in detail grouping by main uncertainty sides of supply chain. Problem areas that uncertainties occur are classified. Used techniques to define and handle uncertainties in papers also are analyzed. Solving methods of problems which contain uncertainties in supply chain are showed. In conclusion part, suggestions for future researches are provided based on review.

1. Introduction

Supply chain is a complex, organized and large scale structure which contains many organizations and structures which are connected to each other's with material, information, labor and services flows. Effective supply chain management depends on external and internal many factors from organization and its environment. Supply chain management is difficult because of complex structure of supply chain. Uncertainties arise from this complex structure of supply chain. For effective supply chain management, uncertainties emerge as key issues that must be managed [77].

In this review, literature is analyzed for uncertainties and handling methods in supply chain to support decision makers for solving problems which contains uncertainties as real life problems and to provide classification of literature and new directions for researchers. As review methodology, "supply chain" and "uncertainty" keywords are searched in literature. 116 articles published between 1998 and 2012 in 23 reputable international journals are reviewed and classified. Articles are classified by their analyzed uncertainties, handling techniques of uncertainty and solving methods for problems which contain uncertainty in supply chain environment. Based on the review, future research recommendations are oriented.

2. Uncertainties, handling and management

2.1. *Uncertainties analyzed in researches*

Uncertainties which are analyzed in literature and their published paper statistics by uncertainty source are given in figure (1). Some of articles contain more than one uncertainty issue. Most worked on uncertainty issue appears demand side of supply chain. More than half of researches made to solve uncertainty problems in demand. The second most worked issue is supply based uncertainty. Main reason for staying ahead of these issues is; demand and supply are in interfaces of organizations. So, uncertainty arises much in these sources of supply chain. Because of demand based uncertainty papers comes first is; information flows from customers and then is forwarded to all of supply chain and causes new uncertainties.

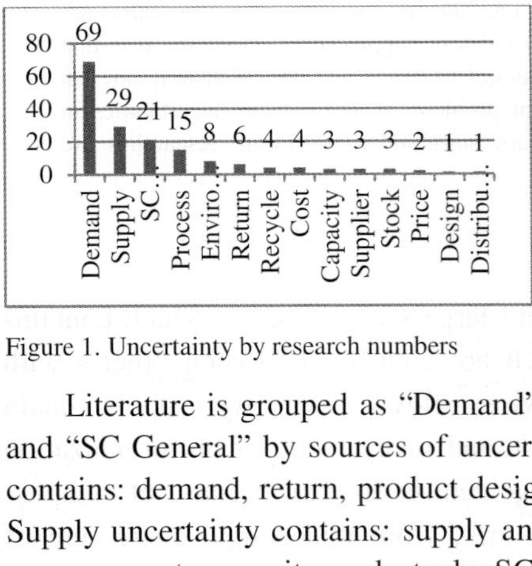

Figure 1. Uncertainty by research numbers

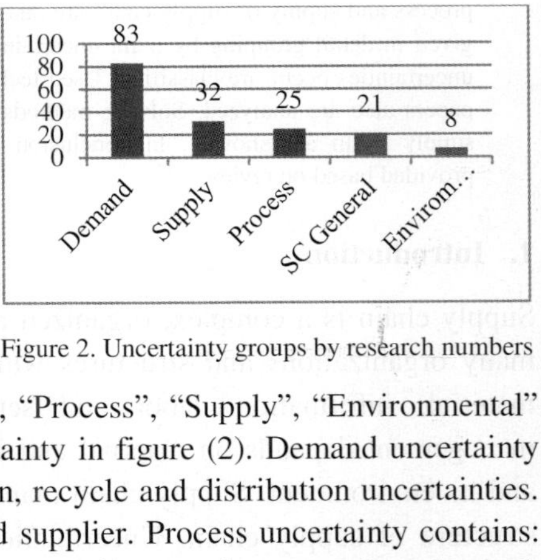

Figure 2. Uncertainty groups by research numbers

Literature is grouped as "Demand", "Process", "Supply", "Environmental" and "SC General" by sources of uncertainty in figure (2). Demand uncertainty contains: demand, return, product design, recycle and distribution uncertainties. Supply uncertainty contains: supply and supplier. Process uncertainty contains: process, cost, capacity and stock. SC general uncertainty contains: demand, supplier, uncertainty management, supply, price, process, environment, capacity, stock, return and recycle uncertainties in overall SC context.

2.2. *Problem areas which contain uncertainty*

Problems which analyzed in papers are given in figure (3). These problems can be grouped by uncertainty sources in SC also. In management of uncertainty in SC, general researches especially deal with uncertainty sources, frameworks and managerial methods. Secondly, in these researches, another problem area is about uncertainties in strategic management, planning. Another working area is supply chain modeling and design under uncertainty. Additionally, some papers offer information systems with case studies to manage and reduce uncertainty in SC. These researches mostly deal with uncertainties at strategic management level. Basic researched issues are reducing uncertainty in SC, performance evaluation of uncertainty, uncertainty management, SC planning, SC behavior, SC uncertainty sources, SC network design and optimization, strengthening of SC, integration of SC, SC strategies, interrelations among SCs, closed loop supply chain (CLSC), SC coordination, SC flexibility and decision making in SC under uncertainty.

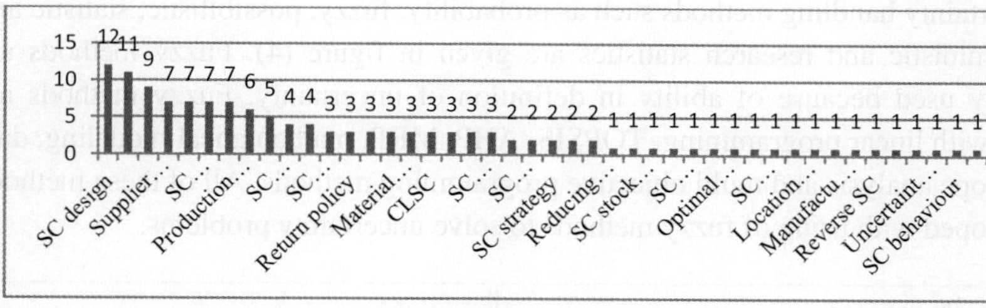

Figure 3. Problem areas which contain uncertainty by research numbers

Demand sourced uncertainty researches can be reviewed in two groups. Most of them deals with only demand uncertainty based problems. This type of researches are made on decision making under demand uncertainty, using information systems to reduce uncertainty, determining suitable demand distributions and effects of demand uncertainty on SC. Other researches deal with return and recycle uncertainty with demand uncertainty, also. SC network return policy under demand uncertainty, closed loop SC network design under demand, return and recycle uncertainty issues are worked. Basic researched issues are optimal order level, return policy, product design, retail decisions, recycles policy, distribution locations determining, delivery policy, and price policy under uncertainty and demand uncertainty, price uncertainty.

Process sourced uncertainty researches are made on raw material, labor and machine uncertainties in production. One part of these articles handles with demand uncertainty in production environment. Others handle with production sourced uncertainties addition to demand stock and supply uncertainty. Uncertainties in process are demand, return, production, capacity, cost, price and stock. Basic researched issues are aggregated production planning, master production scheduling, material requirements planning, capacity planning, production planning, stock management, cell loading, production and reproduction, production planning in CLSC, production stock planning and production planning coordination with uncertainty.

Supply sourced uncertainty researches are made on demand uncertainties effect on supply, supplier uncertainty, supply uncertainty, process uncertainties effect on supply and return uncertainty to supplier. A group of researches handle with supplier selection in uncertainty and others are about order level in uncertainty. Uncertainties in supply are demand, supply, supply level, supplier and cost. Basic researched issues are order amount modeling, supplier selection, stock controlling, stock policies, stocking systems.

2.3. *Uncertainty handling and management methods*

Uncertainty handling methods such as probability, fuzzy, possibilistic, statistic and deterministic and research statistics are given in figure (4). Fuzzy methods are mostly used because of ability in definition of uncertainty. Fuzzy methods are used with linear programming, TOPSIS, AHP, MRP, mathematical modeling, data envelope analysis and multi objective programming methods. All of these methods developed with using of fuzzy methods to solve uncertainty problems.

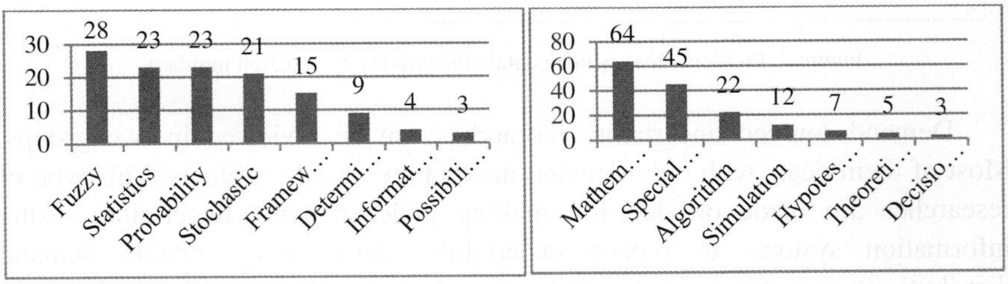

Figure 4. Problem handling methods by research numbers

Highlights of the methods used are uncertainty circle approach, fuzzy set theory based intelligent agents, real choices theory based problem solving methods and uncertainty frameworks. In addition to these framework articles, mixed integer programming models, fuzzy linear programming models and robust programming models are also used other researches like network design in uncertainty problems.

Additionally, mathematical models with fuzzy variables are developed for modeling of problem and objective function in demand sourced uncertainties. At demand uncertainty, mostly used methods are mixed integer programming and genetic algorithm methods. Besides, other solution methods like TOPSIS, game theory are proposed to decision makers. Additionally, mathematical models with fuzzy variables are developed for modeling of problem and objective function, also. Used solution methods are especially mathematical modeling problems with fuzzy, probabilistic and stochastic variables. Mixed integer programming and stochastic programming, Markova dynamic programming models are used. New MRP methods also provided like integrated MRP, fuzzy MRP. Mostly fuzzy based methods are used in supply sourced uncertainties. Some of approaches are fuzzy multi objective linear programming; fuzzy priority based fuzzy AHP, fuzzy SWOT, fuzzy linear programming, and fuzzy optimization model methods.

Using uncertainty handling methods, problems which contain uncertainty in SC are attempted to solve by researches. Proposed methods by researches are given in right side of figure (4). In some other papers developed new frameworks to handle with uncertainties and some others use information technologies to reduce uncertainty.

3. Conclusion and research implications

As seen in figure (1) and (2), most researched uncertainty is demand, and then supply comes. In process side, new researches may be made in future. In figure (3), most analyzed problems are SC wide problems like SC design, planning then sub part of SC problems analyzed like sales, supply and process parts. Sub parts of SC researches may be made in future researches to deeply analyze uncertainty in SC. In figure (4), most used handling technical for uncertainty is fuzzy techniques, and then others come. In figure (4), most used problem solving method is mathematical programming, then algorithms are used, then simulation comes.

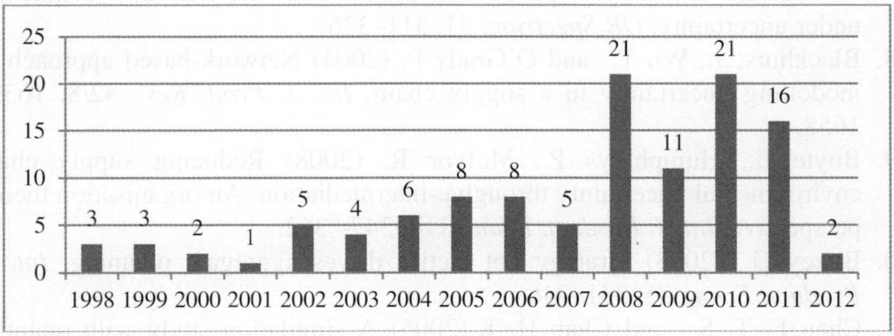

Figure 5. Research statistics by year.

In figure (5), researches by year can be seen. As seen in figure (6), researches in uncertainty in SC area are ascending at last years. Uncertainty in SC research area takes more attentions at last years. Future researches in that area should be made.

In theoretical analyses, researches are generally made under ideal conditions, deterministic criterias and parameters. In real life, uncertainties come from unpredictable situations. Researchers can use this review to develop new solution methods for uncertainties in SC.

References

1. Al-e-hashem S, Malekly H., Aryanezhad M.B. (2011) A multi-objective robust optimization model for multi-product multi-site aggregate production planning in a supply chain under uncertainty, *Int. J. Product. Econ.*, 134, 28–42.
2. Acar Y., Kadipasaoglu S. and Schipperijn P. (2010) A decision support framework for global supply chain modelling: an assessment of the impact of demand, supply and lead-time uncertainties on performance, *Int. J. Prod. Res.,* 48/11, 3245–3268.
3. Amin S.H., Razmi J., Zhang G., (2011) Supplier selection and order allocation based on fuzzy SWOT analysis and fuzzy linear programming, *Expert Syst Appl*, 38, 334–342.
4. Awasthi A, Chauhanb S.S. and Goyal S.K., (2011) A multi-criteria decision making approach for location planning for urban distribution centers under uncertainty, *Math Comput Model*, 53, 98–109.
5. Azadeh A. and Alem S.M., (2010) A flexible deterministic, stochastic and fuzzy Data Envelopment Analysis approach for supply chain risk and vendor selection problem: Simulation analysis, *Expert Syst Appl,* 37, 7438–7448.
6. Baskır M. Bahar and Türksen I. Burhan, (2010) An Uncertainty Analysis of Supplier Selection by Fuzzy Logic, *IEEE.*
7. Bihlmaier R., Koberstein A. and Obst R., (2009) Modeling and optimazing of strategic and tactical production planning in the automotive industry under uncertainty, *OR Spectrum,* 31, 311–336.
8. Blackhurst J., Wu T. and O'Grady P. (2004) Network-based approach to modelling uncertainty in a supply chain, *Int. J. Prod. Res.,* 42/8, 1639–1658.
9. Boyle E., Humphreys P., McIvor R. (2008) Reducing supply chain environmental uncertainty through e-intermediation: An organisation theory perspective, *Int. J. Product. Econ.*, 114, 347– 362.
10. Buxey G., (2003) Strategy not tactics drives aggregate planning, *Int. J. Product. Econ.,* 85, 331–346.
11. Chan F. T. S. and Chan H. K.(2006) A simulation study with quantity flexibility in a supply chain subjected to uncertainties, *Int J Comp Integ M,* 19/2, 148 – 160.
12. Chen C., Lee W. (2004) Multi-objective optimization of multi-echelon supply chain networks with uncertain product demands and prices, *Comput Chem Eng,* 28, 1131–1144.
13. Chen C.L., Yuan T.Y., Chang C.Y., Lee W.C. and Ciou Y.C. (2006) A Multi-criteria Optimization Model for Planning of A Supply Chain Network under Demand Uncertainty, *16th ESCAPE.*

14. Chen X., Huang Y. and Murata T., (2008) Optimization Method of Short Life Cycle Product Supply Chain Network with Recycle Flow and Dispersed Markets, *IEEE.*
15. Childerhouse P. and Towill D.R, (2004) Reducing uncertainty in European supply chains, *JMTM* , No: 15/7, 585-598.
16. Crnkovic J., Tayi G.K., Ballou D.P. (2008) A decision-support framework for exploring supply chain tradeoffs, *Int. J. Product. Econ.,* 115, 28– 38.
17. Dabbenea F., Gayb P. and Saccoc N. (2008) Optimisation of fresh-food supply chains in uncertain environments, Part I: Background and methodology, *Biosyst Eng*, 99, 348 – 359.
18. Dabbenea F., Gayb P. and Saccoc N. (2008) Optimisation of fresh-food supply chains in uncertain environments, Part II: A case study, *Biosyst Eng*, 99, 360 – 371.
19. Das Sanchoy K. and Abdel-Malek L., (2003) Modeling the flexibility of order quantities and lead-times in supply chains, *Int. J. Product. Econ.*, 85, 171–181.
20. Díaz-Madroñero M., Peidro D. and Vasant P., (2010) Vendor selection problem by using an interactive fuzzy multi-objective approach with modified S-curve membership functions, *Comput Math App*, 60, 1038-1048.
21. Ding Y., Vue Y. and Wang X., (2008) Optimal Vendor Selection By Fuzzy Optimization, *IEEE.*
22. Dolgui A., Prodhon C. (2007) Supply planning under uncertainties in MRP environments: A state of the art, *Annual Reviews in Control,* 31, 269–279.
23. Dong J., Zhang D. and Yan H., Nagurney A., (2005) Multitiered Supply Chain Networks: Multicriteria Decision-Making Under Uncertainty, *ANOR*, 135, 155–178.
24. Easwaran G. and Üster H., (2010) A closed-loop supply chain network design problem with integrated forward and reverse channel decisions, *IIE Transactions,* 42, 779–792.
25. Enns S.T., (2002) Production, Manufacturing and Logistics : MRP performance effects due to forecast bias and demand uncertainty, *Eur J Oper Res*, 138, 87-102.
26. Fynes B., De Burca S., Marshall D. (2004) Environmental uncertainty, supply chain relationship quality and performance, *JPSM,* 10, 179–190.
27. Gaffeo E., Scorcu A.E. and Vici L., (2008) Demand distribution dynamics in creative industries: The market for books in Italy, *Inf Econ Policy,* 20, 257–268.
28. Gaoa T., Sirgyb M.J. and Birdb M.M. (2005) Reducing buyer decision-making uncertainty in organizational purchasing can supplier trust, commitment, and dependence help, *J Bus Res*, 58, 397– 405.
29. Georgiadis M.C., Tsiakis P., Longinidis P., Sofioglou M.K. (2011) Optimal design of supply chain networks under uncertain transient demand variations, *Omega*, 39, 254–272.

30. Gomes M.I., Zeballos L.J., Barbosa-Povoa A.P., Novais A.Q. (2011) Optimization of Closed-Loop Supply Chains under Uncertain Quality of Returns, *ESCAPE 21*.
31. Grabot B., Geneste L., Reynoso-Castilloi G. ve Phieve S., (2005) Integration of uncertain and imprecise orders in the MRP method, *J Intell Manuf*, 16, 215–234.
32. Gupta A., Maranas C.D. (2003) Managing demand uncertainty in supply chain planning, *Comput Chem Eng,* 27, 1219-1227.
33. Gupta A, Maranas C.D., McDonald C.M.. (2000) Mid-term supply chain planning under demand uncertainty: customer demand satisfaction and inventory management, *Comput Chem Eng,* 24, 2613–2621.
34. Guilln G., Mele F.D., Espufia A. and Puigjaner L. (2006) Addressing the design of chemical supply chains under demand uncertainty, *ESCAPE 16*..
35. Guillen-Gosalbez G., Grossmann I.E. (2009) Optimal Design and Planning of Sustainable Chemical Supply Chains Under Uncertainty, *AIChE*, 2009, 55/1, 99-121.
36. Halikas J., Virolainen M. and Tuominen M. (2002) Understanding risk and uncertainty in supplier networks transaction cost approach, *Int. J. Prod. Res.*, 40/15, 3519-3531.
37. Handfield R., Warsing D., Wub X.(2009) (Q,r) Inventory policies in a fuzzy uncertain supply chain environment, *Eur J Ope Res*, 197, 609–619.
38. He Y., Zhao X. (2011) Coordination in multi-echelon supply chain under supply and demand uncertainty, *Int. J. Product. Econ.*, doi:10.1016/j.ijpe.2011.04.021.
39. Ho C., Chi Y. and Tai Y. (2005) A Structural Approach to Measuring Uncertainty in Supply Chains, *Int J Electron Comm*, 9/3, 91–114.
40. Hsieh C., Wu C. (2008) Capacity allocation, ordering, and pricing decisions in a supply chain with demand and supply uncertainties, *Eur J Ope Res,* 184, 667–684.
41. Hua Z. and Li S., (2008) Impacts of demand uncertainty on retailer's dominance and manufacturer-retailer supply chain cooperation, *Omega*, 36, 697 – 714.
42. Hua Z., Li S., Liang L. (2006) Impact of demand uncertainty on supply chain cooperation of single-period products, *Int. J. Product. Econ.*, 100,268–284.
43. Iyer K., Germain R. and Frankwick G.L., (2004) Supply chain B2B e-commerce and time-based delivery performance, *IJPDLM*, 34/8, 645-661.
44. Jung J.Y., Blau G., Pekny J.F. (2004) A simulation based optimization approach to supply chain management under demand uncertainty, *Comput Chem Eng*, 28, 2087–2106.
45. Jung J.Y., Blau G., Pekny J.F., Reklaitis G.V., Eversdyk D. (2008) Integrated safety stock management for multi-stage supply chains under production capacity constraints, *Comput Chem Eng,* 32, 2570–2581.

46. Kannabiran G., Bhaumik S., (2005) Corporate turnaround through effective supply chain management: the case of e leading jewellery manufacturer in India, *Supply Chain Management*, 10/5, 340-348.
47. Kara Selin S. and Onut S. (2010) A two-stage stochastic and robust programming approach to strategic planning of a reverse supply network: The case of paper recycling, *Expert Syst Appl,* 37, 6129–6137.
48. Kaur P, Verma R. and Mahanti N. C., (2010) Selection of vendor using analytical hierarchy process based on fuzzy preference programming, *Opsearch*, 47/1, 16–34.
49. Kenné J., Dejax P. and Gharbi A., (2010) Production planning of a hybrid manufacturing-remanufacturing system under uncertainty within a closed-loop supply chain, *Int. J. Product. Econ.,* doi:10.1016/j.ijpe.2010.10.026.
50. Kim J., Lee Y., Moon I. (2008) Optimization of a hydrogen supply chain under demand uncertainty, *Int J Hydrogen Energ,* 33, 4715–4729.
51. Kwak J.K, Gavirneni S. (2011) Retailer policy, uncertainty reduction and supply chain performance, *Int. J. Product. Econ.,* 132, 271–278.
52. Kwakkel J. and Cunningham S., (2008) Techniques and Methods for Uncertainty Management, *PICMET 2008 Proceedings*.
53. Kwon O., Im G., Lee K.C. (2007) MACE-SCM: A multi-agent and case-based reasoning collaboration mechanism for supply chain management under supply and demand uncertainties, *Expert Syst Appl*, 33, 690–705.
54. Lau A., Lau H. and Willett K. (2000) Demand uncertainty and returns policies for a seasonal product: An alternative model, *Int. J. Product. Econ.*, 66, 1-12.
55. Li H. and Azarm S., (2002) An Approach for Product Line Design Selection Under Uncertainty and Competition, *J Mech Design,* 124, 385-392.
56. Li L. and Zabinsky Zelda B., (2009) Incorporating uncertainty into a supplier selection problem, *Int. J. Product. Econ.*, 10.1016.
57. Li J., Huang P. and Ge J., (2007) Demand Uncertainty and Supplier's Buybacks in the Price Dependent Newsvendor Model, *ICMSE.*
58. Li C., Liu F., Cao H. and Wang Q., (2009) A stochastic dynamic programming based model for uncertain production planning of re-manufacturing system, *Int. J. Prod. Res.,* 47/13, 3657–3668.
59. Lin C., Wang T. (2011) Build-to-order supply chain network design under supply, *Transportation Research,* 45-B, 1162–1176.
60. Lo W., Hong T., Jeng R., (2008) A framework of E-SCM multi-agent systems in the fashion industry, *Int. J. Product. Econ.,* 114, 594–614.
61. Longinidis O., Georgiadis M.C. (2011) Integration of financial statement analysis in the optimal design of supply chain networks under demand uncertainty, *Int. J. Product. Econ.*, 129, 262–276.
62. Lu M.M., Irajpour A., Hamidi N., Mavi R.K. (2011) Designing Excellent Supply Chain Network in Uncertainty Environment with Fuzzy Logic, *Int J Bus Manag,* 6/4, 241-151.

63. Mahnam M., Yadollahpour M.R., Famil-Dardashti V. and Hejazi S.R., (2009) Supply chain modeling in uncertain environment with bi-objective approach, *Comput Ind Eng*, 56, 1535–1544.
64. Mantrala Murali K. and Raman K., (1999) Demand uncertainty and supplier's returns policies for a multi-store style-good retailer, Eur J Ope Res, 115, 270-284.
65. Mason-Jones R. and Towill, D.R., (1998) Shrinking the supply chain uncertainty circle, Control, Eylül-1998, 17-22.
66. Matsui K. (2011) Resale price maintenance for supply chains distributing products with demand uncertainty, *Int. J. Product. Econ.*, 134, 375–387.
67. McCaulley-Bell P., (1999) Intelligent agent characterization and uncertainty management with fuzzy set theory: a tool to support early supplier integration, *J Intell Manuf,* 10, 135-147.
68. Mele F.D., Guillen G., Espuna A., Puigjaner L. (2007) An agent-based approach for supply chain retrofitting under uncertainty, *Comput Chem Eng,* 31, 722–735.
69. Merschmann U., Thonemann U.W. (2011) Supply chain flexibility, uncertainty and firm performance: An empiric alanalysis of German manufacturing firms, *Int. J. Product. Econ.,* 130, 43–53.
70. Min H., Ko C.S. and Ko J.H., (2006) The spatial and temporal consolidation of returned products in a closed-loop supply chain network, *Comput Ind Eng*, 51, 309–320.
71. Mishra P.K., Shekhar B.R. (2011) Impact of Risks and Uncertainties on Supply Chain: A Dairy Industry Perspective, J Manag Research, 3/2, 1-18.
72. Mitra K., Gudi R.D., Patwardhan S.C., Sardar G. (2009) Towards resilient supply chains: Uncertainty analysis using fuzzy mathematical programming, *Chem Eng Res Des*, 87, 967–981.
73. Mula J., Peidro D., Poler R. (2010) The effectiveness of a fuzzy mathematical programming approach for supply chain production planning with fuzzy demand, *Int. J. Product. Econ.,* 128, 136–143.
74. Mula J., Peidro D., Poler R. (2010) Fuzzy linear programming for supply chain planning under uncertainty, *Int J Inf Tech Decis,* 9/3, 373–392.
75. Mula J., Poler R. and Garcia J.P. (2006) MRP with flexible constraints: A fuzzy mathematical programming approach, *Fuzzy Set Syst*, 157, 74–97.
76. Pan F. and Nagi R. (2009) Robust supply chain design under uncertain demand in agile manufacturing, *Comput Oper Res*, 37, 668--683.
77. Paulraj A. and Chen I., (2007) Environmental Uncertainty and Strategic Supply Management : A resource dependence perspective and performance implications, *JSCM,* Summer 2007, 29-42.
78. Peidro D., Mula J., Jiménez M. and Botella M., (2010) A fuzzy linear programming based approach for tactical supply chain planning in an uncertainty environment, *Eur J Ope Res,* 205, 65–80.

UNCERTAINTY MODELLING IN SUPPLY CHAIN MANAGEMENT: THE TREND IN THE USE OF FUZZY SET THEORY

ÖZGÜR KABAK

Industrial Engineering Department, Istanbul Technical University, Macka Istanbul 34367, Turkey

Supply chain management (SCM) has been considered as one of the most important operations strategy for improving competitiveness of organizations in the last decades. While managing the supply chain (SC) effectively one of the essential problems is the uncertainties that should be treated properly. Fuzzy set theory allows for a conceptual and theoretical framework for dealing with uncertainties in SC problems. In this paper we investigate the trend in the SC articles that use fuzzy set theory through the years, and the focus of fuzzy set theory in SC modeling. We find that the number of related articles growing exponentially and it is mostly due to the rising interest of the fuzzy environment researchers to SCM. Additionally we find that the demand is the most studied type of uncertainty in fuzzy SC modeling.

1. Introduction

Supply chain management (SCM) has been considered as the most popular operations strategy for improving organizational competitiveness in the twenty-first century. SCM, in general, is a synthesis of planning, controlling and designing of material and information flow through the process starting from procurement of the raw material to the customer delivery. SCM has attracted the attention of the researchers and practitioners since the beginning of 1990s because it is broadly defined to include all classical business functions and theoretic management disciplines impacting this ever emerging field.

Uncertainty is one of the most challenging and important problems in a supply chain (SC) (Sabri and Beamon, 2000). The SC problems are due to uncertainties like other real-life problems. The SC is a dynamic network of several business entities that involve a high degree of imprecision. The complex nature and dynamics of the relationships among the different actors in an SC imply an important degree of uncertainty in SCM decisions. Especially in specific problems that necessitate future projections, such as new product design or strategic planning, there will be several parameters that cannot be estimated deterministically (Kabak and Ulengin, 2011).

The sources of the uncertainties in the SC may be environmental or originated from the system itself. In all circumstances, neglected uncertainties in the models may result in inaccurate solutions.

In the literature, the basic models for modelling SC under uncertainty are stochastic models, simulation approaches, hybrid approaches (i.e., deterministic and stochastic), and fuzzy set theory based models. SC uncertainties are modelled with probability distributions in the first three approaches. Probability distributions are predicted from historical data using statistical approaches. When the data is unavailable fuzzy set theory or the possibility theory may be better choices (Piedro et al., 2009). The fuzzy set theory is an important tool to model those types of uncertainties encountered in SC problems (Stanciulescu, 2003).

In this paper we conducted a literature survey to find out the trend in the use of fuzzy set theory in modelling SC uncertainties, and the focus of fuzzy set theory in SC modelling. Next two sections give the details of the analysis, respectively. The paper concludes with the fourth section.

2. Trends in the use of fuzzy set theory

In this study we investigate the trends in the use of fuzzy set theory in modelling SC under uncertainty. For this purpose we made a search of articles at Web of Science (webofknowledge.com) database. We designed five different searches to reveal the trends (see Table 1.)

Table 1. Searches designed in the study

Search ID	Search Keywords
1	Topic=(supply chain) AND Topic=(fuzzy)
2	Topic=(supply chain) AND Topic=(uncertain OR fuzzy OR stochastic OR uncertainty)
3	Topic=(uncertain OR fuzzy OR stochastic OR uncertainty)
4	Topic=(supply chain)
5	Topic=(fuzzy)

The first search (ID#1) gives the SC articles that use fuzzy set theory. The total number of the results is 357. As can be seen in Table 2, the related articles are growing through the years, and increases to 79 in 2012 from 4 in 2002. This increase may be because of the increase in the fuzzy theory based articles, or increase in the SC articles or increase in the uncertainty modelling articles. Therefore other four searches are designed.

In the second search (ID#2) SC articles for uncertainty modelling are listed. In order to get the uncertainty modelling articles four keywords are issued: uncertain, fuzzy, stochastic, and uncertainty. The third search (ID#3) includes articles for uncertainty modelling. The fourth search (ID#4) gives SC articles. And finally the fifth search (ID#5) gives the articles having fuzzy set theory. The results of the all searches through the years are presented in Table 2.

Each of the research result indicates an increase through years. This shows that the importance of and the interest in SCM, uncertainty modelling and fuzzy modelling is growing. The most significant increase is realized in the number of the SC articles that use fuzzy set theory (search ID#1) as it jumps by 19.75 times from 2002 to 2011. The number of SC articles for uncertainty modelling (search ID#2) also increased very high by 7.55 times from 2002 to 2011.

Table 2. Number of articles in years as a result of the searches

Years \ Search ID	1	2	3	4	5
2002	4	51	13988	545	1959
2003	9	72	15396	681	2309
2004	13	92	16739	765	2689
2005	8	107	18859	893	3174
2006	28	143	21093	1106	3548
2007	27	212	22069	1309	3170
2008	45	253	24718	1598	3850
2009	61	311	26837	1774	4356
2010	61	338	28839	2141	4483
2011	79	385	29105	2094	4634
2002/2011	19.75	7.55	2.08	3.84	2.37
Total	357	2141	349574	15714	51665

In the next step we get the proportional number of the SC articles that use fuzzy set theory (search ID#1) with respect to results of the other searches (See Table 3). These proportions show whether the result of the first search is significant without the effect of the results of other searches.

Table 3. Proportion of the articles in the searchers

Years \ Proportions	(1/2)	(1/3)	(1/4)	(1/5)
2002	7.84%	0.03%	0.73%	0.20%
2003	12.50%	0.06%	1.32%	0.39%
2004	14.13%	0.08%	1.70%	0.48%
2005	7.48%	0.04%	0.90%	0.25%
2006	19.58%	0.13%	2.53%	0.79%
2007	12.74%	0.12%	2.06%	0.85%
2008	17.79%	0.18%	2.82%	1.17%
2009	19.61%	0.23%	3.44%	1.40%
2010	18.05%	0.21%	2.85%	1.36%
2011	20.52%	0.27%	3.77%	1.70%
r^2	0.6038	0.9166	0.8364	0.9228
Yearly effect	0.32%	0.17%	1.24%	0.03%

We build a linear regression model ($Y= ax + b$; x: years) for each proportion to detect pure increase in the number of fuzzy SC articles. According to the results shown in the last two row of the Table 4, all the regression models are significant (r^2 is greater than 0.5). According to the yearly affect coefficients (coefficient of a in the regression model), the number of fuzzy SC articles (search ID#1) increases 0.32% higher than SC uncertainty articles (search ID#1) and 1.24% higher than the SC articles (search ID#4). Results also indicate that the number of fuzzy SC articles (search ID#1) is very parallel to the number of fuzzy articles (search ID#5), due to the yearly affect is very low (i.e., 0.03%). Therefore we can conclude that the trend in the fuzzy SC studies is more related to trend in fuzzy studies. This may be because of rising interest of fuzzy environment researches to SCM.

3. Focus of fuzzy set theory in supply chain modeling

We investigate the literature in details to find out the focus of the fuzzy set theory in supply chain modelling. For this purpose we made a search on sciencedirect database (www.sciencedirect.com) with key words "fuzzy" and "supply chain" for year 2009 until today. We found 43 articles. We analysed these articles according to structure of SC (i.e., single stage SC, two stage SC, serial SC, and network SC) and the uncertain parameters (i.e., demand, supply, price, internal, and environmental).

According the SC structures considered in the articles, majority of the articles are dealing with network supply chains. Serial and two stage SCs modelled less frequently while SCs with single stage are preferred in only 3 articles (see Table 4). This result may be due to the fact that the use of network structure simplifies the representation of SC units and/or functions as well as the interrelationships among them. Consequently, the network structures could be easily transformed to mathematical models. This property is important for the improvement of a SC system as well as for taking the operational and strategic decisions.

Table 4. Classification of articles according to structure of SC

Structure of SC	Number of articles	Percentage
Network SC	25	58.14%
Serial SC	9	20.93%
Two stage SC	6	13.95%
Single stage SC	3	6.98%

The results related to the uncertain parameters show that demand is most studied uncertain parameter (see Table 5.). This result is very consistent with the other literature such as Kabak and Ulengin (2011) that indicates the demand to be the most important and extensively studied source of uncertainty. Given the fact that effectively meeting customer demand is what mainly drives most SCM initiatives, it is appropriate to emphasize incorporating demand uncertainty into planning decisions. Furthermore, because demand fluctuations affect the production system and suppliers over time, demand is the main source of uncertainties.

Other frequently studied parameters are internal, supplier related, environmental parameters and price (see Table 5). This is because these types of uncertainty are also inherent in the system and in the supply side. For instance, internal uncertainty may result from the unreliability of the production process because of machine breakdowns and supplier related uncertainties can be because of the variability of the suppliers' performance due to late deliveries (Petrovic, 2001).

Table 5. Classification of articles according to uncertain parameters

Uncertain parameter	Number of articles	Percentage
Demand	24	55.81%
Internal parameters	11	25.58%
Supplier related parameters	9	20.93%
Environmental parameters	7	16.28%
Price	3	6.98%

4. Conclusions

In this study we, first, investigate trend in the number of SC articles under uncertainty and using fuzzy set theory based models. For this purpose we made five difference article searches on Web of Science database. The results indicate that the number of fuzzy SC articles, SC under uncertainty articles, articles tackling with uncertainty, SC articles and fuzzy articles are increasing through years. Fuzzy SC articles are particularly growing more dramatically than the other types of articles in consideration. Results show that growth in the fuzzy SC articles is very parallel to the increase in the fuzzy articles. This may be referred as the interest of fuzzy environment researchers to SC management rises in years. On the other hand SC researchers' interest to uncertainty modelling or fuzzy modelling is not growing significantly.

Secondly we analysed some SC articles that uses fuzzy set theory in details to find out the related SC structures and uncertain parameters. Results show that fuzzy environment researchers mostly studies network supply chains. Furthermore, the most studied type of uncertainty is the demand. This result underlines a need for some future works related to other types of uncertain parameters such as internal, supplier related, environmental parameters and price.

References

1. Ö. Kabak, F. Ülengin,. European Journal of Operational Research, 209, 253-264 (2011)
2. D. Peidro, J. Mula, R. Poler, J.-L. Verdegay, Fuzzy Sets and Systems, 160, 2640–2657 (2009)
3. D. Petrovic, International Journal of Production Economics, 71, 429– 438 (2001).
4. E.H. Sabri, B.M. Beamon, Omega, 28, 581–598 (2000)
5. C. Stanciulescu, P. Fortemps, M. Installe, V. Wertz,. European Journal of Operational Research, 149, 654–675 (2003)

COLLABORATIVE PARTNER EVALUATION WITH MULTI-CRITERIA DECISION MAKING TECHNIQUES*

JBID ARSENYAN
Industrial Engineering, Bahçeşehir University, Çırağan Cad.
Istanbul, 34100, Turkey

GÜLÇIN BÜYÜKÖZKAN†
Industrial Engineering, Galatasaray University, Çırağan Cad.
Istanbul, 34357, Turkey

Collaborative product development (CPD) can be defined as two or more firms joining resources for the same goal, where the first phase is the partner evaluation process. The choice of partner affects the performance of the whole CPD, and therefore it must be conducted carefully. This paper presents criteria set defined specifically for CPD partner evaluation and proposes an integrated fuzzy multi-criteria decision methodology for the selection of the right partner. The technique enables decision makers to evaluate criteria weights and the potential partners separately.

1. Introduction

Collaborative product development (CPD), by definition, consists of partners from different organizations joining forces for a common goal. Therefore, partner selection is the most fundamental decision of the collaboration process, after the collaboration decision. Selection of an adequate partner ensures the success of the collaboration [1]. This study offers a new set of three-dimensional partner evaluation criteria based on the CPD model introduced by Arsenyan and Büyüközkan [2]. The literature suggests that many multi-criteria decision making (MCDM) methods for the partner evaluation problem. Additionally, the subjective nature of the decision-making problem and the uncertain environment which surrounds CPD and partner evaluation problems are analyzed under fuzzy environment. Consequently, this study employs Fuzzy Analytical Hierarchy Process (AHP) to determine criteria weights and Fuzzy Axiomatic Design

* This work is supported by TÜBİTAK (Project no: 109M147), Bahçeşehir University, and Galatasaray University Research Fund (Project no: 12.402.003).

† gulcin.buyukozkan@gmail.com.

(Fuzzy AD) in order to evaluate and select partners. Evaluation criteria are defined, and potential candidates are identified. Moreover, experts' judgments are employed in order to validate evaluation criteria and determine criteria weights, as well as evaluate partners.

This study aims to provide a criteria set able to cover all aspects of the CPD partner selection process as well as a integrated fuzzy MCDM methodology which can respond to the contingency of the decision-making process. The next section presents partner evaluation criteria and then fuzzy MCDM methodology is introduced. Subsequently, the case study is presented to illustrate the advantages of the three dimensional criteria set as well as the fuzzy techniques, followed by the concluding remarks.

2. Partner selection criteria

Partner selection is the key process in the establishment of a CPD project as highlighted by many authors, and it is critical in finding organizational matches that can effectively relate with each other [3].

Table 1 Partner Evaluation Criteria

	Criteria	*Sources*
C_1	*Partner focused criteria*	
C_{11}	Financial health	Glaister and Buckley [4]
C_{12}	Market position	Emden et al. [3]
C_{13}	Geographical position	Wang et al. [5]
C_{14}	Reputation	Glaister and Buckley [4]
C_{15}	Experience (Business, Market, Sector, Management)	Nguye and Shanks [6]
C_{16}	Network	Glaister and Buckley [4]
C_2	*Collaboration focused criteria*	
C_{21}	Trust (Technical, Personal)	Glaister and Buckley [4]
C_{22}	Compatibility (Managerial, Cultural, IT and communication)	Emden et al. [3]
C_{23}	Complementarity (Knowledge, Resource)	Emden et al. [3]
C_{24}	Willingness to share information	Hou et al. [7]
C_{25}	Flexibility	Hou et al. [7]
C_{26}	Integration ability	
C_{27}	Collaboration experience	Glaister and Buckley [4]
C_3	*Product/Software development focused criteria*	
C_{31}	Technical expertise	Emden et al. [3]
C_{32}	Technological capability	Lederer-Antonucci et al. [1]
C_{33}	Product/Software development experience	Glaister and Buckley [4]
C_{34}	R&D resources	Lederer-Antonucci et al. [1]
C_{35}	Innovative capability	Wang et al. [5]
C_{36}	Project expectations (Risk, Investment, Development time, Quality)	Hou et al. [7]

Given that incompatibility between partners is one of the most important reasons why co-development partnerships fail, selecting a partner becomes a strategic issue for the firm [6]. Individual attributes of firms as well as collaborative experience possess a value in the evaluation of potential partners.

The existing literature offers many different approaches to categorize partner selection criteria. These various approaches are collected and analyzed to define partner selection criteria hierarchy. Partner selection criteria set, summarized in Table 1, is identified through these various aspects influencing partner selection process, which enables the elaboration of the three main dimensions of CPD.

3. Proposed methodology

Partner evaluation literature often refers to MCDM techniques to handle the problem. This paper employs MCDM techniques in a fuzzy environment. The approach consists of Fuzzy AHP for the determination of criteria weights and Fuzzy AD for the evaluation of the alternatives. Fuzzy Set Theory [8] is employed to deal with this imprecision and subjectivity of the decision makers.

AD principles provide a powerful tool to measure how well alternatives respond to requirements [9]. Decision makers employ a "design range" to define the requirements. Then the "system ranges" of the alternatives are assessed against this design range to analyze their suitability. As the study includes incomplete information with subjective judgments, Fuzzy AD is preferred to conventional AD in order to operate in a fuzzy environment. Fuzzy AD has proven to be a powerful MCDM tool since first introduced [10]. This paper employs the methodology of hierarchical Fuzzy AD, introduced in [11]. The design range concept gives Fuzzy AD a dynamic character, given that for each specific case, a new set of requirements can be defined on the criteria set, and redundant criteria may be dismissed without changing the criteria set. However, Fuzzy AD fails to translate functional requirements into criteria weights. Therefore, another technique is required to determine criteria weights. AHP is particularly useful for evaluating complex multi-attribute alternatives involving subjective criteria [12]. As the independence axiom of AD guarantees the independence of the FRs, AHP is a suitable technique as the independence of evaluation criteria is also assured. However, the vagueness in decision makers' judgments makes the crisp pairwise comparison too imprecise to capture the solid judgments. Fuzzy logic in the pairwise comparison of the AHP, which is called Fuzzy AHP, makes up for this deficiency in the conventional AHP. The technique is adapted from [12].

4. Application

A partner evaluation criteria set is tested within a case study. The case study aims to validate partner evaluation criteria and to test the usability of the proposed methodology. Our expert is a Management Engineer and is currently in software industry. He is provided with an evaluation form, which includes an evaluation set of Fuzzy AHP for the determination of criteria weights and another evaluation set of Fuzzy AD system and design ranges for the evaluation of potential partners. The expert employs linguistic variables provided in the form for system and design ranges, whereas he makes pairwise comparisons according to the fuzzy scale. The linguistic terms are then translated into TFNs, through which criteria weights and ICs are calculated.

The case study aims to observe the validity and the managerial advantage the presented methodology provides. The expert supports the significance of the criteria set, however he attributes different importance to the criteria. Fuzzy AHP becomes essential in this phase. Pairwise judgments of the expert are collected in fuzzy scale, and these judgments are translated into criteria weights through Fuzzy AHP. On the other hand, each project type requires different features and some criteria are prioritized over others in each case. The design range of Fuzzy AD serves this purpose by defining a requirement set on each criterion. Both techniques combine in presenting a final evaluation and ranking of the alternatives according to the judgments of the expert.

Table 2 Project evaluation of Expert I

	C_1	C_{11}	C_{12}	C_{13}	C_{14}	C_{15}	C_{151}	C_{152}	C_{153}	C_{154}	C_{16}				
FR	LM	LL	LVL	LMG	LM	LM	LFG	LM	LFL	LM	LM				
A_1	M	VL	VL	M	M	M	G	ML	M	ML	ML				
A_2	G	G	M	VG	M	ML	MG	ML	L	MG	FG				
A_3	ML	FL	L	G	FL	MG	G	G	M	ML	M				
	C_2	C_{21}	C_{211}	C_{212}	C_{22}	C_{221}	C_{222}	C_{223}	C_{23}	C_{231}	C_{232}	C_{24}	C_{25}	C_{26}	C_{27}
FR	LFG	LG	LG	LG	LG	LG	LFG	LG	LFG	LFG	LFG	*LVG*	LFG	LFG	LFG
A_1	G	G	FG	VG	G	FG	G	FG	G	G	G	G	G	MG	MG
A_2	G	G	G	G	G	G	MG	G	G	G	FG	G	G	G	FG
A_3	G	G	G	G	G	G	G	G	G	G	G	G	FG	G	FG
	C_3	C_{31}	C_{32}	C_{33}	C_{34}	C_{35}	C_{36}	C_{361}	C_{362}	C_{363}	C_{364}				
FR	LFG	LG	LFG	LFG	LM	LMG	LG	LMG	LFL	LG	*LVG*				
A_1	MG	G	MG	VG	ML	M	G	G	L	G	VG				
A_2	G	G	VG	MG	G	G	G	G	G	VG	VG				
A_3	FG	FG	MG	G	FG	G	G	G	M	G	G				

The company is about to start a new software development project and three alternatives are considered for the venture. The evaluations on the design and system ranges are presented in Table 2. The highest requirements are set on willingness to share information and expected project quality.

Final evaluations of the expert are presented in Table 3, depicting computed criteria weights and IC's as well as final evaluation. The expert clearly favours collaboration capabilities of the partner. Development focused criteria is considered as of second importance. Partner focused criteria are not considered as important as the others.

Table 3 Partner scores of the Expert

	I_{c1}	I_{c2}	I_{c3}	*Level 1 scores*	*Level 2 scores*	*Level 3 scores*		
w	0.096	*0.653*	0.251					
A_1	0.174	0.292	0.834	1.300	1.858	1.284		
A_2	0.017	0.131	0.373	***0.421***	***1.225***	1.034		
A_3	0.366	0.059	0.493	0.990	1.226	***0.958***		
	I_{c11}	I_{c12}	I_{c13}	I_{c14}	I_{c15}	I_{c16}		
w	0.092	0.037	0.092	0.226	*0.461*	0.092		
A_1	0.397	0.120	0.330	0.409	0.833	0.350	2.438	
A_2	0.007	0.018	0.006	0.409	1.754	0.039	2.233	
A_3	0.126	0.068	0.024	0.862	0.406	0.166	*1.651*	
	I_{c21}	I_{c22}	I_{c23}	I_{c24}	I_{c25}	I_{c26}	I_{c27}	
w	0.220	0.220	0.073	0.073	0.073	0.073	*0.265*	
A_1	0.220	0.220	0.033	0.190	0.033	0.244	0.882	1.822
A_2	0.220	0.220	0.033	0.190	0.033	0.033	0.351	*1.080*
A_3	0.220	0.220	0.033	0.190	0.097	0.033	0.351	1.144
	I_{c31}	I_{c32}	I_{c33}	I_{c34}	I_{c35}	I_{c36}		
w	*0.214*	0.071	*0.214*	0.071	*0.214*	*0.214*		
A_1	0.214	0.237	0.021	0.272	0.768	0.214	1.727	
A_2	0.214	0.007	0.712	0.013	0.056	0.214	*1.216*	
A_3	0.643	0.237	0.096	0.031	0.056	0.214	1.277	

Alternative evaluation presents an interesting analysis. When only the first level criteria are considered, A_2 is clearly the most suitable alternative. As no alternative is eliminated at the first level, sub-criteria are considered as well. A_2 is still the most suitable alternative; however it can be observed that A_3 possesses the minimum information content on partner focused criteria. The low importance of this dimension fails to capture this characteristic of A_3. However, the ranking changes once all levels of hierarchy are considered. A_2 fails dramatically on *experience* sub-criteria, while it is not the best responder to compatibility and complementarity sub-criteria. Given the importance of collaboration focused criteria, A_3 comes out as the most suitable alternative when all hierarchical levels are considered.

5. Concluding Remarks

This study presented an integrated fuzzy multi-criteria decision making methodology for CPD partner evaluation. A three-dimensional criteria set is introduced for partner evaluation. Fuzzy AHP is applied to define criteria weights and Fuzzy AD is applied to evaluate partners. The integrated methodology enables the decision maker to express both his overall judgments and the project-based judgments on the three-dimensional criteria. As the case study shows, the methodology is proven to be easy and effective in reflecting the decision maker's evaluation.

References

1. S.H. Chen, P.W. Wang, C.M. Chen, and H.T. Lee, "An analytic hierarchy process approach with linguistic variables for selection of an R&D strategic alliance partner ," Computers & Industrial Engineering, 58, 278–287, (2010).
2. J. Arsenyan and G. Buyukozkan, "Modelling Collaborative Software Development Using Axiomatic Design Principles," International Journal of Computer Science, 36 (3), (2009).
3. Z. Emden, R.J. Calantone, and C. Droge, "Collaborating for New Product Development: Selecting the Partner with Maximum Potential to Create Value," Journal of Product Innovation Management, 23(4), 330-341, 2006.
4. G. Büyüközkan, O. Feyzioğlu, and E. Nebol, "Selection of the strategic alliance partner in logistics value chain ," International Journal of Production Economics , 113, 148-158, (2008).
5. K.W. Glaister and P.J. Buckley, "Task-related and Partner-related Selection Criteria in International Strategic Alliances," British Journal of Management, 8(3), 199-222, (1997).
6. C. Wu and D. Barnes, "Formulating partner selection criteria for agile supply chains: A Dempster–Shafer belief acceptability optimization approach," International Journal of Production Economics, 125, 284–293, (2010).
7. L. Hou, D. Han, and Z. Lin, "Research on Supplier Selection for Inter-firm Product Collaborative Development," in Proceedings of the 6th World Congress on Intelligent Control and Automation, Dalian, China, (2006).
8. W.Y. Wu, H.-A. Shih, and H.-C. Chan, "The analytic network process for partner selection criteria in strategic alliances," Expert Systems with Applications, 36, 4646–4653, (2009).
9. L.A. Zadeh, "The concept of a linguistic variable and its applications to approximate reasoning," Information Sciences, pp. 199-249 (I); 301-357 (II), (1975).
10. N.P. Suh, The Principles of Design. New York: Oxford University Press, (1990).
11. C. Kahraman and O. Kulak, "Fuzzy multi-attribute decision making using an information axiom based approach," in Fuzzy Multi-Criteria Decision Making. New York: Springer US, 265-278, (2005).
12. C. Kahraman and S. Cebi, "A new multi-attribute decision making method: hierarchical fuzzy axiomatic design," Expert Systems with Applications , 36(3), 4848–4861, (2009).
13. C.K. Kwong and H. Bai, "Determining the importance weights for the customer requirements in QFD using a fuzzy AHP with an extent analysis approach," IIE Transactions, 35, 619-626, (2003).

PART 3

FOUNDATIONS OF COMPUTATIONAL INTELLIGENCE

THE REAL ALGEBRAIC NORMAL FORM: FUZZY BOOLEAN LOGIC SATISFYING $\Delta f = 0$

MICHAEL VIELHABER
Hochschule Bremerhaven, FB2, An der Karlstadt 8, D-27568 Bremerhaven, Germany
Universidad Austral de Chile, Instituto de Matemáticas, Casilla 567, Valdivia, Chile
vielhaber@gmail.com

We define real extensions $f : \mathbb{R}^n \supset [0,1]^n \to [0,1] \subset \mathbb{R}$ of Boolean functions $f^{\mathbb{B}}$ that satisfy $\Delta f = 0$ on the whole domain $[0,1]^n$, *i.e.* potential theory as first principle.

We introduce the Real Algebraic Normal Form RANF $\in \mathbb{Z}[x_1, \ldots, x_n]$ of f, a generalization of the Algebraic Normal Form ANF of $f^{\mathbb{B}}$, and show how the RANF is related to the Disjunctive Normal Form DNF by the Reed-Muller Transform, or the Inclusion-Exclusion Principle.

Keywords: Fuzzy logic; potential theory; Real Algebraic Normal Form; ANF; DNF; Reed-Muller transform.

1. Introduction

This paper was motivated by the two statements "... none of the elementary FCA [fuzzy cellular automata] ... exhibited chaos [in DNF]" and "a large class of CNF-fuzzy rules ... exhibit chaos" in[1]. One and the same Boolean function fuzzyfied thus is tame or chaotic, depending on the representation. This state-of-affaires seems to be odd, in particular in view of safety-critical applications like medical expert systems, or in nuclear science.

So, starting with exactly the 2^n Boolean truth values (the function's truth table) and nothing more (in particular no representation by logic operators), what should a real-valued interpolation yield?

Here is *Nature*'s interpolation rule, the (stationary) heat equation:

$$\Delta(f(x_1, \ldots, x_n)) = \nabla^2(f(x_1, \ldots, x_n)) = 0.$$

The physical interpretation is a unit cube $[0,1]^3$ of heat-conducting material, set to specific temperatures at the corners, *e.g.* to 0°C for corners with $0_{\mathbb{B}}$ and to 100°C for corners with $1_{\mathbb{B}}$, and isolated at its sides. The resulting heat distribution in the cube's interior satisfies the heat equation:

A temperature of *e.g.* 54.4°C is then equivalent to a fuzzy value 0.544 at the corresponding position.

2. Strange Effects of (Wrong) Fuzzification

(*i*) $A \wedge A = A \vee A = A$ will not be satisfied for $\wedge \equiv \cdot_{\mathbb{R}}$, $\vee \equiv +_{\mathbb{R}}$, or if $x \vee y \equiv x + y - xy$.

(*ii*) Using min and max, we get $x \oplus y = (x \vee y) \wedge (\overline{x} \vee \overline{y}) = \min(\max(x, y), \max(1 - x, 1 - y))$:

x	0.0	0.2	0.4	0.6	0.8	1.0
$x \oplus 0.3$	0.3	0.3	**0.4**	0.4	0.3	0.3

; an unjustified inner maximum.

(*iii*) $\wedge \equiv \cdot_{\mathbb{R}}$ and $A \vee B \vee C \equiv \min(1, A + B + C)$ produces chaos[1].

3. Axioms

Let $\mathbb{B} = \{0, 1\}$ be the set of Boolean truth values (0 = false, 1 = true), $\mathbb{I} = [0, 1] \subset \mathbb{R}$ the real unit interval. We immerse $\mathbb{B}$ into $\mathbb{I}$ by identifying $0_{\mathbb{B}} = 0_{\mathbb{I}} = 0_{\mathbb{R}}$ and $1_{\mathbb{B}} = 1_{\mathbb{I}} = 1_{\mathbb{R}}$. We consider a given Boolean function $f^{\mathbb{B}}$ defined on $\mathbb{B}^n$ and ask for its fuzzy extension f to $\mathbb{I}^n$. Starting with the real calculus viewpoint $\Delta f = 0$, we define three basic axioms, our desiderata:

Axiom B (Boole)

The Boolean function $f^{\mathbb{B}}$ and the real function f coincide on $\{0, 1\}^n$.

Axiom L (Laplace)

The real function f has vanishing Laplacian everywhere, *i.e.*

$$\Delta f = \nabla^2 f = \sum_{k=1}^{n} \frac{\partial^2 f}{\partial x_k^2} = 0, \forall x \in \mathbb{I}^n.$$

Axiom S (Subspaces)

Let $f_{1,2}^{\mathbb{B}}$ be defined on $D_i, i = 1, 2$ with $\mathbb{B}^k \cong D_1 \subset D_2 \cong \mathbb{B}^n$, $\iota^{\mathbb{B}} : D_1 \to D_2$ the corresponding injection, and $f_1^{\mathbb{B}}(x) = f_2^{\mathbb{B}}(\iota^{\mathbb{B}}(x))$ for all $x \in D_1$. Let $f_{1,2}$ be the respective extensions to $\mathbb{I}^k, \mathbb{I}^n$, with $\iota : \mathbb{I}^k \to \mathbb{I}^n$ the injection. Then $f_1(x) = f_2(\iota(x)), \forall x \in \mathbb{I}^k$.

Remark 3.1. Axiom **B** is the obvious way to get a faithful representation.

Axiom **L** assures continuity, smoothness (no "sharp edges"), and absence of inner extremal points, unjustified by $f^{\mathbb{B}}$.

Axiom **S** guarantees faithfulness of projections, and defines the full boundary, starting from the corner points to define edges with $\Delta = 0$, then faces, then the 3 –, . . . , (n–1) – dimensional simplices, always building up one more dimension according to Axiom **L**.

4. Uniqueness

Theorem 4.1. *Uniqueness of the solution to the heat equation with boundary condition*

Let $c : \partial\mathbb{I}^n \to \mathbb{R}$ be a boundary condition, a twice differentiable function defined on the boundary of the n-dimensional hypercube.

Let $f, g : \mathbb{I}^n \to \mathbb{R}$ be two solutions of the heat equation with boundary condition c, i.e. $\Delta f(x) = \Delta g(x) = 0$ for $x \in \mathbb{I}^n$, and $f(x) = g(x) = c(x)$ on $\partial\mathbb{I}^n$. Then $f \equiv g$.

Proof. By the maximum principle. In more detail: We consider the difference $h := f - g$, which is zero on $\partial\mathbb{I}^n$ and has vanishing Laplacian everywhere. If $h \equiv 0$, then $f \equiv g$, and we are done.

Assume otherwise. Then h assumes a maximum $M > 0$ somewhere on $\mathbb{I}^n$ (or a minimum < 0, then consider $-h := g - f$), since h is bounded. Let $x_M \in \mathbb{I}^n$ be some argument with $h(x_M) = M$, hence $h(x_M) \geq h(x), \forall x \in \mathbb{I}^n$.

Therefore, $\forall k : \partial^2 h/\partial x_k^2(x_M) \leq 0$, and $\Delta h(x_M) \leq 0$.

Amongst all such arguments x_M there is (at least) one, say $x_{\overline{M}}$, with $\exists E > 0, \forall 0 < \varepsilon < E : h(x_{\overline{M}} + \varepsilon \cdot (1, 0, \ldots, 0)) < M$, since h returns continuously to the boundary value 0.

Here $\partial^2 h/\partial x_1^2(x_{\overline{M}}) < 0$ (strict inequality) and thus $\Delta h(x_{\overline{M}}) < 0$ as well, a contradiction to $\Delta h \equiv 0$. Thus no such difference h apart from the constant function zero may exist, $f \equiv g$, and the solution is unique. □

5. Minterms

Definition 5.1. *Minterm* A Boolean function, which is zero except for a single truth table entry, is called a minterm.

In particular, the AND-ing together of all variables, and each conjunction term in the Disjunctive Normal Form DNF, is a minterm.

Theorem 5.1. *The Boolean* AND *of n variables, $f_\wedge^{\mathbb{B}}(x_1, \ldots, x_n) = x_1 \wedge \cdots \wedge x_n$, has a unique real extension to $\mathbb{I}^n$, compatible with axioms **B**, **L**, and **S**, which is*

$$f_\wedge : \mathbb{R}^n \supset \mathbb{I}^n \ni (x_1, \ldots, x_n) \mapsto \prod_{k=1}^{n} x_k \in \mathbb{I} \subset \mathbb{R}.$$

Proof. Axiom **B** is satisfied by inspection.

The partial derivatives of $f_\wedge$ are $\partial f_\wedge/\partial x_i = \prod_{k=1..n, k\neq i} x_k$ and $\partial^2 f_\wedge/\partial x_i^2 = 0$. Therefore, $\Delta f_\wedge = 0$, satisfying Axiom **L**.

For Axiom **S**, we distinguish two cases:

(*i*) $(1,\ldots,1) \notin D_1$. Then both f_1 and $f_2 = f_\wedge$ are constant zero on D_1.

(*ii*) $(1,\ldots,1) \in D_1$. Then $D_1 = \langle x_i, i \in I\rangle + (c_1,\ldots,c_n)$ with $c_i = \begin{cases} 0, i \in I, \\ 1, i \notin I, \end{cases}$ for some $I \subset \{1,\ldots,n\}$. Hence both $f_\wedge$ and its restriction to D_1 evaluate to the product $\prod_{i\in I} x_i$.

Uniqueness follows from Theorem 4.1, where starting from $f_\wedge^{\mathbb{B}}$ on $\mathbb{B}^n$, we build up the full boundary condition c on $\partial\mathbb{I}^n$ following Remark 3.1 by extending to the k – dimensional simplices for each $k = 1, 2, \ldots$ up to $k = n-1$ in turn. □

6. Symmetry: The Reflection Principle

Theorem 6.1. *Let $f_1^{\mathbb{B}}, f_2^{\mathbb{B}}$ be given such that*

$$f_2^{\mathbb{B}}(x_1,\ldots,x_n) = f_1^{\mathbb{B}}(x_1,\ldots,x_{k-1},\overline{x}_k,x_{k+1},\ldots,x_n)$$

*and let f_1 be the extension of $f_1^{\mathbb{B}}$ satisfying Axioms **B,L,S**. Then*

$$f_2(x_1,\ldots,x_n) = f_1(x_1,\ldots,x_{k-1},1-x_k,x_{k+1},\ldots,x_n).$$

Proof. Axiom **B** is valid by construction.

We obtain f_2 by reflecting at the hyperplane $x_k = \frac{1}{2}, x_k \leftrightarrow (1-x_k)$. Therefore, $\partial f_2/\partial x_i = \partial f_1/\partial x_i$, $\partial^2 f_2/\partial x_i^2 = \partial^2 f_1/\partial x_i^2$ for $i \neq k$, while $\partial f_2/\partial x_k = -\partial f_1/\partial x_k$, $\partial^2 f_2/\partial x_k^2 = (-1)^2\partial^2 f_1/\partial x_k^2$, thus $\Delta f_2 = \Delta f_1 = 0$.

Let $D_1 \subset D_2$ be the subspaces for $f_2^{\mathbb{B}}$ in Axiom **S**. Set $\tilde{D}_2 = \{x \in \mathbb{I}^n \mid (x_1,\ldots,\overline{x}_k,\ldots,x_n) \in D_2\}$ and $\tilde{D}_1 = \{x \in \tilde{D}_2 \mid (x_1,\ldots\ \overline{x}_k,\ldots,x_n) \in D_1\}$. Then Axiom **S** is valid for f_2 on the extensions of $D_1 \subset D_2$, since it is satisfied for f_1 on the extensions of $\tilde{D}_1 \subset \tilde{D}_2$ by assumption. □

7. Superposition Principle: The Real Algebraic Normal Form

Finally, let us recall the following fact:

Theorem 7.1. ***Superposition Principle***

Let $c_{1,2} : \partial\mathbb{I}^n \to \mathbb{R}$ be two boundary conditions, and $f_{1,2} : \mathbb{I}^n \to \mathbb{R}$ be two functions satisfying $\Delta f_i = 0$ on $\mathbb{I}^n$, and $f_i = c_i$ on $\partial\mathbb{I}^n$.

Then the sum function $f := f_1 + f_2$ satisfies $\Delta f = 0$ on $\mathbb{I}^n$ and the sum boundary condition $c_1 + c_2$ on $\partial\mathbb{I}^n$.

To sum up, we obtain the extension for any Boolean function with a single nonzero truth table entry by starting from $f_\wedge^{\mathbb{B}} = \wedge_k x_k \mapsto f_\wedge = \prod_k x_k$

with the 1 at $(1,\ldots,1)$ (Theorem 5.1), and repeatedly reflecting to move the single entry to any desired position by Theorem 6.1.

Repeating this for each minterm of a DNF, we have the extensions for each term of $f^{\mathbb{B}}$, which we now sum up – by the Superposition Principle 7.1 – to the extension f of $f^{\mathbb{B}}$, which is furthermore unique by Theorem 4.1. We thus have the general procedure to fuzzify any Boolean truth table into a real function, the Real Algebraic Normal Form RANF:

Theorem 7.2. *Real Algebraic Normal Form*
Let

$$f^{\mathbb{B}}(x_1,\ldots,x_n) = \bigvee_{I\subset\{1,\ldots,n\}} (d_I \bigwedge_{i\in I} x_i \bigwedge_{h\in\overline{I}} \overline{x}_h) \qquad DNF$$

be the standard Disjunctive Normal Form of $f^{\mathbb{B}}$ with $\overline{I} = \{1,\ldots,n\}\backslash I$ and $d_I \in \mathbb{B}$ the truth table entries. Then its unique extension with $\Delta f = 0$ is

$$f(x_1,\ldots,x_n) = \sum_{I\subset\{1,\ldots,n\}} (d_I \prod_{i\in I} x_i \prod_{h\in\overline{I}} (1-x_h)) = \sum_{J\subset\{1,\ldots,n\}} a_J \prod_{j\in J} x_j. \; RANF$$

Here $\sum_J a_J \prod_{j\in J} x_j$ is the Real Algebraic Normal Form of f, a polynomial with integer coefficients a_J of absolute value at most 2^{n-1}.

8. The Real Algebraic Normal Form and the Reed-Muller Transform

The Boolean function $f^{\mathbb{B}}$ can systematically be transformed to its unique real extension, the RANF, by means of the Reed-Muller Transform, applying the Inclusion-Exclusion-Principle to I, J, see[3].

The coefficients (a_J) of the RANF in n dimensions (2^n terms) are obtained from the (d_I) of the DNF by multiplying with the Reed-Muller transform matrix RMT_n:

$$(a_J) = RMT_n \cdot (d_I), \text{with } RMT_n = \bigotimes_{k=1}^{n} \begin{pmatrix} 1 & 0 \\ -1 & 1 \end{pmatrix}$$

Reducing the RANF mod 2, we obtain the standard ANF in $\mathbb{Z}/2\mathbb{Z}[x_1,\ldots,x_n] \cong \mathbb{F}_2[x_1,\ldots,x_n]$, see[2,3].

$$\begin{array}{lll} & \textbf{DNF} \text{ (truth table)} & \xleftrightarrow{\text{Duality Principle}} \textbf{CNF} \\ & \swarrow \updownarrow \; RMT_{\mathbb{Z}} & \\ RMT_{\mathbb{F}_2} \downarrow & \textbf{RANF} \in \mathbb{Z}[x_1,\ldots,x_n] & \\ & \searrow \downarrow \quad \text{mod } 2 & \\ & \textbf{ANF} \in \mathbb{F}_2[x_1,\ldots,x_n] & \end{array}$$

Example 8.1. The 4 minterms $f^{\mathbb{B}}$ on 2 variables and their coefficients:

$f^{\mathbb{B}}$	f	1	x_1	x_2	x_1x_2	
$f_{00}^{\mathbb{B}} = \overline{x}_2\overline{x}_1$	$f_{00} = (1-x_2)(1-x_1)$	1	−1	−1	1	
$f_{01}^{\mathbb{B}} = \overline{x}_2 x_1$	$f_{01} = (1-x_2)x_1$		1		−1	$= (RMT_2)^T$
$f_{10}^{\mathbb{B}} = x_2\overline{x}_1$	$f_{10} = x_2(1-x_1)$			1	−1	
$f_{11}^{\mathbb{B}} = x_2 x_1$	$f_{11} = x_2 x_1$				1	

The XOR $f_{\oplus}^{\mathbb{B}} := x_1 \oplus x_2 = f_{01}^{\mathbb{B}} \vee f_{10}^{\mathbb{B}}$, extends to $f_{\oplus} = x_1 + x_2 - 2x_1x_2$, with $a_{00} = a_{\{\}} = 0, a_{01} = a_{\{1\}} = 1, a_{10} = a_{\{2\}} = 1, a_{11} = a_{\{1,2\}} = -2$.

Example 8.2. Similarly, in 3 variables, we have the matrix RMT_3:

		d_{000}	d_{001}	d_{010}	d_{011}	d_{100}	d_{101}	d_{110}	d_{111}
a_{000} :	1	**1**							
a_{001} :	x_1	−1	1						
a_{010} :	x_2	−1		1					
a_{011} :	x_2x_1	**1**	−1	−1	**1**				
a_{100} :	x_3	**−1**				1			
a_{101} :	x_3x_1	**1**	−1			−1	**1**		
a_{110} :	x_3x_2	**1**		−1		−1		1	
a_{111} :	$x_3x_2x_1$	**−1**	1	1	**−1**	1	**−1**	−1	1

$f_{000}^{\mathbb{B}} \vee f_{011}^{\mathbb{B}} \vee f_{101}^{\mathbb{B}}$ (columns in bold), *i.e.* $f^{\mathbb{B}} = \overline{x}_3\overline{x}_2\overline{x}_1 \vee \overline{x}_3 x_2 x_1 \vee x_3\overline{x}_2 x_1$, extends to $f = 1 \cdot 1 + (-1)x_1 + (-1)x_2 + 2x_2x_1 + (-1)x_3 + 2x_3x_1 + 1x_3x_2 + (-3)x_3x_2x_1 = 1 - x_1 - x_2 - x_3 + 2x_1x_2 + 2x_1x_3 + x_2x_3 - 3x_1x_2x_3$

Conclusion

We have shown that from the standpoint of modeling phenomena in nature or engineering, in particular satisfying Laplace's equation, there is exactly one way to fuzzify a Boolean function. This real extension, the Real Algebraic Normal Form, a polynomial over $\mathbb{Z}$, can be obtained from the Boolean truth table of $f^{\mathbb{B}}$ by the Reed-Muller transform.

Acknowledgement

Partially funded by HS Bremerhaven through a sabbatical leave.

References

1. D. Forrester and P. Fiocchini, *Chaos in Fuzzy Cellular Automata in Conjunctive Normal Form*, Proc. Automata 2011, p. 253-262, Santiago, Chile.
2. M. A. Thornton, D. M. Miller, and R. Drechsler, *Transformations amongst the Walsh, Haar, Arithmetic and Reed-Muller Spectral Domains*, Proc. 4th Int. Workshop on Appl. of Reed-Muller Exp. in Circuit Design, p. 215, 2001
3. M. Vielhaber, *Speeding up AIDA, the Algebraic IV Differential Attack, by the Fast Reed-Muller Transform*, Proc. ISKE 2009, Intelligent Systems and Knowledge Engineering, p. 504-513, Heerlen, Belgium.

MULTI-EXPERT DECISION MAKING USING LOGICAL AGGREGATION

ANA POLEDICA
Faculty of Organizational Science, University of Belgrade, Jove Ilica 154
Belgrade, 11000, Serbia

ALEKSANDAR RAKICEVIC
Faculty of Organizational Science, University of Belgrade, Jove Ilica 154
Belgrade, 11000, Serbia

DRAGAN RADOJEVIC
Mihajlo Pupin Institute, Volgina 15
Belgrade, 11000, Serbia

This paper proposes logical aggregation based on interpolative Boolean algebra as an aggregation tool for problems of multi-expert decision making. Expert opinions are unlikely to be identical, and they are usually either close or conflicting to various degrees, or sometimes include different forms of interaction. In addition, there are many different types of decisions which fit the broad circumstances of multi-expert decision making and decision maker's point of view. Logical aggregation enables multi-expert aggregation based on linguistic requirements described by logical expressions. Logical expressions are uniquely transformed into corresponding generalized Boolean polynomials. Calculation of generalized Boolean polynomial values provides a decision maker with final alternative assessments by multiple experts.

1. Introduction

Analysis of complex problems requires the efforts and opinions of many experts. In some decision problems, the decision maker considers a set of alternatives that are assessed by a pool of experts regarding a set of criteria. The comparison of alternatives, and consequently the selection of the best alternative, must be done considering all this information. This problem is known as multiple-expert multiple-criteria decision making (ME-MCDM). This problem is generally approached as a two-stage process: an alternative ranking from each expert, and experts' rankings aggregation to get an overall score [1]. In this paper, it is assumed experts' opinions for every alternative are previously described by scores that are from [0,1] interval. The focus is not on the methods or reasoning that experts use to evaluate set of alternatives (multi-criteria decision making), but on the appropriate way to aggregate experts' opinions into single value (multi-expert decision making) that will adequately represent their judgments and/or different forms of interactions.

2. Multi-Expert Decision Making

The increasing complexity of the socio-economic environment nowadays makes it less and less possible for a single decision maker to consider all relevant aspects of a problem [9]. As decision making requires multiple perspectives of different people, most organizational decisions are made in groups. An appealing reason for using the opinions of several experts when solving a problem is that a group approach may produce better solutions to complex problems.

In general there are two main approaches for decision making with multiple experts: group decision making (GDM) and multi-expert decision making (MEDM). GDM usually target achieving a group consensus on the relative importance of the different criteria via discussion and negotiation between the group members. In general, the main disadvantage of the group-based approach is that the role of the individual decision maker in the decision process is minimized and a final decision is only possible if a consensus between the experts is attained. On the contrary, the MEDM pursues development of decision systems that support the employment of individual multi-criteria (MC) models, one for each expert. Experts asses the alternatives using their individual MC models and decision maker (DM) uses some aggregation function in order to aggregate experts' opinions into one final estimate of every alternative analyzed.

2.1. *Aggregation Functions*

Aggregation is a procedure of joining the multiple criteria (a criteria refers to a feature, variable, symptom, partial characteristic, etc.) values of an object or action into one value, which reflects all of these values in a way relevant for the analyzed problem [4]. In MEDM aggregation is used to join experts' opinions in one final decision.

There are numerous approaches to the aggregation of the various scores of the criteria (to a combined score for each alternative) or the scores of the various experts (to a combined group score) [1]. A popular approach uses quasi arithmetic means as a family of algebraic aggregation methods. This family of means is can be divided into four types of means [8]: root-power or generalized mean, geometric mean, harmonic mean, arithmetic mean. In order to average values with different importance the quasi-linear means are introduced. Tsiporkova and Boeva [7] identified the weighted mean approach as one of the most frequently used for aggregation. Another very popular approach is the Ordered Weighted Average (OWA) aggregation method, introduced by Yager [1,3,8]. There are also other approaches such as Choquet integral [3,8], Sugeno integral [8], Leximin ordering [8] and others.

3. Logical Aggregation

Logical aggregation (LA) is a consistent and transparent procedure based on one realization of [0,1]-valued logic, which is compatible with classical logic [5]. Since aggregation is a generic intellectual activity and the process of thinking is based on logic, aggregation can be treated as logical process – logical aggregation [5]. Interaction between attributes is treated as logical interaction.

In this paper, LA is based on Interpolative Boolean Algebra (IBA). IBA [6] is a consistent [0,1]-valued realization of finite Boolean Algebra (BA) in a sense that it preserves, on a value level, all the laws on which BA relies. IBA is technically based on Generalized Boolean Polynomials (GBP). Any element of BA can be uniquely transformed into its' corresponding GBP and the rules for transformation are explained in [6].

The proposed method starts with a finite set of alternatives $A=\{A_1,\ldots,A_n\}$ assessed by various experts $E=\{E_1,\ldots,E_n\}$. Logical aggregation is used to aggregate different expert opinions about alternatives into one final estimate that is used for decision making. LA has two steps [3]:

1. Normalization of attributes' values $\|\cdot\|: \mathrm{A} \to [0,1]$ and
2. Aggregation of normalized values of features into one resulting value by logical or pseudo-logical function as a LA operator $Aggr:[0,1]^n \to [0,1]$.

Operator of logical aggregation in a general case is a pseudo-logical function, a linear convex combination of generalized Boolean polynomials (GBP). Logical aggregation operator depends on the chosen measure of aggregation and operator of generalized product [3].

4. Logical Aggregation in Multi-Expert Decision Making

In general, decision making process depends on the perception of a decision maker (DM). In complex decision making environment, DM hires experts to participate in decision making process because they hold knowledge and expertise about the subject matter. Experts use their own MCDM models to evaluate alternatives based on their characteristics. DM is interested in experts' final judgments not questioning their expertise neither their models.

In this paper, we propose logical aggregation that is based on IBA, as a tool for aggregation of multiple experts' opinions in order to help DM. Different case scenarios concerning MEDM are analyzed. Table 1 represents the data that are used for evaluation of the proposed approach.

Table 1. Experts' opinions (Ei) on five alternatives (Ai).

	A_1	A_2	A_3	A_4	A_5
E_1	0.86	0.9	0.3	0.7	0.44
E_2	0.61	0.65	0.95	0.7	1.0
E_3	0.7	0.55	0.85	0.7	0.75

Case 1: Suppose all three experts come from different backgrounds so they see problem from different aspects. For this reason all experts' opinions are important to DM. In order to evaluate the set of alternatives he uses the following logical expression

$$e_1 \cap e_2 \cap e_3, \quad e_1 \otimes e_2 \otimes e_3 \tag{1}$$

The operator of generalized product $\otimes$ is product $(\otimes := *)$ because there is no interaction between experts' opinions.

Case 2: Experts come from similar background so their view on the problem is from the same aspect. DM wants to take into account all experts' opinions. He can use the same logical expression (6) as in Case 1. In this case, the operator of generalized product $\otimes$ is min $(\otimes := min)$, because we assume there is interaction between experts' opinions.

Case 3: Experts come from similar background and from DMs' point of view one alternative is good if at least one of the experts evaluate it as good. DM uses the following logical expressions and its corresponding GBP where $\otimes := min$:

$$e_1 \cup e_2 \cup e_3, \quad e_1 + e_2 + e_3 - e_1 \otimes e_2 - e_1 \otimes e_3 - e_2 \otimes e_3 + e_1 \otimes e_2 \otimes e_3 \tag{2}$$

Case 4: Experts come from different background. In this case alternative is acceptable for DM if it is evaluated as good enough by two out of three experts. DM uses the following logical expressions and its corresponding GBP where $\otimes := *$:

$$\begin{gathered}(e_1 \cap e_2) \cup (e_1 \cap e_3) \cup (e_2 \cap e_3), \\ (e_1 \otimes e_2) + (e_1 \otimes e_3) + (e_2 \otimes e_3) - 2(e_1 \otimes e_2 \otimes e_3)\end{gathered} \tag{3}$$

Case 5: In this case two experts come from the same field of expertise and the third has different background. For DM good alternative is the one that has been evaluated as good by at least one out of two experts that have the same expertise background and in the same time to be evaluated as good by third expert. DM can use one of the following logical expressions depending on which two experts come from the same field of expertise:

$$(e_1 \cup e_2) \cap e_3, \quad (e_1 \cup e_3) \cap e_2, \quad (e_2 \cup e_3) \cap e_1 \tag{4}$$

The corresponding GBP expressions are:

$$(e_1 + e_2 - e_1 \otimes e_2) \otimes e_3, \quad (e_1 + e_3 - e_1 \otimes e_3) \otimes e_2, \quad (e_2 + e_3 - e_2 \otimes e_3) \otimes e_1 \tag{5}$$

Aggregation of opinions of two experts from the same field of expertise is done with min operator $(\otimes := min)$. In order to aggregate the resulting value with the opinion of the third expert, product is used as operator of $(\otimes := *)$.

Case 6: Suppose that from DMs' point of view it's important that an alternative is evaluated as good by E1 and E2 in particular, or in case it is not evaluated as good by E1, then DM takes into consideration opinion of E3 along with opinion of E2. This can be described by the following logical expression and its corresponding GBP where $\otimes := *$:

$$(e_1 \cap e_2) \cup (\bar{e}_1 \cap e_2 \cap e_3), \quad e_1 \otimes e_2 + e_2 \otimes e_3 - e_1 \otimes e_2 \otimes e_3 \tag{6}$$

Final results for these six cases are presented in Table 2.

Table 2. Alternative (Ai) rankings for different case scenarios.

	Case 1		Case 2		Case 3		Case 4		Case 5		Case 6	
	Value	Rank	Value	Rank	Value	Rank	Value	Rank	Value	Rank	Value	Rank
A_1	0.367	1	0.61	2	0.86	4	0.819	3	0.525	4	0.584	5
A_2	0.322	4	0.55	3	0.9	3	0.794	4	0.585	3	0.621	4
A_3	0.242	5	0.3	5	0.95	2	0.863	1	0.808	1	0.85	2
A_4	0.343	2	0.7	1	0.7	5	0.784	5	0.49	5	0.637	3
A_5	0.330	3	0.44	4	1.0	1	0.86	2	0.75	2	0.86	1

It is shown that alternative ranks differ so that the most suitable alternative will be chosen based on: interaction or no interaction between experts' opinions (Case 2, 1); decision makers' point of view and preferences (Case 3, 6); different forms of interaction between experts' experience and expertise (Case 4, 5).

5. Conclusion

The focus of this paper is aggregation of experts' opinions in multiple expert multiple criteria decision making. In complex decision making environment additional experts participate in decision making process as they hold knowledge and expertise about a problem analyzed. Logical aggregation based on IBA treats aggregation as logical process and is proposed as natural, adequate and transparent method for multi-expert aggregation. The proposed approach recognizes forms of interaction and/or similar nature between experts' opinion depending on experts' background and field of expertise. Furthermore, it is possible to include decision makers' point of view and preferences as particular demands in the process of multi-expert aggregation. The presented approach aims at supporting complex decisions and assists decision makers to improve multi-expert decision making process.

Future research includes comparison of aggregation operators with the proposed approach and its application in decision making in various fields.

References

1. D. Ben-Arieh, *Computers & Industrial Engineering* **48**, 289-309 (2005).
2. G. Canfora and L. Troiano, *Mathware & Soft Computing* **11**, 67-82 (2004).
3. M. Mirkovic, J. Hodolic, D. Radojevic, *Yugoslav Journal of Operational Research* **16 (2)**, 177-188 (2006).
4. D. Radojevic, *Int. Journal of Uncertainy, Fuzziness and Knowledge-Based Systems* **7 (6)**, 577-588 (1999).
5. D. Radojevic and B. Petrovic, *EUROFUSE Workshop on Preference Modelling and Applications*, Granada, Spain, 61-66 (2001).
6. D. Radojevic, In: *From Natural Language to Soft Computing: New Paradigms in Artificial Intelligence* (Eds. L. Zadeh, D. Tufis, F. Filip and I. Dzitac), Editing House of Romanian Academy, Bucharest, 150-171 (2008).
7. E. Tsiporkova and V. Boeva, *Information science* **176**, 2673-2697 (2006).
8. R. Smolikova, M. Wachoviack, *Fuzzy Sets and Systems* **131**, 23–34 (2002).
9. Z. Yue, *Knowledge-Based Systems* **24 (1)**, 146–153 (2011).

ON EVALUATIONS OF PROPOSITIONAL FORMULAS WHOSE RANGE IS A SUBSET OF SOME FIXED COUNTABLE ORDERED FIELD

DRAGAN DODER
University of Belgrade, Faculty of Mechanical Engineering, Kraljice Marije 16, 11000 Belgrade, Serbia

ZORAN OGNJANOVIĆ
Mathematical Institute of Serbian Academy of Sciences and Arts, Kneza Mihaila 36, 11000 Belgrade, Serbia

ALEKSANDAR PEROVIĆ
University of Belgrade, Faculty of Transportation and Traffic Engineering, Vojvode Stepe 305, 11000 Belgrade, Serbia

MIODRAG RAŠKOVIĆ
Mathematical Institute of Serbian Academy of Sciences and Arts, Kneza Mihaila 36, 11000 Belgrade, Serbia

Let L be a countable first-order language such that its set of constant symbols $Const(L)$ is countable. We provide a complete infinitary propositional logic (formulas remain finite sequences of symbols, but we use inference rules with countably many premises) for description of C-valued L-structures, where C is an infinite subset of $Const(L)$. The main goal is to provide a formal framework for reasoning about $\mathbb{F}$-valued evaluations of propositional formulas, where $\mathbb{F}$ is some countable ordered field. The prime examples of $\mathbb{F}$ are the field of rational numbers $\mathbb{Q}$, its real closure $\overline{\mathbb{Q}}$ and the field of fractions $\mathbb{Q}(\varepsilon)$, where ε is a positive infinitesimal.

1. Introduction

In the wider context, we may think of the concept of fuzziness as a multi-valued (which should be read as infinitely many valued) estimation of certain properties. Once we decide to make things operational (even automatized, if possible), the notion of a property boils down to the notion of a formula in some underlying formalism. Some classification problems arousing in relational databases, risk assessment etc can be reduced to the classification of the finite set of $\mathbb{Q}$-valued evaluations of a certain classical (or fuzzy) propositional formula. In particular, one needs formal apparatus for computation of the value of the criterion formula and ordering of computed values. There is a plethora of research papers providing formal environments for reasoning about probability, fuzzy logic, possibility and necessity relations and distributions, nonmonotonic inference relations and so on. The purpose of this note is to provide a general propositional framework, which can be roughly described as a propositional theory of countable structures.

2. Syntax and Semantics

Recall that a first-order language L is a (possibly empty) set of pairwise distinct symbols that are categorized in three groups: symbols for constants, which are the elements of $Const(L)$; symbols for relations, which are the elements of $Rel(L)$; symbols for functions, which are the elements of $Fun(L)$. Note that each of these three sets can be empty. In addition, foe each relation and function symbol is provided its arity (the number of arguments). The arity of a symbol S will be denoted by $ar(S)$. From now on, we will consider only countable languages with countably many symbols for constants.

Let L be such a language. The shortest description of our syntax is the following one: our formulas are Σ_0 L-sentences. However, it seems prudent to present the extended version of definition as well.

Definition 2.1. The set $Term$ of terms is the smallest superset of $Const(L)$ that is closed for all $F \in Fun(L)$, i.e. if $F \in Fun(L)$ and $f_1, \dots, f_{ar(F)} \in Term$, then $F(f_1, \dots, f_{ar(F)}) \in Term$. □

Terms will be denoted by f, g and h, indexed if necessary.

Definition 2.2. Atomic formulas are formulas of the form $f = g$ and $R(f_1, \dots, f_{ar(R)})$, where $f, g, f_i \in Term$ and $R \in Rel(L)$. □

Definition 2.3. The set For of formulas is the smallest set containing all atomic formulas that is closed under negation and conjunction. □

Formulas will be denoted by ϕ, ψ and θ, indexed if necessary. The other connectives (disjunction, implication, equivalence) are defined as usual. We conclude the introduction of our syntax by the convention that C would from now on represent some fixed infinite subset of $Const(L)$.

Concerning semantics, our models are so called C-valued L-structures, i.e. L-structures of the form $\mathbb{A} = (A, \dots)$, where each $a \in A$ is an interpretation of some $c \in C$. Our satisfiability relation coincides with the restriction of the first-order satisfiability relation on Σ_0-sentences. Precise definitions can be found in any graduate textbook of Mathematical logic or some more specialized branches of it such is Model theory, see for instance [2].

3. Complete Axiomatization

The basic axiom system $\mathcal{AS}$ that we are going to introduce here is fairly simple - it contains only two groups of axioms and two types of inference rules.

Propositional axioms

A1 Substitutional instances of classical tautologies;

Equality axioms

A2 $f = f$;

A3 $f = g \to (\phi(\dots, f, \dots) \to \phi(\dots, g, \dots))$;

Inference rules

R1 Modus ponens: from ϕ and $\phi \to \psi$ infer ψ;

R2 f-rule: from the set of premises $\{\phi \to f \neq c \mid c \in C\}$ infer $\phi \to f \neq f$.

Note that R2 is a schemata rule - one instance for each $f \in Term$. The purpose of R2 is to ensure that value of each term lies in the set of interpretations of C. The implicative form of premises and conclusion in R2 is the standard technical detail that provides an easy proof of the deduction theorem.

As it is usual, sets of formulas will be called theories. Due to the presence of infinitary inference rules, the notion of an inference is somewhat different.

Definition 3.1. Suppose that T is a theory and that ϕ is a formula. We say that ϕ is syntactical consequence of T, or that ϕ can be deduced or inferred from T, if there exists a countable ordinal ξ and a $\xi + 1$-sequence of formulas $\phi_0, \phi_1, \dots, \phi_\xi$ such that $\phi_\xi = \phi$ and, for all $i \leq \xi$, ϕ_i is an instance of an axiom, or $\phi_i \in T$, or ϕ_i can be derived from some previous members of the sequence by some inference rule. □

Note that the length of an inference is always at most countable successor ordinal. The notions of satisfiability, finite satisfiability, validity, consistency etc. are defined as usual. As it is usual, $T \vdash \phi$ reads "ϕ is syntactical consequence of T", while $T \vDash \phi$ reads "ϕ is semantical consequence of T".

Theorem 3.1. The compactness theorem fails for $\mathcal{AS}$.

Proof. Suppose that $d \in Const(L) \setminus C$. Let $T = T_1 \cup T_2$, where $T_1 = \{d \neq c \mid c \in C\}$ and $T_2 = \{c_1 \neq c_2 \mid c_1, c_2 \in C \text{ and } c_1 \neq c_2 \text{ (syntactically)}\}$. Furthermore, let $\mathbb{A} = (A, \dots)$ be any L-structure that satisfies T_2. Since all elements of A are interpretations of some elements of C, it is easy to see that T_1 is not satisfiable in A. On the other hand, $\mathbb{A} \vDash T_2$ implies that A is countable, so we can use $\mathbb{A}$ to construct a model for any finite subset of T_1. Hence, T is both unsatisfiable and finitely satisfiable, so compactness theorem fails for $\mathcal{AS}$. □

Deduction theorem for $\mathcal{AS}$ can be shown by induction on the length of the inference similarly as in [4], where the variant of the R2 rule is present. In any case we have

Theorem 3.2 (Deduction theorem). Let T be a theory and $\phi, \psi \in For$. Then, $T, \phi \vdash \psi$ iff $T \vdash \phi \to \psi$. □

Lemma 3.1. Suppose that T is a consistent theory and that f is an arbitrary term. Then, there exists $c \in C$ such that $T, f = c$ is consistent.

Proof. Contrary to the statement of the lemma, suppose that $T, f = c$ is inconsistent for all $c \in C$. Let ϕ be any syntactical consequence of T. Then, for all $c \in C$, $T, f = c \vdash \neg\phi$, so by Theorem 3.2 we obtain $T \vdash f = c \to \neg\phi$, which is equivalent to $T \vdash \phi \to f \neq c$. This is true for all $c \in C$, so by R2, $T \vdash \phi \to \bot$, i.e. $T \vdash \neg\phi$. Hence, T is inconsistent - a contradiction. □

Lemma 3.2 (Lindenbaum's lemma). Every consistent theory can be maximized, i.e. extended to a complete consistent theory.

Proof. Suppose that T is a consistent theory. Since L is countable, sets For, $Term$ and C are all countable. Let $For = \{\phi_n \mid n \in \omega\}$, $C = \{c_n \mid n \in \omega\}$ and $Term = \{f_n \mid n \in \omega\}$. We define an ω-sequence $T_0, T_1, T_2 \ldots$ of theories as follows: (1) $T_0 = T$; (2) If $T_{2n} \cup \{\phi_n\}$ is consistent, then $T_{2n+1} = T_{2n} \cup \{\phi_n\}$. Otherwise, $T_{2n+1} = T_{2n} \cup \{\neg\phi_n\}$; (3) $T_{2n+2} = T_{2n+1} \cup \{f_n = c_m\}$, where m is the least natural number such that $T_{2n+1} \cup \{f_n = c_m\}$ is consistent (the existence of m is provided by Lemma 3.1).

Let $T^* = \bigcup\{T_n \mid n \in \omega\}$. By (2), T^* is complete, i.e. $T^* \vdash \phi$ or $T^* \vdash \neg\phi$ for all $\phi \in For$. In order to see that T^* is consistent, it is sufficient to show that T^* is deductively closed. This fact can be shown by induction on the length of inference similarly as in [4]. □

Theorem 3.3 (Completeness theorem). Every consistent theory is satisfiable.

Proof. Suppose that T is a consistent theory. By Lemma 3.2, there exists a complete consistent theory T^* that extends T. Similarly as in the Henkin's construction, let $c \sim d$ iff $T^* \vdash c = d$, where $c, d \in C$. The domain A of the canonical model $\mathbb{A}$ is defined as the quotient set $C_{/\sim}$. The corresponding interpretation I of L is defined as follows:

(1) If c\in C, then $I(c) =_{\text{def}} c_{/\sim}$;

(2) If $d \in Const(L) \setminus C$, then $I(d) =_{\text{def}} c_{/\sim}$, where $c \in C$ satisfies condition $T^* \vdash d = c$. The existence of such c is provided by the construction of T^*, see the proof of Lemma 3.2;

(3) If f is a functional symbol of arity n and $c_1, \ldots, c_n \in C$, then let $I(f)(I(c_1),$

$\ldots, I(c_n)) =_{\text{def}} I(c)$, where $c \in C$ satisfies condition $T^* \vdash f(c_1, \ldots, c_n) = c$;

(4) If R is a relational symbol of arity n and $c_1, \ldots, c_n \in C$, then let $I(R)(I(c_1), \ldots, I(c_n))$ iff $T^* \vdash R(c_1, \ldots, c_n)$.

The correctness of the definition of $\mathbb{A}$ as well as the fact that $\mathbb{A} \models T^*$ can be proved similarly as in the case of the first-order predicate logic. □

4. Some Examples

As a first and only relatively detailed example of application of $\mathcal{AS}$ in formalization of reasoning about evaluations of $\mathbb{F}$-valued formulas, we will show representation of $\mathbb{Q}$-valued probabilities. Suppose that For_{cl} is some countable set of classical propositional formulas. The elements of For_{cl} will be denoted by α, β and γ, indexed if necessary. Let $Const(L) = \mathbb{Q} \cup For_{cl}$ and let $P, +, -, \cdot, \leq \in L$, where P and $-$ are unary function symbols, $+$ and $\cdot$ binary function symbols and $\leq$ binary relation symbol. We define an $\mathcal{AS}$-theory T as follows:

(1) Diagram of $(\mathbb{Q}, +, -, \cdot, \leq)$;

(2) $f + g = g + f$; $f + (g + h) = (f + g) + h$; $f + 0 = f$; $f + (-f) = 0$;

(3) $f \cdot g = g \cdot f$; $f \cdot (g \cdot h) = (f \cdot g) \cdot h$; $f \cdot 1 = f$;

(4) $f \cdot (g + h) = (f \cdot g) + (f \cdot h)$; $f \leq f$; $f \leq g \wedge g \leq f \rightarrow f = g$;

(5) $f \leq g \wedge g \leq h \rightarrow f \leq h$; \$$f \leq g \vee g \leq f$;

(7) $f \leq g \rightarrow f + h \leq g + h$; $f \leq g \wedge 0 < h \rightarrow f \cdot h \leq g \cdot h$;

(8) $P(\alpha) = 1$ for every tautology α; $P(\neg\alpha) = 1 + \big(-P(\alpha)\big)$;

(9) $P(\alpha \vee \beta) = P(\alpha) + \Big(P(\beta) + \big(-P(\alpha \wedge \beta)\big)\Big)$;

(10) $P(\alpha) = P(\beta)$ for equivalent α and β.

It is easy to see that T is consistent theory. Every model of T provides a $\mathbb{Q}$-valued probability evaluation of r_{cl} . As it was shown by Dragan Radojević (see for instance [3]), the multi-agent case (several probability operators) can be used for solving well known Grabisch's example from [1].

Similarly as above, we can formulate $\mathcal{AS}$-theories for various $\mathbb{F}$-valued evaluations in various fuzzy logics and possibility logics. As it was shown in [4], $\mathbb{Q}(\varepsilon)$-valued conditional probabilities can be used for representation of default reasoning.

References

1. M. Grabisch, Ch. Labreuche. Bi-capacities for decision making on bipolar scales. EUROFUSE Workshop in Information Systems, Varena, Italy: 185-190 (2002).
2. D. Marker. Model Theory: An Introduction. Springer, 2002.
3. A. Perović, Z. Ognjanović, M. Rašković, D. G. Radojević. Finitely additive probability measures on classical propositional formulas definable by Gödel's t-norm and product t-norm. Fuzzy Sets and Systems 169(1): 65-90 (2011).
4. M. Rašković, Z. Marković, Z. Ognjanović. A logic with approximate conditional probabilities that can model default reasoning. Int. J. Approx. Reasoning 49(1): 52-66 (2008).

LOGICAL AGGREGATION FOR MODULAR NEURAL NETWORKS IN CURRENCY EXCHANGE RATE FORECASTING

IVAN NESIC

Faculty of Organizational Science, University of Belgrade, Jove Ilica 154
Belgrade, 11000, Serbia

ALEKSANDAR RAKICEVIC

Faculty of Organizational Science, University of Belgrade, Jove Ilica 154
Belgrade, 11000, Serbia

BRATISLAV PETROVIC

Faculty of Organizational Science, University of Belgrade, Jove Ilica 154
Belgrade, 11000, Serbia

DRAGAN RADOJEVIC

Mihajlo Pupin Institute, Volgina 15
Belgrade, 11000, Serbia

In this paper fields of research are modular neural networks and consistent real-valued logic as a technique for aggregation of the modules. Aggregation is done using logical and pseudo-logical aggregation based on interpolative Boolean algebra (IBA). IBA is technically realized through generalized Boolean polynomials. The proposed technique expands the existing base of modular neural network models and shows respectable results in forecasting currency exchange rates

1. Introduction

Since late 1980s modularization of neural networks has been introduced into common use. Some of its advantages can be attributed to improved robustness and efficiency, better structure interpretation and clear biological analogy. Based on recent researches, it has been concluded that modular design is key to brain efficiency and intelligence. Neurosciences state that brain does not work as a whole, but rather like a set of modules, where each subset can be further modularized. Unified functioning is made possible through feedback connections. For example, in visual cortex of one specific monkey type there are around 30 specialized areas with approximately 300 interconnections. Mice have only 3-5 such areas and people above 100. It is concluded that intelligence rises with increment of modules and not strictly by brain volume growth.

Concept of modularity can be defined by following characteristics [1]:

- Modules are limited to one area of the problem and have structure for solving specific subtask of a complete task;
- Each module works independently from other modules and there are no interactions between them;
- Modules have simpler architectures compared to the system as a whole, thus enabling better efficiency;
- Responses from modules are simple and need integration unit for obtaining a complete model response.

One characteristic of such architectures is that modules do not have to be artificial neural networks; they can be any other type of learning algorithm. This also enables hybridization with other techniques.

Including data into modular neural network structure is also possible. By including it we get a coherent model whose intelligence part is not separated from its knowledge. Integration unit is one of the key points in modular neural network design. Vast number of papers use some sort of weighted sum aggregation including combining it with fuzzy techniques. Weighted sum approach is additive and for all effects of interest that are not additive in their nature is not adequate [2]. When utilizing fuzzy integral as an aggregation method, problem of monotonicity emerges. In general, such requirement is sometimes too much restrictive, since many of logical functions are not monotone in their nature. Usage of logical aggregation that is based on interpolative Boolean algebra does not suffer from such drawback; it is an adequate tool for aggregation in a general case.

2. Logical Aggregation Based on Interpolative Boolean Algebra

Interpolative Boolean algebra provides a frame for consistent realization of all possible elements of finite Boolean algebra. It is consistent in a sense that it preserves, on a value level, all the laws on which Boolean algebra relies.

The task of logical aggregation is the fusion of primary attributes' values into one resulting globally representative value using logical tools [2,4]. Logical aggregation has to steps:

1. Normalization of primary attributes' values and
2. Aggregation of normalized values into one resulting value using pseudo-logical function as a logical operator.

Pseudo-logical function is a linear convex combination of generalized Boolean polynomials [3]. Generalized Boolean polynomial (GBP) $\varphi_{\mu}^{\otimes}(\|a_1\|,...,\|a_n\|)$ in logical aggregation has a role of logically combined attribute [4]. GBP enables calculation of any logical expression. Every element of Boolean algebra has its own unique corresponding GBP, in which it is translated using rules explained in [4].

Operator of logical aggregation in a general case is [2,4]:

$$Agg_{\mu}^{\otimes}(\|a_1\|,...,\|a_n\|) = \sum_{S \in P(\Omega)} \mu(S) \sum_{C \in P(\Omega,S)} (-1)^{|C|} \bigotimes_{a_i \in S \cup C} \|a_i\| . \quad (1)$$

Logical operator depends on the measure of aggregation and on the operator of generalized product $(\otimes)$.

Aggregation measure is a set function $\mu : P(\Omega) \rightarrow [0,1]$, which in general case is not monotone function defined by the following expression [2,4]:

$$\mu(S) = \sum_{i=1}^{m} w_i \sigma_{\varphi_i}(S), \quad S \in P(\Omega), \quad \varphi_i \in BA(\Omega)$$
$$\sum_{i=1}^{m} w_i = 0, \quad w_i \geq 0, \quad i = 1,...,m \quad (2)$$

3. Application of Logical Aggregation in Modular Neural Networks

As an application example, five modules structure of an artificial modular neural network is used in this paper. First module is an artificial neural network which predicts value of a currency exchange rate for the next (T+1) time period. The other four are neural networks that predict whether the currency exchange rate is going to rise or fall, but not the exact future value of that currency. These neural networks are trying to predict exchange rate trend. They are used as an error rectifying tool for the first module. All modules have limited prediction capabilities in terms of success rate and the idea is to combine them in order to try to overcome individual weaknesses.

The second and the third modules use data from a given time period to predict rise or fall of price in the next period. The fourth and the fifth do the same, with a difference that they take two successive periods as input variables.

Incorporation of time series data set into structure is recommended since data is used from previous (T-1) time period. Model can be extended with ease by adding new modules and in such cases larger data set may be needed.

The proposed model of modular neural network has two sub-models. First that incorporates only module 1, which represents the classical approach in using neural networks, and the second one that include all other four modules (2-5) where modules are aggregated using logical aggregation method:

$$\varphi(a_2, a_3, a_4, a_5) = (a_2 \cap Ca_3) \cup (a_4 \cap Ca_5) \quad . \tag{3}$$

The corresponding generalized Boolean polynomial is:

$$\varphi^{\otimes}(a_2, a_3, a_4, a_5) = a_2 + a_4 - a_2 \otimes a_4 - a_2 \otimes a_3 - a_4 \otimes a_5 \\ + a_2 \otimes a_4 \otimes a_5 + a_2 \otimes a_3 \otimes a_4 - a_2 \otimes a_3 \otimes a_4 \otimes a_5 . \tag{4}$$

Lukasieviczs' T-norm is used as operator of generalized product for aggregation of a_2 and a_3 ($\otimes := \max(a_1 + a_2 - 1, 0)$) and also for a_4 and a_5 ($\otimes := \max(a_4 + a_5 - 1, 0)$). All other operators of generalized products in (4) are min ($\otimes := \min$), since they are correlated.

Two sub-models are aggregated using pseudo-logical aggregation operator:

$$Agg^{\otimes}(a_1, a_2, a_3, a_4, a_5) = \frac{1}{2}a_1 + \frac{1}{2}\varphi^{\otimes}(a_2, a_3, a_4, a_5) \tag{5}$$

Table 1 represents 15 days forecast, which is the part of the resulting data .

Table 1. Forecasting results.

Day	**Real market data**	**Sub-model 1** Module 1	**Sub-model 2** Module 2-5	**The proposed approach**
T+1	0.22118	-0.12656	0.07350	0.27355
T+2	0.33175	0.01829	0.18534	0.35238
T+3	0.11058	-0.21780	-0.06559	0.08662
T+4	0.16587	0.21780	0.10659	-0.00463
T+5	-0.05529	0.18852	0.03852	-0.11148
T+6	-0.33173	0.07354	-0.02808	-0.12970
T+7	0.33174	0.07876	0.04979	0.02083
T+8	0.00000	-0.24001	-0.16775	-0.09549
T+9	0.11058	0.18959	0.03059	-0.12841
T+10	0.05529	0.00253	-0.07869	-0.15991
T+11	-0.05529	0.06681	-0.05276	-0.17233
T+12	-0.05529	0.01696	-0.08203	-0.18103
T+13	-0.11057	0.04480	-0.06828	-0.18136
T+14	-0.16586	0.05408	-0.03725	-0.12858
T+15	-0.11058	-0.09909	-0.07488	-0.05067

Values in the table are daily price changes normalized to [0,1] interval. First column are real market data. Second column represents partial results that are obtained using only sub-model 1 (classical neural network approach with only module 1). Partial results from sub-model 2 (logical aggregation of modules 2-5) are presented in third column. Finally, the last column represents results that are obtained using the proposed approach (pseudo-logical aggregation of sub-models).

Figure 1 is a graphical presentation of the data showed in Table 1. Data are presented in form of bars, where each bar represents price change for one single day. Bars are grouped and each four successive bars represent data for one day. First figure in every group of four represents the real price change that happened on the market for that day. Second and third figures represent the forecasts obtained by sub-model 1 and sub-model 2. Finally, the fourth figure represents forecast that are the result of the proposed approach

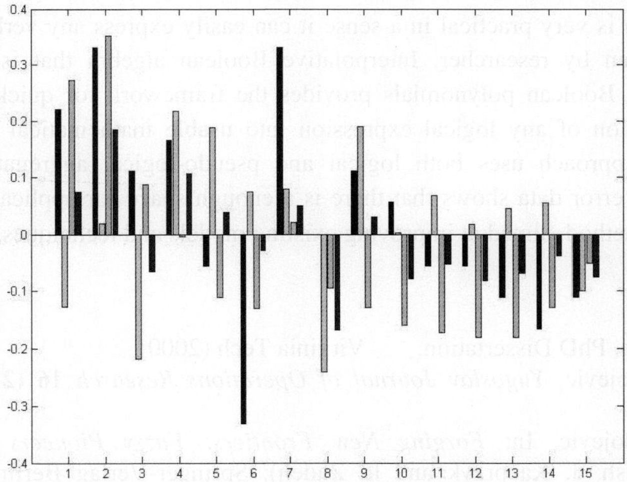

Figure 1. Graphical representation of the results.

Table 2 gives the information about forecasting errors of the proposed model and its submodles.

Table 2. Forecasting errors.

Method	MSE	RMSE	Correct predict. rate	Bayes UP	Bayes DOWN
Sub-model 1	0.1374	0.3707	0.45	0.5839	0.3077
Sub-model 2	0.2231	0.4723	0.53	0.4732	0.6539
Proposed approach	0.2415	0.4914	0.55	0.5923	0.5216

MSE and RMSE columns in Table 2 give the information about mean square error and root mean square error. Sub-model 1 or simple neural network gives the best result, but MSE and RMSE may be misleading. When we look at the correct prediction rate which shows what percentage of our predictions about the trend were correct than we see that previous results about MSE and RMSE were indeed misleading. Column Bayes UP tells us what is the probability that analyzed method is right when it gives signal that price is going to rise. Column Bayes DOWN gives the probability that analyzed method is right when it gives signal that price is going to fall.

4. Conclusion

This paper shows that logical aggregation can be successfully used for aggregation of different modules in modular neural networks. Logical aggregation is very practical in a sense it can easily express any verbal request that is given by researcher. Interpolative Boolean algebra that is based on generalized Boolean polynomials provides the framework for quick and easy transformation of any logical expression into usable mathematical form. The proposed approach uses both logical and pseudo-logical aggregation. Final results and error data shows that there is a enough space for application of the proposed method, aimed at improving existing models and techniques.

References

1. F. Azam, PhD Dissertation, Virginia Tech (2000)
2. D. Radojevic, *Yugoslav Journal of Operations Research* **16 (2)**, 177-188 (2006)
3. D. Radojevic, In: *Forging New Frontiers: Fuzzy Pioneers* (Eds. M. Nikravesh, J. Kacprzyk and L. Zadeh), Springer-Verlag Berlin, 295-318 (2008).
4. D. Radojevic, *Mathware & Soft Computing* **15 (1)**, 125-141 (2008).

EXTENDING AHP WITH BOOLEAN CONSISTENT FUZZY LOGIC AND ITS APPLICATION IN WEB SERVICE SELECTION

IVANA DRAGOVIĆ*, NINA TURAJLIĆ

Faculty of Organizational Sciences, University of Belgrade, Jove Ilića 154
11000 Belgrade, Serbia

DRAGAN RADOJEVIĆ

Mihajlo Pupin Institute, Volgina 15
11000 Belgrade, Serbia

When making a decision on the basis of multiple criteria the traditional AHP method does not take into account the fact that they may be interrelated. It is proposed that the traditional AHP method could be extended by applying fuzzy logic for aggregating the criteria into a single objective function. The application of consistent fuzzy logic is preferred since no conventional fuzzy set theory is in the Boolean frame. The basis of this approach is the Interpolative realization of Boolean algebra (IBA). The proposed approach is illustrated on the web service selection problem.

1. Introduction

Decision making is an integral part of every aspect of life and involves choosing an alternative on the basis of one or more criteria. The problem is to choose the alternative which best fulfills the entire set of goals.

With the proliferation of web services offering similar functionality the problem of how to choose the best one according to multiple non-functional Quality of Service (QoS) attributes has become increasingly important.

One of the most widely used methods for multi-criteria decision making is the Analytic Hierarchy Process method (AHP). The AHP method structures the original complex problem into a set of hierarchically organized sub-problems which are easier to address. The main idea is to evaluate the elements one level at a time and determine their relative priorities with respect to the element above them in the hierarchy. Ultimately, the weighted relative priorities of every level in the hierarchy are aggregated into a single objective function (encompassing the entire set of goals) thus enabling the ranking of the given alternatives.

* ivana.dragovic@fon.rs

Fuzzy logic is often used to address some of the drawbacks of the traditional AHP method. The pairwise comparisons allow decision makers to bring their own valuable experience into the decision making process. Since experience is better represented by linguistic means than by the customary numerical rankings, fuzzy logic has been used to extend the traditional AHP method to more adequately incorporate this experience (Fuzzy AHP) [1,2].

Another challenge arises from the fact that the goal can be, and usually is, a logical demand formulated to reflect the fact that the criteria upon which the decision should be made may be interrelated or conflicting. A number of existing approaches have considered the use of fuzzy logic for taking this correlation into account. Specifically, for the problem of web service selection several approaches have been presented [3,4,5].

The traditional AHP method lacks support for this type of problem. Bearing in mind that correlation can only be adequately described by logical functions it is proposed that AHP should be extended by using fuzzy logic when defining the objective function. A similar technique has been proposed in [6]. However, since no conventional fuzzy set theory (fuzzy logic, theory of fuzzy relation) is in the Boolean frame [7] it is proposed that consistent fuzzy logic, introduced in [8], should be used instead. The basis of this approach is the Interpolative realization of Boolean algebra (IBA) which treats logical functions as a Generalized Boolean polynomial (GBP).

The paper is structured as follows: in Section 2 an overview of the AHP method is given, Section 3 gives the basic concepts of Boolean consistent fuzzy logic and in Section 4 the proposed solution is outlined and a detailed example for the web service selection problem is given. The example points out that the conventional and consistent fuzzy approaches can lead to the choice of different alternatives. Finally, Section 5 concludes the paper and discusses future work.

2. AHP Method

The Analytic Hierarchy Process (AHP) is a multi-criteria decision making approach, introduced by Saaty [9]. It uses a hierarchical structure of goals, criteria, subcriteria and alternatives and assumes that elements of a certain level are dependent on the elements in the previous level. The traditional AHP method (for a simplified problem which includes only goals, criteria and alternatives) involves the following steps:

1. Defining the problem *i.e.* constructing the hierarchical model.
2. Determining the relative importance or weight of the criteria with respect to the goal as well as the alternatives with respect to each

criterion. This is accomplished through pairwise comparisons. The decision maker will assign a number for every pair of elements (according to Saaty's 9-point scale) thereby indicating his degree of preference. The results are presented in comparison matrices.

3. Evaluating the alternatives with respect to the criteria and the criteria with respect to the goal by means of relative weights. The relative weights are estimated by finding the principal right eigenvector of each comparison matrix.
4. Finally, a composite vector is created, representing the product of relative weight vectors of every level. The results are the relative priorities of every alternative with respect to the goal, which ultimately enables their ranking.

One of the main benefits of AHP is that it enables decision making based on both quantitative and qualitative data regardless of their measurement units.

3. Boolean Consistent Fuzzy Logic

Fuzzy logic, introduced by Lofti Zadeh [10] is envisaged to extend binary values 0 and 1 representing strict presence or absence of an entity to any other value in between, indicating an entity's relative presence or absence. It is especially useful for working with linguistic variables which can be mapped to this interval by corresponding membership functions. Another advantage is its ability to take into consideration the correlation existing among elements which should be expressed by logical functions. To this end the use of the Interpolative realization of Boolean algebra (IBA) is proposed since no conventional fuzzy set theory (fuzzy logic, theory of fuzzy relation) is in the Boolean frame [7]. IBA is an illustrative name for the real-valued and/or [0, 1]-valued realization of Boolean algebra [8].

Any logical function can be uniquely transformed into a corresponding generalized Boolean polynomial (GBP) using IBA [8]. Only once the transformations have been conducted and the final structure established will the values be introduced and computed. This is the main difference between the conventional and Boolean consistent approaches which can, in certain cases, lead to different results, as will be shown in the example in Section 4.1.

The GBP maps a corresponding element of Boolean algebra into its value from the real unit interval [0, 1] on the value level, so that a partial order on the value level is preserved contrary to other many-valued (MV) and/or fuzzy approaches [11]. The GBP is a polynomial whose variables are elements of the Boolean algebra and the operators are standard +, standard – and generalized product $\otimes$. The generalized product is a subclass of the conventional fuzzy T norm satisfying the non-negativity axiom [12].

The main goal of Logical aggregation is to combine the initial criteria into a single global criterion using a logical function as a logical aggregation operator. GBP in logical aggregation has the role of a logical combined element [11,12].

4. Proposed Solution

The main weakness of the weighting technique used in the traditional AHP method, as well as the Fuzzy AHP method is that it only allows linear relationships among the criteria. Given that certain criteria may influence or be influenced by other criteria as well as the fact that the importance of the criteria may vary based upon the demonstrated level of other criteria, it is necessary to take this into account while choosing the best alternative. Therefore, it is proposed that a logical function is used for defining the importance of the criteria with respect to the goal instead of comparison matrices. The objective function which takes into account the correlation among the criteria should be defined within the Boolean frame. The proposed Logical Aggregation approach can be applied to both the traditional AHP method (based on crisp numbers) and the Fuzzy AHP method (where preferences are defined as fuzzy numbers).

4.1. *Web Service Selection*

Web services are software components, accessible on the Internet, which provide certain functionality. Multiple services can provide the same functionality hence they could be grouped into classes. The problem consists of choosing a web service from a given class (*i.e.* the class of services that achieve the desired functionality) on the basis of certain criteria. Often the selection is made on the basis of Quality of Service (QoS) attributes. QoS refers to the non-functional properties of a service such as: performance (measured in terms of throughput, response time, latency *etc.*), security, reliability, availability, scalability, integrity *etc.* [13]. The criteria selected for this example are presented in Table 1.

Table 1. Selected QoS attributes

QoS Attribute	Description	Attribute Type	Unit	Type
Response time	Time taken to send a request and receive a response.	Quantitative	millisecond	min
Availability	The probability that a service is operating.	Quantitative	percent	max
Reliability	Related to the ratio of error messages to total messages.	Quantitative	percent	max
Security	The level and kind of security a service provides.	Qualitative	Excellent, Very good, Good, Adequate, Poor	

Generally, certain constraints regarding the required QoS attributes are also given. It is assumed that a previous selection of services satisfying these constraints has been made so that only the remaining services are analyzed. The values of the attributes for the remaining services are given in Table 2.

Table 2. Values of the selected QoS attributes

	Response time	Availability	Reliability	Security
WS_1	1000	98	90	Poor
WS_2	1250	96	95	Excellent
WS_3	2000	92	98	Good

Following the traditional AHP method the first step is to create a hierarchy model. The goal is to achieve the best QoS, the criteria are the chosen QoS attributes and the alternatives are the web services. Next, the comparison matrices representing the relative importance of the alternatives regarding each criterion are constructed (Table 3).

Table 3. Comparison matrices

	Response Time			Availability			Reliability			Security		
	WS_1	WS_2	WS_3	WS_1	WS_2	WS_3	WS_1	WS_2	WS_3	WS_1	WS_2	WS_3
WS_1	1	3	9	1	4	8	1	1/5	1/8	1	1/8	1/4
WS_2	1/3	1	6	1/4	1	5	5	1	¼	8	1	3
WS_3	1/9	1/6	1	1/8	1/5	1	8	4	1	4	1/3	1

The matrices are then processed and the principal right eigenvector of each comparison matrix is calculated (Table 4).

Table 4. Rankings

	Response Time		Availability		Reliability		Security	
WS_1	0.6583	1	0.6893	1	0.0669	3	0.0738	3
WS_2	0.2819	2	0.2438	2	0.2438	2	0.6690	1
WS_3	0.0598	3	0.0669	3	0.6893	1	0.2572	2

The next step in the traditional AHP method would be to repeat the same procedure for the criteria with regard to the goal and finally calculate the composite vector.

However, a decision maker might want to trade off one attribute against another *e.g.* Availability (**A**) is typically related to Reliability (**R**) so this should be taken into consideration: if a service has a high level of **A** then it may be more important to take into consideration the value of the Response Time (**RT**) and if it is low than **R** is more important. On the other hand, the decision maker might prefer a more secure service (**S**) over a less secure one regardless of the value of the other attributes. These requests can only be expressed by a logical function.

$$\left((A \wedge RT) \vee (\neg A \wedge R)\right) \wedge S \tag{1}$$

This objective function could be evaluated by applying the rules of conventional fuzzy logic, and for comparison purposes the results are given in Table 5. Nevertheless, the proposed approach prefers the use of consistent fuzzy logic for reasons previously explained in Section 3. To transform a Boolean function into a GBP the first step is to assess its structure. A detailed description of this transformation is given in [11]. In this particular case the following transformation steps have been taken:

$$\begin{aligned}
\left(\left((A \wedge RT) \vee (\neg A \wedge R)\right) \wedge S\right)^{\otimes} &= \left((A \wedge RT) \vee (\neg A \wedge R)\right)^{\otimes} \otimes S = \\
&= \left((A \wedge RT)^{\otimes} + (\neg A \wedge R)^{\otimes} - \left((A \wedge RT) \wedge (\neg A \wedge R)\right)^{\otimes}\right) \otimes S = \\
&= \left(A \otimes RT + \left((1-A) \wedge R\right)^{\otimes} - (A \wedge RT)^{\otimes} \otimes (\neg A \wedge R)^{\otimes}\right) \otimes S = \\
&= \left(A \otimes RT + (1-A) \otimes R - A \otimes RT \otimes \left((1-A) \otimes R\right)\right) \otimes S = \\
&= \left(A \otimes RT + R - A \otimes R - A \otimes RT \otimes R + A \otimes RT \otimes A \otimes R\right) \otimes S = \\
&= A \otimes RT \otimes S + R \otimes S - A \otimes R \otimes S
\end{aligned} \tag{2}$$

After the transformation has been accomplished focus is transferred to the value level by choosing the standard product as an adequate operator for the generalized product and then inputting the values (which have already been normalized as a result of the previous AHP steps). Table 5 depicts the results.

Table 5. Result comparison

	Consistent Fuzzy Logic		Conventional Fuzzy Logic	
WS_1	0.0350	3	0.0343	3
WS_2	**0.1693**	**1**	0.1608	2
WS_3	0.1664	2	**0.1658**	**1**

As expected the results of these two approaches differ. WS_2 is obviously the better choice regarding the defined requirements.

5. Conclusion

When choosing the best alternative on the basis of a set of criteria using the AHP method, the fact that the criteria may be interrelated or conflicting should be taken into account. This correlation can only be adequately described by logical functions. Logical functions (*i.e.* Logical Aggregation) result in a new structure of the components as opposed to the weighted sum approach. It is proposed that the traditional AHP method, which lacks support for this type of problem, could be extended to cover this situation by applying fuzzy logic when defining the objective function.

Even though, conventional fuzzy logic could be applied to this problem it is shown that, in certain cases, it could lead to the choice of a worse alternative (the ranking: $WS_3 \succ WS_2 \succ WS_1$ instead of the ranking: $WS_2 \succ WS_3 \succ WS_1$ which is the result of the Boolean consistent fuzzy approach). This stems from the fact that no conventional fuzzy set theory is in the Boolean frame. Foremost, because negation is treated differently in these two approaches. But also because the Boolean consistent approach requires that a set of structural transformations be executed before the values can be introduced.

Further work would be aimed at the use of Pseudo Logical Aggregation [11] to allow for several different goals to be defined and fused into one globally representative goal by choosing an adequate measure of aggregation. AHP could also be used for evaluating the weights of the partial goals if the weighted sum is chosen as the aggregation tool.

References

1. D.-Y. Chang, European Journal of Operational Research, **95**, 649 (1996).
2. V. Tran, H. Tsuji, In Proc. of NWeSP , 77 (2008).
3. M. Lin, J. Xie, H. Guo, and H. Wang, In Proc. of EEE, 9 (2005).
4. P. Wang, K.-M. Chao, C.-C. Lo, C.-L. Huang, Y. Li, In Proc. of ICEBE, 585 (2006).
5. I. Sora, D. Todinca, C. Avram, In Proc. of SACI, 497 (2009).
6. J.-J. Shuai, In Proc. of ICICIC, 1452 (2009).
7. D. Radojevic, Int. J. of Comput. Commun., **3**, 121 (2008).
8. D. Radojevic, Studies in Fuzziness and Soft Computing, **218**, 295 (2008).
9. T. Saaty, The Analytic Hierarchy Process, **McGraw-Hill**, (1980).
10. L. A. Zadeh, Information and Control, **8**, 338 (1965).
11. D. Radojevic, In Proc. of FLINS, 511 (2010).
12. D. Radojevic, *Mathware & Soft Computing*, **15**, 125 (2008).
13. W3C, "QoS for Web Services: Requirements and Possible Approaches", http://www.w3c.or.kr/kr-office/TR/2003/ws-qos/ (2003).

CUT SYSTEMS IN Ω-SETS AND REFLECTIONS

JIŘÍ MOČKOŘ*
University of Ostrava Institute for Research and Applications of Fuzzy Modeling
30. dubna 22, 701 03 Ostrava 1, Czech Republic

Let Ω be a complete residuated lattice. Let $\mathbf{SetR}(\Omega)$ be the category of sets with similarity relations with values in Ω (called Ω-sets), which is an analogy of the category of classical sets with relations as morphisms. A cut system in an Ω-set (A,δ) in the category $\mathbf{SetR}(\Omega)$ is a system $(C_\alpha)_{\alpha\in\Omega}$ where C_α are subsets in $A\times\Omega$ with some special properties. It is well known that in the category $\mathbf{SetR}(\Omega)$, there is a close relation between cut systems in an Ω-set on one hand and fuzzy sets in the same Ω-set, on the other hand. Moreover, there exists an extension procedure such that any system $(C_\alpha)_\alpha$ of subsets from $A\times\Omega$ can be extended onto a cut system $(\overline{C_\alpha})_\alpha$ In the paper we prove that the extension procedure is, in some sense, the best possible. This will be expressed by the theorem which states that the category of such subsets is a full reflective subcategory in the category of cut systems.

1. Introduction

In our previous papers [1,4] we introduced a notion of fuzzy sets in sets with similarity relation (A,δ) (the so called Ω-sets), where values of a similarity relation $\delta: A\times A\to\Omega$ are from a residuated lattice Ω. Ω-sets then represent objects in various categories $\mathbf{K}$ with differently defined morphisms. Among these categories we are interested in two special categories: the category $\mathbf{Set}(\Omega)$ with morphisms $(A,\delta)\to(B,\gamma)$ defined as special maps $A\to B$ and the category $\mathbf{SetR}(\Omega)$ with morphisms defined as special relations $A\times B\to\Omega$. A notion of a fuzzy set in (A,δ) then depends on a category $\mathbf{K}$, i.e. f is a fuzzy set in an Ω-set (A,δ) in a category $\mathbf{K}$ (shortly, $f \subseteq_{\mathbf{K}} (A,\delta)$), if $f:(A,\delta)\to(\Omega,\leftrightarrow)$ is a morphism in $\mathbf{K}$, where $\leftrightarrow$ is the biresiduation operation in Ω (= special similarity relation in Ω). This formal extension of classical fuzzy sets enables us to develop a fuzzy set theory in any category of

* E-mail: Jiri.Mockor@osu.cz

Ω -sets, with a lot of properties similar to those of classical fuzzy sets. In papers [2,3] we proved that fuzzy sets in the categories $\mathbf{Set}(\Omega)$ and $\mathbf{SetR}(\Omega)$ can be represent by some cut systems. Namely, any fuzzy set , $f \subseteq_{\sim \mathbf{Set}(\Omega)} (A,\delta)$ can be represented by the so called f-cut $\mathbf{C}=(C_\alpha)_{\alpha\in\Omega}$ where C_α are subsets in A with some special properties, and, any fuzzy set , $g \subseteq_{\sim \mathbf{SetR}(\Omega)} (A,\delta)$ can be analogously represented by an f-cut system $\mathbf{D}=(D_\alpha)_{\alpha\in\Omega}$ where D_α are subsets in $A\times\Omega$ also with some special properties. We also proved that these representation theorems can be expressed as natural isomorphisms between two special functors from the corresponding category $\mathbf{K}=\mathbf{Set}(\Omega)$ or $\mathbf{SetR}(\Omega)$ into the category § of classical sets.

In the present paper we want to show that any cut system $\mathbf{C}=(C_\alpha)_\alpha$ of subsets in A or in $A\times A$, respectively, can be extended onto an f-cut $\overline{\mathbf{C}}$ in an Ω -set (A,δ) in the category $\mathbf{Set}(\Omega)$ or $\mathbf{SetR}(\Omega)$ respectively. Moreover, we want to prove that this extension procedure $\mathbf{C}\mapsto\overline{\mathbf{C}}$ is, in some sense, the best possible. To describe the effectiveness of extension procedures we will use a category theory language. Namely, we will use a notion of a *reflective subcategory* which represents a universal construction how to enhance properties of an object in the most effective way. The principal result of the paper are theorems which state that categories of extensions $\overline{\mathbf{C}}$ are reflective subcategories in categories of cuts $\mathbf{C}$.

2. Notations and preliminary results

In the section we will introduce several notations which will be used in the paper and we present several results from previous papers which can be useful for full understanding of our results. A principal structure used in the paper is a *complete residuated lattice* (see e.g. [5]), i.e. a structure $\Omega=(L,\wedge,\vee,\otimes,\rightarrow,0,1)$ such that $(L,\wedge,\vee)$ is a complete lattice, $(L,\otimes,1)$ is a commutative monoid with operation $\otimes$ isotone in both arguments and $\rightarrow$ is a binary operation which is residuated with respect to $\otimes$, i.e.

$$\alpha\otimes\beta\le\gamma \quad \text{iff} \quad \alpha\le\beta\rightarrow\gamma.$$

Recall that a set with similarity relation (or Ω -set) is a couple (A,δ) where $\delta : A\times A\rightarrow\Omega$ is a map such that

(a) $(\forall x\in A)\quad \delta(x,x)=1$

(b) $(\forall x,y\in A)\quad \delta(x,y)=\delta(y,x)$

(c) $(\forall x,y,z\in A)\quad \delta(x,y)\otimes\delta(y,z)\le\delta(x,z)$ (generalized transitivity).

In the paper we will use two principal categories with Ω -sets as objects and with differently defined morphisms. A morphism $f:(A,\delta)\to(B,\gamma)$ in the first category $\mathbf{Set}(\Omega)$ is a map $f:A\to B$ such that $\gamma(f(x),f(y))\geq\delta(x,y)$ for all $x,y\in A$. The other category $\mathbf{SetR}(\Omega)$ is an analogy of the category of sets with relations between sets as morphisms. Objects of the category $\mathbf{SetR}(\Omega)$ are the same as in the category $\mathbf{Set}(\Omega)$ and morphisms $f:(A,\delta)\to(B,\gamma)$ are maps $f:A\times B\to\Omega$ (i.e. Ω -valued relations) such that

(a) $(\forall x,z\in A)(\forall y\in B)\quad \delta(z,x)\otimes f(x,y)\leq f(z,y)$

(b) $(\forall x\in A)(\forall y,z\in B)\quad f(x,y)\otimes\gamma(y,z)\leq f(x,z)$

As we mentioned in Introduction, a fuzzy set f in an Ω -set (A,δ) in a category $\mathbf{K}$ (shortly, $f\subseteq_{\mathbf{K}}(A,\delta)$) is a morphism $f:(A,\delta)\to(\Omega,\leftrightarrow)$, where $\leftrightarrow$ is a biresiduation operation in Ω defined such that $\alpha\leftrightarrow\beta=(\alpha\to\beta)\wedge(\beta\to\alpha)$.

Any classically defined fuzzy set X in a set A with values in Ω can be defined equivalently by a system of level sets X_α, where $\alpha\in\Omega$, where $X_\alpha=\{a\in A:X(a)\geq\alpha\}$. Conversely, any (nested) system $(Y_\alpha)_\alpha$ of subsets of A such that for any $a\in A$ the set $\{\alpha\in\Omega:a\in Y_\alpha\}$ has the greatest element defines a fuzzy set Y such that $Y(a)=\bigvee_{\{\beta:a\in Y_\beta\}}\beta$. In our previous papers [2,3] we proved that analogously any fuzzy set in (A,δ) in the category $\mathbf{Set}(\Omega)$ or in the category $\mathbf{SetR}(\Omega)$ can be defined equivalently by a system of some special subsets in A or $A\times\Omega$, respectively, called **f-cut system**, which is defined as follows:

Definition 2.1. Let (A,δ) be an Ω -set. Then a system $\mathbf{C}=(C_\alpha)_\alpha$ of subsets in A is called an **f-cut** in (A,δ) in the category $\mathbf{Set}(\Omega)$ if

(a) $\forall a,b\in A,\quad a\in C_\alpha\Rightarrow b\in C_{\alpha\otimes\delta(a,b)}$

(b) $\forall a\in A,\forall\alpha\in\Omega,\quad \bigvee_{\{\beta:a\in C_\beta\}}\beta\geq\alpha\Rightarrow a\in C_\alpha$.

Definition 2.2. Let (A,δ) be an Ω -set. Then a system $(C_\alpha)_{\alpha\in\Omega}$ is called an **f-cut** in (A,δ) in the category $\mathbf{SetR}(\Omega)$ if

(a) $C_\alpha\subseteq A\times\Omega$, for any $\alpha\in\Omega$,

(b) $\forall a,b\in A,\quad (a,\beta)\in C_\alpha\Rightarrow(b,\beta)\in C_{\alpha\otimes\delta(a,b)}$

(c) $\forall a\in A,\forall\gamma\in\Omega,\quad \bigvee_{\{\beta:(a,\gamma)\in C_\beta\}}\beta\geq\alpha\Rightarrow(a,\gamma)\in C_\alpha$

(d) $\forall a\in A,\forall\alpha,\gamma\in\Omega,\quad (a,\alpha)\in C_\beta\Rightarrow(a,\gamma)\in C_{\beta\otimes(\alpha\leftrightarrow\gamma)}$

In the next parts we will use the following notations.

$$C(A,\delta)=\{\mathbf{C}:\mathbf{C}=(C_\alpha)_{\alpha\in\Omega}, C_\alpha\subseteq A\}$$
$$C_R(A,\delta)=\{\mathbf{C}:\mathbf{C}=(C_\alpha)_{\alpha\in\Omega}, C_\alpha\subseteq A\times\Omega\},$$
$$C_f(A,\delta)=\{\mathbf{C}:\mathbf{C} \text{ is an f-cut in } (A,\delta) \text{ in } \mathbf{Set}(\Omega)\},$$
$$C_{R,f}(A,\delta)=\{\mathbf{C}:\mathbf{C} \text{ is an f-cut in } (A,\delta) \text{ in } \mathbf{SetR}(\Omega)\}.$$

In the papers [2,3] it was proved that there exist functors $C_f:\mathbf{Set}(\Omega)\to\mathbf{Set}$ and $C_{R,f}:\mathbf{SetR}(\Omega)\to\mathbf{Set}$ with the above mentioned object functions.

Moreover, in the same papers it was proved that some extensional procedures can be used for object functions C. In fact, for any Ω-set (A,δ) there exist maps

$$^{-}: C(A,\delta)\to C_f(A,\delta), \tag{1}$$

$$^{-}: C_R(A,\delta)\to C_{R,f}(A,\delta), \tag{2}$$

such that $\mathbf{C}=(C_\alpha)_\alpha \mapsto \overline{\mathbf{C}}=(\overline{C_\alpha})_\alpha$ where

$$\overline{C_\alpha}=\{a\in A: \bigvee_{\{(x,\beta):x\in C_\beta\}} \beta\otimes\delta(a,x)\geq\alpha\}, \text{ for any } (C_\alpha)_\alpha, C_\alpha\subseteq A$$

$$\overline{C_\alpha}=\{(a,\beta)\in A\times\Omega: \bigvee_{\{(x,\tau,\rho):(x,\tau)\in C_\rho\}} \rho\otimes\delta(a,x)\otimes(\tau\leftrightarrow\beta)\geq\alpha\}$$

for any $(C_\alpha)_\alpha$ $C_\alpha\subseteq A\times\Omega$ (see papers [2,3]).

3. A category of cut systems in SetR(Ω)

The principal result of the paper are theorems which state that the extension procedures (1) and (2) are , in some sense, the most effective extensions. To describe those properties of some procedures we will use a category theory language. Namely, we will use a notion of a *reflective subcategory* which represents a universal construction how to enhance properties of an object in the most effective way (for more details see [6]). Recall that a full subcategory $\mathbf{K}\ \mathrm{O}\ \mathbf{L}$ of a category $\mathbf{L}$ is a *full reflective subcategory* in $\mathbf{L}$, if

(a) there exists a functor $G:\mathbf{L}\to\mathbf{K}$ (called a reflection),

(b) for any object a of $\mathbf{L}$ there exists a morphism $u_a: a\to G(a)$,

(c) for any other object b of $\mathbf{K}$ and any other morphism $f:a\to b$ there exists the unique morphism $\hat{f}:G(a)\to b$ such that the diagram commutes:

$$\begin{array}{ccc} a & \xrightarrow{u_a} & G(a) \\ f\big\downarrow & & \big\downarrow\widehat{f} \\ b & = & b \end{array}$$

In that case a subcategory $\mathbf{K}$ contains objects with enhanced (or completed) properties with respect to the category $\mathbf{L}$ and from some point of view it represents the most effective extension of objects from the original category $\mathbf{L}$. Many classical examples of well known extension (or completion) procedures can be expressed as reflective subcategories. Recall a completion of a metric space, a completion of a lattice to a complete lattice, a construction of Abelian groups from noncommutative groups etc. In all these examples full subcategories of objects with these enriched properties are full reflective subcategories in categories of original structures.

We will start with definitions of two categories of cut systems.

Definition 3.1. The objects of the category **Cut** will be $(A,\delta,\mathbf{C})$ where (A,δ) is a Ω -set and $\mathbf{C}$ is an element from $C(A,\delta)$ Then $f:(A,\delta,\mathbf{C})\to(B,\gamma,\mathbf{D})$ is a morphism in **Cut** if

(a) $f:(A,\delta)\to(B,\gamma)$ is a morphism in the category $\mathbf{Set}(\Omega)$

(b) $f(\mathbf{C})\subseteq\mathbf{D}$, i.e. $f(C_\alpha)\subseteq D_\alpha$ for all $\alpha\in\Omega$.

Let $\mathbf{Cut}_f$ be a full subcategory in **Cut** with objects $(A,\delta,\mathbf{C})$ such that $\mathbf{C}$ is an f-cut system in (A,δ) in the category $\mathbf{Set}(\Omega)$

Theorem 3.1. A subcategory $\mathbf{Cut}_f$ O **Cut** is a full reflective subcategory in the category **Cut**.

An analogical investigation for the category $\mathbf{SetR}(\Omega)$ is more complicated. We will start with the following definition of $\mathbf{Cut}_R$.

Definition 3.2. Objects of $\mathbf{Cut}_R$ will be $(A,\delta,\mathbf{C})$ where (A,δ) is an Ω -set and $\mathbf{C}$ is an element from $C_R(A,\delta)$ Moreover, $f:(A,\delta,\mathbf{C})\to(B,\gamma,\mathbf{D})$ is a morphism in $\mathbf{Cut}_R$, if

(a) $f:(A,\delta)\to(B,\gamma)$ is a morphism in the category $\mathbf{SetR}(\Omega)$ and

(b) $f(\mathbf{C})\subseteq\mathbf{D}$, i.e. $f(C_\alpha)\subseteq D_\alpha$ for all $\alpha\in\Omega$, where

$$f(C_\alpha)=\{(b,\beta)\in B\times\Omega:\bigvee_{\{(a,\tau):(a,\beta)\in C_\tau\}} f(a,b)\otimes\tau\geq\alpha\}.$$

Let us mention that if **C** is an f-cut, then $f(\mathbf{C})$ can be also expressed by the functor $C_{R,f}$ In fact, for any morphism $f:(A,\delta)\rightarrow(B,\gamma)$ in **SetR**(Ω) and any $\mathbf{C}\in C_{R,f}(A,\delta)$ we have

$$f(\mathbf{C})=C_{R,f}(f)(\mathbf{C}),$$

as follows directly from the definition of the functor $C_{R,f}$ (see [3]).

Proposition 3.1. $\mathbf{Cut}_R$ *is a category.*

By $\mathbf{Cut}_{R,f}$ we denote the full subcategory of the category $\mathbf{Cut}_R$ with objects $(A,\delta,\mathbf{C})$ where **C** is an object from $C_{R,f}(A,\delta)$ i.e. an f-cut.

Theorem 3.2. The subcategory $\mathbf{Cut}_{R,f}$ O $\mathbf{Cut}_R$ is a full reflective subcategory in $\mathbf{Cut}_R$.

For the proof of the theorem we need some technical lemmas and the following proposition.

Proposition 3.2. Let $f:(A,\delta)\rightarrow(B,\gamma)$ be a morphism in the category **SetR**(Ω) and let $\mathbf{C}\in C_R(A,\delta)$ Then we have

$$f(\overline{C})\subseteq\overline{f(C)}.$$

Acknowledgments

This work was supported by the European Regional Development Fund in the IT4Innovations Centre of Excellence project (CZ.1.05/1.1.00/02.0070).

References

1. Močkoř, J., Fuzzy Sets in Categories of Sets with Similarity Relations, Computational Intelligence, Theory and Applications. Heidelberg, Springer Verlag (2006), 677–682.
2. Močkoř, J., Cut systems in sets with similarity relations, Fuzzy Sets and Systems **161(24)**(2010), 3127–3140.
3. Močkoř, J., Fuzzy sets and cut systems in a category of sets with similarity relations, Soft Computing (to appear).
4. Močkoř, J., Morphisms in categories of sets with similarity relations, Proceedings of IFSA Congress/EUSFLAT Conference. Lisabon (2009), 560–568
5. Novák,V., Perfilijeva, I., Močkoř, J., Mathematical principles of fuzzy logic Kluwer Academic Publishers, Boston, Dordrecht, London (1999).
6. Mac Lane, S., Categories for the Working Mathematician, Springer-Verlag, Berlin, Heidelberg, New York (1971).

DUALITY PRINCIPLES IN RESIDUATED LATTICES

IRINA PERFILIEVA*

Centre of Excellence IT4Innovations, division of the University of Ostrava
Institute for Research and Applications of Fuzzy Modeling
30. dubna 22, 702 00 Ostrava, Czech Republic

We investigate residuated lattices as structures with dualities. We formulated the *Principle of Duality* for residuated lattice which divides basic properties of this structure into two groups of dual ones.

1. Introduction

The notion of commutative residuated ℓ-monoid has been introduced in [1] with the goal "to outline a common framework for a diversity of monoidal structures which constitute the basis of various papers in fuzzy set theory". This notion is a bit more general than the earlier introduced [2] notion of "residuated lattice" for an integral, residuated, commutative ℓ-monoid. Both papers became fundamental in the literature related to mathematical fuzzy logic [3–5], algebraic foundations of fuzzy systems [6], algebraic foundations of triangular norms [7], etc. Moreover, they had significant impact on intensive development of various residuated algebraic structures such as residuated ℓ-groupoids, residuated ℓ-semigroups, etc. (see, e.g. [8]). From the point of view of applications, residuated structures play an important role, for example, in mathematical morphology, modeling of linguistic semantics, approximate reasoning, and elsewhere.

In the proposed contribution, we investigate how residuation manifests itself through duality. We will see that existence of a pair of dual semimodules of a commutative semiring implies that its monoidal reduct is a residuated lattice. This result allows to formulate the *Principle of Duality* for residuated lattice which divides basic properties of this structure into two groups of dual ones.

* E-mail:Irina.Perfilieva@osu.cz

2. Residuated, Commutative ℓ-monoid as a Pair of Dual Semimodules

In this section, we will see how a residuated, commutative ℓ-monoid can be characterized without any reference to the property of residuation. To characterize residuated operations, we will use two dually ordered semimodules. Moreover, the existence of two dually ordered semimodules is a part of the criterion under which a monoidal reduct of a lattice-ordered commutative semiring is a residuated, commutative ℓ-monoid. We will formulate and illustrate the *Principle of Duality* for residuated lattice.

2.1. *Semimodules in residuated, commutative ℓ-monoid*

Let us recall that a *semiring* is an algebraic structure with two associative operations which are connected by distributive laws (cf.[6,9,10]). In more details, a semiring $R = (R, +, \cdot, 0, e)$ is an algebraic structure with the following properties:

- $(R, +, 0)$ is a commutative monoid,
- $(R, \cdot, e)$ is a monoid,
- $a \cdot (b + c) = a \cdot b + a \cdot c$ and $(a + b) \cdot c = a \cdot c + b \cdot c$ hold for all $a, b, c \in R$.

A semiring is called *commutative* if $(R, \cdot, e)$ is a commutative monoid. An example of a commutative semiring, which will be used in the sequel, is as follows.

Example 2.1. Let $\mathcal{L} = (L, \leq, *)$ be a residuated, commutative ℓ-monoid. Then $\mathcal{R}_{\vee}^{\mathcal{L}} = (L, \vee, *, \perp, e)$, where $\perp$ is a bottom element of $(L, \leq)$ and e is the unit with respect to $*$ is a semiring reduct of $\mathcal{L}$. This easily follows from the fact that for every $a \in L$, the map $a * (\cdot) : L \to L$ is residuated and therefore, it preserves the join operation, i.e. for all $x, y \in L$, $a * (x \vee y) = (a * x) \vee (a * y)$ holds.

Let us give a definition of a (left) R-*semimodule*.

Definition 2.1. Let R be a semiring. A (left) R-semimodule is an algebra $M = (M, +, 0, (h_a)_{a \in R})$, where $(M, +, 0)$ is a commutative monoid and each $h_a : M \to M$ is a unary operation so that the following properties are fulfilled:

$$h_a(x + y) = h_a(x) + h_a(y) \tag{1}$$

$$h_{a+b}(x) = h_a(x) + h_b(x) \tag{2}$$

$$h_{a \cdot b}(x) = h_a(h_b(x)). \tag{3}$$

Let us remark that the above definition is different from those in [6,9] where there are additional requirements on zero elements of R and M.

The unary operation h_a is usually considered as a (left) *scalar multiplication* by a. An R-semimodule is called *unital* (or *unitary*) if the unit element $e \in R$ determines the identical map on M, i.e. for all $x \in M$, $h_e(x) = x$.

Example 2.2. It is easy to see that if $R = (R, +, \cdot, 0, e)$ is a semiring then $M_R = (R, +, 0, (a \cdot (\cdot))_{a \in R})$ is a unital semimodule where $a \cdot (\cdot) : x \mapsto a \cdot x$. We will say that the semimodule $M_R = (R, +, 0, (a \cdot (\cdot))_{a \in R})$ is induced by a semiring $R = (R, +, \cdot, 0, e)$.

Example 2.3. Let $\mathcal{L} = (L, \leq, *)$ be a residuated, commutative ℓ-monoid and $\mathcal{R}_{\vee}^{\mathcal{L}} = (L, \vee, *, \perp, e)$ its semiring reduct. By Example 2.2, $(L, \vee, \perp, (a * (\cdot))_{a \in L})$ is a $\mathcal{R}_{\vee}^{\mathcal{L}}$-semimodule induced by the respective semiring.

Two R-semimodules M_1 and M_2 are called *isomorphic* if there is a bijection $\varphi : M_1 \to M_2$ which establishes an isomorphism between respective monoids $(M_1, +, 0)$ and $(M_2, +, 0)$ and which commutes with every unary operation h_a, i.e. for all $a \in R$ and for all $x \in M_1$,

$$\varphi(h_a(x)) = h_a(\varphi(x)).$$

Similarly, a homomorphism between M_1 and M_2 can be defined.

In the below given theorems, we show that any residuated, commutative ℓ-monoid can be split into two unital left semimodules such that one of them is a homomorphic image of the other one. As a converse statement, we will prove that existence of two (dual) unital left semimodules is a criterion under which a monoidal reduct of a lattice-ordered commutative semiring is a residuated, commutative ℓ-monoid.

Theorem 2.1. Let $\mathcal{R}_{\vee}^{\mathcal{L}} = (L, \vee, *, \perp, e)$ be a lattice-ordered commutative semiring where $(L, \vee, \wedge, \perp, \top)$ is a bounded lattice and e is the unit with respect to $*$. Then the monoidal reduct $(L, *, e)$ of $\mathcal{R}_{\vee}^{\mathcal{L}}$ is a residuated, commutative ℓ-monoid if and only if there exists a set of unary operations $\{h_a : L \to L \mid a \in L\}$ such that $(L, \wedge, \top, (h_a)_{a \in L}))$ is a $\mathcal{R}_{\vee}^{\mathcal{L}}$-semimodule and for all $a, b \in L$,

$$a \leq b \Leftrightarrow e \leq h_a(b). \tag{4}$$

Corollary 2.1. Let $\mathcal{L} = (L, \leq, *)$ be a residuated, commutative ℓ-monoid and $\mathcal{R}_{\vee}^{\mathcal{L}} = (L, \vee, *, \perp, e)$ its semiring reduct. Then

(i). there are two $\mathcal{R}_{\vee}^{\mathcal{L}}$*-semimodules:* $\mathcal{L}_{\vee,*} = (L, \vee, \perp, (a * (\cdot))_{a \in L})$ *and* $\mathcal{L}_{\wedge,\to} = (L, \wedge, \top, (a \to (\cdot))_{a\text{\inL}})$*, that split* $\mathcal{L}$*,*

(ii). both $\mathcal{R}_\vee^\mathcal{L}$-semimodules $\mathcal{L}_{\vee,*}$ and $\mathcal{L}_{\wedge,\to}$ are unital,

(iii). the condition (4) in Theorem 2.1 is equivalent to the fact that $e = \top$, i.e. $\mathcal{L}$ is integral.

In the sequel, we will assume that a given residuated, commutative ℓ-monoid is integral. For its shorter name we will use the name "*residuated lattice*". It will be denoted by $\mathcal{L} = (L, \leq, *)$ with additional operations as follows: $\to$ as a residual of $*$, and $\bot$ and $\top$ as respective bottom and top elements.

Theorem 2.2. *Let $\mathcal{L} = (L, \leq, *)$ be a residuated lattice with $\mathcal{R}_\vee^\mathcal{L} = (L, \vee, *, \bot, \top)$ as a semiring reduct where $\top$ is the unit with respect to $*$. Then the left $\mathcal{R}_\vee^\mathcal{L}$-semimodule $\mathcal{L}_{\wedge,\to} = (L, \wedge, \top, (a \to (\cdot))_{a\in R}))$ is a homomorphic image of the left $\mathcal{R}_\vee^\mathcal{L}$-semimodule $\mathcal{L}_{\vee,*} = (L, \vee, \bot, (a * (\cdot))_{a\in L})$ under the homomorphism $\varphi : L \to L$, given by*

$$\varphi(x) = x \to \bot.$$

2.2. *Principle of duality for residuated lattice*

Let $\mathcal{L} = (L, \leq, *)$ be a residuated lattice and $\mathcal{R}_\vee^\mathcal{L} = (L, \vee, *, \bot, \top)$ its semiring reduct. By Corollary 2.1, $\mathcal{L}$ splits into two unital $\mathcal{R}_\vee^\mathcal{L}$-semimodules $\mathcal{L}_{\vee,*} = (L, \vee, \bot, (a * (\cdot))_{a\in L})$ and $\mathcal{L}_{\wedge,\to} = (L, \wedge, \top, (a \to (\cdot))_{a\in R}))$. The operations $\vee$ in $\mathcal{L}_{\vee,*}$ and $\wedge$ in $\mathcal{L}_{\wedge,\to}$ determine dual orderings of the same support L of these semimodules:

$$x \leq y - \quad x \vee y = y,$$
$$x \geq y - \quad x \wedge y = y.$$

Below, we will refer to semimodules $\mathcal{L}_{\vee,*}$ and $\mathcal{L}_{\wedge,\to}$ as to dual ones. Because both semimodules are representatives of the same algebraic structure, everything which can be proved for one of them can be proved for the other one. This is used in the below formulated *Principle of Duality* which also uses the fact that both semimodules are parts of a residuated lattice.

Principle of Duality in residuated, commutative ℓ-monoids

To every theorem that concerns a residuated lattice $\mathcal{L} = (L, \leq, *)$ and is formulated in a language of one of its $\mathcal{R}_\vee^\mathcal{L}$-semimodules there is a corresponding theorem that concerns its dual semimodule. This is obtained by replacing symbols of operations $\vee$, $\bot$ and $(a * (\cdot))_{a\in L}$ of the semimodule $\mathcal{L}_{\vee,*}$ by the respective dual symbols $\wedge$, $\top$ and $(a \to (\cdot))_{a\in L}$ of the semimodule $\mathcal{L}_{\wedge,\to}$, and vice versa.

The following list contains some pairs of dual statements where the elements of the semiring $\mathcal{R}_{\vee}^{\mathcal{L}}$ are denoted by characters a, b to be distinguished from the elements of the respective semimodules denoted by characters x, y.

$$a*(x \vee y) = (a*x) \vee (a*y),\ a \rightarrow (x \wedge y) = (a \rightarrow x) \wedge (a \rightarrow y) \tag{5}$$

$$(a \vee b)*x = (a*x) \vee (b*x),\ (a \vee b) \rightarrow x = (a \rightarrow x) \wedge (b \rightarrow x); \tag{6}$$

$$(a*b)*x = a*(b*x), \qquad (a*b) \rightarrow x = a \rightarrow (b \rightarrow x) \tag{7}$$

Remark 2.1. It is worth stressing that the *Principle of Duality* is applicable only to statements in the language of respective semimodules and only to semimodules operations. It is not applicable to operations of the semiring $\mathcal{R}_{\vee}^{\mathcal{L}}$. The following example illustrates this remark.

Example 2.4. In two dual properties (6),

$$(a \vee b)*x = (a*x) \vee (b*x), \quad (a \vee b) \rightarrow x = (a \rightarrow x) \wedge (b \rightarrow x),$$

the symbol $\vee$ in the term $(a \vee b)$ (left-hand side of the left equality) does not correspond to an operation of $\mathcal{L}_{\vee,*}$ (it corresponds to the semiring operation) and thus, it is not changed in the right (dual) equality. However, the same symbol $\vee$ in the expression $(a*x) \vee (b*x)$ (right-hand side of the left equality) does correspond to an operation of $\mathcal{L}_{\vee,*}$ and by the *Principle of Duality*, it is replaced by $\wedge$. Moreover, symbols $*$ in both sides of the left equality correspond to (unary) operations of $\mathcal{L}_{\vee,*}$, and therefore, they are replaced by the respective dual symbols $\rightarrow$.

3. Conclusion

In this paper, we investigated residuated lattices as structures with dualities. we proved that prove that existence of two (dual) unital left semimodules is a criterion under which a monoidal reduct of a lattice-ordered commutative semiring is a residuated, commutative ℓ-monoid. We formulated the Principle of Duality for residuated lattices which divides basic properties of these structures into two groups of dual ones.

Acknowledgment

The author acknowledges that she prepared this paper in connection with the project IT4Innovations Centre of Excellence, reg. no. CZ.1.05/1.1.00/02.0070.

References

1. U. Höhle, Commutative residuated l-monoids, in Non-Classical Logics and Their Applications to Fuzzy Subsets. A Handbook of the Mathematical Foundations of Fuzzy Set Theory, eds. U. Höhle and E. P. Klement (Kluwer, Dordrecht, 1995) pp. 53–106.
2. R. P. Dilworth and M. Ward, Trans. Amer. Math. Soc. 45, 335 (1939).
3. P. Hájek, Metamathematics of Fuzzy Logic (Kluwer, Dordrecht, 1998).
4. V. Novák, I. Perfilieva and J. Močkoř, Mathematical Principles of Fuzzy Logic (Kluwer, Boston, 1999).
5. F. Esteva and L. Godo, Fuzzy Sets and Systems 124, 271 (2001).
6. A. Di Nola, A. Lettieri, I. Per_lieva and V. Nov_ak, Fuzzy Sets and Systems 158, 1 (2007).
7. E. P. Klement, R. Mesiar and E. Pap, Triangular Norms (Kluwer, Dordrecht, 2000).
8. N. Galatos, P. Jipsen, T. Kowalski and H. Ono, Residuated Lattices: an algebraic glimpse at substructural logicsStudies in Logics and the Foundations of Mathematics, Studies in Logics and the Foundations of Mathematics (Elsevier, Amsterdam, 2007).
9. J. S. Golan, Semirings and their Applications (Kluwer Academic Pulishers, Dordrecht, 1999).
10. M. Gondran and M. Minoux, Graphs, Dioids and Semirings (Springer-Verlag, New York, 2008).

INTERPOLATIVITY OF "AT LEAST–AT MOST" MODELS OF MONOTONE FUZZY RULE BASES: MULTIPLE-INPUT CASE

MARTIN ŠTĚPNIČKA

Centre of Excellence IT4Innovations - Division of the University of Ostrava - IRAFM, 30. dubna 22, 701 03 Ostrava, Czech Republic
Martin.Stepnicka@osu.cz

BALASUBRAMANIAM JAYARAM

Department of Mathematics, Indian Institute of Technology Hyderabad, Yeddumailaram-502205, INDIA
jbala@iith.ac.in

Among the many desirable properties of fuzzy inference systems not all of them are known to co-exist. For instance, a system based on a monotone fuzzy rule base need not be monotonic and interpolative simultaneously. Recently, Štěpnička and De Baets have investigated and shown the co-existence of the above two properties in the case of a fuzzy relational inference systems and the single-input-single-output (SISO) rule bases. An extension of these results to the multiple-input-single-output (MISO) case is not straight-forward owing to the lack of a natural ordering in higher dimensions. In this work, we study the MISO case and show that similar results are available when the monotone rule base is modeled based on at-most and at-least modifiers.

1. Introduction

A fuzzy inference mechanism can be viewed as a mapping which - with the help of a model of a given fuzzy rule base - gives a meaningful output from basically imprecise inputs modelled by fuzzy sets. Among many types of inference mechanisms and fuzzy rule based models proposed in fuzzy logic we highlight the fuzzy relational approach that is very common in the literature. The basic principles of such an approach consist in using an appropriate fuzzy relation to model a given fuzzy rule base and in modeling the fuzzy inference mechanism by an appropriate image of a fuzzy set under a fuzzy relation that is derived from a fuzzy relational composition.[3]

Particularly, a fuzzy rule base containing rules of the form:

$$\textbf{IF } X \textbf{ is } A_i \textbf{ THEN } Y \textbf{ is } B_i\,, \tag{1}$$

with X a variable in a universe U, Y a variable in a universe V, and A_i and B_i $(i = 1, \ldots, n)$ fuzzy sets in U and V, respectively, is modelled by a fuzzy relation on $U \times V$.

Let us fix a residuated lattice[9] $\mathcal{L} = \langle [0,1], \wedge, \vee, *, \rightarrow, 0, 1\rangle$, where $*$ is a left-continuous t-norm and $\rightarrow$ its adjoint residuum (also residual fuzzy implication[1]), as the basic algebraic structure for the whole paper and let us denote the set of all fuzzy sets on U by $\mathcal{F}(U)$. Then fuzzy rule base (1) is either modeled by the fuzzy relation $\hat{R} \in \mathcal{F}(U \times V)$ defined by

$$\hat{R}(x,y) = \bigwedge_{i=1}^{n} (A_i(x) \rightarrow B_i(y))\,, \tag{2}$$

which reflects the conditional nature of the rules or by a popular alternative fuzzy relation $\check{R} \in \mathcal{F}(U \times V)$ defined by

$$\check{R}(x,y) = \bigvee_{i=1}^{n} (A_i(x) * B_i(y))\,. \tag{3}$$

As recalled above, within the fuzzy relational framework, the fuzzy inference mechanism is nothing else but a way of computing an *image* of a fuzzy set under a fuzzy relation that is derived from *fuzzy relational composition.* Zadeh's Compositional Rule of Inference (CRI) denoted by $\circ$ (based on the direct image) is the most often used and well-established relational fuzzy inference mechanism. However, Pedrycz[10] proposed to use the Bandler-Kohout Subproduct (BKS)[2] denoted by $\lhd$ (based on the subdirect image) and Štěpnička with Jayaram[19] showed that this choice is fully comparable with the CRI.

We recall that given a fuzzy set $A \in \mathcal{F}(U)$ and a fuzzy relation $R \in \mathcal{F}(U \times V)$, these images are the fuzzy sets in V defined by

$$A \circ R(y) = \bigvee_{x \in U} A(x) * R(x,y)\,, \qquad y \in V\,, \tag{4}$$

$$A \lhd R(y) = \bigwedge_{x \in U} A(x) \rightarrow R(x,y)\,, \qquad y \in V\,. \tag{5}$$

Many real-world applications are fully automated and therefore require some defuzzification strategy to be employed at the end of the inference process. In case of crisp inputs we then obtain the so-called *resulting function,* which maps the crisp input from U to the crisp output from V obtained after the defuzzification.

1.1. *Desirable properties - motivation for this work*

There are many desirable properties of the fuzzy relational inference systems, e.g., continuity of the resulting function[18] or robustness[7,19] of the system. Perhaps the most important property is the interpolativity of the inference mechanism which pertains to the solvability of the fuzzy relational equations:

$$B_i = A_i @ R, \qquad i = 1, \dots n. \tag{6}$$

where $@ \in \{\circ, \lhd\}$, $A_i \in \mathcal{F}(U)$, $B_i \in \mathcal{F}(V)$ and the goal is to find an appropriate $R \in \mathcal{F}(U \times V)$ that solves the system above. In other words, we search for a model R of a given fuzzy rule base (1) such that if an input is equal to one of the antecedents, it enforces the deduced output to be equivalent to the respective consequent. A tremendous job has been done in the filed of solvability of fuzzy relational equations and thus, we only refer to some relevant sources.[5,6,12] In general, it is sufficient for the antecedent fuzzy sets to form a Ruspini partition to ensure interpolativity,[17] which is also the naturally arising context in practical settings.

Recently another important issue with regards to fuzzy inference systems has been discussed - monotonicity of the resulting function. Given a monotone rule base which captures the monotonicity of the underlying system function, the resulting function is expected to preserve this monotonicity. The monotonicity problem was introduced by Broekhoven and De Baets.[20,21] The investigation was motivated by the observation that having an obviously monotone fuzzy rule base, the resulting function may not always be monotone. Even worse, under generally common settings ($\check{R}$ and the COG defuzzification), in many cases monotonicity cannot be guaranteed at all. Results obtained with the use of $\hat{R}$ were even less optimistic. However, the application of the *at-least* (ATL) and *at-most* modifiers (ATM)[13] to both the antecedent and consequent fuzzy sets of the implicative model $\hat{R}$, foreshadowed promising results. Convincing results were obtained recently.[14,15]

Unfortunately, applying the above ATL and ATM modifiers to antecedent fuzzy sets leads us out of the practical setting, as the modified antecedent fuzzy sets then no longer form a Ruspini partition, which, though not necessary, is a sufficient condition for interpolativity. Recently, Štěpnička and De Baets[16] have studied the interpolativity with these modified fuzzy rules, and hence, modified implicative models. While their study deals with the single-input-single-output (SISO) case, in this work we revisit the same issue but for the multiple-input-single-output (MISO) case.

Let us only briefly recall, that in the MISO case, antecedent fuzzy sets A_i are defined on the Cartesian product of universes $U = U_1 \times \cdots \times U_m$ and composed as follows:

$$A_i(x) = A^1_{i_1}(x_1) \wedge \ldots \wedge A^m_{i_m}(x_m)$$

where $A^k_{i_k} \in \mathcal{F}(U_k)$ for $k = 1, \ldots m$. Thus the rules in fuzzy rule base (1) may be expanded as follows

$$\textbf{IF } X_1 \textbf{ is } A^1_{i_1} \textbf{ AND } \cdots \textbf{ AND } X_m \textbf{ is } A^m_{i_m} \textbf{ THEN } Y \textbf{ is } B_i\,, \tag{7}$$

and modelled by a fuzzy relation $R \in \mathcal{F}(U_1 \times \cdots \times U_m \times V)$.

2. Interpolativity and Monotonicity

2.1. *Interpolativity*

We have recalled that the Ruspini partition formed by the antecedent fuzzy sets is a sufficient condition for the solvability of (6). In,[11] we may find a weaker solvability condition that also imposes requirements on antecedent fuzzy sets only. It is called *finitary condition* because it has to be fulfilled only on a finite number of nodes.

Definition 2.1.[11] Let A_i for $i = 1, \ldots, n$ be normal fuzzy sets on U. Then we say that they fulfill the *finitary condition* if

$$A_i(x_j) = 0\,, \tag{8}$$

where x_j is such that $A_j(x_j) = 1$ and where $i \neq j$.

I. Perfilieva[11] has proved that assuming the finitary condition for antecedent fuzzy sets is a sufficient condition for the solvability of System (6) with $@ \in \{\circ, \triangleleft\}$. Note that originally this fact has been proven in[11] for discrete universes, i.e., for fuzzy sets represented by vectors and fuzzy relations represented by matrices. However, the result holds generally even in case of continuous universes as shown below.

Theorem 2.1. *Let A_i, $i = 1, \ldots, n$, be normal and fulfill the finitary condition. Then System (6) with $@ \in \{\circ, \triangleleft\}$ is solvable.*

Sketch of the proof: Using the properties $a * (a \to b) \leq b$ and $a \to (a * b) \geq b$ we may prove that $A_i \circ \hat{R} \subseteq B_i$ and $A_i \triangleleft \hat{R} \supseteq B_i$, respectively. Furthermore, under the finitary condition $A_j(x_i) \leq B_i(y) \to B_j(y)$ as well as $A_j(x_i) \leq B_j(y) \to B_i(y)$ hold which are sufficient conditions in order to prove $A_i \circ \hat{R} \supseteq B_i$ and $A_i \triangleleft \hat{R} \subseteq B_i$, respectively. □

It has to be stressed that assuming the normality of antecedent fuzzy sets leads to sufficient conditions of the solvability that are even weaker than the finitary condition and that this observation (in case of $@ \equiv \circ$) has been published already in Moser and Navara.[8] However, the weaker conditions are given by inequalities that are not as easy to interpret in the practical setting of a fuzzy rule-based system and thus, the finitary conditions is more of a luxury from this point of view.

2.2. *Monotonicity*

We first recall an appropriate partial order relation $\leq_f$ on fuzzy sets based on the ordering of their α-cuts. For a fuzzy set C on a universe U, its α-cut $[C]_\alpha$ is the subset of U defined as

$$[C]_\alpha = \{x \in U \mid C(x) \geq \alpha\}, \quad \alpha \in]0,1],$$

If U is a linear vector space over the real numbers $\mathbb{R}$ we can define the notion of a *convex fuzzy set* in a standard way.[9,18] Then, we can define the ordering of normal convex fuzzy sets.

Definition 2.2. Given two normal convex fuzzy sets C_1 and C_2 on a universe $U \subseteq \mathbb{R}$, $C_1 \leq_f C_2$ if for any $\alpha \in]0,1]$ it holds that

$$\inf[C_1]_\alpha \leq \inf[C_2]_\alpha \quad \text{and} \quad \sup[C_1]_\alpha \leq \sup[C_2]_\alpha .$$

Definition 2.2 is a natural generalization of the well-known ordering of closed real intervals. Requiring this ordering to hold for all α-cuts then obviously leads to a partial order relation on normal convex fuzzy sets. It can be further generalized to non-normal or non-convex fuzzy sets.[4]

Now, we recall the notion of a monotone fuzzy rule base[14,15] that slightly generalizes the original one.[20] We only address the monotonicity of the increasing type here.

Definition 2.3. Let the antecedent fuzzy sets of the fuzzy rule base (7) be linearly ordered on each of the axes. The fuzzy rule base (7) is called *monotone* if for any $s \in \{1,\ldots,m\}$ and any two rules:

$$\textbf{IF } X_1 \textbf{ is } A^1_{i_1} \ldots \textbf{AND } X_s \textbf{ is } A^s_{i_s} \textbf{ AND} \ldots X_m \textbf{ is } A^m_{i_m}, \textbf{THEN } Y \textbf{ is } B_i ,$$
$$\textbf{IF } X_1 \textbf{ is } A^1_{i_1} \ldots \textbf{AND } X_s \textbf{ is } A^s_{j_s} \textbf{ AND} \ldots X_m \textbf{ is } A^m_{i_m}, \textbf{THEN } Y \textbf{ is } B_{i'}$$

whenever we have that $A^s_{i_s} \leq_f A^s_{j_s}$, then it also holds that $B_i \leq_f B_{i'}$.

The *at-least* ($\uparrow$ ATL) and *at-most* ($\downarrow$ ATM) modifiers have been proposed by Bodenhofer *et.al.* [4,13] For a fuzzy set C on a universe $U \subseteq \mathbb{R}$, the fuzzy sets $C^\uparrow, C^\downarrow \in \mathcal{F}(U)$ are defined by

$$\begin{aligned} C^\uparrow(x) &= \sup\{C(t) \mid t \in U, t \leq x\}\,, \\ C^\downarrow(x) &= \sup\{C(t) \mid t \in U, t \geq x\}\,. \end{aligned}$$

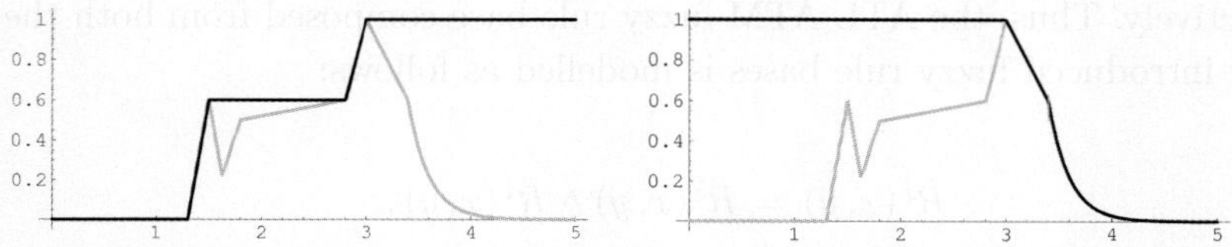

(a) The result of applying the "at-least" modifier (black line) to a fuzzy set (grey line).

(b) The result of applying the "at-most" modifier (black line) to the fuzzy set (grey line).

Fig. 1. Graphical illustration of the application of the "at-least" and "at-most" modifiers to a fuzzy set.

Motivated by,[14,15] we claim that every rule

$$\textbf{IF } X_1 \textbf{ is } A^1_{i_1} \textbf{ AND } \cdots \textbf{ AND } X_m \textbf{ is } A^m_{i_m}\,, \textbf{ THEN } Y \textbf{ is } B_i$$

in a monotone fuzzy rule base[a] may be understood as an *at-least* rule

$$\textbf{IF } X_1 \textbf{ is } (A^1_{i_1})^\uparrow \textbf{ AND } \cdots \textbf{ AND } X_m \textbf{ is } (A^m_{i_m})^\uparrow \textbf{ THEN } Y \textbf{ is } (B_i)^\uparrow\,,$$

and/or an *at-most* rule

$$\textbf{IF } X_1 \textbf{ is } (A^1_{i_1})^\downarrow \textbf{ AND } \cdots \textbf{ AND } X_m \textbf{ is } (A^m_{i_m})^\downarrow \textbf{ THEN } Y \textbf{ is } (B_i)^\downarrow\,,$$

that are naturally modelled by the following fuzzy relations

$$\begin{aligned} \hat{R}^\uparrow_{(i_1,\ldots,i_m)}(x_1,\ldots,x_m,y) &= ((A^1_{i_1})^\uparrow(x_1) \wedge \cdots \wedge (A^m_{i_m})^\uparrow(x_m)) \to B^\uparrow_i(y)\,, \\ \hat{R}^\downarrow_{(i_1,\ldots,i_m)}(x_1,\ldots,x_m,y) &= ((A^1_{i_1})^\downarrow(x_1) \wedge \cdots \wedge (A^m_{i_m})^\downarrow(x_m)) \to B^\downarrow_i(y)\,. \end{aligned}$$

[a] Note that in case of a decreasing relationship, the use of the modifiers would be mixed, in the sense that antecedent and consequent fuzzy sets would be subjected to different modifiers.

respectively; the ATL (ATM) fuzzy rule base is modelled as follows

$$\hat{R}^{\uparrow}(x_1,\ldots,x_m,y) = \bigwedge_{i_1=1}^{n_1}\cdots\bigwedge_{i_m=1}^{n_m} \hat{R}^{\uparrow}_{(i_1,\ldots,i_m)}(x_1,\ldots,x_m,y)\,,$$

$$\hat{R}^{\downarrow}(x_1,\ldots,x_m,y) = \bigwedge_{i_1=1}^{n_1}\cdots\bigwedge_{i_m=1}^{n_m} \hat{R}^{\downarrow}_{(i_1,\ldots,i_m)}(x_1,\ldots,x_m,y)\,,$$

respectively. Thus, the ATL-ATM fuzzy rule base composed from both the newly introduced fuzzy rule bases is modelled as follows:

$$\hat{R}^{\updownarrow}(x,y) = \hat{R}^{\uparrow}(x,y) \wedge \hat{R}^{\downarrow}(x,y)\,.$$

3. Interpolativity of the New Models

As mentioned above, antecedent fuzzy sets after the application of ATL and/or ATM modifiers can no longer form a Ruspini partition that is a sufficient interpolativity condition. What is more alarming is that even the finitary condition may not be fulfilled. However, this does not mean that the interpolativity of these new models is not possible anymore. This is the crucial question that has been opened in[16] and that we discuss in this section.

In other words, we study whether $\hat{R}^{\updownarrow}$ is a proper model of the original fuzzy rule base and thus, whether $\hat{R}^{\updownarrow}$ can be a solution of the system (6), which is illustrated in Fig. 2.

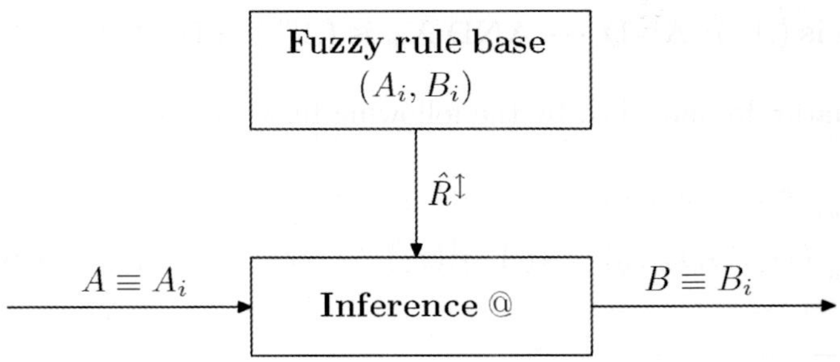

Fig. 2. Illustration of the studied problem.

Lemma 3.1. *For arbitrary $A \in \mathcal{F}(U_1 \times \cdots \times U_m)$*

$$A \circ \hat{R}^{\updownarrow} = \left(A \circ \hat{R}^{\uparrow}\right) \wedge \left(A \circ \hat{R}^{\downarrow}\right), \tag{9}$$

$$A \lhd \hat{R}^{\updownarrow} = \left(A \lhd \hat{R}^{\uparrow}\right) \wedge \left(A \lhd \hat{R}^{\downarrow}\right). \tag{10}$$

Sketch of the proof: The proof of (9) uses the fact that $a * (b \wedge c) = (a * b) \wedge (a * c)$ and the proof of (10) uses the fact that $a \to (b \wedge c) = (a \to b) \wedge (a \to c)$. □

Lemma 3.2. *For arbitrary antecedent fuzzy set $A_i \in \mathcal{F}(U_1 \times \cdots \times U_m)$*

$$A_i \circ \hat{R}^{\uparrow} \subseteq B_i^{\uparrow} \text{ and } A_i \circ \hat{R}^{\downarrow} \subseteq B_i^{\downarrow}.$$

Moreover, if A_i is normal then

$$A_i \lhd \hat{R}^{\uparrow} \subseteq B_i^{\uparrow} \text{ and } A_i \lhd \hat{R}^{\downarrow} \subseteq B_i^{\downarrow}.$$

Sketch of the proof: The proof uses the monotonicity of $*$ and the fact that $a * (a \to b) \leq b$. For $\lhd$ the proof follows from the isotonicity of $\to$, the fact that and $a \to (b \to c) = (a * b) \to c$ and the normality. □

Lemmata 3.1-3.2 hold generally for any dimension and moreover, they do not impose any conditions, e.g., on the convexity of involved fuzzy sets or on the monotonicity of the modelled fuzzy rule base. If we add some of these conditions, we may get further useful results.

Corollary 3.1. *Let B_i be convex and let A_i be normal for any i. Then*

$$A_i \,@\, \hat{R}^{\updownarrow} \subseteq B_i, \quad \text{where } @ \in \{\circ, \lhd\}. \tag{11}$$

Sketch of the proof: The proof is obtained by employing Lemma 3.1, Lemma 3.2 and finally the fact that $B_i^{\uparrow} \cap B_i^{\downarrow} = \mathrm{CVX}(B_i)$ where $\mathrm{CVX}(B_i)$ denotes the convex hull of B_i. □

In the case of $@ \equiv \circ$ we may also prove the opposite inequality to (11) for monotone fuzzy rule bases with valid finitary condition. It is formulated in the following result.

Theorem 3.1. *Let the fuzzy rule base (7) be monotone. Let the antecedents A_{i_k} fulfill the finitary condition for any $k \in \{1, \ldots, m\}$. Then*

$$A_i \circ \hat{R}^{\uparrow} \supseteq B_i^{\uparrow} \text{ and } A_i \circ \hat{R}^{\downarrow} \supseteq B_i^{\downarrow},$$

where $i = (i_1, \ldots, i_m)$ such that $1 \leq i_k \leq n_k$ for any $k \in \{1, \ldots, m\}$.

Sketch of the proof: The proof is a complete analysis of all possible situations and is based on the normality of antecedent fuzzy sets (that is assumed as we require monotonicity of a fuzzy rule base) and the properties of ATL and ATM modifiers. □

This immediately leads to the following corollary that is the crucial result for viewing $\hat{R}^{\updownarrow}$ as a proper model of a monotone fuzzy rule base if we use the $\circ$ inference mechanism.

Corollary 3.2. *Let the fuzzy rule base (7) be monotone. Let the antecedents A_{i_k} fulfill the finitary condition for any $k \in \{1, \ldots, m\}$. Then*

$$A_i \circ \hat{R}^{\updownarrow} = B_i$$

where $i = (i_1, \ldots, i_m)$ such that $1 \leq i_k \leq n_k$ for any $k \in \{1, \ldots, m\}$.

A similar result holds for $\lhd$ if we narrow our focus on a restricted class of t-norms for which $a * (1 - a) = 0$ holds for arbitrary $a \in (0, 1)$.

4. Conclusion

We have described our motivation stemming from the crucial mathematical properties of fuzzy inference systems in detail. The suggested ATL-ATM fuzzy rule base and its model $\hat{R}^{\updownarrow}$ may be safely used as a model of an original monotone fuzzy rule base because it helps to ensure not only the monotonicity of a resulting function (when using MOM defuzzification) but it is also a solution to the adjoint system of fuzzy relation equations. Let us stress that the finitary conditions played a crucial role in our investigation.

Acknowledgments

This paper has been elaborated in the framework of the IT4Innovations Centre of Excellence project, reg. no. CZ.1.05/1.1.00/02.0070 supported by Operational Programme 'Research and Development for Innovations' funded by Structural Funds of the European Union and state budget of the Czech Republic.

References

1. M. Baczyński and B. Jayaram, *Fuzzy Implications*, Vol. 231, Studies in Fuzziness and Soft Computing, (Springer-Verlag, Heidelberg 2008).
2. W. Bandler and L.J. Kohout, Semantics of implication operators and fuzzy relational products. *Internat. J. Man-Mach. Stud.*, **12**,89–116, (1979).

3. L. Běhounek and M. Daňková, Relational compositions in Fuzzy Class Theory, *Fuzzy Sets and Systems*, **160**, 1005–1036, (2009).
4. U. Bodenhofer, Orderings of fuzzy sets based on fuzzy orderings. Part I: the basic approach, *Mathware & soft computing*, **15**, 201–218, (2008).
5. A. Di Nola, S. Sessa, W. Pedrycz and E. Sanchez, *Fuzzy Relational Equations and their Applications to Knowledge Engineering*, (Kluwer, Boston, 1989).
6. B. De Baets, Analytical solution methods for fuzzy relation equations *The Handbook of Fuzzy Set Series Vol. 1.*, D. Dubois and H. Prade Eds., 291–340, (Kluwer, Boston 2000).
7. F. Klawonn and J.L. Castro, Similarity in fuzzy reasoning, *Mathware and Soft Computing*, **2**, 197–228, (1995).
8. B. Moser and M. Navara, Fuzzy controllers with conditionally firing rules, *IEEE Trans. Fuzzy Systems*, **10**, 340–348, (2002).
9. V. Novák, I. Perfilieva and J. Močkoř, *Mathematical Prinicples of Fuzzy Logic*, (Kluwer, Boston 1999).
10. W. Pedrycz, Application of fuzzy relational equations for methods of reasoning in presence of fuzzy data, *Fuzzy Sets and Systems*, **16**, 163–175, (1985).
11. I. Perfilieva, Finitary Solvability Conditions for Systems of Fuzzy Relation Equations, *Information Sciences*, doi: 10.1016/j.ins.2011.04.035, in press.
12. I. Perfilieva and L. Nosková Systems of fuzzy relation equation with $inf \rightarrow$ composition: complete set of solution, *Fuzzy Sets and Systems*, **159**, 2256–2271,(2008).
13. U. Bodenhofer, M. De Cock and E.E. Kerre, Openings and closures of fuzzy preorderings: theoretical basics and applications to fuzzy rule-based systems, *International Journal of General Systems*, **32**, 343-360, (2003).
14. M. Štěpnička and B. De Baets, Monotonicity of implicative fuzzy models, in *Proc. of the FUZZ-IEEE 2010*, 2334-2340, (Barcelona, Spain, 2010).
15. M. Štěpnička and B. De Baets, Implication-based models of monotone fuzzy rule bases, *Fuzzy Sets and Systems*, submitted.
16. M. Štěpnička and B. De Baets, Interpolativity of at-least and at-most models of monotone single-input single-output fuzzy rule bases, *Information Sciences*, doi:10.1016/j.ins.2011.03.005.
17. M. Štěpnička, B. De Baets and L. Nosková, Arithmetic fuzzy models, *IEEE Trans. on Fuzzy Syst.*, **18**, 1058–1069, (2010).
18. M. Štěpnička, U. Bodenhofer, M. Daňková and V. Novák, Continuity issues of the implicational interpretation of fuzzy rules, *Fuzzy Sets and Systems*, **161**, 1959–1972, (2010).
19. M. Štěpnička and B. Jayaram, On the suitability of the Bandler-Kohout subproduct as an inference mechanism, *IEEE Trans. on Fuzzy Syst.* **18**, 285–298, (2010).
20. E. Van Broekhoven and B. De Baets, Monotone MamdaniAssilian models under mean of maxima defuzzication, *Fuzzy Sets and Systems*, **159**, 2819-2844, (2008).
21. E. Van Broekhoven and B. De Baets, Only smooth rule bases can generate monotone Mamdani-Assilian models under center-of-gravity defuzzifcation, *IEEE Trans. on Fuzzy Syst.*, **17**, 1157-1174, (2009).

A METHOD FOR SOLVING FUZZY RELATION EQUATIONS IN COMPLETE BROUWERIAN LATTICES*

XIAOBING QU

College of Mathematics and Information Science, Leshan Normal University, Leshan, Sichuan 614004, People's Republic of China

XUEPING WANG

College of Mathematics and Software Science, Sichuan Normal University, Chengdu, Sichuan 610066, P.R. China

FENG SUN

College of Mathematics and Information Science, Leshan Normal University, Leshan, Sichuan 614004, People's Republic of China

In this paper, we give a new method for solving a system of fuzzy relation equations. First, we get a derived equation from a system of fuzzy relation equations. Then the relationship between the solution sets of a system of fuzzy relation equations and its corresponding derived equation is investigated, and some necessary and sufficient conditions for the solution sets of these equations to be equal are presented.

1. Introduction

The theory of fuzzy relation equations has been intensively investigated both from the point of view of theory and applications (see [2,10]) after Sanchez [14] first introduced it. It plays important roles in many applications, such as intelligence technology [2], fuzzy inference systems [5], coding/decoding images and videos [4,6], and so on. Sanchez [14] showed that the solution set of fuzzy relation equations is a join-semilattice with the greatest solution if it is not empty. After Sanchez results, many works about solving fuzzy relation equations were published (see [1,3,8,9,11,12,13,15,16,17]).

A generic version of fuzzy relation equations arises in the form

* This work is supported by the National Natural Science Foundation of China(No. 11171242), Doctoral Fund of Ministry of Education of China (No. 20105134110002) and the Basic Research Programs of Sichuan Province of China (No.2011JY0133

$$A \odot X = B \quad or \quad \bigvee_{j \in J} (a_{ij} \wedge x_j) = b_i \tag{1}$$

for all $i \in I$. In the above expression, $A = (a_{ij})_{I \times J}$ is a coefficient matrix, $B = (b_i)_{i \in I}$ is a constant column vector (I and J are index sets), $\odot$ denotes the sup-inf composite operation, and all b_i, x_j, a_{ij}'s are in a complete Brouwerian lattice L. An X which satisfies Eq.(1) is called a solution of (1), the solution set of Eq.(1) is denoted by $\mathcal{X} = \{X : A \odot X = B\}$. Moreover, if $\mathcal{X} \neq \varnothing$, then it has the greatest element denoted by X^*, and $X^* = (\bigwedge_{i \in I} (a_{ij} \alpha b_i))_{j \in J}$.

Set $\mathcal{X}^0 = \{X \in \mathcal{X} : X$ is a minimal element of $\mathcal{X}\}$.

For Eq.(1), there are many methods to describe its solution set. In this paper, from Eq.(1), we get a derived equation, and discuss the relationship between the solution sets of Eq.(1) and its derived equation. Furthermore, some necessary and sufficient conditions for the solution sets of these equations to be equal are presented.

This paper is organized as follows. In section 2, some preliminary results on fuzzy relation equations and lattices are given. Section 3 is due to describe the relationship between the solution sets of Eq.(1) and its corresponding derived equation on complete Brouwerian lattices. Conclusions are given in Section 4.

2. Preliminaries

In this section, we give some previous results for the sake of convenience.

Lemma 2.1 [2] Let X_1, $X_2 \in \mathcal{X}$ and $X_1 \leq X_2$. Then $X \in \mathcal{X}$ for any X such that $X_1 \leq X \leq X_2$.

Lemma 2.2 [17] Equation $\bigvee_{j \in J} (a_j \wedge x_j) = b$ has a solution iff $b \leq \bigvee_{j \in J} a_j$.

Lemma 2.3 [15] Let $X_* = (x_{j*})_{j \in J}$ be a minimal solution of equation $\bigvee_{j \in J} (a_j \wedge x_j) = b$. Then $b = \bigvee_{j \in J} x_{j*}$, and for all $j \in J$, $x_{j*} \leq a_j$.

Lemma 2.4[7] Let L be a lattice, a, b, $c \in L$. Then

(i) $(a \wedge b) \alpha c = a \alpha (b \alpha c)$. In particular, $a \alpha (a \alpha b) = a \alpha b$.

(ii) if $a \leq b$, then $a \alpha c \geq b \alpha c$, $c \alpha a \leq c \alpha b$.

Let L be a lattice. For any $a \in L$, denote $M_a = \{S \subseteq L : a = \vee S$, *and for all* $s \in S$, *there is no element* $t \in L, t < s$ *such that* $a = \vee(S - \{s\}) \vee t\}$.

3. Main results

In Eq.(1), denote $\bigvee_{i\in I} b_i = b$, $\bigvee_{i\in I} a_{ij} = a_j$ for all $j \in J$, $Y = \{X = (x_j)_{j\in J}$: $b = \bigvee_{j\in J}(a_j \wedge x_j)\}$ and $Y^0 = \{X \in Y : X$ is a minimal element of $Y\}$. If $Y \neq \varnothing$, then $Y^* = (a_j \alpha b)_{j\in J} = ((\bigvee_{i\in I} a_{ij})\alpha(\bigvee_{i\in I} b_i))_{j\in J}$ is the greatest element of Y. It is easy to see that for all $X = (x_j)_{j\in J} \in X$, we have $X \in Y$, then $X \subseteq Y$. It follows that $X^* \leq Y^*$. If $X \neq \varnothing$, then $Y \neq \varnothing$, but the inverse is not true. In this section, we will discuss the relation between X and Y. Denote $G(b) = \{j \in J : a_j \geq b\}$.

First we suppose that $|J| = 1$, and discuss the relation between X and Y. Consider the system of equations $a_{i1} \wedge x = b_i$, $i \in I$, and let $\bigvee a_{i1} = a_1$.

Lemma 3.1 [3] $X \neq \varnothing$ iff for all i, $k \in I$, $a_{i1} \geq b_i$, and $a_i \wedge b_k = b_i \wedge b_k$.

Theorem 3.1 If $X \neq \varnothing$, then $X = Y$.

Proof. We only need to show that $Y \subseteq X$. Since $X \neq \varnothing$, it follows that $a_{i1} \geq b_i$ for all $i \in I$, and $a_{i1} \wedge b_k = b_i \wedge b_k$ for all $k \in I$. Let $x \in Y$. Then $x \geq b = \bigvee_{i\in I} b_i$, and $x \geq b_i$ for all $i \in I$.

By $a_1 \wedge x = (\bigvee_{i\in I} a_{i1}) \wedge x = \bigvee_{i\in I}(a_{i1} \wedge x) = b = \bigvee_{i\in I} b_i$ we have for all $k \in I$,

$$a_{k1} \wedge x = (a_{k1} \wedge x) \wedge \bigvee_{i\in I} b_i = \bigvee_{i\in I}((a_{k1} \wedge x) \wedge b_i)$$

$$= (a_{k1} \wedge x \wedge b_k) \vee \bigvee_{i\in I, i\neq k}(a_{k1} \wedge x \wedge b_i)$$

$$= b_k \vee \bigvee_{i\in I, i\neq k}(a_{k1} \wedge b_i) = b_k \vee \bigvee_{i\in I, i\neq k}(b_k \wedge b_i) = b_k .$$

So $x \in X_i$ for all $i \in I$, that is, $x \in X$. Then $Y \subseteq X$. Therefore, $X = Y$.

Remark 3.1 If $X = \varnothing$, then $X = Y$ need not be true. For example, in lattice $L = [0,1]$, let $a_{11} = 0.6$, $a_{21} = 0.8$, $b_1 = 0.5$, and $b_2 = 0.7$. Then $a_1 = a_{11} \vee a_{21} = 0.8$ and $b = b_1 \vee b_2 = 0.7$. Obviously, $Y = \{0.7\} \neq \varnothing$, but $X = \varnothing$.

If $|J| > 1$, then Theorem 3.1 fails, as the following example demonstrates:

Example 3.1 In lattice $L = [0,1]$, let $a_{11} = 0.6$, $a_{12} = 0.2$, $a_{21} = 0.8$, $a_{22} = 0.8$, $b_1 = 0.5$ and $b_2 = 0.7$. Then $a_1 = a_{11} \vee a_{21} = 0.8$, $a_2 = a_{12} \vee a_{22} = 0.8$ and $b = b_1 \vee b_2 = 0.7$. Obviously, $X^* = (0.5, 0.7) \in X$, i.e., $X = \varnothing$, but $Y^* = (0.7, 0.7) \notin X$. So $X \neq Y$.

Now we discuss the relation between X and Y under the general case ($|J| \geq 1$).

By Lemma 2.3 and the definition of minimal solutions, we have the following two theorems.

Theorem 3.2 If $X^0 \neq \varnothing$, $X=(x_j)_{j\in J} \in X^0$, then $X \in Y^0$ iff $\{x_j : j \in J_0\} \in M_b$.

Theorem 3.3 If $Y^0 \neq \varnothing$, $Y=(y_j)_{j\in J} \in Y^0$, then $Y \in X^0$ iff $Y \in X$.

Theorem 3.4 If $X = \varnothing$, then $X^* = Y^*$ iff $X=(b)_{j\in J} \in X$.

Proof. Assume that $X^* = Y^*$, for $X = \varnothing$, we have $Y \neq \varnothing$ and $X=(b)_{j\in J} \in Y$. Then $X \le Y^* = X^*$. So $X_i = (b_i)_{j\in J} \le X \le X^* \le X_i^*$ for all $i \in I$, i.e., $X \in X_i$ for all $i \in I$. Then $X \in X$.

Conversely. If $X=(b)_{j\in J} \in X$, then $X \le X^*$, that is, $b = \bigvee_{i\in I} b_i \le \bigwedge_{i\in I}(a_{ij}\alpha b_i)$ for all $j \in J$. Thus $\bigvee_{i\in I} b_i \le a_{kj}\alpha b_k$ for all $k \in I$. Since for all $j\in J$, $k \in I$,

$$(\bigvee_{i\in I} a_{ij})\alpha(\bigvee_{i\in I} b_i) \le a_{kj}\alpha(\bigvee_{i\in I} b_i) \le a_{kj}\alpha(a_{kj}\alpha b_k) = a_{kj}\alpha b_k,$$

we have $(\bigvee_{i\in I} a_{ij})\alpha(\bigvee_{i\in I} b_i) \le \bigwedge_{i\in I}(a_{kj}\alpha b_k)$ for all $j \in J$. For the fact that $X^* \le Y^*$, we have $\bigwedge_{i\in I}(a_{kj}\alpha b_k) \le (\bigvee_{i\in I} a_{ij})\alpha(\bigvee_{i\in I} b_i)$ for all $j \in J$. So $(\bigvee_{i\in I} a_{ij})\alpha(\bigvee_{i\in I} b_i) = \bigwedge_{i\in I}(a_{kj}\alpha b_k)$ for all $j \in J$. Therefore, $X^* = Y^*$.

It is easy to show that if for any $X \in Y$, there exists some solution $X_* \in Y^0$ such that $X_* \le X$, then $X = Y$ iff $X^* = Y^*$, and $Y^0 \subseteq X$. In particular, if L is a complete linear lattice, we have the following conclusions:

Theorem 3.5 Let L be a complete linear lattice. If $X = \varnothing$, then $X = Y$ iff the following conditions hold:

(i) $X = (b)_{j\in J} \in X$;

(ii) for all $k \in G(b)$, $i \in I$, if $b_i < b$, then $a_{ik} = b_i$; if $b_i = b$, then $a_{ik} \ge b$.

Proof. Straightforwardly, by the construct of element in Y^0.

Example3.2 Let $L = [0,1]$. Consider a matrix equation $A \odot X = B$, where

$$A = \begin{pmatrix} 0.6 & 0.5 & 0.6 \\ 0.8 & 0.4 & 0.7 \end{pmatrix}, B = \begin{pmatrix} 0.6 \\ 0.7 \end{pmatrix}.$$

Then its derived equation is $(0.8 \wedge x_1) \vee (0.5 \wedge x_2) \vee (0.7 \wedge x_3) = 0.7$. Obviously, $X^* = (0.7,1,1) = Y^* \in X$, and $X^0 = \{(0.7,0,0),(0,0,0.7)\} = Y^0$. So $X = Y$.

Theorem 3.6 Let L be a complete linear lattice. If $X = (x_j)_{j\in J} \in X^0$, then $X \in Y^0$ iff $|J_0| = 1$.

Proof. The necessity is easy to show. Now we show the sufficiency. Assume that $|J_0|=1$, let $J_0=\{k\}$. Then for all $i\in I$, $a_{ik}\wedge x_k=b_i$, and $x_k\geq b_i$. So $x_k\geq\bigvee_{i\in I}b_i$. Define $Y=(y_j)_{j\in J}$ as follows, for all $j\in J$, if $j=k$, set $y_k=\bigvee_{i\in I}b_i$; if $y\neq k$, set $y_j=0$. Then $Y\in X$, and $Y\leq X$. Since $X\in X^0$ we have $Y=X$. Hence, $x_k=\bigvee_{i\in I}b_i$. Therefore, $X\in Y^0$.

In the following, we assume that each element has irredundant completely join irreducible decomposition (or finite irredundant join irreducible decomposition) in complete Brouwer lattice L. Now suppose that b_i has an irredundant completely join irreducible decomposition $b_i=\vee Q_i$ for all $i\in I$, and b has an irredundant completely join irreducible decomposition $b=\vee P$. Set $Q=\bigcup_{i\in I}Q_i$, and $Q^*=\{q\in Q: q$ is a maximal element of $Q\}$. Then $P=Q^*$. First, for all $p\in P$, $p\leq b=\bigvee_{i\in I}b_i$, then there is an element $k\in I$ such that $p\leq b_k=\vee Q_k$. Thus there exists an element $q\in Q_k$ such that $p\leq q$. Since $q\leq b_k\leq b=\vee P$, it follows that there is $p_0\in P$ such that $q\leq p_0$. So $p\leq q\leq p_0$. Then $p=q=p_0$ for $b=\vee P$ is irredundant, that is, $p=q\in Q_k\subseteq Q$. Therefore, $P\subseteq Q$. Furthermore, if there exists an element $q_0\in Q$ such that $q_0\geq p$, then there is an element $p_1\in P$ such that $q_0\leq p_1$ by $q_0\leq b=\vee P$. Thus $q_0=p$ for $b=\vee P$ is irredundant, that is, $p\in Q^*$. So $P\subseteq Q^*$. In the other case, let $q_1\in Q^*$. By $q_1\leq b=\vee P$ and $b=\vee P$ is irredundant, there is an element $p_2\in P\subseteq Q$ such that $q_1\leq p_2$. Then $q_1=p_2$ for $q_1\in Q^*$, therefore $q_1\in P$. Thus $Q^*\subseteq P$. So $P=Q^*$. In reference [15], if $Y_*=(y_{j*})_{j\in J}\in Y^0$, then there is a partition P_j, $j\in J_0$, of P such that $y_{j*}=\vee P_j$, and $\vee P_j\leq a_j$. If $X\neq\varnothing$, then for all $i\in I$, $q\in Q_i$, there is a $j\in J$ with $q\leq a_{ij}$, and $p\in P$ with $q\leq p$ by $b_i\leq b$. For each $i\in I$, $j\in J$, $p\in P$, set $A_{ij}=\{q\in Q_i: q\leq a_{ij}\}$, $S_p^i=\{q\in Q_i: q\leq p\}$.

Theorem 3.7 If $X=\varnothing$, $X=(x_j)_{j\in J}\in X^0$, then $X\in Y^0$ if and only if there is a partition P_j, $j\in J_0$, of P such that $x_j=\vee P_j$ and $\vee P_j\leq a_j$ for all $j\in J_0$.

Proof. Straightforwardly, by the construct of element in Y^0.

Theorem 3.8 If $X = \varnothing$, $Y = (y_j)_{j\in J} \in Y^0$, $Y \le X^*$, then $Y \in X$ if and only if for the partition P_j, $j \in J_0$, of P, where $y_j = \vee P_j$, we have $\bigcup_{j\in J_0} (A_{ij} \cap (\bigcup_{p\in P_j} S_p^i)) = Q_i$ for all $i \in I$.

Proof. Since $Y = (y_j)_{j\in J} \in Y^0$, by the construct of element in Y^0, there is a partition P_j, $j \in J_0$, of P such that $y_j = \vee P_j$ and $\vee P_j \le a_j$ for all $j \in J_0$. By $Y \in X$, we have $\bigvee_{j\in J} (a_{ij} \wedge y_j) = b_i$ for all $i \in I$. Then there exists a $k \in J_0$ with $q \le a_{ik} \wedge y_k$ for all $q \in Q_i$. Thus $q \le a_{ik}$ and $q \le y_k = \vee P_k$. Moreover, there is a $p \in P_k$ such that $q \le p$, that is, $q \in S_p^i$. So $q \in A_{ik} \cap (\bigcup_{p\in P_k} S_p^i)$. Therefore, $Q_i \subseteq \bigcup_{j\in J_0} (A_{ij} \cap (\bigcup_{p\in P_j} S_p^i))$. It is easy to see that $\bigcup_{j\in J_0} (A_{ij} \cap (\bigcup_{p\in P_j} S_p^i)) \subseteq Q_i$. Then $\bigcup_{j\in J_0} (A_{ij} \cap (\bigcup_{p\in P_j} S_p^i)) = Q_i$.

Conversely. Since $\bigcup_{j\in J_0} (A_{ij} \cap (\bigcup_{p\in P_j} S_p^i)) = Q_i$ for all $i \in I$, it follows that there is a $j \in J_0$ such that $q \in A_{ij} \cap (\bigcup_{p\in P_j} S_p^i)$ for all $q \in Q_i$. Then $q \in A_{ij}$, $q \in \bigcup_{p\in P_j} S_p^i$, that is, $q \le a_{ij}$, $q \vee P_j = y_j$. Thus $q \le a_{ij} \wedge y_j$. By $Y \le X^*$, we have

$$b_i = \vee Q_i \le \bigvee_{j\in J} (a_{ij} \wedge y_j) \le \bigvee_{j\in J} (a_{ij} \wedge x_j^*) = b_i$$

for all $i \in I$, i.e., $\bigvee_{j\in J} (a_{ij} \wedge y_j) = b_i$. So $Y \in X$.

Theorem 3.9 If $X = \varnothing$, then $X = Y$ if and only if the following conditions hold:

(i) $X = (b)_{j\in J} \in X$;

(ii) for any partition P_j, $j \in J_0 \subseteq J$, of P, where $\vee P_j \le a_j$ for all $j \in J_0$, we have $\bigcup_{j\in J_0} (A_{ij} \cap (\bigcup_{p\in P_j} S_p^i)) = Q_i$ for all $i \in I$.

Example 3.3 Let $L = [0,1]^2$. Consider a matrix equation $A \odot X = B$, where

$$A = \begin{pmatrix} \langle 0.8, 0.4\rangle & \langle 0.3, 0.4\rangle \\ \langle 0.7, 0.4\rangle & \langle 0.5, 0.5\rangle \end{pmatrix}, B = \begin{pmatrix} \langle 0.8, 0.4\rangle \\ \langle 0.7, 0.5\rangle \end{pmatrix}.$$

Then its derived equation is $(\langle 0.8,0.4\rangle \wedge x_1) \vee (\langle 0.5,0.5\rangle \wedge x_2) = \langle 0.8,0.5\rangle$. Obviously, $X^* = (\langle 1.0,1.0\rangle, \langle 1.0,1.0\rangle) = Y^* \in X$, and $X^0 = \{(\langle 0.8,0.0\rangle, \langle 0.0,0.5\rangle)\} = Y^0$. So $X = Y$.

4. Conclusion

In this paper we investigated the relationship between the solution sets of a system of fuzzy relation equations and its corresponding derived equation. Under some special conditions, the solution sets of these equations are equal. So under these cases, we only need to solve a single equation for solving a system of fuzzy relation equations.

References

1. B. De Baets, An order-theoretic approach to solving sup-T equations, in: D. Ruan (Ed.), Fuzzy Set Theory and Advanced Mathematical Applications, Kluwer Academic Publishers, Dordrecht, pp. 67-87 (1995). 40.
2. A. Di Nola, S. Sessa, W. Pedrycz and E. Sanchez, Fuzzy Relation Equations and Their Applications to Knowledge Engineering, Kluwer Academic Publishers, Dordrecht, Boston/London, 1989.
3. S. C. Han, H. X. Li, J. Y. Wang, Fuzzy Sets and Systems 159, 40-46 (2008).
4. K. Hirota, W. Pedrycz, Fuzzy Sets and Systems 126, 325-335 (2002).
5. G. J. Klir, B. Yuan, Fuzzy sets and fuzzy logic: theory and applications, Prentice Hall, 1995.
6. V. Loia, S. Sessa, Information Sciences 171, 145-172 (2005).
7. Z. Ma and W. M. Wu, Inform. Sci. 55, 77-97 (1991).
8. C. Miroslav, S. Aleksandar, I. Jelena, P. Tatjana, Journal of Computer and System Sciences 76, 609-633 (2010).
9. L. Nosková, I. Perfilieva, System of fuzzy relation equations with sup-* composition in semi-linear spaces: minimal solutions, In: Proc. Fuzz-IEEE Conf. on Fuzzy Systems, London, July 23-36, pp. 1520-1525 (2007).
10. K. Peeva, Y. Kyosev, Fuzzy Relational Calculus: Theory, Applications and Software, World Scientific Publishing Company, 2004.
11. I. Perfilieva, V. Novák, Information Sciences 177, 3218-3227 (2007).
12. X. B. Qu, X. P. Wang, Chinese Annals of Mathematics 30, 811-828 (2009).
13. X. B. Qu, X. P. Wang, Fuzzy Sets and Systems 158, 1327 -1339 (2007).
14. E. Sanchez, Information and Control 30, 38-48 (1976).
15. X. P. Wang, Fuzzy Sets and Systems 138, 657-666 (2003).
16. X. P. Wang, X. B. Qu, Science China: Mathematics, 53, 2159-2172 (2010).
17. C. K. Zhao, Fuzzy Sets and Systems 22, 303-320 (1987).

FUZZY TRANSFORM FOR IMAGE RECONSTRUCTION

IRINA PERFILIEVA[1], PAVEL VLAŠÁNEK[2], MICHAELA WRUBLOVÁ[1]
[1]*Centre of Excellence IT4Innovations, Division of the University of Ostrava*
Institute for Research and Applications of Fuzzy Modeling
30. dubna 22, 701 03 Ostrava 1, Czech Republic
[2]*Faculty of Science, Department of Informatics and Computers, University of Ostrava*
30. dubna 22, 701 03 Ostrava, Czech Republic

We propose a new image reconstruction technique which uses approximation properties of the fuzzy (F-)transform and its ability to remove noise. The proposed technique is based on the inverse F-transform and on a fusion which combines the inverse F-transform with the original image on a non-damaged area. We claim the new technique has less computation complexity than the usually applied technique of interpolation.

1. Introduction

The goal of this paper is to show that the inverse F-transform can be used for image reconstruction. By image reconstruction we mean a reconstruction of a damaged image where the damage is everything that the original image does not include. It can be noise, text, scratch, etc. We propose to solve the problem of reconstruction with the help of an approximation technique. This means that we will be looking for an approximating image which is close to the given one and at the same time, does not contain what we recognize as damage.

The problem of approximation is as follows. We consider a metric space with the given metric d and an element f in it. We are looking for an element g in the same metric space such that the distance $d(f,g)$ is the smallest possible. The element g is called the approximation of the element f. For more information see [5].

We propose to reconstruct a damaged image with help of a fuzzy technique, namely the F-transform. We recall that the inverse F-transform approximates the original function with an arbitrary precision [1,3] and by this, can be chosen as an appropriate technique. The other reason is that the F-transform has various applications in image processing including the remove of noise, see e.g., [1-4]. The most relevant application of the F-transform to the above discussed problem is image fusion [2].

The problem of reconstruction is different from the problem of fusion. We assume that only one damaged image is given together with the information which allows to distinguish damaged and non-damaged pixels. The goal is to replace the damaged pixels. There exist solutions to this problem which are based on the interpolation technique [6]. These solutions are at the same time solutions of large systems of linear equations. Therefore, the complexity of this approach is rather high. We propose another approach which consists in applying the F-transform. Our reason is that the F-transform successively combines approximation with noise reduction and moreover, it does not require another numeric methods, i.e., the complexity of this approach is low.

2. Approximation Problem

Let us give some details about the problem of *approximation*, see [5].

Let (X,d) be a metric space and $U \subseteq X$. Both sets X, U are equipped with the metric (distance) d. Given $g \in U$ and $f \in X$, the value $d(g,f)$ determines the deviation $g \in U$ from $f \in X$. The *problem of approximation* is to find an element g from U which is close to (approximates) given $f \in X$ in the sense of the distance d.

If we want to find $g^* \in U$ such that the distance $d(f,g^*)$ is the least possible then the $g^* \in U$ is called the best approximation to $f \in X$ so that

$$d(f,g^*) = \inf\{d(f,g) : g \in U\}.$$

In [5] there are theorems related to existence and uniqueness of the best approximation.

We recall some of the approximation methods: least squares approximation, approximation to periodic functions (trigonometric polynomials), etc. (more details in [5]).

3. The F-transform of Functions of Two Variables

The F-transform is a technique which brings a discrete function of one variable into the space of n dimensional vectors. At the beginning, we recall the definition of a fuzzy partition [3]. Next, we introduce the discrete F-transform of functions of two variables [1,3].

Let us remind that a fuzzy set on $[a,b]$ is identified with its membership function which is a mapping from $[a,b]$ to $[0,1]$

Definition 3.1. Let $x_1 < \ldots < x_n$ be fixed nodes within $[a,b]$, such that $x_1 = a$, $x_n = b$ and $n \geq 2$. We say that fuzzy sets $A_1, \ldots, A_n$, identified with their membership functions $A_1,\dots,A_n$, defined on $[a,b]$ form a *fuzzy partition* of $[a,b]$ if they fulfill the following conditions for $k=1,\dots,n$:

i. $A_k:[a,b]\to[0,1],\ A_k(x_k)=1$;

ii. $A_k(x)=0$ if $x\notin(x_{k-1},x_{k+1})$ where for the uniformity of denotation, we put $x_0=a$ and $x_{n+1}=b$;

iii. $A_k(x)$ is continuous;

iv. $A_k(x),\ k=2,\dots,n,$ strictly increases on $[x_{k-1},x_k]$ and $A_k(x)$, $k=1,\dots,n-1,$ strictly decreases on $[x_k,x_{k+1}]$

v. for all $x\in[a,b]$

$$\sum_{k=1}^{n}A_k(x)=1. \tag{1}$$

The membership functions $A_1,\dots,A_n$ are called *basic functions*

In the sequel, we will use a uniform fuzzy partition, more details are in [1–4].

First of all, let $A_1,\dots,A_n$ and $B_1,\dots,B_m$ be basic functions, $A_1,\dots,A_n:[a,b]\to[0,1]$ be a fuzzy partition of $[a,b]$ and $B_1,\dots,B_m$: $[c,d]\to[0,1]$ be a fuzzy partition of $[c,d]$ Let a function $f:[a,b]\times[c,d]\to\mathbb{R}$ be given at points $(p_i,q_j)\in[a,b]\times[c,d]$ where $i=1,\dots,N,\ n<N$, and $j=1,\dots,M,\ m<M$. The set P of nodes $\{p_1,\dots,p_N\}$ and the set Q of nodes $\{q_1,\dots,q_M\}$ are *sufficiently dense with respect to the chosen partitions*. This means that $(\forall k)(\exists i)\ A_k(p_i)>0$ and $(\forall l)(\exists j)\ B_l(q_j)>0$

We say that the $n\times m$-matrix of real numbers $[F_{kl}]$ is called *the (discrete) F-transform* of f with respect to $\{A_1,\dots,A_n\}$ and $\{B_1,\dots,B_m\}$ if for all $k=1,\dots,n,\ l=1,\dots,m,$

$$F_{kl}=\frac{\sum_{j=1}^{M}\sum_{i=1}^{N}f(p_i,q_j)A_k(p_i)B_l(q_j)}{\sum_{j=1}^{M}\sum_{i=1}^{N}A_k(p_i)B_l(q_j)}. \tag{2}$$

The elements F_{kl} are called *components of the F-transform*.

The *inverse F-transform* of the function f with respect to $\{A_1,\dots,A_n\}$ and $\{B_1,\dots,B_m\}$ is defined as follows:

$$f_{nm}^{F}(p_i,q_j)=\sum_{k=1}^{n}\sum_{l=1}^{m}F_{kl}A_k(p_i)B_l(q_j). \tag{3}$$

The function f_{nm}^{F} approximates the original function at nodes $(p_i, q_j) \in [a,b] \times [c,d]$ with a given precision [1,3].

4. Experiments

Our purpose is to reconstruct a damaged image where the damage is in a restricted area which can be separated from a non-damaged part (it can be done manually with an available precision). The damage can be caused by noise, text, scratch, etc. We assume that the separation into damaged and non-damaged areas is realized in the form of a binary mask. The latter can characterize even larger area than that containing damaged pixels. The proposed reconstruction is based on two techniques: the F-transform, namely the inverse F-transform, and the technique of fusion which combines the inverse F-transform with the original image on a non-damaged area. The F-transform is chosen for reconstruction because it approximates the original image [1,3] and removes noise [4]. We distinguish two cases according to a size of a damaged area.

The original image is heavily damaged (more that 50%), e.g., by noise. In this case, the reconstructed (approximating) image is given by the inverse F-transform without fusion.

The original image is not heavily damaged. Then the algorithm of reconstruction is based on the combination of the inverse F-transform with fusion. In details: if a pixel of the original image is undamaged then it is chosen for the reconstructed image, otherwise the value of intensity function at this pixel is replaced by the value of the inverse F-transform.

We show three examples of damaged images and their reconstructions by the proposed technique.

The easiest way how to measure the quality of reconstruction is to compare the reconstructed image with a non-damaged image (we assume that the latter is at disposal). Then the quality can be measured by the Root Mean Square Error (RMSE). According to this measure, the smaller is the length of a support of a basic function in a fuzzy partition, the better is the quality of reconstruction. On the other hand, the bigger is a damaged area, the bigger length of a support of a basic function is required. In our experiments, the least length which removes a damage was chosen manually. The future research will be focused on an automatic choice of the length of a support of a basic function which removes a certain damage and has an optimal quality.

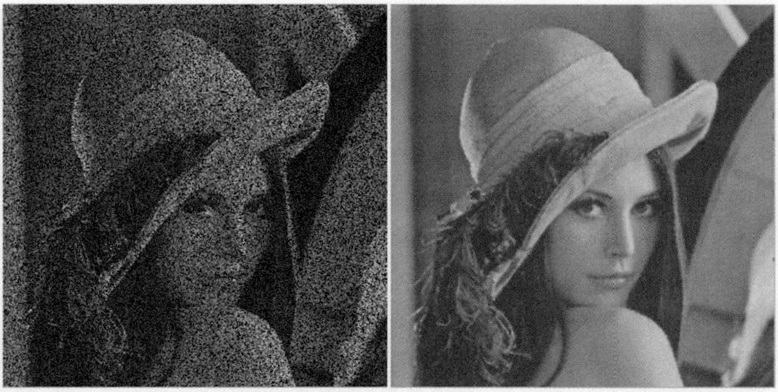

Figure 1. *Left*.The damage is caused by the noise, and the size of the damaged area is above 70%. *Right*.The reconstructed image is obtained by the inverse -transform without fusion.

Figure 2. *Left*.The damage is caused by the "text over image". *Right*.The reconstructed image is obtained by fusion which combines the inverse F-transform with the original image.

5. Conclusion

We have connected the problem of image reconstruction with the approximation problem in classical meaning. We proposed to reconstruct a damaged image with help of a fuzzy technique, namely the F-transform. Actually, the proposed reconstruction is based on two techniques: the inverse F-transform and fusion which combines the inverse F-transform with the original image on a non-damaged area. We showed confirming experimental results and discussed how a quality of reconstruction can be measured.

Figure 3. *Left*.The damage is caused by the scratches or manually painting. *Right*.The reconstructed image is obtained by fusion which combines the inverse F-transform with the original image.

The future research will be focused on an automatic choice of the length of a support of a basic function which removes a certain damage and has an optimal quality. We also plan to make a detailed comparison with other techniques which are used in this area, namely with the RBF-functions interpolation.

Acknowledgments

[1] This paper has been supported by the European Regional Development Fund in the IT4Innovations Centre of Excellence project (CZ.1.05/1.1.00/02.0070).
[2] The research has been supported by the project sgs12/PřF/2012.

References

1. F. Di Martino, V. Loia, I. Perfilieva, S. Sessa, An image coding/decoding method based on direct and inverse fuzzy transforms, *International Journal of Approximate Reasoning*, **48** 110–131, (2008).
2. I. Perfilieva, F-Transforms — Where Fuzzy and Classical Mathematics Meet, *Applied and Computational Mathematics*, **10**, 122--132, (2011).
3. I. Perfilieva, Fuzzy transforms: Theory and applications, *Fuzzy Sets and Systems*, **157,** 993–1023, (2006).
4. I. Perfilieva, R. Valášek, Fuzzy Transforms in Removing Noise, In: *Proc. Fuzzy Days*, 225--234, (Berlin, 2005).
5. M. J. D. Powell, Approximation Theory and Methods, *Cambridge University Press*, 339p., 1981.
6. K. Uhlir, V. Skala, Radial basis function use for the restoration of damaged images. In: *Computer vision and graphics*, 839–844, (Springer, 2006).

REMARKS TO MODEL THEORY IN HIGHER ORDER FUZZY LOGIC

VILÉM NOVÁK
Centre of Excellence IT4Innovations, division of the University of Ostrava
Institute for Research and Applications of Fuzzy Modeling
30. dubna 22, 702 00 Ostrava, Czech Republic

E-mail: *Vilem.Novak@osu.cz*

In this paper, we turn our attention to model theory of higher-order fuzzy logic (fuzzy type theory). This theory generalizes model theory of predicate logic but has some interesting specificities. We will introduce few basic concepts related to homomorphism, isomorphism, submodel, etc. and show some properties of them.

Keywords: fuzzy type theory; model theory; EQ-algebra

1. Introduction

Model theory in fuzzy logic is still in its beginning though few papers have already been published[1–5] (cf. also[6,7]). In this paper, we will focus on model theory in higher-order fuzzy logic. Note that this area is practically non-developed even in classical type theory (cf.[8,9]). Of course, a lot of results from classical model theory of first-order logic can be directly transferred to the latter but still there are some interesting specificities which must be taken into account. Because of lack of space, we refer the reader to all concepts from fuzzy type theory used below to the papers[10,11] and, possibly, the book.[8] This paper presents few basic concepts and results which are in more detail elaborated in.[12]

2. Models in FTT

2.1. *Few introductory concepts*

We will assume that the truth values form an EQ-algebra $\mathcal{E} = \langle E, \wedge, \otimes, \sim, \mathbf{1}\rangle$. Basic syntactical objects of FTT are classical — see,[8] namely *type* and *formula*. The atomic types are ϵ (elements) and o (truth values). General types are denoted by Greek letters $\alpha, \beta, \ldots$. The set of all types is denoted by *Types*.

The *language* of FTT, denoted by J, consists of variables $x_\alpha, \ldots$, special constants $c_\alpha, \ldots$ ($\alpha \in$ *Types*), the symbol λ, and brackets. We will consider the following concrete special constants: $\mathbf{E}_{(oo)o}, \mathbf{E}_{(o\alpha)\alpha}$, $\alpha \in$ *Types* (fuzzy equality), $\mathbf{G}_{o\epsilon}$ (generating function), $\mathbf{C}_{(oo)o}$ (conjunction), $\mathbf{S}_{(oo)o}$ (strong conjunction) $\mathbf{D}_{(oo)}$ (delta operation on truth values), and $\iota_{\epsilon(o\epsilon)}, \iota_{o(oo)}$ (description operators). Formulas are formed of variables, constants (each of specific type), and the symbol λ. Thus, each formula A is assigned a type and we write A_α. A set of formulas of type α is denoted by $Form_\alpha$. The set of all formulas is $Form = \bigcup_{\alpha \in Types} Form_\alpha$. A formula A_o of type o without free variables is called *sentence*. Each formula $A_{\beta\alpha}$ represents a function from some set M_α to M_β.

Let M be a set. A fuzzy equality $\stackrel{\circ}{=}$ on M is a binary fuzzy relation $\stackrel{\circ}{=} \in E^{M \times M}$ which is reflexive, symmetric and transitive w.r.t. the operation $\otimes$. If $m, m' \in M$ then the degree of equality between m and m' will usually be written as $[m \stackrel{\circ}{=} m']$. If objects of M are functions $M_\alpha \longrightarrow M_\beta$ then we introduce a fuzzy equality $\stackrel{\circ}{=}: M_\beta^{M_\alpha} \times M_\beta^{M_\alpha} \longrightarrow E$ which to every two functions $h, h' \in M_\beta^{M_\alpha}$ assigns a truth value of their equality given by

$$[h \stackrel{\circ}{=} h'] = \bigwedge_{m \in M_\alpha} [h(m) \stackrel{\circ}{=}_\beta h'(m)] \qquad (1)$$

where $\stackrel{\circ}{=}_\beta$ is a fuzzy equality on the set M_β.

Definition 2.1. Let $(\mathcal{E}^1, M^1, \stackrel{\circ}{=}^1)$, $(\mathcal{E}^2, M^2, \stackrel{\circ}{=}^2)$ be two sets with fuzzy equalities. We say that a couple of functions (s, f), $s : E^1 \longrightarrow E^2$ and $f : M^1 \longrightarrow M^2$ is a homomorphism between $(\mathcal{E}^1, M^1, \stackrel{\circ}{=}^1)$, $(\mathcal{E}^2, M^2, \stackrel{\circ}{=}^2)$ if the following holds:

(i) s is an algebra-homomorphism preserving all infima and such that $s(a) = \mathbf{1}^2$ iff $a = \mathbf{1}^1$.
(ii) $s([m \stackrel{\circ}{=}^1 m']) = [f(m) \stackrel{\circ}{=}^2 f(m')]$, for all $m, m' \in M^1$.

If (s, f) is a homomorphism between $(\mathcal{E}^1, M^1, \stackrel{\circ}{=}^1)$ and $(\mathcal{E}^2, M^2, \stackrel{\circ}{=}^2)$, then we will write $(s, f) : (\mathcal{E}^1, M^1, \stackrel{\circ}{=}^1) \longrightarrow (\mathcal{E}^2, M^2, \stackrel{\circ}{=}^2)$. If $\mathcal{E}^1 = \mathcal{E}^2$ then we may consider only $f : (M^1, \stackrel{\circ}{=}^1) \longrightarrow (M^2, \stackrel{\circ}{=}^2)$.

Lemma 2.1. *Let* $(s, t) : (\mathcal{E}^1, M^1, \stackrel{\circ}{=}^1) \longrightarrow (\mathcal{E}^2, M^2, \stackrel{\circ}{=}^2)$ *and* $\stackrel{\circ}{=}^1, \stackrel{\circ}{=}^2$ *be separated. Then both* s, f *are injections.*

In the rest of this paper we will consider a fixed language J of FTT. In,[9] it was proved that finite models of type theory must be standard. Using the main idea adopted from this paper we show below that similar property holds also in FTT.

Theorem 2.1. *Let $\mathcal{M}$ be a general model such that M_ϵ is finite. Let α be a type not containing o and β be a type. Then $M_{\beta\alpha} = M_\beta^{M_\alpha}$.*

Corollary 2.1. *Let $\mathcal{M}$ be a general model with finite M_ϵ as well as $\mathcal{E}$. Then it is standard.*

2.2. *Homomorphism of models*

To develop the concepts of embedding, isomorphism, elementary embedding and elementary equivalence, we first need to introduce the concept of commuting triple of functions. Similar concept called "isomorphism condition" was introduced already in classical type theory by Andrews in.[8]

Below, we deal with various sets $M_\alpha, M_\beta, \ldots$. For clarity, we will denote their elements by $m_\alpha, m_\beta \ldots$, respectively. To distinguish more elements from the same set, we may possibly use bars or apostrophes. Note that quite often, M_α is a set of functions and so, its elements $m_\alpha \in M_\alpha$ are functions. For example, if $M_{\beta\alpha} \subseteq M_\beta^{M_\alpha}$ then $m_{\beta\alpha} \in M_{\beta\alpha}$ is a function $m_{\beta\alpha} : M_\alpha \longrightarrow M_\beta$.

Definition 2.2. Let $\mathcal{M}^1, \mathcal{M}^2$ be models. Let α, β be types and $M_\alpha^i, M_\beta^i, M_{\beta\alpha}^i$ for $i = 1, 2$ be corresponding sets, where $M_{\beta\alpha} \subseteq M_\beta^{M_\alpha}$. Let us consider functions $f^\alpha : M_\alpha^1 \longrightarrow M_\alpha^2$, $f^\beta : M_\beta^1 \longrightarrow M_\beta^2$, $f^{\beta\alpha} : M_{\beta\alpha}^1 \longrightarrow M_{\beta\alpha}^2$. Then we say that $\langle f^\alpha, f^\beta, f^{\beta\alpha}\rangle$ is a commuting triple (for the corresponding models $\mathcal{M}^1, \mathcal{M}^2$) if the following diagram commutes:

$$\begin{array}{ccc} M_\alpha^1 & \xrightarrow{f^\alpha} & M_\alpha^2 \\ {\scriptstyle m_{\beta\alpha}^1}\big\downarrow & & \big\downarrow{\scriptstyle m_{\beta\alpha}^2 = f^{\beta\alpha}(m_{\beta\alpha}^1)} \\ M_\beta^1 & \xrightarrow{f^\beta} & M_\beta^2 \end{array}$$

where $m_{\beta\alpha}^i \in M_{\beta\alpha}^i$, $i = 1, 2$ are functions.

Lemma 2.2. *Let f^α, f^β be bijections. Then the function $f^{\beta\alpha}(m_{\beta\alpha}^1) = (f^\alpha)^{-1} \circ m_{\beta\alpha}^1 \circ f^\beta$ is a bijection on $M_\beta^{M_\alpha}$ with the inverse $(f^{\beta\alpha})^{-1}(m_{\beta\alpha}^2) = f^\alpha \circ m_{\beta\alpha}^2 \circ (f^\beta)^{-1}$ and $\langle f^\alpha, f^\beta, f^{\beta\alpha}\rangle$ is a commuting triple. Moreover, if $\langle f^\alpha, f^\beta, q\rangle$ is another commuting triple then $f^{\beta\alpha} = q$.*

Let

$$\mathcal{M}^1 = \langle (M^1_\alpha, \stackrel{\circ}{=}^1_\alpha)_{\alpha \in \mathit{Types}}, \mathcal{E}^1 \rangle$$
$$\mathcal{M}^2 = \langle (M^2_\alpha, \stackrel{\circ}{=}^2_\alpha)_{\alpha \in \mathit{Types}}, \mathcal{E}^2 \rangle$$

be two models and consider a set of functions

$$\mathfrak{f} = \{ f^\alpha : M^1_\alpha \longrightarrow M^2_\alpha \mid \alpha \in \mathit{Types} \} \tag{2}$$

Furthermore, let $p \in \mathrm{Asg}(\mathcal{M}^1)$ be an assignment of elements from $\mathcal{M}^1$ to variables. We will define a related assignment $p \circ \mathfrak{f} \in \mathrm{Asg}(\mathcal{M}^2)$ as a set of compositions of the function p (defined on a set of all variables) and the functions f^α for all $\alpha \in \mathit{Types}$:

$$(p \circ \mathfrak{f})(x_\alpha) = f^\alpha(p(x_\alpha)). \tag{3}$$

Definition 2.3. Let

$$\mathcal{M}^1 = \langle (M^1_\alpha, \stackrel{\circ}{=}^1_\alpha)_{\alpha \in \mathit{Types}}, \mathcal{E}^1 \rangle$$
$$\mathcal{M}^2 = \langle (M^2_\alpha, \stackrel{\circ}{=}^2_\alpha)_{\alpha \in \mathit{Types}}, \mathcal{E}^2 \rangle$$

be two models[a]. Let us consider a set of functions

$$\mathfrak{f} = \{ f^\alpha : M^1_\alpha \longrightarrow M^2_\alpha \mid \alpha \in \mathit{Types} \} \tag{4}$$

such that the following holds:

(i) For all $\alpha, \beta \in \mathit{Types}$, $(f^\alpha, f^\beta, f^{\beta\alpha})$ forms a commuting triple.
(ii) $f^o : E^1 \longrightarrow E^2$ preserves all existing infima.
(iii) For each constant $c_\alpha \in J$, $\alpha \in \mathit{Types}$, it holds that $f^\alpha(\mathcal{M}^1(c_\alpha)) = \mathcal{M}^2(c_\alpha)$.

Then the set (4) is a *homomorphism* of $\mathcal{M}_1$ and $\mathcal{M}_2$ and we will formally write

$$\mathfrak{f} : \mathcal{M}^1 \longrightarrow \mathcal{M}^2.$$

If all functions in (4) are injections then $\mathfrak{f}$ is an *embedding* of $\mathcal{M}^1$ in $\mathcal{M}^2$.

Definition 2.4. Let $\mathfrak{f} : \mathcal{M}^1 \longrightarrow \mathcal{M}^2$ be an embedding.

(i) $\mathcal{M}^1$ is a *submodel* of $\mathcal{M}^2$, in symbols $\mathcal{M}^1 \subset \mathcal{M}^2$, if f^o and f^ϵ are identities and $\mathcal{E}^1$ is a subalgebra of $\mathcal{E}^2$.

[a]Recall that $M^i_o = E^i$, $i = 1, 2$ where E^i is a support of the EQ-algebra $\mathcal{E}^i$.

(ii) Let all functions in (4) be bijections. Then $\mathfrak{f}$ is an *isomorphism* between $\mathcal{M}^1$ and $\mathcal{M}^2$ and we write

$$\mathcal{M}^1 \cong \mathcal{M}^2.$$

Theorem 2.2. *Let $\mathcal{M}_1 \cong \mathcal{M}_2$ be an isomorphism and A_α a formula of type α. Then*

$$f^\alpha(\mathcal{M}^1_p(A_\alpha)) = \mathcal{M}^2_{p\circ\mathfrak{f}}(A_\alpha)$$

for all assignments $p \in Asg(\mathcal{M}^1)$ and the corresponding assignments $p\circ\mathfrak{f} \in Asg(\mathcal{M}^2)$.

Theorem 2.3. *Let $\mathcal{M}_1$, $\mathcal{M}_2$ be standard, $\mathcal{E}^1 \cong \mathcal{E}^2$ and let there be an isomorphism $f^\epsilon : (M^1_\epsilon, \stackrel{\circ}{=}{}^1_\epsilon) \longrightarrow (M^2_\epsilon, \stackrel{\circ}{=}{}^2_\epsilon)$. Let $\mathfrak{f}$ be a set (4) of bijections induced due to Lemma 2.2. If $f^\alpha(\mathcal{M}^1(c_\alpha)) = \mathcal{M}^2(c_\alpha)$ for arbitrary constant c_α, $\alpha \in Types$ then $\mathcal{M}_1 \cong \mathcal{M}_2$.*

Definition 2.5.

(i) Models $\mathcal{M}^1$ and $\mathcal{M}^2$ are elementary equivalent, $\mathcal{M}^1 \equiv \mathcal{M}^2$, if

$$\mathcal{M}^1(A_o) = \mathbf{1}^1 \quad \text{iff} \quad \mathcal{M}^2(A_o) = \mathbf{1}^2 \tag{5}$$

holds for arbitrary sentence $A_o \in Form_o$.

(ii) Models $\mathcal{M}^1$ and $\mathcal{M}^2$ are strongly elementary equivalent, $\mathcal{M}^1 \hat{\equiv} \mathcal{M}^2$, if $\mathcal{E}^1 = \mathcal{E}^2$ and

$$\mathcal{M}^1(A_o) = \mathcal{M}^2(A_o) \tag{6}$$

holds for arbitrary sentence $A_o \in Form_o$.

(iii) An embedding $\mathfrak{f} : \mathcal{M}^1 \longrightarrow \mathcal{M}^2$ is elementary if

$$f^o(\mathcal{M}^1_p(A_o)) = \mathcal{M}^2_{p\circ\mathfrak{f}}(A_o) \tag{7}$$

holds for arbitrary formula $A_o \in Form_o$ and $p \in \mathrm{Asg}(\mathcal{M}^1)$.

(iv) A model $\mathcal{M}^1$ is an elementary submodel of $\mathcal{M}^2$, in symbols $\mathcal{M}^1 \prec \mathcal{M}^2$, if $\mathcal{M}^1 \subset \mathcal{M}^2$ and

$$\mathcal{M}^1_p(A_o) = \mathcal{M}^2_{p\circ\mathfrak{f}}(A_o) \tag{8}$$

holds for arbitrary formula $A_o \in Form_o$ and $p \in \mathrm{Asg}(\mathcal{M}^1)$.

Theorem 2.4.

(a) If $\mathcal{M}^1 \hat{\equiv} \mathcal{M}^2$ then $\mathcal{M}^1 \equiv \mathcal{M}^2$. If $\mathcal{M}^1 \cong \mathcal{M}^2$ then $\mathcal{M}^1 \hat{\equiv} \mathcal{M}^2$.
(b) If $\mathcal{M}^1 \prec \mathcal{M}^2$ then $\mathcal{M}^1 \equiv \mathcal{M}^2$.

(c) If $\mathfrak{f} : \mathcal{M}^1 \longrightarrow \mathcal{M}^2$ *and* $\mathfrak{g} : \mathcal{M}^2 \longrightarrow \mathcal{M}^3$ *are elementary embeddings then* $\mathfrak{f} \circ \mathfrak{g} : \mathcal{M}^1 \longrightarrow \mathcal{M}^3$ *is an elementary embedding.*
(d) If $\mathcal{M}^1 \prec \mathcal{M}^2$ *and* $\mathcal{M}^2 \prec \mathcal{M}^3$ *then* $\mathcal{M}^1 \prec \mathcal{M}^3$.
(e) The relations $\cong, \equiv, \hat{\equiv}$ *are equivalences.*

3. Conclusion

In this paper, we introduced few basic concepts of model theory in fuzzy type theory and some of their properties. We focused on the concepts related to homomorphism, isomorphism, submodel, elementary embedding, etc. The paper is introductory to the topic demonstrating specificities of model theory for fuzzy type theory. In the future, we will focus on questions concerning construction of models and various interrelations among them.

Acknowledgments

This paper has been supported by the European Regional Development Fund in the IT4Innovations Centre of Excellence project (CZ.1.05/1.1.00/02.0070).

References

1. P. Dellunde, *Journal of Logic and Computation* , p. (to appear) (2012).
2. P. Dellunde and F. Esteva, *Archive for Mathematical Logic* , p. (submitted) (2012).
3. P. Murinová and V. Novák, *Mathematical logic quarterly* **52**, 259 (2006).
4. V. Novák, Models and submodels of fuzzy theories, in *Intelligent Systems for Information Processing: From Representation to Applications*, eds. B. Bouchon-Meunier, L. Foulloy and R. R. Yager (Elsevier, Amsterdam, 2003) pp. 363–373.
5. P. Hájek and P. Cintula, *Journal of Symbolic Logic* **71**, 863 (2006).
6. S. Gottwald, *A Treatise on Many-Valued Logics* (Research Studies Press Ltd., Baldock, Herfordshire, 2001).
7. V. Novák, I. Perfilieva and J. Močkoř, *Mathematical Principles of Fuzzy Logic* (Kluwer, Boston, 1999).
8. P. Andrews, *An Introduction to Mathematical Logic and Type Theory: To Truth Through Proof* (Kluwer, Dordrecht, 2002).
9. P. Andrews, *Journal of Symbolic Logic* **37**, 385 (1972).
10. V. Novák, *Fuzzy Sets and Systems* **149**, 235 (2005).
11. V. Novák, *Logic Journal of the IGPL* **19**, 512 (2011).
12. V. Novák, *Fuzzy Sets and Systems* ((to appear)).

INTERVAL-VALUED FUZZY RELATIONAL EQUATIONS WITH MIN-IMPLICATION

QING-QUAN XIONG† AND XUE-PING WANG

College of Mathematics and Software Science,
Sichuan Normal University, Cheng du, Sichuan 610066, P.R. China

This paper introduces the concept of tolerable solution set for interval-valued fuzzy relational equations with min-implication composition on complete Brouwerian lattices. Tolerable solution set interval-valued fuzzy relational equations is shown when the right-hand sides are meet-irreducible. Necessary and sufficient conditions for the existence of solutions are given.

1. Introduction

Sanchez [7] first introduced the fuzzy relational equation with max-min composition. Wu [14] found that in fuzzy relational calculus and reasoning inf-α composition operation is better. Particularly, Di Nola et al. [3] obtained the smallest solution for all fuzzy relational equations with inf-α composition and proved that the solution set of each fuzzy relational equation with inf-α composition is nonempty if and only if the equation has the smallest solution over complete Brouwerian lattices. Li and Wang [5] also discussed inf-α composition fuzzy relational equation on complete Brouwerian lattices, they obtained a necessary and sufficient condition that the solution set is nonempty and described the solution set when the right-hand side has an irredundant finite meet-decomposition. The fuzzy relational equations has no always solutions. Wagenknecht and Hurtmann [8] applied the concept of fuzzy tolerance to allow the relations in interval values. A system of interval-valued min-implication fuzzy relational equations can be represented in the form

$$A^I \circ_{\alpha_T} X^I = b^I \tag{1}$$

where "$\circ_{\alpha_T}$" denotes min-implication operators, $A^I = (a_{ij}^I)_{m\times n}$ is an interval-valued matrix with $a_{ij}^I = [a_{ij}^-, a_{ij}^+]$, and $b^I = (b_i^I)_{i\in I}$ is an interval-valued vector

† xqq309@msn.com

with $b_i^I=[b_i^-,b_i^+]$, a_{ij}^-,a_{ij}^+,b_i^- and $b_i^+\in L$ for all $i\in I, j\in J$. Wang [10] developed an algorithm to solve iterated fuzzy relational equations and allowed the relations in constant and interval forms. Wang and Chang [9] investigated the properties and the solution procedure of composite interval-valued fuzzy relational equations with max-min operator and generalized constant-valued fuzzy relational problems. After that, Li and Fang [6] provided a more efficient approach to find the tolerable solution set for interval-valued fuzzy relational equations with max-min composition. After that, they introduced the concepts of tolerable solution set for interval-valued fuzzy relational equations with min-s-norm composition on the real unite interval. Given a continuous s-norm, it is shown that the tolerable solution set of the solution sets of interval-valued fuzzy relational equations with a min-s-norm composition, if nonempty, is composed of one smallest solution and a finite number of maximal solutions. In fact, min-s composition can be viewed as a dual operation with respect to max-t composition, interval-valued min-s equations can be handled in a completely dual manner. However, these works play on the real unite interval. To our knowledge, no work has been done on a complete Brouwerian lattice and min-implication composition operators for interval-valued fuzzy relational equations. In this paper, we discuss interval-valued fuzzy relational equations on complete Brouwerian lattices with min-implication composition. In the whole paper, unless otherwise stated, $L=(L,\leq,\wedge,\vee)$ always represents a complete Brouwerian lattice with universal bounds 0 and 1. Let $I=\{1,2,\cdots,m\}$, $J=\{1,2,\cdots,n\}$ be two index sets, $A=(a_{ij})_{m\times n}\in L^{m\times n}$ be a matrix, $b=(b_i)_{i\in I}\in L^m$ and $X=(x_j)_{j\in J}\in L^n$ be two column vectors.

2. Preliminaries

We assume that the reader is familiar with the basic order-theoretic notions such as maximal element, (complete) lattice, distributive lattice, meet-irreducible and finite decomposition etc., see [1].

Definition 2.1. [2] A triangular norm (t-norm for short) $T: L\times L\to L$ such that commutativity, associativity, monotonicity and $T(1,a)=a$, $T(a,0)=0$ for every $a,b,c,d\in L$.

Definition 2.2. [12] Let φ be a mapping from $L\times L$ into L. Define I_φ, L_φ as follows:

$$I_\varphi(a,b)=\sup\{x\in L:\varphi(a,x)\leq b\},\forall\ a,b\in L,$$

$$L_\varphi(a,b)=\inf\{x\in L:\varphi(a,x)\geq b\},\forall\ a,b\in L.$$

It is tacitly assumed that the least upper bound of the empty set is 0 and the largest lower bound of the empty set is 1. Obviously, if $\varphi=T$, then $\alpha_T=I_\varphi$.

Lemma 2.1. [12] Let $a,b,c\in L,$ then

(1) If $T \in T(L)$, then $T(a,c) \le b \Leftrightarrow c \le I_T(a,b)$;
(2) If $I_T \in I(L)$, then $I_T(a, L_{I_T}(a,b)) \ge b$.
Denote $A^- = (a_{ij}^-)_{m\times n}$, $A^+ = (a_{ij}^+)_{m\times n}$, $b^- = (b_i^-)_{i\in I}$ and $b^+ = (b_i^+)_{i\in I}$. By extending the lattice order in a component wise manner, A^I and b^I can be denoted as $A^I = [A^-, A^+] = \{A = (a_{ij})_{m\times n} : A^- \le A \le A^+\}$ and $b^I = [b^-, b^+] = \{b = (b_j)_{m\times 1} : b^- \le b \le b^+\}$, respectively. Without loss of generality, we always assume that $A^- \le A^+$ and $b^- \le b^+$.

Definition 2.3. [4] Let $A^I \circ_{\alpha_T} X^I = b^I$ be a system of interval-valued fuzzy relational equations. A column vector $X \in L^n$ is a tolerable solution if for each $A \in A^I$ there exists $b \in b^I$ such that $A \circ_{\alpha_T} X = b$. The set of tolerable solutions for the system is denoted by $X_t(A^I, b^I)$.

3. Some properties of interval-valued fuzzy relational equations

Lemma 3.1. (i) Given a fuzzy matrix $A \in L^{m\times n}$ for all $X^1, X^2 \in L^n$, if $X^1 \le X^2$ then $A \circ_{\alpha_T} X^1 \le A \circ_{\alpha_T} X^2$.

(ii) Given fuzzy matrices $A^-, A^+ \in L^{m\times n}$ and $X \in L^n$, if $A^- \le A^+$ then $A^- \circ_{\alpha_T} X \ge A^+ \circ_{\alpha_T} X$.

Lemma 3.2. (i) Given a vector $X \in L^n$, then for each $b \in [A^+ \circ_{\alpha_T} X, A^- \circ_{\alpha_T} X]$, there exists $A \in [A^-, A^+]$ such that $A \circ_{\alpha_T} X = b$.

(ii) Given a matrix $A \in L^{m\times n}$ and two vectors $X^1, X^2 \in L^n$, if $X^1 \le X^2$, then for each $b \in [A \circ_{\alpha_T} X^1, A \circ_{\alpha_T} X^2]$, there exists $X \in [X^1, X^2]$ such that $A \circ_{\alpha_T} X = b$.

Theorem 3.1. Let A^I be an interval-valued matrix and b^I be an m-dimensional interval-valued column vector. Then

$$X_t(A^I, b^I) = \{X \in L^n : [A^+ \circ_{\alpha_T} X, A^- \circ_{\alpha_T} X] \subseteq b^I\}.$$

4. The tolerable solution set of interval-valued fuzzy relational equations

In order to describe the tolerable solution set, we first consider the following system of fuzzy relational inequalities

$$\begin{cases} A^+ \circ_{\alpha_T} X \ge b^-, \\ A^- \circ_{\alpha_T} X \le b^+, \end{cases} \tag{2}$$

where $A^I = [A^-, A^+]$, $b^I = [b^-, b^+]$ are known.

Theorem 4.1. $X \in X_t(A^I, b^I)$ if and only if X is a solution of (2).

Theorem 4.2. Let $X^I = [X^-, X^+]$. If $X^- \in X_t(A^I, b^I)$ and $X^+ \in X_t(A^I, b^I)$, then $X^I \subseteq X_t(A^I, b^I)$.

Theorem 4.2 suggests further investigation of the structure of $X_t(A^I, b^I)$ by examining the smallest solution and maximal solutions. Let

$$X_*^t \triangleq (x_{j*}^t)_{j \in J} \triangleq \left(\max_{i \in I} T(a_{ij}^+, b_i^-) \right)_{j \in J}. \tag{3}$$

Theorem 4.3. $X_t(A^I, b^I) \neq \varnothing$ if and only if X_*^t is the smallest element of $X_t(A^I, b^I)$.

Theorem 4.4. $X_t(A^I, b^I) \neq \varnothing$ if and only if $A^- \circ_{\alpha_T} X_*^t \leq b^+$.

We suppose that b_i^+ is a meet-irreducible element for all $i \in I$. For any poset P, we denote by P^0 the set of all maximal elements in P.

Lemma 4.1. [1] Any finite nonempty subset of a poset has maximal members.

Theorem 4.5. Let $X \in X_t(A^I, b^I)$ and b_i^+ be a meet-irreducible element for each $i \in I$. Then there exists $j_0 \in J$ such that $\alpha_T(a_{ij_0}^-, x_{j_0}) \leq b_i^+$.

Definition 4.1. For $X = (x_j)_{j \in J} \in X_t(A^I, b^I)$ of inequality system (2), x_{j_0} is an attainable variable if $\alpha_T(a_{ij_0}^-, x_{j_0}) \leq b_i^+$ holds for some $i \in I$.

Lemma 4.2. Let $X_*^t = (x_{j*}^t)_{j \in J}$ be the smallest element of $X_t(A^I, b^I)$, and $X = (x_j)_{j \in J}$ be a solution of (2). If x_j is an attainable variable in the ith inequality, then x_{j*}^t is an attainable variable also. Moreover, if x_{j*}^t is not an attainable variable, then x_j is not an attainable variable for any solution $X = (x_j)_{j \in J} \in X_t(A^I, b^I)$.

Lemma 4.3. Let $X \in X_t(A^I, b^I)$. If x_j is an attainable variable, then $x_j \leq I_{\alpha_T}(a_{i_0 j}^-, b_{i_0}^+)$ for some $i_0 \in I$.

Lemma 4.4. Let $X = (x_j)_{j \in J} \in X_t(A^I, b^I)^0$. If x_k is an attainable variable, then $x_k = \wedge_{i \in I'} I_{\alpha_T}(a_{ik}^-, b_i^+)$, where $I' = \{i \in I : \alpha_T(a_{ik}^-, x_k) \leq b_i^+\}$.

Lemma 4.5. Let $X = (x_j)_{j \in J} \in X_t(A^I, b^I)^0$ and $b^+ < 1$. Then a component x_j is not an attainable component if and only if $x_j = 1$.

By Lemmas 4.3, 4.4 and 4.5, the following result holds.

Theorem 4.6. If $X = (x_j)_{j \in J} \in X_t(A^I, b^I)^0$, then we have either $x_j = 1$ or $x_j = \wedge_{i \in I'} I_{\alpha_T}(a_{ij}^-, b_i^+)$, where $I' = \{i \in I : \alpha_T(a_{ik}^-, x_k) \leq b_i^+\}$.

From Theorem 4.6, we give the following definitions:

Definition 4.2. For each $X = (x_j)_{j \in J} \in X_t(A^I, b^I)$, let

$$M_i(X) \triangleq M_i(X, b_i^-) \triangleq \{j \in J : \alpha_T(a_{ij}^-, x_j) \leq b_i^+\} \quad \text{for all } i \in I \tag{4}$$

and

$$\Gamma(X)=M_1(X)\times M_2(X)\times\cdots\times M_m(X). \quad (5)$$

By Theorem 4.5, $M_i(X)\neq\varnothing$ for all $i\in I$, thus $\Gamma(X)\neq\varnothing$. For each $g\in\Gamma(X)$, let $g(g_1,g_2,\cdots,g_m)$. $\forall\, g\in\Gamma(X)$ and $j\in J$, define $I_j^{(g)}\triangleq\{i\in I: g_i=j\}$.

Definition 4.3. Let $g\in\Gamma(X)$. Define $X^{(g)}=\left(x_1^{(g)},x_2^{(g)},\ldots,x_n^{(g)}\right)'\in L^n$ with

$$x_j^{(g)}=\begin{cases}\wedge_{i\in I_j^{(g)}} I_{\alpha_T}(a_{ij}^-,b_i^+), & \text{if } I_j^{(g)}\neq\varnothing,\\ 1, & \text{if } I_j^{(g)}=\varnothing.\end{cases} \quad (6)$$

Denote $G(X)\triangleq\left\{X^{(g)}: g\in\Gamma(X)\right\}$.

Theorem 4.7. If $X\in X_t(A^I,b^I)$ and $g\in\Gamma(X)$, then $X\leq X^{(g)}$ and $X^{(g)}\in X_t(A^I,b^I)$.

Theorem 4.8. If $X,Y\in X_t(A^I,b^I)$ with $X\leq Y$ and $Y^{(g)}\in G(Y)$ for $g\in\Gamma(Y)$, then $g\in\Gamma(X)$ and $X^{(g)}=Y^{(g)}\in G(X)$.

Theorem 4.9. If $X\in X_t(A^I,b^I)$, then $X\in X_t(A^I,b^I)^0$ if and only if $X^{(g)}=X$ for all $g\in\Gamma(X)$.

Theorem 4.10. The solution set $X_t(A^I,b^I)$ and the finite poset $\left(G(X_*^t),\leq\right)$ contain the same maximal elements, i.e., $X_t(A^I,b^I)^0=G(X_*^t)^0$.

Theorem 4.11. $X\in X_t(A^I,b^I)$ if and only if there exists an $X^0\in X_t(A^I,b^I)^0$ such that $X^0\geq X\geq X_*^t$.

Theorem 4.12. If b_i^+ is a meet-irreducible element $\forall i\in I$, then the tolerable solution set $X_t(A^I,b^I)$ is the union of a finite number of intervals in L^n:

$$X_t(A^I,b^I)=\cup_{X^0\in G(X_*^t)^0}\left[X_*^t,X^0\right]. \quad (7)$$

5. Conclusions

This paper introduces the concept of tolerable solution set for interval-valued fuzzy relational equations with min-implication composition on complete Brouwerian lattices and discusses the tolerable solution set under some special conditions. In further we will discuss the tolerable solution set interval-valued fuzzy relational equations under the more general conditions.

Acknowledgments

This was supported by National Natural Science Foundation of China (No. 11171242), Doctoral Fund of Ministry of Education of China (No.

20105134110002) and Sichuan Youth Fund (No. 2011JQ0055) and Scientific Research Fund of Sichuan Provincial Education Department (No. 11ZB084).

References

1. G. Birkhoff, Lattice theory, vol, XXV, 3rd ed., American Mathematical Society Colloquium Publications, Providence, R.I., 1979.
2. B. De Baets, An order-theoretic approach to solving sup-T equations, in: D. Ruan (Ed.), Fuzzy Set Theory and Advanced Mathematical Applications, Kluwer Academic Publishers, Dordrecht, 1995, pp. 67-87.
3. A. Di Nola, S. Sessa, W. Pedrycz, E. Sanchez, Fuzzy relation equations and their applications to knowledge engineering, Kluwer Academic Publishers, Dordrecht, 1989.
4. P. K. Li, S. C. Fang, A note on solution sets of interval-valued fuzzy relational equations, Fuzzy Optim Decis Making, 8 (2009) 115-121.
5. Y. M. Li, X. P. Wang, The solution sets of α -Fuzzy relational equations in finite domains and on a complete Brouwerian lattice. Indian J. pure appl. Math. 34 (2003) 1249-1257.
6. G. Z. Li, S. C. Fang, Solving interval-valued fuzzy relation equations, IEEE Transactions on Fuzzy Systems, 6 (1998) 321-324.
7. E. Sanchez, Resolution of composite fuzzy relation equations, Information and Control 30 (1976) 38-48.
8. M. Wagenknecht, K. Hartmann, Fuzzy modelling with tolerances, Fuzzy Sets and Systems, 20 (1986) 325-332.
9. H. F. Wang, Y. C. Chang, Resolution of composite interval-valued fuzzy relation equations, Fuzzy Sets and Systems, 44 (1991) 227-240.
10. H. F. Wang, An algorithm for solving iterated composite relation equations, in : Proc. NAFIPS 88 (1988) 242-249.
11. S. M. Wang, S. C. Fang, H. L. W. Nuttle, Solution sets of interval-valued min-s-norm fuzzy relational equations, Fuzzy Optimization and Decision Making, 4 (2005) 331-349.
12. Z. D. Wang, Y. D. Yu, Pseudo-t-norms and implication operations on a complete Brouwerian lattice. Fuzzy Sets and Systems 132 (2002) 113-124.
13. Z. D. Wang, F. M. Dai, The sets of solutions of L-relation equations $T(a,x)=b$ and $I(a,x)=b$. Journal of Yangzhou University 6 (2003) 8-10 (in Chinese).
14. W. M. Wu, The inf-α composition of fuzzy relations, Journal of Shanghai Normal University (Natural). 2 (1985) 1-7 (in Chinese).
15. Q. Q. Xiong, X. P. Wang, Solution sets of inf-α_T fuzzy relational equations on complete Brouwerian lattices, Information Sciences 177 (2007) 4757-4767.

FUZZY LATTICE REASONING (FLR) CLASSIFIER FOR HUMAN FACIAL EXPRESSION RECOGNITION

S. E. PAPADAKIS
Department of Business Planning and Information Systems, TEI of Crete P.O Box 128, GR-72100, Agios Nikolaos, Greece}

V. G. KABURLASOS*, G. A. PAPAKOSTAS
Human-Machines Interaction (HMI) Laboratory, TEI of Kavala GR-65404 Agios Loukas, Kavala, Greece

We deal with the problem of human facial expression recognition. An image is preprocessed for feature extraction using moment descriptors; then, it is represented in the product lattice $(\mathbb{F}_1^{100},\leq)$ of *Intervals' Numbers* (*INs*). Both learning and generalization are pursued in $(\mathbb{F}_1^{100},\leq)$ by a *Fuzzy Lattice Reasoning* (*FLR*) classifier based on an inclusion measure function $\sigma:\mathbb{F}_1^{100}\times\mathbb{F}_1^{100}\rightarrow[0,1]$. Preliminary, experiments on a benchmark set have been promising.

1. Introduction

Pattern recognition is an important application domain, which, typically, includes data preprocessing for feature extraction and pattern representation. Dealing "in principle" (in a well-defined sense) with uncertainty and/or ambiguity is of high interest in pattern recognition applications.

This work considers the problem of recognizing human facial expressions from digital images represented by Intervals' Numbers (INs) induced from image-moments-based preprocessing [1,2]. Decision-making is carried out by Fuzzy Lattice Reasoning (FLR) based on an inclusion measure function.

This work is organized as follows. Section 2 presents the mathematical background. Section 3 describes a fuzzy lattice reasoning (FLR) classifier. Section 4 demonstates, comparatively, a human facial expression recognition application. Section 5 concludes by summarizing our contribution.

* *vgkabs@teikav.edu.gr*

2. Mathematical Background

This section summarizes notions and notation introduced elsewhere [3,4].

2.1. *General Lattice Theory Preliminaries*

Consider the following definition.

Definition 2.1. An *inclusion measure* in a lattice $(\mathbb{L},\leq)$ is a function $\sigma:\mathbb{L}\times\mathbb{L}\to[0,1]$, which satisfies the following conditions:

C1. $\sigma(x,x)=1$.

C2. $x\wedge y<x\Rightarrow\sigma(x,y)<1$.

C3. $u\leq w\Rightarrow\sigma(x,u)\leq\sigma(x,w)$.

Any employment of an inclusion measure for decision-making is called Fuzzy Lattice Reasoning (FLR) [5,6]. Next, we focus on complete lattices.

Theorem 2.1. Let function $v:\mathbb{L}\to[0,+\infty)$ be a positive valuation[†] on a complete lattice $(\mathbb{L},\leq)$ such that both $v(O) = 0$ and $v(I)<+\infty$. Let functions sigma-meet $\sigma_\wedge:\mathbb{L}\times\mathbb{L}\to[0,1]$ and sigma-join $\sigma_\vee:\mathbb{L}\times\mathbb{L}\to[0,1]$ be defined as follows; first, $\sigma_\wedge(O,y)=1$, $\sigma_\wedge(x,y)=\dfrac{v(x\wedge y)}{v(x)}$ for $x>O$; second, $\sigma_\vee(x,y)=1$ for $x\vee y=O$, $\sigma_\vee(x,y)=\dfrac{v(y)}{v(x\vee y)}$ for $x\vee y>O$. Then, both functions $\sigma_\wedge(.,.)$ and $\sigma_\vee(.,.)$ are inclusion measures.

2.2. *Intervals' Numbers (INs)*

Take the set of real numbers $\mathbb{R}$ and the usual order $\leq$. Then $(\mathbb{R},\leq)$ is a totally ordered, noncomplete lattice. Lattice $(\mathbb{R},\leq)$ can be extended to a complete lattice by including both "$-\infty$" and "$+\infty$". Any strictly increasing function $v:\overline{\mathbb{R}}\to[0,+\infty)$ is a positive valuation on $(\overline{\mathbb{R}},\leq)$.

Consider the lattice $(\mathbb{I}_1,\leq)$ of *Type-1 intervals* ordered by set inclusion ($\subseteq$), where $\mathbb{I}_1\doteq\{[a,b]:-\infty\leq a\leq b\leq\infty\}\cup\{\varnothing\}$. The supremum operation, denoted by $\cup$, is computed by $[a,b]\cup[c,d]=[a\wedge c,b\vee d]$.

Definition 2.2. A *mass function* $v_1:\mathbb{I}_1\to[0,+\infty)$ is given by $v_1([a,b])=v(\theta(a))+v(b)$, where $v:\overline{\mathbb{R}}\to[0,+\infty)$ is a strictly increasing function and $\theta:\overline{\mathbb{R}}\to\overline{\mathbb{R}}$ is a bijective, strictly decreasing function.

[†] *Positive valuation in a lattice* $(\mathbb{L},\leq)$ *is a real function* $v:\mathbb{L}\to\mathbb{R}$ *that satisfies both* $v(x)+v(y)=v(x\wedge y)+v(x\vee y)$ *and* $x<y\Rightarrow v(x)<v(y)$

Consider the functions

$$v(x;A,\lambda,\mu)=\frac{A}{1+e^{-\lambda\cdot(x-\mu)}} \text{ and } \theta(x;\mu)=2\mu-x, \quad (1)$$

where $\lambda>0$. Then, it is easy to confirm that $v_1([x,x])=A$.

Another lattice of interest is the lattice of *Type-1 Intervals' Numbers*, or *INs* for short, defined next. Consider a function $E:[0,1]\to\mathbb{I}_1$ which has the additional property $h_1\geq h_2 \Rightarrow E(h_1)\subseteq E(h_2)$. We denote the class of all such functions by $\mathbb{F}_1$ and we equip it with the following order $\leq$: For every pair $F,G\in\mathbb{F}_1$ we define: $F\leq G \Leftrightarrow (\forall h\in[0,1]: F(h)\subseteq G(h))$. Then, $(\mathbb{F}_1,\leq)$ is a complete lattice [7], namely *lattice of (Type-1) INs*.

2.3. *Inclusion Measures*

We present two inclusion measures on $(\mathbb{I}_1,\subseteq)$.

Proposition 2.1. Let $v_1:\mathbb{I}_1\to[0,+\infty)$ be a mass function on $(\mathbb{I}_1,\subseteq)$ satisfying both $v_1(O_1)=0$ and $v_1(I_1)<+\infty$. Then, functions $\sigma_\cap:\mathbb{I}_1\times\mathbb{I}_1\to[0,1]$ and $\sigma_{\dot\cup}:\mathbb{I}_1\times\mathbb{I}_1\to[0,1]$ defined, respectively, by

$$\sigma_\cap([a,b],[c,d])=\begin{cases}1, & \text{if } [a,b]=O_1\\ \dfrac{v_1([a,b]\cap c,d])}{v_1([a,b])}, & \text{if } [a,b]\supset O_1\end{cases}, \text{ and}$$

$$\sigma_{\dot\cup}([a,b],[c,d])=\begin{cases}1, & \text{if } [a,b]\dot\cup[c,d]=O_1\\ \dfrac{v_1([c,d])}{v_1([a,b]\dot\cup[c,d])}, & \text{if } [a,b]\dot\cup[c,d]\supset O_1\end{cases},$$

are inclusion measures on $(\mathbb{I}_1,\subseteq)$.

We extend the aforementioned two inclusion measures to $(\mathbb{F}_1,\leq)$, next.

Proposition 2.2. Let mass function $v_1:\mathbb{I}_1\to[0,+\infty)$ satisfy $v_1(O_1)=0$ and $v_1(I_1)<+\infty$. Let functions $v_1(I_1)<+\infty$ and $\sigma_{\dot\cup}$ be as in Proposition 2.1.. Then functions $\sigma_\curlywedge:\mathbb{F}_1\times\mathbb{F}_1\to[0,1]$ and $\sigma_\curlyvee:\mathbb{F}_1\times\mathbb{F}_1\to[0,1]$ defined, respectively, by

$$\sigma_\curlywedge(F,G)=\int_0^1\sigma_\cap(F(h),G(h))dh \quad\text{and}\quad \sigma_\curlyvee(F,G)=\int_0^1\sigma_{\dot\cup}(F(h),G(h))dh \quad\text{are}$$

inclusion measures.

3. A Fuzzy Lattice Reasoning (FLR) Classifier

Let (labeled) data in $\mathbb{F}_1^N$ be partitioned, first, in n_c classes and, second, in 2 disjoint data sets for *training* and *testing*, respectively.

Learning by the FLR classifier is carried out by Algorithm 3.1 for "structure indentification"; whereas, Algorithm 3.2 describes FLR generalization.

We remark that the function $\ell : \mathbb{F}_1^N \to \mathbb{L}$ assigns a class label to an IN. Moreover, a class $c_j, j \in \mathbb{L} = \{1,\ldots L\}$ in our preliminary experiments here was represented by a single IN per training datum.

Algorithm 3.1: FLR learning (for structure identification)

1: **for** $j=1$ to $j=n_c$ **do**

2: Induce labeled INs $\overrightarrow{\mathbb{W}}_j \in \mathbb{F}_1^N$ from the training data in class c_j.

3: **end for**

Algorithm 3.2: FLR generalization

1: Assume a set $\bigcup_{j\in\{1,\ldots L\}} \{c_j\}$ of labeled classes $c_j = \{\overrightarrow{\mathbb{W}}_j\}$.

2: **for** $i=1$ to $i=n_{tst}$ **do**

3: Consider the next testing datum $(\vec{\mathbb{S}}_i, \ell(\vec{\mathbb{S}}_i)) \in \mathbb{F}_1^N \times \mathbb{L}$.

4: Let $J = \underset{j\in\{1,\ldots,L\}}{\texttt{argmax}}\{\sigma_\curlyvee(\vec{\mathbb{S}}_i \le \overrightarrow{\mathbb{W}}_j)\}$.

5: If $J = \ell(\vec{\mathbb{S}}_i)$ add on to the number of correct classifications.

6: **end for**

4. Computational Experiments

We considered the problem of human facial expression recognition using the JAFFE [8] benchmark dataset. The classes' names and the number of images per class (within parentheses) are: Angry(30), Disgust(29), Fear(32), Happy(31), Neutral(30), Sadness(31), Surprise(30). An image representative from each class in shown in Fig. 1.

4.1. *Data Pre-processing and Feature Extraction*

Irrelevant facial expression content was removed using the Viola-Jones face detector [9] followed by an ellipse-based face masking. Next, feature extraction was carried out using orthogonal moments. In particular, from each image, we induced 100 moments regarding Zernike, Pseudo-Zernike, Fourier-Mellin, Legendre, Tchebichef and Krawtchouk moments [10,11], respectively.

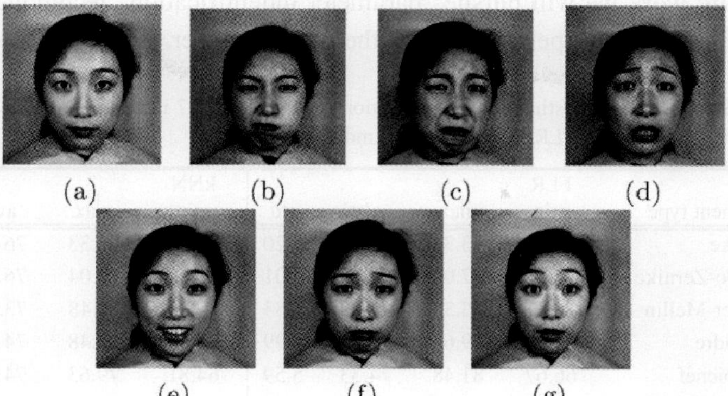

Fig. 1. Seven different facial expressions including (a) "neutral", (b) "angry", (c) "disgust", (d) "fear", (e) "happy", (f) "sadness", (g) "surprise".

4.2. *Experimental Results and Discussion*

We partitioned, randomly, the data in every class in two mutually disjoint sets: one set for training (including around 75% of the data) and another one for testing (including the remaining data). We repeated the above partition 7 times per moment type, and carried out 7 different computer experiments per classifier kNN and FLR using the same data sets for training/testing.

For the kNN classifier we selected both $k=1$ and the $L2$ distance in the Euclidean space $\mathbb{R}^{100}$. For the FLR classifier a (normalized) training datum was represented by a trivial IN; moreover, identical functions $v(x;1,0.5,0.5)$ and $\theta(x;0.5)$ were employed in all 100 data dimensions.

Table 1 displays the statistics of the testing data classification success (%) for both the FLR and the kNN classifiers. Table 1 indicates that the FLR classifier typically performed (marginally) better than the kNN for all types of moments. The best performance was recorded for the Zernike moments, where the FLR on the average performed 3.17 percentage points higher than the kNN. We attributed the superior performance of FLR to the capacity of the inclusion measure. A further improvement is reasonably expected by "parameter indentification", during training, in a future work.

5. Conclusion

This work detailed the FLR classifier for training and testing. Then, the FLR was applied, comparatively with kNN, to a benchmark human facial expression recognition problem. Preliminary experimental results have been encouraging.

In a future work we will pursue "parameter indentification" techniques towards further improving the performance of the FLR classifier.

Table 1. Statistics of the testing data classification success (%) in 7 random computer experiments using classifiers kNN and FLR on six different moment types.

Moment type	FLR min	max	ave	std	kNN min	max	ave	std
1) Zernike	74.07	83.33	79.62	3.20	70.37	83.33	76.45	4.1
2) Pseudo-Zernike	72.22	87.04	77.77	5.01	70.37	87.04	76.45	5.31
3) Fourier-Mellin	70.37	83.33	76.19	4.83	70.37	81.48	73.54	3.81
4) Legendre	66.67	79.63	74.86	4.99	64.81	81.48	74.33	5.49
5) Tchebichef	66.67	81.48	74.33	5.59	64.81	79.63	74.07	4.78
6) Krawtchouk	59.26	83.33	66.41	7.88	55.56	72.22	66.4	5.38

Acknowledgement

This research has been co-funded by the European Union (Social Fund) and Greek national resources under the framework of the "Archimedes III: Funding of Research Groups in TEI of Athens" project of the "Education & Lifelong Learning" Operational Programme.

References

1. D. G. Sim, H. K. Kim and R. H. Park, *Image & Vis. Comput.* **22**, 331 (2004).
2. G. A. Papakostas, E. G. Karakasis and D. E. Koulouriotis, *Pattern Recognition* **43**, 58 (2010).
3. V. G. Kaburlasos and A. Kehagias (submitted).
4. V. G. Kaburlasos, S. E. Papadakis and A. Amanatiadis, *Journal of Mathematical Imaging and Vision* **42**, 118 (2012).
5. V. Kaburlasos, *Towards a Uni_ed Modeling and Knowledge-Representation Based on Lattice Theory* (Springer, Heidelberg, Germany, 2006).
6. V. G. Kaburlasos, I. N. Athanasiadis and P. A. Mitkas, *International Journal of Approximate Reasoning* **45**, 152 (2007).
7. A. Kehagias, *Information Sciences* **181**, 1863 (2011).
8. M. J. Lyons, S. Akamatsu, M. Kamachi and J. Gyoba, Coding facial expressions with gabor wavelets, in *Proc. IEEE Intl. Conference on Automatic Face and Gesture Recognition*, (Nara, Japan, 1998).
9. P. Viola and M. J. Jones, *Intl. Journal of Computer Vision* **57**, 137 (2004).
10. R. Mukundan and K. Ramakrishnan, *Moment functions in image analysis* (World Scienti_c, Singapore, 1998).
11. G. A. Papakostas, D. E. Koulouriotis and E. G. Karakasis, *Information Sciences* **179**, 3619 (2009).

FUZZY COSETS OF FUZZY SUBGROUPS RELATIVE TO FUZZY SUBGROUPS*

LAN SHU, ZHAOHAO WANG

School of Mathematical Sciences, University of Electronic Science and Technology of China, Chengdu, Sichuan, P.R.China 611731

The notion of a fuzzy coset of a fuzzy subgroup relative to a fuzzy subgroup is introduced, and some of their properties are investigated. In particular, it is shown that a normal fuzzy subgroup of a fuzzy subgroup can be characterized in terms of fuzzy cosets of a fuzzy subgroup relative to a fuzzy subgroup.

1. Introduction

The notion of coset is one of the basic concepts in the group theory. Since Resenfeld [1] introduced the concept of fuzzy subgroups of a group, many authors were interested in generalizing the notion of coset in the fuzzy setting [8][9][5]. Mukherjee and Bhattacharya defined the notion of a fuzzy coset and investigated the properties of the notion. Sidky and Mishref gave an analysis for the definition of a fuzzy coset to explain the concept of left and right fuzzy cosets. Clearly, they only defined the notion of coset of a fuzzy subgroup relative to original group. It is well know that there is the notion of coset of subgroup relative to subgroup in the group theory. Therefore we can give a more general definition of fuzzy coset, that is, we introduce the concept of a fuzzy coset of a fuzzy subgroup relative to a fuzzy subgroup and do some researches on the notion.

2. Preliminaries

We give some basic definitions and results which will be used later on. We write $\wedge$ for minimum or infimum and $\vee$ for maximum or supremum.

Definition 2.1 [1] A fuzzy subset μ of a group G is called a fuzzy subgroup of G if μ satisfy (1) $\mu(xy) \geq \mu(x) \wedge \mu(y)$, $x, y \in G$, and (2) $\mu(x) \geq \mu(x^{-1})$, $x \in G$.

* This work is supported by National Natural Science Foundation of China (No.11071178).

Theorem 2.1 [1] Let G be a group. If μ is a fuzzy subgroup of G, then $\mu(x^{-1})=\mu(x)$ and $\mu(e)\geq\mu(x)$ for all $x\in G$, where e denotes the identity of G.

Theorem 2.2 [3] Let G be a group and μ be a fuzzy subset of G. Then μ is a fuzzy subgroup of G if and only if the level subset μ_t, for every $t\in[0,\mu(e)]$, is a subgroup of G, where e is the identity of G.

Definition 2.2 [5] Let μ,ν be fuzzy subgroups of a group G and $\mu\subseteq\nu$. Then μ is called a normal fuzzy subgroup of the fuzzy subgroup ν, written $\mu\lhd\nu$, if $\mu(xyx^{-1})\geq\mu(y)\wedge\nu(x)$, for all $x,y\in G$.

The set of all fuzzy subsets of X is denoted by $FP(X)$.

Definition 2.3 [5] We define the binary operation " $\circ$ " on $FP(X)$ as follows: for all $\mu,\nu\in FP(X)$ and $x\in G$, $(\mu\circ\nu)(x)=\vee\{\mu(y)\wedge\nu(z)|\ y,z\in G,\ yz=x\}$. We call $\mu\circ\nu$ the product of μ and ν.

Theorem 2.3 [6] Let G be a group and μ,ν,η be fuzzy subsets of G. Then $\mu\subseteq\nu\Rightarrow\eta\circ\mu\subseteq\eta\circ\nu,\mu\circ\eta\subseteq\nu\circ\eta$.

Definition 2.4 [7] A fuzzy subset μ of G of the form

$$\mu(y)=\begin{cases}t, & y=x,\\ 0, & y\neq x.\end{cases},\text{ where } t\in[0,1],$$

is said to be a fuzzy point with support x and value t and is denoted by x_t.

3. Main results

A fuzzy point (Definition 2.4) related to a fuzzy subgroup can be considered a element of the fuzzy subgroup. Therefore, in the following, we present the notion of fuzzy coset by means of the notion of fuzzy point.

Definition 3.1 Let μ, ν be fuzzy subgroups of a group G, $\mu\subseteq\nu$ and $x\in G$. Then $x_{\nu(x)}\circ\mu$ ($\mu\circ x_{\nu(x)}$) is called a left (right) fuzzy coset of μ relative to ν.

Remark 3.1 It is easy to check that when we take $\nu=G$ in Definition 3.1 the notion of a fuzzy coset is consistent with the definition of a fuzzy coset which was given in [3,8,9].

Example 3.1 Let $G=S_3$. We define that

$$\mu(x)=\begin{cases}1, & x=e,\\ 0.6, & x\in\{(123),(132)\},\\ 0, & otherwise.\end{cases}\quad \nu(x)=\begin{cases}1, & x=e,\\ 0.8, & x\in\{(123),(132)\},\\ 0, & otherwise.\end{cases}$$

By Theorem 2.2, it is easy to see that μ and ν are fuzzy subgroups of G, and $\mu\subseteq\nu$. Then, by Definition 3.1, $(123)_{\nu((123))}\circ\mu=(123)_{0.8}\circ\mu$ is a left fuzzy coset of μ relative to ν.

We first give some properties about the product of fuzzy points and fuzzy sets, which will help us explore the notion of fuzzy coset.

Proposition 3.1 Let μ be a fuzzy subset of a group G. Then, for all $x\in G$ and $t\in[0,1]$, the following assertions hold:
(1) $(x_t\circ\mu)\circ e_t=e_t\circ(x_t\circ\mu)=x_t\circ\mu$. (2) $(\mu\circ x_t)\circ e_t=e_t\circ(\mu\circ x_t)=\mu\circ x_t$. (3) $x_t\circ x_t^{-1}=x_t^{-1}\circ x_t=e_t$. (4) $x_t\circ e_t=e_t\circ x_t=x_t$.

Proof. (1) Let $z\in G$. Then

$$((x_t\circ\mu)\circ e_t)(z)=(x_t\circ\mu)(z)\wedge t=t\wedge\mu(x^{-1}z)\wedge t=t\wedge\mu(x^{-1}z);$$

$$(e_t\circ(x_t\circ\mu))(z)=t\wedge(x_t\circ\mu)(z)=t\wedge\mu(x^{-1}z)\wedge t=t\wedge\mu(x^{-1}z).$$

Hence $((x_t\circ\mu)\circ e_t)(z)=(e_t\circ(x_t\circ\mu))(z)=t\wedge\mu(x^{-1}z)=(x_t\circ\mu)(z)$. Thus $(x_t\circ\mu)\circ e_t=e_t\circ(x_t\circ\mu)=x_t\circ\mu$. Similarly, (2) holds.

(3) For any $z\in G$, if $z=e$, then by Definition 2.3 we have that $(x_t\circ x_t^{-1})(e)=x_t(x)\wedge x_t^{-1}(x^{-1})=t\wedge t=t$. If $z\neq e$, we can have $x_t(z_1)=0$ or $x_t^{-1}(z_2)=0$ for any $z_1z_2=z$, $z_1,z_2\in G$. Hence,

$$(x_t\circ x_t^{-1})(z)=\vee\{x_t(z_1)\wedge x_t^{-1}(z_2)\,|\,z_1,z_2\in G,\ z_1z_2=z\}=0.$$

Thus, by Definition 2.4 that $x_t\circ x_t^{-1}=e_t$. Similarly, $x_t^{-1}\circ x_t=e_t$.

(4) For any $z\in G$, if $z=x$, then by Definition 2.3 we have that

$$(x_t\circ e_t)(z)=(x_t\circ e_t)(x)=x_t(x)\wedge e_t(e)=t\wedge t=t,$$

then, for any $z_1z_2=z$, $z_1,z_2\in G$, we can have $x_t(z_1)=0$ or $e_t(z_2)=0$ for any $z_1z_2=z$, $z_1,z_2\in G$. Hence,

$$(x_t\circ e_t)(z)=\vee\{x_t(z_1)\wedge e_t(z_2)\,|\,z_1,z_2\in G,\ z_1z_2=z\}=0.$$

It follows from Definition 2.4 that $x_t\circ e_t=x_t$. Similarly, $e_t\circ x_t=x_t$.

In the following, we shall present the conditions under which two fuzzy cosets are identical in the fuzzy group theory, and establishes the relationship between the left cosets and the right cosets.

Theorem 3.1 Let μ, ν be fuzzy subgroups of a group G and $\mu \subseteq \nu$. Then, for all $x, y \in G$, the following assertions hold:

(1) $x_{\nu(x)} \circ \mu = y_{\nu(y)} \circ \mu$

$\Leftrightarrow \nu(x) \wedge \mu(x^{-1}y) = \nu(y) \wedge \mu(e)$ *and* $\nu(y) \wedge \mu(y^{-1}x) = \nu(x) \wedge \mu(e)$.

(2) $\mu \circ x_{\nu(x)} = \mu \circ y_{\nu(y)}$

$\Leftrightarrow \mu(xy^{-1}) \wedge \nu(y) = \mu(e) \wedge \nu(x)$ *and* $\mu(yx^{-1}) \wedge \nu(x) = \mu(e) \wedge \nu(y)$.

(3) $x_{\nu(x)} \circ \mu = y_{\nu(y)} \circ \mu \Leftrightarrow \mu \circ x^{-1}_{\nu(x)} = \mu \circ y^{-1}_{\nu(y)}$.

Proof. (1) "$\Rightarrow$" For any $z \in G$, we have

$$\left(x_{\nu(x)} \circ \mu\right)(z) = \nu(x) \wedge \mu(x^{-1}z) \text{ and } \left(y_{\nu(y)} \circ \mu\right)(z) = \nu(y) \wedge \mu(y^{-1}z).$$

Let $x_{\nu(x)} \circ \mu = y_{\nu(y)} \circ \mu$. Then $\nu(x) \wedge \mu(x^{-1}z) = \nu(y) \wedge \mu(y^{-1}z)$. Taking $z = x$, we conclude that $\nu(y) \wedge \mu(y^{-1}x) = \nu(x) \wedge \mu(e)$; and taking $z = y$, we conclude that $\nu(x) \wedge \mu(x^{-1}y) = \nu(y) \wedge \mu(e)$.

"$\Leftarrow$" For any $z \in G$, on the one hand, we have $\left(x_{\nu(x)} \circ \mu\right)(z)$

$= \nu(x) \wedge \mu(x^{-1}z) = \nu(x) \wedge \mu(x^{-1}yy^{-1}z) \geq \nu(x) \wedge \mu(x^{-1}y) \wedge \mu(y^{-1}z)$.

By the condition $\nu(x) \wedge \mu(x^{-1}y) = \nu(y) \wedge \mu(e)$, we conclude that

$\nu(x) \wedge \mu(x^{-1}y) \wedge \mu(y^{-1}z) = \nu(y) \wedge \mu(e) \wedge \mu(y^{-1}z) = \nu(y) \wedge \mu(y^{-1}z)$.

Hence

$$\left(x_{\nu(x)} \circ \mu\right)(z) \geq \nu(y) \wedge \mu(y^{-1}z) = \left(y_{\nu(y)} \circ \mu\right)(z).$$

On the other hand, we have $\left(y_{\nu(x)} \circ \mu\right)(z) = \nu(y) \wedge \mu(y^{-1}z)$

$= \nu(y) \wedge \mu(y^{-1}xx^{-1}z) \geq \nu(y) \wedge \mu(y^{-1}x) \wedge \mu(x^{-1}z)$.

So by the condition $\nu(y) \wedge \mu(y^{-1}x) = \nu(x) \wedge \mu(e)$, we conclude that

$\nu(y) \wedge \mu(y^{-1}x) \wedge \mu(x^{-1}z) = \nu(x) \wedge \mu(e) \wedge \mu(x^{-1}z) = \nu(x) \wedge \mu(x^{-1}z)$.

Thus $\left(y_{\nu(y)} \circ \mu\right)(z) \geq \nu(x) \wedge \mu(x^{-1}z) = \left(x_{\nu(x)} \circ \mu\right)(z)$. Therefore, $x_{\nu(x)} \circ \mu = y_{\nu(y)} \circ \mu$. Similar reasoning shows that (2) holds.

(3) For any $x, y \in G$, applying (1), (2) and Theorem 2.1, we conclude that

$\mu \circ x^{-1}_{\nu(x)} = \mu \circ y^{-1}_{\nu(y)}$

$\Leftrightarrow \mu(x^{-1}y) \wedge \nu(y^{-1}) = \mu(e) \wedge \nu(x^{-1})$ *and* $\mu(y^{-1}x) \wedge \nu(x^{-1}) = \mu(e) \wedge \nu(y^{-1})$

$\Leftrightarrow \mu(y^{-1}x) \wedge \nu(y) = \mu(e) \wedge \nu(x)$ and $\mu(x^{-1}y) \wedge \nu(x) = \mu(e) \wedge \nu(y)$

$\Leftrightarrow \mu(y^{-1}x) \wedge \nu(y) = \mu(e) \wedge \nu(x)$ and $\mu(x^{-1}y) \wedge \nu(x) = \mu(e) \wedge \nu(y)$

$\Leftrightarrow \nu(x) \wedge \mu(x^{-1}y) = \nu(y) \wedge \mu(e)$ and $\nu(y) \wedge \mu(y^{-1}x) = \nu(x) \wedge \mu(e)$

$\Leftrightarrow x_{\nu(x)} \circ \mu = y_{\nu(y)} \circ \mu$.

Lemma 3.1 Let μ be a fuzzy subgroup of a group G. Then, for any $x \in G$, $e_{\nu(x)} \circ \mu \circ e_{\nu(x)} = e_{\nu(x)} \circ \mu = \mu \circ e_{\nu(x)}$.

Proof. It is follows directly from Proposition 3.1.

Now we give the characterizations of a normal fuzzy subgroup.

Theorem 3.2 Let μ, ν be fuzzy subgroups of a group G and $\mu \subseteq \nu$. Then the following assertions are equivalent:

(1) μ is a normal fuzzy subgroup of the fuzzy subgroup ν.

(2)$\forall x \in G, x_{\nu(x)} \circ \mu \circ x^{-1}_{\nu(x)} \subseteq e_{\nu(x)} \circ \mu$. (3)$\forall x \in G, x_{\nu(x)} \circ \mu \circ x^{-1}_{\nu(x)} \supseteq e_{\nu(x)} \circ \mu$.

(4)$\forall x \in G, x_{\nu(x)} \circ \mu \supseteq \mu \circ x_{\nu(x)}$. (5)$\forall x \in G, x_{\nu(x)} \circ \mu \subseteq \mu \circ x_{\nu(x)}$.

(6)$\forall x \in G, x_{\nu(x)} \circ \mu = \mu \circ x_{\nu(x)}$. (7)$\forall x \in G, x_{\nu(x)} \circ \mu \circ x^{-1}_{\nu(x)} = e_{\nu(x)} \circ \mu$.

Proof. (1) $\Rightarrow$ (2) Let $x \in G$. For any $z \in G$, since $\mu \lhd \nu$, it follows by Definition 2.2 that $\mu(x(x^{-1}zx)x^{-1}) \geq \nu(x) \wedge \mu(x^{-1}zx)$. Hence

$$\left(e_{\nu(x)} \circ \mu\right)(z) = \nu(x) \wedge \mu(z) = \nu(x) \wedge \mu(x(x^{-1}zx)x^{-1})$$

$$\geq \nu(x) \wedge \nu(x) \wedge \mu(x^{-1}zx) = \nu(x) \wedge \mu(x^{-1}zx) = \left(x_{\nu(x)} \circ \mu \circ x^{-1}_{\nu(x)}\right)(z).$$

Therefore $x_{\nu(x)} \circ \mu \circ x^{-1}_{\nu(x)} \subseteq e_{\nu(x)} \circ \mu$.

(2) $\Rightarrow$ (3) Let $x \in G$. By assertion (2), we have $x^{-1}_{\nu(x^{-1})} \circ \mu \circ (x^{-1})^{-1}_{\nu(x^{-1})} \subseteq e_{\nu(x^{-1})} \circ \mu$, that is, $x^{-1}_{\nu(x)} \circ \mu \circ x_{\nu(x)} \subseteq e_{\nu(x)} \circ \mu$ (By Theorem 2.1, $\nu(x^{-1}) = \nu(x)$). Hence, by Theorem 2.3, we have

$$x_{\nu(x)} \circ \left(x^{-1}_{\nu(x)} \circ \mu \circ x_{\nu(x)}\right) \circ x^{-1}_{\nu(x)} \subseteq x_{\nu(x)} \circ \left(e_{\nu(x)} \circ \mu\right) \circ x^{-1}_{\nu(x)}.$$

Applying Proposition 3.1 and Lemma 3.1, we conclude that

$$x_{\nu(x)} \circ \left(x^{-1}_{\nu(x)} \circ \mu \circ x_{\nu(x)}\right) \circ x^{-1}_{\nu(x)} = \left(x_{\nu(x)} \circ x^{-1}_{\nu(x)}\right) \circ \mu \circ \left(x_{\nu(x)} \circ x^{-1}_{\nu(x)}\right)$$

$$= e_{\nu(x)} \circ \mu \circ e_{\nu(x)} = e_{\nu(x)} \circ \mu;$$

$$x_{\nu(x)} \circ \left(e_{\nu(x)} \circ \mu\right) \circ x^{-1}_{\nu(x)} = \left(x_{\nu(x)} \circ e_{\nu(x)}\right) \circ \mu \circ x^{-1}_{\nu(x)} = x_{\nu(x)} \circ \mu \circ x^{-1}_{\nu(x)}.$$

Hence $e_{\nu(x)} \circ \mu \subseteq x_{\nu(x)} \circ \mu \circ x^{-1}_{\nu(x)}$.

(3) $\Rightarrow$ (4) Let $x \in G$. By assertion (3) and Theorem 2.3, we have $\left(x_{\nu(x)} \circ \mu \circ x^{-1}_{\nu(x)}\right) \circ x_{\nu(x)} \supseteq \left(e_{\nu(x)} \circ \mu\right) \circ x_{\nu(x)}$. Applying Proposition 3.1,

$$\left(x_{\nu(x)} \circ \mu \circ x^{-1}_{\nu(x)}\right) \circ x_{\nu(x)} = \left(x_{\nu(x)} \circ \mu\right) \circ \left(x^{-1}_{\nu(x)} \circ x_{\nu(x)}\right) = \left(x_{\nu(x)} \circ \mu\right) \circ e_{\nu(x)} = x_{\nu(x)} \circ \mu;$$

$$\left(e_{\nu(x)} \circ \mu\right) \circ x_{\nu(x)} = e_{\nu(x)} \circ \left(\mu \circ x_{\nu(x)}\right) = \mu \circ x_{\nu(x)}.$$

Hence $x_{\nu(x)} \circ \mu \supseteq \mu \circ x_{\nu(x)}$.

(4) $\Rightarrow$ (5) The proof is similar to that proof of (2) $\Rightarrow$ (3).

(5) $\Rightarrow$ (6) Similar to the proof of (2) $\Rightarrow$ (3), we can prove that for all $x \in G$, $x_{\nu(x)} \circ \mu \supseteq \mu \circ x_{\nu(x)}$. Thus $x \in G$, $x_{\nu(x)} \circ \mu = \mu \circ x_{\nu(x)}$.

(6) $\Rightarrow$ (7) Let $x \in G$. By assertion (6), we have $x_{\nu(x)} \circ \mu = \mu \circ x_{\nu(x)}$. Hence, by Theorem 2.3, we have $\left(x_{\nu(x)} \circ \mu\right) \circ x_{\nu(x)}^{-1} = \left(\mu \circ x_{\nu(x)}\right) \circ x_{\nu(x)}^{-1}$.

Applying Proposition 3.1 and Lemma 3.1, we conclude that $\left(\mu \circ x_{\nu(x)}\right) \circ x_{\nu(x)}^{-1} = \mu \circ \left(x_{\nu(x)} \circ x_{\nu(x)}^{-1}\right) = \mu \circ e_{\nu(x)} = e_{\nu(x)} \circ \mu$. Hence $x_{\nu(x)} \circ \mu \circ x_{\nu(x)}^{-1} = e_{\nu(x)} \circ \mu$.

(7) $\Rightarrow$ (1) We shall prove that $\mu \lhd \nu$. For any $x, y \in G$, by assertion (7), $x_{\nu(x)} \circ \mu \circ x_{\nu(x)}^{-1} = e_{\nu(x)} \circ \mu$ and so $\left(x_{\nu(x)} \circ \mu \circ x_{\nu(x)}^{-1}\right)(y) = \left(e_{\nu(x)} \circ \mu\right)(y)$.

Thus, $\nu(x) \wedge \mu(x^{-1}yx) = \left(x_{\nu(x)} \circ \mu \circ x_{\nu(x)}^{-1}\right)(y) = \left(e_{\nu(x)} \circ \mu\right)(y) = \nu(x) \wedge \nu(y)$, So $\mu(x^{-1}yx) \geq \nu(x) \wedge \mu(x^{-1}yx) = \nu(x) \wedge \nu(y)$. We have proved that $\forall x, y \in G$, $\mu(x^{-1}yx) \geq \nu(x) \wedge \nu(y)$. Thus, by Definition 2.2 , $\mu \lhd \nu$.

References

1. A. Rosenfeld, Fuzzy groups, J. Math. Anal.Appl. 35(1971)512-517.
2. L.A.Zadeh, Fuzzy sets, Inform. and Control 8(1965)338-353.
3. P.S. Das, Fuzzy groups and level subgroups, J. Math. Anal. Appl. 84(1981)264-269.
4. W. Wu, Normal fuzzy subgroups, Fuzzy Math 1(1981)21-30.
5. J.N. Mordeson, K.R. Bhutani, A. Rosenfeld, Fuzzy Group Theroy, Spinger, Berlin, Heidelberg, 2005.
6. B. Yao, Fuzzy Group Theory and Fuzzy Ring Theory, Science Press, Beijing, 2008.
7. P. Pu, Y. Liu, Fuzzy topology I: Neighbourhood structure of a fuzzy point and Moore-Smith convergence, J. Math. Anal.Appl. 76(1980)571-599.
8. N.P. Mukherjee and P. Bhattacharya, Fuzzy normal subgroups and fuzzy cosets, Inform. Sci. 34(1984)225-239.
9. F.I. Sidky and M. A. Mishref, Fuzzy cosets and cyclic and Abelian fuzzy subgroups, Fuzzy Sets and Systems, 43(1991)243-250.

TWO TOPOLOGICAL OPERATORS OF FUZZY FINITE AUTOMATA*

ZHIWEN MO†, NA FENG

College of Mathematics and Software Science, Sichuan Normal University Chengdu, P.R.China, 610066

In this paper, we first introduce some important concepts in fuzzy finite automata (FFA). And then we define bifuzzy source and bifuzzy successor operators of FFA, as well as discuss their basic properties and their essential relationships. Afterwards, we study bifuzzy topologies derived from bifuzzy source and bifuzzy successor operators .

1. Introduction

The notion of automaton was first fuzzified by Wee[6] in 1967. Thereafter, there were many authors, such as Santos[5], Lee and Zadeh[11] et al. who removed made significant contribution to this field. For details, we refer to Refs.[1,2] and especially the recent ones like Malik and Mordeson[3,4]. Also, Daowen Qiu has done much useful work in this field[8], he used representation of fuzzy logic, it is worth pointing out that a similar method was earlier utilized in [7] for dealing with the so-called fuzzifying topology. For the convenience of readers who are unfamiliar with fuzzy logic, he rewrote the successor by using its truth valuation, and he defined the bifuzzy source and successor operators. Then , Zhiwen Mo and Qian Chen[9,10] studied basic properties and essential relationships of two operators, and the bifuzzy topologies derived from them. We defined a series of operators which used the removed method of Daowen Qiu[8], and they have very similar properties to the bifuzzy source and successor operators. In this paper, our purpose is to investigate common properties of these operators in FFA, and to supplement and consummate the systems of algebra and fuzzy topology.

The basic structure of the paper is as follows. In Section 2, we introduce some basic concepts and their properties on FFA. In Section 3, we advance bifuzzy source and successor operators and discuss their properties. In Section 4,

* This work is supported by the National Natural Science Foundation of China(No.11071178).

† email: mozhiwen@263.net

we study the bifuzzy topologies derived from the bifuzzy source and successor operators.

2. Preliminaries

Definition 2.1[8] A triple $M = (Q, \Sigma, \mu)$ is called an FFA , where Q is a nonempty finite set of states, Σ is a finite set of input alphabets and μ is a mapping μ: $Q\times\Sigma\times Q\rightarrow[0,1]$ called the transition function.

Definition 2.2[8] Let (Q, Σ, μ) be an FFA . For any $p, q\in Q$, any $x\in\Sigma^*$ and $a\in\Sigma$, we define μ^*: $Q\times\Sigma^*\times Q\rightarrow[0,1]$ as follows :

$$\mu^*(q,\varepsilon,p)=\begin{cases}1, & q=p\\ 0, & \text{otherwise}\end{cases},$$

and $\mu^*(q,ax,p)=\sup_{r\in Q}\min\{\mu(q,a,r), \mu^*(r,x,p)\}$.

Lemma 2.3[8] Let (Q, Σ, μ) be an FFA . Then

$$\mu^*(q,xy,p)=\sup_{r\in Q}\min\{\mu^*(q,x,r),\mu^*(r,y,p)\},$$

for any $p, q\in Q$ and $x, y\in\Sigma^*$.

3. Some properties of bifuzzy source and successor operators

Definition3.1 Let $M=(Q, \Sigma, \mu)$be an FFA, F(Q) is a fuzzy power set on Q,$\bar{A}\in$F(Q), SO* and SU* are called bifuzzy source and bifuzzy successor operators from F(Q) to F(Q), respectively, they are defined as follows:

$$SO^*(\bar{A})(p)=\sup_{q\in Q}\sup_{x\in\Sigma^*}\mu^*(p,x,q)\oplus\bar{A}(q)$$

$$SU^*(\bar{A})(p)=\sup_{q\in Q}\sup_{x\in\Sigma^*}\mu^*(q,x,p)\oplus\bar{A}(q)$$

For any $a, b\in[0,1]$, $k\in[0,1]$, the symbol $\oplus$ satisfies $a\oplus b=\max\{0, ka+b-k\}$.
Remark, when $x=\varepsilon$, $p=q$, $SO^*(\bar{A})(p) = SU^*(\bar{A})(p) = \bar{A}(p)$, So that

$$SO^*(\bar{A})(p)\geq\bar{A}(p), SU^*(\bar{A})(p)\geq\bar{A}(p).$$

Theorem 3.1 Let $M=(Q, \Sigma, \mu)$ be an FFA, and let SO* and SU* be the operators in Definition 3.1. For any $\bar{A}_i\in F(Q)$, $i\in I$, any $p\in Q$, then

(1) $SO^*(\varnothing) = SU^*(\varnothing) = \varnothing$, $SO^*(Q) = SU^*(Q) = Q$.

(2) $SO^*(\cap_{i\in I}\bar{A}_i(p))\leq\inf_{i\in I}SO^*(\bar{A}_i)(p)$, $SU^*(\cap_{i\in I}\bar{A}_i(p))\leq\inf_{i\in I}SU^*(\bar{A}_i)(p)$,

(3) $SO^*(\cup_{i\in I}\bar{A}_i(p))=\sup_{i\in I}SO^*(\bar{A}_i)(p)$, $SU^*(\cup_{i\in I}\bar{A}_i(p)) = \sup_{i\in I}SU^*(\bar{A}_i)(p)$.

Proof. (1) Immediately $SO^*(\varnothing) = SU^*(\varnothing) = \varnothing$, $SO^*(Q)=SU^*(Q) = Q$.

(2) $SO^*(\cap_{i\in I}\bar{A}_i(p)) = \sup_{q\in Q}\sup_{x\in\Sigma^*}\max\{0, k\mu^*(p,x,q)+\inf_{i\in I}\bar{A}_i(q)-k\}$

$$= \sup_{q\in Q} \sup_{x\in \Sigma^*} \inf_{i\in I} \max \{0, k\mu^*(p,x,q)+\tilde{A}_i(q)-k\}$$

$$\leq \inf_{i\in I} \sup_{q\in Q} \sup_{x\in \Sigma^*} \max \{0, k\mu^*(p,x,q) + \tilde{A}_i(q) - k\}$$

$$= \inf_{i\in I} \mathrm{SO}^*(\tilde{A}_i)(p)$$

Similarly, $\mathrm{SU}^*(\cap_{i\in I}\tilde{A}_i(p)) \leq \inf_{i\in I} \mathrm{SU}^*(\tilde{A}_i)(p)$.

(3) $\mathrm{SO}^*(\cup_{i\in I}\tilde{A}_i(p)) = \sup_{q\in Q} \sup_{x\in \Sigma^*} \max \{0, k\mu^*(p,x,q) + \sup_{i\in I} \tilde{A}_i(q) - k\}$

$$= \sup_{i\in I} \sup_{q\in Q} \sup_{x\in \Sigma^*} \max \{0, k\mu^*(p,x,q) + \tilde{A}_i(q) - k\}$$

$$= \sup_{i\in I} \mathrm{SO}^*(\tilde{A}_i)(p)$$

Using similar method, we can obtain $\mathrm{SU}^*(\cup_{i\in I}\tilde{A}_i(p)) = \sup_{i\in I} \mathrm{SU}^*(\tilde{A}_i)(p)$.

Theorem3.2 $\mathrm{SU}^*(\mathrm{SO}^*(\tilde{A})) = \mathrm{SO}^*(\mathrm{SU}^*(\tilde{A}))$.

Proof. Let

$$M_1 = \mathrm{SU}^*(\mathrm{SO}^*(\tilde{A}))(p)$$

$$= \sup_{q\in Q} \sup_{x\in \Sigma^*} \sup_{r\in Q} \sup_{y\in \Sigma^*} \mu^*(q,x,p) \oplus (\mu^*(q,y,r) \oplus \tilde{A}(r)) \quad (1)$$

$$M_2 = \mathrm{SO}^*(\mathrm{SU}^*(\tilde{A}))(p)$$

$$= \sup_{q_1\in Q} \sup_{x_1\in \Sigma^*} \sup_{r_1\in Q} \sup_{y_1\in \Sigma^*} \mu^*(p,x_1,q_1) \oplus (\mu^*(r_1,y_1,q_1) \oplus \tilde{A}(r_1)) \quad (2)$$

Firstly , we proof $M_1 \leq M_2$, it means that when $M_1 > t$ then $M_2 > t$. When $M_1 > t$, there exists $q_0, r_0 \in Q$, $x_0, y_0 \in \Sigma^*$, satisfying

$$\mu^*(q_0,x_0,p) \oplus (\mu^*(q_0,y_0,r_0) \oplus \tilde{A}(r_0)) > t \quad (3)$$

Suppose $M_2 > t$ does not hold, then $M_2 \leq t$.Then

$$\mu^*(p, x_1,q_1) \oplus (\mu^*(r_1,y_1,q_1) \oplus \tilde{A}(r_1)) \leq t \quad (4)$$

for any $q_1, r_1 \in Q$, $x_1, y_1 \in \Sigma^*$. When $q_1 = p$, $x_1 = \varepsilon$,

$$\mu^*(p,\varepsilon,p) \oplus (\mu^*(r_1,y_1,p) \oplus \tilde{A}(r_1)) \leq t, \text{ so } \mu^*(r_1,y_1,p) \oplus \tilde{A}(r_1) \leq t \quad (5)$$

When $r_1 = r_0, y_1 = y_0x_0$, then $\mu^*(r_1,y_1,p) = \sup_{s\in Q} \min \{\mu^*(r_0,y_0,s), \mu^*(s, x_0,p)\} \geq \max \{\mu^*(r_0,y_0,q_0), \mu^*(q_0,x_0,p)\} \geq \mu^*(q_0, x_0, p)$.

From (5) we have $\mu^*(q_0, x_0, p) \oplus \tilde{A}(r_0) \leq t$, which contradicts (3). So $M_1 \leq M_2$. On the other hand, we can proof that $M_2 \leq M_1$ similarly. Hence, the theorem holds.

4. Bifuzzy topologies derived from bifuzzy source and bifuzzy successor operators

Bifuzzy topology is defined by Daowen Qiu in [8]. It is as follows:

Definition 4.1[8] Let X be a nonempty set, and let $\mathcal{T}\in F(F(X))$. Then $\mathcal{T}$ is called a bifuzzy topology on X if and only if the following conditions are fulfilled:

(i) $\mathcal{T}(X)=\mathcal{T}(\varnothing)=1$,

(ii) For any $\tilde{A}$, $\tilde{B}\in F(X)$,$\min(\mathcal{T}(\tilde{A}),\mathcal{T}(\tilde{B}))\leq \mathcal{T}(\tilde{A}\cap\tilde{B})$.

(iii) For any $\tilde{A}_i\in F(X),i\in I$, $\inf_{i\in I}\mathcal{T}(\tilde{A}_i)\leq \mathcal{T}(\cup_{i\in I}\tilde{A}_i)$.

Lemma 4.1[8] Let X be a nonempty set. For any $\tilde{A},\tilde{B},\tilde{A}_i,\tilde{B}_i\in F(Q)$, any $x\in X$, $i\in I$ (I is an index set), define v as follows:

(i) $\nu(\tilde{A}\subseteq\tilde{B})=\inf_{x\in X}\min(1,1-\tilde{A}(x)+\tilde{B}(x))$,

(ii) $\nu(\tilde{A}\equiv\tilde{B})=\min(\nu(\tilde{A}\subseteq\tilde{B}),\nu(\tilde{B}\subseteq\tilde{A}))$,

So we have $\inf_{i\in I}\nu(\tilde{A}_i\equiv\tilde{B}_i)\leq\min(\nu(\cup_{i\in I}\tilde{A}\equiv\cup_{i\in I}\tilde{B}),\nu(\cap_{i\in I}\tilde{A}\equiv\cap_{i\in I}\tilde{B}))$.

Definition 4.2 Let $M=(Q,\Sigma,\mu)$ be an FFA. Utilizing the definition of SU^* and SO^* we shall establish bifuzzy topologies $\mathcal{T}_i\in F(F(Q))(i=1,2,\cdots\cdots,7)$ in the following. For any $\tilde{A}\in F(Q)$,define

(1) $\mathcal{T}_1(\tilde{A})=\nu(SU^*(\tilde{A})\equiv\tilde{A})$,

(2) $\mathcal{T}_2(\tilde{A})=\nu(SO^*(\tilde{A})\equiv\tilde{A})$,

(3) $\mathcal{T}_3(\tilde{A})=\nu(SU^*(SO^*(\tilde{A}))\equiv\tilde{A})$,

(4) $\mathcal{T}_4(\tilde{A})=\nu(SO^*(SU^*(\tilde{A}))\equiv SU^*(\tilde{A}))$,

(5) $\mathcal{T}_5(\tilde{A})=\nu(SU^*(SO^*(\tilde{A}))\equiv SO^*(\tilde{A}))$,

(6) $\mathcal{T}_6(\tilde{A})=\nu(SU^*(\tilde{A})\equiv SO^*(\tilde{A}))$,

(7) $\mathcal{T}_7=\nu(SU^*(SO^*(\tilde{A}))\equiv SO^*(SU^*(\tilde{A}))=1$.

In the following theorem, we prove that $\mathcal{T}_i$ are bifuzzy topologies on Q.

Theorem 4.1 Let $M=(Q,\Sigma,\mu)$ be an FFA. $\mathcal{T}_i\in F(F(Q))$ $(i=1,2,\cdots\cdots,7)$ are defined as Definition 4.2 , then $\mathcal{T}_i$ are bifuzzy topologies on Q.

Proof . By Theorem 3.1 and Lemma 2.6 we can obtain that $\mathcal{T}_1$, $\mathcal{T}_2$, $\mathcal{T}_3$ are bifuzzy topology on Q. Now we prove $\mathcal{T}_4$ is a bifuzzy topology on Q.

(1) By Theorem 3.1we have $\mathcal{T}_4(\varnothing)=1$, $\mathcal{T}_4(Q)=1$

(2) Suppose $\inf_{i\in I}\mathcal{T}_4(\tilde{A}_i)\leq\mathcal{T}_4(\cap_{i\in I}\tilde{A}_i)$ does not hold. There exits I_0,

$$\inf_{i\in I_0}\mathcal{T}_4(\tilde{A}_i)>\mathcal{T}_4(\cap_{i\in I_0}\tilde{A}_i) \tag{6}$$

$$\mathcal{T}_4(\cap_{i\in I}\tilde{A}_i)=\inf_{x\in Q}\min\{1,1-SO^*(SU^*(\cap_{i\in I_0}\tilde{A}_i))(x)+SU^*(\cap_{i\in I_0}\tilde{A}_i)(x)\}.$$

Thus, there exists $m\in Q$,

$$\mathcal{T}_4(\cap_{i\in I}\mathcal{A}_i) = 1 - SO^*(SU^*(\cap_{i\in I_0}\mathcal{A}_i))(m) + SU^*(\cap_{i\in I_0}\mathcal{A}_i)(m) \quad (7)$$

Since $\inf_{i\in I_0}\mathcal{T}_4(\mathcal{A}_i) = \inf_{i\in I_0}\inf_{x\in Q}\min\{1, 1 - SO^*(SU^*(\mathcal{A}_i))(x) + SU^*(\mathcal{A}_i)(x)\}$, for any $i\in I_0$, $x\in Q$ we have $1 - SO^*(SU^*(\mathcal{A}_i))(x) + SU^*(\mathcal{A}_i)(x) > \mathcal{T}_4(\cap_{i\in I_0}\mathcal{A}_i)$. Since i is arbitrarily chosen, we know ,for any i, $j \in I_0$,$x \in Q$, we have

$$1 - SO^*(SU^*(\mathcal{A}_j))(x) + SU^*(\mathcal{A}_j)(x) > \mathcal{T}_4(\cap_{i\in I_0}\mathcal{A}_i) \quad (8)$$

For any $i, j \in I_0$, when $x = m$, from (7) we know

$$1 - SO^*(SU^*(\mathcal{A}_j))(m) + SU^*(\mathcal{A}_j)(m) > \mathcal{T}_4(\cap_{i\in I_0}\mathcal{A}_i) \quad (9)$$

from (6) (7) (9)we know $1-SO^*(SU^*(\mathcal{A}_j))(m)+SU^*(\mathcal{A}_j)(m) > 1-SO^*(SU^*(\cap_{i\in I_0}\mathcal{A}_j))(m) +SU^*(\cap_{i\in I_0}\mathcal{A}_j)(m)$. It means that for any $j\in I_0$,

$$SU^*(\mathcal{A}_j)(m)-SO^*(SU^*(\mathcal{A}_j))(m) > SU^*(\cap_{i\in I_0}\mathcal{A}_i)(m)-SO^*(SU^*(\cap_{i\in I_0}\mathcal{A}_i))(m) \quad (10)$$

Let $M = SU^*(\mathcal{A}_j)(m) - SO^*(SU^*(\mathcal{A}_j))(m)$
$=\sup_{r_2\in Q}\sup_{x_2\in \Sigma^*}\mu^*(r_2,x_2,m)\oplus\mathcal{A}_j(r_2)-\sup_{r_1\in Q}\sup_{x_1\in \Sigma^*}\mu^*(m,x_1,r_1)\oplus SU^*(\mathcal{A}_j)(r_1)$

Therefore exists $r_0 \in Q$,$x_0 \in \Sigma^*$,$SU^*(\mathcal{A}_j)(m) = \mu^*(r_0,x_0,m) \oplus \mathcal{A}_j(r_0)$, for any $j \in I_0$,$r_1 \in Q$,$x_1 \in \Sigma^*$,

$$M \le \mu^*(r_0,x_0,m) \oplus \mathcal{A}_j(r_0) - \mu^*(m,x_1,r_1) \oplus SU^*(\mathcal{A}_j)(r_1) \quad (11)$$

Let $N = SU^*(\cap_{i\in I_0}\mathcal{A}_i)(m) - SO^*(SU^*(\cap_{i\in I_0}\mathcal{A}_i))(m) = \sup_{p_2\in Q}\sup_{y_2\in \Sigma^*}\mu^*(p_2,y_2,m)$

$$\oplus(\cap_{i\in I_0}\mathcal{A}_i)(p_2)-\sup_{p_1\in Q}\sup_{y_1\in \Sigma^*}\mu^*(m,y_1,p_1)\oplus SU^*(\cap_{i\in I_0}\mathcal{A}_i)(p_1)$$

There exists $p_0\in Q$,$y_0\in \Sigma^*$,$SO^*(SU^*(\cap_{i\in I_0}\mathcal{A}_i))(m) = \mu^*(m,y_0,p_0) \oplus SU^*(\cap_{i\in I_0}\mathcal{A}_i)(p_0)$. Thus, for any $p_2\in Q$,$y_2 \in \Sigma^*$,

$$N > \mu^*(p_2,y_2,m) \oplus (\cap_{i\in I_0}\mathcal{A}_i)(p_2) - \mu^*(m,y_0,p_0) \oplus SU^*(\cap_{i\in I_0}\mathcal{A}_i)(p_0) \quad (12)$$

For any $j\in I_0$, r_1, $p_2\in Q$, x_1, $y_2\in \Sigma^*$, by (10)(11)(12) we have,

$$\mu^*(r_0,x_0,m)\oplus\mathcal{A}_j(r_0)-\mu^*(m,x_1,r_1)\oplus SU^*(\mathcal{A}_j)(r_1) >\ldots$$
$$\ldots\mu^*(p_2,y_2,m)(\cap_{i\in I_0}\mathcal{A}_i)(p_2) - \mu^*(m,y_0,p_0) \oplus SU^*(\cap_{i\in I_0}\mathcal{A}_i)(p_0) \quad (13)$$

For any $j \in I_0$, let $x_1 = y_0$, $r_1 = p_0$, $p_2 = r_0$, $y_2 = x_0$, we have

$$\mathcal{A}_j(r_0) - SU^*(\mathcal{A}_j)(p_0) > (\cap_{i\in I_0}\mathcal{A}_i)(r_0) - SU^*(\cap_{i\in I_0}\mathcal{A}_i)(p_0) \quad (14)$$

Let $(\cap_{i\in I_0}\tilde{A}_i)(r_0) = \tilde{A}_0(r_0)$. Since j is arbitrarily chosen, let $\tilde{A}_j = \tilde{A}_0$, and it follows that,

$$SU^*(\cap_{i\in I_0}\tilde{A}_i)(p_0) > SU^*(\tilde{A}_0)(p_0) \quad (15)$$

By $\cap_{i\in I_0}\tilde{A}_i \subseteq \tilde{A}_0$, we can obtain that $SU^*(\cap_{i\in I_0}\tilde{A}_i) \subseteq SU^*(\tilde{A}_0)$, Thus, for any r $\in$ Q, $SU^*(\cap_{i\in I_0}\tilde{A}_i)(r) \le SU^*(\tilde{A}_0)(r)$, which is a contradiction , So theorem is hold.
(3)By Lemma 2.1 we can obtain $\inf_{i\in I}\mathcal{T}_4(\tilde{A}_i) \le \mathcal{T}_4(\cup_{i\in I}\tilde{A}_i)$, So $\mathcal{T}_4$ is a bifuzzy topology of FFA.

Similarly, $\mathcal{T}_5$ is a bifuzzy topology of FFA can be obtained.

By Lemma 2.1, Theorem 3.1 and Theorem 3.6 in literature [9], we can obtain $\mathcal{T}_6$ is a bifuzzy topology of FFA.

Immediately $\mathcal{T}_7$ is a bifuzzy topology.

Example 4.1 Let $M = (Q, \Sigma, \mu)$ be an FFA , where $Q=\{p,q\}$,$\Sigma=\{a\}$.$\mu\{p,a,p\}= 0.5$,$\mu\{p,a,q\}=0.6$,$\mu\{q,a,p\}=0.9$,$\mu\{q,a,q\}=0.8$. Take $\tilde{A}(p)=0.7$, $\tilde{A}(q)=0.5$, $\tilde{B}(p)=0.5$, $\tilde{B}(q)=0.95$, $k=1$, $\tilde{A}_i(p)=0.7$, $\tilde{A}_i(q)=0.5(i\in I)$. Then we have $SU^*(\tilde{A})(p)=0.7$, $SU^*(\tilde{B})(p)=0.85$, $SU^*(\tilde{A})(q)=0.5$, $SU^*(\tilde{B})(q)=0.95$, then $\mathcal{T}_1(\tilde{A})=1$, $\mathcal{T}_1(\tilde{B})=0.65$, $\mathcal{T}_1(\tilde{A}\cap\tilde{B})=1$, $\mathcal{T}_1(\tilde{A}_i)=1$. So (i) $\mathcal{T}_1(Q)=\mathcal{T}_1(\varnothing)=1$, (ii)$\min(\mathcal{T}_1(\tilde{A}),\mathcal{T}_1(\tilde{B})) \le \mathcal{T}_1(\tilde{A}\cap\tilde{B})$, (iii)$\inf_{i\in I}\mathcal{T}_1(\tilde{A}_i) \le \mathcal{T}_1(\cup_{i\in I}\tilde{A}_i)$, so we can obtain that $\mathcal{T}_1$ is bifuzzy topology on Q.

Theorem 4.2 $\mathcal{T}_1(SO^*(\tilde{A}))=\mathcal{T}_5(\tilde{A})$, $\mathcal{T}_2(SU^*(\tilde{A}))=\mathcal{T}_4(\tilde{A})$.
Immediately, the theorem holds.

5. Conclusions

In this paper, we introduce the concepts of bifuzzy source and bifuzzy successor operators of FFAS, then discuss their properties. Finally, the operators can derive topologies of FFA. To a certain extent, these results generalize the previous conclusions in literature [8].

References

1. M.M. Gupta, G.N. Saridis, B.R. Gaines. Fuzzy Automata and Decision Processes. *North-Holland*, New York, 1977,pp.133–175.
2. A. Kandel, S.C. Lee. Fuzzy Switching and Automata. Theory and Applications, Arnold, London, 1980.
3. D.S. Malik, J.N. Mordeson. Fuzzy Discrete Structures, *Physica-Verlag*, New York, 2000.
4. J.N. Mordeson, D.S. Malik. Fuzzy Automata and Languages. Theory and Applications, Chapman & Hall/CRC, BocaRaton, London, 2002.
5. E.S. Santos. Maximin automata, *Inform.* Control 1968,13, 363–377.

6. W.G. Wee. On generalizations of adaptive algorithm and application of the fuzzy sets concept to pattern classification,*Ph. D. Thesis*, Purdue University, June, 1967.
7. M.S. Ying. A new approach for fuzzy topology (I), *Fuzzy Sets and Systems*,1991,39:303–321.
8. Qiu D W. Characterizations of fuzzy finite automata[J]. *Fuzzy Sets and Systems*, 2004, 141: 391–414.
9. Mo Zhi-wen,Chen Qian.Several Properties of Two Operators of Fuzzy Finite Automata[J]. *Fuzzy Systems and Mathematics*,2007, 21(1): 75-81.
10. Chen Qian,Mo Zhi-wen,Wan Mei-kai. Characterization of Bifuzzy Topology Based on Fuzzy Finite Automata.*Chinese Journal of Engineering Mathematics*, 2009,26(1):17–22.
11. E. T. Lee and L. A. Zadeh. Note on fuzzy languages Information Science, 1969, 1: 403–419.

ON FUZZY ROUGH CONCEPT LATTICES*

CHANG SHU [a,b]

[a]College of Mathematics and Software Science, Si Chuan Normal University, Chengdu, P.R.China,610066

ZHI WEN MO [b]

[b]College of Management Science, Chengdu University of Technology, Chengdu, P.R.China, 610059

Formal concept analysis and rough set theory provide two different methods for data analysis and knowledge processing. Given a fuzzy context $(U, A, \tilde{R})$, one can get the fuzzy rough concept lattice $L_\delta(U, A, \tilde{R})$ in Will's sense. Further, we apply the attribute reduction method for the concept lattices to the reduction of the fuzzy rough concept lattice. Finally, the experimental results show that the reduction method for the fuzzy rough concept lattice is effective in the reduction of the multiple rules and an important example is given.

1. Introduction

In this paper, we will introduce fuzzy rough concept lattice based on fuzzy sets, RST and FCA. Some sufficient condition for a fuzzy rough concept lattice to be algebraic are given. A sufficient and necessary condition for fuzzy rough concept lattice to be completely distributive is obtained and some subtle examples are constructed. We also introduce the concept of definable sets for a context and study the special sublattice of the fuzzy rough concept lattice consisting of definable sets[1].

* This work is supported by National Natural Science Foundation of China (No.11071178).

2. Preliminaries [2-12]

2.1. *Galois Connections and Concept Lattice*

Definition 1 We first recall the Galois connection between powersets of two sets X and Y (Shortly:Galois connection between X and Y), a pair $(\uparrow,\downarrow)$ of mapping $\uparrow: 2^X \to 2^Y, \downarrow: 2^Y \to 2^X$,satisfying :

1) $A_1 \subseteq A_2$ implies $A_2^\uparrow \subseteq A_1^\uparrow$, 2) $B_1 \subseteq B_2$ implies $B_2^\downarrow \subseteq B_1^\downarrow$, 3) $A \subseteq \left(A^\uparrow\right)^\downarrow$, $B \subseteq \left(B^\downarrow\right)^\uparrow \left(A, A_1, A_2 \in 2^X, B, B_1, B_2 \in 2^Y\right)$

For $A \subseteq X, B \subseteq Y$,let

$A^\uparrow = \{y | \forall x \in A : (x, y) \in I\}, B^\downarrow = \{x | \forall y \in B : (x, y) \in I\}$ (I is the L-relationship between X and Y).

Then $A^\uparrow$ is the set of all attributes from Y shared by all objects of A, and $B^\downarrow$ is the set of all objects from X sharing attributes from B.Therefore,a formal concept is a pair $\langle A, B\rangle$ for which $A^\uparrow = B$ and $B^\downarrow = A$.

Definition 2 If $B\langle X,Y,I\rangle$ denotes the set of all formal concepts i.e., $B\langle X,Y,I\rangle$ defined by $\langle A_1, B_1\rangle \le (A_2, B_2)$ if and only if $A_1 \subseteq A_2$ $(or, equivalently, B_1 \supseteq B_2)$,then the structure $(B\langle X,Y,I\rangle,\le)$ is called the formal concept lattice of $\langle X,Y,I\rangle$.

Lemma 1 [8] For a context $\langle X,Y,I\rangle$, the concept lattice $(B\langle X,Y,I\rangle,\le)$ is a complete lattice .

3. Fuzzy Rough Concept Lattice

We often encounter the fuzzy formal context $(U,V,\tilde{R})$in the application, where U is a non-empty finite set of objects, V is a non-empty finite set of attributes and $\tilde{R}$is a fuzzy relation of Cartesian product $U \times V$. The next, we will study the relationship of the fuzzy formal context and concept lattice.

Table 1. Data table with fuzzy attributes Example 1[13]

	1	2	3	4	5	6	7
1 Czech	0.4	0.4	0.6	0.2	0.2	0.4	0.2
2 Hungary	0.4	1.0	0.4	0.0	0.0	0.4	0.2
3 Poland	0.2	1.0	1.0	0.0	0.0	0.0	0.0
4 Slovakia	0.2	0.6	1.0	0.0	0.2	0.2	0.2
5 Austria	1.0	0.0	0.2	0.2	0.2	1.0	1.0
6 France	1.0	0.0	0.6	0.4	0.4	0.6	0.6
7 Italy	1.0	0.2	0.6	0.0	0.2	0.6	0.4
8 Germany	1.0	0.0	0.6	0.2	0.2	1.0	0.6
9 UK	1.0	0.2	0.4	0.0	0.2	0.6	0.6
10 Japan	1.0	0.0	0.4	0.2	0.2	0.4	0.2
11 Canada	1.0	0.2	0.4	1.0	1.0	1.0	1.0
12 USA	1.0	0.2	0.4	1.0	1.0	0.2	0.4

Definition 3 Let $(U, A, \tilde{R})$ be a fuzzy formal context. $R = \begin{cases} 1, \tilde{R} \geq \delta \\ 0, \tilde{R} \langle \delta \end{cases}$ (δ is a precision and $\delta \in [0,1]$)

How to choose the optimal precision? We introduce a method to choose the optimal precision by F statistics.

Let P is the number of schemes, obviously $P \leq n$. Because the two schemes are not much meaning in practical application which all samples are into their classes or into a class, there are P-2 schemes to choose in the fact.

F statistics:

$$F = \frac{\sum_{i=1}^{r} n_i \sum_{k=1}^{m} (\bar{x}_{ik} - \bar{x}_k)^2 \Big/ (r-1)}{\sum_{i=1}^{r} \sum_{j=1}^{n_i} \sum_{k=1}^{m} (x_{ik} - \bar{x}_{jk})^2 \Big/ (n-r)} \sim F(r-1, n-r)$$

Example 2 In Table 1, $n = 12, P = 5$, there are three significant scheme. Let confidence variables $\alpha = 0.05$ and we calculate the F-value of each scheme listed in the following table:

Table 2

Classification number	5	8	9
δ	0.2	0.4	0.6
F	3.98	5.67	8.9
$F_{0.05}$	4.12	6.09	8.85

$F \succ F_{0.05}$ ($\delta = 0.6$)

$\delta = 0.6$ is best precision, so fuzzy equivalence matrix is given.

Table 3

	1	2	3	4	5	6	7
1 Czech	0	0	1	0	0	0	0
2 Hungary	0	1	0	0	0	0	0
3 Poland	0	1	1	0	0	0	0
4 Slovakia	0	1	1	0	0	0	0
5 Austria	1.0	0.0	0	0	0	1.0	1.0
6 France	1.0	0.0	1	0	0	1	1
7 Italy	1.0	0	1	0.0	0	1	0
8 Germany	1.0	0.0	1	0	0	1.0	1
9 UK	1.0	0	0	0.0	0	1	1
10 Japan	1.0	0.0	0	0	0	0	0
11 Canada	1.0	0	0	1.0	1.0	1.0	1.0
12 USA	1.0	0	0	1.0	1.0	0	0

Definition 4 Let $(U, A, \tilde{R})$ be a fuzzy formal context. Then the following approximation operators are defined :(R is defined in Definition 4)

$\nabla': P(U) \to P(V), A \mapsto \{y \in V | R^{-1}(y) \subseteq A\}$ $\nabla': P(V) \to P(U), B \mapsto \{x \in U | R(x) \subseteq B\}$

$\Delta': P(V) \to P(U), B \mapsto \{x \in U | R(x) \cap B \neq \Phi\}$ $\Delta': P(U) \to P(V), A \mapsto \{y \in V | R^{-1}(y) \cap A \neq \Phi\}$

A pair of sets (A,B) is called a fuzzy rough concept of $(U,V,\tilde{R})$, if $A \subseteq U, B \subseteq V$ and $A^{\nabla'} = B, B^{\Delta'} = A$, where A is called the extent of (A,B) and B is called the intent. A fuzzy rough concept $(B,A) \in (P(V) \times P(U))$ is dually defined. The set of all fuzzy rough concepts of $(U,V,\tilde{R})$ is ordered by $(A_1,B_1) \leq (A_2,B_2) \Leftrightarrow A_1 \subseteq A_2$ (which is equivalent to $B_2 \subseteq B_1$). The set of all fuzzy rough concepts of $(U,V,\tilde{R})$ with order $\leq$ is called the fuzzy rough concept lattice of $(U,V,\tilde{R})$ and denoted as $L_O{}^{\delta}(U,V,\tilde{R})$. Similarly, we denote the fuzzy rough concept lattice of concept as $L_p{}^{\delta}(U,V,\tilde{R})$.

Theorem 1 For a fuzzy context $(U,V,\tilde{R})$, the fuzzy rough concept lattice $L_O{}^{\delta}(U,V,\tilde{R})$ is a complete lattice and the infimum and supremum are given by

$$\bigwedge_{i\in I}(A_i,B_i) = \left(\left(\bigcap_{i\in I} A_i\right)^{\nabla'\Delta'}, \bigcap_{i\in I} B_i\right), \bigvee_{i\in I}(A_i,B_i) = \left(\bigcup_{i\in I} A_i, \left(\bigcup_{i\in I} B_i\right)^{\Delta'\nabla'}\right).$$

Proof. Let a partial order relation

$(A_1,B_1) \leq (A_2,B_2) \Leftrightarrow A_1 \subseteq A_2 (B_2 \subseteq B_1)$ and the infimum and supremum operation $\bigwedge_{i\in I}(A_i,B_i) = \left(\left(\bigcap_{i\in I} A_i\right)^{\nabla'\Delta'}, \bigcap_{i\in I} B_i\right)$,

$\bigvee_{i\in I}(A_i,B_i) = \left(\bigcup_{i\in I} A_i, \left(\bigcup_{i\in I} B_i\right)^{\Delta'\nabla'}\right)$

operation is closed in $L_O{}^{\delta}(U,V,\tilde{R})$, then $L_O{}^{\delta}(U,V,\tilde{R})$ is a complete lattice.

Theorem 2 Let $(U,A,\tilde{R})$ be a fuzzy formal context. Then $L_O{}^{\delta}(U,V,\tilde{R}^c)$ is dually isomorphic to $L_p{}^{\delta}(U,V,\tilde{R}^c)$. Other case also has a similar relationship.

Proof. By definition of ∇' and Δ, we have equations:

$A^{\nabla'} = A^{C\Delta' C}, B^{\Delta'} = B^{C\nabla' C}$, and then it is obvious that $L_O{}^{\delta}(U,V,\tilde{R}^c)$ is dually isomorphic to $L_p{}^{\delta}(U,V,\tilde{R}^c)$ and

$$\{(A^C,B) | (A,B) \in L_O^{\delta}(U,V,\tilde{R}^C)\} = \{(A,B^C) | (A,B) \in L_P^{\delta}(U,V,\tilde{R}^C)\}$$

In the sequel, we mainly discuss properties of fuzzy object-oriented concept. For the fuzzy rough concepts $L_O{}^{\delta}(U,V,\tilde{R})$, the set of all extends ordered with set-inclusion relation is denoted as $L_{O_e}{}^{\delta}(U,V,\tilde{R})$ and the set of all intends ordered with set-inclusion relation is denoted as $L_{O_i}{}^{\delta}(U,V,\tilde{R})$. Then we have

$$L_O{}^{\delta}(U,V,\tilde{R}) \cong L_{O_e}{}^{\delta}(U,V,\tilde{R}) \cong L_{O_i}{}^{\delta}(U,V,\tilde{R})$$

To study properties of $L_O^{\delta}(U,V,\tilde{R})$, we may consider those of $L_{O_e}^{\delta}(U,V,\tilde{R})$ or $L_{O_i}^{\delta}(U,V,\tilde{R})$ instead . Since (∇',Δ') is a Galois connection , we have

$$L_{O_e}^{\delta}(U,V,\tilde{R})=\{B^{\Delta'} | B \subseteq V\}=\{A^{\nabla'} | A \subseteq U\}=\{A \subseteq U | A = A^{\nabla'\Delta'}\}$$

$$L_{O_i}^{\delta}(U,V,\tilde{R})=\{A^{\nabla'} | A \subseteq U\}=\{B^{\Delta'\nabla'} | B \subseteq V\}=\{B \subseteq V | B = B^{\Delta'\nabla'}\}$$

The infimum and supremum in $L_{O_e}^{\delta}(U,V,\tilde{R})$ are given by

$$\wedge_{i\in I} A_i = \left(\bigcap_{i\in I} A_i\right)^{\nabla'\Delta'}, \vee_{i\in I} A_i = \bigcup_{i\in I} A_i .$$

The infimum and supremum in $L_{O_i}^{\delta}(U,V,\tilde{R})$ are given by

$$\wedge_{i\in I} B_i = \bigcap_{i\in I} B_i, \vee_{i\in I} B_i = \left(\bigcup_{i\in I} B_i\right)^{\Delta'\nabla'} .$$

4. Attribute Reduction of Fuzzy Rough Concept Lattice

We put forward the methods of attribute reduction based on identified attribute matrix.

There are eight concepts $(\{5,11\},\{1,6,7\})$, $(\{11,12\},\{1,4,5\})$, $(\{5,8,11\},\{1,6\})$, $(\{5,6,7,8,9,10,11,12\},\{1\})$, $(\{3,4\},\{3\})$ $(\{2,3\},\{2\})$,$(U,\Phi),(\Phi,V)$ denoted $FC_i(i=1,2,\cdots,8)$.

There are two attribute reductions , $D_1=\{1,2,3,4,6,7\}$ and $D_2=\{1,2,3,5,6,7\}$. 1,2,3,6,7 are core attributes and 4,5 are relative necessary attribute.

Definition 5 Let $(U,A,\tilde{R})$ be a fuzzy formal context, $(X_i,B_i),(X_j,B_j)\in L_\delta(U,V,\tilde{R})$. $dis_{FC}((X_i,B_i),(X_j,B_j))=B_i\cup B_j-B_i\cap B_j$ is denoted identified attribute set.

$\wedge_{FC}=(dis_{FC}((X_i,B_i),(X_J,B_j)),(X_i,B_i),(X_j,B_j)\in L_\delta(U,V,\tilde{R}))$ is denoted identified attribute matrix.

Example 3 By table 3, identified attribute matrix is following.

Table 4

	FC1	FC2	FC3	FC4	FC5	FC6	FC7	FC8
FC1	Φ							
FC2	4567	Φ						
FC3	7	456	Φ					
FC4	67	45	6	Φ				
FC5	1367	1345	36	13	Φ			
FC6	1267	1245	26	12	32	Φ		
FC7	167	145	16	1	3	2	Φ	
FC8	234589,10,11,12	U-145	U-16	U-1	U-3	U-2	V	Φ

The results of attribute reduction is the same as the results by rough set . In example, core attributes are high cross domestic product per capita, high consumer price index, high unemployment rate, high export per capita , high import per capita.

5. Conclusions

Given a fuzzy context $(U, A, \tilde{R})$,one can get the fuzzy rough concept lattice $L_{\delta}(U, A, \tilde{R})$ in Will's sense.At last ,we apply the attribute reduction method for the concept lattices to the reduction of the fuzzy rough concept lattice and the experimental results show that the reduction method for the fuzzy rough concept lattice is effective in the reduction of the multiple rules.

References

1. Lingyun Yang and Luoshan Xu.On Rough Concept Lattices.Electronic Notes in Theoretical Computer Science 257(2009)117-133.
2. G.-Q.Zhang, G.Shen, Approximable concepts ,Chu spaces, and information systems, Theory and Applications of Categories 17(5)(2006) 80-102.
3. R.Belohlavek,Lattices of fixed points of fuzzy Galois connections.Math, Logic Quart.47(2001)111-116.
4. R.Belohlavek.Fuzzy relational systems, Foundations and Principles, Kluwer Academic\Plenum Publishers, New York,2002.
5. R.Belohlavek, Concept lattice and order in fuzzy logic, Ann. Pure Appl. Logic 128(2004)227-298.
6. G.Georgescu , A. Popescu , Non-dual fuzzy connections, Arch.Math.Logic 43(2004)1009-1039.
7. Hongliang Lai, Dexue Zhang.Concept lattices of fuzzy contexts:Formal concept analysis vs.rough set theory ,international journal of Approximate Reasoning 50(2009)695-707.
8. B.Ganter,R.Wille,Formal Concept Analysis,Springer-Verlag,Berlin,1999.
9. I.Duntsch, G.Gegida, Mondal-style operators in qualitative data analysis, in:Proc.of the 2002 IEEE Int.Conf.on Data Mining (2002),155-162.
10. Y.B.Lei, M.K.Luo, Rough concept lattices and domains, Annals of Pure and Applied Logic, 159(2009)333-340.
11. Y.Y.Yao, A comparative study of formal concept analysis and rough set theory in data analysis, Lecture Notes in Artificial Intelligence 3066(2004)59-68.
12. Y.Y.Yao, Concept lattices in rough set theory , in: Proceedings of 2004 Annual Meeting of the North American Fuzzy Information Processing Society (2004) 796-801.
13. Radim Belohlavek , Jan Outrata, and Vilem Vychodil.Direct Factorization by Similarity of Fuzzy Concept lattices by Factorization of Input Data. Springer-Verlag Berlin Heidelberg 2008. S. Ben Yahia et al. (Eds.) : CLA2006, LNAI 4923, pp. 68-79, 2008.

STUDY ON THE N-INEQUALITY IN THE LATTICE IMPLICATION ALGEBRA*II*

LONG XIQING
School of Maths, YiBin University, YiBin 644000,China.

XU YANG
Department of Mathematics, Southwest Jiaotong University, Chengdu 610031, China.

HE XINGXING
Intelligent Control Development Center, Southwest Jiaotong University, Chengdu 610031, China.

In this paper, the n-inequality is studied in the lattice implication algebra. The necessary and sufficient conditions for existence of solution for the n-inequality will be presented respectively. In the case, that b has decomposition of irreducible finite union, the all minimal elements of solutions and detailed solution sets are given.

1. Introduction

In order to research the many-valued logical system whose propositional value is given in a lattice, in 1993 Xu[1,2] proposed the concept of lattice implication algebras and discussed its properties. Hu[3] studied the properties of residual lattice. Pan[4] proposed the concept of lattice implication algebraic equation in lattice implication algebra and discussed solution sets. Lattice implication algebraic inequalities as the development of the lattice implication algebraic equation, the study of inequalities are great significance in the study of the lattice-valued logic. The authors[5,6,7] investigated the simple lattice implication algebraic inequalities and Long[8,9] studied the N-inequality in the lattice implication algebra. In this paper, the n-inequality is studied in the lattice implication algebra. The necessary and sufficient conditions for existence of solution for the n-inequality will be presented respectively. In the case, that b

* This work is supported by the National Natural Science Foundation of china (Grant No. 61175055) and by the Yibin University hospital-level key projects (Grant No. 2010Z04).

has decomposition of irreducible finite union, the all minimal elements of solutions and detailed solution sets are given. It is hoped that the above work would serve as a foundation for further study lattice-valued logic system based on linguistic truth values lattice implication algebras and develop corresponding applied methods of uncertainty reasoning and resolution automated reasoning with linguistic terms.

2. Preliminaries

Definition 2.1[1] Let $(L,\vee,\wedge,', 0, I)$ be a bounded lattice with the universal boundaries 0(the least element) and I(the greatest element) respectively, and "$'$" an order-reversing involution. If a mapping $\rightarrow: L\times L\rightarrow L$ satisfies for any $x, y, z\in L$,

(1) $x\rightarrow(y\rightarrow z)=y\rightarrow(x\rightarrow z)$;
(2) $x\rightarrow x=I$;
(3) $x\rightarrow y=y'\rightarrow x'$;
(4) $x\rightarrow y=y\rightarrow x=I\Rightarrow x=y$;
(5) $(x\rightarrow y)\rightarrow y=(y\rightarrow x)\rightarrow x$;
(6) $(x\vee y)\rightarrow z=(x\rightarrow z)\wedge(y\rightarrow z)$;
(7) $(x\wedge y)\rightarrow z=(x\rightarrow z)\vee(y\rightarrow z)$,

then $(L,\vee,\wedge,',\rightarrow,0,I)$ is called a lattice implication algebra.

In a lattice implication algebra, we can define a partial ordering "$\leq$" by $x\leq y$ if and only if $x\rightarrow y=I$.

For more details of lattice implication algebras we refer the readers to [3].

Definition 2.2[3] Let L be a lattice. For any $x, y\in L$, if $b=x\vee y$ implies $b=x$ or $b=y$, then b is called an extract irreducible element.

Definition 2.3[3] Let L be a lattice. For any $b\in L$, if there are k extract irreducible elements $p_1,p_2,\cdots,p_k$ such that $b=\bigvee_{i=1}^{k}p_i$, then the b is called that has decomposition of irreducible finite union. And if for any $1\leq j\leq k, b\neq\bigvee_{i=1,i\neq j}^{k}p_i$, then the b is called that has Irreducible and decomposition.

In the following, for any $n\in N$, the symbol $\underline{n}$ means that the number set $\{1,2,3,\cdots,n\}$.

Theorem2.4[2] Let $(L,\vee,\wedge,', 0, I)$ be a lattice implication algebra, then $(L,\vee,\wedge)$ is a distributive lattice.

Lemma2.5[10] If p is an extract irreducible element of a distributive lattice L, then $p\leq\bigvee_{j=1}^{k}x_j$ implies that there is $j\in\underline{k}$ such that $p\leq x_j$.

3. Solutions of the n-inequality

Definition3.1[5]In a lattice implication algebra L, inequalities that included variable and combined by the operators ($\vee$, $\wedge$, $'$, $\rightarrow$) are the lattice implication algebraic inequalities.

In the following, the symbol L is a lattice implication algebra.

For any $a_i, b \in L$, we will find $X = (x_1, x_2, \cdots x_n) \in \overbrace{L \times L \times \cdots \times L}^{n}$ such that $\vee_{i=1}^{n}(a_i \wedge x_i) \geq b$, the symbol $S = \{(x_1, x_2, \cdots, x_n) | \vee_{i=1}^{n}(a_i \wedge x_i) \geq b\}$ is the solution of the inequality $\vee_{i=1}^{n}(a_i \wedge x_i) \geq b$.

If $b = 0$, then $S \neq \varnothing$ and $S = \overbrace{L \times L \times \cdots \times L}^{n}$. In the following, we always suppose that $b \neq 0$.

Theorem3.2[9] For any $b \in L$, $S \neq \varnothing \Leftrightarrow \vee_{i=1}^{n} a_i \geq b$.

Theorem3.3[9] If $S \neq \varnothing$,then $I = (I, I, \cdots, I) \in S$.

Theorem3.4[9] If $S \neq \varnothing$ and b is an extract irreducible element of L, then for any $X \in S$,there is a minimal element X_j^* of S such that $X \geq X_j^*$.

Theorem3.59] If $S \neq \varnothing$ and b is an extract irreducible element of L, then for each minimize X^* has the form $(0, \cdots 0, \overset{\overset{j}{\uparrow}}{b}, 0, \cdots, 0)$ and $|G(b)|$ minimal elements in S, here $G(b) = \{i \in \underline{n} | a_i \geq b\}$.

Theorem3.6[9] If $S \neq \varnothing$ and b is an extract irreducible element of L, then $S = \bigcup_{j \in \{i \in \underline{n} | a_i \geq b\}} [X_j^*, I]$

Theorem3.7 If b has decomposition of irreducible finite union, for any $j \in \underline{k}$, $G(p_j) = \{i \in \underline{n} | a_i \geq p_j\}$ then $S \neq \varnothing \Leftrightarrow \forall j \in \underline{k}, G(p_j) \neq \varnothing$.

Proof. If $S \neq \varnothing$, then there is a element $X = (x_1, x_2, \cdots, x_n) \in S$ such that $\vee_{i=1}^{n}(a_i \wedge x_i) \geq b$. Because that b has decomposition of irreducible finite union, then $b = \vee_{i=1}^{k} p_i$. So for any $j \in \underline{k}$, there is $p_j \leq b \leq \vee_{i=1}^{n}(a_i \wedge x_i)$ and p_j is an extract irreducible element of L, then **there is** $i_j \in \underline{n}$ such that $p_j \leq a_{i_j} \wedge x_{i_j}$, hence $p_j \leq a_{i_j}$ and $G(p_j) \neq \varnothing$.

If for any $j \in \underline{k}$, $G(p_j) \neq \varnothing$, then $\forall j \in \underline{k}$, there is at least one $i \in \underline{n}$ such that $a_i \geq p_j$, then $b = \vee_{i=1}^{k} p_i \leq \vee_{i=1}^{n} a_i$, so $S \neq \varnothing$.

Theorem 3.8 If $S \neq \varnothing$ and b has decomposition of irreducible finite union, then S have minimizes.

Proof. We suppose that $A_1=\{p_i\in P|a_1\geq p_i\}$, $A_2=\{p_i\in P|a_2\geq p_i\}\setminus A_1$, $\cdots$, $A_j=\{p_i\in P|a_j\geq p_i\}\setminus(A_1\cup A_2\cup\cdots\cup A_{j-1})$, ..., $A_n=\{p_i\in P|a_n\geq p_i\}\setminus(A_1\cup A_2\cup\cdots\cup A_{n-1})$, so $A_1\cup A_2\cup\cdots\cup A_n=P=\{p_1,p_2,\cdots,p_k\}$ and for any $i\neq j$, $A_i\cap A_j=\varnothing$, hence $A_1,A_2,\cdots,A_n$ is a partition of P. We define the $X^*=(x_1^*,x_2^*,\cdots,x_n^*)$ as the follows, for any $j\in\underline{n}$, $x_j^*=\begin{cases}\bigvee_{p_i\in A_j}p_i, \text{if } A_j\neq\varnothing\\ 1, \text{else}\end{cases}$, so,

$$\bigvee_{i=1}^n(a_i\wedge x_i^*)=\bigvee_{k\in\underline{n},A_k\neq\varnothing}(a_k\wedge x_k^*)=\bigvee_{k\in\underline{n},A_k\neq\varnothing}(a_k\wedge(\bigvee_{p_i\in A_k}p_i))=\bigvee_{k\in\underline{n},A_k\neq\varnothing}(\bigvee_{p_i\in A_k}(a_k\wedge p_i))$$
$$=\bigvee_{k\in\underline{n},A_k\neq\varnothing}(\bigvee_{p_i\in A_k}p_i)=\bigvee_{p_i\in P}p_i=b\geq b, \text{ then } X^*\in S.$$

In the following, we will proof that X^* is the minimize of S. It means that if $X=(x_1,x_2,\cdots,x_n)\in S$ and $X\leq X^*$, we will proof $X=X^*$. Of course we will only proof that if $x_j^*\neq 0$, then $x_j=x_j^*$. Because $X\in S$ and b has decomposition of irreducible finite union, then $b=\bigvee_{j=1}^k p_j\leq\bigvee_{i=1}^n(a_i\wedge x_i)$, so $\forall p_i\in A_j$, $p_i\leq b\leq\bigvee_{i=1}^n(a_i\wedge x_i)$, for p_i is an extract irreducible element of L, then there is $k_i\in\underline{n}$ such that $p_i\leq(a_{k_i}\wedge x_{k_i})\leq x_{k_i}$. Actually, $k_i=j$. Because that if $k_i\neq j$, by the $x_{k_i}^*\geq x_{k_i}\geq p_i$, we will obtain that $x_{k_i}^*\neq 0$, so $A_{k_i}\neq\varnothing$, then $p_i\leq x_{k_i}^*=\bigvee_{p_i\in A_{k_i}}p_i$, but p_i is an extract irreducible element of L, then there is $p_{t_0}\in A_{k_i}$ such that $p_i\leq p_{t_0}$, if $p_i<p_{t_0}$, then $b=\bigvee_{i=1}^k p_i=\bigvee_{i=1,i\neq t_0}^k p_i$. This is a contradiction with the b has decomposition of irreducible finite union. So $p_i=p_{t_0}\in A_{k_i}$, then $p_i\in A_{k_i}\cap A_j$, but $k_i\neq j$, This is a contradiction with the $A_1,A_2,\cdots,A_n$ is a partition of P. So $k_i=j$, then $\forall p_i\in A_j$, $p_i\leq x_j$, hence $x_j^*=\bigvee_{p_i\in A_j}p_i\leq x_j$, but $x_j^*\geq x_j$, so $x_j^*=x_j$. Hence $X=X^*$ and X^* is the minimize of S.

Theorem3.9 If $S\neq\varnothing$ and b has decomposition of irreducible finite union, then for each $X\in S$, there is an X^* is the minimize of S such that $X\geq X^*$.

Proof. We suppose that $X=(x_1,x_2,\cdots,x_n)\in S$, then $\bigvee_{i=1}^n(a_i\wedge x_i)\geq b$, because b has decomposition of irreducible finite union, then $b=\bigvee_{j=1}^k p_j$, so $\forall j\in\underline{k}$, $p_j\leq b\leq\bigvee_{i=1}^n(a_i\wedge x_i)$, for p_i is an extract irreducible element of L, so there is $i\in\underline{n}$ such that $p_j\leq a_i\wedge x_i$. We suppose that $P_1=\{p_i\in P|a_1\wedge x_1\geq p_i\}$,

$P_2=\{p_i\in P|a_2\wedge x_2\geq p_i\}\backslash P_1$, ..., $P_j=\{p_i\in P|a_j\wedge x_j\geq p_i\}\backslash(P_1\cup P_2\cup\cdots\cup P_{j-1})$, $\cdots$, $P_n=\{p_i\in P|a_n\wedge x_n\geq p_i\}\backslash(P_1\cup P_2\cup\cdots\cup P_{n-1})$, so $P_1\cup P_2\cup\cdots\cup P_n=P=\{p_1,p_2,\cdots,p_k\}$ and if $i\neq j$, then $P_i\cap P_j=\varnothing$. Hence $P_1,P_2,\cdots,P_n$ is a partition of P. We define the $X^*=(x_1^*,x_2^*,\cdots,x_n^*)$ as the follows, for any $j\in \underline{n}$, $x_j^*=\begin{cases}\vee_{p_i\in A_j}p_i, \text{if } P_j\neq\varnothing\\ 1,\text{else}\end{cases}$, of course $X^*\in S$ and $X^*\leq X$. As the same way of the Theorem3.8, we will obtain that X^* is the minimize of S.

Theorem3.10 If $S\neq\varnothing$ and b has decomposition of irreducible finite union, then S have $\prod_{i\in \underline{k}}|G(p_i)|$ minimizes.

Proof. By the Theorem3.7, we know that $S\neq\varnothing\Leftrightarrow\forall j\in\underline{k}$, $G(p_j)\neq\varnothing$. Firstly we proof that S have $\prod_{i\in\underline{k}}|G(p_i)|$ minimizes. We suppose that $G(p_j)=\{j_1,\ j_2,\cdots,j_{\varphi(i)}\}$, $j_1,j_2,\cdots,j_{\varphi(i)}\in\underline{n}$, we will construct the subset of the cluster of P as follows, $\forall i\in\underline{k}$, then p_i belongs and only belongs to one of $A_{j_1},A_{j_2},\cdots,A_{j_{\varphi(i)}}$, if $\exists j\in\underline{n}$, there is not $p_i\in P$ such that $p_i\leq a_j$, then make the $A_j\neq\varnothing$, so we will construct $\prod_{i\in\underline{k}}|G(p_i)|$ different subsets of the cluster($A_1,A_2,\cdots,A_n$) of P, of course $A_1\cup A_2\cup\cdots\cup A_n=P$ and $\forall i,j\in\underline{n}$, if $i\neq j$, then $A_i\cap A_j=\varnothing$, so each cluster is a subset of a partition of P. For any the subset of the cluster ($A_1,A_2,\ \cdots,A_n$), we will define $X^*=(x_1^*,x_2^*,\cdots,x_n^*)$ as follows, for any $j\in\underline{n}$, $x_j^*=\begin{cases}\vee_{p_i\in A_j}p_i, \text{if } A_j\neq\varnothing\\ 1,\text{else}\end{cases}$, as the same way of the Theorem3.9, we will obtain that X^* is the minimize of S, because b has decomposition of irreducible finite union, it is different that by the different subsets defined by clusters ($A_1,A_2,\ \cdots,A_n$) of X^*, so at least there are $\prod_{i\in\underline{k}}|G(p_i)|$ different minimizes in S.

In the following, we will proof that each minimize of S is one of the $\prod_{i\in\underline{k}}|G(p_i)|$ different minimizes. We suppose that $X=(x_1,x_2,\cdots,x_n)$ is the minimize of S, then $X\in S$, as the same way of the Theorem 3.12, we will construct $X^*=(x_1^*,x_2^*,\cdots,x_n^*)$ as follows, for any $j\in\underline{n}$, $x_j^*=\begin{cases}\vee_{p_i\in A_j}p_i, \text{if } P_j\neq\varnothing\\ 1,\text{else}\end{cases}$, of course $X^*\in S$ and $X^*\leq X$; For X is the minimize of S, hence $X^*=X$. Because that $P_1\cup P_2\cup\cdots\cup P_n=P=\{p_1,p_2,$

$\cdots, p_k\}$ and if $i \neq j$, then $P_i \cap P_j = \varnothing$, so each minimize(X) corresponds to the a partition of P($P_1, P_2, \cdots, P_n$), it is clear that the partition of the resulting cluster is a subset of $\prod_{i \in \underline{k}} |G(p_i)|$ different subsets of the cluster one.

Theorem3.11 If $S \neq \varnothing$, then $S = \bigcup_{X^* \in S^*} [X^*, I]$, S^* is the set of all minimal elements of S.

References

1. Xu Yang, "Lattice implication algebra". Journal of Southwest Jiaotong University, Sum. 28(1), 20-27, 1993.
2. Y. Xu, D. Ruan, K.Y, Qin, J. Liu, "Lattice-Valued Logic", Springer-verlag, Berlin, 2003.
3. Hu Changliu , Song Zhengming . The foundation of lattice .1990
4. Pan Xiaodong, Xu Yang , Zhang Qing. Lattice implication algebraic equations in lattice implication algebras, Journal of Southwest Jiaotong University, Vol 40 No. 6, Dec.2005.
5. Lai jiajun, Xu Yang. Lattice implication algebraic inequalities, Journal of Jiangnan University, No.3, June, 2007
6. Long Xiqing. Study on the two classes of simple inequalities in the lattice implication algebras. Journal of Yibin University, Vol 10 No.12, Dec.2010.
7. Long Xiqing. Study on the solution sets of inequalities in the lattice implication algebras. Journal of Neijiang Normal College, Vol 26 No.2, Feb.2011.
8. Long Xiqing. Study on the N-inequality in the Lattice Implication Algebra1. Journal of Mianyang Normal College, Vol 30 No.2, Feb.2011.
9. Long Xiqing. Study on the N-inequality in the Lattice Implication Algebra*II. (Accepted)
10. Birkhoff G.Lattice theory(vol.XXV,3rd ed.)[M]. Providence, RI: American Mathematical Society Colloquium Publications, 1967.

THE PRODUCT OF QUANTUM AUTOMATA[1]

MO ZHI-WEN

College of Mathematics and Software Science, Sichuan Normal University Chengdu, P.R.China ,610066

CHEN QING-LEI

College of Mathematics and Software Science, Sichuan Normal University Chengdu, P.R.China,610066

LIU JUN

Department of Mathematics, Sichuan University for Nationalities Kangding, Sichuan, P.R.China, 626001

Abstract : In this paper, we want to establish the multiplication theory of quantum automata(orthomodular lattice-valued automata). According this theory,we kan construct complex quantum automata.First we give the definition of direct product,cascading product and ring product.Then we discuss the basic properties of them.At last ,we give the form of the language that can be recognized by product quantum automata.

1. Introduction

Quantum finite automata can be seen as a kind of quantum computer model.The essential problem is automata theory that based on quantum logic. In 1936, Birkhoff and Neumann proposed quantum logic when they studied the logical basis of quantum mechanics[1].Because quantum mechanical systems can be described by the closed sub-space of Hilbert Space and all closed sub-space of Hilbert Space construct a orthomodular lattice,so orthomodular lattices is the algebraic form of quantum logic[2].Based on this, in literature[3-4] , complete orthomodular lattice-valued logic was defined as quantum logic ,and automata theory was established based on quantum logic, namely l-valued automata. Iterature[5] mainly discuss grammar theory based on quantum logic. For orthomodular lattice don't necessarily satify the distributive law,some general conclusions

1 This work is supported by The National Science Foundation of Chian (No. 11071178) and Scientific Research Fund of SiChuan Provincial Education Department (No. 10ZB027)

are not correct in quantum automata theory.How to establish the multiplication theory of quantum automata?How to construct complex quantum automata through simple quantum automata?

2. preliminaries

Definition1[4]. Complete orthomodular lattice is a 7-tuple l=<L, $\leq, \wedge, \vee, \perp$, 0,1>,where:

(1)<L,$\leq, \wedge, \vee, \perp$,0,1> is a complete lattice;0 and 1 is the least and greatest elements of L;$\leq$ is the partial ordering in L; and for any $M \subseteq L$,$\wedge M$ and $\vee M$ stand for the greatest lower bound and the least upper bound of M respectively.

(2)$\perp$ is a unary operation on L, called orthocomplement,and required to satisfy the following conditions:

(2.1) $a \wedge a^{\perp}$=0,$a \vee a^{\perp} = 1$.

(2.2) $a^{\perp\perp}$=a.

(2.3) $a \leq b$ implies $b^{\perp} \leq a^{\perp}$.

(2.4) $a \wedge (a^{\perp} \vee (a \wedge b)) \leq b$.

Definition 2[4]. Let l=<L,$\leq, \wedge, \vee, \perp$,0,1> be an orthomodular lattice, and $\sum$ be a finite alphabet.Then a l-valued (quantum) automaton over $\sum$ is a quadruple R=$< Q, I, T, \delta >$,where：

Q is a finite set whose elements are called states.

$I \subseteq Q$ and states in I are said to be initial.

$T \subseteq Q$ and states in T are said to be terminal.

δ is a l-valued subset of $Q \times \sum \times Q$, i.e.,a mapping from $Q \times \sum \times Q$ into L, and called the l-valued (quantum) transition relation of R;and intuitively,for any $p, q \in Q$ and $\sigma \in \sum$, $\delta(p, \sigma, q)$stands for the truth value (in quantum logic) of the proposition that input σ causes state p to become q.

Using transiton function δ, we can describe the recognition of the language over $\sum$. Any quantum automaton R=$< Q, I, T, \delta >$ over $\sum$ decides a unary orthomodular lattice value predicate rec_R over $\sum$. It is defined as : for any $\sigma_1 \sigma_2 \cdots \sigma_k \in \sum^*$,

$$rec_R(\sigma_1 \sigma_2 \cdots \sigma_k) \stackrel{\text{def}}{=}$$
$$(\exists q_0 \in I, q_1 \cdots q_{k-1} \in Q, q_k \in T) path_R(q_0 \sigma_1 q_1 \cdots q_{k-1} \sigma_k q_k)$$

where,

$path_R(q_0 \sigma_1 q_1 \cdots q_{k-1} \sigma_k q_k) \stackrel{\text{def}}{=} \wedge_{i=0}^{k-1}((q_i, \sigma_{i+1}, q_{i+1}) \in \delta)$.

Intuitively,$rec_R(\sigma_1 \sigma_2 \cdots \sigma_k)$ expresses a proposition that the quantum automaton R can recognize $\sigma_1 \sigma_2 \cdots \sigma_k$, Its true value is:

$[rec_R(\sigma_1 \sigma_2 \cdots \sigma_k)] = \vee_{q_0 \in I, q_1 \cdots q_{k-1} \in Q, q_k \in T} \wedge_{i=0}^{k-1} \delta(q_i, \sigma_{i+1}, q_{i+1})$.

3. The main results

Definition 3[6]. Let$(X,\leq_X)$, $(Y,\leq_Y)$be two partial order sets, $T = X \times Y$ (Cartesian product) , in *T*, define a binary relation $\leq$: $(x,y) \leq (x',y') \Leftrightarrow x \leq_X x'$ and $y \leq_Y y'$, (for $\forall(x,y)$, $(x',y') \in T$) . It is easy to verify that $\leq$ is a partial order on *T*. We can say the partial order set $(T, \leq)$ is the basic product of partial order sets *X* and *Y* (referred to as product).Recorded as $T = X \times Y$ or $T = XY$.To put it easily,we also record $(x,y) \in T$ as $x \cdot y \in T$.

Theorem 1. $(X,\leq_X)$, $(Y,\leq_Y)$are two lattices , if and any if $T = XY$ is a lattice , and for any $a_1, a_2 \in X, b_1, b_2 \in Y$, $(a_1 \wedge a_2)\cdot(b_1 \wedge b_2) = (a_1 \cdot b_1) \wedge (a_2 \cdot b_2)$, $(a_1 \vee a_2)\cdot(b_1 \vee b_2) = (a_1 \cdot b_1) \vee (a_2 \cdot b_2)$.

Theorem 2. $\prod_{i\in I} L_i$ is a complete orthomodular lattice$\Leftrightarrow$ $L_i(\forall i \in I)$ are complete orthomodular lattices.

Proof. It is easy to prove that $\prod_{i\in I} L_i$ is a complete lattice $\Leftrightarrow$ $L_i(\forall i \in I)$ are complete lattices, $\prod_{i\in I} L_i$ is a modle lattice$\Leftrightarrow$ $L_i(\forall i \in I)$ are modle lattices , $\prod_{i\in I} L_i$ is an orthomodular lattice $\Leftrightarrow$ $L_i(\forall i \in I)$ are orthomodular lattices, so $\prod_{i\in I} L_i$ is a complete orthomodular lattice$\Leftrightarrow$ $L_i(\forall i \in I)$ are complete orthomodular lattices.

Definition 4. Let $R_1 =< Q_1, I_1, T_1, \delta_1 >$, $R_2 =< Q_2, I_2, T_2, \delta_2 >$ be quantum automatas over $\sum_1$ and $\sum_2$ severally , their value lattices are L_1 and L_2 , $R_1 \times R_2 \overset{\text{def}}{=} < Q_1 \times Q_2, I_1 \times I_2, T_1 \times T_2, \delta_1 \times \delta_2 >$,we say $R_1 \times R_2$ is the direct product of $R_1$and R_2,where:

$$\delta_1 \times \delta_2((p_1,q_1),(x,y),(p_2,q_2)) = \delta_1(p_1,x,p_2)\cdot\delta_2(q_1,y,q_2).$$

Because L_1 and L_2 are complete orthomodular lattices, and by the theorem 2, $L_1 \times L_2$ is a complete orthomodular lattice also, obviously $R_1 \times R_2$ is a quantum automata over $\sum_1 \times \sum_2$,and its value lattice is $L_1 \times L_2$.

Theorem 3. Let $R_1 =< Q_1, I_1, T_1, \delta_1 >$, $R_2 =< Q_2, I_2, T_2, \delta_2 >$ be quantum automatas, $R_1 \times R_2 =< Q_1 \times Q_2, I_1 \times I_2, T_1 \times T_2, \delta_1 \times \delta_2 >$ is the direct product of $R_1$and R_2, then for any $x_1y_1x_2y_2\cdots x_ky_k \in (\sum_1 \times \sum_2)^*$, $\lceil rec_{R_1\times R_2}(x_1y_1x_2y_2\cdots x_ky_k)\rceil =$ $\lceil rec_{R_1}(x_1x_2\cdots x_k)\rceil \cdot \lceil rec_{R_2}(y_1y_2\cdots y_k)\rceil$

Proof. $\lceil rec_{R_1\times R_2}(x_1y_1x_2y_2\cdots x_ky_k)\rceil$

$=\bigvee_{(p_0,q_0)\in I_1\times I_2,(p_1,q_1)\cdots(p_{k-1},q_{k-1})\in Q_1\times Q_2,(p_k,q_k)\in T_1\times T_2} \bigwedge_{i=0}^{k-1} \delta_1 \times \delta_2((p_i,q_i),x_{i+1}y_{i+1},(p_{i+1},q_{i+1}))$

$=\bigvee_{(p_0,q_0)\in I_1\times I_2,(p_1,q_1)\cdots(p_{k-1},q_{k-1})\in Q_1\times Q_2,(p_k,q_k)\in T_1\times T_2} \bigwedge_{i=0}^{k-1} \delta_1(p_i,x_{i+1},p_{i+1})\cdot \delta_2(q_i,y_{i+1},q_{i+1})$

$=\bigvee_{(p_0,q_0)\in I_1\times I_2,(p_1,q_1)\cdots(p_{k-1},q_{k-1})\in Q_1\times Q_2,(p_k,q_k)\in T_1\times T_2}\left(\bigwedge_{i=0}^{k-1} \delta_1(p_i,x_{i+1},p_{i+1})\right)\cdot \left(\bigwedge_{i=0}^{k-1} \delta_2(q_i,y_{i+1},q_{i+1})\right)$

$=(\vee_{p_0 \in I_1, p_1 \cdots p_{k-1} \in Q_1, p_k \in T_1} \wedge_{i=0}^{k-1} \delta_1(p_i, x_{i+1}, p_{i+1})) \cdot$
$(\vee_{q_0 \in I_2, q_2 \cdots q_{k-1} \in Q_2, q_k \in T_2} \wedge_{i=0}^{k-1} \delta_2(q_i, y_{i+1}, q_{i+1})$
$= \lceil rec_{R_1}(x_1 x_2 \cdots x_k) \rceil \cdot \lceil rec_{R_2}(y_1 y_2 \cdots y_k) \rceil$

Definition 5. Let $R_1 = < Q_1, I_1, T_1, \delta_1 >$, $R_2 = < Q_2, I_2, T_2, \delta_2 >$ be quantum automatas over $\sum$, Their value lattice is L, $R_1 \times R_2 \overset{\text{def}}{=} < Q_1 \times Q_2, I_1 \times I_2, T_1 \times T_2, \delta_1 \times \delta_2 >$,we say $R_1 \times R_2$ is the Cartesian product of R_1 and R_2 where $\delta_1 \times \delta_2((p_1, q_1), x, (p_2, q_2)) = \delta_1(p_1, x, p_2) \cdot \delta_2(q_1, x, q_2)$.

Corollary 1. Let R_1=$< Q_1, I_1, T_1, \delta_1 >$, R_2=$< Q_2, I_2, T_2, \delta_2 >$ be quantum automatas,$R_1 \times R_2$=$< Q_1 \times Q_2, I_1 \times I_2, T_1 \times T_2, \delta_1 \times \delta_2 >$ is the Cartesian product of R_1 and R_2 ,then for any $x_1 x_2 \cdots x_k \in \sum^*$, $\lceil rec_{R_1 \times R_2}(x_1 x_2 \cdots x_k) \rceil = \lceil rec_{R_1}(x_1 x_2 \cdots x_k) \rceil \cdot \lceil rec_{R_2}(x_1 x_2 \cdots x_k) \rceil$

Be noted that in literature[4],it has been clearly proved that if we define $\delta_1 \times \delta_2((p_1, q_1), x, (p_2, q_2)) = \delta_1(p_1, x, p_2) \wedge \delta_2(q_1, x, q_2)$, wanting to get the conclusion similar to Corollary 1, i.e.:

$$\lceil rec_{R_1 \times R_2}(x_1 x_2 \cdots x_k) \rceil = \lceil rec_{R_1}(x_1 x_2 \cdots x_k) \rceil \wedge \lceil rec_{R_2}(x_1 x_2 \cdots x_k) \rceil,$$

then we require its true value lattice L must satisfy the distributive.

Theorem 4. Let $R = < Q, I, T, \delta >$ be a quantum automata over $\sum$, its value lattice is L, if $|Q| = m \times n$, $L = L_1 \times L_2$, then there are two quantum automatas R_1=$< Q_1, I_1, T_1, \delta_1 >$, R_2=$< Q_2, I_2, T_2, \delta_2 >$ over $\sum$ and their value lattices are L_1 and L_2 respectively.For any $x_1 x_2 \cdots x_k \in \sum^*$, $\lceil rec_R(x_1 x_2 \cdots x_k) \rceil = \lceil rec_{R_1 \times R_2}(x_1 x_2 \cdots x_k) \rceil$, and satisfy $|Q_1| + |Q_2| = m + n$.

Proof. Let Q_1, Q_2, Q be pairwise disjoint sets of states, and $|Q_1| = m$, $|Q_2| = n$, then $|Q_1 \times Q_2| = m \times n$, so there is a mapping $f: Q \to Q_1 \times Q_2$.Let g_1, g_2 be projections from $Q_1 \times Q_2$ to Q_1, Q_2, for any $q_1, q_2 \in Q, x \in \sum$,let $\delta(q_1, x, q_2) = a \in L$,then there is a unique $b \in L_1, c \in L_2$, making $a = b \cdot c$ .Let $R_1 = < Q_1, g_1 f(I), g_1 f(T), \delta_1 >$, $R_2 = < Q_2, g_2 f(I), g_2 f(T), \delta_2 >$, $\delta_1(g_1 f(q_1), x, g_1 f(q_2)) = b$, $\delta_2(g_2 f(q_1), x, g_2 f(q_2)) = c$, by corollary 1 ,it is easy to show that for any $x_1 x_2 \cdots x_k \in \sum^*$, $\lceil rec_R(x_1 x_2 \cdots x_k) \rceil = \lceil rec_{R_1 \times R_2}(x_1 x_2 \cdots x_k) \rceil$, and satisfy $|Q_1| + |Q_2| = m + n$, so the proof of the theorem.

Theorem 5[6]. If L is a model lattice, then L can be decomposed into a finite product of series of simple model lattices.

The conclusion of theorem 5 will set up also in the case of a complete orthomodular lattice , then we can obtain the following corollary :

Corollary 2. Let $R = < Q, I, T, \delta >$ is a quantum automata over $\sum$, its value lattice is L, if L can be decomposed into product of n simple complete orthomodular lattices, $|Q| = m_1 \times m_2 \cdots m_n$, $m_i \geq 2$,then there are n quantum automatas $R_1 \cdots R_n$, for any $x_1x_2 \cdots x_k \in \sum^*$, $[rec_R(x_1x_2 \cdots x_k)] = \lceil rec_{R_1 \times R_2 \cdots R_n}(x_1x_2 \cdots x_k)\rceil$, and satisfy $|Q_i| = m_i$.

Definition 6. Let $R_1 =< Q_1, I_1, T_1, \delta_1 >$, $R_2 =< Q_2, I_2, T_2, \delta_2 >$ be quantum automatas over $\sum_1$ and $\sum_2$ severally, their value lattices are L_1 and L_2, $R_1\omega R_2 \overset{\text{def}}{=} < Q_1 \times Q_2, I_1 \times I_2, T_1 \times T_2, \delta_1\omega\delta_2 >$,we say $R_1\omega R_2$ is the cascading product of $R_1$and R_2 ,where: $\omega: Q_2 \times \sum_2 \to \sum_1$, $\delta_1\omega\delta_2((p_1, q_1), y, (p_2, q_2)) = \delta_1(p_1, \omega(q_1, y), p_2) \cdot \delta_2(q_1, y, q_2)$.

Theorem 6. $R_1 =< Q_1, I_1, T_1, \delta_1 >$, $R_2 =< Q_2, I_2, T_2, \delta_2 >$ be quantum automatas, $R_1\omega R_2 = < Q_1 \times Q_2, I_1 \times I_2, T_1 \times T_2, \delta_1\omega\delta_2 >$ is the cascading product of R_1 and R_2, then for any, $y_1y_2 \cdots y_k \in \sum_2^*$, $\lceil \mathrm{rec}_{R_1\omega R_2}(y_1y_2 \cdots y_k)\rceil = \{\vee_{q_0 \in I_2, q_2 \cdots q_{k-1} \in Q_2, q_k \in T_2} \lceil \mathrm{rec}_{R_1}(\omega(q_0, y_1)\omega(q_1, y_2) \cdots \omega(q_{k-1}, y_k))\rceil\} \cdot \lceil \mathrm{rec}_{R_2}(y_1y_2 \cdots y_k)\rceil$

Proof. $\lceil rec_{R_1\omega R_2}(y_1y_2 \cdots y_k)\rceil$

$=\vee_{(p_0,q_0)\in I_1\times I_2,(p_1,q_1)\cdots(p_{k-1},q_{k-1})\in Q_1\times Q_2,(p_k,q_k)\in T_1\times T_2} \wedge_{i=0}^{k-1} \delta_1\omega\delta_2((p_i, q_i), y_{i+1}, (p_{i+1}, q_{i+1}))$

$=\vee_{(p_0,q_0)\in I_1\times I_2,(p_1,q_1)\cdots(p_{k-1},q_{k-1})\in Q_1\times Q_2,(p_k,q_k)\in T_1\times T_2} \wedge_{i=0}^{k-1} \delta_1(p_i, \omega(q_i, y_{i+1}), p_{i+1}) \cdot \delta_2(q_i, y_{i+1}, q_{i+1})$

$=\vee_{(p_0,q_0)\in I_1\times I_2,(p_1,q_1)\cdots(p_{k-1},q_{k-1})\in Q_1\times Q_2,(p_k,q_k)\in T_1\times T_2}(\wedge_{i=0}^{k-1} \delta_1(p_i, \omega(q_i, y_{i+1}), p_{i+1}))\,(\wedge_{i=0}^{k-1} \delta_2(q_i, y_{i+1}, q_{i+1}))$

$=\{\vee_{(p_0,q_0)\in I_1\times I_2,(p_1,q_1)\cdots(p_{k-1},q_{k-1})\in Q_1\times Q_2,(p_k,q_k)\in T_1\times T_2} \wedge_{i=0}^{k-1} \delta_1(p_i, \omega(q_i, y_{i+1}), p_{i+1})\}\{\vee_{(p_0,q_0)\in I_1\times I_2,(p_1,q_1)\cdots(p_{k-1},q_{k-1})\in Q_1\times Q_2,(p_k,q_k)\in T_1\times T_2} \wedge_{i=0}^{k-1} \delta_2(q_i, y_{i+1}, q_{i+1})\}$

$=\{\vee_{q_0\in I_2,q_2\cdots q_{k-1}\in Q_2,q_k\in T_2} \vee_{p_0\in I_1,p_1\cdots p_{k-1}\in Q_1,p_k\in T_1} \wedge_{i=0}^{k-1} \delta_1(p_i, \omega(q_i, y_{i+1}), p_{i+1})\} \cdot \{\vee_{q_0\in I_2,q_2\cdots q_{k-1}\in Q_2,q_k\in T_2} \wedge_{i=0}^{k-1} \delta_2(q_i, y_{i+1}, q_{i+1})\}$

$=\{\vee_{q_0\in I_2,q_2\cdots q_{k-1}\in Q_2,q_k\in T_2} \lceil rec_{R_1}(\omega(q_0, y_1)\omega(q_1, y_2) \cdots \omega(q_{k-1}, y_k))\rceil\} \cdot \lceil rec_{R_2}(y_1y_2 \cdots y_k)\rceil$.

Definition 7. Let $R_1 =< Q_1, I_1, T_1, \delta_1 >$, $R_2 =< Q_2, I_2, T_2, \delta_2 >$ be quantum automatas over $\sum_1$ and $\sum_2$ severally, their value lattices are L_1 and L_2, $R_1 \circ R_2 \overset{\text{def}}{=} < Q_1 \times Q_2, I_1 \times I_2, T_1 \times T_2, \delta_1 \circ \delta_2 >$,we say $R_1 \circ R_2$ is the ring product of $R_1$and R_2 .For $\forall(p_1, q_1)$, $(p_2, q_2) \in Q_1 \times Q_2$, $f \in \sum_1^{Q_2}$, $y \in \sum_2$, $\delta_1 \circ \delta_2((p_1, q_1), (f, y), (p_2, q_2)) = \delta_1(p_1, f(q_1), p_2) \cdot \delta_2(q_1, y, q_2)$, whether, $\sum_1^{Q_2} = \{f: Q_2 \to \sum_1\}$ is a set of map.

Theorem 7. Let $R_1 =< Q_1, I_1, T_1, \delta_1 >$, $R_2 =< Q_2, I_2, T_2, \delta_2 >$ be quantum automatas, $R_1 \circ R_2 =< Q_1 \times Q_2, I_1 \times I_2, T_1 \times T_2, \delta_1 \circ \delta_2 >$ is the ring product of R_1 and R_2, then for any $f_1y_1f_2y_2 \cdots f_ky_k \in \left(\sum_1^{Q_2} \times \sum_2\right)^*$, $\lceil rec_{R_1 \circ R_2}(f_1y_1f_2y_2 \cdots f_ky_k)\rceil = \{\vee_{q_0\in I_2,q_2\cdots q_{k-1}\in Q_2,q_k\in T_2} \lceil rec_{R_1}(f_1(q_0)f_2(q_1) \cdots f_k(q_{k-1}))\rceil\} \cdot \lceil rec_{R_2}(y_1y_2 \cdots y_k)\rceil$.

Proof. $\lceil rec_{R_1\circ R_2}(f_1y_1f_2y_2\cdots f_ky_k)\rceil$

$$=\vee_{(p_0,q_0)\in I_1\times I_2,(p_1,q_1)\cdots(p_{k-1},q_{k-1})\in Q_1\times Q_2,(p_k,q_k)\in T_1\times T_2}\wedge_{i=0}^{k-1}\delta_1\circ \delta_2((p_i,q_i),f_{i+1}y_{i+1},(p_{i+1},q_{i+1}))$$

$$=\vee_{(p_0,q_0)\in I_1\times I_2,(p_1,q_1)\cdots(p_{k-1},q_{k-1})\in Q_1\times Q_2,(p_k,q_k)\in T_1\times T_2}\wedge_{i=0}^{k-1}\delta_1(p_i,f_{i+1}(q_i),p_{i+1})\ \delta_2(q_i,y_{i+1},q_{i+1})$$

$$=\vee_{(p_0,q_0)\in I_1\times I_2,(p_1,q_1)\cdots(p_{k-1},q_{k-1})\in Q_1\times Q_2,(p_k,q_k)\in T_1\times T_2}(\wedge_{i=0}^{k-1}\delta_1(p_i,f_{i+1}(q_i),p_{i+1}))\ (\wedge_{i=0}^{k-1}\delta_2(q_i,y_{i+1},q_{i+1}))$$

$$=\{\vee_{(p_0,q_0)\in I_1\times I_2,(p_1,q_1)\cdots(p_{k-1},q_{k-1})\in Q_1\times Q_2,(p_k,q_k)\in T_1\times T_2}\wedge_{i=0}^{k-1}\delta_1(p_i,f_{i+1}(q_i),p_{i+1})\}\cdot\{\vee_{(p_0,q_0)\in I_1\times I_2,(p_1,q_1)\cdots(p_{k-1},q_{k-1})\in Q_1\times Q_2,(p_k,q_k)\in T_1\times T_2}\wedge_{i=0}^{k-1}\delta_2(q_i,y_{i+1},q_{i+1})\}$$

$$=\{\vee_{q_0\in I_2,q_2\cdots q_{k-1}\in Q_2,q_k\in T_2}\vee_{p_0\in I_1,p_1\cdots p_{k-1}\in Q_1,p_k\in T_1}\wedge_{i=0}^{k-1}\delta_1(p_i,f_{i+1}(q_i),p_{i+1})\}\cdot\{\vee_{q_0\in I_2,q_2\cdots q_{k-1}\in Q_2,q_k\in T_2}\wedge_{i=0}^{k-1}\delta_2(q_i,y_{i+1},q_{i+1})\}$$

$$=\{\vee_{q_0\in I_2,q_2\cdots q_{k-1}\in Q_2,q_k\in T_2}\lceil rec_{R_1}(f_1(q_0)f_2(q_1)\cdots f_k(q_{k-1}))\rceil\}\cdot\lceil rec_{R_2}(y_1y_2\cdots y_k)\rceil.$$

4. Conclusions

In this paper,we established the multiplication theory of quantum automata without distributive law,using direct product,cascading product and ring product, we can construct complex quantum automata through simple quantum automata.At last ,we can give the form of the language that can be recognized by product quantum automata.

References:

1. Birkhoff G,von Neumann J.The logic of quantum mechanics.*Annals of Mathematics*, **37(4)**:823~843,(1936).
2. Shor PW.Polynomial-Time algorithms for prime factorization and discrete logarithms on a quantum computer.*SIAM Journal on Computing*, **26(5)**:1484~1509,(1997).
3. Ying MS.Automata theory based on quantum logic(I).*International Journal of Theoretical Physics*,**39(4)**:891~991,(2000).
4. Ying MS.Automata theory based on quantum logic(II).*International Journal of Theoretical Physics*, **39(11)**:2545~2557,(2000).
5. Qiu DW. Automata and grammar theory based on quantum logic . *Journal of Software*, **14(1)**:23—27,(2003).
6. Chen Jie.Elementary lattice theory[M]. Inner Mongolia: *Inner Mongolia University Press*,(1990).

IMAGE MATCHING BASED ON MEDIUM SIMILARITY MEASURE*

NINGNING ZHOU†

College of Computer, Nanjing University of Posts and Telecommunications, Nanjing, 210003, PRC
State Key Laboratory of Software Development Environment, Beihang University, Beijing, 100191,PRC

LONG HONG

College of Computer, Nanjing University of Posts and Telecommunications, Nanjing, 210003, PRC

Similarity measure is an important part of the image matching methods.This paper introduces the medium mathematic system to image matching. It establishes the novel image medium similarity measure between two pixels and that of between two image sets based on the measure of medium truth degree. A new image matching algorithm that is governed by the medium similarity measure is discussed in this paper. The experimental results show that the proposed method is effective.

1. Introduction

Image matching technology has the wide range of applications. As an important image research field, it attracts many scholars and experts. They have presented various matching algorithms from different aspects such as: measure method of the matching degree, effective search strategy, increasing the matching speed, improving robustness of algorithm and so on.

Image matching can be regarded as a process which decides the similarity of the corresponding objects in the different images. Similarity measure which is defined as a certain cost function or a distance function is the important part of the image matching method.Correction function, covariance function, the sum of squared residuals, the Hausdorff distance are the common similarity measures used in image matching. In order to improve the matching precision and speed,

* This work is Supported by the National Natural Science Foundation of China under Grant No.61170322 and by the Open Fund of the State Key Laboratory of Software Development Environment under Grant No. SKLSDE-2011KF-04, Beihang University.

† email:zhounn@njupt.edu.cn

some recent mathematical and computational methods such as neural network[1], fuzzy sets[2], genetic algorithm[3][4], wavelet transformation[5][6] etc, have been introduced into image matching.

This paper introduces MMTD(measure of medium truth degree, MMTD) into image matching. It establishes the medium similarity measure and describes a new grey-based image matching algorithm that is governed by the medium similarity measure. The experimental results show that the proposed method is effective.

2. Medium Similarity Measure

Medium principle was established by Wu-jia Zhu and Xi-an Xiao in 1980s who devised medium logic tools to build the medium mathematics system [7]. In medium mathematics system[8], predicate is represented by P, any variable is denoted as x, with x completely possessing quality P being described as P(x). The "╕"symbol stands for inverse opposite negative and it is termed as "opposite to". The inverse opposite of predication is denoted as ╕P. Then the concept of a pair of inverse opposite is represented by both P and ╕P. Symbol "~"denotes fuzzy negative which reflects the medium state of "either-or" or "both this-and that" in opposite transition process.

2.1. *Medium Similarity Measure of Two Pixels*

Given: $X=[x(i,j)]_{M\times N}$ denotes a reference template of M×N pixels, *x(i,j)* represents the gray level of the pixel located at *(i,j)* in reference template. $Y=[y(k,l)]_{P\times Q}$ denotes an input image of P×Q pixels, where M<P, N<Q. Select a sub-image of M×N pixels in the input image Y ,denoted as $F=[f(i,j)]_{M\times N}$, $F\subset Y$. *f(i,j)* represents the gray level of the pixel located at *(i,j)* in this sub image F, as shown in Figure. 1.

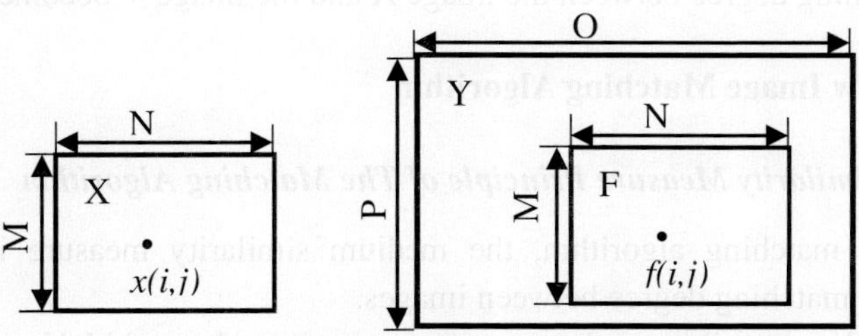

Figure1. The reference template and the input image

To measure the similarity degree between *x(i,j)* and *f(i,j)* ,let predicate *S(x,f)* represents that *x(i,j)* is similar to *f(i,j)*, ╕*S(x,f)* represents that *x(i,j)* is different from *f(i,j)*, and ~ S*(x,f)* transition, as shown in Figure.2. The similarity degree between *x(i,j)* and *f(i,j)* can be calculated by the distance ratio function *hh(i,j)*. According to figure.2., we can get the medium similarity measure between two pixels as follow:

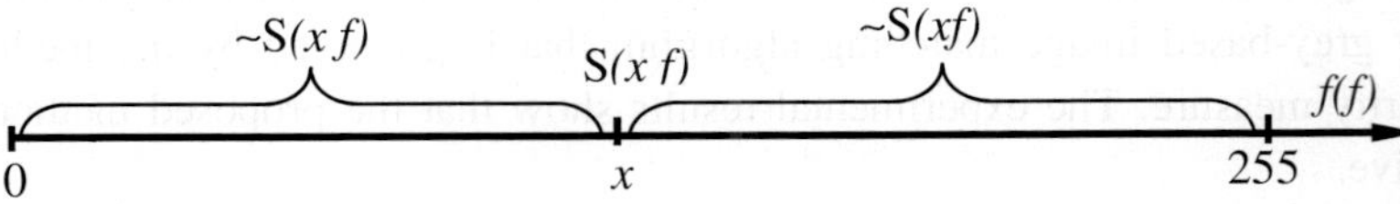

Figure.2. Relation between then gray level of pixel x(i,j) , f(i,j) and predicate similarity

Definition 1. For pixel *x(i,j)* and pixel *f(i,j)* , define

$$hh(i,j)=\begin{cases}\dfrac{d(f(i,j),-1)}{d(x(i,j),-1)} & f(i,j)<x(i,j)\\ 1 & f(i,j)=x(i,j)\\ \dfrac{d(f(i,j),256)}{d(x(i,j),256)} & f(i,j)>x(i,j)\end{cases} \quad (1)$$

as the medium similarity measure of two pixels. The larger the value is, the higher the similarity degree between *x(i,j)* and *f(i,j)* becomes.

2.2. *Medium Similarity Measure of Two Images*

Definition 2. For an image X= $[x(i,j)]_{M\times N}$ of M×N pixels and an image F= $[f(i,j)]_{M\times N}$ of M×N, define

$$s=\frac{1}{M\times N}\sum_{i=1}^{M}\sum_{j=1}^{N}hh(i,j) \quad (2)$$

as the medium similarity measure of two images, where *hh(i,j)* is the distance ratio function as show in expression (1).The larger the value of *s* is, the higher matching degree between the image X and the image Y becomes.

3. The New Image Matching Algorithm

3.1. *The Similarity Measure Principle of The Matching Algorithm*

In the new matching algorithm, the medium similarity measure is used to evaluate the matching degree between images.

Given: X= $[x(i,j)]_{M\times N}$ denotes a reference template of M×N pixels, *x(i,j)* represents the gray level of the pixel located at *(i,j)* in reference template. Y= $[y(k,l)]_{P\times Q}$ denotes an input image of P×Q pixels, where M<P , N<Q. According to expression (2), the medium similarity measure of the reference

image and the sub-image of M╳N which is located at the experiment position *(k,l)* of the input image can be expressed as follows:

$$s(k,l)=\frac{1}{M\times N}\sum_{i=1}^{M}\sum_{j=1}^{N}hh(k+i,l+j) \tag{3}$$

Where the definition of *hh(k+i ,l+j)* is as expression (1), *(k+i, l+j)* represents the gray level of the pixel *(i,j)* in the sub-image which is located at the experiment position *(k,l)* of the input image and *k+i≤P, j+k≤Q*.

Note: *S* is the set of medium similarity measure of reference image X and all the sub-images of M╳N pixels in the input image Y.

$$S=\left[s(k,l)\right]_{(P-M)\times(Q-N)} \tag{4}$$

The larger the value of s*(k,l)* is, the higher matching degree of the reference image X and the sub-image which is located at the experiment position *(k,l)* in the input image Y becomes. Let *max(S)* represents the maximum of the medium similarity measure set S. The value of *max(S)* reflects the matching degree of input image Y and the reference image X. Then set a threshold T, here 0<T<1.The criterions are as follows:

1. When *max(S)*=1, it shows the sub-image located at the position that we can get the maximum of the medium similarity measure can perfectly match the matching template X.
2. When *T<max(S)<1*, it shows the sub-image located at the position that we can get the maximum of the medium similarity measure is highly similar to the matching template. They can match basically.
3. When *max(S)<T*, it shows there isn't any sub-image in the input image Y can match the matching template.

3.2. *The Medium Matching Algorithm*

The details of this matching algorithm are elaborated as follows:

1. Select the gray level, the edge or other eigenvalues of the matching image to establish the feature space of the image.
2. Calculate the medium similarity measure s*(k,l)* of the reference image X and a sub-image in the input image Y.
3. Traverse the input image Y, get the medium similarity measure set *S* of reference image X and all the sub-images in the input image Y.
4. Locate the position *(row,col)* that we can get the maximum value S_{max}(row,col) in the medium similarity measure set *S*.

$$S_{\max}(row,col)=\max(S) \tag{5}$$

5. Set the threshold *T*, where *0<T<1*.
6. According to the criterions mentioned above, complete the matching.

3.3. *Experiment Result Analysis*

Sub-image of Lena image is chosen as reference template to verify the matching algorithm described in this paper. SSDA, IM, Hausdorff are used in comparison in the experiment in order to evaluate the performance of the new algorithm. The results of the experiments are shown in Figure.3-5 and Table 1-3.

Figure. 3 Matching of Standard Lena image

Table. 1 Matching precision and matching rate of Standard Lena image

	Medium	SSDA	IM	Hausdorff
Error	(0,0)	(0,0)	(0,0)	(0,0)
time(s)	0.187	0.113	25.450	1.501

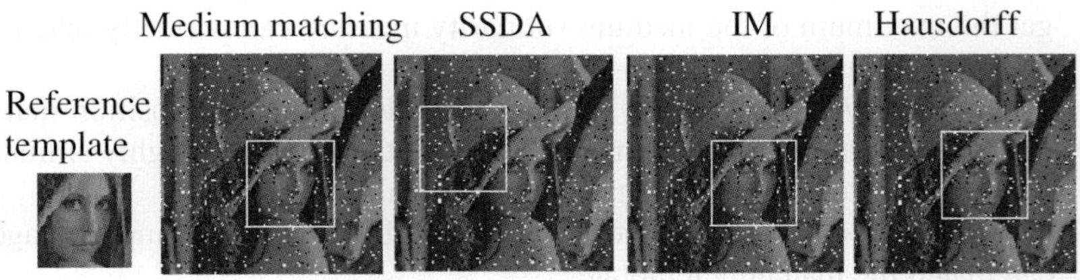

Figure. 4 Matching of Lena image with "salt and pepper" noise (σ=0.05)

Table 2. Matching precision and matching rate of Lena image with noise(σ=0.05)

	Medium	SSDA	IM	Hausdorff
Error	(0,0)	(20,35)	(-1,0)	(5,0)
time(s)	0.193	0.110	26.561	1.673

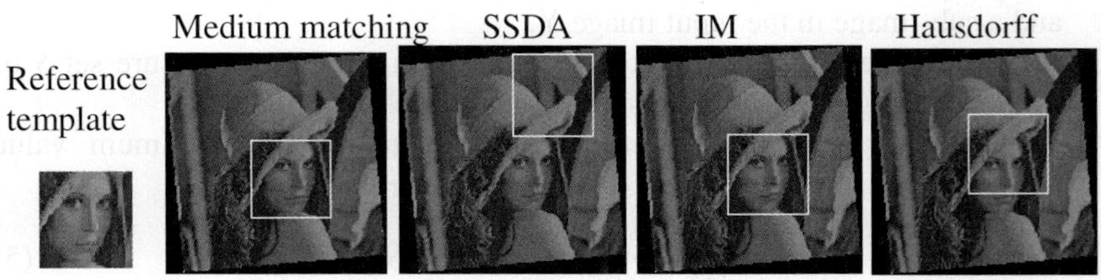

Figure. 5 Matching of Lena image rotated 50

Table 3. Matching precision and matching rate of Lena image (rotated 50)

	Medium	SSDA	IM	Hausdorff
Error	(0,0)	(50,-15)	(-1,-2)	(15,-5)
time(s)	0.309	0.151	35.786	2.346

The experiment results as shown in Figure.3 and Table 1 reveal that SSDA, IM, Hausdorff and medium matching algorithm all can match precisely under the condition of no noise and no distortion. The SSDA has the highest matching rate and the medium matching has a slightly lower rate which is higher than Hausdorff by a magnitude and higher than IM by two magnitudes.

As exhibited in Figure 4,5, Table 2 and 3, when the original image is polluted by noise or rotated 5^0, the matching precision of SSDA is worse, while the IM, the Hausdorff and the medium matching remain the high matching precision. But the matching rate of the medium matching is conspicuously higher than the other two. In SSDA, threshold value has strong influence on the matching precision and matching speed. Improper threshold will result in low matching precision.

The experiments show that the results from this method match a template better than that of the classic image matching algorithms, and it exhibits good anti-noise, anti-distortion and high matching precision and speed.

4. Conclusion

This paper presents a new image matching algorithm based on measuring of medium truth degree. Conclusions about the new algorithm can be drawn as follows. (1)Establish the novel image medium similarity measure between two pixels and that of between two image sets. (2)Design and realize a new algorithm governed by the medium similarity measure. The experiment results show that the new algorithm is effective. We plan to work on the methodology further to promote it in terms of the accuracy and the speed.

References

1. M.G.Li and H. Du, *Journal of Image and Graphics*, **V12**,2357(2008).
2. H.B.Shen, J.Yang, X.J.Liu and N.J.Chen, *Journal of Shanghai Jiaotong University*,**V40**,466(2006).
3. X.F.Leng, J.Y.Liu and Z.Xiong, *Journal on Communications*, **V29,** 17(2008).
4. J.Wang, L.S.Zhou, L.Y.Zhang and Y.J.Zhu, *Journal of Image and Graphics*, **V12**,695(2007).
5. L.Liu and M.Su. *Journal of Image and Graphics*, **V13**,1171(2008).
6. Y.Dong, *University of Electronic Science and Technology of China*,2008
7. X.A.Xiao and W.J.Zhu, *SCIENTIA SINICA*. **V11**, 1320 (1988).
8. L.Hong, X.A.Xiao and W.J.Zhu, *Chinese Journal of Computers*, **V29**, 2186 (2006).

FEATURE-BASED OFF-LINE RECOGNITION OF DEGRADED NUMERAL CHARACTERS *

WEIQING CHENG†, LONG HONG, FEIER ZHU, XIANG LI, JIAN LI

College of Computer, Nanjing University of Posts and Telecommunications, Nanjing, 210003, PRC

State Key Laboratory of Software Development Environment, Beihang University, Beijing, 100191,PRC

This paper proposes an integrated scheme for identifying segmented degraded numeral characters in gray images. The scheme consists of distinguishing foreground from background, wiping off mottles, cutting margins, calculating features, and recognizing numerals by the classifier established based on the measure of medium truth degree and several features selected by logistic regression. The experimental results show that the proposed approach performs well on recognition of degraded numeral characters.

1. Introduction

Off-line recognition of numerals is a classical problem that has been studied for many years. Many methods for off-line recognition of printed or handwritten numeral characters have been proposed [1]-[5]. However, only a few of them have taken degraded numerals into account, and most methods need a large number of samples for training. This paper presents an integrated scheme for identifying segmented degraded numeral characters in gray images. The scheme consists of such steps as distinguishing foreground color from background color, wiping off mottles in an image, cutting margins of an image, calculating features, and recognizing numerals by the classifier established based on the measure of medium truth degree (MMTD) [6][7] and several features selected by an embedded feature selection method using the logistic regression.

---

* This work is supported by the National Natural Science Foundation of China under Grant No.61170322, by the Open Fund of the State Key Laboratory of Software Development Environment under Grant No. SKLSDE-2011KF-04, Beihang University, and by the Fund of NUPT STITP under Grand No. Y2011069.

† chengweiq@njupt.edu.cn

2. Feature extraction

A challenge in feature extraction from a degraded character in a gray image is to locate the character therein. In this paper, we assume that there is only one character in an image and the character size is variable. The other challenge is to determine which features are suitable for classification.

2.1. *Distinguishing between foreground and background*

To determine the character location in a gray image, the image should be firstly converted to a black-and-white image. The pixels in the foreground are generally less than those in the background; thus we design the following algorithm to distinguish between foreground and background:

1) Read the gray image, and store the pixels in the array *pixs* with *pixs*(0,0) and *pixs*(*height*-1,*width*-1) representing the pixels at the top left and the bottom right corners; Get the frequency (i.e. number of pixels) for each color value in the image;

2) Find the color with the maximum number of pixels, denoted as c_1; if the image has only this color then return error;

3) Try to find the color with the maximum number of pixels, denoted as c_2, among colors with the gaps between those and c_1 bigger than a threshold, e.g. the total number of all colors (e.g. 256 for 8 bit pixel) divided by 7; if there is no such color, then try to find the color with the maximum number of pixels among colors except c_1, also denoted as c_2;

4) If $c_1 > c_2$, then the pixels with their colors$\geq (c_1+c_2)/2$ are set to the unique background color (white), and other pixels are set to the unique foreground color (black); If $c_1 < c_2$, then the pixels with their colors$\leq (c_1+c_2)/2$ are set to the background color, and other pixels are set to the foreground color.

2.2. *Wiping off mottles*

The original gray image may be blurry, thus this step is to wipe off mottles in the derived monochrome image. The foreground of the monochrome image is regarded as a set of connected subimages. If a pixel and one of its eight neighbors are both in the foreground, then the two pixels are called connected. Any pixel in a connected subimage is not connected to any pixel outside the subimage. We deem that the character in the image consists of one or few of big connected subimages, other subimages containing few pixels are regarded as mottles. We design the following algorithm to wipe off mottles: 1) find all connected subimages and get the pixel number of every subimage; 2) a subimage with its pixel number less than a threshold, e.g. 0.07*the total number of foreground pixels, is taken as a mottle; 3) a subimage with its pixel number less than a threshold, e.g. 0.25*the pixel number of the biggest subimage, is taken as a mottle; 4) set the pixels in the mottles to the background color.

2.3. *Cutting margins*

This step is to cut margins of the new monochrome image. In fact, we do not actually cut the margins, and just try to obtain the coordinates for the top left corner and the bottom right corner of the foreground.

2.4. *Calculating features*

Fig. 1 shows two original images, and the resulted images after processed by above steps. The images in the rectangles are the normalized images, whose features are to be calculated. 27 statistic features and one structural feature are defined here. The last feature (X_{27}) is defines as the number of rows divided by the number of columns of a normalized image that is logically divided into 3×3 grids averagely, as shown in Fig. 2, to calculate 27 features as follows:

$$X_i = \begin{cases} n_{r,c} / n_r, & if\ (i\%9)\%3 = 0 \\ n_{r,c} / n_c, & if\ (i\%9)\%3 = 1 \\ n_{r,c} / n_{all}, & if\ (i\%9)\%3 = 2 \end{cases}, \tag{1}$$

where i=0,1,2,…,26, r=i/9, c=(i%9)/3, i/9 and i%9 mean the integral quotient and the arithmetical compliment of i divided by 9 respectively, $n_{r,c}$ denotes the number of foreground pixels in grid (r,c), n_r and n_c denote the number of foreground pixels in the row r and that in the column c respectively, n_{all} denotes the total number of all foreground pixels.

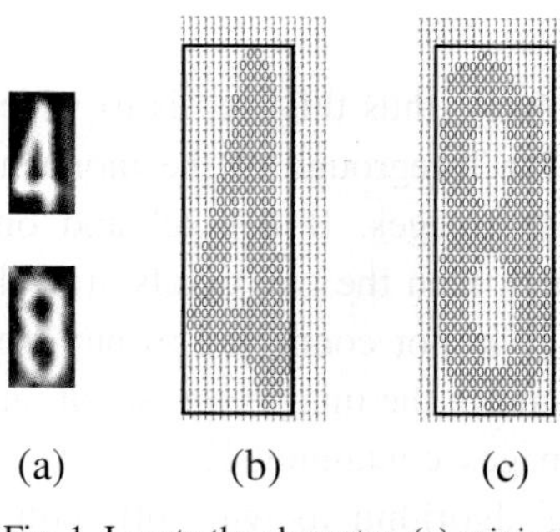

(a) (b) (c)

Fig. 1. Locate the character. (a) original, (b)cleaned, (c)cleaned

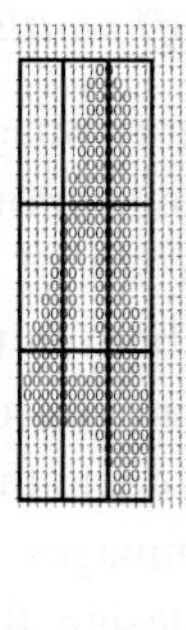

Fig. 2. An image divided into 9 grids

2.5. *Feature selection based on logistic regression*

Feature selection by focusing on a smaller number of features can help to generate more reliable estimates by excluding noises and provide faster and more efficient models [8]. We adopt an embedded feature selection method using the logistic regression.

First, some numeral images are selected as training samples, and 28 features for each sample are calculated according to above steps. In this study, 65 training samples with 6 or 7 ones for each digit are used for feature selection. Next, let X_0~X_{27} be the independent variables, and Y that takes values 0,1,2,..,9 be the response variable; construct a total of the number of response levels (denoted as c) minus 1 cumulative logistic regression models

$$\mathrm{logit}\,(\sum_{i=0}^{k}\pi_i) = \ln(\frac{\sum_{i=0}^{k}\pi_i}{1-\sum_{i=0}^{k}\pi_i}) = \alpha_k + \sum_{j=0}^{m-1}\beta_j X_j, \tag{2}$$

where m=28, c=10, k=0,1,2,…,c-2, i denotes a response level, m denotes the number of features, c denotes the number of response levels or classes, $\pi_i = P(Y = i|X)$, $\sum_{i=0}^{c-1}\pi_i = 1$, $\alpha_0,\alpha_1,\ldots,\alpha_{c-2}$ are intercept parameters, β_j is termed a logistic regression coefficient, X_j is an independent variable and logit is a dependent variable.

Finally, let alpha be 0.05, and then maximum likelihood estimate of the coefficient of a feature having a p-value of 0.05 or less is considered to be statistically significant. Each feature with statistically significant parameter estimate should be selected. In this study, selected features consist of X_0~X_3, X_5~X_8, X_{11}, X_{14}, X_{16}, X_{17}, X_{20}, X_{23}, X_{26}, and X_{27}.

3. Classification based on MMTD (CBM)

We propose a classification method on the basis of MMTD [6] to recognize degraded numeral characters.

3.1. *Measure of medium truth degree*

Definition 1. Assume f_1: $X\rightarrow\boldsymbol{R}$, f_2: $X\rightarrow\boldsymbol{R}$ are two types of one-dimensional quantification mappings of set X of objects. Let $P(x)$ be a predicate, and two distance ratio functions relative to P, that is, h_{T1}: $f_1(X)\rightarrow\boldsymbol{R}$ and h_{T2}: $f_2(X)\rightarrow\boldsymbol{R}$ are defined as below:

$$h_{T1}(y=f_1(x))=\begin{cases}\dfrac{-2.0d(y,\alpha_F-\varepsilon_F)}{d(\alpha_T-\varepsilon_T,\alpha_F-\varepsilon_F)}, & y<\alpha_F-\varepsilon_F\\ 0, & \alpha_F-\varepsilon_F\le y\le\alpha_F+\varepsilon_F\\ \dfrac{d(y,\alpha_F+\varepsilon_F)}{d(\alpha_T-\varepsilon_T,\alpha_F+\varepsilon_F)}, & \alpha_F+\varepsilon_F<y<\alpha_T-\varepsilon_T\\ 1, & \alpha_T-\varepsilon_T\le y\le\alpha_T+\varepsilon_T\\ \dfrac{d(y,\alpha_F+\varepsilon_F)}{d(\alpha_T+\varepsilon_T,\alpha_F+\varepsilon_F)}, & y>\alpha_T+\varepsilon_T\end{cases} \quad (3)$$

$$h_{T2}(y=f_2(x))=\begin{cases}1-12y, & \varepsilon_T=0\\ 1-\dfrac{d(y,\alpha_T+\varepsilon_T)}{d(\alpha_T+\varepsilon_T,\alpha_T-\varepsilon_T)}, & \varepsilon_T\neq0, y>\alpha_T+\varepsilon_T\\ 1, & \varepsilon_T\neq0,\alpha_T-\varepsilon_T\le y\le\alpha_T+\varepsilon_T\\ 1-\dfrac{1.5d(y,\alpha_T-\varepsilon_T)}{d(\alpha_T+\varepsilon_T,\alpha_T-\varepsilon_T)}, & \varepsilon_T\neq0, y<\alpha_T-\varepsilon_T\end{cases} \quad (4)$$

where every $d(a,b)$ denotes Euclidean distance between a and b.

Definition 2. Ten predicates used for numeral character recognition are defined as: $A_i(x)$: the numeral a_i is in the image x, $0\leqslant i\leqslant 9$, where $x\in X$, and the object set X={segmented degraded images}, a_0~a_9 denote 10 numeral characters (0~9) to be identified.

Definition 3. The truth degree function of X relative to $A_i(x)$, $0\leqslant i\leqslant 9$ defined in Def. 2, that is, $g_{nT\text{-}M\text{-}i}: X\to \boldsymbol{R}$ is defined as:

$$g_{nT-M-i}(x)=\left(\sum_{k=0}^{27}w_{i-k}h_{Ti-k}(m_k(x))\right)\Bigg/\left(\sum_{k=0}^{27}w_{i-k}\right) \quad (5)$$

where $w_{i\text{-}0}$~$w_{i\text{-}27}$ denote the weights of features m_0~m_{27} (denoted as X_0~X_{27} in sections 2.4 and 2.5) for identifying whether the numeral character a_i is in the image x, $m_k(x)$ denotes the value of the feature m_k for the image x, $h_{Ti\text{-}k}(m_k(x))$ denotes the distance ratio of $m_k(x)$ relative to $A_i(x)$, and $h_{Ti\text{-}k}$ is h_{T1} or h_{T2} (defined in Def. 1), which depends on the type (1 or 2) of m_k for identifying the numeral character a_i.

3.2. *Parameter setting*

It is obviously that the numeral 1 differs greatly from other numerals in the feature X_{27}(i.e. m_{27}), thus only this feature is chosen to identify the numeral 1. Features in the set S_{subf} are used for identifying other numerals:

$$S_{subf}=\{X_0\text{~}X_3, X_5\text{~}X_8, X_{11}, X_{14}, X_{16}, X_{17}, X_{20}, X_{23}, X_{26}\} \quad (6)$$

If a feature is chosen, then its weight in Eq. (5) is simply set to 1, otherwise 0. Thus $w_{1\text{-}27}$ is set to 1, and $w_{1\text{-}k}$ (k=0,1,2,…26) are set to 0; $w_{i\text{-}0}$~$w_{i\text{-}3}$, $w_{i\text{-}5}$~$w_{i\text{-}8}$, $w_{i\text{-}11}$, $w_{i\text{-}14}$, $w_{i\text{-}16}$, $w_{i\text{-}17}$, $w_{i\text{-}20}$, $w_{i\text{-}23}$ and $w_{i\text{-}26}$ (i=0,2,3,4,…9) are set to 1, and other $w_{i\text{-}k}$ are set to 0.

The type for m_{27} is 1, while types for other features are all 2. Four parameters in Eq.(3) must be set for calculating $h_{T1\text{-}27}(m_{27}(x))$. Two parameters in Eq.(4) must be set for calculating each $h_{Ti\text{-}k}(m_k(x))$ where $i \neq 1$ and $k \neq 27$. Also 65 training samples mentioned in section 2.5 are used to determine parameters for the distance ratio functions as follows:

For each numeral *i*, do {
    For each feature m_j of this numeral, do {
        Sort values of m_j for samples of the same numeral *i* in increasing order and obtain an ordered population;
        Let *p0, p1, p2* be 20th, 50th, 80th percentile of the population respectively, and *p3* = max(0.12**p1*, min(*p2* - *p1*, *p1* - *p0*));
        If the numeral is 1 and the feature is m_{27}, then set α_T, ε_T, α_F, ε_F for Eq.(3), to *p1*, *p3*, *p1**0.4, *p3**0.4 respectively;
        If the numeral is not 1 and the feature is in the set S_{subf} defined in Eq.(6), then set α_T, ε_T for Eq.(4), to *p1*, *p3* respectively;
    }
}

3.3. *Recognition algorithm*

The recognition algorithm is described as below:

1) compute features of an image *x* according to section 2.1~2.4;

2) compute ten truth degrees $g_{nT\text{-}M\text{-}i}$ $(0 \leqslant i \leqslant 9)$ according to Eq.(5);

3) find the highest truth degree denoted as $g_{nT\text{-}M\text{-}k1}$ and the second highest one denoted as $g_{nT\text{-}M\text{-}k2}$;

4) if *k*1 is equal to 1 and $g_{nT\text{-}M\text{-}k1}$ is no less than 0.1, then *obj*=1;
else if *k*1 is not 1, then *obj*= *k*1;
else *obj*= *k*2;

5) assert *x* to be numeral *obj*;

4. Experimental results

We have implemented a numeral character recognition system according to the proposed scheme, and adopt 400 degraded gray images from [9] as the test dataset, where only a total of 65 more legible images are taken as training samples. Comparisons of recognition accuracy using our method (CBM) and the template matching method (TMM) [4] implemented in [9] are given in Table 1. Experimental results show that CBM recognition rates of numerals 4,5,7,9 are notably higher than the TMM, and the average recognition rate is increased by 8%.

But in some cases our method is not better than TMM, especially the case with numeral 8 that may be misclassified as numeral 5 or 4. The reasons for the misclassification errors mainly attribute to: (1) false distinguishing foreground from background; (2) cutting margins without adapting to the leaning character in an image; (3) features not enough for discriminating the numeral pairs "8-5", "4-6", "8-4".

5. Conclusions

The contribution of this paper is fourfold. First, an effective approach is proposed to locate the character in a degraded gray image. Secondly, this work has proved that using logistic regression to obtain a reduction in the number of features is effective. Thirdly, this work has used both statistic and structural features, and experimental results highlight that integrating various features of objects for recognition is usually a good idea. Finally, this work indicates again the measure of n-dimensional medium truth degree is suitable for multi-class classification for its scalability [10]. We plan to work on the proposed scheme further to promote it in terms of the recognition performance.

Table 1 Comparisons of recognition accuracy

Numeral	Total	Recognition errors	
		Template matching method	Our method
0	40	0	0
1	40	2	3
2	41	6	8
3	40	5	6
4	40	17	5
5	40	11	1
6	39	2	3
7	40	13	3
8	40	6	10
9	40	10	2
Average recognition rate		82%	90%

References

1. Adnan Amin, *Pattern Recognition*. **31**, 517(1998).
2. L.M. Lorigo, V. Govindaraju, *IEEE Transactions on Pattern Analysis and Machine Intelligence*. **28**, 712(2006).
3. N. Arica and F.T. Yarman-Vural, *IEEE Transactions on Systems, Man, and Cybernetics, Part C: Applications and Reviews*. **31**, 216(2001).
4. Ivind Due Trier, Anil K. Jain and Torfinn Taxt, *Pattern Recognition*. **29**, 641(1996).
5. V. Vaidya, V. Deshpande, D. Gada and B. Shirole, *Proceedings of the Fifth International Conference on Document Analysis and Recognition*. 273(1999).

6. Hong L, Xiao X A and Zhu W J, *Chinese Journal of Computers*. **29**, 2186 (2006).
7. Xiao X A and Zhu W J, *Scientia Sinica*. **11**, 1320 (1988).
8. Ma S G and Huang J, *Briefings in Bioinformatics*. **9**, 392 (2008).
9. http://www.verysource.com/handwrittenfiguresid-16142.html (2011).
10. Cheng W Q, Gong J, Ding W and Sun Z X, *The Journal of China Universities of Posts and Telecommunications*. **17**, 72(2010).

ON MEASURE METHOD TO THE TWO-PHASE UNCERTAINTY AND ITS APPLICATION *

YULONG DENG †

College of Computer, Nanjing University of Posts&Telecommunications,Nanjing,210003,P.R.China

LONG HONG

College of Computer, Nanjing University of Posts&Telecommunications,Nanjing,210003,P.R.China

In this paper, the uncertainty problem, including randomness and fuzziness as well as the medium mathematics background are introduced firstly. Then the two-phase uncertainty problem is described and two-phase degree (TPD), a measure method to the two-phase uncertainty, is introduced. Next, the SMS (Short Message Service) Spam in the mobile communication is studied as a two-phase uncertainty problem and a TPD of spam based algorithm is applied to make the spam filtering. The algorithm is combined the Naïve Bayes based algorithm with the measure to the n-dimensional distance ratio mean function. The experiment result indicates that the TPD based algorithm is better than the classical Bayes based one and has practical value in current technical conditions.

1. Introduction

In nature and the human society, there are large numbers of problems characterized by uncertainty. Generally, uncertainty is characterized by two different attributes, the randomness and the fuzziness. The uncertainty contained in an event which has a determined definition but may not occur is named as "randomness", meanwhile, the uncertainty contained in an event which has occurred but can not be defined exactly is named as "fuzziness" [1].

The Probability Theory lays the foundation for research on the randomness in the uncertainty problems and there has been lots of research work to deal with fuzziness. In 1965, L A Zadeh brought about the concept of fuzzy sets [2] and built the fuzzy mathematics, which measured and processed the fuzzy

* This work is supported by the National Natural Science Foundation of China under Grant No.61170322 and by the Open Fund of the State Key Laboratory of Software Development Environment under Grant No. SKLSDE-2011KF-04, Beihang University

† email: DYL@njupt.edu.cn

phenomenon based on the membership function and achieved success in both theory and application. In 1982, Z Pawlak made the concept of rough sets [3], which was used to describe the incomplete and indeterminate objects, find out implied principle and knowledge behind the data. Zhu Wujia and Xiao Xi-an proposed medium rule [4] and established medium mathematics (MM) whose primary contents were medium axiomatic set theory (MS) using the medium logic created by themselves as inferential tools in 1980s. In the meaning of pure mathematical foundation, MM has extended the research objects of mathematics from precision to fuzziness and widened the logic and set foundation of classic mathematics.

The randomness and fuzziness are two aspects of one thing. For example, in this sentence "The weather of tomorrow will be warm", the object is the tomorrow's temperature, the warmth is a fuzzy concept but the whole sentence is a random event [5]. For completely understanding objects, it is required to make research on the both sides of the uncertainty. Li Deyi brought about the concept of membership clouds, which described the character of this and that of fuzzy problems and the randomness of the membership as a whole [2]. Hong Long proposed the two-phase uncertainty which depicted the randomness and fuzziness of objects and gave its measure method in macro-level [5]. For describing the uncertainty in more detail, Two-Phase Degree, a measure method to two-phase uncertainty will be discussed in this paper, which depicts the two-phase uncertainty in micro-level.

2. The Measure of Medium Truth Degree (MMTD) and the Measure of the Two-Phase Uncertainty

The theory of the measure of medium truth degree (MMTD) is built on MM, describing the relationship between numerical value interval and truth-value of predicate as well as measuring the numerical degree of the logic truth. [6].

Some discussions in this paper will be based on the foundation of MM and MMTD. For more information, see [6][7].

The uncertainty including only the randomness or fuzziness is named as single-phase uncertainty and the uncertainty including both the randomness and the fuzziness is named as two-phase uncertainty. Hong Long has proposed to use the hybrid entropy [5] to measure the two-phase uncertainty.

3. The Two-Phase Degree

The hybrid entropy describes the two-phase uncertainty degree in macro-level. For studying the two-phase uncertainty degree in micro level, we propose the two-phase degree (TPD), which can be used to measure the hybrid degree which combines the possibility degree of one of inverse opposite properties of an event occurring, with the medium truth value degree relative to one property.

Definition 1: Assume (Ω, *B*, *U*) be the two-phase uncertainty space. Let Ω be the sample space and B be the uncertainty event range, $B \subset P(\Omega)$. Use ordered pair (*p*, *h*) to describe the two-phase degree (TPD) of event x in the sample space. Let *p* be the probability of event x, $p \in [0, 1]$, h be the distance ratio function relative to one of the two inverse opposite properties. Let *h* be h_T or h_F, $h \in \mathbf{R}$.

Definition 2: To two-phase degree of x, assume *S* (*p*, *h*) be the modulus value, *T* (*p*, *h*) be the tendency value and *A* (*p*, *h*) be the area value, then

$$S(p,h) = \sqrt{p^2 + h^2} \ (1) \quad T(p,h) = tg((p,h)) = \frac{p}{h} \ (2) \quad A(p,h) = p * h \ (3)$$

We can discuss TPD of x and its three values in area I~VIII of the coordinated plane, see figure 1. Let P be a predicate and ~P means event *x* has property P partially. Denote the inverse opposite side of a predicate P as ╕P. Hence P and ╕P represent a pair of the inverse opposite concepts. The super truth is expressed in symbol as $^+$, so $^+$P is more than P and ╕$^+$P is more than ╕P. The x-axis represents the value of distance ratio mean function corresponding to P which is used to evaluate the medium truth degree of event x. The y-axis represents the probability of event x.

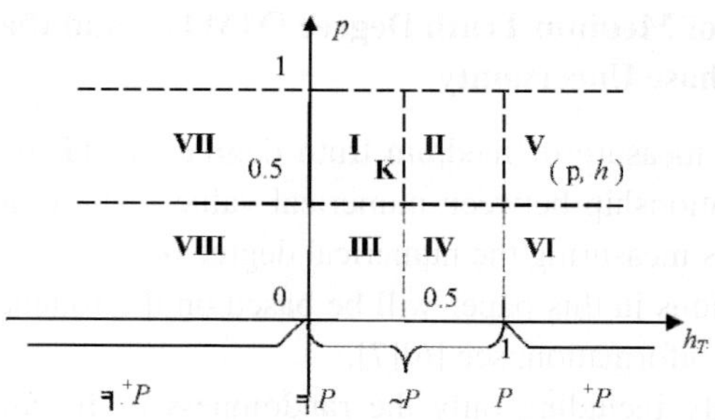

Figure 1. TPD of x in the coordinated plane

4. The Anti-SMS-Spam Algorithm Based on the TPD of Spam

Let SM_x be a short message(SM), The SMS Spam filtering is a dichotomy problem in nature. It gives the value of smc_i of SM_x, $smc_i \in \{smc_0, smc_1\}$, smc_0 means SM_x is spam SM, smc_1 means SM_x is a normal SM.

It is clear that the spam and normal is a pair of inverse opposite properties of SM and meets the medium rule. Let R be a predicate describing "is spam", then $R\ (SM_x)$ denotes that SM_x is spam and $\neg R\ (SM_x)$ denotes that SM_x is normal. $\sim R\ (SM_x)$ describes the transition between spam and normal.

Supposed that the rules provided by the SMS receiver are unchanged, the result to mark SM_x as spam (or normal) in the sample space will demonstrate some kind of statistical distribution. It means the spam filtering has randomness. Therefore, the SMS Spam filtering has both randomness and fuzziness so it is a two-phase uncertainty problem.

We propose an algorithm based on the TPD of spam, which combined the rule based method and the Bayesican based method. So we can calculate the medium truth value of the SM and evaluate $R\ (SM_x)$ in the numerical value interval, and the Bayesican based method will give the posterior probability of the SM_x classification which can be used to evaluate the spam property of SM_x.

4.1. *Spam Filtering Based on the Medium Truth Degree of SM Spam*

Let $f(SMx)$ be the n-dimensional quantification mapping of the SM set. Based on experiential data, we give three rules, $f(SMx)=(f_1, f_2, f_3)$. $h_{3T\text{-}M}(f)$ is the distance ratio mean function corresponding to R which is used to evaluate the medium truth degree of spam for SMx.

$$h_{3T\text{-}M}(f)= \frac{1}{3}\sum_{i=1}^{3} f_i(SMx) \tag{4}$$

4.2. *Spam Filtering Based on the Posterior Probability Evaluation*

We use the classical Naïve Bayes algorithm to evaluate the probability of spam for SMx, it is based on the Multi-Variate Bernoulli Event Model (MBM) and gives the posterior probability. The content of the SMx is described as c_x, $c_x = a_1a_2\ldots a_m$, $a_i \in V$, $1 \leq i \leq m$, $P\ (smc_i \mid c_x)$ can be calculated as follow:

$$P\ (smc_i \mid c_x) = \frac{P(smc_i)P(c_x \mid smc_i)}{P(c_x)} \tag{5}$$

4.3. *The Measure to the Two-Phase Degree (TPD) of SM Spam*

Based on the definition 1 in section 3, the TPD of spam of SMx is denoted by $(P(smc_0 \mid c_x), h_{3T\text{-}M}(f))$, We can set the threshold θ depended on $A(P(smc_0 \mid c_x),$

$h_{3T\text{-}M}(f)$) in the algorithm and evaluate the level of TPD which can be applied to spam filtering.

5. Experiments of the Anti-SMS-Spam Algorithm Based on the TPD of Spam

We build a SM corpus which contains 528 SMs, 259 of SMs are spam and the rest are normal. Similar with Email spam filtering, Confusion Matrix [8] is used as the evaluation tool in experiments. The r (recall ratio) can evaluate the detection ratio to the spam and the p (precision ratio) can evaluate the precision ratio to the spam.

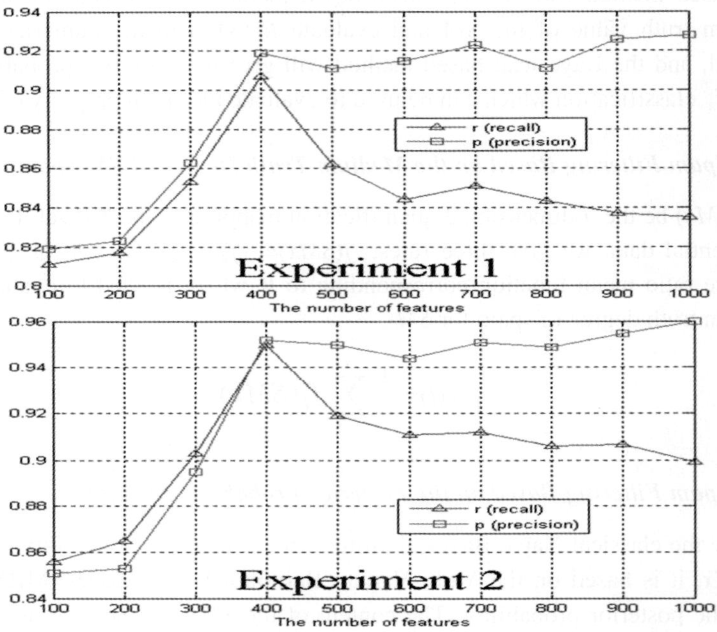

Figure 2. The results of experiment 1 and 2

Experiment 1: The Naïve Bayes based Anti-SMS-Spam algorithm. The features are extracted using information gain (IG) and the posterior probability is given using Eq(5). The posterior threshold is set to 0.5. The relation among the r, p and the number of features is depicted in figure 2.

Experiment 2: The TPD based Anti-SMS-Spam algorithm. The number of keywords is set to 700, which is chosen based on the rule based Anti-SMS-Spam algorithm using Eq(4), the features are extracted by the same method in experiment 1 and the same posterior threshold is set. The threshold for TPD area value is set to 0.25. The relation among the r, p and the number of features is depicted in figure 2.

6. Discussions on the Experiment Results and the Evaluation of the Anti-SMS-Spam Algorithm Based on the TPD of Spam

The variation of the r and p in experiment 2 is similar with experiment 1, but the r and p are higher at the same number of features because of combination of two algorithms in experiment 2. In experiment 2, the decline of r becomes slower after it reaches its peak, which caused by the compensation from the rule based algorithm. Compared with the classical Naïve Bayes based Anti-SMS-Spam algorithm in experiment 1, the TPD based algorithm demonstrates the better effect. But its complexity is higher and requires more computing time. In real application, it can be ignored because of the powerful process ability of today's mobile communication system. The result indicates the TPD based algorithm has practical value and the TPD measure method is reasonable in solving some uncertainty problems as well.

References

1. Li Deyi, Meng Haijun and Shi Xuemei, *Computer R&D*. **V32**,15(1995)
2. L. Zadeh, Fuzzy sets, *Information and Control*. **V8**, 338(1965)
3. Z. Pawlak, Rough sets, *International Journal of Information and Computer Science*. **V11**, 341(1982)
4. Zhu Wujia and Xiao Xian, *Nature Magazine*. **V7**, 723(1984)
5. Hong Long, *Ph.D Thesis in Nanjing University of Aeronautics and Astronautics*. (2006)
6. Hong Long, Xiao Xi-An, and Zhu Wu-Jia, *Chinese Journal of Computers*. **V29,** (2186)2006
7. Hong Long, Xiao Xi-An, and Zhu Wu-Jia, *Chinese Journal of Computers*. **V30**, (1551)2007
8. Gómez,J.M., *SAC'02 Proceedings of the ACM Symposium on Applied Computing, ACM, New York, USA*, (615)2002

FILTERS AND DEDUCTIVE SYSTEMS ON RESIDUATED MULTILATTICES

I. P. CABRERA*, P. CORDERO, G. GUTIÉRREZ, J. MARTÍNEZ, M. OJEDA-ACIEGO
Dept Applied Mathematics. University of Málaga, Spain

We continue the study of the residuated operations in the framework of hyperstructures. We focus on the case of a multilattice as underlying algebraic structure and introduce the notions of filter and deductive system. They differ from the analogous concepts in a pocrim due to the connection to congruence relations. Finally, we prove that the set of filters of a residuated multilattice is a complete lattice.

1. Preliminary definitions

Residuation has a prominent role in the algebraic study of logical systems, which usually are modeled as partially ordered sets together with some operations reflecting the properties of the connectives.

This work is related to the use of residuated implication in the framework of hyperstructures and fuzzy logic reasoning.

Although the most used structure in this context is that of residuated lattice, there are reasons for weakening some of its properties, thus leading to a more general class of algebraic structures for computation.

A commonly considered algebraic structure is that of partially ordered commutative residuated integral monoid[1].

Definition 1.1. A tuple $\mathcal{A} = (A, \leq, *, \to, \top)$ is said to be a *partially ordered commutative residuated integral monoid*, briefly a **pocrim**, if, for every $a, b, c \in A$, the following properties hold:

- $(A, *, \top)$ is a commutative monoid with neutral element $\top$
- $(A, \leq)$ is a partially ordered set with maximum $\top$.
- the operations $*$ and $\to$ satisfy the adjointness condition, that is $a * c \leq b$ if and only if $c \leq a \to b$.

A pocrim $\mathcal{A}$ is said to be a **residuated lattice** if $(A, \leq)$ is a lattice.

* *ipcabrera@uma.es*

Monotonicity of the ordering relation $\leq$ wrt $*$ follows as a consequence of the definition. Specifically, it holds that $x \leq y$ implies $x * z \leq y * z$

It is well-known that residuated lattices are considered to be the algebraic structures of substructural logics~\cite{ono-50sl}, which are logics without some of the structural rules of logic: weakening, contraction, or associativity.

Concerning applications in logic, the notions of filter and deductive system, closely related to *modus ponens*, deserve to be studied in depth.

Definition 1.2. Given $\mathcal{A} = (A, \leq, *, \rightarrow, \top)$ a pocrim, a non-empty subset $F \subseteq A$ is said to be a **filter** if the following conditions hold:

i. if $a, b \in F$, then $a * b \in F$
ii. if $a \leq b$ and $a \in F$, then $b \in F$.

On the other hand, F is said to be a **deductive system** if

i. $\top \in F$ and
ii. $a \rightarrow b \in F$ and $a \in F$ imply $b \in F$.

It is not difficult to see that both definitions are equivalent. In [3] several kinds of filters and congruence relations in pocrims have been introduced.

2. Congruences and filters on multilattices

We focus here on some extensions of the previously defined notions to the framework of multilattices. These are partially-ordered sets together with two non-deterministic operations (hyperoperations) which generalize the supremum and the infimum by weakening the restrictions imposed on a (complete) lattice. Specifically, a *multisupremum of* a and b is defined as a minimal element of the set of upper bounds of a and b, we denote as $a \sqcup b$ *the set of all the multisuprema of* a *and* b; whereas $a \sqcap b$ is introduced similarly. Now, we can proceed with the formal definition of multilattice and related structures.

Definition 2.1. A poset $(M, \leq)$ is said to be a **multilattice** if for all $a, b, x \in M$ with $a \leq x$ and $b \leq x$, there exists[†] $z \in a \sqcup b$, such that $z \leq x$; and, similarly, for all $a, b, x \in M$ with $a \geq x$ and $b \geq x$, there exists $z \in a \sqcap b$, such that $z \geq x$. A multilattice is said to be **full** if $a \sqcup b \neq \varnothing$ and $a \sqcap b \neq \varnothing$ for all $a, b \in M$.

The authors have studied in [5,6] different properties of multilattices in relation to non-determinism and coalgebras; in the rest of the paper, we focus on the aspects related to residuation, filters, and congruences.

[†] Note that the definition is consistent with the existence of two incomparable elements *without* any multisupremum. In other words, $a \sqcup b$, and also $a \sqcap b$, can be empty. Likewise, a poset satisfying $a \sqcup b \neq \varnothing \neq a \sqcap b$ needs not be a multilattice, see [4].

Now, we will introduce a notation which will be useful hereafter. Let $\mathcal{R}$ be a binary relation in M and $X,Y \subseteq M$, then $X\widehat{\mathcal{R}}Y$ denotes that, for all $x \in X$, there exists $y \in Y$ such that $x\mathcal{R}y$ and for all $y \in Y$ there exists $x \in X$ such that $x\mathcal{R}y$.

Definition 2.2. Let $(M,\sqcup,\sqcap)$ be a multilattice, a **congruence** on M is any equivalence relation $\equiv$ such that, for all $a,b,c \in M$, if $a \equiv b$, then $a \sqcup c \mathrel{\hat{\equiv}} b \sqcup c$ and $a \sqcap c \mathrel{\hat{\equiv}} b \sqcap c$.

It is well-known that, for every set A, the set of equivalence relations on A, $Eq(A)$, with the inclusion ordering (in the powerset of $A \times A$) is a complete lattice in which the infimum is the meet, and the supremum is the transitive closure of the join. In [7] the authors proved that the set of congruences on a multilattice is also a complete lattice.

Theorem 2.1. *The set of the congruences in a multilattice M, $Con_m(M)$, is a sublattice of $Eq(M)$. Furthermore, $Con_m(M)$ is a complete lattice w.r.t. the inclusion ordering.*

There exist several ways to give a definition for the notion of filter of a multilattice.

In this section, we introduce the one which is more suitable for extending the classical results about congruences and homomorphisms.

Definition 2.3. Let $(M,\sqcup,\sqcap)$ be a multilattice. A non-empty set $F \subseteq M$ is said to be a **filter** if the following conditions hold:

i. $i, j \in F$ implies $\varnothing \neq i \sqcap j \subseteq F$.

ii. $i \in F$ implies $i \sqcup a \subseteq F$ for all $a \in M$.

iii. For all $a,b \in M$, if $(a \sqcup b) \cap F \neq \varnothing$ then $a \sqcup b \subseteq F$.

Theorem 2.2. *If $(M,\sqcup,\sqcap)$ is a full multilattice then $(\mathcal{F}(M),\subseteq)$, the set of filters in M, is a complete lattice.*

Theorem 2.3. *Let $(M,\sqcup,\sqcap)$ be a multilattice with top element $\top$, and let $\equiv$ be a congruence relation. Then the equivalence class $[\top]$ is a filter of M.*

3. Filters and deductive systems on residuated multilattices

Definition 3.1. A **residuated multilattice** is a pocrim whose underlying poset is a multilattice. If, in addition, there exists a bottom element, the residuated multilattice is said to be **bounded**.

That is, a residuated multilattice is a hyperalgebra $(M,\sqcup,\sqcap,*,\to,\top)$ such that $(M,\sqcup,\sqcap)$ is a multilattice and $(M,\leq,*,\to,\top)$ is a pocrim where $\leq$ is the natural ordering on the multilattice. Observe that, as a consequence of the definition of pocrim, $\top$ is the top element in the multilattice. To ease notation, from now on, every residuated multilattice will be denoted just by M, when no confusion arises.

Notice that every residuated multilattice is full: for all $a,b \in M$ we have that $a,b \leq \top$ and, therefore, $a \sqcup b \neq \varnothing$. Furthermore, $a * b \leq a$, and $a * b \leq b$, hence $a \sqcap b \neq \varnothing$.

It is convenient to remark that any finite poset is actually a multilattice, hence the only proper examples of pocrims not being multilattices have infinitely many elements.

Due to the fact that any residuated multilattice combines the structures of multilattice and pocrim, it is possible to use the notion of filter on the multilattice, filter on the pocrim, or give a new definition that combines both. These three notions are not equivalent, as we will show with an example below. To distinguish them, we will write **p-filter** to denote a filter of the pocrim and **m-filter**, a filter of the multilattice. The notion that we are interested in throughout this work, will be called **r-filter**.

Definition 3.2. Let M be a residuated multilattice. A non-empty subset $F \subseteq M$ is said to be an **r-filter** if it is a deductive system and the following conditions hold: (1) $a,b \in F$ implies $\varnothing \neq a \sqcap b \subseteq F$, and (2) for all $a,b \in M$, if $(a \sqcup b) \cap F \neq \varnothing$ then $a \sqcup b \subseteq F$.

Therefore, F is an r-filter if and only if it is an p-filter and an m-filter.

The set of r-filters in M is denoted by $\mathcal{F}_r(M)$. Since M is a full multilattice, the intersection of a family of r-filters is always nonempty, hence it is also an r-filter. As a consequence, $(\mathcal{F}_r(M), \subseteq)$ is a complete lattice, as it is a complete inf-semilattice with top element.

Example 3.1. Let M be the multilattice given by the bounded poset $(M, \leq)$ described in Fig. 1. Let $A = \{a_i \mid 0 \leq i \leq 5\}$, $B = \{b_i \mid 0 \leq i \leq 5\}$ and $C = \{b_i \mid 2 \leq i \leq 5\}$ and the operations $*$ and $\rightarrow$ defined as follows:

$$x * y = \begin{cases} x & \text{if } y = 1 \\ y & \text{if } x = 1 \\ b_2 & \text{if } x, y \in C \\ b_0 & \text{if } x \in B \setminus C, y \in B \\ b_0 & \text{if } x \in B, y \in B \setminus C \\ a_0 & \text{otherwise.} \end{cases} \qquad x \rightarrow y = \begin{cases} 1 & \text{if } x \leq y \\ y & \text{if } x = 1 \\ a_5 & \text{if } x \in B, y \in A \\ b_1 & \text{if } x \in C \\ & \text{and } y \in B \setminus C \\ b_5 & \text{otherwise.} \end{cases}$$

It is routine calculation that $(M, \leq, *, \rightarrow, 1)$ is a residuated multilattice.

The subset $C \cup \{1\}$ is a p-filter but it is not an r-filter because $b_3 \sqcap b_4 = \{b_1, b_2\} \not\subseteq C$. Likewise, $\{b_5, 1\}$ is an m-filter that is not an r-filter because $b_5 * b_5 = b_2 \notin \{b_5, 1\}$. Finally, $B \cup \{1\}$ is a p-filter and an m-filter, therefore, it is an r-filter.

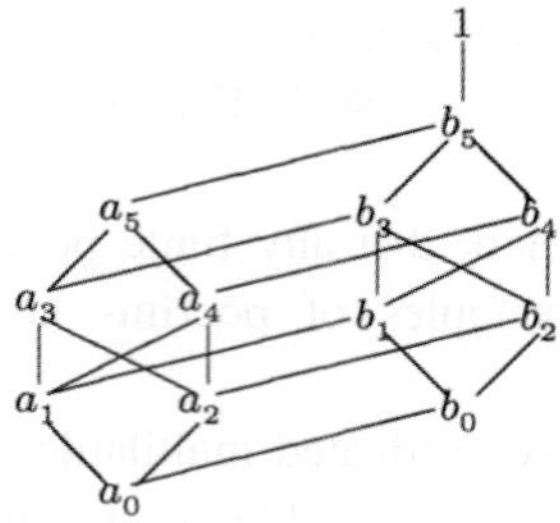

Fig. 1.

Definition 3.3. Let $(M, \leq, *, \rightarrow, \top)$ be a residuated multilattice, a **congruence** on M is any equivalence relation $\equiv$ such that if $a \equiv b$, then $a \sqcup c \mathrel{\hat{=}} b \sqcup c$, $a \sqcap c \mathrel{\hat{=}} b \sqcap c$, $a * c \equiv b * c$, $a \rightarrow c \equiv b \rightarrow c$ and $c \rightarrow a \equiv c \rightarrow b$, for all $a, b, c \in M$.

Example 3.2. In the residuated multilattice given in Example 1, the partition $\{A, B \cup \{1\}\}$ defines a non-trivial congruence relation in which the class $B \cup \{1\}$ is an r-filter.

Theorem 2.1 ensures that the set of congruences in a multilattice, $Con_m(M)$, is a sublattice of the complete lattice $(Eq(A), \subseteq)$. It is not difficult to see that the set of the congruences in a residuated multilattice M, $Con_r(M)$, is a sublattice of $Con_m(M)$.

Definition 3.4. Let $h: M \rightarrow M'$ be a map between residuated multilattices, h is said to be an **homomorphism** if, for all $a, b \in M$, $h(a \sqcup b) \subseteq h(a) \sqcup h(b)$, $h(a \sqcap b) \subseteq h(a) \sqcap h(b)$, $h(a * b) = h(a) * h(b)$ and $h(a \rightarrow b) = h(a) \rightarrow h(b)$.

Observe that $h(\top) = \top$, for all homomorphism h between residuated multilattices.

It is also remarkable that $h(a \sqcup b) = \big(h(a) \sqcup h(b)\big) \cap h(M)$ and $h(a \sqcap b) = \big(h(a) \sqcap h(b)\big) \cap h(M)$, since M is a full multilattice [7].

The relation among filters, congruences and homomorphisms is stated below and follows from the convenient definition of r-filter just introduced.

Theorem 3.1. *Let $h: M \rightarrow M'$ be an homomorphism between residuated multilattices. Then, the relation $a \equiv b$ if and only if $h(a) = h(b)$ is a congruence, and $h^{-1}(\top) = \{x \in M \mid h(x) = \top\}$ is an r-filter of M. On the other hand, given a congruence relation $\equiv$ on M, the mapping $p: M \rightarrow M/_{\equiv}$ such that $p(x) = [x]$ is a surjective homomorphism of residuated multilattices and, as a consequence, the equivalence class $[\top]$ is an r-filter of M.*

4. Conclusions

We have generalized the notions of filter and deductive system considering a residuated multilattice as underlying algebraic structure. These two concepts differ from those based on a pocrim due to the connection to congruence relations. Finally, we have proved that the set of r-filters of a residuated multilattice is a complete lattice, and that the usual relation between r-filters, congruences, and homomorphisms generalizes as expected.

Acknowledgements

Partially supported by projects TIN2009-14562-C05-01 and TIN2011-28084 (Science Ministry of Spain), and P09-FQM-5233 (Junta de Andalucí a).

References

1. W. J. Blok and J. G. Raftery, *Journal of Algebra* 190, 280 (1997).
2. H. Ono, Substructural logics and residuated lattices—an introduction, in *Trends in logic: 50 years of Studia Logica*, (Kluwer, 2003) pp. 177–212.
3. R. Halaš and M. Botur, *Soft Computing* 13, 1063 (2009).
4. J. Medina, M. Ojeda-Aciego and J. Ruiz-Calviño, *Fuzzy Sets and Systems* 158, 674 (2007).
5. I. Cabrera, P. Cordero, G. Gutiérrez, J. Martínez and M. Ojeda-Aciego, *Information Sciences* 180, 4323 (2010).
6. I. Cabrera, P. Cordero, G. Gutiérrez, J. Martínez and M. Ojeda-Aciego, *Applied Mathematics and Computation* (2012), Accepted. doi:10.1016/j.amc.2011.10.081.
7. I. P. Cabrera, P. Cordero, G. Gutiérrez, J. Martínez and M. Ojeda-Aciego, On congruences, ideals and homomorphisms over multilattices, in *EUROFUSE Workshop Preference Modelling and Decision Analysis*, pp. 299–304, 2009.

α-MULTI ARY SEMANTIC RESOLUTION METHODS BASED ON LATTICE-VALUED PROPOSITIONAL LOGIC LP(X)*

YI LIU[1,2], XIAOYAN QIN[3], YANG XU[2]

[1]*College of Mathematics and Information Science, Neijiang Normal Univeristy*
Neijiang, Sichuan 641112, P.R.China
[2]*Intelligent Control Development Center, Southwest Jiaotong Univeristy*
Chengdu, Sichuan 610031, P.R.China

[3]*College of Mathematics and Computer Science, Shanxi Normal Univeristy*
Linfen, Shianxi, 041000, P.R.China

This paper focuses on α-multi ary semantic resolution automated reasoning method based on the general form of α-generalized resolution principle for lattice-valued logic. The definitions of the α-multi ary semantic resolution and α-multi ary semantic resolution deduction in LP(X) are given, the soundness and completeness of this ground case are gotten. This work will provide a theoretical foundation for the more efficient resolution based automated reasoning in lattice-valued logic

1. Introduction

Resolution was introduced by Robinson in his landmark paper [2] in 1965 as a mechanizable method for detecting the unsatisfiability of a given set of formulae in classical first-order logic, it revolutionized the field of automated reasoning, and since then, many refinements of resolution have been proposed by researcher to cut down the search space and increase efficiency.

One of the most important refinements of resolution is semantic resolution [3], which was introduced by Slagle in 1967. In semantic resolution method, it utilize the technology that restraining the type of clauses and the order of literals participated in resolution procedure to reduce the redundant clauses, and can increase the efficiency of reasoning, the order of predicate symbol in semantic resolution is a coarse order, because there can exist more than one predicate symbols in one clause in first-order logic, among these identical predicate

* This work is supported by National Natural Science Foundation of P. R. China (Grant no. 61175055) and Scientific Research Fund of SiChuan Provincial Education Department(11ZB023) and the Sichuan Key Technology Research and Development Program (2011FZ0051)

symbols, there are no order, so the resolution literals can not be selected solely. In order to select the resolution literals solely in resolution procedure and reduce the new clause, in 1971, Reiter, Slagle and Norton presented the ordered clause semantic resolution—OI-resolution [4], the efficiency of OI-resolution is very high in many cases. Unfortunately, it is incomplete. In 1979, Liu applies the idea of indexing the literals in lock resolution method presented by Boyer, and introduced a new semantic resolution method——IDI-resolution [5], IDI-resolution possess the advantages of PI-resolution and OI-resolution, and it is complete for clause set in first-order logic.

Lattice-valued logic, as one of the most important many-valued logics, extends the chain type truth-valued field to general lattice in which the truth-values are incompletely comparable with each other. Aiming to deal with the uncertainty especially for incomparability in the intelligent information processing, Xu et al. proposed lattice implication algebra (LIA)[11], lattice-valued logic [14] based on LIA, uncertainty reasoning and automated reasoning [14] in lattice-valued logic based on LIA, and applied them in many areas, such as rule bases, decision making, natural language processing, etc. For the automated reasoning aspect, Xu et al [12-15] developed α-resolution principle in lattice-valued logic, gave the soundness and weak completeness of this principle. Xu and Zhong [16] investigated the general form of α-resolution principle based on lattice-valued propositional logic LP(X), and gave the theorems of both soundness and weak completeness of this general form of α-resolution principle. This general form of α-resolution principle is extended to the corresponding lattice-valued first-order logic LF(X) and its soundness theorem, lifting lemma and weak completeness theorems are also proved. In order to improve the reasoning efficiency of α-resolution principle in lattice-valued propositional logic LP(X), Zhang [17-18] investigated α-semantic resolution method by applying semantic strategy on α-resolution principle. The soundness and weak completeness of α-group semantic resolution are also established.

2. Preliminaries

In the following, we will introduce some elementary concepts and conclusions of lattice-valued logic with truth-value in lattice implication algebras and the general form of α-resolution principle. We refer the readers to [16, 17, 18] for more details.

Definition 2.1 [11] Let $(L, \vee, \wedge, O, I)$ be a bounded lattice with an order-reversing involution $'$, I and O the greatest and the smallest element of L

respectively, and $\rightarrow$: $L \times L \rightarrow L$ be a mapping. $(L, \vee, \wedge, ', \rightarrow, O, I)$ is called a lattice implication algebra if the following conditions hold for any $x, y, z \in L$:

(I_1) $x \rightarrow (y \rightarrow z) = y \rightarrow (x \rightarrow z)$,

(I_2) $x \rightarrow x = I$,

(I_3) $x \rightarrow y = y' \rightarrow x'$,

(I_4) $x \rightarrow y = y \rightarrow x = I$ implies $x = y$,

(I_5) $(x \rightarrow y) \rightarrow y = (y \rightarrow x) \rightarrow x$,

(l_1) $(x \vee y) \rightarrow z = (x \rightarrow z) \wedge (y \rightarrow z)$,

(l_2) $(x \wedge y) \rightarrow z = (x \rightarrow z) \vee (y \rightarrow z)$.

Definition 2.2 [15] The set $\mathcal{F}$ of formulas of lattice-valued propositional logic LP(X) is the least set Y satisfying the following conditions:

(1) $X \subseteq Y$,

(2) $L \subseteq Y$,

(3) if $p, q \in Y$, then $'(p), \rightarrow (p, q) \in Y$,

where X is the set of propositional variables, L is the set of constants.

In the following, we denote $'(p)$ as p' and $\rightarrow (p, q)$ as $p \rightarrow q$.

Definition 2.3 [15] A mapping v: LP(X) $\rightarrow L$ is called a valuation of lattice-valued propositional logic LP(X), if it is a T-homomorphism.

Definition 2.4 [15] Let $G \in \mathcal{F}$ and $\alpha \in L$. If $v(G) \leq \alpha$ for any valuation v of lattice-valued propositional logic LP(X), we say G is always less than or equal to α (or G is α-false), denoted by $G \leq \alpha$.

Definition 2.5 [16] Let $C_i = p_{i1} \vee \ldots \vee p_{im_i}$ be generalized clauses of LP(X), $H_i = \{p_{i1}, \ldots, p_{im_i}\}$ the set of all disjuncts occurring in C_i, $i = 1, 2, \ldots, m$, $\alpha \in L$. For any $i \in \{1, 2, \ldots, m\}$, if there exist generalized literals $x_i \in H_i$ such that $x_1 \wedge x_2 \wedge \ldots \wedge x_m \leq \alpha$, then

$$C_1(x_1 = \alpha) \vee C_2(x_2 = \alpha) \vee \ldots \vee C_m(x_m = \alpha)$$

is called an α-resolvent of $C_1, C_2, \ldots, C_m$, denoted by $R_{p(g\text{-}\alpha)}(C_1(x_1), C_2(x_2), \ldots, C_m(x_m))$, $x_1, x_2, \ldots, x_m$ are called an α-resolution group.

3. α-multi Ary Semantic Resolution Methods Based on LP(X)

Definition 3.1 Let V be a valuation in lattice-valued propositional logic $LP(X)$, $\alpha \in L$. $N, E_1, E_2, \cdots, E_q (q \geq 1)$ is a generalized clauses sequence in $LP(X)$, $\mathcal{G}$ is an order of generalized literals occurring in these clauses. The finite clauses sequence

$$(\{N\}, \{E_{1_1}, E_{1_2}, \cdots, E_{1_{i_1}}\}, \{E_{2_1}, E_{2_2}, \cdots, E_{2_{i_2}}\}, \cdots, \{E_{s_1}, E_{s_2}, \cdots, E_{s_{i_s}}\}), \quad (*)$$

where $E_{j_{i_j}} \in \{E_1, E_2, \cdots, E_q\} (j = 1, \cdots, s)$, is called an α-$\mathcal{G}v$ semantic clash w. r. t. V and $\mathcal{G}$, if (*) satisfy the following conditions:

(1) $v(E_{j_{i_j}})\le\alpha, j=1,2,\cdots,s$;

(2) Let $R_1=N_1=N$,for any $\{j_1,\cdots,j_{i_j}\}\subseteq\{1,2,\cdots,q\}$, $j=1,2,\cdots,s$,there exists an α -multi ary resolution formula

$$R_{j+1}=(N_j,E_{j_1},E_{j_2},\cdots,E_{j_{l_j}}),$$

where $N_j=R_{j-1}\bigcup N_j^*, N_j^*\subseteq N\bigcup\{R_1,\cdots,R_{j-1}\}$,the resolution generalized literal in $E_{j_1},E_{j_2},\cdots,E_{j_{l_j}}$ are all leftmost generalized literals in $E_{j_1},E_{j_2},\cdots,E_{j_{l_j}}$;

(3) $v(R_{s+1})\le\alpha$.

R_{s+1} is called α - $\mathcal{G}v$ resolvent of this clash, N is called the core and $\{E_{1_1},E_{1_2},\cdots,E_{1_{i_1}}\},\{E_{2_1},E_{2_2},\cdots,E_{2_{i_2}}\},\cdots,\{E_{s_1},E_{s_2},\cdots,E_{s_{i_s}}\}$

are called α -electrons group.

Definition3.2 Let $S=C_1\wedge C_2\wedge\cdots\wedge C_m$, where $C_1,C_2,\cdots,C_m$ are generalized clauses in lattice-valued propositional logic $LP(X)$, v be a valuation in $\mathcal{V}$ and $\alpha\in$ L, $\mathcal{G}$ is an order of generalized literals occurring in these clauses. An α-resolution deduction from S to Φ_t: $\Phi_1,\Phi_2,\cdots,\Phi_t$,is called an α - $\mathcal{G}v$ resolution deduction from S to Φ_t, if it satisfies the following conditions:

(1) $\Phi_i\in\{C_1,C_2,\cdots,C_m\}(i=1,\cdots,t)$ or

(2) Φ_i is an α - $\mathcal{G}v$ resolvent, where the core and electrons of Φ_t are composed of Φ_j $(j<i)$ or generalized clauses occurring in S.

Example 3.3 In lattice-valued propositional logic $L_9P(X)$, let $\alpha=a_6$, generalized clause set $S=\{N,E_1,E_2\}$,where

$N=(x\to y)\vee(s\to t)$, $E_1=(y\to z)\vee(w\to t)\vee(p\to q)$

$E_2=(x\to z)'\vee(s\to q)$.

Define a valuation of $L_9P(X)$ as follows :

$$v(x)=v(y)=v(s)=v(w)=v(p)=I\ , v(z)=v(t)=v(q)=a_4\ ,$$

then $v(E_1)=a_4\le a_6, v(E_2)=a_4\le a_6$.

Let $\mathcal{G}$ be an order of generalized literal in N,E_1,E_2 and $R_1=N$. Since the leftmost generalized literals are $y\to z$, $(x\to z)'$ in E_1,E_2, respectively, and

$$(x\to y)\wedge(y\to z)\wedge(x\to z)'\le\alpha=a_6,$$

we have

$$R_2(N,E_1,E_2)=\alpha\vee(s\to t)\vee(w\to t)\vee(p\to q)\vee(s\to q)\text{, and}$$

$$v(R_2(N,E_1,E_2))=v(\alpha\vee(s\to t)\vee(w\to t)\vee(p\to q)\vee(s\to q))=a_6\le\alpha$$

Therefore $(\{N\},\{E_1,E_2\})$ is an α - $\mathcal{G}v$ semantic clash w. r. t. v and $\mathcal{G}$, $\{E_1,E_2\}$ is α-electron group and N is α-core of this clash, R_2 is an α - $\mathcal{G}v$ resolvent.

Theorem 3.4 Let $S=C_1\wedge C_2\wedge\cdots\wedge C_m$, where $C_1,C_2,\cdots,C_m$ are generalized clauses in lattice-valued propositional logic $LP(X)$, v be a valuation in v and $\alpha\in$ L, $\mathcal{G}$ is an order of generalized literals occurring in these clauses. If there exists an α - $\mathcal{G}v$ semantic clash w. r. t. v and $\mathcal{G}$, R_{s+1} is an α - $\mathcal{G}v$ resolvent of this clash, then $C_1\wedge C_2\wedge\cdots\wedge C_m\le R_{s+1}$.

Proof The proof is straightforward from [16].

Theorem 3.5 Let $S=C_1\wedge C_2\wedge\cdots\wedge C_m$, where $C_1,C_2,\cdots,C_m$ are generalized clauses in lattice-valued propositional logic $LP(X)$, v be a valuation in v and $\alpha\in L$, $\mathcal{G}$ is an order of generalized literals occurring in these clauses. There exists an α - $\mathcal{G}v$ resolution deduction from S to Φ_t

$$\Phi_1,\Phi_2,\cdots,\Phi_t,$$

and Φ_t is α -empty clause, then $S\le\alpha$.

Proof According to the soundness of the general form of α -resolution principle in lattice-valued propositional logic LP(X), we can obtain the result easily.

Theorem 3.6 Let $S=C_1\wedge C_2\wedge\cdots\wedge C_m$, where $C_1,C_2,\cdots,C_m$ are generalized clauses in lattice-valued propositional logic $LP(X)$, v be a valuation in $LP(X)$ and $\alpha\in L$, $\mathcal{G}$ is an order of generalized literals occurring in these clauses. If the following conditions hold:

(1) $S\le\alpha$;

(2) If $S^*\subseteq S$ and generalized literal x_i doesn't occur in S^*, then S^* and x_i are independent each other ;

(3) If g is the leftmost generalized literal of $\mathcal{G}$,then $v(g)=\alpha$;

Then there exists an α - $\mathcal{G}v$ resolution deduction from S to α -empty clause.

References

1. K.Y. Qin, Y. Xu, Z.M. Song, Lattice-valued propositional logic LP(X) (I), Fuzzy Sys. Math. **11 (4)** (1997) 5-10.
2. L. Wos, J. A. Robinson, D. F. Carson, Efficiency and completeness of the set of support strategy in theorem proving, J. ACM, **12(4)** (1965) 536-541.

3. Slagle, J. R. Automatic Theorem Proving with Renamable and Semantic Resolution, J. ACM, **14**(4)(1967) 687-697.
4. Reiter, R. Two Results on Ordering for Resolution with Merging and Linear Format, J. ACM, **18** (4)(1971)630-646 .
5. X.H.Liu,Type-3 IDI-resolution principle, J. Jilin University, (**1**)(1979)47-52.
6. X.H.Liu,Type-3 IDI-resolution principle, J. Jilin University, (**1**)(1979)47-52.
7. X. H. Liu, Lock-semantic resolution principle with lemmas—LI-resolution principle, Journal of Jilin University (Science Edition), **4** (1979) 129-136.
8. X. H. Liu, Input semi-lock resolution method based on Horn sets, Chinese Science Bulletin, **16** (1985) 1201-1202.
9. X. H. Liu, The compatibility among three resolution principles, Journal of Software, **2** (1992) 60-64.
10. X. X. He, Y. Xu, et al., α-Satisfiability and α-Lock Resolution for a Lattice-Valued Logic LP(X), The 5th International Conference on Hybrid Artificial Intelligence Systems, Jun.23-25, 2010, San Sebastián, Spain, pp. 320-327
11. Y. Xu, Lattice implication algebras, J. Southwest Jiaotong Univ. **1**(**2**)(1993) 123-128.
12. Y. Xu, D. Ruan, E.E. Kerre, J. Liu, α-Resolution principle based on lattice-valued propositional logic LP(X). Inform. Sci., **130**(2000) 195-223.
13. Y. Xu, D. Ruan, E.E. Kerre, J. Liu, α-Resolution principle based on lattice-valued first-order logic LF(X), Inform. Sci., **132**(2001) 221-239.
14. Y. Xu, D. Ruan, K.Y. Qin, J. Liu, Lattice-Valued Logic, Springer, 2003.
15. Y. Xu, J. Liu, D. Ruan et al , Determination of α-resolution in lattice-valued first-order logic LF(X), Inform. Sci.,**181**(2011)1836-1862.
16. Xiaomei Zhong, Jun Liu, Shuwei Chen, Yang Xu. α-Group semantic resolution method in lattice-valued first-order logic system LP(X) (finished)
17. Jiafeng Zhang, Yang Xu. Semantic resolution method on lattice-valued propositional logic LP(X), Journal of Liaoning Technical University (Natural Science) , **29**(**5**) (2010)767-770.
18. Jiafeng Zhang, Yang Xu, Xingxing H. Lattice-valued semantic resolution reasoning Method, Computer Science, **38**(**9**)(2011)201-203+210.

AN OPTIMAL SPECTRUM ALLOCATION SCHEME BASED ON STACKELBERG GAME IN MACRO/FEMTOCELL HIERARCHICAL NETWORKS *

PENG XU†, YANG XU
College of Maths, Southwest Jiaotong University,
Chengdu, 610031, China

Femtocells widely deployed in a macrocell forms hierarchical cell networks, has been regarded as one of the most promising approach. However owing to the absence of coordination between the macro and femtocells, and among femtocells, decentralized spectrum allocation between macro and femtocell users become technically challenging. In this paper, an optimal spectrum allocation scheme based on Stackelberg game is proposed, in which femto and macrocell base stations are players, and the same spectrum is the resource players will choose to assign users to minimize the affected interference among each other. The comparison results show that the proposed scheme might be a solution for efficiently allocating the spectrum in hierarchical cell networks, as the improvement in terms of outage probability and spectrum efficiency had been achieved.

1. Introduction

The recent development of hierarchical macro/femtocell networks is a realistic way of providing better quality of service to indoor mobile users [1]. The placement of femtocells has a critical effect on the performance of a pre-existing macrocell network, One of the complicated issues yet to be solved is that femto and macrocells could suffer severe co-channel interference from each other if spectrum management is not appropriately considered .

Generally it has been considered more feasible to introduce femtocells on orthogonal frequency than macrocells due to coexistence problems by reducing the mutual interference. However that will lead low spectrum efficiency, and recent studies have shown that deploying co-channel femtocells with minor impact on the macrocell performance is possible [2]. One conventional approach

---

*This work is supported by National Natural Science Foundation of China (Grant No. 61175055), Sichuan Key Technology Research and Development Program of China (Grant No.2011FZ0051), and Ministry of Industry and Information Technology of China, Radio Regulatory Bureau (Grant No.[2011]146).

† pengxup@my.swjtu.edu.cn

to improving spectrum efficiency in static manner is to reuse the frequency band in multiple geographical areas or cells. However inter-cell interference will be inevitably incurred, when in adjacent cells share the same spectrum [3]. However preplanned frequency assignments cannot real-timely satisfy the variety of femtocells deployed in ad-hoc manner. So the methods of dynamic spectrum allocation are needed [4]. In the meantime, game theory as a mathematical tool has been widely used in radio resource allocation, such as in [5]. The Stackelberg game was given and proved particularly in [5], but spectrum allocation was not discussed concretely.

In this paper an optimal frequency allocation scheme based on Stackelberg game is proposed, through which femto and macro BSs will choose the optimal frequency in shared frequency to assign associated users, where the optimal is defined as minimal interference introduced to users. The scheme indicates superior performance gain over the conventional method with high SINR.

2. System Model

As the femtocells are largely deployed randomly, the focus is concentrated in one macrocell with some femtocells randomly deployed, and some users are distributed among hierarchical networks.

The process of spectrum allocation between m-BS and f-BSs accords with a dynamic game, for example Stackelberg game. Stackelberg game also known as the leader-follower game, is an extension of non-cooperative dynamic game in which there is a group of players, called leaders that have the privilege of making the first move, while the remaining players, called the followers, make their move after the leaders. That is motivated by the fact cross-tier interference by deploying f-BSs has to be minimized, more priority is given at the m-BS. In Stackelberg game-based spectrum allocation scheme, the m-BS is the leader and the f-BSs are the followers, which is rationale that the m-BS being the primary networks is preexisted whereas f-BSs are later deployed.

Then the Stackelberg game can be defined as:

$$G=\{N,S,U\} \tag{1}$$

where $N=\{M,F\}$ denotes the set of players, $M=\{mBS\}$ and $F=\{fBS_1, fBS_2,\ldots,fBS_n\}$ denote the macro base station and some femto base stations in one macrocell, respectively. $S=\{f_1,f_2,\ldots,f_m\}$ denotes the frequency that *mBS* and *fBS* will allocate, as the strategy set. $U=\{U_M, U_F\}$ represents the relevant utility set for the set *M* and *F*.

3. Stackelberg Game Formulation

To simplify the analysis, assume that m-BS and all other existed femtocells have adaptively assigned frequencies, and the suitable power control is adapted. Then the scenario is changed into that a new femtocell is deployed, some new users are also distributed in macro and femtocell. Only spectrum allocation of the new femtocell is considered in this part.

The utility set $\tilde{U} = \{\tilde{U}_M, \tilde{U}_F\}$ can be written as:

$$\tilde{U}_M(mBS_i) = \{\tilde{u}_i^m \mid i \in 1,2,\cdots,m\} \tag{2}$$

$$SINR_{i,m} = \frac{P_{i,m}}{\sum_{m' \in M'-M} P_{i,m'} + \sum_{f \in F} P_{i,f} + N} \tag{3}$$

$$\tilde{u}_i^m : \min(\sum_{x \in X^m} P'_{i,x} + \sum_{y \in Y^f} P'_{i,y}) \tag{4}$$

$$\tilde{U}_F(fBS_j) = \{\tilde{u}_{j(i^*)}^f \mid j \in 1,2,\cdots,m\} \tag{5}$$

$$SINR_{j(i^*),f} = \frac{P_{j(i^*),f}}{\sum_{m \in M'} P_{j(i^*),m} + \sum_{f' \in F'} P_{j(i^*),f'} + N} \tag{6}$$

$$\tilde{u}_{j(i^*)}^f : \min(\sum_{x \in X^m} P'_{j(i^*),x} + \sum_{y \in Y^f} P'_{j(i^*),y}) \tag{7}$$

where $SINR_{i,m}$ denotes the SINR of the macro users when communicating with m-BS on sub-channel *i*, $SINR_{j(i^*)}, f$ denotes the SINR of the femto users when communicating with f-BS on sub-channel *j* after the m-BS allocating frequency *i**. $P_{i,m}$ represents the received power of the macro users on sub-channel *i*, *Pj(i*), f* denotes the received power of the femto users on sub-channel *j* after the m-BS allocating frequency *i**. The set of m-BSs is denoted by *M'*, and *M'-M* denotes the set except the serving m-BS. The set of f-BSs except the serving f-BS is denoted by *F'*. The existed user sets for macro and femtocell are represented as x∈Xm and y∈Yf, respectively. When the *i*th sub-channel is allocated to a new macro user *m*, the interference introduced is mainly for existed macro users and femto users as described in Eq. (4), which is also as the game utility of m-BS. Similar for f-BS, Eq. (7) is also given.

The Stackelberg game process described above is known as a dynamic game model with complete and perfect information [6]. Backward induction [7] is applied to obtain the subgame perfect Nash equilibrium.

4. Numerical Results

In this part, the scheme based on Stackelberg game is evaluated in terms of outage probability and spectrum efficiency, compared to the other schemes based on conventional frequency reuse with reuse factor (RF) 3, fractional frequency reuse (FFR) with RF=3 as in [5], and conventional manner only with SINR, respectively. The simulation parameters are listed in Table 1 [8]. As described in previous part, the focus is the performance of the proposed scheme when a new femtocell deployed in a macrocell. The number of femto users is varied in one macrocell coverage. The macro and femto users are randomly distributed in the overall network at the ratio 4:1.

Table 1. Parameters setting

Parameters	Values
Channel Bandwidth	5Mhz
Inter macrocell distance	800m
Radius of a femtocell	30m
Number of cells	One tier(7 macrocells)
Macro BS transmit power	20W(43dBm)
Femto Bs transmit power	20mW(13dBm)
Number of used sub-carrier	300
Channel model (Path loss, PL)	$PL_{outdoor}$=28+35$\log_{10}$(d) PL_{indoor}=38.5+20$\log_{10}$(d)+L_{walls} L_{walls}: 7dB, d∈(0,10]; 10dB, d∈(10,20]; 15dB, d∈(20,30]
Log-Normal fading	8dB(outdoor), 4dB(indoor)
BER	10^{-6}

The highest spectrum efficiency is obtained by the Stackelberg game scheme from Fig.1. Compared to other three schemes, the maximum increments are 3.8% for FFR, 122% for conventional reuse, 426% for conventional manner. The main reason is that the co-channel interference is reduced to minimum.

Fig.2 depicts the outage probability of total users according to the SINR threshold, when one femtocell deployed in a macrocell and there are 50 new users distributed in the system. In a given SINR threshold, the proposed scheme indicates lower outage probability than other schemes. This implies that the proposed scheme effectively supports more users through cross-tier interference avoidance, even though the interference is severe.

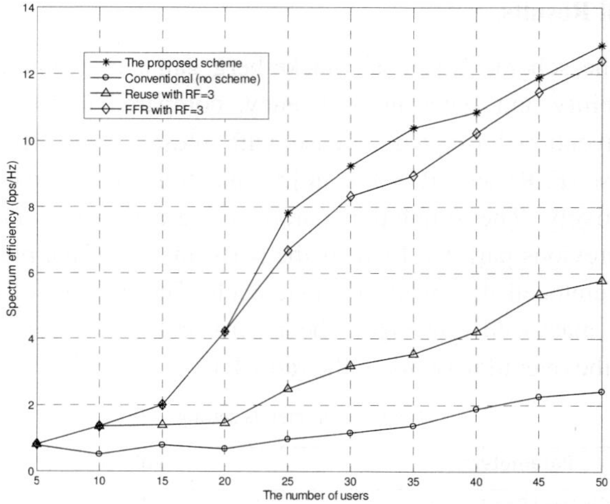

Figure 1. Spectrum efficiency versus users

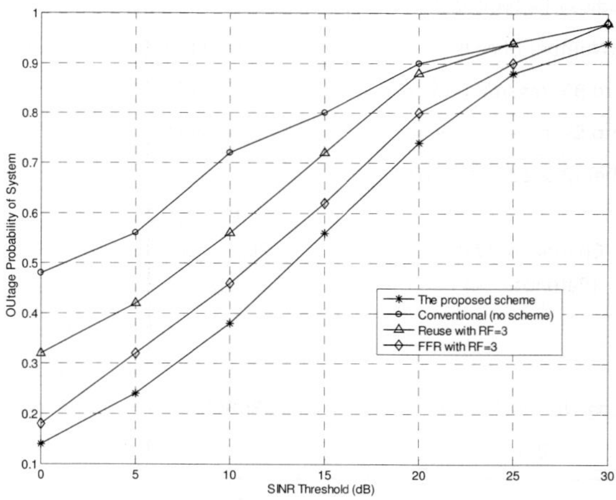

Figure 2. Outage probability vs SINR threshold

5. Conclusions

Through the proposed scheme, the BSs will allocate frequencies according with two principles. The first is to allocate the orthogonal frequencies to macro and femto users, if the spectrum resource is enough. The second is to allocate the frequencies causing the minimum co-channel interference. Although the analysis is only as to one femtocell, it is not limited and can be expanded to more femtocells. Based on the same mechanism the scheme will perform as good as the one femtocell, and the similar results will be obtained.

References

1. Claussen H. Performance of macro- and co-channel femtocells in a hierarchical cell structure. IEEE International Symposium on Personal, Indoor and Mobile Radio Communications, Athens, Greece, September 3-7, 2007:1–5
2. Ho L T W, Claussen H. Effects of user-deployed, cochannel femtocells on the call drop probability in a residential scenario. In Proceedings of the 18th International Symposium on Personal, Indoor and Mobile Radio Communications (PIMRC), Athens, Greece, September 3-7, 2007:1-5
3. Chang R Y, Tao Z, Zhang J, et al. A graph approach to dynamic fractional frequency reuse (FFR) in multi-cell OFDMA networks. In Proc. IEEE ICC, Dresden, Germany, June 14-18 , 2009:1-6
4. Jabbari B, Pickholtz R, Norton M. Dynamic spectrum access and management. IEEE Wireless Communications, August 2010,17(4):6-15
5. Bennis M, Debbah M, Chair A. On spectrum sharing with underlaid femtocell networks. Personal, Indoor and Mobile Radio Communications Workshops, Istanbul, Turkey, 26-30 Sept. 2010:185-190.
6. V.H.Stackelberg. Marketform und Gleichgewicht. Oxford, University Press, 1934
7. D.Fundenberg and J.Tirole. Game Theory, Cambridge:The MIT Press, 1991
8. Femto Forum. Interference Management in OFDMA Femtocells, [March 2010]. http://www.femtoforum.org/femto/pdfs01.php

A STUDY ON FUZZY D-IMPLICATIONS*

A.G. HATZIMICHAILIDIS, G.A. PAPAKOSTAS AND V.G. KABURLASOS†
Human-Machines-Interaction (HMI) Laboratory, TEI of Kavala
GR-65404 Agios Loukas, Kavala, Greece

A first study of some properties of the fuzzy D-implications is presented in this paper. Compared to the *S*-, *R*-, and *QL* - implications, D-implications are less used and reported in the literature. This fact constitutes a particular motivation to investigate and analyze some of their properties by providing analytical proofs. Such analysis is of high importance since it constructs a concrete theoretical framework, which can be used in a latter step for applying the D-implications in real life applications.

1. Introduction and Preliminaries

Mas et al. [5,6] presented the fuzzy implications of type $I(a,b)= S(T(n(a),n(b)),b)$, $\forall a,b\in[0,1]$, where T is a t-norm, S is a t-conorm and n is a fuzzy negation, on $[0,1]$. The triple $<T,S,n>$ is required to satisfy the De Morgan laws.

In order to study the properties and characteristics of the fuzzy D-implications, some concepts and basic notation of fuzzy implications that are more or less known in the literature, are need to be recalled. The definitions and notations that are going to be used in this paper can be found in [2-4].

A function $n:[0,1]\to[0,1]$ is called a fuzzy *negation*, if it is non-increasing, i.e. $a\geq b\Rightarrow n(a)\leq n(b)$, $\forall a,b\in[0,1]$.Moreover $n(0)=1$, $n(1)=0$.

The intersection of two fuzzy sets *A* and *B* is a function of the form: $i:[0,1]\times[0,1]\to[0,1]$ and $i(0,0)=i(0,1)=i(1,0)=0$, $i(1,1)=1$. Properties that a function i must satisfy to be acceptable as an intersection of fuzzy sets are exactly the same properties of functions that are known as *t-norms*.

A *t-norm* is a binary operation $T:[0,1]\times[0,1]\to[0,1]$, that satisfying at least the following conditions $(\forall a,b,c\in[0,1])$:

---

* This research has been co-funded by the European Union (Social Fund) and Greek national resources under the framework of the "Archimedes III: Funding of Research Groups in TEI of Athens" project of the "Education & Lifelong Learning" Operational Programme.

† *e-mail:* vgkabs@teikav.edu.gr

(i) $T(a,1)=a$, *boundary condition*
(ii) $b \le c \Rightarrow T(a,b) \le T(a,c)$, *monotony*
(iii) $T(a,b)=T(b,a)$, *commutative*
(iv) $T(T(a,b),c)=T(a,T(b,c))$, *associative*

The union of two fuzzy sets A and B is a function of the form: $u:[0,1]\times[0,1]\rightarrow[0,1]$ and $u(1,1)=u(0,1)=u(1,0)=1$, $u(0,0)=0$. Properties that a function u must satisfy to be acceptable as a union of fuzzy sets are exactly the same properties of functions that are known as *t-conorms*.

A *t-conorm* is a binary operation $S:[0,1]\times[0,1]\rightarrow[0,1]$, that satisfying at least the following conditions $(\forall a,b,c \in [0,1])$:

(i) $S(a,0)=a$, *boundary condition*
(ii) $b \le c \Rightarrow S(a,b) \le S(a,c)$, *monotony*
(iii) $S(a,b)=S(b,a)$, *commutative*
(iv) $S(S(a,b),c)=S(a,S(b,c))$, *associative*

A *fuzzy implication* I is a function of the form: $I:[0,1]\times[0,1]\rightarrow[0,1]$ where for any possible truth values a,b, of given fuzzy propositions p,q, respectively, it defines the truth value, $I(a,b)$, of the conditional proposition "if p then q". The *implication operator* of the classical logic is a mapping: $m:\{0,1\}\times\{0,1\}\rightarrow\{0,1\}$, which satisfies the conditions: $m(0,0)=m(0,1)=m(1,1)=1$ and $m(1,0)=0$. These conditions are the least ones that we can demand from a fuzzy implication operator. In other words, fuzzy implications collapse to the classical implication, when the truth values are restricted to 0 and 1; i.e. $I(0,0)=I(0,1)=I(1,1)=1$ and $I(1,0)=0$.

One way of defining m in classical logic is to use the logic formula $\forall a,b \in \{0,1\}: m(a,b)=\bar{a} \vee b$.

Another two different Boolean expressions for the implication function m are shown next.

$m(a,b)= \max\{x \in \{0,1\} \mid a \wedge x \le b\}$ and
$m(a,b)=\bar{a} \vee (a \wedge b)$

The extensions of theses equations in fuzzy logic are $\forall a,b \in [0,1]$, respectively,

$$I_S(a,b)=S(n(a),b) \tag{I}$$

$$I_R(a,b)=\sup\{x \in [0,1] \mid T(a,x) \le b\} \tag{II}$$

$$I_{QL}(a,b)=S(n(a),T(a,b)) \tag{III}$$

Where T is a t-norm, S is a t-conorm, n is a fuzzy negation, on $[0,1]$ and the triple $<T,S,n>$ is required to satisfy the De Morgan laws.

The fuzzy implications that are obtained from (I) are usually referred to in the literature as *S-implications*. The fuzzy implications that are obtained from (II) are called *R-implications*. The fuzzy implications that are obtained from (III) are called *QL-implications* [4]. In the following, when we would be reported in *S-*,*R-*, and *QL* - implications we will mean fuzzy implications with types (I), (II), (III) respectively.

2. Fuzzy Implications of Type $I(a,b)=S(T(n(a),n(b)),b)$

In addition to the above three classes of fuzzy implications (*S-*, *R-*, *QL-*implications), which are predominant in the literature, other fuzzy implications are possible 4.

The formula $m(a,b)=\bar{a}\vee b$, $\forall a,b\in\{0,1\}$ may also be rewritten, due to the law of absorption of negation in classical logic, as

$$m(a,b)=(\bar{a}\wedge\bar{b})\vee b\,,\forall a,b\in\{0,1\}.$$

The extension of this equation in fuzzy logic is:

$$I(a,b)=S(T(n(a),n(b)),b),\forall a,b\in[0,1]. \qquad \text{(IV)}$$

Where S and T are dual with respect to n. Function $I(.,.)$ with the sense of (IV) is a fuzzy implication. It collapses to the classical implication when truth values are restricted to 0 and 1, i.e. $I(0,0)=I(0,1)=I(1,1)=1$ and $I(1,0)=0$.

In the literature, these fuzzy implications are called *D*–Implications 4. . It is important to point out that hereinafter the fuzzy implication with the sense of (IV), they will be symbolized $I_D(.,.)$ and is called an I_D*-implication*

The following Propositions introduce properties and characteristics of I_D-implications. These properties are required 1. , for some *S-*, *R-* and *QL-*implications and could be important in some applications.

Proposition 1. Let the fuzzy implication I_D, then the following properties hold $\forall a,b,x\in[0,1]$.

i) $[a\le b\Rightarrow I_D(a,x)\ge I_D(b,x)]\Leftrightarrow[I_D(a\wedge x,b)=\vee(I_D(a,b),I_D(x,b))]$.

ii) $[a\le b\Rightarrow I_D(a,x)\ge I_D(b,x)]\Leftrightarrow[I_D(a\vee x,b)=\wedge(I_D(a,b),I_D(x,b))]$.

iii) $[a\le b\Rightarrow I_D(x,a)\le I_D(x,b)]\Leftrightarrow[I_D(a,x\wedge b)=\wedge(I_D(a,x),I_D(a,b))]$.

iv) $[a\le b\Rightarrow I_D(x,a)\le I_D(x,b)]\Leftrightarrow[I_D(a,x\vee b)=\vee(I_D(a,x),I_D(a,b))]$.

Proof:

i) $(\Rightarrow)$

We have two different cases:

Case $a \le x$: From (i) it holds that $a \le x \Rightarrow I_D(a,b) \ge I_D(x,b)$ (1)

Moreover we have $I_D(a \wedge x,b) = I_D(a,b)$ and from (1)

$\vee(I_D(a,b), I_D(x,b)) = I_D(a,b)$.

Therefore $I_D(a \wedge x,b) = \vee(I_D(a,b), I_D(x,b))$ and property (i) has been proved.

Case $a \ge x$: From (i) it holds that $a \ge x \Rightarrow I_D(x,b) \ge I_D(a,b)$ (2)

Moreover we have $I_D(a \wedge x,b) = I_D(x,b)$ and from (2)

$\vee(I_D(a,b), I_D(x,b)) = I_D(x,b)$ .Therefore

$I_D(a \wedge x,b) = \vee(I_D(a,b), I_D(x,b))$ and property (i) has been proved.

($\Leftarrow$)

We have two different cases:

Case $a \le x$: From (i) it holds that $I_D(a \wedge x,b) = \vee(I_D(a,b), I_D(x,b))$ (3)

Moreover we have $I_D(a \wedge x,b) = I_D(a,b)$ and from (3)

$I_D(a,b) = \vee(I_D(a,b), I_D(x,b))$.

Therefore $I_D(a,b) \ge I_D(x,b)$ and property (i) has been proved.

Case $a \ge x$: We have $I_D(a \wedge x,b) = I_D(x,b)$ and from (3)

$I_D(x,b) = \vee(I_D(a,b), I_D(x,b))$.

Therefore $I_D(x,b) \ge I_D(a,b)$ and property (i) has been proved.

ii) This statement is being proven in a similar way.

iii) ($\Rightarrow$)

We have two different cases:

Case $x \le b$: From (iii) it holds that $x \le b \Rightarrow I_D(a,x) \le I_D(a,b)$ (4)

Moreover we have $I_D(a, x \wedge b) = I_D(a,x)$ and from (4)

$\wedge(I_D(a,x), I_D(a,b)) = I_D(a,x)$. Therefore

$I_D(a, x \wedge b) = \wedge(I_D(a,x), I_D(a,b))$ and property (iii) has been proved.

Case $x \ge b$: From (iii) it holds that $b \le x \Rightarrow I_D(a,b) \le I_D(a,x)$ (5)

Moreover we have $I_D(a, x \wedge b) = I_D(a,b)$ and from (5)

$\wedge(I_D(a,x), I_D(a,b)) = I_D(a,b)$. Therefore

$I_D(a, x \wedge b) = \wedge(I_D(a,x), I_D(a,b))$ and property (iii) has been proved.

($\Leftarrow$)

We have two different cases:

Case $x \le b$:From (iii) it holds that $I_D(a, x \wedge b) = \wedge(I_D(a,x), I_D(a,b))$ (6)

Moreover we have $I_D(a, x \wedge b) = I_D(a,x)$ and from (6) $I_D(a,x) = \wedge(I_D(a,x), I_D(a,b))$

Therefore $I_D(a,x) \le I_D(a,b)$ and property (iii) has been proved.

Case $x \ge b$:We have $I_D(a, x \wedge b) = I_D(a,b)$ and from (6)

$I_D(a,b) = \wedge(I_D(a,x), I_D(a,b))$.

Therefore $I_D(a,x) \ge I_D(a,b)$ and property (iii) has been proved.

iv) This statement is being proven in a similar way.

Proposition 2. Let the fuzzy implication $I_D(a,b) = \vee(\wedge(n_S(a), n_S(b)), b)$, where $n_S(a) = 1 - a$, then the following properties hold $\forall a,b,c \in [0,1]$.

i) $I_D((a \wedge b), c) \ge I_D((a \vee b), c)$.

ii) $\wedge(I_D(a,c), I_D(b,c)) \le I_D(\wedge(a,b), c) = \vee(I_D(a,c), I_D(b,c))$.

iii) $\wedge(I_D(a,c), I_D(b,c)) = I_D(\vee(a,b), c) \le \vee(I_D(a,c), I_D(b,c))$.

iv) $\vee(a, I_D(a,b)) = I_D(a, \vee(a,b))$.

Proof:

We prove the property (i). The rest of the properties are proven in a similar way.

i) We have two different cases:

Case $a \le b$:It holds that $I_D(\wedge(a,b), c) = I_D(a,c) = \vee(\wedge(1-a, 1-c), c)$ (7)

and $I_D(\vee(a,b), c) = I_D(b,c) = \vee(\wedge(1-b, 1-c), c)$ (8)

From (7) and (8) we have $I_D(\wedge(a,b), c) \ge I_D(\vee(a,b), c)$.

Case $a \ge b$: The statement is proven in a similar way.

The property (i) has been proved.

Proposition 3. Let the fuzzy implication $I_D(a,b) = \vee(\wedge(n_S(a), n_S(b)), b)$, $a,b,c \in [0,1]$ where $n_S(a) = 1 - a$, then the following properties hold.

i) $\wedge\left(I_D(a,c), I_D(b,c)\right) \leq I_D(a,b)$, for $a \leq c \leq b$

ii) $\wedge\left(a, I_D(a,b)\right) \leq b$, for $a \leq b$

Proof:

i) From Proposition1 we have $a \leq b \Rightarrow I_D(a,x) \geq I_D(b,x)$. Therefore $\wedge\left(I_D(a,c), I_D(b,c)\right) = I_D(b,c) = S\left(T\left(n(b), n(c)\right), c\right) \leq S\left(T\left(n(a), n(b)\right), b\right) =$

$I_D(a,b)$. The statement has been proved.

ii) This statement is proven in a similar way.

3. Conclusion

Some very important properties of the D-implications have been discussed and proved in the previous sections. The above analysis has shown that the D-implications satisfy the same requirements as S-, R- and QL-implications and common properties hold. Apart from the previously analyzed theoretical properties of the D-implications, the authors intend to use them in constructing distance and similarity measures, to apply them to pattern recognition problems and to compare their recognition performance with that of other implications from the literature.

References

1. M. Baczynski, Residual implications revisited. Notes on the Smerts-Magrez Theorem, Fuzzy Sets and Systems 145 (2004) 267-277.
2. J.C. Fodor and T. Keresztfalvi, Non-standard Conjunctions and Implications in Fuzzy Logic, International Journal of Approximate Reasoning 12 (1995) 69-84.
3. J.C. Fodor, M. Roubens, Fuzzy preference modelling and multicriteria decision support, Kluwer Academic Publishers, Netherlands, 1994.
4. G. Klir and Bo Yuan, Fuzzy Sets and Fuzzy Logic Theory and Applications, Prentice Hall P T R Upper Saddle River, New Jersey, USA, 1995.
5. M. Mas, M. Monserrat, J. Torrens, and E. Trillas, A survey on fuzzy implication functions, IEEE Transactions on Fuzzy Systems, 15 (2007) 1107–1121.
6. M. Mas, M. Monserrat and J. Torrens, QL-Implications versus D-Implications, Kybernetika 42 (2006) 351-366.

RESEARCH ADVANCES ON RESOLUTION AUTOMATED REASONING IN LATTICE-VALUED LOGIC BASED ON LATTICE IMPLICATION ALGEBRA*

YANG XU, XIAOMEI ZHONG†, XINGXING HE

School of Mathematics, Southwest Jiaotong University
Chengdu, Sichuan 610031, P.R. China

JUN LIU

School of Computing and Mathematics, University of Ulster
Northern Ireland, UK

This paper provides a systematic review on the research development of α-resolution automated reasoning theories, methods, algorithms and prototype in lattice-valued propositional logic and first-order logic based on lattice implication algebra (LIA), linguistic truth-valued lattice-valued propositional logic and first-order logic based on LIA, respectively. In addition, some important problems needed to be further investigated are proposed.

1. Introduction

Incomparability is an important type of uncertainty often associated with humans' intelligent activities in the real world, and it exists not only in the processed object itself, but also in the course of the object being processed. It is a kind of overall uncertainty of objects due to the complexity of objects themselves, associated with many factors and the inconsistency among those factors.

It is caused by the complex degree of an object. The more complex for certain object, the more for attribute of certain, and the larger for incomparability. This fact implies an overall uncertainty of objects, which can be due to missing information, ambiguity or conflicting evaluations.

* This work is partially supported by the National Natural Science Foundation of China (Grant No. 60875034, 61175055); the project TIN-2009-0828; Sichuan Key Technology Research and Development Program of China (Grant No. 2011FZ0051); Wireless Administration of Ministry of Industry and Information Technology of China ([2011]146)

† E-mail: zhongxm@126.com

To provide a logical foundation under incomparability in the uncertainty inference and automated reasoning, we started the corresponding research work from 1980's. Xu (1993) proposed the concept of LIA by combining lattice and implication algebra [10]. In 1993, Xu and Qin [11] proposed the lattice-valued propositional logic LP(X) based on LIA. In 1997, we further proposed the lattice-valued first-order logics LF(X) on LIA [12, 13].

Automated reasoning is an important field of intelligent information processing; where resolution automated reasoning is one of important research directions. Since the introduction of resolution principle in 1965 by Robinson [6], resolution automated reasoning based on classical logic has been extensively investigated and developed due to its effective inference ability for certainty information.

Based on the academic idea of resolution automated reasoning in classical logic, Xu *et al.* proposed α-resolution principles in lattice-valued propositional logic and first-order logic respectively and further established resolution-based automated reasoning theories, methods, algorithms and prototype in these two lattice-valued logics. The paper provides a systematic overview of those research works in last few years.

2. α-Resolution Automated Reasoning in Lattice-Valued Propositional Logic LP(X) Based on LIA [16]

In 2000, Xu *et al.* [14] proposed a kind of gradational resolution principle, i.e., α-resolution principle in LP(X), and established its soundness and completeness. 2001 to 2003, Xu *et al.* [16] designed the resolution automated reasoning demo prototype in LP(X) using VB. In 2003, Liu *et al.* [3] discussed the validity of some logical formulae that can be equivalently transformed into certain kind of general normal forms in lattice-valued propositional logic LP(X), and a α-resolution method which can be used to judge that a lattice-valued logical formula is always α-false (i.e. is always less than α) was established, as well as its soundness and completeness. In view of the limitation of the assumption of implication connective in classical logic, they further proposed a resolution-like procedure that was aimed at implication connective in LP(X), and its soundness was also obtained. In 2007, Ma *et al.* [5] proposed a filter-based resolution method in LP(X), and proved that it was sound and weak complete for two special generalized clauses. 2010 to 2011, Xu *et al.* [21-23] established the general form of α-resolution principle, α-generalized resolution principle and the general form of α-generalized resolution principle in LP(X), and obtained their soundness and completeness. 2010 to 2011, Xu and some other co-authors

[1, 2, 7, 9, 25-27] proposed α-generalized semantic resolution, α-lock resolution, α-generalized lock resolution, α-ordered linear resolution, α-generalized linear resolution, α-quasi-lock semantic resolution methods and α-group resolution principle in LP(X), and established their soundness and completeness respectively.

3. α-Resolution Automated Reasoning in Linguistic Truth-Valued Lattice-Valued Propositional Logic Based on LIA [20]

When humans deal with uncertainty, they often use linguistic truth values to represent its truth value, such as Slightly true, Somewhat true, Rather true [17]. Hence, the truth-valued field of LP(X) can be the LIA with linguistic truth values as its element, i.e., linguistic truth-valued LIA. Therefore, we can obtain various linguistic truth-valued lattice-valued propositional logic systems. On this basis, many researchers have investigated a variety of resolution automated reasoning in linguistic truth-valued propositional logic based on lattice-valued propositional logic. For example, in 2007, Xu *et al.* [18] established the weak completeness of α-resolution principle for linguistic truth-valued lattice-valued propositional logic based on linguistic truth-valued lattice implication algebra L-LIA. 2010 to 2011, Zhong and He *et al.* [2, 27] concluded that the general form of α-resolution principle, α-group resolution principle, α-quasi-lock semantic resolution, α-generalized lock resolution in linguistic truth-valued lattice-valued propositional logic $\mathcal{L}_{V(n\times2)}$P(X) can be equivalently transformed into that in lattice-valued propositional logic $\mathcal{L}_{Vn}$P(X) respectively. 2009 to 2011, Xu and He *et al.* [24] proposed the algorithms of α-generalized semantic resolution, α-generalized linear resolution and α-generalized lock resolution automated reasoning in $\mathcal{L}_{V(n\times2)}$P(X), and designed the corresponding demo prototypes.

4. α-Resolution Automated Reasoning in Lattice-Valued First-Order Logic LF(X) Based on LIA [16]

In 2001, Xu *et al.* [15] proposed α-resolution principle and strong α-resolution principle in LF(X), and further established the completeness of the former and the latter's soundness. In 2003, Li *et al.* [4] established lattice-valued modal first-order logic LMF(X), and further proposed α-resolution principle in LMF(X). In 2010, Xu *et al.* [22, 23] proposed the general forms of α-resolution principle and α-generalized resolution principle in LF(X), and further obtained their soundness and completeness respectively. In 2011, Xu, Zhong, and He *et al.* [2, 8, 9, 25, 27] established α-generalized semantic resolution, α-ordered linear resolution, α-generalized linear resolution, α-generalized lock resolution,

α-quasi-lock semantic resolution methods and α-group resolution principle in LF(X), and also proved their soundness and weak completeness respectively.

5. α-Resolution Automated Reasoning in Linguistic Truth-Valued Lattice-Valued First-Order Logic Based on LIA [20]

In lattice-valued first-order logic LF(X), if the truth-valued LIA of interpretation is linguistic truth-valued LIA, then we can obtain linguistic truth-valued lattice-valued first-order logic system. On this basis, many researchers have investigated a variety of linguistic truth-valued first-order logic resolution automated reasoning based on lattice-valued first-order logic. Such as, in 2007, α-resolution in LF(X) was equivalently transformed into that in lattice-valued propositional logic LP(X) by Xu *et al.* [19]. And they further proved that α-resolution for lattice-valued propositional logic $(L_n \times L_2)P(X)$ can be equivalently transformed into that in lattice-valued propositional logic $L_nP(X)$ based on chain-type LIA L_n. 2010 to 2011, Zhong and He *et al.* [2, 27] concluded that the general form of α-resolution principle, α-group resolution principle, α-quasi-lock semantic resolution and α-generalized lock resolution in linguistic truth-valued lattice-valued first-order logic $\mathcal{L}_{V(n\times 2)}F(X)$ can be equivalently transformed into that in lattice-valued propositional logic based on chain-type LIA. 2009 to 2011, Xu and He *et al.* [24] gave the algorithms of α-generalized semantic resolution, α-generalized linear resolution and α-generalized lock resolution automated reasoning in $\mathcal{L}_{V(n\times 2)}F(X)$, and designed the corresponding demo prototypes.

6. Conclusions and Further Researches

The theory, methods, algorithms and prototype program of resolution automated reasoning in lattice-valued logic based on LIA have been extensively studied and established. Although a lot of progresses have been made, to form an effective means of resolution automated reasoning in lattice-valued logic based on LIA, the following important questions need to be further investigated:

(1) When the level of resolution is given, which generalized literals can be resolution literals, especially for more complex generalized literals, how to effectively determine whether some generalized literals can be resolution literals?

(2) For the linear resolution, when the level of resolution is given, how to choose the top clause and side clause to improve the efficiency of resolution?

(3) For the lock-semantic resolution, when the level of resolution is given, in order to improve the efficiency of resolution, how to choose the interpretation

and whether the interpretation could be changed for different bloc semantic resolution of the resolution deduction?
(4) For the lock resolution, when the level of resolution is given, how to equip each generalized literal with a lock to improve the efficiency of resolution?
(5) About all kinds of resolution deductions, more effective resolution deductive methods, algorithms should be further proposed and implemented.
(6) Resolution automated reasoning in lattice-valued logic based on LIA should be further used to solve practical problems with incomparability.

References

1. He X.X., Xu Y., Liu J., Ruan D., α-Lock resolution method for a lattice-valued first-order logic. Engineering Applications of Artificial Intelligence, 24, 1274-1280 (2011).
2. He X.X., Xu Y., Liu J., α-Generalized lock resolution method in linguistic truth-valued lattice-valued logic, to appear.
3. Liu J., Ruan D., Xu Y., Song Z. M., A resolution-like strategy based on a lattice-valued logic, *IEEE Trans. on Fuzzy Systems,* 11(4), *560-567* (2003).
4. Li W.J., Xu Y., Ma J., Semantics on lattice-valued modal first-order logic, Proc. of the 2nd Asian Workshop on Foundations of Software (Dec. 2-3, 2003, Nanjing, China), 53-55 (2003).
5. Ma J., Li W.J., Ruan D., Xu Y., Filter-based resolution principle for lattice-valued propositional logic LP(X), Information Sciences, 177, 1046-1062 (2007).
6. Robinson J.P., A Machine-oriented logic based on the resolution principle. J. ACM, 12, 23-41 (1965).
7. Xu W.T., Xu Y., α-Generalized linear resolution method based on lattice-valued propositional logic LP(X), 8th Int. Conf. on Fuzzy Systems and Knowledge Discovery, July 26-28, Shanghai, China, 1413-1416 (2011).
8. Xu W.T., Xu Y., α-Generalized linear resolution method based on lattice-valued first-order logic system, The IEEE Joint International Information Technology and Artificial Intelligence Conference, Aug. 20-22, Chongqing, China, 217-221 (2011).
9. Xu W.T., Xu Y., α-Ordered linear resolution method for lattice-valued logic system based on lattice implication algebra, Int. J. of Applied Management Science, accepted and in press.
10. Xu Y., Lattice implication algebras. J. Southwest Jiaotong University, 89(1), 20-27 (1993). (in Chinese)
11. Xu Y., Qin K.Q., Lattice-valued propositional logic (I). J. Southwest Jiaotong University, 1(2), 123-128 (1993).
12. Xu Y., Qin K.Q., Song Z.M., Syntax of first order lattice valued logic system fm. Chinese Science Bulletin, 42 (16), 1337-1340 (1997).

13. Xu Y., Qin K.Y., On semantic of first-order lattice-valued logic systems FM. System Analysis, Modeling and Simulation, Germany (1997).
14. Xu Y., Ruan D., Kerre E.E., Liu J., α-Resolution principle based on lattice-valued propositional logic LP(X), Information Sciences, 130, 195-223 (2000).
15. Xu, Y., Ruan D., Kerre E.E., Liu J., α-Resolution principle based on first-order lattice-valued logic LF(X), Information Sciences, 132, 221-239 (2001).
16. Xu Y., Concluding report of National Natural Science Foundation of China—Study on uncertain reasoning in lattice-valued logic (Grant No. 60074014) (2004). (in Chinese)
17. Xu Y., Chen S. and Ma J., Linguistic truth-valued lattice implication algebra and its properties, in: Proc. CESA'06, 1123-1129 (2006).
18. Xu Y., Chen S.W., Liu J., Ruan D., Weak Completeness of Resolution in a Linguistic Truth-Valued Propositional Logic, Proc. IFSA2007, June 18-21, 2007, Cancun, Mexico, 358-366 (2007).
19. Xu Y., Li X.B., Liu J., Ruan D., Determination of α-Resolution for Lattice-Valued First-Order Logic Based on Lattice Implication Algebra, Proc. International Conference on Intelligent Systems and Knowledge Engineering (ISKE2007), Oct 15-16, Chengdu, China, 1567-1574 (2007).
20. Xu Y., Concluding report of National Natural Science Foundation of China—Study on linguistic truth-valued resolution automated reasoning in lattice-valued logic (Grant No. 60474022) (2007). (in Chinese)
21. Xu Y., Xu W.T., Zhong X.M., He X.X., α-Generalized resolution principle based on lattice-valued propositional logic LP(X), Proc. of the 9th Int. FLINS Conf. (FLINS2010), Emei, Chengdu, China, 66-71 (2010).
22. Xu Y., Zhong X.M., Liu J., Chen S.W., General form of α-resolution principle based on lattice-valued logic with truth-value in lattice implication algebras, Information Sciences, under review.
23. Xu Y., He, X.X., General form of α-generalized resolution principle based on lattice-valued logic. Int. J. of Computational Intelligence Systems, under review.
24. Xu Y., Concluding report of National Natural Science Foundation of China—Study on linguistic truth-valued α-generalized resolution automated reasoning in lattice-valued logic (Grant No. 60875034) (2011). (in Chinese)
25. Xu Y., Zhang J.F., Zhong X.M., α-Generalized semantic resolution method in lattice-valued logic (in Chinese), to appear.
26. Zhong X.M., Xu Y., α-Group resolution method based on lattice-valued propositional logic LP(X), Proc. of the 8th Int. Conf. on Fuzzy Systems and Knowledge Discovery, Shanghai, China, July 26-28, 1418-1422 (2011).
27. Zhong X.M., Study on α-quasi-lock semantic resolution automated reasoning based on lattice-valued logic, Southwest Jiaotong University Doctor Degree Dissertation.

GENERAL FORM OF α-LINEAR RESOLUTION METHOD BASED ON LATTICE-VALUED PROPOSITIONAL LOGIC SYSTEM LP(X)*

WEITAO XU†1, DEXIAN ZHANG[1], HUA ZHU[2], YANG XU[3]

[1]*College of Information Science and Engineering, Henan University of Technology, Zhengzhou 450001, P.R. China*

[2]*Department of Maths, Zhengzhou University, Zhengzhou 450001, P.R. China*

[3]*College of Maths, Southwest Jiaotong University, Chengdu 610031, P.R. China*

In this paper we propose a novel resolution method for theorem automated proving. General form of α-linear resolution method is investigated in lattice-valued propositional logic system LP(X) based on lattice implication algebra. Both soundness and weak completeness theorems are established. The contribution of this method is to provide a scientific and extended resolution method for automated reasoning based on lattice-valued logic system. It also provides a foundation for constructing an α-linear resolution automated reasoning algorithm.

1. Introduction

Automated reasoning is an important research realm in artificial intelligence. Resolution-based automated reasoning is one of the research branches. It provides a strong tool for intelligent information processing.

Since the resolution principle based on classical logic has been presented by Robinson in 1965[4], resolution-based automated reasoning becomes an active realm and pays any attention by many scholars. Some resolution methods have also been studied. Particularly, there are three typical resolution methods based on the resolution principle, such as semantic resolution, lock resolution and linear resolution[7]. All these methods are devoted to reducing the redundant clause and improving the resolution efficiency in the course of resolution.

With the development and application of non-classical logic, resolution automated reasoning based on these logic systems have also been focused. As an important non-classical logic, lattice-valued logic system based on lattice

---

* This work is supported by the National Natural Science Foundation of P.R. China(No. 61175055, 61100046), the Sichuan Key Technology Research and Development Program(No. 2011FZ0051).

† hnxmxwt@163.com, hnxmxwt@haut.edu.cn.

implication algebra provides a strict and reasonable foundation for automated reasoning. Xu established α-resolution principle in LP(X) and applied them to automated reasoning in 2000[9]. Xu, et al., gave the determination of α-resolution for generalized literals in LF(X) in 2011[10]. However, two generalized clauses were always selected in one resolution process. This can confine the resolution efficiency to a certain extent.

In the present paper, we focus on a new resolution method aimed to provide a scientific and efficient automated reasoning method for mechanical theorem proving. It will be a great extension with respect to the formerly proposed methods.

The paper is organized as follows. In section 2, we briefly recall some essential concepts. In section 3, general form of α-linear resolution method for lattice-valued propositional logic system LP(X) based on lattice implication algebra is presented. The soundness and weak completeness theorems are established. Finally, the conclusions and future works are also presented.

2. Preliminaries

In this section we review α-resolution principle for lattice-valued propositional logic based on lattice implication algebra, and suppose $\mathcal{L} = (L, \vee, \wedge, ', \rightarrow, O, I)$ is a lattice implication algebra.

Definition 2.1.[9] (α-resolution principle) In lattice-valued propositional logic system LP(X), let $\mathcal{L}$ be a lattice implication algebra, $\alpha < I$, $\alpha \in L$, G_1 and G_2 two generalized clauses of the form

$$G_1 = g_1 \vee \cdots \vee g_i \vee \cdots \vee g_m,$$

$$G_2 = h_1 \vee \cdots \vee h_j \vee \cdots \vee h_n,$$

where $g_i (i = 1, 2, \cdots, m)$ and $h_j (j = 1, 2, \cdots, n)$ are generalized literals in G_1 and G_2 respectively. If $g_i \wedge h_j \leq \alpha$, then

$$g_1 \vee \cdots \vee g_{i-1} \vee g_{i+1} \vee \cdots \vee g_m \vee h_1 \vee \cdots \vee h_{j-1} \vee h_{j+1} \vee \cdots \vee h_n$$

is called an α-resolvent of G_1 and G_2, denoted by $R_\alpha(G_1, G_2)$, and (g_i, h_j) is called an α-resolution pair, denoted by $(g_i, h_j) - \alpha$.

Definition 2.2.[6] Let $C = g_1 \vee g_2 \vee \cdots \vee g_i \vee \cdots \vee g_n$ be a generalized clause. If there exist some equivalent generalized literals in C, then we remain the leftmost generalized literal and delete the other equivalent generalized literals. The above process is called left-combination rule.

Definition 2.3.[6] A generalized clause C is called an ordered generalized clause if the following criterions hold:

(r1) C has been simplified by using the left-combination rule,
(r2) align by the length of the generalized literals from left to right and put the longest generalized literal on to the leftmost position in C, where the length means the order of the generalized literal (if the order is equivalent, they can be arbitrary arrange),
(r3) especially, put the generalized literals containing the constants in front of the constants in C.

Definition 2.4.[11] In LP(X), let C be a generalized clause, g a generalized literal, $A(C)$ and $A(g)$ the atom sets of C and g respectively. C and g are independent if $A(C) \cap A(g) = \varnothing$.

Definition 2.5.[11] In LP(X), $\alpha \in L$. The generalized literals $h_1,\ldots,h_i,\ldots,h_k$ are called collaborative normal if $h_1 \wedge \cdots \wedge h_i \wedge \cdots \wedge h_k \leq \alpha$ such that $v(\wedge_{j=1, i\neq j}^{k} h_j) = I$ under a valuation v.

We refer the readers to [8, 9] for more details about the concepts and properties about lattice implication algebra and lattice-valued propositional logic system and α-resolution principle based on LP(X).

3. General form of α-linear resolution method in lattice-valued propositional logic system LP(X)

In this section, α-resolution method is extended in lattice-valued propositional logic system LP(X) based on lattice implication algebra.

Definition 3.1. Let LP(X) be a lattice-valued propositional logic system, C_0 an ordered generalized clause, $C_i(i=1,\cdots,m)$ the generalized clauses, H_i a set of all the generalized literals in C_i, $\alpha \in L, \alpha < I$. If there exist the leftmost generalized literal h_0 in C_0 and generalized literal $h_i \in H_i (i=1,\cdots,k)$ such that

$$h_0 \wedge h_1 \wedge \cdots \wedge h_i \wedge \cdots \wedge h_k \leq \alpha,$$

then

$$\vee_{i=0}^{k} C_i(h_i = \alpha)$$

is called anα-ordered resolvent of C_0 and $\{C_1,\cdots,C_i,\cdots,C_k\}$, denoted by $R_{\alpha-LG}^{p}(C_0,\{C_1,\cdots,C_i,\cdots,C_k\})$, where the resolvent is implemented the ordered combination rule, $\{h_0,h_1,\cdots,h_i,\cdots h_k\}$ is called α-resolution generalized literals set.

Theorem 3.1. In LP(X), let C_0 be an ordered generalized clause, $C_i(i=1,\cdots,m)$ the generalized clauses, H_i a set of all the generalized literals in C_i, $\alpha \in L$. If there exist the leftmost generalized literal h_0 in C_0 and generalized literal

$h_i \in H_i (i=1,\cdots,k)$ such that $h_0 \wedge h_1 \wedge \cdots \wedge h_i \wedge \cdots \wedge h_k \le \alpha$, $R^p_{\alpha-LG}(C_0,\{C_1,\cdots,C_i,\cdots,C_k\})$ is an α-ordered resolvent of C_0 and $\{C_1,\cdots,C_i,\cdots,C_k\}$, then

$$C_0 \wedge C_1 \wedge \cdots \wedge C_i \wedge \cdots \wedge C_k \le R^p_{\alpha-LG}(C_0,\{C_1,\cdots,C_i,\cdots,C_k\}).$$

Definition 3.2. Let S be a set of generalized clauses in LP(X), an ordered generalized clause $C_0 \in S$, $\alpha \in L$. An α-linear resolution deduction from S to an ordered generalized clause D_t is obtained as follows (see Figure 3.1):

(1)C_0 is a top ordered generalized clause, T_i is a set of side generalized clauses with respect to D_i, where $i=0,1,\cdots,t-1$, D_0=C_0;

(2) $D_i = R^p_{\alpha-LG}(D_{i-1},T_{i-1}), i=1,\cdots,t$;

(3)For any side generalized clause set T_i, then $T_0 \subseteq S-\{C_0\}$, $T_i \subseteq S \cup \{D_0,\cdots,D_{i-1}\}, i=1,\cdots,t$.

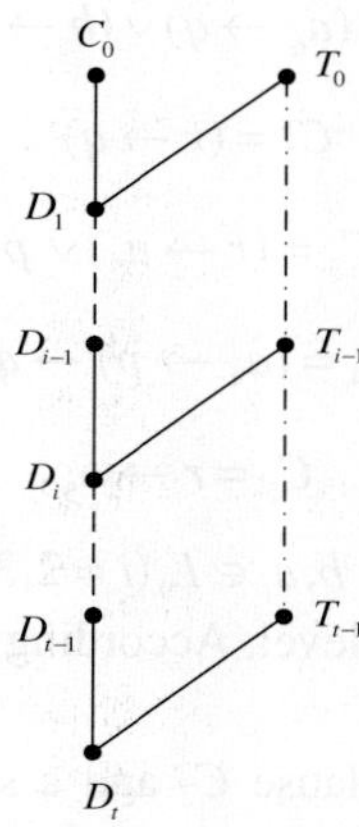

Figure 3.1 α-linear resolution deduction from S to D_t with the top generalized clause C_0.

Remark 3.1. In an α-linear resolution deduction, every resolvent D_i is a resolution result of the ordered generalized clause D_{i-1} and all the side generalized clauses in T_{i-1}, where $i=1,\cdots,t$.

Theorem 3.2. (Soundness) In LP(X), suppose S is a set of generalized clauses, an ordered generalized clause $C_0 \in S$, $\alpha \in L$, $\alpha < I$. $\omega = \{C_0, D_1, \cdots, D_t\}$ is an α-linear resolution deduction from S to an ordered generalized clause D_t with the top generalized clause C_0. If D_t=α-◎, then $S \le \alpha$, i.e., if $D_t \le \alpha$, then $S \le \alpha$.

Theorem 3.3. (Weak Completeness) In LP(X), suppose S is a set of generalized clauses, an ordered generalized clause $C_0 \in S$, $\alpha \in L$, $\alpha < I$. If

(1) $S \le \alpha$,

(2) S-$\{C_0\}$ is α-satisfiable,

(3)There exist the leftmost generalized literal g in C_0 and generalized literal h_i in $C_i\,(i=1,\cdots i,\cdots,t)$ such that $g_0,h_1,\cdots,h_i,\cdots,h_t$ satisfy the following conditions:

① $g_0\wedge h_1\wedge\cdots\wedge h_i\wedge\cdots\wedge h_t\le\alpha$,

② $g,h_1,\cdots,h_i,\cdots,h_t$ are independent with S^* respectively, where H_S=H_{S^*}-{g, $h_1,\cdots,h_i,\cdots,h_t$ }, H_S and H_{S^*} are the generalized literal sets of generalized clauses in S and S^*,

③ $g,h_1,\cdots,h_i,\cdots,h_t$ are collaborative normal,

then there exists an α-linear resolution deduction from S to α-◎ with the top generalized clause C_0.

Example 3.1. Let $L_9P(X)$ be a lattice-valued propositional logic system based on lattice implication algebra $(L_9,\vee,\wedge,',\rightarrow,a_1,a_9)$, S={C_1,C_2,C_3,C_4,C_5} a set of generalized clauses, where

$$C_1=(a_6\rightarrow q)\vee(b\rightarrow p)',$$

$$C_2=(r\rightarrow q)',$$

$$C_3=(r\rightarrow a_3)\vee p,$$

$$C_4=(a_4\rightarrow p)\rightarrow q',$$

$$C_5=r\rightarrow a_2,$$

p, q, r are the literals in $L_9P(X)$, $b,a_i\in L_9\,(i=2,3,4,6)$.

Take $\alpha=a_5$ as a resolution level. According to Definition 3.2, the resolution processes are as follows:

Step 1. Set a top generalized clause C_2 and a set {C_3, C_5} of side generalized clauses such that $(r\rightarrow q)'\wedge(r\rightarrow a_3)\wedge(r\rightarrow a_2)\le\alpha$, then α-ordered resolvent

$$D_1=R^p_{\alpha-LG}(C_2,\{C_3,C_5\})=p\vee\alpha\,.$$

Step 2. Set a center generalized clause D_1 and a set {C_1, C_4} of side generalized clauses such that $p\wedge(a_6\rightarrow q)\wedge((a_4\rightarrow p)\rightarrow q')\le\alpha$, then α-ordered resolvent

$$D_2=R^p_{\alpha-LG}(D_1,\{C_1,C_4\})=(b\rightarrow p)'\vee\alpha\,.$$

Step 3. Set a center generalized clause D_2 and a set {D_1} of side generalized clauses such that $(b\rightarrow p)'\wedge p\le\alpha$, then the α-ordered resolvent

$$D_3=R^p_{\alpha-LG}(D_2,\{D_1\})=\alpha\,.$$

4. Conclusions

In this paper, an α-resolution method is extended for resolution automated reasoning based on lattice-valued propositional logic system LP(X). General form of α-linear resolution method is presented in lattice-valued propositional logic system LP(X) based on lattice implication algebra. In a resolution process, we select a set of side generalized clauses with respect to a center generalized clause, and obtain their resolvent. Hence, this method improves the resolution efficiency to some extent. On the basis of the present method, an automated reasoning algorithm for general form of α-linear resolution method based on lattice-valued propositional logic system LP(X) can be constructed.

References

1. C.L. Chang, R.C.T. Lee, Symbolic Logic and Mechanical Theorem Proving, New York: Academic Press (1973).
2. D.W. Loveland, Automated Theorem Proving: A logical Basis, North-Holland, Amsterdam (1978).
3. D. Luckham, Refinements in resolution theory, Proceeding of IRIA Symposium Automatic Demonstration, Versailes, France: Spring-Verlag. 163-190(1970).
4. J. A. Robinson, A machine-oriented logic based on the resolution principle, Journal of the Association for Computing Machinery, **12(1)**, 23-41(1965).
5. Jun Liu, Da Ruan, Yang Xu, Zhengming Song, A resolution-like strategy based on lattice-valued logic, IEEE Transactions on Fuzzy Systems. **11(4)**, 560-567 (2003).
6. Weitao Xu, Yang Xu, α-ordered linear resolution for lattice-valued logic system based on lattice implication algebra, Int. J. Applied Management Science, in press (2012).
7. X. H. Liu, Resolution-based Automated Reasoning, Beijing: Science Press.1994 (in Chinese).
8. Y. Xu, Lattice implication algebra, J. Southwest Jiaotong University. **28(1)**, 20-27(1993).
9. Y. Xu, D. Ruan, K.Qin and J. Liu, Lattice-Valued Logic: An alternative approach to treat fuzziness and incomparability, Berlin: Springer-Verlag (2003).
10. Y. Xu, J. Liu, D. Ruan, et al., Determination of α-Resolution in Lattice-Valued First-order Logic LF(X), *Information Sciences*. **181(10)**, 1836-1862 (2011).
11. Y. Xu, et al., General form of α-resolution principle in lattice-valued logic system, in press (2012).

N-TRUTH DEGREES OF FORMULAE IN TWO-VALUED PREDICATE LOGIC*

QIN XIAOYAN†1,2, LIU YI3, XU YANG1
1 *Intelligent Control Development Center, Southwest Jiaotong University, Chengdu,Sichuan 610031, China*
2 *College of Mathematics and Computer Science, Shanxi Normal University, Linfen, Shanxi 041004, China*
3 *Key Lab of Numerical Simulation in Sichuan Provincial College, Neijiang Normal University, Neijiang, Sichuan 641112, China*

Great achievement have been obtained in the research on the truth degrees of formulae in propositional logic, but now the investigation on the truth degrees of formulae in predicate logic is very lack. Based on this situation, the n-truth degrees of formulae in two-valued predicate logic is proposed in this paper, and some important properties are studied. This paper enriches the content of the theory of quantitative logic, and provides a kind of framework for further approximate reasoning.

1. Introduction

Early in 1952 Rosser and Turquett[1] first proposed the idea of distinguishing truth degrees between logic formulas, and this initial idea was improved by many scholars later on, among them Pavelka[2], DeGlas[3], and Ying[4] made important contributions. In 2009, Wang proposed the quantitative logic[5] which combined the mathematical logic and the computation mathematics to provide a graded approach to mathematical logic. There have been many research findings of quantitative propositional logic[5-9].As to quantitative predicate logic, it's hard to establish the theory of the truth degree of first-order formulae because of the complexity of the interpretations of the first-order language. On the basis of such situations, the author proposed the theory of the quasi-truth degree of first-order formulae[10].However, the quasi-truth degrees of first-order formulae was only a preliminary test far from satisfactory because it defined truth degrees of predicate formulae simply by taking the average of the values obtained from two extreme cases.

*This work is supported by National Natural Science Foundation of China (No. 61175055), and Sichuan Key Technology Research and Development Program under Grant No.2011FZ0051.

† lisaqin1981@126.com

Based on the above circumstances, we will grade all the finite interpretations by the cardinalities of their interpretation domains, and then average all the relative truth degrees of predicate formulae in the finite interpretations at every level. Thus, we will obtain a kind of more precise truth degrees of predicate formulae, i.e., n-truth degree. Furthermore, some related basic properties are discussed and it is proved that MP Rule, HS Rule and Gen Rule in two-valued predicate logic is preserved by n-truth degrees of first-order formulae. This paper enriches the content of theory of quantitative predicate logic, and provides a kind of possible framework for approximate reasoning in predicate logic.

2. n-truth degrees of first-order formulae

Suppose that the first order language $\mathcal{L}$ contains no function symbols (it is proved reasonable in Ref. [11]), $\mathbf{I} = (I, (r_P)_P, (m_c)_c)$ is an interpretation of the first order language $\mathcal{L}$, where the domain *I* of the interpretation is an nonempty set, r_P is the corresponding relation on *I* to the predicate symbol *P*, and m_c is an element in *I* corresponding to the individual constant *c*. The valuation $v : \mathcal{T} \to I$ of $\mathcal{L}$ in $\mathbf{I}$ is a mapping from the set $\mathcal{T}$ of terms of $\mathcal{L}$ to the domain *I* of the interpretation $\mathbf{I}$, and satisfies that $v(c) = m_c, v(x) \in I$, in which *c* is an individual constant, x is a variable, and every t in $\mathcal{T}$ is an individual constant or a variable. In the following, We will denote the set of all first-order formulae by $\mathcal{F}$, the set of all valuations in $\mathbf{I}$ by $\Omega_{\mathbf{I}}$, the set of all variables in $\mathcal{L}$ by S, and the class of all interpretations with nonempty finite domains by $\mathcal{I}_f$. For a valuation v in $\Omega_{\mathbf{I}}$ and a formula *A* in $\mathcal{F}$, we denote that $\|A\|_{\mathbf{I},v} = 1$ if v satisfies *A*, and $\|A\|_{\mathbf{I},v} = 0$ if not. And we will denote the cardinality of the nonempty finite domain *I* by $|I|$.

Definition 1[12] Suppose that $(X_n, \mathcal{A}_n, \mu_n)$ is a probabilistic measure space where μ_n is a probability measure on X_n, and $\mathcal{A}_n$ is the family consisting of all μ_n-measurable subsets of $X_n (n = 1, 2, \cdots)$. Assume that $X = \prod_{n=1}^{\infty} X_n$, then $\prod_{n=1}^{\infty} \mathcal{A}_n$ generates on X a σ-algebra $\mathcal{A}$, and there exists on *X* a unique measure μ such that (i) $\mathcal{A}$ is the family of all μ-measurable subsets of *X*; (ii) $E \times \prod_{n=m+1}^{\infty} X_n$ is μ-measurable and

$$\mu(E \times \prod_{n=m+1}^{\infty} X_n) = (\mu_1 \times \mu_2 \times \cdots \times \mu_m)(E), m = 1, 2, \cdots \quad (1)$$

whenever E is a measurable subset of $\prod_{n=1}^{m} X_n$. μ is called the infinite product of $\mu_1, \mu_2, \cdots$. The space $(X, \mathcal{A}, \mu)$ will often be simplified as $\mathbf{X}$.

Suppose that $\mathbf{I} \in \mathcal{I}_f$, $X_{n,\mathbf{I}} = I, \mu_{n,\mathbf{I}}$ is the evenly distributed probability measure on $X_n (n = 1, 2, \cdots)$. Let $X_\mathbf{I} = \prod_{n=1}^{\infty} X_{n,\mathbf{I}}$, $\mu_\mathbf{I}$ is the infinite product of $\{\mu_{n,\mathbf{I}}\}_{n=1}^{\infty}$. For the reason that the valuation v of $\mathcal{L}$ in $\mathbf{I}$ is completely decided by the restriction $v|S$ of v on S, suppose that $v \in \Omega_\mathbf{I}, v(x_k) = v_k (k = 1, 2, \cdots)$, then there is an unique corresponding element $\mathbf{v} = (v_1, v_2, \cdots, v_n, \cdots)$ in $X_\mathbf{I}$; Conversely, if $\mathbf{v} = (v_1, v_2, \cdots, v_n, \cdots) \in X_\mathbf{I}$, then there is an unique corresponding valuation v in $\Omega_\mathbf{I}$. Hence there exists a one-to-one correspondence φ between $\Omega_\mathbf{I}$ and $X_\mathbf{I}$, where $\varphi(v) = \mathbf{v}$.

Definition 2 Suppose that $\mathbf{I} \in \mathcal{I}_f$ and $A \in \mathcal{F}$. Define $[A]_\mathbf{I}$ and $\tau_\mathbf{I}(A)$ as follows

$$[A]_\mathbf{I} = \{\mathbf{v} \in X_\mathbf{I} | v \in \Omega_\mathbf{I}, \|A\|_{\mathbf{I},v} = 1\}, \tau_\mathbf{I}(A) = \mu_\mathbf{I}([A]_\mathbf{I}), \tag{2}$$

then $\tau_\mathbf{I}(A)$ is called the relative truth degree of A in $\mathbf{I}$.

Example 1 Suppose that $\mathbf{I} = (I, r_P, (m_c)_c) \in \mathcal{I}_f$, $A = P(x) \to (\forall x)P(x)$. Calculate $\tau_\mathbf{I}(A)$.

Solution It follows from Definition 2 that

$$\begin{aligned} \tau_\mathbf{I}(A) &= \mu_\mathbf{I}(\{\mathbf{v} \in X_\mathbf{I} | v \in \Omega_\mathbf{I}, \|P(x) \to (\forall x)P(x)\|_{\mathbf{I},v} = 1\}) \\ &= 1 - \mu_\mathbf{I}(\{\mathbf{v} \in X_\mathbf{I} | v \in \Omega_\mathbf{I}, \|P(x) \to (\forall x)P(x)\|_{\mathbf{I},v} = 0\}) \\ &= 1 - \mu_\mathbf{I}(\{\mathbf{v} \in X_\mathbf{I} | v \in \Omega_\mathbf{I}, \|P(x)\|_{\mathbf{I},v} = 1, \|(\forall x)P(x)\|_{\mathbf{I},v} = 0\}) \end{aligned}$$

When $r_P = I$ or $r_P = \emptyset$, obviously $\tau_\mathbf{I}(A) = 1$. Otherwise, for all valuation v in $\Omega_\mathbf{I}$, we have that $\|(\forall x)P(x)\|_{\mathbf{I},v} = 0$, then it follows from the above formula that $\tau_\mathbf{I}(A) = 1 - \frac{|r_P|}{|I|}$. Consequently, we have

$$\tau_\mathbf{I}(A) = \begin{cases} 1, & \text{if } r_P = I \text{ or } \emptyset, \\ 1 - \frac{|r_P|}{|I|}, & \text{else.} \end{cases}$$ □

Note that, if the cardinalities of two sets I and I' are equal, i.e., there exists a one-to-one correspondence $h : I \longrightarrow I'$, then the interpretation $\mathbf{I} = (I, (r_P)_P, (m_c)_c)$ with the domain I can be naturally changed into the interpretation $\mathbf{I}' = (I', (r_P)_P, (m_c)_c)$ with the domain I', where $x \in r_P$ if and only if $h(x) \in r'_P$, and $m'_c = h(m_c)$. For this reason, in the following we will think of those interpretations having the same structures but different carriers as identical ones. Especially, without any loss of generality, we will take the set $I_n = \{a_1, a_2, \cdots, a_n\}$ as a standard universe of discourse for all the finite interpretations of which the cardinalities of their domains are all equal to n. In the sequel, we denote by $\mathcal{I}_n$ the set of all interpretations possessing standard domains and suppose that $\mathcal{I}_f = \bigcup_{n=1}^{\infty} \mathcal{I}_n$. Moreover, because I_n is a finite set, its power set is also finite. And due to $r_P \subseteq I_n^k$ for any k-ary predicate symbol P, the amount of all the interpretations with its domain I_n is finite for a specific formula, i.e., $|\mathcal{I}_n|$ is a finite number.

Definition 3 Suppose that $A \in \mathcal{F}$. For all positiveintegers n, define $\tau_n(A)$ as follows:

$$\tau_n(A) = \frac{1}{|\mathcal{I}_n|} \sum \{\tau_{\mathbf{I}}(A) | \mathbf{I} \in \mathcal{I}_n\}, \tag{3}$$

then we call $\tau_n(A)$ the n-truth degree of A.

Example 2 Let $A = P(x), B = (\forall x)P(x), C = P(x) \to (\forall x)P(x)$. Calculate the n-truth degree of A, B and C respectively.

Solution Consider an arbitrary finite interpretation $\mathbf{I} = (I_n, (r_P)_P, (m_c)_c)$ where $I_n = a_1, a_2, \ldots, a_n$. Note that the predicate symbol P is unary and r_P is a subset of I. If $|r_P| = k$, then there are altogether C_n^k subsets of I that can be taken as the corresponding relation r_P to P. Furthermore we can take no account of $(m_c)_c$ since no individual constants occur in A , B or C. For example, we can fix $(m_c)_c$ to be a_1. Thus $|\mathcal{I}_n| = 2^n$.

(i) It's easy to prove that $\tau_{\mathbf{I}}(A) = \frac{|r_P|}{n}$, so

$$\tau_n(A) = \tfrac{1}{|\mathcal{I}_n|} \sum \{\tau_{\mathbf{I}}(P(x)) | \mathbf{I} \in \mathcal{I}_n\} = \tfrac{1}{2^n} \sum_{k=0}^{n} \tfrac{k}{n} C_n^k = \tfrac{1}{n2^n} \sum_{k=0}^{n} kC_n^k.$$

According to the fact that $\sum_{k=0}^{n} kC_n^k = n2^{n-1}$, we have that ¿n (A) = $\frac{1}{2}$:

(ii) Obviously, $\tau_{\mathbf{I}}(B) = 1$ holds if and only if $r_P = I$, and otherwise $\tau_{\mathbf{I}}(B) = 0$. So

$$\tau_n(B) = \tfrac{1}{|\mathcal{I}_n|} \sum \{\tau_{\mathbf{I}}((\forall x)P(x)) | \mathbf{I} \in \mathcal{I}_n\} = \tfrac{1}{2^n}(1 + 0 + \cdots + 0) = \tfrac{1}{2^n}.$$

(iii) It follows from Definition 3 and Example 1 that

$$\begin{aligned} \tau_n(C) &= \tfrac{1}{|\mathcal{I}_n|} \sum \{\tau_{\mathbf{I}}(P(x) \to (\forall x)P(x)) | \mathbf{I} \in \mathcal{I}_n\} \\ &= \tfrac{1}{2^n}(1 + C_n^1(1 - \tfrac{1}{n}) + C_n^2(1 - \tfrac{2}{n}) + \cdots + C_n^{n-1}(1 - \tfrac{n-1}{n}) + 1). \end{aligned}$$

We can prove that

$$1 + C_n^1(1 - \frac{1}{n}) + C_n^2(1 - \frac{2}{n}) + \cdots + C_n^{n-1}(1 - \frac{n-1}{n}) + 1 = 2^{n-1} + 1.$$

Therefore, $\tau_n(C) = \frac{1}{2^n}(2^{n-1} + 1) = \frac{1}{2} + \frac{1}{2^n}$.□

Remark We can easily prove that the quasi-truth degrees of A, B and C are all equal to $\frac{1}{2}$. However from example 1, $\tau_n(B) \leq \tau_n(A) \leq \tau_n(C)$ holds for any positive integer n. It fully shows that n-truth degree can more precisely describe the truth degree of formulae than the quasi-truth degree.

Proposition 1 Suppose that $A, B \in \mathcal{F}$, $n = 1, 2, \cdots$.
(i) $0 \le \tau_n(A) \le 1$;(ii) $\tau_n(A) + \tau_n(\neg A) = 1$;(iii) If A is a logically efficient formula, then $\tau_n(A) = 1$;(iv) If the formulae A and B are logically equivalent to each other, then $\tau_n(A) = \tau_n(B)$.
Theorem1 Suppose that $A, B, C \in \mathcal{F}$, $n = 1, 2, \cdots$.

(i) $\tau_n(A \vee B) + \tau_n(A \wedge B) = \tau_n(A) + \tau_n(B)$; (4)
(ii) If $\tau_n(A) \ge \alpha, \tau_n(A \to B) \ge \beta$, then $\tau_n(B) \ge \alpha + \beta - 1$;
(iii) If $\tau_n(A \to B) \ge \alpha, \tau_n(B \to C) \ge \beta$, then $\tau_n(A \to C) \ge \alpha + \beta - 1$.

Proof (i) Because the definition of $\tau_n(A)$ given by (3) is actually obtained by taking the average of all possible values of $\tau_{\mathbf{I}}(A)$, so in order to prove (4), it suffices to prove the following equality:

$$\tau_{\mathbf{I}}(A \vee B) + \tau_{\mathbf{I}}(A \wedge B) = \tau_{\mathbf{I}}(A) + \tau_{\mathbf{I}}(B), \forall \mathbf{I} \in \mathcal{I}_f.$$

In fact, we have

$$\begin{aligned} \tau_{\mathbf{I}}(A \vee B) &= \mu_{\mathbf{I}}([A \vee B]_{\mathbf{I}}) = \mu_{\mathbf{I}}([A]_{\mathbf{I}} \cup [B]_{\mathbf{I}}) \\ &= \mu_{\mathbf{I}}([A]_{\mathbf{I}}) + \mu_{\mathbf{I}}([B]_{\mathbf{I}}) - \mu_{\mathbf{I}}([A]_{\mathbf{I}} \cap [B]_{\mathbf{I}}) \\ &= \tau_{\mathbf{I}}(A) + \tau_{\mathbf{I}}(B) - \tau_{\mathbf{I}}(A \wedge B). \end{aligned}$$

(ii) Since it follows from (4) that

$$\begin{aligned} \tau_n(A \to B) &= \tau_n(\neg A \vee B) = \tau_n(\neg A) + \tau_n(B) - \tau_n(\neg A \wedge B) \\ &= 1 - \tau_n(A) + \tau_n(B) - \tau_n(\neg A \wedge B), \end{aligned}$$

we have $\tau_n(B) = \tau_n(A) + \tau_n(A \to B) + \tau_n(\neg A \wedge B) - 1$.
Then it follows from $\tau_n(A) \ge \alpha$, $\tau_n(A \to B) \ge \beta$, and the above formula that

$$\tau_n(B) \ge \alpha + \beta - 1.$$

(iii) The proof of part (iii) can be done in the same way and hence is omitted.□
Corollary 1 Suppose that $A, B, C \in \mathcal{F}$, $n = 1, 2, \cdots$.
(i) If $\tau_n(A) = 1, \tau_n(A \to B) = 1$, then $\tau_n(B) = 1$;
(ii) If $\tau_n(A \to B) = 1, \tau_n(B \to C) = 1$, then $\tau_n(A \to C) = 1$.
Theorem2 Suppose that $A \in \mathcal{F}$, $n = 1, 2, \cdots$.
(i) If $\tau_n(A) = 1$, then $\tau_n((\forall x)A) = 1$;(ii) $\tau_n((\forall x)A) \le \tau_n(A)$.
According to Definition 2 and Definition 3, it follows the theorem that
Theorem 3 Suppose that $A \in \mathcal{F}$.
(i) $\tau(A) = 1$ holds if and only if $\tau_n(A) = 1$ holds for any positive integer n;
(ii) $\tau(A) = 0$ holds if and only if $\tau_n(A) = 0$ holds for any positive integer n.

3. Conclusions

In this paper, the n-truth degrees of formulae are proposed, and some related basic properties are studied. In the following work, we will further consider the distribution of the n-truth degrees of formulae, study the relation between the n-truth degrees and the quasi-truth degrees of formulae, and discuss the theory of approximate reasoning, which will enrich further the content of the theory of quantitative predicate logic.

References

1. J.B. Rosser and A.R. Turquette, Many-Valued Logics. Amsterdam: North-Holland, 1952.
2. J. Pavelka, On fuzzy logic I,II,III. Zeitschr f Math Logik u Grundlagen d Math., 25(1979): 45-52,119-134,447-464.
3. M. De Glas, Knowledge Representation in a Fuzzy Setting. Technical Report 89/48. Universite' Paris VI, Laforia,1989.
4. M.S. Ying, A logic for approximate reasoning. J Symbolic Logic, 59(1994): 830-837.
5. G.J. Wang and H.J. Zhou, Quantitative logic. Inf Sci, 179 (2009): 226-247.
6. G.J. Wang and L. Fu, Theory of truth degrees of propositions in two-valued logic. Sci. China A 11(2001): 998-1008.
7. G.J. Wang and Y. Leung, Integrated semantics and logic metric spaces. Fuzzy Sets Syst. 136 (2003) : 71-91.
8. H.J. Zhou and G.J. Wang, A new theory consistency index based on deduction theorems in several logic systems. Fuzzy Sets Syst. 157(2006): 427-443.
9. H.J. Zhou and G.J. Wang, Generalized consistency degrees of theories w.r.t. formulas in several standard complete logic systems. Fuzzy Sets Syst. 157(2006):2058-2073.
10. G.J.Wang, X.Y. Qin and X.N. Zhou, Theory of quasi-truth degrees of formulasin two-valued predicate logic. Journal of Shaanxi Normal University(Natural Science Edition),1(2005): 1-6(in Chinese).
11. Petr Hájek, Metamathematics of Fuzzy Logic. London: Kluwer Academic publishers, 1998.
12. Paul Richard Halmos, Measure Theory. New York: Springer Verlag, 1974.

AUTO-ASSOCIATIVE MEMORIES BASED ON COMPLETE INF-SEMILATTICES*

PETER SUSSNER† and CARLOS RENATO MEDEIROS

Dept. of Applied Math., MICI Group‡, University of Campinas, Rua Sérgio Buarque de Holanda 651 Campinas - SP, 13083-859, Brazil

In the mid 1990's, the morphological associative memory (MAM) was introduced as a distributive associative memory model that performs certain morphological operations defined in the mathematical theory of minimax algebra. MAM models come in two different versions that are tolerant to different types of noise in the input patterns. To overcome this drawback, we resort to the more recent theory of mathematical morphology (MM) on inf-semilattices whose elementary operators are self-dual and we define an associative memory (AM) model in this framework.

1. Introduction

MAMs including fuzzy morphological associative memories (FMAMs) have been extensively analyzed and applied to a variety problems such as pattern recognition and classification [1,2], prediction [3], vision-based self-localization in robotics [4,5], image compression [6], color image segmentation [7], and hyperspectral image analysis [8]. The characteristics of the autoassociative morphological memories (AMMs) W_{XX} and M_{XX} include optimal absolute storage capacity and one-step convergence [1,9]. The AMMs W_{XX} and M_{XX} constitute fully-connected models whose weight matrices have values in $\mathbb{R}_{\pm\infty}$ i.e., the extended real numbers. Note that $\mathbb{R}_{\pm\infty}$ represents a complete 1-group extension [10,11]. The disadvantages of the aforementioned AMMs models are a restricted error correction capability and a large number of spurious memories [1]. Since the output produced by either W_{XX} or M_{XX} depends on the partial order in the complete lattice $\mathbb{R}^n_{\pm\infty} = (\mathbb{R}_{\pm\infty})^n$, a change in partial order may lead to improvements in the tolerance with respect to noise [12]. Based on this observation, we resort to MM on complete inf-semilattices (cisls) in order to

* This work was partially supported by CAPES, by FAPESP under grant no. 2006/06818-1, and by CNPq under grant no. 306040/2006-9.

† sussner@ime.unicamp.br

‡ www.milab.ime.unicamp.br

devise a new AM model that relies on a non-standard partial ordering of patterns. Recall that a cisl can be constructed from a certain partially ordered mathematical structure by defining a reference element representing the least element in the cisl [13]. Special care has to be taken because the supremum does not always exist in this framework. The paper is organized as follows: We begin by providing some basic concepts of mathematical morphology on complete lattices and inf-semilattices. After considering the original MAM models in a more general algebraic setting than usual, section 4 introduces MAMs on inf-semilattices. Finally, section 5 provides some experimental results concerning the reconstruction of gray-scale images.

2. Some Lattice-Theoretical Background of Mathematical Morphology

Lattice theory is a well-established branch of pure and applied mathematics that grew out of propositional logic [14]. Lattice theory was also independently discovered by Dedekind during his research concerning ideals of algebraic numbers [15]. Since then, lattice theory has found applications in diverse areas such as mathematical morphology [16,17], fuzzy set theory [18], computational intelligence [1,2,11,19,20], automated decision making [21], and formal concept analysis [22]. Recall that originally MM was conceived as a theory for the processing and the analysis of objects, i.e. images or signals, using operators and functions based on topological and geometrical concepts [23]. The formalization of MM by means of lattice theory is due to Matheron and Serra who perceived that complete lattices provide for an appropriate mathematical framework for binary and gray-scale MM [16]. The more recent theory of MM on complete inf-semilattices has some additional advantages concerning the self-duality of operators [13,24]. A partially ordered set $\mathbb{L}$ is called a *lattice* if every finite, non-empty subset of $\mathbb{L}$ has an infimum and a supremum in $\mathbb{L}$ [10]. In particular, every totally ordered set or *chain* such as $\mathbb{R}$ and $\mathbb{Z}$ is a lattice. For any $X \subseteq \mathbb{L}$, we denote the infimum of X using the symbol $\bigwedge X$ and the supremum of X using the symbol $\bigvee X$. If $X = \{x_j \in \mathbb{L} : j \in J\}$ then we write $\bigwedge_{j\in J} x_j$ instead of $\bigwedge X$ and $\bigvee_{j\in J} x_j$ instead of $\bigvee X$. A lattice $\mathbb{L}$ is *complete* if every subset of $\mathbb{L}$ has an infimum and a supremum in $\mathbb{L}$. A lattice $\mathbb{L}$ is called *conditionally complete* if every bounded subset of $\mathbb{L}$ has an infimum and a supremum in $\mathbb{L}$. In particular, the *set of finite elements* of a complete lattice $\mathbb{L}$, i.e., $\mathbb{L} \setminus \{\bigwedge \mathbb{L}, \bigvee \mathbb{L}\}$ is conditionally complete. If every finite, non-empty subset of a partially ordered set $\mathbb{L}$ has an infimum in $\mathbb{L}$ then $\mathbb{L}$ constitutes an *inf-semilattice*. If every subset of $\mathbb{L}$ has an infimum in $\mathbb{L}$ then $\mathbb{L}$ is called a *complete inf-semilattice* or *cisl*. If $\mathbb{L}$ is a (complete) lattice or an inf-semilattice then the direct product $\mathbb{L}^n$ is also a (complete) lattice or an inf-semilattice, respectively. These ideas can be further extended by considering $\mathbb{L}^X$, the class of functions from a set $X \neq \varnothing$ to $\mathbb{L}$. A function $\phi : \mathbb{L} \to \mathbb{M}$, where $\mathbb{L}$ and $\mathbb{M}$ are cisls, is called a *cisl homomorphism* or (algebraic) *erosion* if we have the following equation for all index sets J and all $x_j \in \mathbb{L}$ (the concept of (algebraic) *dilation* is defined in a similar way):

$$\phi(\bigwedge_{j \in \mathbb{L}} x_j) = \bigwedge_{j \in \mathbb{L}} \phi(x_j). \tag{1}$$

The chains $\mathbb{R}$ and $\mathbb{Z}$ also constitute examples of *lattice-ordered groups*, for short l-*groups*, i.e., lattices that also form a group in which every group translation $x \mapsto a + x + b$ is isotone [10]. If $\mathbb{G}$ is a complete lattice whose set of finite elements forms a group then we refer to $\mathbb{G}$ as a *complete* l-*group extension* [11].Of course, a conditionally complete lattice $\mathbb{F}$ can form a group at the same time and, in this case, $\mathbb{F}$ is simply called a *conditionally complete* l-*group* [10]. For example, the l-groups $\mathbb{R}$ and $\mathbb{Z}$ are conditionally complete. Note that a conditionally complete l-group $\mathbb{F}$ induces conditionally complete l-groups $\mathbb{F}^n$ and $\mathbb{F}^{\mathbf{X}}$. Let $\mathbb{G}$ be a complete l-group extension. In a recent paper, we proved that erosions and dilations $\mathbb{G}^n \to \mathbb{G}^m$ can be defined via matrix products in minimax algebra [11]. For the purposes of this paper, it suffices to consider the following situation where $\mathbb{F}$ denotes the set of finite elements of $\mathbb{G}$ For $A \in \mathbb{F}^{m \times p}$ and $B \in \mathbb{G}^{p \times n}$, the matrix $\mathrm{C} = \mathrm{A} \boxed{\vee} \mathrm{B}$ and the matrix $\mathrm{D} = \mathrm{A} \boxed{\wedge} \mathrm{B}$, called respectively called the *max-product* and the *min-product* of A and B, are defined by:

$$c_{ij} = \bigvee_{k=1}^{p} (a_{ik} + b_{kj}), \quad d_{ij} = \bigwedge_{k=1}^{p} (a_{ik} + b_{kj}). \tag{2}$$

The following operators $\varepsilon_A, \delta_A : \mathbb{G}^n \to \mathbb{G}^m$ represent respectively an erosion and a dilation between complete lattices for every $A \in \mathbb{F}^{m \times n}$ [11].

$$\varepsilon_A(\mathbf{x}) = A \boxed{\wedge} \mathbf{x}, \quad \delta_A(\mathbf{x}) = A \boxed{\vee} \mathbf{x}. \tag{3}$$

Apart from these operators, the original MAM models also employ the operation of conjugation [1,25]. If $x \in \mathbb{F} \subseteq \mathbb{G}$ then its conjugate x^* is simply given by $-x$ and $X^* \in \mathbb{F}^{m \times n}$ is given by $(X^*)_{ij} = -x_{ji}$ for all $X \in \mathbb{F}^{n \times m} \subseteq \mathbb{G}^{n \times m}$. Note that $\mathbb{F} = \mathbb{G} \setminus \{+\infty, -\infty\}$ constitutes a conditionally complete l-group. In an upcoming paper, we will show that cisls can be constructed from complete l-group extensions or from conditionally complete l-groups. From now on, let $\mathbb{F}$ stand for an arbitrary conditionally complete l-group. The *cone* $\mathbb{F}^+$ is defined as $\{x \in \mathbb{F} : 0 \leq x\}$ where 0 denotes the neutral element of the group $\mathbb{F}$ Note that $(\mathbb{F}^+, \leq)$ represents a cisl. Every $x \in \mathbb{F}$ satisfies $x = x^+ - x^-$, where $x^+ = x \vee 0$ and $x^- = -(x \wedge 0)$ [13].

Theorem 2.1. Let $r \in \mathbb{F}$, where $\mathbb{F}$ is a conditionally complete 1-group. A cisl arises via the following definition of " $_r$ " which constitutes a partial order on $\mathbb{F}$.

$$x \;_r\; y \Leftrightarrow x \vee r \leq y \vee r \text{ and } y \wedge r \leq x \wedge r. \tag{4}$$

The resulting cisl $(\mathbb{F}, \;_r)$ *can be denoted using the symbol* $\mathbb{F}_r$.

3. MAMs in Complete 1-Group Extensions

MAMs have been defined in spaces of vectors with entries in the complete 1-group extension $\mathbb{R}_{\pm\infty} = \mathbb{R} \cup \{+\infty, -\infty\}$ A closer look at MAMs reveals that $\mathbb{R}_{\pm\infty}$ can be replaced by an arbitrary complete 1-group extension $\mathbb{G}$ whose set of finite elements is denoted using the symbol $\mathbb{F}$. Suppose that we want to record k pairs of patterns $(\mathbf{x}^1, \mathbf{y}^1), \ldots, (\mathbf{x}^k, \mathbf{y}^k) \in \mathbb{F}^n \times \mathbb{F}^m$ using a MAM. Let X denote the matrix in $\mathbb{F}^{n\times k}$ whose columns are given by the vectors $\mathbf{x}^\xi \in \mathbb{F}^n$ and let Y denote the matrix in $\mathbb{F}^{m\times k}$ whose columns are given by the vectors $\mathbf{y}^\xi \in \mathbb{F}^m$, where $\xi = 1, \ldots, k$. Two types of matrix associative memories arise from the following two matrices W_{XY} amd M_{XY} of synaptic weights:

$$W_{XY} = Y \boxed{\wedge} X^*, M_{XY} = Y \boxed{\vee} X^* \in \mathbb{F}^{n\times m}. \tag{5}$$

For an input pattern $\mathbf{x} \in \mathbb{G}^n$, the outputs $\mathbf{y}$ and $\mathbf{z}$ of W_{XY} and M_{XY} are respectively computed as follows:

$$\mathbf{y} = W_{XY} \boxed{\vee} \mathbf{x}, \quad \mathbf{z} = M_{XY} \boxed{\wedge} \mathbf{x}. \tag{6}$$

4. MAMs in Cisls Based on Conditionally Complete 1-Groups

Let $\mathbb{F}$ be an arbitrary conditionally complete 1-group $\mathbb{F}$ In the following, we present a new auto-associative memory on the cisl derived from $\mathbb{F}^n$. To this end, we consider k patterns $\mathbf{x}^1, \mathbf{x}^2, \ldots, \mathbf{x}^k \in \mathbb{F}^n$. Let $X_r^\pm$ denote the matrix whose ξth column is given by $(\mathbf{x}^\xi - r(\mathbf{x}^\xi))^+$ and whose $(\xi + k)$h column is given by $(\mathbf{x}^\xi - r(\mathbf{x}^\xi))^-$ for all $\xi = 1, \ldots k$. Furthermore, let $r : \mathbb{F}^n \to \mathbb{F}^n$ and $M_{XX}^r = M_{X_r^\pm X_r^\pm}$. We define an AM model as follows:

$$\mathcal{M}_r(\mathbf{x}) = M_{XX}^r \boxed{\wedge} (\mathbf{x} - r(\mathbf{x}))^+ - M_{XX}^r \boxed{\wedge} (\mathbf{x} - r(\mathbf{x}))^- + r(\mathbf{x}).$$

Theorem 4.1. *Let* $\mathcal{M}_r : \mathbb{F}^n \to \mathbb{F}^n$ *be defined as in Equation 7. The function* $\mathcal{M}_r$ *describes an associative memory model that is guaranteed to yield perfect recall for an arbitrary set of patterns* $\{\mathbf{x}^1, \ldots, \mathbf{x}^k\} \subseteq \mathbb{F}^n$ *and an arbitrary number of patterns* $k \in \mathbb{N}$. *Formally, we have*

$$\mathcal{M}_r(\mathbf{x}^\xi) = \mathbf{x}^\xi \; \forall \mathbf{x}^\xi \in \mathbb{F}^n. \tag{7}$$

Fig 1. Fundamental memories.

For a given input pattern $\mathbf{x} \in \mathbb{F}^n$, *the output pattern* $\mathcal{M}_r(\mathbf{x}) \in \mathbb{F}^n$ *satisfies*

$$r(\mathbf{x}) \quad {}_{r(\mathbf{x})}\mathcal{M}_r(\mathbf{x}) \quad {}_{r(\mathbf{x})}\mathbf{x}. \tag{8}$$

If $r(\mathbf{x}) = \mathbf{r} \in \mathbb{F}^n \; \forall \mathbf{x} \in \mathbb{F}$ *then* $\mathcal{M}_r : \mathbb{F}_0^n \to \mathbb{F}_0^n$ *represents an erosion.*

5. Some Experimental Results on Gray-Scale Image Reconstruction

In this section, we considered the eight gray-scale images of size 128×128 (resized versions of images of size 256×256 that are depicted in Figure 1. We converted these images into eight patterns in $\mathbb{R}^{16384}$ by applying the row-scan method. For each of the image $\mathbf{x}$ let $r(\mathbf{x}) = \overline{\mathbf{x}}$ be the median with window size 3×3 We confirmed that an application of $\mathcal{M}_r$ to the eight original images resulted in perfect recall in accordance to Theorem 4.1 (Equation 8 is obviously satisfied for all $\mathbf{x}^\xi$ since $\mathcal{M}_r(\mathbf{x}^\xi) = \mathbf{x}^\xi$). Subsequently, we corrupted the original images by introducing the following types of noise: salt & pepper (0.1 density) and Gaussian (0 mean and 0.01 variance).

For each type of noise, we conducted a set of 100 simulations for each original image. Figure 2 illustrates these simulations by providing visual examples for each type of noise. This figure also includes the outputs produced by the median operator for each input image. Our new AM model produced slightly better results that the median operator alone. For example, the mean over all images of the relative quadratic errors produced by $\mathcal{M}_r z$ and the median operator yielded 0.0731 versus 0.0911 (salt & pepper) and 0.1156 versus 0.1236 (Gaussian).

Fig. 2. From left to right: original image, noisy version, and outputs of median filtering and of $\mathcal{M}_r$.

References

1. P. Sussner and M. E. Valle, *IEEE Transactions on Neural Networks* **17**, 559 (2006).
2. M. E. Valle and P. Sussner, *Fuzzy Sets and Systems* **159**, 747 (2008).
3. M. E. Valle and P. Sussner, *Neural Networks* **24**, 75 (2011).
4. P. Sussner, E. Esmi, I. Villaverde andM. Graña, *Journal of Mathematical Imaging and Vision* **42**, 134 (2012).
5. B. Raducanu, M. Graña and X. F. Albizuri, *Journal of Mathematical Imaging and Vision* **19**, 113 (2003).
6. E. Guzman, O. Pogrebniak, C. Yañes and L. P. Sanchez Fernandez, *EURASIP Journal on Advances in Signal Processing* **2008** (2008).
7. G. Urcid, J.-C. Valdiviezo-N. and G. X. Ritter, *Journal of Mathematical Imaging and Vision* **42**, 150 (2012).
8. G. X. Ritter and G. Urcid, *Inf. Sci.* **181**, 1787 (2011).
9. G. X. Ritter, P. Sussner and J. L. D. de Leon, *IEEE Trans. on Neural Networks* **9**, 281 (1998).
10. G. Birkhoff, *Lattice Theory*, 3rd edn. (AMS, Providence, NJ, 1993).
11. P. Sussner and E. Esmi, *Information Sciences* **181**, 1929 (2011).
12. M. E. Valle, *Information Sciences* **180**, 4136 (2010).
13. H. J. A. M. Heijmans and R. Keshet, *Journal of Math. Im. and Vision* **17**, 55 (2002).
14. G. Boole, *An Investigation of the Laws of Thought, on Which are Founded the Mathematical Theories of Logic and Probabilities* (Walton and Maberley, London, 1854).
15. R. Dedekind, *Mathematische Annalen* **53**, 371 (1900).
16. J. Serra, *Image Analysis and Mathematical Morphology, Volume 2: Theoretical Advances* (Academic Press, New York, 1988).
17. P. Sussner, M. Nachtegael, T. Mélange, G. Deschrijver, E. L. Esmi and E. E. Kerre, *Journal of Mathematical Imaging and Vision* **43**, 50 (2012).
18. J. Goguen, *Journal of Mathematical Analysis and Applications* **18**, 145 (1967).
19. V. G. Kaburlasos, *IEEE Transactions on Systems Man and Cybernetics, Part B - Cybernetics* **34**, 1017 (2004).
20. V. G. Kaburlasos, *Towards a Unified Modeling and Knowledge-Representation Based on Lattice Theory* (Springer, Berlin, Germany, 2006).
21. Y. Xu, D. Ruan, K. Qin and J. Liu, *Lattice-Valued Logic* (Springer, Heidelberg, Germany, 2003).
22. B. Ganter and P. Wille, *Formal Concept Analysis* (Springer, Berlin, Germany, 1999).
23. J. Serra, *Image Analysis and Mathematical Morphology* (Ac. Press, London, 1982).
24. R. Keshet, *Journal of Mathematical Imaging and Vision* **22**, 309 (2005).
25. R. Cuninghame-Green, *Minimax Algebra: Lecture Notes in Economics and Mathematical Systems 166* (Springer-Verlag, New York, 1979).

AN EXTENDED VARIABLE CONSISTENCY MODEL UNDER DOMINANCE-BASED ROUGH SETS*

DUN LIU†

School of Economics and Management, Southwest Jiaotong University, Chengdu, 610031, China

TIANRUI LI

School of Information Science and Technology, Southwest Jiaotong University, Chengdu, 610031, China

Different variants of variable consistency dominance-based rough set approach (VC-DRSA) have been proposed by considering the ordinality and inconsistency of information systems simultaneously. These approaches focus mainly on the criterion selection of rule generation, but rarely consider the effectiveness of decision results. In this paper, an extended dominance relation is induced into the variable-precision rough set (VPRS) and an extended variable consistency model under dominance-based rough sets is proposed. An illustration validates the feasibility of the proposed method.

1. Introduction

Rough set theory (RST), introduced by Pawlak in 1982, is an important approach to deal with incomplete, inexact and insufficient information [4]. The lower and upper approximations in Pawlak's RST are defined by an equivalence relation, that is, the lower approximation is the union of those equivalence classes that are completely included in a finite set and the upper approximation is the union of those equivalence classes that have a non-empty intersection with the set. However, such definitions proposed by Pawlak do not allow any tolerance of errors, and the inconsistencies just like errors and outliers may affect the validity of decision results, particularly when the information system is large. Based on the above-mentioned analysis, probabilistic rough set models are induced by allowing certain acceptable level of errors and a pair of threshold

* This work is partially supported by the National Science Foundation of China (Nos. 61175047, 60873108, 61100117, 71090402/G1), the Youth Social Science Foundation of the Chinese Education Commission (11YJC630127).

† E-mail:newton83@163.com.

parameters is used to redefine the lower and upper approximations. Nevertheless, in many multi-criteria decision problems, we also need to consider the order relation in an information system. Greco et al. suggested dominance principles should be taken into account and proposed the Dominance-based Rough Set Approach (DRSA) to deal with the dominance-based problems [1-3].

In this paper, we both consider the ordinality and inconsistency in an information system. The remainder of the paper is organized as follows. Section 2 provides basic concepts of RST and its extensions. In Section 3, two new equivalence relations are constructed, and an extended variable consistency model under DRSA is proposed. Then, a case is given to illustrate our approach in Section 4. The paper ends with conclusions and further research topics in Section 5.

2. Preliminaries

Definition 2.1: A complete information system is defined as a 4-tuple $S=(U,A,V,f)$, where $U = \{x_1, x_2, \dots, x_m\}$ is a non-empty finite set of objects. A stands for an attribute set and $A = C\cup\{CL\}$. $C = \{a_1, a_2, \dots, a_k\}$ denotes the set of condition attributes and CL denotes the set of decision attributes. $V=\bigcup_{a\in A} V_a$, where V_a is a domain of a, $f: U\times V\rightarrow V_a$ is an information function such that $f(x,a)\in V_a$ for every $x\in U$, $a\in A$.

$\forall x,y\in X$, $xO_a^{\succeq} y$ if x is at least as good as y with respect to criterion a, and $xO_a^{\succ} y$ if x is better than y with respect to criterion a. Furthermore, x dominates y if $xO_a^{\succeq} y$ for every $a\in C$. Similarly, the concepts of $xO_a^{\preceq} y$ and $xO_a^{\prec} y$ are defined respectively as follows..

Definition 2.2: Given $P\subseteq C$ and $X\subseteq U$, we say that x dominates y with respect to P, denoted by $xD_P y$, if $\forall q\in P$, $xO_q^{\succeq} y$ holds. Let

$$D_P^+(x)=\{y\in U : yD_P x\}=\{y\in U \mid \forall a\in P\subseteq C, f(y,a)\succeq f(x,a)\};$$

$$D_P^-(x)=\{y\in U : xD_P y\}=\{y\in U \mid \forall a\in P\subseteq C, f(y,a)\preceq f(x,a)\}.$$

represent P-dominating set and P-dominated set with respect to $x\in U$. Obviously, the dominance relation D_P is reflexive and transitive.

Definition 2.3: Suppose the decision attribute $\{CL\}$ induces a partition of U into n classes. Let $T = \{1, 2, \dots, n\}$ be the domain of decision criterion. V_{CL} consists of n elements and $CL=\{CL_t, t\in T\}$, where $CL_t=\{x\in U, f(x,CL)=t\}$. Each object $x\in U$ is assigned to one and only one class CL_t. The classes are preference-ordered according to an increasing order of class indices, *i.e.*, for $\forall r,s\in T$ such that $r > s$, the objects from CL_r are strictly preferred to the objects from CL_s. The upward and downward unions of classes are defined respectively as:

$CL_t^{\succeq} = \bigcup_{s \ge t} CL_s$; $CL_t^{\preceq} = \bigcup_{s \le t} CL_s$; $t \in T$.
$CL_1^{\succeq} = CL_n^{\preceq} = U$; $CL_1^{\preceq} = CL_1$; $CL_n^{\succeq} = CL_n$.
$CL_t^{\succeq} = U - CL_{t-1}^{\preceq}$; $CL_t^{\preceq} = U - CL_{t+1}^{\succeq}$.

Definition 2.4: Given a consistency level *l*, the *P*-lower approximations, *P*-upper approximations and *P*-boundaries of VC-DRSA are defined respectively as follows.

$$\underline{P}^l(X_i^{\succeq}) = \{x \in X_i^{\succeq} : |D_P^+(x) \cap X_i^{\succeq}| / |D_P^+(x)| \ge l\};$$
$$\underline{P}^l(X_i^{\preceq}) = \{x \in X_i^{\preceq} : |D_P^-(x) \cap X_i^{\preceq}| / |D_P^-(x)| \ge l\};$$
$$\overline{P}^l(X_i^{\succeq}) = U - \underline{P}^l(X_{i-1}^{\preceq}); \overline{P}^l(X_i^{\preceq}) = U - \underline{P}^l(X_{i+1}^{\succeq});$$
$$BND_P^l(X_i^{\succeq}) = \overline{P}^l(X_i^{\succeq}) - \underline{P}^l(X_i^{\succeq}); BND_P^l(X_i^{\preceq}) = \overline{P}^l(X_i^{\preceq}) - \underline{P}^l(X_i^{\preceq}).$$

3. An Extended Variable Consistency Approach under DRSA

By Definition 2.2, the relation D_P^+ and D_P^- are reflexive and transitive. However, it is better to let these relations become more strict in dealing with real decision problems. For example, consider information system $S = (U, A, V, f)$ and $X \subseteq U$. $\forall x, y \in X$, x strictly dominates y if $\exists a' \in C$, $xO_{a'}^{\succ} y$ and for each $a \in C - \{a'\}$, $xO_a^{\succeq} y$ holds; y strictly dominates x if $\exists a'' \in C$, $xO_{a''}^{\prec} y$ and for each $a \in C - \{a''\}$, $xO_a^{\preceq} y$ holds. Therefore, we need to extend the dominance relation D_P into a strict dominance relation D_P^* .

Definition 3.1: Given $P \subseteq C$ and $x \in U$, we say that x strictly dominates y with respect to $P \subseteq C$, denoted by $xD_P^* y$, if $\forall a \in P$, $xO_a^{\succeq} y$ holds; and x strictly dominated y with respect to $P \subseteq C$, denoted by $yD_P^* x$, if $\forall a \in P$, $xO_a^{\preceq} y$ holds. Let

$$D_P^{+*}(x) = \{y \in U \mid \forall a \in P, f(y,a) \succeq f(x,a) \wedge \exists a' \in P, f(y,a') \succ f(x,a')\};$$
$$D_P^{-*}(x) = \{y \in U \mid \forall a \in P, f(y,a) \preceq f(x,a) \wedge \exists a'' \in P, f(y,a'') \prec f(x,a'')\}.$$

represent *P*-strict dominating set and *P*-strict dominated set with respect to $x \in U$, respectively. The extended dominance relation D_P^* is anti-reflexive and transitive.

Definition 3.2: $\forall x \in U$, we define $[x]_{R^+}$ and $[x]_{R^-}$ as the extended dominating equivalence class and extended dominated equivalence class as follows, and R^+ and R^- induce a partition of the universe, respectively.

$$[x]_{R^+} = \{x' \mid x' \in U, D_P^{+*}(x') = D_P^{+*}(x)\}; [x]_{R^-} = \{x'' \mid x'' \in U, D_P^{-*}(x'') = D_P^{-*}(x)\}.$$

Definition 3.3: Let $S = (U, A, V, f)$ be a complete information system, $U / R^+ = \{[x]_{R^+} \mid x \in U\} = \{X_1^+, X_2^+, \cdots, X_n^+\}$, $CL = \{CL_t, t \in T\}$, $CL_t^{\succeq} = \bigcup_{s \ge t} CL_s$,

$\forall X^+ \subseteq U/R^+$, $\forall Y \subseteq CL_i^{\succeq}$, the relative error for X^+ with respect to Y is defined as:

$$E(X^+,Y)=\begin{cases}1-|X^+\cap Y|/|X^+| & |X^+|>0\\ 0 & |X^+|=0\end{cases}$$

Let $0 \le \beta < 0.5$, we define the β inclusion relation for X^+ with respect to Y as: $X^+ \subseteq_\beta Y \Leftrightarrow E(X^+,Y) \le \beta$.

Definition 3.4: Let $S=(U,A,V,f)$ be a complete information system. $\forall X^+ \subseteq U/R^+$, $Y \subseteq CL_i^{\succeq}$, the lower approximation, upper approximation and boundary region are defined as follows:

$\underline{R}^\beta(X^+)=\bigcup\{x\in U/R^+ : E([x]_{R^+},Y)\le\beta\}$;

$\overline{R}^\beta(X^+)=\bigcup\{x\in U/R^+ : E([x]_{R^+},Y)<1-\beta\}$;

$BND_\beta(X^+)=\bigcup\{x\in U/R^+ : \beta< E([x]_{R^+},Y)<1-\beta\}$;

Similarly, we denote $U/R^-=\{[x]_{R^-}\mid x\in U\}=\{X_1^-,X_2^-,\cdots,X_{n'}^-\}$, $CL=\{CL_t, t\in T\}$, $CL_i^{\preceq}=\bigcup_{s\le t}CL_s$. For $\forall X^-\subseteq U/R^-$, $Y'\subseteq CL_i^{\preceq}$, we can define $E(X^-,Y')$, $\underline{R}^\beta(X^-)$, $\overline{R}^\beta(X^-)$ and $BND_\beta(X^-)$, respectively. In the following, we mainly illustrate our proposed approach by an practical illustration.

4. An Illustration

Let us illustrate the above concepts on a didactic example. The example refers to the salary of teachers in a university. In the information system given in Table 1, $U = \{x_1, x_2,\ldots, x_{14}\}$ stands for the 14 kinds of teacher's classifications. We choose "length of service", "educational background", "research level" and "teaching level" as condition attributes and described by the set $C = \{a_1, a_2, a_3, a_4\}$. The level of salary is considered as the decision attribute $D =\{d\}$. All criteria are to be maximized, and *Num* stands for the cardinal number of objects. The analyzed data is presented in Table 1. Obviously, the values of the attribute sets perform a preference-order as: $1 \prec 2 \prec 3 \prec 4$.

Table 1. The information table about the salary of teachers

U	a_1	a_2	a_3	a_4	d	*Num*	U	a_1	a_2	a_3	a_4	d	*Num*
x_1	1	1	1	1	1	10	x_8	2	3	3	2	3	45
x_2	1	2	1	1	1	30	x_9	3	2	2	3	1	10
x_3	1	2	1	1	2	5	x_{10}	2	4	3	3	1	3
x_4	1	2	1	2	3	4	x_{11}	2	4	3	3	3	45
x_5	1	2	2	1	2	25	x_{12}	3	4	3	3	3	5
x_6	1	2	2	1	4	3	x_{13}	3	4	3	3	4	35
x_7	2	2	2	2	3	5	x_{14}	3	4	4	4	4	10

From the values of the decision attribute in Table 1, the universe is divided into four parts: $CL_1=\{x_1,x_2,x_9,x_{10}\}$, $CL_2=\{x_3,x_5\}$, $CL_3=\{x_4,x_7,x_8,x_{11},x_{12}\}$, $CL_4=\{x_6,x_{13},x_{14}\}$. Then, we calculate the upward and downward unions according to Definition 2.3:

$CL_1^{\succeq}=U$; $CL_2^{\succeq}=CL_2\bigcup CL_3\bigcup CL_4=\{x_3,x_4,x_5,x_6,x_7,x_8,x_{11},x_{12},x_{13},x_{14}\}$

$CL_3^{\succeq}=CL_3\bigcup CL_4=\{x_4,x_6,x_7,x_8,x_{11},x_{12},x_{13},x_{14}\}$; $CL_4^{\succeq}=CL_4=\{x_6,x_{13},x_{14}\}$.

$CL_1^{\preceq}=CL_1=\{x_1,x_2,x_9,x_{10}\}$; $CL_2^{\preceq}=CL_1\bigcup CL_2=\{x_1,x_2,x_3,x_5,x_9,x_{10}\}$; $CL_3^{\preceq}$ $=CL_1\bigcup CL_2\bigcup CL_3=\{x_1,x_2,x_3,x_4,x_5,x_7,x_8,x_9,x_{10},x_{11},x_{12}\}$; $CL_4^{\preceq}=U$.

According to Definition 3.1, we can compute $D_P^{+*}(x_i)$ and $D_P^{-*}(x_i)$ for $x_i\in U$. Then we have $U/R^+=\{\{x_1\},\{x_2,x_3\},\{x_4,x_5,x_6\},\{x_7\},\{x_8\},\{x_9,x_{10},x_{11}\},\{x_{12},x_{13}\},\{x_{14}\}\}$. We denote $U/R^+=\{X_1^+,X_2^+,X_3^+,X_4^+,X_5^+,X_6^+,X_7^+,X_8^+\}$. Similarly, $U/R^-=\{\{x_1\},\{x_2,x_3\},\{x_4,x_5,x_6\},\{x_7\},\{x_8,x_9\},\{x_{10},x_{11}\},\{x_{12},x_{13}\},\{x_{14}\}\}$. We denote $U/R^+=\{X_1^-,X_2^-,X_3^-,X_4^-,X_5^-,X_6^-,X_7^-,X_8^-\}$.

Furthermore, we calculate $E(X^+,Y)$ for every $X^+\subseteq U/R^+$ and $Y\subseteq U/CL_j^{\succeq}$. Analogously, we calculate $E(X^-,Y')$ for every $X^-\subseteq U/R^-$ and $Y'\subseteq U/CL_j^{\preceq}$, which is shown in Table 2.

Table 2. The calculation of $E(X^+,Y)$ and $E(X^-,Y')$

$E(X^+,Y)$	$CL_1^{\succeq}$	$CL_2^{\succeq}$	$CL_3^{\succeq}$	$CL_4^{\succeq}$	$E(X^-,Y')$	$CL_1^{\preceq}$	$CL_2^{\preceq}$	$CL_3^{\preceq}$	$CL_4^{\preceq}$
X_1^+	0	-	-	-	X_1^-	0	0	0	0
X_2^+	0	0.857	-	-	X_2^-	0.143	0	0	0
X_3^+	0	0	0.718	0.906	X_3^-	-	0.219	0.094	0
X_4^+	0	0	0	-	X_4^-	-	-	0	0
X_5^+	0	0	0	-	X_5^-	0.818	0.818	0	0
X_6^+	0	0.224	0.224	-	X_6^-	0.938	0.938	0	0
X_7^+	0	0	0	0.125	X_7^-	-	-	0.875	0
X_8^+	0	0	0	0	X_8^-	-	-	-	0

According to Table 2, when $\beta\geq 0.224$, we have $\{x_1,x_2,x_3\}\subseteq CL_1^{\succeq}$, $\{x_4,x_5,x_6\}\subseteq CL_2^{\succeq}$, $\{x_7,x_8,x_9,x_{10},x_{11}\}\subseteq CL_3^{\succeq}$, $\{x_{12},x_{13},x_{14}\}\subseteq CL_4^{\succeq}$. Analogously, when $\beta\geq 0.219$, we have $\{x_1,x_2,x_3\}\subseteq CL_1^{\preceq}$, $\{x_4,x_5,x_6\}\subseteq CL_2^{\preceq}$, $\{x_7,x_8,x_9,x_{10},x_{11}\}\subseteq CL_3^{\preceq}$, $\{x_{12},x_{13},x_{14}\}\subseteq CL_4^{\preceq}$. Therefore, when $\beta\geq\max(0.224,0.219)=$

0.224, we have the consistent results as: $\{x_1, x_2, x_3\} \subseteq_{0.224} CL_1$, $\{x_4, x_5, x_6\} \subseteq_{0.224} CL_2$, $\{x_7, x_8, x_9, x_{10}, x_{11}\} \subseteq_{0.224} CL_3$, $\{x_{12}, x_{13}, x_{14}\} \subseteq_{0.224} CL_4$.

5. Conclusions

By considering the ordinality and inconsistency in real information system simultaneously, we proposed an extended dominance relation based on probabilistic rough sets and DRSA in this paper. The *P*-strict dominating set and *P*-strict dominated sets were defined to construct an extended dominated equivalence relations R^+ and R^-, which is utilized to present the new approach. Finally, we validated the rationality and validity of the proposed method by a case study. However, the work here was only based on the complete information system. Our future research work will focus on the robustness and stability of the rules generation when the parameter is changed. It is certainly worthwhile to explore if the proposed approach can be extended to incomplete information system.

References

1. M. Beynon, MJ. Peel.: Variable precision rough set theory and data discretisation: an application to corporate failure prediction. Omega, 29, 581-576.(2001)
2. J. Blaszczynski, S. Greco, R. Slowinski, M. Szelag.: Monotonic Variable Consistency Rough Set Approaches. Int J Approx Reason, 50, 979-999. (2009)
3. S. Greco, B. Matarazzo, R. Slowinski.: Rough approximation of a preference relation by dominance relations. Eur J Oper Res, 117, 63-83. (1999)
4. Z. Pawlak: Rough sets. Int J Comput Inf Sci 11, 341-356. (1982)

LIFTING QUASI-FILTERS OF LATTICE IMPLICATION ALGEBRAS*

XIAODONG PAN†1, XIAOHONG LIU[2]

[1]*School of Mathematics, Southwest Jiaotong University,Chengdu, Sichuan, 610031, P.R. China*

[2]*College of Management, Southwest University for nationalities, Chengdu, 610041, P.R. China*

Filters play a significant role in studying of structure and properties of partially ordered algebraic structures, in particular in the algebraic semantics of formal fuzzy logics and, moreover, it is also very important for the proof of the completeness of these logics. The aim of this paper is to introduce a kind of quasi-filters in lattice implication algebras, which are the algebraic abstractions of the sets of provable formulae and are closed with respect to lifting rules (These rules were defined by Pavelka in Zeitschr. f. math. Logik und Grundlagen d. Math. 25: 447-464, 1979) in the corresponding formal fuzzy logic, and we examine their mutual dependencies and the connections with other different kinds of filters in lattice implication algebras.

1. Introduction

As we know, algebraic logic can be regarded as a structuralist approach to the subject matter of logic, which sets up a duality theory between the world of logics and the world of classes of algebras. As the algebraic counterpart, lattice implication algebras (LIA's for short) [1] were invented by Xu in 1993 in order to study the many-valued logical system whose propositional values are given in a bounded distributive lattice. LIA theory play an important role in helping and improving, clarifying the corresponding logic. Until now, this logical algebra has been extensively investigated by several researchers (see [2-11]). In [2], Xu and Qin introduced the notions of filters and implicative filters in lattice implication algebras and investigated some of their properties. Filters can be used in the algebraic semantics of lattice-valued logics based on LIA's, and are

* The work was supported by the National Natural Science Foundation of China (Grant No. 61100046, 61175055), the application fundamental research plan project of Sichuan Province (Grant No. 2011JY0092) and the Fundamental Research Funds for the Central Universities (Grant No. SWJTU12CX054).

† xdpan1@163.com (X.D Pan).

of crucial importance in the proof of the completeness of these logics. In 1998, Jun et al. [3] introduced the notions of positive implicative filters and associative filters in lattice implication algebras, and investigated some of their properties. Also Jun defined fuzzy positive implicative and fuzzy associative filters of lattice implication algebras (see [4]). But in 2011, Masoud et al. proved that there is no associative filter in LIA (see [12]).

From a logical point of view, filters can be regarded as sets of provable formulae, which are closed with respect to Modus Ponens. In [13], Pavelka established a kind of fuzzy propositional logic, in which he defined a kind of infer rule, called (*L*, *a*)-lifting rule, using this rule one can infer $a \Rightarrow \varphi$ from any logical formula φ. In order to study the deductive system with respect to this kind of infer rules, in this paper, we intend to introduce a kind of quasi-filters in lattice implication algebras, which can be regarded as the algebraization of sets of the corresponding provable formulae.

2. Preliminary

In the following we review some basic notions about lattice implication algebras and filter according to Xu et al.(for details, we prefer readers to refer to [14]).

Definition 2.1. A bounded lattice $(L,\vee,\wedge,O,I)$ with order-reversing involution ' and a binary operation $\rightarrow$ is called a lattice implication algebra if it satisfies the following axioms: for all $x, y, z \in L$,

(I_1) $x \rightarrow (y \rightarrow z) = y \rightarrow (x \rightarrow z)$,

(I_2) $x \rightarrow x = I$,

(I_3) $x \rightarrow y = y' \rightarrow x'$,

(I_4) $x \rightarrow y = y \rightarrow x = I \Rightarrow x = y$,

(I_5) $(x \rightarrow y) \rightarrow y = (y \rightarrow x) \rightarrow x$,

(L_2) $(x \wedge y) \rightarrow z = (x \rightarrow z) \vee (y \rightarrow z)$,

(L_1) $(x \vee y) \rightarrow z = (x \rightarrow z) \wedge (y \rightarrow z)$.

Definition 2.2. Let *L* be a lattice implication algebra and *F* is a non-empty subset of *L*, *F* is called a filter (MF for short) if it satisfies the following conditions: (1) $I \in F$, (2) for all $x, y \in L$, if $x \in F$ and $x \rightarrow y \in F$, then $y \in F$.

Theorem 2.1. Let $(L,\vee,\wedge,',\rightarrow,O,I)$ be a lattice implication, then $(L,\vee,\wedge)$ is a distributive lattice.

In a lattice implication algebra *L*, define the following operations $\oplus$ and $\otimes$: for all $a,b \in L$, $a \oplus b = a' \rightarrow b$, $a \otimes b = (a \rightarrow b')'$, then

Theorem 2.2. Let *L* be a lattice implication algebra. Then $(L,\oplus,\leq)$ and $(L,\otimes,\leq)$ are all lattice ordered semigroups.

In what follows, unless otherwise stated, L always denotes any given lattice implication algebra. a^n is an abbreviation of $\overbrace{a\otimes a\otimes\cdots\otimes a}^{n\ times}$, na is an abbreviation of $\overbrace{a\oplus a\oplus\cdots\oplus a}^{n\ times}$ for any $n\in \mathrm{N}^+$, a^0 and $0a$ are defined as the greatest element I and the least element O in L.

Xu et al. [14] proved the following

Proposition 2.1. Let L be an LIA. For all elements $x, y, z\in L$,
(1) $x\to y\geq x'\vee y$,
(2) $x\to y=I$ if and only if $x\leq y$,
(3) $x\vee y=I$ if and only if $x\to y=y$, $x\wedge y=O$ if and only if $x\to y=x'$,
(4) $x\to(y\to z)=(x\otimes y)\to z$.

3. Lifting quasi-filter

In this section, we introduce the notion of lifting quasi-filter and present its some properties.

Definition 3.1. Let $\alpha\in L\backslash\{O,I\}, A\subset L\backslash\{O,I\}$. A non-empty subset F of L is called a lifting quasi-filter with respect to α (α-LQF for short) if it satisfies: (1) $I\in F$, (2) if $x\in F$, then $\alpha\to x\in F$. F is called a lifting quasi-filter with respect to D (D-LQF for short) if it satisfies: (1) $I\in F$, (2) if $x\in F$, then for any $\alpha\in D$, we have $\alpha\to x\in F$.

From the definition above, it is easy to show that F is a D-LQF if and only if F is an α-LQF for any $\alpha\in D$. If we let α =O or I, then all subsets of L are α-LQF's.

Example 3.1. Let $L=\{O,a,b,c,d,I\}$, the Hasse diagram of L, its implication operation $\to$ and order-reversing involution ' are defined as follows:

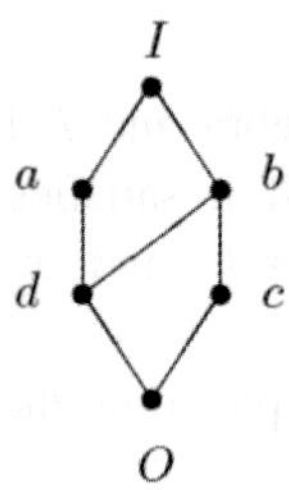

x	x'
O	I
a	c
b	d
c	a
d	b
I	O

$\to$	o	a	b	c	d	I
o	I	I	I	I	I	I
a	c	I	b	c	b	I
b	d	a	I	b	a	I
c	a	a	I	I	a	I
d	b	I	I	b	I	I
I	o	a	b	c	d	I

then $(L,\vee,\wedge,',\to,O,I)$ is a lattice implication algebra (see [14]). It is easy to show that $\{I,a\}$ is a filter, and is also a lifting quasi-filter of L with respect to a, b, c, d. {I,c} is a lifting quasi-filter with respect to a, c, but is not lifting quasi-filter with respect to *b, d*. This is because that $b \to c = d \to c = b$ is not included in the set.

Proposition 3.1. Let $F \subset L, \alpha \in L$. If *F* is a filter, then *F* is also an α-*LQF*.

Proof. Assume that F is a filter. For any $x \in F$, then $x \to (\alpha \to x) = I \in F$, this follows that $\alpha \to x \in F$. Hence, F is an α-*LQF*.

The converse is not always true. For example, in example 3.1, $\{I, c\}$ is a *c*-*LQF* but not a filter.

Theorem 3.1. Let L_1 and L_2 be lattice implication algebras, f an onto implication homomorphism from L_1 to L_2, $\alpha \in L_1, \beta \in L_2$.

(i) If F_1 is an α-LQF of L_1, then $f(F_1)$ is a $f(\alpha)$-LQF of L_2.

(ii) If F_2 is a β-LQF of L_2, then $f^{-1}(F_2)$ is a $f^{-1}(\beta)$-LQF of L_1.

Proof. Let I_1 and I_2 be the greatest elements of L_1 and L_2, respectively. It is easy to show that $f(I_1) = I_2$.

(i) Let F_1 be an α-LQF of L_1. First, $I_2 \in f(F_1)$. Secondly, for any $y \in f(F_1)$, there exists $x \in F_1$ such that $f(x) = y$ and $\alpha \to x \in F_1$, which follows that $f(\alpha) \to y = f(\alpha) \to f(x) = f(\alpha \to x) \in f(F_1)$. Hence, $f(F_1)$ is a $f(\alpha)$-LQF of L_2.

(ii) Let F_2 be a β-LQF of L_2. It follows from the discussion above that $I_1 \in f^{-1}(F_2)$. For any $x \in f^{-1}(F_2)$, then $f(x) \in F_2$ and $\beta \to f(x) \in F_2$. Since f is an epimorphism, $f^{-1}(\beta) \neq \varnothing$; for any $z \in f^{-1}(\beta)$, we have $f(z \to x) = \beta \to f(x)$, thus $z \to x \in f^{-1}(F_2)$. Hence $f^{-1}(F_2)$ is a $f^{-1}(\beta)$-LQF of L_1.

Proposition 3.2. Let $F, F_1, F_2 \subset L$ and $\alpha \in L, D \subset L$. The following conclusions hold:

(1) If both F_1 and F_2 are α-LQF's (*D*-LQF's), then both $F_1 \cup F_2$ and $F_1 \cap F_2$ are also α-LQF's (*D*-LQF's).

(2) If $I \in F$ and for any $x \in F, x \geq \alpha$, then *F* is an α-LQF.

(3) If *F* is nonempty and closed with respect to $\to$, then for any $\alpha \in F$, *F* is an α-LQF, and thus is *D*-LQF for any $D \subset F$ as well.

(4) Let *F* be an α-LQF. If $\alpha \wedge \beta = O$ and $\beta \in F$, then $\alpha' \in F$.

(5) Let *F* be a *D*-LQF and $\alpha_1, \alpha_2 \in D$, then *F* is also an $(\alpha_1 \otimes \alpha_2)$-LQF.

Proof. (1), (2) and (3) are obvious. The remaining we only need to prove (4), (5).

If $\alpha \wedge \beta = O$, by Proposition 2.1 we have $\alpha \to \beta = \alpha'$ and $\beta \in F$. Since *F* is an α-LQF, thus $\alpha' \in F$. Thus (4) is proved.

$I \in F$ is obvious. For any $x \in F$, since *F* is a *D*-LQF, thus is α_1-LQF and is α_2-LQF as well. It follows that $\alpha_1 \to x \in F$, thus $\alpha_2 \to (\alpha_1 \to x) = (\alpha_1 \otimes \alpha_2) \to x \in F$. Hence (5) hold.

As we have known, the intersection of LQF's of L is an α -LQF. Suppose $A \subseteq L$, the least filter containing A is called the lifting quasi-filter with respect to α generated by A and denoted by $[A]_\alpha$. It is trivial to verify that

$$[A]_\alpha = \bigcap \{B \mid A \subseteq B \subseteq L, B \text{ is an } \alpha - LQF\}.$$

Theorem 3.2. Let A be a nonempty subset of L, and $\alpha \in L$. Then

$$[A]_\alpha = \{x \mid x \in L, \text{there exist } n \in \mathrm{N} \text{ and } y \in A \text{ s.t., } x = \alpha^n \to y\} \cup \{I\}.$$

Proof. Let $B = \{x \mid x \in L, \text{there exist } n \in \mathrm{N} \text{ and } y \in A \text{ s.t., } x = \alpha^n \to y\} \cup \{I\}$. Firstly, for any $a \in A$, since $\alpha^0 \to a = I \to a = a$, thus $a \in B$, it follows that $A \subseteq B$. Secondly, let $x \in B$, if $x = I$, then $I = \alpha \to I \in B$; otherwise there exist $n \in \mathrm{N}$ and $y \in A$ such that $x = \alpha^n \to y$. Since $\alpha \to x = \alpha \to (\alpha^n \to y)$ $= \alpha^{n+1} \to y$, so we have $\alpha \to x \in B$. Hence B is an α -LQF. Finally, assume that $A \subseteq C$ and C is an α -LQF. For any $x \in B$, if $x = I$, then $x \in C$; otherwise there exist $n \in \mathrm{N}$ and $y \in A$ such that $x = \alpha^n \to y$. By $y \in A \subseteq C$ and C is an α -LQF, thus $x \in C$ and hence $B \subseteq C$. To sum up, B is an α -LQF generated by A, that is, $B = [A]_\alpha$.

Corollary 3.1. Let $a, \alpha \in L$. Then

$$[\{a\}]_\alpha = \{x \mid x \in L, \text{there exist } n \in \mathrm{N} \text{ s.t., } x = \alpha^n \to a\} \cup \{I\}.$$

By Theorem 2.2, we know that $(L, \oplus)$ is a semigroup and O is the identity element. In semigroup notations, F is an α -LQF if and only if $I \in F$ and $\alpha' \oplus x \in F$ for any $x \in F$. In what follows, we study the properties of LQF's from the viewpoint of semigroup. We omitted the proofs.

Proposition 3.3. Let F be a subset of L. If F is a subsemigroup of $(L, \oplus)$, then $[F]_\alpha$ is also a subsemigroup of $(L, \oplus)$.

Corollary 3.2. Let F be a subset of L. If F is a submonoid of $(L, \oplus)$, then $[F]_\alpha$ is also a submonoid of $(L, \oplus)$.

Proposition 3.4. Let F be a subset of L. If F is an ideal of $(L, \oplus)$, then $[F]_\alpha$ is also an ideal of $(L, \oplus)$.

4. Conclusion

The paper introduced a kind of quasi-filters in lattice implications algebras, some important properties are presented. It could serve as a fundamental algebraic theory for characterizing the corresponding deductive system. Moreover, we desire to see that the theory and method developed in this paper can be applied to characterize other logical algebras, such as MTL algebras,

BCK/BCI algebras etc. Next, we need to investigate the relationship between this quasi-filters with other several filters, consequently, we can make clear the relationship among several inference rules. Meanwhile, it is also an important topic to consider the fuzzification of this quasi-filters.

References

1. Y. Xu, Lattice implication algebras, *J. Southwest Jiaotong Univ.* 28, 1(1993) 20-27.(in Chinese)
2. Y. Xu, K.Y. Qin, On filters of lattice implication algebras, *Fuzzy Math.* 1(1993) 251-260.
3. Y.B. Jun, Y. Xu, K.Y. Qin, Positive implicative and associative filters of lattice implication, *Bull Korean Math Soc.* 35, 1(1998) 53-61.
4. Y.B. Jun, Fuzzy positive implicative and fuzzy associative filters of lattice implication algebras, *Fuzzy Sets Syst.*, 121(2001) 353-357.
5. Y.B. Jun, Implicative filters of lattice implication algebras, *Bull Korean Math Soc.* 34(1997) 193-198.
6. Y.B. Jun, Topological aspects of filters in lattice implication algebras, *Bull Korean Math Soc.* 43, 2(2006) 227-233.
7. J. Liu, Y. Xu, On the prime filters and decomposition theorem of lattice implication algebras, *Fuzzy Math.* 6(1998) 1001-1008.
8. S. Mokhtari, A. Kordi, A. Moussavi et al., On LI-ideals of lattice implication algebras, *J Math Appl.* 32(2010) 67-74.
9. X.D. Pan, Y. Xu, Q. Zhang, Lattice Implication Algebraic Equation in Lattice Implication Algebras, *J. Southwest Jiaotong Univ.* 40, 6(2005) 842-845.
10. X.D. Pan, Y. Xu, Lattice implication ordered semigroups, *Information Sci.* 178(2008) 403-413.
11. Y.Q.Zhu, W. B. Tu, A note on lattice implication algebras, *Bull Korean Math Soc.* 38,1(2001) 191-195.
12. M.Haveshki, M. Mohamadhasani, Lack of associative filters in lattice implication algebras, *Soft computing*. (2011) DOI: 10.1007/s00500-011-0783-3.
13. J. Pavelka, On fuzzy logic I: Many-valued rules of inference, II: Enriched residuated lattices and semantics of propositional calculi, III: Semantical Completeness of some many-valued propositional calculi. *Zeitschr F Math Logik Und Grundlagend Math*, 25(1979) 45-52; 119-134; 447-464.
14. Y. Xu, D. Ruan, K.Y. Qin, J. Liu, Lattice-Valued Logic-An Alternative Approach to Treat Fuzziness and Incomparability, *Springer-Verlag Berlin Heidelberg New York Press* (2003).

INTUITIONISTIC FUZZY PROPOSITIONAL SYSTEM BASED ON LINGUISTIC TRUTH-VALUED INTUITIONISTIC FUZZY LATTICE

LI ZOU*

Informedia Electronic Co.,Ltd., Dalian 116001, China
School of Computer and Information Technology,Liaoning Normal University, Dalian 116029, China

XIN LIU

School of Mathematics, Liaoning Normal University, Dalian 116029, China

DENGEN HUANG

School of Computer Science and Technology, Dalian University of Technology, Dalian 116024, China

A kind of intuitionistic fuzzy logic system LP(S) based on linguistic truth-valued intuitionistic fuzzy algebra is proposed in this paper. The linguistic truth-valued intuitionisitic fuzzy algebra comes from the linguistic truth-valued implication algebra which is fit to express both comparable and incomparable information. The method can deal with the uncertain problem which has both positive evidence and negative evidence at the same time. Some axioms and reasoning rules of linguistic truth-valued intuitionistic fuzzy propositional logic are discussed. The proofs and theorems are also obtained.

1. Introductions

In many decision makings under uncertain environments, linguistic terms rather than probabilistic values are taken into account the managing of uncertainty in decision processes. A linguistic value is consisted of atomic linguistic value and linguistic hedge. In computing with words (CWW), semantic of very true is expressed by a fuzzy set on $[0,1]$ Espinilla, et al. presented several computational approaches to manage multigranular linguistic scales in decision making problems [3]. There are many numeric aggregation operators and linguistic aggregation operators to aggregating them [4,6,9]. Xu, et al. proposed

* zouli@lnnu.edu.cn

linguistic truth-valued lattice implication algebra to deal with linguistic truth inference [5,7,8]. Zou [10-12] proposed a framework of linguistic truth-valued propositional logic and developed its reasoning method.

Intuitionistic fuzzy sets (A-IFSs) concentrate on expressing advantages and disadvantages, pros and cons [1,2]. Formally, intuitionistic fuzzy set is defined as $A = \{(x, \mu_A(x), \nu_A(x)) \mid x \in U\}$, where U is a discourse, $\mu_A(x)$ and $\nu_A(x)$ are the membership degree and nonmembership degree of the object $x \in U$ belonging to $A \subseteq U$ which satisfied with $0 \le \mu_A(x) + \nu_A(x) \le 1$ for any $x \in U$.

Inspired by A-IFSs, we discuss the linguistic truth-valued intuitionistic fuzzy lattice in linguistic truth-valued lattice implication algebra. we construct a kind of intuitionistic fuzzy logic system. Some axioms and reasoning rules of linguistic truth-valued intuitionistic fuzzy propositional logic are discussed. The proofs and theorems are also obtained.

2. Linguistic truth-valued intuitionistic fuzzy lattice $\mathcal{LI}_{2n}$

Based on $2n$ linguistic truth-valued lattice implication algebra $\mathcal{L}_{V(n\times 2)}$ we can construct linguistic truth-valued intuitionistic fuzzy lattice. Formally, denote $\mathcal{LI}_{2n} = (LI_{2n}, \cup, \cap)$ as a linguistic truth-valued intuitionistic fuzzy lattice where $((h_n, t), (h_n, f))$ and $((h_1, t), (h_1, f))$ are the greatest element and the least element of $\mathcal{LI}_{2n}$ respectively(Figure 1).

Definition 2.1. In the linguistic truth-valued intuitionistic fuzzy lattice $\mathcal{LI}_{2n} = (LI_{2n}, \cup, \cap)$, for any $((h_i, t), (h_j, f)), ((h_k, t), (h_l, f)) \in LI_{2n}$ $((h_i, t), (h_j, f)) \le ((h_k, t), (h_l, f))$ if and only if $i \le k$ and $j \le l$, also

$$((h_i, t), (h_j, f)) \cup ((h_k, t), (h_l, f)) = ((h_{max(i,k)}, t), (h_{max(j,l)}, f)), \quad (1)$$

$$((h_i, t), (h_j, f)) \cap ((h_k, t), (h_l, f)) = ((h_{min(i,k)}, t), (h_{min(j,l)}, f)). \quad (2)$$

3. Linguistic truth-valued intuitionistic fuzzy propositional logic and its axiom

The language of Linguistic truth-valued intuitionistic fuzzy propositional logic are as follows:

(1) Let $S = \{p_1, p_2, \cdots\}$ be countable propositional variables set and $\{\overline{0}, \overline{u}\}$ be constant set, where for any $p_i \in S$, p_i is a distinct proposition or a uncertain atomic propositional;

(2) The logical connectives are T, F, $\vee$, $\wedge$, & and $\rightarrow$, where T and F are unary logical connectives. $\vee$, $\wedge$, & and $\rightarrow$ are binary logical connectives. The advantage of the connectives. The advantage is : unary

logical connectives are prior to binary logical connectives, binary logical & connectives are prior to $\vee$, $\wedge$ and $\to$, $\vee$ and $\wedge$ are prior to $\to$;

(3) auxiliary symbols: "(" and ")".

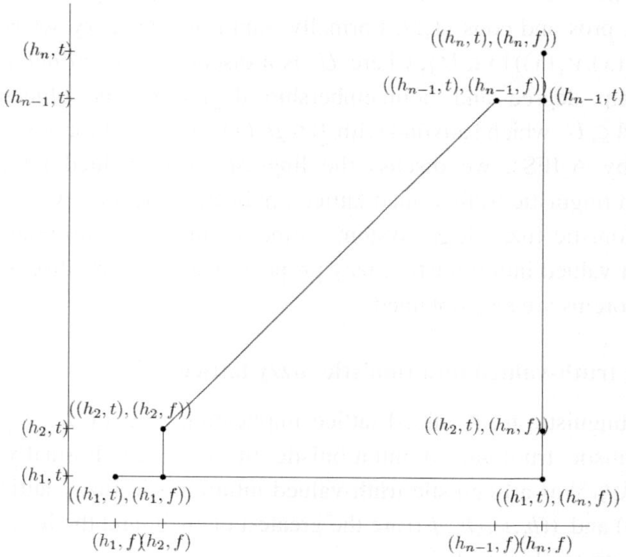

Figure 1. Structure Diagrams of $\mathcal{LI}_{2n}$

Definition 3.1. Well-formed formula of Linguistic truth-valued intuitionistic fuzzy propositional logic or formula for short are defined recursively as follows:

(1) For any $p_i \in S$, $\overline{0}$ and $\overline{u}$ are formulae;

(2) If ϕ and ψ are formulae, then $T\phi$, $F\phi$, $\phi \vee \psi$, $\phi \wedge \psi$, $\phi \& \psi$ and $\phi \to \psi$ are formulae;

(3) No expression is a formula unless it is compelled to be one by (1) and (2).

In formally, let $P(S)$ be a set of all the formulae of $LTV - IFP$, then P(S) is a free algebra with the type of $\{T, F, \vee, \wedge, \&, \to\}$ from $\{\overline{0}, \overline{u}\} \cup S = \{\overline{0}, \overline{u}, p_1, p_2, \cdots\}$ For any $\phi, \psi \in P(S)$ we denote as follows:

$$\overline{1} = \overline{0} \to \overline{0};$$

$$\neg\phi = \phi \to \overline{0};$$

$$\phi \leftrightarrow \psi = (\phi \to \psi) \wedge (\psi \to \phi).$$

Definition 3.2. We get the axioms of $P(S)$ as follows:

(1) $(\phi \to \psi) \to ((\psi \to \chi) \to (\phi \to \chi))$

(2) $\phi \to (\phi \vee \psi)$

(3) $\psi \to (\phi \vee \psi)$

(4) $(\phi \to \chi) \to ((\psi \to \chi) \to ((\phi \vee \psi) \to \chi))$

(5) $(\phi \wedge \psi) \to \psi$;

(6) $(\phi \wedge \psi) \to \phi$;

(7) $(\phi \& \psi) \to \psi$;

(8) $(\phi \& \psi) \to (\psi \& \phi)$

(9) $(\phi \to \psi) \to ((\phi \to \chi) \to (\phi \to (\psi \wedge \chi)))$

(10) $(\phi \to (\psi \to \chi)) \to ((\phi \& \psi) \to \chi)$

(11) $((\phi \& \psi) \to \chi) \to (\phi \to (\psi \to \chi))$

(12) $\overline{0} \to \phi$;

(13) $T\phi \to \phi$, $\phi \to F\phi$;

(14) $T\phi \to T(T\phi)$, $F(F\phi) \to F\phi$;

(15) $T\phi \wedge T\varphi \to T(\phi \wedge \psi)$, $F\phi \wedge F\varphi \to F(\phi \wedge \psi)$;

(16) $T(\phi \vee \psi) \to T\phi \vee T\psi$, $F(\phi \vee \psi) \to F\phi \vee F\psi$;

(17) $T\overline{1}$, $\neg F\overline{0}$

(18) $\neg T\overline{u}$ $F\overline{u}$;

(19) $F\phi \to T(F\phi)$, $F(T\phi) \to T\phi$;

(20) $T(\phi \to \psi) \to (T\phi \to T\psi)$;

(21) $(T\phi \leftrightarrow T\psi) \& (F\phi \leftrightarrow F\psi) \to (\phi \leftrightarrow \psi)$

(22) $(T\phi \to T\psi) \to T(T\phi \to T\psi)$

Definition 3.3. There exist three reasoning rules in $P(S)$:

(1) modus ponens rule: ψ deduces From ϕ and $\phi \to \psi$;

(2) $T\phi$ deduces from ϕ;

(3) $F\phi \to F\psi$ deduces from $\phi \to \psi$.

The linguistic truth-valued intuitioinistic fuzzy propositional logic system is consists of $P(S)$, Axiom 1-22 and Reasoning rules 1-3, for short $\mathcal{LP}(S)$

Definition 3.4. A proof of $\mathcal{LP}(S)$ is a finite sequence $\phi_1, \phi_2, \cdots, \phi_n$ of formulae in $P(S)$. For every $\phi_i (i \le n)$ ϕ_i is an axiom in $\mathcal{LP}(S)$ or there exist $\phi_j (j < i)$ and $\phi_k (k < i)$ such that ϕ_i is a formula from ϕ_j and ϕ_k using Reasoning rule 1, 2 or 3. They are all called proofs of ϕ_n denote as $\vdash \phi_n$ and ϕ_n is called theorem.

Definition 3.5. Assume $\Gamma \subset P(S)$ $\phi \in P(S)$ A deduce from Γ to ϕ is a finite sequence $\phi_1, \phi_2, \cdots, \phi_n = \phi$ of $P(S)$. For every $\phi_i (i \le n)$ ϕ_i is an axiom of $\mathcal{LP}(S)$ or $\phi_i \in \Gamma$, or there exist $\phi_j (j < i)$ and $\phi_k (k < i)$ such that ϕ_i is a formula from ϕ_j and ϕ_k using Reasoning rule1, 2 or 3, it is called the formula ϕ can deduce from Γ, denote as $\Gamma \vdash \phi$.

Obviously, ϕ is a theorem if and only if $\varnothing \vdash \phi$.

Theorem 3.1. Syllogism rule: let $\Gamma = \{\phi \to \psi, \psi \to \chi\}$ then

$$\Gamma \vdash \phi \to \chi.$$

Theorem 3.2. For any $\phi, \psi \in P(S)$

(1) $\vdash \phi \to \phi$;

(2) $\vdash \phi \to (\psi \to \phi)$

(3) $\vdash (\phi \to (\psi \to \chi)) \to (\psi \to (\phi \to \chi))$

(4) $\vdash (\phi \to \psi) \to (\neg\psi \to \neg\phi)$

4. Conclusion

We construct a kind of linguistic truth-valued intuitionisic fuzzy propositional logic system. The linguistic truth-valued intuitionisitic fuzzy algebra comes from the linguistic truth-valued implication algebra which is fit to express both comparable and incomparable information. The method can deal with the uncertain problem which has both positive evidence and negative evidence at the same time. More properties will be discussed in the future work.

Acknowledgements

This work is partly supported by National Nature Science Foundation of China (Grant No. 61105059, 61175055, 61173100), Liaoning Excellent Talents in University(LJQ2011116), Sichuan Key Technology Research and Development Program under Grant No.2011FZ0051, Sichuan Key Laboratory of Intelligent Network Information Processing(SGXZD1002-10) and Key Laboratory of Radio Signals Intelligent Processing (Xihua University) (XZD0818-09).

References

1. K.T. Atanassov, Answer to D. Dubois, S. Gottwald, P. Hajek, J. Kacprzyk and H. Prade's paper "Terminological difficulties in fuzzy set theorythe case of `Intuitionistic Fuzzy Sets'", Fuzzy Sets and Systems, Vol.156, 496-499,2005.
2. G. Deschrijver, E.E. Kerre,On the position of intuitionistic fuzzy set theory in the framework of theories modelling imprecision,Information Sciences, Vol.177,1860-1866,2007.

3. M. Espinilla, J. Liu, L. Martinez, An extended hierarchical linguistic model for decision-making problems, Computatioinal Intelligence, Vol. 27(3),489-512,2011.
4. L. Martinez, D. Ruan, F. Herrera, Computing with Words in Decision support Systems: An overview on Models and Applications, International Journal of Computational Intelligence Systems, Vol.3(4), 382-395,2010.
5. Z. Pei, D. Ruan, J. Liu, Y. Xu, Linguistic Values based Intelligent Information Processing: Theory, Methods, and Application, Atlantis Computational Intelligence Systems-Vol. 1 (Atlantis press & World Scienti¯c, 2009.
6. Z. Pei, Y. Xu, D. Ruan, K. Qin, Extracting complex linguistic data summaries from personnel database via simple linguistic aggregations, Information Sciences. Vol.179, 2325-2332,2009.
7. Y. Xu, D. Ruan, E.E. Kerre and J. Liu, α-resolution principle based on first-order lattice-valued propositional logic $LF(X)$, *Information Sciences*, Vol.132, 195-223,2001.
8. Y. Xu, J. Liu, D. Ruan and T.T. Lee, On the consistency of rule bases based on lattice-valued ¯rst-order logic $LF(X)$, International Journal of Intelligent Systems, Vol.21, 399-424,2006.
9. Z.S. Xu, Intuitionistic fuzzy aggregation operators, IEEE Transactions on Fuzzy Systems, Vol.15(11), 1179{1187,2007.
10. L. Zou, D. Ruan, Z. Pei, and Y. Xu, A linguistic truth-valued reasoning approach in decision making with incomparable information, Journal of Intelligent and Fuzzy Systems, Vol.19(4-5), 335-343,2008.
11. L. Zou, Z. Pei, X. Liu, Y. Xu, Semantics of linguistic truth-valued intuitionistic fuzzy proposition calculus, International Journal of Innovative Computing, Information and Control,Vol.5(12), 4745-4752,2009. Vol.7(4), 387-397,2008.
12. L. Zou, D. Ruan, Z. Pei, Y. Xu, A Linguistic-Valued Lattice Implication Algebra Approach for Risk Analysis. Journal of Multi-Valued Logic and Soft Computing. Vol.17, 293-303,2011.

FUZZY MEMBERSHIP FUNCTION CONSTRUCTION BASED ON MULTI-VALUED EVALUATION

VASILE PATRASCU *

Department of Informatics Technology, Tarom
Bucharest-Otopeni, Romania

This paper presents a fuzzy set construction method by transforming a membership multi-valued representation into a scalar one. The method has two steps: firstly, a bifuzzy set is constructed using an aggregator operator from the family of averaging operators; secondly, the bifuzzy set obtained in the first step is transformed into a fuzzy one, preserving some indices, like the net truth or the proportional truth. Furthermore, the paper uses the method described above to create new formulae for entropy calculus of intuitionistic fuzzy sets and then, new formulae for color luminosity calculus.

1. Introduction

The problem of fuzzy membership function generation is of fundamental importance [1], [5], [12], [13], [8]. Numerous and various methods are proposed for the construction of values of membership function. These include the histogram method, transformation of probability distribution to possibility distribution, based on Zadeh's possibility theory [14], and methods based on fuzzy c-means algorithm capable of generating fuzzy membership function during the clustering process [4]. The neural network can also be utilized to generate membership functions [11]. Having many methods, we can obtain many variants for the membership function. Here, we will discuss a way of generation of the membership function in the context of a multiple evaluation.

2. Determining Membership Function using Multi-valued Evaluation

A fuzzy set A is defined by the membership function $w_A: E \to [0,1]$ where E is a fixed universe. Let us have k different methods $M_1, M_2, \dots, M_k$ of fuzzy membership estimation for the element $x \in E$. In [3] these methods are called estimators. Let the estimations are collected in the vector $V = (v_1, v_2, \dots, v_k)$. Using the traditional statistics we can define the value of membership function considering the mean of the V components, namely:

* patrascu.v@gmail.com.

$w_A = (v_1 + v_2 + \cdots + v_k)/k$. In the following, we will construct other calculus formulae for the membership function w_A. We consider the ideal vector for the belongingness, $T = (1,1,\ldots,1)$ and the ideal vector for the non-belongingness, $F = (0,0,\ldots,0)$. Now, using the *p-norm* we compute the following distances:

$$D_p(V,T) = \left(\frac{(1-v_1)^p + (1-v_2)^p \ldots + (1-v_k)^p}{k}\right)^{1/p} \tag{1}$$

$$D_p(V,F) = \left(\frac{(v_1)^p + (v_2)^p \ldots + (v_k)^p}{k}\right)^{1/p} \tag{2}$$

Having these two distances, we will define a bifuzzy set, thus:
The membership function:

$$\mu = D_p(V,F) \tag{3}$$

The non-membership function:

$$\nu = D_p(V,T) \tag{4}$$

For a bifuzzy set, we can consider *the net truth* defined by:

$$\tau = \mu - \nu \tag{5}$$

and the *proportional truth* defined by:

$$\rho = \frac{\mu}{\nu} \tag{6}$$

In [2], Atanassov proposed two methods for transforming a bifuzzy set to an intuitionistic fuzzy set. The first method preserves the proportional truth and the second preserves the net truth. Similarly, in this paper, we will transform the bifuzzy set defined by (3) and (4) into a fuzzy set. For a fuzzy set defined by the membership function w the formulae (5) and (6) become:

$$\tau = 2w - 1 \tag{7}$$

$$\rho = \frac{w}{1-w} \tag{8}$$

Preserving the *net truth*, from equations (5) and (7) it results:

$$w = \frac{1 + \mu - \nu}{2} \tag{9}$$

Preserving the *proportional truth*, from equations (6) and (8), it results:

$$w = \frac{\mu}{\mu + \nu} \tag{10}$$

The formulae (9) and (10) represent two variant for fuzzy membership function construction. In the following we will consider the well known three particular values for the parameter p: 1, 2 and ∞.

Case $p = 1$. From formulae (1) and (2) we obtain: $D_p(V,T) = 1 - v$, $D_p(V,F) = v$, where v represents the mean of V components. Then, from formulae (9) and (10) it results the same formula:

$$w = \frac{v_1 + v_2 + \cdots + v_k}{k} \tag{11}$$

Case $p = 2$. From formulae (1) and (2) we obtain: $D_p(V,T) = \sqrt{(1-v)^2 + \sigma^2}$ and $D_p(V,F) = \sqrt{v^2 + \sigma^2}$, where v represents the mean defined by (11) and σ^2 represents the variance, defined by:

$$\sigma^2 = \frac{(v_1 - v)^2 + (v_2 - v)^2 + \cdots + (v_k - v)^2}{k} \tag{12}$$

Then, from formulae (9) and (10) it results:

$$w = \frac{1 + \sqrt{v^2 + \sigma^2} - \sqrt{(1-v)^2 + \sigma^2}}{2} \tag{13}$$

$$w = \frac{\sqrt{v^2 + \sigma^2}}{\sqrt{v^2 + \sigma^2} + \sqrt{(1-v)^2 + \sigma^2}} \tag{14}$$

Case $p = \infty$. From formulae (1) and (2) we obtain: $D_p(V,T) = 1 - \min(v_1, v_2, \ldots, v_k)$ and $D_p(V,F) = \max(v_1, v_2, \ldots, v_k)$. Then, from formulae (9) and (10) it results:

$$w = \frac{M + m}{2} \tag{15}$$

$$w = \frac{M}{M + 1 - m} \tag{16}$$

where $m = \min(v_1, v_2, \ldots, v_k)$ and $M = \max(v_1, v_2, \ldots, v_k)$.

3. The Entropy of Intuitionistic Fuzzy Set

An intuitionistic fuzzy set is defined by the membership function μ and the non-membership function ν, where $\mu + \nu \leq 1$ [2]. In addition, it is defined the index of incompleteness $\pi = 1 - \mu - \nu$. We consider the vector $V = (\mu, 1 - \nu)$. Using one of the formulae (11), (13), (14), (15) and (16), we can transform the intuitionistic fuzzy set in to a fuzzy set. We obtain the following four variants:

from (11) and (15) it results the same formula:

$$w = \mu + \frac{\pi}{2} \tag{17}$$

from (13) it results:

$$w = \frac{1 + 0.5\left|\sqrt{(2\mu+\pi)^2+\pi^2} - \sqrt{(2\nu+\pi)^2+\pi^2}\right|}{2} \tag{18}$$

from (14) it results:

$$w = \frac{\sqrt{(2\mu+\pi)^2+\pi^2}}{\sqrt{(2\mu+\pi)^2+\pi^2} + \sqrt{(2\nu+\pi)^2+\pi^2}} \tag{19}$$

from (16) it results:

$$w = \frac{\mu+\pi}{1+\pi} \tag{20}$$

Using the entropy formula proposed by Kaufmann [6] for the membership functions defined by (18) and (19) one obtains the following formulae:

$$e = 1 - 0.5\left|\sqrt{(2\mu+\pi)^2+\pi^2} - \sqrt{(2\nu+\pi)^2+\pi^2}\right| \tag{21}$$

$$e = 1 - \frac{\left|\sqrt{(2\mu+\pi)^2+\pi^2} - \sqrt{(2\nu+\pi)^2+\pi^2}\right|}{\sqrt{(2\mu+\pi)^2+\pi^2} + \sqrt{(2\nu+\pi)^2+\pi^2}} \tag{22}$$

Using the entropy formula proposed by Kaufman [6] or Kosko [7] for the membership function defined by (20), one obtains the following entropy formulae:

$$e = 1 - \frac{|\mu-\nu|}{1+\pi} \tag{23}$$

$$e = \frac{1-|\mu-\nu|+\pi}{1+|\mu-\nu|+\pi} \tag{24}$$

Formula (24) was proposed by Szmidt and Kacprzyk [10] while (21), (22) and (23) represent new formulae for intuitionistic fuzzy entropy calculus.

4. The Color Luminosity

Considering the color $V = (r, g, b) \in [0,1]^3$, we define:
the maximum

$$M = max(r, g, b) \tag{25}$$

the minimum

$$m = min(r, g, b) \tag{26}$$

the mean

$$v = \frac{r + g + b}{3} \tag{27}$$

the saturation

$$s = \sqrt{\frac{(r - v)^2 + (g - v)^2 + (b - v)^2}{3}} \tag{28}$$

From formulae (11), (13), (14), (15) and (16) it results the following formulae for luminosity calculus:

$$L = \frac{r + g + b}{3} \tag{29}$$

$$L = \frac{1 + \sqrt{v^2 + s^2} - \sqrt{(1 - v)^2 + s^2}}{2} \tag{30}$$

$$L = \frac{\sqrt{v^2 + s^2}}{\sqrt{v^2 + s^2} + \sqrt{(1 - v)^2 + s^2}} \tag{31}$$

$$L = \frac{M + m}{2} \tag{32}$$

$$L = \frac{M}{M + 1 - m} \tag{33}$$

The formulae (29) and (32) represents two standard variants, the formula (33) was used in [9] for an HSL system definition in the framework of color clustering algorithms while (30) and (31) represent new formulae for color luminosity calculus.

5. Conclusion

Having a multi-valued representation of a fuzzy membership function, we constructed a transformation for a scalar representation. This transformation from a vector to a scalar uses an intermediary status: bifuzzy set. We have obtained five scalar variants. The first (11) uses the statistical mean, the second (13) and the third (14) use the mean and the standard deviation, while the forth (15) and the fifth (16) use the maximum and the minimum functions. The presented method was used for conversion of intuitionistic fuzzy set into a fuzzy one. Then, computing the entropy of the resulted fuzzy set, there were obtained some entropy formulae for intuitionistic fuzzy set.

The second application of the presented method is related to the color luminosity calculus.

For both cases, the intuitionistic fuzzy entropy and color luminosity, among of the obtained formulae, we discover some formulae that already exist in the specialty literature. This fact proves the effectiveness of our approach.

References

1. K. Atanassov, E. Szmidt and J. Kacprzyk, On some ways of determining membership and non-membership functions characterizing intuitionistic fuzzy sets, Sixth International Workshop on IFSs, Banska Bystrica, Slovakia, 10 Oct. 2010, *NIFS 16*, **4,** 26 (2010).
2. K.T.Atanassov, Remark on a Property of the Intuitionistic Fuzzy Interpretation Triangle, *Notes on Intuitionistic Fuzzy Sets* **8**, 34 (2002).
3. J. Buckley, Fuzzy Statistics, Springer, Berlin, (2004).
4. J. C. Bezdek, Pattern recognition with Fuzzy Objective Function Algorithms. Plenum Press, New York, (1981).
5. A. Kaufmann, Introduction a la theorie des soub-ensembles flous, Paris, Masson, (1977).
6. A. Kaufmann, Introduction to the Theory of Fuzzy Subsets, academic Press, New York (1975).
7. B. Kosko, Fuzzy entropy and conditioning, *Information sciences* **40**, 165 (1986).
8. S. Medasani, J. Kim and R. Krishnapuram, An overview of membership function generation techniques for pattern recognition, *International Journal Approx. Reason*, **19**, 391 (1998).
9. V. Pătraşcu, New fuzzy color clustering algorithm based on hsl similarity, *Proceedings of the International Fuzzy Systems Association World Congress* (IFSA 2009), Lisbon, Portugal, (2009).
10. E. Szmidt and J. Kacprzyk, A new measure of entropy and its connection with a similarity measure for intuitionistic fuzzy sets, *Proceedings of the EUSFLAT-LFA 2005 Conference*, Barcelona, Spain, 461 (2005).
11. C. C. Yang and N. K. Bose, Generating fuzzy membership function with self-organizing feature map, Elsevier Press, *Pattern Recognition Letters* **27**, 356 (2006).
12. L. Zadeh, Fuzzy sets, *Information and Control, Vol.* **8**, 338 (1965).
13. L. Zadeh, The concept of a linguistic variable and its application to approximate reasoning, American Elsevier Publ. Co. New York, (1973).
14. L. Zadeh, Fuzzy sets as a basis for a theory of possibility, *Fuzzy Sets System* **1**, 3 (1978).

INTERVAL IDENTIFICATION OF UNCERTAIN NONLINEAR SYSTEMS USING BELIEF RULE-BASED SYSTEMS *

YU-WANG CHEN, DONG-LING XU†, JIAN-BO YANG

Decision and Cognitive Sciences Research Centre, MBS, The University of Manchester Manchester M15 6PB, United Kingdom

This paper combines the methodology of belief rule-based (BRB) systems with nonlinear min-max optimisation techniques to develop a novel interval identification method, in which the outputs of an uncertain nonlinear system are represented by belief structures. On a finite set of measured data, the l_∞-norm is used as the optimisation criterion of minimising the identification errors. This identification method is capable of modelling the optimal lower and upper bounds simultaneously, which can be used to approximate the interval output of an uncertain nonlinear system. The proposed method is tested with a family of uncertain nonlinear functions, and numerical results validate the functionality of the proposed identification method for uncertain nonlinear systems.

1. Introduction

Identification of uncertain nonlinear systems is concerned with characterising an unknown nonlinear system on the basis of measured input-output data in an uncertain environment. It is of fundamental importance in control, fault diagnosis, signal processing and information sciences since most real-life systems are nonlinear, and are often associated with uncertainties due to uncertain physical parameters, unpredictable disturbance and measurement noise, unexpected variation in external surrounding, or incomplete knowledge [1][3][5]. Over past few decades, a large number of studies have been conducted for effectively identifying uncertain nonlinear systems, especially with the advent of neural network and fuzzy rule-based system techniques [4]. For nonlinear systems with parametric uncertainty, Ackermann [1] extensively studied various analytical state space and transfer function modelling techniques. On the basis of fuzzy identification methodology, Škrjanc et al. [7][8] presented an interval fuzzy model (INFUMO) to model a class of

* This work forms part of the project supported by the UK Engineering and Physical Science Research Council under Grant No.: EP/F024606/1.

† Ling.Xu@mbs.ac.uk.

nonlinear systems with interval parameters. It results in a lower and upper fuzzy model or a fuzzy model with lower and upper parameters. Linear programming techniques are used to find the set of lower and upper parameters using the l_∞-norm as the optimality criterion. The idea behind the INFUMO is to find optimal lower and upper bound fuzzy systems that define a confidence band which encloses all the measured input-output data [9]. However, in INFUMO the lower and upper bound fuzzy models are independent, and they may generate invalid lower and upper bounds if no measured data are available in some regions of input space [6][8].

In this paper, we combine the methodology of belief rule-based (BRB) systems with nonlinear min-max optimisation techniques, and propose a novel interval identification method, in which the outputs of uncertain nonlinear systems are represented with belief structures [12][13]. The rest of the paper is organised as follows. In Section 2 nonlinear system modelling with belief rules is introduced. In Section 3 the new belief rule-based system identification method is investigated by using l_∞-norm as the optimality criterion. Section 4 shows a numerical example on a family of uncertain nonlinear functions. Finally, this paper is concluded in Section 5.

2. Uncertain nonlinear system modelling with belief rules

A typical BRB system is given in the form of belief rules as follows.

$$\begin{aligned}
&\text{IF } x_1 \text{ is } A_1^k \wedge x_2 \text{ is } A_2^k \wedge \cdots \wedge x_{M_k} \text{ is } A_{M_k}^k, \\
R_k:\ &\text{THEN } \{(D_1,\beta_{1,k}),(D_2,\beta_{2,k}),\ldots,(D_N,\beta_{N,k})\}, \left(\sum_{n=1}^{N}\beta_{n,k} \le 1\right), \\
&\text{with rule weight } \theta_k, \text{ and weight of varibles } \delta_{1,k},\delta_{2,k},\ldots,\delta_{M_k,k}, k\in\{1,\ldots,K\}.
\end{aligned} \quad (1)$$

where $x_1, x_2,\ldots, x_{M_k}$ denote the antecedent variable in the kth rule, and these variables belong to the whole set of input variables $\boldsymbol{x} = \{x_i; i = 1,\ldots, M\}$. $A_i^k\ (i = 1,\ldots, M_k)$ is the referential value taken by the ith antecedent variable in the kth rule and $A_i^k \in A_i$. $A_i = \{A_{i,j}; j = 1,\ldots, J_i\}$ denotes the set of referential values for the ith antecedent variable and J_i is the number of the referential values. $\beta_{n,k}\ (n = 1,\ldots, N; k = 1,\ldots, K)$ represents the belief degree to which the consequent element D_n is believed to be the consequent, given the logical relationship of the kth rule IF x_1 is $A_1^k \wedge x_2$ is $A_2^k \wedge \cdots \wedge x_{M_k}$ is $A_{M_k}^k$. Note that the element D_n in the set of consequent elements $D = \{D_n; n = 1,\ldots, N\}$ can either be a conclusion or an action and a subset of elements can also be part of the consequent [11]. If

$\sum_{n=1}^{N}\beta_{n,k}=1$, the kth rule is said to be complete; otherwise, it is incomplete. θ_k is the relative weight of the kth rule, and $\delta_{M_k,k}$ represents the relative weight of variables in the kth rule.

Under the unified scheme as defined above, a belief rule can represent a functional mapping between antecedent inputs and output possibly with uncertainties. Once a belief rule base is established, the knowledge embedded in all belief rules can be used to perform inference for a specific input vector. Given an input vector $x=\{x_i, i=1,...,M\}$, an utility interval can be established for representing the interval output of uncertain nonlinear system, which is characterised by the maximum, minimum, and average utilities of $S(f(x))$ defined as follows, provided that $u(D_{n+1})\geq u(D_n)$:

$$f_{\max}(\mathbf{x})=\sum_{n=1}^{N-1}\beta_n(\mathbf{x})u(D_n)+(\beta_N(\mathbf{x})+\beta_{\mathbf{D}}(\mathbf{x}))u(D_N) \tag{2a}$$

$$f_{\min}(\mathbf{x})=(\beta_1(\mathbf{x})+\beta_{\mathbf{D}}(\mathbf{x}))u(D_1)+\sum_{n=2}^{N}\beta_n(\mathbf{x})u(D_n) \tag{2b}$$

where the belief degrees on the inference output can be generated through the aggregation of all activated belief rules using the analytical evidential reasoning (ER) approach [10]. It is evident that $f(\mathbf{x})=f_{\max}(\mathbf{x})=f_{\min}(\mathbf{x})=f_{aver}(\mathbf{x})=\dfrac{f_{\max}(\mathbf{x})+f_{\min}(\mathbf{x})}{2}$ if $\beta_D(x)=0$.

3. Belief rule based-system identification using l_∞-norm

In this section, we describe how BRB systems can be used to identify nonlinear systems with uncertain physical models. The l_∞-norm is used as the optimality criterion of minimising the identification errors. Suppose that a finite set $\boldsymbol{X}=\{\boldsymbol{x}_1,\boldsymbol{x}_2,...,\boldsymbol{x}_T\}$ can be constructed to represent the measured input data, which are collected from a compact set $\boldsymbol{U}\subset\boldsymbol{R}^m$. A set of corresponding outputs is also defined as $\boldsymbol{Y}=\{y_1,y_2,...,y_T\}$, which can be represented as the result of mapping the input data on a member of a family of nonlinear functions [7][8].

$$\mathbf{G}=\{g:\mathbf{U}\rightarrow\mathbf{R}\,|\,g(\mathbf{x})=g_{\text{norm}}(\mathbf{x})+\Delta g(\mathbf{x})\} \tag{3}$$

where $g_{\text{norm}}(\boldsymbol{x})$ denotes the nominal function and the uncertainty $\Delta g(\boldsymbol{x})$ satisfies $\sup_{x\in U}|\Delta g(\boldsymbol{x})|\leq c, c\in\boldsymbol{R}$. If we consider a nonlinear function $g(\boldsymbol{x})$ that is a member of the class $g\in\mathbf{G}$, the corresponding set of measured outputs over the set of input vectors X can also be represented as $y_t=g(\mathbf{x}_t), g\in\mathbf{G}, \mathbf{x}_t\in\mathbf{X}, t=1,...,T$. The idea of the BRB system identification method for nonlinear systems with uncertain physical models is to find a BRB system satisfying

$$f_{\min}(\boldsymbol{x}) \le g(\boldsymbol{x}) \le f_{\max}(\boldsymbol{x}) \tag{4}$$

This implies that the outputs of an uncertain nonlinear function from the class $\boldsymbol{G}$ can always be found in the utility interval defined in Eqs. (2a) and (2b). The main aim of designing a BRB system is to generate the utility interval on the whole input space $\boldsymbol{U}$ as narrow as possible in accordance with relevant constraints. This can be realised by solving the following *min-max* optimisation problem.

$$\min_{P} \max_{x_t \in X} \left(\left| y_t - f_{\min}(\boldsymbol{x}_t) \right|, \left| f_{\max}(\boldsymbol{x}_t) - y_t \right| \right) \tag{5}$$

subject to

$$y_t - f_{\min}(\boldsymbol{x}_t) \ge 0, \quad t = 1,...,T \tag{5a}$$

$$y_t - f_{\max}(\boldsymbol{x}_t) \le 0, \quad t = 1,...,T \tag{5b}$$

Constraints of BRB system parameters (5c)

The *min-max* optimization problem (5) can be solved as the nonlinear programming problem of minimizing λ subject to the linear and nonlinear constraints (5a)-(5c) and the following inequalities.

$$y_t - (\beta_1(\mathbf{x}_t) + \beta_{\mathbf{D}}(\mathbf{x}_t))u(D_1) - \sum_{n=2}^{N} \beta_n(\mathbf{x}_t)u(D_n) \le \lambda, \quad t = 1,...,T \tag{6a}$$

$$-y_t + \sum_{n=1}^{N-1} \beta_n(\mathbf{x}_t)u(D_n) + (\beta_N(\mathbf{x}_t) + \beta_{\mathbf{D}}(\mathbf{x}_t))u(D_N) \le \lambda, \quad t = 1,...,T \tag{6b}$$

$$\lambda \ge 0 \tag{6c}$$

4. A numerical example

There is a class of $\boldsymbol{G}$ with the nominal function $g_{\text{norm}}(x) = \cos(x)\sin(x)$ and the uncertainty $\Delta g(x) = \gamma \cos(8x), 0 \le \gamma \le 0.2$. The function from the class is defined in the domain $U = \{x | -1 \le x \le 1\}$ and the set of measured input set is $\boldsymbol{X} = \{x_t | x_t = 0.021t, t = -47, -46, ..., 47\} \subset \boldsymbol{U}$ [7].

In this example, the dimensionality of the input space is $m = 1$. To construct a BRB system, 5 initial referential values for the input variable x are uniformly selected from the interval [-1, 1], and they are {-1, -0.5, 0, 0.5, 1}. The outputs of the nonlinear function with the parameter γ = 0, 0.05, 0.1, 0.15, and 0.2 can be calculated with the analytical model. Through observing the measured outputs on these referential inputs, the following referential utilities can be defined for measuring the consequent elements of belief rules.

$$\{D_1, D_2, D_3, D_4, D_5\} = \{-0.8, -0.4, 0, 0.4, 0.8\}$$

Correspondingly a set of initial belief rules can be constructed in Table 1.

Table 1. Initial BRB system

R_k	θ_k	x	Consequents $\{D_1, D_2, D_3, D_4, D_5\}$={-0.8, -0.4, 0, 0.4, 0.8}
1	1	-1	$\{(D_1, 0.2), (D_2, 0.7), (D_3, 0), (D_4, 0), (D_5, 0), (\boldsymbol{D}, 0.1)\}$
2	1	-0.5	$\{(D_1, 0.3), (D_2, 0.5), (D_3, 0), (D_4, 0), (D_5, 0), (\boldsymbol{D}, 0.2)\}$
3	1	0	$\{(D_1, 0), (D_2, 0), (D_3, 0.5), (D_4, 0.4), (D_5, 0), (\boldsymbol{D}, 0.1)\}$
4	1	0.5	$\{(D_1, 0), (D_2, 0), (D_3, 0.1), (D_4, 0.6), (D_5, 0.1), (\boldsymbol{D}, 0.2)\}$
5	1	1	$\{(D_1, 0), (D_2, 0), (D_3, 0), (D_4, 0.8), (D_5, 0.1), (\boldsymbol{D}, 0.1)\}$

In Table 1, the sum of the belief degree on the individual consequent element D_n is less than 1, and the belief degree on the whole set of consequent elements $\boldsymbol{D}$ is not equal to 0 as usual. This implies that the output of the nonlinear system represented by the belief structures is uncertain. To obtain a utility interval to enclose all the uncertain outputs of the nonlinear system, the measured input-output data set above can be used to train the initial belief rules. The optimisation model of combining a BRB system with the optimisation criterion of using l_∞-norm is discussed in Section 3. Using the nonlinear optimization solver *fmincon* in the Optimization Toolbox of Matlab, the simulation results are presented in Fig. 1.

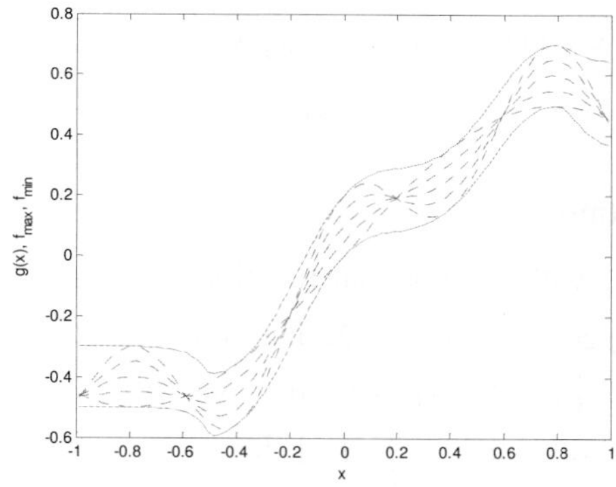

Figure 1. Data, lower and upper bounds for the BRB system identification model

In Fig. 2, the dotted lines are the outputs of the nonlinear function with the above 5 different parameters, and the solid lines are the lower and upper bounds of the uncertain outputs respectively, which are generated by the trained BRB system identification model. It is obvious from Fig. 2 that the trained BRB system can generate very tight boundaries around the family of the nonlinear functions. In the trained BRB system, the referential values of the input variable

x are updated to be {-1.0000, -0.4521, 0.3202, 0.7118, 1.0000}. Table 2 lists the updated parameters in the trained belief rules.

Table 2. Trained BRB system

R_k	θ_k	x	Consequents $\{D_1, D_2, D_3, D_4, D_5\}$={-0.800, -0.550, 0.165, 0.439, 0.800}
1	1.000	-1.000	$\{(D_1, 0.000), (D_2, 0.811), (D_3, 0.000), (D_4, 0.000), (D_5, 0.061), (D, 0.127)\}$
2	0.250	-0.452	$\{(D_1, 0.735), (D_2, 0.000), (D_3, 0.000), (D_4, 0.000), (D_5, 0.137), (D, 0.127)\}$
3	0.425	0.320	$\{(D_1, 0.000), (D_2, 0.000), (D_3, 0.575), (D_4, 0.295), (D_5, 0.000), (D, 0.131)\}$
4	0.461	0.712	$\{(D_1, 0.000), (D_2, 0.002), (D_3, 0.164), (D_4, 0.000), (D_5, 0.690), (D, 0.144)\}$
5	0.629	1.000	$\{(D_1, 0.059), (D_2, 0.041), (D_3, 0.000), (D_4, 0.004), (D_5, 0.707), (D, 0.189)\}$

In this example, the identification results show that the BRB system is appropriate and effective in the modelling of the uncertainties for the family of nonlinear functions. Even better approximation performance can be obtained if we increase the number of referential values for the input variable according to the universal approximation property of BRB systems [2].

5. Conclusion

In this paper, we combined the methodology of belief rule-based (BRB) systems with nonlinear min-max optimisation techniques, and proposed a novel uncertain nonlinear system identification method, in which the outputs of uncertain nonlinear systems are represented by the belief structures. The l_∞-norm is used to measure the optimality of the identification model on a finite set of measured data. The proposed method was tested with a family of uncertain nonlinear functions, and numerical results validated the functionality of the proposed BRB system identification method and showed that it has great potential in system identification in uncertain environments. Further work will explore effective ways of constructing initial belief rules and identifying optimal parameters.

References

1. Ackermann J., 1993. Robust control: systems with uncertain physical parameters. Springer-Verlag, Berlin Heidelberg.
2. Chen Y.W., Yang J.B., Xu D.L., 2011. On the inference and approximation properties of belief rule based systems. Information Sciences (under review).
3. Isermann R., Münchhof M., 2011. Identification of dynamical systems: an introduction with applications. Springer-Verlag, Berlin Heidelberg.

4. Nelles O., 2001. Nonlinear system identification: from classical approaches to neural networks and fuzzy models. Springer-Verlag, Berlin Heidelberg.
5. Reinelt W., Ljung L., 2003. Robust control of identified models with mixed parametric and non-parametric uncertainties. European Journal of Control 9(4), 373–380.
6. Senthilkumar D., Mahanta C., 2010. Identification of uncertain nonlinear systems for robust fuzzy control. ISA Transactions 49, 27–38.
7. Škrjanc I., Blažič S., Agamennoni O., 2005a. Interval fuzzy model identification using l∞-norm. IEEE Transactions on Fuzzy Systems 13(5), 561–567.
8. Škrjanc I., Blažič S., Agamennoni O., 2005b. Identification of dynamical systems with a robust interval fuzzy model. Automatica 41, 327–332.
9. Škrjanc I., 2009. Confidence interval of fuzzy models: an example using a waste-water treatment plant. Chemometrics and Intelligent Laboratory Systems 96, 182–187.
10. Wang Y.M., Yang J.B., Xu D.L., 2006. Environmental impact assessment using the evidential reasoning approach. European Journal of Operational Research 174(3), 1885–1913.
11. Xu D.L., Yang J.B., Wang Y.M., 2006. The ER approach for multi-attribute decision analysis under interval uncertainties. European Journal of Operational Research 174(3), 1914–1943.
12. Yang J.B., Liu J., Wang J., Sii H.S., Wang H.W., 2006. A belief rule-base inference methodology using the evidential reasoning approach – RIMER. IEEE Transactions on Systems, Man, and Cybernetics – Part A 36(2), 266–285.
13. Yang J.B., Liu J., Xu D.L., Wang J., Wang H.W., 2007. Optimization models for training belief rule based systems. IEEE Transactions on Systems, Man, and Cybernetics – Part A 37(4), 569–585.

CONSTRUCTION OF STRONG EQUALITY INDEX FROM IMPLICATION OPERATORS

H. BUSTINCE*, J. FERNANDEZ, J. A. SANZ, D. PATERNAIN

Departmento de Automatica y Computacion, Universidad Publica de Navarra, Pamplona, 31006, Spain

M. BACZYSKI

Institute of Mathematics, University of Silesia, ul. Bankowa 14, Katowice, 40-007, Poland

G. BELIAKOV

School of Information Technology, Deakin University, Burwood 3125, Australia

R. MESIAR

Slovak University of Technology, Radlinskeho 11, Bratislava, Slovakia, Centre of Excellence IT4Innovations, Division University of Ostrava, IRAFM, Ostrava, 70103, Czech Republic

We present the concept of strong equality index, starting from the definition of strong inclusion given by Dubois and Prade in 1980. We also present a construction method based on the use of implication operators and two specific properties of the implications.

1. Introduction

It is a well-known fact that inclusion of a fuzzy set A in another fuzzy set B can be defined basically in two ways:

(A) The most usual one, proposed by Zadeh in 1965[1]: $A \leqslant B$ if and only if $\mu_A(x_i) \leqslant \mu_B(x_i)$ for all $x_i \in X$.

(B) A stronger one, proposed by Dubois and Prade in 1980[2], which considers that all elements belonging to A must be prototypes of B; that is,A S B if and only if $\mu_A(x_i) = 0$ or $\mu_B(x_i) = 1$ for all $x_i \in X$.

* E-mail: bustince@unavarra.es

In this work, based on these ideas, we propose a definition of a "strong equality function" that allows to build strong inclusions. Following the ideas of Bandler and Kohout [3], B. De Baets [4] and others [5, 6], we characterize this strong equality functions in terms of implications satisfying appropriate conditions.

The structure of this paper is the following. We start with some Preliminaries. In Section 3 we introduce the concept of strong equality function and how we can characterize it in terms of implications. We finish with some conclusions and references.

2. Preliminaries

We assume the reader is familiar with the concepts of negation, t-norm and t-conorm. Regarding the concept of implication, although also a very well-known, we prefer to review it here briefly.

Definition 2.1. [7]: A fuzzy implication is a function $I : [0, 1]^2 \rightarrow [0, 1]$ that satisfies the following properties:

i. The first place antitonicity
$\forall x1, x2, y \in [0, 1]\ (x1 \leqslant x2 \Rightarrow I(x1, y) \geqslant I(x2, y))$, (FA)

ii. The second place isotonicity
$\forall x, y1, y2 \in [0, 1]\ (y1 \leqslant y2 \Rightarrow I(x, y1) \leqslant I(x, y2))$, (SI)

iii. The dominance of falsity of antecedent
$I(0, y) = 1, y \in [0, 1]$, (DF)

iv. The dominance of truth of consequent
$I(x, 1) = 1, x \in [0, 1]$, (DT)

v. The boundary condition
$I(1, 0) = 0$. (BC)

In this work, we will always use the word implication to mean fuzzy implication operators in the sense of Definition 2.1. In other case, we will only speak of mappings $I : [0, 1]^2 \rightarrow [0, 1]$.

3. Strong equality functions

The condition $A \subseteq_S B$ and $B \subseteq_S A$ if and only if $\mu_A(x_i) = \mu_B(x_i) \in \{0, 1\}$, for all $x_i \in X$, has suggested us the following definition, which we will use later to construct strong equality indexes between fuzzy sets.

Definition 3: A function ψ: $[0, 1]^2 \rightarrow [0, 1]$ is called a strong equality function if it satisfies the following conditions:

i. $\psi(x, y) = 1$ if and only if $x = y \in \{0, 1\}$;

ii. $\psi(x, y) = 0$ if and only if $(x, y) \in \{(0, 1), (1, 0)\}$;

iii. ψ(x, y) = (y, x) for all x, y ∈ [0, 1].

Example 3.1: The function ψ (x, y) = max(xy, (1 – x)(1 – y)) is a strong equality function.

Our key result is the following. In order to present it we present the following two properties:

$$I(x, y) = 1 \text{ if and only if } x = 0 \text{ or } y = 1; (1)$$

$$I(x, y) = 0, \text{ if and only if } x = 1 \text{ and } y = 0. (2)$$

Theorem 3.1: For a function ψ : [0, 1]2 → [0, 1] the following statements are equivalent:

i. ψ is a strong equality function

ii. There exists a function I : [0, 1]2 → [0, 1] which satisfies (1), (2) and such that

$$\psi (x, y) = \min(I(x, y), I(y, x)), x, y \in [0, 1]. (3)$$

Proof: (ii)=>(i) We will show that defined by (3) satisfies all axioms from Definition 3.1. Axiom (i). Observe that

$$\psi (x, y) = 1 \Longleftrightarrow \min(I(x, y), I(y, x)) = 1 \Longleftrightarrow I(x, y) = I(y, x) = 1$$

$$\Longleftrightarrow (x = 0 \text{ or } y = 1) \text{ and } (x = 1 \text{ or } y = 0)$$

$$\Longleftrightarrow (x = 0 \text{ and } y = 0) \text{ or } (x = 1 \text{ and } y = 1)$$

$$\Longleftrightarrow x = y \in \{0, 1\}.$$

Axiom (ii). Observe that

$$\psi (x, y) = 0 \Longleftrightarrow \min(I(x, y), I(y, x)) = 0 \Longleftrightarrow I(x, y) = 0 \text{ or } I(y, x) = 0$$

$$\Longleftrightarrow (x = 1 \text{ and } y = 0) \text{ or } (x = 0 \text{ and } y = 1)$$

$$\Longleftrightarrow (x, y) \in \{(0, 1), (1, 0)\}.$$

Axiom (iii) is obvious.

(i) ⟹ (ii)

Suppose that ψ is a strong equality function and define a binary operation I by

I(x, y) = 0, if x = 1 and y = 0;

I(x, y) = 1, if x = 0 or y = 1;

I(x, y) = ψ (x, y), otherwise.

We will show that such defined I satisfies axioms (1) and (2).

Axiom (1). Suppose that I(x, y) = 1. There are two possibilities:

- x = 0 or y = 1, and the result will be proved.

- ψ (x, y) = 1, then by the property (i) in Definition 3.1 we get x = y $\in\{0, 1\}$, so x = y = 0 or x = y = 1; a contradiction.

The opposite implication in (1) is obvious.

Axiom (2). Suppose that I(x, y) = 0. Then there again are two possibilities:

- x = 1 and y = 0, and the result will be proved.
- ψ (x, y) = 0, then by the property (ii) in Definition 3.1 we get (x, y) $\in\{(0, 1), (1, 0)\}$; a contradiction.

The opposite implication in (2) is obvious.

Finally, we must show that ψ is given by (3). Notice that the values of I(x, y) for x, y $\neq \{0, 1\}$ are irrelevant. In case x, y $\in \{0, 1\}$, there are four possibilities:

a) x = 0 and y = 1. Clearly, we have min(I(0, 1), I(1, 0)) = 0 = ψ (0, 1);
b) x = 1 and y = 0. Then min(I(1, 0), I(0, 1)) = 0 = ψ (1, 0);
c) x = y = 0. Then min(I(0, 0), I(0, 0)) = I(0, 0) = 1 = ψ (0, 0);
d) x = y = 1. Then min(I(1, 1), I(1, 1)) = I(1, 1) = 1 = ψ (1, 1).

Observe that Theorem 3.1 states that each strong equality function is, in fact, a special biimplication operator. Also notice that in the case of residual implications linked to additively generated continuous t-norms (continuous Archimedean t-norms) with an additive generator f : [0, 1] $\rightarrow[0,\infty]$ (with f continuous, strictly decreasing and f(1) = 0, and T(x, y) = f^{-1}(min(f(0), f(x) + f(y)))), the residual implication is given by

$$I(x, y) = f^{-1}(\max(0, f(y) - f(x))),\ x, y \in [0, 1],$$

and the corresponding biimplication is then BI(x, y) = $f^{-1}(|f(x)-f(y)|)$, i.e., it is an isomorphic transformation of the classical distance of real numbers from [0, 1]. But residual implications do not satisfy (1) and thus BI in this case is not a strong equality function.

Example 3.2: Consider Kleene-Dienes implication operator, which is given by I_{KD}(x, y) = max(1 - x, y). This operator satisfies properties (1) and (2). By Theorem 3.1 we get that

$$\psi (x, y) = \min(\max(1 - x, y),\max(1 - y, x)),\ x, y \in [0, 1],$$

is a strong equality function.

Observe that Theorem 3.1 can be extended as follows.

Theorem 3.2: For a function $\psi : [0, 1]^2 \rightarrow [0, 1]$ the following statements are equivalent:

i. ψ is a strong equality function
ii. There exist two mappings

- $M: [0, 1]^2 \rightarrow [0, 1]$ such that M is symmetric, $M(x, y) = 1$ if and only if $x = y = 1$, and $M(x, y) = 0$ if and only if $x = 0$ or $y = 0$, and
- $I : [0, 1]^2 \rightarrow [0, 1]$ which satisfies properties (1) and (2) such that

$$\psi(x, y) = M(I(x, y), I(y, x)), x, y \in [0, 1].$$

The proof of this result is as that of Theorem 3.1. Notice that any aggregation function without divisors of zero and without multipliers of one and such that $M(1, 0) = M(0, 1) = 0$ can be used as M in Theorem 3.2; in particular, any t-norm T with no zero divisors can be used as M.

Of course these results make the following question arise:

Which conditions on ensure that the mapping I in Theorem 3.1 (or in Theorem 3.2) is in fact an implication operator in the sense of Definition 2.1?

We can state the following:

Theorem 3.3. Let I be a fuzzy implication which satisfies properties (1) and (2). Then the mapping

$$\psi(x, y) = I(\max(x, y),\min(x, y)), x, y \in [0, 1], (4)$$

is a strong equality function.

Proof. Just observe that, if I is a fuzzy implication and if we define ψ by (4), due to the monotonicity properties of I, we have that

$$I(\max(x, y),\min(x, y)) \leq I(\min(x, y),\max(x, y)), x, y \in [0, 1],$$

so in fact, we have $\psi(x, y) = I(\max(x, y),\min(x, y)) = \min(I(x, y), I(y, x))$.

By Theorem 3.1 we get the proof.

Theorem 3.1 leads us to study under which conditions a mapping $I : [0, 1]^2 \rightarrow [0, 1]$ satisfies properties (1) and (2).

4. Conclusions

We have introduced the concept of strong equality function and we have shown how we can characterize these functions by means of functions that satisfy appropriate conditions. We intend to develop the study of such conditions in future works.

Acknowledgments

The first four authors were supported by project TIN2010-15055 of the Spanish Government. R. Mesiar was supported by grant VEGA 1/0171/12 and the project IT4Innovations Centre of Excellence, reg. no. CZ.1.05/1.1.00/02.0070 supported by Research and Development for Innovations Operational Programme financed by Structural Founds of Europe Union and from the means of state budget of the Czech Republic. M Baczynski was supported by the Polish Ministry of Sciences and Higher Education Grant Nr N519 384936.

References

1. L. A. Zadeh, Fuzzy sets, Inform. Control 8, 338 (1965).
2. D. Dubois and H. Prade, Fuzzy Sets and Systems: Theory and Applications (Academic Press, New York, 1980).
3. W. Bandler and L. Kohout, Fuzzy power sets and fuzzy implication operators, Fuzzy Sets Syst. 4, 13 (1980).
4. B. D. Baets and E. Kerre, Fuzzy inclusion and the inverse problem,, in Proc. 2nd European Congress on Intelligent Techniques and Soft Computing EU-FIT'94 , (Aachen, Germany, 1995).
5. H. Bustince, V. Mohedano, E. Barrenechea and M. Pagola, Definition and construction of fuzzy DI-subsethood measures, Inform.Sci. 176, 3190 (2006).
6. H. Bustince, M. Pagola and E. Barrenechea, Construction of fuzzy indices form DI-subsethood measures: Application to the global comparison of images,Inform.Sci. 177, 906 (2007).
7. J. Fodor and M. Roubens, Fuzzy Preference Modelling and Multicriteria Decision Support (Kluwer Academic Publishers, Dordrecht, 1994).

ON DERIVATIONS OF LATTICE IMPLICATION ALGEBRAS*

HUA ZHU, JIANBIN ZHAO†

Department of Mathematics, Zhengzhou University, Zhengzhou 450001, P.R. China

WEITAO XU

College of Information Science and Engineering, Henan University of Technology, Zhengzhou 450001, P.R. China

YANG XU

Intelligent Control Center, Southwest Jiaotong University, Chengdu 610031, China

The notion of derivation in a lattice implication algebra is introduced, and some related properties are discussed. The properties of the isotone derivations are mainly researched. Using the concept of derivations as defined, a class of lattice implication algebras is constructed.

1. Introduction

Lattice implication algebra which plays a significant role in lattice value logic was introduced by Xu Yang firstly[12]. Several researchers have investigated this logic algebra, and many beautiful results have been obtained [4-6,8-9,13-14]. The notion of derivation which came from the analytic theory, is useful to research the structure and property of algebraic system. Several authors have studied derivations in rings and near-rings[1-2]. In[7], Jun and Xin introduced the notion of derivation in BCI-algebras and investigated some of its properties. In [11], Xin and Li applied the notion of derivation to lattices and described the structure of lattice by using the properties of derivation. In [3], Li introduced the notion of derivation in λ-lattice and obtained some important results. In this paper, we apply the concept of derivation to lattice implication algebras and discuss some of its properties. Especially we research the properties of the isotone derivations in lattice implication algebras. We construct a class of lattice implication algebras by using the notion of derivations as defined.

---

*This work is supported by the Natural Science Foundation of China (Grant No.61175055).

† zhaojianbin@zzu.edu.cn.

2. Preliminaries

Definition 2.1.[12] By a lattice implication algebra we mean a bounded lattice $(L,\vee,\wedge,0,1)$ with order-reversing involution $'$, I and O the greatest and the smallest element of L respectively, and a binary operation $\to$ satisfying the follow axioms: for all $x, y, z\in L$,

(I_1) $x\to(y\to z)=y\to(x\to z)$;

(I_2) $x\to x=I$;

(I_3) $x\to y=y'\to x'$;

(I_4) if $x\to y=y\to x=1$, then $x=y$;

(I_5) $(x\to y)\to y=(y\to x)\to x$;

(L_1) $(x\vee y)\to z=(x\to z)\wedge(y\to z)$;

(L_2) $(x\wedge y)\to z=(x\to z)\vee(y\to z)$.

If a lattice implication algebra L satisfies

$$\forall x, y, z\in L, x\vee y\vee((x\wedge y)\to z)=1 ,$$

then we call it a lattice H implication algebra.

A mapping $f: L_1\to L_2$ from lattice implication algebras L_1 to L_2 is called a lattice implication homomorphism if it satisfies: $\forall x, y\in L_1$

$$f(x\to y)=f(x)\to f(y), f(x\vee y)=f(x)\vee f(y) ,$$

$$f(x\wedge y)=f(x)\wedge f(y), f(x')=(f(x))' .$$

A one-to-one and onto lattice implication homomorphism is called a lattice implication isomorphism.

For more details of lattice implication algebras, we refer readers to Reference[12].

3. The derivations in lattice implication algebras

Definition 3.1. Let L be a lattice implication algebra and $d: L\to L$ be a function on L. If $\forall x, y\in L, d(x\wedge y)=(dx\wedge y)\vee(x\wedge dy)$, then d is called a derivation on L.

Example 3.2. Let L be a lattice implication algebra, and define a function d on L by any $x\in L, dx=x$. Then d is a derivation on L, which is called the identity derivation.

Proposition 3.3. Let d be a derivation on L, the following hold:

(1) $dx\le x$, $d0=0$ and $d1\le 1$;

(2) $dx\wedge dy\le d(x\wedge y)\le dx\vee dy$;

(3) $dx=(x\wedge d1)\vee dx$;

(4) $dx = (x \wedge d(x \vee y)) \vee dx$.

Proof. (1)By definition 3.1, then

$$dx = d(x \wedge x) = (dx \wedge x) \vee (x \wedge dx) = x \wedge dx,$$

so we can get $dx \leq x$. Clearly $d0 = 0$, $d1 \leq 1$.

(2) By definition 3.1and (1), then

$$d(x \wedge y) = (dx \wedge y) \vee (x \wedge dy) \leq dx \vee dy,$$

$$dx \wedge dy = (dx \wedge dy) \vee (dx \wedge dy) \leq (dx \wedge y) \vee (x \wedge dy) = d(x \wedge y).$$

(3) $\forall x \in L$, we obtain

$$dx = d(x \wedge 1) = (dx \wedge 1) \vee (x \wedge d1) = (x \wedge d1) \vee dx.$$

(4) $\forall x \in L$, we obtain

$$dx = d(x \wedge (x \vee y))$$

$$= (dx \wedge (x \vee y)) \vee (x \wedge d(x \vee y)) = (x \wedge d(x \vee y)) \vee dx.$$

Example 3.4. Let $L = \{0, a, b, c, d, 1\}$, $0' = 1$, $a' = c$, $b' = d$, $c' = a$, $d' = b$, $1' = 0$, the Hasse diagram of L be defined as Fig.1 and its implication operator be defined as Table 1, then $(L, \vee, \wedge, ', \rightarrow)$ is a lattice implication algebra[12].

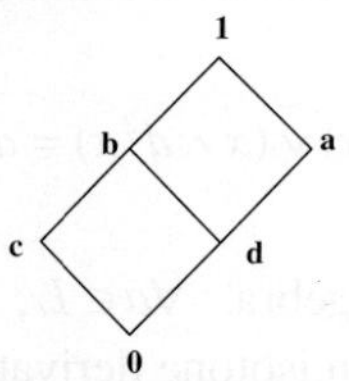

Fig. 1 Hasse Diagram of L

Table 1 Implication Operation of L

$\rightarrow$	0	a	b	c	d	1
0	1	1	1	1	1	1
a	c	1	b	c	b	1
b	d	a	1	b	a	1
c	a	a	1	1	a	1
d	b	1	1	b	1	1
1	0	a	b	c	d	1

Define a function d on L by $$dx = \begin{cases} a & x = b \\ b & x = a \\ x & x = 0, c, d, 1 \end{cases},$$

Then d is not a derivation on L . Because $d(a \wedge b) = d(d) = d$, but $(da \wedge b) \vee (a \wedge db) = b \vee a = 1$.

Definition 3.5. Let d be a derivation on L. If $x \leq y$ implies $dx \leq dy$, then d is called an isotone derivation.

In the follows, we research the properties of the isotone derivation on L.

Theorem 3.6. Let d be a derivation on L. Then the following conditions are equivalent:

(1) d is an isotone derivation on L;

(2) $dx = x \wedge d1$;

(3) $d(x \wedge y) = dx \wedge dy$;

(4) $d(x \vee y) = dx \vee dy$.

Proof. (1)$\Rightarrow$ (2) Assume d is an isotone derivation on L, We have $dx \leq d1$, and we obtain $dx \leq x \wedge d1$ by proposition 3.3(3).

(2)$\Rightarrow$ (3)Assume (2) holds, we obtain

$$dx \wedge dy = (x \wedge d1) \wedge (y \wedge d1) = x \wedge y \wedge d1 = d(x \wedge y).$$

(2)$\Rightarrow$ (4) Assume (2) holds, we obtain

$$dx \vee dy = (x \wedge d1) \vee (y \wedge d1) = (x \vee y) \wedge d1 = d(x \vee y).$$

(3)$\Rightarrow$ (1)If $x \leq y$, then $x = x \wedge y$. And we obtain $dx = d(x \wedge y) = dx \wedge dy \leq dy$. Hence, d is an isotone derivation on L.

(4)$\Rightarrow$ (1) If $x \leq y$, then $y = x \vee y$. And we obtain $dy = d(x \vee y) = dx \vee dy \geq dx$. Hence, d is an isotone derivation on L.

Remark 3.1. Let d be an isotone derivation on L, define

$$A_d(L) = \{x \in L : dx = x\}.$$

Theorem 3.7. Let d be an isotone derivation on L. Define a function d on L $\forall x \in L, d^2x = d(dx)$, then $dx \in A_d(L)$.

Proof. $\forall x \in L, d^2x = d(dx) = d(x \wedge dx) = (dx \wedge dx) \vee (x \wedge d^2x) = dx$, thus we obtain $dx \in A_d(L)$.

Example 3.8. Let L be a lattice implication algebra. $\forall a \in L$, define a function $d_a : L \to L$ by $d_a x = x \wedge a$, then $d_a : L \to L$ is an isotone derivation on L

.

Proof. $\forall x, y \in L$, we have $d_a(x \wedge y) = (x \wedge y) \wedge a$ and

$$(d_a x \wedge y) \vee (x \wedge d_a y) = (x \wedge a \wedge y) \vee (x \wedge y \wedge a) = (x \wedge y) \wedge a,$$

these show that $d_a(x \wedge y) = (d_a x \wedge y) \vee (x \wedge d_a y)$.

By definition 3.1, we have $d_a : L \to L$ is a derivation on L.

On the other side, $\forall x, y \in L$, if $x \leq y$, then $d_a x = x \wedge a \leq y \wedge a = d_a y$.

Thus, we can get $d_a : L \to L$ is an isotone derivation on L.

By some well-defined operators, we can get a class of lattice implication algebras.

Example 3.9. Let L be a lattice implication algebra. Define a set $D(L)=\{d_a \mid a\in L\}$, then $D(L)=\{d_a \mid a\in L\}$ is a lattice implication algebra by the following defined operators. $\forall d_a, d_b \in D(L), x\in L$, define

$$(d_a \vee_1 d_b)x = d_a x \vee d_b x,\quad (d_a \wedge_1 d_b)x = d_a x \wedge d_b x$$

$$(d_a x)' = d_{a'} x,\quad (d_a \to_1 d_b)x = d_a x \to d_b x.$$

Firstly, we prove that the above operators are well-defined.

By $d_a x = x\wedge a$, we have

$$(d_a \vee_1 d_b)x = d_a x \vee d_b x = (x\wedge a)\vee(x\wedge b) = x\wedge(a\vee b) = d_{a\vee b}x,$$

and $(d_a \wedge_1 d_b)x = d_a x \wedge d_b x = (x\wedge a)\wedge(x\wedge b) = x\wedge(a\vee b) = d_{a\wedge b}x$.

Because $d_a x \vee d_{a'} x = d_1 x, d_a x\wedge d_{a'} x = d_0 x = 0$, then we have $(d_a x)' = d_{a'} x$ and d_1 is a maximal element of $D(L)=\{d_a \mid a\in L\}$.

Because

$$(d_a \to_1 d_b)x = d_a x \to d_b x = (x\wedge a)\to(x\wedge b)$$
$$= (x\wedge a)\to b \ge a\to b \ge x\wedge(a\to b)$$

and

$$(d_a \to_1 d_b)x = d_a x \to d_b x = (d_{a'} x)' \to d_b x$$
$$= (x\wedge a')' \to (x\wedge b) = (x'\vee a)\to(x\wedge b)$$
$$=(x'\to(x\wedge b))\wedge(a\to(x\wedge b))$$
$$\le (x'\to x)\wedge(a\to b)\le x\wedge(a\to b)$$

We have $(d_a \to_1 d_b)x = x\wedge(a\to b) = d_{a\to b}x$.

Hence, the above operations are well-defined. It is not difficult to prove that $(D(L),\wedge_1,\vee_1,\to_1,',d_1,d_0)$ is a lattice implication algebra.

Theorem 3.10. Let L be a lattice implication algebra, $D(L)=\{d_a \mid a\in L\}$, then L is isomorphic to $D(L)$.

Proof. Define a function $\phi: L\to D(L)$ by $\phi(a)=d_a$, then it is easily verified that $\phi: L\to D(L)$ is one-to one and onto.

4. Conclusions

In this paper, we introduced the notion of derivations in lattice implication algebra and investigated some of the related properties. We hope that these will supply foundations for further research in lattice implication algebras, even for lattice-valued logic.

References

1. H. E. Bell, G. Mason, On derivations in rings, near-rings and near-fields, North-Holland, Math. Studies. **137**: 31-35(1987).
2. H. E. Bell, L. C. Kappe, Rings in which derivations satisfy certain algebraic conditions, Acta Math Hungar. 53: 339-346(1989).
3. Y. J. Li, X. L. Xin, Derivation of λ-lattice, Computer Engineering and Applications. **46(6)**: 46-47(2000).
4. J. Liu, Y. Xu, Filters and structure of lattice implication algebra, Chinese Sci. Bull. **42(18)**: 1517-1520(1997).
5. Y. L. Liu, S. Y. Liu, Y. Xu, et al, ILI-ideals and prime LI-ideals in lattice implication algebras, Information Sci. **155**: 157-175(2003).
6. Y.B. Jun, Y. Xu, J.Liu, Quotient lattice implication algebras induced by fuzzy LI-ideals, J. Fuzzy Math. **7(4)**, 1013-1020 (1999).
7. Y. B. Jun, X. L. Xin, On derivations of BCI-algebras, Information Sciences. **159**: 167-176(2004).
8. J. Ma, S.W. Chen, Y. Xu, Fuzzy logic from the viewpoint of machine intelligence, Fuzzy sets and systems, **157**, 628-634 (2006).
9. X. D. Pan, Y. Xu, Lattice implication ordered semigroups, Information Science. **178(2)**: 403-413(2008).
10. X. Y. Xie, An Introduction to Ordered Semigroup, Publishing Company for Science, Peking. 2001(in Chinese).
11. X. L. Xin, T. Y. Li, J. H. Lu, On derivations of lattices, Information Sciences. **178(2)**: 307-316(2008).
12. Y. Xu, D. Ruan, K. Y. Qin, et al, Lattice-valued logic, Germary: Springer-Verlag, 2003.
13. H. Zhu, S. W. Chen, J. B. Zhao, The ideal's radical in lattice implication algebras, The 9th International FLINS Conference River Edge: World Science Press. 117-123 (2010).
14. Hua Zhu, S.W. Chen, J.B. Zhao, The Extended filter in Lattice Implication Algebras, Journal of Fuzzy Mathematics and Systems Association of the Systems Engineering Society of China. **25(5)**, 50-53 (2011).

SOME REMARKS ON A NUMERICAL REPRESENTATION THEOREM BY NEWMAN AND READ

J. C. CANDEAL

Departamento de Análisis Económico, Universidad de Zaragoza, 50005, Zaragoza, Spain

E. INDURÁIN*

Departamento de Matemáticas, Universidad Pública de Navarra, 31006, Pamplona, Spain

G. B. MEHTA

Department of Economics, University of Queensland, Brisbane, 4072, Australia

Throughout this note, a classical construction by Newman and Read, that arises in Utility Theory, is revisited and analyzed from a modern perspective, relating it to several key principles on representability of ordinal structures.

1. Introduction

As a tool for Decision Making, and also to build models to deal with Uncertainty by means of theoretical approaches that convert qualitative scales into quantitative ones, the numerical representation theory of ordinal structures translates *qualitative* preferences, or hierarchies between objects, into *quantitative* comparisons of real numbers. This setting is classical in Economics (Utility Theory) among many other disciplines. In the corresponding mathematical theory, there are some abstract results that characterize the existence of a certain numerical representation. However, some of the typical theorems do not provide an *explicit* construction of the representation involved. These explicit constructions do not abound.

Our aim is to re-interprete an interesting classical representation result by Newman and Read [1] in the light of recent developments in utility theory [2].

* steiner@unavarra.es

In an important paper written in 1957, Mckenzie [3] introduces the expenditure function into demand analysis, and claims that this new approach provides a more efficient way to analyze problems in consumer theory in as much as the expenditure function obviates the need for the use of the utility function. In a comment on this paper, Newman and Read [1] demonstrate that the assumptions used by Mckenzie in fact imply the existence of a numerical representation. In our opinion, the *explicit* construction about the existence of a continuous utility function on certain subsets of $\mathbb{R}^n$, made by Newman and Read, and perhaps undeservedly neglected, should be considered in justice among the classical ones in utility theory, at least at the same level of importance as other well-known ones as, e.g., the Arrow-Hahn approach [4] or the Eilenberg-Richter method (see section 4.4. in Ref. 5).

2. Preliminaries

Throughout the paper, $\precsim$ will denote a total preorder (or *preference*) defined on a (nonempty) set X, i.e., a binary relation on X which is *reflexive* ($x \precsim x$ for all $x \in X$), *transitive* ($x \precsim y$, , $y \precsim z$ implies $x \precsim z$), and *total* (either $x \precsim y$ or $y \precsim x$ for every $x, y \in X$). We also say that $(X, \precsim)$ is a *totally preordered set*. Two elements $x, y \in X$ are said to be *indifferent* if $x \precsim y$ and $y \precsim x$ (briefly, $x \sim y$). If, in addition, $\precsim$ is anti-symmetric ($x \precsim y$ and $y \precsim x$ implies $x = y$), then $\precsim$ is said to be a *total order*.

Given a total preorder $\precsim$, we define its *asymmetric binary relation* $\prec$ on X as: $x \prec y$ if and only if no $(y \precsim x)$. The preorder $\precsim$ is said to be *non-trivial* if there are $x, y \in X$ such that $x \prec y$. We also consider the *symmetric binary relation* $\sim$ on X given by $x \sim y$– $(x \precsim y) \wedge (y \precsim x)$, $x, y \in X$. Let $(X, \precsim)$ be a totally preordered set. A real-valued function $u : X \to \mathbb{R}$ is said to be *order-preserving* (or a *utility function* or *numerical representation*) for $\precsim$, if $x \precsim y$– $u(x) \le u(y)$, for every $x, y \in X$. The total preorder $\precsim$ is said to be *representable* if it admits a numerical representation.

Let (X, τ) be a topological space (i.e.: a nonempty set X with a topology τ). A total preorder $\precsim$ on X is τ *-continuous* if for each $x \in X$ the sets $U(x) = \{y \in X : x \precsim y\}$ and $L(x) = \{z \in X : z \precsim x\}$ are both τ -closed.

Let now X be a nonempty subset of the space $\mathbb{R}^n$. On X we consider the relativized Euclidean topology. A total preorder (preference) on X is *locally non-satiated* if for each $x \in X$ and each neighbourhood V of x there is $y \in V$ such that $x \prec y$. For vectors $x = (x_1, \ldots, x_n); y = (y_1, \ldots, y_n) \in X \subseteq \mathbb{R}^n$ denote the vector inequality by $x < y$ if and only if $x_i \le y_i$ for all $i \in \{1, \ldots, n\}$ and there is $j \in \{1, \ldots, n\}$ such that $x_j < y_j$. Denote also $x << y$ if and only if $x_i < y_i$ for all

$i \in \{1,\ldots,n\}$. A total preorder $\precsim$ on $X \subseteq R^n$ is *monotone*† if $x < y$ implies that $x \prec y$, for every $x, y \in X$. Also, $\precsim$ is *anti-monotone* if $x < y$ implies that $y \prec x$, for every $x, y \in X$.

3. Newman and Read approach revisited

We start by observing that Newman and Read [1] list the following four assumptions which they claim are among those made by Mckenzie: 1) the commodity space X is a subset of some finite-dimensional Euclidean space $\mathbb{R}^n$ and not all goods are necessarily divisible; 2) there is a preference relation $\precsim$ on the commodity space that is a total preorder; 3) the preference relation $\precsim$ is continuous on the commodity space; 4) the consumer is not satiated in some good which is divisible. Under these assumptions, Newman and Read [1] result reads as follows.

Theorem 3.1. *Let X be a commodity space endowed with a preference relation $\precsim$. Assume that the assumptions 1 to 4 above are all accomplished. Then there exists a continuous utility function on X that represents $\precsim$.*

It can be argued that the stated assumptions are not clear. As an example, let us consider the *non-satiation* assumption. Newman and Read claim that it is needed to ensure that all income is spent. Unfortunately, however, there is an ambiguity in assumption 4). From the Newman and Read [1] description it is not clear what kind of non-satiation is intended. It should be observed that neither local non-satiation nor global satiation is sufficient to ensure that all income is spent. Indeed, it is easy to see that global satiation is not enough for this purpose: To put an example, consider two commodities (whose quantities are measured by x and y). Let p and q respectively be its unitary prices. Let $\mathcal{I}$ denote income. Suppose also that $p,q,\mathcal{I} > 0$ and $x, y \geq 0$. Under these conditions it is claimed that a locally non-satiated (even continuous) preference relation may have an optimal bundle in which not all income is spent (i.e. $p.x + q.y < \mathcal{I}$). Suppose, for example, that $\precsim$ is an anti-monotone locally non-satiated preference relation on the positive orthant of $\mathbb{R}^2$, representable by the continuous function $f(x, y) = -x.y;\ x, y \in \mathbb{R}$. Then this preference $\precsim$ is continuous and locally non-satiated. But at the utility maximum $(0,0)$ not all income is spent since $p.0 + q.0 = 0 < \mathcal{I}$. To ensure that all income is spent the more restrictive condition of strong monotonicity is required. However, it is clear from the first paragraph of section 4 in Ref. 1 (p. 198) that some kind of

† There is a related assumption of weak monotonicity. A total preorder $\precsim$ on $X \subseteq \mathbb{R}^n$ is *weakly monotone* if $x < y$ implies that $x \precsim y$ for every $x, y \in X$. The significance of the difference between monotonicity and weak monotonicity is explained in Ref. 4. There are also other kinds of monotonicity assumptions that one can make.

strong monotonicity is the desired connotation. This is because the real-valued order-preserving (utility) function constructed there on the non-negative real numbers is given by $U(x) = \frac{x}{1+x}$ which is strictly increasing implying some kind of monotonicity of the preference relation. We conclude that Newman and Read are implicitly assuming that the preference relation satisfies a strong monotonicity condition with respect to the divisible good.

We now briefly describe the Newman and Read [1] ideas about the construction of a real-valued order-preserving (utility) function. Since there are only finitely many divisible goods it follows that we may assume that the commodity space is given by the product of the non-negative real numbers with the non-negative integers. Actually, if there are *more than one divisible good* as well as *more than one* non-divisible good, then the commodity space is not (a priori) $\mathbb{R} \times \mathbb{N}$ but, instead, $\mathbb{R}^p \times \mathbb{N}^q$ for some $p, q > 1 \in \mathbb{N}$. The *discrete* factor $\mathbb{N}^q$ $(q > 1)$ can easily be handled as $\mathbb{N}$ without loss of generality, since $\mathbb{N}^q$ $(q > 1)$ and $\mathbb{N}$ share the same cardinality and inherit the discrete topology. However, if $p > 1$, the *continuous* factor $\mathbb{R}^p$ $(p > 1)$ is, from the topological point of view, quite different from $\mathbb{R}$. However, even in the case of being $p > 1$, some important topological result concerning continuous total preorders defined on the Euclidean space $\mathbb{R}^n (n \geq 1)$ comes in our rescue. Thus, we can prove [6] that whenever $X \subseteq \mathbb{R}^n$ is a connected closed subset endowed with a continuous total order, then X is homeomorphic to a subset of the real line $\mathbb{R}$. Using this result, we can definitively assume without loss of generality that our commodity space is $\mathbb{R} \times \mathbb{N}$, and go ahead with Newman and Read [1] proof. Intuitively, we may think of the commodity space as being given by a series of horizontal lines (or *stripes*) $S_n : n \in \mathbb{N}$ in the non-negative quadrant. On S_1, as we have said above, monotonicity of preferences implies that $U(x) = \frac{x}{1+x}$ is the desired utility function which takes values in the unit interval $[0,1]$. We now consider the line S_2 and take an arbitrary point b on S_1. Then continuity and monotonicity of the preference relation imply that S_2 consists of three subsets (perhaps empty) L, M, N consisting respectively of those commodity bundles on S_2 that are all inferior to b, those that are all superior or preferred to b and those that are neither in L or N. Using the fact that any two intervals of real numbers are order-homeomorphic, Newman and Read then give an explicit formula for the utility function. This utility formula maps the subsets L and N onto intervals that are adjacent to $[0,1]$. The utility values of points of M are defined to be equal to the utilities of points on S_1 to which they are indfferent.

The resulting function is easily seen to be continuous on S_1 and S_2 Proceeding inductively, a continuous utility function is constructed on the whole commodity space.

4. Alternative proofs of the Newman and Read theorem

It should be observed that the Newman and Read [1] proof is ad hoc and complicated. Furthermore, it is not based on or related to other theoretical principles in utility theory. Another objective of the present note is to try to integrate this result with some fundamental principles of utility theory [2].We now offer an alternative shorter proof which is based on those key principles in utility theory. At first, we retain all the assumptions made by Newman and Read. The idea of our proof is based on Debreu's Gap Lemma (see section 3.1 in Ref. 5) that states (roughly) that if there a utility function on the consumption set X then there also exists a utility function that is continuous with respect to any topology that is finer than the order topology generated by the total preorder $\precsim$ on X. Indeed it is sufficient to prove the existence of a utility function since continuity follows from the Debreu's Gap Lemma. The proof below is similar to but much easier than that of Newman and Read and proceeds as follows. On S_1 we argue as in Newman and Read to get a utility function U with values in the interval $[0,1]$. We then consider S_2 and define our utility function on the set M as in Newman and Read [1]. For the subsets L and N of S_2 we need not choose adjacent intervals or any particular complicated formula to define utility as in Ref. 1. Instead, we map the subsets L and N onto any bounded interval of the form $[a,b]$ which is disjoint from $[0,1]$ and has the same order relation to $[0,1]$ that the subsets L and N have to points on S_1. Proceeding inductively, we extend this utility function which may not be continuous to all the horizontal lines. Hence, we have a utility function on the commodity space and it follows from Debreu's Gap Lemma that there is a continuous utility function. In this way, we have obtained a new, simple and shorter proof of the existence of a continuous utility function which is related to other fundamental ideas in utility theory.

Newman and Read [1] result can be generalized by replacing their strong monotonicity assumption by the weak monotonicity assumption with respect to the divisible good. Now instead of using the formula $U(x)=\dfrac{x}{1+x}$ as in Ref. 1, which is possible only if the preference relation is monotonic and is not valid under the weaker assumption of weak monotonicity of preferences, we use Theorem 1 in Ref. 4 to conclude that there is a utility function on S_1 and so on.

Then Debreu's Gap Lemma applies and the rest of the proof proceeds as described above and retains the simplicity of the original argument. Furthermore, by using Theorems 1 and 2 in Ref. 4 one can even avoid Debreu's Gap Lemma and use instead the simpler construction of Wold [4]. Actually, the following result can also straightforwardly be proved. As a matter of fact, this result is a corollary of a more general one that states that every continuous total preorder defined on a locally connected and separable topological space admits a continuous utility representation [2].

Theorem 4.1. *Every continuous total preorder on* $\mathbb{R}\times\mathbb{N}$ *has a continuous utility representation.*

5. Conclusion

Explicit constructions of representations to convert qualitative scales into quantitative ones, building mathematical models of preferences, are quite scarce. Moreover, they are dispersed in papers corresponding to a wide sort of miscellaneous applications (e.g.: Economic Theory, Measurement in Psychology, Social Choice, Entropy Theory in Physics [7], etc.) instead of having been issued through papers devoted to Pure and Applied Mathematics. It seems necessary to revise these classical results under a modern perspective.

Acknowledgments

This work has been partially supported by the Research Projects MTM2007-62499, ECO2008- 01297 and MTM2009-12872-C02-02 (Spain).

References

1. P. Newman and R. Read, Demand theory without a utility index - a comment, *Rev. Econom. Stud.* **25**, 197-199 (1958).
2. J.C. Candeal, E. Indur_ain and G. B. Mehta, Utility functions on locally connected spaces, *J. Math. Econom.* **40**, 701-711 (2004).
3. L. Mckenzie, Demand theory without a utility index, *Rev. Econom. Stud.* **24**, 185-189 (1957).
4. A.F. Beardon and G. B. Mehta, The utility theorems of Wold, Debreu and Arrow-Hahn, *Econometrica* **62**, 181-186 (1994).
5. D.S. Bridges and G.B. Mehta, *Representations of preference orderings*, Springer, Berlin (1995).
6. J.C. Candeal, E. Indur_ain and G. B. Mehta, Further remarks on totally ordered representable subsets of Euclidean space, *J. Math. Econom.* **25**, 381-390 (1996).
7. J.C. Candeal, J.R. De Miguel, E. Indur_ain and G.B. Mehta, Utility and entropy, *Econom. Theory* **17**, 233-238 (2001).

CONDITIONAL CHOQUET EXPECTATION*

HONGXIA WANG, SHOUMEI LI†, JUAN GAO

College of Applied Science, Beijing University of Technology,
Beijing, 100124, P. R. China

The article aims at discussing conditional Choquet expectations with respect to capacities. We introduce the concept of conditional Choquet expectation with respect to submodular capacity, and prove some properties. Moreover, we state that C_r inequality, Schwarz inequality, Hölder's inequality, Minkowski inequality and Jensen's inequality still hold for this kind of conditional Choquet expectations.

1. Introduction

It is well known that the concepts of capacity and the Choquet integral [3] are reasonable generalizations of probability measure and the mathematical expectation. Capacity is non-additive measure and special monotone set function. The Choquet integral can be considered as one kind of nonlinear expectations with respect to capacities. Many papers developed the Choquet theory and its applications (e.g. [1, 4, 5, 6, 7]).

In capacity theory several conditional capacities have been proposed, but there are only a few papers to have discussions on conditional Choquet expectation. Chateauneuf et al. [2] defined the conditional expectation with respect to capacity in two cases: when information, represented as a random variable Y, is comonotone with the random variable X or when it is anti-monotone with X. How to define the conditional Choquet expectation in more general case is an open problem so far. In 1994, Denneberg [5] studied the conditional capacities in the submodular and supermodular case. Based on this kind of conditional capacities, we shall define the conditional Choquet expectation under the condition that the capacity is submodular. And we also state some properties of this kind of conditional Choquet expectations.

On the other hand, the inequalities for conditional expectations, such as C_r inequality, Schwarz inequality, Hölder's inequality, Jensen's inequality and

* This work is supported by NSFC (No. 11171010), PHR (No. 201006102) and Beijing Natural Science Foundation (Stochastic Analysis with Uncertainty and Applications in Finance).

† lisma@bjut.edu.cn.

Minkowski inequality, play an important role in probability theory. Then it is natural to wonder whether those classical integral inequalities still hold for the conditional Choquet expectations with respect to capacities. In this paper, we shall show that these inequalities still hold for the conditional expectations with respect to submodular capacities.

The rest of this paper is organized as follows. Section 2 is for preliminaries and notations of capacity, the Choquet integral and conditional capacity that will be used later. In Section 3, we firstly state the concept and present some properties of conditional Choquet expectation; secondly we present some inequalities for this kind of conditional Choquet expectations.

Due to the page limit, we have to omit all proofs.

2. Preliminaries and Notations

Throughout of the paper, assume that $(\Omega, \mathcal{F})$ is a measurable space, $\mathcal{P}$ is the set of all probabilities on $\mathcal{F}$ and $\mathbb{R}$ is the set of all real numbers.

We first recall some concepts and some elementary results of capacity and the Choquet integral.

Definition 2.1 [3] A set function μ: $\mathcal{F}\rightarrow [0, 1]$ is called a capacity if it satisfies

- $\mu(\emptyset)=0$, $\mu(\Omega)=1$;
- $\mu(A) \le \mu(B)$ for any $A \subseteq B$ and $A, B \in \mathcal{F}$.

A triplet $(\Omega, \mathcal{F}, \mu)$ is called a capacity space.

The conjugate of μ, $\overline{\mu} : \mathcal{F} \rightarrow [0, 1]$ is defined as

$$\overline{\mu}(A) = 1 - \mu(A^c), for\ all\ A \in \mathcal{F}.$$

A capacity μ is called submodular (or concave) if $\mu(A \cup B) + \mu(A \cap B) \le \mu(A) + \mu(B), A, B \in \mathcal{F}$. If the above inequality is reversed, μ is called supermodular (or convex). μ is called modular if it is sub- and supermodular, and at the moment, μ is a finite additive measure.

We define for a capacity μ on $\mathcal{F}$

$$C(\mu) = \{P \in \mathcal{P}: \overline{\mu}\,(A) \le P(A) \le \mu(A)\ for\ all\ A \in \mathcal{F}\},$$

and call it the core of μ. We have C $(\mu) \neq \emptyset$ if the capacity μ is submodular [5]. Obviously, if the capacity μ is supermodular, then the core of its conjugate $\overline{\mu}$ is

$$C(\,\overline{\mu}\,) = \{P \in \mathcal{P}: \mu(A) \le P(A) \le \overline{\mu}\,(A)\ for\ all\ A \in \mathcal{F}\},$$

and $C(\overline{\mu}) \neq \emptyset$.

A random variable f on Ω is a (Borel) measurable function f: $(\Omega, \mathcal{F}) \rightarrow (\mathbb{R}, \mathcal{B}(\mathbb{R}))$, where $\mathcal{B}(\mathbb{R})$ is the Borel σ-field of $\mathbb{R}$. The family of all the bounded random variables is denoted as L^∞. The concept of the integral with respect to a capacity was introduced by Choquet [3].

Definition 2.2 Assume that $f \in L^\infty$. The Choquet integral of f with respect to the capacity μ on A $\in \mathcal{F}$ is defined by

$$(C)\int_A f d\mu = \int_0^{+\infty} \mu(A \cap \{f \geq t\})dt + \int_{-\infty}^0 [\mu(A \cap \{f \geq t\}) - 1]dt,$$

where the integrals in the right hand are taken in the sense of Lebesgue's.

Remark 2.3 Instead of $(C)\int_\Omega f d\mu$, we shall write $(C)\int f d\mu$. We say that f is Choquet integrable, if we have $-\infty < (C)\int f d\mu < \infty$.

Proposition 2.4 [4] Assume that μ is a capacity and $f, g \in L^\infty$.

- $(C)\int 1_A d\mu = \mu(A), A \in \mathcal{F}$.
- (positive homogeneity) For all $\lambda \in \mathbb{R}^+$, we have $(C)\int \lambda \cdot f d\mu = \lambda \cdot (C)\int f d\mu$
- (translation invariance) $(C)\int (f + c)d\mu = (C)\int f d\mu + c$ for all $c \in \mathbb{R}$.
- (monotonicity in the integrand) $(C)\int f d\mu \leq (C)\int g d\mu$ for $f \leq g$.

(monotonicity in the set function) $(C)\int f d\mu \leq (C)\int f d\nu$ for $\mu \leq \nu$.

- (subadditivity) $(C)\int (f + g)d\mu \leq (C)\int f d\mu + (C)\int g d\mu$ if μ is submodular.

(superadditivity) $(C)\int (f + g)d\mu \geq (C)\int f d\mu + (C)\int g d\mu$ if μ is supermodular.

We say two functions $f, g \in L^\infty$ are comonotonic if for any pair ω, $\omega' \in \Omega$, we have $(f(\omega) - f(\omega'))(g(\omega) - g(\omega')) \geq 0$.

Proposition 2.5 [4] (comonotonic additivity) If μ is a capacity and $f, g \in L^\infty$ are comonotonic, then $(C)\int (f + g)d\mu = (C)\int f d\mu + (C)\int g d\mu$.

In the following subsection, we shall list some preliminaries about the conditional capacities [5]. Assume that $(\Omega, \mathcal{F}, \mu)$ is a capacity space, $\mathcal{A}$ is the sub-σ algebra of $\mathcal{F}$.

Definition 2.6 The conditional capacity of the event $A \in \mathcal{A}$ under the given event B is defined as

$$\mu_{|B}(A) = \frac{\mu(A \cap B)}{\mu(A \cap B) + \overline{\mu}(A^C \cap B)}.$$

Theorem 2.7 If μ is submodular and $\overline{\mu}_{|B} > 0$, then $\mu_{|B}$ and $\overline{\mu}_{|B}$ are well defined on $\mathcal{A}$.

Proposition 2.8 For all B $\in \mathcal{F}$, we have:

- If $\mu(B) > 0$, we have $\mu_{|B}(\Omega) = \mu_{|B}(B) = 1$.
- $\forall A_1, A_2 \in \mathcal{A}$, if $\mu_{|B}(A_i), i = 1,2$ exist, then we have

$$A_1 \subseteq A_2 \Rightarrow \mu_{|B}(A_1) \le \mu_{|B}(A_2).$$

- If ν is another capacity on $\mathcal{A}$, and $\forall A \in \mathcal{A}$, $\mu_{|B}(A), \nu_{|B}(A)$ exist, then

$$\mu \le \nu \Rightarrow \mu_{|B}(A) \le \nu_{|B}(A).$$

- $\overline{\mu_{|B}} = \overline{\mu}_{|B}$. In particular, $\overline{\mu}_{|B}(A)$, $A \in \mathcal{A}$ is defined if and only if $\mu_{|B}(A^C)$ is defined.

Theorem 2.9 If μ is submodular and $\overline{\mu}(B) > 0$, then $\mu_{|B}, \overline{\mu}_{|B}$ is well defined on $\mathcal{A}$, and we have $\mu_{|B}(A) = \max\limits_{P \in C(\mu)} P_{|B}(A), \overline{\mu}_{|B}(A) = \min\limits_{P \in C(\mu)} P_{|B}(A), \forall A \in \mathcal{A}$.

Theorem 2.10 If μ is submodular and $\overline{\mu}(B) > 0$, then $\mu_{|B}$ is submodular and $\overline{\mu}_{|B}$ is supermodular.

Theorem 2.11 If μ is submodular and for given B, C $\in \mathcal{F}$, $\overline{\mu}(B \cap C) > 0$, then $\mu_{|B\cap C} \le (\mu_{|B})_{|C} \le (\mu_{|B})_{|B\cap C}$.

3. Conditional Choquet Expectation

We firstly define the concept of conditional Choquet expectation.

Definition 3.1 Assume that $f \in L^\infty$ and μ is submodular. The conditional expectation of f with respect to μ under the σ- algebra $\mathcal{A}$ is defined as

$$\mathbb{E}^{\mu}(f|\mathcal{A}) = \max_{P \in C(\mu)} \mathbb{E}_P(f|\mathcal{A}).$$

Remark 3.2 From above definition, we conclude that:

- If μ is a probability, then $\mathbb{E}_{\mu}(f|\mathcal{A})$ is the classical conditional expectation.
- Conditional Choquet expectation is generalization of conditional capacity. In fact, the conditional capacity of the event A under the given event B may be the special case of the conditional expectation, i.e.

$$\mathbb{E}_{\mu}\left(1_A \middle| \mathcal{A}\right) = \max_{P \in C(\mu)} \mathbb{E}^{P}\left(1_A \middle| \mathcal{A}\right) = \max_{P \in C(\mu)} P_{|B}(A) = \mu_{|B}(A).$$

- The above definition and the following theorems hold almost everywhere with respect to μ, abbreviated a. e. $[\mu]$.
- If μ is a supermodular capacity, then its conjugate $\overline{\mu}$ is a submodular capacity. So we can define the conditional expectation under the supermodular capacity as

$$\mathbb{E}^{\mu}(f|\mathcal{A}) = \min_{P \in C(\overline{\mu})} \mathbb{E}_P(f|\mathcal{A}).$$

The following results summarize the elementary properties of conditional Choquet expectation.

Theorem 3.3 Assume that $f, g \in L^{\infty}$ and μ, ν are submodular. If conditional expectations $(f|\mathcal{A})$, $\mathbb{E}_{\mu}(g|\mathcal{A})$, $\mathbb{E}_{\nu}(f|\mathcal{A})$ exist, then we have:

1. $\mathbb{E}_{\mu}(a|\mathcal{A}) = a$, for all $a \in \mathbb{R}$.
2. (positive homogeneity) $\mathbb{E}_{\mu}(af|\mathcal{A}) = a\,\mathbb{E}_{\mu}(f|\mathcal{A})$, for all $a \in \mathbb{R}^{+}$.
3. (translation invariance) $\mathbb{E}_{\mu}(f + a|\mathcal{A}) = \mathbb{E}_{\mu}(f|\mathcal{A}) + a$, for all $a \in \mathbb{R}$.
4. (subadditivity) $\mathbb{E}_{\mu}(f + g|\mathcal{A}) \le \mathbb{E}_{\mu}(f|\mathcal{A}) + \mathbb{E}_{\mu}(g|\mathcal{A})$.
5. (monotonicity) If $f \le g$, then $\mathbb{E}_{\mu}(f|\mathcal{A}) \le \mathbb{E}_{\mu}(g|\mathcal{A})$.
6. (monotonicity in capacity) If $\mu \le \nu$, then $\mathbb{E}_{\mu}(f|\mathcal{A}) \le \mathbb{E}_{\nu}(f|\mathcal{A})$.

In the following, we shall present some inequalities for the conditional Choquet expectations. In the following inequalities, we always assume that $f, g \in L^{\infty}$ and μ is submodular.

Theorem 3.4 If conditional expectations $\mathbb{E}^{\mu}\left(|f|\middle|\mathcal{A}\right)$, $\mathbb{E}^{\mu}\left(|g|\middle|\mathcal{A}\right)$ exist, then $\mathbb{E}^{\mu}\left(|f+g|\middle|\mathcal{A}\right) \le \mathbb{E}^{\mu}\left(|f|\middle|\mathcal{A}\right) + \mathbb{E}^{\mu}\left(|g|\middle|\mathcal{A}\right)$.

Theorem 3.5 (C_r inequality) If conditional expectations $\mathbb{E}_{\mu}\left(|f|^{r}\middle|\mathcal{A}\right)$, $\mathbb{E}_{\mu}\left(|g|^{r}\middle|\mathcal{A}\right)$, $r > 0$ exist, we have

$$\mathbb{E}^{\mu}\left(|f+g|^{r}\middle|\mathcal{A}\right) \le C_r \cdot [\,\mathbb{E}^{\mu}\left(|f|^{r}\middle|\mathcal{A}\right) + \mathbb{E}^{\mu}\left(|g|^{r}\middle|\mathcal{A}\right)],$$

where $C_r = \max\{1, 2^{r-1}\}$

Theorem 3.6 (Hölder's inequality) If conditional expectations $\mathbb{E}_{\mu}\left(|f|^{p}\middle|\mathcal{A}\right)$, $\mathbb{E}_{\mu}\left(|g|^{q}\middle|\mathcal{A}\right)$, $p, q > 1$, $\frac{1}{p} + \frac{1}{q} = 1$ exist, then we have

$$\mathbb{E}^{\mu}\left(|fg|\middle|\mathcal{A}\right) \le \left(\mathbb{E}^{\mu}\left(|f|^{p}\middle|\mathcal{A}\right)\right)^{\frac{1}{p}} \cdot \left(\mathbb{E}^{\mu}\left(|g|^{q}\middle|\mathcal{A}\right)\right)^{\frac{1}{q}}.$$

Corollary 3.7 (Schwarz inequality) If conditional expectations $\mathbb{E}_\mu\left(|f|^2 \middle| \mathcal{A}\right)$, $\mathbb{E}_\mu\left(|g|^2 \middle| \mathcal{A}\right)$ exist, then $\mathbb{E}^{\mu}_{2}\left(|fg| \middle| \mathcal{A}\right) \leq \mathbb{E}^{\mu}\left(|f|^2 \middle| \mathcal{A}\right) \cdot \mathbb{E}^{\mu}\left(|g|^2 \middle| \mathcal{A}\right)$.

Corollary 3.8 If conditional expectations $\mathbb{E}_\mu\left(|f|^l \middle| \mathcal{A}\right)$, $\mathbb{E}_\mu\left(|g|^k \middle| \mathcal{A}\right)$ exist, and $1 < l, k, r < \infty$ with $\frac{1}{l} + \frac{1}{k} = \frac{1}{r}$, then

$$\left(\mathbb{E}^{\mu}\left(|fg|^r \middle| \mathcal{A}\right)\right)^{\frac{1}{r}} \leq \left(\mathbb{E}^{\mu}\left(|f|^l \middle| \mathcal{A}\right)\right)^{\frac{1}{l}} \left(\mathbb{E}^{\mu}\left(|g|^k \middle| \mathcal{A}\right)\right)^{\frac{1}{k}}.$$

Theorem 3.9 (Minkowski inequality) If conditional expectations $\mathbb{E}_\mu\left(|f|^r \middle| \mathcal{A}\right)$, $\mathbb{E}_\mu\left(|g|^r \middle| \mathcal{A}\right)$, $r \geq 1$ exist, we have

$$\left(\mathbb{E}^{\mu}\left(|f+g|^r \middle| \mathcal{A}\right)\right)^{\frac{1}{r}} \leq \left(\mathbb{E}^{\mu}\left(|f|^r \middle| \mathcal{A}\right)\right)^{\frac{1}{r}} + \left(\mathbb{E}^{\mu}\left(|g|^r \middle| \mathcal{A}\right)\right)^{\frac{1}{r}}.$$

Theorem 3.10 (Jensen inequality) If $\Phi : \mathbb{R}^+ \to \mathbb{R}^+$ is a convex function and the conditional expectations $\mathbb{E}_\mu\left(|f| \middle| \mathcal{A}\right)$, $\mathbb{E}_\mu\left(\Phi \circ |f| \middle| \mathcal{A}\right)$ exist, then

$$\Phi\left(\mathbb{E}^{\mu}\left(|f| \middle| \mathcal{A}\right)\right) \leq \mathbb{E}^{\mu}\left(\Phi \circ |f| \middle| \mathcal{A}\right).$$

Corollary 3.11 If conditional expectation $\mathbb{E}_\mu\left(|f|^p \middle| \mathcal{A}\right)$, $1<p<\infty$ exists, then

$$\left(\mathbb{E}^{\mu}\left(|f| \middle| \mathcal{A}\right)\right)^p \leq \mathbb{E}^{\mu}\left(|f|^p \middle| \mathcal{A}\right).$$

Corollary 3.12 (Moment inequality) For $0 \leq \alpha < \beta$, we have

$$\left(\mathbb{E}^{\mu}\left(|f|^\alpha \middle| \mathcal{A}\right)\right)^{\frac{1}{\alpha}} \leq \left(\mathbb{E}^{\mu}\left(|f|^\beta \middle| \mathcal{A}\right)\right)^{\frac{1}{\beta}}.$$

Theorem 3.13 (Jensen inequality in 2 dimensions) If f, g are comonotonic, conditional expectations $\mathbb{E}^{\mu}\left(|f| \middle| \mathcal{A}\right)$, $\mathbb{E}^{\mu}\left(|g| \middle| \mathcal{A}\right)$, $\mathbb{E}^{\mu}\left(\Phi \circ (|f|, |g|) \middle| \mathcal{A}\right)$ exist and $\Phi: \mathbb{R}^+ \times \mathbb{R}^+ \to \mathbb{R}^+$ is a convex function, then

$$\Phi(\mathbb{E}^{\mu}\left(|f|\,\middle|\,\mathcal{A}\right),\mathbb{E}^{\mu}\left(|g|\,\middle|\,\mathcal{A}\right))\leq\mathbb{E}^{\mu}\left(\Phi\circ(|f|,|g|)\,\middle|\,\mathcal{A}\right).$$

References

1. M. Allais, Le comportement de l'homme rationnel devant le risque: Critique des postulates et axiomes de l'ecole americaine, *Econometrica.* **21**, 503-546 (1953).
2. A. Chateauneuf, R. Kast, Conditioning Capacities and Choquet Integrals: The Role of comonotony, *Theory and Decision*. **51**, 367-386 (2001).
3. G. Choquet, Theory of capacities, *Ann Inst. Fourier.* **5,** 131-295 (1953).
4. D. Denneberg, Non-Additive Measure and Integral, *Kluwer Academic Publishers, Boston.* (1994).
5. D. Denneberg, Conditioning (updating) non-additive measures, *Annals of Operations Research*. **52,** 21-42 (1994).
6. D. Schmeidler, Subjective probability and expected utility without additivity, *Econometrica.* **57**, 571-587 (1989).
7. Z. Wang and J. Yan, A selective overview of applications of Choquet integrals, *Advanced lectures in Mathematics*. 484-515 (2007).

ASCRIBING STRONG CAUSES FROM OBSERVATIONAL AND INTERVENTIONAL DATA UNDER THE BELIEF FUNCTION FRAMEWORK

IMEN BOUKHRIS[1,2 †], SALEM BENFERHAT[1] AND ZIED ELOUEDI[2]

[1] *CRIL-Université d'Artois-Faculté Jean Perrin-Lens-France*

[2] *LARODEC-Université de Tunis-ISG-Tunis-Tunisie*

Ascribing causality amounts to detect causally related events. In this paper, we consider that these events are uncertain and expressed with the belief function formalism. They are either results of observations or interventions which are external actions forcing the variable to take a specific value. We show how the proposed concepts of strong accept/reject can be used, instead of changes in uncertainty, to discriminate between potential causes.

1. Introduction

Causal discovery is a challenging task especially when mere correlations are confused with causal links. One solution is to use aside from observational data, interventional data[1]. Interventions are external actions disturbing the spontaneous behavior of the system. These latter allow to see if manipulating some potential causes variables affect the distribution of some effects variables.

Unlike standard diagnosis problems, causality ascription[2,3] is a problem of identifying what elements in a set of observed or intervened events in a sequence are causally related, on the basis of some background knowledge about the normal course of things. In fact, diagnosing amounts to make an inspection to find the history of the faults that explain the observations. It provides an explanation for what components failed and when they did[4]. For instance, let us consider that I make my presentation, I have poured water on my computer and accordingly it crashes. A diagnosis problem consists to find which component has burned and considered as faulty while ascribing causality amounts to describe as causal the link among the series of reported events, namely, I have made the presentation, someone came, someone went out, I have poured some water, etc.

---

† imen.boukhris@hotmail.com.

Causal networks are simple and efficient tools to graphically represent the agent's background knowledge. In fact, we indicate for each variable what its directed causes are. Structural equations[5] may be appropriate. However, we do not specify exact equations for each variable. We just need to specify for each variable which variables are involved in its associated equation.

Since imperfection is ubiquitous in real world, it seems impossible for the agent to model the information that he has in a precise and certain form. The theory of belief functions is a general framework for reasoning with uncertainty. In fact, it has connections to other frameworks such as probability, possibility and imprecise probability theories. Besides, it is the most appropriate theory to model ignorance situations. Accordingly, we will use in this paper belief function causal networks[6] to represent the causality ascribing agent's background knowledge.

In this paper, we propose a model that an intelligent agent will use to ascribe causality from a sequence of observations or interventions occurring in his environment under the belief function framework. In fact, he will look for the causes of an exceptional event. To model such changes, we use the concepts of acceptance and rejection instead of changes in uncertainty distributions. This paper introduces definitions of acceptance and rejection allowing the categorization of causes according to their strength. Accordingly, we show that aside from favoring abnormal events, we may use the idea of the strength of the causes to discriminate between potential causes.

The rest of the paper is organized as follows: In Section 2, we briefly sketch the basics of the belief function theory. Section 3 is dedicated to our definitions of the concepts of acceptance and rejection. Section 4 presents our model for ascribing strong causes. Section 5 concludes the paper.

2. Belief Function Theory

2.1. *Background*

In the following we recall the basics of the belief function theory[7]. A finite set of events denoted by $\boldsymbol{\Theta}$ is referred to the frame of discernment. Beliefs are expressed on propositions belonging to the power set of $\boldsymbol{\Theta}$. The mapping m: $\mathbf{2^{\Theta}}$ → [0,1] is the basic belief assignment (bba) such that: $\sum_{A\subseteq\Theta} \boldsymbol{m(A)} = \mathbf{1}$, where m(A) is the basic belief mass (bbm) representing the part of belief exactly committed to the event A. A bba is said to be normalized, if m(∅)=0.

A bba can equivalently be written as a plausibility function Pl: $\mathbf{2^{\Theta}}$ → [0,1] where: $\mathrm{Pl(A)} = \sum_{B\cap A\neq\emptyset} \boldsymbol{m(B)}$.

Conditioning allows to update beliefs originally defined on A, following the disposal of a new piece of information B. The mass initially allocated to A will be transferred to A∩B. m(A|B) denotes the degree of belief of A in the context where B holds with $A, B \subseteq \Theta$. The Dempster rule of conditioning is computed by: $m(A|B) = \frac{\sum_{C, C\cap B=A} m(C)}{Pl(B)}$.

2.2. *Belief Causal Networks*

A belief causal network[6] is a compact representation of a mass distribution. It consists of two components:

- A graphical component: It is a directed acyclic graph (DAG), where nodes represent variables and arcs follow the causal process, i.e. from causes towards effects. A subset from the set of the parents of A_i, $PA(A_i)$, is denoted by $Pa(A_i)$. A subset from one parent node, $PA_j(A_i)$, is denoted by $Pa_j(A_i)$.
- A numerical component: It is the set of bba's associated to each node in the graph. For each root node A_i having a frame of discernment Θ_{A_i}, an a priori mass m(a) is defined on the powerset $2^{\Theta_{A_i}}$. For other nodes, a conditional bba $m(a|Pa_j(A_i))$ is specified.

An intervention on a variable A_i forces it to take the specific value a_{ij}. This action makes the initial causes of the manipulated variable no more responsible of its state. Graphically, it amounts to remove the arcs linking the variable of interest to its parents (mutilated graph). The effect of an intervention, denoted by $m(.|do(a_{ij}))$, corresponds to observing the value a_{ij} on this graph.

3. On The Use of Acceptance/Rejection to Ascribe Causality

3.1. *Role of Acceptance and Rejection in Ascribing Causality*

The causality-ascribing agent will use the concepts of acceptance, rejection to affirm that an event truly causes another one. An advantage of working with belief functions compared with the qualitative models[3], is the possibility to represent events with n-ary variables.

An agent will identify a causal link from these components: (1) his non-causal uncertain background knowledge about the natural course of things represented with a belief function causal network, (2) a sequence of observations or interventions occurring in his environment, (3) an event that contradicts his judgment about the normal course of things: an abnormal event.

Unlike the approach proposed by Shafer[8], the notion of abnormality is central in our model. The agent will have to identify the causes of the abnormal event. He will use the concepts of acceptance and rejection instead of changes in uncertainty distributions to model such transformations. To ascribe causality, the notion of acceptance is used to describe abnormal events and that of rejection is devoted to causes events. To reduce the set of potential causes, abnormal events from the set of observed/intervened events are privileged. In the following we will denote events by e_i, $e_i \subseteq \Theta_{E_i}$.

3.2. *Levels of Acceptance and Rejection*

An event e_i has different possible statuses, in particular it said to be[9]:

- very strongly accepted: if the confidence in this event is strictly greater than the confidence in its complement: $Pl(e_i) > Pl(\overline{e_i})$;

- very strongly rejected: if the confidence in this event is strictly less than the confidence in its complement: $Pl(e_i) < Pl(\overline{e_i})$;

- ignored: if the confidence in this event is the same than the confidence in its complement: $Pl(e_i) = Pl(\overline{e_i})$.

In this paper, we propose alternative definitions, namely:

- strongly accepted: if an event is *exclusively* in $Argmax_{e_x \in \Theta_{E_i}} Pl(e_x)$, it is said to be strongly accepted. Formally, an event e_i is strongly accepted if it is the only one that has the highest plausibility (the best confidence):

- $Pl(e_i) \in Argmax_{e_x \in \Theta_{E_i}} Pl(e_x)$.
- $\nexists\, e_k \neq e_i$ where $Pl(e_k) \in Argmax_{e_x \in \Theta_{E_i}} Pl(e_x)$.

- strongly rejected: if an event is exclusively in $Argmin_{e_x \in \Theta_{E_i}} Pl(e_x)$, it is said to be strongly rejected. Formally, an event e_i is strongly rejected if it is the only one that has the lowest plausibility (the worst confidence):

- $Pl(e_i) \in Argmin_{e_x \in \Theta_{E_i}} Pl(e_x)$.
- $\nexists\, e_k \neq e_i$ where $Pl(e_k) \in Argmin_{e_x \in \Theta_{E_i}} Pl(e_x)$.

If all the events are in $Argmin_{e_x}$ or in $Argmax_{e_x}$, then it is a situation of ignorance.

Example 3.1

Let us consider the events e_1,e_2,e_3. If m({e_1})=0.3, m({e_1,e_2})=0.25, m({e_3})=0.45 then Pl({e_1})=0.55, Pl({e_1,e_2})=0.7, Pl({e_2})=0.25 and Pl({e_3})=0.45.Thus, Pl({e_1}) < Pl({e_1,e_2}); Pl({e_1}) > Pl({e_2}) and Pl({e_1}) > Pl({e_3}). Hence, e_1 is a strongly accepted event.

4. Ascribing Strong Relations

4.1. *Ascribing Strong Causes*

Two events are perceived as causally related, if the agent at time *t*, starts believing that one of them is strongly rejected and at time *t+1*, after observing or acting on the other one, he changes his beliefs and strongly accepts it.

Definition 4.1 If an event e_i is strongly rejected, and after observing an event e_j, it becomes strongly accepted, then e_j is said to be a strong cause of e_i. Namely,

$$Pl(e_i) = Argmin_{e_x}(Pl(e_x)),\ Argmax_{e_x}(Pl(e_x|\ e_j)) = Pl(e_i|e_j) \quad (1)$$

Definition 4.2 If an event e_i is strongly rejected, and after acting on an event e_j, it becomes strongly accepted, then the intervention do(e_j) is said to be a strong cause of e_i. Namely,

$$Pl(e_i) = Argmin_{e_x}(Pl(e_x)), Argmax_{e_x}(Pl(e_x,|do(e_j))) = Pl(e_i|do(e_j)) \quad (2)$$

4.2. *Confirmation and Attenuation*

Acceptance, rejection can be confirmed or attenuated upon observing or intervening on some variables which is not the case for qualitative models[3,10].

Definition 4.3 Observing e_j (or acting on it), is said to confirm another event e_i if the plausibility of observing e_i after observing (acting) on e_j is greater than the plausibility of observing e_i alone, i.e.

$$Pl(e_i|do(e_j)) > Pl(e_i). \quad (3)$$

Definition 4.4 Observing e_j (or acting on it), is said to attenuate e_i if the plausibility of observing e_i after observing (acting) on e_j is smaller than the plausibility of observing e_i alone, i.e.

$$Pl(e_i|do(e_j)) < Pl(e_i). \quad (4)$$

5. Conclusion

In this paper, we proposed a causal model that an agent will use to identify the causes of an abnormal event. Interventions allow him to distinguish between mere correlations and causal relations. Levels of acceptance and rejection control the size of the set of potential causes. As future work, we plan to explore alternative definitions of causality ascription by combining different levels of acceptance and rejection.

References

1. J. Pearl, *Causality: Models, Reasoning and Inference* (Cambridge University Press, 2000).
2. S. Benferhat, J.-F. Bonnefon, P. Chassy, R. D. S. Neves, D. Dubois, F. D. de Saint-Cyr, D. Kayser, F. Nouioua, S. Nouioua-Boutouhami, H. Prade and S. Smaoui, A comparative study of six formal models of causal ascription, in *SUM*, 2008.
3. J. Bonnefon, R. Da Silva Neves, D. Dubois and H. Prade, Int. *J. Approx. Reasoning* **48**, 752 (2008).
4. M. Otero and R. P. Otero, Using causality for diagnosis, in *Int. Workshop on Principles of Diagnosis*, DX-80, 2000.
5. J. Halpern and J. Pearl, Causes and explanations: A structural-model approach – part I: Causes, in *UAI*, 2001.
6. I. Boukhris, Z. Elouedi and S. Benferhat, Modeling interventions using belief causal networks, in *FLAIRS*, 2011.
7. G. Shafer, *A Mathematical Theory of Evidence* (Princeton Univ. Press. Princeton, NJ, 1976).
8. G. Shafer, *The Art of Causal Conjecture* (The MIT Press, 1997).
9. I. Boukhris, S. Benferhat and Z. Elouedi, Ascribing causality from interventional belief function knowledge, in *BELIEF*, 2012.
10. S. Benferhat and S. Smaoui, Quantitative possibilistic networks: Handling Interventions, in *MICAI*, 2008.

SMOTE-FRST: A NEW RESAMPLING METHOD USING FUZZY ROUGH SET THEORY*

E. RAMENTOL
Department of Computer Science, University of Camagüey, Cuba

N. VERBIEST†
Department of Applied Mathematics and Computer Science, Ghent University, Belgium

R. BELLO
Department of Computer Science, Universidad Central de las Villas, Cuba

Y. CABALLERO
Department of Computer Science, University of Camagüey, Cuba

C. CORNELIS
Department of Computer Science, University of Camagüey, Cuba
Department of Computer Science and AI, University of Granada, Spain

F. HERRERA
Department of Computer Science and AI, University of Granada, Spain

In this paper, we introduce a new hybrid preprocessing method for editing imbalanced data. The algorithm we propose first resamples the training data using the Synthetic Minority Oversampling Technique (SMOTE) method, and subsequently applies an editing technique based on fuzzy rough set theory to the balanced training set. We evaluate the performance of our algorithm in an experimental study, using the C4.5 classifier as the learning algorithm. Statistical tests show the superiority of our method over state-of-the-art resampling methods.

1. Introduction

An important data mining task is the problem of classifying imbalanced data. These are datasets for which one of the classes is highly under- or

* This work was partially supported by the Spanish Ministry of Science and Technology under project TIN2011-28488, and the Andalusian Research project TIC-06858
† Nele.Verbiest@UGent.be

overrepresented. In case of two classes, the problem we discuss in this paper, this means that there is one class with significantly more instances than the other class. The Imbalance Ratio (IR), defined by the number of instances in the majority class divided by the number of instances in the minority class, expresses to which extent a dataset is imbalanced: a dataset with IR equal to 1 is perfectly balanced, the higher the IR, the more imbalanced the dataset. Imbalanced datasets occur in many research fields, fraud detection, oil spill detection, text classification and medical applications are a few examples.

Standard classification methods are often biased towards the majority class, because they attempt to perform well w.r.t. global quantities such as classification accuracy, not taking the data distribution into account. As a result, examples from the majority class are generally correctly classified, whereas examples from the minority class are more often misclassified. In order to evaluate if a method is able to classify both instances from the minority and majority class well, the Receiver Operating Characteristic (ROC) curve that plots the rate of correctly classified minority instances against the rate of incorrectly classified minority instances can be used. The Area Under the ROC curve (AUC[2]) reflects the trade-off between correctly classified instances in the minority class and a high classification accuracy of instances in the majority class.

Several techniques have emerged to improve a classifier's performance under the data imbalance assumption. They can be divided into two classes depending on the level on which they operate, viz. learning algorithms level, and data level. In this paper, we work with the latter type: data level algorithms mainly focus on resampling instances, i.e. either generating synthetic instances from the minority class, removing examples from the majority class or both. Specifically, we focus on the Synthetic Minority Oversampling Technique (SMOTE[4]) methodology. As its name implies, this method generates synthetic instances, belonging to the minority class, to balance the training set.

In this paper, we attempt to improve SMOTE's performance by monitoring the quality of the generated synthetic instances. In an iterative process, after applying SMOTE, we remove synthetic minority instances, as well as original majority instances, that have a small membership degree to the fuzzy positive region. Such instances are considered to be noisy and hence are filtered out from the training data. The whole process (SMOTE instance generation and fuzzy-rough instance removal) is repeated until the training set is perfectly balanced.

The remainder of this paper is structured as follows. In Section 2, we review basic preliminaries on fuzzy rough set theory, while in Section 3 we detail the methodology of our proposal, which is called SMOTE-FRST. In Section 4, an experimental study is set up to compare SMOTE-FRST to leading resampling techniques, using the well-known C4.5 classifier as the learning algorithm, and evaluating their performance on a large corpus of 44 datasets obtained from the KEEL dataset repository. Finally, in Section 5 we conclude.

2. Fuzzy Rough Set Theory

Rough set theory[6] provides a methodology for data analysis based on the approximation of concepts in a decision system $(X, A \cup \{d\})$ in which X is a set of instances, A is a set of conditional attributes and d is the decision or class attribute. The theory revolves around the notion of (in)discernibility: the ability to distinguish between instances, based on their attribute values. When fuzzy rough sets are used, indiscernibility is typically modeled by means of a fuzzy tolerance relation R in X. In this paper, R is defined as, for x and y in X,

$$R(x, y) = T_L(\underbrace{R_a(x, y)}_{a \in A}) \tag{1}$$

where T_L represents the Lukasiewicz t-norm given by $T_L(a, b) = \max(0, a + b - 1)$ for $a, b \in [0,1]$, and $R_a(x, y)$ is given by

$$R_a(x, y) = \max\left(1 - \frac{|a(y) - a(x)|}{\sigma_a}, 0\right) \tag{2}$$

if a is a quantitative (real) attribute, and

$$R_a(x, y) = \begin{cases} 1 & \text{if } a(x) = a(y) \\ 0 & \text{otherwise} \end{cases} \tag{3}$$

when a is discrete. Given R, the fuzzy-rough positive region of X is defined as the fuzzy set POS in X defined in this paper by

$$POS(x) = \min_{y \in X} I_L(R(x, y), R_d(x, y)) \tag{4}$$

where I_L is the Lukasiewicz implicator, given by $I_L(a, b) = \min(1, 1 - a + b)$ for $a, b \in [0,1]$, and $R_d(x, y) = 1$ if x and y belong to the same class, and 0 otherwise. The idea of the fuzzy-rough positive region is that instances on the border of a class (i.e., for which there exists a similar instance in another class) will have a small membership value to POS compared to instances in the center of a class. This makes the fuzzy-rough positive region suitable to measure the quality of an instance as a typical representative of its class.

3. Methodology

In this section, we introduce SMOTE-FRST, our new hybrid editing method for imbalanced datasets. The algorithm consists of two stages, which are repeated until the training set is balanced, or until a maximum number of iterations is reached (10 in our experimental study):

1. Apply SMOTE to introduce new synthetic minority class instances to the training set.
2. Remove synthetic instances, or majority class instances, for which the membership to the fuzzy-rough positive region of the training set falls below a given threshold (= 0.8 in our experimental study).

Our hypothesis is that some of the introduced synthetic minority class instances may not be suitable for use in the learning phase, and hence should be eliminated. Similarly, majority class instances in the original training data that do not sufficiently belong to the fuzzy-rough positive region are removed. Note that we do not apply this procedure to the original minority class instances, since they are relatively sparse and are better left untouched.

4. Experimental Study

In this section we compare SMOTE-FRST to six well known preprocessing algorithms based on SMOTE:

1. The **SMOTE** algorithm itself
2. SMOTE with data cleaning using Tomek Links (**SMOTE-TL**[1])
3. SMOTE with data cleaning using the Edited Nearest Neighbour technique (**SMOTE-ENN**[1])
4. Borderline SMOTE, where only border instances are resampled (**SMOTE-BL1**[5])
5. A variation on SMOTE-BL1 where the synthetic instances are closer to the minority class (**SMOTE-BL2**[5])
6. SMOTE weighting the minority instances according to their safe-level (**SMOTE-SL**[3])

We apply these algorithms and the SMOTE-FRST algorithm to 44 datasets from the KEEL dataset repository[‡]. The datasets and their IR (we only use datasets with an IR higher than 9) are listed in Table 4. For each experiment, we follow a 10 fold cross validation strategy: we divide the data in 10 folds and

‡ http://sci2s.ugr.es/keel/datasets.php

classify the instances of each fold using the remaining folds as training data. In Figure 1, we show the average AUCs over all datasets for each method[§].

SMOTE-FRST has the best average score. The improvement over SMOTE is small, so we must check if the differences are significant. We apply the Wilcoxon's signed ranks statistical test[7] to compare SMOTE-FRST against all other considered SMOTE methods. This is a non-parametric pairwise test that aims to detect significant differences between two sample means; that is, the behavior of the two implicated algorithms in the comparison. For each comparison we compute the sum of ranks of the Wilcoxon's test in favor of SMOTE-FRST R+, the sum of ranks in favor of the other methods R- and also the p-value obtained for the comparison. The observed values of the statistics are listed in Table 2. The p-values are all lower than 0.10, which means that SMOTE-FRST outperforms the other SMOTE approaches at the 10 percent significance level.

Table 1: Datasets used for the experimental study and their IRs.

Dataset	IR	Dataset	IR	Dataset	IR
Yeast2vs4	9.08	Yeast2vs8	23.1	Ecoli01vs235	9.17
Yeast05679vs4	9.35	Yeast4	28.1	Ecoli0267vs35	9.18
Vowel0	9.98	Yeast1289vs7	30.57	Glass04vs5	9.22
Glass016vs2	10.29	Yeast5	32.73	Ecoli0346vs5	9.25
Glass2	11.59	Ecoli0137vs26	39.14	Ecoli0347vs56	9.28
Shuttle0vs4	13.87	Yeast6	41.4	Ecoli067vs5	10
Yeast1vs7	14.3	Abalone19	129.44	Ecoli0147vs2356	10.59
Glass4	15.47	Ecoli034vs5	9	Led7digit02456789vs1	10.97
Ecoli4	15.8	Ecoli067vs35	9.09	Glass06vs5	11
Pageblocks13vs4	15.86	Ecoli0234vs5	9.1	Ecoli01vs5	11
Abalone9-18	16.4	Glass015vs2	0.12	Glass0146vs2	11.06
Glass016vs5	19.44	Yeast0359vs78	9.12	Ecoli0147vs56	12.28
Shuttlec2vsc4	20.5	Yeast02579vs368	9.14	Cleveland0vs4	12.62
Yeast1458vs7	22.1	Yeast0256vs3789	9.14	Ecoli0146vs5	13
glass5	22.78	Ecoli046vs5	9.15		

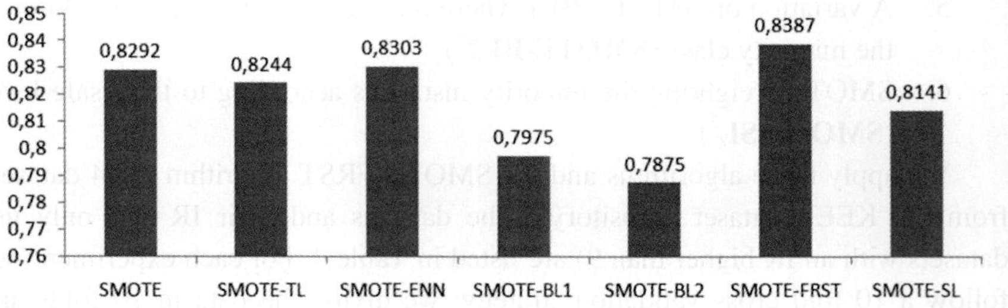

Figure 1: Average AUC over all datasets

§ Due to space constraints, we cannot list all results in this paper, but they can be found on our website: http://users.ugent.be/~nverbies

Table 2: Wilcoxon test comparing SMOTE-FRST to the other SMOTE methods w.r.t. AUC.

SMOTE-FRST vs.	R+	R-	p-value
SMOTE	554	266	0.056
SMOTE-TL	667	236	0.007
SMOTE-ENN	572	289	0.066
SMOTE-BL1	833	113	0.00
SMOTE-BL2	805	98	0.00
SMOTE-SL	843	147	0.00

5. Conclusion

In this paper, we proposed a hybrid preprocessing technique for imbalanced datasets, which iteratively applies SMOTE to balance the dataset and uses fuzzy rough techniques to carry out editing on the synthetical and majority instances. A preliminary experimental study shows that our proposed technique outperforms the existing SMOTE based techniques significantly.

References

1. G.E.A.P.A. Batista, R.C. Prati, and M.C. Monard. A study of the behavior of several methods for balancing machine learning training data. SIGKDD Explorations, 6(1):20-29, 2004.
2. Andrew P. Bradley. The use of the area under the roc curve in the evaluation of machine learning algorithms. Pattern Recognition, 30:1145-1159, 1997.
3. C. Bunkhumpornpat, K. Sinapiromsaran, and C. Lursinsap. Safe-level-smote: Safe-level-synthetic minority over-sampling technique for handling the class imbalanced problem. In Pacic-Asia Conference on Knowledge Discovery and Data Mining (PAKDD09), volume 5476, pages 475-482, 2009.
4. N.V. Chawla, K.W. Bowyer, L.O. Hall, and W.P. Kegelmeyer. Smote: Synthetic minority over-sampling technique. Journal of Artificial Intelligence Research, 16:321-357, 2002.
5. Hui Han, Wenyuan Wang, and Binghuan Mao. Borderline-smote: A new oversampling method in imbalanced data sets learning. In ICIC (1), pages 878-887,2005.
6. Z. Pawlak. Rough sets. International Journal of Computer and Information Science, 11:341-356, 1982.
7. F. Wilcoxon. Individual comparisons by ranking methods. Biometrics Bulletin, (6):80-83, 1945.

DEALING WITH NATURAL LANGUAGE INTERFACES IN A GEOLOCATION CONTEXT[1]

MOHAMMED-AMINE ABCHIR[†]
Deveryware, 43 rue Taitbout
Paris, 75009, France
LIASD – EA4383, Université Paris 8, 2 rue de la Liberté
Saint-Denis, 93526, France

ISIS TRUCK, ANNA PAPPA
LIASD – EA4383, Université Paris 8, 2 rue de la Liberté
Saint-Denis, 93526, France

Geolocation applications and software are one of the fields where the configuration of the tools is a very complex task. An interesting challenge is to try to automatize it in proposing natural language interfaces able to interact and configure the low-level devices. This paper shows how such interfaces and fuzzy models can be mixed to facilitate the elicitation of business-level objectives in the geolocation context. Two 2-tuple linguistic representation models are considered and compared in a simple use case.

1. Introduction

The geolocation applications with the mobile market are rapidly growing. Needs in geolocation software, mostly concern the problems of delivery rounds (optimization problems), fleet, vehicle and personal tracking (for people suffering from Alzheimer's disease) or child location. Usually there is one server (a kind of *hub*) that coordinates geo-information on a single platform in order to track devices (vehicles or people, they all considered as *devices* in this paper) and to give them orders knowing their position (*e.g.* what to do in case of traffic jam, or accident on the road, or if the vehicle has broken down, etc.).

Among the problems to face there are those concerning the configuration of the hub, which is often unique and has to interact with all the tracked devices. The interaction is done through low-level messages (written in the Forth

[1] This work is partially supported by the French National Research Agency (ANR) under grant number ANR-09-SEGI-012.
[†] maa@ai.univ-paris8.fr.

language, for example) in "push" or "pull" mode. In such cases, we distinguish two roles: (1) the enterprise who configures the hub to satisfy the client and (2) the client who expresses his need (his fleet tracking, for example).

Ideally, clients should configure it by themselves either through a Web interface or directly by phone, having a conversation with a virtual assistant. It is still hard to do since it implies an important expertise able to translate the needs expressed in a natural language into a set of Forth scripts and programs written in other languages. The fuzzy logic and its various methods and tools are of great help for such needs, especially the ones dealing with linguistic variables.

The paper is organized as follows: we briefly expose a linguistic fuzzy model which deal with imprecisions. Then we explain our approach that permits to mix NLP with fuzzy tools in order to translate human needs in business processes. Finally a use case show the interest of the approach as well as what is missing and what shall be continued.

2. State of the art

The first techniques (1960-1980) to deal with natural language use sets of hand-written rules [1,2], or machine learning algorithms proposing models for understanding and disambiguating sentences [3,4].

However these techniques don't deal with imprecision or vagueness in the meaning. The first approaches come from Zadeh when he introduced in 1965 the fuzzy set theory, the fuzzy logic and the concept of linguistic variables [5]. We distinguish the 2-tuple fuzzy linguistic model [6]. The idea is to deal only with words or linguistic expressions in translating them into a linguistic pair (s_i,α) where s_i is a triangular-shaped fuzzy set and α a symbolic translation.

If α is positive then s_i is reinforced else s_i is weakened. If the information is perfectly balanced (*i.e.* the distance between words is exactly the same, then all the s_i values are equally distributed on the axis). But if not – that may happen when talking about distance, for example, "almost arrived" and "close to" are closer to each other than "near" and "out of the route" – the s_i values may not be equally distributed on the axis. That led to a model called multigranular linguistic information [7]. It has been shown that despite its advantages, the 2-tuple model or unbalanced linguistic term sets doesn't fit our needs especially when one (or more) linguistic expression is very far away from its next neighbor [8,9].

We propose a new model that fully takes advantage of the symbolic translations α that become a very important element to generate the data set. Our 2-tuples are twofold. In fact, except the first and the last part of the partition, they are all composed by two half 2-tuples: an upside and a downside 2-tuple.

The choice of our 2-tuple model is relevant since the linguistic terms used in the geolocation context are usually unbalanced.

3. How a linguistic 2-tuple model can help NLP?

The methodology we use to deal with imprecision is inspired by the Parts of Speech (PoS) recognition and tagging, using simply semantic tags because the context (geolocation) is known. As an example, let us take the following preference (translated in English for the paper): "I want to create an alert when the truck gets very close to the warehouse" (see below).

```
<tokens>
        <token gram="PRON">I</token>
        ...
        <token gram="NOUN" sem="ALERT">alert</token>
        ...
        <token gram="VERB" sem="ZONE_ENTRY">gets</token>
        <token gram="ADV" sem="FUZZY_MODIF_+">very</token>
        <token gram="ADJ" sem="DISTANCE">close</token>
        ...
</tokens>
```

A tree using a simplified tree-adjoining grammar (TAG)-based is then created, where each leaf node represents the semantic tag of a token from the lexicon, *e.g.*:

```
ALERT=TYPE,MOBILE,PLACE,NOTIFICATION
TYPE=ZONE_ENTRY|ZONE_EXIT|CORRIDOR
...
PLACE=TOWN|ADDRESS|POI|ZOI
...
```

Upstream a fuzzy partition with our 2-tuple model has been performed by experts for each imprecise piece of data. For example, the expert gives 5 terms: *A, B, C, D, E* that are by default uniformly distributed on their axis. But, looking for synonyms (*e.g.* http://www.crisco.unicaen.fr/cgi-bin/cherches.cgi that exhibits a French dictionary) and the distance between the terms, our algorithm permits to obtain 5 synonym bags. Resemblance rates between bags are then computed in order to obtain new positions of *A, B, C, D, E* on the axis. Downstream the semantic tokens (*e.g.* "close") are expressed through 2-tuples and compared to the fuzzy partition. Notice that quantitative adverbs such as "very" modify their associated 2-tuple through the α value.

4. Use case and comparison of the fuzzy models

The chosen use case is about an alert where the frequency of the positions sending depends on the distance between the device and the arrival. The more the mobile gets near to the destination, the more the device must send its current position in order to notify the client as accurately as possible . The software is written in Java, using jFuzzyLogic with the extension we have proposed [9] and the FCL (Fuzzy Control Language) specification (IEC 61131-7). The FCL script below uses three different partitions for the input Distance.

```
FUNCTION_BLOCK Corridor
VAR_INPUT    Distance_V1:REAL; Distance_V2:LING; Distance_V3:LING;
END_VAR
VAR_OUTPUT
             Frequency:REAL;
END_VAR
FUZZIFY Distance_V1
             TERM InTheCenter := trian 0.0 0.0 50.0;
             ...
             TERM OutOfRoute := trian 80.0 300.0 300.0;
END_FUZZIFY
FUZZIFY Distance_V2
        TERM S := ling InTheCenter VeryCloseTo Near | Far | OutOfRoute,
extreme extreme;
END_FUZZIFY

FUZZIFY Distance_V3
        TERM S := pairs (InTheCenter, 0.0) (VeryCloseTo, 50.0) (Near, 60.0)
(Far, 80.0) (OutOfRoute, 300.0);
END_FUZZIFY
DEFUZZIFY Frequency // 5 linguistic terms
        TERM Min := trian 2.0 2.0 5.0;
        ...
        TERM Max := trian 25.0 60.0 60.0;
        METHOD : LM;              // 'Center Of Gravity'
END_DEFUZZIFY
RULEBLOCK Rules
        RULE 1: IF Distance_V1 IS InTheCenter THEN Frequency IS Min;
        ...
        RULE 6: IF Distance_V2 IS InTheCenter THEN Frequency IS Min;
                 ...
        RULE 11: IF Distance_V3 IS InTheCenter_R THEN Frequency IS Min;
        RULE 12: IF Distance_V3 IS VeryCloseTo_L
                 OR Distance_V3 IS VeryCloseTo_R THEN Frequency IS Weak;
        ...
        RULE 15 : IF Distance_V3 IS OutOfRoute_L THEN Frequency IS Max;

END_RULEBLOCK

END_FUNCTION_BLOCK
```

Figure 1 shows the three different partitions for the input: the ideal partition (Distance_V1, without 2-tuples, i.e. the subsets correspond exactly to the user's wishes but they don't fit any linguistic model), the 2-tuple partition from Herrera *et al.* (Distance_V2) and our 2-tuple partition (Distance_V3); and the partition for the output. The comparison between the three models is highlighted in Figure 2 where Herrera's partition (in green) is far away from the ideal one, especially for a distance between 55 and 120 meters and between 180 and 240. This is due to the fact that the unbalancement is weaker in this model than in the other ones. Generally speaking, our partition (in red) fits better but is less efficient after 230 meters.

The lesson to learn is that the reasoning process (GMP: *generalized modus ponens*) probably is not convenient for the 2-tuple models (whatever model).

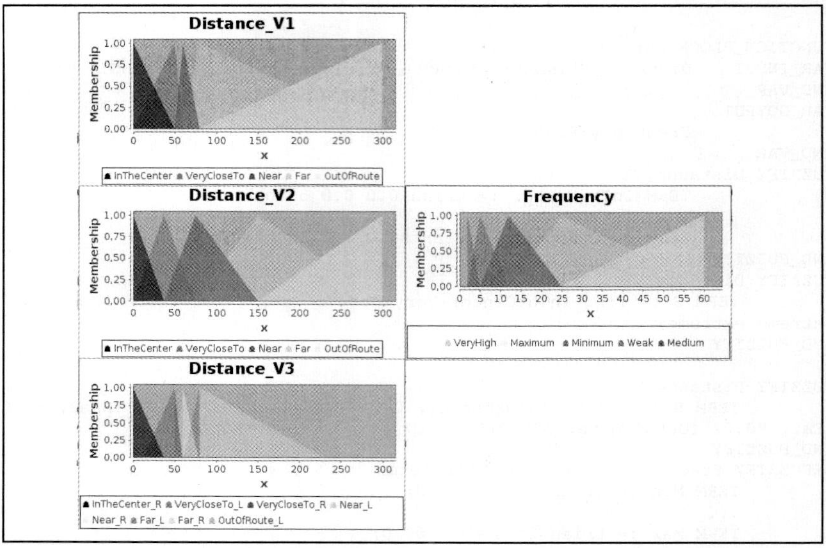

Figure 1. Left, the 3 partitions for Distance and right, the partition for the output Frequency.

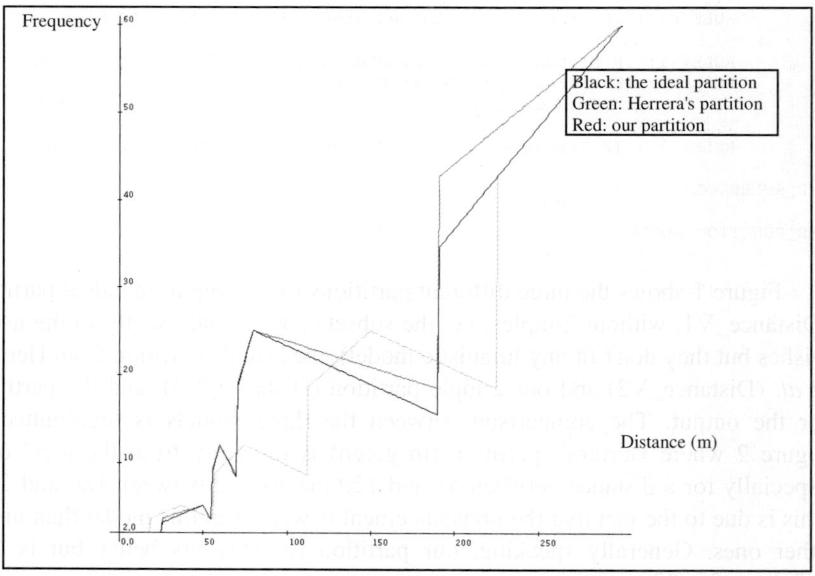

Figure 2. Comparison between the 3 different partitions in our example.

With the GMP, abscissa and ordinates are used in the computations. That shouldn't be the case using 2-tuple models because only s_i terms and α translations should be used. A membership value μ to the s_i term has no meaning in such models, since only the reinforcement or the weakening of the term is considered (through α). In other words, we have to conceive a new reasoning process in order to take into account this particularity.

5. Conclusions

In this paper we have presented a methodology to deal with natural language interfaces when data are vague. We mix NLP techniques with a 2-tuple representation model to express data with their imprecision. In a second part, we compare 3 representation models and we show that ours fits better reality in some cases. It appears that a major direction should be considered in future works: defining a new reasoning process dedicated to 2-tuple models (ours and Herrera's one).

References

1. T. Winograd, "Procedures as a Representation for Data in a Computer Program for Understanding Natural Language", MIT AI Technical Report 235, (1971).
2. J. Weizenbaum, "ELIZA — A Computer Program For the Study of Natural Language Communication Between Man And Machine", Communications of the ACM 9 (1): 36–45 (1966).
3. R.S. Michalski, J. Carbonell and T. Mitchell (Eds.), chapter in the book, Machine Learning :An artificial Intelligence Approach. Tioga Publishing Co., pp 41-81, Palo Alto (1983).
4. R. Kuhn and R. De Mori, "A Cache-Based Natural Language Model for Speech Recognition". IEEE Transactions on Pattern Analysis and Machine Intelligence 12 (6): 570–583 (1990).
5. L. A. Zadeh, "Fuzzy sets", Information and Control 8 (3): 338–353 (1965).
6. F. Herrera and L. Martínez, "A 2-tuple fuzzy linguistic representation model for computing with words". IEEE Transactions on Fuzzy Systems, 8(6):746–752 (2000).
7. F. Herrera, E. Herrera-Viedma, and L. Martínez, "A fuzzy linguistic methodology to deal with unbalanced linguistic term sets." IEEE Transactions on Fuzzy Systems, 354–370 (2008).
8. L. Martínez, D. Ruan, and F. Herrera, "Computing with words in decision support systems: An overview on models and applications". International Journal of Computational Intelligence Systems, 3(4):382–395 (2010).
9. M.-A. Abchir and I. Truck, "Towards a New Fuzzy Linguistic Preference Modeling Approach for Geolocation Applications". In Proc. of the EUROFUSE Workshop, 413–424 (2011).

INDUCED AND HEAVY AGGREGATION OPERATORS*

JOSÉ M. MERIGÓ†1, YEJUN XU[2]
[1]Department of Business Administration, University of Barcelona, Av. Diagonal 690, Barcelona, 08034, Spain
[2]Business School , Hohai University, Jiangning, Nanjing 210098, PR China

We present some new induced and heavy aggregation operators in a unified framework between the weighted average and the ordered weighted average. We introduce the induced heavy ordered weighted averaging weighted average operator. It is an aggregation operator that provides a parameterized family of aggregation operators between the minimum and the total operator. It considers complex reordering processes assessed with order inducing variables. We study some of its main properties and particular cases.

1. Introduction

The aggregation operators are becoming more popular in the scientific community [1,6]. They provide better tools for representing the information and have been applied in a wide range of areas. A very common an aggregation operator is the ordered weighted averaging (OWA) operator [7,10]. It provides a parameterized family of aggregation operators between the minimum and the maximum. It can be generalized by using order inducing variables that represent complex reordering processes obtaining the induced OWA (IOWA) operator [5,9]. It can also be extended allowing the weighting vector to range from the minimum to the total operator forming the heavy OWA (HOWA) operator [3,4,8]. Recently, Merigó and Casanovas have presented the induced heavy OWA (IHOWA) operator as a generalization of the previous approaches.

Further interesting aggregation operators are those that use the weighted average in the analysis. In this context, Merigó [2] has presented the induced OWA weighted average (IOWAWA) operator as a unified model between the weighted average and the OWA operator that considers the degree of importance that each concept has in the analysis.

The objective of this paper is to introduce a more general framework to the previous approaches that represent the information in a more efficient way. We

* This work is supported by the University of Barcelona through the project 099311.
† jmerigo@ub.edu

present the induced heavy OWA weighted average (IHOWAWA) operator. It is an aggregation operator that unifies the weighted average and the IOWA operator in the same formulation and allowing the weighting vector to range from the minimum to the sum of all the arguments. Moreover, it uses order inducing variables that deals with complex environments. We study some of its main properties and particular cases and we see that the IHOWAWA operator includes a wide range of particular cases.

2. Preliminaries

In this Section we briefly review the HOWA operator, the IOWA operator and the OWAWA operator.

2.1. *The Heavy OWA Operator*

The HOWA operator [8] is an extension of the OWA operator that allows the weighting vector to sum up to n. Thus, we are able to include the total operator in the aggregation. It can be defined as follows:

Definition 1. A HOWA operator is a mapping HOWA: $R^n \rightarrow R$ that has an associated weighting vector W with $w_j \in [0, 1]$ and $1 \leq \sum_{j=1}^{n} w_j \leq n$, such that:

$$HOWA\ (a_1, \ldots, a_n) = \sum_{j=1}^{n} w_j b_j\ , \tag{1}$$

where b_j is the jth largest of the a_i and a_i is the argument variable.

As we can see, if $\sum_{i=1}^{n} w_i = 1$, then, we get the usual weighted average and if $\sum_{i=1}^{n} w_i = n$, the total operator.

2.2. *The Induced OWA Operator*

The IOWA operator [9] is an extension of the OWA operator. Its main difference is that the reordering step is not carried out with the values of the arguments a_i. In this case, the reordering step is developed with order inducing variables that reflects a more complex reordering process. It can be defined as follows.

Definition 2. An IOWA operator of dimension n is a mapping IOWA: $R^n \rightarrow R$ that has an associated weighting vector W of dimension n such that the sum of the weights is 1 and $w_j \in [0, 1]$, then:

$$IOWA\ (\langle u_1, a_1\rangle, \langle u_2, a_2\rangle, \ldots, \langle u_n, a_n\rangle) = \sum_{j=1}^{n} w_j b_j\ , \tag{2}$$

where b_j is the a_i value of the IOWA pair $\langle u_i, a_i\rangle$ having the jth largest u_i, u_i is the order inducing variable and a_i is the argument variable.

2.3. *The OWAWA Operator*

The ordered weighted averaging – weighted averaging (OWAWA) operator [2] is an aggregation operator that unifies the WA and the OWA operator in the same formulation considering the degree that each concept has in the analysis. It can be defined as follows.

Definition 3. An OWAWA operator of dimension n is a mapping OWAWA: $R^n \rightarrow R$ that has an associated weighting vector W of dimension n such that $w_j \in [0, 1]$ and $\sum_{j=1}^{n} w_j = 1$, according to the following formula:

$$OWAWA\ (a_1, \ldots, a_n) = \sum_{j=1}^{n} \hat{v}_j b_j\ , \qquad (3)$$

where b_j is the jth largest of the a_i, each argument a_i has an associated weight (WA) v_i with $\sum_{i=1}^{n} v_i = 1$ and $v_i \in [0, 1]$, $\hat{v}_j = \beta w_j + (1-\beta) v_j$ with $\beta \in [0, 1]$ and v_j is the weight (WA) v_i ordered according to b_j, that is, according to the jth largest of the a_i.

3. The Induced Heavy OWAWA Operator

The IHOWAWA operator is an aggregation operator that uses induced and heavy aggregation operators in a unified framework between the OWA operator and the weighted average. Thus, it provides a parameterized family of aggregation operators between the minimum and the total operator. It also deals with complex environments where the information is assessed with order-inducing variables. Its main advantage is that it can deal with subjective information and with the complex attitudinal character of the decision maker in the same formulation. It can be defined as follows.

Definition 4. An IHOWAWA operator of dimension n is a mapping IHOWAWA: $R^n \times R^n \rightarrow R$ that has an associated weighting vector W of dimension n such that $w_j \in [0, 1]$ and $1 \le \sum_{j=1}^{n} w_j \le n$, according to the following formula:

$$IHOWAWA\ (\langle u_1,a_1\rangle, \langle u_2,a_2\rangle, \ldots, \langle u_n,a_n\rangle) = \sum_{j=1}^{n} \hat{v}_j b_j\ , \qquad (4)$$

where b_j is the a_i value of the IHOWAWA pair $\langle u_i, a_i\rangle$ having the jth largest u_i, u_i is the order inducing variable, each argument a_i has an associated weight (WA)

v_i with $1 \le \sum_{i=1}^{n} v_i \le n$ and $v_i \in [0, 1]$, $\hat{v}_j = \beta w_j + (1-\beta) v_j$ with $\beta \in [0, 1]$ and v_j is the weight (WA) v_i ordered according to b_j, that is, according to the jth largest of the a_i.

Note that it is also possible to formulate the IHOWAWA operator separating the part that strictly affects the IHOWA operator and the part that affects the HWA.

$$IHOWAWA\ (\langle u_1,a_1\rangle, \langle u_2,a_2\rangle, \ldots, \langle u_n,a_n\rangle) = \beta \sum_{j=1}^{n} w_j b_j + (1-\beta) \sum_{i=1}^{n} v_i a_i\ , \quad (5)$$

where b_j is the a_i value of the IHOWAWA pair $\langle u_i, a_i\rangle$ having the jth largest u_i, u_i is the order inducing variable and $\beta \in [0, 1]$.

In the following, we are going to give a simple example of how to aggregate with the IHOWAWA operator.

Example 1. Assume the following arguments in an aggregation process: (40, 20, 70, 50). Assume the following order inducing variables U = (7, 4, 9, 2), weighting vector W = (0.7, 0.6, 0.4, 0.3) and the following weighting vector V = (0.4, 0.4, 0.5, 0.7). Note that the WA has a degree of importance of 60% while the weighting vector W of the OWA a degree of 40%. If we want to aggregate this information by using the IHOWAWA operator, we get the following:

$$\hat{v}_1 = 0.4 \times 0.7 + 0.6 \times 0.5 = 0.58\ ; \ \hat{v}_2 = 0.4 \times 0.6 + 0.6 \times 0.4 = 0.48$$

$$\hat{v}_3 = 0.4 \times 0.4 + 0.6 \times 0.4 = 0.4\ ; \ \hat{v}_4 = 0.4 \times 0.3 + 0.6 \times 0.7 = 0.54$$

Then, we calculate the aggregation process as follows:

IHOWAWA = 0.58×70 + 0.48×40 + 0.4×20 + 0.54×50 = 94.8.

Note that with Eq. (5) we also get the same result.

It is possible to distinguish between descending (DIHOWAWA) and ascending (AIHOWAWA) orders by using the transformation $w_j = w^*_{n-j+1}$.

The IHOWAWA is monotonic, idempotent and bounded. Note that this operator is not commutative because the weighted average is not commutative. It is monotonic because if $a_i \ge e_i$, for all i, then, $IHOWAWA\ (\langle u_1,a_1\rangle, \ldots, \langle u_n,a_n\rangle) \ge IHOWAWA\ (\langle u_1,e_1\rangle, \ldots, \langle u_n,e_n\rangle)$. It is idempotent because if $a_i = a$, for all a_i, then, $IHOWAWA\ (\langle u_1,a_1\rangle, \ldots, \langle u_n,a_n\rangle) = a$. It is bounded because the IHOWAWA aggregation is delimitated by the minimum and the total operator. That is, $\mathrm{Min}\{a_i\} \le IHOWAWA\ (\langle u_1,a_1\rangle, \ldots, \langle u_n,a_n\rangle) \le \sum_{i=1}^{n} a_i$.

A further interesting issue is to analyze the measures for characterizing the weighting vector W following the methodology used for the HOWA operator [5,14,16]. Different types of IHOWAWA operators are found by using a different manifestation in the weighting vector or in the coefficient β. For example, we can obtain the following cases.

- The IOWAWA operator: When the sum of the weights W and V is 1.
- The total operator: When the sum of the weights W and V is n.
- The total weighted average operator: When $W = n$ and $V = 1$.
- The total heavy weighted average operator: When $W = n$.
- The total IOWA operator: $V = n$ and $W = 1$.
- The total heavy IOWA operator: $V = n$.
- The weighted heavy IOWA operator: $V = 1$.
- The IOWA heavy weighted average: $W = 1$.
- The IHOWA operator: If $\beta = 1$.
- The heavy weighted average: If $\beta = 0$.
- The heavy average: If $w_j = |W|/n$ and $v_i = |W|/n$, or $w_j = |W|/n$ and $\beta = 1$, or $v_i = |W|/n$ and $\beta = 0$.
- The heavy arithmetic weighted average: If $w_j = |W|/n$.
- The heavy arithmetic IOWA operator: If $v_i = |W|/n$.
- The arithmetic mean: If $w_j = 1/n$ and $v_i = 1/n$, for all i.
- The minimum: When $w_n = 1$, $w_j = 0$, for all $j \neq n$ and $\beta = 1$.
- The heavy weighted minimum: When $w_n = 1$, $w_j = 0$, for all $j \neq n$.
- Heavy weighted push up allocation: We use $w_j = (1 \wedge (|W| - (j-1))) \vee 0$.
- Heavy weighted push down allocation: In this case, $w_{n-j+1} = (1 \wedge (|W| - (j-1))) \vee 0$.
- Heavy weighted step-IHOWAWA allocation: We assume $b = \text{Min}[(K-1), (n-K)]$, and we allocate the weights for j to b as $w_K = 1$ and $w_{K+j} = w_{K-j} = [1 \wedge ([(|W|-1) - 2(j-1)]/2)] \vee 0$.

Note that it is possible to develop other types of IHOWAWA operators by using other manifestations of the weighting vector of the WA and the IOWA.

4. Conclusions

We have presented the IHOWAWA operator. It is an aggregation operator that provides a parameterized family of aggregation operators between the minimum and the total operator and in a unified framework between the weighted average and the IOWA operator. We have studied some of its main properties and particular cases including the heavy weighted average, the HOWA operator, the IOWAWA operator, the HOWAWA operator and many others. In future research we expect to develop further developments by adding new concepts in the analysis such as the use of generalized aggregation operators and distance measures.

References

1. G. Beliakov, A. Pradera and T. Calvo, *Aggregation functions: A guide for practitioners* (Springer-Verlag, Berlin-Heidelberg, 2007).
2. J.M. Merigó, A unified model between the weighted average and the induced OWA operator, *Expert Syst. Applic.* **38**, 11560 (2011).
3. J.M. Merigó and M. Casanovas, Induced and heavy aggregation operators with distance measures, *J. Syst. Engineering and Electronics* **21**, 431 (2010).
4. J.M. Merigó and M. Casanovas, Induced and uncertain heavy OWA operators, *Computers & Industrial Engineering* **60**, 106 (2011).
5. J.M. Merigó and A.M. Gil-Lafuente, The induced generalized OWA operator. *Inform. Sciences* **179**, 729 (2009).
6. Z.S. Xu and Q.L. Da, An overview of operators for aggregating information. *Int. J. Intelligent Syst.* **18**, 953 (2003).
7. R.R. Yager, On ordered weighted averaging aggregation operators in multi-criteria decision making, *IEEE Trans. Syst. Man Cybern. B* **18**, 183 (1988).
8. R.R. Yager, Heavy OWA operators. *Fuzzy Optim. Decision Making*, **1**, 379 (2002).
9. R.R. Yager and D.P. Filev, Induced ordered weighted averaging operators, *IEEE Trans. Syst. Man Cybern. B* **29**, 141 (1999).
10. R.R. Yager, J. Kacprzyk and G. Beliakov. *Recent developments on the ordered weighted averaging operators: Theory and practice*. (Springer-Verlag, Berlin-Heidelberg, 2011).

PART 4

STATISTICS, DATA ANALYSIS AND DATA MINING

FORECASTING HIGH FREQUENCY FINANCIAL DATA: STATISTICAL AND FUZZY LOGIC RBF ANN APPROACH*

DUSAN MARCEK

Research Institute of the IT4I Centre of Excellence, Faculty of Philosophy and Science Silesian University in Opava, Bezruc. Square 13, 746 01 Opava, Czech Republic

Forecast accuracy of economic and financial processes is a popular measure for quantifying the risk in multi-criteria decision making. In this paper, we develop forecasting models based on the statistical (stochastic) methods, sometimes called as hard computing, and on the soft methods using granular computing. To illustrate the forecasting performance of these approaches the learning aspects of RBF networks are presented and an application is included. We show a new approach of function estimation for nonlinear time series model by means of a granular fuzzy logic neural network based on Gaussian activation function with cloud concept. In a comparative study is shown that the presented approach is able to model and predict the high frequency data with reasonable accuracy and more efficient than the statistical methods.

1. Introduction

Over the past ten years academics of computer science have developed new soft techniques based on the latest information technologies such as soft, neural and granular computing to help to predict the future values of high frequency financial data. At the same time, the field of financial econometrics has undergone various new developments, especially in finance models, stochastic volatility, and software availability. This paper discusses and compares the forecasts from volatility models which are derived from competing statistical and Radial Basic Function (RBF) neural network (NN) specifications. Our motivation for this comparative study lies in both the difficulty for constructing of appropriate statistical Autoregressive/Generalised Conditionally Heteroscedastic (ARCH-GARCH) models (so called hard computing) to forecast the volatility even in ex post simulations, and the recently emerging problem-solving methods that exploit tolerance for impression to achieve low

* This paper has been elaborated in the framework of the IT4Innovations Centre of Excellence project, reg. no. CZ.1.05/1.1.00/02.0070 supported by Operational Programme 'Research and Development for Innovations' funded by Structural Funds of the European Union and state budget of the Czech Republic.

solution costs (soft computing). The aim of the paper is to explain, and to show achieved aspects of both statistical and soft approach for quantifying the forecast accuracy applied to the daily closed prices for AUD (Australian dollar) against the US dollar (denoted AUD/USD) which is, according to definition by Zivot and Wang [1], the high frequency data sequence. This paper is organized in the following manner. In Section 2 we briefly describe the basic methodology of ARCH-GARCH models. In Section 3 we present the data, conduct some preliminary statistical analysis of the time series and demonstrate the forecasting abilities of the ARCH-GARCH models of an application. Section 4 is devoted to the granular computing approach applied for fuzzy logic RBF neural networks. This approach is suggested as an alternative to the approach analyzed in Section 3. In Section 5 we put an empirical comparison. Section 6 briefly concludes.

2. Some ARCH-GARCH Models for Financial Data

ARCH models are considered as some specific nonlinear time series models, which allows for quite exhaustive study of the underlying dynamics. ARCH-GARCH models are designed to capture certain characteristics that are commonly associated with financial time series. They are among others: fat tails, volatility clustering, persistence, mean-reversion and leverage effect. The first model that provides a systematic framework for volatility modelling is the ARCH model introduced by Engle [2]). Bollerslev [3] proposes a useful extension of Engle´s ARCH model known as the generalized ARCH (GARCH) model for time sequence $\{ y_t \}$ in the following form

$$y_t = \varepsilon_t \sqrt{h_t} \qquad . \tag{1}$$
$$h_t = \alpha_0 + \sum_{i=1}^{m} \alpha_i y_{t-i}^2 + \sum_{j=1}^{s} \beta_j h_{t-j}$$

where $\{ \varepsilon_t \}$ is a sequence of iid random variables with zero mean and unit variance, α_i a β_i are the ARCH and GARCH parameters, h_t represents the conditional variance of time series conditional on all the information to time t -1.

In the literature several variants of basic ARCH-GARCH model (1) has been derived. Nelson [4] proposed the exponential GARCH model abbreviated as EGARCH to allow for leverage effects in the form

$$\log h_t = \alpha_0 + \sum_{i=1}^{p} \alpha_i \frac{\left|\hat{\varepsilon}_{t-i}\right| + \gamma_i \hat{\varepsilon}_{t-i}}{\sigma_{t-i}} + \sum_{j=1}^{q} \beta_j h_{t-j}, \tag{2}$$

where $\hat{\varepsilon}_{t-i}$ are the residuals, γ_i denotes the coefficient of the leverage effects. Note if $\hat{\varepsilon}_{t-i}$ is positive or there is “good news”, the total effect of $\hat{\varepsilon}_{t-i}$ is $(1+\gamma_i)\hat{\varepsilon}_{t-i}$. However contrary to the “good news”, i.e. if $\hat{\varepsilon}_{t-i}$ is negative or there is "bad news", the total effect of $\hat{\varepsilon}_{t-i}$ is $(1-\gamma_i)\left|\hat{\varepsilon}_{t-i}\right|$. Bad news can have a larger impact on the volatility. Then the value of γ_i would be expected to be negative.

Ding and *et al.* [5] extended the basic GARCH model to allow for leverage effects. This model has the form as follows

$$h_t^d = \alpha_0 + \sum_{i=1}^{p} \alpha_i \left(\left|\hat{\varepsilon}_{t-i}\right| + \gamma_i \hat{\varepsilon}_{t-i}\right)^d + \sum_{j=1}^{q} \beta_j h_{t-j}^d, \quad (3)$$

where d is a positive exponent (see also [1]).

A model which is also able to generate the leverage effect is the TGARCH (threshold) model. This model is also called as GJR-GARCH because Glosten, Jagannathan, Runkle [6] proposed the same model in the following form

$$h_t = \alpha_0 + \sum_{i=1}^{p} \alpha_i \varepsilon_{t-i}^2 + \sum_{i=1}^{p} \gamma_i S_{t-i} \varepsilon_{t-i}^2 + \sum_{j=1}^{q} \beta_j h_{t-j},$$

where $S_{t-i} = 1$, if $\varepsilon_{t-i} < 0$ and $S_{t-i} = 0$, if $\varepsilon_{t-i} \geq 0$. In this model the exceeded value of the leverage effect is always positive.

Another ARCH-GARCH models as the ARCH-GARCH regression and ARCH-GARCH mean model can be found in [7].

3. An Application of ARCH-GARCH Models

We illustrate the ARCH-GARCH methodology on the developing of forecast model for the time series of daily exchange rates AUD currency against US dollar. We would like to develop a prediction model for one daily ahead forecast of the changes in the AUD/USD-exchange rates (see Figure 1 left). Firstly, the sample period for analysis from January 2, 2001 to October 30, 2010, (denoted as training data set) was defined, i.e. the period over which the forecasting model can be developed and the ex post forecast period from November 1, 2010 to December 31, 2010 (denoted as validation or ex-post data set). After some experiments, the following AR(0) + TGARCH (1,1,1) model was quantified.

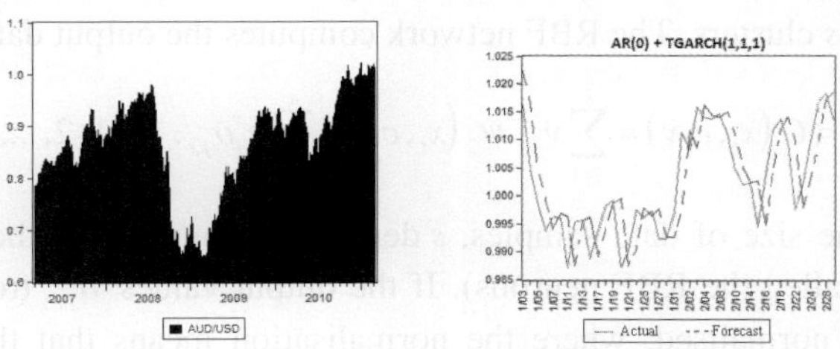

Figure 1. The time series of the daily exchange rates AUD currency against the US dollar (2007 – 2011) (left.). Actual (solid line) and the forecast (dotted line) values of the AUD/USD exchange rate generated from model AR(0)+ TGARCH (1,1,1) with Gaussian error distribution (right).

$$y_t = 0{,}00035547 + \varepsilon_t \cdot \qquad (4)$$

$$h_t = 1{,}247174 10^{-6} + 0{,}048885\varepsilon_{t-1}^2 + 0{,}055831\varepsilon_{t-1}^2 S_{t-1} + 0{,}908330 h_{t-1} \cdot \qquad (5)$$

where $S_{t-i} = 1$ if $\varepsilon_{t-i} < 0$, and $S_{t-i} = 0$ if $\varepsilon_{t-i} \geq 0$.

4. Models based on RBF NNs

In this section we show an approach of function estimation based on RBF NNs. RBF NNs are described by their architectures (see Figure 2). In Figure 2(a) each circle or node represents the neuron. This neural network consists an input layer with input vector **x** and an output layer with the output value $\hat{y}_t$.

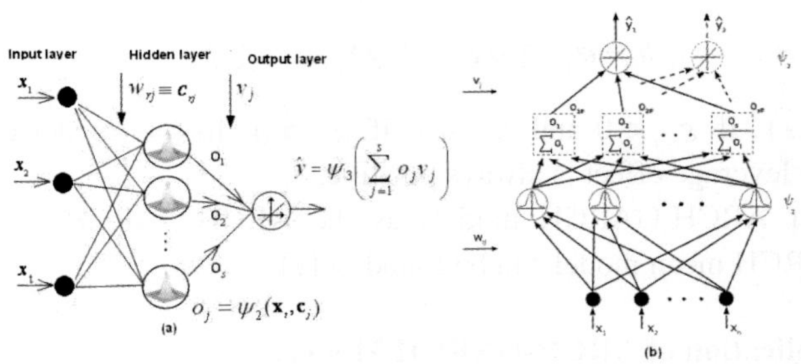

Figure 2. Classic RBF NN architecture (a), fuzzy logic RBF NN (b) architecture (b).

The output signals of the hidden layer are

$$o_j = \psi_2\left(\left\|\mathbf{x} - \mathbf{w}_j\right\|\right), \cdot \qquad (6)$$

where x is a k-dimensional neural input vector, $\mathbf{w}_j$ represents the hidden layer weights, ψ_2 are radial basis (Gaussian) activation functions. Note that for an RBF network, the hidden layer weights $\mathbf{w}_j$ represent the centers $\mathbf{c}_j$ of the activation functions ψ_2. To find the weights $\mathrm{w_j}$ or centers of activation functions we used the adaptive (learning) version of K-means clustering algorithm for s clusters. The RBF network computes the output data set as

$$\hat{y}_t = G(x_t, c, v) = \sum_{j=1}^{s} v_{j,t}\, \psi_2(x_t, c_j) = \sum_{j=1}^{s} v_j o_{j,t},\ t = 1, 2, ..., \mathrm{N} \cdot \qquad (7)$$

where N is the size of data samples, s denotes the number of the hidden layer neurons (so called the RBF neurons). If the output values $o_{j,t}$ from the hidden layer will be normalised, where the normalisation means that the sum of the outputs from the hidden layer is equal to 1, then the RBF network will compute the "normalised" output data set $\hat{y}_t$ as follows

$$\hat{y}_t = G(x,c,v) = \sum_{j=1}^{s} v_{jt} [\psi_2(x_t, c_j)] / \left[\sum_{j=1}^{s} \psi_2(x_t, c_j) \right], t = 1, 2, ..., \mathrm{N} \quad (8)$$

The network with one hidden layer and the normalised output values $o_{j,t}$ is the fuzzy logic model or soft RBF network [8]. In our case, the subjects of learning are the weights $v_{j,t}$ only. These weights can be adapted by the error back-propagation algorithm. The learning rule has the form

$$v_{j,t} \leftarrow v_{j,t} + \eta\, o_{j,t} e_t, j = 1, 2, ..., \mathrm{s},\ t = 1, 2, \ldots, \mathrm{N} \quad (9)$$

where the term η, $\eta \in (0,1)$ is a constant called the learning rate, e_t is the error given by $e_t = y_t - \hat{y}_t$.

5. Empirical Comparison

The RBF NN was trained using the variables and data sets as the ARCH-GARCH models (4), (5). In the granular RBF neural network framework, the non-linear forecasting function *f*(**x**) was estimated according to the expressions (8) with cloud Gaussian RB function $\psi_2(./.)$ [9]. The forecasting ability of particular networks was measured by the RMSE (Root Mean Square Error) criterion of ex post forecast periods (validation data set). The detailed computational algorithm for calculation the forecast accuracy values and the weight update rule for the granular network is shown in [9]. The results of this application for various architectures of granular RBF networks are shown in Table 1. The forecast accuracy measured by RMSE of neural networks is compared with AR(0)+TGARCH (1,1,1) model given by equations (4) and (5). A direct comparison between the statistical (RMSE = 0.006205) and neural network models (RMSE= 0.00531) shows that the neural network approach is better than the statistical one. The achieved ex post accuracy of the AR-TGARCH model is still reasonable and acceptable for use in forecasting systems that routinely predict the values of variables important in decision processes. As we could see, the RBF NNs have such attributes as computational efficiency, simplicity, and ease adjusting to changes in the process being forecast. ARCH-GARCH models require more costs of development, installation and operation in a management system, management comprehension and cooperation, and often a lot of computational time.

6. Conclusion

This paper has presented the granular RBF neural network based on Gaussian activation function modeled by cloud concept for solving approximation and prediction problems of real financial and economic processes. The neural network is suggested as an alternative to widely used statistical and econometric techniques in time series modeling and forecasting. The power of the granular

RBF NN was tested against some nonlinear high frequency data. A comparative analysis of two empirical studies was executed in order to evaluate forecast accuracies for both approaches. The presented neural network, or soft computing approach, is applied on real data and time series with different models. It is able by using input-output data to find a relevant functional structure between the input and the output. The proposed methodology is believed to be helpful in future research and its applications.

Table 1. Ex post forecast RMSEs for various granular RBF NNs (validation data set).

Number of RBF neurons	Classic RBF NN	Fuzzy logic with cloud concept RBF NN
1	0,01355	0,013572
2	0,009171	0,008118
3	0.009654	0.007232
4	0.011336	0.006066
5	0.012981	0.005762
6	0.016152	0.00638
7	0.015543	**0.00531**
8	0.016474	0.00584
10	0.017914	0.005692

References

1. E. Zivot, E., J. Wang, J., Modeling Financial Time Series with S-PLUS®, NY: Springer Verlag, NY, (2005).
2. R.F. Engle, Autoregressive Conditional Heteroscedasticity with Estimates of the Variance of United Kingdom Inflation. *Econometrica* 50 (4), 987-1007 (1982).
3. D. Bollerslev, Generalized Autoregressive Conditional Heteroscedasticity, *Journal of Econometrics* Vol. 31, 307-327 (1986)
4. D. B. Nelson, Conditional Heteroskedasticity in Asset Returns: A New Approach, *Econometrica* 59 (2), 347-370 (1991).
5. Z. Ding, C. W. Granger and R. F. Engle, A Long Memory Property of Stock Market Returns and a New Model, *Journal of Empirical Finance*, 1, 83-106, (1993).
6. L.R. Glosten; R. Jaganathan, and D.E. Runkle: On the relation between the expected value and the volatility of the nominal excess returns on stocks, Journal of Finance 48, 1779-1801, (1993).
7. M. Marcek, Multiple Statistical Data Analysis and Time Series Models in Economics, The University Press of Opava (2009).
8. V. Kecman, Learning and soft computing: support vector machines, neural networks, and fuzzy logic, Massachusetts: *The MIT Press*, (2001).
9. M. Marcek, L. Pancikova and D. Marcek, Econometrics & Soft Computing, The University Press, Zilina, (2008).

STATISTICAL MODELS, SV REGRESSION MODELS AND SOFT NEURAL NETWORKS: AN APPLICATION TO THE INPUT-OUTPUT FUNCTION ESTIMATION OF SALES PROCESSES*

DUSAN MARCEK†

Research Institute of the IT4Innovations Centre of Excellence, Institute of Computer Science, Silesian University in Opava, Bezruc. Square 13,746 01 Opava, Czech Republic

This article discusses statistical, SVM models and fuzzy logic neural network as new managerial tools for finding the input/output function of sales processes for industrial companies. The study reviews experimental results, discusses, analytically and numerically demonstrates the quality and interpretability of the obtained results.

1. Introduction

In economics there are many situations where quantities under investigation are functionally related or related in a more obscure manner. Very often it is difficult, or even impossible, to distinguish the independent variables. Traditionally, the formulation of multivariate models based on the econometric approach requires the following steps [1]: (i) choice of independent variables; (ii) specification of a functional form of the model; (iii) analysis of a data set appropriate to the theoretical and functional form of the model; (iv) model estimation, statistical testing and evaluation of the model´s forecasting capabilities. Any multivariate econometric (regression) model must first be based on a theoretical foundation for the multivariate relationship. In economic and finance, where our understanding of real phenomena is very poor and incomplete, it seems to be more realistic and more useful, instead of making unrealistic mathematical assumptions about functional dependency, to take the data as they are and try to represent the relationships among them in such a way that as much information as possible would be preserved. Very frequently, in

* This paper has been elaborated in the framework of the IT4Innovations Centre of Excellence project, reg. no. CZ.1.05/1.1.00/02.0070 supported by Operational Programme 'Research and Development for Innovations' funded by Structural Funds of the European Union and state budget of the Czech Republic.

† email: Dusan.Marcek@fri.uniza.sk

such cases more sophisticated approaches are considered. These approaches are based on the human experience knowledge. They consist of series linguistic expressions each of which takes the form of an´ if ... then ...´ fuzzy rule and they are well known under the common name fuzzy controllers. But also, an expert is usually unable linguistically describe the behaviour of economic processes in particular situations. Hence, most recent researches in the fuzzy controllers design for deriving of linguistically interpreted fuzzy rules have been centered on developing automatic methods to build these fuzzy rules using a set of numerical input-output data. In [2], it is shown how to use and incorporate both fuzzy sets theory and computational networks to determine the fuzzy relational equations from the database for inflation modelling. For applying these methods it is supposed that a database describing previous input-output behaviour of a system is available. Majority of these models and data-driven techniques relies on the use of Takagi-Sugeno type controllers [3] and clustering/fuzzy-clustering [4].

This contribution discusses the determining of appropriate causal (statistical) models to find an appropriate input/output function for a sales process. We use these models as a tool to compare their fit to the data with those obtained using SVM´s methods and models based on fuzzy logic RBF NN. The paper is organised as follows. In section 2, we briefly describe the framework of SVM´s methods (SV regression model). In section 3 the performance statistical, SV regression models and models based on RBF NN are investigated and evaluated. Section 4 briefly concludes.

2. SVMs for Functional Approximation and Prediction

This section quickly presents a relatively new type of learning machine – the SVM´s applied in the regression (functional approximation and prediction) problems. For details we refer to [5]. The general regression learning task is set as follows. The learning machine is given *n* training data, from which it attempts to learn the input-output relationship $y = f(x)$, where $\{ x_i, y_i \in \Re^n \times \Re, i = 1,2,...,n \}$ consists of *n* pairs $\{ y_i, x_i \}_{i=1}^{n}$. The x_i denotes the *i*-th input and y_i is the *i*-th output. The SVM´s considers the regression functions of two (primal and dual) representations [6]. The first one has the form

$$f(\mathbf{x}, \mathbf{w}) = \sum_{i=1}^{n} w_i \varphi_i(\mathbf{x}) + b \tag{1}$$

and the second one is

$$f(\mathbf{x}) = \sum_{\substack{i=1 \\ j=1}}^{n} (\alpha_i - \alpha_i^*)\psi(\mathbf{x}_i, \mathbf{x}_j) + b \tag{2}$$

where α_i, α_i^*, w_i are the adaptable weights, b is a real constant, $\psi(./.), \varphi_i(./.)$ are the kernel functions. The admissible kernels have the following forms: $\psi(\mathbf{x}_i, \mathbf{x}_j) = \mathbf{x}_i^T \mathbf{x}_j$ (the linear SVM´s) $\psi(\mathbf{x}_i, \mathbf{x}_j) = (\mathbf{x}_i^T \mathbf{x}_j + 1)^d$ (the polynomial SVM´s of degree d), $\psi(\mathbf{x}_i, \mathbf{x}_j) = \exp\left(-\theta \left\| \mathbf{x}_i - \mathbf{x}_j \right\|_2^2\right)$ (the radial basis SVM´s), where θ is a positive real constant and other (spline, B-spline, exponential RBF, etc.), $\varphi(.)$ is a non-linear function (kernel) which maps the input space into a high dimensional feature space.

The SV regression approach is based on defining a loss function that ignores errors that are within a certain distance of the true value. There are different error (loss) functions in use and that each one results from a different final model. Figure 1 shows the typical shapes of three loss functions [7].

Formally this results from solving the following Quadratic Programming problem

$$\min_{\mathbf{w},b,\xi,\xi^*} R(w, \xi, \xi^*) = \frac{1}{2}\mathbf{w}^T\mathbf{w} + C\sum_{i=1}^{n}(\xi_i + \xi_i^*) \tag{3}$$

where C is the value of capacity degrees,

$$\text{subject to} \quad \begin{cases} y_i - \mathbf{w}^T\varphi(\mathbf{x}) - b \le \varepsilon + \xi_i & i = 1,2,\ldots,n, \\ \mathbf{w}^T\varphi(\mathbf{x}) + b - y_i \le \varepsilon + \xi_i^* & i = 1,2,\ldots,n, \\ \xi_i, \xi_i^* \ge 0 & i = 1,2,\ldots,n. \end{cases} \tag{4}$$

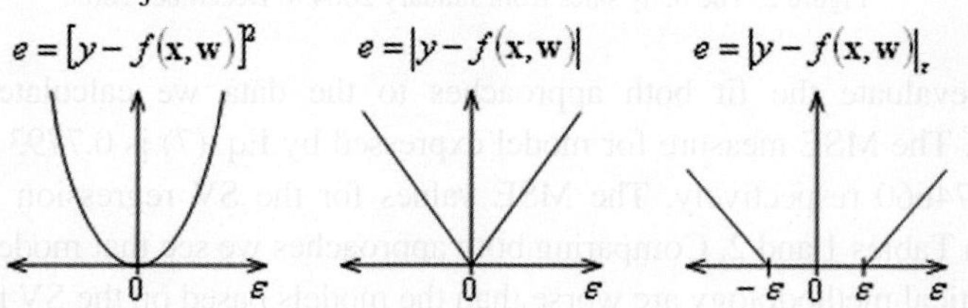

Figure 1. The error (loss) functions. Left: the quadratic 2- norm. Middle: the absolute error 1-norm. Right: the Vapnik´s ε-insensitive loss function.

3. Data, Results and an Empirical Comparison

In this section of the paper the performance of the SV regression models and models based on the RBF NN are investigated. Both types of models will be developed with the same database. The time plot of the data set used in these applications (the 724 daily sales for Hansa Flex company (denoted as y_t) are shown in Figure 2 (so called the sample data set).

After some experimentation using the Statgraphics procedures the statistical models were chosen as follows

$$y_t = \phi_1 y_{t-7} + \varepsilon_t \tag{5}$$

or

$$y_t - y_{t-7} = \theta_1 \varepsilon_{t-7} + \varepsilon_t \tag{6}$$

Both statistical models have typical seasonal behaviour with the seventh lag. The fitted models have the following forms

$$\hat{y}_t = -0.1248 y_{t-7} \tag{7}$$

or

$$y_t - y_{t-7} = 0.93868 \varepsilon_{t-7} \tag{8}$$

The usual diagnostic checking procedures according to Box & Jenkins [1] do not reveal any inadequacies in these models

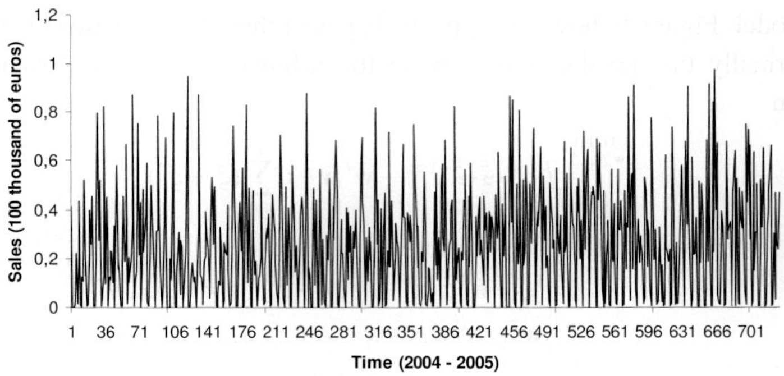

Figure 2. The daily sales from January 2004 to December 2005.

To evaluate the fit both approaches to the data we calculate the MSE statistics. The MSE measure for model expressed by Eq. (7) is 0.7793 and by Eq. (8) is 0.74660 respectively. The MSE values for the SV regression models are shown in Tables 1 and 2. Comparing both approaches we see that models based on the statistical methodology are worse than the models based on the SV regression.

In [8], the approximation ability of sales time series $\{y_t\}_{t=1}^{724}$ was analysed by various types of the RBF NN (classic, soft, granular). The MSE values were calculated analogously as in the case of SV regression models. Their values for various RBF´s networks related to the different number of clusters (RBF neurons) are shown in Table 3. From Table 3 we see that the MSE measures for models based on the fuzzy logic RBF networks and SV regression are approximately identical. Moreover, as we could also see, the RBF NN has such attributes as computational efficiency, simplicity, and ease adjusting to changes in the process being forecast. SV regression models require more costs of development, installation and operation in a management system, management comprehension and co-operation, and often a lot of computational time. Another disadvantage of SV regression models is that there is not a convenient way to modify or update the estimates of the model parameters as each new observation becomes available.

Table 1. The SV regression representation for model (5): The SV regression results of three different choice of the kernels and the results of the dynamic model on the training set. In the last column, the fit to the data and forecasting performance respectively are analysed.

SVM MODEL	KERNEL FUNCTION	STANDARD DEVIATION σ	DEGREE -d	LOSS FUNCTION	MSE
Causal	RBF	1		Quadratic	0.5411
Causal	RBF	0.9		Insensitive	0.5745
Causal	Polynomial		4	Iinsensitive	0.5745

Table 2. The SV regression representation for model (6): The SV regression results of three different choices of the kernels and the results of the dynamic model on the training set. In the last column, the fit to the data and forecasting performance respectively are analysed.

SVM MODEL	KERNEL FUNCTION	STANDARD DEVIATION σ	DEGREE -d	LOSS FUNCTION	MSE
Causal	RBF	0.8		Quadratic	0.6425
Causal	RBF	0.9		Insensitive	0.6723
Causal	Polynomial		4	Quadratic	0.6521

Table 3. The approximation results of various RBF's networks related to the different number of clusters (RBF neurons).

Neural network architecture:	Gaussian (classic RBF network)	Gaussian with normalised outputs (soft or fuzzy logic RBF network	Gaussian (classic RBF network with normal cloud concept	Gaussian soft RBF network (with normal cloud concept)
Number of RBF neurons		MSE		
		RBF network representations for model (5)		
3	1.439	0.698	1.503	0.729
5	0.729	0.693	0.817	0.716
10	0.687	0.675	0.671	0.678
15	0.697	0.681	0.681	0.678
		RBF network representations for model (6)		
3	0.783	0.646	0.786	0.647
5	0.810	0.632	0.803	0.630
10	0.607	0.571	0.607	0.571
15	0.582	0.563	0.582	0.563

4. Conclusion

In this article we have applied the fuzzy logic neural network to find the I/O function of sales process for Hansa Flex company. The performance of classic, fuzzy logic RBF neural networks was compared with models based on SVM approaches by the summary statistical measure, the mean square error. The experimental results show that the SVM approaches, soft RBF neural networks outperforms the ARIMA methodology that is widely used by managers as forecasting tool.

References

1. G. E. P. Box, and G. M. Jenkins, Time Series Analysis, *Forecasting and Control.* San Francisco, CA : Holden-Day, (1970).
2. D. Marcek. Determination of Fuzzy Relations for Economic Fuzzy Time Series Models by Neural Networks. *Computing and Informatics,* Vol. 22, SAS Bratislava, 457-471 (2003).
3. T. Takagi; M. Sugeno. Fuzzy identification of systems and its applications to modelling and control. *IEEE trans. System Man. Cybernet,* 16 yr, 116-132 (1985).
4. J.,Q., Chen, Z., G., Xi, and Z., J Zhang, A clustering algorithm for fuzzy model identification, *Fuzzy Sets and Systems.* Vol. 98, 319-329 (1998).
5. V. Kecman, Learning and Soft Computing – Support Vector Machines, Neural Networks and Fuzzy logic Models. *Massachusetts,* Institute of Technology (2001).
6. N. Cristianini, J. Shawe-Taylor, An introduction to support vector machines, *Cambridge University Press* (2000)
7. V. Vapnik: The support vector method of *function.* In: Nonlinear Modeling: Advanced Black-Box Techniques, J.A.K. Suykens,. J. Vondewalle, (Eds.), Kluwer Academic Publishers, Boston, 55-85 (1998).
8. M. Marcek, D. Marcek, *RBF* Neural Network Implementation of Fuzzy Systems: Application to Time Series Modelling. In: Aijun An et al Eds.: RSFDGrC 2007 – Rough Sets, Fuzzy Sets, Data Mining and Granular Computing, LNAI 4482, Springer-Verlag Berlin, 500 – 507 (2007).

A HYBRID APPROACH TO SELECTING INFORMATIVE CONSTRAINTS FOR SEMI-SUPERVISED CLUSTERING *

XIANHUA NI
School of Information Science & Technology, Southwest Jiaotong University
Chengdu 610031, P.R China

YAN YANG†
School of Information Science & Technology, Southwest Jiaotong University
Chengdu 610031, P.R China

Instance-based Must-Link and Cannot-Link constraints are two forms of background knowledge famously used in semi-supervised clustering. In this work we demonstrate that it is possible to step-up the quality and quantity of constraint sets by adopting a method that combines automated and active constraint selection. This method capitalizes on the cluster feature that most of the data objects in the cluster are actually core points and just small parts are border points. Therefore, considering Must-Link constraints between core points and Cannot-Link constraints between border points adds more effective information because border points would separate each cluster clearly. Our experiments show that our approach provides a competitive edge in identifying informative constraints that would enhance the accuracy of clustering solutions.

1. Introduction

In Semi-Supervised Clustering, Must-Link (ML) and Cannot-Link (CL) constraints on data instances have been successfully used to guide various clustering algorithms [1-2], towards more accurate results, reduced execution speed or even to capture important semantics of the dataset. While the use of pair-wise supervision has the potential to improve clustering accuracy, research has empirically shown that the achieved improvement strongly depend on the set of constraints chosen [3, 4].

Conventionally, constraints are randomly derived from already labeled data. In some semi-supervised clustering tasks[3, 5], constraints are obtained from co-

* This work is partially supported by the National Science Foundation of China (Nos. 61170111, 61003142 and 61152001) and the Fundamental Research Funds for the Central Universities (No. SWJTU11ZT08).

† You can contact the author via e-mail: yyang@swjtu.edu.cn.

associations between pairs of objects in a clustering ensemble, while in others, selecting constraints is active in nature where the constraint oracle actually takes the form of a human expert; some limitations do exist within these approaches; such as failure to represent the whole population and need for constant interaction with the user. A mechanism is therefore necessary that can combine both automated and active approaches to build optimum constraint sets. In this paper, we proposed a "*hybrid constraint selector*"; a mechanism that extends the must-link constraint set by using a threshold on the co-associations between a pair of *core points* in an ensemble of clusterings. On the other hand it extends the cannot-link set by considering a pair of *border points* and query the oracle if they belong to the same cluster and if not; the pair is added to the cannot-link set.

The rest of the paper is organized as follows: In section 2, we discuss the basic notions of semi-supervised clustering, clustering ensemble as well as the crucial concepts of *Density-Based Clustering* which strongly makes the background of our hybrid approach. Section 3 describes the underlying framework on which the *hybrid constraint selector* is built. In section 4 we present the experimental results. Finally, we draw our conclusions and future work in section 5.

2. Background

2.1. *Semi-Supervised Clustering*

Given a data set $X = \{x_1, \ldots, x_n\}$, a pair-wise constraint set pertaining to X can be incorporated into conventional partition clustering algorithms by adapting the objective function not to violate the constraints. Semi-Supervised clustering algorithms put to use background knowledge to guide the clustering process towards more accurate results. Pair-wise instance constraints are a popular form of background knowledge be-cause in most domains, constraints on which instances ML or CL re-side together in the same cluster are either known or are computable automatically from background knowledge.

2.2. *Ensemble Clustering*

A clustering ensemble is a paradigm that seeks to best combine the outputs of several clustering algorithms with a decision fusion function to achieve a more accurate and stable final output. Ensemble clustering typically involves two phases: a ***generation*** phase where a collection of base clusterings is produced and an ***integration*** phase where an aggregation function is applied to the ensemble members to produce a consensus solution. The most frequently employed integration strategy has been to use the information provided by an ensemble to determine the level of association between pairs of objects in a dataset [6, 7]. In practice, these pair-wise co-associations are represented using a symmetric co-association matrix and a consensus solution is consequently recovered by applying a similarity-based algorithm [3] to the matrix.

2.3. *Density-Based Clustering*

Density-based clustering locates regions of high density that are separated from one another by regions of low density.

Definition 1[8]: (Eps-neighborhood) The *Eps-neighborhood* of a point *p*, denoted by $N_{Eps}(p)$, is defined by $N_{Eps}(p) = \{q \in D \mid dist(p,q) \leq Eps\}$. The main idea is that for each point of a cluster the neighborhood of a given radius has to contain at least a minimum number of points (MinPts). However, this would be unrepresentative of the cluster since it has two categories of point (core & border) point. A point in a cluster is ***Core*** if it has more than a specified number of points (MinPts) within Eps Where as a point is a ***Border*** if it has fewer than MinPts within Eps, but is in the neighborhood of a core point.

3. The Hybrid Approach to Constraint Selection

Our approach focuses on the data sets (core objects set and border object set) separated from the original data. As some algorithms mentioned in Section 1, we manage the constraint selection procedure from core objects by automatic ensemble method to avoid the impossibility of querying an oracle to identify the constraints on a mass of objects. However when processing the border objects set which have a small amount of instances but contains more effective information to help separating each cluster we use the active constraint selection method to collect more reliable and effective constraints by querying the oracle.

3.1. *Automatic Approach to Constraint Selection on Core Objects*

Following the approach described in [3] to generate constraints, it computes the co-association matrix by the results of K-Means algorithm, and the final matrix **A** represent the relation between each object, the entity $\mathbf{A_{ij}}$ indicates the confidence that two data points $\mathbf{x_i}$ and $\mathbf{x_j}$ located in the same cluster. In our approach, we run the K-Means algorithm π times using the objects that is not the whole data set but the core objects set $\mathbf{P_c}$, and finally got π partitions € from $\mathbf{P_c}$. The main steps are given in Alg.1.

Algorithm 1. Must-Link Constraints Selection Algorithm from Core Objects

Input : π partitions of r data objects from $\mathbf{P}_c$(core objects set) ;
the threshold k_m ;
the initialized Must-Link set ML ;
1. Initialize **A** as the r×r empty(0) co-association matrix for $\mathbf{P}_c$
2. For t =1 to π:
For each pair (x_i, x_j) assigned to the same cluster in $€_t$,
update **A**: $A_{ij} = A_{ij} + 1$
3. Compute the final co-association matrix **A** from the updated **A**:$A_{ij} = A_{ij}/\pi$
4. Impute constraint set ML' based on **A** using a threshold k_m which indicate that if $A_{ij} \geq k_m$ then add <x_i,x_j> to ML'
5. Construct constraint sets ML = ML $\cup$ ML'

Output: The resulting constraint set ML

3.2. *Active approach to constraint selection on border objects*

Most of the time, the border objects cannot be separated from each other clearly, but the cannot-link constraints between these objects will help semi-supervised clustering algorithm to distinguish the boundary of each cluster. So we choose the Active Constraint Selection method to deal with the identifying constraints problem for border objects set $\mathbf{P}_b$ which owns obvious characteristics of holding a small proportion in the whole data set. The main steps are given in Alg.2.

Algorithm 2. Constraints Selection Algorithm from Border Objects

Input : the border objects set $\mathbf{P}_b$ with size s;
the updated Must-Link set ML and initialized Cannot-Link set CL ;
For t = 1 to s:
For q = t+1 to s:
1) query the oracle if the tth and qth object in $\mathbf{P}_b$ can be in the same cluster?
2) if "YES" then add this pair to ML with replacement
else if "NO" then add this pair to CL with replacement
else "UNCERTAINTY" query the next

Output: The resulting constraint sets (ML,CL)

4. Evaluation

We experiment on the data sets of different areas, and evaluate the effectiveness of the algorithm according to the experimental results. In our experiment we used six real data sets (Iris [3 classes, 4 attributes, 150 instances], Ionosphere [2 classes, 33 attributes, 239 instances], Wine [3 classes, 13 attributes, 178 instances], Ecoli [8 classes, 8 attributes, 336 instances], Breast [2 classes, 9 attributes, 478 instances] and Letter (A, B) [2 classes, 16 attributes, 1555 instances]) from UCI.

We firstly run our hybrid constraint selection approach to get the constraint sets ML and CL in each data set. We choose the parameters ε and MinPts in our approach with the method used in [9] which determines the two parameters automatically via analysis of the statistical characteristics off the data set. In the ensemble phase we use different initial center running to generate the base clustering, as an oracle we answer the query by natural classification supplied for the data set.

To evaluate our hybrid constraint selection algorithm, we compare it with two constraint selection strategies. The first is the random constraint selection method which generates a set of ML and CL constraints by randomly choosing pairs of objects and computing the constraint with the labels, another approach is advanced in [3] and we named it as "EA-IIC". We employ Cop-Kmeans with the cosine similarity measure as our semi-supervised clustering algorithm, and set the value of *k* to the number of classes in the data set. For each data set we averaged the Rand Index value by conducting 25 runs because of the nondeterminacy.

For each data set, the Fig. 1 reports a comparison of the mean RI scores with the growth of 15 constraints.

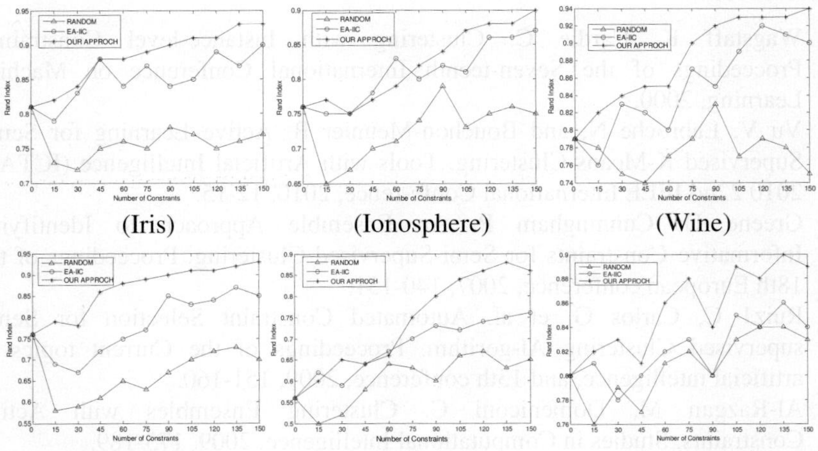

Fig.1. Comparison of mean RI scores for constraint selection strategies

We can clearly see from Fig. 1 that both the EA-IIC and our approach get a better effect than the random selection method. One general phenomena caused by the useless and wrong constraints is that as the number of constraint increasing the RI scores may be cut back, but we can notice the most striking contrast is the count of such phenomena between our approach and the others. And our approach can get a better performance in most of time from the result. The consequence demonstrates the constraints selected by our approach are more useful and correct base on the fact that the must-link constraint between core objects and the cannot-link constraint between the border points contain more valuable information.

5. Conclusion

In this work, we proposed a hybrid approach of constraint selection for semi-supervised clustering algorithm. We focus on ML constraints in core objects and CL in border objects, and select ML constraints with high reliability from the co-association matrix generated by different clustering results and select ML or CL constraints by query an oracle from the less border objects. Experimental results show that our approach can provide a better solution for selecting informative constraints. In the future, our work will focus on the constraint selection strategy on the massive and high dimensional data sets.

References

1. Wagstaff K, Cardie C. Clustering with Instance-level Constraints. Proceedings of the Seven-teenth International Conference on Machine Learning, 2000.
2. Vu V, Labroche N, and Bouchon-Meunier B. Active Learning for Semi-Supervised K-Means Clustering. Tools with Artificial Intelligence (ICTAI), 2010 22nd IEEE International Confe-rence, 2010, 12-15.
3. Greene D, Cunningham P. An Ensemble Approach to Identifying Informative Constraints for Semi-Supervised Clustering. Proceedings of the 18th European conference, 2007, 140-151.
4. Ruizl C, Carlos G, et al. Automated Constraint Selection for Semi-supervised Clustering Al-gorithm. Proceedings of the Current topics in artificial intelligence, and 13th conference, 2009, 151-160.
5. Al-Razgan M, Domeniconi C. Clustering Ensembles with Active Constraints. Studies in Computational Intelligence, 2009, 175-189.
6. Al-Shaqsi J, Wang W J. A Clustering Ensemble Method for Clustering Mixed Data. Neural Networks (IJCNN), The 2010 International Joint Conference, 2010, 1-8.
7. Mimaroglu S, Erdil E. Combining multiple clusterings using similarity graph. Pattern Recog-nition, 2011, 694-703.
8. Ester M, Kriegel H P, Sander J, Xu X W. A Density-Based Algorithm for Discovering Clus-ters in Large Spatial Databases with Noise. Proceedings of 2nd International Conference on Knowledge Discovery and Data Mining (KDD-96), 1996.
9. Xia L N, Jing J W. SA-DBSCAN: A self-adaptive density-based clustering algorithm[J]. Journal of the Graduate School of the Chinese Academy of Sciences, 2009(4):530-538. (In Chinese)

ANALYSIS OF THE IMPACTS OF THE DATA LENGTH ON THE UNCERTAINTY OF THE ARCHIMEDEAN COPULA MODELING*

TONG XIN

Department of Hydrosciences, School of Earth Sciences and Engineering, Nanjing University, Nanjing 210093, P. R. China

WANG DONG †

Department of Hydrosciences, School of Earth Sciences and Engineering, Nanjing University, Nanjing 210093, P. R. China

CHEN YUANFANG

College of Water Resources and Environment, Hohai University, Nanjing 210098, P. R. China

Copulas have been frequently used in solving issues about modeling multivariate variables for the significant advantage that it can capture the multivariate dependence of the variables irrespective of their marginal distributions. The parametric families of Archimedean copulas are widely used in hydrologic frequency analysis due to their simple and closed form expressions. This paper mainly deals with the impacts of the length of hydrological data on the uncertainty of the modeling results of the Archimedean copulas. Several estimation methods are implemented including the inference from marginals (IFM) method, canonical maximum likelihood estimator (CMLE) and the estimation based on Kendall's tau. The results show that the length of data set will affect the choice of the best fitted copula and it is more reliable to use CMLE and estimation via Kendall' tau when the length of the data set changes.

1. Introduction

There have been many issues in hydrologic frequency analysis that involve multivariate variables modeling, such as modeling the dependence between flood volume, duration and magnitude. Copula is a relatively new method that

* This work is supported by National Natural Science Fund of China (No. 41071018, 41030746, 40725010 and 40730635), the Skeleton Young Teachers Program and Excellent Disciplines Leaders in Midlife-Youth Program of Nanjing University.

† wangdong@nju.edu.cn.

surpasses the traditional multivariate models such as bivariate normal, log-normal, and extreme-value distributions and so on. The main advantage of copula is that it can capture the dependence of variables irrespective of their marginal distributions. It is widely used in hydrologic frequency analysis involving the following aspects: (1) flood frequency analysis e.g. Zhang and Singh [1] derived bivariate flood frequency distribution using several copulas. (2) Drought variables modeling e. g. Kao and Govindaraju [2] used copulas to capture the joint behaviors of drought variables. (3) Modeling the dependence structure between storm or rainfall characteristics e.g. Vandenberghe et al. [3] applied some commonly used bivariate copulas and techniques to fit the dependence between storm variables. (4) Using copulas to simulate streamflow. e. g. Lee, T. and Salas, J. D. [4] introduced copula method to stochastic streamflow simulation.

When using copula in the hydrologic analysis, sufficient length of data set is an indispensible element. Long-term data set can make the model more reliable and the parameter calibration process easier. But in reality, such sufficient data can be very difficult to get. Different lengths of data set will intuitively bring different parameter estimating results. So it is in need to analyze the uncertainty in the results of copula modeling when employing various length of data set. Actually, there have been some researches referring the uncertainties in copula modeling. Genest et al. [5] investigated the effect of the sample size and the strength dependence on the power of the blanket test for various copulas. But few researches have been done exploring the uncertainty that the length of data set bring to the Archimedean modeling results.

This paper mainly discusses the uncertainty of the modeling results of three Archimedean copulas when using different lengths of the annual maximum flood magnitude (AMFM) sets from two stations of Yangtze River. The second Section will briefly review the Archimedean copulas and parameter estimating methods and the simulating steps. The case study will be in Section 3. Section 4 and Section 5 give the discussion and conclusions respectively

2. Methodology

2.1. *Archimedean copulas*

The concept of copulas is functions that link multivariate probability distributions and details about copulas can be found in the book of Nelsen [6].

Archimedean copulas have a wide range in hydrology for the following reasons that: (1) they can be constructed easily; (2) there exist a great variety of families of copulas in Archimedean copulas; (3) Archimedean copulas possess many good properties as mentioned by Nelsen [6]. Gumbel-Hougaard (G-H) copula, Clayton copula and Frank copula are three common used families from Archimedean copulas in hydrology as stated by Genest and Favre [7]. Table 1 shows the relationships between Kendall's tau and the parameter θ and the corresponding functions of the three copulas.

Table1. Relationship between Kendall's tau and the parameter θ and the corresponding functions of three Archimedean copulas.

Family	Kendall's tau	Function of copula
G-H	$1-1/\theta$	$\exp\left\{-[(-\ln u)^{\theta}+(-\ln v)^{\theta}]^{1/\theta}\right\}$
Clayton	$\theta/(\theta+2)$	$(u^{-\theta}+v^{-\theta}-1)^{-1/\theta}$
Frank	$1-4/\theta+4D_1(\theta)/\theta$	$\frac{1}{\theta}\ln\left[1+\frac{[\exp(\theta u)-1][\exp(\theta v)-1]}{\exp(\theta)}\right]$

2.2. *Parameter estimation method and goodness-of-fit*

This paper implements a nonparametric, a semiparametric and a fully parametric estimation method to estimate θ, namely the estimation form Kendall's tau, the canonical maximum estimator (CMLE) and the inference from marginals (IFM).

The estimation from Kendall's tau is intended to estimate θ based on ranks exclusively. The relationship between Kendall's tau and θ is shown in Table 1 from which θ is easy to calculate. Actually, CMLE method has been proved to be superior compared to the IFM method in most situations when the marginals are unknown. The calculation steps of the two methods and their comparison can be found in the work of Kim *et al.* [8]. But in this paper, the IFM method is still applied because several lengths of data set will be discussed under the assumption that the marginals are already known.

In order to measure how the variation of the length of the data set will influence the modeling results of the copulas. Two criterions will be applied, including the AIC value and ordinary least square (OLS) value:

$$OLS=\sqrt{\frac{1}{n}\sum_{i=1}^{n}(p_{ei}-p_i)^2} \tag{1}$$

$$AIC=n\log\left[\frac{1}{n}\sum_{i=1}^{n}(p_{ei}-p_i)^2\right]+2m \tag{2}$$

Where n is the length of the data. m is the number of the parameters, p_i is the theoretical joint probability, p_{ei} is the empirical joint probability. The modeling results reach best when the corresponding OLS and AIC values are the minimum.

2.3. *Simulation methods*

In this paper, the impacts of the length of the data set on the modeling results will be analyzed. The paper collects 3 different lengths of the data set of the AMFMs from two stations and discusses the uncertainty which the variation of the data length bring to the modeling results. For each length of data set, the main steps are listed as following:

1. Calculate the parameter θ of the three families of Archimedean copulas under the assumption that the marginal distribution of the AMFMs are known, using the estimation via Kendall's tau and the IFM method.
2. Calculate the parameter θ under the assumption that the marginal distribution of the AMFMs are unknown, using the estimation from Kendall's tau and the CMLE method.
3. Use the calculated θ from step 1and 2 to model the dependence between the variables and get the AIC and OLS values via Eq. (1) an Eq. (2).

Then the goodness-of-fit results of the three kinds of Archimedean copula models for all the lengths of data sets are compared and analyzed.

3. Case study

The paper collected 109 year, 70 year and 40 year of the AMFM data sets from two stations of the Yangtze River, namely the Yichang station and Cuntan station as shown in Table 2.

Table 2. Goodness-of-fit results of the copula models of different lengths

data length		With assumed marginals					
		IMF			via Kendall's tau		
		G-H	Clayton	Frank	G-H	Clayton	Frank
109	AIC	-854.5128	-789.3112	**-868.5147**	-849.5023	-861.3731	**-864.977**
	OLS	0.0199	0.0265	**0.0184**	0.0201	0.0191	**0.0187**
70	AIC	-511.2132	-489.3071	**-514.4668**	**-514.8973**	-508.3137	-513.7329
	OLS	0.0256	0.0299	**0.0250**	**0.0249**	0.0261	0.0251
40	AIC	-266.2034	-262.8750	**-273.0859**	-268.9675	-269.8415	**-275.1969**
	OLS	0.0350	0.0365	**0.0321**	0.0338	0.0334	**0.0313**
data length		Marginals are unknown					
		CMLE			via Kendall's tau		
		G-H	Clayton	Frank	G-H	Clayton	Frank
109	AIC	-936.119	-781.5964	**-990.7597**	-960.3654	-840.541	**-993.9589**
	OLS	0.0135	0.0275	**0.0105**	0.0121	0.021	**0.0104**
70	AIC	-570.3349	-532.1200	**-584.8775**	-581.2847	-534.1154	**-589.0667**
	OLS	0.0168	0.0220	**0.0151**	0.0155	0.0217	**0.0147**
40	AIC	**-314.7653**	-294.1719	-301.5955	**-309.6580**	-291.5056	-304.3962
	OLS	**0.0191**	0.0247	0.0225	**0.0203**	0.0255	0.0217

The dependence structures between the three different lengths of AMFMs of the two stations are modeled with the three Archimedean copulas, i.e. G-H copula, Clayton copula and Frank copula. The marginal distributions of the AMFMs are assumed to be P-III distribution. The goodness-of-fit results are represented by the values of AIC and OLS which are shown in Table 2.

4. Discussion

From the case study in the last section, the impacts of the length of data set on the modeling results can be analyzed. It can be seen from Table 2 that in general, when the data length increases, the values of AIC and OLS decrease, which indicate the copula models fit better when data length is longer. So it is recommended that the longer the data set can be the better for the modeling effect.

When the data length is the fixed, the AIC and OLS values under the condition that the marginal distributions of the AMFMs are known are bigger than that under the assumption that the marginals are unknown, which means the latter gives better modeling results. It may be explained by the fact that the dependence structure between the variables has nothing to do with the characteristics of the variable itself; the numbers in bold in Table 2 show the best modeling results among the three types of copulas. It can be perceived that the differences between the modeling results of the CMLE and estimation via Kendall's tau are a slightly more significant than the difference between the modeling results of the IFM and estimation via Kendall's tau.

Under the condition that the marginal distributions are known, when the length of the data set changes, the best fitted copulas are all Frank copula for the IFM method, while the best fitted copula changes for the estimation via Kendall's tau; On the other hand, the best fitted copula varies both for the CMLE method and the estimation via Kendall's tau when the data length changes under the assumption that the marginals are unknown. That is to say, the change of data length will affect the choice of the best fitted copula. But for a fixed data length, the CMLE and estimation via Kendall's tau always get the same best fitted type of copula under the condition that the marginals are unknown no matter how long the data set is, which is not the case when the marginal distribution are assumed. This may be explained by the fact that when the marginals are unknown, the CMLE and estimation via Kendall's tau both substitute the empirical probability for the marginal distribution while the assumed marginals used in IFM may somehow be inaccurate when the data length changes. That is to say, it is more reliable to use the CMLE and estimation when the data length changes.

5. Conclusion

The paper discusses the uncertainty that the data length varying brings to the modeling effect of Archimedean copulas. Three different lengths of the AMFM data sets from two stations of the Yangtze River are collected for analysis. The paper uses G-H copula, Clayton copula and Frank copula to construct the dependence structures between the AMFMs of the two stations under two conditions that the marginal distributions are P-III distribution and the marginals are unknown. Parametric, semiparametric and nonparametric estimation methods are implemented. Results show that it is better to use long data set. The data length will affect the choice of the best fitted copula and it is more reliable to use CMLE and estimation via Kendall's tau when the data length changes.

References

1. L. Zhang and V. P. Singh, *J. Hydrol. Eng.* **11**(2), 150– 164 (2006).
2. S.-C. Kao and R. S. Govindaraju, *J. Hydrol.* **380**, 121-134 (2010).
3. S. Vandenberghe, N. E. C. Verhoest and B. De Baets, *water Resour. Res.* **40**,W01512 (2010).
4. T. Lee and J. D. Salas, *Hydrol. Res.* **42**(4), 318-330 (2011).
5. Genest, B. Rémillard and D. Beaudoin, *Insur. Math. Econ.* **44**(2), 199-213 (2009).
6. R. B. Nelsen, An introduction to copulas. Springer, New York (1999).
7. C. Genest and A. –C. Favre, *J. Hydrol. Eng.* **12**(4), 347– 368 (2007).
8. G. Kim, M. Silvapulle J. and P. Silvapulle, *Comput. Stat. data Annual.* **51**, 2836-2850 (2007).

AN INITIAL RESEARCH TO SEGMENT A CROWDSOURCING PROJECT PARTICIPANTS USING TEXT MINING

BAŞAR ÖZTAYŞI*
Industrial Engineering Department, Istanbul Technical University, Beşiktaş, İstanbul 34367, Turkey

LARA CEMCEM
Industrial Engineering Department, Istanbul Technical University, Beşiktaş, İstanbul 34367, Turkey

Crowd creation is a branch of crowdsourcing that focus on creation of activities such as asking individuals to film TV commercials, perform language translation or solve challenging scientific problems. BurtonStory is a crowd creation project by Tim Burton that was run over twitter between November 20th and December 6th of 2010. The author started a new story with a single sentence, and the rest of the story is created by the crowd. The aim of this paper is to analyze participants based on their profile information using text mining approach.

1. Introduction

Open innovation is the use of purposive inflows and outflows of knowledge to accelerate internal innovation, and expand the markets for external use of innovation, respectively. In general, customer integration into innovation activities stands for an important competitive strategy, known as open innovation (Chesbrough,2003). Crowd sourcing, proposed by Jeff Howe (Howe,2009), can be described as the act of outsourcing a task to a large group of people or community. With this definition crowd sourcing is highly related to open innovation since it corporates with outer sources. The mostly known crowd sourcing projects are Linux operating system and Wikipedia open encyclopedia. Both systems are generated by the interested people working willingly without any task assignment from a higher level of management.

Howe (2009) describes the four primary types of crowdsourcing; (i) Crowd Creation is the best known forms of crowdsourcing are "creation" activities such asking individuals to film TV commercials, perform language translation or

* oztaysib@itu.edu.tr

solve challenging scientific problems. (ii) Crowd Voting utilizes the crowd's judgment to organize, filter and stack-rank content such as newspaper articles, music and movies. (iii)Crowd Wisdom principle attempts to attach many people's knowledge or expectations in order to solve problems or predict future outcomes or help direct corporate strategy. (iv) Crowd Funding is a funding system where the crowd corporates to finance new projects with individual micro funding. The terms crowdsourcing and open innovation has much in common and the specially crowd creation is the type of crowdsourcing that can directly be used as an open innovation tool.

Intelligent Systems has developed powerful tools to analyze and understand text data. Using these tools, businesses can gain business intelligence from many sources of text data such as web feedback, e-mail & websites and documents. Leveraging techniques from machine learning, natural language processing, and statistics, these tools can help understand and organize text data for trend analysis, summarization, classification, taxonomy generation, content management and business intelligence.

In this study, a crowd creation project, "BurtonStory" is investigated and the specific groups of participants are identified based on self-description text and text mining techniques. The following of the paper is organized as follows; section 2 gives a brief literature review about text mining. In section 3 the data collection and preparation activities are summarized. Section 4 the analysis process is explained and the results are discussed and the last section gives the conclusions and suggestions for further research.

2. Text Mining

Text Mining is the discovery of new, previously unknown information, by automatically extracting information from different written resources. With text mining, unstructured data generated within a company or over internet has become a source for analytical analyzes (Turban et al. 2011).

The most popular text mining applications can be given as; information extraction, topic tracking, summarization, categorization, clustering, concept linking and question answering. Information or knowledge extraction is the identification of key phrases and relationships with in text. The second application, topic tracking is the process of predicting documents of interest to a user based on user profile and other supplementary data. Summarization is the text mining application automatically summarizes the documents without information loss. Another important text mining application is categorization. Categorization is the process that starts with identifying the main themes of a document and places the documents to predefined set of categories. Clustering, just like categorization, aims to group documents but in clustering there are no predefined categories. The aim of clustering is to group similar documents together. Concept linking is the text mining process that finds related documents by identifying their shared concepts. And finally, text mining is used to automatically find the most suitable answer to a question which is known as "question answering". The relevant studies about the listed text mining applications are given in Table 1.

Table 1: Review of text mining applications

Application	Purpose	Reference
Information Extraction	Identification of key phrases and relationships with in text	Tilak et al. (2011)
Topic Tracking	Predict documents of interest	Qiu et al. (2010)
Summarization	Automatically summarize the documents	Bhattacharya et al. (2011)
Categorization	Place the documents to predefined set of categories	Trappey et al.(2009)
Clustering and Concept Linking	Group similar documents together.	Liu and Liu (2011)
Question answering	Find the best answer to a given question.	Dizier and Moens (2011),

3. Data Collection

Burtonstory project is a crowd writing project designed by famous film director Tim Burton. After writing the first sentence, the director of the project waited for the crowd (the twitter followers) to send alternatives for the following sentences and selected a suitable one among the alternatives. As a selection is done, it is published to the crowd and the alternatives for the next sentence are collected.

The project was online over burtonstory.com and twitter.com. Twitter is a microblogging website which users can write messages not longer than 140 characters. All the messages are published on the website so in order to define the messages about a special topic or project a tag is defined. The users use this tag within their messages so that when searched with the tag all the related messages can be reached. In the project "#BurtonStory" is used as a tag to follow the project. The data that is used in this study belongs to the messages sent to the project. The project was run over twitter between November 20th and December 6th of 2010. The initial data collection for the research is accomplished on 5th January 2011.

Data collection for the research is accomplished in three phases. In the first phase, all the messages that contain "#BurtonStory" tag are downloaded. In the second phase, the messages are investigated to distinguish the informative messages from the contributive messages. Contributive messages are the ones that are directly written as an alternative story part. On the other hand informative messages are the ones that are written about the project but not aim to be a part of the story. As the contributive messages are identified, the profile information about the authors of these messages is collected in the third phase. As a result description text of 1366 authors are collected and used for the text mining activities.

4. Analysis and Results

In this study Rapid Miner 5.02 is used to cluster the description text of the project participants. In order to reach the adequate results, a series of applications are accomplished in two phases. The first phase, the process documents from data phase, the non letter text are tokenized, stop words and the words shorter than 3 characters are eliminated. Finally, TF-IDF (term frequency–inverse document frequency) matrix is generated with the remaining set of words. The tf*idf weight is a numerical statistic which reflects how important a word is to a document in a collection. The tf-idf value increases proportionally to the number of times a word appears in the document, but is offset by the frequency of the word in the corpus, which helps to control for the fact that some words are generally more common than others. A part of the matrix is presented in Figure 1.

Row ...	text	artist	academic
388	child artist problem staying artist grow	0.533	0
330	digital media producer photographer artist	0.438	0
242	comic book artist colorist penciler inker	0.433	0
369	editor artist husband father	0.413	0
127	amphibias friend lillypad twitter pond cyber:	0.410	0
129	art culture lover traveler aspirant artist seth	0.400	0
147	artist author musician etc	0.400	0
1351	writer singer actress director artist	0.387	0
1275	vincitrice del concorso letterario streghe vai	0.374	0
476	graphic artist schmuck drunk dad necessa	0.363	0

Figure 1: Sample TF-IDF scores

In the second phase, k-means clustering method is used to cluster the description texts, based on the IT-IDF matrix calculated in the previous phase. k-means clustering is a method of cluster analysis which aims to partition observations into k clusters in which each observation belongs to the cluster with the nearest mean. Given a set of observations (x1, x2, …, xn), where each observation is a *d*-dimensional real vector, *k*-means clustering aims to partition the n observations into k sets ($k \leq n$) $\mathbf{S} = \{S_1, S_2, \ldots, S_k\}$ so as to minimize the within-cluster sum of squares (WCSS):

$$\underset{\mathbf{S}}{\arg\min} \sum_{i=1}^{k} \sum_{\mathbf{x}_j \in S_i} \left\| \mathbf{x}_j - \boldsymbol{\mu}_i \right\|^2 \tag{1}$$

where μi is the mean of points in Si.

The results of the k-means clustering where k is set to six provide six clusters as shown in Table 2. The clusters are given descriptive names based on the centroid values.

Table 2. Descriptive names and number of items in each cluster.

Descriptive Names	Number of Items
Students & Journalists	130
Hobby Owner	96
Artists	82
Social Media Professionals	69
Bloggers	62
Others	927

Table 3 presents the resulting centroid table. The centroid of a cluster represents the most typical case. From the table we can see the top five typical word of each cluster.

Table 3. Top 5 typical values of each cluster.

Cluster1	Cluster2	Cluster3	Cluster4	Cluster5
Film (0.10)	love (0.180)	writer (0.22)	social (0.18)	com (0.21)
student (0.1)	life (0.143)	actor (0.03)	media (0.16)	http (0.20)
world (0.06)	music (0.045)	photographer (0.03)	marketing (0.13)	www (0.13)
director (0.04)	movies (0.038)	editor (0.02)	digital (0.05)	tumblr (0.04)
art (0.042)	people (0.025)	Director (0.02)	advertising (0.04)	writer (0.02)

5. Conclusion

Crowdsourcing is becoming an important source for organizations to empower open innovation and reducing costs. In this initial study, data of a crowdsourcing project is investigated to define the groups of participants. A total of 1366 authors' self description text is clustered using k-means cluster analysis. The results indicate that students, journalists, hobby owners, artists, social media professionals and bloggers are the different groups of participants. A limitation of the study is that the number of classes is identified as six, in the further studies different number cluster numbers can be used to define other groups of participant. Also other clustering methods can be used and the results of the clusters can be compared.

References

1. Chesbrough, H.W. 2003, Open Innovation: The new imperative for creating and profiting from technology. Boston: Harvard Business School Press.
2. Turban E., Sharda R. and Delen D. *Decision Support and Business Intelligence Systems*, Prentice Hall, New Jersey, 2011.
3. Tilak O., Hoblitzell A., Mukhopadhyay S., You Q., Fang S., Xia Y., Bidwell J., Multilevel text mining for bone biology, *Concurrency and Computation: Practice and Experience*, 23/17, pp. 2355–2364, 2011.
4. Qiu J., Liao L. and Li P., Enhancing Topic Tracking for Chinese News Web Pages with Temporal Information and Key Web Contexts , *International Journal of Innovative Computing, Information and Control,* 6(1), pp. 399-408, 2010.
5. Bhattacharya S., Thuc V. and Srinivasan P., MeSH: a window into full text for document summarization,*Bioinformatics*, 27 (13), pp. 120-128, 2011.
6. Trappey A.,Trappey C., Wu C.Y., Automatic patent document summarization for collaborative knowledge systems and services, *Journal of Systems Science and Systems Engineering*, 18(1), pp. 71-94, 2009.
7. Liu Y., Liu C.W., Research of fast SOM clustering for text information, Expert Systems with Applications, 38(8), pp. 9325–9333, 2011.
8. Dizier P.S., and Moens M.F., Knowledge and reasoning for question answering: Research perspectives, *Information Processing and Management*, 47(6), pp. 899–906,2011.
9. Howe, J.(2009) Crowdsourcing: Why the Power of the Crowd Is Driving the Future of Business, Crown Business

SPARE PART DEMAND FORECASTING WITH BAYESIAN MODEL

SINAN APAK*

Industrial Engineering Department, Istanbul Arel University
Istanbul, Turkey

Continuous change in technology and differentiation in product models in industrial market have an indispensable impact on forecasting demand for spare parts. Inventory managers periodically update their predictions of future demand rates for products. A Bayesian model, using a prior probability distribution for the demand rate, was used to obtain optimal inventory levels over several periods assuming a known cost for surplus and shortage. However, its performance has not been examined under various demand rates such as intermittent demand. Study examines such conditions with a research question.

1. Introduction

Spare part demand forecast is the process of prediction of the future amount to be requested by consumers. This is the basis for the determination of the production level of the business. The demand forecasts are often interpreted by which product in which demand for this product's spare part would date on the probability for the realization of this demand. In today's economy, spare part market has not only become the primary point of contact between customers and business but also a major investment for many organizations. According to some recent automotive industry estimates, the market value will be 230 billion Euros in 2020 in Europe [16]. Due to magnitude of the market, proper management of such production estimations arrivals is the most difficult of the primitives to assess.

In different situations forecasting demand requires some historical demands which are unavailable or invalid for new consumption parts [3]. Some spare parts may be selling fast and others may be selling slowly due to randomness in customer buying patterns. This approach is used to mitigate the effects of insufficient data in predicting future demand rates. The study examines both the performance of the Bayes model approach and the Poisson model approach in determining the optimal level of inventory.

* E-Mail: sinanapak@arel.edu.tr

Snyder [12] discusses the difficulty in forecasting demand for high-end challenge for automobile parts. Accurate forecasting of demand is important in inventory control, but there are three main difficulties in forecasting the demand of spare parts [4]. First one is, demand of spare part is often intermittent and the nature of demand makes forecasting difficult [5, 15]. Second one is historical data of spare part demand are usually very limited. Thirdly, in some industries, spare parts inventory level is largely a function of how equipment is used and how it is maintained [5].

The study provides information about the performance of proposed inventory managements tools and the conditions under which these tools are reliable and useful in inventory decision making. The study addresses the situation in which a manager may need an estimate of the future demand rate of products.

The paper organized as follows. Second chapter conducts previous studies about moving inventory forecasting techniques. Third chapter examines the methodology employed by the research. Chapter four presents key finding and discussions finally last chapter presents conclusions, limitations and describes areas for further investigation.

2. Methods used to forecast spare part inventory demand

Many modern inventory management and control software packages assume that future demand follows a normal distribution and thus include analytic models believed to have sufficient forecasting accuracy [14]. Most stochastic inventory models use either periodic or continuous review [10]. The periodic review approach determines the amount of inventory at intermittent intervals to compute the order quantity required to bring it back to a desired level, which is popular with small retailers. The continuous system approach, in contrast, updates inventory instantly and submits an order for a fixed quantity to minimize the total cost whenever it reaches a predetermined minimum level.

Strijbosch et al. [11], Razi and Tarn [8], and Dolgui and Pashkevich [2] propose models for managing spare parts, which are important in ensuring adequate availability of items, thus providing organizations with a competitive advantage. Syntetos and Boylan [13] empirically compare simple moving average, single exponential smoothing, Croston's method, and a modified Croston method proposed by Boylan and Syntetos [1] on spare parts from the automotive industry with a focus on intermittent demands.

An inventory management system for slow-moving and fast-moving products is proposed by Leven [6]. His model is based on the Croston method and sporadic demand is assumed to have an Erlang distribution utilizing a Bayesian approach for spare parts. Shale et al. [9] adapted standard testing methodology is presented which overcomes the complications arising from the necessity of sampling from time series data. They found that the frequency of orders is well represented by a Poisson process. Viswanathan et al. [15] evaluate the performance of top-down and bottom-up forecasting strategies in estimating the aggregate data series when the sub-aggregate time series components are intermittent and for forecasting the aggregate demand using top-down forecasting, the simple exponential smoothing technique outperformed Croston's method in a case.

3. Methodology

Using Bayesian approach methodology, a study is conducted to compare optimal levels of inventory with respect to the costs for shortages and surpluses to a traditional approach based on the Poisson distribution. The application of this alternative method requires a user-supplied prior distribution for the demand rate, which is treated as a random variable having a gamma distribution.

This study assessed the robustness of the reliability of both one-sided and two-sided prediction intervals for the future demand rate of products with no sales history as well as those with no more than one sale. The performance of the Bayesian approach is evaluated using intermittent sales data.

Popovic [7] states that the optimal inventory level in the Ith unit time interval I_i can be computed by considering the ratio of the cost of surplus C_1 and the cost of shortage C_2 of an item as well as the demand described by the a posteriori distribution. As additional information is accumulated, the posteriori distribution of demand can be updated to improve the accuracy of the parameter estimates. The optimal inventory level r_i^* for product i will be such that it satisfies one of the inequalities in (1,2).

Bayesian Model Approach:

$$\sum_{k=0}^{r_i^*-1}\binom{\alpha+k-1}{k}\left(\frac{\beta}{\beta+1}\right)^{\alpha}\left(\frac{1}{\beta+1}\right)^{k}<\frac{C_2}{C_1+C_2}\leq\sum_{k=0}^{r_i^*}\binom{\alpha+k-1}{k}\left(\frac{\beta}{\beta+1}\right)^{\alpha}\left(\frac{1}{\beta+1}\right)^{k} \quad (1)$$

Poisson Model Approach:

$$\sum_{k=0}^{r_i^*-1}e^{-\lambda t}\frac{(\lambda t)^k}{k!}<\frac{C_2}{C_1+C_2}\leq\sum_{k=0}^{r_i^*}e^{-\lambda t}\frac{(\lambda t)^k}{k!} \quad (2)$$

Application that use the ratio of C_1 and C_2 to determine optimal inventory levels include style goods and perishable items. Finding appropriate costs for shortage and surplus goods for inventory with intermittent demand may not always be practical and, thus, could potentially be a limitation to this approach. Popovic [7] describes an application of this approach to the ordering of spare parts from a warehouse at the beginning of each month and explains how the equalities in (1) can be used for ordering optimal stock levels.

4. Research application

This Bayesian model will be compared with a traditional maximum likelihood approach, referred to as the Poisson model, in which the rate parameter of a Poisson distribution is estimated from the previous data. A Monte Carlo simulation of demand rates over ten time periods is performed to assess the effectiveness of each of these two methods. The term, moderate-demand products, is used in this research question to mean products that are not intermittent. Let the underlying unknown demand rate of a product be λ. The estimate of this rate could be zero for a product that has no demand or no recorded history of demand. Since the observed period of time may not be long enough to allow for an accurate prediction of future demand, a well chosen prior distribution for the demand rate may provide a more reasonable forecast than a forecast of zero.

Form of the cost expression for inequality (1) requires that only the ratio of the surplus cost to shortage cost be known. That is, a shortage cost of \$5 and a surplus cost of \$1 will yield the same results as a shortage cost of \$50 and a surplus cost of \$10. A pool of 100 products is used in the simulation study to compare the costs based on the Bayesian model and the Poisson model. Demand rates for this pool of products are generated from a gamma distribution with a selected value of the mean equal to 3, which allow some products to be slow moving provided the standard deviation is not too small. Four standard deviations were selected: 5.5 (high), 1.7 (moderately high), 0.55 (low), and 0.17 (very low). Two parameters of the gamma distribution determine these values: α, the shape parameter, and β, the scale parameter. The mean for the gamma distribution is equal to α/β and the variance is equal to α/β^2. The Standard deviations resulted from the following four pairs of α and β values: $\alpha = 0.3$ and $\beta = 0.1$, $\alpha = 3$ and $\beta = 1$, $\alpha = 30$ and $\beta = 10$, and $\alpha = 300$ and $\beta = 100$. Once the demand rate λ is selected, actual product sales are generated using a Poisson process. The number of units sold for each product is simulated 10 times, representing sales over a period of 10 time units. This process is replicated 20 times and the average costs per time unit are recorded. The time units could be days, weeks, months or something else. Expected costs from holding this amount of inventory is then determined using the underlying distribution of demand for that product. The performances of the two models will be judged to determine which model has the lower expected cost.

There is a noticeable gap between the total costs for the Bayesian model and the Poisson model for the first two time periods. As mentioned previously, the total cost per period using either model improves dramatically after only three periods. There is an advantage to using the Bayesian model particularly for the first couple of time periods when the variance of the product demand rates is moderately high. The ratio of the surplus cost to shortage cost affects the scale of the graphs, but the relative performance of the model does not change much.

5. Conclusion and limitations

Inventory managers desire to stock products that customers want to purchase. This research addresses one aspect of this challenge: predicting the future demand rate of products. The Bayesian methodology is appropriate when a prior distribution of the demand rate is available and when surplus and shortage costs are known. Furthermore, the standard deviation of the products' demand rate will determine if the Bayesian or the Poisson model would be better suited.

The Bayesian model and the Poisson model proposed in this research make several assumptions, which may not be applicable in practice. A Poisson process was assumed for the demand of goods. This was a general assumption often made in the literature.

Alternative multi-period inventory methodology for optimizing inventory levels should be explored. The Bayesian approach may be extended to more complex inventory management problems to account for inventory that has a limited life span or allows for replenishment in the middle of the single period.

References

1. J. E. Boylan, and A. A. Syntetos, *Spare parts management: a review of forecasting research and extensions.* **IMA Journal of Management Mathematics**, 21 (3) 227-237 (2010).
2. A. Dolgui and M. Pashkevich, On the performance of binomial and beta-*binomial models of demand forecasting for multiple slow-moving inventory items*. **Computers and Operations Research** 13 (8) 112-129 (2007).
3. B. Gohodrati, P. A. Akrsten, and U. Kumar, *Spare part estimation and risk assessment conducted at Choghart Iron Ore Mine.* **Journal of Quality in Maintenance Engineering**, 13 (4) 353-363 (2007).

4. Z. S. Hua, B. Zhang, J. Yang, and D. S. Tan, *A new approach of forecasting intermittent demand for spare parts inventories in the process industries.* **Journal of the Operational Research Society**, 58 (1) 52-61 (2007).
5. W.J. Kennedy, J. W. Patterson, and L. D. Fredendall, *An overview of recent literature on spare parts inventories.* **International Journal of Production Economics**, 76 201–215 (2002).
6. E. Leven and A. Segerstedt, *Inventory control with a modified Croston procedure and Erlang distribution.* **International Journal of Production Economics**, 90 (3) 361-367 (2004).
7. J.B. Popovic, *Decision making on stock levels in cases of uncertain demand rate.* **European Journal of Operational Research**, 32 (2) 276-290 (1987).
8. L.A. Razi and J.M. Tarn, *An applied model for improving inventory management in ERP systems.* **Logistics Information Management**, 16 (2) 114-124 (2003).
9. E. A. Shale, J. E. Boylan, and F. R. Johnston, *Demand Forecasting for Inventory Management: Characterizing the frequency of orders received by a stockiest.* **IMA Journal of Management Mathematics** 19 (2) 137-143 (2008).
10. W. J. Stevenson, **Operations management** (9th ed.). St. Louis: McGraw-Hill/Irwin, (2007).
11. J. W. G. Strijbosch, R. J. M. Heuts, and E. H. M. Schoot, *A combined forecast-inventory control procedure for spare parts.* **Journal of the Operational Research Society**, 51 (10) 1184-1192 (2000).
12. R. D. Snyder, *Forecasting sales of slow and fast moving inventories.* **European Journal of Operational Research**, 140 (3) 684-699 (2002).
13. A. A. Syntetos and J. E. Boylan, *On the stock-control performance of intermittent demand estimators.* **International Journal of Production Economics**, 103 (1) 36-47 (2006).
14. A. Vereecke, A. and P. Verstraeten, *An inventory management model for an inventory consisting of lumpy items, slow movers and fast movers.* **International Journal of Production Economics**, 35(1/3) 379-389 (1994).
15. S. Viswanathan, H. Widiarta, and R. Piplani, Forecasting aggregate time series with intermittent subaggregate components: top-down versus bottom-up forecasting. **IMA Journal of Management Mathematics**, 19 (3) 275-287 (2008).
16. A. Yılmaz, **Yedek parça piyasası**, http://www.subconturkey.com/2010/Mart/koseyazisi-Yedek-parca-piyasasi.html (available on 28.12.2011).

DECISION MAKING IN COMPLEX ENVIRONMENTS WITH UNCERTAIN GENERALIZED UNIFIED AGGREGATION OPERATORS*

JOSÉ M. MERIGÓ†

Department of Business Administration, University of Barcelona, Av. Diagonal 690, Barcelona, 08034, Spain

A new decision making approach is presented where the available information is uncertain and can be assessed with interval numbers. We consider complex environments where the available information is formed by using different sources of information that have to be considered in the analysis. We introduce a new aggregation operator adapted for these situations that we call the uncertain generalized unified aggregation operator (UGUAO). It includes a wide range of classical aggregation operators in the same formulation.

1. Introduction

Decision making is a fundamental task that individuals have to correctly assess in order to maximize their benefits. When dealing with the real world, we find a lot of heterogeneities that come from different sources of information such as subjective and objective information, particular attitudes of the decision maker in the specific problem considered different opinions and so on. Usually, when dealing with decision problems we use a wide range of mathematical and statistical techniques to efficiently manage the problem. A very useful tool in this assessment is the aggregation operator. In the literature, we find a wide range of aggregation operators [1] such as the weighted average and the ordered weighted average (OWA) [8,10].

Some aggregation operators are focused on providing a general framework that includes a wide range of particular cases. We call them generalized aggregation operators [4,9]. Recently, Merigó [2] has suggested a deeper approach that is able to unify other aggregation operators in the same formulation and considering the degree of importance that each concept has in the analysis. He called it the unified aggregation operator (UAO).

* This work is supported by the University of Barcelona through the project 099311.

† jmerigo@ub.edu.

In the previous approaches it is implicitly assumed that the available information is known and can be assessed with the classical exact numbers. However, in real world problems it is very common to find situations where the information is vague or imprecise. Thus, we need to use aggregation operators that are able to deal with this imprecision [3,6].

The aim of this paper is to introduce a new approach that uses UAO operators in uncertain environments that can be assessed with interval numbers and by using generalized means. We call it the uncertain generalized UAO (UGUAO) operator. Its main advantage is that it can use different sources of information in the analysis including subjective and objective information and the attitudinal character of the decision maker. Thus, it is able to represent the information in a more complete way.

2. Preliminaries

In this Section we briefly review the interval numbers, the uncertain generalized aggregation operators and the unified aggregation operator.

2.1. *Interval Numbers*

The interval numbers [7] are a very useful and simple technique for representing the uncertainty. They have been used in an astonishingly wide range of applications. They can be expressed in different forms. For example, if we assume a 4-tuple (a_1, a_2, a_3, a_4), that is, a quadruplet; we could consider that a_1 and a_4 represents the minimum and the maximum of the interval number, and a_2 and a_3, the interval with the highest probability or possibility, depending on the use we want to give to the interval numbers. Note that $a_1 \leq a_2 \leq a_3 \leq a_4$. If $a_1 = a_2 = a_3 = a_4$, then, the interval number is an exact number; if $a_2 = a_3$, it is a 3-tuple known as triplet; and if $a_1 = a_2$ and $a_3 = a_4$, it is a simple 2-tuple interval number.

In the following, we are going to review some basic interval number operations as follows. Let A and B be two triplets, where $A = (a_1, a_2, a_3)$ and $B = (b_1, b_2, b_3)$:

1) $A + B = (a_1 + b_1, a_2 + b_2, a_3 + b_3)$
2) $A - B = (a_1 - b_3, a_2 - b_2, a_3 - b_1)$
3) $A \times k = (k \times a_1, k \times a_2, k \times a_3)$; for $k > 0$.

Note that other operations could be studied [7] but in this paper we will focus on these ones.

2.2. *Uncertain Generalized Aggregation Operators*

Uncertain generalized aggregation operators are those aggregation operators that use generalized means or quasi-arithmetic means with information represented in the form of interval numbers. For example, the uncertain generalized OWA (UGOWA) operator [5] can be defined as follows.

Definition 1. Let Ω be the set of interval numbers. An UGOWA operator of dimension n is a mapping $UGOWA$: $\Omega^n \rightarrow \Omega$ that has an associated weighting vector W of dimension n with $\sum_{j=1}^{n} w_j = 1$ and $w_j \in [0, 1]$, such that:

$$UGOWA\ (\tilde{a}_1, \tilde{a}_2, \ldots, \tilde{a}_n) = \left(\sum_{j=1}^{n} w_j b_j^{\lambda} \right)^{1/\lambda} \quad (1)$$

where b_j is the jth largest of the $\tilde{a}_i$, λ is a parameter such that $\lambda \in (-\infty, \infty)$ and $\tilde{a}_i$ is the argument variable represented in the form of an interval number.

In a similar way we could consider other types of uncertain generalized aggregation operators such as the uncertain generalized weighted average (UGWA), the uncertain generalized probabilistic aggregation (UGPA) and the uncertain generalized mean (UGM).

2.3. *The Unified Aggregation Operator*

The unified aggregation operator (UAO) [2] is an aggregation operator that unifies a wide range of other aggregation operators in the same formulation and considering the degree of importance that each concept has in the aggregation. Its main advantage is that it can represent the information in a more complete way because it can deal with a wide range of aggregations that consider several sources of information. It can be defined as follows.

Definition 2. A unified aggregation operator of dimension m is a mapping UAO: $R^m \times R^n \rightarrow R$, that has an associated weighting vector C of dimension m representing concepts with a degree of importance of C_h, such that:

$$UAO\ (a_1, \ldots, a_n) = \sum_{h=1}^{m} \sum_{i=1}^{n} C_h w_i^h a_i \ , \quad (2)$$

where C_h is the degree of importance that each concept has in the aggregation with $C_h \in [0, 1]$ and $\sum_{h=1}^{m} C_h = 1$, w_i^h is the ith weight of the hth weighting vector W with $w_i^h \in [0, 1]$ and $\sum_{i=1}^{n} w_i^h = 1$.

3. The Uncertain Generalized Unified Aggregation Operator

The UGUAO operator is an aggregation operator that unifies any common aggregation operator in a general framework that includes it as a particular case and considering the degree of importance that has in the analysis. Moreover, it also deals with uncertain information that can be assessed with interval numbers. Furthermore, it uses generalized means that permit to aggregate different types of aggregations in the same framework such as arithmetic and quadratic aggregations. It can be defined as follows.

Definition 3. Let Ω be the set of interval numbers. A unified aggregation operator of dimension m is a mapping $UGUAO$: $\Omega^m \times \Omega^n \rightarrow \Omega$, that has an associated weighting vector C of dimension m representing concepts with a degree of importance of C_h, such that:

$$UGUAO\ (\tilde{a}_1, \tilde{a}_2, \ldots, \tilde{a}_n) = \sum_{h=1}^{m} C_h \left(\sum_{i=1}^{n} w_i^h \tilde{a}_i^{\lambda} \right)^{1/\lambda} \tag{3}$$

where C_h is the degree of importance that each concept has in the aggregation with $C_h \in [0, 1]$ and $\sum_{h=1}^{m} C_h = 1$, w_i^h is the ith weight of the hth weighting vector W with $w_i^h \in [0, 1]$, $\sum_{i=1}^{n} w_i^h = 1$, λ is a parameter such that $\lambda \in (-\infty, \infty)$ and $\tilde{a}_i$ is the argument variable represented in the form of an interval number.

The UGUAO operator is monotonic, bounded and idempotent. It is idempotent because if $\tilde{a}_i = \tilde{a}$, for all i, then, $UGUAO\ (\tilde{a}_1, \tilde{a}_2, \ldots, \tilde{a}_n) = \tilde{a}$. It is monotonic because if $\tilde{a}_i \geq \tilde{a}$, for all i, then, $UGUAO\ (\tilde{a}_1, \tilde{a}_2, \ldots, \tilde{a}_n) \geq UGUAO\ (\tilde{e}_1, \tilde{e}_2, \ldots, \tilde{e}_n)$. It is bounded because the UGUAO aggregation is delimitated by the minimum and the maximum.

A further interesting issue is to consider different particular types of UGUAO operators. First, we will consider some basic cases such as the uncertain generalized POWAWA (UGPOWAWA) operator. The UGPOWAWA operator is found if we simplify the GUAO operator by using the UGWA, the UGPA and the UGOWA in the same formulation. That is:

$$UGPOWAWA\ (\tilde{a}_1, \tilde{a}_2, \ldots, \tilde{a}_n) = C_1 UGOWA + C_2 UGWA + C_3 UGPA\,. \tag{4}$$

As we can see, the UGPOWAWA operator also includes many other particular cases such as:

- If $C_1 = 1$, we get the UGOWA operator.
- If $C_2 = 1$, we get the UGWA.
- If $C_3 = 1$, we get the UGPA.

- If $C_1 = 0$, we form the uncertain generalized probabilistic weighted average (UGPWA).
- If $C_2 = 0$, we form the uncertain generalized probabilistic OWA (UGPOWA) operator.
- If $C_3 = 0$, we form the uncertain generalized OWAWA (UGOWAWA) operator.

Other interesting aggregations that could be studied in the UGUAO operator are those formed by using a different expression in the parameter λ. Note that in Definition 3 we assumed that λ is equal in all the situations but often we may have an aggregation process that uses a different value λ for each subaggregation. For example:

- If $\lambda = 1$, we get the uncertain UAO (UUAO) operator.
- If $\lambda \to 0$, we get the geometric UUAO operator.
- If $\lambda = 2$, we get the quadratic UUAO operator.
- If $\lambda = -1$, we get the harmonic UUAO operator.
- If $\lambda = 3$, we get the cubic UUAO operator.
- If $\lambda 1 = 1$, $\lambda 2 = 2$, $\lambda 3 = 3$, we get the uncertain generalized probabilistic ordered weighted quadratic averaging weighted cubic averaging (UGPOWQAWCA) operator.

Next, we could analyze more complex constructions that use several UGOWAs, UGWAs and UGPAs in the same formulation. For example, we could look to the UGP2GOWA3GWA2 operator. That is, an aggregation operator that uses two generalized probabilistic vectors, two GWAs and three GOWAs. Following this methodology, we could form more and more complex aggregation operators by using more aggregations in the analysis such as UGP4GOWA6GWA5, UGP7GOWA6GWA8, and so on. Finally, we could imagine extremely complex situations where we consider thousands of aggregations in the analysis like a UGP3000GOWA6000GWA4000 operator. Note that these aggregations are very common in the real world because the real high quality analysis that represents the world should be assessed with this type of constructions.

4. Conclusions

We have introduced the UGUAO operator as an aggregation operator that unifies a wide range of aggregation operators under an uncertain environment that can be assessed with interval numbers. The main advantage of the UGUAO is that it can represent different types of information in the same problem such as objective and subjective information and the attitudinal character of the decision maker. In future research we expect to develop further developments by adding new characteristics in the analysis such as the use of mixture and multi-person aggregation operators.

Acknowledgments

This work is supported by the University of Barcelona through the project 099311.

References

1. G. Beliakov, A. Pradera and T. Calvo, *Aggregation functions: A guide for practitioners* (Springer-Verlag, Berlin-Heidelberg, 2007).
2. J.M. Merigó, Decision making with probabilities, weighted averages and OWA operators, *Proceedings of the IEEE-SSCI 2011*, Paris, France, pp. 122-129.
3. J.M. Merigó, The uncertain probabilistic weighted average and its application in the theory of expertons, *African J. Business Manag.* **5**, 6092 (2011).
4. J.M. Merigó and M. Casanovas, The uncertain induced quasi-arithmetic OWA operator, *Int. J. Intelligent Syst.* **26**, 1 (2011).
5. J.M. Merigó and M. Casanovas, The uncertain generalized OWA operator and its application in financial decision making, *Int. J. Information Techn. Decision Making* **10**, 211 (2011).
6. J.M. Merigó and G.W. Wei, Probabilistic aggregation operators and their application in uncertain multi-person decision making, *Techn. Econ. Develop. Economy* **17**, 335 (2011).
7. R.E. Moore, *Interval analysis*, (Prentice Hall, Englewood Cliffs, NJ, 1966).
8. R.R. Yager, On ordered weighted averaging aggregation operators in multi-criteria decision making, *IEEE Trans. Syst. Man Cybern. B* **18**, 183 (1988).
9. R.R Yager, Generalized OWA aggregation operators, *Fuzzy Optim. Decision Making* **3**, 93 (2004).
10. R.R. Yager, J. Kacprzyk and G. Beliakov. *Recent developments on the ordered weighted averaging operators: Theory and practice*. (Springer-Verlag, Berlin-Heidelberg, 2011).

SOFTWARE COST ESTIMATION BY FUZZY ANALOGY FOR ISBSG REPOSITORY

ALI IDRI*, FATIMA AZZAHRA AMAZAL

Department of Software Engineering, ENSIAS, Mohamed V University, Rabat, Morocco

Software cost estimation is one of the most important and complex tasks in software project management. As a result, several techniques for estimating development effort have been suggested. Fuzzy Analogy is one of these techniques suggested to estimate project effort when it is described either by linguistic or numerical values. Based on reasoning by analogy and fuzzy logic, this technique uses fuzzy representation of software project attributes by using expert knowledge or clustering techniques. From this work, we evaluate the accuracy of this approach to estimate the software effort using the International Software Benchmarking Standards Group (ISBSG) repository.

1. Introduction

Accurate estimation for software projects is one of the greatest challenges for software developers. Hence, numerous software cost estimation techniques have been suggested. A review of the literature revealed that these techniques may be grouped into two major categories: (1) algorithmic models, which are derived from the statistical or numerical analysis of historical projects data [1], and (2) non-algorithmic models, which are based on a set of artificial intelligence techniques such as artificial neural networks, genetic algorithms, regression trees, rule-based induction and analogy-based reasoning [3,7,8,9]. Case-based reasoning is one of the most promising techniques in the software cost estimation field due to its ability to model complex relationships between the dependent variable (such as cost) and the independent variables (cost drivers). Itis also an intuitive method that can be understood and explained easily to users (as opposed to black-box approaches like neural networks) [7,9].

Case-based reasoning is an attractive technique for software cost estimation [7,9]. However, it is limited by its inability to manage uncertainties and imprecision when estimating the development effort of the new project. To overcome this limitation we have developed a new approach referred to as Fuzzy Analogy. This approach, which is based on estimation by analogy, uses fuzzy sets theory to deal with linguistic values [3].

* Email: idri@ensias.ma

Fuzzy analogy was validated, in two earlier works, by using the COCOMO'81[1]and the Tukutuku [4] datasets. The results of these validation show that Fuzzy Analogy generates more accurate estimates than other techniques and it handles correctly the imprecision and the uncertainty when describing a software project [3,4].The aim of this study is to validate fuzzy analogy using the International Software Benchmarking Standards Group (ISBSG) repository, which contains more than 2000 software projects, gathered from different countries.

The remaining of this paper is organized as follows. In the second section, we present the principle of the Fuzzy Analogy Approach. The third section describes the ISBSG repository and briefly how to build fuzzy sets and their membership functions using the Fuzzy C-Means clustering technique and a Real Coded Genetic Algorithm. The last section presents and discusses the results obtained from the empirical experiments on the ISBSG repository.

2. Estimation by Fuzzy Analogy

As for the classical analogy procedure, Fuzzy Analogy is composed of three steps: identification of cases, retrieval of similar cases and case adaptation [4].

Identification of cases

The objective of this step is to describe each project by a set of attributes, which can be, measured either by numerical or linguistic values. The goal of our Fuzzy Analogy approach is to deal with linguistic values by using fuzzy sets rather than classical intervals.

Retrieval of similar cases

To measure the similarity between two projects, which are described by linguistic values such as ‘low’, and 'high', we have proposed a set of newmeasures based on fuzzy logic [3,4]. Using these measures, two steps are required for evaluating similarity:

- *First step*: We calculate the similarity between two projects P_1 and P_2according to each individual attribute V_j describing P_1 and P_2, $d_{v_j}(P_1, P_2)$
- *Second step*: We calculate the overall distance $d(P_1,P_2)$ by aggregating the individual similarities $d_{v_j}(P_1, P_2)$ using Regular Increasing Monotone(RIM) linguistic quantifiers such as all, most, many, or there exists. The choice of the appropriate RIM linguistic quantifier, Q, depends on the characteristics and the needs of each environment.

Case adaptation

The aim of this step is to generate an estimate for the new project by using the real costs of the most similar projects. Two steps are required in the case adaptation. First, we have to decide on the number of analogies that will be used to estimate the cost of the new project. The second step is to adapt the chosen analogies in order to generate an estimate for the new project.

3. Dataset Description

The ISBSG repository is a multi-organizational dataset, which contains more than 2000 software projects gathered from different organizations in different countries [5]. Each project in this dataset is described by several numerical and categorical attributes from which we identified 12 numerical attributes (table1) to characterize an ISBSG project and its development process.

In order to have relevant data, we have taken into account projects with the same development type (new development), having a data quality and UFP rating of A or B. Furthermore, we have selected projects that use the same count approach (IFPUG) so as to be sure we make appropriate analogies. Since many projects have missing values, only 148 projects have been retained. As for the web dataset, field descriptions were insufficient to empirically build their fuzzy representation. As a consequence we have used Fuzzy C-Means clustering technique (FCM) and a Real Coded Genetic Algorithm (RCGA) to build fuzzy sets and their membership functions [3].

Table 1.Examples of software attributes for the ISBSG repository

Software attribute	Description
Value Adjustment factor	The adjustment to the function points
Resource Level	Level of people whose time is included in the work effort data reported
Max Team Size	The maximum number of people that worked at any time on the project
User-Base Business units	Number of business units that the system services

In the first step, we use Fuzzy C-Means clustering technique (FCM) and the Xie-Beni validity criterion to decide on the number of clusters.Table 2 shows the classification obtained according to the Xie-Beni index.

Table 2. Examples of classification generated by the FCM algorithm for the minimum value of the Xie-Beni index

Attributes	#fuzzy sets	Attributes	#fuzzy sets
Value Adjustment factor	6	Max Team Size	4
Resource Level	3	User-Base Business units	4

Once the clusters and their partition matrix have been generated, the next step is to build membership functions for these clusters by using the RCGA. Membership functions can be trapezoidal, triangular or Gaussian.

Figure 1 shows three different shapes of membership functions associated with the fuzzy sets of the Resource level.

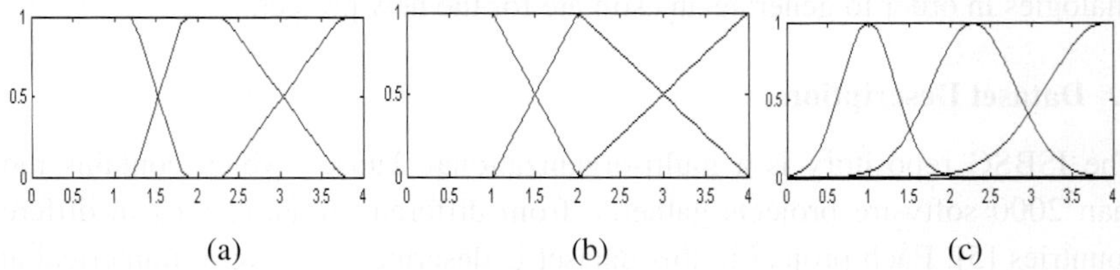

Figure 1.Membership functions associated with the fuzzy sets of the Resource Level attribute: (a) Trapezoidal for Resource Level; (b) Triangular for Resource Level; and (c) Gaussian for Resource Level.

4. Empirical Results

The proposed software cost estimation model using Fuzzy Analogy has been validated through a series of experiments using the ISBSG Repository. The calculations were made using a software prototype developed with Matlab 7.0.

Measuring the accuracy of the proposed estimation model is accomplished by usingtheprediction level *Pred* which is calculated as:

$$\Pr ed(p) = \frac{k}{N} \quad (1)$$

Where N is the total number of projects, k is the number of projects with an MRE (Magnitude Relative Error) less than or equal to p. In our evaluation, we use p equal to 0.20.

The prototype developed allows us to try applying various α-RIM linguistic quantifiers (Q) to the ISBSG historical data. Usually, each environment must define an appropriate quantifier by studying its features and its requirements. In this illustration, we used various α-RIM linguistic quantifiers to calculate the overall distances since the lack of knowledge concerning the appropriate quantifier for the environment from which the dataset was collected. The α-RIM linguistic quantifiers used are defined by:

$$Q(r) = r^{\alpha} \quad \alpha > 0 \quad (2)$$

Figure 2 shows the results of the validation of the fuzzy Analogy approach using the minimum value of the Xie-Beni index. From the results obtained, we conclude that the accuracy of the estimates is, in general, monotonous increasing according to α. Indeed, when α tends towards zero, we obtain lower accuracy because the overall similarity takes into account fewer attributes among all those describing software projects. The minimum number of attributes to consider is one. For the other α-RIM linguistic quantifiers ($0<\alpha<\infty$), the accuracy increases with α because additional attributes will be considered in the evaluation of the overall similarity. As a maximum we may consider all attributes.

By comparing the accuracy of the Fuzzy Analogy when using different shapes of the membership functions (Figure 2), we noticed that Fuzzy Analogy performs better when trapezoidal membership functions are used than when triangular membership functions are used. In addition, it performs better when triangular membership functions are used than when Gaussian membership functions are used.

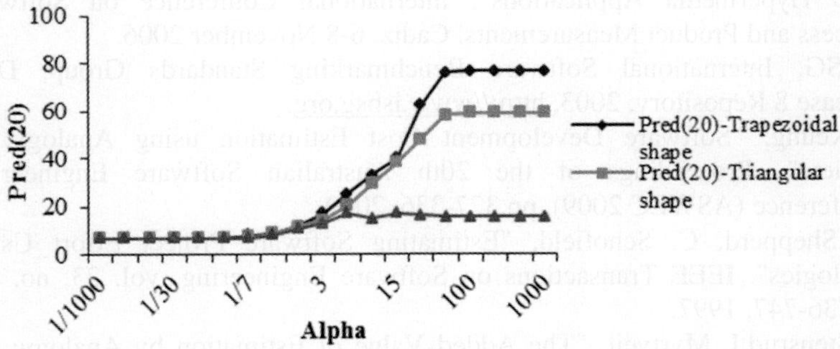

Figure 2. Relationship between α and the accuracy of Fuzzy Analogy for the three shapes of membership functions using the minimum value of the Xie-Beni index.

5. Conclusion

In this paper, we have described briefly the Fuzzy Analogy approach for software cost estimation. This approach is based on reasoning by analogy, fuzzy logic and linguistic quantifiers. Such an approach can be used when the software projects are described by linguistic and/or numerical values. The main benefit of this model is its ability to handle correctly the imprecision and the uncertainty when describing a software project. Also, we have validated the Fuzzy Analogy approach on the ISBSG repository. The descriptions given of the ISBSG attributes were insufficient to empirically build their fuzzy representations. Hence, for each attribute and its associated linguistic values, corresponding fuzzy sets were defined by using the Fuzzy C-Means clustering technique (FCM). To build membership functions for these fuzzy sets we have used the

Real Coded Genetic Algorithm (RCGA). These membership functions can be trapezoidal, triangular or Gaussian. The results obtained show that Fuzzy Analogy performs better and generate accurate predictions when using the trapezoidal fuzzy representation.

References

1. B.W. Boehm, "Software Engineering Economics". Prentice-Hall, 1981.
2. A. Idri, A. Abran., T. Khoshgoftaar, "Estimating Software Project Effort by Analogy Based on Linguistic Values ". 8th IEEE International Software Metrics Symposium, pp. 21-30, Ottawa, June 2002.
3. A. Idri, A. Zahi, A. Abran, "Generating Fuzzy Term Sets for Software Project Attributes using Fuzzy C-Means and Real Coded Genetic Algorithms", ICT4M, pp.120-127, Malaysia, November 2006.
4. A. Idri, A. Zahi, A. Abran, "Software Cost Estimation by Fuzzy Analogy for Web Hypermedia Applications", International Conference on Software Process and Product Measurements, Cadiz, 6-8 November 2006.
5. ISBSG, International Software Benchmarking Standards Group, Data Release 8 Repository, 2003, http://www.isbsg.org.
6. J. Keung, "Software Development Cost Estimation using Analogy: A Review". Proceedings of the 20th Australian Software Engineering Conference (ASWEC 2009), pp.327-336, 2009.
7. M. Shepperd, C. Schofield, "Estimating Software Project Effort Using Analogies". IEEE Transactions on Software Engineering, vol. 23, no. 12, pp.736-747, 1997.
8. E. Stensrud,I. Myrtveit, "The Added Value of Estimation by Analogy: An Industrial Experiment". Proceedings of the European Software Measurement Conference (FESMA 98), pp. 549–556, Germany, May,1998.
9. S. Vicinanza, M. J. Prietolla, "Case-Based Reasoning in Software Effort Estimation". Proceedings of the 11th International Conference on Information Systems, 1990.

PROBABILISTIC KERNEL RANKING

BERND-JÜRGEN FALKOWSKI*

School of Business Administration, University of Applied Sciences Stralsund, Zur Schwedenschanze 15, D-18435 Stralsund, Germany

A solution of the ranking problem using kernels is described.. As a consequence an elegant probabilistic ranking algorithm is obtained from the pocket algorithm. This removes a restriction from a former paper and simplifies proofs considerably. In addition results concerning class probabilities are exhibited. Some preliminary experimental results are reported and briefly discussed.

1. Introduction

Recently the importance of ranking functions has been recognized in different contexts. As the most important example so-called *scoring systems* should be mentioned (information retrieval using the vector space model provides another one): At least since the beginning of the world-wide financial crisis it has been realized that the objective creditworthiness rating of individual banking customers is a major problem. Its solution even now is mainly achieved by exploiting bank records to compute a linear discriminant dividing customers into "good" and "bad" ones. Although some results obtained by this method are reasonable, certain difficulties indicate that a finer distinction (more preference classes) would be advantageous, cf. e.g. [3], [13]. Of course, for the ranking of countries several rating classes have been used for quite some time, cf. e.g. [8]. However, as the present Euro crisis shows, these methods still lack the necessary accuracy.

Although the ranking problem may be solved using general classifiers, cf. e.g. [1], special ranking functions have been investigated for some time. An approach via Bayesian statistics and logistic regression may be found in [9]. There the modelling of the probabilities introduced some assumptions that could be hard to justify in practice. A somewhat different method was employed in [12], where support vector machines were utilized. Unfortunately a restriction concerning the monotonicity of the *cut-offs* was apparently neglected. In [11] support vector machines led to a somewhat complicated algorithm. In [6] a new

* bernd.falkowski@fh-stralsund.de

online learning algorithm based on perceptron learning, cf. [10], and a special but fairly wide class of popular kernels was described, since former perceptron learning algorithms lead to certain disadvantages (e.g. an unnecessarily large Novikoff bound and a somewhat obscure correctness proof), cf. e.g. [12], [5]. It is the purpose of this paper to remove the restriction on the kernels, simplify the correctness proof in [6], and generalize the algorithm to fault tolerant learning. Moreover some results concerning class probabilities will be exhibited. Some preliminary experimental results will be reported and briefly discussed.

2. The Ranking Problem and its Solution

Definition: A map $K: X \times X \to \mathfrak{R}$, where X is a set and $\mathfrak{R}$ is the real line is called a *positive semi-definite kernel* if for all choices $x(1), x(2), \ldots, x(l) \in X$ and $r_1, r_2, \ldots, r_l \in \mathfrak{R}$ the inequality

i. $\Sigma_{ij}\, r_i r_j K(x(i),x(j)) \geq 0$ holds, and
ii. K is symmetric, i.e. $K(x,y) = K(y,x)$ for all $x, y \in X$.

2.1. *The Problem*

Suppose that a finite set $X := \{x(1), x(2), \ldots, x(m)\}$ is given. Assume further that $C_1, C_2, \ldots, C_k$ form a partition of X (i.e. each x(i) belongs to precisely one C_j) and that a preference relation $C_1 \rangle C_2 \rangle \ldots \rangle C_k$ is defined. For definiteness think of classes describing the creditworthiness of bank customers (here membership of C_1 would indicate the highest creditworthiness). Then it is required to find a map $m_\mathbf{w}: X \to \mathfrak{R}$ that preserves this preference relation, where of course **w** denotes a weight vector. More precisely one must have

$$x(j) \rangle x(k) \Rightarrow m_\mathbf{w}(x(j)) > m_\mathbf{w}(x(k)).$$

If a positive semi-definite kernel K on X is given, then a special solution to the problem is obtained with a weight vector **w** and *cut-offs* $c_1 > c_2 > \ldots > c_{k-1}$ such that

$$x \in C_1 \Rightarrow \quad \Sigma_i\, w(i)K(x(i),x) > c_1$$
$$x \in C_\ell \Rightarrow \quad c_{\ell-1} \geq \Sigma_i\, w(i)K(x(i),x) > c_\ell \quad \text{for } \ell = 2, 3, \ldots, k-1$$
$$x \in C_k \Rightarrow \quad c_{k-1} \geq \Sigma_i\, w(i)K(x(i),x),$$

since then defining $m_\mathbf{w}$ by $m_\mathbf{w}(x) := \Sigma_i\, w(i)K(x(i),x)$ provides a solution.

Of course, in practice an ideal solution may not be possible and hence some (as few as possible) violations of the above implications may occur However, the monotonicity of the cut-offs must definitely be guaranteed.

2.2. *Solution*

Denote by $\mathbf{e}_i$ the i-th unit vector in $\mathfrak{R}^{k-1}$ (considered as column vector). Suppose further that ψ: X → H, the *feature map* from X to a Hilbert space H, for details see e.g. [12], p. 61, is given and satisfies $K(x(i),x(j)) = <\psi(x(i)),\psi(x(j))>_H$, where $<.,.>_H$ denotes the Hilbert space inner product. Then embed the set X as a set Y into the Cartesian product $H\times\mathfrak{R}^{k-1}$ as follows. Y contains the k-2 vectors $(\mathbf{0}, \mathbf{e}_i-\mathbf{e}_{i+1})$ for i = 1, 2, ..., k-2, where **0** denotes the zero vector in H. Denote this subset of Y by R (rules). In addition, for x in C_1 respectively C_k Y contains a vector $(\psi(x), -\mathbf{e}_1)$ respectively $(-\psi(x), \mathbf{e}_{k-1})$ and for x in C_ℓ with $1 < \ell < k$ it contains the *two* vectors $(\psi(x), -\mathbf{e}_\ell)$ and $(-\psi(x), \mathbf{e}_{\ell-1})$.

Now an application of the pocket algorithm with rules, cf. [7], gives the required result: First of all one obtains a discriminant $D((h,\mathbf{v})) = <\Sigma_j\beta(j)h_j,h>_H + (\Sigma_j\beta_j\mathbf{v}_j)^T\mathbf{v}$, where an arbitrary but fixed enumeration of the vectors in Y has been assumed. Since all h_j are of the form $\alpha\psi(x(i))$ with $\alpha = +1$ or -1 this may be rewritten as $D((h,\mathbf{v})) = <(\Sigma_i w(i)\psi(x(i)),\mathbf{c}),(h,\mathbf{v})>$, where the summation now runs from 1 to m and **c** has been equated to $(\Sigma_j\beta_j\mathbf{v}_j)^T$. This discriminant is strictly positive on the set of rules R and thus one obtains the monotonicity of the cut-offs as required. Moreover even the conditions

$$x \in C_1 \;\Rightarrow\; \Sigma_i\, w(i)K(x(i),x) > c_1$$

$$x \in C_\ell \;\Rightarrow\; c_{\ell-1} > \Sigma_i\, w(i)K(x(i),x) > c_\ell \quad \text{for } \ell = 2, 3, \ldots, k\text{-}1$$

$$x \in C_k \;\Rightarrow\; c_{k-1} > \Sigma_i\, w(i)K(x(i),x)$$

hold. For example, if $x\in C_1$, then $<(\Sigma_i w(i)\psi(x(i)),\mathbf{c}),(\psi(x), -\mathbf{e}_1)> \;> 0$implies $\Sigma_i w(i)K(x(i),x) > c_1$ (summation from 1 to m here), as desired.

3. Exploiting the Kernel Trick

A little consideration shows that an explicit use of the feature map is not needed, since the computations to be performed involve only Hilbert space inner products, which can be entirely described in terms of kernels and thus the following algorithm is seen to be correct.

3.1. *Ranking Algorithm (Kernel Version)*

Input: A training set X:= {x(1), x(2), ... , x(m)} as above and a positive semi-definite kernel K on X.

Output: A weight vector **p** and *cut-offs* $c_1 > c_2 > ...> c_{k-1}$ (collected in a cut-off vector $\mathbf{c_p}$) satisfying the inequalities in 2.1 (or, if this is not possible, a maximal number of the inequalities).

Note: An element x from X will be said to be correctly classified using **w** and $\mathbf{c}_w$ if the implications in 2.1 hold.

Step 1 (temporary data):

w (discriminant weight vector), **p** (pocket weight vector), $\mathbf{c}_w$ (cut-off vector for **w**), $\mathbf{c}_p$ (cut-off vector for **p**), run_**w** (number of successive correct classifications using the discriminant weight vector and $\mathbf{c}_w$), run_**p** (number of successive correct classifications using the pocket weight vector and $\mathbf{c}_p$), num_ok_**w** (number of all correct classifications using the discriminant weight vector and $\mathbf{c}_w$), num_ok_**p** (number of all correct classifications using the pocket weight vector and $\mathbf{c}_p$)

Step 2 (initialization):

Fix the number of required iterations iter of the algorithm and set

w:= **0**, **c**:= **0**; run_**w** = run_**p** = num_ok_**w** = num_ok_**p** = 0.

Step 3:

```
Repeat  iter  times:
If (there exist  i+1,  i such that  cw(i+1)  ≥  cw(i) )
Then
        cw(i):= cw(i) + 1;
        cw(i+1) := cw(i+1) - 1;
Else
        randomly select an element x(i) from X.
        If   x(i)  is correctly classified using w and cw
        Then
                run_w = run_w + 1;
                If  run_w > run_p
                Then
                        compute num_ok_w;
                        If ((num_ok_w > num_ok_p)
                        Then
                                p = w;
                                cp = cw;
                                run_p = run_w;
                                num_ok_p = num_ok_w;
        Else
                If (x(i) ∈ Cℓ, 1≤ ℓ ≤k-1and Σi w(i)K(x(i),x)  ≤ cℓ )
                Then
                        w[i] = w[i] + 1;
                        cw [ℓ]:= cw [ℓ] - 1;
                If (x(i) ∈ Cℓ, 2≤ ℓ ≤k and Σi w(i)K(x(i),x) ≥ cℓ-1 )
                Then
                        w[i] = w[i] - 1;
```

$$c_w[\ell\text{-}1] := c_w[\ell\text{-}1] + 1;$$
$$\text{run\_}\mathbf{w} = \text{num\_ok\_}\mathbf{w} = 0;$$

Step 4: Return **p**, c_p

3.2. *Class Probabilities*

In [3] it is shown that under certain assumptions, where in particular X is supposed to be a subset of some Euclidean space (which is not necessary in general), the output of the pocket algorithm with rules may be interpreted in terms of class probabilities as follows:

$$P(C_1|\mathbf{x}) := g(<\psi(\mathbf{x}), \mathbf{w}> - c_1),$$
$$P(C_\ell|\mathbf{x}) := g(<\psi(\mathbf{x}), \mathbf{w}> - c_\ell) - g(<\psi(\mathbf{x}), \mathbf{w}> - c_{\ell-1}) \text{ for } \ell = 2, 3, \ldots, k\text{-}1;$$
$$P(C_k|\mathbf{x}) := 1 - g(<\psi(\mathbf{x}), \mathbf{w}> - c_{k-1}),$$

where $P(C_\ell|\mathbf{x})$ denotes the probability of C_ℓ conditioned on **x** for $\ell = 1, 2, 3, \ldots, k\text{-}1$ and g is the logistic activation function $g(y) := [1+\exp(-y)]^{-1}$. Hence in this case the algorithm of 3.1 yields analogous class probabilities as

$$P(C_1|\mathbf{x}) := g(\Sigma_i w(i)K(\mathbf{x}(i),\mathbf{x}) - c_1),$$
$$P(C_\ell|\mathbf{x}) := g(\Sigma_i w(i)K(\mathbf{x}(i),\mathbf{x}) - c_\ell) \quad - g(\Sigma_i w(i)K(\mathbf{x}(i),\mathbf{x}) \quad - c_{\ell-1}), \quad 1 < \ell < k\text{-}1,$$
$$P(C_k|\mathbf{x}) := 1 - g(\Sigma_i w(i)K(\mathbf{x}(i),\mathbf{x}) - c_{k-1}).$$

4. Functionality and Performance Tests

Only a small number of real life examples (about 360) were available for functionality tests. The data provided described bank customers and were partitioned into 4 preference classes. The number of iterations needed for faultless learning using a Radial Basis Function kernel was just 5000 (7.4 sec CPU time).

For performance and generalization tests two (real life) sets with 5164 data vectors containing 52 attributes each describing "good" or "bad" banking cus tomers were used. Both sets contained 123 "bad" customers. One set was used for training and the other for validation purposes. The above algorithm was implemented using Java concurrent programming as described in [4]. 50 000 iterations of the main loop were run on a QuadCore PC (2 GB RAM) using about 3 hours of CPU time, where a Radial Basis Function kernel was employed as well. This resulted in faultless (or near faultless) learning, as was to be expected, cf. [2], [13]. Unfortunately, due to overfitting the validation results proved disappointing even when using a smoothing parameter. Since with a cubic polynomial kernel good generalization results were obtained, cf. [6], it seems that there were just not sufficiently many data vectors available.

5. Conclusion

An elegant ranking algorithm based on the pocket algorithm and utilizing the kernel trick has been described. In addition class probabilities have been computed in terms of kernels. This approach generalizes the algorithm described in [6] by allowing arbitrary kernels and fault tolerant learning. In order to evaluate the practical use of such kernels further experiments are needed.

References

1. Bishop,C.M.: Pattern Recognition and Machine Learning, Springer, (2006)
2. Cover, T.M.: Geometrical and Statistical Properties of Systems of Linear Inequalities with Applications in Pattern Recognition. IEEE Trans. on Electronic Computers, Vol. 14, (1965)
3. Falkowski, B.-J.: On a Ranking Problem Associated with Basel II. Journal of Information & Knowledge Management, Vol. 5, No. 4, (2006)
4. Falkowski,B.-J.: Parallel Implementation of Certain Neural Network Algorithms. In: Computational Intelligence in Decision and Control, Proceedings of the 8th International FLINS Conference, Eds.: Ruan, D.; Montero, J.; Lu, J.; Martinez, L.; D'hondt, P.; Kerre, E., World Scientific, (2008)
5. Falkowski, B.-J.; Appelt, M.; Finger, C.; Koch, S.; van der Linde, H.: Parallel Java Implementation of a Kernel Ranking Algorithm. In: Proc. of the 12th IASTED Conference on AI and Soft Computing, ACTA Press, (2008)
6. Falkowski, B.-J.: Ranking and Perceptron Learning. Proceedings of the 9th International FLINS Conference, Emei, Chengdu, China, 2. – 4. August 2010
7. Gallant, S.I.: Perceptron-based Learning Algorithms. IEEE Transactions on Neural Networks, Vol. I, No. 2, (1990)
8. Morgan, J.P.: CreditMetrics^{TM}-Technical Document. First Edition, New York, (1997)
9. Mathieson, M.: Ordinal Models for Neural Networks. Neural Networks in Financial Engineering, Proceedings of the 3rd International Conference on Neural Networks in the Capital Markets, World Scientific, (1996)
10. Minsky, M.L.; Papert, S.: Perceptrons. MIT Press, (Expanded Edition 1990)
11. Shashua, A.; Levin, A.:Ranking with Large Margin Principle: Two Approaches, NIPS 14, (2003)
12. Shawe-Taylor, J.; Cristianini, N.: Kernel Methods for Pattern Analysis. CUP, (2004)
13. Vapnik, V.N.: Statistical Learning Theory. John Wiley & Sons, (1998)

PROBABILISTIC OPERATION MANAGEMENT FOR A MICROGRID BASED ON POINT ESTIMATE METHOD AND MODIFIED GRAVITATIONAL SEARCH ALGORITHM

MASOUD JABBARI

Department of Electrical Engineering, Marvdasht Branch, Islamic Azad University, Marvdasht , Iran

FARANAK GOLESTANEH, TAHER NIKNAM

Electronic and Electrical Department, Shiraz University of Technology, Shiraz, Iran

This paper proposes a probabilistic approach for energy management of MicroGrids (MGs) under uncertain environment. The proposed framework includes $2m+1$ point estimate method for covering the existing uncertainties in the MGs and a modified optimization algorithm based on the Gravitational Search Algorithm (GSA) to determine the economic assessment of MGs. This paper takes account of uncertainties in load demand, market prices and the available electrical power of wind farms and photovoltaic systems. The weibull and normal distributions are employed to model the input random variables. Moreover, the Gram-Charlier expansion is used to find an accurate distribution of the total operational cost of MGs for the next day-ahead. The effectiveness of the proposed method is validated on a typical grid-connected MG including energy storage and different power generating units.

1. Introduction

MGs are the integration of electrical and/or thermal loads and the Distributed Energy Resource (DER) units which can work in both grid-connected and autonomous mode [1].

Several studies have been done on optimizing the operation management of MGS. A smart energy management system based on the matrix real-coded genetic algorithm is presented in [2] to optimize the operation of MG. The load demand and generation of two MGs, including wind farms is modeled in [3] by using optimal power flow and Particle Swarm Optimization (PSO) algorithm. Chedid et al. used a linear programming to minimize the average production cost of electric power in a hybrid solar-wind MG while environmental factors were considered [4].

One of the drawbacks with the previous works on the OM of the MGs is because of neglecting the uncertainty of the generation, load demand and market price patterns.

This paper proposes $2m+1$ Point Estimate Method (PEM) as a probabilistic one to cover the uncertainties in load demand, market prices and available output power of PV and WT units. The normal and weibull Probability Density Function PDF are used to model the variations of Input Random Variables (IRVs). Moreover, the Gram-Charlier expansion is employed to provide more accurate probability distribution of total operational cost. Finally, a modified Gravitational Search Algorithm (MGSA) is devised to optimize the OM of the MGs.

2. Operation Management of a MicroGrid

2.1. *Objective function*

$$Min\ \widetilde{f(X)} = \sum_{t=1}^{NT} Cost^t = \sum_{t=1}^{NT} \{ \sum_{i=1}^{N_g} [u_i^t\, p_{Gi}^t B_{Gi}^t + S_{Gi}\left|u_i^t - u_i^{t-1}\right|] + \sum_{j=1}^{N_s} [u_j^t\, p_{sj}^t B_{sj}^t + S_{sj}\left|u_j^t - u_j^{t-1}\right|] + p_{Grid}^t B_{Grid}^t \} \tag{1}$$

where NT is total number of hours. N_g, N_s are total number of generation and storage units, respectively. B_{Gi}^t, B_{sj}^t, B_{Grid}^t are bid of the ith Distributed Generation (DG) source, the jth Distributed Storage (DS) unit and utility at hour t, respectively. p_{Gi}^t, p_{sj}^t and p_{Grid}^t are active power output of the generator i and the storage device j and utility at time t, respectively. S_{Gi}, S_{sj} are start-up/shut-down costs for the generation unit i and the storage device j, respectively. u_i^t is status of unit i at hour t and $\mathbf{X} = [X^1\, X^2 \ldots X^t \ldots X^{NT}]$, X^t is state variables vector including active powers of units and their related states that can be described as follows:

$$X^t = [p_{G1}^t, p_{G2}^t, \ldots, p_{GNg}^t, p_{s1}^t, p_{s2}^t, \ldots, p_{sNs}^t, u_1^t, u_2^t, \ldots, u_{Ns+Ng}^t] \tag{2}$$

2.2. *Constraints*

Power Balance:

$$\sum_{i=1}^{N_g} p_{Gi}^t + \sum_{j=1}^{Ns} p_{sj}^t + p_{Grid}^t = \sum_{D=1}^{N_D} P_{L_D}^t \tag{3}$$

Real Power Generation Capacity:

$$p_{i,\min}^t \le p_i^t \le p_{i,\max}^t \tag{4}$$

3. 2m Point Estimate Method

This paper uses the Hong's $2m+1$ PEM to cover the uncertainty in load demand market prices and available output power of WT and PV units [5]. Mostly, the $2m$ PEM has been employed in power system research. The $2m+1$ PEM can provide more accurate solutions in comparison with the $2m$ because of two reasons: the accuracy of the $2m+1$ PEM doesn't depend on the number of IRVs. Moreover, the $2m+1$ PEM employs both skewness and kurtosis of IRVs while the $2m$ scheme doesn't use the second one. It is worth noting that, the $2m+1$ PEM needs just one evaluation more than $2m$ one.

The $2m+1$ PEM finds the statistical information of the output random variables using the first few central moments of the IRVs. The details can be found in [5].

The probability density function of the output random variable can be approximated and plotted using the calculated mean and standard deviation by the $2m+1$ PEM and Gram-Charlier series approach [6].

4. Modified GSA

4.1. *Original GSA*

The GSA is a new powerful optimization algorithm which is inspired by the Newton's law of gravity [7]. In the GSA, the individuals are a set of the objects and the solutions are positions of them. Moreover, the mass of the objects are defined by the fitness value associated with corresponding solution. The GSA expresses the gravitational force as follows.

$$F_{je}^{k,t} = G^k \times (M_j^k \times M_e^k) \times (X_j^{k,t} - X_e^{k,t}) / (R_{je}^k + \varepsilon) \tag{5}$$

where t, k are the time interval and iteration index, respectively, M_j^k, M_e^k are the gravitational mass related to particles *j* and *e*, R_{je}^k is the Euclidian distance between two particles *j* and *e*. ε is small constant and G^k is the gravitational constant at the kth iteration.

The value of mass of each particle is computed as :

$$m_j^k = (Fit(\mathbf{X}_j^k) - worst^k) / (best^k - worst^k) \tag{6}$$

$$M_j^k = m_j^k / (\sum_{e=1}^{N_{swrm}} m_e^k) \tag{7}$$

where $Fit(\mathbf{X}_j^k)$ is the fitness value of the jth particle. N_{swrm} is the total number of the bees in the swarm. $worst^k$ and $best^k$ are the maximum and the minimum fitness values at iteration k (in minimization problems).

The overall gravitational forces, which exerts on the particle j, is calculated as:

$$F_j^{k,t} = \sum_{e=1,e\neq j}^{N_{swrm}} r_e \times F_{je}^{k,t} \quad (8)$$

The acceleration, velocity and the new position of each particle are defined as follows:

$$a_j^{k,t} = F_j^{k,t} / M_j \quad (9)$$

$$V_{new,j}^{k,t} = a_j^{k,t} + r \times V_{old,j}^{k,t} \quad (10)$$

$$X_{new,j}^{k,t} = X_{old,j}^{k,t} + V_{new,j}^{k,t} \quad (11)$$

where $V_{new,j}^{k,t}$, $V_{old,j}^{k,t}$ are the new and old velocity of the jth particle, respectively. $X_{new,j}^{k,t}$, $X_{old,j}^{k,t}$ are the new and old position of the jth particle, respectively and r is a random number with normal distribution between 0 and 1.

4.2. *Mutation operator*

The MGSA algorithm proposes this mutation to enhance convergence characteristics of the original GSA and escape from local optima and premature convergence. This mutation tries to employ the information found by the other members of the swarm more efficient. In this approach, for the jth object, two objects are selected randomly as $m_1 \neq m_2 \neq j$ thereafter, the new solution is obtained as:

$$\begin{cases} if \quad Fit(\mathbf{X}_{m_1}) \leq Fit(\mathbf{X}_{m_2}) \quad \text{then } X_{mut,j}^{k,t} = X_{new,j}^{k,t} + r \times (X_{m_1}^{k,t} - X_{m_2}^{k,t}) \\ else \quad X_{mut,j}^{k,t} = X_{new,j}^{k,t} + r \times (X_{m_1}^{k,t} - X_{m_2}^{k,t}) \end{cases} \quad (12)$$

5. Simulation Results

The case study of this paper is a grid-connected MG consists of different DER units such as the MT, a low temperature fuel cell (PAFC), PV, WT and also NiMH-Battery. System data can be found in [1].

Firstly, to verify the effectiveness of the proposed MGSA, deterministic OM is solved by it. Table 1 offers a comparative study in terms of the some statistical indices between the proposed MGSA and some of the other optimization techniques reported in [1]. Table 1 shows the efficiency of the

MGSA in comparison with the other cited method. As can be seen in Table 1, although at each iteration, the number of OFEs using the MGSA is more than GSA, the MGSA converges to optimum solutions after fewer numbers of iterations. Furthermore, the small value of the standard deviation using the MGSA shows the reliability and robustness of the proposed method. Moreover, both the GSA and MGSA need much fewer numbers of the objective function evaluation in comparison with the other cited method.

Then, the $2m+1$ PEM is used to model the uncertainty in load demand, market price and the available output power of the WT and PV units in a day with 24 time interval. Using the optimal mean and the standard deviation values and also Gram-charlier expansion the distribution of the total operational cost can be found. Fig. 1 shows the PDF and cumulative Density function (CDF) of the operational cost for the next day ahead. From this figure it can be found that the proposed probabilistic optimization method can find the distribution of the cost efficiently.

Finally, to verify the impact of the considering uncertainty in the decision making, the Value of the Stochastic Solution (VSS) is computed [8]. The VSS for the test system is calculated as 6.1 (€ct).

Table 1 Comparison of different optimization algorithms for first scenario out of 50 trials

Method	Best solution (€ct)	Average (€ct)	Standard deviation (€ct)	Mean number OFEs
GA [1]	277.7444	290.4321	13.4421	Up to16000
AMPSO-T [1]	274.4317	274.5643	0.0921	Up to16000
GSA	275.5369	277.8021	2.9283	1920
MGSA	269.7600	269.8000	0.0901	1024

Table 2 Comparison of Cases I, II and III of first scenario

Case	Expected cost (€ct)	σ_s	VSS (€ct)
Case I	273.000	12.4122	3.24
Case II	449.448	13.78	4.12
Case III	673.377	14.6	4.81

Table 2 is offered to qualify the relative impact of uncertain sources penetration on the decision making. In Case I, the maximum installed WT and PV units are as given in [1] while in cases II and III the capacity of the WT and PV units are twice and third times larger than what are noted in [1]. Considering Table 2, by increasing the penetration of the uncertain WT and PV units:

1. The expected cost increases because WT and PV units have higher bids than the other units.

2. The uncertainty in problem increases accordingly, so the standard deviation of the operational cost distributions (σ_s) will rise.

3. The VSS increases i.e. the system operator should spend more to know the future realization of the system.

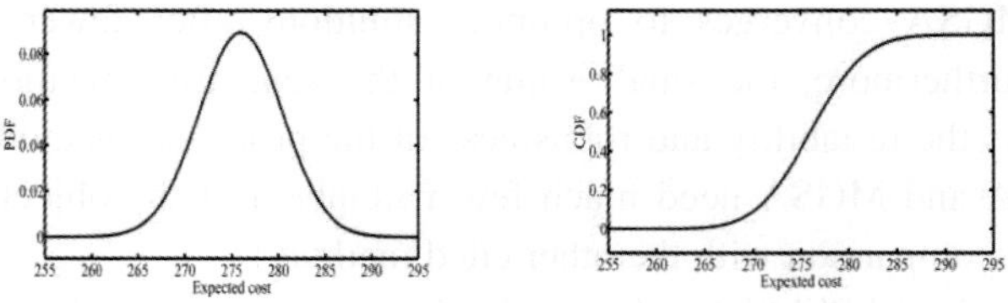

Fig. 1 PDF and CDF of expected cost

6. Conclusion

This paper proposed a probabilistic optimization technique for operation management of the MGs under uncertain environments. The proposed framework consists of the $2m+1$ PEM to model the uncertainty and a modified optimization technique named MGSA. The proposed technique was implemented to find optimal operational cost for a typical MG. Firstly; deterministic analyses were done which showed superiority of the proposed MGSA against some of the other methods. After that, $2m+1$ PEM along with the MGSA method were employed to find the optimal distribution of the total operational cost for the next day ahead. Simulation results demonstrated that the proposed method could successfully find the distribution of the output random variable. Moreover, to validate the impact of the considering uncertainty in the decision making, the VSS was computed. Simulation results showed that the value of the VSS increases by increasing the penetration of the uncertain sources.

References

1. A. Moghaddam, A. Seifi, T. Niknam, M.R. Pahlavani, *Energy* **36**, 6490 (2011).
2. C. Chen, S. Duan, T. Cai, B. Liu, G. Hu, *IET Renew. Power Gener.* **5**, 258 (2011).
3. E. Sortomme, M.A. El-Sharkawi, *IEEE/PES Power Systems Conf.*, 1(2009).
4. R. Chedid, S. Raiman, *IEEE Trans. Energy Convers.* **12**, 79 (1997).
5. JM. Morales, J. Perez-Ruiz, *IEEE Trans. Power Syst.* **22**, 1594 (2007).
6. P. Zhang, ST. Lee, *IEEE Trans. Power Syst.* **19**, 676 (2004).
7. E. Rashedi, H. Nezamabadi-pour, S. Saryazdi, *Information Sciences* **179**, 2232 (2009).
8. JM. Morales., *Ph.D thesis* . dissertation. Dept. Electr. Eng., Univ. Castilla-La Mancha, Ciudad Real, Spain; (2010).

EFFECTIVENESS EVALUATION OF COMPLEX NETWORKS: A SIMULATION-BASED ALGORITHM

WUHONG WANG*, FUGUO HOU, RONGJIE LIN, FANG LI

Department of Transportation Engineering,, Beijing Institute of Technology
Beijing 100081, P. R. China

As the effectiveness of complex networks mainly depend on probability distribution of time between failure and time to repair that involves system operations in different routes, all combinations of system operations are quantitatively described by various running and repairing states. In this paper, we describe a sampling method of random variable's probability distributions of failures and repairs time according to priority and event sequencing. Based on various states of system operations, a detailed simulation procedure has been established through identifying different failure and repair states. Finally, we have carried out the reliability and maintainability evaluation of system operations for a special complex line using Monte Carlo simulation as a case study. The simulation results demonstrate that the reliability and maintainability simulation algorithm is adaptable to measure the effectiveness of system service whose system operations follow to arbitrary failure and repair distributions.

1. Introduction

Reliability and maintainability evaluation of special complex network is playing increasingly important role in designing routes, creating timetables, scheduling system to trips and assigning drivers. Since failures and repairs can occur at any time and even simultaneously during the system operation duration, in-operate repair or replacement that subjects to random times to failure must be made before the maximum system downtime is exceeded. Also, in-operate maintenance must take system failure and repair rate into account. Now, modeling of this process forms the cornerstone for system service assessment including analytic method and simulation modeling.

As the system in special complex lines could be regarded as a complex repairable system, it is usually supposed that system's failure rates and repair rates are constant while general reliability theory is used to reliability and maintainability evaluation of complex repairable system [1]. Although such assumptions are convenient for theory analysis, the analytic method is often too complicated to be

*wangwuhong@bit.edu.cn

reached within budge and time constraints; in particular, the analysis results are not accurate under some circumstances [2–8]. Moreover, it is difficult to describe all combinations of system states and to determine the distributions of failure time and repair time. Therefore, it is necessary to investigate the reliability and maintainability of complex repair system through simulation.

With the rapid development of high-speed computers and large capacity storage, the computational cost is likely to become less a concern. As a result, Monte Carlo methods will become a more practical and power tool to describe the succession of system operation complex system. There are a number of research papers in the literature that suggest alternative strategies for the reliability and maintainability problem [8–12]·Wang analyzed and summarized the typical exiting Monte Carlo reliability, availability and maintainability simulation procedures, variance reduction methods, and random vitiate generation algorithms, discussed the advantages, drawbacks, accuracy computer time of Monte Carlo simulation in evaluating reliability, availability and MTBF of a complex network. Although many problems about traffic system reliability and safety(i. e. traffic accident) have resolved with Fault Tree Analysis (FTA), it is very difficult and complex to construct fault tree [13].

Using the priority and event sequencing for system operations, this paper presents a reliability and maintainability evaluation method for effective of system service using priority-based Monte Carlo simulation, the simulation results clearly demonstrated the potential and feasibility of the method is able to describe the service succession of system operations with maintainability, Mean Time To Restore Functionality (MTTRF) and Mean Time Between Critical Failure (MTBCF) parameters.

2. Algorithm of reliability and maintainability simulation

Herewith the simulation procedures is summarized and described as follows:

Step 1: As for setting & reading of basic data, the basic data includes the characteristic parameters of the subsystems, the number of the simulation, the mission time, and the initialization of the correlative parameters and so on. At the beginning of the simulation, the basic data is inputted with user interface, and then the computer program saves them.

Step 2: The current number of the simulation increases by 1, and then S is compared with M. If $S>M$, the simulation is finished, then go to the step (7); if $S<M$, the simulation will be continue, then go to the step (3).

Step 3: For the simulation at a time, all kinds of variables will be set the initial values. The initial failure time of each subsystem obtains from the failure distribution before the simulation carries out.

If the initial failure time of subsystem i is T_{Sj1}, the failure time is writing as

$$T_{S11}, T_{S21}, \cdots, T_{Sj1}, \cdots, T_{SN1}$$

Step 4: Select the earliest event in time from the event calendar and let it as the current event. If the current event is the kth time's failure of subsystem i, the failure time is T_{Sik}. The current time of the simulation is given as $St = T_{Sik}$

If $St > T_m$, the failure can't occur again, the current mission will finish and succeed. Then set $T_{Sr} = T_{Sik}$, determine the values of A_S, R_S, and implement step (2). If $St \leq T_m$, the current mission is continued, then the failure will be repaired. Finally, Generate the repair time of subsystem i from its repair distribution function $M_i(t)$, then set the value T_{Sik} is greater than T_m.

If PR_i=1, the failure is critical failure, then go to step (5). If $PR_i \neq 1$, the sub priority must be used to judge the critical failure. If the failure doesn't cause the critical failure, the program updates the event calendar and selects the earliest event in time from the event calendar again, then returns to step (4). Otherwise, the critical failure will occur, the program select the subsystems with critical failure is to repair, then go to step (5).

Step 5: Suppose that the critical failure is the rth time's critical failure, so we have $T_{Sr} = T_{Sik}$, $\tau_{Sr} = \tau_{Srk}$ .According to step (5), determine if failure can be repaired, then go to step (6), if not, go back step (4).

Step 6: If the critical failure can't be repaired, the system mission is aborted and the program carries out the nextsimulation after recording the value of R_S, then the simulation goes back to step (2). Otherwise, the critical failure will be repaired. Generate the next fault time of subsystem i from its failure distribution function $F_i(t)$ after the repair finished, then update the event calendar and select the earliest event in time from the event calendar again, then returns to step (4).

Step 7: Completes all statistical outputs and summaries the simulation data. During the course of the simulation, all statistical outputs and summaries are saved. When the simulation is finished, the program compiles and outputs these variables such as the reliability and the maintainability.

3. Application of simulation algorithm to system operation

A simple network that three sub-systems that are running is as case study for application of simulation algorithm, it was assumed that the service mission time of the network is 24 hours, every sub-system will run 8 hours. Data of sub-system dynamics operation (from sub-system 1 to sub-system 3) is as follows: Sub-system 1 density: Weibull, $\hat{\theta}$=2000h, K=3; Sub-system 2 density: Exponential, $\hat{\theta}$=1212.12h; Sub-system 3 density: Gamma, $\hat{\lambda}$=0.02/h, a=3.

When the service mission time is 24h, the number of the simulation is 2400 and the max mission repair time is 0.5 hours. The program output can provide the entire history of the mission, which save as a run-information text document, and a summary of pertinent statistics save as a result text document file. Fig 1 gives the simulation output and 95% confidence limit of a case study of network operations.

Table 1 Reliability of each system

System	Distribution	Reliability Formula	Mission Time	Reliability
1	Weibull	$R(t) = \exp\left[-\left(\frac{t}{\eta}\right)^m\right]$	24	0.99936
2	Exponential	$R(t) = \exp(-\lambda t)$		0.9726
3	Gamma	Simulation Method		0.9687

In order to validate the performance of simulation algorithm and availability of simulation sampling, we calculate the reliability of each system according to analytic method, the result is shown in table 1.

Consequently, the reliability of service concerning system line is

$$R_z=0.99936\times0.\ 9726\times0.9687 = 0.9416$$

So we can obtain the error between the simulation result and the analytic result. The error is

$$E=0.9416\text{-}0.9391=0.\ 0025$$

From Fig. 1 and Table 1, we can find that the error is very small with 2400 simulation replications, and the analytic result is in the 95% confidence intervals of the simulation result.

4. Conclusion remarks and future work

There are some difficulties in evaluating the effective of system service and its relative reliability, maintainability of system operations in complex system by using classical statistics or analytic method. However, it is convenient to obtain parameter values concerning designing system routes, creating timetables, scheduling system to trips and assigning drivers with Monte Carlo simulation technique.

In this study, we run the reliability and maintainability simulation algorithm for analysis of system operations in a simple specific system line consisted of 3 sub-systems as an example to validate the correctness by comparing with an analytic method. If the number of simulations is adequate, the accuracy of reliability and maintainability parameter is much more suitable of dynamic scheduling and operation evaluation for complex system line, especially system service effective assessment.

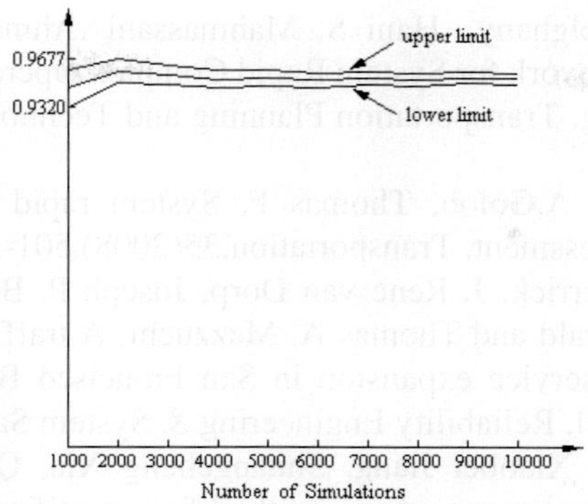

Fig. 1 Reliability in different number of simulations

On the basis of identifying the system's failure state and repair states, we can determine if the system service could satisfy all requirement of performance of system operations. Based on proposed priority principals, the reliability and maintainability evaluation of system operations using priority-based Monte Carlo simulation has been conducted, which is able to use to estimate the reliability, maintainability parameters of system service whose system follows to arbitrary failure and repair distributions.

Acknowledgments

This research was supported in part by the Programme of Introducing Talents of Discipline to Universities under grant B12022 and National Nature Science Foundation of China under Grant 51110305060.

References

1. Wuhong Wang, Yan Mao, Jing Jin, Xiao Wang, Driver's various information process and multi-ruled decision-making mechanism: a fundamental of intelligent driving shaping model, International Journal of Computational Intelligence Systems, Vol.4, No. 3 (May, 2011)297-304.
2. Christos Valouxis, Efthymios Housos, Combined system and driver scheduling, Computers & Operations Research 29 (2002) 243-259.
3. Ali Haghani, Mohamadreza Banihashemi , Kun-Hung Chiang, A comparative analysis of system complex vehicle scheduling models, Transportation Research Part B 37 (2003) 301–322.
4. Chen XM, Yu L, Zhang YS, Analyzing urban system service reliability at the stop, route, and network levels, Transportation research part a-policy and practice ,43(2009),722-734 .

5. Khaled F. Abdelghany, Hani S. Mahmassani ,Ahmed F. Abdelghany,A Modeling Framework for System Rapid Complex Operations Evaluation and Service Planning, Transportation Planning and Technology, 30(2007),571 – 591.
6. Hensher, David A,Golob, Thomas F, System rapid complex systems: a comparative assessment, Transportation,35(2008),501-518.
7. Jason R. W. Merrick, J. Rene van Dorp, Joseph P. Blackford, Gregory L. Shaw, Jack Harrald and Thomas A. Mazzuchi, A traffic density analysis of proposed ferry service expansion in San Francisco Bay using a maritime simulation model, Reliability Engineering & System Safety,81(2),p119-132.
8. Wuhong Wang, Xiaobei Jiang, Shuangcheng Xia, Qi Cao, Incident tree model and incident tree analysis method for quantified risk assessment: an in-depth accident study in traffic operation, Safety Science, 2010 48(10),1248-1262
9. S.J.Kamat, M.W.Riley, Determination of reliability using event-based Monte Carlo simulation. IEEE Transactions on reliability, vol. R-24, 1976.
10. Albert H.Moore, H.Heon Harter, Robert C.Snead, Comparison of Monte Carlo Techniques for Obtaining System-Reliability Confidence Limits, IEEE Transactions on Reliability, 29(1980), 327~p331.
11. William H. Widwsky, Reliability and Maintainability Parameters Evaluated with Simulation, IEEE Transactions on Reliability, Vol. R-20, No.3, August 1971, p158~p164.
12. Hongzhou Wang and Hoang Pham, Survey Reliability and Availability Evaluation of Complex Networks Using Monte Carlo Techniques, Micro electron Reliability, Vol.37, No.2, 1997, p187~p209.
13. Kuzminski P, Improvement of Highway safety I: Identification of causal factors through fault-tree modeling, Risk Analysis, 15 (1995) 293-312.

A NEW "IMPLICIT" PARAMETER ESTIMATION FOR CONDITIONAL GAUSSIAN BAYESIAN NETWORKS

AIDA JARRAYA* [1,2] AFIF MASMOUDI [1] AND PHILIPPE LERAY [2]
[1] *Laboratory of Probability and Statistics*
Faculty of Sciences of Sfax, Sfax University. B. P. 1171 Sfax. Tunisia
[2] *LINA Computer Science Lab UMR 6241*
Knowledge and Decision Team, University of Nantes, France.

Among existing Bayesian network (BN) parametrizations, conditional Gaussian are able to deal with discrete and continuous variables. Bayesian estimation of conditional Gaussian parameter needs to define several a priori parameters which are not easily understandable or interpretable for users. The approach we propose here is free from this priors definition. We use the Implicit estimation method which offers a substantial computational advantage for learning from observations without prior knowledge and thus provides a good alternative to Bayesian estimation when priors are missing. We illustrate the interest of such estimation method by first giving the Bayesian Expectation A Posteriori estimator (EAP) for conditional Gaussian parameters. We then describe the Implicit estimator for the same parameters. One experimental study is proposed in order to compare both approaches.

1. Introduction

In this paper, one simplest component of Bayesian networks (BN) learning, parameter learning, is addressed. We use the Implicit method recently proposed [1] to overcome some of the shortcomings associated with Bayesian estimation related to the choice of the prior distribution parameters. This approach was already applied for learning parameters in discrete BN [2,3]. We propose here new theoretical results concerning Implicit learning in conditional Gaussian Bayesian networks (CGBNs). To take back the classical notations used in the domain, we inspire from Murphy's work [4] devoted to Maximum Likelihood (ML) and Bayesian Maximum A Posteriori (MAP) estimation for conditional Gaussian models and enlarge them with Expectation A Posteriori (EAP) and Implicit approaches. The remainder of this paper is organized as follows: in section 2, we briefly present the Implicit method and recall the principles of

* aidajarraya@yahoo.fr

Implicit estimation. In section 3 we discuss the problem of parameter learning in CGBNs by using Bayesian estimation. In section 4 we present parameter estimation in CGBNs using Implicit approach and we compare the results found with both approaches. Finally, we conclude with perspectives for future work.

2. Implicit estimation

2.1. *Principle*

The Implicit estimation is an alternative to Bayesian estimation which does not need to specify any prior for the parameters. In the context of the Bayesian theory, the unknown parameter θ is assumed to be a random variable with a known prior distribution. Alternatively, the concept of Implicit distribution previously proposed by [1], can be described as a kind of posterior distribution of a parameter given data. To explain the principle of Implicit distribution, we suppose that the distribution $p_\theta(dx) = p(x,\theta)\nu(dx)$ is defined on $\mathbb{R}^k$.

The Implicit distribution is given by $Q_x(d\theta) \propto p(x,\theta)\sigma(d\theta)$. The distribution $Q_x(d\theta)$ plays the role of a posterior distribution of θ given x in the Bayesian method. For instance, let $X_1,\ldots,X_n$ be n independent random variables identically distributed according to the distribution with density $p(x,\theta)$ and let $P_\theta(\underline{x}) = \prod_{i=1}^n p(x_i,\theta)\nu(dx_i)$ denotes the joint distribution of $\underline{X} = (X_1,\ldots,X_n)$. Suppose that the sample Implicit distribution $Q_{\underline{x}}(d\theta)$ of the parameter θ exists, then one may define the Implicit estimator of θ corresponding to a given cost function $\varphi(\theta,t) = (\theta - t)^2$ by $T_n = E(\theta \mid \underline{X}) = \int_\Theta \theta Q_{\underline{x}}(d\theta)$.

2.2. *Example with variance estimation*

Let $X_i \sim N(0,\sigma^2)$ be a centered Gaussian random variable with unknown variance $\sigma^2 \in]0,+\infty[$, the likelihood of σ^2 for n independent observations $\underline{x} = (x_1,\ldots,x_n)$ is $l(\underline{x},\sigma^2) = (2\pi\sigma^2)^{-n/2}\exp\{-\frac{1}{2\sigma^2}\sum_{i=1}^n x_i^2\}$. By a standard calculation, we show that the Implicit estimator of σ^2 is $(\hat{\sigma}^2)^{Imp} =$ $E(\sigma^2 \mid X_1,\ldots,X_n) = \frac{1}{n-4}\sum_{i=1}^n X_i^2$ with $n > 4$, which is different from the ML estimator for which the normalizing factor is $1/n$. We can compare this result with the Bayesian estimation obtained by EAP approach (Expectation a posteriori) $(\hat{\sigma}^2)^{Bay} = \frac{\sum_{i=1}^n X_i^2 + 2b}{n+2a-2}$, where the prior distribution of σ^2 is an Inverse-Gamma $IG(a,b)$ with shape parameter a and scale parameter b.

We can firstly see that the Implicit estimator does not need to tune any hyper-parameters such as a and b in EAP approach. We can also notice that the Bayesian variance estimator should be equal to the Implicit one for $a=-1$ and $b=0$, which is not possible since a and b, parameters of an Inverse Gamma distribution, have to be positive. To compare the performance of Implicit and Bayesian method, we start in a first stage by simulation of data by using Matlab software, and in a second stage we will validate our results by comparing them with Bayesian method. We generated 1000 observations from the Gaussian model, we compare the Bayesian and Implicit estimator in terms of mean squared errors (MSE) for different true parameter values of σ^2 We replicate the process 10000 times and compute the average estimates and the MSE. We perform two Bayesian simulation studies based on two different prior density for the parameter σ^2 Results are reported in Table 1.

Table 1. Estimators of σ^2 obtained by Implicit method, Bayesian method with true a priori (Bay^* and different one (Bay^{**} , for several prior parameters (a and b) and σ^2

σ^2	priors	$(\widehat{\sigma^2})^{Imp}$	MSE(Imp)	$(\widehat{\sigma^2})^{Bay^*}$	$MSE(Bay^*)$	$(\widehat{\sigma^2})^{Bay^{**}}$	$MSE(Bay^{**})$
0.1	a=2, b=0.1	0.1	$2 \,.\, 10^{-6}$	0.1	$2 \,.\, 10^{-6}$	0.11	10^{-5}
0.05	a=2, b=0.05	0.05	$4.99 \,.\, 10^{-7}$	0.05	$5 \,.\, 10^{-7}$	0.053	10^{-6}
1	a=2, b=1	1.0002	$1.99 \,.\, 10^{-4}$	1.0006	$1.99 \,.\, 10^{-4}$	1.019	10^{-4}
1.5	a=2, b=1.5	1.5	$4.45 \,.\, 10^{-4}$	1.5006	$4.46 \,.\, 10^{-4}$	1.52	0.0013
2	a=2, b=2	2	$8.03 \,.\, 10^{-4}$	2.0007	$8.05 \,.\, 10^{-4}$	2.039	0.0024

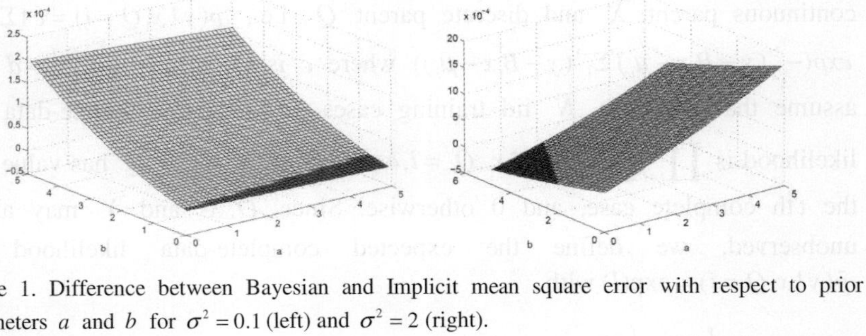

Figure 1. Difference between Bayesian and Implicit mean square error with respect to prior parameters a and b for $\sigma^2=0.1$ (left) and $\sigma^2=2$ (right).

Bayesian estimation Bay* is obtained by specifying true values as prior parameters. Bayesian estimation Bay** correspond to parameter estimated by using prior different from true values. The comparison of MSE obtained by Bay* and Bay** proves the sensibility of the parameter estimation with respect to the choice of the prior distribution. This result proves the fragility of the Bayesian estimation. Biased result can be obtained if we fix a bad choice for the prior distribution. Results obtained by Implicit approach (without need of any prior information) and Bay* approach (based on true values as priors) are close to the true value. We see a very good concordance between both approaches. However, we point out a better precision of parameter estimated by Implicit method than the corresponding one for Bay** method (with priors different from true values). These results also are illustrated in Fig. 1. The yellow part correspond to the area where the MSE Implicit is smaller than the MSE Bayesian with respect to the values of the prior coefficients a and b. Whatever the value of σ^2 (low or high value), Implicit MSE is in majority lower than the Bayesian one.

3. Conditional Gaussian Bayesian network parameter estimation

3.1. *Definitions and notations*

In this paper, we develop metrics for domains containing continuous variables or a mixture of discrete and continuous variables, under the assumption that continuous data is sampled from a multivariate normal (Gaussian) distribution. Consider a finite set $X = \{X_1, \ldots, X_n\}$ of random variables. A Bayesian network (BN) is a directed acyclic graph G and a set of conditional probability distributions which represent a joint probability distribution [5].

In order to simplify future equations, we summarize below the notations initially proposed by Murphy [4]. We will consider the problem of finding the estimators of the parameters of a conditional Gaussian variable Y with continuous parent X and discrete parent Q, i.e., $p(y \mid x, Q = i) = c \mid \Sigma_i \mid^{-1/2}$ $exp(-\frac{1}{2}(y - B_i x - \mu_i)' \Sigma_i^{-1} (y - B_i x - \mu_i))$ where c is a constant and $|y| = d$. We assume that we have N iid training cases $\{e_t\}$ so the complete-data log-likelihood is $\prod_{t=1}^{N} \prod_{i=1}^{|Q|} p(y_t \mid x_t, Q_t = i, e_t)^{q_t^i}$ where $q_t^i = 1$ if Q has value i in the t'th complete case, and 0 otherwise. Since Q, X and Y may all be unobserved, we define the expected complete-data likelihood Let $\tilde{p}(y \mid x, Q = i) = \exp(l)$ with

$$l = -\frac{1}{2} \sum_t E \left(\sum_i q_t^i \log |\Sigma_i| + q_t^i (y_t - B_i x_t - \mu_i)' \Sigma_i^{-1} (y_t - B_i x_t - \mu_i) \mid e_t \right).$$

We can write $E(q_t^i x_t x_t' \mid e_t) = E(q_t^i \mid e_t) E(x_t x_t' \mid Q_t = i, e_t) = w_t^i E(XX')$ where the weights $w_t^i = p(Q = i \mid e_t)$ are posterior probabilities and $E_{ti}(XX')$ is a conditional second moment. We obtain,

$$l=-\frac{1}{2}\sum_t\sum_i w_t^i \log|\Sigma_i|-\frac{1}{2}\sum_t\sum_i w_t^i E_{ti}\left((y_t-B_ix_t-\mu_i)'\Sigma_i^{-1}(y_t-B_ix_t-\mu_i)\,|\,e_t\right).$$

$$w_i=\sum_t w_t^i\ ,\ S_{XX',i}=\sum_t w_t^i E_{ti}(XX') \text{ and } S_{X,i}=\sum_t w_t^i E_{ti}(X).$$

In this section we inspire from Murphy's work [4] devoted to MAP and ML estimation for parameters of conditional Gaussian models and enlarge them with EAP estimation.

3.2. *Bayesian parameter estimation*

We assume that the prior for the regression matrix B_i is a multivariate normal distribution with mean a and covariance matrix V. Hence

$$\widehat{B_i}^{Bay}=(\Sigma_i^{-1}(S_{YX',i}-\mu_i S_{X',i})+aV^{-1})\,(\Sigma_i^{-1}S_{XX',i}+V^{-1})^{-1}.$$

For the parameter μ_i mean of the conditional Gaussian distribution, the prior density is a multivariate normal distribution with parameters [5]. Then $\widehat{\mu_i}^{Bay}=(\psi^{-1}m+\Sigma_i^{-1}(S_{Y,i}-B_iS_{X,i}))\,(w_i\Sigma_i^{-1}+\psi^{-1})^{-1}$.

We classically assume that the prior for the covariance matrix Σ_i is an Inverse-Wishart distribution with α degrees of freedom and positive definite precision matrix V. Thus $\widehat{\Sigma_i}^{Bay}=\frac{1}{w_i+\alpha-d-1}(A_i+V)$ where

$$A_i=S_{YY',i}-S_{YX',i}B_i'-S_{Y,i}\mu_i'-B_iS_{XY',i}+B_iS_{XX',i}B_i'+B_iS_{X,i}\mu_i'-\mu_iS_{Y',i}+\mu_iS_{X',i}B_i'+\mu_i\mu_i').$$

3.3. *Implicit parameter estimation*

In the Implicit estimation context,

$$\tilde{p}(B_i\,|\,y_t,x_t,\Sigma_i,\mu_i)\propto\exp(-\tfrac{1}{2}\sum_t w_t^iE_{ti}(y_t-B_ix_t-\mu_i)'\Sigma_i^{-1}(y_t-B_ix_t-\mu_i)).$$

which gives us the Implicit estimator of the regression matrix $\widehat{B_i}^{Imp}=(S_{YX',i}-\mu_iS_{X',i})(S_{XX',i})^{-1}$ We also have

$$\tilde{p}(\mu_i\,|\,y_t,x_t,\Sigma_i,B_i)\propto\exp(-\tfrac{1}{2}\sum_t w_t^iE_{ti}(y_t-B_ix_t-\mu_i)'\Sigma_i^{-1}(y_t-B_ix_t-\mu_i))$$

which gives us the Implicit estimator of the mean μ_i $\widehat{\mu_i}^{Imp}=\frac{S_{Y,i}-B_iS_{X,i}}{\sum_t w_t^i}$.

$\tilde{p}(\Sigma_i\,|\,y_t,x_t,B_i,\mu_i)$ is an Inverse-Wishart distribution with (w_i-d-1) degrees of freedom and positive definite precision matrix $\sum_t w_t^iE_{ti}(y_t-B_ix_t-\mu_i)$

$(y_t - B_i x_t - \mu_i)'$. Then the Implicit estimator of the covariance matrix Σ_i is given by $\widehat{\Sigma}_i^{Imp} = \frac{1}{w_i - 2d - 2}(A_i)$

3.4. *Comparative study*

We first remark that there is a difference between the Implicit estimator and the classical ML one (described in [4]) in the estimation of the covariance matrix, whereas it coincides for the mean and regression parameters. Concerning the comparison between Implicit and Bayesian estimation, we point out that parameters estimated by Bayesian approach depend on the prior parameters and as we know, the choice of prior information has always been problematic. As Implicit and ML approaches coincide for the estimation of mean parameter μ and the parameter of regression B , we only have to compare the estimations of the covariance matrix. We point out that the estimator of $\hat{\Sigma}$ by using Implicit method corresponds to the Bayesian one by taking $V = 0$ and $\alpha = d - 1$ This situation is not possible because V is a positive definite precision matrix. This situation is similar to the one described in example 2.2 for σ^2 estimation. Our Implicit estimation will then give more robust results than the Bayesian one, particularly if priors used in Bayesian estimation are far away from the true ones.

4. Conclusion and perspectives

In this paper, we introduce the notion of Implicit approach for the estimation of parameters in conditional Gaussian Bayesian Networks. This method of estimation is similar to the Bayesian one, but happens in a natural context without specifying any prior parameters. Bayesian estimation with priors far from the true values can lead to poor results. the implicit (prior free) estimators proposed here are then very attractive to avoid such situations. This preliminary theoretical work has now to be consolidated with an experimental validation of these generic properties of Implicit estimation. This Implicit approach can also be used to learn the network structure. Structure learning is a much harder issue than parameter learning. Within this framework, our future work is to propose an Implicit score function on which network structure inference can be based without determining any priors.

References

1. A. Hassairi, A. Masmoudi, and C Kokonendji. Implicit distributions and estimation. *Communications in Statistics - Theory and Methods*, 34(2):245 – 252, 2005.
2. Hanen Ben Hassen, Afif Masmoudi, and Ahmed Rebai. Causal inference in biomolecular pathways using a bayesian network approach and an implicit method. *Journal of Theoretical Biology*, 253(4):717 – 724, 2008.
3. Hanen Ben Hassen, Afif Masmoudi, and Ahmed Rebai. Inference in signal transduction pathways using em algorithm and an implicit algorithm: Incomplete data case. *Journal of Computational Biology*, 16(9):1227 – 1240, 2009.
4. Kevin P. Murphy. Fitting a conditional linear gaussian distribution. Technical report, UC Berkeley, Computer Science Division, 2003.
5. R. E. Neapolitan. *Learning Bayesian Networks*. Prentice Hall, 2003.

DISSIMILARITY-BASED BIPOLAR SUPERVISED CLASSIFICATION[1]

J. TINGUARO RODRÍGUEZ, JAVIER MONTERO, BEGOÑA VITORIANO
Faculty of Mathematics, Complutense University of Madrid, Plaza de Ciencias 3, 28040 Madrid, Spain

Frequently, the set of classes of a supervised classification problem presents an structure related to the specific features of each application context. However, standard classification models does not use to consider such an structure in their learning and reasoning processes. By means of the introduction of a bipolar approach, this paper proposes a revision of the basic notions of supervised classifiers, aimed to extend their generalization power and adaptation to problems with an structured set of classes.

1. Introduction

This work proposes a bipolar approach (see [1] for an introduction to the notion of *bipolarity*) to supervised classification (see [3] for instance). Therefore, this proposal is focused on the replication, in the context of supervised classification, of the human ability of learning and reasoning on the basis of information with positive and negative character [2]. It is important to remark that this work does not present a new sophisticated learning or reasoning method (though the proposed approach admits a later application of adaptative techniques, for example), but it rather poses a revision of the conceptual basis of the supervised classification framework, in order to extend its generalization power and adaptation to different application contexts.

The main improvement enabled by the proposed approach is the capability of dealing with classification problems in which the set of classes presents a relevant *structure*, i.e. in which classes are not independent, unrelated items, but rather they are somehow linked and conform an structured whole. Usually, this structure comes from and depends on both the operative requirements and the specific semantics of the classes in each application context. For example, in disaster-severity assessment an ordinal structure is usually considered, in the sense of the possible consequences (classes) being ordered from the smallest (e.g. no casualties) to the largest (lot of casualties). However, due to

[1] This work has been partially supported by grant TIN2009-07190.

underestimation-risk avoidance requirements, a certain asymmetry of the ordinal structure has to be considered (see [5,7]). Similarly, in a fault detection context, relationships between the different faults could be non-trivial, and asymmetry assumptions are imposed to avoid underestimation of grave faults. Also, in e-marketing, as some products have moral resonances some customers should not be spammed with certain adds (e.g. children and adult movies).

Therefore, in order to met the requirements of each specific context and make use of all the available information, thus enabling the adaptation of the classification models to different application contexts, it is relevant (if not crucial) both to consider the underlying structure of the set of classes and to introduce it in the learning and reasoning processes of the classifiers. As this work shows, this last is achievable through the notion of semantic antagonism (or dissimilarity) introduced in [6], which allows to define negative information in such a way that it reflects the opposition or antagonism relationships between the classes, and leads to a bipolar classification framework that enables positive and negative information to interact in the learning and reasoning stages so the relevant features of the structure of the classes are taken into account.

2. Fundamentals

Recall that a *classification problem* consists of assigning a *class* C_j, from a predefined set of N_C classes $\zeta=\{C_1,\ldots,C_{N_C}\}$, to an *object* (or instance, example, etc.) described by a vector $x=(x_1,\ldots,x_n)\in U_{X_1}\times\ldots\times U_{X_n}=U_X$ of features or *attributes* $X_1,\ldots,X_n$, U_{X_i} being the range of the attribute X_i, $i=1,\ldots,n$. In a supervised learning setting, a set of *training patterns* $\{x^p=(x_1^p,\ldots,x_n^p;C^p), p=1,\ldots,m\}, C^p\in\zeta\ \forall p$, is available and the objective is to *learn* an appropriate mapping $C:U_X\to\zeta$ that generalizes the training sample in an optimal way with respect to certain criteria $\varepsilon(C)$ measuring the classifier's accuracy. Usually, the objective is to obtain a classifier that assigns classes to the objects from a test dataset with the lowest possible validation error. Different methodologies can be applied to develop a classifier: fuzzy classifiers, neural networks, classification trees, support vector machines, etc.

Roughly speaking, in a fuzzy classification context, the standard measure of evidence for the classification of a query x in each class C_j, j=1,..,N_C, is given by the *confidence degree*, basically obtained as

$$cf(C(x)=C_j)=\mu_{N(x)}(x)\cdot\frac{\sum\limits_{p=1,\ldots,m/C^p=C_j}\mu_{N(x)}(x^p)}{\sum\limits_{p=1}^{m}\mu_{N(x)}(x^p)}, \tag{1}$$

where $N(x) \subset U_X$ is a neighbourhood of the query x, and $\mu_{N(x)}(x^p)$ (resp. $\mu_{N(x)}(x)$) is the compatibility of each training pattern x^p (resp. the query x) with $N(x)$. For example, in case-based learning, $N(x)$ could represent the *K*-nearest neighbourhood (*K-NN*) of x, so that $\mu_{N(x)}(x)=1$ and $\mu_{N(x)}(x^p)$ can be understood as the *proximity* degree of each training pattern to x (only the *K*-nearest patterns having positive *proximity*). Similarly, in a rule-based context, $N(x)$ could be associated to the set of premises $A_1,\ldots, A_q$ that are compatible with the query x, so that $\mu_{N(x)}(x^p) = (\mu_{A_1}(x^p),\ldots,\mu_{A_q}(x^p))$ and $\mu_{N(x)}(x) = (\mu_{A_1}(x),\ldots,\mu_{A_q}(x))$. Finally, once these confidence degrees are obtained for each class, the query x is then normally assigned to the class with maximum confidence degree.

Therefore, it is easy to see that $cf(C(x)=C_j)$, as an approximation of the conditional probability $P(C_j \mid N(x))$, only depends on the proportion of training patterns of class C_j that occurs in $N(x)$ out of the total number of examples in $N(x)$. In this sense, $cf(C(x)=C_j)$ does not depend on how are the remaining not-in-C_j patterns distributed in the other classes. For example, consider $N_C=4$ and $K=10$ in the previous illustration of a *K-NN* framework. If we assume proximity degrees to be 1 for the ten nearest examples and 0 for the others, and four instances of class C_1 are found among the first group, then $cf(C(x)=C_1)=\frac{4}{10}$ *independently* of the behaviour of the six remaining *K*-nearest neighbours. However, precisely because of this *independence*, the confidence degree (alone) is unable to represent the relevant structural relationships between classes that can arise in each application context. Similarly, by assigning x to the class with maximum confidence, we further neglect these relationships, as only quantitative differences are considered.

Thus, it is easy to understand the difficulties found by the standard fuzzy (as well as the non-fuzzy) classification framework in order to face problems with an structured set of classes. As it has been shown, it is not a matter of methodologies or lacks of some specific models, but simply that it is usually impossible to represent such an structure by only relying in a single[2] value as the confidence degree. Thus, it is the implicit knowledge representation framework which in fact has to be expanded in order to enable an adequate representation of the relevant structure of each problem. In this paper, we propose a bipolar approach to this question, in which the *positive* description given by the

[2] It can be argued that the description given by the *confidence* $P(C_j \mid N(x))$ is usually completed by the *support* $P(N(x))$. However, when we restrict to training examples in $N(x)$, support becomes uninformative, as it is $P(C_j \mid N(x)) = P(C_j \cap N(x)) / P(N(x))$, i.e. its influence is identical for each class C_j, $j=1,\ldots,N_C$.

confidence degree is accompanied of a *negative* description given by the opposition relationships between classes.

3. Dissimilarity-based bipolar classification

It is interesting to study the case in which complementation (or negation) is used to construct such a negative description, that is then given by

$$cf(C(x)=(C_j)^c)=P((C_j)^c \mid N(x))=1-P(C_j \mid N(x))=1-cf(C(x)=C_j). \quad (2)$$

Since without loss of generality it can be assumed that the sum of the $P(C_j \mid N(x))$ for j=1,…,N_C is 1, we then get that

$$cf(C(x)=(C_j)^c)=1-cf(C(x)=C_j)=\sum_{l=1,\dots,N_C / C_l \neq C_j} cf(C(x)=C_l), \quad (3)$$

i.e. the negative description $cf(C(x)=(C_j)^c)$ is just the sum of the evidences for the N_C–1 remaining classes. Therefore, a negative description given by the complementation is associated to a trivial structure in which all classes are considered as equally (and totally) opposite between them. Notice also that, as the former is just the negation of the latter, $cf(C(x)=(C_j)^c)$ does not add any information to that already contained in $cf(C(x)=C_j)$ (and see [4]).

As discussed above, such an structure of total opposition between classes could be too restrictive in many application contexts. In practice, what we found is that opposition relationships are gradual and heterogeneous, some classes being strongly opposite and some others being only weakly opposite (or even not opposite at all). In this sense, between the extreme cases of total and void opposition between all classes, there exists a wide range of possibilities for opposition relations, that can be modelled in the framework of *antagonism* [6].

For the purposes of this paper, it is enough to recall that antagonism constitutes a generalization of complementation allowing to define opposite predicates. Particularly, given a set of predicates $\zeta=\{C_1,\dots,C_{N_C}\}$, in [4] it was shown that it is possible to define a *dissimilarity structure* on ζ by means of a *dissimilarity matrix* $\Delta=(d_{ij})_{i,j=1,\dots,N_C}$ that reflects the pairwise opposition relationships between the predicates in ζ, i.e. $d_{ij}\in[0,1]$ states the degree up to which the predicate C_j is opposite to (i.e. constitutes a source of negative information against) the predicate C_i (here we assume that $d_{ii}=0\ \forall i$). This allows to construct, for each predicate C_j, the *opposite predicate* ∂C_j.

Therefore, if a dissimilarity matrix Δ representing the structure of a set of classes ζ is given, then for each class C_j, the opposite predicate ∂C_j represents the opposition relationships regarding C_j, and can be used to construct the negative description $cf(C(x)=\partial C_j)$ associated to these relationships, defined as

$$cf(C(x)=\partial C_j)=\mu_{N(x)}(x)\cdot\frac{\sum_{p=1,\ldots,m/C^p=C_l\neq C_j} d_{jl}\cdot\mu_{N(x)}(x^p)}{\sum_{p=1}^{m}\mu_{N(x)}(x^p)}. \quad (4)$$

Notice the resemblance between this definition of $cf(C(x)=\partial C_j)$ and that of $cf(C(x)=C_j)$ in (1). Also, it is easy to see that

$$cf(C(x)=\partial C_j)=\sum_{l=1,\ldots,N_C/C_l\neq C_j} d_{jl}\cdot cf(C(x)=C_l), \quad (5)$$

showing that antagonism effectively generalizes the description (3) obtained through the complementation.

Thus, given a class C_j and a query x, the pair $\langle cf(C(x)=C_j), cf(C(x)=\partial C_j)\rangle$ constitutes a *bipolar* description of the evidence for the classification of x in C_j contained in the training dataset, the first element being a positive description that informs of the presence of examples of class C_j in a neighbourhood $N(x)$ of x, while the second element informs of the presence in $N(x)$ of examples structurally opposite to C_j. This allows to capture the structure of the set of classes and to introduce it in the learning and reasoning process of classifiers attending to context requirements.

For example, consider an e-marketing context, in which an intelligent agent has to classify the preferences of the incoming customers x in order to decide what add-up it has to show them. Suppose that two of the possible classes are C_C = *child looking for toys* and C_A =*adult looking for adult movies*. There is an evident danger in classifying a query actually belonging to class C_C in class C_A. Class C_C's semantics is strongly opposite or antagonistic (from a moral point of view) to that of C_A. Therefore, the presence of examples from class C_C in $N(x)$ should reduce the *confidence* in classifying x under class C_A. This requirement can be addressed by considering an opposition relationship between classes C_C and C_A, e.g. such that $d_{AC}=1$, leading to obtain the bipolar evidence $\langle cf(C(x)=C_A), cf(C(x)=\partial C_A)\rangle$. These values can then be aggregated in a final evidence degree, for example by means of the expressions

$$t_A(x)=\max\{cf(C(x)=C_A)-cf(C(x)=\partial C_A),0\} \text{ or } v_A(x)=\begin{cases}0 & \text{if } cf(C(x)=\partial C_A)>0\\ cf(C(x)=C_A) & \text{else}\end{cases}.$$

By using $t_A(x)$, the confidence for classifying x in class C_A could be significantly reduced, therefore making it more difficult to assign this class to x. By using $v_A(x)$ instead, class C_A is avoided whenever there is evidence for any of its opposite classes (particularly C_C).

4. Conclusions

This last example shows how the bipolar approach proposed in this paper can be useful for addressing classification problems in which the set of classes presents a relevant structure, enabling the introduction of such structure in the learning and reasoning processes of the classifiers in order to adapt them to the requirements of each application context. As it has been shown, this bipolar approach is mainly concerned with expanding the underlying knowledge representation framework by complementing the usual confidence degree with a negative description that capture the relevant opposition or antagonism relationships between classes

References

1. D. Dubois, H. Prade (2008) An introduction to bipolar representations of information and preference. Int Jour of Intellig Syst 23 866-877.
2. J.T. Cacioppo, W.L. Gardner, G.G. Bernston (1997) Beyond bipolar conceptualizations and measures: The case of attitudes and evaluative space, Personality and Social Psychology Review 1 3-25.
3. E. Hullermeier (2005) Fuzzy methods in machine learning and data mining: Status and prospects, Fuzzy Sets and Systems, 156 (3) 387-406
4. Montero, J., Gomez, D., Bustince, H. (2007) On the relevance of some families of fuzzy Sets, Fuzzy Sets and Systems, 158 (22) 2429-2442
5. J.T. Rodríguez (2010) Classification of disasters and emergencies with bipolar knowledge representation. PhD Thesis. Complutense University of Madrid
6. J.T. Rodríguez, C.A. Franco, B. Vitoriano, J. Montero (2011) An axiomatic approach to the notion of semantic antagonism. Proceedings IFSA-AFSS Conference FT104-1/6.
7. J.T. Rodríguez, J. Montero, B. Vitoriano (2009) A general methodology for data-based rules building and its application to natural disaster management. Computers and Operations Research 39 (4) 863-873

USING GROUP METHOD OF DATA HANDLING FOR BRAIN COMPUTER INTERFACE

RAVIKANTH VARAGANTI
Institute of Research in Applied Computing, University of Bedfordshire, UK, Ravikanth.Varaganti@beds.ac.uk

VITALY SCHETININ
Institute of Research in Applied Computing, University of Bedfordshire, UK, Vitaly.Schetinin@beds.ac.uk

Electroencephalograms (EEG) contain the information important for brain-computer interface (BCI), despite that their earlier use is limited by variations caused by background brain activity of operators as well as by artifacts. Thus achieving high accuracy of EEG classification is an important task for BCI systems. In this paper, we explore possible ways of improving the classification performance by using Group Method of Data Handling (GMDH). Specifically, we aim to find the most suitable selection criteria to be used in GMDH. The experimental results conducted on EEG representing the motor imagery indicates that the proposed criteria are capable of improving the classification performance as compared with Radial Basis networks.

1. Introduction

Brain Computer Interface (BCI) is a communication system, which enables the user to control special computer applications by using only his or her thoughts. A BCI system is a direct communication pathway between a brain and an external artificial device. BCI systems are aimed at assisting, augmenting or repairing human cognitive or sensory-motor functions. Different research groups have examined and used different methods to achieve this. Almost all of them are based on electroencephalography (EEG) recorded from the scalp. The EEG is measured and sampled while the user imagines different things. Depending on the BCI, particular preprocessing and feature extraction methods are applied to the EEG sample of certain length. It is then possible to detect the task-specific EEG signals or patterns from EEG samples with a certain level of accuracy [13]. A BCI should be able to detect the user's wishes and commands while the user remains silent and immobilized. In order to do this, the brain activity must be monitored.

EEG is a method used in measuring the electrical activity of the brain. This activity is generated by billions of nerve cells, called *neurons*. Each neuron is connected to thousands of other neurons. Some of the connections are *excitatory* while others are *inhibitory*. The signals from other neurons sum up in the receiving neuron. When this sum exceeds a certain potential level called *threshold*, the neuron fires *nerve impulse*. The electrical activity of a single neuron cannot be measured with scalp EEG. However, EEG can measure the combined activity of millions of neurons [14, 15]. There exists various properties in EEG, which can be used as a basis for a BCI e.g. Event Related Potentials (ERPs). ERP is a common title for the potential changes in the EEG that occur in response to a particular "event" or a stimulus. These changes are so small that in order to reveal them, EEG samples have to be averaged over many repetitions. This removes the "random" fluctuations of the EEG, which are not stimulus-locked. The most commonly studied ERP is P300 also known as P3. The electrical activity of the brain is recorded non-invasively, i.e. from the surface of the scalp using normally small metal plate electrodes [15, 16]. While the number of electrodes varies from study to study, they are usually arranged according to international 10-20 system.

2. Review of Literature

Earlier work on BCI EEG is done by using other classification algorithms which are non-adaptive and non-self-organizing. Group Method of Data Handling (GMDH) was applied in a great variety of areas for data mining and knowledge discovery, forecasting and systems modeling, optimization and pattern recognition. Inductive GMDH algorithms give possibility to find automatically interrelations in data, to select optimal structure of model or network and to increase the accuracy of existing algorithms. This self-organizing approach substantially has inductive nature as it finds the best solution by sorting-out possible variants [3, 4]. By sorting of different solutions, GMDH networks aims to minimize the influence of the author on the results of modeling. The algorithm itself finds the structure of the model and the laws which act in the system. GMDH is a set of several algorithms for different problems solution. This inductive approach is based on sorting out of gradually complicated models and selection of the best solution by minimum of external criterion characteristic [3, 4, 9]. The optimal complexity of model structure is found, adequate to level of noise data sample. The number of layers and neurons in hidden layers, model structure and other optimal parameters are determined automatically. It guarantees that the most accurate or unbiased models will be found. As input variables are used any non-linear functions or features, which can influence the output variable. It automatically finds interpretable relationships in data and selects effective input variables. These algorithms are rather simple for programming [3, 4, 11]. The method uses information directly from data sample and minimizes influence of a-priori author assumptions about results of modeling. Following is the state of art literature review of the two variants of GMDH i.e. Classical and Abductive networks.

2.1. *Classical GMDH Algorithm*

In mid 1960's, *GMDH*, a method was introduced, that allows the researcher to build models of complex systems without making assumptions about the internal workings. The idea is to have the computer construct a model of optimal complexity based only on data and not and not on any preconceived ideas of researcher; that is, by knowing only simple input-output relationships of the system, *GMDH* algorithm will construct a self-organizing model that can be used to solve prediction, identification, control synthesis, and other system problems [5, 6, 7, 12]. A complex multidimensional decision hyper-surface can be approximated by a set of polynomials in the input signals (properties) which contain information about the hyper-surface of interest. The hyper-surface is usually described by a number of experimental (vector) points and simple functions of their coordinates. The approach taken in this paper to approximating the decision hyper-surface, and hence the input-output relationship of a complex system, is to fit a high-degree multinomial to the input properties using a multilayered perceptron like network structure. Thresholds are employed at each layer in the network to identify those polynomials which best fit into the desired hyper-surface [6, 7, 8, 10].

Only the best combinations of the input properties are allowed to pass to succeeding layers, where more complex combinations are formed. Each element in each layer in the network implements a nonlinear function of two inputs. The coefficients of each element are determined by a regression technique which enables each element to approximate the true outputs with minimum mean-square error. The experimental data is divided into training and testing set. The training set is used to obtain the element coefficients, and the testing set is used to determine the utility of a given element in the network and to control over-fitting of the experimental data. The latter is termed as "*decision regularization*" [7]. In contrast to the statistical decision theoretic approach which is single layered, it is argued that the type of multilayered structure presented should be used to solve complex problems for four primary reasons: (1) a smaller training set is of data required; (2) the computational burden is reduced; (3) the procedure automatically filters out input properties which provide little information about the location and shape of the decision hyper-surface; and (4) a multilayered structure is a computationally feasible way to implement multinomial of very high degree [6, 7, 10, 12].

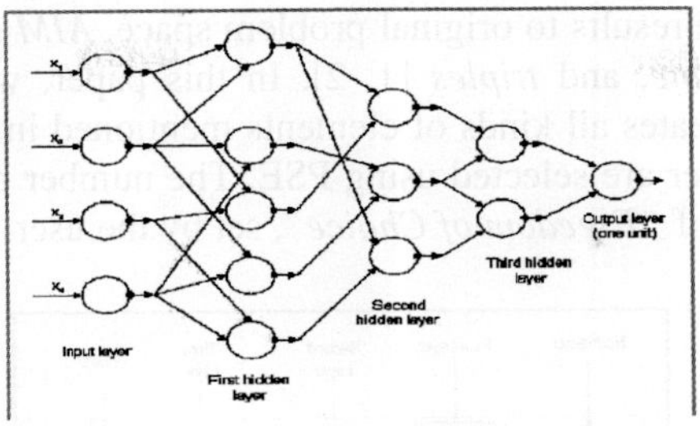

Figure 1: Classical GMDH Polynomial Neural Network

2.2. *Abductory Inductive Modeling (AIM) GMDH Algorithm*

AIM is a supervised inductive machine learning tool for automatically synthesizing abductive network models from a database of inputs and outputs representing a training set of solved examples. As a *GMDH* algorithm, the tool can automatically synthesize adequate models that embody the inherent structure of complex and highly non-linear systems. Automation of model synthesis not only lessens the burden on the analyst but also safeguards the model generated against influences by human biases and misjudgments [1, 2]. The *GMDH* approach is a formalized paradigm for iterated polynomial regression capable of producing a high-degree polynomial model in effective predictors. To prevent exponential growth and limit model complexity, the algorithm selects only relationships having good predicting powers within each phase. Iteration is stopped when the new generation regression equations start to have poorer prediction performance than those of the previous generation, at which point the model starts to become overspecialized and therefore unlikely to perform well with new data. A number of *GMDH* methods have been proposed which operate on the whole training dataset thus eliminating the need for a dedicated selection set.

The *ALN* (*Adaptive Learning Network*), *AIM* being an example, uses the *predicted squared error* (*PSE*) criterion for selection and stopping to avoid model over-fitting, thus solving the problem of determining when to stop training in neural networks. The criterion minimizes the expected error that would be obtained when the network is used for predicting new data [1, 2]. *AIM* builds networks consisting of various types of polynomial functional elements. The network size, element types, connectivity, and coefficients for the optimum model are automatically determined using well-proven optimization criterion, thus reducing the need for user intervention compared to neural networks. This simplifies model development and considerably reduces the learning/ development time and effort. The models take the form of layered feed-forward abductive networks of functional elements. In addition to the main layers of the network, an input layer of *normalizers* convert the input variables into an internal representation as *Z scores* with *zero mean* and *unity variance*, and an *unitizer* restores the results to original problem space. *AIM* supports elements as *whites*, *single*, *double*, and *triples* [1, 2]. In this paper, we propose the AIM algorithm which creates all kinds of elements mentioned in this section, and the models for next layer are selected using PSE. The number of models selected is based on the value of *"Freedom of Choice"*, set by the user.

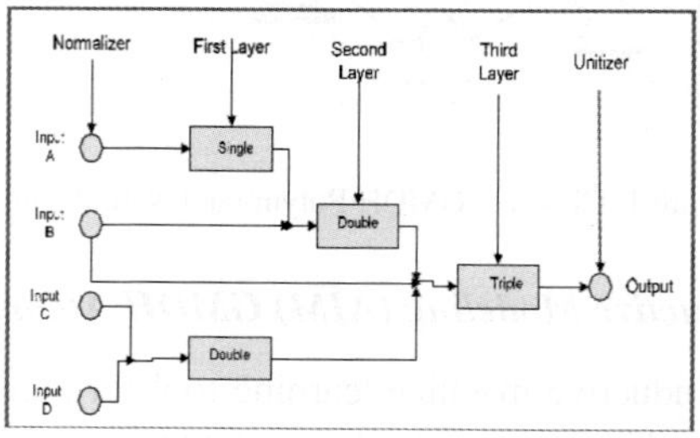

Figure 2: AIM Network showing various types of elements

3. Experiments

The data set is provided by University of Tubingen, Germany, Department of Computer Engineering for BCI Competition III [17]. It is the time series of the electrical brain activity of a subject picked up during trials. The train data consists of 278 trials and test data consists of 100 trials. Each sample is 64 by 300 matrix which is reduced to 64 by 1 vectors by taking the standard deviation of the recorded potentials over all the 64 channels. Hence, the train set becomes 278 by 64 and test set 100 by 64 matrices. PCA is applied to reduce the dimension from 64 to 15. The data is then standardized using z-scores. The train set is used to train models. Following are the accuracies and performances over train and test sets. The statistics are described using the notation $\mu\pm2\sigma$ (95% confidence interval) over 5 fold cross-validation where 'μ' and 'σ' represent mean and standard deviation. Standard deviation represents the possible spread over the performance distribution.

Method	Accuracy ($\mu\pm2\sigma$)	Performance ($\mu\pm2\sigma$)
Radial Basis Networks	65 ± 5 (%)	55 ± 8 (%)
Classical GMDH	70 ± 6 (%)	70 ± 7 (%)
AIM GMDH	75 ± 2 (%)	73 ± 7 (%)

Figure 3: Accuracy and Performance over Train and Test Data

In the above Figure, we could see, both classical and AIM models outperforming RBF networks in terms of classification accuracy and performances. Hence, Classical and AIM models are selected for comparison using other statistics. We could see that, AIM models generalize well on Test data.

True	P (μ±2σ)	N (μ±2σ)
P	35±3 (%)	15±5 (%)
N	16±3 (%)	35±5 (%)

Figure 4: Confusion rates for Classical GMDH classifier over 5-folds

True	P (μ±2σ)	N (μ±2σ)
P	37±4 (%)	12±3 (%)
N	13±4 (%)	40±4 (%)

Figure 5: Confusion rates for AIM GMDH classifier over 5-folds

In the above Figures, the true positive rate and true negative rate of AIM models outperforms that of the Classical model. Overall error rate for the classical model(s) is 31±6, whereas for the AIM model(s) it is 25±1. So, there is also decrease in the overall error rate in AIM model(s) over the Classical model(s). Here, P and N denote the class labels 1 and -1, one corresponding to tongue movement and other to finger movement in the motor imagery experiments.

Model	Sensitivity (μ±2σ)	Specificity (μ±2σ)
Classical GMDH	69±5 (%)	70±7 (%)
AIM GMDH	74±4 (%)	76±4 (%)

Figure 6: Sensitivity and Specificity for both Models over 5-folds

In the above Figure, both *Sensitivity* and *Specificity* rates are higher for AIM models than that of Classical models. Also, the difference of both *Sensitivity* and *Specificity* is ~ 2%, which means, the AIM models are capable of handling imbalance data if used in ensembles.

4. Conclusion

For various applications of the BCI systems, it is necessary that the EEG feature related to BCI were to be analyzed. The paper aimed at exploring the performance of the specified classifiers trained for discriminating the cognitive states such as finger movement and tongue movement. The paper describes a modification of GMDH proposed for classification of EEG for BCI. As found in our experiments the AIM method has outperformed the conventional GMDH as well as RBF neural networks.

References

1. Abdel-Aal, R. E. "Abductive Network Committees for Improved Classification of Medical Data". Methods Inf Med 2/2004, vol. 43, no. 2, pp. 192-201, 2004.

2. Abdel-Aal, R. E. "Improved Classification of Medical Data Using Abductive Network Commitees Trained on Different Feature Subsets". Journal of Computer Methods and Programs in Biomedicine, vol. 80, no. 2, pp. 141-153, November 2005.
3. Farlow, S. J. "The GMDH Algorithm of Ivakhnenko". The American Statistician, vol. 35, no. 4, pp. 210-215, 1981.
4. Farlow, S. J. "Self-Organising Methods in Modelling: GMDH Type Algorithms". New York: Marcel Dekker, 1984.
5. G. A. Ivakhnenko, A. G. Ivakhnenko. The Review of Problems Solvable by Algorithms of the Group Method of Data Handling. Pattern Recognition and Image Analysis, vol. 5, no. 4, pp. 527-535, 1995.
6. Ivakhnenko, A. G; Ivakhnenko, G. A. "Problems of Further Development of the Group Method of Data Handling Algorithms". IEEE Transactions on Pattern Recognition and Image Analysis, vol. 10, no. 2, pp. 187-194, 2000.
7. Ivakhnenko, A. G. "Polynomial Theory of Complex Systems". IEEE Transactions on Systems, Man and Cybernetics, pp. 364-378, 1971.
8. G. A. Ivakhnenko, A. G. Ivakhnenko, E. A. Savchenko, D. Wunsch. Problems of Further Development of GMDH Algorithms: Part 2. Mathematical Theory of Pattern Recognition, 2002.
9. Ivakhnenko, A. G; Ivakhnenko, G. A; Muller, J. A. "Self-Organization of Neural Networks with Active Neurons". Journal of Pattern Recognition and Image Analysis, vol. 4, no. 2, pp. 185-196.
10. J.-A. Mueller, F. Lemke, A. G. Ivakhnenko. "GMDH Algorithm for Complex Systems Modeling". Mathematical Modelling of Systems, 1998.
11. Madala, H. R; Ivakhnenko, A. G. "Inductive Learning Algorithms for Complex Systems Modelling". CRC Press, 1994.
12. Nikolay, Y. Nikolaev and Hitoshi Iba. "Adaptive Learning of Polynomial Networks". Springer-Verlag, 2006
13. Nikolay, C; Nikolay, V. M; Adrien, C; Johan, A. K. S; Marc, M. V. H. "An Application of Feature Selection to Online P300 Detection in Brain Computer Interface". Proceedings of IEEE Conference, 2009.
14. Tugce Balli, Ramaswamy Palaniappan. "A Combined Linear & Nonlinear Approach for Classification of Epileptic EEG Signals". Proceedings of the 4th International IEEE EMBS Conference on Neural Engineering, Antalya, Turkey, pp. 714-717, April 29 - May 2, 2009.
15. Turnip, A; Hong, Keum-Shik. "Adaptive Neural Network Classifier for EEG Signals of Six Mental Tasks". International Conference on Control, Automation, and Systems, pp. 331-336, Oct. 26-29, 2011, Gyeonggi-do, Korea.
16. Wu, J; Majoj, T; Cuntai, G. "Robust Classification of EEG Signal for BCI". IEEE Transactions on Neural Systems and Rehabilitation Engineering, Vol. 14, no. 1, March 2006.
17. Wudman, G; Schroder, M; Hill, J; Rosentiel, W; Elger, C; Scholkopf, B; Neils, B; Thilo, H; Lal, T. Methods Towards Invasive Human Brain Computer Interfaces. *Advances in Neural Information Processing Systems (NIPS)*, 2004.

UAPRIORI: AN ALGORITHM FOR FINDING SEQUENTIAL PATTERNS IN PROBABILISTIC DATA

METANAT HOOSHSADAT*, SAMANEH BAYAT, PARISA NAEIMI, MAHDIEH S. MIRIAN, OSMAR R. ZAÏANE

Computing Science Department, University of Alberta, Canada

Uncertainty in various domains implies the necessity for data mining techniques and algorithms that can handle uncertain datasets. Many studies on uncertain datasets have focused on modeling, query ranking, discovering frequent patterns, classification models, clustering, etc. However despite the existing need, not many studies have considered uncertainty in sequential data. This paper introduces UAprioriAll, a method to mine frequent sequences in the presence of uncertainty in transactions. UAprioriAll scales linearly in time relative to the size of the dataset.

1. Introduction

Statistical studies on uncertain data have recently attracted significant attention due to the fact that data produced and collected in modern applications are often uncertain or noisy. Uncertainty happens because of the limitations in the equipment, privacy reasons, information conversion or extraction, etc.

In a probabilistic transactional database, an item in a transaction can have a probability attached to it indicating its existential uncertainty of appearing in the transaction. This is a common form of uncertainty, called existential uncertainty. As an example of this assume a health-related database in which data is extracted from hand-written or text-based medical records using machine learning approaches [1]. Each attribute in this database describes a fact about the patient, and may be inaccurate for several reasons including inaccuracy in the information extraction method.

Sequential Pattern Mining or *SPM* is a well known and important problem in data mining and has been addressed by many studies [2-5]. However mining frequent sequences from uncertain datasets is still an open problem. In this paper, we propose a solution for the aforementioned problem by introducing a new algorithm called UAprioriAll.

---

* {hooshsad, samaneh, naiemi,mirianho,zaiane}@cs.ualberta.ca

UAprioriAll is designed to enable mining frequent sequences in datasets with existential uncertainty. The proposed algorithm uses *expected support* as the measure of frequentness of transactions and sequences. Expected support is a metric that measures the expected frequency of an itemset/sequence in uncertain datasets. It is memory efficient and very fast to compute [6].

In this paper, we briefly overview the studies in the field of uncertain datasets in Section 2, then introduce and describe our proposed algorithm in Section 3. We discuss the experiments designed to evaluate the proposed algorithm in Section 4.

2. Related Works

Frequent sequential pattern mining or *SPM* [2-5] deals with datasets in which each transactions is considered to be associated with an id. Each id may be associated with a sequence of transactions. The definition of the problem is as follows.

Given a set of sequences where each sequence consists of a list of elements and each element consists of a set of items, and given a user-specified minimum support threshold, sequential pattern mining is to find all of the frequent subsequences, i.e., the subsequences whose occurrence frequency in the set of sequences is no less than the minimum support.

Uncertain datasets have attracted much attention recently [6]. Some studies have addressed the SPM problem for specific and limited types of uncertain datasets [2,3]. The main difference between related works and our study is the nature of the problem and the data modeling. In our model, each item has a probability of existence in each transaction, which implies that the spaces of the models in the previous work are subspaces of our model.

The data model that we propose for a realistic capture of uncertainty in sequential datasets is existential probabilistic datasets. In such datasets, each item exists within a transaction with a probability. Equation 1 shows the general form of our datasets. Each dataset D with size $|D|$ is a set of $|D|$ sequences S_i, where each sequence of size S_i contains S_i transactions $t_{i,j}$.

$$\begin{aligned} D &= \{S_i: i = 1..|D|\} \\ S_i &= < t_{\{i,j\}}: j = 1..|S_i| > \\ t_{\{i,j,k\}} &= (it_{i,j,k}, pr_{\{i,j,k\}}): k = 1..|t_{i,j}| \end{aligned} \quad (1)$$

Our novel algorithm, UAprioriAll, his algorithm has three phases: a) U-Litemset; b) U-Transformation; c) U-Sequence.

2.1. *UItemset: Mining Single Sequences*

In this phase, the sequences of size 1 (containing only one transaction) are evaluated. Each single sequence (itemsets) is marked as a candidate, if its expected support is above the minimum threshold. In the probabilistic datasets *D* the expected support is computed by Equation 2.

$$E\big(s(x)\big) = \Sigma_{\{S \in D\}} P(x \in S)$$
$$P(x \in S) = 1 - \Pi_{\{T \in S\}}\big(1 - p(x \in T)\big) \quad (2)$$
$$P(x \in T) = \prod_{\{i \in x\}} P(i \in T)$$

We mine the *probabilistic frequent patterns* using a UApriori based technique [7]. These itemsets are put in a set called L_1. Next, each of the patterns in set L_1 is mapped to a unique integer number and L_1 is transformed using this map. Set L_1 is the output of this phase.

2.2. *U-Transformation: Simplifying the Dataset*

In this phase, we transform the sequential dataset based on set L_1. The transformed dataset has two major differences with the original dataset. First, all the infrequent itemsets of the original dataset, that is the ones that are not contained in L_1, are removed from the transformed dataset. Second, the frequent itemsets are mapped into integer numbers as in L_1.

UC-Sequence: Mining the Sequences

In this phase, we mine the frequent sequences from the transformed dataset (output of U-Transformation) using an UApriori-like algorithm. At each step, the new candidate set (C_k) is filled up based on the frequent sequences of the previous level (L_{k-1}) by evaluating $L_{k-1} \bowtie L_{k-1}$ for all two tuples that have *k-2* items in common and then removes those that have infrequent subsets. The procedure starts from L_1 which was calculated in phase 2. We continue till the set L_k is empty.

To form L_k, we choose *c* from C_k if the expected support of *c* is greater than the minimum value. To calculate the expected support based on Equation 2, we need to calculate the value of $P(x \in S)$ for *x* which size is greater than 1.The probability by which the first *k* items of candidate $c \in C_m$ (*m*>*k*) appears at least once within the first *j* items of the sequence $s \in D$ is denoted by $P_{\{j,k\}}(c,s)$ and computed by a recursive approach presented in Equation 3. The recursive equation allows us to benefit from the dynamic programming scheme.

$$
\begin{aligned}
P\{c \subseteq s\} &= P_{\{|c|,|s|\}}(c \subseteq s) \\
P_{\{j,k\}}(c \subseteq s) &= P_{\{j-1,k-1\}}(c \subseteq s) * p(c[k] \subseteq s[j]) + \\
&\quad P_{\{j-1,k\}}(c \subseteq s) * (1 - p(c[k] \subseteq s[j])) \\
P_{\{j,k\}}(c \subseteq s) &= 0 \; if \; k > j \\
P_{\{j,1\}}(c \subseteq s) &= 1 - \Pi_{\{t_l \subseteq s\}}^{\{l=1..j\},}(1 - p(c[1] \subseteq t_l))
\end{aligned} \tag{3}
$$

The recursive formula is achieved by dividing the problem into two mutually exclusive states. State *a* is when s_j contains c_k and state *b* is otherwise. The probability value is the addition of the probabilities of the two states. State *a* requires two events to happen, both $c[k] \subseteq s[j]$ and *s[1..j-1]* (the *j-1* first elements of *s*) should contain at least one appearance of *c[1..k-1]* (the first *k-1* elements of *c*). State *b* also requires two events, *c[k]∉s[j]* and *c* being a subset of *s[1..j-1]*. It is evident that states *a* and *b* are mutually exclusive. The total number of computations required can be assessed by Equation 4.

$$
\begin{aligned}
DC_{\{(c,s)\}} &= \{P_{\{j,k\}(c,s)} : |c| - k > |s| - j\} \\
Computation[P(c \subseteq s)] &= |s|.|c| - \frac{\{|c|^2 - |c|\}}{2} - \frac{\{|c|^2 - |c|\}}{2} \\
&= |c|.(|s| - |c| + 1)
\end{aligned} \tag{4}
$$

Relying on the mathematical meaningfulness of the expected support, the purpose of UAprioriAll is to find all sequences that have higher expected support than the predefined threshold. It is easily verifiable that our algorithm is sound. An induction using the downward closure lemma [7] can prove the algorithm completeness. In addition, the algorithm terminates when the set L_k is empty, that is worst case happens at the K_{max}-th level, where K_{max} is the maximum length of the sequences. Therefore, total correctness can be proven for UAprioriAll.

3. Experiments

As there are no uncertain sequential datasets publicly available similar to other studies on uncertain data (such as in [8]), we used synthetic datasets in our experiments. Each generated synthetic data is characterized by two parameters: the number of sequences *L* and the total number of items *I*. Our goal is to investigate the effect of *L* on the time consumption, so we set *I* to 20 and *L* is varied from 50 to 10000.

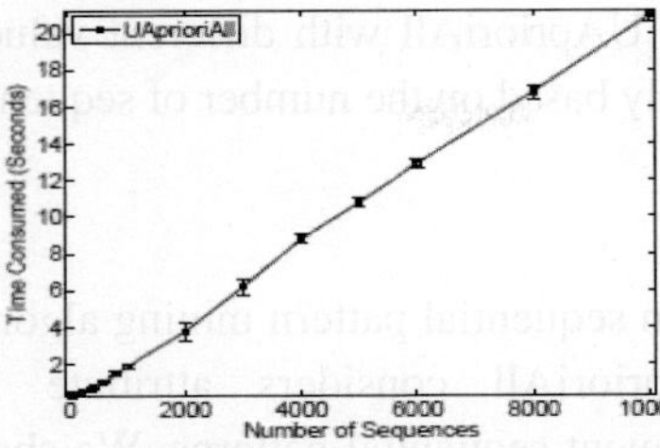

(a) Time consumption of *UAprioriAll* when minimum support = 0.5

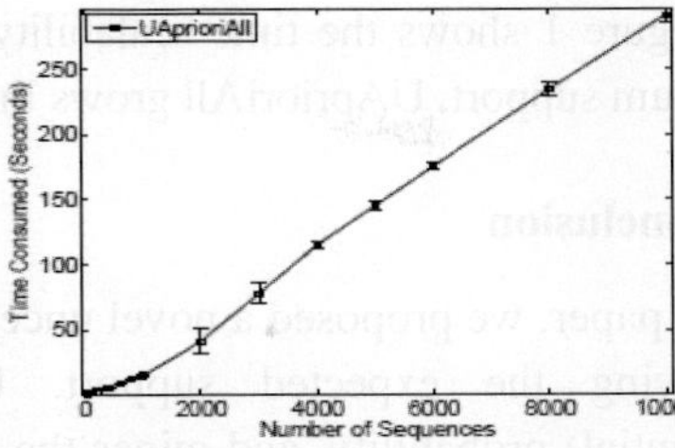

(b) Time consumption of *UAprioriAll* when minimum support = 0.33

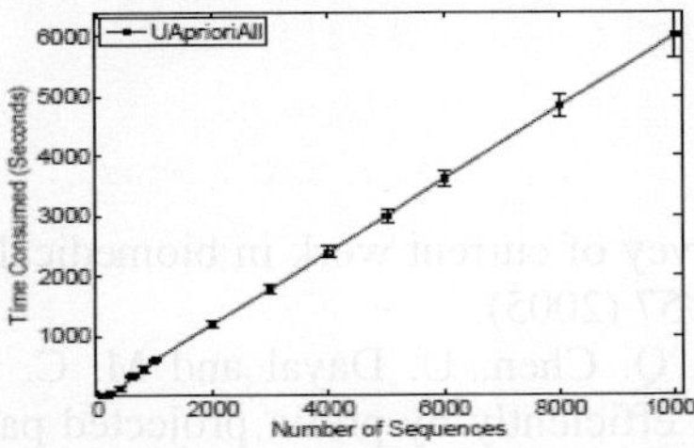

(c) Time consumption of *UAprioriAll* when minimum support = 0.25

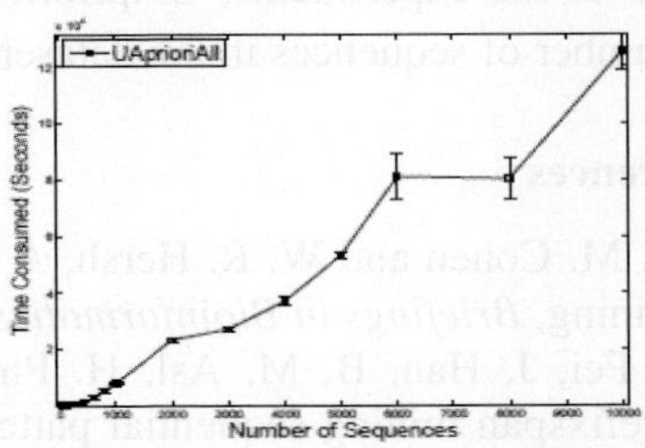

(d) Time consumption of *UAprioriAll* when minimum support = 0.10

Figure 1-Time consumption of UAprioriAll with different support values

The number of transactions in each sequence is a monotonically distributed pseudo-random number from 1 to 0. We randomly select the items within the transaction, where all items have equal chances. The existence probability attached to each item within the transaction is a randomly generated number between 0 and 1.

To increase the reliability of the results, for each value of *L*, 5 datasets were generated and the method was applied 5 times to each dataset, and then averaged. In this process, we used the *Scaling Method*. This method scales down a large dataset by randomly eliminating some transactions, to get a smaller dataset with lower number of sequences.

The experiments were carried out on a machine with 2.66 GHz clock speed and 8 GB of RAM. In the implementation of the algorithm we adopted a free online Java implementation of Apriori [9].

The experiments include the time consumption of the algorithm based on the minimum support. This measure is important because decreasing it may cause a dramatic drop in performance and may affect the consistency in the behavior of UAprioriAll. Setting the minimum support in real world applications depends greatly on the domain.

Figure 1 shows the time scalability of UAprioriAll with different values of minimum support. UAprioriAll grows linearly based on the number of sequences.

4. Conclusion

In this paper, we proposed a novel uncertain sequential pattern mining algorithm employing the expected support. UAprioriAll considers attribute level (existential) probability, and mines the frequent sequential patterns. We showed the feasibility of our new algorithm in terms of time consumption. Based on the results of the experiments, UAprioriAll's runtime is linearly scalable based on the number of sequences in the dataset.

References

1. A. M. Cohen and W. R. Hersh, A survey of current work in biomedical text mining, *Briefings in Bioinformatics* **6**, 57 (2005).
2. J. Pei, J. Han, B. M. Asl, H. Pinto, Q. Chen, U. Dayal and M. C. Hsu, Prefixspan mining sequential patterns efficiently by prefix projected pattern growth, in *ICDE '01:Proceedings of the 17th International Conference on Data Engineering*, (Heidelberg, Germany, 2001).
3. J. Yang, T. J. Watson, W.Wang, P. S. Yu and J. Han, Mining long sequential patterns in a noisy environment, in *SIGMOD '02: Proceedings of the 2002 ACM SIGMOD international conference on Management of data*, (Madison, Wisconsin, USA, 2002).
4. M. J. Zaki, SPADE: An Efficient Algorithm for Mining Frequent Sequences, *Mach.Learn.* **42**, 31 (2001).
5. R. Agrawal and R. Srikant, Mining sequential patterns, in *Data Engineering, 1995.Proceedings of the Eleventh International Conference on*, (Taipei, Taiwan, 1995).
6. C. C. Aggarwal and P. Yu, A Survey of Uncertain Data Algorithms and Applications,*IEEE Transactions on Knowledge and Data Engineering* **21**, 609(May 2009).
7. C. C. Aggarwal, Y. Li, J. Wang and J. Wang, Frequent pattern mining with uncertain data, in *KDD '09: Proceedings of the 15th ACM SIGKDD international conference on Knowledge discovery and data mining*, (Paris, France, 2009).
8. Q. Zhang, F. Li and K. Yi, Finding frequent items in probabilistic data, in *SIGMOD '08: Proceedings of the 2008 ACM SIGMOD international conference on Management of data*, (Vancouver, Canada, 2008).
9. P. Fournier-Viger, Algorithms/frequent itemset mining algorithms (2010), url: http://www.philippe-fournier-viger.com/spmf/index.php.

SEASONAL VARIATION OF TURKEY: A CONSENSUS CLUSTERING APPROACH

AYCA YETERE KURSUN*

Department of Scientific Computing, Institute of Applied Mathematics, Middle East Technical University, METU Üniversiteler Mah. Dumlupınar Blv. No:1 Cankaya Ankara, 06800, Turkey

CEM IYIGUN

Department of Industrial Engineering, Middle East Technical University, METU Üniversiteler Mah. Dumlupınar Blv. No:1 Cankaya, Ankara, 06800, Turkey

CEYLAN YOZGATLIGIL

Department of Statistics, Middle East Technical University, METU Üniversiteler Mah. Dumlupınar Blv. No:1 Cankaya, Ankara, 06800, Turkey

INCI BATMAZ

Department of Statistics, Middle East Technical University, METU Üniversiteler Mah. Dumlupınar Blv. No:1 Cankaya, Ankara, 06800, Turkey

A "consensus clustering" methodology is applied to long-term (1950-2010) Turkey's meteorological data (temperature, humidity and precipitation) in order to analyze the seasonal variations. The consensus clustering analysis applied is based on the methodology of disturbing the original data using resampling techniques, proposed by Monti et al [2]. We employed four different clustering algorithms (Agglomerative Nesting, Divisive Analysis, Partitioning Around Medoids and K-means) in order to form a consensus solution among different algorithms. Results indicated that Turkey is experiencing longer winter and summer, and rather short spring and fall seasons than usual.

1. Introduction

In general meteorologists identified seasons as groupings of three whole months (Spring, Summer, Fall and Winter) with spring starting from 1st of March for Northern hemisphere. Yet a season is also defined as a sub-division of the year, marked by changes in weather. All over the world different regions experience different weather conditions throughout a year. So the main goal of this study is

* e112987@metu.edu.tr

to answer the questions like how many seasons Turkey go through within a year and how we can define those seasons. For this reason the Turkey's meteorological data between years 1950 and 2010 is analyzed using consensus clustering methodology. Consensus clustering ensembles multiple clusterings of a dataset into one consolidated clustering. Consensus clustering methodologies offers benefits such as improved quality of solution, improved robustness against wide ranges of datasets, elimination of the model selection process, knowledge reuse, distributed clustering and effective consolidation of clusters depending on different views of data having multiple features [1].

2. Methodology

The consensus methodology developed by Monti et al. proposes a resampling-based approach for class discovery and clustering validation. The proposed methodology provides a method to represent the consensus across multiple runs of clustering algorithms [2]. In this study we used this proposed approach to investigate the seasonal variation of Turkey.

2.1. *Procedure*

For a selected bootstrapping technique with different clustering algorithms and number of clusters;

1. Resample the dataset
2. Apply the clustering algorithm to the resampled data
3. Compute the consensus clustering matrix using all the runs for the same algorithm
4. Combine the consensus clustering matrices obtained for each algorithm using weights in order to form a merged matrix
5. Use the merged matrix as a similarity matrix and solve for the final clustering
6. Choose the most suitable number of cluster.

3. Dataset

For the investigation of seasonal variation of Turkey, dataset used contains the meteorological data between years 1950 and 2010. It comprises 244 measurement stations from all over Turkey. Dataset contains five variables, monthly maximum temperature, monthly minimum temperature, monthly average temperature, monthly average humidity, and monthly total precipitation. Further details for the dataset can be found in Kartal et al [3]. In order to form a unique dataset for feeding into the clustering algorithms, the data is standardized for each variable and 14 geographical regions seperately.

4. Results and Discussion

Tables 1, 2, 3 and 4 summarize the results of the consensus clustering algorithm for different number of clusters (K=2-5). 500 bootstrap resamples are used for each clustering algorithm employed, which are Agglomerative Nesting (Agnes), Divisive Analysis Clustering (Diana), Partitioning Around Medoids Clustering (Pam), K-means Clustering.

Table 1. Results of consensus clustering for K=2

	Agnes	Diana	Pam	K-means	Merged
January	1	1	1	1	1
February	1	1	1	1	1
March	1	1	1	1	1
April	2	1	1	2	2
May	2	2	2	2	2
June	2	2	2	2	2
July	2	2	2	2	2
August	2	2	2	2	2
September	2	2	2	2	2
October	2	2	2	2	2
November	1	1	1	1	1
December	1	1	1	1	1
Area Under CDF	0.5434	0.5578	0.5632	0.5528	0.5499

Table 2. Results of consensus clustering for K=3

	Agnes	Diana	Pam	K-means	Merged
January	1	1	1	1	1
February	1	1	1	1	1
March	1	1	1	1	1
April	2	1	2	2	2
May	2	2	2	2	2
June	3	3	3	3	3
July	3	3	3	3	3
August	3	3	3	3	3
September	3	3	3	3	3
October	2	2	2	2	2
November	1	1	1	1	1
December	1	1	1	1	1
Area Under CDF	0.7520	0.7249	0.7324	0.6433	0.7054
Δ(K)	0.3838	0.2995	0.3004	0.1637	0.2828

Table 3. Results of consensus clustering for K=4

	Agnes	Diana	Pam	K-means	Merged
January	1	1	1	1	1
February	1	1	1	1	1
March	2	2	2	2	2
April	3	3	3	3	3
May	3	3	3	3	3
June	4	4	4	4	4
July	4	4	4	4	4
August	4	4	4	4	4
September	4	4	4	4	4
October	3	3	3	3	3
November	2	2	2	2	2
December	1	1	1	1	1
Area Under CDF	0.8253	0.8187	0.8214	0.6193	0.7672
Δ(K)	0.0975	0.1295	0.1216	-0.0373	0.0876

Table 4. Results of consensus clustering for K=5

	Agnes	Diana	Pam	K-means	Merged
January	1	1	1	1	1
February	1	1	1	1	1
March	2	2	2	2	2
April	3	3	3	3	3
May	3	3	3	3	3
June	4	4	4	4	4
July	4	4	4	4	4
August	4	4	4	4	4
September	4	4	4	4	4
October	3	3	3	5	3
November	2	2	2	2	2
December	5	5	5	1	5
Area Under CDF	0.8676	0.8663	0.8613	0.5996	0.7952
Δ(K)	0.0513	0.0581	0.0485	-0.0318	0.0366

For solving the consensus matrices and the merged consensus matrix, Agglomerative Nesting algorithm is utilized. The Euclidian distance measure is used for all algorithms. For Agglomerative Nesting, Average Linkage method is selected, and Hartigan-Wang method is employed for the K-means clustering algorithm.

The consensus matrixes and the merged consensus matrix are obtained from the R package 'clusterCons'. This package contains functions that generate cluster membership based on generating matrices from bootstrapped clustering experiments in which a random proportion of rows of the dataset are used in each individual clustering [4].

The results obtained revealed that Turkey have seasonal differences within a year. As this is a known fact, the basic questions that we are trying to answer are: how many seasons there are within a year, and how we can define those seasons.

As can be seen from the results, different clustering algorithms produced different results. For K=2 Agnes and K-means algorithms and Diana and Pam algorithms produced the same results respectively. For K=3 and K=5 there is a more unified result with respect to K=2, with only Diana and K-means producing a different result from the other three algorithms, respectively. As we analyze the results for K=4, we observed that there is a total consensus among the results. But this does not mean that K=4 is the right number of clusters that we are looking for.

In order to select the appropriate number of clusters for Turkey's seasonal variation, we can look into the concept of cumulative distributions (CDF) and proportion increase in the CDF's as described by Monti et al [2]. Monti et al. proposed that the inspection of the CDF progression is to select the largest K that induces a large enough increase in the area under the corresponding CDF (which is represented with $\Delta(K)$ in the results table). Figure 1 plots the value of $\Delta(K)$ for different number of clusters, that is, K's. As shown, the $\Delta(K)$ for K=3 is greater than the other number of clusters. Yet we should consider whether the differences between $\Delta(K)$'s are significant enough to declare K=3 as the right number of clusters, K_{true}. K=4 could also be considered as K_{true} for algorithms other than K-means. For K-means it is obvious that the CDF is even getting smaller for K>3. The result for the merged consensus matrix also suggests that K=4 can also be K_{true}.

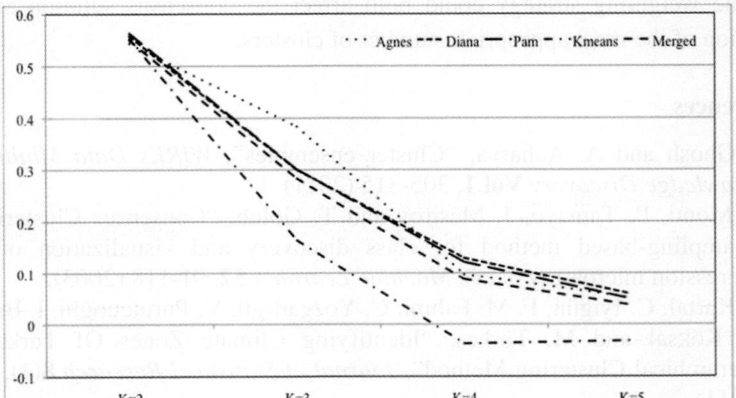

Figure 1. Proportion increase $\Delta(K)$ in the area under the CDF.

As a result, within the scope of this study, depending on the out comes of the merged consensus matrix, we can select K=4 as K_{true} (with Winter: December, January and February; Spring: April and May; Summer: June, July, August and September; Fall: October). By taking account the results of K=3 we can also conclude that Turkey is experiencing longer winter and summer seasons and rather short springs, with fall being one month only.

5. Conclusion

In this study we analyzed the seasonal variation of Turkey employing a consensus clustering algorithm by perturbing the original data with bootstrapping technique. The results show that the four clusters are appropriate to define the seasonal variation. This result suggests that Turkey experiences winter during December, January and February. Summer is defined by the months June, July, August and September. April and May stands out as spring, with October being the fall. But this identification made depends on the semantics that mankind had used so far and the idea of consecutive months defines a season. One should keep in mind that the results suggest that April, May and October are identified as having the same characteristics by being in the same cluster. March and November is standing out as transition seasons between our regular season definitions.

We have used different clustering algorithms and the merged consensus matrix is formed by weighting each consensus matrix (for different algorithms). In this study as we have selected to form the merged matrix with equal weights, the idea of the weighting the matrices does not really reflected into the study. The weighting methodologies are a point to explore further as we have observed that the weighting strategy could both affect the consensus solution and the selection of the most appropriate number of clusters.

References

1. J. Ghosh and A. Acharya, "Cluster ensembles", *WIREs Data Mining and Knowledge Discovery* **Vol.1**, 305-315 (2011)
2. S. Monti, P. Tamayo, J. Mesirov and T. Golub, "Consensus Clustering: A resampling-based method for class discovery and visualization of gene expression microarray data", *Machine Learning* **52**, 91-118 (2003).
3. E. Kartal, C. İyigün, F. M. Fahmi, C. Yozgatlıgil, V. Purutçuoğlu, İ. Batmaz, G. Köksal and M. Türkeş, "Identifying Climate Zones Of Turkey By Hierarchical Clustering Method", *Journal of Statistical Research* **8(1)**, 13-25 (2011)
4. T. I. Simpson, "R Package 'clusterCons", (8 January 2011).

SPATIAL ANALYSIS BASED METHOD FOR DETECTION OF DATA TRAFFIC PROBLEMS IN COMPUTER NETWORKS*

GRZEGORZ KOŁACZEK†

Faculty of Computer Science and Management,
Wroclaw University of Technology, Wyb..Wyspianskiego 27,
50-370 Wroclaw, Poland

The paper presents an original approach to the analysis of network data flows with the application of statistical spatial analysis methods. An illustrative example has been presented how to identify the correlations in network nodes behaviour and how to support the process of detection of network traffic related problems. The proposed method of data traffic analysis can be used to support various network management and security related tasks for example anomaly detection. The presented example shows how spatial correlations reflect the changes in communication patterns and so how they may be used for the detection of ongoing attacks, data link failures or network traffic routing problems.

1. Introduction

1.1. *Network Traffic and Network Security Level Evaluation*

Network traffic shows some quantitative and topological features that appear to be invariant and characteristic for given network [1] [2] Moreover, general rules underlying those features are the same for almost any network of remarkable size. These distinct features concern topology of network communication, considered as origin-destination flows graph, the distribution of data volumes sent between destinations and the in/out ratio of data sent between nodes/subnets and outside world.

For example distributed interactive multimedia environments (DIME) are characterized by a large number of cooperating multimedia devices [4]. Dependencies of these type of software components over a time-span imply also very strong correlations at the communication level. With respect to these properties, wide range of network problems and attacks can be detected by observation of communication patterns and comparison existing under normal

* Work partially supported by the European Union within the European Regional Development Fund program no. POIG.01.03.01-00-008/08

state of the network to new ones which occur under data link failure or attack. For example, in case of Internet worm attacks, within a network there could be scanning and attack flows which differ substantially from normal network activity. Moreover, total scanning rate into the sub-network (or given set of nodes) is a function of the number of all infected nodes in the network. Another invariant for a long time periods and different scales (subnet sizes) or traffic types (protocols) is proportion between a number of internal (Fan-in) and external (Fan-out) data flows [5].

1.2. *Spatial Analysis*

Spatial analysis is the process of discovering potentially useful patterns from spatial datasets. Extracting interesting patterns from spatial datasets could be more difficult than extracting the corresponding patterns from traditional numeric and categorical data due to the complexity of spatial data types, spatial relationships, and spatial autocorrelation.

The first source of difficulties that appears in spatial analysis is the data input. Spatial analysis requires some additional information that is very often more complex than the input of classical data analysis approaches. The input for spatial analysis must include some extended objects such as points, lines or polygons. There are two distinct data types inputs of spatial data: non-spatial attribute and spatial attribute. Non-spatial attributes are used to characterize non-spatial features of objects, such as a node's name, incoming traffic level, etc. They are the same as the attributes used in the data inputs of classical data analysis. Spatial attributes are used to define the spatial location and relations among spatial objects. The spatial attributes most often include information related to spatial locations, e.g., longitude, latitude and elevation, as well as shape [6].

The paper focuses on a statistical model to represent observations of network traffic in data network. The spatial relationships which are often used as a synonymous with geographical distance between objects, in this case have been redefined as the logical distance between network nodes. This means that the distance between network nodes is measured in number of network hops on communication path. This model has been used for estimation and description of spatial correlations among network nodes.

To the author's best knowledge, the method of computer network related problems detection proposed in this paper is the first one that benefits from spatial analysis approach.

2. Models and Assumptions

Data flow in communication networks, as well as anomaly detection, can be considered at several different levels of abstraction. The 'low level' is where one focuses on the physical links between network nodes. The example where this low level point of view could be convenient is analysis of wide area backbone network where vertices of communication graph are routers or network gateways. The other level of data flow analysis is 'logical links level'. At this level one does not need to be aware of the physical connection between communicating parts but only identifies the source and destination of the transmitted data.

The aim of the proposed approach is to benefit from the knowledge about the structural and logical relationships among communicating network nodes and to provide a convenient tool for identification of communication problems e.g. anomaly detection in data networks.

2.1. *Application of Spatial Statistics Methods to Diagnosis Computer Network Data Flow*

During this research the possibility to detect the abrupt changes in communication patterns or network structure has been investigated. The two types of network topologies have been generated with simulated data flow between each pair of network nodes (see Figure 1). For all cases the statistical spatial analysis have been conducted.

The first network topology that has been analyzed was a ring, where each node is connected only to two other network nodes. The ring has been composed by 20 nodes. To investigate the influence of the communication structure change to the spatial statistics the next two cases were considered. These cases introduce modifications to the ring structure. The first variant was a starting ring network with addition of one communication link. The additional link was put at the diameter of the original network. The second variant was a starting ring network modified by addition of two communication link at two distinct ring's diameters. The results obtained while analyzing ring-based network topologies were compared to network with a random topology. This network has been composed of 30 distinct nodes with random connections among them. During analysis the same data flow level has been used for all networks.

The spatial analysis has been performed in the following steps:

1. Moran-I test

   It is the most common test for the existence of spatial autocorrelation. Moran coefficient calculates the ratio between the product of the variable of interest and its spatial lag, with the product of the variable of interest, adjusted for the spatial weights used.

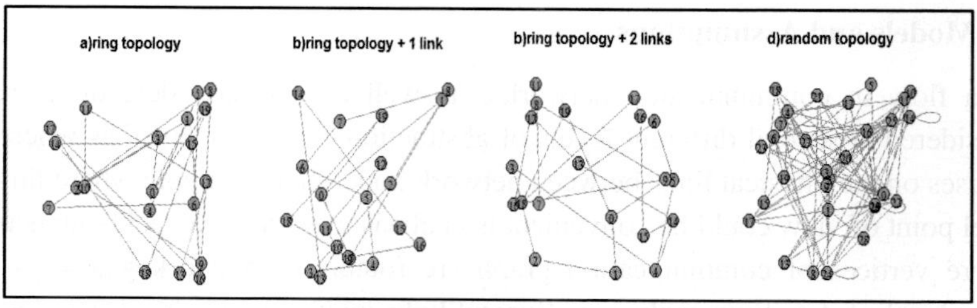

Figure 1 Network topologies used in experiments

$$I = \frac{n\sum_i^n \sum_j^n \omega_{ij}(x_i - \bar{x})(x_j - \bar{x})}{\sum_i^n \sum_j^n \omega_{ij} * \sum_i^n (x_i - \bar{x})^2} \tag{1}$$

Where x is the value of a variable for the ith observation, $\bar{x}$ is the mean of x and the ω_{ij} is the spatial weight of the connection between i^{th} an jth node

Values range from -1 (dispersion) to +1 (correlation). A zero value indicates a random spatial pattern.

2. Local Moran test

A local Moran's test for the unit i can be constructed as one of the n components which comprise the global test:

$$I_i = \frac{(x_i - \bar{x})\sum_j^n \omega_{ij}(x_j - \bar{x})}{\frac{\sum_i^n (x_i - \bar{x})^2}{n}} \tag{2}$$

As with global statistics, it is assumed that the global mean $\bar{x}$ is an adequate representation of the variable of interest. As before, local statistics can be tested for divergence from expected values, under assumptions of normality.

3. Spatial correlogram

Spatial autocorrelation measures the degree to which observations are correlated to itself in space. Spatial correlogram allows to tests and visualize whether the observed value of a variable at one location is independent of values of that variable at neighboring locations. Positive spatial autocorrelation indicates that similar values appear close to each other, or cluster, in space while negative spatial autocorrelation indicates that neighboring values are dissimilar or that similar values are dispersed. Null spatial autocorrelation indicates that the spatial pattern is random. In this research the correlogram of input/output flows of the corresponding network nodes has been investigated.

3. Experiments and Results of Analysis

Due to space limitation in this section the overview of the main results of the spatial analysis applied to data flow in computer networks will be briefly presented.

1. Moran-I test

Table 1. Moran-I test results for examined topologies

	Network topology			
	Ring	Ring +1 link	Ring +2 links	Random
Moran I statistic	0.62191729	0.70500961	0.26689018	-0.065609756
p-value	0.000787	9.222e-05	0.05888	0.6412
Expectation	-0.05263158	-0.05263158	-0.05263158	-0.03448276
Variance	0.04554737	0.04105034	0.04172549	0.01899098

The results presented in Table 1. shows that there is a high spatial correlation in ring-based networks while the random communication topology destroys this type of correlation.

2. Local Moran test

Table 2. Loca Moran-I test results summary

	Network topology			
	Ring	Ring +1 link	Ring +2 links	Random
Median(I_i)	0.26489	0.4809227	0.19972	-0.05212
Mean(I_i)	0.62192	0.7050096	0.26689	-0.06561
Mean(Pr(z>0))	2.870e-01	2.276e-01	0.33687	0.5209676

Additionally it could be noticed that in the ring type network the most significant results (the greatest spatial correlation) obtains the nodes with the highest degree (with additional links) while in random network the highest Local Moran test values get the nodes with the lowest degree.

3. Spatial correlogram

The figures 2a,2b,2c show that the simple communication topologies are very good represented in spatial log charts. Additionally it is also visible that any small change in the data routing between nodes induces changes in correllograms. Contrary, the diagram of spatial lag for random network shows randomness between data flows measured at subsequent distances from the source node.

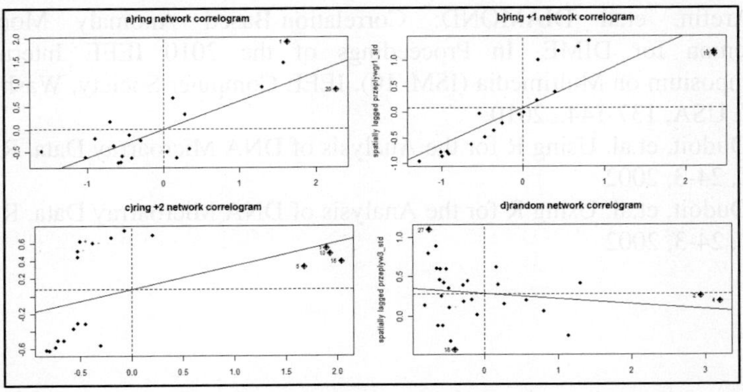

Figure 2 Spatial lag correlogram

4. Conclusions and Future Work

The paper presented an original approach to the analysis of network data flows with the application of statistical spatial analysis methods. The proposed method of data traffic analysis can be used to support various network security related tasks for example anomaly detection. The obtained results are very promising, so the further research in this field has been planned.

For example a presented spatial analysis based method can be used for detection of spatial outliers in data networks. A spatial outlier is a spatially referenced object whose non-spatial attribute values differ significantly from those of other spatially referenced objects in its spatial neighborhood. The network nodes which would be classified as spatial outliers could be a local instability center in communication network. The identification of such spatial outliers can become a vital element of network management and security providing processes. The further works are intended to apply spatial analysis to real world data and to extend the analysis with some additional methods to enable reliable classification of network events e.g. DDoS attack, attack against routing protocol, communication link failure, etc.

References

1. V. Chandola, A. Banerjee, V. Kumar: Anomaly Detection: A Survey. ACM, 2009.
2. R. Fontugne, P. Borgnat, P. Abry, K. Fukuda: MAWILab: Combining Diverse Anomaly Detectors for Automated Anomaly Labeling and Performance Benchmarking. ACM, 2010.
3. G. Kołaczek, A.Prusiewicz, "*Social network approach to anomaly detection in network* systems", w: Intrusion detection systems, Rijeka, str. 81-96, InTech, 2011
4. A.Arefin, et.al DIAMOND: Correlation-Based Anomaly Monitoring Daemon for DIME. In Proceedings of the 2010 IEEE International Symposium on Multimedia (ISM '10). IEEE Computer Society, Washington, DC, USA, 137-144., 2010
5. S. Dudoit, et.al. Using R for the Analysis of DNA Microarray Data, R News, 2, 1, 24-3, 2002
6. S. Dudoit, et.al. Using R for the Analysis of DNA Microarray Data, R News, 2, 1, 24-3, 2002

UNCERTAINTY AND A NEW MEASURE FOR CLASSIFICATION UNCERTAINTY

HALUK DAMGACIOĞLU
Industrial Engineering Department, Middle East Technical University Ankara, 06531, Turkey

CEM İYİGÜN
Industrial Engineering Department, Middle East Technical University Ankara, 06531, Turkey

Ben-Israel and Iyigun ([1] and [2]) presents a new clustering method which is probabilistic distance clustering (P-D Clustering). In this method, the probability of assignment to cluster for each point is inversely proportional to distances between data point and centers of clusters according to given number of clusters and their centers. In this paper, we study on new uncertainty measure for classification using the assignment probabilities of P-D Clustering. Moreover, the relationship of the new measure with Kullback – Liebner divergence is discussed.

1. Introduction

Information is separated into two concepts with respect to scope of information as singular and non-singular. While singular information gives the exact solution or answer of the problem, in nonsingular information there is uncertainty since the answer is not clear. Uncertainty is appeared in such that information is linguistic based, there is a lack of information and it does not need singular information [6]. Therefore it can be said that uncertainty is an attribute of information (see [3], [4] and [5] for more details).

In clustering problems, there are two important questions whose answers are nonsingular information. The first question is that what "right" number of clusters for a given data set is.

The other question is the assignment of each data point. When clustering methods in data mining are used to determine the assignment of a point $\mathbf{x}_i$, $\{\mathbf{x}_i \in$ D, D is the dataset$\}$, to cluster C_k, $\{k \in \overline{1,K}\}$; the idea of clustering method is assigning the "close" points to the same cluster with respect to a distance function $d_k(\mathbf{x}_i)$, that shows the distance between data point $\mathbf{x}_i$ and center of cluster C_k. The logic defining "close" is a kind of linguistic information and as it is stated in [6], linguistic information is one of the reasons of uncertainties. Therefore this uncertainty should be modeled with the answers of two basic questions: what does mean "close" and what is the "right number of clusters" along with determining the assignments of the points?

In this paper, a new classification uncertainty measure for clustering is stated. P-D clustering is briefly explained in section 2. Classification Uncertainty Function (CUF) is introduced in section 3. Its relation with Kullback – Leibner Divergence is discussed in Section 4. The paper concludes with the conclusion and future work in Section 5.

2. Probabilistic Distance Clustering

Consider a data set denoted by D with N data points $\mathbf{x_i}$ ($i \in \overline{1,N}$) and K clusters,

$$D = C_1 \cup C_2 \cup \ldots \cup C_K, \; C_k \cup C_l = \emptyset \quad \forall k \neq l.$$

Data points have n attributes, $\mathbf{x_i} \in \mathbb{R}^n$. Center of kth- cluster C_k is shown by $\mathbf{c_k}$, and a distance function $d(\mathbf{x_i}, \mathbf{c_k})$ measures the distance between data point $\mathbf{x_i}$ and $\mathbf{c_k}$, such as Euclidean distance,

$$d(\boldsymbol{x_i}, \boldsymbol{c_k}) = \|x_i, -\boldsymbol{c_k}\|.$$

In probabilistic distance clustering, membership of data points to clusters is not deterministic, probabilities are used. Membership probability that data point $\mathbf{x}$ belongs to k^{th}- cluster C_k is denoted by $p_k(\mathbf{x})$, $\forall\, k: \overline{1,K}$. It depends on the distance and its assignment probability would be greater if data point is closer to center of a cluster. This relation can be modeled as,

$$p_k(x) d_k(x, c_k) = constant, \quad \forall x \tag{1}$$

Let $M(\mathbf{x})$ shows the constant in equation (1), and the probabilities considered in the model satisfy the following,

$$\textstyle\sum_{k=1}^{K} p_k(x) = 1\,; \qquad \forall x. \tag{2}$$

Then from equation (1) and (2) the assignment probabilities are

$$p_k(x) = \frac{\prod_{j \neq k} d_j(x, c_k)}{\sum_{j \in \overline{1,K}} \prod_{i \neq j} d_i(x, c_i)}, \quad k \in \overline{1,K}, \tag{3}$$

and the constant

$$M(x) = p_k(x) d_k(x, c_k) = \frac{\prod_{k=1}^{K} d_k(x,c_k)}{\sum_{i=1}^{K} \prod_{j \neq i} d_j(x,c_i)} \quad (4)$$

$M(\mathbf{x})$ measures the distance of $\mathbf{x}$ from all cluster centers, the constant, *M(D)*, for the whole data set is the sum of (4) over all points,

$$M(D) = \sum_{i=1}^{N} \frac{\prod_{k=1}^{K} d_k(x,c_k)}{\sum_{i=1}^{K} \prod_{j \neq i} d_j(x,c_i)} \quad (5)$$

The solution of the optimization problem (P) defined for a single point is the probabilities in equation (3) and the optimal value is $D(\mathbf{x})$ in equation (4)

(P)

$$Min \sum_{k=1}^{K} p_k^2(x) d_k(x, c_k)$$

$$s.t \sum_{k=1}^{K} p_k(x) = 1;\ p_k(x) \geq 0, \forall k = 1, \dots, K.$$

The optimization problem can be extended to the whole dataset for a given K clusters,

$$Min \sum_{k=1}^{K} \sum_{i=1}^{N} p_k^2(x_i)\, d_k(x_i, c_k)$$

$$s.t \sum_{k=1}^{K} p_k(x_i) = 1, \qquad \forall i = 1 \dots N$$

$$p_k(x_i) \geq 0;\ \forall i = 1 \dots N \text{ and } \forall k = 1 \dots K. \quad (6)$$

The optimal centers in (6) are the convex combination of the data points,

$$c_k = \sum_{i=1}^{N} w_{ki} x_i \quad (7)$$

where

$$w_{ki} = \frac{\frac{p_k^2(x_i)}{d_k(x_i, c_k)}}{\sum_{j=1}^{N} \frac{p_k^2(x_i)}{d_k(x_i, c_k)}}, \quad \forall i = 1, N, \forall k = 1, K$$

Algorithm 1 uses the ideas in equations (3-6) for data clustering. It calculates the distances, probabilities and centers in an iterative fashion until the cluster centers settle down.

Algorithm1	
Initilize	for a given data set D and for a given number of clusters K
	determine centers of clusters randomly
	$\alpha > 0$
Step 1	Calculate the distances, $d_k(\mathbf{x_i}, \mathbf{c_k})$
Step 2	Update the centers, $\mathbf{c'_k}$ using (7)
Step 3	IF $\sum_{k=1}^{K} \|(c'_k - c_k)\| < \alpha$ STOP, ELSE return step 1

3. Classification Uncertainty Function (CUF)

This section defines a new measure for classification uncertainty of a point, named as classification uncertainty function (CUF) based on assignment probabilities in equation (3) and given number of clusters, K.

Definition 1: Let $E(\mathbf{x})$ shows the classification uncertainty of data point $\mathbf{x}$. It is the harmonic mean of the distances $\{d(\mathbf{x}, \mathbf{c_k}) : k \in \overline{1,K}\}$ divided by their geometric mean.

From definition, the CUF is calculated by using only probabilities calculated in the Algorithm 1:

$$E(x) = K\left(\prod_{k=1}^{K} p_k(x)\right)^{\frac{1}{K}} \quad (8)$$

$$0 \le E(x) \le 1$$

To interpret the CUF, consider the two special cases. First, if a point $\mathbf{x}$ is the center of C_k, according to equation (3), $p_k(\mathbf{x}) = 1$, which means that $\mathbf{x}$ is assigned exactly to C_k. In this situation, there is a certainty because there is no another probable cluster for point $\mathbf{x}$, and $E(\mathbf{x}) = 0$. When $p_k(\mathbf{x})$ values are equal for all $k = \overline{1,K}$, there is totally uncertainty since there is no information about the assignment of that data point, and $E(\mathbf{x}) = 1$. As a result, uncertainty about assignment of any point is greater when CUF takes larger values.

Definition 2: Let $E_K(D)$ denotes the CUF for a given data set and for a given number of clusters, K. The CUF for data set is the mean of the CUF values of all data points.

$$E_K(D) = \frac{1}{N} \sum_{i=1}^{N} E(\boldsymbol{x}_i),\ where\ D = \{\boldsymbol{x}_i : i \in \overline{1,N}\} \quad (9)$$

$E_K(D)$ value depends on number of clusters, K, since when K is changed, probabilities and then, CUF values for each data point also alters. Trivially, greater the number of clusters, the smaller uncertainty about assignment of data points to clusters. For example, if N data points is clustered into N clusters then each data point will be cluster center, denoting that $E(\mathbf{x}) = 0, \forall i = \overline{1, N}$ and $E_N(D) = 0$. On the other hand, CUF for the whole data set, $E_1(D)$ will be equal to 1 if all data points are clustered to only one cluster. Figure 1 shows the results of CUF on a two cluster case of using Algorithm 1.

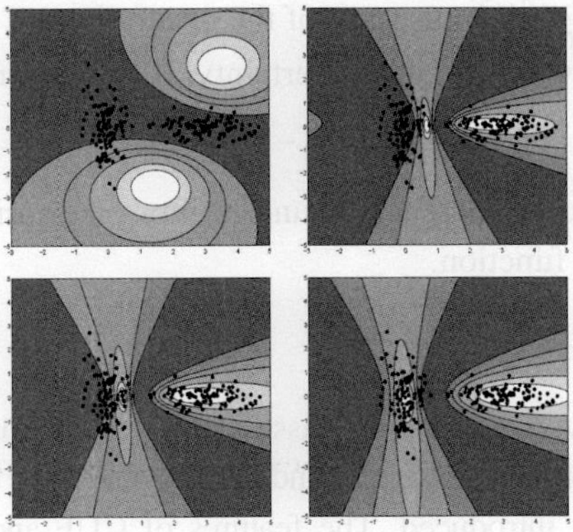

Figure 1. The level sets of the evolving classification uncertainty function at iteration 0 (top left), iteration 2 (top right), iteration 3 (bottom left) and iteration 14 (bottom right)

Original data and the level sets of CUF are shown at iteration 0 (top left) and each colored area denotes the level sets of classification uncertainty. While darker regions correspond to larger values of CUF, lighter regions show smaller values of CUF. After each iteration, data points are located on the lighter areas, where the CUF values getting smaller. The algorithm stops at iteration 14 (bottom right).

4. CUF and Kullback – Leibner Divergence

In this section, we show the relationship between CUF and Kullback-Leibner (KL) divergence.

The CUF in equation (8) can be written as

$$E(\boldsymbol{x}) = \left(\prod_{k=1}^{K}\left(\frac{p_k(\boldsymbol{x})}{1/K}\right)\right)^{1/K}$$

and then the logarithm of the E($\mathbf{x}$) is

$$-\log E(\boldsymbol{x}) = \sum_{k=1}^{K}(1/K)\left(\log\left(\frac{p_k(\boldsymbol{x})}{1/K}\right)\right) = \boldsymbol{I}\left(p(\boldsymbol{x}),\frac{1}{K}\mathbf{1}\right) \quad (10)$$

Then the function in (10) is the KL divergence between the distributions $\mathrm{p}(\mathbf{x}) = (\mathrm{p}_1(\mathbf{x}), \mathrm{p}_2(\mathbf{x}) \ldots \mathrm{p}_K(\mathbf{x}))$ and $(\frac{1}{K}\ \mathbf{1}) = \left(\frac{1}{K}, \frac{1}{K} \cdots \frac{1}{K}\right)$ where the KL divergence is defined as

$$I(p, q) = \sum_{i=1}^{K} p_i \log\left(\frac{p_i}{q_i}\right)$$

The distribution $(\frac{1}{K}\ \mathbf{1})$ is of maximal uncertainty and consequently the divergence $\boldsymbol{I}\left(p(\boldsymbol{x}),\frac{1}{K}\mathbf{1}\right)$is a measure of the uncertainty of the distribution $p(\boldsymbol{x})$. Its smaller values show the greater uncertainty. If (10) is written as,

$$E\ (\boldsymbol{x}) = exp\left\{-\boldsymbol{I}\left(p(\boldsymbol{x}),\frac{1}{K}\mathbf{1}\right)\right\} \quad (11)$$

It shows an entropic measure of uncertainty of classification, which is an increasing monotonic function.

5. Conclusion

We introduced a new measure for classification uncertainty CUF. Probabilistic distance clustering approach, definition and the calculation of CUF for a single point and whole dataset are introduced. The features of CUF and its relation with the entropy (KL Divergence) are discussed. The proposed uncertainty function may be used as a clustering validity criterion; it will be studied as a future work.

References

1. Ben-Israel A. and Iyigun C. Iyigun, 2008. Clustering, classification and contour approximation of data, RUTCOR{Rutgers Center for Operations Research, Rutgers University, 640 Bartholomew Road, Piscataway, NJ 08854-8003, USA
2. Ben-Israel A. and Iyigun C. Iyigun, 2008. Probabilisic D-Clustering, Journal of Classification 25:5-26.
3. Dubois D., Possibility Theory and Statistical Reasoning, Didier Dubois Institut de Recherche en Informatique de Toulouse Universit´e Paul Sabatier, 118 route de Narbonne, 31062 Toulouse Cedex 4, France
4. Uffink J.B.M., 1990. Measures of uncertainty and the uncertainty principle, 12-21.
5. Yager R.R., 2002. Uncertainty representation using fuzzy measures, Systems, Man, and Cybernetics, Part B: Cybernetics, IEEE Transactions on volume 32, 13-20.
6. Zadeh L.A., 2005. Toward a generalized theory of uncertainty (GTU)—an outline, Information Sciences 172, 1–40

A NEW PROCESS MONITORING CONTROL CHART

SU-FEN YANG*

Department of Statistics, National Chengchi University, Teipei, 116, Taiwan

CHUNG-MING YANG

Department of Insurance and Risk Management, Ling-Tung University, Taichung, 408, Taiwan

Using a single chart to monitor simultaneously the process mean and variability if found would cut down the time and effort. Some researches have been done in finding such charts. Moreover, some researches have showed that the adaptive control charts could detect process shifts faster than the traditional Shewhart charts. A much easier loss chart with variable design parameters is proposed here to monitor both the deviation from the process target and the shift in process variability. Furthermore, a more efficient optimal variable design parameters loss chart is proposed and performs better than the loss chart with fixed design parameters and the joint $\overline{\overline{X}}$ and S charts.

1. Introduction

When dealing with a variables quality characteristic, it is more advantageous to use a single chart to monitor process mean and variability simultaneously. A single chart is simple to use and interpret; it would cut down the time, resource, manpower, money and effort. In the last ten years, quite a bit of researches in this area have been done, see Grabov and Ingman (1996), Sullivan and Woodall (1996), Chen and Cheng (1998), Cheng and Spiring (1998), Wu and Tian (2006), Yang et. al (2012) and Yang and Arnold (2012). Cheng and Thaga (2006) gave an overview on many of the single charts developed in 2006.

In industry, quality of products and loss of productivities are important factors among competitive companies. Loss function is widely used in industry to measure the loss due to poor quality (Spiring and Yeung (1998)). From Taguchi's philosophy (Gopalakrishnan, Jaraiedi, Iskander and Ahmad (2007)), the target value is a vital process measurement. Sullivan (1984) gave examples that stress the importance of monitoring the target value and variability. In practice, the in-control process mean may not be the process target. Hence, a

* yang@nccu.edu.tw.

single control chart is effective if it could detect the deviation from the process target and the shift in process variability with smallest loss. Many papers on the performance of adaptive control charts had suggested using adaptive control schemes instead of the fixed control scheme. But, only a few researches proposed the double adaptive charts or a single adaptive chart to simultaneously monitor the process mean and the variance, like Chengular, Arnold, Reynolds (1989), Costa (1998), Reynolds and Stoumbos (2001), Zhang and Wu (2006), Costa and Magalh (2007), Costa and Rahim (2008), Yang and Chen (2009), Yang and Yang (2009), Yang and Yu. (2009), Wu, Wang and Wang (2009) and Prajapati and Mahapatra (2009). So far, no single variable design parameters (VP) control chart detecting both the deviation from the process target and the shift in process variability has ever been proposed. It seems reasonable to consider such a chart. A single VP loss chart is thus proposed to improve the performance of the single loss chart with fixed design parameters (FP), and/or the combined $\overline{X}-S^2$ charts.

2. The Distribution of Taguchi Loss Function

Taguchi quadratic loss function is defined as

$$L = k'E(X-T)^2 \tag{1}$$

where L = loss, k' = coefficient of the loss, X = the quality variable of interest and T = the process target. Let $X\sim N(\mu,\sigma^2)$ when process is in-control. When μ and σ^2 are unknown, the average loss is estimated by

$$\hat{AL} = \frac{\sum_{i=1}^{n}(X_i-T)^2}{n} \tag{2}$$

where n = the sample size. Without lose generalization, let k=1. The in-control distribution of the statistic $\hat{AL}$ is derived as

$$\hat{AL} \sim \frac{\sigma^2\chi^2_{n,\lambda_0}}{n} \tag{3}$$

where χ^2_{n,λ_0} is a non-central chi-square distribution with n degrees of freedom and non-centrality $\lambda_0 = n\delta_3^2$.

Let $X\sim N(\mu+\delta_1\sigma, \delta_2^2\sigma^2)$, $\delta_1 \neq 0$ and $\delta_2 > 1$, when the process is out-of-control. The out-of-control distribution of the estimated average loss is derived as

$$\hat{AL} \sim \frac{\delta_2^2\sigma^2\chi^2_{n,\lambda_1}}{n} \tag{4}$$

where non-centrality $\lambda_1 = n[(\mu+\delta_1\sigma-T)/\sigma]^2 = n[(\delta_1+\delta_3)/\delta_2]^2$.

3. Design of the VP Loss Control Chart

To quickly detect deviation from the process target and variance simultaneously, a VP $\hat{AL}$ chart is designed. The design parameters - the sample size (n_q), the sampling interval (t_q), the warning factor (w_q) and the control factor (k_q) or the false alarm rate (α_q) are all variables. The limits of the VP $\hat{AL}$ chart are:

$$UCL = k_q\sigma^2 \qquad UWL = w_q\sigma^2 \qquad LCL = 0,$$

where, k_q = the control factor of UCL and $k_q = \dfrac{\chi^2_{n,\lambda_0,1-\alpha_q}}{n_q}$, w_q = the warning factor of upper warning limit (UWL), $0 < w_q < k_q$ and α_q = the false alarm rate, $q = 1, 2$.

Two variable combinations of the design parameters, (n_1, t_1, w_1, k_1) and (n_2, t_2, w_2, k_2), are adopted. The small sample size (n_1), long sampling interval (t_1), the corresponding warning factor (w_1) and the corresponding control factor (k_1) or the corresponding false alarm rate (α_1) are adopted when the plotted point fell into the interval (LCL, UWL); the large sample size (n_2), the short sampling interval (t_2), the corresponding warning factor (w_2) and the corresponding control factor (k_2) or the corresponding false alarm rate (α_2) are adopted when the plotted point fell into the interval (UWL, UCL), where $n_1 < n_2$, $t_2 < t_1$, $w_1 < k_1$ and $w_2 < k_2$.

Assuming that the process is in-control at the beginning, the first sample size (n_q) is randomly taken from the process. During the in-control period all samples including the first one, should have a probability of p_{01} to adopt the parameter combination (n_1, t_1, w_1, k_1) and a probability of p_{02} to adopt (n_2, t_2, w_2, k_2), where $p_{01} + p_{02} = 1$ and

$$p_{01} = P\left(\hat{AL} < w_q\sigma^2 \mid \hat{AL} < k_q\sigma^2, \delta_1 = 0, \delta_2 = 1\right) = \frac{F_{\chi^2_{n_q,\lambda_0}}(n_q w_q)}{F_{\chi^2_{n_q,\lambda_0}}(n_q k_q)}, q = 1,2 \qquad (5)$$

where $F_{\chi^2_{n_q,\lambda_0}}(\bullet)$ = the cumulative probability function of a non-central chi-square distribution with n_q degree of freedom and non-central parameter $\lambda_0 = n\delta_3^2$, $p_{02} = 1 - p_{01}$.

To evaluate the performance of the FP and the VP $\hat{AL}$ charts, the adaptive control schemes should be compared under equal conditions; that is, they should demand the same average sample size, the average sampling interval of time and the false alarm rate under the in-control period. Hence, we let

$$n_1 p_{01} + n_2 p_{02} = n_0 \; ; \; t_1 p_{01} + t_2 p_{02} = t_0 \text{ and } \alpha_1 p_{01} + \alpha_2 p_{02} = \alpha_0 \tag{6}$$

Specify the values of μ, σ, δ_1, δ_2, δ_3, T_r, T_f n_0, t_0, k_0 (or α_0) and n_1, n_2, t_2, α_1(or α_1), the k_2, w_1, w_2 and t_1 can be obtained, and the VP $\hat{AL}$ chart are constructed.

4. Performance Measurement of the VP $\hat{AL}$ chart

For a VP chart, the detection speed is measured by the average time from either mean or variance or both shifting until $\hat{AL}$ chart signals, which is known as the adjusted average time to signal (AATS). That is, the AATS is the mean time that the process remains out-of-control.

5. Performance Comparisons Among VP $\hat{AL}$, FP $\hat{AL}$ and $\overline{X} - S^2$ Charts

Considering the specific parameters are δ_1 = (0.5, 1.0, 1.5) , δ_2 = (1.2, 1.5, 2.0), γ_1 = (0.01, 0.05, 0.1), δ_3 = (0, 1.5, 2.5), (T_r, T_f) = ((0.5, 1.0), (1.0, 0.5), (2.0,1.0)), $n_0 = 4$, $t_0 = 1$ and $\alpha_0 = 0.0054$. 27 combinations of the parameters are based on orthogonal array $L_{27}(3^{13})$ table. Results provide the AATS of the optimal VP, FP $\hat{AL}$ charts and $\overline{X} - S^2$ chart under the 27 combinations of the parameters. The optimization technique, applying Fortran IMSL BCONF subroutine, is used to minimize AATS and obtain the optimal design parameter (n_q, t_q, w_q, k_q), q = 1, 2, satisfying the constraints, $2 \le n_1 \le n_0 \le n_2 \le 30$, $0.1 \le t_2 \le t_0 \le t_1 \le 2$, $0.0027 \le \alpha_1 \le \alpha_0$.

Comparing the AATS between the optimal VP and FP $\hat{AL}$ charts, we found that the optimal VP $\hat{AL}$ chart detects the small to medium shifts ($0.5 \le \delta_1 \le 1.5$, $1.2 \le \delta_2 \le 2.0$ and $0 < \delta_3 \le 2.5$) better and saved detection time between 3.01% and 58.71% more than the FP $\hat{AL}$ chart. Comparing the performance between the optimal VP $\hat{AL}$ chart and FP $\overline{X} - S^2$ charts, it can be seen that the optimal VP $\hat{AL}$ chart detects the small to medium shifts ($0.5 \le \delta_1 \le 1.5$ and $1.2 \le \delta_2 \le 2.0$) in process means and variance better and saved detection time between 3.96% and 58.03% more than the $\overline{X} - S^2$ charts. Hence, we recommend the usage of the optimal VP $\hat{AL}$ chart.

6. Conclusions

The proposed VP scheme to monitor the deviation from the process target and variance using a single chart improves the performance of the FP scheme by increasing the speed with small to medium shifts in the mean and variance. We found that the optimum VP $\hat{AL}$ chart always works better (in the cases examined) than the FP $\hat{AL}$ chart and the combined $\overline{X} - S^2$ charts.

We recommend to use the optimum VP scheme in monitoring a process when quality engineers were unable to specify the VPs. This paper proposed a single chart. However, the studies of the variables control scheme could be extended to study CUSUM/EWMA charts or the multiple steps of cascade processes.

Acknowledgements

Support for this research was provided in part by the National Science Council of the Republic of China, grant no. NSC-96-2118-M-004-001-MY2, Commercial College at National Chengchi University, Taiwan, and NCTS, Taiwan.

References

1. Chen, G. and Cheng, S. W. (1998) Max chart: combining X-bar chart and S chart. *Statistica Sinica*, 8(1), 263-272.
2. Cheng, S. W. and Spiring. F.A. (1998) An alternate variable control chart: the univariate and multivariate cases. *Statistica Sinica*, 8(1), 273-287.
3. Cheng, S. W. and Thaga, K. (2006). "On single variables control charts - an overview". *Quality and Reliability Engineering International*, 22(7), 811-820.
4. Chengular, I., Arnold, J. and Reynolds, M. R. (1989) Variable Sampling Intervals for Multiparameter Shewhart Charts. *Communications in Statistics — Theory and Methods* 18, 1769-1792.
5. Costa, A. B. (1998) Joint X and R charts with variable parameters. *IIE Trans* 30(6):505–514
6. Costa, A. B. and Magalh, M. S. (2007) An adaptive chart for monitoring the process mean and variance. *Qual. Reliab. Eng. Int,.* 23(7), 821-831.
7. Costa, A. F. B. and Rahim, M.A. (2008) EWMA chart for monitoring process mean and process variance. *Quality Technology and Quantitative Management* , 3(3), 295-305.
8. Gopalakrishnan, B., Jaraiedi, M., Iskander, W.H., Ahmad, A. (2007) Tolerance synthesis based on Taguchi philosophy. *International Journal of Industrial and Systems Engineering*, 2(3), 311 – 326.

9. Grabov P. and Ingman D. (1996) Adaptive control limits for bivariate process monitoring. *J Qual Technol* 28:320–330.
10. Prajapati, D. R. and Mahapatra, P. B. (2009) Control charts for variables to monitor the process mean and dispersion: a literature review. *International Journal of Productivity and Quality Management.* 4 (4), 476 – 520.
11. [11]. Reynolds, M.R., Jr. and Stoumbos, Z. G. (2001) Monitoring the process mean and variance using individual observations and variable sampling intervals. *Journal of Quality Technology* 33, 181-205.
12. Sullivan, L.P. (1984) Reducing variability: A new approach to quality. *Quality Progress*. July, 15-21.
13. Spiring, F. A. and Yeung, A. S. (1998) A general class of loss functions with individual applications. *J Qual Technol*, 30, 152–162.
14. Sullivan, J. H., and Woodall, W. H. (1996) A comparison of multivariate quality control charts for individual observations. *Journal of Quality Technology*, 265 -278
15. Wu, Z. and Tian, Y. (2006) Weighted-loss-function control charts. *International Journal of Advanced Manufacturing Tech.*, 31, 107-115.
16. Wu, Z., Wang, P. and Wang, Q. (2009) A loss function-based adaptive control chart for monitoring the process mean and variance. *Int. J. Adv. Manuf. Technol.* , 40:948–959.
17. Yang, S. and Chen, Y. (2009) Controlling over-adjusted process means and variances using VSI cause selecting control charts. *Expert System with Application*, 36, 7170-7182.
18. Yang, C and Yang S. (2009) Adaptive control scheme for a process with incorrect adjustment. First Asian Conference on Intelligent Information and Database Systems, Dong hoi, Quang binh, Vietnam , 404-409.
19. Yang, S and Yu, Y. (2009) Monitoring cascade processes. *Journal of Chemometrics*, 23, 449-462.
20. Zhang, S. and Wu, Z. (2006) Monitoring the process mean and variance using a weighted loss function CUSUM scheme with variable sampling intervals. *IIE Transactions*, 38(4), 377 – 387.
21. Yang, S., Cheng, C., Hung, Y. and Cheng, S. (2012) A new chart for monitoring service process mean. Quality and Reliability Engineering. Published online.
22. Yang, S. and Arnold, B. (2012) Signal detection for process with unknown distribution. Proceedings of 2012 International Conference on Frontiers of Manufacturing Science and Measuring Technology, Xi'An, China, 2012.

FORECASTING FUTURE ENROLLMENTS TO UNIVERSITIES OF NORTH CYPRUS THROUGH TREND CENTRALITY BASED ON FUZZY TIME SERIES

EBRU TURANOGLU*

Department of Industrial Engineering, Selcuk University, Selcuklu, Konya, 42031, Turkey

OZLEM SENVAR

Department of Industrial Engineering, Marmara University, Göztepe, Istanbul, 34722,Turkey

In this study, student enrollments to universities of North Cyprus are examined from 2000 to 2009. Future values are forecasted using trend centrality of actual and forecasted enrollments that are obtained by using fuzzy time series.

1. Introduction

Higher education seemed to be a large part of the answer to economic isolation of Turkish Republic of Northern Cyprus (TRNC). It is observed that there is an incremental trend in student numbers in TRNC.

Today, there are more than 40,000 students, which are studying at various universities in TRNC. These students are more than 25.000 from Turkey and more than 4000 from 65 different nationalities.

This paper is organized as follows. Section 2 presents Fuzzy Time Series. Section 3 summarizes the profile of TRNC's higher education. Section 4 includes an application of forecasting enrollments to universities of TRNC using fuzzy time series. The last section gives the conclusion and suggestion for further research.

2. Fuzzy Time Series

In this section, the basic concept of fuzzy time series and the forecasting methods are viewed [1-2]. Let U be the universe of discourse with U={x1, x2, …, xn}. A fuzzy set Ai is of U is defined as:

$$A_i = \mu_{A_i}(x_1)/x_1 + \mu_{A_i}(x_2)/x_2 + \mu_{A_i}(x_3)/x_3 + \cdots + \mu_{A_i}(x_n)/x_n \qquad (1)$$

* eturanoglu@selcuk.edu.tr.

where μ_{A_i} is the membership function of the fuzzy set A_i, such that $\mu_{A_i}: U \to [0,1]$ and $\mu_{A_i}(x_j)$ represents the grade of membership of x_j in A_i, $\mu_{A_i}: \in \to [0,1]$.

Definition 2.1 Let $X(t)$ a subset of $\Re$, be the universe of discourse on which fuzzy sets, $f_i(t)(i = 1,2,\ldots)$ are defined $F(t)$ is the collection $f_i(t)$, then $F(t)$ is defined as fuzzy time series on $X(t)(t =\ldots,0,1,2,\ldots)$. Here, $F(t)$ be considered as a linguistic variable and $f_i(t)(i = 1,2,\ldots)$ as possible linguistic values of $F(t)$. As at different times, the values of $F(t)$ can be different, $F(t)$ is a function of time and a similar way, the universe of discourse *X(t)* may be different at different times.

Definition 2.2 If for any $f_j(t) \in F(t)$ and $f_j(t-1) \in F(t-1)$ there exist a fuzzy relation $R_{ij}(t,t-1)$ such that

$$f_i(t) = f_i(t-1) \circ R_{ij}(t,t-1) \tag{2}$$

where "∘" is the max-min composition operator, then $F(t)$ is said to be caused only by $F(t-1)$ and is denoted by $f_j(t-1) \to f_j(t)$ or by $F(t-1) \to F(t)$. This relation is called first order relation.

Definition 2.3 If for any $f_j(t) \in F(t)$ and $f_j(t-1) \in F(t-1)$ there exist a fuzzy relation $R_{ij}(t,t-1)$ such that $f_i(t) = f_i(t-1) \circ R_{ij}(t,t-1)$ and let $R(t,t-1) = \cup_{ij} R_{ij}(t,t-1)$, where $\cup$ is union operator. Then $R(t,t-1)$ is called the fuzzy relation between $F(t)$ and $F(t-1)$ and define as a fuzzy relation equation

$$F(t) = F(t-1) \circ R(t,t-1) \tag{3}$$

Definition 2.4 Suppose $F(t)$is caused only by $F(t-1)$ or $F(t-2)$ or …or $F(t-m)$ (m > 0), then this relation can be expressed as fuzzy relational equation

$$F(t) = F(t-1) \circ R(t,t-m) \tag{4}$$

or

$$F(t) = (F(t-1) \cup F(t-2) \cup \ldots \cup F(t-m)) \circ R(t,t-m) \tag{5}$$

then (4) or (5) is called first order model.

Definition 2.5 Suppose $F(t)$is caused only by $F(t-1)$ or $F(t-2)$ or …or $F(t-m)$ (m > 0) simultaneously, then relation can be expressed in fuzzy relation equation as

$$F(t) = (F(t-1) \times F(t-2) \times \ldots \times F(t-m)) \circ R_m(t,t-m) \tag{6}$$

Eq. (6) is called the *m*th order model $F(t)$.

Definition 2.6 Suppose $F(t)$, a fuzzy time series and $R(t,t-1)$ be a first order model of $F(t)$ such that $R(t,t-1) = R(t-1,t-2)$ for any time t, the $F(t)$ is called time-invariant fuzzy time series. But if $R(t, t-1)$ is time dependent i.e. $R(t, t-1)$ may be different from $R(t-1, t-2)$ for any time t then $F(t)$ is called time-variant fuzzy time series.

Some studies about forecasting methods based on fuzzy time series are proposed by Song and Chissom [3]; Wang and Hsu [4] and Li et al. [5].

3. Higher Education in TRNC

Private universities began to spring up around North Cyprus in turn. Girne American University (established in 1985), Near East University (established in 1988), European University of Lefke (established in 1989), and Cyprus International University (established in 1997). In addition to these universities, there are private campuses of Turkish state universities. Middle East Technical University North Cyprus campus (METU-NCC) established in 2002. In 2006, Istanbul Technical University (ITU) has signed a protocol with the TRNC Ministry of Education and Culture to open campuses in TRNC. Also, Gazi University is performing the initiatives of opening campus that provides education in healthcare.

4. An Application

In this section, it is employed the forecasting model proposed by Hwang et al.'s [6] to deal with the forecasting enrollments to universities of TRNC. This method is now presented as follows:

Step 1: From the historical enrollment data of universities of TRNC, it is computed the variations of the enrollments between any two continuous years. The variation of this year is the enrollment of this year minus the enrollment of last year and the obtained the variations of the enrollments between any two years shown in Table 1. After it is found the minimum increase Dmin and maximum increase Dmax, it is defined the universe of discourse U, in this case U can be represented as U=[-5500, 6500].

Table 1. Actual enrollments and variations of historical data

Years	Actual Enrollments	Variations	Years	Actual enrollments	Variations	Years	Actual Enrollments	Variations
1995	16651	2611	2000	25765	556	2005	41865	-3066
1996	19262	1582	2001	26321	1427	2006	38799	-3066
1997	20844	654	2002	27748	2857	2007	43021	4222
1998	21498	2433	2003	30605	4868	2008	45634	2613
1999	23931	1834	2004	35473	6392	2009	40431	-5203

Step 2: It is separated the universe of discourse U into six intervals, where u1=[-5500, -3500], u2=[-3500, -1500], u3=[-1500, -500], u4=[-500, 2500], u5=[2500, 4500], and u6=[4500, 6500].

Step 3: It is determined some linguistics values represented by fuzzy sets to describe the degree of variation between two continuous years. In this study it is considered six fuzzy sets which are A_1=(big decrease), A_2=(decrease), A_3=(no change), A_4=(increase), A_5=(big increase), A_6=(too big decrease). It is defined fuzzy sets $A_1, A_2, \ldots, A_6$ on the universe of discourse U as follows:

$$A_1 = {}^{1}/_{u_1} + {}^{0.5}/_{u_2} + {}^{0}/_{u_3} + {}^{0}/_{u_4} + {}^{0}/_{u_5} + {}^{0}/_{u_6}, A_2 = {}^{0.5}/_{u_1} + {}^{1}/_{u_2} + {}^{0.5}/_{u_3} + {}^{0}/_{u_4} + {}^{0}/_{u_5} + {}^{0}/_{u_6}, A_3 = {}^{0}/_{u_1} + {}^{0.5}/_{u_2} + {}^{1}/_{u_3} + {}^{0.5}/_{u_4} + {}^{0}/_{u_5} + {}^{0}/_{u_6}, A_4 = {}^{0}/_{u_1} + {}^{0}/_{u_2} + {}^{0.5}/_{u_3} + {}^{1}/_{u_4} + {}^{0.5}/_{u_5} + {}^{0}/_{u_6}, A_5 = {}^{0}/_{u_1} + {}^{0}/_{u_2} + {}^{0}/_{u_3} + {}^{0.5}/_{u_4} + {}^{1}/_{u_5} + {}^{0.5}/_{u_6}, A_6 = {}^{0}/_{u_1} + {}^{0}/_{u_2} + {}^{0}/_{u_3} + {}^{0}/_{u_4} + {}^{0.5}/_{u_5} + {}^{1}/_{u_6} \quad (7)$$

Step 4: It is fuzzified the values of historical data. If the number of variation of the enrollment of year i is p, where $p \in u_j$, and if there is a value represented by fuzzy set A_k in which the maximum membership value occurs at u_j, then p is translated to A_k.

Step 5: It is chosen a suitable window basis, it is calculated the output from the operation matrix $O^w(t)$ and the criterion matrix $C(t)$, where t is the year for which wanted to forecast the enrollment, and then calculated the relation matrix $R(t)$ by $R(t)[i,j] = O^w(t)[i,j] \times C(t)[j]$ where $1 \le i \le w$, and $1 \le j \le m$. The fuzzified forecasted variations from 2000 to 2009 can be calculated and all results are listed in Table 2.

Table 2. Forecasted variations with the window basis w=5

Years	Membership functions of forecasted variations						Years	Membership functions of forecasted variations					
	u_1	u_2	u_3	u_4	u_5	u_6		u_1	u_2	u_3	u_4	u_5	u_6
2000	0	0	0.25	1	0.5	0	2005	0	0	0	0	0.5	0.5
2001	0	0	0.25	1	0.5	0	2006	0	0	0	0	0.5	1
2002	0	0	0.25	1	0.25	0	2007	0	0	0.25	0	0	0
2003	0	0	0.25	1	0.25	0	2008	0	0	0	0.25	1	0.5
2004	0	0	0	0.5	0.5	0	2009	0	0	0	0.25	1	0.5

Step 6: It is defuzzified the fuzzy forecasted variations derived in Step 5.

Step 7: It is calculated the forecasted enrollments which are forecasted variation plus the number of actual enrollment of last year. The results of the forecasted enrollment of the universities of TRNC and the following error of each year under the window basis w=5 is also shown in Table 3.

Table 3. Forecasted results using the proposed by Hwang et al.'s fuzzy time series method with window

Years	Actual Enrollments	Forecasted Enrollments	Errors	Years	Actual Enrollments	Forecasted Enrollments	Errors
2000	25765	27265	5.82	2005	41865	46365	10.75
2001	26321	27821	5.70	2006	38799	44299	14.18
2002	27748	29248	5.41	2007	43021	42521	1.16
2003	30605	32105	4.90	2008	45634	49134	7.67
2004	35473	37973	7.05	2009	40431	43931	8.66

From Table 4 we can see that the average forecasting errors for different window bases range from 7.86% down to 5.86%. The best average forecasting error value is obtained when the window basis equals to 4.

Table 4. Forecasting errors with different window bases

	Window Bases							
	w2	w3	w4	w5	w6	w7	w8	w9
Average Forecasting Errors	7.10	6.43	5.86	6.90	7.02	7.18	7.43	7.86

Step 8: Future values are forecasted using trend centrality of actual and forecasted enrollments that are obtained by using fuzzy time series. The forecasted enrollments of universities of TRNC from 2010 to 2015 are shown in Table 5.

Table 5. Forecasting enrollments from 2010 to 2015

	Years					
	2010	2011	2012	2013	2014	2015
Forecasting Enrollments	49900	50024	49581	48362	46286	44190

5. Conclusion

The technique used in this study give good prediction values. For future studies, fuzzy regression and fuzzy exponential smoothing techniques can be used.

References

1. S.M. Chen, Forecasting enrollments based on fuzzy time series, Fuzzy Sets and Systems, 81 (1996) 311–319.
2. L.A. Zadeh, The concept of a linguistic variable and its application to approximate reasoning Part I, Information Sciences, 8 (1975) 199–249.
3. Q. Song, and B.S. Chissom, Forecasting enrollments with fuzzy time series – Part II, Fuzzy Sets and Systems, 62 (1994) 1–8.
4. C-H. Wang and L-C. Hsu, , Constructing and applying an improved fuzzy time series model: Taking the tourism industry for example, Expert Systems with Applications, 34 (2008) 2732-2738.
5. S-T. Li, S-C. Kuo, Y-C. Cheng, & C-C. Chen, A vector forecasting model for fuzzy time series, Applied Soft Computing, 11 (2011) 3125-3134.
6. J-R. Hwang, S-M. Chen, Y-C. & C-H. Lee, Handling forecasting problems using fuzzy time series, Fuzzy Sets and Systems 100 (1998) 217-228.

FUZZY EXPERT SYSTEM FOR FORECASTING RETURN AMOUNT IN REVERSE LOGISTICS

GÜL TEKİN TEMUR*

Department of I Management Engineering, Istanbul Technical University, Macka, Istanbul, 34367, Turkey

MUHAMMET BALCILAR

Department of Computer Engineering, Yıldız Technical University, Davutpasa, Istanbul, 34220, Turkey

BERSAM BOLAT

Department of Management Engineering, Istanbul Technical University, Macka, Istanbul, 34367, Turkey

Depending on decrease in natural resources and increase in perception of people on environment, governments have begun to attach much importance to legal regulations for protecting environment. In consequence of legal enforcements and customer perception, all actors in the business area are forced to operate recovery activities for getting value from used or scrapped products. In reverse logistics (RL) which includes the control of backward move of goods, for the actors it becomes hard to make scheduling regarding to amount and time of returned products. Because in the uncertain nature of RL, it is not easy to know how much products will return at a given time. Therefore, in this study a fuzzy expert system is proposed for forecasting return amount. An application at an electrical and electronic equipment recycling (e-recycling) facility is conducted for clarifying how the method is used in a real decision process.

1. Introduction

Environmental regulations which force companies to organize the backward move of used or scrapped products have become a major concern in plans of organizations. By following the business strategy called as reverse logistics (RL), companies collect used or scrapped products from consumers in order to recover them for increasing sustainability, total profit and productivity [1]. In various industries ranging from electronics to beverage, the importance of RL

* temurg@itu.edu.tr.

approaches has been increased [2]. But in practical, RL implementation has a hard nature, because there are many unknown parameters such as product return quantity, quality and time. In order to accomplish difficulties resulted from uncertainty, the factors affecting return amount in RL should be determined and an effective forecasting system that provides realistic results for decision makers should be improved. Thus, in this study, a forecasting model is proposed by using fuzzy expert system. A case study with real data is applied in an electrical and electronic equipment facility. This study differs from other forecasting studies in RL literature in that it uses fuzzy expert system successfully and is applied to a real case study.

2. Product Return Amount Forecasting

A quick look at RL literature reveals that current studies [3,4,5] usually do not address forecasting return amount from a comprehensive viewpoint. It is usually supposed that product return is a function of one or a few factors. But interviewing with managers and looking at factors complied from the waste of electrical and electronic equipments directive [6] and different studies in RL literature show that there are many quantitative and qualitative factors. This study contributes to the literature by focusing on return amount forecasting that considers more different factors compared to current studies.

3. Fuzzy Expert System Methodology

Besides methods such as least square [7] and partial least square [8], there are artificial neural network (ANN) based systems [9] for developing forecasting system which works depending on generalizing gathered data. However, these methods can be unsuccessful at giving satisfactory results if the train set contains less data. In order to deal with disadvantage of less data, expert systems are used [10]. In this study, a fuzzy expert system which models fuzzy expressions of experts is used.

3.1. *Fuzzy Logic*

In real life, it is not possible to evaluate human-based preferences with exact values. In 1965, Zadeh [11] proposed fuzzy set theory to deal with the uncertainty resulting from human-based judgments. Let be an element in a universe set and is a membership function that defines the membership degree of element to set. In classical set theory, the membership function gets the value 0 or 1, but in fuzzy set theory, the membership function can get a value between 0 and 1 as shown in Eq. (1).

$$0 \leq \mu_A(x) \leq 1 \quad x \in X \tag{1}$$

3.2. *Fuzzy Rule Base*

In fuzzy set literature, two types of fuzzy rule models are defined which are called the Mamdani and Sugeno models [12]. The output of the Mamdani rule model is determined by the fuzzy set, while the output of the Sugeno rule model is determined by the combinations of linear and crisp functions. Let k be the number of rules in the rule base, n is the number of input parameters, A_i^n is the fuzzy set which belongs to ith rules of nth input parameter, B_i is the fuzzy set of ith rule's output, x_i is the ith input parameter and y is the output of fuzzy rule. The Mamdani rule base is given in Eq. (2).

$$If\ x_1 \in A_i^1\ and\ x_2 \in A_i^2\ and \ldots..x_n \in A_i^n\ then\ y \in B_i \quad i = 1..k \tag{2}$$

3.3. *Fuzzy Inference Method*

A crisp result of fuzzy expert systems is produced by a fuzzy inference system. In a case where the rule base is Mamdani type; Mamdani, Larsen or Tsukamoto inference systems are preferred. On the other hand, when the rule base is Sugeno, the Takagi-Sugeno-Kang (TSK) inference method is used for output. Let the rule base be such as Eq.(2), where α is the minimum membership value of the antecedent fuzzy set, C_i is the output fuzzy set of ith rule depending on input parameters and z is the crisp output. The Mamdani inference method is given in Eq. (3,4,5).

$$\alpha_i = \min\left(\mu_{A_i^1}(x_1), \mu_{A_i^2}(x_2) \ldots \mu_{A_i^n}(x_n)\right) \quad i = 1..k \tag{3}$$

$$\mu_{C_i}(x) = \min(\mu_{B_i}(x), \alpha_i) \quad i = 1..k \tag{4}$$

$$z = \sum_{i=1}^{k} \int x.\mu_{C_i}(x).dx / \sum_{i=1}^{k} \int \mu_{C_i}(x).dx \tag{5}$$

The Centre of Gravity (COG) defuzzification step is shown in Eq. (5). Also other defuzzification methods such as indexed center of gravity, basic defuzzification distributions, and maxima methods are used.

3.4. *Learning*

In most cases, the rule of experts does not directly generalize the training set. For this reason, the values of parameters which define membership functions must be adjusted. To change the membership functions definition to converge to the desired output value and predicted value, derivative-based optimization methods such as gradient descent, Levenberg-Marquardt [13] and evolutionary-based methods [14] such as genetic algorithms are used. Let X^i be i^{th} input vector which consists of input scalars, θ is a vector that consists of the membership function's adjustable parameter, and $f(X^i, \theta)$ is the fuzzy expert system's output, mean square error of prediction E is shown as in Eq. (6). The aim of the learning step is to find θ which minimizes error function.

$$E(\theta) = \sum_{i=1}^{k}\left(y^i - f(X^i,\theta)\right)^2 \qquad (6)$$

3.5. *Validation and Performance Metrics*

The most well-known validation methods are holdout validation, k-fold cross validation and leave-one-out cross validation methods. According to the k-fold cross validation method, n number data are divided randomly by k number set in the form of each cluster having equal elements, and each cluster is labeled from 1 to k, as given in Eq. (7).

$$v_i \in \{1,2,..k\} \quad i = 1..n \qquad (7)$$

Firstly the data whose v_i value is 1 are taken to the test set and data whose value is different from 1 are taken to the training set. The suggested method is trained with the training set and examined with the test set. Then data whose v_i value is 2 are sent to the test set, data whose v_i value is different from 2 are sent to the training set. The method is retrained with the new training set and examined with the new test set. This process continues until all folds are tested [15]. Correlation coefficient (R) and mean square error (MSE) are used as a metric that shows the success of the proposed approach.

4. Application and Results

A case study considering real data is conducted in an e-recycling facility which plans to extend its recycling operations to different locations. It is assumed that by realizing enlargement strategy, the facility will collaborate with 19 new cities. The aim is to forecast return amount of new cities. The most important factors are defined by conducting interviews with managers of the firm as: *population density (pop), electronic waste quantity per inhabitant (ew), investment on environment per inhabitant (inv), income (gdp),* and *take-back price (prc)* for four different electronic product groups. It is assumed that the return amount of each product group has no interaction with each other. Furthermore, models are explored for each product group separately. There are 15 existing cities, and then vector X which defines inputs can be given as Eq. (8).

$$X^i = \{pop^i, ew^i, inv^i, gdp^i, prc^i\} \quad i = 1 \dots 15 \tag{8}$$

The return amount of ith city and j[th] product type can be shown as y_j^i.

The Mamdani type rule base and inference method was used in Eq. (5). The COG method was used as the defuzzification method in application. The used rules which were established by reverse logistics experts by the way of interview are shown in Table 1. All antecedent components in this rule are connected each other with the "and" conjunction.

Table 1 Mamdani Type Rule Base

Rules	pop^i	ew^i	inv^i	gdp^i	prc^i	y^i
R1	Low	Low	Low	Low	Low	Low
R2	Medium	Medium	Medium	Medium	Medium	Medium
R3	High	High	High	High	High	High

All inputs are normalized on [0 1] interval. Gaussian curve function is used for membership in Eq. (9) and standard deviation of all membership functions are fixed as 0.1. At the beginning mean value parameters of membership functions $\theta_{Low}, \theta_{Medium}, \theta_{High}$ are set 0, 0.5, and 1 respectively.

$$\mu_\theta = e^{\frac{-(x-\theta)}{2\,0.1^2}} \tag{9}$$

Table 2 Results of Fuzzy Expert Forecasting Approach

City	pop	ew	inv	gdp	prc_1	f_1	f_2	prc_3	f_3	prc_4	f_4
1	19.0	3.5	92.5	1158.4	0.0	0.0	183.0	0.3	12923.6	0.2	1073.6
2	26.0	5.9	67.4	2012.3	0.0	0.0	147.9	0.3	700.5	0.2	0.0
3	30.0	2.8	94.3	1061.3	0.0	0.0	152.1	0.3	12904.6	0.2	2719.7
4	33.0	10.2	18.4	4215.6	0.0	0.0	146.0	0.3	31895.2	0.2	0.0
5	38.0	5.5	27.8	1951.3	0.0	0.0	146.8	0.3	25903.7	0.2	0.0
6	52.0	6.5	38.7	2116.9	0.0	0.0	151.3	0.3	10862.4	0.2	0.0
7	52.0	8.0	34.7	2584.4	0.0	0.0	149.2	0.3	34492.0	0.2	0.0
8	55.0	8.9	117.6	2512.6	0.0	487.0	355.5	0.3	869.5	0.2	0.0
9	61.0	9.5	74.2	2724.7	0.0	487.0	263.0	0.3	2126.7	0.2	0.0
10	65.0	5.3	101.1	1704.3	0.0	0.0	162.1	0.3	8550.5	0.2	0.0
11	72.0	6.3	44.4	1806.3	0.0	0.0	411.0	0.3	674.5	0.2	0.0
12	80.0	6.8	48.8	2132.8	0.0	63.8	161.2	0.3	1199.2	0.2	0.0
13	81.0	5.5	60.3	1215.7	0.0	0.0	96.4	0.3	645.3	0.2	1973.4
14	102.0	3.7	102.7	1313.4	0.0	0.0	152.0	0.3	12954.6	0.2	1335.9
15	105.0	7.4	48.2	2458.6	0.0	50.9	147.8	0.3	36739.1	0.2	0.0
16	106.0	8.3	65.6	2452.0	0.0	487.0	211.1	0.3	1116.4	0.2	0.0
17	138.0	4.4	68.6	1679.7	0.0	0.0	146.4	0.3	663.1	0.2	146.2
18	188.0	7.6	48.0	2969.5	0.0	47.8	146.0	0.3	36708.8	0.2	208.7
19	249.0	5.6	80.7	1592.5	0.0	0.0	532.9	0.3	702.7	0.2	1800.6

In application there are 18 fuzzy set (6 different set for each of low, medium and high set in Table 1). Hence, adjustable parameters vector of membership function θ have 18 scalar numbers. Genetic algorithm was used to determine θ

parameters which minimize Eq. (6). The selected values of genetic algorithm are double vector type population, 200 population sizes, rank scaling function, roulette selection function, scattered crossover function, uniform mutation function, 0.01 mutation rate, hybrid search with Nelder-Mead method and maximum 200 epochs. 5-fold cross validation was applied for validation stage. R value of test component was calculated as 0.8118, and MSE value was calculated as 0. 0072. All results can be seen at Table 2. The results can help managers check their capacities and plan their future operations by considering results of fuzzy expert forecasting approach.

5. Conclusion

This study proposes a return amount forecasting system using fuzzy expert system in order to decrease uncertainty in RL. By training data of existing cities, the return amounts of new cities are forecasted. The performance metrics of the proposed model are found as satisfactory. That means, decision makers can benefit from this system as a support tool when forecasting return amount of selected new areas. In the future, the proposed model can be used for forecasting other uncertainty parameters such as return quality and return time. In addition to this type-2 fuzzy sets can be used for increase prediction success.

References

1. S. Dowlatshahi, *Interfaces,* **30**, 143-155 (2000).
2. M.P. De Brito and R. Dekker, Rev.log.: quant. models for clsc (2004).
3. M. Klausner and C. Hendrickson, *Interfaces*, **30**, 3, 156–165 (2000).
4. B.L. Toktay, E.A. van der Laan, and M.P. De Brito, Rev.log.: quant. models for clsc (2004).
5. J. Hanafi,, S. Kara and H. Kaebernick. Proceedings of IEEE International Symposium on Electronics and the Environment, 245-250, (2007).
6. European Parliament and Council, Directive 2002/96/EC on WEEE (2003).
7. H. W. Bode and C. E. Shannon, IRE Proceedings, **38**, 4, 417-425 (1950).
8. Y. Zhou, J. Wu, and F. Qin, Int. Cont. and Aut. Congress, 2777 (2006).
9. S. Ekinci, U. B. Celebi, M. Bal, M. F. Amasyali, and U. K. Boyaci, *Applied Soft Computing*, **11**, 2, 2356-2366, (2011).
10. W. Siler and J. J. Buckley, Fuzzy Expert Systems and Fuzzy Reasoning (2005).
11. Zadeh, L. A Fuzzy sets. *Information Control*, **8**, 338–353 (1965).
12. J. S. R. Jang, C. T. Sun, and E. Mizutani, Neuro-Fuzzy and Soft Computing (1997).
13. J. Nocedal and S. J. Wright, Numerical Optimization (1999).
14. M. Gabryel and L. Rutkowski, *Learning*, 354-359 (2006).
15. E. Alpaydin, *Introduction to Machine Learning*, **56**, 2, 187 (2004).

PART 5

INTELLIGENT INFORMATION PROCESSING

RAINFALL-RUNOFF MODELING USING ARTIFICIAL NEURAL NETWORKS AND ADAPTIVE NEURO-FUZZY INFERENCE SYSTEM MODELS

GHOLAMALI GHAFARI

Department of Watershed Management Engineering, Faculty of Natural Resources, Tarbiat Modares University, P.O.Box 46417-76489, Noor, Mazandaran Province, Iran

MEHDI VAFAKHAH*

Department of Watershed Management Engineering, Faculty of Natural Resources, Tarbiat Modares University, P.O.Box 46417-76489, Noor, Mazandaran Province, Iran

In this study, rainfall-runoff modeling was carried out in Hajighoshan watershed using artificial neural network (ANN) and adaptive neuro-fuzzy interface system (ANFIS) with different inputs(current day rainfall; current rainfall and pervious day rainfall; current rainfall, pervious day rainfall and two previous day) methods. Root mean squared error (RMSE), mean absolute error (MAE) and correlation coefficient (R) statistics are employed to evaluate the performance of the ANNs and ANFIS models for forecasting runoff. Comparison of the obtained results reveals that the ANFIS model outperforms the ANN models. Based on the results of test stage, ANFIS with RMSE=7.11 m3 s-1, MAE=2.18m3 s-1 and R=0.60 is superior to rainfall-runoff modeling than the ANN with RMSE=6.03m3 s-1, MAE=1.97 m3 s-1 and R=0.39.

1. Introduction

Short-term and long-term forecasting of streamflow is very significant for developing operational flood forecasting systems, reliable operation of flood mitigation reservoir systems, prediction of sediment amounts carried by rivers to reservoirs and optimally managing water resources. Due to the abilities of time series, ANN and ANFIS models in modeling complex nonlinear systems, successful applications of these methods in rainfall-runoff modeling have been extensively reported (Ahmat Nor[6]; Sajjad Khan and Coulibaly[7]; Dorum *et al.* [8]; Nourani *et al.* [9]). A comprehensive review of the application of ANNs to hydrology can be found in the ASCE Task Committee [4, 5]. However, soft computing applications in streamflow forecasting still has been advancing to

* E-mail: vafakhah@modares.ac.ir

provide novel and robust. The aim of this paper is to investigate the performance of ANN and ANFIS models for rainfall-runoff modeling.

2. Description of the Models

The most commonly used neural network structure is feed-forward back propagation network (FFBP) [3]. The feed-forward back propagation network (FFBP) details can be found in Hagan and Menhaj [2]. The methodology used for adjusting the weights of the ANN model was Broyen-Fletcher-Golofrab-Shanno (BFGS). Sigmoid, hyperbolic tangant and linear activation functions were used for the hidden and output node (s), respectively. The ANFIS used in the study is a fuzzy inference model of Sugeno type, and is a composition of ANNs and fuzzy logic approaches [1]. The root mean squared error (RMSE), mean absolute error (MAE) and correlation coefficients(R) were considered as statistical performance evaluation.

3. Description of the Study Area and the Data

The time series of daily streamflow data collected from station of Hajighoshan (station no: 12-063, 55° 21′E, 37° 24′ N) on the Gorgan River operated by Iranian Water Research Institute were used in this study. The used data spans a period of 26 years from 1982 to 2008(9497 days) for the mentioned station. The rainfall data were comprised the observations belonging to four meteorological stations (Houtan, Maravehtapeh, Tamar, Cheshmehkhan and Tangrah climatologic stations). The average rainfall of Hajighoshan watershed was computed using Thiessen polygon.

In the modeling process, the data sets of streamflow and rainfall were scaled to the range between 0.1 and 0.9 for ANN and ANFIS models as follow:

$$N_i = 0.8 \times \left(\frac{x_i - x_{\min}}{x_{\max} - x_{\min}} \right) + 0.1 \qquad (1)$$

Where N_i is the normalized value, x_i is the original data and $x_{\min}$, $x_{\max}$ are, respectively, the minimum and maximum of streamflow and rainfall. After examining the data and noting the periods in which there were gaps in the streamflow and rainfall variables, the periods for training and testing were chosen. The 19 year rainfall and streamflow data are used to train the ANFIS and ANN models and the remaining 7 year records are used for testing. For the Hajighoshan Station, the daily flow statistics of training, test and entire data set are presented in Table1.

Table 1. The statistical characteristics of the daily rainfall and streamflow data

Variable	Data set	Numbers of data	Average	Standard deviation	Maximum amount	Minimum amount	Skewness
Rainfall (mm)	Training	6940	1.44	4.12	53.37	0	4.78
	Test	2557	1.69	4.79	46.04	0	4.60
	Entire	9497	1.51	4.31	5.37	0	4.75
Streamflow ($m^3 s^{-1}$)	Training	6940	1.94	6.64	248	0	17.78
	Test	2557	2.12	8.74	267	0	19.13
	Entire	9497	1.99	7.27	267	0	18.89

It can be seen from table that the rainfall and streamflow data show significantly high skewed distribution.

4. Results and Discussions

In this study, feedforward back-propagation (FFBP) and ANFIS models was accomplished using STATISTICA software and algorithm written in MATLAB, respectively. Hidden layer unit number was found separately for each of the input layer scenarios. The partial correlation analyses of the data of rainfall are employed for selecting appropriate input vectors in ANN and ANFIS runoff estimation models. The partial and cross correlation statistics from lag 0 to lag10 are estimated for the rainfall and rainfall-streamflow (Table 2).

Table 2.The partial auto-correlation function (PACF) and cross-correlations

Lag	0	1	2	3	4	5
Rainfall(mm)	-	0.22	-0.05	0.003	-0.0009	0.013
Rainfall(mm)-Streamflow($m^3 s^{-1}$)	0.249	0.027	0.009	0.001	-0.002	-0.003

It is obvious from Table 2 that the partial auto-correlation function (PACF) indicated significant correlation at lag-1 and lag-2 after lag-2, fell within the confidence limits. The partial autocorrelation coefficients suggested incorporating daily rainfall values up to 2 day-lag in input vector to the ANN and ANFIS models. Considering the correlation analyses, two input combinations based on previous daily rainfall were evaluated to estimate current flow value. The input combinations evaluated in the study were; (i) Rt, (ii) Rt, Rt-1 and Rt-2, (iii) Rt, Rt-1 and Rt-2. In all cases, the output was the streamflow Qt for the current. The RMSE and R statistics of ANN models in training and test periods as well as the optimum ANN structures are given in Table 3.

By studying Table 3, it is obvious that the ANN (3-8-1) model used at the Hajighoshan station whose inputs were the rainfalls of two previous days (input combination iii), has the smallest RMSE, MAE and the highest R. Here, the

ANN (3-8-1) denotes an ANN model comprising 3 inputs, 8 hidden and 1 output nodes. The training parameters of ANFIS model are given in Table 4.

Table 3. The RMSE, MAE and R statistics of ANN models

Model inputs	ANN structure	Activation function		Training period			Test period		
		Hidden layer	Output layer	RMSE (m3/s)	MAE	R	RMSE (m3/s)	MAE	R
Rt	1-10-1	hyperbolic tangant	Linear	5.15	1.80	0.33	10.38	2.15	0.27
Rt,Rt-1	2-4-1	hyperbolic tangant	Linear	6.62	1.82	0.44	6.60	1.87	0.31
Rt,Rt-1,Rt-2	3-8-1	hyperbolic tangant	Linear	**6.94**	**2.01**	**0.39**	**6.03**	**1.97**	**0.39**

Table 4.The training parameters of the ANFIS

Parameters	Hajighoshan Station
Membership function	trimf, gbellmf, gaussmf, gauss2mf
AND method	Prod
Or method	Maximum
Imp. method	Prod
Aggr. method	Maximum
Defuzzification method	wtaver

Table 5 The RMSE, MAE and R statistics of ANFIS models

Model inputs	Numbers of the rules	Membership function	Training period			Test period		
			RMSE (m3/s)	MAE	R	RMSE (m3/s)	MAE	R
Rt	2	trimf	6.87	1.90	0.00	8.46	2.32	0.25
		gbellmf	6.90	1.91	0.00	8.47	2.32	0.25
		gaussmf	6.89	1.90	0.00	8.46	2.32	0.25
		gauss2mf	6.90	1.91	0.00	8.48	2.32	0.25
Rt,Rt-1	4	trimf	6.70	1.86	0.18	8.07	2.24	0.20
		gbellmf	6.74	1.87	0.18	12.45	251	0.03
		gaussmf	6.73	1.87	0.18	9.33	2.39	0.16
		gauss2mf	6.71	1.87	0.18	10.60	2.45	0.05
Rt,Rt-1,Rt-2	8	trimf	**6.62**	**1.82**	**0.22**	**7.11**	**2.18**	**0.60**
		gbellmf	6.65	1.81	0.22	13.90	2.62	0.00
		gaussmf	6.64	1.81	0.22	10.91	2.51	0.00
		gauss2mf	6.60	1.81	0.23	10.27	2.42	0.10

The RMSE and R statistics of ANFIS models in training and test periods are given in Table 5. By studying Table 5, we can see that the ANFIS model used at the Hajighoshan station whose inputs were the rainfalls of two previous days (input combination iii) and member function of trimf, has the smallest RMSE, MAE and the highest R. Also, by studying Tables 3 and 5, it is obvious that the ANFIS models seem to be more accurate than the ANN models in test period.

Figure 1 demonstrates the streamflow forecasts of the ANN and ANFIS models in the test period. There, the ANN and ANFIS models predict the maximum peak as $14.22 m^3s^{-1}$ and $137.07\ m^3s^{-1}$ instead of measured $141 m^3s^{-1}$ and $267\ m^3s^{-1}$ with underestimations of 89.90% and 48.66%. However, the ANN and ANFIS prediction of the second maximum peak $133 m^3s^{-1}$ and $208 m^3s^{-1}$ are $3.88 m^3s^{-1}$ and $7.02 m^3s^{-1}$, respectively with underestimations of 97% and 96.62%, respectively. The ANN and ANFIS seem to be the worst at forecasting peak flows.

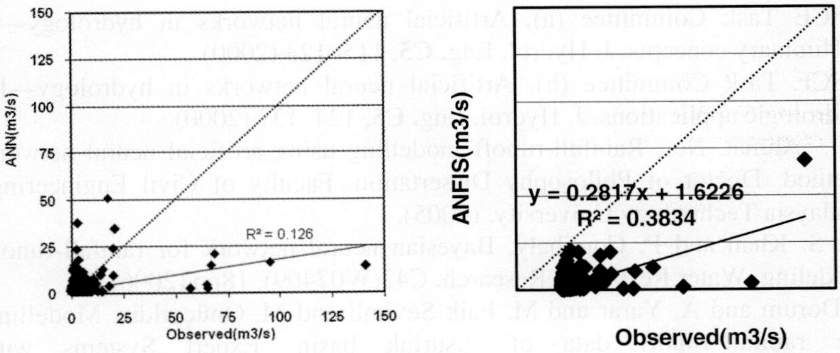

Figure 1. The scatterplot of streamflow forecasts of the ANN (3, 8, 1) and ANFIS models in test period

5. Conclusions

The performances of the ANN and ANFIS techniques in rainfall-runoff modeling have been investigated in this study. The ANN and ANFIS models were tested by using different input combinations of daily rainfall data of the Hajighoshan station on the Gorgan River in the northeast Caspian Sea region of Iran. Comparison results indicated that the ANFIS model performed better than ANN models. The combination of two models using pervious rainfall as inputs proved to be an excellent alternative to perform high quality daily runoff prediction. It was also seen that the partial auto-correlation function (PACF) values provide quite valuable information about the structure of the ANN and ANFIS input layer. The examination of partial auto-correlation table enables the ANN and ANFIS user to select the most appropriate number of input nodes preventing time losses due to the testing of several input layer alternatives. It can be said that, in general, the ANFIS model provides a superior alternative to ANN models for developing input–output simulations and for rainfall-runoff modeling. The results of the study are highly encouraging and suggest that an ANFIS approach is viable for rainfall-runoff modeling.

References

1. J. S. R. Jang, ANFIS: adaptive-network-based fuzzy inference system. IEEE Trans. System Management in Cybernetics. **23(3),** 665–685 (1993).
2. M.T. Hagan and M. B. Menhaj, Training feed forward networks with the Marquardt algorithm. IEEE Transactions on Neural Networks. **C5,** 989–993 (1994).
3. P. Coulibaly and F. Anctil and e. B. Bobeó, Preó vision hydrologique parreó seaux de neurones artificiels: eó tatdeló art. Can. J. Civ. Eng. **C26(3),** 293-304 (1999).
4. ASCE Task Committee (a), Artificial neural networks in hydrology—I: preliminary concepts. J. Hydrol. Eng. **C5,** 115–123 (2000).
5. ASCE Task Committee (b), Artificial neural networks in hydrology—II: hydrologic applications. J. Hydrol. Eng. **C5,** 124–137 (2000).
6. N.I. Ahmat. Nor, Rainfall-runoff modelling using artificial neural network method, Doctor of Philosophy Dissertation. Faculty of Civil Engineering, Malaysia Technology University. (2005).
7. M. S. Khan and P. Coulibaly, Bayesian neural network for rainfall-runoff modeling. Water Resources Research. C42(W07409), 18pp (2006).
8. A.Dorum and A. Yarar and M. Faik Sevimli and M. Onüçyildiz, Modelling the rainfall–runoff data of susurluk basin. Expert Systems with Applications. **C37 (9),** 6587-6593 (2010).
9. V. Nourani and Ö. Kisi and M. Komasi, Two hybrid Artificial Intelligence approaches for modeling rainfall–runoff process. Journal of Hydrology. **C402 (1-2),** 41-59 (2011).

FUZZY RULE-BASED DEMAND FORECASTING FOR DYNAMIC PRICING

ÖZLEM COŞGUN†
Fatih University, Department of Industrial Engineering, Istanbul, 34500, Turkey

YELIZ EKINCI
Dogus University, Department of Industrial Engineering, Istanbul, 34722, Turkey

SEDA UĞURLU
Istanbul Technical University, Department of Industrial Engineering, Istanbul, 34367, Turkey

In this study, the pricing problem of a transportation service provider company is considered. Our goal is to find optimal prices by using probabilistic dynamic programming. A fuzzy rule-based expert system is used to identify the demand levels under different price levels and other characteristics of the journey. The results obtained by optimal price policies show that the revenue levels and the capacity utilization increase by applying dynamic pricing policy instead of fixed pricing. Thus, the diversification of price policies under different conditions is advantageous for the company.

1. Introduction

Dynamic pricing, often referred to as individual-level price discrimination, has become common in the last decade [1]. Companies try to maximize their profits via pricing strategies. The bulk of the literature deals with the revenue management and dynamic pricing problems of especially airline, freight, and tourism markets. To the best of our knowledge, there has been no study conducted for maritime public transportation industry. The aim of this study is to estimate demand via FRBS and to find optimal prices via probabilistic dynamic programming.

2. Literature Review

Some revenue management models are reviewed below but for a detailed review of pricing models please refer to [2] or [3], and for a survey of the literature that considers both pricing and inventory decisions, see [4] and [5]. In the revenue management literature, dynamic pricing problems for a fixed stock of a single item sold in a finite selling horizon have attracted considerable attention. Gallego and Ryzin [6] formulate an intensity control model of the dynamic pricing problem and derive several structural properties. Zhao and Zheng [7] consider a similar problem with non-homogeneous demand. Zhang and Cooper [5] develop a Markov decision process formulation of a dynamic pricing problem for multiple substitutable flights. Kuyumcu and Diaz [8] develop a new analytical procedure for joint pricing and seat allocation problem which considers demand forecasts, number of fare classes, and aircraft capacities. Some of the previous studies in the literature consider the case of competition for the problem of dynamic pricing in transportation industry. However, in the case of this study the market is monopolistic similar to the most of the research on dynamic pricing such as [6], [9], [10], [7].

Characterization of demand is required in order to model the dynamic pricing problem. Gamma, Normal shapes and Poisson function have been proposed to characterize demand distributions in the previous studies [8], [11], [12], [5]. Some of the studies make assumptions for modeling demand under different price levels based on previous data [11], [13], [14], [12], some use price elasticity functions derived from previous data [15], [16]. Price elasticity which characterizes consumer demand has also been derived using market research techniques such as conjoint analysis [17]. In this study, there is no previous data which show the relation between different prices and demand. Moreover demand is related to some other criteria, such as weather, day of week. Thus, the fuzzy set theory proposed by [18] and fuzzy-rule based system (FRBS) may be useful for modeling this kind of imprecise problem.

3. Dynamic Pricing Model

We use probabilistic dynamic programming (DP) to solve the price optimization problem. The aim of this stochastic optimization problem is to determine the optimal policy given in (1).

$$\max_{\pi} E\left[\sum_{t=1}^{T} V^{\pi}(S_t, a_t^{\pi})\right] \tag{1}$$

where V is the value function for state S_t. The system state S_t is defined with number of seats at time *t*. The actions a_t are the prices and the decision a_t^* at time *t* will be given according to these actions that will be evaluated at time *t*. In this case the possible action set $A(S_t)$ of the S_t state is $A(S_t) = \{a_{1t}, a_{2t}, ..., a_{kt}\}$

if there are k prices observed. Then, the policy π is the set of decisions given for each period, $\pi = \{a_1^*, a_2^*, ..., a_T^*\}$. Demand is estimated under different prices together with some other criteria through fuzzy rule based systems.. When demand D_t is appeared, the system state S_t which is the number of seats changes into $S_{t+1} = \max(0, S_t - D_t)$. The optimality function is as follows which is known as Bellman Equation:

$$V_t(S_t) = \max_{a_t \in A(S_t)} \left(C(S_t, a_t) + \gamma \sum_{S_{t+1} \in S} p(S_{t+1} \mid S_t, a_t) V_{t+1}(S_{t+1}) \right) \quad (2)$$

$V_t(S_t)$ is the maximum expected discounted reward that can be earned during t periods if the state of the current period is S_t, $C(S_t, a_t)$ is the cost (or revenue) function of state S_t under decision a_t and $p(.)$ denotes the probability of next period's state S_{t+1} given the current state is S_t. Moreover, γ is the discount factor between two periods.The value of being in state S_t in any period t, denoted by $V_t(S_t)$, is formed from two parts: immediate reward $C(S_t, a_t)$ and discounted cost-to-go value function $\gamma \sum_{S_{t+1} \in S} p(S_{t+1} \mid S_t, a_t) V_{t+1}(S_{t+1})$ shown in (2). Then dynamic programming algorithm tries to find optimal action for each state S_t by using (3).

$$a_t^* = \arg\max_{a_t \in A(S_t)} \left\{ C_t(S_t, a_t) + \gamma \sum_{S_{t+1} \in S} p(S_{t+1} \mid S_t, a_t) \bar{V}_t(S_{t+1}) \right\} \quad (3)$$

Furthermore, the reward function $C(S_t, a_t)$ is the profit obtained from the difference of the sales and the cost of empty seats.

4. Fuzzy Rule Based Demand Forecasting System

A fuzzy rule-based system is a systematic reasoning methodology to model the complex behaviour of a system by using fuzzy set theory. FRBS express the input–output relationship of the process by a collection of 'if-then' rules [19].

The two types of fuzzy logic inference systems are the Sugeno type and Mamdani type. The Sugeno model is a data driven approach where membership functions and rules are developed using a training data set. When training data is not available, Mamdani model which relies on expert knowledge is useful. Since Mamdani approach is not reliant on a data set, with sufficient expertise on the system a generalized model for effective future predictions can be obtained [21].

Fuzzy inference process of Mamdani approach includes five steps: (i) fuzzification of the input variables, (ii) application of the fuzzy operator (AND or OR) in the antecedent, (iii) implication from the antecedent to the consequent, (iv) aggregation of the consequents across the rules, and (v) defuzzification.

In this study, Mamdani fuzzy inference system is implemented with α-cuts as implication and centroid as defuzzification method which is expressed in (4):

$$Y = \frac{\sum_{r=1}^{R} A^{\alpha r} C_{A^{\alpha r}}}{\sum_{r=1}^{R} A^{\alpha r}} \tag{4}$$

where $A^{\alpha r}$ is the area of the output fuzzy subset covered by α membership value using rth rule and $C_{A^{\alpha r}}$ is the center distance of the area $A^{\alpha r}$. In order to identify the variables in the FRBS and to fuzzify the variables, we have used domain expert opinion. The fuzzy inference system offered by MATLAB 7.6.0 fuzzy logic toolbox is used in this study. Using the IF-THEN rules, the demand level for a scheduled ferry is estimated based on the input variables of weather condition, day, time and the ticket price. Finally, the numerical value of "occupancy percentage" is obtained as the output. The membership functions of some variables are shown in Figure 1. An example of the rules is as follows:
Rule1 : IF (weather is warm) AND (day is late) and (time is late) and (price is low) THEN (percentage is veryHigh)

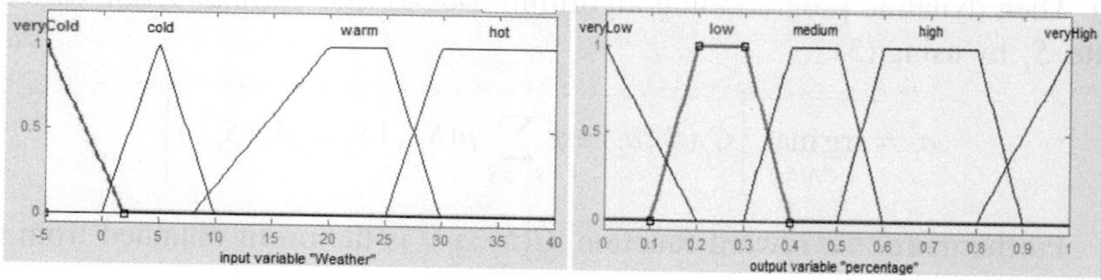

Figure 1. Membership functions of some input and output variables

5. Application

In this study, we analyzed a vehicle-passenger ferry line, namely Istanbul-Bandırma. The dynamic pricing model was run for 4 cases. The demand estimation is conducted with the FRBS. The output value is the expected occupancy percentage of the seats based on the weather, time and price of the trip which are given in Table 1 for the four cases. Table 2 shows the seven days dynamic pricing model results for a pre-determined demand path; 9, 11, 25, 38, 45, 57, 65 seats.

Table 1. Expected occupancy percentage of the seats

	Temperature	Day of Week	Time of Day	Price	Occupancy %
Case	26 ℃	Saturday	16:30	110	54.4%
Case	26 ℃	Saturday	16:30	90	91.2%
Case	26 ℃	Saturday	16:30	70	92.5%
Case	7 ℃	Sunday	16:30	110	45.0%
Case	7 ℃	Sunday	16:30	90	50.0%
Case	7 ℃	Sunday	16:30	70	90.7%
Case	35 ℃	Thursday	14:00	110	70.0%
Case	35 ℃	Thursday	14:00	90	92.8%
Case	35 ℃	Thursday	14:00	70	93.0%
Case	19 ℃	Saturday	08:00	110	50.0%
Case	19 ℃	Saturday	08:00	90	90.9%
Case	19 ℃	Saturday	08:00	70	93.4%

Table 2. Optimal dynamic pricing policies

			Optimal Policies			
Time	Demand	Stock	Case1	Case2	Case3	Case4
1	9	200	110	70	90	70
2	11	191	110	70	110	110
3	25	180	110	110	110	110
4	38	155	110	110	110	110
5	45	117	110	110	110	110
6	57	72	110	110	110	110
7	65	15	110	110	110	110

Comparing Case 1 and 2, it is seen that the day of the week is late and time of day is mid-late, but weather differs among them. In case 1, the weather is warm and in case 2, the weather is cold. Since Bandırma is a city which is visited mostly in good weathers for spending holidays, the demand is higher, therefore prices are higher. In case 1, the optimal policy is to apply the highest prices while in Case 2, the optimal policy is to apply the lowest in first 2 days. As in Case 1 and Case 2, when Case 3 is compared with Case 2, similar results are obtained. In Case3 since the weather is hot and in case 2 the weather is cold, the optimal prices in Case 3 are higher than Case 2. Finally, Case 1 and 4 can be compared since the only difference between them is time of the day. In Case 1, time of the day is mid-late while in Case 4, it is early. The demand is expected to be higher in late hours, therefore the optimal prices are higher in Case 1. The prices are 110 in all days in Case 1 while the price in the first day is 70 and 110 in the other days in Case 4.

6. Conclusion

In this study, we consider a dynamic pricing problem of a transportation service provider company. Demand is estimated under different prices together with the other criteria such as day and time of trip and weather condition through fuzzy rule based systems. We have resorted probabilistic dynamic programming to solve the optimization problem. The optimal policies give us an idea about the necessity of applying dynamic pricing policy instead of fixed pricing and show us the diversification of optimal policies under different conditions.

References

1. K.L. Haws and W.O.Bearden, *Jr. of Consumer Research*. **33**, 304 (2006).
2. G. Bitran and R. Caldentey, *Manuf. and Serv. Oper. Man.* **5 (3)**, 203 (2003).
3. K. Talluri and G. van Ryzin, *Kluwer Academic Publishers,* Boston (2004).
4. W. Elmaghraby and P. Keskinocak, *Management Science*. **49 (10)**, 1287 (2003).
5. D. Zhang and W.L. Cooper, *Eur. Jr. of Oper. Res.* **197**, 848 (2009).
6. G. Gallego and G. van Ryzin, *Management Science*. **40 (8)**, 999 (1994).
7. W. Zhao and Y.S. Zheng, *Management Science*. **46 (3)**, 375 (2000).
8. A. Kuyumcu and A. Garcia-Diaz, *Comp. and Ind. Eng.* **38**, 375 (2000).
9. R E. Chatwin, *Eur. Jr. of Oper. Res.* **125**, 149 (2000).
10. Y. Feng and B. Xiao, *Management Science*. **46(5)**,644 (2000).
11. X Yong-bo, C. Jian, and L. Xiao-ling, J*SETP*. **28(1),** 46 (2008).
12. L. Li and P. Ji-hua, *SETP*, **27(11)**, 15 (2007).
13. T.L. Friesz, R. Mookherjee, J. Holguin-Veras, and M.A. Rigdon, *Transportation Research Part B*. **42,** 305 (2008).
14. B. Maddah, L. Moussawi-Haidar, M. El-Taha, and H. Rida, *Eur. Jr. of Oper. Res.* **207,** 445 (2010).
15. R. Bharill and N. Rangaraj, *Trans. Research Part A*. **42**, 1195 (2008).
16. T. Gorin and P. Belobaba, *Jr. of Air Transport Management*. **10**, 259 (2004).
17. S. Ugurlu, O. Cosgun, Y. Ekinci, *Int. Jr of Comp. Int. Sys* (article in press).
18. L.A. Zadeh, Fuzzy sets, *Information and Control.* **8**, 338 (1965).
19. K. Tahera, R.N. Ibrahim, *Exp. Sys. with App.* **34**, 2630–2638, (2008).
20. A.K. Nandi and J.P Davim, J. P., *Mechatronics*, **19(2)**, 218–232, (2009).
21. D.R. Keshwani, D.D. Jones, G.E. Meyer, and R.M. Brand, *Applied Soft Computing*, **8(1)**, 285–294, (2008).

NEURAL NETWORK AND PARAMETER PRODUCTION MODELING: METAHEURISTICS CLUSTERING METHODS*

A. F. BERNATE LARA† AND F. DUGARDIN AND F. YALAOUI
Laboratoire d'Optimisation des Systèmes Industriels (LOSI), University of Technology of Troyes, 12 Rue Marie-curie, Troyes, 10000, France

F. ENTZMANN
OSIRIS, SOUFFLET Group, Rue de la poterne à sel Nogent sur Seine, 10400, France

This article presents a real case application of a neural network used in a production criterion modeling. The goal here is to determine the most adapted configuration of a neural network in a complete predictive model construction for the criterion. Since the evaluated criterion cannot be measured in real time and its influence on the performance of the production system could be important, a model is required to estimate the criterion value as correctly as possible. The analyzed criterion is supposed to be related to a set of independent variables. The employed neural network is a Radial Basis Function Network. Moreover metaheuristics have been used to increase the performance of the proposed method, by modifying the neural network composition.

1. Introduction

Radial Basis Function Networks has been used for different kinds of mathematical approximations as interpolation, functions approximation and pattern recognition, in presence of noise [4].

As remained in Wang et al [5], two mainly stages can be cited about the Radial Basis Function Networks (RBFN): initialization and parameters optimization. Initialization stage provides initial values for the different RBFN parameters, as centers and weights. Optimization phase finds optimal or near optimal values for those parameters. This article is about the implementation of both stages in a real problem, of a production plant. Neural networks were chosen because of their adaptability to solve problems with noise on the incoming data [6].

---

* This work is supported by SOUFFLET Group
† andres.bernate_lara@utt.fr.

The present article is declined in the next sections: first, the problem is presented. Then, the used neural network is explained. The most relevant elements are exposed. After, a RBFN is adapted. Results and conclusion close the article.

2. Incoming Data

This study was proposed in an biochemical production plant. This project aims to reduce the number of non-controlled production criteria, including those that are not easy to measure on the product or over the production line. In the production system, the real criterion, that we call dependent variable, is actually measured by a time consuming process. The goal of this study is to estimate the evaluation criterion from other criterion that is shorter to measure. This second criterion is called the independent variable. Both criterion are measured on samples taken directly from the production line. The independent variable is a spectral non-destructive measure of the sample. The dependent variable is a physical characteristic destructive measure of the same sample. The relation is established with a model of dependent and independent variables. To model the criterion, the dependence between variables must be assumed[1].

The performance of the proposed method, in term of difference from the real value, cannot exceed 2%, and has to provide results in a very short time. The independent variable will be represented by X, with $\vec{x}_i$ the set of independent variable data for the sample i, and $x_{i,j}$ the component j of the sample i of the same variable, with $i \in \{1,...P\}$ and $j \in \{1,...R\}$, where P is the number of samples, and R the number of independent variables by sample. The studied criterion will be symbolized by $\vec{Y}$, with y_i as the value of the dependent variable for the sample i. To build the predictive model, data were collected for more than 2 years.

3. The Radial Basis Function Network

Radial Basis Function Networks are neural networks which main feature is the use of a single hidden layer, centers and a radius. The figure 1 includes the graphical representation of a RBFN. The incoming layer contains the X, variables. The hidden layer aims is the reduction of noise with the use of a Radial Basis Function (RBF). The final layer is the answer layer, where the estimated $\vec{Y}$ value is calculated. This structure depends on many parameters, whose values are in the following sections.

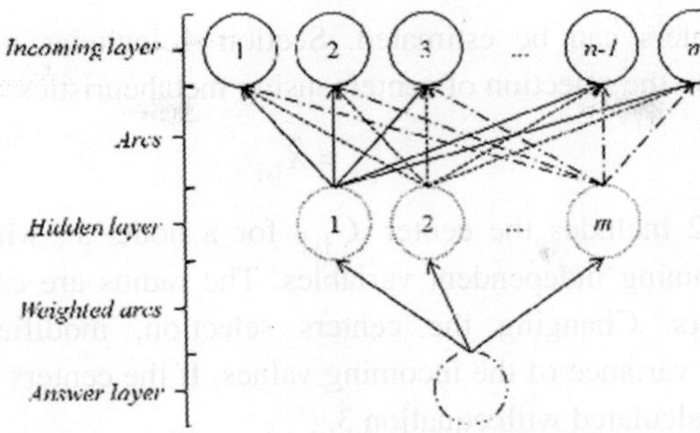

Figure 1: Radial Basis Function Network

Where n and m are the number of incoming nodes and hidden nodes respectively.

3.1. *Radial Basis Function*

Two types of parameters have to be fixed for the RBF: the centers and the radius. Centers are depends on the central value of the incoming data and they are calculated for each of the nodes of the hidden layer. A radius is also calculated for each hidden node. Equation 1 describes the exponential radial basis function used, for one of the hidden nodes.

$$h_{\{i\}}(X) = \exp\frac{-\left\|X - C_{\{i\}}\right\|}{r_{\{i\}}^2} \qquad (1)$$

Where $h_{\{i\}}(X)$ denotes the value of the RBF for a node i, with the set of variables X, and the two parameters $C_{\{i\}}$ the center of the values of the node, and $r_{\{i\}}$ the radius.

3.2. *Nodes and Arcs*

The number of incoming nodes is defined by the number of variables X. Each node on the incoming layer figures a $x_{i,j}$ variable. The nodes on the hidden layer are defined by the structure of the neural network. Each hidden node is connected to a set of incoming nodes. Those connections are the arcs on fig. 1.

3.3. *Centers and Radius*

A center is calculated for each node. The objective is to calculate the center of values of the connected incoming nodes. If the range of received values is undefined, centers can be estimated. Section 4 includes a description of a improvement on the selection of centers, using metaheuristics.

$$C_{\{s\}} = \bar{X}_{\{s\}} \tag{2}$$

Equation 2 includes the center $C_{\{s\}}$ for a node s, with $X_{\{s\}}$ the set of connected incoming independent variables. The radius are calculated with the selected centers. Changing the centers selection, modifies the radius. It symbolizes the variance of the incoming values. If the centers are estimated, the radius can be calculated with equation 3.

$$r_{\{s\}} = E\left[\left(X_{\{s\}} - C_{\{s\}}\right)^2\right] \tag{3}$$

3.4. *Weighted Arcs*

The weights are the relative impact of the hidden nodes on the answer nodes. After identifying the centers and radius, weights can be found with a least square minimization [3] with equation 4.

$$\vec{W} = \left(H^T H + \lambda I_m\right)^1 H^T \vec{Y} \tag{4}$$

With $\vec{W}$ the vector of weights, H the array of RBF values for all the nodes for all the samples, λ a coefficient to avoid dominant weights.

4. Industrial Case

The explained neural network was adapted to solve the real case problem. Parameters n and m were fixed as: $m = n = R$. About the structure of the network, we have chosen to allow all the arcs between the incoming and the hidden layer. Optimization phase has been concentrated on selecting the centers. The answer layer is composed of only one node.

4.1. *Center Selection*

Since the current problem is the construction of a predictive model, the range of incoming data is undefined. We are proposing a genetic algorithm which objective is to find the most appropriated set of centers to minimize the error. The next sub-sections present the genetic algorithm used [2].

4.1.1. *Solution representation*

The solution representation is a vector with integers numbers. Each number correspond to a sample used to compute a center. No sample can be taken twice.

4.1.2. *Crossover and mutation*

Crossover and mutation are the movements chosen to produce the offspring. Selecting parents is made according to the fitness function value (section 4.1.3). The crossover involves two parents. The first half of a new element is copied from the Parent 1. The new element is completed with elements from Parent 2. The mutation use one parent to create one new element. A random number selects how many elements to copy from the parent. Solutions are randomly completed. New elements are verified to avoid selecting twice the same sample.

4.1.3. *Fitness function*

The evaluation is made with the sum of the square difference between the real and the estimated values of the studied criterion (equation 5). Where $\{C_o\}$ is the set of centers of the solution O. $\vec{Y}$ and $\hat{\vec{Y}}$ are the real and the estimated value of the studied parameter for the used samples.

$$e^2_{\{C_o\}} = \left(\vec{Y} - \hat{\vec{Y}}\right)^2 \quad (5)$$

5. Results

The proposed methods were coded in PYTHON according Table 1. The proposed method is compared to a single least square minimization regression. The performance criteria is the sum of the square error. The performance is measured over a set of data not included during the modeling phase, so performance evalues predicting. Results are given in Table2.

Table 1. Genetic Algorithm parameters.

Parameter	Value
Initial population size	10
Number of generations	100
Crossover/Mutation probability	80%/20%
Time limit	1000 Seconds

Table 2. Comparative results.

R	*AG-RBFN*	*RND-RBFN*	*MinLsmr*
100	2198	3018	2329
200	1730	2887	2028
300	1939	4120	3478

For example, the first line provides the performance of the proposed method ($AG-RBFN$), the random selection process ($RND-RBFN$), and the single least square minimization regression ($MinLsmr$), in predicting the value of the studied criterion in samples with $R=100$ (the number of independent variables by sample).

6. Conclusions

The present article presents a RBFN to build a predictive model for an industrial criterion. The proposed network was improved with a genetic algorithm. Results show that the genetic algorithm increases the performance measured over a set of new samples. RBFN seems to be an appropriated solution method. Improvements could be done modifying the allowed arcs between the incoming and hidden layer.

References

1. A. Bernate-Lara, F. Entzmann, F. Yalaoui and F. Dugardin, International conference on industrial engineering and manufacturing. (2012).
2. A. Bernate-Lara and F. Yalaoui, International Conference on Metaheuristics and Nature Inspired Computing. (2012).
3. S. Chen, C. Cowan and P. Grant, IEEE Transactions on Neural Networks. **2**, 302 (1991).
4. M. Paliwal and U. Kurmar, Experts Systems with Applications. **36**, 2 (2009).
5. D. Wang, X. Zeng and J. Keane, Neurocomputing. **77**, 144 (2012).
6. B. Warner and M. Misra, The American Statistician. **50**, 284 (1996).

IN-CORE SENSORS READINGS DIAGNOSTICS BASED ON NEURO-FUZZY TECHNIQUES*

STEFAN FIGEDY†

Department of diagnostic systems, VUJE,a.s., Okruzna 5, Trnava 91864, Slovak Rep.

In this work an alternative method of in-core sensor validation is outlined. Instead of using direct sensor readings, it is based on neuro-fuzzy modeling of residuals between the experimental values and their theoretical counterparts obtained from reactor core simulator calculations. Throughout the fuel cycle, the neural networks are subject to the incremental training on data prior to the current monitoring period. While for thermocouples the method shows very good sensitivity, for self powered neutron detectors the results are not satisfactory. The purpose of this paper is to present the newest results from considerably long time effort in development of this computational intelligence based approach.

1. Introduction

The purpose of this paper is to present the newest results from considerably long time effort in development of and give some recommendations to new in-core sensor validation methods for both self-powered neutron detectors (SPNDs) and reactor core outlet temperature thermocouples (TCs). This approach, which might be used as alternative and supplementary technique to the classical ones implemented in reactor core surveillance, is based on computational intelligence techniques.

For in-core sensors signal validation, the OECD Halden Reactor Project's neuro-fuzzy system PEANO [1-2] was chosen. The signal validation model is based on the set of artificial neural networks (ANNs), each driven by a classification algorithm. Originally, PEANO was developed for various process parameters validation like power, pressure, temperature, flow, water level etc. It is a challenging task to use it for signal validation of in-core sensors due to the ever changing fuel load patterns and burn up of both fuel and SPND emitters. To

* This paper is the result of implementation of the project: Increase of Power Safety of the Slovak Republic (ITMS: 26220220077) supported by the Research & Development Operational Program funded by the ERDF.

† Email: figedy@vuje.sk

overcome this difficulty, the idea arose to train PEANO to residuals, i.e. the differences between core simulator physics code results and their experimental counterpart values. The author hopped that the system trained to residuals can be used in the next fuel cycles without having to be retrained. The aim of this work was to verify experimentally if this assumption holds.

2. Diagnostics Based on Neuro-Fuzzy System Trained to Residuals

Direct validation of SPND readings in [μA] encounters a problem of their dependence on ever changing fuel load patterns as well as burn-up of both fuel and SPND emitters. The training would normally have to be done on simulated data prior to each fuel cycle. As mentioned above, the idea arose to train PEANO to residuals instead. Because both the SPND and TC readings are not directly calculated by a core simulator, the SPND residuals are the differences of linear power calculated from the 3D power reconstruction by reactor core physics code and linear power derived by formula considering measured SPND readings, burn-up of both fuel and detector emitter, temperature, power, boron acid concentration, uneven power distribution across fuel rods inside the fuel assembly, and vicinity of the fuel assembly, i.e. the fuel load pattern. The TC residuals are differences of fuel assembly power calculated from the 2D power reconstruction and fuel assembly power calculated from the enthalpy difference related to fuel assemblies outlet and inlet temperatures. Because reactor core physics parameters such as fuel load pattern and both fuel and SPND emitters burn up are involved in calculation of both factors in the difference, the residuals are expected to incorporate them implicitly. At the same time the residuals are supposed to depend on effective days values, control rods position, power, and maybe few other parameters. It was supposed and hopped that the system would be enough to train once on data from one fuel cycle and then to use it in the next fuel cycles.

2.1. *Experiences from testing in the successive campaigns*

PEANO was trained with data from one fuel cycle covering various operating situations and then tested with data from the successive fuel cycle [3,4]. Due to inaccuracies in both calculations and measurements, residuals have some systematic values during the given fuel cycle. In contrast to the situation when training and testing was performed with data from the same fuel cycle and validation performance was excellent, in the new campaign the PEANO system did not model the residuals for some sensors properly. Because new fuel loading may have certain influence on accuracy of calculations which resulted in residual patterns different to those PEANO was trained to, it reflected in the validation performance deterioration. Whatever the cause of the change of residuals in the other than the training fuel cycle might be, the conclusion is that training to all possible operating situations in one fuel cycle prior to its use in the successive ones is not a feasible solution.

2.2. *Incremental mini-batch training and testing within the same campaign*

Because the validation performance was excellent within the same fuel cycle, the proposed solution was to use the incremental mini-batch training in which training was performed with data from operation situations prior to the current one, yet from the same cycle. In the beginning of the fuel cycle after reactor start-up, data were first acquired for about two weeks period of time after which PEANO was trained and then used for signal validation. This process continued as the fuel cycle proceeded, each time PEANO being retrained with the new batch of data added to the previous one.

Regarding the SPND sensors, in most cases prediction of residual values was very good. Yet, there were situations when the accuracy deteriorated to five or even ten percent of linear power at the detector location. It is to note that even though we tried to tune the PEANO system by proper adjusting of clustering and neural networks topology parameters during the training, yet we did not strive deliberately for the best possible solution. We wanted to model the operational staff or user behavior, who is not expected to act like this in the real life and wants to have a system functioning relatively simply. This may perhaps be responsible for the deterioration of the signal predictions, which appeared from time to time.

Summarizing the above mentioned experience with SPND sensors diagnostics brings up the conclusion that the usage of a system would be to a certain extent cumbersome when constant attention by the repeated and not quite easy retraining throughout the fuel cycle is to be given by the user while the demanded accuracy may not be guaranteed. We therefore are not inclined to recommend this approach for validation of SPNDs, which still are of less importance to control the reactor comparing to TC readings.

Similarly to SPNDs, for validation of TCs the additional measured parameters like reactor power, main control rods positions, and overall flow rate through reactor were added to the model as inputs. The number of middle hidden layer neurons was chosen close to the number of principal components in data files. The topology of the neural networks was kept unchanged during the whole process of retraining and testing. The so called robust training was applied when different noise and faults were superimposed on input signals while the networks were trained to get the denoised signals on output. This forced the neural networks to act as a noise filter. Figure 1 shows the typical prediction of TC signal when all monitored signals to be validated were correct. A certain problem related to incremental training can appear when a new operational situation, not having been included during the previous training, is encountered. The system then cannot model signal (residual) behavior very correctly before this situation data are added to training files.

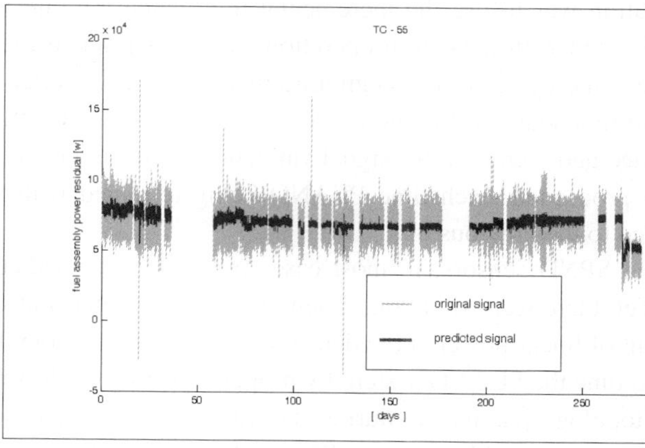

Figure 1. The typical prediction of the correct TC residual

The other problem is a drift in input signals. If the drift is small and not apparent on the first look, the corrupted signals are used in the next training phase, which impairs the ability to recognize deviations in the drifted signals. Five TCs were simulated to drift simultaneously, each with the rate of 0.1°C/month immediately after the initial two weeks training data gathering period. The deviation of the drifted signals (residuals) at the fuel cycle end corresponds to 0.96°C deviation of the fuel assembly outlet temperature. As supposed, after some time the prediction performance degraded. For this reason further testing was stopped.

To solve this problem, the retraining style was changed by replacing the drifted signals by their predicted values. It resulted in much better elimination of drift as can be seen from Figure 2. The initial part of prediction line represents the period of time of approx. one month before we started "as if" to suspect the signals of drifting. This was how we tried to model the real life use of the system. It roughly corresponds to the method sensitivity, which was about 0.1°C deviation or fault in the measured temperatures. Below this we cannot be sure whether it is the prediction error. After this period the replacement of drifted values by their predictions in the retraining process began. It is to note that due to the periodical retraining of the system, the result of which are always new neural models, the predictions are not smooth and the step breaks can be seen periodically. We expect the better results might have been achieved if more effort had been put into the training. For the same reasons as already mentioned, it was not done.

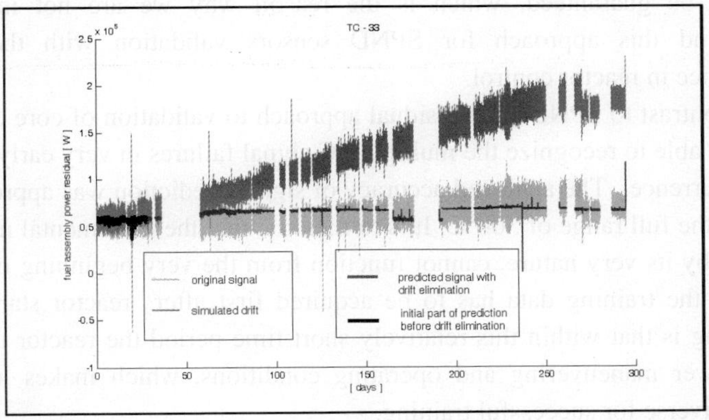

Figure 2. The prediction of the drifted TC residual with the drift elimination

Let us answer the question why we prefer the more complicated residual approach, which requests core physics simulator data in addition to measured quantities, to the simpler way to use only the sensor readings. The reason is that the residual approach comprises burn-up and other reactor physics characteristics. The various operating conditions are involved incrementally in data coming from previous time periods. The incremental training is supposed to enable the use of trained system for longer periods. The other reason for using residuals is that the system can be more sensitive and reveal even very small deviations in sensor readings in spite of modeling the signal full range.

3. Conclusions

The in-core sensors validation method we were examining is based on neuro-fuzzy modeling of residuals instead of raw measured values. Residuals are differences between core simulator physics code results and their experimental counterpart values. We found that the original idea to train the neuro-fuzzy system to residuals acquired during one fuel cycle and then using the trained system in the following cycles failed. The remedy to this can be the incremental training, the process of which is as follows: In the beginning of fuel cycle after the reactor start-up data were first acquired for about two weeks period of time after which the system was trained and then used for signal validation. This process continued as the fuel cycle proceeded, each time the system being retrained with new batch of data added to the previous one. The price the user would pay for such system would be its relatively cumbersome usage when constant attention by the repeated and not quite easy retraining throughout the fuel cycle is to be given. In addition to this, the demanded validation accuracy may not be guaranteed, which is the reason why we are not inclined to recommend this approach for SPND sensors validation with their lesser significance in reactor control.

In contrast to SPNDs, the residual approach to validation of core outlet TC signals is able to recognize the multiple TC signal failures in very early stages of their occurrence. The achieved accuracy of signal prediction was approximately 0.1°C at the full range of 300°C. It is to mention that the incremental mini-batch training, by its very nature, cannot function from the very beginning of the fuel cycle, as the training data has to be acquired first after reactor start-up. The good thing is that within this relatively short time period the reactor undergoes more power maneuvering and operating conditions, which makes input data enough diverse for successful training.

References

1. P.F.Fantoni, S. Figedy,B. Papin, A neuro-fuzzy model applied to full range signal validation of PWR nuclear power plant data. *Second OECD Specialist Meeting on Operator Aids for Severe Accident Management (SAMOA-2)*, Lion, (1997)
2. P. Fantoni, M. Hoffmann, R. Shankar, E. Davis, "On-Line Monitoring of Instrument Channel Performance in Nuclear Power Plant Using PEANO", *Progress in Nuclear Energy*, Vol. 43, No. 1-4, (2003)
3. S.Figedy, Outline of New Approaches to In-Core Sensor Validation Based on Modern Signal Processing and Neuro-Fuzzy Techniques, *NPIC&HMIT 2009*, Knoxville, Tennessee, Apr.5-9,(2009)
4. S.Figedy, Experiences with Residual Approach to Validation of In-Core Sensors by PEANO Neuro-Fuzzy System, *NPIC&HMIT 2010*, Las Vegas, Nevada, Nov.7-11,(2010)

AN APPROACH FOR ATTRIBUTE REDUCTION IN SET-VALUED ORDERED INFORMATION SYSTEMS BASED ON INFORMATION ENTROPY*

CHUAN LUO, TIANRUI LI†, HONGMEI CHEN, JUNBO ZHANG

School of Information Science and Technology, Southwest Jiaotong University
Chengdu, 610031, China

Rough sets theory is an important mathematic tool for data mining. Feature selection is a main step in the course of data mining. In this paper, we focus on the attribute reduction by using rough set theory in Set-valued Ordered Information Systems (SOIS) from the view of information theory. Firstly, we introduce the information entropy and relative information quantity to SOIS. Then, we define attribute reduction by means of information entropy. Furthermore, an algorithm for computing reducts is proposed based on the information entropy. Finally, an illustrative example is employed to show the effectiveness of the proposed method for attribute reduction in SOIS.

1. Introduction

Attribute reduction is a major research direction of Rough Set theory (RST) which is an extended version of the classical set theory to deal with imprecise, uncertain, and incomplete data [1]. Many researchers have done a lot of research in the area of attribute reduction based on RST [2, 3, 4]. Recently, some researchers introduced information entropy to measure the uncertainty of an information system, and adopted it as the heuristic metrics for attribute reduction based on RST [5].

In real-life applications, some objects in information systems may have more than one value for an attribute, and attributes inside the information systems are sometimes with preference-ordered domains which the ordering properties of the considered attributes play a crucial role. For this issue, a so-called Set-valued Ordered Information Systems (SOIS) is proposed to deal with these situations [6]. In this paper, we study the problem of attribute reduction in

* This work is supported by the National Science Foundation of China (Nos. 60873108, 61175047, 61100117), the Youth Social Science Foundation of the Chinese Education Commission (No. 11YJC630127), and the Fundamental Research Funds for the Central Universities (SWJTU11ZT08).

† trli@swjtu.edu.cn.

SOIS from the viewpoint of information theory. Firstly, we introduce the information entropy to SOIS. Then we define attribute reduction of SOIS by means of information entropy. Finally, we propose an algorithm to calculate the reduct in SOIS.

This paper is organized as follows: In Section 2, we review some basic notions in SOIS. In Section 3, we investigate the definition of information entropy and relative information quantity in SOIS and study the attribute reduction based on it. In section 4, we proposed an algorithm for attribute reduction from the view of information theory. At the end, an example is given to verify the effectiveness of our method in the Section 5.

2. Preliminaries

In this section, some basic concepts and preliminaries of SOIS are introduced firstly.

A set-valued information system is a quadruple (U, AT, V, f) where $U=\{x_1, x_2, \ldots x_n\}$ is a nonempty finite set of objects, $AT=\{a_1, a_2, \ldots, a_l\}$ is a finite set of attributes, V is the set of attributes values, $V=\{V_{a_1}, V_{a_2}, \ldots, V_{a_l}, a_i \in AT\}$, $V_{a_i}\,(i=1,2,\ldots,l)$ is domain of attribute $a_i\,(a_i \in AT)$. f is a mapping from $U \times AT$ to V such that $f: U \times AT \to 2^V$ is a set-valued mapping.

In a set-valued information system, if the domain of an attribute is ordered according to a decreasing or increasing preference, then the attribute is a criterion.

Definition 1 A set-valued information system is called a SOIS if all attributes are criterions.

Definition 2 Let (U, AT, V, f) be a SOIS, and $A \subseteq AT$, the dominance relation in terms of A is defined as:

$$R_A^{\supseteq} = \{(x, y) \in U \times U \mid f(y,a) \supseteq f(x,a) (\forall a \in A)\} \tag{1}$$

The dominance class induced by the dominance relation $R_A^{\supseteq}$ are the set of objects dominating x, *i.e.*,

$$[x]_A^{\supseteq} = \{y \in U \mid f(y,a) \supseteq f(x,a) (\forall a \in A)\} \tag{2}$$

The set of objects dominated by x is denoted as follows.

$$[x]_A^{\subseteq} = \{y \in U \mid f(y,a) \subseteq f(x,a) (\forall a \in A)\} \tag{3}$$

where $[x]_A^{\supseteq}$ and $[x]_A^{\subseteq}$ are called the $A-$dominating set and $A-$dominated set with respect to $x \in U$, respectively.

3. Attribute reduction of SOIS based on information entropy

In the following, we introduce an information measure for the discernibility power of knowledge in SOIS firstly, and then, the attribute reduction of SOIS from the viewpoint of information entropy is studied as following:

Definition 3 Let $S=(U,AT,V,f)$ be a SOIS, and $A \subseteq AT$. The information entropy of knowledge A is defined as:

$$I(A)=\frac{1}{|U|}\sum_{i=1}^{|U|}\left(1-\frac{\left|[x_i]_A^{\supseteq}\right|}{|U|}\right)=1-\frac{1}{|U|^2}\sum_{i=1}^{|U|}\left|[x_i]_A^{\supseteq}\right| \tag{4}$$

where $|\cdot|$ denotes the cardinality of a set.

Definition 4 Let $S=(U,AT,V,f)$ be a SOIS, and A , $B \subseteq AT$, the relative information quantity of B to A be defined as follows:

$$I(A/B)=I(B)-I(A)=\frac{1}{|U|^2}\sum_{i=1}^{|U|}\left(\left|[x_i]_B^{\supseteq}\right|-\left|[x_i]_A^{\supseteq}\right|\right) \tag{5}$$

From the above definitions, the following properties in SOIS can be obtained.

Property 1 Let $S=(U,AT,V,f)$ be a SOIS, and $A,B \subseteq AT$, we have:

1. If $A \supseteq B$, then $I(A) \geq I(B)$, $I(A/B) \geq 0$;
2. If $A \subseteq B$, then $I(A) \leq I(B)$, $I(A/B) \leq 0$.

Now we study the attribute reduction of SOIS from the viewpoint of information entropy.

Theorem 1 Let $S=(U,AT,V,f)$ be a SOIS, and $A \subseteq B \subseteq AT$. If $I(A)=I(B)$, then $U/R_A^{\supseteq}=U/R_B^{\supseteq}$.

Theorem 2 Let $S=(U,AT,V,f)$ be a SOIS. $a \in AT$ is redundant if and only if $I(AT/AT-\{a\})=0$.

Corollary 1 Let $S=(U,AT,V,f)$ be a SOIS, and $a \in AT$ is indispensable if and only if $I(AT/AT-\{a\})>0$.

By Corollary 1 and the definition of the attribute core, we have the following result.

Theorem 3 Let $S=(U,AT,V,f)$ be a SOIS. The core of attribute set AT is: $Core(AT)=\{a\in AT \mid I(AT/AT-\{a\})>0\}$.

Corollary 2 Let $S=(U,AT,V,f)$ be a SOIS, $A\subseteq AT$ is a reduct of AT if and only if $I(A)=I(AT)$ and $I(AT/AT-\{a\})>0$.

Definition 5 Let $S=(U,AT,V,f)$ be a SOIS, $A\subseteq AT$. $\forall a\in AT-A$, the significance of attribute a in attribute set A is defined by $Sig_A(a)=I(A\cup\{a\})-I(A)$.

4. An algorithm for attribute reduction in SOIS based on information entropy

As mentioned above, the approach shows that by using the information entropy we can indeed acquire the reduction in a SOIS. Based on the information view of point, we present the process (See Algorithm 1) of attribute reduction in SOIS.

Algorithm 1: *An approach to attribute reduction in SOIS based on information entropy.*

Input: The SOIS $S=(U,AT,V,f)$.

Output: An attribute reduction R of SOIS $S=(U,AT,V,f)$.

```
Begin:
Let Core(AT) = ∅;
Calculate I(AT);
For each a ∈ AT
    Calculate I(AT/AT − {a});
    If I(AT/AT − {a}) > 0 then
        Core(AT) = Core(AT) ∪ {a};
    Endif
Endfor   // find the attribute core of S: Core(AT).
Calculate I(Core(AT));
If I(AT) = I(Core(AT)) then
    Return R = Core(AT)   // Core(AT) is the smallest reduction.
Else
    Let R = Core(AT);
Endif
While I(R) ≠ I(AT) do
```

For each $a \in AT - R$
    Calculate $Sig_R(a)$
Endfor
If $Sig_R(a) = \max\{Sig_R(a_i) \mid a_i \in AT - R\}$ then
    $R = R \cup \{a\}$ // select the attribute with maximum significance
Endif
Calculate $I(R)$
Endwhile
Return R // find reduction

5. An illustrative example

Consider a SOIS $S = (U, AT, V, f)$ as follows (see Table 1): where $U = \{x_1, x_2, x_3, x_4, x_5, x_6, x_7, x_8, x_9, x_{10}\}$, $AT = \{a_1, a_2, a_3, a_4\}$ = {Audition, Spoken language, Reading, Writing} and V = {E, F, G} = {English, French, German}. Then, we compute the reduct of the SOIS in Table 1 by using the above algorithm.

Firstly, from the Definition 6, we can calculate the information entropy as follows: $I(AT) = 81/100$; $I(AT/AT - \{a_1\}) = 0$; $I(AT/AT - \{a_2\}) = 0$; $I(AT/AT - \{a_3\}) = 11/100 > 0$; $I(AT/AT - \{a_4\}) = 5/100 > 0$. Then we can obtain the core of the attribute set in information system by Theorem 4: $Core(AT) = \{a_i \in AT \mid I(AT - \{a_i\}) > 0\} = \{a_3, a_4\}$. Secondly, we calculate the information entropy of the core: $I(Core(AT)) = I(\{a_3, a_4\}) = 69/100 \neq I(AT)$. So $Core(AT)$ is not a reduct of Table 1.Thirdly, let $R = Core(AT) = \{a_3, a_4\}$, we calculate the significance of attribute a, $a \in AT - R = \{a_1, a_2\}$: $Sig_R(a_1) = I(R \cup \{a_1\}) - I(R) = 3/25$; $Sig_R(a_2) = I(R \cup \{a_2\}) - I(R) = 3/25$. Then we select the attribute $a \in AT - R$, which has the maximal value of significance, and add it to the attribute set R. We observe that the significance of attribute a_1 is equal to the significance of attribute a_2. So neither a_1 nor a_2 can be added to R. Here we choose attribute a_1 to add to R, and set $R = R \cup \{a_1\}$. Then we calculate the information entropy of R: $I(R) = 81/100$. Since $I(R) = I(AT)$, so we obtain that $R = \{a_1, a_3, a_4\}$ is a reduct of the SOIS in Table 1.

6. Conclusions

Attribute reduction is a main research direction of RST. In this paper, we considered the problem of attribute reduction in SOIS. Firstly, we introduced the concepts of information entropy and relative information quantity in SOIS. Then we used the information entropy as the heuristic information to design and implement an algorithm to compute the reduct of SOIS. Finally, the example shows that the algorithm is feasible. In our further work, we will verify the efficiency of the proposed algorithm, and develop new incremental approaches for attribute reduction in SOIS.

Table 1. A SOIS about language ability.

U	Audition	Spoken language	Reading	Writing
x_1	{E}	{E}	{F,G}	{F,G}
x_2	{E,F,G}	{E,F,G}	{F,G}	{E,F,G}
x_3	{E,G}	{E,F}	{F,G}	{F,G}
x_4	{E,F}	{E,G}	{F,G}	{F}
x_5	{F,G}	{F,G}	{F,G}	{F}
x_6	{F}	{F}	{E,F}	{E,F}
x_7	{E,F,G}	{E,F,G}	{E,G}	{E,F,G}
x_8	{E,F}	{F,G}	{E,F,G}	{E,G}
x_9	{F,G}	{F,G}	{F,G}	{F,G}
x_{10}	{E,F}	{E,G}	{F,G}	{E,F}

Reference

1. Z. Pawlak, Rough Sets: Theoretical Aspects of Reasoning about Data, *Kluwer Academic Publishers*. (1991).
2. A. Skowron, Extracting laws from decision tables: a rough set approach, *Computational Intelligence*. **11**(2): 371-388 (1995).
3. M. Dash and H. Liu, Consistency-based search in feature selection, *Artificial Intelligence*. **151**: 155-176 (2003).
4. T. R. Li, D. Ruan, W. Geert, J. Song and Y. Xu, A rough sets based characteristic relation approach for dynamic attribute generalization in data mining, *Knowledge-Based Systems*. **20**(5): 485-494 (2007).
5. D. Q. Miao and G. R. Hu, A heuristic algorithm for reduction of knowledge. *Chinese Journal of Computer Research and Development*. **36**(6): 681-684 (1999).
6. Y. H. Qian, C. Y. Dang, J. Y. Liang and D. W. Tang, Set-valued ordered information systems, *Information Science*. **179** (16): 2809-2832 (2009).

HYBRID SIMULATED ANNEALING/GENETIC ALGORITHM APPROACH TO BUFFER ALLOCATION PROBLEM IN UNRELIABLE PRODUCTION LINES*

SIMGE YELKENCI KOSE†, OZCAN KILINCCI

Department of Industrial Engineering, Dokuz Eylul University, Izmir, Turkey

The buffer allocation problem, i.e. how to distribute a certain amount of buffers among the buffer locations of a production line, is an NP-hard combinatorial optimization problem. In this study, a hybrid approach combining simulated annealing and genetic algorithms is presented to find optimal buffer sizes for unreliable production lines. The objective is to improve the capacity of production line by implementing proposed hybrid approach. To evaluate the capacity of the line, a detailed simulation model is developed. Numerical examples are given for evaluating the performance of the proposed heuristic, traditional genetic algorithm, and simulated annealing procedures. The results show that the proposed hybrid approach gives good solutions in reasonably short computation times.

1. Introduction

Efficient management of buffer size allocation is a key concern in manufacturing systems. The basic structure of allocation is a decision on how to distribute a certain amount of buffers among the buffer locations of a production line. This problem is well-known as a buffer allocation problem (BAP) in literature and arises in a wide range of manufacturing systems.

In the relevant literature, solution approaches for BAP involve applying a generative method and an evaluative method in an iterative manner. In our proposed hybrid heuristic, simulated annealing and genetic algorithms are used as generative methods, and simulation is preferred as evaluative method. Therefore we focus on recent studies related with simulated annealing, genetic algorithm, and simulation as solution approaches. In recent times, simulated annealing [1], [2] and genetic algorithms [3-6], are widely used to solve BAP. For evaluative method, simulation is chosen in recent studies [3], [6-11].

---

* This research was done during the PhD study of the first author at the Graduate School of Natural and Applied Sciences, Dokuz Eylul University.

† E-mail: simge.yelkenci@deu.edu.tr

Considering the advantages of hybridization, in this study a hybrid genetic algorithm-simulated annealing, namely Genetic Annealing Algorithm (GAA) approach is proposed to deal with the BAP in unreliable production line. This is the first study that attempts to solve the BAP using the proposed genetic annealing algorithm approach.

The rest of this paper is organized as follows. Section 2 explains the bufffer allocation problem. The details of the proposed hybrid approach is given in Section 3. Section 4 presents numerical examples and computational results about the performance of the proposed approach. Finally, the last section summarizes the findings of this study and some future research directions.

2. Buffer Allocation Problem

The BAP is an NP-hard combinatorial optimization problem and has been considered in the literature with respect to different optimality criteria [12] such as production rate, inventory cost, and total buffer capacity. Our optimality criterion is the average production rate. Also, the BAP in an unreliable serial production line is addressed in this study. The considered system consists of M machines, where $M-1$ buffer stages are to be determined in between. A fixed nonnegative integer, K, denoting the total buffer space available in the system which has to be allocated among the M-1 buffer locations. For environmental reasons such as layout limitations, each buffer size varies from a lower value to an upper value. Each part goes through all the machines in exactly the same order from the first machine to the last machine. The first machine is never starved, i.e. input is always available, and the last machine is never blocked, i.e. there is always space to dispose of the output. The main aim is to maximize the average production rate of the line, subject to a given total buffer capacity.

3. Proposed Hybrid Approach

In this section, the proposed hybrid genetic annealing algorithm approach has been introduced. As mentioned before, the solution approaches to solve BAP involve applying a generative method and an evaluative method in an iterative manner. As an evaluative tool, a detailed simulation model of the production line is used by estimating the fitness value of each candidate buffer configurations. As a generative tool, the proposed GAA is used to search the solution space and determine optimal buffer sizes. While the flow of GAA and its general structure are similar to the traditional GA , its performance has been enhanced with the adoptation of acceptance test in SA as the replacement test in the genetic algorithm. For the new population, chromosomes are replaced by new chromosomes generated either after the crossover or the mutation operation. The replacement test is designed with the change in cost, ΔE, and the acceptance probability which is determined by a temperature control parameter that decreases during the SA procedure. The proposed approach is described step by step as follows:

Step 0: Initialize SA and GA parameters
Step 1: Initial population generation
Step 2: Evaluation of initial population using simulation module
Step 3: Selection of parents for crossover
Step 4: Crossover and mutation
Step 5: Evaluation of offsprings using simulation module
Step 6: Replacement of the individuals to next generation using SA algorithm strategy
Step 7: Lower the annealing temperature
Step 8: Termination check

4. Computational Study

In this study, the BAP in a serial production line is addressed. The considered system consists of five machines, where four buffer stages are to be determined in between. The total buffer space available in the system which has to be allocated among four buffer locations is considered to be at most 50. Moreover, it should be noted that an upper bound, i.e. 15, and a lower bound, i.e. 1, are considered for each buffer location.

As an evaluative method, discrete event simulation is used to estimate the expected value of the production rate of this stochastic combinatorial optimization problem. The detailed simulation model of this line has been developed using simulation language Arena V10.0, assuming that the process times, the failure times and the repair times of each machine are stochastic and exponentially distributed. The system has been simulated around 50000 time units with ten independent replications.

As a generative tool, the GAA has been coded in Matlab V7.6 software product using a 2 Ghz Pentium (R) 4 CPU processor. GAA parameters to be used in proposed hybrid approach for Pc (crossover probability), Pm, (mutation probability), population size, generation number, initial temperature (T0), and cooling constant for exponential schedule are set to 0.80, 0.033, 50, 20, 0.5 and 0.9, respectively. The results of the proposed hybrid approach to solve the BAP are given in Table 1. The columns in this table show the genaration number, the buffer size configurations, maximum production rate, average production rate and the relative difference which is found as by dividing (maximum fitness – average fitness) to maximum fitness.

Table 1. Results of the genetic annealing algorithm procedure for the five-station serial production system

Gen. number	*Buffer Configuration* B_1	B_2	B_3	B_4	*Total Buffer Size*	*Max Production Rate*	*AVG Production Rate*	*Relative Difference*
1	2	14	7	2	25	0.09106	0,089658	0.015396
2	15	10	10	2	37	0.09128	0.089658	0.017770
3	3	10	11	7	31	0.09150	0.089644	0.020284
4	13	14	15	6	48	0.09150	0.089646	0.020262
5	9	15	9	13	46	0.09150	0.089650	0.020219
6	13	14	15	6	48	0.09150	0.089646	0.020262
......	......	.....	....	....	....	....	....	....
8	11	6	10	15	42	0.09150	0.089646	0.020262
9	6	14	13	10	43	0.09150	0.089650	0.020219
10	3	11	13	14	41	0.09150	0.089650	0.020219
11	7	14	13	14	48	0.09150	0.089646	0.020262
12	3	14	9	11	37	0.09150	0.089656	0.020153
......	......	.....	....	....	....	....	....	....
16	3	14	9	7	33	0.09150	0.089656	0.020153
17	3	14	9	10	36	0.09150	0.089656	0.020153
18	3	14	9	11	37	0.09150	0.089656	0.020153
19	**5**	**14**	**11**	**2**	**32**	**0.09128**	**0.089662**	**0.017726**
......	......	.....	....	....	....	....	....	....

It has been noted that the best average production rate is found after nineteen iterations as 4084 products in the specified time with the total minimum buffer sizes. As a result, with the total number of buffer size of 32, B_i={5,14,11,2} is identified as best buffer configuration in 43.859 minutes as a computational time.

In order to evaluate the performance of the hybrid procedure, traditional SA and GA based simulation optimization approaches have been employed for the specified BAP. First, SA based simulation optimization approach has been employed with exactly the same parameters values proposed in [1] and these parameters used in SA based algorithm for *maximum trials at given temperature, maximum successes (U) at given temperature, initial temperature* (T_0), *cooling constant* for exponential schedule and final temperature (T_f) are set to 500, 50, 0.5, 0.90 and 0.001, respectively. The second approach for comparison with hybrid approach is the simulation optimization method in which traditional genetic algorithm in conjunction with simulation model works as an optimizer in search of an optimal buffer allocation for the average production rate. GA parameters used in GA-based simulation optimization approach for *Pc*, *Pm*, *population size*, and *generation number* are set to 50, 20, 0.80 and 0.033 respectively. Unlike the replacement strategy used for the proposed hybrid approach, elitism replacement scheme is used for GA-based simulation optimization approach.

Comparing the proposed GAA based approach with traditional SA-simulation and traditional GA-simulation approaches, the results in Table 2 show that the proposed hybrid approach gives good solutions with the average production rate and total required buffer size in reasonably short computation times.

Table 2. Comparison of SA, GA and GAA based approaches

Approach	B_1	B_2	B_3	B_4	Max Production Rate	AVG Production Rate	Total Buffer Size	Computational Time (mins)
SA-simulation	4	15	8	9	0.09150	0.089650	36	176.516
GA-simulation	13	8	14	2	0.09134	0.089666	37	61.990
Hybrid GAA-simulation	**5**	**14**	**11**	**2**	**0.09128**	**0.089662**	**32**	**43.859**

5. Conclusion

This study proposes a hybrid GAA based simulation optimization approach combining the key advantages of simulation, metaheuristics and also hybridization to find optimal buffer sizes in unreliable production system. The proposed method combines the GA and SA which includes some problem specific features with simulation in a closed loop configuration. The capability of the proposed hybrid approach is evaluated with numerical examples. Results show that the proposed approach has great potential to improve the system in terms of total buffer size cost in reasonably short computation times.

In a further study, this study can be solved for large-sized problems and extended to search for what new features can be added to the proposed hybrid GAA so that the capacity of the system can be improved. Moreover, the problem can be also studied in a multi-objective procedure considering different optimality criteria such as inventory cost, buffer acquisition cost minimization.

References

1. D. D. Spinellis and C. T. Papadopoulos, A simulated annealing approach for buffer allocation in reliable production lines. *Annals of Operations Research*. 93, 373-384 (2000).
2. N. Nahas, D. Ait-Kadi and M. Nourelfath, A new approach for buffer allocation in unreliable production lines. *International Journal of Production Economics*, 103, 873-881 (2006).
3. A. A. Bulgak, Analysis and design of split and merge unpaced assembly systems by metamodelling and stochastic search. *International Journal of Production Research*, 44,18-19, 4067-4080 (2006).
4. J. A. Qudeiri, H. Yamamoto, R. Ramli and K.R. Al-Momani, Development of production simulator for buffer size decisions in complex production systems using genetic algorithms. *Journal of Advanced Mechanical Design, Systems, and Manufacturing*, 1, 3, 418-429 (2007).
5. H. Yamamoto, J. A. Qudeiri and E. Marui, Definition of FTL with bypass lines and its simulator for buffer size decision. *International Journal of Production Economics*, 112, 18-25 (2008).
6. M.Amiri and A. Mohtashami, Buffer allocation in unreliable production lines based on design of experiments, simulation, and genetic algorithm. *International Journal of Advanced Manufacturing* (2011).
7. I. Sabuncuoglu, E. Erel and Y. Gocgun, Analysis of serial production lines: characterization study and a new heuristic procedure for optimal buffer allocation. *International Journal of Production Research*. 44, 13, 2499-2523 (2006).
8. I. Um, H. Lee and H. Cheon, Determination of Buffer Sizes in Flexible Manufacturing System by Using the Aspect-oriented Simulation. *International Conference on Control, Automation and Systems,* 1729-1733 (2007).
9. F. Altiparmak, B. Dengiz and A. A. Bulgak, Buffer allocation and performance modeling in asynchronous assembly system operations: An artificial neural network metamodeling approach. *Applied Soft Computing*, 7, 946-956 (2007).
10. M. Hasama, T. Ito and S. Matsuno, Optimization of Buffer-size Allocation Using Dynamic Programming, *International Journal of Mathematical Models and Methods in Applied Sciences*, 5, 1, 125-132 (2011).
11. K. Sörensen and G.K. Janssens, Simulation Results on Buffer Allocation in a Continuous Flow Transfer Line with Three Unreliable Machines, *Advances in Production Engineering & Management,* 6, 1, 15-26 (2011).
12. A. Dolgui, A. Eremeev, A. Kolokolov and V. Sigaev, A genetic algorithm for the allocation of buffer storage capacities in a production line with unreliable machines. *Journal of Mathematical Modelling and Algorithms.* 1, 89-104 (2002).

A SUBSTITUTE KEYWORD SUGGESTION METHOD FOR SEARCH ENGINE ADVERTISING*

QIANG WEI, JIN ZHANG†, GUOQING CHEN, XUNHUA GUO

School of Economics and Management, Tsinghua University
Beijing 100084, China

Keyword suggestion methods are important to effectively match the keywords chosen by the advertisers and the queries in search engine advertising. Existing keyword suggestion methods can only suggest keywords with high complementary relevance based on concurrence analysis which cannot support competitive keyword advertising. This paper proposes a notion of substitute relevance as well as the corresponding definition of the substitute keyword. Moreover, a novel substitute keyword suggestion method named SKS is designed to help advertisers find more appropriately competitive keywords. Examples and experiments demonstrate the efficiency and effectiveness of the proposed method.

1. Introduction

Over the past few years, search engine advertising has been one of the most popular forms of online advertising. In search engine advertising, advertisers create ads beforehand and bid on the relevant keywords. Then the ads are displayed alongside the search results when the users query using the bid keywords. Search engine advertising occupies more than 1/3 Internet Ads market share, due to its high targeting, low cost, high flexibility and high variability. However, only a small fraction of the searching keywords are used by advertisers, though the total number of the searching keywords is estimated to exceed a billion [1]. In addition, as for the volume of searching keywords, a long tail distribution is also observed, which means that only a small part of keywords occupy the vast majority of search traffic, resulting in fierce competition and high auction price among a few keywords [2]. Therefore, it makes sense to automatically explore more relevant keywords with low competition, cheap price and similar traffic by

* This work was partly supported by the National Natural Science Foundation of China (70890083/71072015/71110107027), and the Research Center for Contemporary Management in Tsinghua University.

† Email: zhangj5.09@sem.tsinghua.edu.cn

leveraging keyword suggestion methods to help advertisers increase advertising performance and reduce the cost.

Existing keyword suggestion methods used in today's search engine market usually only suggest keywords with high complementary relevance based on concurrence/association analysis of query logs [1, 3, 4]. For example, Google's Adword Tools [3] and Overture's Keyword Selection Tool [4] recommend keywords using concurrence analysis with the seed keywords in popular queries. These methods only generate keywords that contain the original seed keywords and fail to take into consideration the keywords with substitution characteristics which are deemed important in competitive marketing and could help seize market share of competitors [5]. For example, for a "BMW" advertiser, existing keyword suggestion tool may suggest "BMW Car", "BMW repair", "BMW discount", etc., but could hardly suggest "Audi Car", "Toyota", etc., as an aggressive marketing strategy.

Targeting on this problem, this paper is to propose the notions of substitute relevance and substitute keywords from the perspective of substitution in economics [6], along with a novel substitute keyword suggestion method named SKS which is developed to generate substitute keywords for a seed keyword given by the advertisers.

2. Substitute Relevance and Substitute Keywords

Given the keyword set K containing all the keywords used in search engine advertising and the seed keyword s, all the queries that contain s constitute the set labeled $\{s\}$ and the search volume is marked as $|\{s\}|$. For another keyword k, there are two cases to consider. The first case is that k and s frequently co-occur in a same query together, which means that the volume of the queries containing both k and s, i.e., $|\{sk\}|$, is large enough or more than a specified threshold. The combination of keywords s and k reveals that these two keywords complement each other from the viewpoint of economics [6]. For example, users may frequently query "BMW" with the keyword "Car". The keyword "BMW" is complementarily relevant to "Car". Moreover, the degree of complementary relevance of k with respect to s, denoted as $d^C(k, s)$, could be defined as follows:

$$d^C(k,s) = |\{sk\}| / |\{s\}| \tag{1}$$

The degree of complementary relevance in Equation (1) is to describe the occurring probability of k in all the queries containing s. The larger the degree is, the more semantically complementary k to s. Notably, current mainstream keyword suggestion methods focus on suggesting complementary keywords.

For the second case, k and s do not frequently occur in a same query and have low complementary relevance. However, low degree of complementary relevance does not mean that there is no semantic link between these two words. To describe this, the notion of substitute relevance is introduced.

In economics, the so-called substitutability refers that different products may replace each other in the specific function [6]. For example, various brands of cars could be approximately considered to replace each other. Since keywords could be regarded as special products, the notion of substitutability can be extended to the field of keyword advertising from the semantic perspective, which is called substitute relevance. In many situations, substitute relevance reflects competitiveness, which is considered useful for competitive intelligence analysis [5], such as the keywords "Pepsi" vs. "Coca Cola", "BWM" vs. "Audi", "iPad" vs. "Tablet", etc. Therefore, the substitute relevance between different keywords, which is evaluated based on the query logs, may provide an effective way for the competitive keyword suggestion.

The substitute relevance between keywords k and s reflects three properties: a) Keywords k and s do not frequently occur in a same query; b) Keywords k and s frequently occur in the queries with the same intermediate keyword a respectively; c) The more the intermediate keywords, the higher the degree of substitute relevance of k and s.

Based on the above properties, the degree of the substitute relevance can be described in a probability-based framework in a similar sprit of the complementary relevance. Given an intermediate keyword a, the degree of substitute relevance k with respect to s, denoted as $d^S{}_a(k, s)$, is defined as follows:

$$d_a^S(k,s) = |\{ka\}| / |\{a\} - \{sa\}| \tag{2}$$

Given an intermediate keyword set K_M, there often exist more intermediate keywords between k and s in K_M, e.g., a_1, a_2, …, a_m intermediate keywords for the seed s and k. In other words, the overall degree of the substitute relevance of k with respect to s may be consisted of the $d^S{}_*(k, s)$ on the m intermediate keywords. For an intermediate keyword a, the corresponding weight of k, denoted as $w^S{}_a(k)$, is defined as:

$$w_a^S(k) = |\{sa\}| / |\{s\}| \tag{3}$$

After aggregating all the degree of substitute relevance $d^S{}_a$, the total degree of the substitute relevance of k with respect to s, denoted as $d^S(k, s)$, is defined as:

$$d^S(k,s) = \sum_{i=1}^{m} \{ w_{a_i}^S(k) \times d_{a_i}^S(k,s) \} \tag{4}$$

Through Equations (2)-(4), the degree of the substitute relevance could be calculated to reflect the intensity of competition between *k* and *s*. In this paper, the keywords with high degree of substitute relevance are called substitute keywords that could be leveraged to achieve competitive intelligence analysis and support the corresponding keyword marketing strategy.

3. A Substitute Keywords Suggestion Method - SKS

Based on the above discussions, a keyword suggestion method named SKS (Substitute Keyword Suggestion) is proposed, which is to, based on search query log, automatically generate sufficient substitute keywords for a given seed keyword. The pseudo-codes of the algorithm are given in Figure 1.

```
Initialization: Search log: List = {l1, l2,..., ln};
li is represented by (Ki={k1,k2,...,km} , Volumei), Volumei is the corresponding search volume;
Set of suggested keywords: KM = ∅; Set of intermediate keywords: IK = ∅;
Seed keyword s; Threshold Volume Volume;
Step 1: For ∀ li ∈ List {            /* obtain the set of IK and the volumes of s and sa */
  If s ∈ Ki then {
   Volume_of_s += Volumei;         /* compute the search volume of s*/
   If (a ∈ Ki) && (Volumei ≥ Volume) then {Volume_of_sa += Volumei; IK = IK ∪ {a};}
  }
}
Step 2: For ∀ li ∈ List {            /* obtain the set of KM and the volumes of a */
  For ∀ a ∈ IK { If a ∈ Ki then Volume_of_a += Volumei;
   For ∀ k ∈ Ki {
    If (SearchVolume(sk) < Volume) && (SearchVolume(ak) ≥ Volume ) then KM = KM ∪ {k};
  }
}
Step 3: For ∀ a ∈ IK {               /* compute the value of wSa(k) and dSa(k, s) */
  Calculate wSa(k); For ∀ k ∈ KM {Calculate dSa(k, s)}
}
Step 4: For ∀ k ∈ KM {               /*compute the value of dS(k, s) */
  For ∀a ∈ IK { dS(k, s) = dS(k, s) + wSa(k)×dSa(k, s); }
  Update KM with dS(k, s);}
Step 5: Rank(KM); Output(KM);   /* rank the value of dS(k, s) */
```

Figure 1. The pseudo-codes of SKS method.

In this algorithm, the first step is to obtain the intermediate keyword set *IK* and the search volumes of *s* and *sa*. In Step 2, the keyword set of K_M and the search volumes of the keywords *a* in *IK* are obtained. According to *IK* and K_M, the value of $w^S{}_a(k)$ and $d^S{}_a(k, s)$ can be computed in Step 3. Step 4 computes all the substitute relevance values for the keywords in K_M. Then the substitute relevance values are ranked by a sorting algorithm in Step 5 and the keywords in K_M with the first *l* largest substitute values are provided to the advertisers.

4. Examples and Experiments

To further examine the efficiency and effectiveness of SKS proposed in this paper, this section conducts efficiency experiments on synthetic data of different scales and introduces an example. The synthetic data was constructed randomly. In other words, the queries with the seed *s* were selected randomly and the keywords in a given query are generated randomly. The data experiment environment was a Windows XP system on a PC with Intel E8400 CPU and 1GB RAM. The programs were implemented with the basic routines in VC 6.0.

For the time and space complexity, the experiments were conducted by finding substitute keywords with the proposed method SKS from 10^3 to 10^4 scales of query logs. Figure 2 shows the running time of different scales, revealing a polynomial complexity. Figure 3 shows the maximum memory of different scales. The results reveal a nearly linear trend of the space complexity.

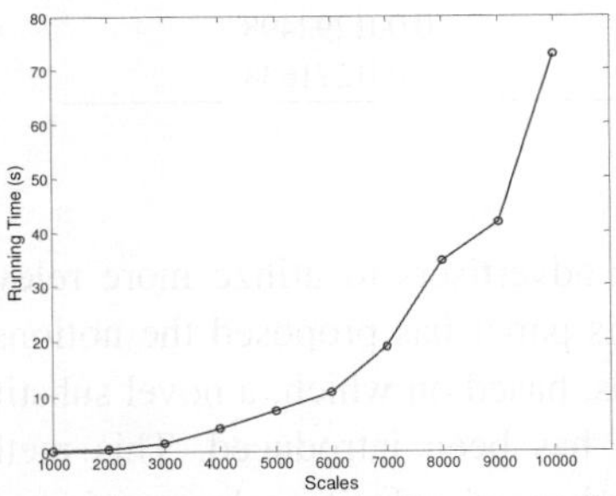

Figure 2. Running time of different scales.

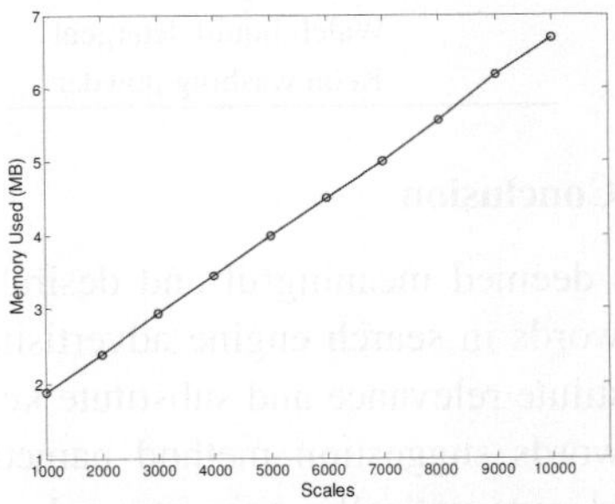

Figure 3. Memory used of different scales.

To explain the effectiveness of SKS, a real-world example is given by setting the seed keyword to "Tide" that is a well-known P&G detergent brand in China. By processing and analyzing the query logs crawled from Google's AdWord Tool, and filtering the keywords with low search volumes, the intermediate keywords of "Tide" were obtained, such as "washing powder", "liquid detergent", "dioxane" and so on. Moreover, the keywords "production equipment", "environmental formula", "OMO", "Fresh Hygiene", "Green Umbrella", "Keon", "Ariel" and some others were found out to be the candidates of the substitute keywords. By computing the substitute relevance values with Equations (2)-(4), the top ten keywords with highest substitute relevance values were recommended to the advertisers and listed in Table 1.

In Table 1, the keywords "OMO", "Fresh Hygiene", "Green Umbrella", "Keon", "Ariel", "Pahmi", "Walch", and "IEKK" are also detergent brands in China, which are competitive to "Tide". The recommendation results in Table 1 show that the substitute keywords suggested by SKS were very competitive and could support the advertisers to achieve competitive intelligence and the corresponding marketing strategy. On the other hand, the suggested keywords and the seed keyword were sufficiently relevant. This provides a chance to advertisers to leverage more keywords with low bidding prices, which is very helpful for advertisers to reduce the budget of online advertising.

Table 1. The recommendation results of the seed keyword "Tide".

Substitute Keywords	Substitute Relevance Values d^S
Detergent production equipment	0.003884276
OMO liquid detergent	0.003788358
Fresh Hygiene liquid detergent	0.003788358
Green Umbrella liquid detergent	0.002027886
environmental formula washing powder	0.001965259
Ariel liquid detergent	0.001626765
IEKK washing powder	0.001618448
Pahmi liquid detergent	0.001292498
Walch liquid detergent	0.001292498
Keon washing powder	0.001271638

5. Conclusion

It is deemed meaningful and desirable for advertisers to utilize more relevant keywords in search engine advertising. This paper has proposed the notions of substitute relevance and substitute keywords, based on which, a novel substitute keywords suggestion method named SKS has been introduced. This method could automatically generate a large number of substitute keywords to be candidates for search engine advertising, which is considered useful to do competitive intelligence analysis and reduce the budget of advertising.

References

1. K. Bartz, V. Murthi and S. Sebastian, Logistic regression and collaborative filtering for sponsored search term recommendation, In *Second Workshop on Sponsored Search Auctions*, 2006.
2. V. Abhishek and K. Hosanagar, Keyword generation for search engine advertising using semantic similarity between terms, In *Proceedings of the ninth international conference on Electronic commerce*, pp. 89-94, 2007.
3. Google AdWords Keyword Tool, https://adwords.google.com/select/KeywordToolExternal
4. Overture Keyword Selection Tool, http://inventory.overture.com/d/searchinventory/suggestion/
5. L. Kahaner, Competitive Intelligence: How to Gather, Analyze and Use Information to Move your Business to the Top, Touch stone, NY.1998
6. M. Nicholson, Microeconomic Theory, The Dryden Press, 1998.

SUPPORTING SITUATION AWARENESS USING NEURAL NETWORK AND EXPERT SYSTEM

MOHSEN NADERPOUR*, JIE LU

Decision Systems and e-Service Intelligence Laboratory
Centre for Quantum Computation and Intelligent Systems (QCIS)
School of Software, Faculty of Engineering and Information Technology
University of Technology, Sydney, P.O. Box 123, Broadway, NSW 2007, Australia

Situation awareness (SA) is a critical factor for human decision making and performance in dynamic environments. Actually SA is a mental model of the current state of the environment and includes many types of complex systems such as safety supervisory systems. The current paper employs two focus areas including neural network and expert system for maintaining SA in a safety supervisory system. The neural network components provide adaptive mechanisms for perception, and the expert system offers the ability to support comprehension and projection.

1. Introduction

Situation awareness (SA) involves being aware of what is happening in the vicinity to understand how information, events, and actions will impact the goals and objectives. Lacking or inadequate SA has been identified as one of the primary factors in accidents attributed to human error. Thus, SA is especially important in work domains where the information flow can be quite high, and poor decisions may lead to serious consequences. The idea did not receive much attention until the late 1980s, but has become a hot topic ever since.

The primary research came from the aviation industry but a similar review in other domains, such as nuclear power showed that this is not a problem limited to aviation, but one faced by many complex systems where combining and presenting vast amounts of data now available from many technological systems in order to provide true SA is a big challenge (whether it is to a pilot, a physician, a business manager, or an automobile driver) [1].

Despite having its roots in aviation, it has been suggested that the concept is equally applicable to human supervisory control for land based industries. It is

* Mohsen.Naderpour@student.uts.edu.au

also argued that problems in human supervisory control may be due to poor situational awareness [2], such as: 1) failure to detect critical cues regarding the state of the system; 2) failure to interpret the meaning of information perceived via SCADA (supervisory control and data acquisition) technology; 3) failure to understand individual task responsibilities and the responsibilities of others; 4) failure to communicate with other operators in the team; and 5) failure to communicate with other teams.

This paper considers the applicability of SA concepts to safety in the control of complex systems. Safety supervisory systems continue to increase in degree of automation and complexity as operators are decreasing. As a result, each operator must be able to comprehend and respond to an ever increasing amount of available risky status and alert information. In our previous work, we proposed an innovative conceptual model that assesses situations and helps the decision maker to take appropriate action in hazardous situations [3]. In this paper, we will describe how neural network and expert system can support SA in a complex environment.

2. Basic Concepts

2.1. *Situation Awareness*

One of the earliest and most widely used definitions of SA describes it as the "perception of the elements in the environment within a volume of time and space, the comprehension of their meaning and the projection of their status in the near future". Based on this definition, SA is comprised of three levels: (1) perception, (2) comprehension, and (3) projection.

Level 1 SA, perception, involves the sensory detection of significant environmental cues. For example, operators need to be able to see relevant displays or hear an alarm sound. Comprehending the meaning or significance of that information in relation to operator's goals is also important. Operators with good Level 2 SA are able to understand the immediate impact of an outage on other parts of the system. Projection, the highest level of SA, consists of extrapolating information forward in time to determine how it will affect future states of the operating environment [1].

2.2. *Neural Networks*

Neural network or artificial neural network (ANN) is the interconnection of artificial neurons that tends to simulate the nervous system of a human brain. ANNs are commonly used in machine learning and utilize supervised, unsupervised, and reinforcement learning approaches to achieve predictive properties based on example data. Unsupervised learning can be effective when ground-truth is not available with a dataset. Supervised learning provides a means to use experience to correctly classify yet unseen situations. Reinforcement learning offers promise for machine learning in difficult learning environments by taking advantage of feedback about the performance of a system. In particular, they can be embedded into appropriate sensors for operation at the lowest levels of information fusion with effective, but low-complexity designs. Although neural networks have been applied to sensor fusion, their use in SA has been limited, possibly because of the lack of rich training data for this problem [4].

2.3. *Expert Systems*

The basic idea behind expert systems (ES) is simply that expertise, which is the vast body of task-specific knowledge, is transferred from a human to a computer. This knowledge is then stored in the computer and users call upon the computer for specific advice as needed. The computer can make inferences and arrive at a specific conclusion. Then like a human consultant, it gives advices and explains, if necessary, the logic behind the advice. A rule-based ES is defined as one, which contains information obtained from a human expert, and represents that information in the form of rules, such as IF–THEN. The rule can then be used to perform operations on data to inference in order to reach appropriate conclusion. These inferences are essentially a computer program that provides a methodology for reasoning about information in the rule base or knowledge base, and for formulating conclusions [5].

2.4. *Supporting Situation Awareness using NN and ES*

In an attempt to effectively provide SA for a decision maker, we have brought together two focus areas of research. An overview of these areas is depicted in Figure 1. For determining the aspects that are important for an operator's SA in a safety supervisory system, we use the goal-directed task analysis which was introduced in [1]. The results are showed in Table 1.

In a complex and dynamic environment, a casualty requires immediate response. There are often multiple actions required in a very short time and the trend toward smaller operators' size and increasing automation has resulted larger volumes of more and more complex information in a dynamic environment. Therefore, it is necessary to assist the operators in quickly understanding the available information by providing that data in a format that allows them to act quickly and correctly. The information provided for situational awareness must be more than just information gathering. This implies collecting the right multi-domain information across a net-centric environment for shared awareness and presenting results for the human to understand and make quick decisions. Any new approach must efficiently bring together the human operator, sensor equipment data, and real world events to provide a subset of actionable information [4].

Table 1. Safety goals and decisions

1.0 Eliminate or reduce the risks to a level that is as low as reasonably practicable (ALARP)
1.1 Hazards identification
1.1.1 Causes
1.1.2 Consequences
1.2 Risks reduction
1.2.1 Practical options

For the safety supervisory system, critical factors in the environment (SA level 1) relate to the actual state of the system and identifying the hazards. In this step we rely on neural networks. The historical database that includes information related with normal and abnormal situations can be used to train the ANN structure. The on-line or off-line measurements from the plant are the ANN's inputs. The ANN's outputs are the signals of different suspected hazards. These outputs are a subset of the set of ES's inputs.

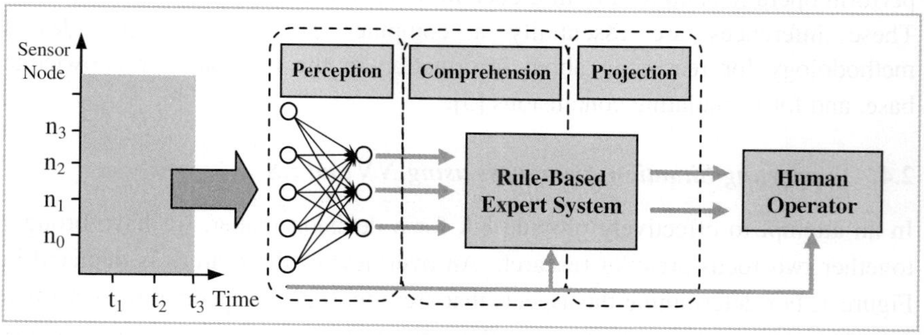

Figure 1.Situation awareness diagram

SA level 2 relates to the operator's understanding of the system as a whole, and emergent events. At this level, it is also necessary to understand the causality and consequences of the hazards. For SA level 3, the system and operator should understand future required actions. This is not only limited to responding to changing engineering conditions, it also addresses the changing tactical situation to eliminate or reduce the risks. For both these levels we utilize an expert system. The HAZOP analysis and FTA are useful to build the knowledge base (KB).

This base relies on the knowledge of the operators and engineers about the process and allows formulating artificial intelligence algorithms. Our approach still relies on the effectiveness of the operator. If the approach is focused solely on the automated features, then the operator can become more disconnected from the tools and resources needed to assess situations and make objective and effective decisions.

3. Functional Demonstration

Chemical plants are one of the complex environments that the supervision tasks have increased considerably. A decision support system is needed to assist operators in understanding and assessing status and responding quickly to risky situation. For example, in the explosion of Texas City, TX BP Amoco Refinery, that 15 workers were killed and 170 injured, the key problem was the poor operator's SA in a complex, fast moving environment [6]. For functional demonstration, an ethylbenzene process plant, involving two reactors and two distillation columns, is chosen. The safety system includes eleven sensors which describe in Table 2. For perception phase we use BP neural network that has eleven inputs ($I_1,\ldots,I_{11}$) and 6 outputs ($O_1,\ldots,O_6$) that show the possible hazards for this plant as shown in Table 2. Using steady state data from simulation or from experimental data, when available, the ANN is trained and the outputs are useful for the next step. The activation functions are sigmoidals in the hidden layer and linear in the output.

Table 2. Sensors and hazards

Zone	Sensor	Description	Hazard	Description
R1	TCR1	Temperature controller	H1	High Temperature in R1
	LCR1	Level controller	H2	High Temperature in R2
R2	LCR2	Level controller	H3	High Pressure in R1
D1	TCD1	Temperature controller	H4	High Level in D1
	PCD1	Pressure controller	H5	High Level in D2
	LC1D1	Level controller	H6	High Pressure in D2
	LC2D1	Level controller		
D2	TCD2	Temperature controller		
	PCD2	Pressure controller		
	LC1D2	Level controller		
	LC2D2	Level controller		

For comprehension and projection phases, a set of if–then rules has been defined to build the knowledge base of expert system. The antecedents are usually the outputs of the ANN with or without some measured/calculated variables. For example the rule which is relevant to H1, is shown in Table 3. To simplify and accelerate the performance of expert system, forward reasoning that starts from the fact, reasoning towards target is used.

Table 3. Set of rules

Rule		Antecedent	Causes	Consequences	Suggestions
Rule 1	IF	H1			
	THEN		Poor cooling or input low level	shut-down, release and explosion	switch to redundancy pump in cooling system and administrative checks
...					

4. Conclusion and Future Work

As safety supervisory systems continue to increase in degree of automation and complexity, the task of providing actionable information for SA becomes more difficult and costly to achieve. The proposed approach employs neural network and expert system to support SA in a safety supervisory system. Reinforcement learning is a general area of research worth pursuing in the area of SA, so in future work we will attempt to create principles for employing reinforcement learning to support SA.

References

1. M. R. Endsley, "Situation Awareness," in *Handbook of Human Factors and Ergonomics*, G. Salvendy, Ed.: John Wiley and Sons Press 2006, pp. 528-542.
2. N. A. Stanton, P. R. G. Chambers, and J. Piggott, "Situational awareness and safety," *Safety Science,* vol. 39, pp. 189-204, 2001.
3. M. Naderpour, J. Lu, and E. Kerre, "A Conceptual Model for Risk-Based Situation Awareness," in *International Conference on Intelligent Systems and Knowledge Engineering (ISKE)*, Shanghai, China, 2011, pp. 297-306.
4. N. G. Brannon, J. E. Seiffertt, T. J. Draelos, and D. C. Wunsch Ii, "Coordinated machine learning and decision support for situation awareness," *Neural Networks,* vol. 22, pp. 316-325, 2009.
5. L. Shu-Hsien, "Expert system methodologies and applications: a decade review from 1995 to 2004," *Expert Systems with Applications,* vol. 28, pp. 93-103, 2005.
6. J. L. Pridmore, "Designing for the improvement of operator situation awareness in automation systems," *Ph.D. Thesis*, United States-Alabama: Auburn University, 2007.

MAKING INTELLIGENT DECISIONS ON NOISY IMAGES

GONZALO FARIAS
Escuela de Ingeniería Eléctrica, Pontificia Universidad Católica de Valparaíso (PUCV), Valparaíso, 2362804, Chile.

SEBASTIÁN DORMIDO-CANTO
Departamento de Informática y Automática, ETSI-Informática, UNED, Madrid, 28040, Spain.

MATILDE SANTOS
Departamento de Arquitectura de Computadores y Automática
Facultad de Informática, Universidad Complutense de Madrid, Madrid, 28040, Spain.

JESÚS VEGA
Asociación EURATOM/CIEMAT para fusión, Avenida Complutense 22, Madrid, 28040, Spain.

Making successful decisions on image classification can be influenced by the level of present noise. The noise on data can be reduced applying many classical and advanced techniques such as low pass filters or wavelets. However, in some cases the presence of the noise is not global but located only in some particular regions of the image. In these situations the application of global filters over the entire image is not a suitable option since the noise and the information are equally reduced. This article describes an intelligent approach to eliminate noise and avoid affecting the overall image.

1. Introduction

Classification data is a relevant problem in pattern recognition, especially if the classification has to be done in massive databases [1-4]. Usually, previous data processing is needed in order to improve the successful rate of classifiers. One of the most important pre-processing tasks is noise reduction. Real data is acquired by sensors with non-free error measurements and is highly recommended to use filters in order to reduce the impact of such error measurements on the images. The range of techniques that can be used for this purpose goes from the classical low pass filters to some advanced techniques such as wavelets [5]. However many of such algorithms are globally applied to all components of the data, reducing in the same proportion noise and significant information. This kind of pre-processing is particularly inadequate on images with noise located only on specific regions. Such local noise can be produced by faulty optics, flare effects or straight light [6]. Thus, searching and identification of the noise can be a suitable solution for this type of problem, reducing successfully local noise without eliminating relevant data on the image.

In past decades, most object recognition approaches were based on the sliding window scanning [7-8], localizing the bounding box of an object. This technique is computationally expensive and is mostly appropriate for object classes that are well-approximated by a rectangle, but not for objects with irregular borders (i.e., with noise). Recently, there has been a lot of interest in using results of generic image segmentation algorithms [9-10]. Segmentation is used to subdivide an image into a set of regions. These regions do not have predefined shapes like rectangular patches and their boundaries are more likely to overlap with the boundaries of other regions in the image. Segmentation provide a usually much smaller set of hypothesis compared to the sliding window, and more close to reality scales are obtained through segmentation.

This paper is organized as follows. First the problem of local noise is described in Section 2. Section 3 briefly describes the sliding window method. Section 4 introduces the segmentation algorithm and the proposed approach to reduce local noise on images. Section 5 summarizes the main conclusions.

2. Local noise on images

We can use two different approaches in order to remove some specific regions associated to the noise: 1) exhaustive detection and, 2) region extraction with connected objects. To explain these approaches let us consider a toy example (Figure 1a). The whole image has 10x12 pixels. The goal is to remove de noise which is defined by the sets of pixels on the left part of the image (9,4), [(2,3), (2,4), (3,3), (4,3)] and [(6,2), (6,3), (7,2), (8,2)] (orange). The background color in the image corresponds with the white color, for example pixel (1,1).

3. Exhaustive detection

This method is based on the sliding window scanning throughout the image. It is necessary to move the template associated to the noise in order to find matches. After that, all coincidences are replaced by background color.

Figure 1 shows two matches for the selected noise template and how they are replaced by background. As the template moves through the image we must decide if what is inside is noise or not. In this example, it is sufficient to consider that in the template there are four pixels correctly oriented. It is easy to see some problems that can occur with this technique: What happens if there are connected pixels that lie outside the template? What determines whether pixels within the template are noise or not? The technique of extraction regions with connected components tries to answer all these questions.

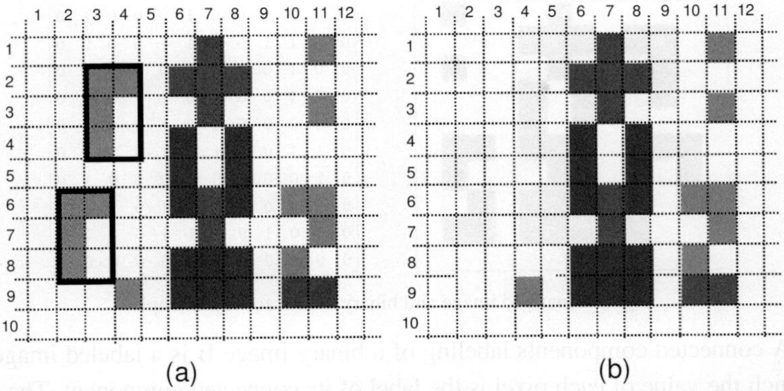

Figure 1. (a) Original image with two matches for a noise template; (b) Matches eliminated.

4. Extraction of regions with connected components

The extraction of regions with connected components is based on the segmentation theory. Segmentation refers to the process of partitioning a digital image into multiple segments (sets of pixels). More precisely, image segmentation is the process of assigning a label to every pixel in an image so pixels with the same label share certain visual characteristics. In general, a segmentation of an image is a partition into connected sub-images (regions) R_1, R_2,..., R_n such that all regions are disjointed, and the union of all of them makes up the image. Each sub-image satisfies a predicate such as all pixels in any sub-image R_i must not differ by more than Δx gray levels, all pixels in any sub-image R_i must be joined by a connectivity factor, etc.

The procedure proposed to reduce noise by using image segmentation is shown in Figure 2. Each step of the process is explained below.

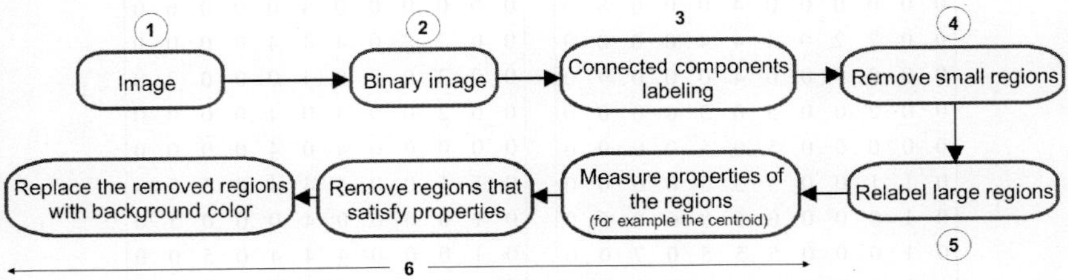

Figure 2. Flowchart for region extraction with connected components.

A binary image can be obtained from a gray scale or color image by thresholding, which selects a subset of the image pixels as foreground pixels and leaves the rest as background. Figure 3 shows the binarized image and the binary matrix corresponding to Figure 1a.

$$\begin{bmatrix} 0&0&0&0&0&0&1&0&0&0&1&0\\ 0&0&1&1&0&1&1&1&0&0&0&0\\ 0&0&1&0&0&0&1&0&0&0&1&0\\ 0&0&1&0&0&1&0&1&0&0&0&0\\ 0&0&0&0&0&1&0&1&0&0&0&0\\ 0&1&1&0&0&1&1&1&0&1&1&0\\ 0&1&0&0&0&0&1&0&0&0&1&0\\ 0&1&0&0&0&1&1&1&0&1&0&0\\ 0&0&0&1&0&1&1&1&0&1&1&0\\ 0&0&0&0&0&0&0&0&0&0&0&0 \end{bmatrix}$$

Fig. 3. Binarized image and binary matrix for toy example.

A connected components labeling of a binary image B is a labeled image, LB, in which the value of each pixel is the label of its connected component. The pixels of a binary image are 0's and 1's; the 1's stand for foreground pixels and the 0's background ones. The term B[r,c] denotes the value of the pixel located at row r, column c of the binary matrix of the image. A binary matrix has M rows and N columns. Thus B[1,1] refers to the value of the upper leftmost pixel and B[M,N] refers to the value of the lower rightmost pixel. The connected components labeling is associated to a connectivity factor between neighbors. The two most common areas of neighbors are the four-neighbors and the eight-neighbors of a pixel. The four-neighborhood N4(r,c) of pixel (r,c) includes pixels (r-1,c), (r+1,c), (r,c-1) and (r,c+1), which are often referred to as its north, south, west and east neighbors, respectively. The eight-neighborhood N8(r,c) of pixel (r,c) includes each pixel of the four-neighborhood plus the diagonal neighbor pixels (r-1,c-1), (r-1,c+1), (r+1,c-1) and (r+1,c+1). Figure 4 shows the labeled matrices (LB) with respect to both the four-neighborhood and the eight-neighborhood. There are nine and seven objects (regions) for N4 and N8 respectively.

$$\begin{bmatrix} 0&0&0&0&0&0&4&0&0&0&8&0\\ 0&0&2&2&0&4&4&4&0&0&0&0\\ 0&0&2&0&0&0&4&0&0&0&9&0\\ 0&0&2&0&0&5&0&5&0&0&0&0\\ 0&0&0&0&0&5&0&5&0&0&0&0\\ 0&1&1&0&0&5&5&5&0&6&6&0\\ 0&1&0&0&0&0&5&0&0&0&6&0\\ 0&1&0&0&0&5&5&5&0&7&0&0\\ 0&0&0&3&0&5&5&5&0&7&7&0\\ 0&0&0&0&0&0&0&0&0&0&0&0 \end{bmatrix} \quad \begin{bmatrix} 0&0&0&0&0&0&4&0&0&0&6&0\\ 0&0&2&2&0&4&4&4&0&0&0&0\\ 0&0&2&0&0&0&4&0&0&0&7&0\\ 0&0&2&0&0&4&0&4&0&0&0&0\\ 0&0&0&0&0&4&0&4&0&0&0&0\\ 0&1&1&0&0&4&4&4&0&5&5&0\\ 0&1&0&0&0&0&4&0&0&0&5&0\\ 0&1&0&0&0&4&4&4&0&5&0&0\\ 0&0&0&3&0&4&4&4&0&5&5&0\\ 0&0&0&0&0&0&0&0&0&0&0&0 \end{bmatrix}$$

(a) (b)

Figure 4. Connected component labeling (a) N4, (b) N8

The next step is to remove from LB all regions that have fewer than P pixels (small regions), and then the remaining regions are relabeled. After that, for each labeled region in the label matrix (LB) it is possible to calculate certain properties of the regions such as the area, centroid, circularity, elongation, etc. The regions which meet the desired properties are replaced by background color and small regions that were deleted are restored. In our example the parameters, properties and conditions have been the following: N8, P=2 and elimination of regions whose centroid is on the left half plane. The final result is exactly the same as figure 1.b. Figure 5 shows the state of the image processing after each stage of the procedure described in figure 2.

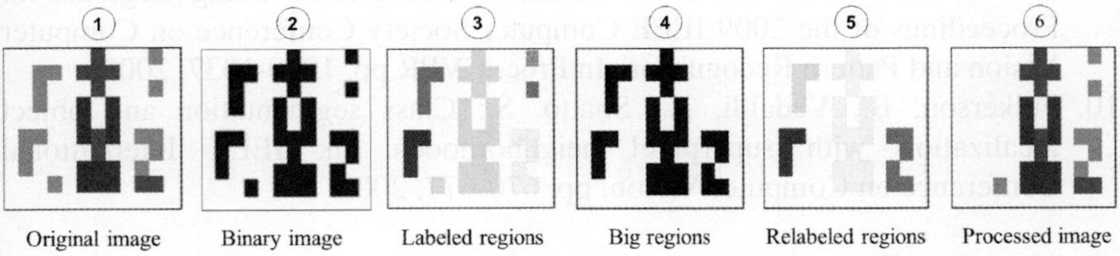

Figure 5. Step by step image processing through the algorithm proposed.

5. Conclusions

In this work a simple but efficient intelligent approach to eliminate local noise on images is proposed and described. The application of classical methods, such filters, is not adequate to face this kind of noise. Extraction regions with connected components can be used to separate significant information from the noise on images. The noise can be identified by using some region properties. After this pre-processing task, performance of classification methods should be enhanced.

References

1. Theodoridis, S. and Koutroumbas, K., Pattern recognition. 2nd ed., Academic Press, 2004.
2. Lakovidis, D. K. et al., A Pattern Similarity Scheme for Medical Image Retrieval, IEEE Transactions on Information Technology in Biomedicine,vol.13, no.4, pp. 442-450, 2009.
3. Farias, G., Santos, M., López, V. Making Decisions on Brain Tumour Diagnosis by Soft Computing Techniques, Soft Computing, ISSN 1432-7643, Vol. 14, No. 12, pp. 1287-1296, 2010.
4. Makili, L., Vega, J., Dormido-Canto, S., Pastor, I., Pereira, A., Farias, G., Portas, A., Pérez-Risco, D., Rodríguez-Fernández, M.C., Busch, P. Upgrade of the automatic analysis system in the TJ-II Thomson Scattering

diagnostic: New image recognition classifier and fault condition detection. Fusion Engineering and Design 85 (2010) 415–418.
5. Mallat, S., A wavelet tour of signal Processing. 2nd ed, Academic Press, 2001.
6. R. P. Breault, Control of Stray Light in Handbook of Optics, Vol. I, McGraw-Hill, Inc., New York, 1995.
7. Rowley, H., Baluja, S., Kanade, T.: Human face detection in visual scenes. Advances in Neural Information Processing Systems, pp. 875-881, 1996.
8. Ferrari, V., Fevrier, L., Jurie, F., Schmid, C.: Groups of adjacent contour segments for object detection. IEEE Transactions on Pattern Analysis and Machine Intelligence 30, pp. 36-51, 2008
9. Gu, C., Lim, J., Arbelaez, P., Malik, J.: Recognition Using Regions. In: Proceedings of the 2009 IEEE Computer Society Conference on Computer Vision and Pattern Recognition, In Proc. CVPR pp. 1030-1037, 2009.
10. Fulkerson, B., Vedaldi, A., Soatto, S.: Class segmentation and object localization with superpixel neighborhoods. In: IEEE International Conference on Computer Vision, pp. 670-677, 2009.

SME CLUSTERING BASED ON INNOVATION NEEDS

AYCA ALTAY, DIDEM CINAR, GULGUN KAYAKUTLU

ITU Dept. of Industrial Engineering, Macka
Istanbul, 34376, Turkey

IREM DUZDAR

Arel University, Dept. of Industrial Engineering, Tepekent
Istanbul, Turkey

Innovation is a competitive obligation for the SMEs but the needs vary by region. In this study, Turkish SMEs in Thrace region are clustered using Fuzzy C-Means which is a strong clustering tool for qualitative data, in order to compare the past and current innovational needs of SMEs. Similar questions are surveyed in 2007 and 2011and the results are evaluated to emphasize the innovation based improvements.

1. Introduction

Innovation is a competitive obligation for the small and medium size companies (SME). They are known to be agile in change, but fragile in facing the economic fluctuations (Levy and Powell, 1998). This study is realized to evaluate the improvements in a regional approach for innovation, which is a risky field. Clustering is used to define the natural similarities and differences in the innovation based needs of SME groups. SME managers allow uncertainties in defining the needs for innovation; that is why Fuzzy C-Means is chosen as the clustering method. This paper will present the clusters structured as a result of 2011 survey and compares them with the similar study run by authors in 2007.

The paper is so organized that the next section provides a brief literature review on Fuzzy C-Means clustering and innovation of SMEs. The application is described and discussed in the third section. The last section is reserved for the conclusion and suggestions.

2. SME Clustering

SME is defined as an enterprise with less than 250 employees, a turnover less than 50 million € and a balance less than 43 million € (European Comission, 2003). SME innovation studies include clustering of the pre-generated classes, as industrial sector, geography or company size. Lazoi et al. (2011) propose aerospace industry SME clusters in Italy to collaborate only with bigger companies. Braun (2003) analyzes tourism industry and offers clusters with focus on IT Systems. SME characteristics are recognized and precautions are proposed cluster-wise as in Chen and Can (2006). Carpinetti et al. (2007) proposes a performance measure for each cluster. Min (2010) encourages a cooperative game theory model for sustainable innovation. SME Clustering for the regional strategies is introduced by Cinar& Kayakutlu (2009).

3. Fuzzy C-Means

Fuzzy sets can be used to handle vagueness in SME clustering. Each data point belongs to all determined clusters with a different degree of membership instead of one certain cluster as in the crisp case (Yang and Shih, 2001). Baraldi and Blonda (1999) give a theoretical background on fuzzy clustering.

Fuzzy C-Means algorithm is one of the most widely used fuzzy clustering algorithms (Wu and Yang, 2002). Clustering the GSM users to investigate market success (Karahoca and Karahoca, 2011), bankruptcy forecasting (De Andres et al., 2011), medical diagnosis (Chaira, 2011; Lim et al., 2005), clustering fire risks in a forest (Iliadis et al., 2011) are some of the recent engineering applications that use Fuzzy C-Means clustering. Cinar and Kayakutlu (2009) cluster SMEs with respect to their need for performance improvement by using hard c-means and fuzzy c-means methods. Fuzzy c-means is found to be the most successful one for this case. The superiority against crisp clustering and vague environment of the problem are the reasons to apply fuzzy c-means algorithm to cluster the SMEs according to their needs and investigate the changes of needs in 4 year period in this study.

Fuzzy c-means clustering is based on minimizing the distances between objects and the center of related clusters. Since each data sample has m coordinates, each cluster center also has m coordinates. j^{th} coordinate of ith cluster center is calculated by

$$v_{ij} = \sum_{k=1}^{n} \mu_{ik}^{m'} \cdot x_{kj} \,/\, \sum_{k=1}^{n} \mu_{ik}^{m'} \tag{1}$$

where μ_{ik} is the membership value that the kth data point has in the ith class and m' is a weighting parameter with a range of $[1,\infty)$. Then the objective function, J_m, of fuzzy clustering is formulated as following:

$$J_m\left(\underset{\sim}{U}, v\right) = \sum_{k=1}^{n} \sum_{i=1}^{c} \left(\mu_{ik}\right)^{m'} \left(d_{ik}\right)^2 \tag{2}$$

where $\underset{\sim}{U}$ is a partition matrix in which the membership values of all data points for all classes are ranked and d_{ik} is the distance of point k to the center of the class i. The fuzzy c-means algorithm can be summarized by the following steps (Bezdek, 1981):

1. The number of classes (c), weighting parameter (m') and level of accuracy (ε_L) are determined.
2. Partition matrix $\underset{\sim}{U}$ is initialized.
3. The center of clusters (Equation 2) and Euclidean distances are computed.
4. $\underset{\sim}{U}$ is updated as follows:

$$\mu_{ik} = \left[\sum_{j=1}^{c} (d_{ik} / d_{jk})^{2/(m-1)} \right]^{-1} \tag{3}$$

If $d_{ik} = 0$ then $\mu_{ik} = 1$.

5. If $\|U_{new} - U_{old}\| \leq \varepsilon_L$ the algorithm is terminated, else returned to step 3.

4. Case Study

In this study the change of SME innovation needs in Thrace Turkey from 2007 to 2011 is investigated. Our objective is to support the innovative performance in our region. A survey is run with 48 SMEs in 2007 and with 51 SMEs in 2011 in the same region. Randomly chosen SMEs are distributed in a variety of industries as in Table 1.

Table 1. Distribution of SMEs according to industries (%)

Industry	2007	2011	Industry	2007	2011
Food	24.0	17.6	Service	0.0	15.7
Clothing & Textile	28.0	13.7	Health	0.0	5.7
Machinery & Electronics	21.0	19.6	IT & Communication	0.0	7.8
Automotive	7.0	2.0	Construction	2.0	2.0
Chemical & Pharmaceutical	9.0	0.0	Furniture	0.0	2.0
Plastics	7.0	2.0	Metal	0.0	2.0
Publishing	2.0	2.0	Miscellaneous	2.0	7.9

The questionnaires are prepared for the regional SME issues as given in Table 2.

Table 2. Needs considered in the survey

Criteria	Need	Criteria	Need
P1	An improved system for business support and advice	P6	Physical infrastructure and better equipped premises
P2	Funds for R&D and knowledge related activities	P7	Improvement on supply chain and logistics
P3	Training and workforce development	P8	Attraction of high value foreign investments
P4	Entering and accessing new markets	P9	Creation of startup companies
P5	Relations with universities and other R&D organizations		

Clustering the responses of the year 2007, regarding 9 criteria, Table 3 is obtained. Best reflections and explanations are provided by 3-clusters with dissimilar need structures. The first cluster which contains majority of the SMEs every need is mentioned. Clustering results indicate that all needs can be classified as vital, but the most vital needs are "system for business support" and "the supply chain and logistics" which are needed by the overwhelming majority of SMEs in this cluster. On the other hand, the firms in the second cluster are in only in vital need of "funds for R&D and knowledge-based activities" whereas the other needs are noted as "critical but not vital". As for the third cluster, only "critical but not vital" needs are observed as "accessing new markets", "an improved physical infrastructure" and "creation of new start-up companies". Consequently, in 2007, the majority of SMEs showed discontent with the company assets and funds for innovation.

Table 3. Cluster of SMEs in the year 2007

	2 Clusters	3 Clusters	4 Clusters
Size of the 1st Cluster	41	30	36
Size of the 2nd Cluster	7	7	5
Size of the 3rd Cluster	N/A	11	4
Size of the 4th Cluster	N/A	N/A	3

Clustering data of the year 2011, regarding the same 9 criteria, Table 4 is obtained. Analyzing the cluster data, the optimum cluster number is found to be 4. All clusters are in vital need of "entering new markets". Three clusters are in vital need of "improved business support", "relations with scientific foundations", and "creation of start-up companies". One cluster vitally needs "workforce development opportunities" in addition to the other needs.

The basic changes within the firms from 2007 to 2011 are highly profound. The critical needs in the year 2007 are revealed as funding R&D activities and foreign investment. Whereas in the year 2011, the critical need is revealed as entering new markets. Other less critical needs involve relations with scientific foundations and improved business support. These requirements indicate that, the needs have evolved into an expansion related dimension.

Table 4. Cluster of SMEs in the year 2011

	2 Clusters	3 Clusters	4 Clusters
Size of the 1st Cluster	38	15	6
Size of the 2nd Cluster	13	25	25
Size of the 3rd Cluster	N/A	11	11
Size of the 4th Cluster	N/A	N/A	9

5. Conclusions

It is observed that awareness for innovation has risen among the SMEs. Collaboration with similar SMEs has become compulsory with the global crisis. In this study, Turkish SMEs in Thrace region are clustered to investigate the past and current innovation states of SMEs. The emphasized needs will enlighten further studies on regional innovation and collaboration strategies.

The case study results indicate that the needs for innovation have changed drastically. In the year 2007, the companies have shown grand need for each innovation resource, especially funding. On the contrary, in the year 2011, these SMEs are in need of growth strategies, new markets and scientific knowledge. Regional policies designed for the basic needs 2007 onwards are to be changed for more collaboration and new market oriented strategies.

Further research may include building the SME clusters that maximizes the innovation capacities and minimizes the research risks to select the collaboration partners and alternative regional policies.

References

1. Baraldi, A. and Blonda, P., 1999, A Survey of Fuzzy Clustering Algorithms for Pattern Recognition—Part I, IEEE Transactions on Systems, Man, and Cybernetics—Part B: Cybernetics, 29(6), 778-785.
2. Bezdek J. C., 1981, Pattern Recognition with Fuzzy Objective Function Algoritms, Plenum Press, New York.
3. Braun, P., 2003, SME Networks: Clustering for Regional Innovation Purposes, 16th Annual Conference of Small Enterprise Association of Australia & New Zealand, Ballarat, Victoria, September 28- October
4. Carpinetti, L.C.R., Gerolamo, M.C. and Galdamez, E.V.C., 2007, Continuous Innovation and Performance Management of SME Clusters, Creativity and Innovation Management, 16(4), 376-385.

5. Chaira, T., 2011, A novel intuitionistic fuzzy C means clustering algorithm and its application to medical images, Applied Soft Computing, 11(2), 1711-1717.
6. Chen, X. and Cao, L., 2006, SME Clusters in China: One Way build up Innovation Capacities, Tech Report, July-August 2006, 38-43.
7. Cinar, D. and Kayakutlu, G., (2009) "Clustering the Open Ended Future Needs", PICMET 2009, Portland, USA.
8. De Andres, J., Lorca, P., de Cos Juez, F. J., Sanchez-Lasheras, F., 2011, Bankruptcy forecasting: A hybrid approach using Fuzzy c-means clustering and Multivariate Adaptive Regression Splines (MARS), Expert Systems With Applications, 38(3), 1866-1875.
9. European Comission 2003. The New SME Definition: User Guide and Model Declarations, Enterprise and Industry Publications.
10. Hair, J.F., Black, W.C., Babin,B.J., Anderson, R.E. and Tatham, R.L., 2006, Multivariate Data Analysis, Pearson Prentice Hall, New Jersey.
11. Iliadis, L. S., Vangeloudh, M. and Spartalis, S., 2010, An intelligent system employing an enhanced fuzzy c-means clustering model: Application in the case of forest fires, Computers and Electronics In Agriculture,70(2),276-284.
12. Karahoca, A. and Karahoca, D., 2011, GSM churn management by using fuzzy c-means clustering and adaptive neuro fuzzy inference system, Expert Systems With Applications, 38(3), 1814-1822.
13. Lazoi, M., Ceci, F., Corallo, A. and Secundo, G., 2011, Collaboration in an Aerospace SMEs Cluster: Innovation and ICT Dynamics, International Journal of Innovation and Technology Management, 8(3), 393-414.
14. Levy, M. and Powell, P., 1998, SME Flexibility and the Role of Information Systems, Small Business Economics, 11(8), 183-196.
15. Lim, C., Kuan, M. and Harrison, R., 2005, Application of fuzzy ARTMAP and fuzzy c-means clustering to pattern classification with incomplete data, Neural Computing & Applications, 14(2), 104-113.
16. Min, L., 2010, Research on Enterprises' Cooperation Innovation Behavior in Yiwu SME clusters, 2nd ICIFE Conference, Changqing, China, 405-408, September 17-19.
17. Valente de Oliveira, J. and Pedrycz, W. (eds), 2007, Front Matter, in Advances in Fuzzy Clustering and its Applications, John Wiley & Sons, Ltd, Chichester, UK. doi: 10.1002/9780470061190.fmatter.
18. Wu, K.L. and Yang, M.S., 2002, Alternative c-means clustering algorithms, Pattern Recognition, 35, 2267 – 2278.
19. Xu, R. and Wunsch, D., 2005, Survey of Clustering Algorithms, IEEE Transactions on Neural Networks, 16(3), 645-678.
20. Yang, M.S. and Shih, H.M., 2001, Cluster analysis based on fuzzy relations, Fuzzy Sets and Systems, 120, 197–212.

ARTIFICIAL NEURAL NETWORK FOR DISCRETE MODEL ORDER REDUCTION WITH FREQUENCY SELECTIVITY

OTHMAN M.K. ALSMADI[1*], ZAER. S. ABO-HAMMOUR[2], ADNAN M. AL-SMADI[3]

[1]*Department of Electrical Engineering, University of Jordan, Amman, Jordan*

[2]*Department of Mechatronics Engineering, University of Jordan, Amman, Jordan*

[3]*Department of Computer Science, Al Al-Bayt University, Mafraq, Jordan*

This paper presents a new technique for model order reduction (MOR) of discrete systems with the advantage of selecting some of the desired frequencies. The proposed technique is achieved using the Artificial Neural Network (ANN) prediction. The ANN is implemented for predicting the unknown elements of the reduced order model while selecting the desired frequencies. Prediction of the ANN architecture is based on minimizing the cost FUNCTION obtained by the difference between the actual and desired system behaviour. The proposed ANN-based model order reduction method is compared to recently published work of MOR techniques. Simulation results verify the validity of the new MOR technique.

1. Introduction

For many years, MOR has been an important subject area in control engineering [1]-[7]. Derivation of the mathematical model often leads to detailed description of a complex model in the form of high order differential equations. Hence, it is useful, and sometimes necessary for analysis and design purposes, to search for the possibility of finding some lower order equations that may be considered to adequately reflect the dominant characteristics of the original system.

In model order reduction, maintaining the original substructures in the reduced model yields a meaningful physical interpretation as presented by Reis and Stykel [4]. In spite of the several methods available in literature, each method has advantages and disadvantages when tried on a particular system. In addition to that, no approach always gives the best results for all systems.

Different neural network architectures lately have been proposed for MOR such as neural black-box identification by means of a non-linear auto-regressive with exogenous model, or neural network of linear and nonlinear principal component analysis combined with generalized dimensional analysis as

* othman_mk@yahoo.com

proposed by Yahya Chetouani [7], Bellamine [6]. However, those architectures usually train the estimated model for closest system behaviour response regardless of the dominant critical frequencies. In this paper, a new technique for model order reduction of dynamical control systems is introduced using artificial neural networks with the advantage of frequency selectivity. The proposed method has been compared to different recent MOR techniques, where the superiority of the new method is shown.

2. Problem Formulation

Consider the difference equation (with order n)

$$y(k) + a_1 y(k-1) + a_2 y(k-2) + \cdots + a_n y(k-n) = b_0 u(k) + b_1 u(k-1) + \cdots + b_n u(k-n) \quad (1)$$

where u(k) is the input and y(k) is the output of the system at the kth sampling instant. The corresponding reduced r^{th} order model can be given as

$$G_r(z) = \frac{Y_r(z)}{U(z)} = \frac{\bar{b}_0 z^r + \bar{b}_1 z^{r-1} + \cdots + \bar{b}_r}{z^r + \bar{a}_1 z^{r-1} + \cdots + \bar{a}_r} \quad (2)$$

The desired objectives in obtaining the reduced model (system stability, frequency selectivity, and minimum response deviation) can be achieved by realizing the following state space representation

$$x_r(k+1) = A_r x_r(k) + B_r u(k) \quad (3)$$

$$y_r(k) = C_r x_r(k) + D_r u(k) \quad (4)$$

where $x_r(k)$ is r-state vector, $u(k)$ is the system input, $y_r(k)$ is the reduced order model output, and A_r, B_r, C_r, and D_r are matrices with appropriate dimensions. The state space matrices of Equations (3) and (4) are to be obtained by training an ANN based on minimizing the following function

$$e(k) = y(k) - y_r(k) \quad (5)$$

3. MOR via ANN

3.1. *ANN Presentation*

Consider the system of Equations (3) and (4). The ANN will provide the reduced order model with the advantage of eliminating the feed-forward matrix D_r. The ANN will simultaneously estimate the A_r and B_r matrices. The used ANN is illustrated as shown in Figure 1 for a simple architecture of a 1st order model.

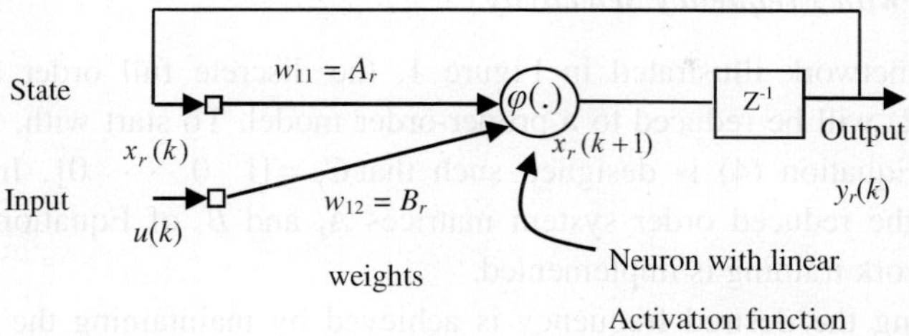

Figure 1. Neural network architecture for 1st order dynamical system

Based on Figure 1, the reduced order model matrices are simply estimated directly by the architecture weights given as [8]

$$w = [A_r \ B_r] \tag{6}$$

Based on the cost function of Equation (5), a modification maybe given by

$$e(k) = y(k) - C_r x_r(k) \tag{7}$$

The instantaneous sum of squared error at time k is

$$\varepsilon(k) = \frac{1}{2} e^2(k) \tag{8}$$

The minimization of the cost function is obtained by summing $\varepsilon(k)$ over time k; that is, $\varepsilon_{\text{total}}(k) = \sum \varepsilon(k)$ [9]. In order to achieve this objective, the gradient of $\varepsilon(k)$ with respect to the weight matrix $w(k)$ is used. That is,

$$\nabla_w \varepsilon_{\text{total}} = \frac{\partial \varepsilon_{\text{total}}}{\partial w} = \sum_k \frac{\partial \varepsilon(k)}{\partial w} = \sum_k \nabla_w \varepsilon(k) \tag{9}$$

For an rth order system, there will be an r number of neurons with w_j being the **weight vector** of neuron j given as

$$w_j = [A_{r,j} \ B_{r,j}], \qquad j = 1, 2, \cdots, r \tag{10}$$

To minimize the function of Equation (7), $\varepsilon(k)$ is differentiated with respect to the weight vector w_j which leads to the following updating weight vector [9]

$$\Delta w_j = -\eta \frac{\partial \varepsilon(k)}{\partial w_j} = \eta C \frac{\partial x_r(k)}{\partial w_j} e(k) \tag{11}$$

where η is the learning-rate parameter.

3.2. *MOR with Frequency Selectivity*

Using the network illustrated in Figure 1, the discrete full order system of Equation (2) will be reduced to a proper-order model. To start with, the output matrix in Equation (4) is designed such that $C_r = [1 \quad 0 \quad \cdots \quad 0]$. In order to determine the reduced order system matrices A_r and B_r of Equations (3), the neural network training is implemented.

Selecting the desired frequency is achieved by maintaining the full order dominant frequencies in the reduced order model. This can be done by restricting the reduced system state matrix A_r to have the following format

$$A_r = \begin{bmatrix} \lambda_1 & a_{12} & a_{13} & a_{14} & & \cdots & & & a_{1r} \\ 0 & \lambda_2 & a_{23} & a_{24} & & \cdots & & & \\ & \ddots & \ddots & \vdots & & & & & \\ . & & 0 & \lambda_{\bar{b}} & & & & & \vdots \\ . & & & 0 & \sigma_1 & \alpha_1 & & & \\ . & & 0 & & -\alpha_1 & \sigma_1 & & & \\ & & 0 & & & 0 & \ddots & & a_{(r-2)r} \\ & & & & & & 0 & \sigma_{\bar{p}} & \alpha_{\bar{p}} \\ 0 & & . & . & . & & 0 & -\alpha_{\bar{p}} & \sigma_{\bar{p}} \end{bmatrix} \tag{12}$$

where the original system eigenvalues are preserved in the diagonal, seen as λ_i, $i = 1, 2, \ldots \bar{b}$ (real) and $\sigma_i \pm \alpha_i$, $i = 1, 2, \ldots \bar{p}$ (complex). The elements to be predicted in Equation (12) are seen as (a_{ij}), where $r = \bar{b} + 2\bar{p}$ and $j = i+1, 2, \ldots r$.

4. Simulation Results

Consider the 8th order transfer function investigated by Ramesh et al. [8] given as:

$$G(z) = \frac{0.1625z^7 + 0.125z^6 - 0.0025z^5 + 0.00525z^4 - 0.02263z^3 - 0.00088z^2 + 0.003z + 0.000413}{z^8 - 0.6307z^7 - 0.4185z^6 + 0.078z^5 - 0.057z^4 + 0.1935z^3 + 0.09825z^2 - 0.0165z + 0.00225}$$

with critical frequencies: [0.8797 ± 0.2442i, -0.0542 ± 0.6558i, -0.5875 ± 0.0959i, 0.0773 ± 0.1078i]. Notice that all of the critical frequencies are complex. Simulating this model for a step input, the ANN was used to train a reduced 2nd order model to provide the same behaviour with minimum deviation. Using the restriction of Equation (12), we have preserved the most dominant critical frequencies given by [0.8797 ± 0.2442i] in the following obtained reduced order model

$$G_r(z) = \frac{0.16234z + 0.082445}{z^2 - 1.7594z + 0.83347}$$

The proposed method has been compared with recently published work [1]. Comparison results are shown in Table 1 and Figure 2 where the advantage of the new method is clearly shown.

Table 1. Model order reduction method comparison.

System	Critical frequencies	Error index, *J*
Original 8th order	0.8797 ± 0.2442i, 0.0773 ± 0.1078i -0.5875 ± 0.0959i, -0.0542 ±0.6558i	----
Proposed 2nd order	0.8797 ± 0.2442i	0.000369
Ramesh (2009)[1] 2nd order	0.8768 ± 0.2484i	0.0011
Selvaganesan (2007)[3] 2nd order	----	0.3397

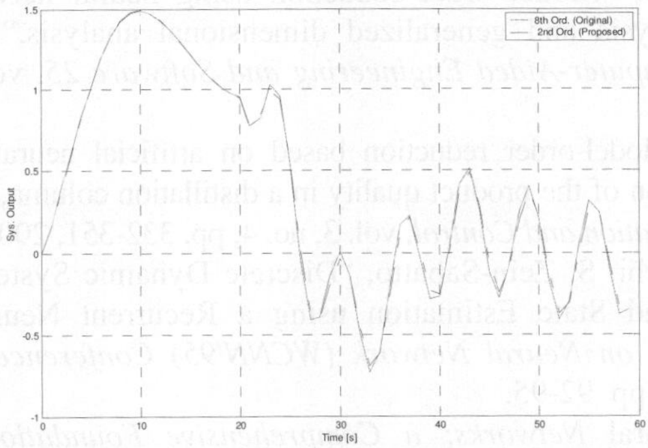

Figure 2. Mixed signal input (step/sinusoidal) responses to the 8th and 2nd order models.

5. Conclusion

A new method of ANN-MOR of dynamical systems with frequency selectivity is presented in this paper. The reduced-order model is obtained by ANN training using a set of input-output data. The convergence process is achieved based on the minimization of the cost function presented by system response deviation. Simulation results show the performance of the proposed ANN-based method compared with other well-known methods in terms of maintaining system stability, preserving the exact dominant frequencies, and providing system response for SISO models. Based on the results presented in Section 4, it is clearly seen that, in general, the proposed method outperforms the other methods.

References

1. K. Ramesh, A. Nirmalkumar, and G. Gurusamy, "Design of Discrete Controller via Novel Model Order Reduction Technique," *International Journal of Electrical and Power Engineering*, vol. 3, no. 3, pp. 163-168, 2009.
2. C. B. Vishwakarma, R. Prasad, "Clustering Method for Reducing Order of Linear System using Pade Approximation," *IETE Journal of Research*, vol. 54, no. 5, pp. 326-330, 2008.
3. N. Selvaganesan, "Mixed Method of Model Reduction for Uncertain Systems," *Serbian Journal of Electrical Engineering*, vol. 4, no. 1, pp. 1-12, 2007.
4. T. Reis, T. Stykel, "Balanced truncation model reduction of second-order systems," *Mathematical and Computer Modelling of Dynamical Systems*, vol. 14, no. 5, pp. 391–406, 2008.
5. G. D. Howitt, R. Luus, "Model reduction by minimization of integral square error performance indices," *Journal of Franklin Inst*, vol. 327, pp. 343-357, 1990.
6. F. H. Bellamine, "Model order reduction using neural network principal component analysis and generalized dimensional analysis," *International Journal for Computer-Aided Engineering and Software 25*, vol. 5, pp. 443-463, 2008.
7. Y. Chetouni, "Model-order reduction based on artificial neural networks for accurate prediction of the product quality in a distillation column," *International Journal of Automation and Control*, vol. 3, no. 4, pp. 332-351, 2009.
8. J. G. Kusckewschi, S. Zein-Sabatto, "Discrete Dynamic Systems Parameter Identification and State Estimation using a Recurrent Neural Network," *World Congress on Neural Network (WCNN'95) Conference*, Washington D.C., July 1995; pp. 92-95.
9. S. Haykin, Neural *Networks: a Comprehensive Foundation.* Macmillan College Publishing Company, New York, 1994.

RESPONSE OF SOIL-STRUCTURE INTERACTION SYSTEM USING ARTIFICIAL NEURAL NETWORK (ANN)

BADREDDINE SBARTAI[†]
Civil Engineering Department, University of 20 Août 1955-Skikda, Route el hadeik Skikda, 21000, Algeria

KAMEL GOUDJIL
Civil Engineering Department, University of 20 Août 1955-Skikda, Route el hadeik Skikda, 21000, Algeria

Artificial Neural Networks (ANN) has seen an explosion of interest over the last few years. Indeed, anywhere that there are problems of prediction, classification or control, neural networks are being introduced. Hence, the main objective of this paper is to develop a model to predict the response of the soil-structure interaction system without using the calculate code based on sophisticate numerical methods by the employment of a statistical approach based on an Artificial Neural Network model (ANN). In this study, a data base which relates the impedance functions to the geometrics characteristic of the foundation and the dynamic properties of the soil is implemented. This leads to develop a neural network model to predict impedances functions (all modes) of a rectangular surface foundation. Then the results are compared with unused data to check the ANN model's validity.

1. Introduction

The major problem that must be addressed during the construction of particular types of structures, such as conventional and nuclear power plants, chemical factories, liquid natural gas tanks, etc., concerns the rigorous safety conditions that must be established to avoid damage caused by various types of excitations to which these structures can be subjected. The strong interest in this problem is not only because of the constant increase in the quality and stability requirements of the constructions in the vicinity of these structures but also because of the need to protect sensitive material. The analysis of the response of these structures and that of the neighbouring soil is a wave propagation problem that leads to consideration of the soil-structure interaction [4]. To face these

[†] This work is supported by grant J0401620100010 of the Algerian Ministry of Higher Education and Scientific Research.

challenges, research in this field has been oriented toward numerical methods because the classical analytical methods, which are generally based on restrictive assumptions about the geometry of the foundation and the elastic properties of the soil, are not adapted to treat the problems of such great complexity. Also, the study of soil-structure interaction remains important and justifies the particular interest that many researchers have shown in it up to the present day. There are two main methods dealing with soil-structure interaction analysis: Direct Method, and Substructure Method. In the direct method, the response of the soil and structure is determined simultaneously by analysing the idealized soil–structure system in a single step. In substructure method the SSI problem is divided into sets of simpler problems, which are solved independently, and the results are then superposed to obtain the response of the structure. The basic step in the substructure approach is to determine the force–displacement characteristics of the soil. This relationship may be in the form of an impedance (stiffness) function, or, inversely, a compliance (flexibility) function. These impedances functions (K_{ij}) are often represented in the form of a sum of two terms depending of the frequency. The first being the real part (k_{ij}) and the second the imaginary part (c_{ij}).

The aim of this work is to avoid the calculation of the impedances functions with the above numerical methods (FEM, BEM, and DEM) because of theirs complexities. Using the Artificial Neural Networks and the gathered data from Wong and Luco tables [5] necessary information can be reached with acceptable engineering approximation. In recent years, different computational solutions using neural networks have been studied, examples of which can be found in the literature where Pala et al. [3] studied the dynamic soil-structure interaction of building by (ANN). In this study, a database which relates the impedance functions (dynamic stiffness) to the geometrics characteristic of the foundation and the dynamic properties of the soil is implemented. This leads to develop a neural network model to predict impedances functions (all modes) of a rectangular surface foundation. Then the results are compared with unused data to check the ANN model's validity.

2. Modeling by (ANN)

Artificial Neural Network (ANN) gives an alternative for mathematical modeling and makes a part of the statistical nonparametric and nonlinear models capable to answering to the problematic of identification, of assistance to decision, of diagnostic, of prediction, etc. The application of this type of model only east appeared at the beginning of 1990 and their advantage resides in their capacity of generalization. Generally, the Artificial Neural Networks are often confused with the genetic algorithms, the cybernetics and the artificial intelligence. They constitute actually a current very precise of the artificial intelligence, of which developmental model is based on idea to copy of the algorithms of training on a simplified brain model. The Artificial Neural Networks are inspired of the biology and represent a mathematical model of the functioning of the biological neuron [2]. Idea is to present to (ANN) the input data and output, and to make him learn the relation between two by training process. This last consists in minimizing error by adjustment of the parameters model. A neural network, in general, has three layers namely, the input, output and hidden layers. The neurons in the input layer receive input from the external environment. The structure of a three-layer neural network is shown in (figure 1). At first we created our database starting with the tables of impedances functions (all modes) of a rigid rectangular foundation resting on a homogeneous visco-elastic semi-infinite soil [5]. These impedances functions were gotten numerically by the code of computation (Continuum Linear Analysis of Soil-Structure Interaction). This database is made of 24 tables which incorporate the values of the impedances according to the frequency a_o, the relative length B/C (B and C are the sizes of the foundation), the coefficient of Poisson's ratios v and the material damping β.

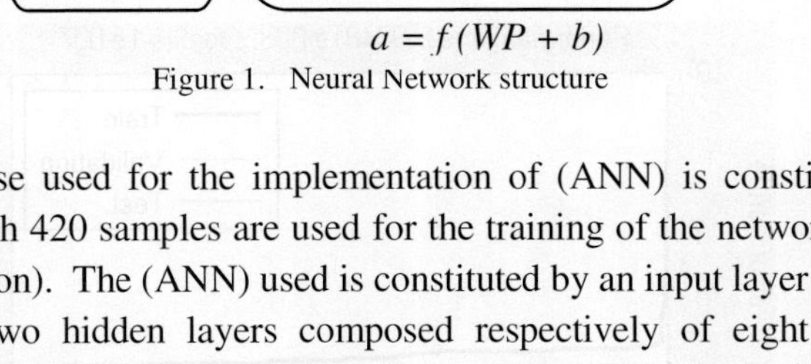

Figure 1. Neural Network structure

The database used for the implementation of (ANN) is constituted of 504 samples of which 420 samples are used for the training of the network and 84 for the test (validation). The (ANN) used is constituted by an input layer consisting of four neurons, two hidden layers composed respectively of eight and sixteen neurons and an output layer consisting of two neurons (4 - 8 16 - 2), Fig 2. The activation function used for all neurons is a sigmoid type. The output parameters of the (ANN), represents the complex impedance functions (K_i) of a rectangular foundation: the real part of the impedance is (k_i), and the imaginary part of impedance is (c_i). (i) Represents the type of mode of the vibrant foundation. After the phase of training, the validation (test) of RNA is used to judge the capability of generalization of the model [1]. This technique consists in testing the model on not used data for the adjustment of the weights. The data of the phase test represent generally 20% of the database. The programming of the different algorithms was carried out in the environment Matlab ®. Before choosing the type of network to make the training, one has shared the database in two: the twenty first tables for training and the last four tables for the test without forgetting to normalize input data and output. After chosen the type of network, we carried out the training. This last stop them if the epoch number were perfect or if the goal (error) east reaches (fig.3). The back propagation method was used for training. It is the best known training algorithm for neural networks, and still one of the most useful. This algorithm is based on the computation of the gradient of error, and aims to minimize error in output network.

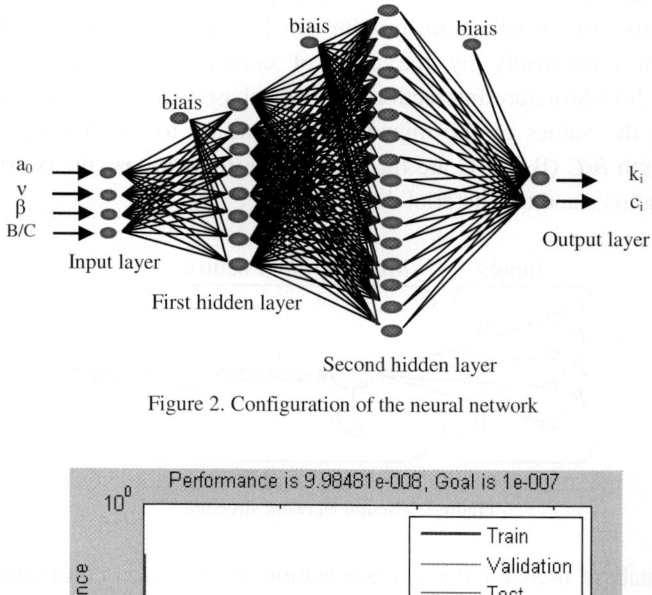

Figure 2. Configuration of the neural network

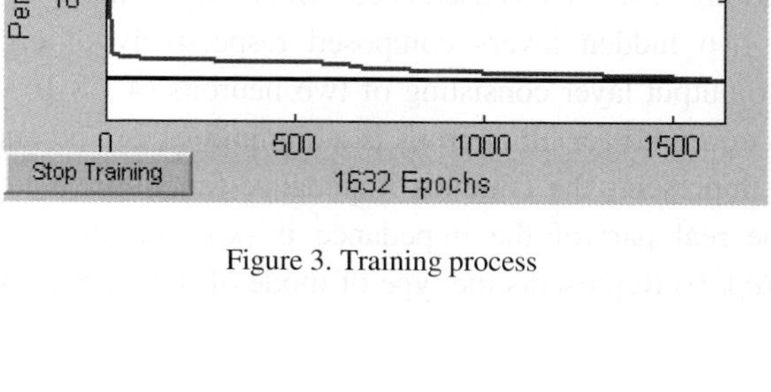

Figure 3. Training process

3. Results

The figure 4a illustrates the correlation between the true horizontal dynamic stiffness k_x of the foundation and predicted by the model for the training process. The correlation is excellent with an absolute error of prediction inferior to 1 %. On the twenty tables, note us that 99, 93 % of the examples were predicted. On the database of validation (testing), the same remarks were observed with however less of precision (fig. 4b). On the four tables that this database composes, we note that 99, 77 % of the examples were predicted. These results testify about the capabilities of generalization of adopted (ANN). On the figure 5a, we show the correlation between true horizontal damping C_x and predicted by the model based on the database of training. It is noticed clearly that the model predict damping with an excellent precision. Indeed, 99, 92 % of the data of the twenty four tables are predicted. On the database of test (fig. 5b), note us that 99.45 % of the data of the four remaining tables were predict but with less precision.

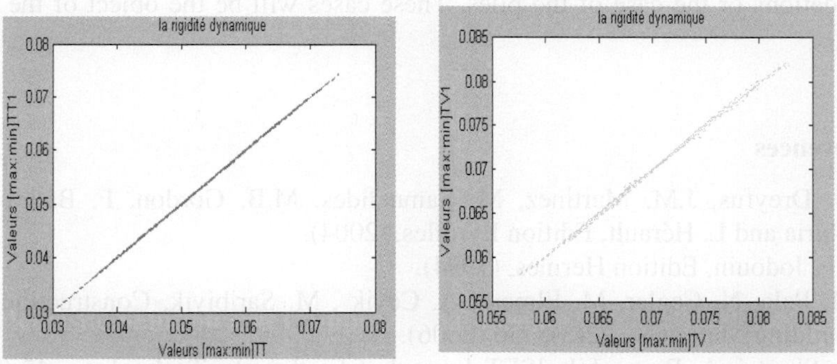

(a) Training (b) Test

Figure 4. Correlation between actual dynamic stiffness and predicted by the ANN Model

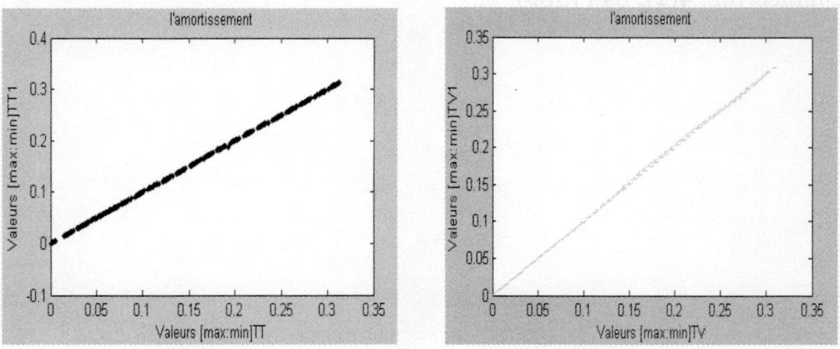

(a) Training (b) Test

Figure 5. Correlation between real damping c_x = TV and predicted of (ANN) model c_{x1} = TV1

4. Conclusion

The purpose of this study is to develop a model able of predicting the analysis of a problem of interaction soil– structure problem without passing by sophisticated numerical methods established in the calculation codes (example: calculation of the impedances functions of a foundation), by using a statistical approach based on the concept of artificial neuronal networks (ANN). For this reason a database was constituted based on the tables of impedances functions by [5], linking up the input parameters (geometry of foundation, mechanical characteristics of the soil, frequency of excitation) with the dynamic impedance function of foundation. At the end of the training, the set up model was considered satisfactory. This model allows predicting the impedance (all modes) of a rigid rectangular foundation resting on the surface of a semi-infinite soil with an excellent precision (± 0.5 %). This model is valid only for a rigid surface foundation, but can be developed to predict the case of embedded foundations or the case of the piles. These cases will be the object of the next papers.

References

1. G. Dreyfus, J.M. Martinez, M. Samuelides, M.B. Gordon, F. Badran, S. Thiria and L. Hérault, Edition Eyrolles, (2004).
2. J.F. Jodouin, Edition Hermès, (1994).
3. M. Pala, N. Caglar, M. Elmas , A. Cevik , M. Saribiyik, Construction and Building Materials, **22(3)**, 330 (2006).
4. B. Sbartai, A. Boumekik, ISET Journal of Earthquake Technology, **45 (3-4)**, 65 (2008).
5. H.L. Wong, J.E. Luco, Journal of Soils Dynamics and Earthquake Engineering, **4(2)**, 64 (1985).

RECOGNITION OF INDIVIDUALS BASED ON HAND GEOMETRY

SILARBI SAMIYA*, BENDAHMANE ABDERRAHMANE, BENYETTOU ABDELKADER

Laboratory signal image speech SIMPA Department of computer science, University of sciences and technology, USTOMB BP 1.505 EL-M'NAOUAR, Oran, Algeria.

Biometrics is an emerging field, where technology enhances our ability to identify an individual. The identification by hand geometry is one of the most widely approved by the general public, it is very easy to use and less expensive than other techniques. Our work is based on the realization of a system that is capable for recognizing an individual based solely on the geometry of his hand, gpdsHAND database was used, where a preprocessing has been applied to extract the relevant characteristics of each image, these will be used in the identification phase. Two classification approaches are implemented, the SVMs (support vector machine) and neural networks PMC (Multilayer Perceptron). We then present our experiments and results and we will fix our perspectives.

1. Introduction

The domain of the automatic identification and authentication in recent years is growing. Indeed, the need for secure access to physical or virtual environments is in our duties, including personalized services that continue to grow. However, the conventional means based on passwords or magnetic cards associated with a personal code have many disadvantages. As an example the risk to be stolen, lost or forgotten. Thus we need to use physiological or biometric measures of the individual (biometrics), such as fingerprint, hand, voice, face, signature etc. they provide a solution to conventional identification systems and present safer and more reliable mediums.

In this article we are interested in recognition systems based on hand geometry, as they are widespread, easy to use, widely accepted by the public and relatively inexpensive. Increasingly this technology has the advantage of being less capricious than the recognition of fingerprints, dirt and small cuts will not prevent the system from stop working. It is also less intrusive and not dangerous as the recognition of the retina or the iris. In addition, it is advocated by the National Commission for Computing and Liberties CNIL.

* silarbi_samiya@hotmail.fr.

The aim of our study is to develop a system that is capable of recognizing individuals by their hands geometry and specifically by the curve. Thus by focusing on the most important part of our biometric system (extraction of pertinent parameters), we performing several processing steps (ACP principal component analysis, Fourier transform TF). We have applied SVMs in the classification and we add the MLP for comparison.

2. Biometric recognition based on hand geometry

The process of a biometric system goes through three stages: acquisition, preprocessing and feature extraction, then classification. Multitudes of studies have been achieved in this research [1] [2] [3] [4].

Feature extraction is a crucial step in any application, which consists of performing several image processing procedures to extract the relevant features to be used later in the verification phase. Several types of geometric parameters can be extracted from an image of a hand, such as: The width of the finger, the finger length, palm width, the length of the palm, the palm, the length of the hand ... etc. In the following sections, we describe the complete processes used in preprocessing phase. The identification by hand geometry will be carried out by learning models on the images of the database. We will use two models for the classification phase the SVM and the MLP that we will be seen in detail later.

3. Support Vector Machine SVM

SVM was introduced by Vladimir Vapnik in1995. It is a classification model by binary supervised learning; and is based on the concept of maximizing the margin and the existence of a linear classifier in an appropriate space [5].

In a classification problem, we can see two cases of separability: the linearly separable problems and non-linearly separable ones. A kernel function (polynomial, Gaussian ...) allows SVMs to solve the second case by projecting the non-linearly separable input space to a linearly separable description space [6].

Fig1 shows two classes linearly separable problem:

- Find a classifier (hyperplane) line that separates the data and maximizes the margin between these two classes.

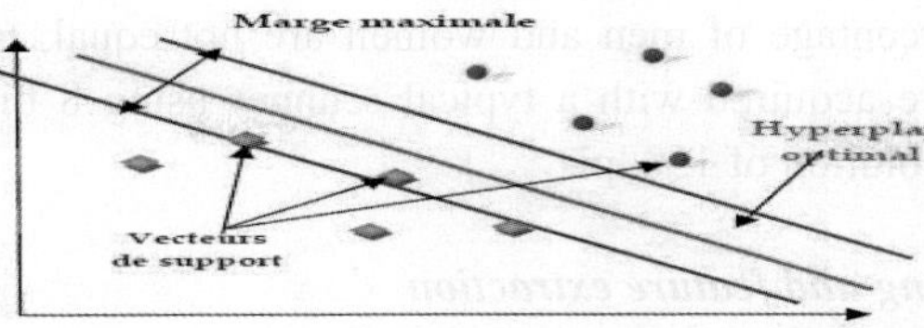

Figure 1. Principe of SVMs

SVMs can be extended to the case of N classes, two solutions are possible:

- One against all (N classifiers) each class is opposed to all others.
- One against one (N (N-1) / 2 classifiers) each class is discriminated from another one.

1.1. Multi Layer Perceptron MLP

A multilayer perceptron (MLP) is a feedforward artificial neural network model that maps sets of input data onto a set of appropriate output. An MLP consists of multiple layers of nodes in a directed graph, with each layer fully connected to the next one. Except for the input nodes, each node is a neuron (or processing element) with a nonlinear activation function. MLP utilizes a supervised learning technique called backpropagation for training the network[7].

Backpropagation learning algorithm 'BP' has two phases:

- Forward pass phase: computes 'functional signal', feedforward propagation of input pattern signals through network
- Backward pass phase: computes 'error signal', *propagates* the error *backwards* through network starting at output units (where the error is the difference between actual and desired output values)

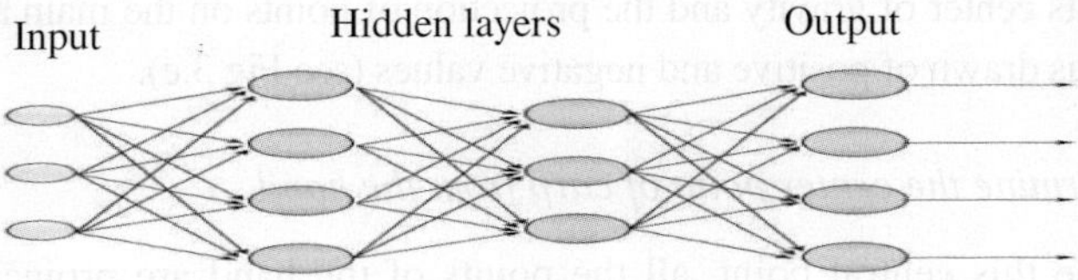

Figure 2. MLP architecture

4. Experimental results

4.1. *Corpus*

The gpdsHAND database is an open access and can be downloaded from [8] it contains pictures of hands of 50 people; the right hand of each person was photographed 10 times; most users are selectively within a range age from 23 to

30 years; the percentage of men and women are not equal, 68% of users are men; images were acquired with a typical scanner using 8 bits by Pixel (256 gray levels), a resolution of 150dpi.

4.2. *Preprocessing and feature extraction*

For the feature extraction phase our approach is based on the estimation of the curve formed by the fingers and palm. To do this, it is necessary to determine the center point of carp of hand and then apply a polarization, shown in Fig1 a-h. The feature extraction of the hand involves several steps:

4.2.1. *Use of the median filter*

Use the median filter with 10x10 mask to reduce the noise shown in Fig 3.b.

4.2.2. *Thresholding of the image*

The application of the thresholding operation on the image, the threshold was chosen after several tests and set to 100 (fig 3.c).

4.2.3. *Determine the inclination by the ACP*

To determine the inclination angle of the hand, first the recovery of coordinates (x, y) of white pixels (pixels> threshold) from the thresholded image , the storage of these coordinates in a matrix on which a principal components analysis was carried out (see Fig 3.d).

After determining the principal axis (the inclination of the hand) it is necessary to determine the position of the fingers and palm of the hand. For this the image was centered on its center of gravity and the projection of points on the main axis then, a histogram was drawn of positive and negative values (see Fig 3.e).

4.2.4. *Determine the center point of carp from the hand*

To determine this central point, all the points of the hand are projected on the main axis, and the space of points was divided into 200 equal intervals, then we calculates the number of points on each interval (see Fig 3.f). The threshold was set at 400 points minimum (in the negative part only) to determine the end of the palm and the beginning of the wrist. We have determined after several tests on several images of the database that 200 intervals with a threshold of 400 points gave us satisfactory results.

4.2.5. *Passage in polar coordinates*

This operation is to transform the points of the hand from the Cartesian coordinates to polar coordinates using the center of the carp of the hand as the origin. To retrieve only the curve of the fingers we have limited the angle of the polarization [-Π / 3 + Π / 3] around the main axis, and we have recovered only the curve of the points by dividing the space abscissa of points in 100 intervals and calculating the number of points in each interval (histogram see fig 3.g).

4.2.6. *Using the FFT*

After the step of polarization we calculated the Fourier transform of the curve of the hand (100 coef). keeping only the first 32 coefficients (complex coef representing low frequencies) which gives us 32 amplitude coefficients + 32 coefficients of phases and therefore real values at all 64 (Fig 3.h).

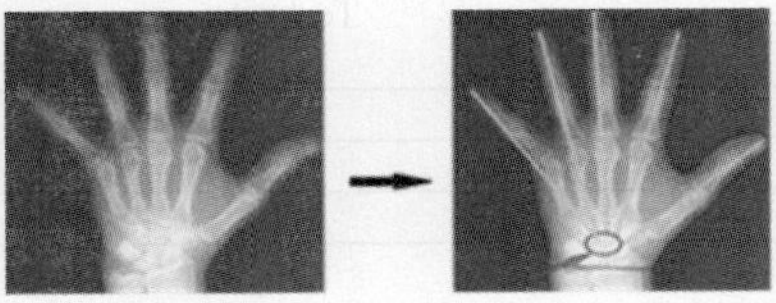

(a) In the red the center point of the carp of the hand, and in blue the curve of the hand

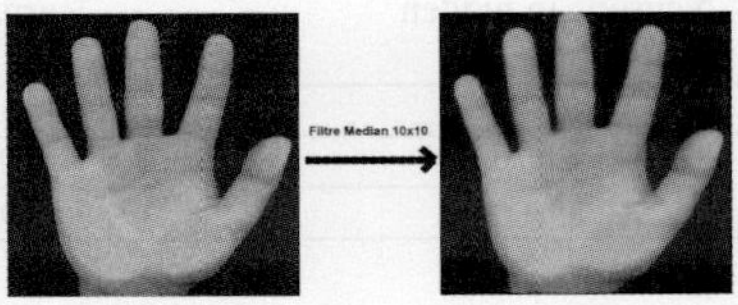

(b) Applying a median filter

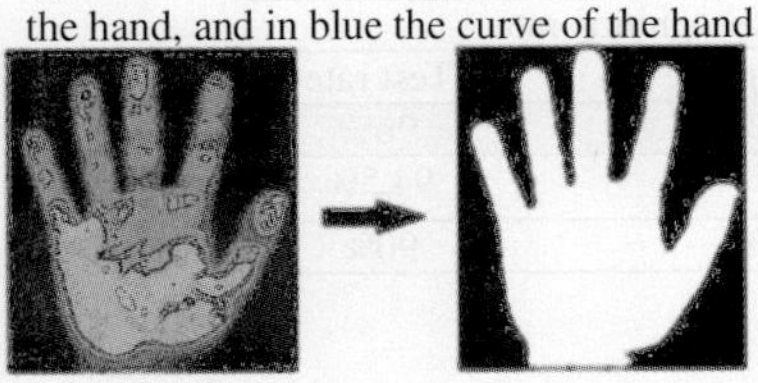

(c) Thresholding of the image

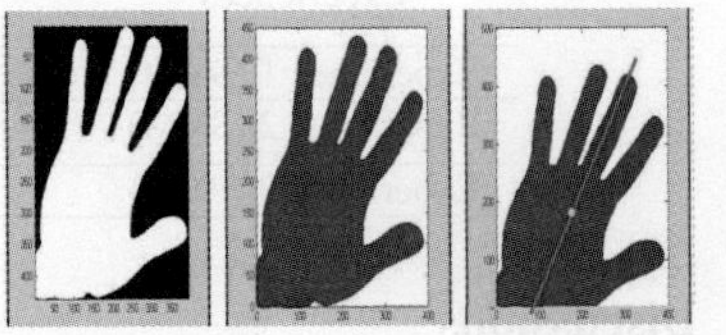

(d) Determination of the main axis by ACP

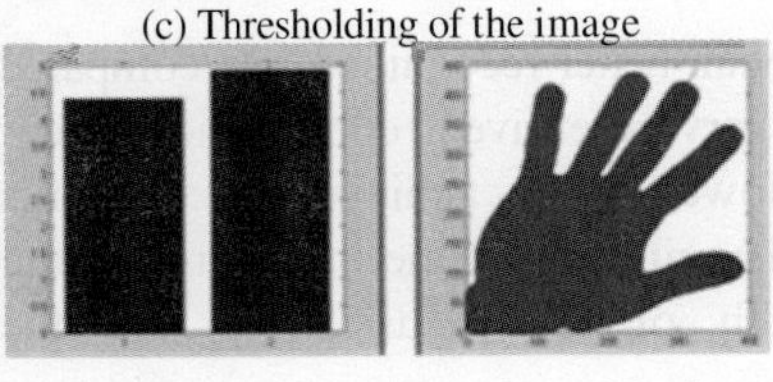

(e) Determination of the position of the fingers

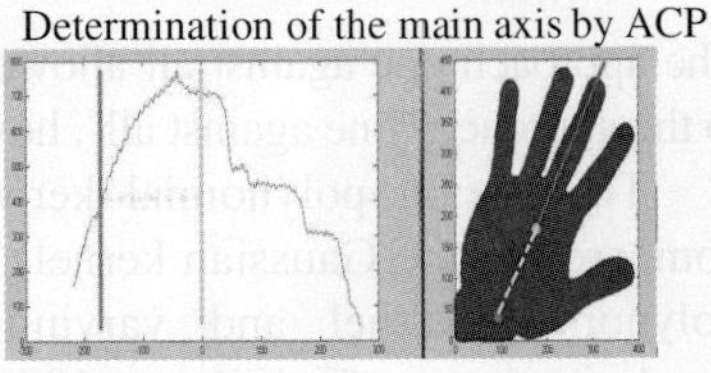

(f) Determination of the position of the center point of carp from the hand

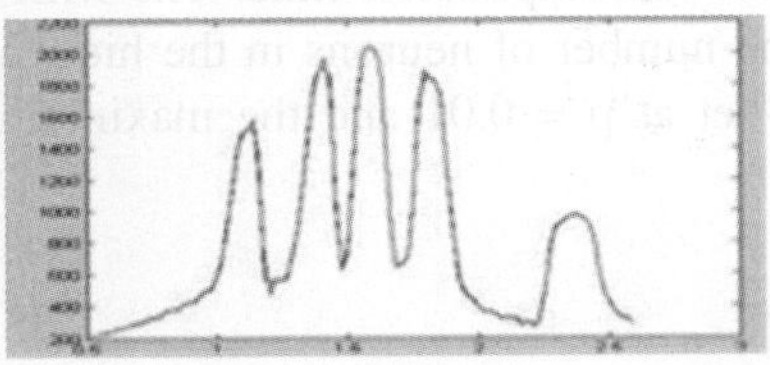

(g) in bleu the curve of the palm, en in red the curve returned by FFT

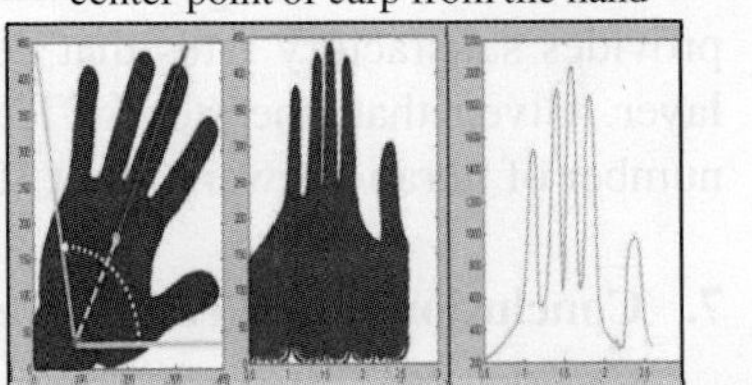

(h) Extracting the palm curve by passing to polar coordinates

Figure 3. Preprocessing phases

5. Classification

For learning we have taken five images of each individual hand, so in total we making 50 images (training set), and five images for the test phase which is in total 50 images (test set). First experiment: SVM (one against one), 2nd experiment: SVM (one against all), 3rd experiment: MLP

Table 1. Results for the two kernels "one against one" and "one against all"

Kernel	Learning rate	Test rate	Kernel	Learning rate	Test rate
Polynomial kernel	98%	96%	Polynomial kernel	96%	94,5%
Gaussian kernel	95,5%	93%	Gaussian kernel	94%	92%

Table 2. Results of the MLP

Neurons in hidden layer	learning rate	Test rate
32	97.50%	30%
64	98%	45%
250	92%	90%

Table3. Comparative results

Experiences	Recognition rate	
	learning rate	Test rate
Experience 1 (SVM)	98%	96%
Experience 2 (SVM)	96%	94,5%
Experience 3 (PMC)	92%	90%

6. Discussions

The approach one against all allowed us to obtain better recognition rate compared to the approach "one against all", however it is very expensive in computation time.

The use of polynomial kernel has allowed us to achieve better results compared to the Gaussian kernel. This by changing each time the degree of the polynomial kernel and varying the width of the Gaussian γ and the regularization coefficient ($c = 1000$).

Identification using the MLP is very fast in computation time. The MLP provides satisfactory rates that depend on the number of neurons in the hidden layer. Given that: the step of learning was set at $p = 0.01$ and the maximum number of iterations were set at 10,000.

7. Conclusions and Perspectives

We have presented in this paper the implementation of a biometric identification system by hand geometry. Several pretreatment steps were performed on the images of hands in order to extract the parameters (64 values) that were used in the classification phase by SVM and MLP.

The use of the FFT has allowed us to move from 100 to 64 values only, thereby reducing input data, and having a more robust than the curve itself, including small changes, can alter the quality of learning, unlike the Fourier coefficients.

The use of SVM returns especially to select a good family of kernel functions and set the parameters of these functions. If the implementation of an SVM is generally inexpensive in time, we must however mention that finding the best parameters may require testing phases quite long.

In future work, several improvements can be developed to improve our results by combining hand geometry with texture or using evolutionary approaches to optimize the classification models used.

References

1. Benamara Djahid, Recognition of individuals by the geometry of the hand, memory for the graduation of engineering in computer science. University of Science and Technology of Oran USTO_LAMOSI, (2007).
2. Bendiabdallah Ghizlène, Biometric identification through hand geometry by the AIS, to obtain memory of engineering in computer science. University of Science and Technology of Oran LAMOSI_USTO, (2008).
3. Raul Sanchez-Reillo et Carmen Sanchez-Avila, RBF Neural Networks for Hand-Based Biometric Recognition, Madrid, Published in the 3rd International Conference in 'Audio and Video-Based Biometric Person Authentication', AVBPA, pp330-335, (2001).

   Ajay Kumar1, David C. M. Wong, Helen C. Shen1, Anil K. Jain, Personal Verification using Palmprint and Hand Geometry Biometric, Hong Kong, Published in the fourth international conference 'Audio and Video-Based Biometric Person Authentication', AVBPA, pp.668-678, (2003).
4. Antoine Cornuéjols, A learning method: The vast margin separators SVM, University of Paris-Sud, Bulletin of the AFIA number51, Orsay, pp.14-23, in June (2002).
5. Olivier Bousquet, Introduction to Support Vector Machines (SVM), Centre of Applied Mathematics, Ecole Polytechnique, Palaiseau Orsay, November (2001).
6. Philippe Leray, Some Typesof backpropagation Neural Networks, Department ASI, PSI Laboratory, INSA Rouen.
7. www.gpds.ulpgc.es the database used gpdsSHAND.

AN NEW BELIEF RULE BASE REPRESENTATION SCHEME AND ITS GENERATION BY LEARNING FROM EXAMPLES *

J. LIU[1†], L. MARTINEZ[2], H. WANG[1], A. CALZADA[1], S.W. CHEN[1]

[1] *School of Computing and Mathematics, University of Ulster, Northern Ireland, UK*

[2] *School of Computing, Dept. of Computer Science, University of Jaén, Jáen, Spain*

A new rule-based representation scheme is proposed with belief degrees embedded in the entire consequent terms and in the entire antecedent terms of each rule, which is shown to be capable of capturing uncertainty and nonlinear causal relationships in an integrated way. The overall representation and inference framework offers a further improvement and extension of the recently developed belief rule base inference methodology (called RIMER). Subsequently, a simple but efficient method for automatically generating such extended belief rule-base from numerical data is proposed involving neither time-consuming iterative learning procedure nor complicated rule generation mechanisms, which is mainly attributed to the new features of the ex-tended rule base with belief structures. Then a case study in oil pipeline leak detection is provided to illustrate the proposed approaches.

1. Introduction

During the last quarter of a century many different types of rule-based systems emerged, certainly including the fuzzy rule-based system, which, as one of the dominant and main framework in rule-based system, has been widely accepted, investigated and applied in many application areas [1]. In recognition of the need to handle hybrid information with uncertainty in human decision making, a new methodology has been proposed recently for modeling a hybrid rule-base using a belief structure and for inference in the belief rule-based system using the evidential reasoning (ER) approach [2], is referred to as RIMER [3], where a rule-base is designed with belief degrees embedded in the entire consequent terms of a rule, called belief rule-base (BRB), is used to capture nonlinear causal relationships as well as uncertainty. The inference of a rule-based system is implemented using the ER approach. RIMER approach has been further investigated and applied to different areas, such as work in [4-6].

---

* This work is partially supported by the research project TIN2009-08286.

† j.liu@ulster.ac.uk.

How to generate the rule base is a fundamental issue in the design and implementation of rule-base system. For RIMER, an optimal modeling has been proposed [5, 7], with some further improved learning algorithms [8-9]. In this paper, to facilitate the more general application cases and more flexible and simpler rule-base generation scheme, the belief rule base in RIMER is extended with belief degrees embedded also in the entire antecedent terms of each rule. A simple but efficient method for automatically generating such extended belief rule-base from numerical data is then proposed involving neither time-consuming iterative learning procedure nor complicated rule generation mechanisms

The rest of this paper is organized as follows. Extended belief rule base framework is proposed in Section 2. In Section 3, we propose extended belief rule generation method, which is then demonstrated in Section 4 using a real world case study in oil pipe leak detection compared with some existing approaches. Conclusions are drawn in Section 6.

2. Extended Belief Rule Base Framework

Take for example the following belief rule in RIMER for safety analysis [4]:

R_k: IF the **failure rate** is *frequent* and the **consequence severity** is *critical* and the **failure consequence probability** is *unlikely,* THEN the **safety estimate** is {(*good,* 0), (*average,* 0), (*fair,* 0.7), (*poor,* 0.3)}, (1)

where the consequent is a belief distribution representation for safety consequent, stating that it is 70% sure that safety level is fair and 30% sure that safety level is poor. This belief rule is extended with belief degrees embedded in the entire possible antecedent terms of each rule as well, e.g.:

R_k: IF the **failure rate** is {(*very low,* 0), (*low,* 0), (*average,* 0), (*frequent,* 0), (*highly frequent,* 1) AND the **consequence severity** is {(*negligible,* 0), (*marginal* , 0), (*moderate* , 0.3), (*critical,* 0.7), (*catastrophic,* 0.7)} AND the **failure consequence probability** is {(*unlikely* , 0.8), (*reasonably unlikely,* 0), (*likely* , 0), (*highly likely,* 0), (*definite,* 0)} THEN the **safety estimate** is {(*Good,* 0), (*Average*, 0.1), (*Fair*, 0.6), (*Poor*, 0.3)} (2)

This kind of extended belief rule is able to not only capture vagueness (with linguistic terms), uncertainty (with beliefs), incompleteness (partially known beliefs in antecedents and/or consequents) and nonlinear relationships (IF-THEN rules) between the three parameters and the safety level but also provide the flexible way to incorporate the hybrid input information and more efficient rule generation scheme (as illustrated in the subsequent sections).

More generally, the kth rule in an EBRB can be represented as follows:

R_k^*: IF $\boldsymbol{U}$ is $(\boldsymbol{A}, \boldsymbol{\alpha}^k)$ THEN $\boldsymbol{D}$ with belief degrees $\boldsymbol{\beta}^k$, with a rule weight θ_k and attribute weights $\delta_1, \delta_2, \ldots, \delta_T$ (3)

where $(\boldsymbol{A}, \boldsymbol{\alpha}^k)$ is the packet antecedents $\{\{(A_{ij}, \alpha_{ij}^k), j=1,\ldots, J_i\}| i=1,\ldots, T\}$ representing antecedent with belief structure; here A_{ij} ($j\in\{1,\ldots, J_i\}$) is the jth referential value of the ith antecedent attribute U_i ($i=1, \ldots, T$), α_{ij}^k the likelihood to which U_i is evaluated to be A_{ij} in the kth rule with $\alpha_{ij}^k \geq 0, \sum_j^{J_i} \alpha_{ij}^k \leq 1$, $k=1,\ldots, L$, L is the total number of all the packet rules in the rule-base. A rule base with rules in the form of (3) is called an extended belief rule base (EBRB) compared with it in RIMER.

3. Generating EBRB from Numerical Data

Based on the feature of new EBRB in (3) and following the similar way used in [10], we propose a simple but efficient and powerful method for automatically generating belief rules from numerical data. The common features shared in both approaches are efficiency. However, the procedures are very different from each other in that our belief rule-base has been a great extension of traditional fuzzy if-then rule-base in terms of representation power, applicability, and flexibility. Here we consider a single output in the T-dimensional input space R^T. Let us assume that the following M input-output pairs are given as training data for constructing an EBRB:

$$\{(\boldsymbol{x}_p; y_p)| p=1, 2, \ldots, M\}, \tag{4}$$

where $\boldsymbol{x}_p=(x_{p1},\ldots, x_{pT})$ is the pth input vector and y_p is the corresponding output. The approach to derive a set of belief rules in the form of (3) from those training data pairs consists of the following steps:

Step 1: Determine the referential set of each antecedent and consequent attribute and their representations. Finally, each A_{ij} will be determined and represented either in fuzzy set or utility values.

Step 2: Generate belief distribution from the given input-output data pairs

This step is equivalent to transforming an input data (output data) into a distribution on referential values using belief degrees.

Step 2-1: Antecedent belief distribution generation from input data

Given the input vector $\boldsymbol{x}_p=(x_{p1},\ldots, x_{pT})$ can be assessed to a distribution representation of the referential values using belief degrees as follows:

$$E(x_{pi})=\{(A_{ij}, \alpha_{ij}^p); j=1,\ldots, J_i\}, i=1,\ldots T; p=1,\ldots, M \tag{5}$$

where α_{ij}^p is the likelihood to which the pth input x_{pi} for U_i belongs to A_{ij} (i=1, 2,…, T), which can be generated using various ways depending on the given representation schemes of the referential values [3].

Step 2-2: Consequent belief distribution generation from output data

Consider the pth output y_p, it can be assessed to a distribution representation of the referential values using belief degrees as follows:

$$E(y_p)=\{(D_s, \beta_{sp}); s=1,\ldots, N\}, p=1, 2,,\ldots M \quad (6)$$

where β_{sp} is the belief degree to which D_s is believed to be the consequent given the pth input vector $\boldsymbol{x}_p$, with $\beta_{sp} \geq 0$ and $\sum_{s=1}^{N} \beta_{sp} \leq 1$ (s=1,…, N; p=1, 2,…, M).

Step 3: Create the extended belief rule base

Based on Step 2, the pth extended belief rule can be generated as follows:

$$R_p^*: \text{IF } U_1 \text{ is } \{(A_{1j}, \alpha_{1j}^p), j=1,\ldots, J_1\} \text{ and } \ldots \text{ and } U_T \text{ is } \{(A_{Tj}, \alpha_{Tj}^p), j=1,\ldots, J_T\}$$
$$\text{THEN } (D_1, \beta_{1p}),\ldots, (D_1, \beta_{Np}) \quad (7)$$

Step 4: Assign the rule and antecedent attribute weights

Since each data pair generates one rule, the rule weights depend on the importance of the data pairs, along with the antecedent weights.

It follows from Step 1 to 4 that this rule generation method is simple and quick to construct in that it is a one-pass build procedure without need of complicated or time consuming training procedure. Once the EBRB is established from the given training data, following the inference procedure, the predictive output can be obtained given the new input data.

4. A Case Study

An example for oil pipeline leak detection in [6] is used here. The difference between inlet mass flow and outlet mass flow in a pipeline, called Flow Difference (FD), and the average pipeline pressure change over time, called Pressure Difference (PD), are the two very important factors in detecting whether there is a leak in the pipeline. They are the two antecedent attributes of the rule base. The consequent attribute is the Leak Size (LS). During the leak trial, samples of 4%, 16% and 25% leak data were collected at the rate of 10s per sample, respectively. In order to generate the EBRB using the proposed algorithm in Section 3, 3733 data pairs are selected as training data, which include 853 from 25% leak, 880 from 16% leak and 2000 from 4% leak data sets. The process of generation and testing the EBRB is implemented using java-based EBRB inference system. For testing the generated EBRB, 1135 data from 25% data set are used. There are a total of 3733 training data pairs leading to 3733 rules in the EBRB. Fig. 1 displays the actual and predicted LS on the time scale. It demonstrates that the predicted outcomes match the real ones very closely, and can detect the leak in a high accurate rate.

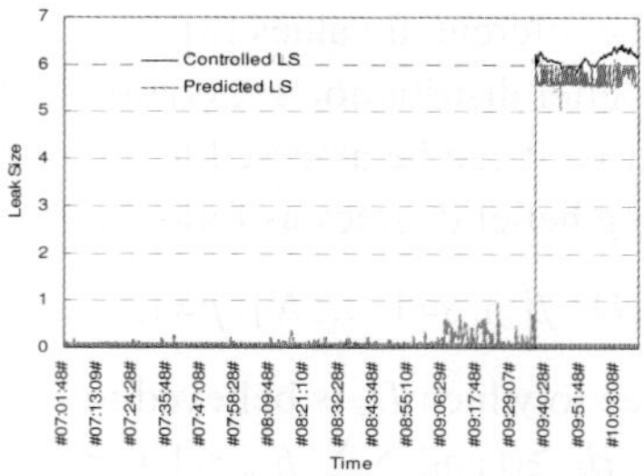

Fig. 1 Testing data and the output by the EBRB system

The widely used performance evaluation criteria mean square error (MSE), mean absolute error (MAE), and root mean squared error (RMSE) was used in this study. In [5] (denoted as M1), [8] (M2), and [9] (M3), the calculated time (in minutes) of three algorithms has been 300 (for M1), 20 (for M2), and 2 minutes (for M3). However, in the proposed method, the time counting for generating the rule for 3733 training data (i.e., the generation time of rules) is less than 1 second, the execution time for the testing for 1135 data pairs are less than 30 seconds, on the PC with Inter® Pentium® 4 CPU @3.40GHz Processor with 2GB RAM. The performances of the proposed EBRB model (denoted as M4), the existing learning models based on RIMER (M1, M2, and M3), and fuzzy rule-based model in [10] (denoted as M5) are summarized in Table 1, from which it is quite clear that the EBRB model achieves much better performance than the other models. Overall, the EBRB model outperforms M1-M3 in terms of performance and efficiency.

Table 1: Performances comparison of several models

Criteria	M1	M2	M3	**M4**	M5
MSE	0.2245	0.4227	0.1732	**0.0829**	0.5292
RMSE	0.4738	0.6502	0.4162	**0.2879**	0.7275

The feature of the proposed method leads to decision attributes definitions, rule-base representation and generation, inference can be implemented in an integrated system, no need link to a standalone training package.

5. Conclusions

A new belief rule base representation, called EBRB, was proposed as a great extension of traditional fuzzy rule based systems and also a further improvement and extension of the belief rule base introduced in [3] in terms of flexibility, applicability and predictive performance. Specially, the major advantage of this EBRB is it offers and facilitates a simple and efficient rule-base generation approach with high performance from the sample data. The case study in a real world application to oil pipe leak detection has shown the high efficiency and better performance compared with some existing approaches. We believe that such more general, flexible and efficient rule-base system is acceptable in more complex system.

References

1. P. Angelov and R. Yager. A simple fuzzy rule-based system through vector membership and kernel-based granulation. In *Proc. of 2010 IEEE Int. Conf. on Intelligent Systems*, pages 349-354, 7-9 July, London, UK.
2. J.B. Yang and D.L. Xu. On the evidential reasoning algorithm for multiple attribute decision analysis under uncertainty, *IEEE Trans. on Syst., Man, and Cyber. -Part A*, 32(3):289-304, 2002.
3. J.B. Yang, J. Liu J, J. Wang J, H-S. Sii, H-W. Wang. Belief rule-base inference methodology using the evidential reasoning approach – RIMER, *IEEE Trans. on Syst., Man, and Cyber.-Part A*, 36(2):266-285, 2006.
4. J. Liu, J.B. Yang, J. Wang, H.S. Sii and Y.M. Wang. Fuzzy rule-based evidential reasoning approach for safety analysis, *International Journal of General Systems*, 33(2-3): 183-204, 2004.
5. J.B. Yang, J. Liu, D.L. Xu, J. Wang, and H.W. Wang. Optimization models for training belief-rule-based systems, *IEEE Trans. on Syst., Man, and Cyber.-Part A*, 37(4): 569-585, 2007.
6. D.L. Xu, J. Liu, J.B. Yang, G.P. Liu, J. Wang, I. Jenkinson, and J. Ren. Inference and learning methodology of BRB expert system for pipeline leak detection, *Expert Systems with Applications*, 32 (1): 103-113, 2007.
7. J. Liu, J.B. Yang, D. Ruan, L. Martinez, and J. Wang. Self-tuning of fuzzy belief rule bases for engineering system safety analysis, *Annals of Operations Research*, 163(1): 143-168, 2008.
8. Z.J. Zhou, C.H. Hu, J.B. Yang, D.L. Xu, and D.H. Zhou, Online updating belief-rule-based system for pipeline leak detection under expert intervention. *Expert Systems with Applications*, 36(3): 7700-7709, 2008.
9. Z.J. Zhou, C.H. Hu, J.B. Yang, D.L. Xu and D.H. Zhou. Bayesian reasoning approach based recursive algorithm for online updating belief rule based expert system of pipeline leak detection, *Expert Systems with Applications*, 38(4): 3937-3943, 2011.
10. L.X. Wang, J.M. Mendel, Generating fuzzy rules by learning from examples, *IEEE Trans. on Syst., Man, and Cyber.* 22:6 (1992) 1414-1427.

ESTIMATION OF PREDICTION INTERVALS OF NEURAL NETWORK MODELS BY A MULTI-OBJECTIVE GENETIC ALGORITHM

RONAY AK[1], YANFU LI[1], ENRICO ZIO[1,2 *]

[1]Chair on Systems Science and the Energetic Challange, European Foundation for New Energy-Electricité de France

Ecole Central Paris, Grande Voie des Vignes, Châtenay-Malabry, 92290 France

SUPELEC, Plateau du Moulon - 3 rue Joliot-Curie, Gif-Sur-Yvette, 91192 France

[2]Department of Energy, Polytechnic of Milan, Via Ponzio 34/3

Milan, 20133 Italy

In this work, we implement a multi-objective genetic algorithm (namely, non-dominated sorting genetic algorithm–II (NSGA-II)) to find the parameters of an ANN which provides in output the lower and upper bounds of the prediction intervals of a quantity of interest. We apply the proposed method on a synthetic case study of literature.

1. Introduction

Artificial neural networks (ANNs) [1]-[3], [5], or simply neural networks (NNs), are powerful regression, prediction and classification models capable of approximating non-linear relationships among numerical variables. However, in practice the accuracy of the point predictions provided by an NN can be significantly affected by the uncertainties in the network structure and in the input data [1]-[3]. For this reason, it is important to estimate NN prediction intervals (PIs).

PIs are comprised of lower and upper bounds within which the actual target is expected to lie with a predetermined probability [1], [2]. Delta, Bayesian, Mean-variance Estimation (MVE) and Bootstrap techniques have been proposed in the literature for estimating PIs for NN model outputs [1].

In this paper, we train a NN to estimate in output the lower bound $L(x)$ and the upper bound $U(x)$ of the PI corresponding to a given input x.

There are two conflicting criteria for assessing the quality of the estimated PIs: coverage probability (CP) and prediction interval width (PIW) [2]. To obtain representative PIs, one should aim at maximizing the CP and minimizing

* enrico.zio@ecp.fr

the PIW, simultaneously. In this work, the multi-objective, non-dominated sorting genetic algorithm–II (NSGA-II) [6] is used for training such NN, i.e. optimizing its parameters (weights and biases). The proposed method is implemented on a synthetic case study.

The rest of the paper is organized as follows. Section 2 formally introduces the concept of PI. Section 3 illustrates the setting of NSGA-II for construction of PIs in NN models. Experimental results are given in Section 4. Finally, Section 5 concludes the paper with a critical analysis of the results obtained.

2. Prediction Intervals

The formal definition of a PI can be presented as follows:

$$\Pr(L(x) < y(x) < U(x)) = 1 - \alpha \tag{2.1}$$

where L(x) and U(x) are the lower bound and upper bound of the PI of the output y(x) corresponding to input x; the confidence level (1-α) refers to the expected probability that the true value of y(x) lies within the PI (L(x), U(x)).

As mentioned in the Introduction, two measures are used to evaluate the quality of the PIs: the coverage probability (CP) and the interval width (IW) [1], [2]. The prediction interval coverage probability (PICP) represents the probability that the set of estimated PI values will contain a certain percentage of the true output values. Prediction interval width (PIW) simply measures the extension of the interval as the difference of the estimated upper bound and lower bound values.

The mathematical formulation of the PICP and PIW measures used in this paper are [2]:

$$\mathrm{PICP} = \frac{1}{n}\sum_{i=1}^{n} c_i \tag{2.2}$$

where n is the number of samples in the training or testing sets, and $c_i = 1$ if $y_i \in [L(x_i), U(x_i)]$ and otherwise $c_i = 0$.

$$\mathrm{NMPIW} = \frac{1}{n}\sum_{i=1}^{n} \frac{(U(x_i) - L(x_i))}{t_{max} - t_{min}} \tag{2.3}$$

where NMPIW is the Normalized Mean PI Width, t_{min} and t_{max} represent the true minimum and maximum values of the targets (i.e., the bounds of the range in which the true values fall).

3. Implementation of NSGA-II for the Construction of the PIs of NN Models

We use NSGA-II for finding the values of the parameters of the NN which minimize the two objective functions PICP (2.2) and NMPIW (2.3) simultaneously, in a Pareto optimality sense (the maximization of PICP has been converted to minimization by subtracting from one).

The practical implementation of NSGA-II on our specific problem can be summarized as follows:

Step 1. Split the input data into training (D_{train}) and testing (D_{test}) subsets.

Step 2. Fix the maximum number of generations and the number of chromosomes Nc in each population; set the generation number n = 1. Initialize the first population P_n of size Nc, by randomly generating Nc chromosomes (individuals); each chromosome consists of M real-valued genes, where M is the total number of parameters in the NN.

Step 3. For each input vector x in the training set, compute the lower and upper bound outputs of the Nc NNs, each one with M parameters.

Step 4. Evaluate the two objectives PICP and NMPIW for the Nc NNs (one pair of values 1-PICP and NMPIW for each of the Nc chromosomes in the population P_n).

Step 5. Rank the chromosomes (vectors of M values) in the population P_n by running the fast non-dominated sorting algorithm [6] with respect to the pairs of objective values, and identify the ranked non-dominated fronts $F_1, F_2, \dots, F_k$ where F_1 is the best front, F_2 is the second best front and F_k is the least good front.

Step 6. Apply to P_n a binary tournament selection based on the crowding distance [6] for generating an intermediate population S_n of size Nc.

Step 7. Apply the crossover and mutation operators to S_n, to create the offspring population Q_n of size Nc.

Step 8. Apply Step 3 onto Q_n and obtain the lower and upper bound outputs.

Step 9. Apply Step 4 onto Q_n.

Step 10. If the maximum number of generations is reached, stop and return P_n. Select the first Pareto front F_1 as the optimal solution set. Otherwise, go to Step 11.

Step 11. Combine P_n and Q_n to obtain a union population $R_n = P_n \cup Q_n$.

Step 12. Apply Steps 3-5 onto R_n and obtain a sorted union population.

Step 13. Select the Nc best solutions from the sorted union to create the next parent population P_{n+1}.

Step 14. Apply Steps 6-9 onto P_{n+1}. Set n=n+1; and go to Step 10.

Finally, the best front in terms of ranking of non-dominance and diversity of the individual solutions is chosen. Once the best front is chosen, then the testing step is performed on the trained NN with optimal parameter values (weights and biases).

4. Experiments and Results on a Synthetic Example

In this Section, the proposed method is applied to a synthetic case study of literature. The synthetic function used is [1]:

$$y = g(x) + \varepsilon, \text{ where } g(x) = x^2 + \sin(x) + 2 \tag{4.1}$$

where the error term ε follows a zero-mean Gaussian distribution with variance $\sigma^2(x)$ equal to $\sigma^2(x) = g(x)/5$

By repeatedly calculating this function, with x values sampled from a uniform distribution between -10 and 10, 500 input/output samples are generated; the first 350 (70%) are used for training and the rest is used for testing. The NN consists of one input, one hidden and one output layers. The number of hidden neurons is set to 7 after a trial-and-error process; the number of output neurons is 2, one for the lower and one for the upper bound values of the PI.

For comparison, the single objective simulated annealing (SOSA) procedure in [2] has been implemented based on the single objective function called coverage width-based criterion (CWC) [2]:

$$\text{CWC} = \text{NMPIW}(1 + \gamma(\text{PICP})\, e^{-\eta(\text{PICP}-\mu)}) \tag{4.2}$$

where η and μ are constants. The role of η is to magnify any small difference between μ and PICP. The value of μ gives the nominal confidence level, which is set to 90% in our example. Then, η and μ are two parameters determining how much penalty is paid by the PIs with low coverage probability. The function $\gamma(\text{PICP})$ is equal to one during training, whereas in the testing of the NN it is given by the following step function:

$$\gamma(\text{PICP}) = \begin{cases} 0, & \text{PICP} \geq \mu \\ 1, & \text{PICP} < \mu \end{cases} \tag{4.3}$$

In the NSGA-II, the population size (Nc) is set to 50 and the number of generations to 300, for a total number of evaluations equal to 15000. For fair comparison, SOSA is configured to have equal number of evaluations: therefore, the maximum number of iterations is set to 15000 as termination condition. In the SOSA, the parameters "μ", "η", "T_{init}", "T_{min}", "geometric cooling factor" and "CWC_{int}" are set to 0.9, 50, 5, 10^{-50}, 0.95 and 10^{80} respectively [2].

The first (best) front found by NSGA-II includes 25 non-dominated solutions. For each of these, the CWC value has been *a posteriori* calculated by combining the individual PICP and NMPIW values. Among 25 Pareto front solutions, the lowest CWC value is 0.226, which improves the CWC value found at convergence by SOSA, which is equal to 0.44.

To account for the randomness in the search procedures by SOSA and NSGA-II, both have been run twenty times. Then, *a posteriori* CWC evaluation for the solutions of the 20 NSGA-II Pareto fronts has been performed. For each Pareto front, the solution with smallest (best) CWC value is selected among all solutions on the front. This leads to obtaining twenty best CWC values, one from each Pareto front. After training, we perform the testing of the trained NNs with fixed optimal parameter values (weights and biases). The testing results reported in Table 1 show that statistics of the NSGA-II CWC values are smaller (better) than or equal to those of SOSA. As to the individual objectives, the mean of the PICP values obtained by NSGA-II is larger and that of NMPIW is smaller, than those obtained by SOSA.

Table 1. Descriptive Statistics for testing results of NSGA-II and SOSA runs

	SOSA			NSGA-II		
	CWC	PICP	NMPIW	CWC	PICP	NMPIW
Mean	0.324	0.933	0.324	0.309	0.954	0.309
Std. Dev.	0.051	0.021	0.051	0.051	0.025	0.051

Finally, an overall best non-dominated Pareto front is obtained from the twenty NSGA-II fronts. Figure 1 illustrates the testing results. In this plot, comparison is done in terms of 1-PICP and NMPIW. The crosses indicate the SOSA values and the diamonds represent the NSGA-II values. We can see that the testing results of NSGA-II are better than those of SOSA.

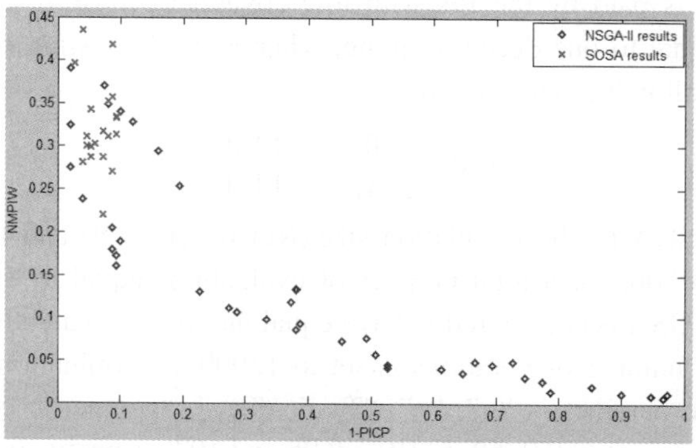

Figure 1. 20 SOSA results versus overall best NSGA-II front (testing set)

5. Conclusion

In this study, we propose an estimation of prediction intervals for NN models within a multi-objective optimization framework, and proceed to its solution by the powerful NSGA-II. The strength of the approach has been demonstrated on a synthetic case study of literature. The results obtained are satisfactory.

References

1. Khosravi, A., Nahavandi, S., Creighton, D. and Atiya, A. F. (2011a). Comprehensive Review of Neural Network-Based Prediction Intervals and New Advances. IEEE Transactions on Neural Networks, Vol. 22, No.9, pp. 1341-1356.
2. Khosravi, A., Nahavandi, S., Creighton, D. and Atiya, A. F. (2011b). Lower Upper Bound Estimation Method for Construction of Neural Network-Based Prediction Intervals. IEEE Transactions on Neural Networks, Vol. 22, No. 3, pp. 337-346.
3. Dybowski, R. and Roberts, S. J. (2000). Confidence Intervals and Prediction Intervals for Feed-forward Neural Networks. Clinical Applications of Artificial Neural Networks, Cambridge University Press.
4. Konak, A. Coit, D. W. and Smith, A. E. (2006). Multi-objective Optimization Using Genetic Algorithms: A Tutorial. Reliability Engineering and System Safety, Vol. 91. pp. 992-1007.
5. Kalagirou, S. A. (1999). Applications of Artificial Neural Networks in Energy Systems: A Review. Energy Conversion and Management, Vol. 40, No. 10, pp. 1073-1087.
6. Deb, K., Pratap, A., Agarwal, S., and Meyarivan, T. (2002). A fast and Elitist Multi-objective Genetic Algorithm: NSGA-II, IEEE Transactions on Evolutionary Computation, Vol. 6, No.2, pp. 182-197.

PART 6

PRODUCTIVITY AND RELIABILITY

LOAD AVERAGE FORECAST: A NEW MONITOR TO ANALYSE INSTABILITY IN COMPUTER SYSTEMS

VICTORIA LOPEZ*, GUADALUPE MIÑANA

Department of Computer Architecture, Complutense University, Madrid 28040, Spain

In this paper a new software monitor is presented. Computer systems must be adequately monitored to ensure stability control, reliability and risks to unexpected problems. Classic monitors are based on the analysis of some key variables which are in change during operation of the systems. In this new monitor, some modeling techniques (statistics, reliability and decision making) are used to provide an easy, automated way for the intervention into a system under risk.

1. Introduction

A computer system is any device or network of devices that are operated by human intervention and in which some software is running. Nowadays there are million applications for mobile devices with Internet connection. These applications are also computer systems, and consistently appear on the market from any source often without any warranty or control. In any of these cases, we can analyze the system to make appropriate decisions to each particular problem.

On the other hand and due to the current economic situation, the researchers recommend developing reliable and quality software. This software must be developed for ease of maintenance and can be updated and reused whenever possible. In this sense [1] proposes government actions on Intellectual Property to encourage the reuse and maintenance of software development. Also several research groups are working in formal engineering, more specifically in reusing of the software [2]. Prof. R. Yager in his plenary talk last ISKE Conference [3] also encouraged programmers and engineers to develop reliable and maintainable software instead of creating new applications again and again.

There is a set of fundamental variables in the modeling and analysis of a computer system. These variables are as follows: the processor load average, the response time of each device, the average time in queue, the mean time to

* vlopez@fdi.ucm.es

failure, and so on. The study of computer systems deals with monitoring these variables and the modeling of the system itself by using suitable techniques. Some previous studies have been developed by the authors of this paper (see [4]). In this and other related work, the authors attempt to jointly assess performance and reliability of computer systems. However, a specific analysis of variables and devices is required to achieve good results. Taking into account the importance of processors in any computer system, in this article we focus on the study of load average of the processors as an essential indicator of stability [5].

This paper is organized as follows. Section 2 shows network monitoring fundamentals. In section 3 definitions and computational features of the new monitor are given. Then, an example is shown in section 4. Finally, conclusions and future work are proposed in section 5.

2. Network Monitoring

Network monitors are software solutions for monitoring computer networks. Usually they allow monitoring in a visual way the status and performance of several parameters from different operating systems, servers, applications and hardware systems such as firewalls, proxies, databases, web servers or routers. Pandora FMS (for Pandora Flexible Monitoring System) [6] is a good example of a commercial network monitor. Network monitors usually get the information by means of small monitors already integrated in operative systems and available to the user. Many of them are system commands very easy to use. The problem is to store and analyze the information. In this sense, network monitors define several indexes and benchmarks more friendly to be understood by the customer.

In this paper, we developed a new monitor for analyzing load of a processor. This monitor outputs a benchmark about stability of the system. This information helps to make decisions about the operation of the system.

3. Formal Definition and Specification

Definition 1. [7] The *LA* triplet is defined by means of the following structure:

$$LA = (l_1, l_2, l_3) : l_i \geq 0 \forall i = \{1, \ldots, 3\} \quad (1)$$

Each number into the triplet is defined in terms of an array of 15 non-negative x_i numbers where x_i is the load of the system (15-i+1) minutes ago. This array is defined as follows:

$$
\begin{aligned}
&\exists(x_1,x_2,\ldots,x_{15}):\\
&\forall i\in\{1\ldots15\}:(x_i\geq 0)\wedge \qquad (2)\\
&(l_1=x_{15})\wedge\left(5l_2=\sum_{i=11}^{15}x_i\right)\wedge\left(15l_3=\sum_{i=1}^{15}x_i\right)
\end{aligned}
$$

Equations (1) and (2) constitute the formal of the LA triplet.

Although it is possible to find the value of the 15 numbers (by making an exhaustive monitoring), this idea constitutes an overhead in monitoring and also because the information that they offer is not too much informative, in fact, these series of numbers are original time series. In addition, the triplet is formed by the moving averages of sizes 1, 5 and 15 respectively. Therefore, these data are more valuable in terms of smooth samples. Instead, we are interested in the most and the less stable samples. These samples can be calculated without monitoring, therefore without adding overhead to the system.

Definition of the most and the less stable sample related is given as follows.

Definition 2. [7] Let $LA = (l_1, l_2, l_3)$ be a load average triplet. Then,

a. the most stable sample for LA is the array defined by the maximum number of identical components, under restrictions (1) and (2). That means its variance is minimum.

$$
\begin{aligned}
&X_+=(x_1,\ldots,x_{15})\equiv(\alpha,\alpha,\alpha,\alpha,\alpha,\alpha,\alpha,\alpha,\alpha,\alpha,\beta,\beta,\beta,\beta,\delta):\\
&where\quad \alpha=\frac{3l_3-l_2}{2},\quad \beta=\frac{5l_2-l_1}{4},\quad \delta=l_1
\end{aligned} \qquad (3)
$$

b. the less stable sample is the array defined by the maximum distance between its components, which means, its variance is maximum.

$$
\begin{aligned}
&X_+=(x_1,\ldots,x_{15})\equiv(\alpha',0,0,0,0,0,0,0,0,0,\beta',0,0,0,\delta'):\\
&where\quad \alpha'=15l_3-5l_2,\quad \beta'=5l_2-l_1,\quad \delta'=l_1
\end{aligned} \qquad (4)
$$

Proof. The expressions (1) and (2) represent three equations and fifteen variables, but in booth case, the most stable and the less stable samples, the number of variables is reduced to three. In these cases is very easy to find the figure of coefficients α, β, δ of the system.

For notation, X_+ is the sample in which components have the load values within the last 15 minutes under more stability. On the contrary, X- contains the sample under less stability. Since the sample X_+ is the only possibility for stability while instability has 40 possibilities under a discrete universe conditions, the distance between both variances of samples seems a good way for defining stability of the processor. Moreover, variance of X_+ works as a smooth parameter over the variance of X-. Although this is a somewhat pessimistic allocation, the result is enough good because it ensures the robustness of the procedure. Next definition is about load average stability and, for extension, about system stability. Taking into account that greater variance means greater instability and lower variance means greater stability, the definition of stability of the load is given by the distance between the two variances as follows.

Definition 3. [7] The load average stability is defined as the distance between the two variances (*Var* is the statistic variance).

$$D=abs(Var(X_+) - Var(X_-)) \qquad (5)$$

Proposition 1. [7] A short way to compute the load average stability is given by

$$D = \frac{1}{7}(45\alpha^2 + 6\beta^2) \qquad 6)$$

Proof. The expressions (6) is obtained by solving the equation (5)

The load average stability is warranted if D is small enough. The new monitor switches an alarm on when the number of events is oversize. The general way of working of the new monitor is shown in Figure 1 below.

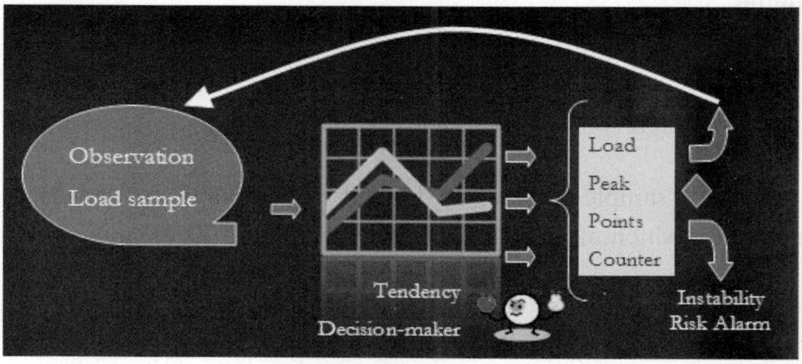

Figure 1. New monitor flow chart.

4. Example

4.1. *Scenario*

We have applied the monitor described in this paper in different computers in order to get data from different sources: Computer 1 (C1) is a personal computer with one processor running Ubuntu 9.04 OS. Computer 2 (C2) is a front-end of the one cluster. The OS is again Ubuntu 9.04. Computer 3 (C3) is a personal computer with dual core and with Debian OS. In each one of the computers the monitor has been getting information about stability of the system during 24 hours at regular intervals of 25 minutes. All data about this information ($Var(X_+)$, $Var(X_-)$, D, …) can be found in [7].

4.2. *Analysis of the information*

Figure 2 shows some graphs obtained by the monitor. Each graph shows the following information: on one hand green and cyan lines represent the values of $Var(X_+)$ and Var(X-) respectively. Red line is the difference between those variances (D). On the other hand, the thin black horizontal line shows the value of the D average (S), and the other black line represents the D Average square root.

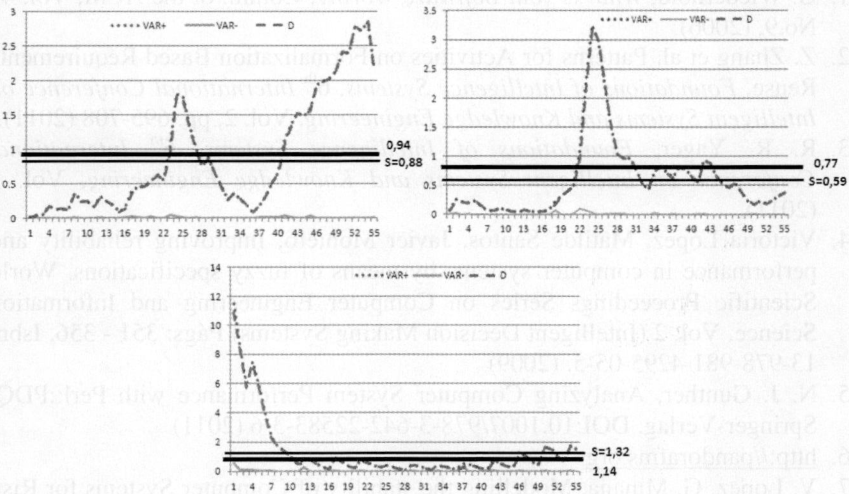

Figure 2. Graphical stability information about computers 1, 2, and 3.

In computer C1, D is increasing. That means instability is increasing. In C3, D is decreasing. That means a good stability of the system. Computer C2 shows more changes along the time but there is no instability tender because D is not increasing, and due to it is a dual core, the threshold for instability decision must be greater than in a simple processor cases.

5. Conclusions and Future Work

This paper shows a new monitor for discovering load instability into a computer system as an aid system for auditing. The procedure is based in the study of load average triplets, the aggregation of the information obtained by them and the study of both features tendency and load peaks. One of the best advantages of this procedure is the optimization of the overhead in the system, due to its simple calculations running in parallel way. Moreover it is adaptable to any system since it permits the selection of a suitable way for aggregation of the information. As future work, we are working in integrating it in the EMSI tool [8, 9] to make some comparative analysis of the load stability regarding to different systems. This issue is a part of several developments about measuring quality of computer systems for promoting the good behavior, easy maintenance and reuse of the systems already in working.

References

1. G. Wiederhold, *What is your Software Worth?,* Comm. of the ACM, Vol. 49 No.9, (2006).
2. Z. Zhang et al. Patterns for Activities on Formalization Based Requirements Reuse, *Foundations of Intelligence Systems*, 6th *International Conference on Intelligent Systems and Knowledge Engineering,* Vol. 2, pp. 695-708 (2011).
3. R. R. Yager, *Foundations of Intelligence Systems, 6th International Conference on Intelligent Systems and Knowledge Engineering*, Vol. 1 (2011).
4. Victoria López, Matilde Santos, Javier Montero, Improving reliability and performance in computer systems by means of fuzzy specifications, World Scientific Proceedings Series on Computer Engineering and Information Science, Vol. 2 (Intelligent Decision Making Systems) Págs: 351 - 356, Isbn: 13-978-981-4295-05-5. (2009)
5. N. J. Gunther, Analyzing Computer System Performance with Perl::PDQ. Springer-Verlag, DOI:10.1007/978-3-642-22583-3-6 (2011)
6. http://pandorafms.org
7. V. Lopez, G. Minana, Modelling the stability of Computer Systems for Risk Prevention, *Int. J.UFKBS* (to be published).
8. G-TeC team research, www.tecnologiaUCM.es, (2010).
9. V. Lopez, M. Santos and T. Rodriguez. An Approach to the Analysis of Performance, Reliability and Risk in Computer Systems, *Foundations of Intelligence Systems*, 6th *International Conference on Intelligent Systems and Knowledge Engineering,* Vol. 2, pp. 653-658 (2011).

COMPUTATION OF COMPONENT IMPORTANCE MEASURES FOR RELIABILITY OF COMPLEX SYSTEMS

[1,2]QAMAR MAHBOOB*, [3]FAISAL MAQBOOL ZAHID, [1]ERIC SCHÖNE, [1]MICHAEL KUNZE, [1]JOCHEN TRINCKAUF

[1]Department of Railway Signalling & Transport Safety Technology, Technische Universität Dresden, Germany

[2]Mechanical Department, Pakistan Railways HQ office Lahore, Pakistan

[3]Statistics Department, GC University, Allama Iqbal Road Faisalabad, Pakistan

The classical method of computing component importance measures (CIM), which are used for decision making on system reliability, is by means of fault tree (FT) analysis. However, the FT method has limitations for complex systems – characterized by the presence of various interrelated aspects like dependencies, uncertainties and knowledge of a reliability problem. The Bayesian Networks (BN) use graph-based representations that define probability distributions over network and over random variables in a concise and flexible way. In this paper, the computation of CIM using BN is investigated for complex railway systems.

1. Introduction

Failure of a component, which is more important for the system reliability, can lead to adverse consequences. The importance of each component towards system reliability can be determined by using importance measure (CIM) [1]. The purpose of CIM is to assist (1) the designers in identifying the components that should be improved (2) maintenance engineers to improve maintenance plans for the important components, and (3) decisions on utilization of engineering budgets and public money. A number of CIM exist [2, 3] that can be used for reliability analysis [4]. The fault tree (FT) uses logical representation of a technical system and is a common technique to compute CIM for reliability analysis. However, the FT method has limitations for modelling complex systems – characterized by the presence of a number of dependencies, uncertainties and knowledge of the problem [5, 6]. Failing to consider such characteristics will lead to over or under estimation of system reliability. This motivates an investigation into the use of Bayesian Networks (BN), which are probabilistic graphical models and have several advantages over FT. The BN define probability distributions over network over random variables and allow combining systematic, expert and factual knowledge of safety problems in a flexible and concise way [7, 8, 9]. In this paper we compute a number of CIM and investigate advantages of the BN over the FT for complex systems.

2. Definitions of Component Importance Measures

To present simple definitions of the reliability related CIM the time t, which is implicitly present in all of the definitions in Table 1, is omitted. Here, $\Pr(F_s)$ and $\Pr(F_i)$ are failure probabilities of the system and component i, respectively.

Table 1. Definitions of component importance measures (CIM)

CIM	Theoretical description	Mathematical definition
Conditional Probability (CP)	CP gives the $\Pr(F_s)$ given the $\Pr(F_i)$	$CP_{(i)} = \frac{\Pr(F_s \cap F_i)}{\Pr(F_i)}$
Risk Achievement Worth (RAW)	RAW measures the worth of component F_i in achieving the present level of system reliability	$RAW_{(i)} = \frac{\Pr(F_s\|F_i)}{\Pr(F_s)}$
Risk Reduction Worth (RRW)	RRW measures the decrease of the system unreliability by increasing the reliability of F_i	$RRW_{(i)} = \frac{\Pr(F_s)}{\Pr(F_s\|\overline{F_i})}$
Diagnostic Importance Factor (DIF)	DIF gives the $\Pr(F_i)$ given the $\Pr(F_s)$	$DIF_{(i)} = \frac{\Pr(F_s \cap F_i)}{\Pr(F_s)}$
Birnbaum's Measure (BM)	BM measures the sensitivity of the system unreliability w.r.t to the changes in the $\Pr(F_i)$	$BM_{(i)} = \frac{\partial \Pr(F_s)}{\partial \Pr(F_i)}$
Fussel-Vesely (FV) Importance	The standard FV failure importance is the contribution of the $\Pr(F_i)$ to the $\Pr(F_s)$	$FV_{(i)} = \frac{\Pr(F_s) - \Pr(F_s\|\overline{F_i})}{\Pr(F_s)}$
Criticality Importance Factor (CIF)	CIF gives the probability that the component F_i has caused system failure given $\Pr(F_s)$	$CIF_{(i)} = \frac{\Pr(F_i)}{\Pr(F_s)} \cdot BM_{(i)}$
Improvement Potential (IP)	IP gives the improvement potential if failed component is replaced by a perfect one	$IP_{(i)} = \Pr(F_i) \cdot BM_{(i)}$

3. Fault Tree Analysis

Fault Tree (FT) uses series of logical symbols to develop scenarios (originated from basic events) leading to the system failure, also so-called Top Event (TE). The system is analyzed in the context of its design, operation, safety requirements and environment to find all combinations of basic as well as intermediate events that can lead to the TE [10].

3.1. *Fault Tree for Train Derailment due to SPAD*

The aim is to show advantages of the BN for complex systems, not to provide a detailed analysis. Therefore, we adopt an informal description of the proposed system, which is sufficient to serve our aim. According to the FT in Fig 1 the train can derail when the SPAD (signal passed at danger) event is followed by a turnout point that is not set or followed by a curve and train speed is too high. SPAD events occur when trains do not stop before a red light.

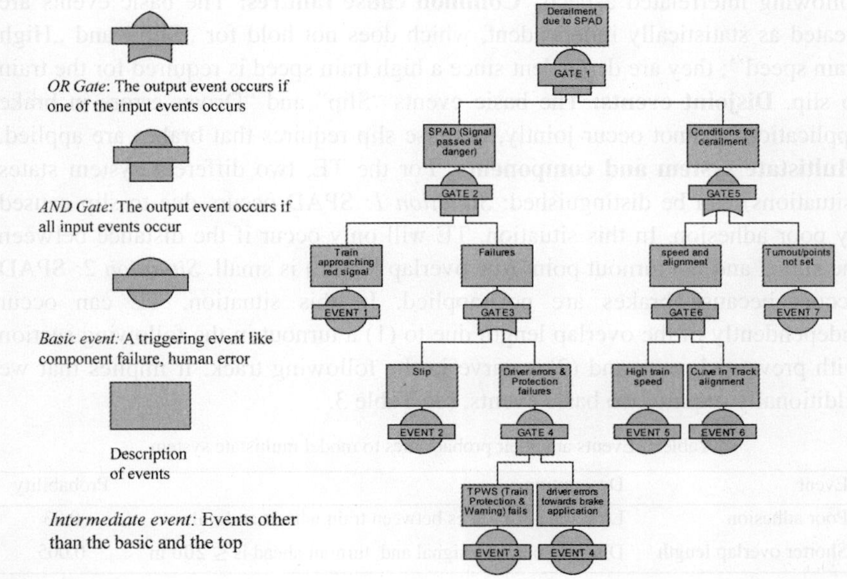

Figure 1. Fault Tree for Train derailment due to SPAD and its graphical symbols

The possible scenarios leading to SPAD are failure to break by the combined failure of TPWS and driver errors and slip between the wheels and the rails. The basic events in the FT are summarized in Table 2 where probabilities are for illustration only. The probability of the TE will be computed as a function of the probabilities of the basic events by identifying (minimal) cut-sets: $E_1E_3E_4E_7 + E_1E_3E_4E_5E_6 + E_1E_2E_7 + E_1E_2E_5E_6$. From the FT, the $\Pr(TE)$ is $2.84E-04$ and the CIM values are presented in Table 4.

Table 2. Events and their probabilities in the FT for train derailment due to SPAD

Event	Description	Probability
Train approaching red signal	Train is running towards a red signal	0.05
Slip	Sliding of train over rails due to poor adhesion	0.05
TPWS fails	TPWS fail on demand such as while passing a signal	0.0001
Driver errors in brake app.	Driver fails to react to a signal in time	0.001
High train speed	Speed of the train is > 60 miles/hour	0.15
Curve in Track alignment	Railway track is not straight after passing a signal	0.1
Turnout/point not set	A turnout/point prevents the route ahead	0.1

4. Interrelated Aspects of Complex System

The FT in Figure 1 assumes many simplifications and therefore fails to address following interrelated aspects. **Common cause failures:** The basic events are treated as statistically independent, which does not hold for „Slip'' and „High train speed''; they are dependent since a high train speed is required for the train to slip. **Disjoint events:** The basic events "Slip" and "Driver errors in brake application" cannot occur jointly, because slip requires that brakes are applied. **Multistate system and components:** For the TE, two different system states (situations) can be distinguished: *Situation 1:* SPAD occurs due to slip caused by poor adhesion. In this situation, TE will only occur if the distance between the signal and the turnout point (the overlap length) is small. *Situation 2:* SPAD occurs because brakes are not applied. In this situation, TE can occur independently of the overlap length due to (1) a turnout in the following section with prevented route and (2) a curve in the following track. It implies that we additionally require two basic events, see Table 3.

Table 3. Events and their probabilities to model multistate system

Event	Description	Probability
Poor adhesion	Low adhesive forces between train wheels and rails	0.03
Shorter overlap length	Distance between signal and turnout ahead is ≤ 200 m	0.005

Failure dependency: The probability of the gate 4 will be higher (positive dependency) if TPWS fails earlier than driver errors. **Functional/failure uncertainty:** The logic (e.g. OR, AND) between the events is uncertain. For example, uncertainty in functional (failure) will arise when there is a track section, which involves both (1) curve in track alignment and (2) turnout/point. **Uncertainty in expert knowledge:** Two experts have different believes on the probability of basic events i.e. *Curve in track alignment.*

Some of the aspects mentioned above can be modeled using advanced FT techniques [5], which lead to an exponential increase in the FT structure and models become nonintuitive and difficult to handle. The functional uncertainty and expert uncertainties cannot be modeled using FT [6]. Moreover, the FT has limitations in modeling multistate and failure dependency. Failing to consider such aspects will lead to over or under estimation of the CIM values.

5. Bayesian Networks

Bayesian Networks (BN) are directed acyclic probabilistic graphical models that define probability distributions over network over random variables. The nodes in BN are random variables and directed links represent the dependence structure. A (discrete) BN with random variables $\mathbf{X} = [X_1,\ldots,X_n]$ will have the probability mass function (PMF) for each random variable as conditional on its parents, i.e., $p(x_1 \mid x_2, \ldots, x_n)$ and the joint PMF $p(\mathbf{x})$ of a set of random variables $\mathbf{X}$ is obtained by multiplying all the conditional probabilities:

$$p(\boldsymbol{x}) = \prod_{i=1}^{n} p[x_i|pa(x_i)] \tag{1}$$

Where $pa(x_i)$ are set of realizations of the parents of variables X_i. BN can predict and estimate the states of the variables using total probability and Bayes theorems, respectively. For further details on BN we refer to [11, 12, 13].

5.1. *Modeling of Interrelated Aspects and Computation of CIM using Bayesian Network*

Any FT can be mapped into an equivalent BN [6, 14]. We model all interrelated aspects by exploiting the use of BN structure and conditional probability tables attached to (child) variables. We use probabilities for increased (failure) tendency as $p(G3 \mid E_3, \overline{E_4}) = 0.05$, functional uncertainty as $p(TE \mid SPAD, E_2, E5, E6) = 0.1$ and for expert knowledge as $p(E_6 \mid expert_1) = 0.1$ and $p(E_6 \mid expert_2) = 0.2$. The reliabilities of the first and second experts are taken 0.55 and 0.45, respectively.

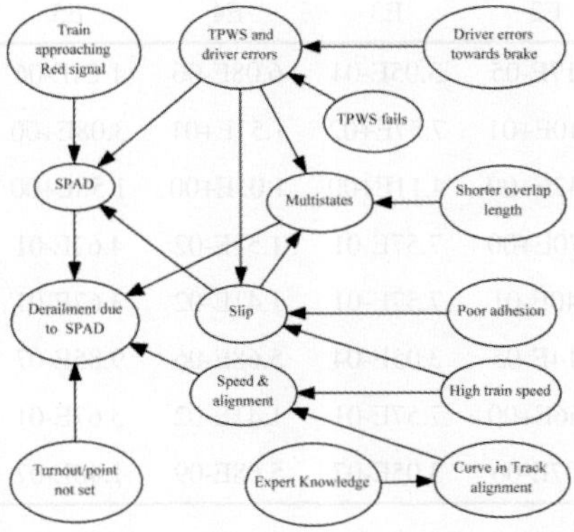

Fig 4. Bayesian Network model after modeling interrelated aspects of a complex system

To answer probability questions in BN we make use of standard Bayesian inference, which is also used to compute the values of the CIM, see Table 5. In the BN, the $\Pr(TE) = 4.02E - 07$, which is lower than the $\Pr(TE)$ in the FT.

6. Results

The BN was able to model a number of interrelated aspects, which was not possible using the FT. Thus, different values of CIM are obtained from both the models that can influence the decision on system reliability.

Table 4 Values of CIM without dependencies and uncertainties in the FT model

CIM	E1	E2	E3	E4	E5	E6	E7
CP	5.75E-03	5.75E-03	2.93E-04	2.93E-04	5.00E-04	6.25E-04	2.54E-03
RAW	2.00E+01	2.00E+01	1.02E+00	1.02E+00	1.74E+00	2.17E+00	8.83E+00
RRW	∞	5.07E+04	1.01E+00	1.01E+00	1.15E+00	1.15E+00	7.67E+00
DIF	1.00E+00	1.00E+00	1.02E-03	1.02E-03	2.61E-01	2.17E-01	8.83E-01
FV	1.00E+00	1.00E+00	1.31E-02	1.31E-02	1.30E-01	1.30E-01	8.70E-01
BM	5.75E-03	5.75E-03	9.50E-06	9.50E-06	2.50E-04	3.75E-04	2.50E-03
CIF	1.00E+00	1.00E+00	3.30E-05	3.30E-05	1.30E-01	1.30E-01	8.70E-01
IP	2.88E-04	2.88E-04	9.50E-09	9.50E-09	3.75E-05	3.75E-05	2.50E-04

Table 5. Values of CIM after modeling interrelated aspects using BN

CIM	E1	E2	E3	E4	E5	E6	E7
CP	8.05E-06	2.17E-05	3.05E-04	6.08E-06	1.24E-06	6.10E-07	3.53E-06
RAW	2.00E+01	5.40E+01	7.57E+02	1.51E+01	3.08E+00	1.52E+00	8.76E+00
RRW	∞	1.32E+00	4.11E+00	1.01E+00	1.58E+00	1.10E+00	7.26E+00
DIF	1.00E+00	2.70E+00	7.57E-01	1.51E-02	4.62E-01	1.52E-01	8.76E-01
FV	1.00E+00	2.40E-01	7.57E-01	1.41E-02	3.67E-01	8.74E-02	8.62E-01
BM	8.05E-06	2.14E-05	3.05E-04	5.68E-06	9.85E-07	2.43E-07	3.47E-06
CIF	1.00E+00	2.66E+00	7.57E-01	1.41E-02	3.67E-01	6.03E-02	8.62E-01
IP	4.02E-07	1.07E-06	3.05E-07	5.68E-09	1.48E-07	2.43E-08	3.47E-07

7. Conclusions

A number of component importance measures (CIM) were computed using Fault Tree (FT) and Bayesian Networks (BN). The BN provided a compact way of modeling interrelated aspects of a complex system and offered a straightforward way to compute the values of the CIM, which was complicated using FT. The BN offered a suitable mechanism for exploiting BN structure not only to model dependencies and uncertainties for quantitative assessment, but also for knowledge management of the specific reliability problem and this question is quite central in model engineering.

Acknowledgement

Financial support from Higher Education Commission Pakistan is acknowledged

Bibliography

1. E. Borgonovo, G. E. Apostolakis, S. Tarantola and A. Saltelli, "Comparison of global sensitivity analysis techniques and importance measures in PSA," Reliability Engineering & System Safety,79(2), pp. 175-185, 2003.
2. Z. W. Birnbaum, "On the importance of different components in a multi-component system. In Multivariable Analysis II, P.R. Kor-ishnaiah, Ed," Academic Press, New York,NY, 1969.
3. B. Fussell, "How to hand-calculate system reliability characteristics," IEEE Trans. on Reliability, p. 24(3), 1975.
4. M. Rausand and A. Hoyland, System Reliability Theory: Models, Statistical Methods, and Applications, Hoboken: John Wiley & Sons, 2004.
5. L. Xing and S. V. Amari, "Fault Tree Analysis," in Handbook of Performability Engineering, London, Springer, 2008, pp. 595-620.
6. N. Khakzad, F. Khan and P. Amyotte, "Safety Analysis in Process Facilities: Comparison of Fault Tree and Bayesian Network Approaches," Reliability Engineering and System Safety, p. doi:10.1016/j.ress.2011.03.012, 2011.
7. D. M. HANEA, "Human Risk of Fire: Building a decision support tool using Bayesian networks," PhD Dissertation, TU Delft, Nov 2009.
8. M. Lampis and J. D. Andrews, "Bayesian Belief Networks for System Fault Diagnostics," Quality and Reliability Engineering International, pp. 409-426, 2009.
9. D. Straub and A. D. Kiureghian, "Bayesian Networks Enhanced with Structural Reliability Methods.Part B: Application," Journal of Engineering Mechanics, pp. Trans. ASCE, in press, 2010.
10. W. Vesely, F. Goldberg, N. Roberts and D. Haasl, Fault Tree Handbook, Washington, D.C: U.S.N.R.C, 1981.
11. D. Koller and N. Friedman, Probabilistic Graphical Models, ISBN-10: 0-262-01319-3 : MIT Press, 2009.
12. F. V. Jensen and T. D. Nielsen, Bayesian Networks and Decision Graphs, Berlin: Springer, 2007.
13. R. Ahmed, Bayesian Network, Rijeka: Sciyo, 2010.
14. A. Bobbio, L. Portinale, M. Minichino and E. Ciancamerla, "Improving the analysis of dependable systems by mapping fault trees into Bayesian networks," Reliability Engineering and Systen Safety 71, pp. 249-260, 2001.

ON THE REVISION TO THE RESULT OF FINANCE EARLY-WARNING LOGIT MODEL FROM THE PERSPECTIVE OF EFFICIENCY

SHUANGJIE LI[1], SHAO WANG[2]

[1] *Economics & Management School, Beijing University of Technology Pingleyuan 100,100124, Chaoyang District, Beijing, China*

[2] *China Security CO, LTD, Beijing, China*

This paper use LOGIT model to study financial early-warning by selected ST and normal companies. The in-sample and out-of-sample prediction rate are respectively 97.1% and 94.1%. With revision and re-judgment to logit model prediction result by efficiency indicators, the study result shows that in-sample and out-of-sample correction rate is respectively 95.8% and 96.2%.

1. Introduction of financial early-warning model study

Foreign study on the financial early-warning model is more mature than that of Chinese domestic study. In China, the study on company's financial early-warning model began in the 1980s, directly using foreign method or modifying their models according to the characteristics of China.

Quantitative financial early-warning models include single-variable financial early-warning model, multiple linear financial early-warning models, the probability of selection models (including linear probability model (LPM), logit model and probit model), artificial neural network (ANN), etc.

2. LOGIT model

2.1. *Sample indicators and data*

This paper selected 17 companies for two consecutive years whose net profit is negative in 2009 which is called ST Company (Special Treatment) and 17 companies in the same industry as the pair's normal. In order to ensure a comprehensive study, we selected three years of data before being ST, i.e. 2006 data. All the financial indicators data are from the CCER database and the financial statements of listed companies.

In order to fully reflect the company's operating performance, this study initially sets six categories of 13 financial indicators into the logit model of the financial early-warning according to "enterprise performance evaluation standard" of the SASAC of the state council. Profitability indicator: X_1—Rate of Return on Common Stockholders' Equity, X_2—Return on assets. Growth indicator: X_3—Net asset growth rate, X_4—Operating income growth rate, X_5—After-tax profit growth rate. Flow indicator: X_6—Liquidity ratio, X_7—Quick ratio. Cash indicator: X_8—Cash to current liability ratio, X_9—Cash debt ratio. Structure indicator: X_{10}—Debt to assets ratio. Sales indicator: X_{11}—Inventory turnover ratio, X_{12}—turnover of account receivable, X_{13}—total assets turnover ratio.

2.2. *Logit model of financial early-warning*

We define the event "the company in 2009 is being ST" as 1, the event "the company is not being ST" as 0. By this definition on the dummy variables for the above six categories 13 financial indicators, we use SPSS17.0 regression model to do automatic filtering variables which has a significant influence into the model, the sample can be modeled to predict the probability values calculated. As a general rule, if the calculated probability is greater than 0.5, the company is considered as the financial crisis company, classifying as a crisis class, the company will be ST after three years; if the calculated probability is less than 0.5, the company is classified as normal classes, the company will not be ST after three years.

Eventually variables entering the model and related statistics in Table 1 as below:

Table 1 logit regression result

variable	Variable name	coefficients	standard deviation	Wald Statistics	Significance level
C	Constant	4.557	2.176	4.384	0.036
X_2	Return on assets	-110.933	48.710	5.187	0.023
X_5	Net profit growth rate	-1.931	0.983	3.858	0.050

Shown by the regression results, in addition to the constant, the original 13 variables remain only two: return on assets X_2 and net profit growth rate X_5.

$$\ln\frac{p}{1-p}=4.557-110.933X_2-1.931X_5 \tag{1}$$

$$p=\frac{e^{4.557-110.933X_2-1.931X_5}}{1+e^{4.557-110.933X_2-1.931X_5}} \qquad 2)$$

Where p is the probability of the company being ST three years later.

2.3. *Test in sample*

Substituting asset profit rate, the growth rate of net profit of the 34 companies in 2006 respectively type in (2), we can obtain the probability of each company of being ST after three years, the result shows the effect of the predict of the model is very good. Among the 34 companies, only one (No. 22, 600485) occurs prediction error. The overall correct prediction rate is 97.1%. Using the mapping function by SPSS17.0, according to the classification of ST group and normal group samples obtained within the above to make a prediction probability histogram (Figure 1), only one normal group company is judged for ST group, the remaining 33 companies all predict correctly.

3. The revision of the result of LOGIT model from the perspective of efficiency

We establish the logit financial early-warning model of samples within predicted high correct rate reached 97.1%.when we apply the model to another company out of the samples, is the prediction (being ST or normal after three years) reliable? We hope to use the company's data to find an indicator, to predict the company's prospects and to revise or to confirm the results of the above financial early-warning model. There are very few examples of solving the financial early warning problems by non-financial indicators in the existing literature. This article use DEA efficiency as a non-financial indicator, determine (amend) the results of the logit model for early warning of financial.

First, calculate output DEA efficiency and value added DEA efficiency. The input-output variables of output DEA efficiency are: Total assets, main business cost, and number of employees, the main business income, net profit. The input-output variables of value added DEA efficiency are: Total assets, number of employees, net profit.

Second, calculate the correlation coefficient of the two Logit model indicators (return on assets, net profit growth rate) and the value of DEA efficiency, as shown in table 2:

Table 2 correlation between efficiencies and two logit model indicators

		output efficiency	value added efficiency
Return on assets	Correlation coefficient	0.738	0.824
	Significance level	0.001	0.001
Net profit growth	Correlation coefficient	0.292	0.488
	Significance level	0.094	0.033

From Table 2, it is suitable to comprehensively evaluate company financial problems with the output efficiency DEA and value added DEA:

First, at 10% significance level, the correlation coefficients are significant. Although DEA efficiency is a non-financial indicator, it still has some relevance with financial indicators. Therefore, from the perspective of significant, the use of efficiency indicator is appropriate.

Second, in theory, if the correlation coefficient is too high, then the results judge by efficiency and the predicted results obtained from the financial early-warning logit model are exactly the same or substantially the same. Although the efficiency indicator itself has a strong predictive, it cannot play the auxiliary judgment function on the financial early warning. If the correlation coefficient is too small, the efficiency indicator itself may have nothing to do with financial problems, it cannot be suitable for auxiliary judgments too. The coefficients of Net profit growth and efficiency are (0.738, 0.824), the coefficients of Net profit growth and efficiency are (0.292, 0.488), they are not too large and not too small. They are suitable for auxiliary judgments.

Here we give the amendment.

(1) Calculate the mean output DEA efficiency of the 17 ST companies and 17 normal companies separately usually there's $Mean_{ST} < Mean_{normal}$. So we get three intervals: [0, $Mean_{ST}$], ($Mean_{ST}$, $Mean_{normal}$), [$Mean_{normal}$, 1]. This paper consider the bankruptcy of the company which fell into the first interval is deterministic, assigned -1, the normal company which fell into a third interval is certain, assigned 1, and the company which fell into the second interval, its financial position is uncertainty, assigned 0.

(2) For each company's value added DEA efficiency, repeat (1), add the two assigned scores of each company, we get the final score, there are five possible values: -2, -1, 0, 1, 2.

(3) According to the principle of "ST Company's score is negative, the normal company's score is positive", we judge the company financial position.

Table 3 amendment and judge the Logit model results according to the efficiency scores

ST company group's mean of output DEA efficiency is 57.94%, normal company group's is 84.70%, The uncertainty range is 57.94%~84.70%.The uncertainty range of value of added DEA efficiency is 12.87%~51.35%. In all the 34 companies, 23 companies of the amendments is in the right direction, accounting for 67.6%; there are 10 companies have no correct conclusion, accounting for 29.4%; One company amendment is in the wrong direction, accounting for 2.9%.

In addition, if we do not take the 10 companies without amendment conclusions into account, there are 24 companies of amendments, 23 companies of amendments is in the right direction, the correct rate is 95.8%.

4. Test out-of-sample

This paper has also selected another 17 companies which have being ST in 2009, and selected another 17 paired companies in accordance with the same rules. We put the assets returns ratio, net profit growth rate of the above 34 companies in 2006 into the financial early-warning logit model, obtaining the predicted probability of corresponding company, and according to this prediction probability to determine whether the model results are revised.

For these 34 out-of-sample companies, logit financial early-warning model still has high prediction accuracy. 34 companies, only two ST companies (NO.13 company, 000036; No.14 company, 600234) were judged to be normal company. Correctly prediction rate is 94.1%. Such a high prediction accuracy of financial early-warning logit model is more effective. In all 34 companies, 25 companies of the amendment is in the right direction, accounting for 73.5%; 8 companies has no correct conclusion, accounting for 23.5%; one company is in the wrong direction, accounting for 2.9%. In addition, if we do not take the 8 companies without amendment conclusions into account, there are 25 companies of amendments in the right direction among the 26 companies, the correct rate is 96.2%. In other words, the published correctly amendments rate was 96.2%.

Finally, we should focus on the two companies of the wrong prediction in the logit financial early warning model. No.13 company,000036, score of -1, which means that, although the company is predicted to be normal, according to the score the prediction results may be questionable; No.14 company, 600234, score of -2, which means that, although the company is predicted to be normal, according to the score the prediction results are doubtful. On both of the prediction error of the companies, although DEA efficiency revisition cannot directly deny the predict results of logit financial early warning model, it reminds us to doubt the results. At this point, people can through other aspects (such as corporate governance, corporate structure, etc.) of the company and get a more comprehensive and integrated evaluation.

5. Conclusion

In this paper, we set a Logit model to do the financial early-warning, the 34 in-sample prediction accuracy was 97.1%, only a normal company misjudged for the ST companies; 34 out-of-sample companies, only two ST companies were misjudged to be normal companies, prediction accuracy was 94.1%.

We use Efficiency to amend and judge the results of financial early-warning logit model, for the 34 in-sample companies, 1 company is in the wrong direction, accounting for 2.9%; For the 34 out-of-sample companies,1 company is in the wrong direction, accounting for 2.9%. Especially for the two companies which have an error in the logit model prediction, given the right amendments, will help us to make another judgment. The traditional approaches in financial analysis can not evaluate enterprise performance from the perspective of efficiency. The paper did the evaluation. This is a new perspective.

References

1. E. I. Altman. Financial Ratios, Discriminant Analysis and the Prediction of Corporate Bankruptcy. Journal of Finance. 9, 589-609 (1968).
2. E. I. Altman, R.G. Haldeman, P. Narayanan, ZETA Analysis: A New Model to Identify Bankruptcy Risk of Corporations. Journal of Banking and Finance. 1(1), 29-54 (1977).
3. W. H. Beaver, Financial Ratios as Predictors of Failure. Journal of Accounting Research(supplement). 4, 71-111(1966).
4. P. Coats, L. Fant, Recognizing Financial Distress Patterns Using a Neural Network Tool. Financial Management. 22(3), 142-155 (1993).
5. P. J. Fitpatrick, A Comparison of Ratios of Successful Industrial Enterprises with Those of Failed Firms. Certified Public Accountant. 6, 589-731 (1932).
6. J. S. Ohlson. Financial, Ratios and the Probabilistic Prediction of Bankruptcy. Journal of Accounting Research. 18(1) , 109-131 (1980).
7. M. E. Zmijewski, Methodological Issues Related to the Estimation of Financial Distress Prediction Models. Journal of Accounting Research. 22, 59-82 (1984).
8. Chen Jing, An Empirical Analysis to the Prediction of Listed Company's Financial Deterioration. Accounting Research. 4, 31-38 (1999).
9. Chen Xiao,Chen Zhihong, The prediction of China Listed Company's Financial Dilemma. China Accounting and Finance Review. 2(3), 55-92 (2000).
10. Wang Zhengmiao, Zhang Ruiwen, The Deficiency and Improvement of Traditional Financial Index System. Technology Economics. 10, 30-50 (2002).
11. Zhou Shouhua, Yang Jihua, Wang Ping, The Analysis on the Early Warning of Financial Crisis—F Score. Accounting Research. 8, 8-11 (1996).
12. The SASAC of the state council, Enterprise performance evaluation standard, Economic science press,2011.

FUZZY AHP/DEA APPROACH FOR RELATIVE EFFICIENCY OF STATE UNIVERSITY IN TURKEY

AHMET ÇALIK*, NIMET YAPICI PEHLIVAN, AHMET PEKGÖR

Department of Statistics, Selçuk University, Campus Konya, 42075, TURKEY

This paper aims at integrating fuzzy analytic hierarchy process (FAHP) and data envelopment analysis (DEA) to evaluate relative efficiencies of 69 state universities in Turkey. Since most similar evaluations are multi-objection problems, which both FAHP and DEA are capable of solving, the integration of these two approaches will be shown to be even more powerful. The proposed integrated FAHP/DEA model can evaluate and rank different alternatives. On the first stage, we apply FAHP for determine weights of criterias. On the second stage, we rank alternatives by DEA of universities. A practical case study demonstrates the effectiveness of the proposed methodology.

1. Introduction

The Higher Education Institutions (HEIs) are very important for individuals and societies all over the world. It is obvious that the effectiveness and presence of HEIs is a prerequisite for a strong and modern society. The higher education sector has characteristics which make it difficult to measure efficiency: it is non-profit making; there is an absence of output and input prices; and higher education institutions produce multiple outputs from multiple inputs [2].

Multiple criteria decision-making (MCDM) methods which is a powerful tool widely used for evaluating problems containing multiple, usually conflicting criteria. In addition, because subjective considerations are relevant to partner evaluation and selection decision, a fuzzy logic approach is adopted. FAHP which is one of the MCDM approaches, is particularly suitable for modeling qualitative criteria and has found comprehensive applications in a wide variety of areas. However, due to the fact that there are tenths of decision making units-alternatives to be evaluated and prioritized, while the FAHP method can only compare a very limited number of decision alternatives, the pair-wise comparison manner is obviously infeasible in this situation [3]. To overcome this difficulty, FAHP and Data Envelopment Analysis (DEA) are combined and an integrated FAHP/DEA methodology is proposed.

---

* ahmetcalik@selcuk.edu.tr

DEA is a linear programming methodology that evaluates the efficiency of a number of decision making units. Efficiency of HEIs was examined a various contries such as the UK, the USA, Canada, Finland, Israel and Australia [5].

In this paper, we applied the integrated FAHP/DEA methodology for the ranking of universities in TURKEY. On the first stage, FAHP effectively represents the vagueness of human evaluations and FAHP method is to determine the weights of criteria, linguistic terms such as Equally importance (EI), Moderate Importance (MI), Strong importance (SI),Very strong importance (VSI) and Extreme importance (EXI) to assess under each criterion. On the second stage, the DEA method measures the relative efficiency of the state universities in TURKEY.

This paper is organized as follows: The second and third sections include FAHP and DEA methodologies, respectively. In Section 4, an application for ranking of universities in TURKEY is presented. Finally, the conclusions and suggestions for the future studies are given in the last section.

2. Fuzzy Analytic Hierarchy Process/Data Envelopment Analysis Methodology

2.1. *Fuzzy Analytic Hierarchy Process*

Fuzzy set theory which was first introduced by Zadeh (1965), is an important method to provide measuring the ambiguity of concepts that are associated with human beings' subjective judgments including linguistic terms, satisfaction degree and importance degree that are often vague [6].

The Analytic Hierarchy Process (AHP) method was first developed by Thomas L. Saaty in the 1980s. AHP assumes that evaluation criteria can be completely expressed in a hierarchical structure. In the AHP method, the pairwise comparison is made by using a ratio scale and it does not take into account the uncertainty associated with the mapping of one's perception (or judgment) to a number [1].

In order to deal with the uncertainty and vagueness from the subjective perception and the experience of humans in the decision-making process. Fuzzy AHP is based on pairwise comparisons and it allows the utilization of linguistic variables. There are the several methods to obtain the priorities in FAHP. Some of these methods are Laarhoven and Pedrycz(1983)'s Fuzzy Extension of Saaty's priority approach, Buckley (1985)'s geometric mean method, Chang (1996)'s synthetic extend analysis method, Csutora and Buckley (2001)'s Lambda-Max method and Mikhailov (2000, 2003)'s fuzzy preference programming. Triangular fuzzy numbers (TFNs) are often used in applications because of their calculation easiness and features. TFNs are used for representing the linguistic variables.

In this study, we considered AHP method of Saaty (1980), the FAHP methods of Chang(1996), Buckley(1985) and Mikhailov(2003) to obtain the weights of criteria. Fuzzy Importance scale for pairwise comparison of FAHP is given in Table 1.

Table 1. Fuzzy Importance scale for pairwise comparison of FAHP

Definition	**Fuzzy Importance scale**
Equal importance (EI)	(1, 1, 1)
Moderate importance (MI)	(2/3, 1, 3/2)
Strong importance (SI)	(3/2, 2, 5/2)
Very strong importance (VS)	(5/2, 3, 7/2)
Extreme importance (EXI)	(7/2, 4, 9/2)

Chang(2006)'s extent analysis method is used to consider the extent of an object to be satisfied for the goal, that is, satisfied extent. In the method, the ‘‘extent’’ is quantified by using a fuzzy number. On the basis of the fuzzy values for the extent analysis of each object, a fuzzy synthetic degree value can be obtained. Fuzzy Importance scale for pairwise comparison of FAHP is given in Table 1. The algorithm of the Chang's extent analysis model is given in detailed in Chang (1996) [7]. Buckley (1985)'s Geometric Mean Method is developed for computing the weights which is easily extended to fuzzy pairwase comparison matrix. The method is given in detailed in [8]. Mikhailov(2003)'s fuzzy preference programming method is able to derive inconsistency value and crisp priorities vector of pairwise comparison matrices concurrently. Mikhailov (2003) use both linear and nonlinear optimization methods, which maximize the decision maker's satisfaction on the derived weight vector. Nonlinear optimization model proposed by Mikhailov (2003) is given as;

$$
\begin{aligned}
& Max \ \ \lambda \\
& (m_{ij} - l_{ij})\lambda w_j - w_i + l_{ij} w_j \leq 0 \\
& (u_{ij} - m_{ij})\lambda w_j + w_i - u_{ij} w_j \leq 0 \\
& \sum_{k=1}^{n} w_k = 1, \ w_k > 0; \ k = 1,2,...,n \\
& i = 1,2,...,n-1; \ j = 2,3,...,n-1; \ j > i
\end{aligned}
\tag{1}
$$

Where; w_j is the weight of criterion j and λ is the consistency index. If $\lambda = 1$ then the fuzzy pairwise comparison matrix is consistent, and if $\lambda < 0$

then the matrix is strongly inconsistent. On the other hand, if λ is close to 1 then the fuzzy pairwise comparison matrix is almost consistent [9].

2.2. *Data Envelopment Analysis*

DEA is a non-parametric performance measurement technique, based on linear programming (LP), for determining the relative efficiencies of DMUs [4]. DEA models use these inputs and outputs to compute an efficiency score for a given DMU when this particular DMU is compared with all the other DMUs considered. An efficiency score of entity shows the highest values of desirable attributes and the lowest values of undesirable attributes. Two assumptions can be made while computing efficiency scores using DEA, namely, constant returns to scale (CCR) and variable returns to scale (BCC). There are two types of CCR models. The first one is the input oriented model which the inputs are maximized. The other one is the output-oriented model which the outputs are maximized. As the focus is on maximizing multiple outputs, this study employs the output-oriented CCR model.

In CCR model the efficiency of the entity evaluated is obtained as a ratio of its weighted output to its weighted input subject to the condition that the ratio for each entity is not greater than 1 and formulated as following:

$$\begin{aligned} \max_{u,v} \quad & u^t y_o \\ & v^t x_o = 1 \\ & u^t y_j - v^t x_j \le 0 \quad (j = 1,\ldots,n) \\ & u, v \ge 0 \end{aligned} \tag{2}$$

Where; the evaluated entities (DMUs) form a reference set and n is the number of DMUs.

3. Application and Conclusion

The integrated FAHP/DEA methodology is simple enough, easy to use, applicable to any number of decision alternatives. Also it is particularly useful and effective to complex MCDM problems with a large number of decision alternatives, where pairwise comparisons are certainly impossible to be made [3].

In this section, we used a practical case study to illustrate the efficiency and effectiveness of the FAHP/DEA methodology. The focus of this study, which integrates AHP and DEA approaches, is to compare state universities in Turkey.

Data are collected from the Council of Higher Education (YÖK) (www.yok.gov.tr). The Council of Higher Education is responsible for the supervision of universities in Turkey. Due to the changing in the universities, it is difficult to obtain series of consistent data. The present study therefore uses data from the rankings published in 2009. In this study, we collected quantitative data related to universities based on such factors as the number of undergraduate students, the number of undergraduate students gruduated from, the number of graduate students, the number of graduate students gruduated from, the number of academic staff, research publications.

FAHP analysis used for obtaining the variable weights which indicate the relative importance of the universities for each criterion. According to considered AHP and FAHP methods, weights are obtained as given in Table 2.

Table 2.Relative weights of criterias

	Weights of the Criteria					
Method	**K1**	**K2**	**K3**	**K4**	**K5**	**K6**
AHP	0.029	0.074	0.248	0.029	0.072	0.548
Chang	0	0	0.1801	0	0	0.8198
Buckley	0.0855	0.1277	0.2325	0.0855	0.1272	0.3588
Mikhailov	0.1218	0.0812	0.2474	0.0812	0.0989	0.3692

Table 2 describes the relative weights of the six criteria obtained from fuzzy AHP process. The results of FAHP showed that publications of academic staff was the most preferred factor at the criteria level. After we obtained the FAHP weights, we normalized the quantitative data with FAHP weights. Then, we applied DEA methodology to the normalized data. In application, at first, we determined the ranking of universities for data set. According to the results, we saw that some of new universities better than the other ones, so, we applied cluster analysis to data. We used SPSS for clustering analysis and we were able to come up with a solution that partitions our data set into three clusters. The clusters are categorized as new founded universities (Cluster 1), devoloping universities (Cluster 2) and advanced universities (Cluster 3). The results of FAHP/DEA shown in Table 3.

According to the FAHP/DEA results, Gaziantep, Akdeniz, Erciyes, ODTU, Hacettepe and Boğaziçi considered to be the best universities of Turkey. However, some of the advanced universities in Turkey couldn't enter the ranking. To be competitive with the others, they need to focus on producing much more quantitative outcomes.

Table 3. FAHP/DEA Results of the application

	AHP/DEA			Chang's FAHP/DEA		
No	**Cluster1**	**Cluster2**	**Cluster3**	**Cluster1**	**Cluster2**	**Cluster3**
1	GYTE	Akdeniz	Boğaziçi	Karabük	Erciyes	ODTÜ
2	Gçantep	Atatürk	Hacettepe			
3	Amasya	19 Mayıs	ODTÜ			
4	Abant iB.	Cumhuriyet	Marmara			
5	Galatasaray	Yıldız Teknik	Selçuk			
	Buckley's FAHP/DEA			**Mikhailov's FAHP/DEA**		
No	**Cluster1**	**Cluster2**	**Cluster3**	**Cluster1**	**Cluster2**	**Cluster3**
1	Karabük	Atatürk	ODTÜ	Sinop	19 Mayıs	Boğaziçi
2	G. antep	Erciyes	Boğaziçi	Gaziantep	Uludağ	Marmara
3	Amasya	Uludağ	Marmara	GYTE	YTU	Hacettepe
4	Sinop	YTU	Selçuk	Karabük	Akdeniz	Selçuk
5	GYTE	Akdeniz	Hacettepe	Amasya	Erciyes	ODTÜ

References

1. G. Büyüközkan, O. Feyzioğlu, E. Nebol, Selection of the strategic alliance partner in logistics value chain, *International Journal of Production Economics*, **113,** 148-158 (2008).
2. J. Johnes, Data envelopment analysis and its application to the measurement of efficiency in higher education, *Economics of Education Review,* **25,** 273-288 (2006).
3. Hadi-Vencheh and A. Mohamadghasemi, A fuzzy AHP-DEA approach for multiple criteria ABC inventory classification, *Expert Systems with Applications*, **38**, 3346-3352 (2011).
4. Charnes, W.W. Cooper, B. Golany, L. Seiford, J. Stutz, Foundations of data envelopment analysis for Pareto-Koopmans efficient empirical production functions, *Journal of Econometrics* **30** (1–2), 91–107 (1985).
5. T. Ahn, V. Arnold, A. Charnes, W.W. Cooper, DEA and ratio efficiency analyses for public institutions of higher learning in Texas. *Research in Governmental and Nonprofit Accounting*, **5**, 165–185 (1989).
6. H.J. Zimmerman, Fuzzy Set Theory and its Applications, *Kluwer Academic Publishers*, (1991).
7. D.Y. Chang, Applications of the extent analysis method on Fuzzy AHP, *European Journal of Operational Reserarch*, **95** 649-655 (1996).
8. J.J. Buckley, Fuzzy Hierarchical Analysis, *Fuzzy Sets and Systems*, **17** 233-247 (1985)
9. L.Mikhailov, Deriving Priorities from fuzzy pairwise comparison judgements, *Fuzzy Sets and Systems*, **134** 365-385 (2003).

IMPROVING THE COMPUTATIONAL COMPLEXITY AND WEIGHTS DISPERSION IN FUZZY DEA

SABER SAATI*

Department of Mathematics, Tehran-North Branch, Islamic Azad University, P.O. Box 19585-936, Tehran, Iran

ADEL HATAMI-MARBINI

LSM, Center of Operations Research and Econometrics (CORE), Universite catholique de Louvain, 34 Voie du Roman Pays, L1.03.01, B-1348 Louvain-la-Neuve, Belgium

One of the prominent features of standard and fuzzy data envelopment analysis (DEA) is the representation of each of the participating decision making units (DMUs) in the best possible light. This causes two problems; first, the different set of factor weights with large number of zeros and second a large number of linear programming models to solve. In this paper, we propose an efficient method to address these two problems. In proposed method by solving just one linear programming a Common Set of Weights (CSW) is achieved in fuzzy DEA. Since resulted efficiencies by the proposed CSW are interval numbers rather than crisp values, it is more informative for decision maker. The proposed model is applied to a numerical example to demonstrate the concept.

1. Introduction

Data envelopment analysis (DEA) [1] is a widely used non-parametric frontier analysis method, implemented in linear programming, for comparing the inputs and outputs of a set of comparable decision making units (DMUs). The flexibility of the DEA models, i.e. the absence of *a priori* assumptions on the production technology, is implemented through a minimal spanning hull. This flexibility in combination with the number of input and output dimensions normally render a relatively high proportion of the DMUs fully technically efficient, in particular for limited datasets.

Between these value-free and value-exact approaches, Thompson et al. [2] used boundary conditions (bounds) for the virtual multipliers in DEA standard models. Solutions to these models identified only one efficient DMU for location of the lab in Texas. A series of possible approaches for setting bounds on factor weights in DEA have been put forward [2-3].

* *Email: s_saatim@iau-tnb.ac.ir*

Imposing bounds on factor weights, limits the flexibility of DEA in assigning individual sets of weights to each of the participating DMUs. In the extreme case, when no flexibility is allowed, a common set of weights (CSW) is applied for the assessment of all DMUs [4-7].

In this paper, we propose an algorithm to reduce the computational complexities and costs as well as overcoming pitfalls in the fuzzy DEA including the zero-value and differing weights and multiple solutions. Initially, we determine the lower and upper bounds of factor weights by comparing the data. We then use a linear programming problem to determine a CSW. Finally, we calculate the efficiency of the DMUs with the obtained CSW as intervals. The proposed models are applied to a numerical example.

The remainder of this paper is organized as follows. In Section 2 we present the basic DEA models. In Section 3 we present the details of our proposed method and Section 4 contains a numerical example. The paper is closed in Section 5 with conclusions and some suggested future research directions.

2. Basic DEA Models

Let us assume that there are n DMUs to be evaluated where each DMU$_j$, $j=1,2,...,n$ convert m inputs into s outputs. Suppose x_{ij} $(i=1,\dots,m,$ $j=1,\dots,n)$ and y_{rj} $(r=1,\dots,s, j=1,\dots,n)$ are the ith input and the rth output of DMU$_j$, respectively. The relative efficiency of DMU$_P$, $p\in\{1,2,...,n\}$, is defined as the maximum value of θ_p and can be obtained by using the following programming model (for constant returns to scale, CRS) [1]:

$$
\begin{aligned}
\max \quad & \theta_p = \sum_{r=1}^{s} u_r y_{rp} \\
s.t. \quad & \sum_{i=1}^{m} v_i x_{ip} = 1 \\
& \sum_{r=1}^{s} u_r y_{rj} - \sum_{i=1}^{m} v_i x_{ij} \le 0, \quad \forall j, \\
& u_r, v_i \ge 0, \quad \forall r, i.
\end{aligned} \tag{1}
$$

The relative efficiency of DMU$_P$ is determined by assigning weights to the inputs and outputs of the DMU and maximizing the ratio of the weighted sum of outputs to the weighted sum of inputs. The only underlying assumption for the weights of the inputs and outputs is non-negativity which is recognized as either a weakness or strength of the traditional DEA models. In addition, with respect to weights flexibility, some of input and output weights take the zero values. In the original DEA and its extensions, all the inputs and outputs assume the form of specific numerical values. In many real-world problems, however, the data can be imprecise or vague or described by qualitative terms. Fuzzy logic and fuzzy sets can represent imprecise or ambiguous data in DEA by formalizing inaccuracy in decision making [8]. The fuzzy CCR model with bounded factor weights is as follows:

$$\begin{aligned} \max \quad & \theta_p = \sum_{r=1}^{s} u_r \tilde{y}_{rp} \\ s.t. \quad & \sum_{i=1}^{m} v_i \tilde{x}_{ip} = 1, \\ & \sum_{r=1}^{s} u_r \tilde{y}_{rj} - \sum_{i=1}^{m} v_i \tilde{x}_{ij} \le 0, \quad \forall j, \\ & U_r^l \le u_r \le U_r^u, \quad \forall r, \quad V_i^l \le v_i \le V_i^u, \quad \forall i. \end{aligned} \tag{2}$$

where, U_r^l, U_r^l, U_r^l and U_r^l are lower and upper bounds on output and input weights, respectively.

Note that in the fuzzy CCR model (2) the right hand sides of the constraints are assumed to be crisp values because they are similar to the original CCR model used for normalization of the value of the efficiency in the objective function.

3. Proposed Model

In this study we use triangular fuzzy numbers to develop our model. However, our model is adaptable to other types of fuzzy numbers. To assess a CSW, an upper bound is determined for each factor. This is done by considering the following problems:

$$\begin{aligned} \max \quad & u_p \quad (or\ v_t) \\ s.t. \quad & \sum_{i=1}^{m} v_i \tilde{x}_{ij} \le 1 \\ & \sum_{r=1}^{s} u_r \tilde{y}_{rj} - \sum_{i=1}^{m} v_i \tilde{x}_{ij} \le 0, \quad \forall j, \\ & u_r, v_i \ge 0, \quad \forall r, i. \end{aligned} \tag{3}$$

The following model, therefore, can be obtained when fuzzy coefficients in model (3) are assumed to be triangular fuzzy numbers denoted as $\tilde{x}_{ij} = (x_{ij}^l, x_{ij}^m, x_{ij}^u)$ and $\tilde{y}_{rj} = (y_{rj}^l, y_{rj}^m, y_{rj}^u)$:

$$
\begin{aligned}
\max \quad & u_p \quad (or\ v_t) \\
s.t. \quad & \sum_{i=1}^{m} v_i(x_{ij}^l, x_{ij}^m, x_{ij}^u) \le 1, \qquad \forall j, \\
& \sum_{r=1}^{s} u_r(y_{rj}^l, y_{rj}^m, y_{rj}^u) - \sum_{i=1}^{m} v_i(x_{ij}^l, x_{ij}^m, x_{ij}^u) \le 0, \quad \forall j, \\
& u_r, v_i \ge 0, \qquad \forall r, i.
\end{aligned} \tag{4}
$$

By introducing α-cuts of constraints, the following model is obtained:

$$
\begin{aligned}
\max \quad & u_p \quad (or\ v_t) \\
s.t. \quad & \sum_{i=1}^{m} [\alpha x_{ij}^m + (1-\alpha)x_{ij}^l, \alpha x_{ij}^m + (1-\alpha)x_{ij}^u] v_i \le 1 \quad \forall j, \\
& \sum_{r=1}^{s} [\alpha y_{rj}^m + (1-\alpha)y_{rj}^l, \alpha y_{rj}^m + (1-\alpha)y_{rj}^u] u_r \\
& - \sum_{i=1}^{m} [\alpha x_{ij}^m + (1-\alpha)x_{ij}^l, \alpha x_{ij}^m + (1-\alpha)x_{ij}^u] v_i \le 0, \quad \forall j, \\
& u_r, v_i \ge 0, \qquad \forall r, i.
\end{aligned} \tag{5}
$$

Since (5) has a special structure, so the upper and lower bounds of weights may achieved by comparing their constraints [9]. Since the weights are restricted to non-negativity, hence the lower bound of weights is zero, that is $V_i^l = 0$ and $U_r^l = 0\ (\forall i, r)$. Considering the first constraints of (6), the upper bounds of input weights are calculated as follows:

$$
V_t^u = \min_{1 \le j \le n} \left\{ \frac{1}{\alpha x_{tj}^m + (1-\alpha)x_{tj}^l} \right\} \tag{6}
$$

And using the second constraints of (6), the upper bounds of output weights are determined as follows:

$$
U_p^u = \min_{1 \le j \le n} \left\{ \frac{\sum_{i=1}^{m} [\alpha x_{ij}^m + (1-\alpha)x_{ij}^u] V_i^u}{\alpha y_{pj}^m + (1-\alpha)y_{pj}^l} \right\} \tag{7}
$$

Starting from bounded model (2), a CSW can be achieved by expressing the deviation from either bound as a fraction of the range between the upper and lower bounds. Assuming the same deviation from bounds across all DMUs and applying Saati et al. [6] method, we get:

$$
\begin{aligned}
\max \quad & \phi \\
s.t. \quad & \sum_{i=1}^{m} \hat{x}_{ij} \le 1 \qquad \forall j, \\
& \sum_{r=1}^{s} \hat{y}_{rj} - \sum_{i=1}^{m} \hat{x}_{ij} \le 0, \qquad \forall j, \\
& [\alpha x_{ij}^{m} + (1-\alpha) x_{ij}^{l}] v_i \le \hat{x}_{ij} \le [\alpha x_{ij}^{m} + (1-\alpha) x_{ij}^{u}] v_i \qquad \forall i, j, \\
& [\alpha y_{rj}^{m} + (1-\alpha) y_{rj}^{l}] u_r \le \hat{y}_{rj} \le [\alpha y_{rj}^{m} + (1-\alpha) y_{rj}^{u}] u_r \qquad \forall r, j, \\
& u_r, v_i \ge 0, \qquad \forall r, i.
\end{aligned} \tag{8}
$$

A CSW is obtained by solving (8), and the interval efficiency of each DMU can be evaluated as follows:

$$
e_j = [e_j^l, e_j^u] = \left[\frac{\sum_{r=1}^{s} u_r^* \left[\alpha y_{rj}^{m} + (1-\alpha) y_{rj}^{l} \right]}{\sum_{i=1}^{m} v_i^* \left[\alpha x_{ij}^{m} + (1-\alpha) x_{ij}^{u} \right]}, \frac{\sum_{r=1}^{s} u_r^* \left[\alpha y_{rj}^{m} + (1-\alpha) y_{rj}^{u} \right]}{\sum_{i=1}^{m} v_i^* \left[\alpha x_{ij}^{m} + (1-\alpha) x_{ij}^{l} \right]} \right] \qquad \forall j \tag{9}
$$

where, $u_r^* \ (r = 1,...,s)$ and $v_i^* \ (i = 1,...,m)$ are optimal values of (8).

4. Numerical Example

In this section, we use the proposed method and demonstrate the applicability of our framework and exhibit the simplicity and efficacy of the procedures with an example. In this illustration, we address a problem with 10 hypothetical DMUs introduced in [7]. The resulted upper bounds of output and input weights are presented in Table 1.

Table 1. The CSW by proposed model.

α	Upper bounds				Factor weights			
	Iu_1	Iu_2	Ou_1	Ou_2	v_1	v_2	u_1	u_2
0.1	0.0581	0.0055	0.0172	0.0016	0.0388	0.0037	0.0044	0.0004
0.3	0.0568	0.0055	0.0164	0.0016	0.0379	0.0036	0.0042	0.0004
0.5	0.0556	0.0054	0.0155	0.0015	0.0370	0.0036	0.0040	0.0004
0.7	0.0543	0.0054	0.0147	0.0014	0.0362	0.0036	0.0037	0.0004
0.9	0.0532	0.0054	0.0140	0.0014	0.0355	0.0036	0.0036	0.0004
1	0.0526	0.0054	0.0136	0.0014	0.0351	0.0036	0.0035	0.0003

Table 2. The corresponding efficiencies by CSW.

DMU	α					
	0.1	0.3	0.5	0.7	0.9	1
A	[0.2467,0.3093]	[0.2479,0.2956]	[0.2491,0.2825]	[0.2503,0.2699]	[0.2514,0.2578]	0.2519
B	[0.4107,0.5099]	[0.4045,0.4778]	[0.3982,0.4480]	[0.3920,0.4204]	[0.3856,0.3947]	0.3825
C	[0.3843,0.5421]	[0.3872,0.5054]	[0.3902,0.4715]	[0.3931,0.4401]	[0.3960,0.4111]	0.3974
D	[0.3037,0.4259]	[0.3053,0.3968]	[0.3069,0.3698]	[0.3085,0.3448]	[0.3100,0.3158]	0.3107
E	[0.7641,0.9735]	[0.7641,0.9226]	[0.7641,0.8743]	[0.7641,0.8285]	[0.7641,0.7745]	0.7641
E	[0.5414,0.6978]	[0.5407,0.6585]	[0.5400,0.6216]	[0.5393,0.5868]	[0.5386,0.5463]	0.5383
G	[0.3343,0.4237]	[0.3315,0.3982]	[0.3288,0.3744]	[0.3260,0.3522]	[0.3231,0.3273]	0.3217
H	[0.5587,0.7863]	[0.5520,0.7178]	[0.5452,0.6564]	[0.5385,0.6012]	[0.5317,0.5415]	0.5282
I	[0.2207,0.2589]	[0.2175,0.2460]	[0.2142,0.2338]	[0.2109,0.2222]	[0.2076,0.2094]	0.2060
J	[0.3045,0.3795]	[0.2987,0.3537]	[0.2930,0.3300]	[0.2874,0.3083]	[0.2817,0.2850]	0.2789

These bounds are introduced in (8) and the CSW is evaluated. This CSW for different α in [0,1] is presented in Table 1. After fixing the input and output prices for all inputs and outputs, the efficiencies of DMUs are evaluated by (9). The interval efficiencies by the CSW are shown in Table 2 for different α in [0,1].

5. Conclusion

The calculation of DEA scores requires a linear program per DMU and generates an individual set of endogenous weights for each performance dimension. Given a large number of DMUs in real applications, the computational and conceptual complexities are considerable with weights that are potentially zero-valued or incommensurable across units. To overcome this problem, a procedure for finding a CSW in fuzzy CCR model is suggested. In contrast to the other methods, the presented procedure is very useful and applicable. The proposed algorithm reduces the computational complexities and costs as well as overcoming pitfalls in the DEA including the zero-value and differing weights and multiple solutions.

References

1. A. Charnes, W.W., Cooper, and E.L. Rhodes, Measuring the efficiency of decision making units, *Eur. J. Operat. Res*, **2(6),** 429(1978).
2. R.G. Thompson, F. Singleton, R. Thrall, and B. Smith, Comparative site evaluations for locating a high-energy physics lab in Texas. *Interfaces*, **16**, 35(1986).
3. M. Khalili, A.S. Camanho, M.C.A.S. Portela and M.R. Alirezaee, The measurement of relative efficiency using data envelopment analysis with assurance regions that link inputs and outputs. *Eur. J. Operat. Res*, **203 (3)**, 761(2010).
4. G.R. Jahanshahloo, F. Hosseinzadeh Lotfi, M. Khanmohammadi, M. Kazemimanesh, and V. Rezaie, Ranking of units by positive ideal DMU with common weights. *Expert Syst Appl.* **37 (12)**, 7483(2010).
5. F.H.F. Liu and H.H. Peng, Ranking of DMUs on the DEA frontier with common weights. *Comput Oper Res,* **35 (5)**, 1624 (2008).
6. S. Saati, Determining a common set of weights in DEA by solving a linear programming. *J Ind Eng Int*, **4 (6)**, 5(2008).
7. S. Saati and A. Memariani, Reducing weight flexibility in fuzzy DEA, *Appl Math Comput,* **161(2)**, 611(2005).
8. A. Hatami-Marbini, A. Emrouznejad and M. Tavana, A taxonomy and review of the fuzzy data envelopment analysis literature: Two decades in the making. *Eur. J. Operat. Res*, **214(3)**, 457(2011).
9. R.E., Moore, B.B. Kearfortt, and M.J. Cloud, *Introduction to interval analysis*, SIAM, Philadelphia, pp. 11(2009).

GENERAL DEA WITH FUZZY DATA

HILDA SALEH[1], MOHSEN ROSTAMY-MALKHALIFEH, ZOHREH MOGHADDAS
Dept. of Mathematics, Science and Research Branch, Islamic Azad University Tehran 14515-775,Iran.

The assumption in DEA is that all inputs and outputs are definite, but in real situation, this assumption is not true. In this study, we suggest a model with fuzzy data called the fuzzy generalized DEA (FGDEA) model, which can treat the stated basic DEA models with fuzzy data in a unified way. In addition, we linearize this model. And finally with a numerical example we evaluate DMUs with fuzzy data by the fuzzy DEA model and the FGDEA model and compare the results of two models.

1. Introduction

Data Envelopment Analysis (DEA)was suggested by Charnes et al. [3], by using the idea of Farrell [5], and then developed by others [2,4]. DEA is a method for measuring the efficiency of decision making units (DMUs) with common inputs and outputs. In real situation the most of data are indefinite and are stated by natural language, so, some researchers have carried out studies in this field and used fuzzy set theory and comparison of fuzzy numbers [6,7,10]. Yun et al. [11] proposed a generalized model for DEA, called GDEA, which can treat basic DEA models, specifically the CCR, BCC and FDH models in a unified way.In this study, we suggest a model with fuzzy data called the fuzzy generalized DEA (FGDEA) model, which can treat the stated basic DEA models with fuzzy data in a unified way. In addition, we linearize this model. This paper consists of the following sections: In section 2 the fuzzy number is defined and in section 3 fuzzy DEA model is introduced. linear GDEA model is presented in section 4, and also introduce a method for solving the fuzzy GDEA model in section 5. Finally, an example with fuzzy data will be give.

2. Fuzzy numbers

A fuzzy set A in X is a set of ordered pairs:

$$A = \{(x, \mu_A(x)) \mid x \in X\}$$

[1] hilda_saleh @yahoo.com.

$\mu_A(x)$ is called the membership function of x in A.

A fuzzy number M is a convex normalized fuzzy set M of real line R such that :

1) there exists exactly one $x \in R$ with $\mu_M(x) = 1$.

2) $\mu_M(x)$ is piecewise continuous.

The (crisp) set that element belong to the fuzzy set A at least to the degree α -cut set:

$$A_\alpha = \{x \in X : \mu_A(x) \geq \alpha\}.$$

The lower and upper endpoints of any α -cut set, A_α, are represented by $\underline{A}(\alpha)$ $\overline{A}(\alpha)$, respectively.

Let u be a fuzzy number and $m(u) = 1/2\int_0^1 (\underline{u}(\alpha) + \overline{u}(\alpha))d\alpha$.

For arbitrary fuzzy numbers u and v

$$\text{u} \geq \text{v} \quad \text{iff} \quad m(u) \geq m(v)$$
$$\text{u} \cong \text{v} \quad \text{iff}\; m(u) = m(v) \qquad [1].$$

Theorem 1. Let u and v be fuzzy numbers therefore, $m(v+v) = m(u) + m(v)$.

Proof. See [8].

Theorem 2. Let u be a fuzzy number and λ is a scaler so $m(\lambda u) = \lambda m(u)$.

Proof. See [8].

3. Linear GDEA model

The GDEA model can evaluate efficiency in several basic models in special cases, as follows:

$$\begin{aligned}
&\max \quad \Delta \\
&s.t. \quad \Delta \leq \tilde{d}_j + \alpha\left(\sum_{r=1}^{s} u_r(y_{ro} - y_{rj}) + \sum_{i=1}^{m} v_i(-x_{io} + x_{ij})\right), \qquad j = 1,2,\dots,n, \\
&\qquad \sum_{r=1}^{s} u_r + \sum_{i=1}^{m} v_i = 1, \\
&\qquad u_r \geq \varepsilon, v_i \geq \varepsilon, \qquad i = 1,2,\dots,m, \quad r = 1,2,\dots,s.
\end{aligned} \tag{3}$$

where α is appropriately the selected according to the give problem, and

$$\tilde{d}_j = \max_{i,r}\{u_r(y_{ro} - y_{rj}), v_i(-x_{io} + x_{ij})\}$$

Definition 1. For given positive number α, DMUo is defined to be α -efficient if and only if the optimal value to the model (GDEA) is equal to zero. Otherwise, DMUo is said to be α -inefficient.

Theorem 3. DMU_o is BCC-efficient if and only if DMU_o is α -efficient for some sufficiently large positive number α .(see [9])

Theorem 4. DMU_o is CCR- efficient if and only if it is α -efficient for some sufficiently large positive number α when the condition $U^tY_o - V^tX_j = 0$ is added to the GDEA model. (see [9])

Let z^j =(-X^j , Y^j), now we change the constraint $\Delta \le \tilde{d}_j + \alpha(U,V)(z_0 - z_j),\ j = 1,\ldots,n$ to $\Delta \le (U,V)(z_0 - z_j),\ j = 1,\ldots,n.$

So we have:

$$\begin{aligned}
\max\quad & \Delta \\
s.t.\quad & \Delta \le (U,V)(z_0 - z_j), \quad j = 1,\ldots,n, \\
& \sum_{r=1}^{s} u_r + \sum_{i=1}^{m} v_i = 1, \\
& u_r \ge \varepsilon, v_i \ge \varepsilon, \qquad r = 1,2,\ldots,s, i = 1,2,\ldots,m,
\end{aligned} \tag{4}$$

$$\begin{aligned}
\max\quad & \Delta \\
s.t.\quad & \Delta \le (U,V)(z_0 - z_j), \quad j = 1,\ldots,n, \\
& \sum_{r=1}^{s} u_r + \sum_{i=1}^{m} v_i = 1, \\
& (U,V)z_0 = 0, \\
& u_r \ge \varepsilon, v_i \ge \varepsilon, \qquad r = 1,2,\ldots,s, i = 1,2,\ldots,m.
\end{aligned} \tag{5}$$

Theorem 5 :

1- DMU_0 is BCC - efficient if and only if it is efficient in model (4).

2- DMU_0 is CCR - efficient if and only if it is efficient in model (5).

Proof(only if part): Let $(U^*, V^*, \Delta^* = 0)$ be the optimal solution of model (4). That is, $0 = \Delta^* \le (U^*, V^*)(z_0 - z_j),\ j = 1,\ldots,n,$ consequently $(U^*, V^*)z_0 - (U^*, V^*)z_j \ge 0,\ j = 1,\ldots,n.$

Let $u_0^* = -(U^*, V^*)z_0$, that is $(U^*, V^*)z_0 + u_0^* \le 0$

Suppose that $V^* X_0 = \beta > 0$, so $(\frac{U^*}{\beta}, \frac{V^*}{\beta}) z_0 + \frac{u_0^*}{\beta} \le 0$, consequently $(\frac{U^*}{\beta}, \frac{V^*}{\beta}, \frac{u_0^*}{\beta})$ is a feasible solution for the BCC model, also the value of the objective function for this solution is equal to 1.

(if part). Suppose that DMU_0 is BCC- efficient and $(\overline{U}, \overline{V})$ is the optimal solution for DMU_0, then we have

$$
\begin{aligned}
&\overline{U}^t Y_0 + \overline{u}_0 = 1, \\
&\overline{V}^t X_0 = 1, \\
&\overline{U}^t Y_j - \overline{V}^t X_j + u_0 \le 0, \quad j = 1,2,\dots, n \\
&\overline{u}_r \ge \varepsilon, \overline{v}_i \ge \varepsilon \qquad r = 1,2,\dots, s, i = 1,2,\dots, m.
\end{aligned}
$$

It is evident that $\bar{U}^t Y_0 - \bar{V}^t X_0 + u_0 = 0, j = 1,2,\dots,n$ or $(\overline{U}, \overline{V}) z_0 + u_0 = 0$

and also, $-(\overline{U}, \overline{V}) z_j - u_0 \ge 0, j = 1,2,\dots,n$

Then $(\overline{U}, \overline{V})(z_0 - z_j) \ge 0, j = 1,\dots,n$. Therefore $(\overline{U}, \overline{V}, \Delta = 0)$ is an optimal solution for model (4)and the proof is complete.

The proof of the second part is similar to the above proof and is hence omitted.

So the GDEA model were transformed a simple LP to calculate and rank the efficiency of the units.

Definition 2. DMUo is defined to be efficient in model 4 and 5 if and only if the optimal value in these models are equal to zero.

4. GDEA model with fuzzy data

We consider the GDEA model with fuzzy data as follows:

$$
\begin{aligned}
&\max \quad \tilde{\Delta} \\
&s.t. \quad \tilde{\Delta} \circ\le \sum_{r=1}^{s} u_r (\tilde{y}_{ro} - \tilde{y}_{rj}) + \sum_{i=1}^{m} v_i (-\tilde{x}_{io} + \tilde{x}_{ij}), \quad j = 1,\dots,n, \\
&\qquad \sum_{r=1}^{s} u_r + \sum_{i=1}^{m} v_i = 1, \\
&\qquad u_r \ge \varepsilon, v_i \ge \varepsilon, \qquad i = 1,\dots,m, \; r = 1\dots,s,
\end{aligned} \tag{6}
$$

In model (6), inequalitys for all of constraints are regarded as fuzzy ineqyality.

By using Theorems (1) and (2) we convert model (6) to model (7) as follows:

$$\begin{aligned} \max\quad & m(\tilde{\Delta}) \\ s.t.\quad & m(\tilde{\Delta}) \le \sum_{r=1}^{s} u_r (m(\tilde{y}_{ro}) - m(\tilde{y}_{rj})) + \sum_{i=1}^{m} v_i(-m(\tilde{x}_{io}) + m(\tilde{x}_{ij})), \quad j=1,\ldots,n, \\ & \sum_{r=1}^{s} u_r + \sum_{i=1}^{m} v_i = 1, \\ & u_r \ge \varepsilon, v_i \ge \varepsilon, \qquad i=1,\ldots,m,\ r=1,\ldots,s, \end{aligned} \tag{7}$$

So we rewrite model 7 as follow:

$$\begin{aligned} \max\quad & \frac{1}{2}\int_0^1 (\underline{\Delta}(\alpha) + \overline{\Delta}(\alpha))d\alpha \\ s.t.\quad & \\ & \frac{1}{2}\int_0^1 (\underline{\Delta}(\alpha) + \overline{\Delta}(\alpha))d\alpha \le + \frac{1}{2}(\sum_{r=1}^{s} u_r [\int_0^1 (\underline{y}_{ro}(\alpha) + \overline{y}_{ro}(\alpha))d\alpha - \int_0^1 (\underline{y}_{rj}(\alpha) + \overline{y}_{rj}(\alpha))d\alpha]) + \\ & \frac{1}{2}(\sum_{r=1}^{s} v_i [-\int_0^1 (\underline{x}_{io}(\alpha) + \overline{x}_{io}(\alpha))d\alpha + \int_0^1 (\underline{x}_{ij}(\alpha) + \overline{x}_{ij}(\alpha))d\alpha]) \qquad j=1\ldots n \\ & \sum_{r=1}^{s} u_r + \sum_{i=1}^{m} v_i = 1, \\ & u_r \ge \varepsilon, v_i \ge \varepsilon \qquad i=1,\ldots,m,\ r=1,\ldots,s. \end{aligned} \tag{8}$$

We consider the following variable changes:

$$\hat{x}_{ij} = \frac{1}{2}\int_0^1 (\underline{x}_{ij}(\alpha) + \overline{x}_{ij}(\alpha))d\alpha,$$

$$\hat{y}_{rj} = \frac{1}{2}\int_0^1 (\underline{y}_{rj}(\alpha) + \overline{y}_{rj}(\alpha))d\alpha,$$

$$\hat{\Delta} = \frac{1}{2}\int_0^1 (\underline{\Delta}(\alpha) + \overline{\Delta}(\alpha))d\alpha.$$

Therefore we have:

$$\begin{aligned} \max\quad & \hat{\Delta} \\ s.t.\quad & \hat{\Delta} \le \sum_{r=1}^{s} u_r (\hat{y}_{ro} - \hat{y}_{rj}) + \sum_{i=1}^{m} v_i(-\hat{x}_{io} + \hat{x}_{ij}), \quad j=1,\ldots,n, \\ & \sum_{r=1}^{s} u_r + \sum_{i=1}^{m} v_i = 1, \\ & u_r \ge \varepsilon, v_i \ge \varepsilon, \qquad i=1,\ldots,m,\ r=1,\ldots,s. \end{aligned} \tag{9}$$

By applying previous method we have:

$$
\begin{aligned}
&\max \quad \hat{\Delta} \\
&s.t. \quad \hat{\Delta} \le \sum_{r=1}^{s} u_r(\hat{y}_{ro} - \hat{y}_{rj}) + \sum_{i=1}^{m} v_i(-\hat{x}_{io} + \hat{x}_{ij}), \quad j = 1,\dots,n, \\
&\qquad \sum_{r=1}^{s} u_r + \sum_{i=1}^{m} v_i = 1, \\
&\qquad \sum_{r=1}^{s} u_r \hat{y}_{ro} - \sum_{i=1}^{m} v_i \hat{x}_{io} = 0, \\
&\qquad u_r \ge \varepsilon, v_i \ge \varepsilon, \qquad i = 1,\dots,m, \; r = 1,\dots,s,
\end{aligned} \tag{10}
$$

5. Conclusion

In this paper FGDEA model was applied in order to evaluated efficiency of DMUs with fuzzy inputs and outputs by solving unified model and the results was compared to other results obtained by evaluating DMUs with fuzzy BCC and CCR models. This method can be used for computing other concept in fuzzy DEA such as ranking and return to scale to name but a few, by solving a FGDEA model. Also in mentioned method all multiplier were regarded as a crisp but we regarded all multiplier as a fuzzy variable and extend this method to general method.

References

1. B. Asadi, A. Zendenam, Ranking fuzzy number by distance minimization, Applied Mathematical Modeling, 31(2007)2589-2598.
2. R. D. Banker, A. Charnes, W.W. Cooper, Some models for estimating analysis, Management Science, 30(1984)1078-1092.
3. A. Charnes, W. W.Cooper, E. Rhodes, Measuring the efficiency of decision-making, European J. Oper. Res. 2.(1978)429-444.
4. D. Deprins, L. Simar, H. Tulkens, Measuring labor-efficiency in post offices, in: M. Marchand, P. Pestieau, H. Tulkens(Eds), The performance of public enterprises: Concepts and Measurements. Amsterdam, (1984)247-263.
5. M. J. Farrell, The measurement of productive efficency, Journal of the Royal Statistical Society A 120 (1957) 253-281.
6. P. Guo, H. Tanaka, Extended fuzzy DEA, in: Proceedings of the 3rd Asian Fuzzy system symposium, (1997)517-521.
7. P. Guo, H. Tanaka, Fuzzy DEA: a perceptual, evaluation method, Fuzzy Sets Syst. 119(2001)149-160.
8. G. R. Jahanshahloo, M. Sanei, M. Rostamy-Malkhalifeh, H.Saleh,
9. G. R. Jahanshahloo, F. Hosseinzadeh Lotfi, M. Rostamy-Malkhalifeh, M. Ahadzadeh Namin, Applied mathematical modling, 33(2009)3237-3244.
10. M. Soleimani-damaneh, Fuzzy upper bounds and their applications, Chaos, solitons and Fractal, 36(2008)217-225.
11. Y. B. Yun, H. Nakayama, T. Tanino, A generalized model for data envelopment analysis, Eur J of Operational Research 157(2004)87-105.

EFFICIENCY ANALYSIS UNDER IMPRECISE INPUTS AND OUTPUTS REDUCTION

KOBRA GHOLAMI*
Department of Science, Bushehr Branch, Islamic Azad University, Bushehr, Iran

FARHAD HOSSEINZADEH LOTFI
Department of Mathematics, Science and Research Branch, Islamic Azad University, Tehran, Iran

ADEL HATAMI-MARBINI, PER J AGRELL
LSM, Center of Operations Research and Econometrics (CORE), Universite catholique de Louvain, 34 Voie du Roman Pays, L1.03.01, B-1348 Louvain-la-Neuve, Belgium

ZAHRA GHELEJ BEIGI
Department of Mathematics, Science and Research Branch,Islamic Azad University, Tehran, Iran

Data envelopment analysis (DEA) is a powerful tool for measuring the relative efficiencies of a set of decision making units (DMU) that consume multiple inputs to produce multiple outputs. DEA can also be used in activity planning and economic regulation of decentralized units under a common management. One important management challenge in such systems (e.g. schools, hospitals, utilities) is how to implement reductions in common resources (budget, staff, raw material) as well as centrally allocated outputs (students, patients, service areas) while maintaining a high operational efficiency of the system. We extend earlier work on DEA activity planning under resource and output restrictions for the case in which the centralized planner is facing an imprecise shrinkage factor, here expressed as interval data. The implementation of the proposed method is applied for a numeric example.

1. Introduction

Performance evaluation is a crucial issue for agents in the present competitive environment. Data envelopment analysis (DEA) was introduced by Charnes et al. [1] and has been extensively used for evaluating the performances of the decision making units (DMUs). In the weight (multiplier) DEA model, the dual weights in the formulation are individual and endogenous. In consequence, these

* E-mail: kobra_gholami@yahoo.com

dual weights may be radically different between units that objectively share a common technology and market. To deal with the total flexibility of weights we can use an endogenous common set of weights (CSW) for providing a desirable evaluation (e.g. [2]).

The classical DEA models have limited to the use of perfect and complete data. However, impreciseness in the input and output data is unavoidable in most real-life problems. Fuzzy and interval DEA approaches are two common approaches for handling this challenging issue [3, 4]. Moreover, the central manager of a set of units may be facing uncertain reductions in the allocation or resources and outputs while still carrying the responsibility for an optimal operating efficiency of the system.

Amirteimoori and Emrouznejad [5] put forward a DEA-based approach to reduce the inputs and output by a given amount while preserving the efficiency. However, the amount of inputs and outputs reduction is often uncertain and imprecise. To deal with this problem, we introduce a new imprecise DEA method with the interval data contraction. This method determines the reduction in the fair way such that the efficiency will not get worse compared to the prior situation.

2. Imprecise inputs and outputs reduction in DEA

Consider n DMUs under the evaluation process that each DMU_j, $j=1,...,n$, use m inputs, $x_{ij} \in R^+$, $i=1,...,m$, to produce s outputs, $y_{rj} \in R^+$, $r=1,...,s$. Suppose that there are two types of inputs $I_1=\{i_1,i_2,...,i_k\}$, and $I_2=\{1,...,m\}-I_1$, and two types of outputs as $O_1=\{r_1,r_2,...,r_t\}$ and $O_2=\{1,...,s\}-O_1$ where I_1 and O_1 are, respectively, the set of selected inputs and outputs for reduction. After reducing the selected data, we aim to avoid decreasing the efficiency of DMUs. To do so, we introduce the interval variables $C_i=[C_i^L,C_i^U]$, $i\in I_1$, and $F_r=[F_r^L,F_r^U]$, $r\in O_1$ as the total reduction in inputs and outputs set, respectively. In this paper, $\overline{c}_{ij}$ and $\overline{f}_{rj}$ are the ith and rth variables of inputs and outputs reduction corresponding to DMU_j, $j=1,...,n$. Note here that $\sum_{j=1}^{n}\overline{c}_{ij}=C_i, i\in I_1$, $\sum_{j=1}^{n}\overline{f}_{rj}=F_r, r\in O_1$. As a result, the following system can be formulated:

$$
\begin{aligned}
&\frac{\sum_{r\in O_2} u_r y_{rj} + \sum_{r\in O_1} u_r (y_{rj} - \overline{p}_{rj})}{\sum_{i\in I_2} v_i x_{ij} + \sum_{i\in I_1} v_i (x_{ij} - \overline{c}_{ij})} \ge \theta_j^*, \quad j = 1,\dots,n, \\
&\frac{\sum_{r\in O_2} u_r y_{rj} + \sum_{r\in O_1} u_r (y_{rj} - \overline{p}_{rj})}{\sum_{i\in I_2} v_i x_{ij} + \sum_{i\in I_1} v_i (x_{ij} - \overline{c}_{ij})} \le 1, \quad j = 1,\dots,n, \\
&\sum_{j=1}^{n} \overline{c}_{ij} = C_i, \qquad i \in I_1, \qquad \sum_{j=1}^{n} \overline{p}_{rj} = P_r, \qquad r \in O_1, \\
&\overline{c}_{ij} \le x_{ij}, \quad i \in I_1, j = 1,\dots,n, \qquad \overline{p}_{rj} \le y_{rj}, \quad r \in O_1, j = 1,\dots,n, \\
&C_i^L \le C_i \le C_i^U, \qquad i \in I_1, \qquad P_r^L \le P_r \le P_r^U, \qquad r \in O_1, \\
&u_r, v_i, \overline{c}_{ij}, \overline{p}_{rj} \ge 0, \qquad r, i, j.
\end{aligned}
\tag{1}
$$

To normalize the inputs and outputs, we insert the multipliers ρ_{ij} and μ_{rj} in the above system as follows:

$$
\rho_{ij} = \frac{x_{ij}}{\sum_{t=1}^{n} x_{it}}, \; j = 1,\dots,n, \; i \in I_1, \quad \mu_{rj} = \frac{y_{rj}}{\sum_{t=1}^{n} y_{rt}}, \; j = 1,\dots,n, \; r \in O_1.
$$

Note that $\sum_{j=1}^{n} \rho_{ij} = \sum_{j=1}^{n} \mu_{rj} = 1$.The resulting model is implemented using goal programming:

$$
\begin{aligned}
\min \quad & \sum_{j=1}^{n} (d_j^- + d_j^+) + \varepsilon\Big[\sum_{j=1}^{n} \sum_{i\in I_1} (\alpha_{ij}^- + \alpha_{ij}^+) + \sum_{j=1}^{n} \sum_{r\in O_1} (\beta_{rj}^- + \beta_{rj}^+)\Big] \\
s.t. \quad & \Big(\sum_{r=1}^{s} u_r y_{rj} - \sum_{r\in O_1} p_{rj} + d_j^+\Big) - \theta_j^* \Big(\sum_{i=1}^{m} v_i x_{ij} - \sum_{i\in I_1} c_{ij} - d_j^-\Big) \ge 0, \quad j = 1,\dots,n, \\
& \Big(\sum_{r=1}^{s} u_r y_{rj} - \sum_{r\in O_1} p_{rj}\Big) - \Big(\sum_{i=1}^{m} v_i x_{ij} - \sum_{i\in I_1} c_{ij}\Big) \le 0, \qquad j = 1,\dots,n, \\
& c_{ij} + \alpha_{ij}^- - \alpha_{ij}^+ = \rho_{ij} \overline{C}_i \qquad i \in I_1, \quad j = 1,\dots,n, \\
& p_{rj} + \beta_{rj}^- - \beta_{rj}^+ = \mu_{rj} \overline{P}_r, \qquad r \in O_1, \; j = 1,\dots,n, \\
& \sum_{j=1}^{n} c_{ij} = \overline{C}_i, \quad i \in I_1, \qquad \sum_{j=1}^{n} p_{rj} = \overline{P}_r, \quad r \in O_1, \\
& c_{ij} \le v_i x_{ij}, \quad i \in I_1, j = 1,\dots,n, \qquad p_{rj} \le u_r y_{rj}, \quad r \in O_1, j = 1,\dots,n, \\
& v_i C_i^L \le \overline{C}_i \le v_i C_i^U, \quad i \in I_1, \qquad u_r P_r^L \le \overline{P}_r \le u_r P_r^U, \quad r \in O_1, \\
& u_r, \; v_i, \; c_{ij}, \; p_{rj}, \alpha_{ij}^-, \alpha_{ij}^+, \; \beta_{rj}^-, \beta_{rj}^+, d_j^-, d_j^+ \ge 0, \qquad \forall \; r, \; i, \; j.
\end{aligned}
\tag{2}
$$

where θ_j^*, as the efficiency of jth unit, is obtained via the CSW method. In the proposed model (6), d_j^-, d_j^+, α_{ij}^-, α_{ij}^+, β_{rj}^-, β_{rj}^+, $\forall j$, $i \in I_1$, $r \in O_1$, are deviation variables and $u_r\overline{p}_{rj} = p_{rj} (\forall j, r \in O_1)$; $v_i\overline{c}_{ij} = c_{ij} (\forall j, i \in I_1)$; $v_iC_i = \overline{c}_i \ (i \in I_1)$ and $u_rP_r = \overline{p}_r (r \in O_1)$.

3. Numerical example

In this section, we illustrate the applicability of the proposed method with a numerical example used in [5]. Table 1 shows the data involving 27 DMUs, three inputs and two outputs. We assume that the reduction of the first input and second output can be changed within [2.5, 3.5] and [1.5, 2.5], respectively. We use model (2) to calculate the optimizing reduction of 1st input and 2nd output among 27 units. The optimal values of $\overline{c}_{1j}$ and $\overline{p}_{2j}$ are presented, respectively, in the seventh and eighth columns of Table 1. The efficiency scores calculated by CSW model before (θ_j^*) and after reduction (θ_j^{*new}) are given in the ninth and tenth columns of Table 1.

Table 1. The data and results for 27 DMUs

DMU	I_1	I_2	I_3	O_1	O_2	$\overline{c}_{ij}$	$\overline{p}_{rj}$	θ_j^*	θ_j^{*new}
1	0.150	0.713	13.300	14.478	0.232	0.0637	0.03	0.57	0.57
2	0.170	1.071	16.900	19.502	0.340	0.0721	0.04	0.61	0.62
3	0.235	1.224	24.000	20.952	0.363	0.0997	0.05	0.47	0.47
4	0.211	0.363	15.600	13.902	0.211	0.0895	0.03	0.45	0.46
5	0.133	0.409	18.485	15.206	0.237	0.0564	0.03	0.48	0.48
6	0.497	5.846	56.420	81.186	1.103	0.2109	0.14	0.76	0.76
7	0.060	0.918	56.420	81.186	1.103	0.0254	0.14	0.97	0.98
8	0.071	1.235	12.000	11.441	0.199	0.0301	0.02	0.54	0.54
9	1.500	18.12	89.510	124.072	1.858	0.6365	0.23	0.60	0.60
10	0.120	1.821	19.800	17.425	0.274	0.0510	0.03	0.50	0.50
11	0.120	1.915	19.800	17.425	0.274	0.0510	0.01	0.50	0.50
12	0.050	0.874	13.100	14.342	0.177	0.213	0.02	0.66	0.67
13	0.370	6.918	12.500	32.491	0.648	0.1570	0.05	0.80	0.80
14	0.440	4.432	41.900	47.653	0.636	0.1867	0.08	0.58	0.58
15	0.431	4.504	41.100	52.63	0.741	0.1828	0.09	0.65	0.65
16	0.110	1.241	14.400	17.493	0.243	0.0467	0.03	0.67	0.67
17	0.053	0.450	7.6000	9.512	0.067	0.0224	0.01	0.71	0.71
18	0.345	5.892	15.500	42.469	1.002	0.1464	0.13	1.00	1.00
19	0.128	0.973	12.600	18.987	0.243	0.0543	0.03	0.79	0.79
20	0.055	0.444	5.6000	7.546	0.153	0.0233	0.02	0.71	0.71
21	0.057	0.508	5.7000	7.595	0.123	0.0242	0.01	0.69	0.70
22	0.098	0.370	14.100	16.906	0.233	0.0416	0.03	0. 69	0.70
23	0.104	0.395	14.600	17.264	0.263	0.0442	0.03	0.68	0.69
24	0.206	2.680	19.600	36.43	0.601	0.0874	0.08	0.93	0.93
25	0.067	0.781	10.500	11.581	0.120	0.0285	0.01	0.63	0.63
26	0.100	0.872	12.100	22.207	0.248	0.0424	0.10	1.00	1.00
27	0.0106	1.757	12.700	20.67	0.253	0.0046	0.03	1.00	1.00
					SUM	2.5001	1.5	18.64	18.71

As shown in Table 1, the efficiency scores of DMU_2, DMU_4, DMU_7, DMU_{12}, DMU_{21}, DMU_{22} and DMU_{23} improve after assigning the optimal reduction. The remaining DMUs keep the previous efficiency score in the proposed method.

4. Conclusions

The integrated use of DEA in performance evaluation and activity planning is an important field for ongoing research and application. In this paper, we extend a recent model for the optimal allocation of resource and output reductions under common endogenous weights for the case with resource reductions given as interval data so as to slightly improve the efficiency scores of DMUs. Given that budgeting, resource and output allocation are normally activities that are performed iteratively between the principal and the planners, we argue that the imprecise representation is realistic and result in more useful information for the planner.

References

1. A. Charnes, W.W. Cooper and E. Rhodes, Measuring the efficincy of decision making units, *Eur. J. Operat. Res*, **2(6),** 429(1978).
2. F. Hosseinzadeh Lotfi, G. R. Jahanshahlooand A. Memariani, A method for finding common set of weights by multiple objective programming in data envelopment analysis. SW J Pure Appl Math, **1**, 44 (2000).
3. A. Hatami-Marbini, A. Emrouznejad and M. Tavana, A taxonomy and review of the fuzzy data envelopment analysis literature: Two decades in the making, *Eur. J. Operat. Res*, **214**, 457 (2011).
4. W.W. Cooper, K.S. Parkand G. Yu, IDEA and AR-IDEA: models for dealing with imprecise data in DEA, *Management Sci.***45**, 597–607(1999).
5. A. Amirteimooriand A. Emrouznejad, Input/output deterioration in production processes. *Expert Syst Appl.* **38 (5)**, 5822(2010).

RESOURCE ALLOCATION AND TARGET SETTING IN INTERVAL DATA ENVELOPMENT ANALYSIS

ZAHRA GHELEJ BEIGI*[1], FARHAD HOSSEINZADEH LOTFI[1], ADEL HATAMI-MARBINI[2], PER J AGRELL[2], KOBRA GHOLAMI[1]

[1]*Department of Mathematics, Science and Research Branch, Islamic Azad University, Tehran, Iran*

[2]*LSM, Center of Operations Research and Econometrics (CORE), Universite catholique de Louvain, 34 Voie du Roman Pays, L1.03.01, B-1348 Louvain-la-Neuve, Belgium*

Data envelopment analysis (DEA) is a commonly used non-parametric technique for performance measurement of decision making units (DMU) that can also be used in normatively in activity planning, resource allocation and target setting. Whereas earlier work in this line have considered completely defined data, prospective use of DEA in activity planning often involves uncertainty with respect to the feasible ranges for allocation of resources and target settings. In this paper, we present an imprecise DEA-based method with interval inputs and outputs as to address this shortcoming. The proposed model corresponds to reasonable managerial objectives concerning the technical efficiency of the subordinate units after implementing resource allocation and target setting.

1. Introduction

Data envelopment analysis (DEA) is a non-parametric method for measuring the efficiency of a set of decision making units (DMUs) with multiple inputs and multiple outputs [1]. Although the conventional DEA models developed based on the perfect and accurate in the real-life problems the data are often uncertain and imprecise. To deal with the uncertain data, some authors used the interval analysis in DEA such as Cooper et al. [2]. Wang et al. [3] presented two DEA models with same technology to calculate the lower and upper efficiency bounds for interval data.

The common set of weights (CSW) is a popular approach to handle the disadvantage of the full weight flexibility (e.g. different weights, zero weights, etc.) in the multiplier DEA model [4]. Our work is related to DEA in activity planning where a centralized planner has some authority over (part) of the DMUs in the reference set with respect to allocated resources and targets

* zahra.ghlichbegi@yahoo.com.

defined for output [5]. Normally, the planner is striving to achieve an allocation of resources and targets such that the expected technical efficiency of the DMUs is maximized [6]. We generalize this approach for the setting where the planner ignores the exact value of inputs and outputs attributable to each DMU, know only as ranges or intervals. Our method obtains the interval efficiency before and after assigning the new inputs and outputs. The upper bound of all the DMUs attains unity (fully technically efficient) while the lower bound efficiency is improved or maintained.

2. Interval DEA preliminaries

Suppose that there are n DMUs to be evaluated, and each DMU using m positive inputs $x_{ij}, \forall i$ to produce s non-negative outputs $y_{rj}, \forall r,$. Assume that $x_{ij} \in \left[x_{ij}^{l}, x_{ij}^{u}\right], \forall i,j,$ and $y_{rj} \in [y_{rj}^{l}, y_{rj}^{u}]$ $\forall r,j$. Based on the imprecise DEA models proposed by Wang et al. [3], the lower and upper bounds of efficiency can be calculated using the CSW approach as follows:

$$\begin{aligned} \theta^{u*} &= \max\left\{\frac{\sum_{r=1}^{s} u_r y_{r1}^{u}}{\sum_{i=1}^{m} v_i x_{i1}^{l}}, \frac{\sum_{r=1}^{s} u_r y_{r2}^{u}}{\sum_{i=1}^{m} v_i x_{i2}^{l}}, \ldots, \frac{\sum_{r=1}^{s} u_r y_{rn}^{u}}{\sum_{i=1}^{m} v_i x_{in}^{l}}\right\} \\ s.t.\quad & \frac{\sum_{r=1}^{s} u_r y_{rj}^{u}}{\sum_{i=1}^{m} v_i x_{ij}^{l}} \le 1,\ \forall j, \qquad u_r, v_i \ge \varepsilon,\ \forall r,i. \end{aligned} \tag{1}$$

$$\begin{aligned} \theta^{l*} &= \max\left\{\frac{\sum_{r=1}^{s} u_r y_{r1}^{l}}{\sum_{i=1}^{m} v_i x_{i1}^{u}}, \frac{\sum_{r=1}^{s} u_r y_{r2}^{l}}{\sum_{i=1}^{m} v_i x_{i2}^{u}}, \ldots, \frac{\sum_{r=1}^{s} u_r y_{rn}^{l}}{\sum_{i=1}^{m} v_i x_{in}^{u}}\right\} \\ s.t.\quad & \frac{\sum_{r=1}^{s} u_r y_{rj}^{u}}{\sum_{i=1}^{m} v_i x_{ij}^{l}} \le 1, \quad \forall j, \quad u_r, v_i \ge \varepsilon \quad \forall r,i. \end{aligned} \tag{2}$$

where ε is a non-Archimedean infinitesimal. Models (1) and (2) are multi-objective fractional programming (MOFP) problems and several solving methods have been reported in the literature. Assume that $\left(u_r^{*}, v_i^{*}\right)$ and $\left(\overline{u}_r^{*}, \overline{v}_i^{*}\right), \forall r,i,$ are optimal solutions of models (1) and (2), respectively. Then the upper and lower efficiency for each unit can be obtained as:

$$\theta_j^{*u} = \sum_{r=1}^{s} u_r^{*} y_{rj}^{u} \Big/ \sum_{i=1}^{m} v_i^{*} x_{ij}^{l} \quad \forall j; \qquad \theta_j^{*l} = \sum_{r=1}^{s} \overline{u}_r^{*} y_{rj}^{l} \Big/ \sum_{i=1}^{m} \overline{v}_i^{*} x_{ij}^{u}, \quad \forall j.$$

where $\theta_j^{*l} \le \theta_j^{*u}$. DMU_o is [technically input-] efficient iff $\theta_o^{*u} = 1$, otherwise, it is [technically input-] inefficient.

3. Resource allocation and target setting for uncertain data

Assume that a central manager of a set of DMUs intends to allocate q new resources among DMUs for which (s) he is only able to determine the upper and lower bounds as $\left[F_k^l, F_k^u\right], F_k^l \in R^+, \forall k,$. As a result, the DM expects to attain p interval target outputs as $\left[G_w^l, G_w^u\right], G_w^l \in R^+, \forall w$. Suppose that $\left[f_{kj}^l, f_{kj}^u\right]$ and $\left[g_{wj}^l, g_{wj}^u\right]$ are non-negative interval input and output variables, respectively, allocated among the DMU_j. Furthermore, the balance constraints $\sum_{j=1}^n f_{kj}^l = F_k^l$, $\sum_{j=1}^n f_{kj}^u = F_k^u$ and $\sum_{j=1}^n g_{wj}^l = G_w^l$, $\sum_{j=1}^n g_{wj}^u = G_w^u$ must be respected. We formulate the following system to find the lower bound of resource allocation and the upper bound of target setting:

$$
\begin{aligned}
&\frac{\sum_{r=1}^{s} u_r y_{rj}^u + \sum_{w=1}^{p} u_{s+w} g_{wj}^u}{\sum_{i=1}^{m} v_i x_{ij}^l + \sum_{k=1}^{q} w_{m+k} f_{kj}^l} = 1, && \forall j,\\
&\sum_{j=1}^{n} f_{kj}^l = F_k^l, \quad \forall k; \qquad \sum_{j=1}^{n} g_{wj}^u = G_w^u, && \forall w,\\
&u_r, v_i, u_{s+w}, w_{m+k} \geq \varepsilon, && \forall r, i, w, k.
\end{aligned}
\tag{3}
$$

By the use of the above system, the upper bound on the efficiency for each DMU reaches unity. To transform the non-linear problem (3) into a linear program, we introduce the alternation variables $u_{s+w} g_{wj}^u = \overline{g}_{wj}^u$ and $w_{m+k} f_{kj}^l = \overline{f}_{kj}^l$. The relative weights of the input and output factors can be expressed with $\lambda_{kj} = (1/m)\sum_{i=1}^m x_{ij}^l \big/ \sum_{t=1}^n x_{it}^l$ and $\mu_{wj} = (1/s)\sum_{r=1}^s y_{rj}^u \big/ \sum_{t=1}^n y_{rt}^u$, where $\sum_{j=1}^n \lambda_{kj} = \sum_{j=1}^n \mu_{wj} = 1\ (\forall k; \forall w)$. For a general $\lambda_{kj} F_k^l$ and $\mu_{wj} G_w^u$ the solution to (3) may be infeasible. To deal with this problem, we use a goal programing approach. Let $\alpha_{kj}^-, \beta_{wj}^-$ and $\alpha_{kj}^+, \beta_{wj}^+$ be the negative and positive deviational variables to achieve the goals $w_{m+k} \lambda_{kj} F_k^l$ and $u_{s+w} \mu_{wj} G_w^u$ to formulate model (3) we minimizing the sum of the deviations as follow:

$$
\begin{aligned}
\min \quad & \sum_{j=1}^{n}\left(\sum_{k=1}^{q}(\alpha_{kj}^- + \alpha_{kj}^+) + \sum_{w=1}^{p}(\beta_{wj}^- + \beta_{wj}^+)\right)\\
s.t. \quad & \sum_{r=1}^{s} u_r y_{rj}^u + \sum_{w=1}^{p} \overline{g}_{wj}^u - \left(\sum_{i=1}^{m} v_i x_{ij}^l + \sum_{k=1}^{q} \overline{f}_{kj}^l\right) = 0, && \forall j,\\
& \overline{f}_{kj}^l + \alpha_{kj}^- - \alpha_{kj}^+ = w_{m+k} \lambda_{kj} F_k^l, && \forall j,\ \forall k,
\end{aligned}
\tag{4}
$$

$$\bar{g}^{u}_{wj} + \beta^{-}_{wj} - \beta^{+}_{wj} = u_{s+w}\mu_{wj}G^{u}_{w}, \qquad \forall j,\ \forall w,$$

$$\sum_{j=1}^{n} \bar{f}^{l}_{kj} = w_{m+k}F^{l}_{k}, \ \forall k; \qquad \sum_{j=1}^{n} \bar{g}^{u}_{wj} = u_{s+w}G^{u}_{w}, \ \forall w,$$

$$u_r, v_i, u_{s+w}, w_{m+k}, \alpha^{-}_{kj}, \alpha^{+}_{kj}, \beta^{-}_{wj}, \beta^{+}_{wj}, \qquad \forall r,i,w,k.$$

Once the optimal solution of (4) $(U^*, V^*, \bar{f}^{*l}, \bar{g}^{*u})$ is computed we can calculate the lower bound of resource allocation (f^{*l}_{kj}) and the upper bound of target setting (g^{*u}_{wj}) as: $g^{*u}_{wj} = \bar{g}^{*u}_{wj} / u^{*}_{s+w}$ and $f^{*l}_{kj} = \bar{f}^{*l}_{kj} / w^{*}_{m+k}$. Similarly, we propose the following system to obtain the upper bounds of resource allocation value and lower bounds of target setting:

$$\frac{\sum_{r=1}^{s} u_r y^{u}_{rj} + \sum_{w=1}^{p} u_{s+w} g^{*u}_{wj}}{\sum_{i=1}^{m} v_i x^{l}_{ij} + \sum_{k=1}^{q} w_{m+k} f^{*l}_{kj}} \le 1, \qquad \forall j,$$

$$\theta^{l*}_{j} \le \frac{\sum_{r=1}^{s} u_r y^{l}_{rj} + \sum_{w=1}^{p} u_{s+w} g^{l}_{wj}}{\sum_{i=1}^{m} v_i x^{u}_{ij} + \sum_{k=1}^{q} w_{m+k} f^{u}_{kj}} \le 1, \qquad \forall j,$$

$$\sum_{j=1}^{n} f^{u}_{kj} = F^{u}_{k}, \quad \forall k; \qquad \sum_{j=1}^{n} g^{l}_{wj} = G^{l}_{w}, \quad \forall w, \tag{5}$$

$$f^{u}_{kj} \ge f^{*l}_{kj}, \quad \forall k; \qquad g^{*u}_{wj} \ge g^{l}_{wj}, \quad \forall w,$$

$$u_r, v_i, u_{s+w}, w_{m+k} \ge \varepsilon, \qquad \forall r,i,w,k.$$

Note that in the setting of the above system θ^{l}_{j} varies within $[\theta^{l*}_{j}$, 1] and all the DMUs remain in the technology. Analogously, we define $u_{s+w} g^{l}_{wj} = \tilde{g}^{l}_{wj}$ and $w_{m+k} f^{u}_{kj} = \tilde{f}^{u}_{kj}$ for estimating $\tilde{g}^{l}_{wj}$ and $\tilde{f}^{u}_{kj}$. Hence, the following linear program is obtaing using the goal programming method:

$$\min \ \sum_{j=1}^{n} \left(\sum_{k=1}^{q} (\alpha^{-}_{kj} + \alpha^{+}_{kj}) + \sum_{w=1}^{p} (\beta^{-}_{wj} + \beta^{+}_{wj}) \right)$$

$$s.t. \ \sum_{r=1}^{s} u_r y^{u}_{rj} + \sum_{w=1}^{p} u_{s+w} g^{*u}_{wj} - \sum_{i=1}^{m} v_i x^{l}_{ij} + \sum_{k=1}^{q} w_{m+k} f^{*l}_{kj} \le 0, \quad \forall j, \tag{6}$$

$$\theta^{l*}_{j} \left(\sum_{i=1}^{m} v_i x^{u}_{ij} + \sum_{k=1}^{q} \tilde{f}^{u}_{kj} \right) - \sum_{r=1}^{s} u_r y^{l}_{rj} + \sum_{w=1}^{p} \tilde{g}^{l}_{wj} \le 0, \qquad \forall j,$$

$$\sum_{r=1}^{s} u_r y_{rj}^l + \sum_{w=1}^{p} \tilde{g}_{wj}^l - \sum_{i=1}^{m} v_i x_{ij}^u + \sum_{k=1}^{q} \tilde{f}_{kj}^u \le 0, \qquad \forall j,$$
$$\tilde{f}_{kj}^u + \alpha_{kj}^- - \alpha_{kj}^+ = w_{m+k} \lambda_{kj} F_k^l, \qquad \forall j, \forall k,$$
$$\tilde{g}_{wj}^l + \beta_{wj}^- - \beta_{wj}^+ = u_{s+w} \mu_{wj} G_w^u, \qquad \forall j, \forall w,$$
$$\sum_{j=1}^{n} f_{kj}^u = w_{m+k} F_k^u, \quad \forall k; \quad \sum_{j=1}^{n} g_{wj}^l = u_{s+w} G_w^l, \qquad \forall w,$$
$$\tilde{f}_{kj}^u \ge w_{m+k} f_{kj}^{*l}, \quad \forall k; \quad u_{s+w} g_{wj}^{*u} \ge \tilde{g}_{wj}^l, \qquad \forall w,$$
$$u_r, v_i, u_{s+w}, w_{m+k} \ge \varepsilon, \ \alpha_{kj}^-, \alpha_{kj}^+, \beta_{wj}^-, \beta_{wj}^+ \ge 0, \quad \forall r, i, w, k.$$

An optimal solution of (6) $(U^*, V^*, \tilde{f}^{*u}, \tilde{g}^{*l})$ gives the upper bounds of resource allocation ($\tilde{f}_{kj}^{*u}$) and lower bounds of target setting ($\tilde{g}_{wj}^{*l}$) from $g_{wj}^{*l} = \tilde{g}_{wj}^{*l} / u_{s+w}^*$ and $f_{kj}^{*u} = \tilde{f}_{kj}^{*u} / w_{m+k}^*$. To illustrate the approach, consider 23 DMUs with three interval inputs $(I_i, i = 1,2,3)$ and two interval outputs $(O_r, r = 1,2,3)$. The data are given in Table 1. Let us distribute a total interval resource [20, 40] among all the DMUs and we accordingly assume that a total produced output changes in [10, 30]. The upper bounds of resource allocation value and lower bounds of target setting are calculated via model (4) and model (6) computes the upper bounds of resource allocation value and lower bounds of target setting. The results of resource allocation and target setting intervals are shown in 8th and 9th colomns of Table 1, respectively. The efficiency before and after resource allocation and target setting are shown in 7th and 10th colomns of Table 1, respectively. From Table 1, the proposed approach leads to the efficiency improvement for all the DMUs after resource allocation and target setting.

4. Conclusion

The applicability of DEA not only to ex post performance evaluation, but to ex ante performance management and activity planning, requires a careful analysis of the preconditions for the decision situation. First, the planner is usually given a limited discretion with respect to allocation of resources (staff, equipment, raw material) as well as in target setting (market allocation, client and area assignments). This could be expressed in form of well-defined constraints for each unit or resource, but in many contexts this is an over-simplification. In a short- or medium-term planning context, the manager is likely facing unit-defined constraints for minimum and maximum resource and target assignment, resulting from consequences of long-term decisions with respect to fixed assets and location. In our model, the interval setting corresponds to this need. Second,

Table 1. The data, efficency and result for 23 DMUs

DMU	I_1	I_2	I_3	O_1	O_2	$[\theta^L, \theta^U]$	$[f_{1j}^{l*}, f_{1j}^{u*}]$	$[g_{1j}^{l*}, g_{1j}^{u*}]$	$[\theta_{new}^L, \theta_{new}^U]$
1	[11, 3]	[12, 6]	[22,28]	[2, 5]	[12, 9]	[0.1,0.3]	[0, 1.2]	[1.4, 1.4]	[0.9,1]
2	[12,18]	[14,19]	[24,30]	[4, 8]	[14,21]	[0.1,0.3]	[0.0]	[1.4, 1.4]	[0.9,1]
3	[13,19]	[16,22]	[26,32]	[6,10]	[16,22]	[0.2,0.4]	[0,0]	[0.8, 1.4]	[0.9,1]
4	[13,17]	[18,27]	[28,40]	[8,12]	[18,25]	[0.2,0.4]	[0,0]	[0.3, 1.4]	[0.9,1]
5	[15,22]	[3, 12]	[29,39]	[10,17]	[20,28]	[0.2,0.5]	[0.6,0.6]	[0.4,0.4]	[0.9,1]
6	[56,79]	[6, 19]	[30,38]	[1, 7]	[22,29]	[0.1,0.2]	[0,2]	[0.3,3.6]	[0.9,1]
7	[22,29]	[8, 19]	[13,25]	[2, 9]	[24,31]	[0.2, 0.4]	[0.5,0.6]	[0.3,0.4]	[0.9,1]
8	[17,29]	[9, 14]	[14,18]	[3,17]	[26,34]	[0.2, 0.7]	[1.7, 1.9]	[0,0]	[0.9,1]
9	[10,28]	[12, 9]	[16,45]	[4,15]	[28,38]	[0.2, 0.9]	[2.3, 2.4]	[0,0]	[0.9,1]
10	[9, 17]	[20, 3]	[11,44]	[5,14]	[30,39]	[0.2, 0.9]	[2.0, 2.1]	[0,0]	[0.9,1]
11	[6, 21]	[22, 1]	[15,30]	[6,13]	[11,23]	[0.1, 0.6]	[0.6, 0.6]	[0.3,0.7]	[0.9,1]
12	[76,45]	[26,29]	[19,26]	[7,18]	[12·24]	[0.1,0.1]	[0, 3.2]	[0.4,5.1]	[0.9,1]
13	[11,19]	[32, 4]	[21,29]	[8,20]	[13,25]	[0.1,0.5]	[0, 1.8]	[0.4,1.2]	[0.9,1]
14	[13,29]	[33, 7]	[22,29]	[9,17]	[14,23]	[0.1,0.4]	[0, 1.9]	[0.4,1.8]	[0.9,1]
15	[12,43]	[31,54]	[34,65]	[14,21]	[15,28]	[0.1,0.4]	[0, 2.2]	[0.4,1.8]	[0.9,1]
16	[21, 2]	[25,43]	[38,54]	[16,26]	[16,32]	[0.1,0.4]	[0, 2.3]	[0.5,1.6]	[0.9,1]
17	[25,33]	[27,44]	[37,55]	[18,25]	[17,33]	[0.1,0.4]	[0, 2.4]	[0.5,2.0]	[0.9,1]
18	[29,36]	[28,37]	[22,38]	[20,28]	[18,39]	[0.2, 0.5]	[1, 1.1]	[0.6,1.2]	[0.9,1]
19	[32,36]	[19,30]	[11,25]	[11,29]	[19,32]	[0.1, 0.5]	[0.8, 0.8]	[0.4,0.4]	[0.9,1]
20	[44,66]	[16,25]	[30,42]	[12,28]	[20,40]	[0.1, 0.3]	[1,1]	[0.6,1.8]	[0.9,1]
21	[55,67]	[20,35]	[32,54]	[13,29]	[31,44]	[0.1, 0.3]	[0.1, 2.8]	[0.6,2]	[0.9,1]
22	[17,25]	[2, 20]	[26,30]	[14,31]	[32,41]	[0.3, 0.9]	[4.2, 4.3]	[0,0]	[0.9,1]
23	[16,19]	[11,23]	[11,24]	[15,32]	[21,36]	[0.3,1]	[4.7, 4.8]	[0,0]	[0.9,1]
SUM							[20,40]	[10,30]	

the planner in her prospective role of activity planner is normally interested and incentivized to ensure the (ex ante) technical efficiency of the units under her control following the planning. Here our approach leading to a range including full technical efficiency is consistent with this behavioral conjecture, as opposed to other work in which the planner does not leave all units with the same planned opportunities for performance provision.

References

1. A. Charnes, W.W. Cooper and E. Rhodes, Measuring the efficincy of decision making units, *Eur. J. Operat. Res.* **2(6),** 429 (1978).
2. W.W. Cooper, K.S. Park and G. Yu, IDEA and AR-IDEA: models for dealing with imprecise data in DEA, *Management Sci.* **45**, 597–607(1999).
3. Y. Wang, R. Greatbanksa and J. Yanga, Interval efficiency assessment using data envelopment analysis, *Fuzzy Set. Syst.* **153**, 347 (2005).
4. F. Hosseinzadeh Lotfi, G. R. Jahanshahloo and A. Memariani. A method for finding common set of weights by multiple objective programming in data envelopment analysis, *SW. J. Pure. Appl. Math.* **1**, 44 (2000).
5. P. Bogetoft. DEA and activity planning under asymmetric information, *J. Productiv. Anal.* **13(1),** 7–48 (2000).
6. A. Amirteimoori and M. Mohaghegh Tabar, Resource allocation and target setting in data envelopment analysis, *Expert Syst. Appl.* **3**, 3036 (2010).

PROJECT RISK MANAGEMENT TAKING INTO CONSIDERATION THE INFLUENCE OF VARIOUS RISK LEVELS BASED ON LINGUISTIC APPROACH*

DOROTA KUCHTA†

Faculty of Computer Science and Management, Institute of Industrial Engineering and Management, Wroclaw University of Technology, Wybrzeze Wyspianskiego 27
50-370 Wroclaw, Poland
Faculty of Management, Wyższa Szkoła Oficerska Wojsk Ladowych we Wroclawiu, ul. Czajkowskiego 109,
51-150 Wroclaw, Poland

DARIUSZ SKORUPKA

Faculty of Management, Wyższa Szkoła Oficerska Wojsk Ladowych we Wroclawiu, ul. Czajkowskiego 109
51-150 Wroclaw, Poland

A new method of project time overrun risk estimation is proposed. The method takes into consideration various risk levels, corresponding to the project environment understood in a more or less wide manner, and interdependencies between them. It is based on expert estimations expressed in linguistic terms, which are flexible and may be defined in each case accordingly to experts preferences. The method allows to insert buffers protecting the project planned deadline, taking into account the complex, interdependent risk factors – in contrast to the existing methods, which usually treat risk factors as independent.

1. Introduction

The problem of project risk management (e.g.[6]) has been studies for many years is by an innumerous number of researchers. It is still unsolved and probably will never be completely solved, because risk and uncertainty is inherent to projects. What is more worrying is the fact that in spite of all the project risk management research apparently no progress has been made and the project successful terminations are rather rare ([7]). Thus is still worth looking

* This work is partially supported by Polish National Science Center grant "The buffer size and location in a proactive and reactive risk management of a project duration overruning" 2011/01/B/HS4/02821

† dorota.kuchta@pwr.wroc.pl.

for new approaches which are substantially different from the existing ones and thus offer a chance to help project managers to be successful with their projects.

It seems that one major problem of most project risk management methods is that they do not take into account interdependencies between risks. One attempt to take these dependencies into account has been made in [1] and [2], where, for construction projects, it is taken into account that the situation on the macro level or on the construction market may influence the risks on the level of the construction project itself. However, in [1] and [2] the probabilistic approach is applied. Here we try to help to solve also another problem linked to practical project risk management: the probabilistic approach is very often rejected as being too "mathematical" (e.g.[5]). Thus here we propose to use in our method the linguistic approach, applied for public projects in [5].

First, we propose the project a new risk estimation method, and then we illustrate the approach with a small example.

2. Method Outline

We assume that we have a project, understood as a set of activities $\{A_i\}_{i=1}^{n}$, with a set precedence relations of the type "Finish-To-Start", imposed on a set of activity couples. We also dispose of basic estimations $\{TB_i\}_{i=1}^{n}$ of activity time duration, given by the experts for the most probable scenario.

Then, applying the risk management process, experts estimate, at the project level, the risk $\{RD_i\}_{i=1}^{n}$that the activities may take more time if a negative scenario occurs. We assume that these risks are expressed in linguistic terms, e.g. taken from the set $D = \{VL.L.M,H,VH\}$, where the abbreviations mean, respectively: very low, low, medium, high, very high. The exact meaning of these terms has to be clarified with the experts each time, here we assume that VL means the possibility of 10% of time overrun with respect to the most probable scenario and the other percentages are, respectively, 25%, 50%, 80% and 100%.

On the basis of expert opinion at the project level, the basic estimations $\{\mathrm{TB_i}\}_{\mathrm{i=1}}^{\mathrm{n}}$ are corrected in such a way that buffers $\{BF_{\mathrm{i}}\}_{\mathrm{i=1}}^{\mathrm{n}}$ are added to the activity duration estimations.

In the second step, other levels, higher than the project level, are taken into account. In [1] and [2], where construction projects are considered, the other levels are: the construction market level and the macro level. In other projects we may consider the situation of the team, company department, the company itself, the region, the branch, the country etc. For simplicity reasons we assume here just one additional level. On this higher level we identify *m* risks: $\{R_i\}_{i=1}^{m}$.

These risks are assumed to take on values for set *D* defined above. Each of these risks may be correlated with each of the project activities. We ask experts to determine the correlation level. Here we assume that there are three possibilities: no correlation (thus no influence of the risk on the higher level on the activity duration), a high (H) correlation (the risk on the higher level reinforces considerably the risk of a longer duration on the activity level) and a very high (VH) correlation (the influence of the risk on the higher level on the risk of activity longer duration is still stronger).

In Table 1 we propose a linguistic expression of this influence. Of course, in each case the linguistic expression may be different and is given by experts.

Table 1. Proposal of translating the influence of the risk degree on a higher level on the risk linked to a project activity depending on the correlation between the higher level risks and the project activities.

	Risk at higher level **VL**		Risk at higher level **L**		Risk at higher level **M**		Risk at higher level **H**		Risk at higher level **VH**	
Risk at activ. level **VL**	C. **H**	C. **VH**	C. **H**	C. **VH**	C. **H**	C. **VH**	C. **H**	C. **VH**	C. **H**	C. **VH**
	R. A.L. **VL**	R. A.L. **L**	R. A.L. **M**	R. A.L. **H**	R. A.L. **VH**	R. A.L. **VH**	R. A.L. **VH**	R. A.L. **VH**	R. A.L. **VH**	R. A.L. **VH**
Risk at activ. level **L**	C. **H**	C. **VH**	C. **H**	C. **VH**	C. **H**	C. **VH**	C. **H**	C. **VH**	C. **H**	C. **VH**
	R. A.L. **L**	R. A.L. **M**	R. A.L. **H**	R. A.L. **VH**	R. A.L. **VH**	R. A.L. **VH**	R. A.L. **VH**	R. A.L. **VH**	R. A.L. **VH**	R. A.L. **VH**
Risk at activ. level **M**	C. **H**	C. **VH**	C. **H**	C. **VH**	C. **H**	C. **VH**	C. **H**	C. **VH**	C. **H**	C. **VH**
	R. A.L. **M**	R. A.L. **H**	R. A.L. **VH**	R. A.L. **VH**	R. A.L. **VH**	R. A.L. **VH**	R. A.L. **VH**	R. A.L. **VH**	R. A.L. **VH**	R. A.L. **VH**
Risk at activ. level **H**	C. **H**	C. **VH**	C. **H**	C. **VH**	C. **H**	C. **VH**	C. **H**	C. **VH**	C. **H**	C. **VH**
	R. A.L. **H**	R. A.L. **VH**	R. A.L. **VH**	R. A.L. **VH**	R. A.L. **VH**	R. A.L. **VH**	R. A.L. **VH**	R. A.L. **VH**	R. A.L. **VH**	R. A.L. **VH**
Risk at activ. level **VH**	C. **H**	C. **VH**	C. **H**	C. **VH**	C. **H**	C. **VH**	C. **H**	C. **VH**	C. **H**	C. **VH**
	R. A.L. **VH**	R. A.L. **VH**	R. A.L. **VH**	R. A.L. **VH**	R. A.L. **VH**	R. A.L. **VH**	R. A.L. **VH**	R. A.L. **VH**	R. A.L. **VH**	R. A.L. **VH**

The dictionary in Table 1 is only a proposal of taking into account the influence of the higher project level risks on the project activities and their risk of time overrun. Table 1 is to be read as follows:

- Abbreviation R.A.L means "Risk on activity level after taking into consideration the influence of the higher level risk"
- Abbreviation C. means "Correlation between the risk on the higher level and the risk at the activity level"
- If e.g. risk at activity level before the correction taking into consideration the influence of the higher level risk (the first cell of the 2nd row) is low (L), the risk at the higher level is also low (2nd column), however, the correlation between the two is very high (VH), we assume that the corrected risk at the activity level is very high – we assume here that the low risk at the higher level, highly correlated with the low risk at the activity level, gives a reinforced, thus a very high risk at the activity level (which is only an assumption, it might be estimated to be low or high too, depending on the experts).

Taking into account Table 1, the risk degree for each activity is corrected, assuming the most pessimistic result from those obtained for the given activity for all the *j*=1,..,*m*. Then the corrected buffers $\{CBF_i\}_{i=1}^{n}$ are inserted into the project network. The resulting network gives a much better estimation possibility of the project time overrun and may be the starting point for the construction of schedules with buffers, using such methods as those presented in [3] or [4].

3. Example

We consider the following example: the project consists of four activities A_1, A_2, A_3, A_4. There are four precedence relations: A_2 and A_3 can start only once A_1 is finished and A_4 can start only once both A_2 and A_3 are finished. Also the basic duration time estimations for individual activities ($TB_1 = 3, TB_2 = 5, TB_3 = 3, TB_4 = 4$) are given.

Taking only the basic estimations of the duration time into account, for the most probable scenario, we will get the estimation of the project completion time equal to the longest path in the project network (the critical path) equal to 12 time units. However, we assume that the project manager has estimated the risk of a longer durations linked to each activity: $RD_1 = VH, RD_2 = H, RD_3 = M, RD_4$=*L*. Using the percentages assumed for those linguistic values in Section 2, we insert into the project network buffers ($BF_1 = 3, BF_2 = 4, BF_3 = 1{,}5, BF_4 = 1$). Now the estimated project duration (when the activities duration times are taken to be equal to $TB_i + BF_i, i = 1, \dots, n$) is 20 time units. This is an estimation of the project duration based on the knowledge of the risk on the project level only.

Let us now assume that project manager wants to take into account also the information about risks on higher levels and that four risks on one higher levels have been identified: R_1, R_2, R_3, R_4. Also, on the basis of expert opinions we assume that A_1 correlated with R_1 and R_3 to a high degree and with R_4 to a very high degree, A_3 is correlated with R_1 to a very high degree and with R_3 and R_4 to a high degree and A_4 is correlated with R_1 to a high degree. A_2 is correlated to no risk from the higher level.

Taking into account Table 1 and assuming the maximal corrected risk for ach activity, we correct the values of the risks of a longer duration for each activity to the following values: $CRD_1 = VH, CRD_2 = H, CRD_3 = VH, CRD_4$=M. When we base our calculations of the activity duration times $TB_i + CBF_i, i = 1, \dots, n$, we get another estimation of the total project completion time (21 time units) and a reduced float of activity A_3. This a more reliable estimation of the project duration, as it takes into account also the situation on higher levels than the project and its team itself.

4. Conclusions

A new method of project risk estimation, taking into account the influence of higher levels (the project environment in the more or less wide meaning) on the project itself, and based on linguistic values, which are easy to use on practice, is proposed. The method has been presented as focused on time overrun risk, but could be also applied to cost overrun risk. Experiments on real world projects are necessary to validate the method.

References

1. D. Skorupka, Identification and Initial Risk Assesment of Construction Projects on Poland, *Journal of Management in Engineering*, ASCE, vol. 3, p.120-27 (2008).
2. D. Skorupka, 2009.Method of planning construction projects, taking into account risk factors, *Operations Research and Decisions*, vol. 3, p. 119-128.
3. E. M. Goldrat, *Critical Chain*, North River Press, Great Barrington, 1997.
4. Kuchta D., Kobylański P. A note on the paper by M. A. Al-Fawzan and M. Haouari about a bi-objective problem for robust resource-constrained project scheduling, *International Journal of Production Economics*, 107(2), s.496-501, 2007.
5. E. Ptaszyńska Ewa, D. Kuchta, The concept of system supporting risk management in European projects, w: Zofia Wilimowska et al. (red.),

Informations Systems Architecture and Technology: Information as the Intagible Asset and Company Values Source, Oficyna Wydawnicza Politechniki Wrocławskiej, Wrocław, 2011, s.43-52;

6. P.S. Royer, *Project Risk Management*, Management Concepts, Inc., Vienna, Virginia, 2002
7. Standish Group, *CHAOS Report*, West Yarmouth, Massachusetts, Standish Group, 2004.

RISK ASSESSMENT OF THE SPANISH NATIONAL RAILWAY SYSTEM*

A. MATEOS, A. JIMENEZ, H.J. PICHARDO

Departamento de Inteligencia Artificial, Universidad Politécnica de Madrid (UPM), 28660 Boadilla del Monte, Madrid, Spain

The principal risks in the railway industry are mainly associated with collisions, derailments and level crossing accidents. An understanding of the nature of previous accidents on the railway network is required to identify potential causes and develop safety systems and deploy safety procedures. Risk assessment is a process for determining the risk magnitude to assist with decision-making. We propose a three-step methodology to predict the mean number of fatalities in railway accidents. The first is to predict the mean number of accidents by analyzing generalized linear models and selecting the one that best fits to the available historical data on the basis of goodness-of-fit statistics. The second is to compute the mean number of fatalities per accident and the third is to estimate the mean number of fatalities. The methodology is illustrated on the Spanish railway system. Statistical models accounting for annual and grouped data for the 1992-2009 time period have been analyzed. After identifying the models for broad and narrow gauges, we predicted mean number of accidents and the number of fatalities for the 2010-18 time period.

1. Introduction

As in all types of transportation there are risks we have to face in the railway networks. The occurrence of an accident in railways infrastructures has a very low probability, but when an accident occurs it gets a lot of attention[3]. Train accidents sometimes lead to a number of casualties and injuries and entail substantial financial costs, such as damage to equipment, increased insurance premiums, legal costs, fines, compensations, and loss of company reputation.

Developing safety systems and deploying safety procedures requires an understanding of the nature of previous accidents on the railway network to identify potential causes.

Many of the railway risk assessment techniques currently used are comparatively mature tools. The results of using these tools are heavily reliant

* The paper was supported by the projects S2009/ESP-1685, TIN 2008-06796-C04-02 and MTM2011-28983-C03-03.

on the availability and accuracy of the risk data[1]. However, data are usually incomplete or there is a high level of uncertainty involved in such data[1].

Most lines in Europe and North America were built to the so-called *standard* or *international gauge* (1435 mm), although there are some exceptions where broader gauges were chosen, like Spain and Portugal (1668 mm). Narrow gauge lines have also been built all around the world, mainly in mountainous areas or for branch lines. The total length of the broad and narrow gauge networks in Spain in 2009 was 13,354 km and 1,2691 km, respectively.

According to EU Safety Directives, EU member states have established independent national accident investigation bodies. In Spain the task was assigned to the *Railway Accident Investigation Committee* (CIAF). CIAF publishes both the *common safety indicators* and the accident reports on its website and in annual reports.

We propose a methodology that helps to improve the safety of railway systems by reducing the risk of fatal accidents and providing for the possibility of applying new strategies.

2. Methodology

2.1. *Phase 1. Prediction of the Mean Number of Accidents*

The number of accidents is predicted by fitting statistical models to historical data. We have analyzed generalized linear models where the dependent variable is linearly related to the independent variables via a specified link function. These models allow for the dependent variable to have a non-normal distribution such as a Poisson, negative binomial, gamma[5] distribution and COM-Poisson[4], while the most commonly used link functions are log and power.

Parameter estimates are obtained using the principle of maximum likelihood. Selecting the model that best fits to data depends on the following goodness-of-fit statistics: *deviance*, *Pearson* χ^2, *log-likelihood*, *Akaike's information criterion* (AIC), *finite sample corrected AIC* (AICC), *Bayesian information criterion* (BIC) and c*onsistent AIC* (CAIC).

Original annual data and three-year grouped data accounting for broad and narrow gauges were originally considered. Generalized linear models accounting for *log* and *power* functions were analyzed. We found that most goodness-of-fit statistics were better for both broad and narrow gauge when data were grouped, see Table 1.

Thus, the statistical models that best fitted three-year grouped data were used to predict the mean number of accidents. Specifically, Poisson Power -1 model was selected for broad gauge, see Table 2, while the Gamma Power -1 was the best one for narrow gauge.

Table 1. Three-year period grouped data.

Years	Broad Gauge			Narrow Gauge		
	Train-kms	Accidents	Fatalities	Train-kms	Accidents	Fatalities
1992-1994	496.5	421	93	22.7	56	41
1995-1997	485.6	304	80	24.1	62	26
1998-2000	501.9	203	58	29.3	96	20
2001-2003	519.7	170	84	30.0	46	20
2004-2006	524.8	137	138	29.8	16	17
2007-2009	540.2	135	135	29.6	16	13

Table 2. Goodness-of-fit statistics for grouped data and broad gauge.

Goodness-of-fit	Poisson Log	Poisson Pow-1	Negative Binomia l Log	Negative Binomia l Pow-1	Gamma Log	Gamma Pow-1	COM-Poisso n
Deviance	30.0	22.4	0.6	0.5	0.6	0.5	34.3
Pearson χ^2	30.8	22.1	0.6	0.5	0.6	0.5	35.2
Log Likel.	69.6	65.8	94.5	94.4	70.8	69.3	70.9
AIC	145.2	137.6	195.0	194.9	149.6	146.5	149.7
AICC	146.9	139.3	196.7	196.6	152.6	149.6	152.8
BIC	147.9	140.3	197.6	197.5	153.1	150.1	153.3
CAIC	150.9	143.3	200.6	200.5	157.1	154.1	157.3

Regarding **broad gauge**, based on the Poisson Power – 1 model, predictions for the next 19 years (2010-2018) can be computed as follows: $E(y_{1t}) = (0.00108867013927t + 8.764125618932E\text{-}0.6tkm_t - 0.00306052078)^{-1}$, where t is the time period and broad gauge train-kilometers for the 2010-18 period have to be previously estimated to predict the number of accidents in such period. A linear regression model from 1992-2009 data was used to estimate train-kms in 2010-2018: $tkm_t = 10.1114285742857142 85t + 476.06$.

For **narrow gauge**, the selected Gamma Power -1 model is used to make predictions for the number of accidents in the coming 9 years: $E(y_{2t}) = (0.0177103395923t - 0.00597228124517tkm_t + 0.132157197445)^{-1}$, where narrow gauge train-kilometers for the 2010-18 period is again previously estimated assuming a linear regression model applied to 1992-2009: $tkm_t = 1.488171428571t + 22.3914$, where t is the time period.

Table 3 shows the corresponding estimation for train-kilometers and the prediction of the mean number of broad and narrow gauge accidents for the

2010-2018 period, respectively. Finally, to predict the total number of accidents in the 2010-18 period we just have to add the predictions for broad and narrow gauges in Table 3.

Table 3. Prediction of the mean number of accidents.

Years	Broad Gauge		Narrow Gauge	
	Train-kms (millions)	Accidents	Train-kms (millions)	Accidents
2010-2012	546.8	106.9	32.8	16.6
2013-2015	557.0	95.0	34.3	14.5
2016-2018	567.1	85.4	35.8	12.8

2.2. *Phase 2. Computation of the Mean Number of Fatalities per Accident*

The mean number of fatalities per accident can vary hugely depending on the characteristics of the system. Therefore, it might be worthwhile finding out how to reduce the number of fatalities, which we do by changing some characteristics. Suppose that we classify the accidents in r groups, according to the characteristics we have determined. Then, c_j is the mean number of fatalities per accident given characteristic j, $j = 1 \ldots r$. Within the j group there could be other sub-characteristics. Suppose we have r_j subgroups within group j. Within each group r_j there could be other subgroups, and so on. The mean number of fatalities per accident in period t is defined as $c_t = \sum_{j=1}^{r} p_{jt} c_{jt}$, where p_{jt} represents the proportion of a selected characteristic that belong to a group of characteristics j in period t and $c_{jt} = \sum_{k=1}^{r_j} p_{jkt} c_{jkt}$, where p_{jkt} is the proportion of the selected characteristic that belong to subgroup jk in period t. c_{jkt} follows the same process, and more c can be calculated until there are no more subgroups to consider. If there are not more subgroups to consider we propose two possible methods to calculate c:

1. If all data are considered equally significant and reliable, then c_j (c_{jk}) is the mean number of fatalities per accident given the selected characteristic in group j (jk) in period t.
2. Recent data are more representative and significant than the older information. With this assumption, c_j (c_{jk}) is computed as a weighted mean of the number of fatalities per accident in group j (jk) in period t. We used the *rank-order centroid function*[2] to weight the values.

Now, we estimate the mean number of fatalities per accident in broad and narrow gauge, denoted by c_1 and c_2 respectively. We consider three possibilities:

- c_1 and c_2 are the mean number of fatalities per broad and narrow gauge accident in the 1992-2009 period ($c_1 = 0.42919708$, $c_2 = 0.469178082$).

- c_1 and c_2 are the weighted mean number of fatalities per broad and narrow gauge accident in the 1992-2009 period. The centroid function was used to weight the different years, i.e., recent years are assigned a higher weight than years further in the past ($c_1 = 0.692831966$, $c_2 = 0.52795136$).
- c_1 and c_2 are computed in the 2004-2009 period, in which accident investigations were more exhaustive ($c_1 = 1.003676471$, $c_2 = 0.9375$).

The mean number of fatalities per accident, c_t, is computed as follows: $c_t = p_t\, c_1 + (1 - p_t)\, c_2$, with p_t as the proportion of broad gauge train-kilometers.

2.3. *Phase 3. Prediction of the Mean Number of Fatalities*

The total number of fatalities, f_t, in the 2010-2018 period are estimated using the following model: $f_t = E(y_t)c_t$. The total predicted number of fatalities per 3-year period is shown in Table 4.We have noticed broad gauge is safer than narrow gauge, and narrow gauge train-kilometers are expected to grow. If we reduce their proportion by 1/1000 and take this to be the proportion for broad gauge train-kilometers, the results would be as follows, see Table 4.

Table 4. Prediction of the mean number of fatalities.

	Mean number of fatalities (f_t)			f_t applying changes		
	Mean	Weighted mean	Mean for the last 6 years	Mean	Weighted mean	Mean for the last 6 years
2010-2012	53.3	84.4	123.5	51.3	81.5	119.2
2013-2015	47.2	74.8	109.4	45.2	71.8	104.9
2016-2018	42.4	67.1	98.2	40.4	64.2	93.7

As we can see, the proposed strategy reduces the number of fatalities. With the different methods, i.e., weighted mean, mean and mean for last six years, we reduce fatalities by a total of 9, 6 and 13 respectively. We consider the last method to be the best for the reasons explained above. The final decision has to be made by the experts, who are the ones to evaluate whether or not the cost of such changes are worth it.

3. Conclusions

We have proposed a methodology to predict the mean number of fatalities in railway network accidents, based on three phases. The methodology has been applied on the Spanish railway system on the basis of annual and grouped data for the 1992-2009 time period. In both cases, we made a distinction between broad and narrow gauges.

The proposed methodology could be a starting point for a more complete analysis of the Spanish railway network, accounting for other variables apart from gauge, since CIAF has conducted fuller investigations in Spain from 2004 on wards. Variables such as the gauge traffic level, manual or automatic block options or centralized traffic control can add meaningful analyses in the future.

References

1. M. An, A Review of Design and Maintenance for Railway Safety the Current Status and Future Aspects in the UK Railway Industry, *World Journal of Engineering* 163, 3, 99-119 (2005).
2. Barron, F., Barret, B. Decision Quality Using Ranked Attribute Weights. Management Science 42 (11), 1515-1523 (1996)
3. A.W. Evans, Fatal Train Accidents on Europe's Railways: 1980-2009, *Accident Analysis and Prevention* 43, 10-23 (2011).
4. D. Lord, S. Guikema and S. Geedipally, Application of the Conway-Maxwell-Poisson Generalized Linear Model for Analyzing Motor Vehicle Crashes, *Accident Analysis and Prevention* 40, 1123-1134 (2008)
5. R. Winkelmann and K. Zimmermann. Recent Developments in Count Data Modeling: Theory and Applications, *J. of Economy Surveys* 9, 1-24 (1995).

PERFORMANCE EVALUATION OF LOT SIZING STRATEGY VIA DISCRETE EVENT SIMULATION

SERDAR BAYSAN

Department of Industrial Engineering, Istanbul Technical University, Istanbul, 34367, Turkey

ÇAĞATAY IRIS[†]

Department of Industrial Engineering, Istanbul Technical University, Istanbul, 34367, Turkey

MEHMET MUTLU YENISEY

Department of Industrial Engineering, Istanbul Technical University, Istanbul, 34367, Turkey

Lot Sizing strategies are based on determining periods where production will take place and quantities to be produced in order to satisfy demand pattern imposed to system as parameter. Most of the studies in the literature aim to minimize total cost of lot sizing strategies. It is very well-known that related type of problem structure is very hard to solve optimally. Hence, a number of heuristic approaches are proposed to find a near optimum solution for pre-determined problem characteristics. In this study, one of recent algorithms proposed for problem is evaluated in respect of robustness. The aim of this paper is to test whether algorithmic structure of lot sizing strategy is adequate to fully characterize fluctuations in parameters. A discrete-event simulation model is established to understand performance outcomes of proposed strategy.

1. Introduction

The capacitated lot sizing problem (CLSP) pertains to the efficient utilization of resources, and provides an economic production schedule with pre-determined parameters. The complexity theory proofs that Capacitated Lot Sizing Problem with Overtime decisions (CLSPO) and extensions derived from fundamental structure are NP-hard problems. Heuristics proposed for dynamic lot sizing problem satisfies inventory balance and capacity constraints [1].

The aim of this study is to test a pure lot sizing heuristic under parameter variability and internal fluctuations. Proposed algorithmic strategy neglects

† ciris@itu.edu.tr

presence of stochasticity in input parameters. Comparison is based on discrete event simulation with factorial design approach. Robustness is tested by considering not only external parameters, but also, algorithmic fluctuations. It is very well known that simulation model is directly triggered by probability distributions (pdf) of input parameters while heuristic algorithm works out with instances (random numbers) derived from pdf.

Discrete event simulation has been extensively used to evaluate performance of planning activities. Habenicht and Mönch have established a comparison structure about batching strategies in a multi-product waterfab via simulation. It is emphasized that increasing number of batches in one family results in worse performance in overall costs. [2]

2. Sources of variability in CLSP heuristic

In proposed CLSP heuristic, there are three sources of variability. One of most influential variability is demand pattern of products. Most of the studies make assumption of normal distribution with predefined parameters. Standard deviation of demand (*STD*) is widely used to test robustness of algorithm. The ratio of average production cost to average holding cost is another source (*C/H*). Due to the fact that, algorithms involve trade-off between production and inventory cost, it is very crucial to define C/H ratio which reflects different product characteristics. There are also some other sources of volatility which depends on metaheuristic used by algorithm. These strategies are used to reach global optima. Metaheuristic algorithms use RND for threshold controlling, crossover, mutation to embed stochasticity in decision making procedure. [3]

Iris and Yenisey [4] is tested by using simulation. Main feature of algorithmic variability is simulated annealing embedded. Global search procedure is capable to improve given initial solution within the feasible solution. The philosophy maintained in improvement heuristic is transferring some of lots from one period to another. The flow of algorithm is as follows;

Step 1: Find periods where lot transfer take place using dominance property

Step 2: After determining periods, identify product that will be backward shifted by using holding and production cost savings

Step 3: Calculate maximum possible amount of lot transfer, determine lot size to be shifted randomly

Step 4: Transfer required production lot between periods, control overall cost change in objective function. If the change in cost is negative, then apply move. Otherwise, calculate simulated annealing parameters

Step 5: If move is rejected, initialize forward shifting procedure, apply step 2 and 3 iteratively, and transfer an amount of production lot to forward periods.

The performance of algorithm is evaluated by using performance measure in Table 1. (Z is the objective function value of given alternative (H: Heuristic, O: Optimal))

Table 1. Results for Class A (4x4) problems and deviation from optimality

	Mean	Std. Dev.	Parameters
GAP_1	3,58	2,85	$[(Z_H-Z_O)/Z_O]$*100 (distance of heuristic from optimal)

3. Robustness tests

Optimality gap is obtained by using 4 items and 4 periods planning horizon. It is very well-known that robustness test may be applied to algorithms which are proved as well performing on optimality gaps. For robustness structures, input parameters variability and algorithmic variability is considered separately.

3.1. *Input factors variability*

Demand pattern of items in planning system is reflected as different levels of coefficient of variation Table 2. gives information about obtained data.

Table 2. Demand fluctuation tests (Three levels of CV)

	Replication	Demand Mean	Demand STD	C/H Mean	C/H STD	Min COST	Iteration number	Avg. Delta_max	Avg. Delta
Scen. 1	50	140	70	7,0	3,0	114628	43,1	53,28	26,331
Scen. 2	50	140	140	7,0	3,0	122992	51,02	55,671	27,929
Scen. 3	50	140	210	7,0	3,0	140390	42	57,939	29,333

It can be understood that increasing values of coefficient of variation (CV) results in higher overall costs. Adversely, number of iteration until termination is fluctuating with higher levels of demand CV. This is quite sensible, because algorithm moves freely between periods with higher fluctuation in data. Additionally, this helps not get stuck in local optima. Average amounts of shifts are also increased with higher fluctuation in demand. For related analysis, three levels of CV are set as; [0.5; 1; 1.5]. Each replication of 50 trials is done by using randomly generated data.

Table.3 describes C/H volatility. Scenario 1 has equal holding and production costs with 0.2 standard deviation. Number of iterations to find solution is very small. This helps us to understand whenever fluctuation with cost parameters are decreased, computational solution quality is increased.

Scenario.2-9 present c/h ratio fixed to seven while standard deviation is changed. It can be interpreted as; increasing values of standard deviation in c/h, results in lower overall cost (summation of inventory holding, overtime, regular time, production costs). This phenomena is quite interesting, although given fluctuation in parameters is increased cost associated is decreased. This is because of capability of algorithm to reflect cost associations with dominance properties proposed. On the other hand, iteration number to find near optimum solution is increased whenever; standard deviation of c/h is increased. This is because of the enlargement of search space which contains optimum results. Average amount of lot which is transferred between periods seems to have a stabilized (very slowly decreasing) characteristic against fluctuations in c/h cost ratio. Final scenario is based on having products whose holding cost is very higher than production costs. For these types of products, iteration number and cost associated is the least of overall test set. Because, ending inventory in each period is most commonly less than production costs. Additionally, algorithmic structure has an efficient search potential to fully reflect variation in holding and production costs in lower levels. For this data set, demand is fixed with optimum performing demand pattern.

Table 3. C/H fluctuation tests (Three levels of Mean and STD)

	Replication	Demand Mean	Demand STD	C/H Mean	C/H STD	Min COST	Iteration number	Avg. Delta_max	Avg. Delta
Scen. 1	50	140	70	1,0	0,2	42486	4,06	50,242	25,202
Scen. 2	50	140	70	7,0	7,0	84513	60,52	53,754	26,702
Scen. 3	50	140	70	7,0	6,0	88704	50,4	53,99	26,874
Scen. 4	50	140	70	7,0	5,0	92992	52,06	53,158	26,654
Scen. 5	50	140	70	7,0	4,0	99732	46,58	52,15	26,497
Scen. 6	50	140	70	7,0	3,0	114628	43,1	53,28	26,331
Scen. 7	50	140	70	7,0	2,0	134888	38,16	52,583	26,01
Scen. 8	50	140	70	7,0	1,0	155347	20,06	51,493	25,497
Scen. 9	50	140	70	7,0	0,0	166141	13,84	51,383	25,632
Scen10	50	140	70	0,1	0,0	24486	1,76	49,611	24,953

After focusing on input based fluctuation, related analyses should be made according to changes due to algorithmic structure. It is very well known that delta amount that is shifted between periods is determined randomly between zero and Delta_max. Determination of Delta value is tested for different types of probability distributions.

3.2. *Algorithmic variability*

Determining the amount of lot to transfer backward or forward between periods is key movement of algorithm proposed by authors. Hence, focusing on sensitivity of results based on decisions given in shifting phase is crucial. Algorithm is simulated in four different levels for distribution characteristics of Delta calculation. The one which is uniform(both), right triangular to Delta_max (forward-backward), right triangular to zero (both shifting direction), and lastly right skewed right triangular for forward shifting and left skewed right triangular for backward shifting. Related tests are done for different levels of demand patterns. Firstly, Table 4. represents a mean c/h ratio of 7 with 3 standard deviation (selected among best scenarios). Related analysis results are shown in Figure 1 with their standard deviations with respect to minimum costs.

Table 4. Delta determination test (Four different probability distributions)

	Delta_pattern	Demand STD	C/H Mean	C/H STD	Min COST	Iteration number	Avg. Delta_max	Avg. Delta
Scen. 1	delta_unif_unif	70	7,0	3,0	114628	43,1	53,28	26,331
Scen. 2	delta_tria_tria_max	210	7,0	3,0	140771	38,12	44,537	29,668
Scen. 3	delta_tria_tria_min	70	7,0	3,0	116956	59,38	66,185	21,777
Scen. 4	tria_min_tria_max	210	7,0	3,0	140771	38,12	44,537	29,459

According to test results, best operating conditions with heuristic algorithm can't be determined exactly. Because, there is an obvious tradeoff between overall cost and iteration number. It is also anticipated that minimum number of runs to obtain these results corresponds to Scenario 2 & 4. For scenario 2, average Delta_max is maximized due to the fact that algorithm iterative shifts lots that are with minimal edge of probability distribution. Figure1. illustrates all of tested results of analysis. It can be interpreted easily with the values of minimum costs and their standard deviation as box plot. Latest scenario set with least c/h ratio have minimum min. cost values. Standard deviations in Figure 1. proofs that minimum cost results moves accordingly with related parameters. Whenever, standard deviation of parameters is increased, standard deviation of minimum cost is also increased in 50 replications. It is also observed that most of symmetric structures in statistics set are preserved.

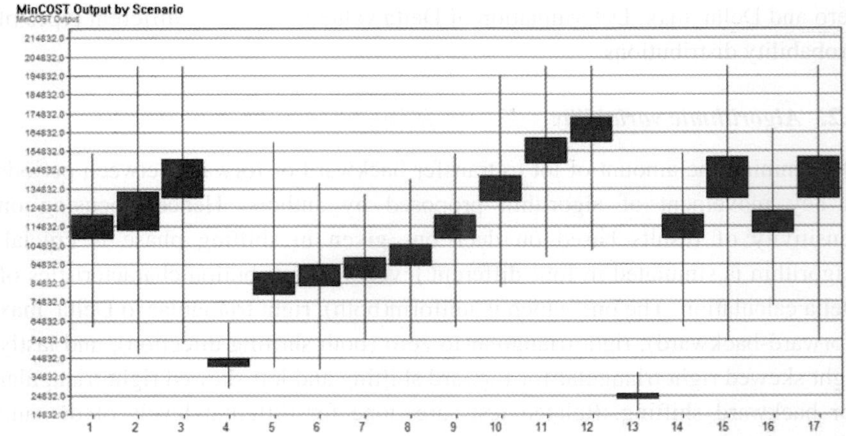

Figure 1. Minimum cost values of robustness test

4. Results and Future Research

In this study, we have focused on robustness test of a proposed heuristic on capacitated lot sizing problem. It is found out that fluctuation due to input variables or algorithm directly effect obtained results. These validations help us to understand characteristics of problem better. Discrete event simulation is used to run related heuristic, and variability is embedded in simulation structure. For future studies, it is planned to integrate aspects that are not covered by algorithm to simulation model (e.g. sequence dependent setup times, backlogs, scheduling etc.). Hence, simulation could help to understand performance of related production environments without including them in algorithm flow.

References

1. D. Quadt, H. Kuhn, "Capacitated Lot Sizing with Extensions". *4OR,* No.(6), pp. 61-83, (2008).
2. I. Habenich and L. Mönch, "Evaluation of Batching Strategies In A Multi-Product waferfab by Discrete-Event Simulation", *16th European Simulation Symposium*, (2004).
3. L. Ozdamar, G. Barbarosoğlu, "An integrated Lagrangean relaxation-simulated annealing approach to the multi-level multi-item capacitated lot sizing problem". *International Journal of Production Economics*, No. 68, pp. 319-331, (2000).
4. Ç. Iris, M. Mutlu Yenisey, "A Two-stage Metaheuristic Approach to Multi-item Multi-period Capacitated Lot sizing with Overtime Decisions", *APMS*, Norway (2011).

PART 7

APPLIED RESEARCH

A HEURISTIC APPROACH FOR PROVIDER SELECTION AND TASK ALLOCATION MODEL IN TELECOMMUNICATION NETWORKS UNDER STOCHASTIC QOS GUARANTIES*

HASAN HUSEYIN TURAN
Department of Industrial Engineering, University of Yalova, Merkez
Yalova, 77100, Turkey

NIHAT KASAP
School of Management, Sabancı University, Tuzla
İstanbul, 34956, Turkey

In this study, we model provider selection and task allocation problem as an expected cost minimization problem with stochastic chance constraints. Two important parameters of Quality of Service (QoS), delay and jitter are considered as random variables to capture stochastic nature of telecom network environment. As solution methodology, stochastic model is converted into its deterministic equivalent and then a novel heuristic algorithm is proposed to solve resulting nonlinear mixed integer programming model. Finally, performance of solution procedure is tested by several randomly generated scenarios.

1. Introduction

We consider an environment in which the firm can acquire network capacity with different service qualities at different prices in order to complete its daily tasks [1]. Day-to-day operations such as video conferencing, voice over IP and data applications are allocated between these acquired capacities by considering QoS requirement of each operation. For telecommunication networks QoS offered by providers are measured in terms of delay, jitter, lost rate and latency [2]. Specifications of these measures are described in Service Level Agreements (SLA) [3]. In this paper, two important parameters of QoS, delay and jitter are considered as random variables to capture stochastic nature of telecom network environment. The resulting optimization problem is nonlinear mixed integer problem with probabilistic constraints which is considered as NP-hard. Therefore, a heuristic algorithm capable of solving large problem instances is developed.

* This research is supported by the Scientific and Technological Research Council of Turkey (TÜBİTAK), Career Development Program (Grant No. 106 K 263).

The rest of the paper is organized as follows. Section 2 presents problem definition, notation and the proposed nonlinear mixed integer programming model with chance constraints [4, 5]. Heuristic algorithm is summarized in Section 3 to unravel the problem. Results of computational studies are also demonstrated in this section. Finally, Section 4 drowns conclusions and describes future goals respectively.

2. Stochastic and Deterministic Mathematical Models

We will use parameters and decision variables given in Table 1 while presenting formulations and solution procedures throughout the text. Each formulation uses a subset of these variables and parameters.

Table 1. Problem notations and parameters

Parameters	
I, J:	The ordered index set of resources, and tasks respectively.
A_T, A_S :	The index set of tasks with fixed transmission time and fixed size respectively where $A_T \cap A_S = \varnothing$
δ_j :	The maximum amount of delay tolerated by task j.
ρ_j :	The maximum amount of jitter tolerated by task j
D_i :	Amount of delay in resource i, which is a random variable and has a probability distribution $F^D{}_i(.)$
R_i	Amount of jitter in resource i, which is a random variable and has a probability distribution $F^R{}_i(.)$
α_i :	Transmission efficiency, calculated as one minus the packet loss rate of resource i.
φ_i :	The maximum allowed overflow probability to assign a task into resource i.
p_j^{δ} , p_j^{ρ} :	Confidence level parameters for delay and jitter respectively for task j.
β_i, L_i:	The bandwidth and duration of resource i. $L_i = min$ (length of contract, planning horizon)
c_i :	The total cost of resource i for specific β_i, L_i.
c_j^o :	The opportunity cost of missing the target transmission rate for task j.
c_j^{δ} :	The opportunity cost of not meeting minimum quality requirement for delay in task j.
c_j^{ρ} :	The opportunity cost of not meeting minimum quality requirement for jitter in task j.
q_i :	The unit overflow cost for going beyond the length of contract period for resource i
$\bar{r}_j^U$, $\bar{r}_j^L$:	Target and the minimum transmission rate of task j at the receiving node, respectively, $j \in A_T$.
Δt_j :	Estimated scheduled transmission time for time-fixed task $j \in A_T$.

$\bar{x}_j$:	The (fixed) length of task $j \in A_s$ in number of bits.
μ_i^D, σ_i^D :	Mean and standard deviation of probability distribution $F_i^D(.)$.
μ_i^R, σ_i^R :	Mean and standard deviation of probability distribution $F_i^R(.)$.
$\psi(a,b,c)$:	Return value of integral $\int_a^\infty x \frac{1}{\sqrt{2\pi c^2}} e^{-\frac{(x-b)^2}{2c^2}} dx$
$\phi(a,b,c)$:	Return value of integral $\int_a^\infty \frac{1}{\sqrt{2\pi c^2}} e^{-\frac{(x-b)^2}{2c^2}} dx$
Decision variables:	
v_i :	1, if resource i is selected 0, otherwise
r_j :	The transmission rate of task j.
y_{ij} :	1 if task j is assigned to resource i, 0 otherwise.
y_{ijt} :	1 if task j is active (transmitting) at time t on resource i, 0 otherwise
t_j :	Start time of task j.

2.1. *Stochastic Problem (SP) Formulation*

We model the capacity and the loss probability requirements explicitly, but formulate delay and jitter as random variables and added stochastic chance constraints associated with marginal probability distributions. We consider the all-you-can-send pricing scheme [1, 6] in which the firm pays a fixed price for a fixed bandwidth available for a fixed duration. We also assume that a real-time task incurs an opportunity cost if its transmission rate falls below a desired level.

$$\text{Minimize} \quad z = \sum_i c_i v_i + \sum_i \sum_{j \in A_T} \left(\bar{r}_j^U - r_j \alpha_i\right) c_j^o y_{ij} + \sum_i \sum_{j \in A_T \cup A_S} y_{ij} \Xi_{D_i}\left[c_j^\delta Max\left(D_i - \delta_j, 0\right)\right]$$
$$+ \sum_i \sum_{j \in A_T \cup A_S} y_{ij} \Xi_{R_i}\left[c_j^\rho Max\left(R_i - \rho_j, 0\right)\right] + \sum_i \sum_{j \in A_T \cup A_S} y_{ij} \Xi_{D_i}\left[q_i Max\begin{pmatrix} t_j + \Delta t_j + \\ D_i - L_i, 0 \end{pmatrix}\right] \tag{1}$$

$$\text{Subject to} \sum_{j \in A_s} \bar{x}_j y_{ij} + \sum_{j \in A_T} \alpha_i \Delta t_j r_j y_{ij} \le \alpha_i \beta_i L_i \quad \forall i \in I \tag{2}$$

$$\Pr\left\{y_{ij} D_i \le \delta_j\right\} \ge p_j^\delta \quad \forall j \in J, \forall i \in I, \quad 0 \le p_j^\delta \le 1 \tag{3a}$$

$$\Pr\left\{y_{ij} R_i \le \rho_j\right\} \ge p_j^\sigma \quad \forall j \in J, \forall i \in I, \quad 0 \le p_j^\rho \le 1 \tag{3b}$$

$$\Pr\left\{(t_j + \Delta t_j + D_i) y_{ij} \le L_i\right\} \ge \varphi_i \quad \forall j \in A_T, \forall i \in I \tag{4}$$

$$\sum_{j \in J} r_j y_{ijt} \le \beta_i \quad \forall i \in I, t = 1..L_i \tag{5}$$

$$\sum_{i\in I} y_{ij} = 1, \quad \forall j \in J \tag{6}$$

$$\sum_{i\in I}\sum_{t\le L_i} y_{ijt} = \Delta t_j, \quad \forall j \in A_T \tag{7}$$

$$\sum_{i\in I}\sum_{t\le L_i} r_j y_{ijt} = \overline{x}_j, \quad \forall j \in A_S \tag{8}$$

$$y_{ij} \le v_i \quad \forall i \in I,\ j \in J \tag{9}$$

$$y_{ijt} \le y_{ij}, \forall i \in I, j \in J, t \le L_i \tag{10}$$

$$r_j \alpha_i \ge y_{ij} \overline{r}_j^L \quad \forall j \in A_T,\ \forall i \in I \tag{11}$$

$$r_j \alpha_i y_{ij} \le \overline{r}_j^U \quad \forall j \in A_T,\ \forall i \in I \tag{12}$$

$$r_j, t_j \ge 0 \tag{13}$$

$$v_i, y_{ij}, y_{ijt} \in \{0,1\} \tag{14}$$

Objective function(1) of problem SP considers the tradeoff between the costs of acquiring capacity, the cost of not meeting target transmission rates in real-time tasks, expected opportunity cost for not meeting minimum quality specifications for delay and jitter, and expected overflow cost due to not completing transmission any task assigned within contract period. We assume that available capacity is purchased at a fixed price for a specific bandwidth and duration (all-you-can-send pricing). Constraint set (2) guarantees that we can use only up to available capacity. Stochastic Constraint sets (3a) and (3b) ensure that the resources satisfy the minimum QoS requirements of the tasks that are assigned to it within determined confidence limits. Constraint (4) ensures with φ_i probability that all time-fixed tasks assigned to resource *i* are completed within the contracted duration of resource *i*. Note that a similar constraint is not needed for size-fixed tasks since constraint set (8) guarantees that a size-fixed task is only assigned if there is enough capacity and can be completed on time since we can arbitrarily vary the transmission rate (i.e. morph its shape). Constraint set (5) prevents using more bandwidth than available at any time (bandwidth dimension). Constraint (6) along with (14) ensures that a task is assigned to only one resource and all tasks are assigned. Constraint sets (7) guarantee that the tasks are actually allocated the required amount of time slices. Constraint set (9) guarantees that a network resource is selected only if at least one task is assigned to it. Constraint set (10) ensures that a task is assigned to a network resource only if it occupies a time slice on it. Constraint set (11) states that transmission rate for a time-fixed task *j* should be high enough to satisfy the minimum transmission rate at the sink node. Constraint set (12) enforces the target transmission limitation for all tasks.

2.2. *Deterministic Equivalent of Stochastic Problem (DESP)*

Terms in expected form in Eq. (1) can be transferred into deterministic domain as following:

$$\Xi_{D_i}\left[Max(D_i-\delta_j,0)\right]=\Psi(\delta_j,\sigma_i^D,\mu_i^D)-\delta_j\,\phi(\delta_j,\sigma_i^D,\mu_i^D)\ \forall i\in I,\forall j\in J$$

$$\Xi_{R_i}\left[Max(R_i-\rho_j,0)\right]=\Psi(\rho_j,\sigma_i^R,\mu_i^R)-\delta_j\,\phi(\rho_j,\sigma_i^R,\mu_i^R)\ \forall i\in I,\forall j\in J$$

$$\Xi_{D_i}\left[Max(t_j+\Delta t_j+D_i-L_i,0)\right]=\Psi(L_i-t_j-\Delta t_j,\sigma_i^D,\mu_i^D)-(t_j+\Delta t_j-L_i)\,\phi(L_i-t_j-\Delta t_j,\sigma_i^D,\mu_i^D)\forall i\in I,\forall j\in J$$

All stochastic equations namely Eq. (3a), (3b) and (4) in SP can be converted in to linear constraints Eq.(15), (16) and (17), respectively:

$$y_{ij}\left(\frac{\delta_j-\mu_i^D}{\sigma_i^D}-Z_{p_j^\delta}\right)\geq 0,\ \forall i\in I,\forall j\in J \tag{15}$$

$$y_{ij}\left(\frac{\rho_j-\mu_i^R}{\sigma_i^R}-Z_{p_j^\rho}\right)\geq 0,\forall i\in I,\forall j\in J \tag{16}$$

$$y_{ij}\left(\frac{L_i-t_j-\Delta t_j-\mu_i^D}{\sigma_i^D}-Z_{\varphi_i}\right)\geq 0,\forall i\in I,\forall j\in J \tag{17}$$

3. Heuristic Solution and Computational Results

Pseudo-code of the heuristic algorithm (HDESP) for DESP is given in Figure 1 and initial results of algorithm are summarized in Table 2.

Sort tasks by descending ρ and δ, then **Sort** tasks by descending p^δ, p^ρ
    Sort above list by decreasing size of tasks
Sort resources by increasing unit cost $\bar{c} = c/(1-\alpha)L\beta$
While all tasks not assigned
    If Eqs. 13-15 is satisfied, assign task to resource
    Else open new resource and make assignments
End while
Sort selected resources by decreasing unused capacity
While enough unused capacity
    While new $z \leq z$
        Swap tasks among selected resources
    End While
Close unused resource
End While

Figure 1. Pseudo-code for HDESP

Table 2.Effect of Pricing Type on Cost Components

Pricing Type	Opportunity Cost	Quality Loss Cost	Capacity Leasing Cost	Total Cost
Random	94.34	35.96	449.11	579.41
Parallel cost curve	128.01	45.14	877.80	1050.95
Intersecting cost curve	129.11	47.51	1072.20	1248.82

4. Conclusion and Future Study

We presented a novel formulation to solve the firm's network resource acquisition problem subject to stochastic QoS parameters, opportunity and quality loss cost. We also proposed a heuristic to handle QoS requirements and chance constraints. The proposed approach is able to solve moderate size problem sets in reasonable time limits which is not possible with standard commercial mathematical programming softwares. As a next step, quality of solutions generated by HDESP will be compared with calculated lower bounds.

References

1. Kasap, N., Aytug, H. and Erenguc, S. S. Provider selection and task allocation issues in networks with different QoS levels and all you can send pricing", *Decision Support Systems*, 43, pp.375 – 38 (2007).
2. Pan W., Yu L., Wang S., Hua G., Xie G, and Zhang J., Dynamic Pricing Strategy of Provider with Different QoS Levels in Web Service, *Journal of Networks*, 4(4), pp.228-234 (2009).
3. Rana O., Warnier M. and Quillinan T. B. (2008), Monitoring and Reputation Mechanisms for Service Level Agreements, The Fifth International Workshop on Grid Economics and Business Models (GECON).
4. Charnes A. and Cooper W. W., Chance-constrained programming, *Management Science*, 6(1), pp.73-79 (1959).
5. Charnes A. and Cooper W. W., Deterministic equivalents for optimizing and satisfying under chance constraints, *Operations Research*, 11(1), pp.18-39 (1963).
6. Courcoubetis, C., Kelly, F. and Weber, R., Measurement-based usage charges in communications networks, *Operations Research*, 48 (4), pp.535–548 (2000).

A NEW ARCHITECTURE FOR ASSESSMENT OF MULTIPLE USERS IN COLLABORATIVE MEDICAL TRAINING ENVIRONMENTS BASED ON VIRTUAL REALITY[1]

RONEI M. MORAES*

Department of Statistics, Federal University of Paraíba
João Pessoa/PB, Brazil, ronei@de.ufpb.br

LILIANE S. MACHADO

Department of Informatics, Federal University of Paraíba
João Pessoa/PB, Brazil, liliane@di.ufpb.br

In collaborative systems (CS), procedures are performed by a team of professionals acting simultaneously, as in surgical rooms in medicine. Using Virtual Reality is possible to create virtual rooms that join people to simulate those medical procedures. In such cases, collaborative features are used to coordinate the interactions of the multiple participants in the environment. To verify if the group performed the procedure correctly, assessment architectures has been proposed. However, network latency can affect that assessment. This paper presents a new architecture for adaptive assessment systems for training based on virtual reality for multiple users, which is able to solve that problem.

1. Introduction

Training systems based on virtual reality (VR) are 3D environments composed of objects with topologies and behaviors similar to real objects, the interaction ways to deal with this world and the possibility of performing the training several times without risk or damage, among others [4]. Systems for different modalities in medicine have been developed [6, 8, 17]. The goal of most of these systems is to provide training, in which devices and techniques explore the human senses. However, most of them can provide training for only one user at a time. Some procedures need to be performed by more than one person, as those performed in surgery rooms. Then, it is necessary a team working together

[1] Work partially supported by grants 312375/2009-3 and 310339/2009-0 of the Brazilian Council for Scientific and Technological Development, CNPq. This work is related to the National Institute of Science and Technology for Medicine Assisted by Scientific Computing also supported by CNPq.
* email: *ronei@de.ufpb.br*

in a collaborative activity and in that one, the dexterity of each member can interfere in the group performance [11].

Collaborative systems (CS) based on VR for complex training have been designed for several purposes. Although the possibilities of training in these environments, they present little value if the trainees do not have any feedback about their performance. Then, the existence of an assessment tool attached to a simulation system based on VR is important to allow knowing users performance and help learning improvement. As example can be mentioned the assessment of psychomotor skills in VR systems that include haptic devices to quantify surgical dexterity with objective metrics [5].

Some systems can already provide methodologies to assess users' performance [1, 8, 10, 12, 17]. Those methodologies are used to assess only one user at a time. However, only three methodologies which monitor multiple users in complex training environments based on virtual reality can be found in literature [7, 11, 13]. All of them were proposed for assessment based on fuzzy methodologies using as input statistical measures. Two of them using local networks and did not taking into account delays in them.

In this paper is proposed a new architecture for multiple users' assessment in CS for any kind of network, which can use a set of fuzzy measures to perform those assessments. In the next section is presented a review of user´s assessment methodologies and a new architecture for multiple user´s assessment is proposed on Section 3. Section 4 brings some considerations about the new architecture and some advantages of the proposed approach.

2. Assessment in Virtual Reality Simulators

The first methodologies for Single User's Assessment System (SUAS) of training based on VR were recently proposed. In general, SUAS works coupled to a virtual reality simulator [9]. Off-line SUAS can be defined as methods which can be or can be not coupled to VR systems, whose assessment results are provided some time (which can be minutes, hours or days) after the end of the VR-based training. On the other hand, on-line SUAS are coupled to the training system and should be capable to monitor user's interactions and compare their performance with classes of performance previously defined by experts to recognize user's level of training. On-line SUAS should provide its assessment immediately at the end of the simulation.

The main differences of training systems based on VR for multiple users are: increase of complexity of the system (use of clusters of computers or a computer capable to generate realistic multiple views, support changes in virtual environments for multiple users and support assessment system) and high speed peer-to-peer or local network for communication among computers without compromising the simulation. Eventually, more than one interaction device (haptics and tracking, as example) must be available on each computer for each user in training. Observing those restrictions, some computational systems for multiple users have been developed [2, 7, 18]. Multi-user training on network requires different approaches for a MUAS: monitor all users in training according to relevant variables to the training; some tasks must be completed by specific users and according to a specific schedule; take measures of specific interactions among users during the time of simulation; take into account the length of assessment; the creation of user and group profiles; present high accuracy level [7].

Despite technological advances, the most common problem in CS (by local network or by Web) is still network latency [14] that may result in inconsistencies and unsynchronized views of the shared workspace. As consequence, can present difficulties to perform specific schedules with desired accuracy. In a real surgical room, this can cause patient's death. As mentioned before, after extensive literature research, only three MUAS has been found [7, 11, 13]. In all architectures, user and group interactions are monitored during collaborative training and their data are collected. In [7] a fuzzy expert system [16] uses as input statistical measures and models from data to create individual user and group profiles. In [11] was used a possibilistic network which use information from statistical measures, models and hypothesis testing from data. In [13] a fuzzy expert system uses as input statistical measures, models and hypothesis testing from data.

3. A New MUAS Architecture

Network latency is unpredictable and does not show constancy all the time. For this reason, an assessment system should be adaptive to consider different network latencies. Then, network latency should not affect the assessment when checking the execution of each user task and the global schedule of the procedure. As in SUAS, MUAS should monitor all variables of interest. However, unlike [7] and [11] designed for local networks, the new MUAS architecture should be decentralized to fulfill the assessment task, as in [13].

In the new MUAS, each user connected receives a copy of the training simulator. During the execution of the simulation, interaction data and scenario changes are exchanged among users. Each user has their own local assessment system (LAS) that locally monitor the users' interactions and perform a pre-assessment. The LAS communicate with the central assessment system (CAS) which receives the individual interactions. CAS must supervise the group activity and identify the time each user need to perform its task locally. This way, CAS can monitor all connections and determine when each user received and performed a movement. The CAS should also correct times of response in order to synchronize the interaction events discounting latency networks. The central system can synchronize users data to perform the final users' assessment and the group assessment from pre-assessment provided by the several LAS. The union of each LAS and the CAS is the new MUAS architecture. From that architecture, MUAS provide Individual and Group Assessment Reports.

The LAS is equivalent to the SUAS in a system with only one user. However, LAS has other tasks to fulfill: generate a report with the timetables of the receipt of system updates, user's position in space, user's interactions, duration of those interactions and a short local user assessment. In fact, LAS does not generate reports for the user. This local assessment is based on a fuzzy expert system [8], whose fuzzy rules are provided by an expert in the specific task, defined according to predefined classes of performance. This way, to assess multiple users in a collaborative task, several experts are necessary, each one with their own expertise. Each LAS contains fuzzy rules to assess specifically tasks and interactions of a single user and recognizes patterns of user's behavior when performing the procedure. The data used by LAS can be the user position in relation to patient and the interaction information, such as force, torque and position of surgical tools. Specific devices must be used to provide that information, such as tracking systems and haptic devices. From that information, fuzzy statistical measures, fuzzy models and fuzzy testing of hypothesis are calculated [3] and used as input for the fuzzy expert system, which analyzes that information to recognize individual user's level of training (Figure 1).

All information collected by the several LAS is sent to the CAS. The CAS should identify the relative positions of the users, time elapsed between the interactions of each user discounting the different latencies of the sub-networks present in the simulation. After that, it should be checked whether the procedures were performed within the correct time, and the correct sequence of tasks and if users do not stumbled into each other in the virtual environment must be observed. CAS should use those information to provide details about user's and group performance. Then, at the end of simulation, CAS will emit two types of report: an individual assessment report for each user and a group assessment report. The first one is about the individual user performance on the training and the second one is about group performance and the interactions among users during training. Those reports can be used to allow skills improvement [15] of individual and group performance in real tasks. The report with interactions among users can be used to correct and improve details in specific procedures, as sequential tasks, simultaneous tasks or collaborative tasks. These tasks are common in surgical rooms, particularly in cases when group's performance is essential for save or sustain patients life.

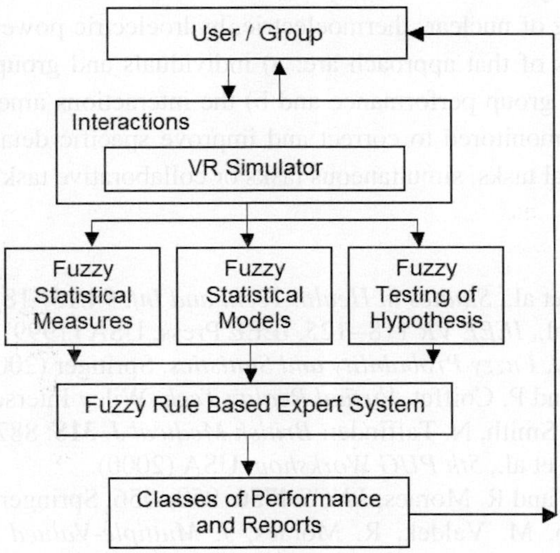

Figure 1. Diagram of new MUAS architecture.

This MUAS can be used for several kind of training in medicine, as procedures in surgical rooms, training of paramedics groups in emergency situations, etc. However, it is a generic architecture and can be used in training systems for other areas, as aeronautical simulators, maintenance group, etc.

4. Conclusions

This paper presents a new architecture for adaptive assessment systems for training based on virtual reality for multiple users (MUAS) with interactions among them. This approach solves the network delays problem, by computing the time used to perform a task after the reception of data. This methodology is based on fuzzy based rules expert system which uses input variables from training, fuzzy statistical measures, fuzzy models and fuzzy testing of hypothesis in order to create measures of assessment for trainees and group performance.

The MUAS can create two types of report: an assessment report for individual user performance, on the training, and an assessment report about group performance and interactions among users during the training. The MUAS can be used in training situations when a group of trainees performs a task together, as in surgical rooms.

This MUAS is adaptive, since it can deal with latency of the several networks present in the simulation. Its architecture is also generic and can be used in training systems for other areas, as aeronautical simulators, maintenance group, managing of nuclear, thermoelectric, hydroelectric power plant, etc. The main advantages of that approach are: a) individuals and group reports can be used to increase group performance and b) the interactions among users during training can be monitored to correct and improve specific details in procedure such as sequential tasks, simultaneous tasks or collaborative tasks.

References

1. M. Alcañiz et al., *Studies in Health Tech. and Inf.* **94**:16--18 (2003).
2. H. Baier et al., *IEEE VR* 118--125. IEEE Press, USA (1999).
3. J. J. Buckley, *Fuzzy Probability and Statistics*. Springer (2006).
4. G. Burdea and P. Coiffet, *Virtual Reality Tech.* Wiley Interscience (2003).
5. A. Darzi, S. Smith, N. Taffinder, *British Medical J.* **318**: 887--888 (1999).
6. L. Machado et al., *5th PUG Workshop*. USA (2000).
7. L. Machado and R. Moraes, *LNCS* **4756**: 950--956. Springer (2007).
8. L. Machado, M. Valdek, R. Moraes, *J. Multiple-Valued Logic and Soft Comp.* **14**: 511--524 (2008).
9. L. Machado and R. Moraes, *Comp. Intelligence in Complex Decision Systems*, chapter 4. Atlantis Press (2010).
10. R. Moraes and L. Machado, *9th Int. FLINS Conf.* 243--248. China (2010).
11. R. Moraes and L. Machado, *Proc. SHEWC'2011*, 46--50. Brazil (2011).
12. R. Moraes and L. Machado, *Virtual Reality* 325--348. Intech (2011).
13. R. Moraes and L. Machado, *Proc. SHEWC'2012*, Brazil (to appear).
14. K-S. Park and R.V. Kenyon, *IEEE VR* 104--111. USA (1999).
15. R.J. Sternberg and E. Grigorenko, *Dynamic Testing: The Nature and Measurement of Learning Potential*. Cambridge University Press (2001).
16. Terano, T, Asai, K., Sugeno, M., *Fuzzy systems theory and it's applications*. Academic Press Inc. (1987).
17. G. Voss et al., *Studies in Health Technology and Inf.* **70**: 359--364 (2000).
18. S. Yoshida, H. Noma, K. Hosaka, *IEEE VR* 269--272. USA (2006).

LARGE-SCALE SPATIAL GRASSLAND FIRE SPREAD SIMULATION BASED ON GIS*

XINGPENG LIU†, JIQUAN ZHANG, ZHIJUN TONG

Natural Disaster Research Institute, College of Urban and Environmental Sciences, Northeast Normal University, Changchun, Jilin, 130024, P. R. China

Grassland fire simulation is complex because of many influence input parameters. Based on the field data of grassland fire behavior, the paper simulated the spread process of the grassland fire applying GIS tools. Applying experimental data, the study established initial grassland fire speed in control environment at first, and then calculated grassland fire spread speed rate in different wind speed, wind direction, and slope. The study obtained the grassland fire spread speed model applying composition of forces. Finally, the study calculated grassland fire spread time in large-scale grassland spatial. Take the severe grassland fire disaster of Dongwuqi in May 16, 2005 as a cases, comparing with actual results, this model showed reliable, and this model can provides decision-making for grassland fire management department.

1. Introduction

The grassland fire is a critical problem in China due to global warming and human activity. It have been the cause of numerous and irreversible damages with deep ecological and socio-economic impacts. The control of fire spreading in an environment is one of the most challenging aspect of environmental research, since fire, of course, heavily affects the quality state of grasslands and the public safety[1]. Grassland fires may represent a very dangerous situation, especially for fire fighters, and are responsible for a large number of injuries during some suppression operations and prescribed burning operations. Therefore, the need for designing and developing effective ways of dealing with grassland fire is constantly increasing as such phenomena appear ever more often. In the latest years, numerous grassland fire spread models have been

---

* This study is supported by the Fundamental Research Funds for the Central Universities (11QNJJ023), the National Natural Science Foundation of China under Grant Nos. 40871236 and 41071326, the National Key Technology R&D Program of China under Grant No. 2011BAD32B00-04, and the National Grand Fundamental Research 973 Program of China under Grant Nos. 2010CB951102 and 2009CB426305.

† email: liuxp912@nenu.edu.cn

proposed. Such models would be also useful in the case of controllable fires which are often used as an important grassland management tool in areas like Savanna or in Australia [2]. They can be grouped into: empirical models, semi-empirical models, and physical models [3]. According to these models, the famous operational tools BEHAVE [4] and FARSITE [5] were developed, and currently used in United States and other countries all around the world.

A model for predicting a grassland fire spread would have to take into account external environmental factors like fuel characteristics, meteorological conditions and terrain. According to literature [6-9], the most important factors that affect the rate of grassland spread are the fuel load and humidity, wind speed and direction, grassland topography (slope), and fuel continuity (vegetation thickness). Although there are many grassland fire behavior models widely used in world, such models all have its own constraints. They cannot directly used in China for limitations of external conditions. To predict the behavior of grassland fire in China, the study firstly proposed a grassland fire spread model using an empirical approach with experimental data obtained for outdoor or indoor fires, and then a simulation process was designed. Finally, simulated a historical fire sample to test the reliability of the model. The purposes of a fire spreading model in this study are to aid the real-time decision support system and to help the determination of strategies for fire control and land use planning. By calculating large-scale grassland fire spread time, it is helpful for fire fighters and fire managers to control grassland fire disaster.

2. Materials and methods

2.1. *Initial fire speed model*

The grassland fire speed is mainly influenced by fuel characteristic, terrain, weather. In order to analyze the grassland fire spread speed in no wind, flat ground, and continuous fuels, many times of indoor experiments were operated by control external influence factors.

In order to establish initial grassland fire speed, the study established the relationship of fuel moisture and air temperature and relative humidity. According to grassland fire behaviors research [10], the grassland fuel moisture (FM, %) is established as follows.

$$FM = 2.18 - 0.14x_1 + 0.34x_2 \qquad r^2 = 0.97, p < 0.01 \tag{1}$$

Where x_1 is temperature, x_2 is relative humidity。

In the indoor laboratory condition and continuous grassland fuel, the experimental results show that the fuel not burning when the grassland fuel moisture content exceed 45.0%, the grassland fuel can be burnt out but don't spread when the grassland fuel moisture content between 35.0% and 45.0%, and grassland fire could spread when the fuel moisture content below 35.0%. The grassland fire spread speed (S_m) has a negative correlation with grassland fuel moisture content, and their relationship could be expressed as follows [11].

$$S_m = 0.30 - 0.67FM \qquad r^2 = -0.93, p < 0.01 \tag{2}$$

As grassland fuels reaching balance moisture content, the experimental results show the grassland fire spread speed up with increased grassland fuel load, and the grassland fire could not spread when the grassland fuel load below 50 g/m2.

$$\begin{cases} S_l = 0.237\ln(FL) - 1.0065, \quad \text{FL} \geq 50 \\ S_l = 0, \ \text{FL} < 50, \ (r^2 = 0.899) \end{cases} \tag{3}$$

Where S_l is grassland fire spread speed (m/s), FL is grassland fuel load (g/m^2). For large spatial grassland area, the study applied normalized difference vegetation index (NDVI) and regress analysis to calculate the grassland fuel load. Through regression analysis, the relationship of grassland NDVI value of vegetation growth period (in the midmonth of August) and dry grass load was calculated using following function.

$$NDVI = (NIR - RED) / (NIR + RED) \tag{4}$$

$$FL = 82.385 * e^{4.815x}, (r^2 = 0.885, \text{P} < 0.001) \tag{5}$$

Where RED and NIR stand for the spectral reflectance measurements acquired in the red and near-infrared regions, FL is surface grassland dry fuel load (kg/hm2). In order to avoid data with cloud contamination we have used ten days composite image that was obtained by using the mean value composite procedure for the decade under consideration. If grassland fire burnt out in spring, the study applies NDVI value of last year. The finally initial grassland fire spread speed (R_0) was established as following function.

$$R_0 = \frac{S_m + S_l}{2} \tag{6}$$

2.2. *Grassland fire spread model*

Suppose initial grassland fire spread speed is 1, the grassland fire gain speed in different slopes and winds conditions were calculated using following functions.

$$R_t = e^{1.005tg(\phi)^{0.6696}}, r^2 = 0.9879 \text{ (Uphill)} \quad (7)$$

$$R_t = e^{0.4601tg(\phi)^{0.383}}, r^2 = 0.9606 \text{ (Downhill)} \quad (8)$$

$$R_w = e^{0.2015v\cos(\varphi)}, r^2 = 0.9842 \text{ (Wind)} \quad (9)$$

where ϕ is intersection angle of terrain and grassland fire direction, and φ is intersection angle of wind and grassland fire.

The grassland fire direction and the spread velocity were determined using of a parallelogram vector superposition principle.

$$R = \begin{cases} cR_0\vec{R}_w & v \neq 0, \varphi = 0 \\ cR_0\vec{R}_t & v = 0, \varphi \neq 0 \\ cR_0\vec{R}_{wt} & v \neq 0, \varphi \neq 0 \end{cases} \quad (10)$$

where R is grassland fire speed (m/min), c is fuel continuous degree, R_0 is Initial velocity (m/min), v is wind speed(m/s), ϕ is slope, $\vec{R}_{wt}$ is synthesized speed of wind speed and terrain, and it expressed as follows.

$$\vec{R}_{wt} = (\vec{R}_w + \vec{R}_t) = \sqrt{R_w{}^2 + R_t{}^2 + 2R_wR_t\cos(\theta/2)} \quad (11)$$

Many researches prove that vegetation grows well and continuous, the NDVI value will high and sparse vegetation have low NDVI value. Therefore, the grassland fuel continuous degree was calculated using NDVI value.

$$c = (NDVI - NDVI_{\min})/(NDVI_{\max} - NDVI_{\min}) \quad (12)$$

Where c is grassland fuel continuous degree which range from 0 to1. $NDVI_{min}$ is normalized difference vegetation index in the area of sparse vegetation, and $NDVI_{max}$ is normalized difference vegetation index in the area of dense vegetation.

2.3. *Data sources and process*

In this study, the land use map, climate data and elevation map were obtained from data sharing infrastructure of earth system science (http://www.geodata.cn). NDVI in northern China was obtained from long term vegetation index dataset of China-SPOT VEGETATION NDVI (Source for this dataset was the VITO (Flemish Inst. Technological Research, Belgium), http://www.vgt.vito), the data process flowchart was showed as Fig.1.

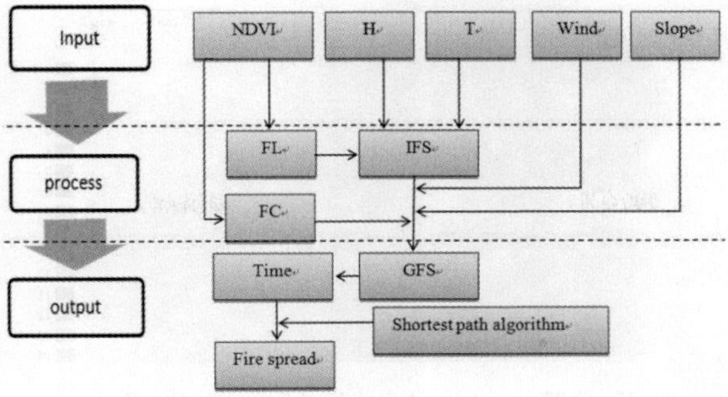

Fig.1 the data process flowchart. H: relative humidity, T: temperature, FL: fuel load, IFS: initial fire speed, FC: fuel continuity, GFS: grassland fire speed.

3. Conclusion and discussion

The proposed methodology was applied for the prediction of the spreading of a grassland fire that occurred on October 16, 2005. The fire cropped up due to unknown causes somewhere near the northeastern of the Dongwuqi grassland and was quickly spread fanned to the southeast by moderate to strong northwest winds (about 17.1m/s). The fire was extinguished around 14 h later, after having burned down a grassland area of almost 10 000 km^2.

Based on GIS, all related factors were imported to database. The grid map of grassland fire spreading speed was calculated using Eq.1-12 and GIS technology, and the grassland fire spread time of every grid also could be calculated using Eq.10. Applying shortest path algorithm, the shortest times from grassland fire point to every spatial grid were calculated (Fig.2). Figure 2 shows that the grassland fire spread quickly in downwind areas, and grassland fire in two flanks spread slower, the burnt area was showed as fan shape. According to calculation, the burnt area account for 15000 km^2. Due to fire preventing work, the calculated burned area could not match actual situation accurately. But it is a useful tool to calculate the reach time of grassland fire in spatial area, and it is helpful for grassland firefighting. It could also use to analyze cost-benefit of firefighting.

Through comparing of actual result, the model was reliability by tested. The applicability of grassland fire spread model will be tested in other grassland area in the next step.

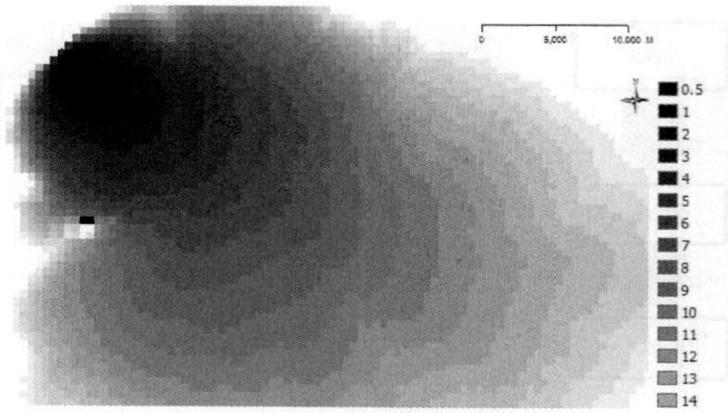

Fig. 2 The contour of grassland fire spreading time

References

1. Joseph Quartieri, Nikos E. Mastorakis, Gerardo Iannone, Claudio Guarnaccia. A Cellular Automata Model for Fire Spreading Prediction. Latest Trends on Urban Planning and Transportation. 2010, 173-179.
2. A. Alexandridis, D. Vakalis, C.I. Siettos, G.V. Bafas, A cellular automata model for forest fire spread prediction: The case of the wildfire that swept through Spetses Island in 1990, Applied Mathematics and Computation, 2008 (204) :191–201.
3. Sullivan, A.L., Wildland surface fire spread modelling, 1990–2007.1: physical and quasi-physical models. Int. J. Wildland Fire 2009,18(4):349–368.
4. P.L. Andrews, C.D. Bevins, Proceedings of the Second International Congress on Wildland Fire Ecology & Fire Management 5th Symposium on Fire &Forest Meteorology, 2003.
5. M. A. Finney, FIRESITE: Fire Area Simulator-Model Development and Evaluation, RMRS-RP-4, Ogden, UT, 1998, 47pp.
6. N.P.Cheney, J.S.Gould, W.R.Catchpole, The influence of fuel, weather and fire shape variables on fire spread in grasslands, Int. J. Wildland Fire 1993, 3(1):31–44.
7. N.P.Cheney, J.S.Gould, Fire growth in grasslands, Int.J. Wildland Fire 1995, 5(4):237–247.
8. N.P.Cheney, J.S.Gould,W.R.Catchpole, Prediction of fire spread in grasslands, Int. J. Wildland Fire, 1998,8(1):1–13.
9. W. Fons, Analysis of fire spread in light forest fuels, J. Agric. Res. 1946, 72:93–121.
10. P. Guo, C. Kang, J. Li, D. Zhou, The Changing Pattern of Fuel Moisture Content of Four Herbaceous Plants while Drying, ACTA AGRESTIA SINICA, 2003, 11(3):251-255.
11. P. Guo, X. Ma, C. Kang, D. Zhou, J. Li, Flammability Regularity of Flammable Materials in *Aneurolepidium* Chinense Fuel, Journal of Jilin University(science Edition), 2002,40(4):426-429.

DECOMPOSING RELATION SCHEMES OF GROUP-BUYING AGGREGATORS IN E-COMMERCE: A REVERSE ENGINEERING APPROACH

MING REN

School of Information Resource Management, Renmin University of China
Beijing 100872, PR China

QIANG WEI *

School of Economics & Management, Tsinghua University, Beijing 100084, PR China

Group buying aggregators collect data from heterogeneous sources on the Internet and fit them into a unified schema. As the amount of data grows, the initial schemes need to be maintained and optimized such as being decomposed into a desired normal form. This paper provides an approach to decompose relation schemes of group buying aggregator in a reverse engineering manner. Functional dependencies with degree of satisfaction that are discovered from massive data are used to decompose relation schemes into 3NF. A preliminary data experiment has been done to illustrate the effectiveness of the approach.

1. Introduction

With the development of electronic commerce, group buying aggregator, a service that collects group-buying data and deals from various sources on the Internet, has grown at a tremendous rate [7.]. For example, price.com finds and compares the best prices of airline tickets, hotel reservations, electronics, office supplies, etc; another popular site, dealsea.com, provides best deals and coupons (mostly on electronics, cosmetics) from various websites. They help individuals to effectively access the growing amount of group-buying data.

To handle the data from various sources, a global schema is usually constructed in terms of the data sources, without knowing either the underlying data model or the distribution of the data, and then data from autonomous and heterogeneous sources are crawled and fit into the unified global schema. However, the initial schema is constructed to be simple and general, sometimes as a 1-NF scheme in the relational database (RDB). In RDB design, 1-NF sets

*email: weiq@sem.tsinghua.edu.cn

very basic rules that no duplicative columns, unique rows, atomic value in the intersection of each column and row [1.]. As the data is accumulated day by day, the initial schema needs to be maintained and adapted to the realities in the purpose of effectively managing the accumulated data, e.g. decomposing the schemes into a desired normal form to eliminate data redundancy and insertion, deletion, and update anomalies. Typically 3-NF is chosen, where partial dependencies and transitive dependencies are eliminated [3.].

Traditionally, functional dependencies (FDs) that are declared and documented beforehand are used for decomposition [1.], while they may be not available to group buying aggregators. In a reverse engineering manner, functional dependencies can be extracted from massive data [2. ,4.], and these extracted functional dependencies reflect valuable knowledge about the structure of the current data. An issue of concern is that in massive database a functional dependency may not hold for a few records (i.e. exceptions), notably in the context of heterogeneous data sources, where there are missing data, noise, inconsistencies, etc. For example, a functional dependency *Merchant Name* →*Address* does not hold due to two records that (SAKS FIFTH AVE, …, 767 Fifth Ave) and (SAKS FIFTH AVE, …, 5th Ave 767), in which inconsistencies between various sources arise. In a traditional manner, such a dependency will not hold, though, it provides valuable knowledge about the current data. To deal with this, researchers have proposed functional dependency with degrees of satisfaction, where a functional dependency $X \rightarrow Y$ holds for a relation at a degree of α [5. ,6.].

This paper adopts functional dependency with degrees of satisfaction for decomposing a relation schema. The 1-NF schemes of a group-buying aggregator are decomposed, after obtaining functional dependency with degrees of satisfaction, into 3-NF which is the best normal form obtained through lossless join and dependency preserving decomposition [3.].

2. Functional Dependencies with Degrees of Satisfaction (FD_d)

Let **R**(**U**) be a 1-NF relation scheme on domains D_1, D_2, … Dn, $X, Y \subseteq \mathbf{U}$, and R be a relation of **R**(**U**), that is, $R \subseteq D_1 \times D_2 \times \ldots \times Dn$, where records t_i, $t_j \in R$ and $t_i \neq t_j$. Then Y is called to functionally depend on X for a record pair (t_i, t_j), if $t_i(X) = t_j(X)$, then $t_i(Y) = t_j(Y)$. It is obvious that the FD for a record pair could be represented in terms of the degree of satisfaction, $d_{(ti,\ tj)}(X \rightarrow Y)$. If $t_i(X) = t_j(X)$ and $t_i(Y) \neq t_j(Y)$, $d_{(ti,\ tj)}(X \rightarrow Y)=0$, otherwise 1. Subsequently, the degree that R satisfies $X \rightarrow Y$ can be defined as [5.]:

$$d_R(X \rightarrow Y) = \frac{\sum_{\forall t_i, t_j \in R, t_i \neq t_j} d_{(t_i, t_j)}(X \rightarrow Y)}{C_n^2} \quad (1)$$

Given a minimum satisfaction threshold θ, 0 ≤ θ ≤ 1, $X \rightarrow Y$ is called a satisfied functional dependency for relation R if $d_R(X \rightarrow Y) \geq \theta$, and will be used in decomposing a scheme. Moreover, the algorithm MFDD has been proposed to discover satisfied functional dependency [5.]. The approach has been proved to be efficient in that it can derive many dependencies by inference from previously discovered ones without scanning databases, and filter out those unsatisfied ones during (rather than after) the generating process.

Once the satisfied FD_d are discovered, schemes can be further decomposed into a desired normal form, e.g. 3-NF.

3. Decomposing Relation Schemes with FD_d

In a 1-NF relation schema, each relation scheme will be decomposed in three steps.

1. Given a relation R, initiate a relation scheme $\mathbf{R}(\mathbf{U}, \mathbf{F}_m)$, where **U** is the set of attributes, $\mathbf{F}_m = \varnothing$;
2. Given a minimum satisfaction threshold θ, discover a minimal set of satisfied functional dependency $\mathbf{F}_m$ through MFDD, and update $\mathbf{R}(\mathbf{U}, \mathbf{F}_m)$;
3. Decompose $\mathbf{R}(\mathbf{U}, \mathbf{F}_m)$ into $\mathbf{R}(\mathbf{U}_i, \mathbf{F}_{mi})$ which is in 3NF.

We illustrate the approach through a toy example. Suppose a relation is shown in Table 1. Classical functional dependencies that hold for R include Deal_ID→Product_Name, Deal_ID→Category. The scheme is in 3NF and does not need to be decomposed. Actually, it makes sense that Deal_ID→Category holds due to Deal_ID→Product_Name and Product_Name→Category. The latter does not hold, unfortunately, due to the null value of Category in a record.

Table 1. An example of relation R

Deal_ID	Product_Name	Category
001	b1	N/A
002	b2	c2
003	b2	c2
004	b1	c1
005	b1	c1
006	b2	c2

In the following, the scheme will be decomposed with FD_d. First, a relation scheme is initiated as $\mathbf{R}(\mathbf{U}, \mathbf{F}_m)$, where **U**={Deal_ID, Product_Name,

Category}, $\mathbf{F}_m=\varnothing$. Second, given the threshold $\theta=0.85$, a minimal set of FD_d is discovered as $\mathbf{F}_m$={Deal_ID→Product_Name, Product_Name→Category}. Finally, a relation scheme is decomposed into $\mathbf{R}(\mathbf{U}_1, \mathbf{F}_{m1})$ and $\mathbf{R}(\mathbf{U}_2, \mathbf{F}_{m2})$, where $\mathbf{U}_1$={Deal_ID, Product_Name}, $\mathbf{F}_{m1}$={Deal_ID→Product_Name}, $\mathbf{U}_2$={Product_Name→Category}, $\mathbf{F}_{m2}$={Product_Name→Category}.

In the optimized schema, each scheme is in 3NF, where data redundancy and update anomalies are eliminated. It can be seen the decomposition process is driven by data, rather than user defined functional dependencies. And the decomposed schemes represent the real structure of the current data, although the extracted dependencies may not be true in a strict sense.

An algorithm has been proposed for the decomposition with FD_d. See Figure 1 for details of the algorithm.

```
FOR EACH Dependency X->A IN R<U,F> AND A=B1B2……Bm
    FOR EACH ItemBi IN A
        Closure=X-Bi
        FOR EACH Dependency U->V IN R
            IF U∈Closure THEN Closure=Closure∪V
        ENDFOR
        IF A∈Closure THEN Replace X with X-Bi
    ENDFOR
ENDFOR
FOR EACH Attribute in R<U,V>
    IF Attribute∉F
    THEN Remove Attribute FROM U TO U0
ENDFOR
Construct Schema with AttributeSet U0 AND Functional_DependencySet∅
FOR EACH Dependency in R<U,V>
   IF X->A∈F AND XA=U
   THEN Result={R,R0}
ENDFOR
FOR EACH Dependency IN R<U,F>
   Sort F into Subsets with same Antecedents And Construct SubSchemas Ri
with Subsets
ENDFOR
FOR EACH Schema Ri IN Result
   FOR EACH Schema Rj IN RESULT
   IF Ui∈Uj    THEN Combine Ri And Rj
ENDFOR
```

Figure 1. An algorithm of the lossless join dependency preserving decomposition with FDd

4. Decomposing Schemes of Group-buying Aggregators

A preliminary data experiment has been done, after crawling real data from tuan800.com, a popular group buying aggregator in China. A relation scheme is initiated according to data sources (attributes shown in Table 2), and 50 pieces group buying information are randomly selected for analysis.

Table 2. Attributes of the Table

Table Name	Attributes (19)
Group Buying Deal	Deal ID, GB Provider, Merchant Name, Region, Address, Tel, Merchant Description, Product Category, Product Subcategory, Product Name, GB Price, Original Price, Discount Rate, Start Time, End Time, Min No of Purchase, Current No of Purchase, Deal URL, Deal Specification

Given the threshold θ=0.95, dozens of FD_d are discovered using MFDD and the dependencies with sufficiently high degrees of satisfaction are shown in Table 3. By using the decomposition algorithm, the initial scheme is decomposed into four schemes, which represent information about Deal, Merchant, Product, and Product Category respectively (Figure 2).

Such a data-driven decomposing process employs knowledge about the structure of the current data and helps to optimize and organize the data as it should be.

Table 3. Selected Results of the FD_d discovered from the data

FD_d	degree	FD_d	degree
Deal ID→GB Provider	1.00	Deal ID→GB Price	1.00
Deal ID→Merchant Name	1.00	Deal ID→Original Price	1.00
Merchant Name→Region	0.95	Deal ID→Discount Rate	1.00
Merchant Name→Address	0.98	Deal ID→Start Time	1.00
Merchant Name→Telephone Number	0.98	Deal ID→End Time	1.00
Merchant Name→Merchant Description	0.98	Deal ID→ Min. No of Purchase	1.00
Product Subcategory→Product Category	0.98	Deal ID→Current No of Purchase	1.00
Product Name→Product Subcategory	0.96	Deal ID→ Deal URL	1.00
Deal ID →Product Name	1.00	Deal ID→ Deal Description	1.00

5. Conclusions and Future Work

This paper has proposed an approach to decompose relation schemes of group buying aggregators in a reverse engineering manner. First functional dependencies with degree of satisfaction are discovered from the data, and then each relation scheme is decomposed with these dependencies into 3NF. A preliminary data experiment has been done based on real data from a popular group buying aggregator in China.

Ongoing research centers on optimization of the approach and more data experiments on large real datasets for selecting suitable thresholds, and evaluating the efficiency.

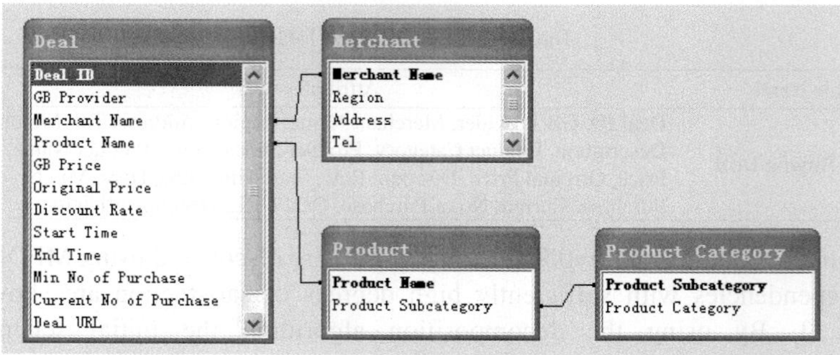

Figure 2. Decomposed schemes in 3NF

Acknowledgements

This work is partly supported by National Natural Science Foundation of China (71072015) and Doctoral Program Foundation of Ministry of Education of China (20100004120005).

References

1. E.F. Codd, Recent investigations in relational database systems. In: *IFIP 74 Conf.* North Holland, 1974, pp. 1017–1021.
2. P.A. Flach and I. Savnik. Database dependency discovery: A machine learning approach. *AI Commun* 1999;12(3):139–160.
3. H. Garcia-Molina, J.D. Ullman, J. Widom, Database Systems: The Complete Book, Prentice Hall, 2009.
4. I. Savnik, and P.A. Flach, Discovery of multivalued dependencies from relations. Technical Report 00135. Marseille: Albert-Ludwigs-Universitaet Freiburg, Institut fuer Informatik; 2000.
5. Q. Wei, and G.Q. Chen, Efficient Discovery of Functional Dependencies with Degrees of Satisfaction, *INT J INTELL SYST*, Vol. 19, 1089-1110, 2004.
6. Q. Wei, and G.Q. Chen, Optimized Algorithm of Discovering Functional Dependencies with Degrees of Satisfaction, In: *Proc the 7th Int'n FLINS Conf*, World Scientific Press, pp. 169 -176, 2006.
7. http://technode.com/2011/09/13/google-china-still-wants-to-be-local-launched-group-buying-aggregator-service/

A BIBLIOMETRIC EXPLORATION OF THE EVOLUTION OF 3D TECHNOLOGY

GİZEM İNTEPE*
Mathematical Engineering Department, Istanbul Technical University , Room 410, Istanbul, Maslak, 34469, Turkey

TUFAN KOÇ
Indstrial Engineering Department, Istanbul Technical University Istanbul, Maçka, Turkey

The purpose of this paper is to use bibliometric analysis to forecast 3D TV technology. Forecasting emerging technologies and identifying the rate of diffusion of products based on these technologies is difficult because of lack of data. For this aim, ISI Web of Science publication database is searched and posed with findings appropriate S-curve is composed. Then technology life cycle of 3D TV technology is obtained and results are explained.

1. Introduction

Technological forecasting applies to all purposeful and systematic attempts to anticipate and understand the potential direction, rate, characteristics and effects of technological change, especially invention, innovation, adoption and use [1]. The main function of TF is "to lead the decision making process towards profitable solutions with minimum uncertainties [2].

Technology forecasting assesses impacts of implementing a new technology on both the firm and its external environment [3]. It is important for predicting as clearly as probable impacts of a rapidly growing technology. Choice between alternative paths may make all the difference in competitive performance

There are hundreds of TF techniques which can be divided into groups. In literature there can be seen different separations. Some are numerical data based, some are judgment based. In general, numeric data-based forecasting extrapolates history by generating statistical fits to historical data. A few numeric methods deal with complex interdependencies. Judgmental forecasting may also be based on projections of the past, but information sources in such models rely on the subjective judgments of experts. Mainly used techniques are;

* E-mail: karakang@itu.edu.tr

trend extrapolation, growth curves, analogy, morphological matrices, patent analysis, system simulation, scenario writing, cross impact analysis, delphi, relevance trees and system dynamics.

In this paper, we use growth curves method for predicting the life cycle of 3D TV technology. 3D TV technology is becoming increasingly popular with each passing day. Due to the rise of popular 3D feature films major television manufacturers began developing three dimensional home TV technology in 2009. Today, the 3D TV is slowly becoming a household essential and soon we would not be able to imagine our lives without it anymore. This study uses the bibliometric analysis to collect useful data for the growth curve model to forecast the development of 3D TV technology.

2. Bibliometric Analysis

The bibliometric analysis, which applied in this paper, can provide an interesting alternative data source for quantitative evidence of R&D activity and text materials [2, 4, 5, 6]. Bibliometric analysis could provide a nicely accessible and cost-effective data or information. It helps to explore, organize and analyze amounts of historical data helping researchers to identify “hidden patterns” that may help researchers in the technology forecasting and decision making process [2, 4].

Bibliometric analysis is defined by Norton [7] as the measurement of literatures and texts. The approach is to capture some of the information inherent in the content and patterning of the literature. Bibliometric analysis uses counts of publications, patents, or citations to measure and interpret technological advances [4]. Historically, bibliometric analyses have been used to trace back academic journal citations. Nowadays, bibliometric analysis can be used to understand the past and even potentially to forecast the future [8, 9, 10].

2.1. *Research method*

Growth Analysis is highly quantitative and requires numerical data and uses S-shaped curves. Most used S-shaped curve is logistic curve, is helpful for estimating the level of technological growth (or decline) at each stage in the life cycle and for predicting when a specific technology will reach a particular stage

The importance placed on research into growth curve models, such as the logistic curve model, for examining technology diffusion and substitution has increased in the last few decades. The growth curve model involves fitting a growth curve to a dataset of technological performance. The growth curve beyond the data range is used to estimate future performance [11].

In the innovation management field the S-Curve illustrates the introduction, growth and maturation of innovations as well as the technological cycles that most industries experience. In the early stages large amounts of money, effort and other resources are expended on the new technology accumulates, progress becomes more rapid. As soon as major technical obstacles are overcome and the innovation reaches a certain adoption level an exponential growth will take place. During this phase relatively small increments of effort ans resources will result in large performance gains. Finally, as the technology starts to approach its physical limit, further pushing the performance becomes increasingly difficult, as the figure below shows.

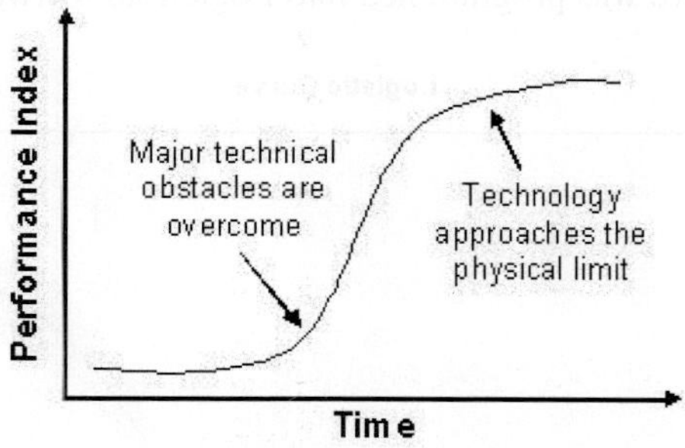

Figure 1. Technology Life Cycle

2.2. *Data collection*

The dataset for this study was constructed from the ISI Web of Science publication database. Scholarly literature and professional publications from 1990 to 2009 are included in the study. We provided these data from article search.

3. Case Study

To test the bibliometric analysis modelling 3D technology will be used. 3D is a very powerful technology which has the ability to immerse us in a viewing experience, and make the action seem to 'pop' from the screen.

3.1. *History*

Employing technology first discovered in the early 1800s, the leading TV manufacturers are using ground-breaking technological skills to give us improved access to stunning 3D imaging - also known as stereoscopy - which basically tricks your brain into creating an illusion of visual depth by displaying two different images. This is known as a parallax view - the two slightly different perspectives of a visual object that each eye sees naturally because they are spaced physically apart. Our brains resolve these two different object views and this is what gives us the perception of depth and three dimensions [12].

3.2. *S-curve implementation*

We utilize growth curve analysis based on bibliometric data to identify trends in publications about 3D TV technology during the period 1981-2011. Using LogletLab software, which can fit growth processes using one or more S-curves and a possible initial displacement, a logistic fit is produced as Figure 2. Logistic growth curve was programmed into LogletLab software.

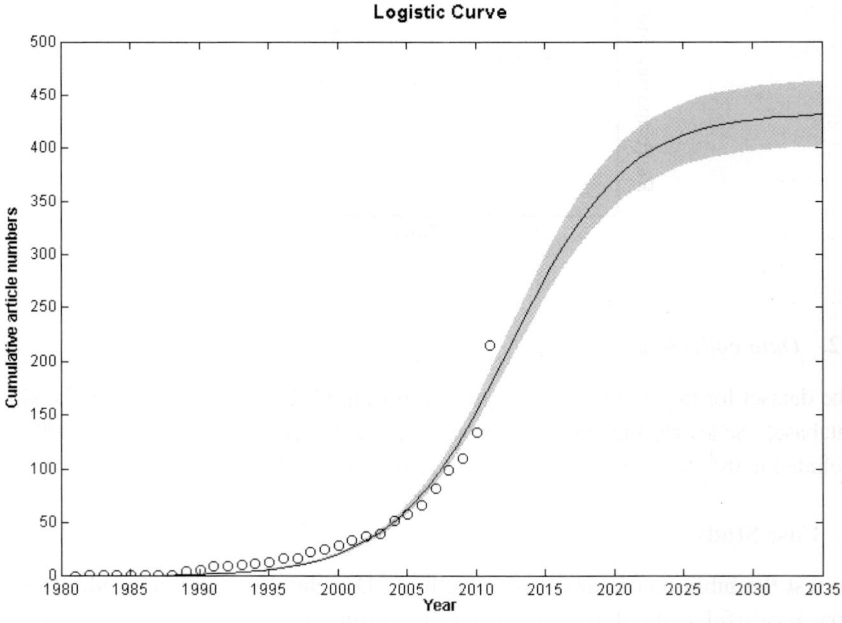

Figure 2. Logistic curve of 3D TV Technology

3.3. *Results and discussion*

We saw that number of the articles issued in 2011 is really far more than the others. When we composed the logistic curve as Figure 2, we can see that, the year 2013 is the inflection point. This means in 2013 the most rapid growth come through, this growth will continue till 2020, then will slow down. After

2026 this technology comes its saturation point, and then gives its place to a new technology. Studies will continue on this new technology after 2026.

4. Conclusion

Along with knowing what is 3D TV, you should also know how this technology is evolving and what you can expect from it in future. For forecasting future life cycle 3D TV technology we inspected scientific developments.

Scientific studies are followed by reformed technologies. Density changes in scientific works could be a good deterministic index for a specific technology which wanted to be forecasted. Hence these changes could be watched closely and carefully. Thus bibliometric analysis was proposed as the simple and efficient to link the science and technology activities and to obtain quantitative and historical data especially in rare historical data available fields. With these data growth curve is produced and life cycle of the technology is predicted.

For future works, new studies could be done using other technological forecasting techniques which will confirm our results.

References

1. Fırat A.K., Woon W. L., Madnick S, Technological forecasting - A review (2008).
2. Bengisu, M., Nekhili, R., Forecasting Emerging Technologies with the Aid of Science and Technology Databases, Technological Forecasting and Social Change, 73 (7), 835-844 (2006).
3. A. L. Porter, A. T. Roper, T. W. Mason, F. A. Rossini, and J. Banks, Forecasting and Management of Technology, John Wiley & Sons, Inc., New York (1991).
4. J. R. Watts, A. L. Porter, Innovation Forecasting, Technological Forecasting and Social Change, 56, 25-47 (1997).
5. A. L. Porter, A. T. Roper, T. W. Mason, F. A. Rossini, J. Banks,. Forecasting and Management of Technology. Wiley, New York (1991).
6. M. Martino, A review of Selected Recent Advantages in Technological Forecasting, Technological Forecasting and Social Change, 70 (8), 719-733 (2002).
7. M. J. Norton, Introductory concepts in information science, New Jersey: ASIS (2001).
8. S. W. Cunningham, The Content Evaluation of British Scientific Research, Ph.D. thesis, UK (1997).

9. Morris S, De Yong C, Wu Z, Salman S, Yemenu D, DIVA: A Visualization System for Exploring Document Databases for Technology Forecasting, Comput. Ind. Eng., 43(4), 841-862 (2002).
10. F. Narin, D. Olivastro, K. A. Stevens, Bibliometrics - Theory, Practice and Problems. Eval. Rev., 18(1): 65.76 (1994).
11. Y. H. Chen, C.Y Chen, S.C Lee, Technology forecasting and Patent Strategy of Hydrogen Energy and Fuel Cell, International Journal of Hydrogen Energy, 36, 6957-6969 (2011).
12. http://www.3dtvguide.org/

ANFIS MODEL FOR VIBRATION SIGNALS OF AN INDUCTION MOTOR BASED ON AGING PROCESS DECREASING THE NONLINEARITY

D. BAYRAM, S. ŞEKER

Electrical Engineering Department, Istanbul Technical University, Maslak, Istanbul, 34469, Turkey

In this study an ANFIS model is aimed to achieve the vibration signals decreasing the nonlinearity with moving average technique. This nonlinearity is observed by an aging process of an induction motor. The aging process is identified by vibration data taken from healthy and aged (faulty) cases of the same induction motor. In order to change the nonlinearity degree of the system, moving average method is applied to the input and output pairs for different lagging factors. The success of the model is attached two conditions; these are the progress of the ANFIS and the characteristics of the model output. The gauges of these conditions are training - checking errors of ANFIS and cross correlation coefficient between Power Spectral Densities of the model output and faulty case data (real data), respectively.

1. Introduction

Induction motors are the backbone of the industrial process since they are low-priced, simple and maintenance - free. Because of these properties, they are constructed at various power levels in order to get the industrial requirements [1]. Despite they have high reliability, they might function improperly [2].

To provide the sustainability of the industrial process, condition monitoring and investigations about the lifetime of induction motors have vital importance. Condition monitoring studies are used to observe the trends of the induction motor parameters. These trends play an important role on determination of induction motor performance analysis and detection of failure potentials. In addition, some surveys and academic studies are executed on the lifetime of induction motors [3]. Therefore, aging process is researched extensively in the literature [4-5]. Aging process has a complicated structure and it contains a lot of unfamiliar components which makes the process highly nonlinear. For this reason, modeling of the aging processes is very difficult.

Different signal types can be used to investigate the aging process of an induction motor. Electrical signals carry hints about aging, whereas mechanical outputs, like torque and vibration, directly show the aging effects. For this reason, torque and vibration signals give an opportunity to define the aging. Therefore vibration signals are the indispensable elements to realize a proper condition monitoring. Vibration signals can be investigated by signal processing techniques in order to find the distinctive failure signatures [6].

Also, in the related literature, number of the studies based upon electric motors and their aging processes in terms of ANFIS application is very low. In this manner, one of them [7] is connected with the motor vibration and current as well as temperature changes.

In this study the ANFIS is used to get the possible model approach for the aging process using vibration signals under the nonlinearity which occurs between healthy and aged motor cases. However, this nonlinearity can be defined considering all aging steps (7 cycles for this study) between healthy and aged motor cases. And, the degree of nonlinearity is changed by Moving Average (MA) method which is applied to the input and output data. Hence, the success of the model is observed for different degrees of nonlinearity depending on the order of MA.

2. Mathematical Background

In this study, three main mathematical approaches are used. These methods are Spectrum Estimations (Power Spectral Density), ANFIS and Moving Average methods.

2.1. *Power Spectral Density Estimation*

Power spectral density (PSD) estimation is a common method to obtain frequency component of a signal. It calculates the density of the energy of the signal as a function of frequency [8].

2.2. *Moving Average*

Moving average can be interpreted as a low pass filter in signal processing. It is employed to smooth data and to decrease the fluctuations by computing the arithmetic mean of data's subsets [9-10].

2.3. *Adaptive Neuro Fuzzy Inference System (ANFIS)*

ANFIS defines rule base and tunes the parameters of the rule base using neural network. In this manner, it incorporates the learning ability of neural network and inference power of fuzzy logic. ANFIS is a good choice to model nonlinear systems because of its intelligence.

Basic ANFIS is an inference system which has a structure similar to multi layer feedback network. The network has adaptive and fixed nodes. The adaptive nodes serve to minimize the error by tuning relevant parameters. The algorithm is constituted by 5 layers, 2 inputs (x, y) and 1 output f. Layer1 includes adaptive nodes with their membership functions. The most popular membership function type is a bell shape function with a maximum value of one and minimum value of zero. The parameters of membership function are named as premise parameters which can change its shape. Then nodes located at this layer become adaptive by this change. Layer 2 combines the input signals to produce an activation (firing) strength. Nodes at this layer are the fixed nodes which multiply its inputs and defining a rule. Layer 3, which contains fixed nodes, is used to normalize the activation (firing) strengths. Layer 4 includes adaptive nodes which compute the consequent parameters. This layer represents the inference ability of the algorithm. Layer 5 is formed by a fixed node which calculates the output of the ANFIS as the sum of all incoming signals [6, 11].

3. Experiment and Data Acquisition System

In this study an artificial aging process is applied to an induction motor. The process is composed by 7 cycles. During each cycle (step) two types of aging techniques are applied to the motor. These are Electrical Discharge Machining (EDM) and the Thermal and Chemical aging actions, respectively. In this paper first and last cycle's vibration data are used to model an aging process. As a result of these aging processes, bearing damage occurs on the test motor.

Two accelerometers are used to record the vibration measurement. The sampling frequency of the recorded data is 12 kHz. Besides, in order to avoid from the high frequency noise interfered in the recorded data, an anti aliasing filter with cut off frequency at 4 kHz is used. The type of the motor is an induction motor of 5 HP, three - phase, four - poles, designed for 60 Hz supply frequency and with the nominal speed 1742 rpm (rotation per minute). Details about the experiment can be found from the PhD thesis of A.S. Erbay [12].

4. Application

Modeling of an aging process is impossible using Fuzzy Inference System (FIS) because defining a rule base for an aging process is very difficult and complicated [13]. For this reason ANFIS is employed to identify this nonlinear process. Highly nonlinear processes are in the need of reducing the nonlinearity to increase the model success. For this reason moving average method is used in this study. The nonlinearity of the data can be tune by moving average parameters. The success of ANFIS is investigated for different moving averaged input – output pairs.

The aging process is aimed to model using the ANFIS. In this study a Sugeno's type ANFIS is used which has 3 inputs and 1 output with 2 bell-shaped membership functions for each input. In other words, system produces 8 rules, 18 premise parameters and 32 consequent parameters. Modeling and identification of a nonlinear system using ANFIS can be done employing some techniques like selecting input variable from a pool of delayed input and output data. Three input variables are selected from the pool which consists of ten delayed input or output data [13].

The vibration data achieved from first and final aging cycles are used in this study. It can be seen obviously that the amplitude of the vibration is increased in Figure 1. Another identifier of aging is the change of their PSDs. As it is well known, aging manipulates the frequency content of a vibration signal (Figure 1).

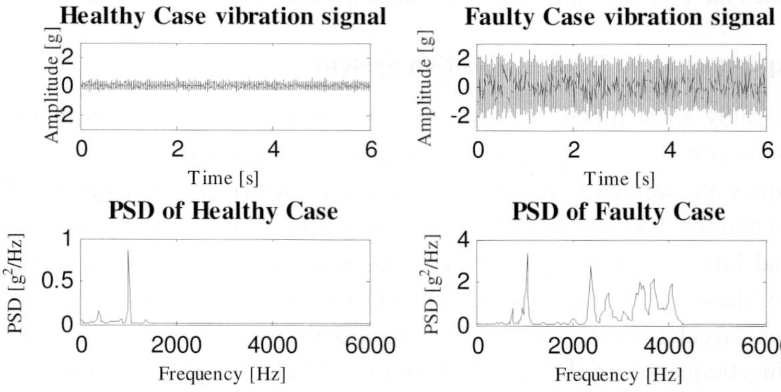

Figure 1. The vibration signals of the electric motor for Healthy - Faulty Cases and their PSDs.

The schematic of an ANFIS model can be seen in Figure 2. An ANFIS, which has 3 inputs comprise delayed healthy or faulty case data, is expected to produce its model's faulty case data.

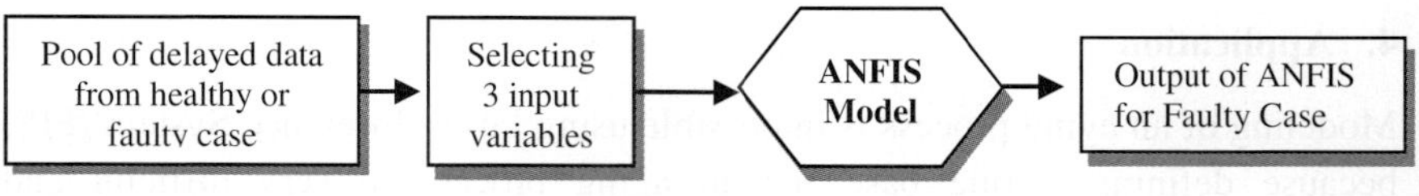

Figure 2. The schematic of an ANFIS model.

Different ANFIS models are designed for moving averaged data pairs. The models are named by their moving average lagging parameters. The highest lagging parameter provides the smoothest data. As seen on the Table 1, model's training and checking errors decrease whereas the model order increases. This is an expected fact because the nonlinearity decreases when the moving average order increases.

Table 1. Training and checking errors of the proposed models.

Model Name	Moving average parameters (*Lagging*)	Errors *Training Error*	*Checking Error*
M 0	0 (Without Moving Average)	0.49043	0.48159
M 5	5	0.1945	0.19669
M 10	10	0.10187	0.10414
M 15	15	0.06663	0.073

However, the similarities of the PSDs must be considered to identify the best model. The success of the system is identified as producing ability of the high frequency components occurred at high frequency region, which can be seen at the Figure 1.

Because of this, calculating the cross correlation coefficient between the model and original signal PSDs is required. The maximum correlation coefficient is the indication for the best model. The cross correlation coefficient between PSDs of signals and its ANFIS models can be seen on the Table 2.

Table 2. The cross correlation coefficient between PSDs of signals and its ANFIS models.

Model Name	Cross Correlation Coefficient
M 0	0.8392
M 5	0.9605
M 10	0.9504
M 15	0.7740

As seen on the Table 2, the last model (M 15) is the worst ANFIS model with low cross correlation coefficient with the faulty case's PSD which means the loss of information on the spectral domain. Model trained with 5 lagging parameters (M 5) is accepted as an optimum ANFIS model, although it has higher training and checking errors than the other models (M 10 and M 15).

5. Conclusion

In this study an aging process is considered as an example of highly nonlinear system. The modeling and identification of aging process is aimed to be realized by an ANFIS using the vibration signals for healthy and faulty cases. The amplitudes and the spectral densities are distinctive properties between the healthy and faulty cases.

In order to reduce the degree of system nonlinearity, moving average method is used for different lagging parameters. The success of the realized models is discussed with two criteria. One of them is the training and checking errors of ANFIS and the other is the cross correlation between the PSDs of the model output and the faulty data.

As a result of this observation, it is noticed that increasing in the moving average order causes the disappearance spectral details which characterize the aging process. Although the highest order moving average operation provides the lowest training and checking error, however it cannot be accepted as a successful aging model. The model named "M 5" is the optimum aging process model because of its highest cross correlation coefficient even though it does not have achievement by means of training and checking errors. In other words, having low ANFIS error levels do not make the model the best of proposed models. In this sense, this study shows an interesting application of ANFIS approach on the motor aging studies.

Acknowledgements

The authors would present their special thanks to Prof. B.R. Upadhyaya from the Maintenance and Reliability Center and Nuclear Engineering Department of University of Tennessee Knoxville, USA, for permission to the use of experimental data.

References

1. A. Siddique, G. S. Yadava, B. Singh, *IEEE Trans. Energy Convers.*, **20/1**, 106 (2005).
2. S. Nandi, H. A. Toliyat, *Proc. Ind. App. Soc.*, **1**, 197(1999).
3. P.J. Tavner, *IET Electric Power App.*, **2/4**, 215 (2008).
4. S. Seker, *IEEE Power Eng. Rev.*, **20/7**, 48 (2000).
5. P. Werynski, D. Roger, R. Corton, J.F. Brudny, *IEEE Trans. Energy Convers.*, **21/3**, 673 (2006).
6. F.A. Sanz, J.M. Ramirez, R. E. Correa, *NAPS 10,* 1 (2010).
7. M.S. Yilmaz, E. Ayaz, *IEEE Eurocon 09,* 1140 (2009).
8. S. V. Vaseghi, "Advanced DSP and Noise Reduction", 4th ed., John Wiley and Sons, UK, 2008.
9. E. Alessio, A. Carbone, G. Castelli, V. Frappietro, *EPJB*, **27**, 197 (2002).
10. Yongzhen Zhuangt, Lei Chen, X.S. Wang, Jie Lian, *ICDCS 07*, 38 (2007).
11. J.-S.R. Jang, *IEEE Trans. Syst., Man, Cybern.,* **23/3**, 665 (1993).
12. A.S. Erbay, PhD Thesis, The University of Tennessee (1999).
13. J.-S.R. Jang, C.-T. Sun, E. Mizutani "Neuro-Fuzzy and Soft Computing", Prentice Hall, NJ (1997).

MODELLING THE SOLID FRACTION TRANSPORT OF A FIBER SUSPENSION WITH CELLULAR AUTOMATA

STEFAN KITTAN

IPM, Zittau/Goerlitz University, Theodor-Koerner-Allee 16
02763 Zittau, Germany

WOLFGANG KAESTNER

IPM, Zittau/Goerlitz University, Theodor-Koerner-Allee 16
02763 Zittau, Germany

The processes of cake forming filtration during a loss of coolant accident in a nuclear power plant are of safety-related importance. Thereby these accretion processes are characterized by high complexity. Thus, precise predications to the operation course and effect are hindered. Hence a dynamic model is reasonable for a broad investigation and understanding of these processes. One submodel is the solid fraction movement in the fluid, which is presented in this paper.

1. Introduction

The solid-liquid separation based on cake filtration is highly relevant in versatile technical applications. For example, this operation is used by the chemical industry or for water treatment [1]. Furthermore, also in the field of reactor safety, cake filtration is a matter of consequence. The process background is a loss of coolant accident. In this scenario is presumed, that insulation material is detached by a leak jet (water or steam) from pipes or other components. As a result, the cooling liquid is suspended by the released insulation material (Barsebaeck scenario [2]). The research project in context of which the presented model was created, is related to this example of reactor safety research.

One aim of the project is to model the dynamic behavior of the accretions. Filter cakes composed of mineral wool are characterized by high porosity and compressibility. Hence a rising differential pressure over the filter bed, caused for example by an increased flow rate, can lead to a porosity drop. This fact causes higher flow resistance and therefore a further rise of differential pressure. The filter cake compression has versatile reasons like fiber deformation or relocation of the fibers [1]. Figure 1 illustrates this dynamic behavior of the accretions by means of experimental data.

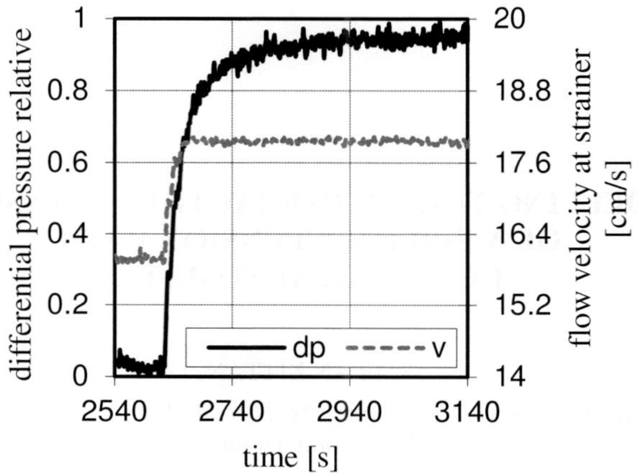

Figure 1. Pressure drop over strainer and accretions after a stepwise rise of the flow velocity

A further goal of the project is to model and simulate the dynamic accretion process itself. Such a dynamic accretion model enables, for example, the investigation of the effectiveness of different sustainer geometries. As one part of this model the movement of the fiber phase is presented in this paper.

2. Theory of Cellular Automata

Cellular Automata (CA) are time- and spatial-discrete systems providing the possibility of simulating emergent systems or processes. Normally, they are one- or two-dimensional. But generally the dimension is not restricted. The cellular grid is built up of limited number equal cells. Mostly common for two-dimensional grids are square cells. But there are also models which are made up of different cell shapes. One example is the FHP-model for simulating gases and fluids [3].

Every cell is characterized by a limited number of states. At every time step, the rules of the CA are checked for each cell. Subsequently, an adjustment of the cell states is occurred. The CA rules can be deterministic (e.g. based on physical laws) as well as stochastic [4]. Furthermore a combination of determined equations with stochastic elements is possible.

For the adjustment of a cell state, the cell neighborhood (NH) is generally included. The most common NH for square cells is the “Moore-NH” and the “Von-Neumann-NH”. At the “Von-Neumann-NH” the reference cell is considered as well as the four edge cells. The “Moore-NH” enhances the “Von-Neumann-NH” by the four cells at the reference cell corners. In addition it is possible to choose a radius for the cell NH [4].

In practice, the dimension of the cellular grid is limited by the calculating capacity. Because of that it is necessary to define a boundary condition for the CA. This means, that the NH of cells at the cellular grid border has to be defined. In [4] the following alternatives are recommended: open, periodic or symmetric conditions. But in principle there are no limitations for the modeler. E.g. an edge of cells with defined or stochastic states around the cellular grid is also imaginable.

3. Cellular Automaton for the Fiber Movement Simulation

3.1. *Design of the Cellular Automaton*

At present, a downward channel with a square sectional area is assumed as flow geometry for experiments (Figure 2). Thus, the basic CA design is occurred on this geometry. Consequently, there is used a three-dimensional cellular grid with cubic cells. Each cell has an edge length of $1mm$. The state of one cell is primarily identified by the mass of the present insulation material. The maximum capacity for the insulation material mass of each cell can be calculated by the insulation material density and the cell volume:

$$m_{CellIsoMat} = \rho_{IsoMat} V_{cell} = 2740 \frac{kg}{m^3} 10^{-9} m^3 \tag{1}$$

Since the maximum mass on fibers at a cell can only arise under pressure and not under normal flow conditions, a limiting factor f_{lim} was introduced to the model.

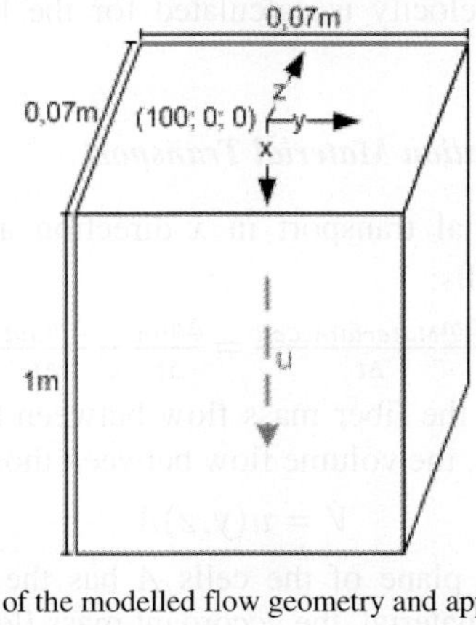

Figure 2. Dimensions of the modelled flow geometry and applied coordinate system

3.2. *Properties of the Fluid Phase*

For the accretion model only very thin and small fibers are relevant. Bigger agglomerations of insulation material do not reach the strainer, because of sedimentation or deposition on plant components. Hence there is no noticeable influence of the fibers on the fluid phase. Therefore, the fluid phase is implemented to the model as a constant and directed flow profile.

The calculation of the Reynolds numbers has shown that a turbulent flow can be expected for a wide range of flow velocities. In literature many different and partly highly specialized approaches for calculating turbulent velocity profiles in thin channels do exist. But for the accretion model a simplified flow treatment is sufficient. So an established power law is used [5]:

$$u(r) = \left(1 - \frac{r}{R}\right)^n \cdot u_{max} \tag{2}$$

With the maximum flow velocity u_{max} and the position dependent velocity $u(r)$ the exponent n is usually between $\frac{1}{8}$ and $\frac{1}{6}$. A lower exponent causes a more compressed velocity profile. Since equation 2 shall be applied for flow channels with a circular sectional area, a modification is required. Therefore, this equation is initially established independently to the y-coordinate and the z-coordinate. Subsequently, one of these equations has to be inserted into the maximum velocity of the other one. Thus, the velocity profile for a square flow channel results in:

$$u(y,z) = \left(1 - \frac{|y|}{0{,}035m}\right)^n \cdot \left(1 - \frac{|z|}{0{,}035m}\right)^n \cdot u_{max} \tag{3}$$

With this equation the velocity next to the channel walls results in zero. Hence all boundary cells of the CA would have no flow velocity. This condition is unrealistic and results for this phenomenon from a rough spatial discretization. Therefore a new flow velocity is calculated for the boundary cells by linear extrapolation.

3.3. *Rules for the Insulation Material Transport*

The rules for the material transport in x-direction are educed by the mass balance for the several cells:

$$\frac{\Delta m_{MaterialAtCell}}{\Delta t} = \frac{\Delta m_{in}}{\Delta t} - \frac{\Delta m_{out}}{\Delta t} \tag{4}$$

For the determination of the fiber mass flow between two adjacent cells at the same y and z coordinates, the volume flow between those cells is needed:

$$\dot{V} = u(y,z)A \tag{5}$$

in which the connection plane of the cells A has the value $1mm^2$. With the density of the insulation material, the accordant mass flow can be calculated by:

$$\frac{\Delta m_{Material}}{\Delta t} = u(y,z) \cdot 1mm^2 \cdot 0{,}00274 \frac{g}{mm^3} \tag{6}$$

Since the flow velocity remains constant in x direction and time, the transported material mass within one time step results by transposition of equation 6:

$$\Delta m_{Material} = u(y,z) \cdot 0{,}00274g \cdot \Delta t \quad (7)$$

It is necessary to set Δt in a way, that the theoretical maximum value for the transported material mass cannot be exceeded. This maximum value corresponds to the maximum capacity for the insulation material mass of a cell. Furthermore the maximum transported mass appears at the maximum flow velocity. Since these values are constant during a simulation, it is reasonable to calculate Δt before simulation start by the respective boundary conditions:

$$\Delta t = \frac{m_{CellIsoMat}}{u_{max} \cdot 1mm^2 \cdot 0{,}00274 \frac{g}{mm^3}} \quad (8)$$

The simulation course is realized in a way that the process is going on parallel at every time step. Hence the rules are executed on the cells against the flow direction at each step. Additionally the cells in one yz-plane are chosen randomly. After selecting a cell, it is tested if any material can be transported to the next cell (the maximum amount is calculated by equation 7). Every material transport is saved for each cell in an auxiliary variable. At the end of a simulation step the current cell mass is calculated by the old mass and the auxiliary variable:

$$\mathrm{m}_{Cell,i} = \mathrm{m}_{Cell,i-i} + \mathrm{m}_{aux} \quad (9)$$

The turbulence influence on the fibers is modeled by dispersion. Dispersion processes can be distinguished into processes with and without outer force effect. A dispersion process which is only caused by the thermic particle energy is called diffusion. If the dispersion of the solid fraction phase in a turbulent flow is considered, the diffusion rate on the particle mixing is negligible [6]. Hence a simple dispersion rule was designed and implemented.

The dispersion step is affiliated to the transportation step. The first action of this step is that a cell is chosen. This selection is occurred plane-by-plane in flow direction. At each plane the single cells are selected randomly. Subsequent to the cell selection, the dispersion process is applied to the respective cell, if the material mass at this cell is greater than zero. This dispersion process can be structured into the following steps:

1. Throw a random number and determine if the dispersion process is carried out.
2. Choose a random neighbor cell (Moore-NH with arbitrary radius).
3. Swapping the material mass with the neighbor cell.

4. Summary and Outlook

In the present paper a CA for simulating the transport of insulation material in a channel flow was introduced.

For the further work, the focus is set on the accretion experiments. For these works the final experimental setup has to be clarified. If essential, the models have to be readjusted to this setup. The realization and instrumentation of the experiments is designed in a way that the several submodels can be parameterized and validated.

Depending on the quality of the simulation results further improvements of the movement model are planned. This is e.g. the consideration of the gravity in a horizontal flow.

Acknowledgements

This work is supported by funds of Federal Ministry of Economics and Technology.
Support code: 1501402

References

[1] C. M. Alles: *Prozeßstrategien für die Filtration mit kompressiblen Kuchen.* Dissertation, Universität Fridericiana Karlsruhe (TH) (2000)

[2] M. Maqua: *Das Barsebäck-Ereignis: Ablauf, Bedeutung und Folgen.* KTG-Fachtagung, Rossendorf (2006)

[3] U. Frisch, B. Hasslacher and Y. Pomeau: *Lattice-Gas Automata for the Navier-Stokes Equation.* Physical Review Letters Vol. 56(14), S. 1505–1508, American Physical Society (1986)

[4] M. Gerhardt and H. Schuster: *Das digitale Universum Zelluläre Automaten als Modelle der Natur.* Vieweg Verlag, Braunschweig, Wiesbaden (1995)

[5] H. Sigloch: *Technische Fluidmechanik.* Springer-Verlag Berlin, Heidelberg (2009)

[6] T. Schmitz: *Dispersion passiver Partikel in der dreidimensionalen Turbulenz.* Diplomarbeit, Westfälische Wilhelms-Universität, Münster (2010)

REDUCING EMI IN MULTIPLE-OUTPUT DC CONVERTERS WITH CHAOS CONTROL*

YUHONG SONG[1, 2], ZHONG LI[1], JUNYING NIU[1, 2], GUIDONG ZHANG[1], AND WOLFGANG HALANG[1], HOLGER HIRSCH[3]

[1]*Faculty of Mathematics and Computer Science, FernUniversität in Hagen, Germany*

[2] *Dept. of Electronic and Information Enginneering, Shunde Polytechnic, China*

[3]*Faculty of Engineering, University of Duisburg-Essen, Germany*

It was proposed in the former research that chaos control can be used to suppress electromagnetic interference (EMI) in DC-DC converters. Here, a practical application in a power supply ATX is to be given to show that inputting an external chaotic signal to a Pulse Width Modulation (PWM) control circuit can reduce EMI by spreading the spectra of switch voltage, transformer current and output inductor current over the whole frequency band.

1. Introduction

Nowadays, due to lower power consummation demand, switch mode power supplies have increasingly wide applications in various electrical and electronic devices in industry and daily life. The switching power makes use of high rates of changes of voltage and current, resulting in the electromagnetic interference (EMI), which impairs other devices' performance and harms human bing's health.

Normally, EMI is handled by appending properly-tuned filters or by the electromagnetic shielding according to the frequency bands. However, both approaches are not perfect solutions of the EMI. Many researchers have proposed some approaches for the EMI problems [1, 2].

Hong Li et al. [3, 6] have contributed to the application of chaos control in DC-DC converters to reducing EMI in DC-DC converters, including the system design, dynamics analysis, simulations, and hardware implementations of chaotic DC-DC converters. There, the models, such as Boost converters, are ideal, lacking a practical application. Therefore, an real example, ATX2.0 power supply, is to be given to verify the effectiveness of chaos control in EMI suppression.

* This work was supported by AiF project under Grant No. 17211N

2. Design of Chaotic PWM in ATX Power Supplies

2.1. *System Design*

Power supply ATX2.0 is a typical accessory of desk computer, which compromises many modules, in which the PWM control circuit is to be driven by a chaotic signal, generated by Chua's circuit. The whole diagram of power supply ATX2.0 with the chaos control is depicted in Fig. 1.

The main circuit consists of a half bridge reverting module and a high frequency (HF) isolating transformer, which can generate multiple DC voltages, such as ±5V, ±12V, etc. Some other peripheral components include the auxiliary power, the signal detection and protection components. The main circuit topology, as shown in the dashed rectangle in Fig. 1, is detailed in Fig. 2.

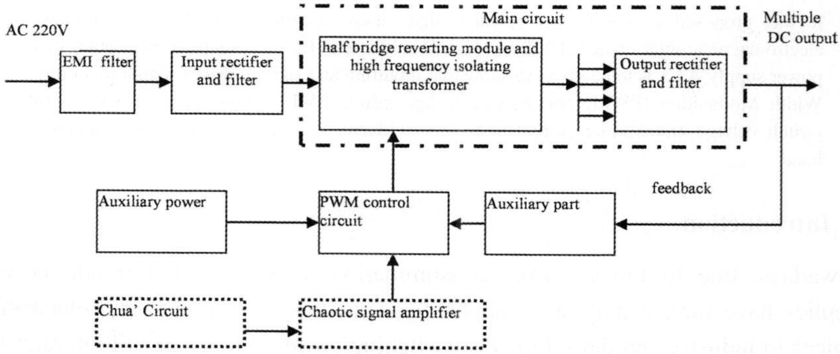

Fig. 1. Schematic diagram of the computer power supply with chaotic control

2.2. *Chaotic PWM*

The above-mentioned PWM control module adopts a special IC TL494[9].The TL494 is a fixed–frequency pulse width modulation control circuit, incorporating the primary building blocks required for the control of a switching power supply, as shown in Fig. 3. An internal linear saw-tooth oscillator is frequency–programmable by two external components, R_T and C_T. The oscillator frequency is determined by

$$f_{osc} = \frac{1.1}{R_T \cdot C_T} \quad (1)$$

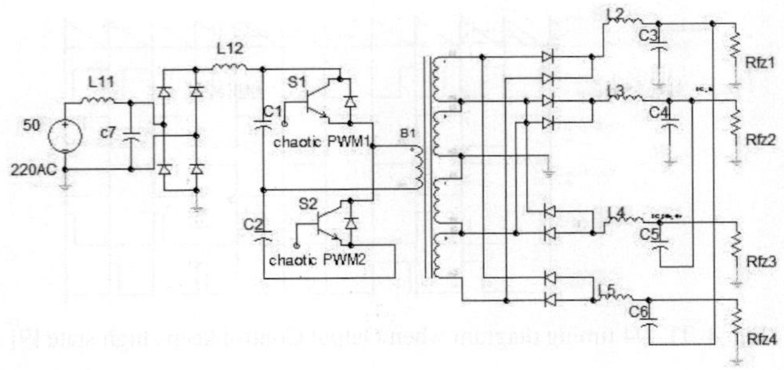

Fig. 2 Main circuit topology

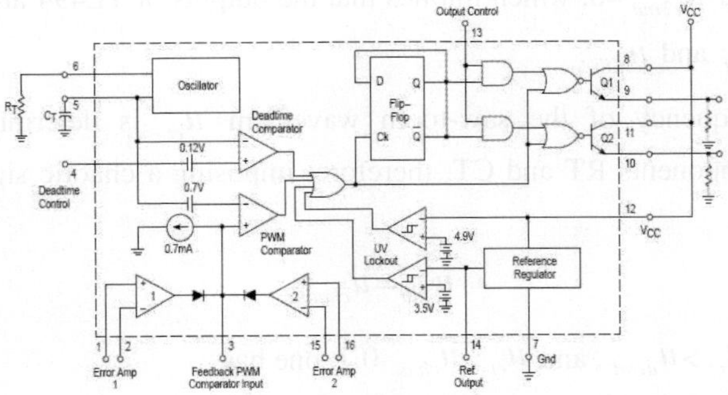

Fig. 3 TL494 Block Diagram

The control signals are external inputs that can be fed into the dead–time control (DTC) through pin 4, the error amplifier inputs through pins 1, 2, 15, 16, or the feedback input through pin 3, as shown in Fig. 3. Output pulse width modulation is accomplished by comparing the positive saw-tooth waveform across capacitor C_T with either of non-inverting inputs of Dead time Comparator and PWM Comparator. When the saw-tooth wave transient value is greater than that of the control signals, the NOR gates are enabled only if the flip–flop clock pulse, denoted by CK, is changed to logic 0. NOR gates drive output transistors Q1 and Q2. Therefore, an increase in control–signal amplitude causes a corresponding decrease of output pulse width.

Assume that the output control at pin 13 keeps at logic 1. Denote the outputs of Error Amplifiers 1 and 2 by u_{e1out} and u_{e2out}, respectively, the signal at pin 5 by u_{ct}, another input of the Dead-time Control Comparator by u_{dead}, and the feedback signal at pin 3 by u_{fdb}.

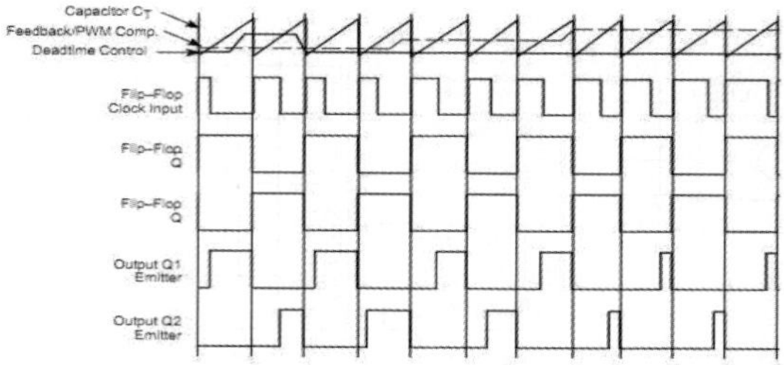

Fig. 4 TL494 timing diagram when Output Control keeps high state [9]

Fix the DTC input u_{dead} to a constant between 0 and 3.3 V and assume that u_{e1out} =0 and u_{e2out} =0, which implies that the outputs of TL494 are determined by u_{dead} ,u_{ct} and u_{fdb} .

The frequency of the saw-tooth waveform u_{ct} is determined by two external components RT and CT, therefore, imposing a chaotic signal to pin 3 results in

$$u_{fdb} = u_{chaotic} . \tag{2}$$

When $u_{ct} > u_{dead}$, and $u_{ct} > u_{chua}$ -0.7, one has

$$CK = 0 \text{ , } = \bar{Qn} \text{ , and } Q2_b = Qn \text{ ,} \tag{3}$$

otherwise,

$$CK = 1 \text{ , } Q1_b = 0, \quad Q2_b = 0, \tag{4}$$

where Qn and $\bar{Qn}$ stand for the outputs of the flip–flop. Qn and $\bar{Qn}$ have the opposite state. if Qn is at logic 1, then $\bar{Qn}$ is at logic 0.

It is remarked that a change in chaotic signal amplitude causes a corresponding change of output pulse width.

2.3. *Analogue Chaotic Signal*

In recent decades, chaotic oscillators have been extensively investigated [4, 5].Chaotic oscillators have been used in the PWM control of DC-DC converters to reduce EMI [1]. Among the existing chaotic oscillators, Chua's, Lorentz's, and Chen's oscillators are well known. In this paper, Chua's oscillator is adopted due to its simplicity and maturity. As shown in Fig. 5, Nr is Chua's diode. Chua's oscillator can be described by the following differential equations,

$$\begin{cases} \dfrac{du_1}{dt} = \dfrac{1}{C_{11}}\left[(u_2 - u_1)G - f(u_1)\right] \\ \dfrac{du_2}{dt} = \dfrac{1}{C_{22}}\left[(u_1 - u_2)G + i_1\right] \\ \dfrac{di_1}{dt} = -\dfrac{1}{L_1}(u_2 + r_0 i_1) \end{cases} \tag{5}$$

where G stands for the reciprocal of Ohm to Chua's diode Nr and r0 stands for resistor of the inductor L1.

The voltage at the feedback pin 3 of TL494, u_{fdb}, is required to vary between 0.5 V and 3.5 V, thus, an amplifier is designed to scale the signal $u_{chua} = u_2$ into [0.5, 3.5]. The circuit output is indicated by $u_{chaotic}$.

3. Simulation Results

A default value to dead-time control is given by connecting the resister to the ground, and inputting fixed values to two error amplifiers results in two zero outputs. Let R_T =12K Ohm (Ω) and C_T=1.5n Fala (F), and the parameters of other components are taken from a real ATX2.0 power supply.

The simulation is conducted with Saber Ver. 2008. Comparisons of the simulation results under the conventional and chaotic PWM controls have been done. EMI reduction is observed by using FFT by spreading the spectra of the critical components signals such as the voltage between the collector and emitter of the HF transistor, the voltage across the HF transformer and the inductor currents of multiple outputs. The simulation results are given in Fig.6.

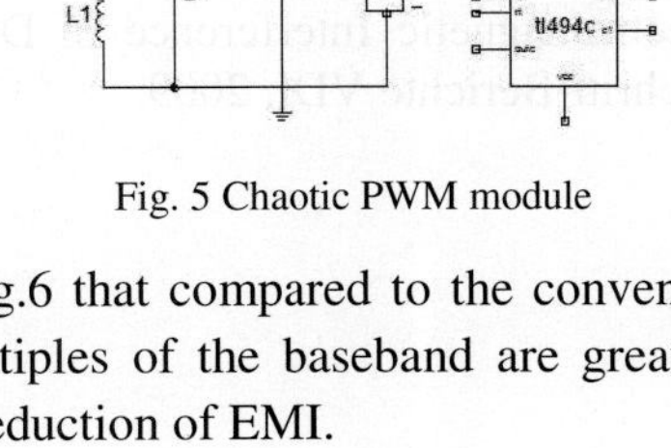

Fig. 5 Chaotic PWM module

It is seen from Fig.6 that compared to the conventional PWM control the amplitudes of the multiples of the baseband are greatly reduced under chaos control, implying the reduction of EMI.

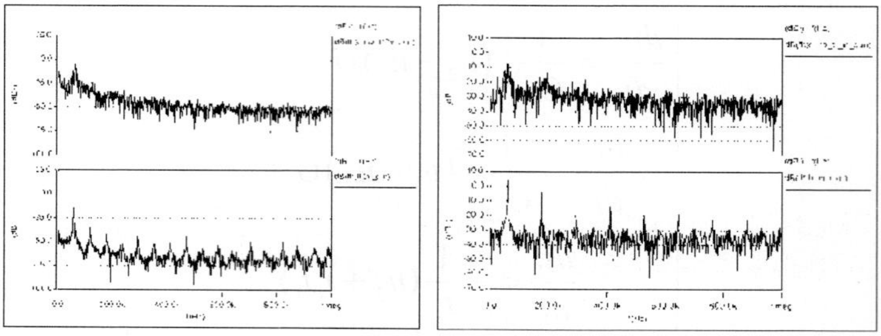

a. Inductor current at +12V output b. primary winding current of HF transformer

Fig. 6. FFT spectra（lower waveform in each figure corresponds to the conventional PWM control, the upper one to the chaotic PWM control）

4. Conclusion

In order to verify the effectiveness of chaos control in suppressing EMI in DC converters, a practical example, i.e. ATX2.0 power supply, was used as a test-bed. FFT spectra of critical signals indicate significant reduction of EMI under the chaotic mode operation.

References

1. M. Balestra, M. Lazzarini, G. Setti, and R. Rovatti. Experimental performance evaluation of a low-emi chaos-based current-programmed dc-dc boost converter. In IEEE Int. Symp. Circuits and Systems, Vol. 2, pp. 1489-1492, May 2005.
2. A. A. Amini and J. Nazarzadeh. Improvement behavior and chaos control of buck converter in current mode controlled. In Proc. IEEE Int. Conference on Industrial Technology, pp. 1-6, April 21-24, 2008.
3. H. Li, Z. Li, B. Zhang, Wallace K.S. Tang and W. HaLang. Suppressing Electromagnetic Interference in Direct Current Converters. IEEE Circuits and Systems Magazine, Vol. 2, pp. 10-28, Fourth Quarter 2009.
4. E. Bilotta, P. Pantano, and F. Stranges. A gallery of chua attractors: Part I. Int. J. Bifurcation and Chaos, Vol.17, pp. 1-60, 2007.
5. G. Chen. Controlling chua's global unfolding circuit family. IEEE Trans. Circuits and Systems, Part I, Vol. 40, pp. 829-832, 1993.
6. H. Li. Reducing Electromagnetic Interference in DC-DC Converters with Chaos Control. Fortschritt-Berichte VDI, 2009.

AN APPLICATION OF FILLED FUNCTION METHOD TO THE HARDNESS PROPERTY OF FE-MN BINARY ALLOYS*

AHMET SAHINER

Suleyman Demirel University, Department of Mathematics, 32260, Isparta, TURKEY

HAVVA GOKKAYA

Suleyman Demirel University, Department of Mathematics, 32260, Isparta, TURKEY

In this paper, we apply the filled function method to solving a global optimization problem which investigates the maximum microhardness of Fe-Mn binary alloys in different layer pairs. A solution algorithm is also proposed.

1. Introduction

The filled function method was first proposed by Ge [3,6]. Ge specifically proposed the following two-parameter filled function which has some disadvantages such as including an exponential term. When the value of the exponential term increases rapidly, changes in $P\left(x,x_1^*,r,\rho\right)$ and $\nabla P\left(x,x_1^*,r,\rho\right)$ become indistinguishable [3].

$$P\left(x,x_1^*,r,\rho\right)=\frac{1}{r+f\left(x\right)}\exp\left(\frac{\left\|x-x_1^*\right\|^2}{\rho^2}\right)$$

We can see different studies on filled function method in the literature trying to tackle these disadvantages [5,7,8].

In this article, by using filled function method we offer an algorithm to solve the physical problem examining the microhardness of $Fe-Mn$ binary alloys.

Definition 1.1. The basin B_k^* at a local minimizer x_k^* of $f\left(x\right)$ is a connected domain which contains x_k^* and in which starting from any point the

* This research is partially supported by Suleyman Demirel University Project Grant, under Grant number 2948-YL-11.

steepest descent trajectory of $f(x)$ converges to x_k^*, but outside which the steepest descent trajectory does not converges to x_k^* [7].

Definition 1.2. A hill of $f(x)$ at x_k^* is the basin of $-f(x)$ at its minimizer x_k^*, if x_k^* is a maximizer of $f(x)$ [7].

Definition 1.3. A local minimizer x_{k+1}^* of $f(x)$ is said to be lower (higher) than x_k^* if

$$f(x_{k+1}^*) < f(x_k^*) \ \left(f(x_{k+1}^*) > f(x_k^*)\right)$$

and in this case the basin B_{k+1}^* of $f(x)$ at x_{k+1}^* is said to be lower (higher) than B_k^* [7].

Definition 1.4. $p(x)$ is called a filled function of the function of $f(x)$ at a local minimizer x_k^* if $p(x)$ has the following properties: (i) x_k^* is a local maximizer of $p(x)$, (ii) $p(x)$ has neither minimizer or saddle point in $W_k^* \setminus \{x_k^*\}$ and the set W_k^* becomes a part of a hill of $p(x)$ at x_k^*, where

$$W_k^* = B_k^* \cup U_k^*$$

and U_k^* is the union of all basins higher than B_k^*, (iii) If $f(x)$ has a basin B_{k+1}^*, lower than B_k^*, then there exists a point $x' \in B_{k+1}^*$ that minimizes $p(x)$ along the ray $x_k^* + \lambda(x' - x_k^*)$, $\lambda > 0$.

The above properties of a filled function ensure that when any of the descent methods, for example the steepest decent method, the Newton method, the conjugate gradient method, etc., is employed to minimize the constructed filled function, the sequence of iteration will not terminate at any point in W_k^*, and when there exists basins of $f(x)$ lower than B_k^*, the sequence will either terminate at a point in a basin lower than B_k^* or generate a point x such that the value of $f(x)$ is less then $f(x_k^*)$ [7].

2. Algorithm FF

This work focuses on the constrained two dimensional physical problem. Accordingly, we have an algorithm as follows:

(1) Choose a constant $\varepsilon > 0$ (set $\varepsilon > 10^{-4}$. Choose lower and upper bound in which the global optimum of $f(x)$ is to be found. Start

from an x_0 point which is in the lower and upper bound. Minimize $f(x)$ by using a local algorithm to find a local minimizer x_1. Use a_0 as the initial value of the weight factor a ($a_0 = 2$).

(2) Construct the filled function *ff* by using x_1 and a. Minimize the *ff* function with x_0 as the starting point to obtain a local minimizer denoted x.

(3)

i. If such a point x is found that it is minimizer of the *ff* function, go to 4

ii. If the iteration sequence reaches the lower bound of D, then stop and the current x is adopted as the global minimizer.

(4) With x_1 as the starting point, minimize $f(x)$ to find another local minimizer $x_2(\neq x_1)$ (if existing).

i. If $f(x_2) < f(x_1)$, x_1 is updated with x_2 but a is unchanged, then go to 2.

ii. If $f(x_2) > f(x_1)$, x_1 is unchanged but a is updated (a replaced by $a+4$), then go to 2.

Although different filled functions can be used, we have used $Q(x,a) = -\left(f(x) - f(x_1^*)\right)e^{a\|x-x_1^*\|^2}$, $a > 0$ filled function which is proposed by Qin and $H_1(x,a) = \dfrac{1}{y\left[f(x) - f(x_1^*)\right]} + \dfrac{a}{\|x - x_1^*\|^2}$, $a > 0$ filled function where $y(.)$ is a one-dimensional real function with several analytical properties which is proposed by Liu [4,5].

3. An Application of Filled Function Method to the Physical Problem: *Fe–Mn* Binary Alloys

$Fe-Mn$ binary alloys are used in many fields such as construction of planes, weapon production, etc. The goal of the optimization procedure is typically to maximize the microhardness. The task of the $Fe-Mn$ binary alloys optimization is to make a specified objective function reach its optimal value while keeping other technical indices within acceptable ranges. $Fe-Mn$ binary alloys having different Mn concentrations were subjected to the process of boriding at different times and different temperatures. Then, microhardnesses of the binary alloys were measured [1]. $Fe-Mn$ binary alloys having different Mn concentrations at $3h$ and $1073-1373K$ temperature were subjected to the process of boriding. After boronizing process, three distinct regions have been identified on the surface of binary alloys: boride layer including borides, transition zone, the matrix which is not affected by boron. The microhardnesses of these regions were measured. For more information about this subject see [1,2].

By using the data in [1,2] we obtain the following maximum microhardness function for different pairs of layers:

$$f(x_k) = a_k + b_k \cos(x_k z_k) + c_k \sin(x_k w_k).$$

Then the resultant maximum microhardness function for $Fe-Mn$ binary alloys is

$$f(x) = \sum_{k=1}^{n} f_k(x_k),$$

where $x = (x_1, x_2, ..., x_n)$. The model also contains the constraints $ls_k \leq x_k \leq us_k$ and $\sum_{k=1}^{n} x_k = B =$ constant.

3.1. *Experiments and Result Analysis*

We have the constants as shown in Table 1 for $Fe-Mn$ binary alloy at 0.42 wt. % Mn. a_1, b_1, c_1, z_1, and w_1 are belongs to boride layer. a_2, b_2, c_2, z_2, w_2 and a_3, b_3, c_3, z_3, w_3 are belongs to transition zone and matrix layer, respectively.

Table 1. Parameters of the example for 0.42 wt.% M006E

a_1	b_1	c_1	z_1	w_1
770.6	-49.1	476.5	0.1	0.01075
a_2	b_2	c_2	z_2	w_2
162.9	-11.19	-24.74	0.02121	0.02121
a_3	b_3	c_3	z_3	w_3
143.6	-19.42	-13.5	0.02026	0.02026

If we use the following constraints

$$x_1 + x_2 = 2746,$$
$$1073 \leq x_1 \leq 1373,$$
$$1073 \leq x_2 \leq 1373.$$

problem's dimension will be reduced from two to one.

It is obvious from Figures 1 and 2 that the objective functions are multimodel. We conclude that $Fe-Mn$ binary alloy at 0.42 wt.% Mn achieved maximum microhardness at 1347.5366K for boride layer and transition zone. $Fe-Mn$ binary alloy at 0.42 wt.% Mn achieved maximum microhardness at 1291.2922K for boride layer and matrix layer and at 1358.1688K for transition zone and matrix layer.

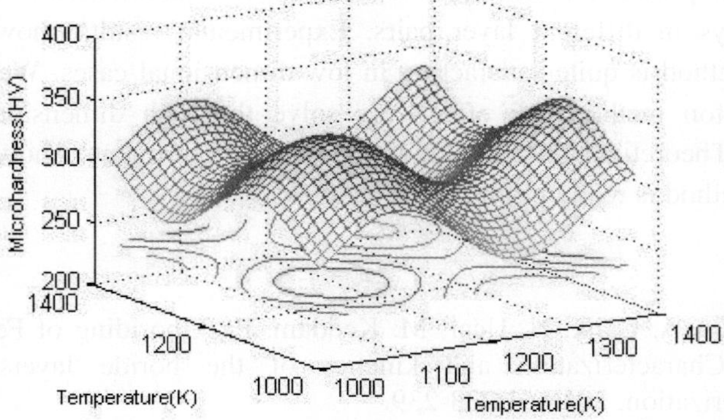

Figure 1. The Microhardness Profile Without Constraint For Transition Zone and Matrix Layers

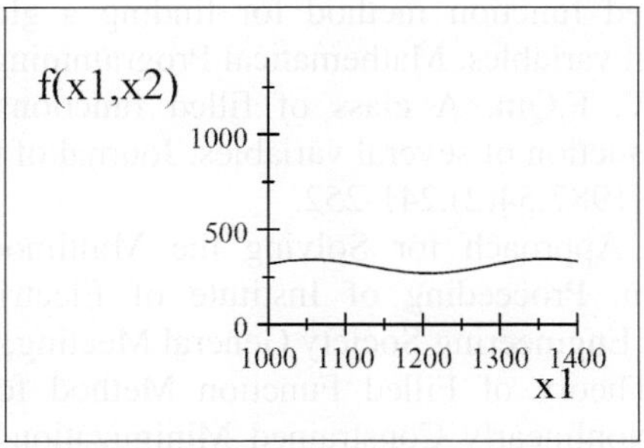

Figure 2. The Microhardness Profile With Constraint For Transition Zone and Matrix Layers

Making similar observations for $Fe-Mn$ binary alloy at 0.76 wt.% Mn and at 0.94 wt.% Mn, we obtained that $Fe-Mn$ binary alloy at 0.94 wt. % Mn achieve maximum microhardness at the upper bound temperature point 1373K for boride layer and transition zone. Also, the other layer pairs (boride layer, the matrix and transition zone, the matrix) achieved maximum microhardness at 1373K $Fe-Mn$ binary alloy at 0.76 wt.% Mn achieved maximum microhardness at 1366.6635K for boride layer and transition zone, at 1398.0226K for boride layer and matrix layer and at the upper bound temperature point 1373K for transition zone and matrix.

4. Conclusions

Although there are many efficient methods for finding a local minimizer of the objective function the filled function method is one of the fewer methods offering a global minimizer. This makes filled function method advantageous compared to other methods.

In this paper, we applied filled function method to solving a global optimization problem which seeks the maximum microhardness of $Fe-Mn$ binary alloys in different layer pairs. Experimental results show that filled function method is quite satisfactory in low dimensional cases. We expect that filled function method can effectively solve the high dimensional physical problems. Theoretical analyses and numerical experiments show that filled function method is really promising.

References

1. M. Bektas, A. Calik, N. Ucar, M. Keddam, Pack-boriding of Fe-Mn binary alloys: Characterization and kinetics of the boride layers, Materials Characterization, 2010;61:233-239.
2. A. Calik, O. Sahin, A. E. Ekinci, N. Ucar, Mechanical Properties of Boronized Fe-0.94% Mn Binary Alloy, Z. Naturforsch 2007;62:545-548.
3. R. P. Ge, A filled function method for finding a global minimizer of a function of several variables, Mathematical Programming, 1990;46:191-204.
4. R. P. Ge, and Y. F.Qin, A class of filled functions for finding global minimizers of a function of several variables, Journal of Optimization Theory and Applications, 1987;54(2):241-252.
5. X. Liu, A New Approach for Solving the Multimodal Economic Load Dispatch Problem, Proceeding of Institute of Electrical and Electronics Engineers, Power Engineering Society General Meeting, 2007.
6. R. P. Ge, The Theory of Filled Function Method for Finding a Global Minimizer of a Nonlinearly Constrained Minimization Problem, Journal of Computational Mathematics, 1987;5(1) :1-9.
7. Z. Xu, H.X. Huang, P. Pardalos, C.X. Xu, Filled Functions For Unconstrained Global Optimization, Journal of Global Optimization, 2001;20:49-65.
8. X. Liu, A class of augmented filled functions, Computational Optimization and Applications, 2006; 33:333-347.

SHEDDING LIGHT ON THE UNDERLYING LONG-RUN PRICE LEADERSHIP: A STUDY OF US GASOLINE PRICE DATA*

TAKAMITSU KURITA†

Faculty of Economics, Fukuoka University, 8-19-1 Nanakuma, Jonan-ku, Fukuoka 814-0180, Japan

This note aims to shed some useful light on the underlying long-run price leadership in market data. A cointegrated vector autoregressive analysis is performed using weekly data for gasoline prices observed in three regions in the US. Statistical evidence revealed in the analysis indicates that one of the prices plays the role of long-run leadership, thus exerting persistent influences on the price determination in the other regions.

1. Introduction

The purpose of this note is to reveal the underlying long-run price leadership in market data. An empirical understanding of such price leadership can be of practical importance for economists analyzing commodity and financial markets. This section briefly reviews cointegration, common stochastic trends and long-run price leadership, then describing the significant points of this note.

Economic and financial time series data often exhibit trending behavior and thus should be seen as integrated processes rather than stationary. A cointegrated vector autoregressive (VAR) system for data integrated of order 1, denoted as I(1) hereafter, is introduced by Johansen [1][2] and has gained a great popularity among economists and econometricians. See also Burke and Hunter [3] as well as Kurita [6] for theories and applications of cointegrated VAR systems. The presence of cointegration in a VAR system implies that common stochastic trends are incorporated in the system. Cointegration and common stochastic trends are therefore regarded as two sides of the same coin (see Johansen [2], Ch.3). The presence of r cointegrating relations in an n-dimensional I(1) cointegrated VAR system (for $r < n$) implies that the system is driven by $n - r$ common stochastic trends.

* I would like to thank John Hunter and anonymous referees for their helpful comments and suggestions. A financial support from JSPS KAKENHI (19830111) is also gratefully acknowledged.

† tkurita@fukuoka-u.ac.jp.

Let us turn to a review of long-run price leadership. The idea of long-run price leadership, which is introduced by Burke and Hunter [8], is closely related to the existence of a single common trend in a cointegrated VAR system. Suppose that there are a group of n prices, $p_{1,t}, \dots, p_{n,t}$, interacting with each other, which are well fitted in a cointegrated VAR system. The prices are thus bundled in a vector $X_t = (p_{1,t}, \dots, p_{n,t})'$, so that a cointegrated VAR(k) system for X_t, based on Johansen [1,2], can be defined as

$$\Delta X_t = \alpha(\beta', \gamma)\binom{X_{t-1}}{1} + \sum_{i=1}^{k-1} \Gamma_i \Delta X_{t-i} + \varepsilon_t, \ for \ t = 1, \cdots, T \qquad (1)$$

where a sequence of innovations ε_t has independent and identical normal $N(0, \Omega)$ distributions conditional on $X_{-k+1}, \cdots, X_0$, the parameters $\alpha, \beta \in \mathbf{R}^{n\times r}$ for $r < n$, $\gamma \in \mathbf{R}^r$ and $\Gamma_i, \Omega \in \mathbf{R}^{n\times n}$ vary freely, and Ω is a positive definite matrix. The parameter α is called adjustment vectors, while β is referred to as cointegrating vectors, so that $\beta' X_{t-1}$ represents cointegrating relationships. An orthogonal complement $\alpha_\perp$ is defined such that $\alpha'\alpha_\perp = 0$ with the matrix $(\alpha, \alpha_\perp)$ being of full rank. A sequence of shocks $\alpha'_\perp \varepsilon_t$ accumulates to the underlying common stochastic trends, which give rise to the observable non-stationary trending behavior of X_t. In addition, let us denote $\beta^{*\prime} = (\beta', \gamma)$ and $X^*_{t-1} = (X_{t-1}', t)'$ for future reference.

Let us conceive a situation where $p_{n,t}$ plays the role of long-run leadership in Eq. (1). Burke and Hunter [8] introduce a set of conditions required for the existence of long-run price leadership. According to them, the definition of long-run price leadership may be given as follows. Suppose $X_t = (p_{1,t}, \dots, p_{n,t})'$ is generated by Eq. (1). The n-th price $p_{n,t}$ is said to play the role of long-run leadership in Eq. (1) if the following conditions are satisfied:

1. there is a single common stochastic trend, *i.e.* $n - r = 1$,
2. the single common stochastic trend derives from the innovation process for $p_{n,t}$, *i.e.* $\alpha_\perp = (\mathbf{0}_{1\times(n-1)}, 1)'$,
3. $n - 1$ cointegrating relations are expressed as

$$\beta' X_{t-1} = \begin{pmatrix} 1 & 0 & \dots & \dots & -1 \\ 0 & 1 & 0 & \dots & -1 \\ 0 & 0 & \ddots & \dots & \vdots \\ 0 & 0 & \dots & 1 & -1 \end{pmatrix} \begin{pmatrix} p_{1,t-1} \\ p_{2,t-1} \\ \vdots \\ p_{n,t-1} \end{pmatrix} = \begin{pmatrix} p_{1,t-1} - p_{n,t-1} \\ p_{2,t-1} - p_{n,t-1} \\ \vdots \\ p_{n-1,t-1} - p_{n,t-1} \end{pmatrix} \qquad (2)$$

The source of the long-run price leadership is accumulated innovations for $p_{n,t}$ solely, rather than those of combined multiple prices, as demonstrated in the second condition above. The concept of long-run price leadership plays a key role in this note.

This note, with a view to revealing the underlying long-run price leadership, analyzes weekly data for gasoline prices in three regions of the US: New York Harbor, Los Angeles and Gulf Coast. Petrochemical is known as one of the dominant industries in the area of Gulf Coast. It is, therefore, possible to conceive the hypothesis that the price set in Gulf Coast plays the role of long-run leadership, affecting the price determination in the other regions, New York Harbor and Los Angeles. An econometric analysis in this note, using the cointegrated VAR methodology, turns out to be in support of the validity of this hypothesis. It is true that price dynamics in gasoline markets have been subjected to empirical explorations; however, the literature seems to center on the investigation of relationships between gasoline prices and crude oil prices (see Chen, Finney and Lai [4], and Kurita [7], *inter alia*). This note, in contrast, focuses on US gasoline markets and aims to cast useful light on their internal long-run price interactions. Consistent with Kurita [7], this study also lends weight to the validity of the approach of Burke and Hunter [8] to the description of long-run price leadership in real-life markets. Finally let us note that all empirical analysis and graphics in this study use OxMetrics/PcGive (Doornik and Hendry [5]).

2. Empirical Investigation of US Gasoline Price Data

This section analyzes weekly data for gasoline prices in the three areas of the US. Thus the dimension of a VAR system is three (n = 3). The analysis starts with the estimation of a general unrestricted VAR system for $X_t = (p_t^{NY}, p_t^{LA}, p_t^{GC})'$, in which the members are the logs of weekly spot prices for conventional regular gasoline (cents per gallon) in New York Harbor, Los Angeles and Gulf Coast, respectively. The sample period runs from the first week in January 2000 to the final week in August 2008. The number of observations amounts to 452, which is so large that one can rely on asymptotic theory for cointegration analysis. The data source is the website of US Energy Information Administration (Oct 2008). The order of 2 is chosen for the lag length of the system based on F tests. The estimation of a VAR(2) system reveals several outliers in its residuals in September and October 2005, caused by two large hurricanes, Katrina and Rita. A set of dummy variables are thus introduced: $D_{1,t} = 1(1\text{w Sep. } 2005)$, $D_{2,t} = 1(2\text{w Sep. } 2005)$, $D_{3,t} = 1(4\text{w Sep. } 2005)$, $D_{4,t} = 1(1\text{w Oct. } 2005)$, and 0 otherwise, in which "1w" denotes the first week, for example. It is then found that the residuals have become free from any outliers and exhibit no strong serial correlation; we are, therefore, justified in performing a likelihood analysis of cointegration using this VAR system.

2.1. *Determination of the Cointegrating Rank*

We are now in a position to examine the hypothesis that the price set in Gulf Coast plays the role of long-run leadership in the gasoline markets. The first question of interest is whether there is a single common stochastic trend in the data. This corresponds to a case where cointegrating rank is judged to be two ($r = n - 1 = 2$). Table 1 presents the log-likelihood ratio (log LR) test statistics for cointegrating rank, in addition to the modulus of the six largest roots of a companion matrix for the VAR system (see Johansen [2]). The log LR tests in the first panel reject the null of $r = 0$ and $r \leq 1$ with their p-values both being zero, and do not reject $r \leq 2$ with its p-value holding a fairly large value, 0.75. The second panel shows modulus (denoted mod) of the three largest eigenvalues of the companion matrix restricted with $r = 2$, or $n - r = 1$. All the eigenvalues, apart from the first one (imposed unit root), appear to be distinct from unit roots. The overall results support the conclusion of $r = 2$, which implies that there is a single common stochastic trend in the data.

Table 1. Determination of the cointegrating rank

	r = 0	r ≤ 1	r ≤ 2
log LR	126.95** [0.00]	60.52** [0.00]	2.15 [0.75]
mod (r = 2)	1	0.72	0.72

Notes: Figures in the square brackets are p-values according to the limiting distribution. ** denotes significance at the 1% level.

2.2. *Restrictions on α and β*

The choice of the cointegrating rank ($r = 2$) enables us to test hypothetical restrictions on the estimates for α and β. According to the definition of long-run price leadership, the next questions of interest are: (i) whether zero restrictions can be imposed on those elements of α which correspond to p_t^{GC} so that we find $\alpha_\perp = (0,0,1)'$ and (ii) whether the cointegrating space is expressible as Eq. (2). The joint restrictions are imposed on the estimates to find the following results:

$$\hat{\alpha}\widehat{\beta}^{*'} X_{t-1}^{*} = \begin{pmatrix} \underset{(0.03)}{-0.25} & \underset{(0.013)}{0.007} \\ \underset{(0.07)}{0.008} & \underset{(0.03)}{-0.23} \\ \underset{(-)}{0} & \underset{(-)}{0} \end{pmatrix} \begin{pmatrix} \underset{(-)}{1} & \underset{(-)}{0} & \underset{(-)}{-1} & \underset{(0.004)}{-0.01} \\ \underset{(-)}{0} & \underset{(-)}{1} & \underset{(-)}{-1} & \underset{(0.01)}{-0.11} \end{pmatrix} \begin{pmatrix} p_{t-1}^{NY} \\ p_{t-1}^{LA} \\ p_{t-1}^{GC} \\ 1 \end{pmatrix} \tag{3}$$

where the figures in parentheses are standard errors. The log LR test statistic for the joint restrictions is 5.15 [0.27], in which the figure in the square bracket is a p-value according to $\chi^2(4)$. Thus the joint hypothesis is not rejected at the 5% level. All the conditions for long-run price leadership being fulfilled, we are justified in concluding that the price set in Gulf Coast drives the prices in New York Harbor and Los Angeles, playing the role of long-run leadership in the gasoline markets.

Furthermore, both of the off-diagonal coefficients for α in Eq. (3) appear to be insignificant, suggesting the possibility that we can reach a more simplified data-representation. Zero-restrictions are thus placed on these coefficients to find that the resultant restricted estimates are given by

$$\hat{\alpha}\,\hat{\beta}^{*\prime}X_{t-1}^{*} = \begin{pmatrix} \underset{(0.03)}{-0.24} & \underset{(-)}{0} \\ \underset{(-)}{0} & \underset{(0.03)}{-0.23} \\ \underset{(-)}{0} & \underset{(-)}{0} \end{pmatrix} \begin{pmatrix} \underset{(-)}{1} & \underset{(-)}{0} & \underset{(-)}{-1} & \underset{(0.004)}{-0.01} \\ \underset{(-)}{0} & \underset{(-)}{1} & \underset{(-)}{-1} & \underset{(0.01)}{-0.11} \end{pmatrix} \begin{pmatrix} p_{t-1}^{NY} \\ p_{t-1}^{LA} \\ p_{t-1}^{GC} \\ 1 \end{pmatrix} \tag{4}$$

The log LR test statistic for the joint restrictions above is 5.50 [0.48], in which its p-value according to $\chi^2(6)$ is given in the square bracket. Again, the joint hypothesis is not rejected at the 5% level. The explicit expression of $\hat{\alpha}\hat{\beta}'X_{t-1}$ is

$$\hat{\alpha}\hat{\beta}'X_{t-1} = \begin{pmatrix} -0.24p_{t-1}^{NY} + 0.24p_{t-1}^{GC} \\ -0.23p_{t-1}^{LA} + 0.23p_{t-1}^{GC} \end{pmatrix} \tag{5}$$

The markets in New York and Los Angeles may, thus, be seen as separated in the sense that there is no long-run interaction between p_{t-1}^{NY} and p_{t-1}^{LA}.

2.3. *Equilibrium Correction Model*

Mapping the data to I(0) space by differencing and using the restricted cointegrating relations, the author estimates a vector equilibrium correction model, then removing insignificant regressors and finally arriving at a preferred parsimonious equilibrium correction model as follows:

$$\widehat{\Delta p_t^{NY}} = \underset{(0.03)}{-0.24}(p_{t-1}^{NY} - p_{t-1}^{GC} - 0.01) - \underset{(0.07)}{0.25}\Delta p_{t-1}^{NY} + \underset{(0.07)}{0.38}\Delta p_{t-1}^{GC} + \underset{(0.05)}{0.39}D_{1,t} - \underset{(0.05)}{0.26}D_{2,t} - \underset{(0.05)}{0.23}D_{4,t}$$

$$\widehat{\Delta p_t^{LA}} = \underset{(0.03)}{-0.23}(p_{t-1}^{LA} - p_{t-1}^{GC} - 0.11) + \underset{(0.04)}{0.24}\Delta p_{t-1}^{LA} + \underset{(0.06)}{0.26}D_{1,t} - \underset{(0.06)}{0.3}\,D_{2,t} - \underset{(0.06)}{0.19}D_{4,t}$$

$$\widehat{\Delta p_t^{GC}} = \underset{(0.08)}{0.42}\Delta p_{t-1}^{GC} - \underset{(0.08)}{0.28}\Delta p_{t-1}^{NY} + \underset{(0.02)}{0.04}\Delta p_{t-1}^{LA} + \underset{(0.05)}{0.37}D_{1,t} - \underset{(0.05)}{0.23}D_{2,t} + \underset{(0.02)}{0.15}D_{3,t} - \underset{(0.05)}{0.31}D_{4,t} \quad (6)$$

The reduced model above shows that there is no short-run interaction between Δp_t^{NY} and Δp_t^{LA}, in line with the structure of the long-run relations in Eq. (5).

3. Concluding Remarks

The analysis of this note has cast useful light on the issue of empirical long-run price leadership. The study also paves the way for further research extensions in the analysis of various price data.

References

1. S. Johansen, Statistical analysis of cointegration vectors, Journal of Economic Dynamics and Control, **12**, 231-254 (1988)
2. S. Johansen, Likelihood-Based Inference in Cointegrated Vector Auto-Regressive Models, 2nd printing, Oxford University Press (1996)
3. S. P. Burke and J. Hunter, Modelling Non-Stationary Economic Time Series - A Multivariate Approach, Palgrave Texts in Econometrics, Palgrave MacMillan (2005)
4. L.H. Chen, M. Finney and K.S. Lai, A threshold cointegration analysis of asymmetric price transmission from crude oil to gasoline prices, Economics Letters, **89**, 233-239 (2005)
5. J.A. Doornik and D.F. Hendry, Modelling Dynamic Systems Using PcGive, Volume 2, Timberlake Consultants LTD (2007)
6. T. Kurita, A dynamic econometric system for the real yen-dollar rate, Empirical Economics, **33**, 115-149 (2007)
7. T. Kurita, Time series analysis of transatlantic market interactions: evidence from crude oil and gasoline prices, International Journal of Business and Economics, **9**, 157-173 (2010)
8. S. P. Burke and J. Hunter, Long-run equilibrium price targeting, Quantitative and Qualitative Analysis in Social Sciences, **5**, 26-36 (2011)

INDIVIDUAL ASSESSMENT MODEL OF THE PUBLIC SPREADING IN THE ENVIRONMENTAL POLLUTION ACCIDENT*

BIN LUO[1], XIAOHONG LIU[2], XIAODONG PAN[3]

[1]*College of Chemistry and environment protection engineering, Southwest University for Nationalities. Chengdu, 610041,China*

[2]*College of Management, Southwest University for Nationalities, Chengdu, 610041, China*

[3]*School of Mathematics, Southwest Jiaotong University, Chengdu, 610031, China*

Environmental pollution accident is an eternal phenomenon in the process of human social activities. Phenomenon of the public spreading often appears after environmental pollution accident and before the government announces the formal environmental conclusion. It will easily result in distorted information and especially magnified harm. This paper establishes an individual assessment model of the public spreading in the Environmental pollution accident based on fuzzy comprehensive evaluation method and a numerical example shows the application of this model.

1. Introduction

Since ancient times, the various kinds of environmental problems caused by human activities are numerous. The complex affinity between humanity and nature has been come into being. Humanity is a production of the environment as well as innovator of the environment. As the world economy has developed rapidly, people pay more attention to environmental pollution accidents and their serious threat to the nature and the people's daily life. In recent years environmental pollution accidents happen frequently in our country. But there are some defects in the accident crime of great environmental pollution due to legislation value. In many countries, the predicament of accident crime of great environmental pollution is very low rates of applicability. In the relevant laws and legislative behind other cases, it is necessary to further understand that environmental pollution accident might have caused panic on the people's

* This work is This work is supported by grant 09SZYZJ05 of the Fundamental Research Funds for the Central Universities, Southwest University for Nationalities, and supported by grant 2010GXS5D253 of National Soft Science of China, and supported by grant 2011XWD-S1201 of Southwest University for Nationalities.

psychology and adverse effects on the people' s daily life, even affect on government's credibility and environmental pollution accident efficiency. Based on the analysis of environmental pollution accident and the concept of delivery of the people, the article suggests basic individual model of crowd spread about pollution accident. Forever, it establishes the comprehensive evaluation model of the public spreading pollution accident using a fuzzy comprehensive approach, offers a theoretical support to explain the environmental pollution accident on the phenomenon of crowd spread and assess application procedures the model through an example.

2. The Concept of Pollution Accident and the Public Spreading

In fact, the whole process of social activities has been accompanied by environmental pollution accidents. But due to the uncertainty of environmental pollution accident and people in the environmental pollution accident relying on the transfer of information, environmental pollution accidents often lead to psychological panic unduly and have great effect on normal work and life. (For example, in the nuclear leakage caused by the Japanese earthquake, salts were out of stock in many parts of the China overnight due to the public spreading. This incident shows that we are still lack a correct understanding of environmental pollution accident. It is necessary to attach importance to the function of public spreading the environmental pollution accident.

Environmental pollution, According to Chinese encyclopedia volumes of environmental science, environmental pollution is mainly environmental quality degradation caused by human activities and is harmful to normal survival and development of human and other living things. From the 1950s, people have widely begun to pay attention to environmental pollution problems. Pollution accident mainly means the environmental pollution results caused by improper behavior in the production process. The public spreading environmental pollution accident means that the natural spread phenomenon of relevant information among relevant crowd after environmental pollution accident have happened and before the government announced formal conclusion. In short, it means that accident is spread among relevant crowd and may lead to the mutual influence behavior. At present, its information natural communication path which is not been prevented completely. Obviously, the information natural communication is the disadvantages over the advantages in the dispose of the environmental pollution accident. It will easily result in distorted information and magnified harm. Therefore, we need to realize basic process of public spreading the environmental pollution accident and effectively describe mechanisms of individual receiving (input) information and sending(output) information in the public spreading. From human sustainable development, we need to focus on the earth's whole environmental systems, namely various environmental factors and their interrelations on the earth's surface. From environmental pollution accidents arising from the direct harm to the human race and other organisms, we need to focus on environmental systems within certain limits. Conversely, individual should not be affected by negative impact due to the public pass information if beyond a certain extent of environmental systems.

3. The Basic Process of Public Spreading Environment Pollution Accident

In general, environmental pollution accident is a public event and is not restricted from law and morality in principle, so it is very easy to appear public spreading information. If the government departments concerned can not release the truth in time or take measures to restrict spread of environmental pollution accident, a negative impact will also be greater. Therefore, relevant government departments should announce pollution accident timely and accurately.

The article proposes an individual model of public spreading environmental pollution accident which is mainly used to interpret the basic process of public spreading environment pollution accident.

Set of individual A, he (she) in certain cultural environment (H), receiving from different layers (reflecting the degree of tightness of differences) relationship between an object (X) about environmental pollution accident of input information, individual A to other people around (Y) then the output of environmental pollution accident information process.

Therefore, using (1) to explain the behavior of individual of environmental pollution accident population transfer:

$$Y_i = f_i(A_i, X_i, H) \qquad (1)$$

In type (1), Y_i, A_i, X_i, H denotes the spread information of individual of environmental pollution accident, individual information, received information and the humanistic environment.

This article mainly discusses the individual whether to spread received information about environmental pollution accidents. In fact, the secondly question is about the object of spread information. Because of the factors of A_i, X_i, H involved in fuzzy contents, therefore, we put forward the application the method of fuzzy comprehensive evaluation establish a decision model that whether individual continue to spread the information of environmental pollution accidents in the population.

4. Individual Comprehensive Evaluation Index Systems of Public Delivery of Environment Pollution Accident

This paper puts forward public spreading process of the environmental pollution accident, the individual comprehensive evaluation indexes are as follows:

Individual differences factors: An individual difference refers that different basic conditions of individual will result in the different spread scope and spread speed of environmental pollution accident. Public spreading environment pollution accident always base on individual, Individual differences decide to the spread continuity of environmental pollution accident and authenticity of dissemination of information. Individual differences factors include mainly environmental consciousness, relationship and the communion habit.

Information receiving factors: Information receiving factors refer that different information that individual has received will result in the different spread scope and spread speed of environmental pollution accident. Information receiving is the main external burst factors. Information receiving factors mainly include the path, speed and scope of public pass etc.

Human environment factors: Human environment factors refers that different external conditions that individuals are within bounds of political, economic, social and technology will result in the different spread scope and spread speed of environmental pollution accident. Human environments are the main external interference factors. The human environment factors mainly include politics and law, economic base, social values and technical conditions.

As shown in table 1, the article regards individual differences, information receiving and human environment as the first evaluating index and proposes the index systems of individual comprehensive evaluation.

5. Fuzzy Comprehensive Evaluation Model

5.1. *Fundamental model*

The two-layer individual evaluation model of the public spreading environment pollution accident based on fuzzy comprehensive evaluation is presented.

1. The definition of evaluation set of the public spreading environment pollution accident is $C = (C_1, C_2, C_3, C_4, C_5)$, where C_1, C_2, C_3, C_4, C_5, stands for the possibility of individual spreading pollution accident, namely respectively.

2. The definition of the 1st fuzzy evaluation factor set of the public spreading environment pollution accident is $T=T_1 \cup T_2 \cup T_3$, whereT_1、T_2、T_3 stands for individual differences, information receiving and human environment

respectively. The weight vectors of the 1st fuzzy evaluation factor set is defined as A= (a_1, a_2, a_3), and a_1+ a_2 +a_3=1.

3.The definition of the 2nd fuzzy evaluation factor set of the public spreading environment pollution accident is T_i={T_{i1}, T_{i2}, …,T_{ik}}, i=1,2,…, k is the secondary indicator number of the i-level first indicators. The weight vectors of the 2nd fuzzy is defined as Ai=(w_{i1}, w_{i2},…, w_{ik},), and w_{i1}+…+w_{ik}=1.

Table1 Individual comprehensive evaluation index systems of public delivery of environment pollution accident

First-class indexes	secondary-class indexes	Explanation about secondary-class indexes
1.individual differences ($\tilde{B}_1$) (weight:a1)	1.1environmental knowledge	Individual mastery of environmental knowledge
	1.2 relationship	Individual in possession of relationships
	1.3communion habit	Individual Communication habit
2.information receiving ($\tilde{B}_2$) (weight:a2)	2.1channels	Environmental pollution accident crowd spread channels
	2.2spread	Environmental pollution accident crowd spread channels
	2.3range	Environmental pollution accident crowd spread range
3.human environment ($\tilde{B}_3$) (weight: a3)	3.1politics and law	within the scope of politics and law
	3.2economic base	within the scope of economic base
	3.3social values	within the scope of social values
	3.4technical conditions	within the scope of technical conditions

5.2. *Matrix of evaluation and its calculation*

Set the result of single factor evaluation is t_{ijk}, i=1,2,3,j=1,2,…,l,k=1,2,…,5,l is the number of corresponding to the two grade index. According to the table 1, ask participators to evaluate the possible of spread the information of environmental pollution accidents in the population using the evaluation comment sets, such as great, large, general, small and small, and then according to the method of the maximum membership degree or other methods to treat the raw data of t_{ijk} ,and get the single factor evaluation matrix $\tilde{R}_i$,

$$\tilde{R}_i = \begin{bmatrix} t_{i11} & t_{i12} & t_{i13} & t_{i14} & t_{i15} \\ \cdots & \cdots & \cdots & \cdots & \cdots \\ t_{il1} & t_{il2} & t_{il3} & t_{il4} & t_{il5} \end{bmatrix} \quad (2)$$

According to the two stages fuzzy set of factors T1, T2, T3, T4, T5 in the weight vector, then a fuzzy comprehensive evaluation of single factor evaluation matrix $\tilde{R}$,

$$\tilde{R} = \begin{pmatrix} \tilde{B}_1 \\ \tilde{B}_2 \\ \tilde{B}_3 \end{pmatrix} = \begin{bmatrix} A_1 \circ \tilde{R}_1 \\ A_2 \circ \tilde{R}_2 \\ A_3 \circ \tilde{R}_3 \end{bmatrix} \tag{3}$$

According to the first level fuzzy evaluation factors set T weight vector, then the fuzzy comprehensive evaluation formula $\tilde{B}$,

$$\tilde{B} = A \circ \tilde{R} \tag{4}$$

For (4), according to the needs, using the principle of maximum degree of membership, or weighted average operator result determination.

6. Conclusion

It is important elements to concern about public spreading environment pollution accident during dealing effectively with the environmental pollution accident. Although the public spreading has obvious negative impact, this phenomenon can also help the government's environmental management and enterprise timely release of information and assume corresponding environmental responsibility. This article mainly analyses whether the individual will continue to disseminate such comprehensive assessment of environmental pollution accident and illuminates individual continue traditional pollution accident is decided by many factors. Through the individual's analysis, you can extend the scope of the crowd by use of arithmetic operators and determines whether it will continue to spread of environmental pollution accident, at the same time, it is necessary to further strengthen environmental education and the quality of nation education

References

1. T .W .Parr, A. R. J.Sier and R. W. Battarbee, Detecting environmental change: science and society perspectives on long term research and monitoring in the 21st century, The Science of the total environment. 310(1/2/3)(2003).
2. F. Jun and W.W Song, The Improvement of Accident Crime of Great Environmental Pollution on the Global Environmental Crisis, Journal of Hebei University (Philosophy and Social Science), Apr. (2010).

NEUTRON POWER CONTROL OF A TRIGA MARK III REACTOR USING STABLE ADAPTIVE FUZZY CONTROL

BENÍTEZ-READ, JORGE S., ROJAS-RAMÍREZ, ERICK,
QUINTANA-CARAPIA, GUSTAVO
Instituto Nacional de Investigaciones Nucleares de Mexico, Carretera México-Toluca s/n, La Marquesa, Ocoyoacac, Estado de México, 52750, México.
Instituto Tecnológico de Toluca, Av. Tecnológico, s/n, Metepec, Estado de México, 52140, México.

An innovative control scheme to regulate the power of the TRIGA Mark III research reactor of the National Nuclear Center of Mexico is proposed. Based on the safety limit value imposed on the power rate of change, a sectionally continuous time profile for the ascent of power from its source level to a desired value is generated. A group of fuzzy systems is used to identify the nonlinear functions of the plant's state-space equations. Using Lyapunov stability theory, the consequent parameters of these fuzzy identifiers are adapted. A supervisor control stage ensures the boundedness of the neutron power error signal. The dynamic behavior of the reactor is obtained from the point kinetics equations given by the Fuch-Hansen model. The controller shows a robust regulation against plant's parameter disturbances.

1. Introduction

A research nuclear reactor of the TRIGA Mark III type is used by the National Nuclear Research Institute of Mexico (ININ) to produce radioisotopes, which are applied in agriculture, industry, and for medical treatments [1]. Throughout its life, the reactor has required major maintenance works and instrumentation & control system modernizations, being the replacement of the operation control, from an analog to a digital type, one of them. Currently, this console can operate the reactor in the manual and automatic modes [2]. In this latter, a PID control algorithm controls the ascent of power up to a pre-established setpoint. The objective is to maintain the power level during long periods of time, commonly of a few days. Both the ascent and regulation of power is attained by the insertion and withdrawal of the control rods. For long period operations, the buildup of Xe-131 in the fuel could eventually lead to loss of efficiency, and in some cases to instability, manifested as oscillations about the power setpoint [5]. Although, in general, the PID algorithm can manage this buildup, ININ's personnel in charge of the reactor's I&C systems have endeavored to find innovative control schemes that can incorporate this phenomena to yield control rules to effectively manage this potential loss of regulation. Thus, a fuzzy controller whose inputs depend only on the power error and the instantaneous period has been reported [2]. Proposals from other groups include, for instance, the use of adaptation of a fuzzy system to control the neutron flux [3]. Likewise, a controller based on recurrent neural networks, complemented with the identification of the dynamics of a TRIGA Mark III reactor has been proposed in [4].

In this work, an innovative stable power tracking scheme is proposed. The time profile for the ascent of power from a source level up to a desired steady state level is generated considering the safety constraint of maintaining the power rate of change within pre-specified limits to prevent any spurious shutdown of the reactor. Likewise, the control scheme also considers the lack of knowledge of the functions describing the reactor's dynamics. Thus, their estimates are used in the control law. These estimated functions are determined by a group of Mamdani-type fuzzy approximators [6], whereas the control law is obtained from the theory of nonlinear input-output linearization. Moreover, a supervisory control, based on a Lyapunov function, guarantees that the power error signal is maintained within a bounded region. The simulation results show the feasibility of using this control scheme to ensure that a power profile can be properly tracked for long periods of time, even in the presence of plant parameter disturbances.

2. Point kinetics model of the TRIGA Mark III reactor

The neutron diffusion equations describe the spatial dynamic behavior of the neutron population in nuclear reactors. However, for small reactors, such as the majority of the research reactors, a simplified model, called the point kinetics model, is obtained from the diffusion equations where the spatial variables are neglected. This model is given by:

$$\dot{x}_1 = \frac{\alpha T_0 - \beta}{\Lambda} x_1 + \lambda x_2 + \frac{\alpha}{\Lambda} x_1 x_3 + \frac{1}{\Lambda} x_1 u \tag{1}$$

$$\dot{x}_2 = \frac{\beta}{\Lambda} x_1 + \lambda x_2 \tag{2}$$

$$\dot{x}_3 = K(x_1 - n_0) - \gamma(x_3 - T_0) \tag{3}$$

where α, β, λ, Λ, γ, and K, are reactor parameters [5], x_1 is the reactor power, x_2 the concentration of delayed neutron precursors, x_3 the fuel temperature, n_0 the initial reactor power, and T_0 the initial fuel temperature. The control input to the system is denoted by u and represents the external reactivity inserted into the reactor core by the proper withdrawal or insertion of the control rods.

3. Application of fuzzy systems in control

A group of Mamdani-type fuzzy systems are used to approximate the reactor's functions that describe its nonlinear dynamics. The estimated functions are further used by the control law, which is denoted by:

$$u_n = \frac{1}{\hat{g}(x)}\{-\hat{f}(x) + \dot{y}_{ref} - ek\} \quad (4)$$

where $e = y_{ref} - x_1$ represents the output error, k denotes the proportional gain. The variable y_{ref} represents the reference trajectory used to guide the ascent of the reactor power. This reference signal complies with the reactor shutdown constraint due to reactor period off-limits (SCRAM). The functions $\hat{f}(x)$ and $\hat{g}(x)$ represent the group of approximators based on fuzzy systems. Equation (1) shows that the input signal is obtained from the first derivative of x_1; thus, the system has a relative degree n=1.

4. Robust differentiator HOSMC

The structure of the control law (4) indicates that it is necessary to guarantee the existence of the reference path derivative. A high order sliding mode differentiator (HOSMC) [8] is used to compute the derivative. The HOSMC is described by the following equations:

$$\dot{z}_0 = v_0; \quad v_0 = -\lambda_0 L^{1/2}|z_0 - y_{ref}(t)|^{1/2} sign\left(z_0 - y_{ref}(t)\right) + z_1 \quad (5)$$

$$\dot{z}_1 = -\lambda_1 L sign(z_1 - v_0); \quad z_1 = \dot{y}_{ref}(t) \quad (6)$$

where z_0 represents the estimation of y_{ref}, z_1 represents its first derivative, λ_0 and λ_1 are constants, and L is the Lipschitz number.

5. Adaptability and the Lyapunov stability criterion

In order to guarantee the stability of the reactor power, we use the Lyapunov stability concept [7]. Let V the error dependent Lyapunov candidate function, denoted as:

$$V_T\left(e, \theta_f, \theta_g\right) = \frac{1}{2}e^2 + \frac{1}{2\psi_1}\varphi_f^T\varphi_f + \frac{1}{2\psi_2}\varphi_g^T\varphi_g \quad (7)$$

The system (7) is stable if and only if the function V is positive definite; $V(0) = 0$ and $\dot{V}(e) \leq 0$ (the derivative of V is negative definite). The supervisory control law is obtained from the first derivative of (7) and guarantees that the error is maintained within a tolerance region.

$$u_s = LU_s sign(e)\frac{1}{g_L}[|\hat{f}| + f^U + (|\hat{g}| + g^U)u_n] \quad (8)$$

where u_s is the supervisory control signal that acts when the system does not comply with the Lyapunov stability criterion, f^U, g^U and g^L are constants that were obtained from the plant operating limits, $\hat{f}$ and $\hat{g}$ are the fuzzy approximator outputs. The adaptability conditions of the consequents of the fuzzy systems are:

$$\dot{\theta}_f = -\psi_1 e\xi \quad \text{and} \quad \dot{\theta}_g = -\psi_2 e\xi u_n \tag{9}$$

where ψ_1 and ψ_2 represent adaptive parameters, and ξ represents the radial basis function of the fuzzy systems.

6. Results

An ascent of power is simulated starting at a source level of $x_0 = 5W$ up to a full power level of $x_f = 1E6W$.

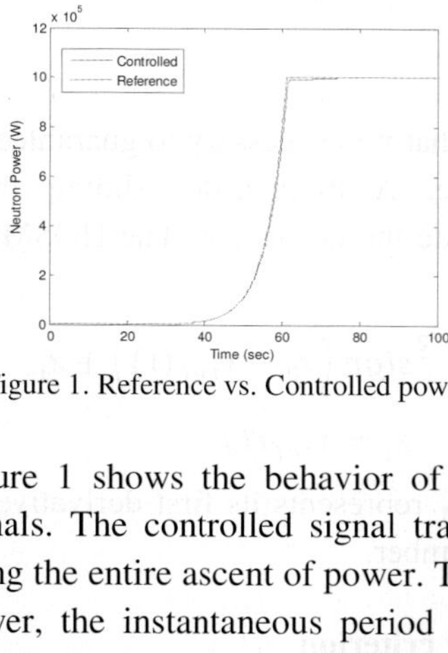

Figure 1. Reference vs. Controlled power

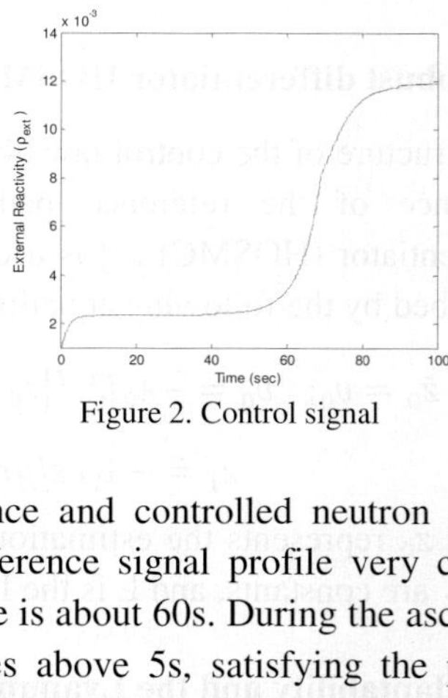

Figure 2. Control signal

Figure 1 shows the behavior of the reference and controlled neutron power signals. The controlled signal tracks the reference signal profile very closely along the entire ascent of power. The rise time is about 60s. During the ascent of power, the instantaneous period takes values above 5s, satisfying the period requirement to avoid spurious reactor shutdown. Figure 2 shows the control signal behavior (external reactivity). The smoothness of the control input shape makes it suitable for its physical implementation.

7. Conclusions

The proposed adaptive fuzzy system is capable of increasing power towards the desired setpoint following a profile that minimizes the rise time but also prevents the occurrence of a shutdown due to an excursion off-limits of the reactor period. The stability of the system is monitored by means of an output-energy Lyapunov function, which guarantees that the reactor power lies inside a region of stability. In case parametric disturbances are present in the plant, threatening the power stability, a supervisory control is fired-up, driving the power back to the stability region. The constraint on the power rate of change, which in nuclear reactor literature is known as reactor period, is always satisfied when using this control method. Thus, the possibility of a shutdown of the plant, during ascent and regulation of power, due to off-limit reactor period is highly reduced. Given its adaptive characteristic, which greatly reduces undesired power fluctuations during long periods of time, this control architecture represents an interesting option for power control in TRIGA type reactors.

Acknowledgements

The authors wish to thank the National Council for Science and Technology of Mexico (Conacyt) for the financial support provided through its scholarship program.

References

1. Mazón R. Ruperto (2008), El reactor nuclear TRIGA Mark III. En Actividad científica y tecnológica en el Instituto Nacional de Investigaciones Nucleares, Innovación Ed. Lagares, México.
2. J. S. Benítez-Read y D. Vélez-Díaz, “Controlling Neutron Power of a TRIGA Mark III Research Nuclear Reactor with Fuzzy Adaptation of the Set of Output Membership Functions,” Fuzzy Systems and Soft Computing in Nuclear Engineering, Editor Da Ruan, Editorial Physica-Verlag. (2000).
3. Marseguerra, M. and Zio, E., Model-free fuzzy tracking control of a nuclear reactor, Annals of nuclear energy, 30, p. 953-981, (2003).
4. Perez-Cruz, JH, Chairez I., Pozniak A. and De Jesus-Rubio, J., Constrained Neural Control for the Adaptive Tracking of Power Prolifes in a TRIGA Reactor, Int. J. Innovative Computing, Information and Control, Vol. 7, No. 7(B), July (2011).
5. Lamarsh, JR and Baratta, AJ., Introduction to nuclear engineering, Third edition, Prentice Hall, New Jersey, (2001).
6. Wang, LX, Adaptive Fuzzy Systems and Control, Prentice Hall, Upper Saddle River, NJ, (1994).
7. Khalil, H.K, Nonlinear systems, third ed., Prentice-Hall, Englewoods, Cliffs, (2004).
8. Levant, Arie (2006), Exact Differentiation of Signals with Unbounded Higher Derivates, Proc. of the 45th IEEE Conference on Decision and Control, San Diego, California, December, pp.14-17.

A FRAMEWORK PROPOSAL UPON THE ESTIMATION OF ELECTRONIC-WASTE PRODUCTS IN ISTANBUL*

TUGBA EFENDIGIL†, VILDAN OZKIR, NIHAN C. DEMIREL, TUFAN DEMIREL, ONDER ONDEMIR

Department of Industrial Engineering, Yildiz Technical University, Barbaros Bulvari, 34349, Besiktas, Istanbul, Turkey

Due to the increase in both consumption and obsolescence rates for new technologies, but shorter life cycles of electronic products cause a significant rise in the volume of electronic waste (e-waste). E-waste practices mostly focus on the processes for a collection system; however, the amount of e-waste critically affects the network design considering the number of facilities and the capacity assignments, as well. We present a framework proposal to predict the amount of e-waste including IT/telecommunication and consumer equipments in Istanbul metropolitan. The main selection reason of Istanbul is the requirement of an urgent intervention to deal with increasing e-waste amount due to its young and crowded population. The result of this study will strongly assist the authorities to configure well-structured strategies for future e-waste management system.

1. Introduction

E-waste is defined as an informal name for electronic products after the end of their "useful life" [1]. Computers, televisions, monitors, laptops, cell phones, stereos, copiers, fax machines are just a few of the common electronic products that fall into the e-waste categories. The e-waste volume has been rapidly growing segment of the municipal solid waste stream over the last few decades. However, as opposed to other municipal waste, e-wastes are hazardous since it contains toxic substances that cause environmental risks and health hazards when inappropriate disposal practices or primitive recycling methods are utilized. Environmental consequences have driven government policies to explore alternative solutions such as the reuse and/or recycling of outdated electronics [16]. 27 European countries have been authorized to collect and recycle electronics under the WEEE (Waste Electrical and Electronic Equipment) Directive of the European Union since 2005. Some countries like

* This work is supported by Office of Scientific Research Project Coordination in Yildiz Technical University.

† Correspondence author: tugbae@yildiz.edu.tr

Switzerland, Norway, Belgium and Japan have earlier enacted legislated national electronics takeback and recycling systems.

There are several benefits in recycling e-waste such as *preserving natural resources by recycling* aims to recover valuable materials from outdated electronics in order to make new products; *protecting the surroundings by safe recycling of outdated electronics* including toxic chemicals like lead, barium, cadmium, and mercury; *creating new business for professionals and markets* for the valuable dismantled components, and followed by *saving landfill space*. It is obviously seemed that, especially, natural resources are conserved and air-water pollution caused by hazardous disposal is avoided by dismantling and providing reuse possibilities. Moreover, recycling *reduces the amount of greenhouse gas emissions* caused by the manufacturing of new products.

In our country, insufficient number of e-waste collection points and the deficiencies in planning of collection and recovering activities prevent benefit from this potential resource. Hence, the main objective of this study is to propose a framework for the estimation of return for optimal product recovery network in order to plan e-waste collection and recovery activities in Istanbul for obsolete mobile phones. Moreover, the framework covers the analysis to define the long-term trends in the generation of mobile phones e-waste. The proposed framework aims to assist decision makers for management of the organizations.

2. Literature Review

The studies about e-waste have started to appear in the literature in the beginning of 2000s. There have been several approaches for predicting e-waste generation. However, to precisely estimate the annual quantity of obsolete e-waste is significantly difficult due to the lack of statistical data on their consumption patterns in developing countries [1, 2].

Three methods have been mostly utilized to predict possible quantities of e-waste as follows; (i) The "*consumption and use method*", used in the Netherlands, (ii) the "*market and supply method*", used by the German Electrical and Electronic Industries Association, (iii) the "*Swiss Environmental Agency*" estimated Swiss e-waste [3]. The most practical and widely used method is the market and supply method which utilizes *past domestic sales data coupled with the average life of electronic applian*ces to estimate the e-waste generation. Hence, it is possible to extrapolate the assumed lifespan backwards in time to predict potential e-waste from past production and sales figure [1]. Alternatively, "*Carnegie Mellon Method*" [4] uses the same assumptions which are based on *sales data* but differs regarding material flows through different end-of-life (EOL) alternatives like reuse, storage, recycling and land filling to estimate current and future quantities of e-waste.

Reference [5] developed an integrated framework for the formulation, analysis, and computation of e-waste management and recycling. Reference [3] presented a case study to predict the amount of obsolete equipment for five main kinds of electronic appliances from urban households and analyzed the flow after the end of their useful phase, due to the consumption growth and the expansion of urbanization with the *annual sales amount* and *yearly obsolete ratio of the product* criteria. Reference [2] modified the EOL model proposed by [4] to predict future outflows of e-waste generation in the Philippines with a focus on televisions, refrigerators, air conditioners, washing machines, and radios. Reference [6] estimated future quantities of e-waste via a time-series material flow analysis by modeling the stages of production, usage, and disposal with four scenarios in California. Reference [7] aimed to characterize long-term trends in the generation of obsolete computers in the U.S. by using the logistic function with the *historical sales data* and *lifespan distribution* criteria. Reference [1] constructed an approach to estimate the future outflows by modeling their usage and disposal considering two scenarios based on user preferences to store or to recycle the e-waste. Reference [8] addressed the generation of obsolete PCs using the logistic models and material flow analysis. Reference [9] collected the data of e-products (televisions, washing machine, air-conditioner, refrigerator, and computer) consumption at household level by a tailor-made questionnaire survey from the public to obtain an e-waste generation estimate. Robinson's approach which requires the minimal data such as *unit weight of e-products*, *number of units of e-products*, *average lifespan* was applied to calculate the quantity of e-waste generated. Reference [10] proposed a model for WEEE generation estimation and realized that the most important variable is the *equipment lifetime* which requires the understanding of consumer behavior. Besides those general e-waste studies, there are few studies focusing on mobile phones in literature. Reference [11] developed a generic remanufacturing plan for mobile phones under the circumstances like changing product, process, and market constraints. Reference [12] presented an overview of cell phone recycling programs currently available in the U.S. Reference [13] examined the economics of cell phone reuse and recycling and showed that cell phone reuse has a healthy profit margin.

So far, Istanbul municipality runs a project LIFE SMILE to announce the importance of collecting e-waste to citizens. The aim of the project is the constitution of a framework for the preparation and execution of the plan, database and pilot projects in Istanbul regarding and Integrated Waste Management System for Electrical and Electronic Materials (including collection systems, re-use, recycling and secure discharge) and improvement of capacity and administrative structure in respect of municipality/governorship levels. However, there is no analytical search for predicting the quantity of e-waste or determining the collecting points or defining the e-waste potential of Istanbul via mathematical modeling. Hence, as an initial step, we suggest building a structure in order to estimate e-waste quantity of mobile phones to plan the rest of the project.

3. Proposed Framework

To date, there has been no study done to estimate e-waste generation in Turkey. For a more concentrated framework, Istanbul metropolitan is selected to search the potential e-waste quantity for mobile phone product group. A flow for mobile phone usage is depicted in Fig.1.

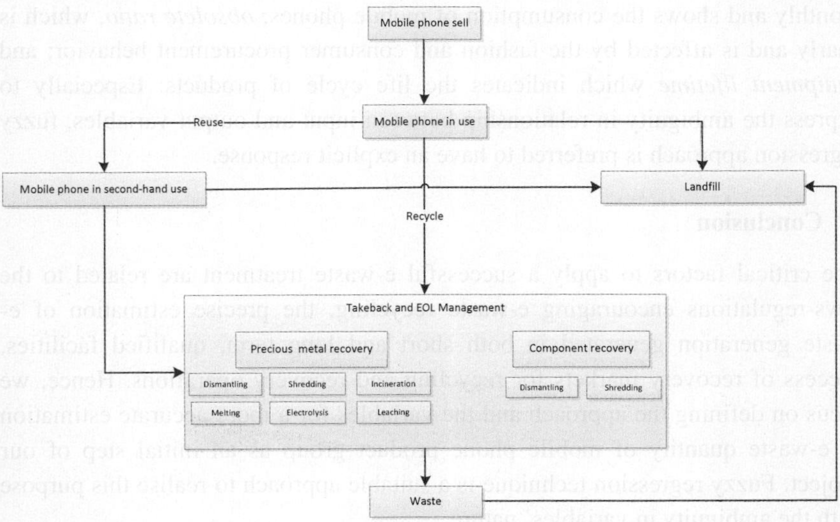

Figure 1. Mobile phone usage flow

Our aim is to estimate the return quantity for optimal product recovery network to plan properly e-waste collection and recovery activities in Istanbul for obsolete mobile phones. For this purpose, fuzzy set theory is utilized in a regression analysis to define the vagueness in the relationship between the variables. Statistical regression has been used for many years for predicting in marketing, sales and management areas; however it is inadequate whether the data set is too small, or there is ambiguity associated with the event or linearity. Fuzzy regression is based on possibility and fuzzy set theory, so deviations between observed values and estimated values are assumed to be due to system fuzziness.

The fuzziness in regression analysis was firstly proposed by Tanaka et al (1982) [14] as a fuzzy linear regression (FLR) model. The parameter estimations were considered under two main factors defined as the degree of the fitting and the vagueness of the model in FLR models. The estimation problems were transformed into linear programming (LP) based on those two factors. Diamond (1988) [15] proposed the fuzzy least-squares approach, which is a fuzzy extension of the ordinary least squares based on a new defined distance on the fuzzy numbers space. Briefly, the fuzzy regression methods can be categorized in those two approaches, namely Tanaka's LP approach and fuzzy least-squares approach.

As mentioned in literature section, there have been some approaches and criteria defined for the estimation of e-waste quantity. In our model we predict the *e-waste quantity* depending on the variables as *historical sales data*, which is monthly and shows the consumption of mobile phones; *obsolete ratio*, which is yearly and is affected by the fashion and consumer procurement behavior; and *equipment lifetime* which indicates the life cycle of products. Especially to express the ambiguity in relationship between input and output variables, fuzzy regression approach is preferred to have an explicit response.

4. Conclusion

The critical factors to apply a successful e-waste treatment are related to the laws-regulations encouraging e-waste recycling, the precise estimation of e-waste generation generated in both short and long term, qualified facilities, success of recovery markets for recycling and recovery operations. Hence, we focus on defining the approach and the variables for a more accurate estimation of e-waste quantity of mobile phone product group as an initial step of our project. Fuzzy regression technique is a suitable approach to realise this purpose with the ambiguity in variables' nature.

References

1. M. Dwivedy and R.K. Mittal, *Waste Management*. 30(3), 483 (2010).
2. G.L. Peralta and P.M. Fontanos, *Journal of Material Cycles & Waste Management*, 8, 34 (2006).
3. X.B. Liu, M. Tanaka and Y. Matsui, *Waste Management & Research*, 24(5), 434 (2006).

4. H.S. Matthews, C.T. Hendrickson, F.C. McMichael and D.J. Hart, *Green Design Initiative Technical Report*, Carnegie Melon University, Pittsburgh (1997).
5. A. Nagurney and F. Toyasaki, *Transportation Research Part E: Logistics and Transportation Review*, 41(1), 1(2005).
6. H.Y. Kang and J.M. Schoenung, *Journal of Hazardous Materials*, 137(2), 1165 (2006)
7. Y.Yang and E. Williams, *Technological Forecasting and Social Change*, 76(8), 1105, (2009).
8. J .Yu , E. W illiams , M. Ju and Y. Yang, *Environmental Science & Technology*, 44(9), 3232, (2010).
9. S.S. Chung, K.Y. Lau and C. Zhang, *Waste Management*, 31(3), 544, (2011).
10. M.G. Araújo, A. Magrini, C. F.Mahler and B. Bilitewski, *Waste Management*, 32(2), 335, 2012.
11. C. Franke, B. Basdere, M. Ciupek, S. Seliger, *Omega-International Journal of Management Science*, 34(6), 562, (2006).
12. G.T.R. Silveira and S.Y. Chang, *Waste Management*, 30(11), 2278, (2010).
13. R.Geyer and V.D. Blass, *International Journal of Advanced Manufacturing Technology*, 47(5-8), 515 (2010).
14. H. Tanaka, S. Uejima and K. Asai, *IEEE. Trans. Systems Man Cybernet* 12, 903 (1982).
15. P. Diamond, *Information Sciences*, 46(3), 141, (1988).
16. http://www.ewastecenter.com

VARIANCE REDUCTION FOR FINANCIAL OPTIONS PRICING

SEMIH YÖN[†]

Department of Industrial Engineering, İstanbul Technical University, Maçka, Beşiktaş İstanbul,34367, Turkey

CAFER ERHAN BOZDAĞ

Department of Industrial Engineering, İstanbul Technical University, Maçka, Beşiktaş İstanbul,34367, Turkey

Options pricing is still an open area for exact results. Monte Carlo integration is the unique solution for especially complicated options. It is desired to control variability while implementing Monte Carlo techniques. In order to supply reliable results variance of simulation trials should be decreased. Importance Sampling is one of the variance reduction techniques commonly used in Monte Carlo applications. This study includes a research to gather the appropriate Importance Sampling density which gives the lowest variance. We illustrate the Importance Sampling method on financial options and calculate the value of options. By the same way, it is possible to calculate any expectation that cannot be calculated analytically. Numerical results indicate that longer tailed proposal distributions provide substantial decrease in the estimated variance.

1. Introduction

An *option* is a contract between two sides about trading an asset at a future time. One side is the *writer* that fixes the terms of the option contract and sells the option. The other side is the *holder* who purchases the option for the market price. Holders may exercise the option till a specified time called *expiration date* and denoted by *T*. An option gives its holder the right (but not the obligation) to trade the *underlying asset* at a *specified price* in the future. The specified price of the underlying is called *strike price* or *exercise price* and denoted by *K*. The underlying asset is typically a stock. Some other general examples of the underlying are stock indices, currencies and commodities. Nevertheless, the underlying need not necessarily to be an asset, weather condition, electricity consumption or even sheep and rams[*] could be the underlying as well. From this point of view, options gain high importance day by day.

[†] yon@itu.edu.tr

[*] The contracts on sheep and rams were included to VOB Izmir in 2011

In the modern economies, firms and individuals try to minimize the risk arises from their transactions. Options allow investors to make proper decisions in their trades. However, value of the option, which is called *premium* and denoted by *V*, should be calculated correctly. *Arbitrage-free pricing* and the *equivalent martingale measures* are the essentials of options pricing, see Neftci (2000). In an efficient market, which refers to no-arbitrage principle, it is assumed that the price of an option (any derivative) can be calculated as the expected value of its discounted payoffs. This expectation is taken under the original (equivalent martingale) probability measure. The literature on the options pricing was started with Fischer Black and Myron Scholes at 1973. They developed very famous Black&Scholes formula for standard European options. Black and Scholes (1973) modeled the stochastic process of underlying as a geometric Brownian motion. Robert Merton (1973) generalized this formula. Although there is a huge literature on options pricing no closed form solutions could be gathered for all kinds of options. Charnes (2000), Boyle, et al. (1997) used Monte Carlo Integration technique to price complicated options. The random number based approach generates a variance phenomena. Bodie (2007), Jackel (2001) deal with variance reduction techniques for Monte Carlo applications. Hesterberg (1995) used Importance Sampling (IS) technique for substantial decrease in the estimated variance. Broadie and Glasserman (1997) used IS by changing the drift of the original distribution. Schepper and Heijnen (2007) used distribution-free options pricing by defining bounds on the option price. We give some background on options in second section. Then we mention Importance Sampling via dealing with some maths and stats. Then we give numerical results in the fourth part. And finally, we set the conclusions at the end.

2. Background

Options allow investors to change the shape of the probability distribution of investment returns. That is why options are essential for risk management. We give commonly used notations at first. List of general variables are set in Table 1.The price of the underlying at time t is denoted by St, for $0 \leq t \leq T$. Options are basically in two types:

- *Call options* give its holder the right to **purchase** the underlying security for the strike price. The *payoff* for a call option with strike price K when it is exercised on date t is $(S_t - K)^+$ where $(X)^+ \equiv \max(X, 0)$.

- *Put options* give its holder the right to **sell** the underlying security for the strike price. The payoff for a put option with strike price K when it is exercised on date t is $(K - S_t)^+$.

Table1. List of important variables

t	current time $0 \le t \le T$, T expiration time
$r > 0$	risk-free interest rate
S, S_t	spot price, current price per share of underlying
μ	annual drift
σ	annual volatility
K	strike price per share
$V(S,t)$	premium, value of an option at time t and underlying price S

Some investors are risk aversion and trade options for hedging their portfolios. Nonetheless speculators play an active role in the options and derivatives markets. So market participants demand different types of options.

2.1. *Types of Options*

Standard options also called *vanilla options* are the first traded options. There are two styles, European and American. European options can only be exercised *on the expiration date* whereas American options can be exercised *any time* till the expiration date. Hence, American options are more complicated than European ones. The premium for European ones can be calculated by closed formulas. American options can just be priced via advanced simulations, see Broadie and Glasserman (1996). Nevertheless options on stocks are mostly American styles.

Exotic Options are in many different styles. For example *Asian options* pay the difference between strike price and the average price of the underlying over a specified period. *Up-and-in barrier options* pay the difference between strike price and spot price at exercise only if the spot price has exceeded barrier level. *Down-and-out barrier* options stay alive as long as the spot price has not gone below barrier level. *Lookback options, Bermuda options, Quanto options* and any other different kinds of agreements are samples of exotic options.

2.2. *Black and Scholes Formula*

According to Black-Scholes model the price of an option can be found using the assumption of no arbitrage principle, see Neftci (2000) for details. Black and

Scholes (1973) modeled the stochastic process as a geometric Brownian Motion, see Jackel (2001). Famously known stochastic differential equation is

$$dS_t = \mu S_t dt + \sigma S_t dW_t \tag{1}$$

where *W(t)* represents a Wiener process, see Seydel (2006) for details. Then Black-Scholes price for a European call option on a nondividend-paying stock trading at time *t* is

$$C_t(S_t, T-t) = S_t N(d_1) - Ke^{-r(T-t)} N(d_2) \tag{2}$$

where *N(.)* is the cumulative distribution value for standard normal random variables with

$$d_1 = \frac{\log(S_t / K) + (r + \frac{1}{2}\sigma^2)(T-t)}{\sigma\sqrt{(T-t)}} \tag{3}$$

and

$$d_2 = \frac{\log(S_t / K) + (r - \frac{1}{2}\sigma^2)(T-t)}{\sigma\sqrt{(T-t)}} = d_1 - \sigma\sqrt{(T-t)} \tag{4}$$

The Black-Scholes price for a European Put option on a non-dividend paying stock trading at time *t* is

$$P_t(S_t, T-t) = -S_t N(-d_1) + Ke^{-r(T-t)} N(-d_2) \tag{5}$$

3. Importance Sampling

The contribution of Importance Sampling (IS) is to introduce definite selection rules to generate the most likely samples and hence to obtain more accurate values of statistical averages, Hesterberg (1995). The basic idea is to compute a correction factor to the IS estimates, based on sample weights accumulated during sampling. With proper weights the correction factor compensates for statistical fluctuations and lead to a lower variance. IS can be carried out as in the following with a mathematical approach:

weight function

$$\theta = E_g[q(X)\frac{f(X)}{g(X)}] = \int_{R^d} q(x)\frac{f(x)}{g(x)} g(x)dx \tag{6}$$

IS density

θ is calculated as in (7), where N is the sample size and $\mathbf{x}$ is a random vector.

$$\theta = \frac{1}{N}\sum_{i=1}^{N}\left\{ q(\mathbf{x}_i)\prod_{j=1}^{d}\frac{f(x_j)}{g(x_j)}\right\} \tag{7}$$

4. Numerical Results

We applied the IS method with gamma distribution and pareto distribution. The results are set in Table 2 and Table 3. When we compare the results the factor of variance reduction is substantial.

Table 2. T=1, r=0.1, K=50 ve σ=0.2 Black-Scholes, Monte Carlo simulation ve Importance Sampling with Gama distribution results

	B&S	Monte Carlo Simulation			IS Gama Distribution				Factor
S	Premium	Premium	Variance	Std Dev	Premium	Variance	Std Dev	Parameter	
30	0.053837	0.053414	0.351507	0.592881	0.053088	0.015098	0.122874	a=0.9 b=3.2	23.282
40	1.394964	1.396355	12.8775	3.588524	1.393550	3.044087	1.744731	a=0.9 b=1.7	4.230
50	6.634841	6.639414	64.8351	8.052025	6.646235	15.62166	3.952425	a=0.9 b=1.2	4.150
60	15.129237	15.13415	133.3456	11.54754	15.13546	34.72188	5.892528	a=0.9 b=1.2	3.840
70	24.815723	24.82018	196.8174	14.02916	24.82394	65.13932	8.070893	a=0.9 b=0.7	3.021

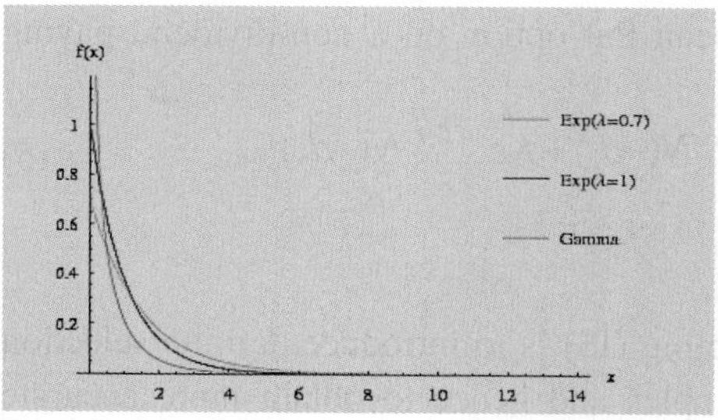

Figure 1. Gama density & Exponential comparison.

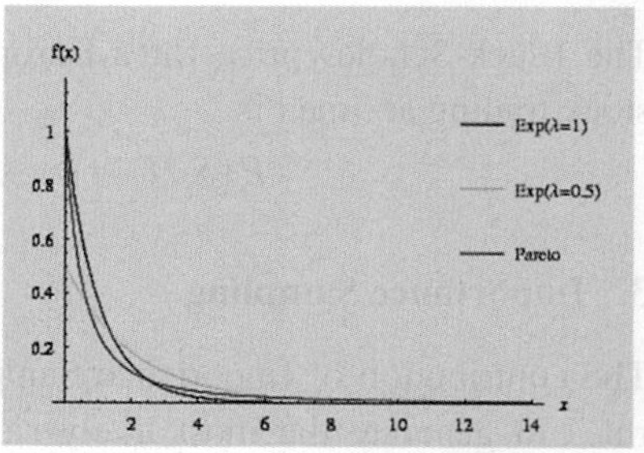

Figure 2. Pareto density & Exponential comp.

As seen in figure 1 and figure 2 higher tailed distributions give substantial variance reduction as proposed by IS.

Table 3. T=1, r=0.1, K=50 ve σ=0.2 Black-Scholes, Monte Carlo simulation ve Importance Sampling with pareto distribution results

	B&S	Monte Carlo Simulation			IS Pareto Distribution				Factor
S	Premium	Premium	Variance	Std Dev	Premium	Variance	Std Dev	Parameter	
30	0.053837	0.053414	0.351507	0.592881	0.053099	0.014712	0.121292	a=10 b=35	23.893
40	1.394964	1.396355	12.8775	3.588524	1.391207	2.784419	1.668658	a=40 b=60	4.625
50	6.634841	6.639414	64.8351	8.052025	6.649338	12.66737	3.559125	a=65 b=80	5.118
60	15.129237	15.13415	133.3456	11.54754	15.42982	28.16991	5.307533	a=60 b=65	4.734
70	24.815723	24.82018	196.8174	14.02916	24.82208	51.57958	7.181893	a=75 b=80	3.816

5. Conclusions

Options are examples of derivatives since the option value is derived from another asset. A key issue in order to achieve small errors on the obtained result is a suitable strategy of sampling the available multidimensional space. Fundamental idea with importance sampling is that the sampling process is distorted in order to take into account the weighting of the underlying distribution. The term "importance sampling" refers to choosing the proposal density so that the sampled values lie in the region that is important for the value of the integral.

Acknowledgments

I would like to gratefully thank to TÜBİTAK for support of the BIDEP 2022.

References

1. Black, F. and M. Scholes 1973. The pricing of options and corporate liabilities. *Journal of Political Economy*81:637–654.
2. Bodie Z, Kane A, Marcus A. J., 2007. *Essentials of Investments*. 6th Edt NY
3. Brandimarte, P. 2006. Numerical methods in Finance and Economics, Wiley,
4. Broadie M. and Glasserman P. 1997. Pricing American style securities using simulation. *Journal of Economic Dynamics and Control* 21:1323–1352.
5. Boyle, P., Broadie, M., and Glasserman, P. 1997. Monte Carlo methods for security pricing. *Journal of Economic Dynamics and Control* 21:1267–1321.
6. Boyle P, Broadie M, and Glasserman P. 1995. Recent advances in sim for security pricing. In *Proceedings of the 1995 Winter Simulation Conference*
7. Charnes, J. M. 2000. Using Simulation for Option Pricing. *Proceedings of the 2000 Winter Simulation Conference* Orlando, FL.
8. Hesterberg, T., 1995. *Weighted Average Importance Sampling and Defensive Mixture Distributions.* Technometrics 37(2):185-194
9. Jackel P., 2001. *Monte Carlo Methods in Finance*.
10. Merton, R.C. 1973. Theory of rational options pricing. *Bell Journall of Economics and Management Science* 4, 141-183.
11. Neftci, S.N 2000. An introduction to the Mathematics of Financial Derivatives, Academic Press Second Edition.
12. Schepper, A.D., Heijnen B. 2007. Distribution-free option pricing. Mathematics and Economics 40, 179-199.
13. Seydel R. U. 2006. *Tools for Computational Finance*. 3rd Edition, Köln: Springer

LEARNING CAPABILITIES AND THEIR EFFECT ON ORGANIZATIONAL CAPABILITIES: AN EMPIRICAL APPROACH

SEZİ ÇEVİK ONAR

Industrial Engineering Department, Istanbul Technical University, Beşiktaş, İstanbul 34367, Turkey

The capability sets of organizations are the key the organization's success. The learning capabilities are crucial for improving capability sets. In literature although this relation has been examined conceptually there are limited studies which show this relation empirically. The objective of this study is to reveal the relation between organizational learning and organizational capabilities. A statistical technique entitled Structural Equation Modeling will be used in order to achieve this objective. Since it reveals the mechanisms behind capability building this model will help managers in their decision making processes.

1. Introduction

In order to survive organizations should quickly adapt to the changes in the environment [1].To be able to adapt new changes can be named as organizational flexibility. This flexibility can be gained through company's capabilities, skills and assets therefore creating new capabilities are crucial for organizations' success; managers should effectively manage capability building and capability leveraging processes. Capability building process can be defined as "any process by which an organization achieves qualitative changes to its existing stocks of assets and capabilities, including new abilities to coordinate and deploy new or existing assets and capabilities, in ways that help organization achieve its goals" [2]. And capability leveraging can be defined as "the process by which the organization exercises its existing strategic options to create and market products" [3]. Learning capability of organizations is the key to creating new capabilities [4]. If an organization is a learning organization, the capabilities of the organization will be higher.

Organizational learning refers to the process by which organizations learn [5]. In literature many scientists highlighted the role of learning capability on capability building process but most of these studies are conceptual; there are little empirical evidences about these conceptual models [3], [6], [7]. There is a lack of quantitative articles in this area [4]. Therefore the aim of this study is to reveal the relation between competence building process and the learning capability.

2. Organizational Capability and Learning

The learning organization concept implies a change in the traditional way of dealing with business management [8]. The analysis of organizational learning is an important area which attracted many scientists from different areas such as psychological, sociological, organizational theory. In strategy since organizational learning is accepted as the key for competition, it has a special importance [9]. For instance Gebauer et al. [10] highlighted the importance of learning processes in strategic innovation; Camison and Lopez [11] claimed that learning capabilities are crucial for competitive advantage, Akgün et al. [12] investigated the impact of learning capability on product innovativeness.

Organizational learning has a multidimensional nature and is a dynamic process. It starts with individual knowledge acquisition with the exchange and integration of this knowledge a collective knowledge will be formed which will form the organizational memory [13]. Organizational memory has a strong effect on individual knowledge acquisition. Gomez et al. [14] claimed that organizational learning has four factors namely Managerial Commitment (Mc), Systems perspective (Sys), Openness and experimentation (Exp), Knowledge transfer and integration (Trans), and created a model measuring organizational learning. In our study we used this measurement model for measuring organizational learning only the managerial commitment factor since it's a part of general management capability is not considered. Managerial commitment shows the degree of managers on supporting learning and it has five dimensions [14]. The first dimension shows whether the managers frequently involve their staff in important decision making processes or not. The second dimension reveals whether employee learning is considered as an investment. The third dimension shows willingness of management to adapt to and/or keep ahead to the changes. Fourth one shows importance given to employee learning capability. Fifth dimension shows whether innovative ideas are supported or not. Managerial commitment itself is a part of general management capability therefore putting this factor in the model would duplicate its effect on the capability set therefore in this model managerial commitment is not considered. Systems perspective is related with bringing organization's members together around a common identity [8], [14]. The first dimension shows whether employees have general knowledge of the company's objectives. The second dimension reveals the awareness of employee's on their role for achieving the overall objectives. The third dimension questions the coordination among organizational parts. In order to create organizational learning openness to new ideas and experimenting will be the key [8], [14]. Openness and experimentation have four dimensions. The first dimension shows whether experimentation considered as way for improving the work processes. The second dimension shows whether the firm adopts the useful techniques in the other firms of the sector. The third dimension shows whether the company is using external expertise or not. The fourth dimension questions employee's ability on expressing their opinions and make suggestions. The last factor is knowledge transfer and integration which shows the ability to transfer of best practices within the firm [14]. Knowledge transfer and integration have four dimensions. The first dimension reveals whether errors and failures discussed and analyzed in this firm. The second dimension shows employees chance on talking about new ideas, programs, and activities. The third dimension shows the level of teamwork in the firm. The last dimension questions the availability of instruments such as manuals, organizational routines.

Capability sets are the main sources of organizations' success; therefore managers should effectively manage capability building and capability leveraging processes [1]. Capability set can be defined as 'bundle of tangible and intangible assets, and skills, that stand alone and also interact with each other to satisfy specific customer needs' [6]. Organizational learning is the key to enhance capability set [15]. Although the relation between learning and capability set has been accepted by many scientists; there are several studies measured this relationship. The main reason behind this scarcity is that it's not easy to measure capabilities and there is not a unique approach for measurement. Measuring functional capabilities such as general management, production, marketing and sales etc. and evaluate the level of these capabilities based on them is one of the approaches for measuring capability sets [16]. In this study capability set is measure via functional capabilities.

3. Methodology

In order to reveal the relation between organizational learning and capability set, the organizational learning level and capability set value of 104 firms quoted to The Istanbul Stock Exchange (IMKB) are evaluated. In this study the functional capabilities and the levels of these capabilities are measured via a questionnaire. The questionnaire is filled out by either the General Manager or the Human

Resources Manager of the firm. In order to evaluate capability levels the level of functional capabilities (namely general management, production, marketing and sales, finance and accounting, human resources, technology management) are asked to the respondents [16]. The respondents are asked to evaluate their firms' functional capabilities with a 1 to 7 Likert scale where 1 indicates very low and 7 indicates very high. Then the relation between organizational learning and capability set is revealed through Structural Equation Modeling (SEM). The advantage of SEM is that it combines principles of factor analysis and multiple regression analysis in one confirmatory. Hair et al. [17] claimed that the software packages utilized to evaluate SEM create robust results.

4. Results

In this study structural equation model is evaluated using the software package LISREL 8.54 [18]. Figure 2 summarizes the structure equation model for organizational capability and learning. GMC refers, General management capability, PC refers production capability, FC refers accounting and finance capability, HRC refers human resources capability, MRC refers marketing and sales capability. These capabilities form the latent variable capability which is labeled as "Cap". Organizational learning factors are coded as follows: Systems perspective (Sys), Openness and experimentation (Exp), Knowledge transfer and integration (Trans). S1, S2, S3 shows the tree dimensions of Systems perspective (Sys); EX1, EX2, EX3, EX4 shows the four dimensions of Openness and experimentation (Exp); TR1,TR2,TR3,TR4 shows the four dimensions of Knowledge transfer and integration (Trans).

NFI is a ratio of the difference in the chi-square value for the fitted model and the null model divided by the chi-square value for the null model. A model with perfect fit would produce an NFI of 1 [17]. In this model the NFI value is 0.87 and shows the model fits well. Also all the relations are significant and positive accept the first dimension of knowledge transfer and integration. The first dimension of knowledge transfer and integration does not have a significant impact on competence building and can be removed from the model but since the model fit indices are well, it can still remain at the model. Figure2 shows the standardized coefficients; all the coefficients are positive which shows a positive relation between variables. The model indicates that organizational learning have a direct positive effect on capability set.

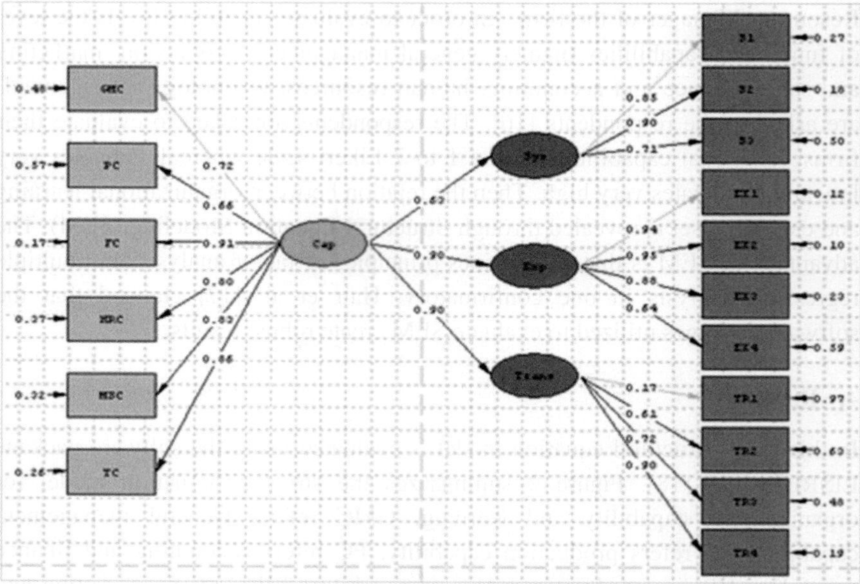

Fig. 2 SEM for organizational capability and learning.

5. Conclusion

This study shows that there is a significant relation between the organizational learning and organizational capabilities and supports the theoretical studies with an empirical research. The firms can have higher capabilities when they manage their organizational learning dimensions. Since it reveals the mechanisms behind capability building this model will help managers in their decision making processes. For further research the number of firms can be increased and the model can be tested on other countries. In a further research the effect of the other factors which are not considered in this study can be evaluated.

References

1. S. Çevik Onar, S. Polat, The factors affecting the relationship between strategic options and competence building process: an empirical examination, in “A focused issue on identifying, building and linking competencies Volume 5 of the Book series: Research in Competence-based management”, pp. 59-77, Ron Sanchez, Aime Heene, Thomas Ede Zimmerman (Editor), Emerald (2010).
2. R. Sanchez, A. Heene and H. Thomas. (Eds.) Dynamics of competence based competition: Theory and practice in the new strategic management. Oxford:Elsevire Pergamon Press(1996).

3. R. Sanchez, A. Heene Reinventing Strategic Management: New Theory and Practice for Competence-based Competition. European Management Journal, 15 (3), 303-317(1997)
4. U. Lichtenthaler, and M. Muethel. *Journal of Engineering and Technology Management,* The role of deliberate and experiential learning in developing capabilities: Insights from technology licensing, 29, 187-209, (2012)
5. J. Alegre, R. Chiva, *Technovation*, Assessing the impact of organizational learning capability on product innovation performance: An empirical test, 28, 315-326 (2008)
6. A. Soyer, S.Çevik Onar, Ö Çavuşoğlu. The Effect of Path Dependency on Capability Building and Leveraging Process, Proceedings of the TUB-ITU Joint Conference, Berlin, Germany (2012)
7. R. Y. Hung, ,B. Yang, B. Y. Lien, G. N. McLean, Y. Kuo, *Journal of World Business,* Dynamic capability: Impact of process alignment and organizational learning culture on performance, 28, (2010)
8. P.M. Senge The fifth discipline: art and practice of the learning organization. New York: Doubleday; (1990)
9. C. K. Prahalad, and G. Hamel, The core competence of the corporation. Harvard Business Review, 33, 79-91, (1990)
10. H. Gebauer, H. Worch ,B. Truffer, *European Management Journal,* Absorptive capacity, learning processes and combinative capabilities as determinants of strategic innovation, 30(1), 187-209, (2012)
11. C. Camisón and A.V. López, Industrial Marketing Management, Non-technical innovation: Organizational memory and learning capabilities as antecedent factors with effects on sustained competitive advantage 40,1294-1304,(2011)
12. A. E. Akgün, H. Keskin, J.C. Byrne, S. Aren, *Technovation*, Emotional and learning capability and their impact on product innovativeness and firm performance,27,501-513,(2007)
13. J. Hubler, and P.X. Meschi, Takeovers and Alliances As Competence Building Decisions a Stock Market-Based View. Advances in Applied Business Strategy, 6(C), 125-146, (2000).
14. P.J. Gomez, J. Cespedes-Lorente, R. Valle-Cabrera, Organizational learning capability: a proposal of measurement. Journal of Business Research, 58, 715– 725, (2005)
15. D.Teece, G. Pisano, and A. Shuen, Dynamic Capabilities and Strategic Management, Strategic Management Journal. 18(7), 509-533(1997).
16. A. Carmeli, A. Tishler, Resources, capabilities, and the performance of industrial firms: A multivariate analysis. Managerial and Decision Economics, 25(6-7), 299-216 (2004)
17. J.F. Hair, R.E.Anderson, R.L. Tatham, and W.C. Black, Multivariate data analysis. New York, Prentice-Hall(1999).
18. K.G. Jöreskog, D. Sörbom, LISREL 8 User's referance guide. Chicago, SSI (1993).

ENSEMBLE OF NEURAL NETWORKS FOR DETECTION AND CLASSIFICATION OF FAULTS IN NUCLEAR POWER SYSTEMS

ROOZBEH RAZAVI-FAR[1,*] , PIERO BARALDI[1] ENRICO ZIO[2,1]

[1] *Dipartimento di Energia, Politecnico di Milano, Milan, Italy*

[2] *Chair on System Science and the Energetic challenge, European Foundation for New energy – Electricite de France, Ecole Centrale, Paris and Supelec, Paris, France*

In this work, an ensemble of neural networks is built by an algorithm called Learn^{++}.NC and applied for fault detection and diagnosis . The algorithm is capable of incrementally learning new classes of faults, while retaining the previously acquired knowledge. The detection of new classes in subsequent data is achieved by thresholding the normalized weighted average of outputs (NWAO) of the neural networks in the ensemble. The unknown classes detected remain unlabeled until their correct labels can be assigned. The proposed method is applied to the identification of simulated faults in the feedwater system of a Boiling Water Reactor (BWR).

1. Introduction

The problem of diagnosing faults in Nuclear Power Plants (NPPs) is safety-critical due to the possible disastrous consequences of an accident. In this respect, a large number of classification methods have been applied for fault diagnosis of NPP sensors and components [1-5].

In real applications, collecting adequate and representative data for the development of the diagnostic system is often a dynamic process of successive data acquisition campaigns. In such dynamic environments, the datasets become available successively, over a period of time. Moreover, datasets collected in successive installments may contain patterns of newly emerged classes of fault that were not present in prior datasets. Then, it is essential for a fault classifier to be trained and incrementally updated, as new data become available without forgetting previously learnt knowledge.

An effective way for performing incremental learning is that of using an ensemble of fault classifiers. In [5], a bagged ensemble algorithm is applied to

---

* E-mail: roozbeh.razavi@mail.polimi.it

diagnose faults in dynamic environments. The ensemble system allows learning the new relationships between the signals in the new operational conditions. However, since in practice it is not feasible to have datasets containing examples of all the possible fault classes, the classifiers are trained only on subsets of the known classes, and, thus, they are doomed to misclassify patterns from classes on which they were not trained.

In this work, an algorithm called Learn^{++}. New Class (NC) [6] is adopted and applied for fault diagnosis. The algorithm is capable of learning incrementally and also allows additional classes to be introduced, while retaining the previously acquired knowledge. However, due to the supervised learning, the original Learn^{++}.NC can classify classes when class labels of all patterns are a priori available. In this work, a novel method is introduced to improve this limitation: unseen classes in subsequent data are detected based on thresholding the normalized weighted average of outputs (NWAO) of the base classifiers of the ensemble. These unknown classes remain unlabeled until their correct labels can be assigned with sufficient robustness.

2. Ensemble based incremental learning

Ensemble-based incremental-learning algorithms have been developed for classification in non-stationary environments. The algorithm 'Learn^{++}' is a keen ensemble approach for incremental learning [7] inspired primarily by the AdaBoost algorithm [8]. Learn^{++} creates an ensemble of (weak) classifiers, each one trained on a subset of the available patterns: in order to focus the training of the classifier on patterns carrying novel information, the training patterns are drawn from a distribution iteratively updated on the basis of the performance of the entire ensemble. Furthermore, as new data arrive, Learn^{++} generates additional classifiers, until the ensemble learns the novel information. This allows Learn^{++} to learn new data, even when previously unseen classes are introduced. Since no classifier is discarded, previously acquired knowledge is retained.

With respect to the combination of the outputs of the individual classifiers, Learn^{++} employs a weighted majority voting strategy [7].

An issue of concern in Learn^{++} is the relatively large number of classifiers required for learning patterns coming from a new concept class. This is because, when a new dataset introduces a previously unseen class, new classifiers are trained to learn the new concept class; however, the existing classifiers continue to misclassify patterns from the new concept class. Therefore, the decisions of latter classifiers that recognize the new class are out-voted by the more numerous previous classifiers that do not recognize the new concept class; this occurs until a sufficient number of new classifiers that recognize the new class is generated. Therefore, the out-voting problem is a natural artifact of the voting procedure, to be rectified by a more strategic classifier combination procedure.

In this respect, Learn^{++}.NC is specifically designed to address the issue of classifier proliferation; this is done by adjusting the voting weights of the classifiers, in proportion to the predicted confidence they have on their decision, which are themselves weighed by the confidence of other classifiers in the ensemble [6]. The classifier combination strategy used in Learn^{++}.NC allows individual classifiers to consult with each other for determining their voting weights for each test pattern. Learn^{++}.NC also uses a new protocol for reinitializing the algorithm when new data arrive.

For a generic test pattern, Learn^{++}.NC compares the class predictions of each classifier and cross-references them with the classes on which they were trained. If a subsequent ensemble overwhelmingly chooses a class it has seen before, then the voting weights of those classifiers not trained with that class are proportionally reduced.

In this work, the MultiLayer Perceptron (MLP) Neural Networks are used due to their ability to discriminate complex and nonlinear decision boundaries and their flexibility of acting as weak classifiers by adequate parameter changing [9]. The MLP architecture of the classifier here adopted is based on three layers, with a number of neurons in the input layer equal to the number of signals used for the diagnosis, a number of neurons in the output layer equal to the number of classes, ω_c, and a number of neurons in the hidden layer properly chosen by the analyst. During the training phase, a pattern of class j is associated to a ω_c-dimensional output vector whose elements are all zeros except for the *j-th* component which takes the value of 1. When the trained MLP is used to classify an incoming pattern, the *j-th* output produces a value between 0 and 1 which can be interpreted as the degree of confidence in the assignment to class j of the pattern under analysis. Finally, the pattern is assigned to the class with the associated highest degree of confidence.

3. Unknown class detection

The proposed diagnostic algorithm learns new data incrementally, including patterns of new concept classes, and discriminates all classes in a supervised way. Therefore, class labels of all patterns should be a priori known. On the other hand, adding a new ensemble of classifiers each time a new dataset becomes available significantly increases the complexity of the system. In this work, to control the proliferation of classifiers, a new ensemble is added only if the new dataset that becomes available introduces new concept classes of faults. This entails the ability to detect a new concept class in the data that becomes available. The detected new concept class would then be discriminated from the other classes as unknown, until a correct label is assigned to them.

The agreement between the classifiers of the ensemble plays an important role to detect an unknown class. A test pattern of an unknown class is expected to be assigned to different classes by the T different base classifier of the ensemble. Taking into account this observation, a heuristic index for new concept class detection is here proposed. The class j Normalized Weighted Average of all the ensemble Outputs $NWAO^j$ is defined by:

$$NWAO^j = \frac{\sum_{t=1}^{T} W_t^j y_t^j}{\sum_{t=1}^{T} W_t^j} \qquad j = 1,\ldots,\omega_c \qquad (1)$$

where y_t^j indicates the *j-th* output of the *t-th* MLP classifier of the ensemble and W_t^j the weight assigned to the same classifier by the algorithm. In order to detect a new class pattern, the $NWAO^j$ values are compared to two predefined thresholds. A pattern belongs to an unknown class if the highest $NWAO^j$ value is lower than the upper threshold and at least one of the other $NWAO^j$ values is higher than the lower threshold.

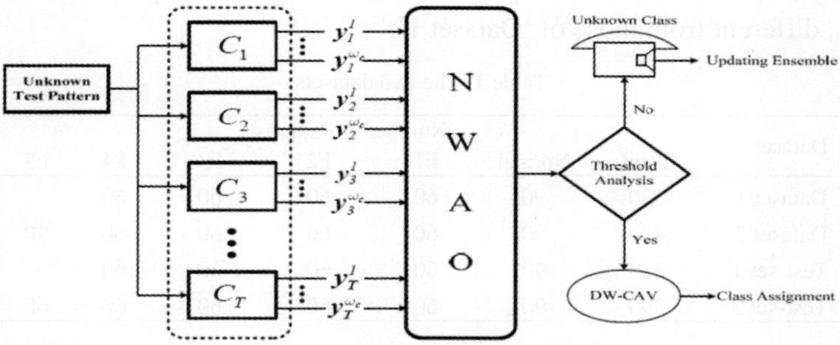

Figure 1: The new concept class fault detection and classification scheme

The overall updating procedure for the new concept class fault detection and classification system is illustrated in Figure 1. A test pattern is sent to the current ensemble of classifiers, and their $NWAO^j$ values are computed for all classes. By comparing these values with the preset thresholds, a decision is made on whether the test pattern belongs to one of the trained faults or to a new concept fault class. In the former case, the test pattern is classified, whereas in the latter case an alarm is activated. After several patterns of the new class become available, it is possible to assign them a label. Then, incremental learning procedure can be applied to add to the previous ensemble new classifiers trained on the new available patterns of the new class.

4. Fault Diagnosis in the feedwater system of a BWR power plant

In this Section, the capability of the proposed procedure is tested with respect to the identification of transients in the feedwater system of a BWR. Five classes of faults have been considered: leakage in the first high pressure preheater to the drain tank (F1); leakage through the second high-pressure preheater (F2); leakage through the first high pressure preheater drain back-up valve to the condenser (F3); leakage through the high-pressure preheaters bypass valve (F4); leakage through the second high-pressure preheater drain back-up valve to the feedwater tank (F5). The corresponding transients have been simulated by the HAMBO simulator of the Forsmark 3 BWR plant in Sweden [10]. The signals used for the fault classification are: Temperature drain 4 before VB3; Water level tank TD1; Temperature feedwater EA2; Temperature feedwater EB2; Position level control valve for EA1 [4-5].

Two datasets have been considered, containing patterns taken from simulated faults (Table 1). 'Dataset 1' includes patterns taken from normal condition and F1-F4 classes. 'Dataset 2' introduces a new class of patterns taken from F5 class and contains also patterns taken from normal operation and F1-F4 classes, different from those of 'Dataset 1'.

Table 1. The two datasets

Dataset	Number of Patterns						
	Total	Normal	F1	F2	F3	F4	F5
Dataset 1	330	90	60	60	60	60	-
Dataset 2	420	90	60	60	60	60	90
Test-set 1	330	90	60	60	60	60	-
Test-set 2	390	90	60	60	60	60	60

Initially, a diagnostic system, *E*1, made by an ensemble of 10 classifiers has been developed. The base classifiers are MLP networks with 40 neurons in the hidden layer, trained using 'Dataset 1'. Then, *E*1 has been tested considering 'Dataset 2' that consequently becomes available. 'Dataset 2' includes patterns of the normal and F1-F5 classes. As expected, the detection module does not give any false alarm on the presence of new concept class transients when it is tested with patterns of classes F1-F4. On the other hand, the NWAO values obtained testing transients of class F5 correctly lead to the detection of new concept classes.

In the next experiment, the diagnostic system *E*1 has been updated by considering the information in 'Dataset 2' which includes patterns of the normal and F1-F5 classes. The obtained diagnostic system, *E*2, formed by the same 10 classifiers of *E*1 and 12 new classifiers, is tested on the same transients considered for the test of *E*1. *E*2 correctly does not give any alarm. This is due to the fact that the classifiers of *E*2 have been trained considering patterns of F1-F4 and F5 classes. Finally, the classification performance of *E*2 is verified with respect to two test datasets, 'Test-set 1' and 'Test-set 2' formed by patterns taken from the same classes of Datasets 1 and 2, respectively (Table 1).

A satisfactory high performance is obtained in the classification of the two test sets: *E*2 correctly classifies 98.4% of the patterns of 'Test-set 1' and 98.7% of the patterns of 'Test-set 2'.

5. Conclusion

In this work, a diagnostic system based on a boosted ensemble of neural networks has been developed by means of the algorithm Learn^{++}.NC and successfully applied to the classification of faults in the feedwater system of a BWR nuclear power plant.

The novelty of the approach stands in its ability to detect the onset of transients caused by new concept classes, which is obtained by thresholding the normalized weighted average of outputs of the neural networks. The Learn^{++}.NC incremental learning procedure is then applied to add new classifiers to the previous ensemble, trained on the newly available data belonging to the new fault class.

References

1. J. Reifman,. Nuclear Technology. **119**, (1997).
2. E. Zio and P. Baraldi, Annals of Nuclear Energy. **32**, (2005).
3. R. Razavi-Far, H. Davilu, V. Palade and C. Lucas, Neurocomputing. **72**, (2009).
4. P. Baraldi, R. Razavi-Far and E. Zio, Annals of Nuclear Energy. **38**, (2011).
5. P. Baraldi, R. Razavi-Far and E. Zio, RESS. **96**, (2011).
6. M.D. Muhlbaier, A. Topalis and R. Polikar, IEEE Transactions on Neural Networks. **20**, **1**, (2009).
7. R. Polikar, L. Udpa, S.S. Udpa, and V. Honavar, IEEE transactions on systems, man, and cybernetics-part c: applications and reviews. **31, 4** (2001).
8. Y. Freund, and R., Schapire, Int. Conference on Machine Learning, (1996).
9. C. Bishop, Oxford press, UK: Oxford University, (1995).
10. E.K. Puska and S. Norrman, Proc. of Enlarged Halden G. Norway, **2** (2002).

AN AREA DEFUZZIFICATION TECHNIQUE AND ESSENTIAL FUZZY RULES FOR DEFUZZIFYING NUCLEAR EVENT FAILURE POSSIBILITIES INTO RELIABILITY DATA

JULWAN HENDRY PURBA[1], JIE LU, GUANGQUAN ZHANG

Decision Systems and e-Service Intelligence Laboratory
Centre for Quantum Computation & Intelligent Systems
School of Software, Faculty of Engineering and Information Technology
University of Technology, Sydney (UTS)
P.O. Box 123, Broadway, NSW 2007, Australia

A fuzzy reliability approach is a good alternative for a probabilistic reliability approach to assess event reliability data when event historical data is unavailable or inadequate for statistical calculation. It applies possibility theory to represent event failure in qualitative natural languages and fuzzy set theory to represent those qualitative natural languages in mathematical form. In this study, we develop a new area defuzzification technique and propose it to assess nuclear event reliability data from their corresponding failure possibilities. To justify the applicability of the new technique, we define five fuzzy rules to be satisfied by the technique and use five sets of fuzzy subsets to mathematically confirm the technique. The results show that the new area defuzzification technique is suitable for assessing nuclear event reliability data in the fuzzy reliability approach.

1. Introduction

Reliability data is essential for probabilistic safety assessment (PSA) by fault tree analysis (FTA). A fuzzy reliability approach is a good alternative for a probabilistic reliability approach to assess system reliability when historical data is unavailable or insufficient for statistical calculation [1-3]. In our previous study, the fuzzy reliability approach can be implemented in a nuclear power plant PSA by an FTA [4-6]. In this study, we develop a new area defuzzification technique and propose five fuzzy rules, which need to be satisfied by the technique. Five sets of fuzzy subsets are then used to mathematically justify the technique. The rest of the paper is organized as follows. Section 2 explains the new area defuzzification technique. The five essential fuzzy rules are given in Section 3. Section 4 presents mathematical examples to justify the new area defuzzification technique. Finally, Section 5 summarizes the study and briefly describes further research directions.

---

[1] Email: Julwan.Purba@student.uts.edu.au

2. An Area Defuzzification Technique

In this section, the new area defuzzification technique, as represented by the grayed area in Fig. 1, is described.

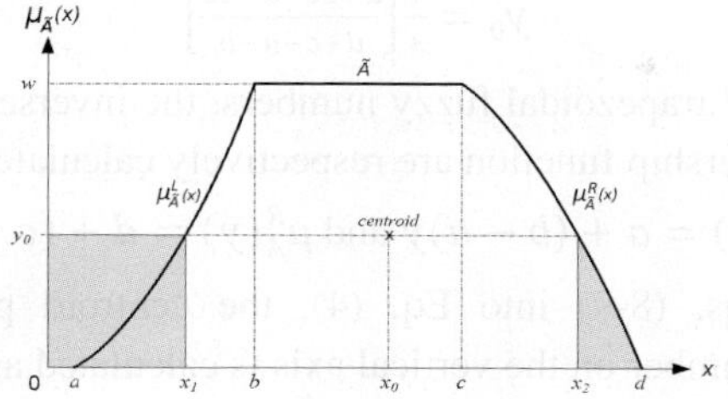

Fig. 1.Area defuzzification technique

Definition 2.1 (Nuclear Event Reliability Score). A nuclear event reliability score (R_s) is a crisp score, which is converted from the membership function of fuzzy numbers using a defuzzification technique, to represent the experts' belief of the most value that a nuclear event might happen.

$$R_s = d\big(\mu_{\tilde{A}}(x)\big) \tag{1}$$

Definition 2.2 (Nuclear Event Fuzzy Failure Rate). A nuclear event fuzzy failure rate (R) is a nuclear event reliability data, which is calculated from the R_s using a logarithmic function proposed by Onisawa [7].

$$R = \begin{cases} \frac{1}{10^m}, & R_s \neq 0 \\ 0, & R_s = 0 \end{cases} \tag{2}$$

where$m = \left[\frac{1-R_s}{R_s}\right]^{1/3} \times 2.301$.

Definition 2.3 (Area Defuzzification Technique). An area defuzzification technique(ADT) utilizes the centroid point of the membership functions on the vertical axis and its intersection with the left and the right membership functions as shown by the grayed area in Fig. 1.

$$ADT = d\big(\mu_{\tilde{A}}(x)\big) = x_1 y_0 + \int_{x_2}^{d} \mu^R_{\tilde{A}}(x)dx \tag{3}$$

where y_0 is the centroid point of the real fuzzy number $\tilde{A}$ on the vertical axis, x_1 is the intersection point between the line y_0 and the left membership function $\mu^L_{\tilde{A}}(x)$ on the horizontal axis, and x_2 is the intersection point between the line y_0 and the right membership function $\mu^R_{\tilde{A}}(x)$ on the horizontal axis. The y_0, x_1 and x_2 are calculated using Eqs. (4-6).

$$y_0 = \frac{\int_0^w y.\mu^R_{\tilde{A}}(y)dy - \int_0^w y.\mu^L_{\tilde{A}}(y)dy}{\int_0^w \mu^R_{\tilde{A}}(y)dy - \int_0^w \mu^L_{\tilde{A}}(y)dy} \tag{4}$$

$$x_1 = \mu_{\tilde{A}}^{L}(y_0) \text{ and } x_2 = \mu_{\tilde{A}}^{R}(y_0) \quad (5)\ (6)$$

Theorem 2.1 *If* $\tilde{A}$ = *(a,b,c,d;1) is a normal trapezoidal fuzzy number, then its centroid point on the vertical axis is shown in Eq. (7)*

$$y_0 = \frac{1}{3}\left[\frac{d+2c-a-2b}{d+c-a-b}\right] \quad (7)$$

Proof. For a normal trapezoidal fuzzy numbers, the inverse functions of the left and the right membership function are respectively calculated as in Eqs. (8-9).

$$\mu_{\tilde{A}}^{L}(y) = a + (b-a)y \text{ and } \mu_{\tilde{A}}^{R}(y) = d + (c-d)y \quad (8)\ (9)$$

By substituting Eqs. (8-9) into Eq. (4), the centroid point of the normal trapezoidal fuzzy number on the vertical axis is calculated as follows.

$$y_0 = \frac{\int_0^1 [y(d+(c-d)y)]dy - \int_0^1 [y(a+(b-a)y)]dy}{\int_0^1 [d+(c-d)y]dy - \int_0^1 [a+(b-a)y]dy} = \frac{1}{3}\left[\frac{2c+d-a-2b}{c+d-a-b}\right]$$

therefore proof is complete.

Theorem 2.2 *If* $\tilde{A}$ = *(a,b,c,d;1) is a normal trapezoidal fuzzy number,* y_0 *is its centroid point on the vertical axis, then the intersection between the line* y_0 *and the left membership function* $\mu_{\tilde{A}}^{L}(x)$ *on the horizontal axis is shown in Eq. (10)*

$$x_1 = \frac{2(a+b)^2 - a(2b+c+2d) - b(2c+d)}{3(a+b-c-d)} \quad (10)$$

Proof. By substituting Eq. (8) into Eq. (5), the intersection point between the line y_0 and the left membership function $\mu_{\tilde{A}}^{L}(x)$ on the horizontal axis is calculated as follows.

$$x_1 = a + (b-a)\left(\frac{1}{3}\left[\frac{2c+d-a-2b}{c+d-a-b}\right]\right) = \frac{2(a+b)^2 - a(2b+c+2d) - b(2c+d)}{3(a+b-c-d)}$$

therefore proof is complete.

Theorem 2.3 *If* $\tilde{A}$ = *(a,b,c,d;1) is a normal trapezoidal fuzzy number,* y_0 *is its centroid point on the vertical axis, then the intersection between the line* y_0 *and the right membership function* $\mu_{\tilde{A}}^{R}(x)$ *on the horizontal axis is shown in Eq. (11)*

$$x_2 = \frac{2(c+d)^2 - c(a+2b) - d(2a+b+2c)}{3(c+d-a-b)} \quad (11)$$

Proof. By substituting Eq. (9) into Eq. (6), the intersection point between the line y_0 and the right membership function $\mu_{\tilde{A}}^{R}(x)$ on the horizontal axis is calculated as follows.

$$x_2 = d + (c-d)\frac{1}{3}\left[\frac{d+2c-a-2b}{d+c-a-b}\right] = \frac{2(c+d)^2 - c(a+2b) - d(2a+b+2c)}{3(c+d-a-b)}$$

therefore proof is complete.

Proposition 2.1 *If* $\tilde{A}$ = *(a,b,c,d;1) is a normal trapezoidal fuzzy number, then its area defuzzification technique (ADT) is shown in Eq. (12)*

$$ADT = \frac{(a+2b-2c-d)\left((2a+2b)^2+(c+d)(-3a+2c-d)-2c(3b+d)-4ab\right)}{18(a+b-c-d)^2} \quad (12)$$

3. Essential Fuzzy Rules to Defuzzify Nuclear Event Failure Possibilities in Fuzzy Reliability Approach

Nuclear event failure possibility distribution is designed within interval [0,1] in the Cartesian coordinate. This means that the lowest failure possibility is set closer to zero (0) and the highest failure possibility is set closer to one (1). Nuclear event reliability data has to increase when the membership function is shifted from the left to the right. It is also very important to note that the calculated reliability data should be closely similar to the real data obtained from operating experiences and/or experiments. In general, nuclear event reliability data, which are directly collected from nuclear power plant operating experiences, are mostly less than 10^{-2} and could be of order 10^{-5} to 10^{-11}[8-11].

Based on those analyses, we define five essential fuzzy rules to be satisfied by the ADT. Let $\tilde{A}_i$ be a normal trapezoidal fuzzy numbers. Let a be the left support, b be the left core, c be the right core, and d be the right support of the $\tilde{A}_i$. Let Δx be the translation range of the $\tilde{A}_i$ in the horizontal axis. Let $R(\tilde{A}_i)$ be the reliability data calculated from the $\tilde{A}_i$.

(1) If $\tilde{A}_1(a,b,c,d) \rightarrow \hat{A}_1(a+\Delta x,b+\Delta x,c+\Delta x,d+\Delta x)$ and $d+\Delta x \le 1$ then $R(\tilde{A}_1)<R(\hat{A}_1)$

(2) If $\tilde{A}_1(a,b,c,d) \rightarrow \hat{A}_1(a+\Delta x,b+\Delta x,c,d)$ and $b+\Delta x \le c$ then $R(\tilde{A}_1)<R(\hat{A}_1)$

(3) If $\tilde{A}_1(a,b,c,d) \rightarrow \hat{A}_1(a,b+\Delta x,c+\Delta x,d)$ and $c+\Delta x< d$ then $R(\tilde{A}_1)<R(\hat{A}_1)$

(4) If $\tilde{A}_1(a,b,c,d) \rightarrow \hat{A}_1(a,b,c+\Delta x,d+\Delta x)$ and $d+\Delta x \le 1$ then $R(\tilde{A}_1)<R(\hat{A}_1)$

(5) If $\tilde{A}=\{\tilde{A}_i \mid i=1,2,\ldots, n\}$ and $R=\{R_i \mid i=1,2,\ldots, n\}$ then $1.0E\text{-}2 \le R_i \le 1.0E\text{-}12$

4. Numerical Examples

We use five sets of fuzzy subsets to mathematically justify the ADT for assessing nuclear event reliability data from failure possibilities. The fuzzy rules, the sets of the fuzzy subsets and their membership function, and the possible reliability data calculated by the ADT are given in Table 1. We can see that the reliability data calculated from the membership function of the five sets of fuzzy subsets meet all the predefined fuzzy rules. This means the ADT is suitable to assess nuclear event reliability data from failure possibilities, which are expressed in qualitative natural languages and mathematically represented by the membership function of fuzzy numbers.

Table 1. Membership functions and the corresponding reliability data

Rule#	$\tilde{A}_i$	$\mu_{\tilde{A}i}$	R_i	
(1)	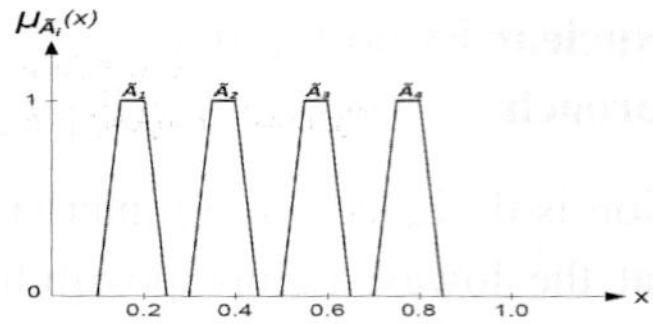	(0.1, 0.15, 0.2, 0.25)	1.1234E-06	$R(\tilde{A}_1)$
		(0.3, 0.35, 0.4, 0.45)	5.7888E-05	$R(\tilde{A}_2)$
		(0.5, 0.55, 0.6, 0.65)	3.1662E-04	$R(\tilde{A}_3)$
		(0.7, 0.75, 0.8, 0.85)	9.3452E-04	$R(\tilde{A}_4)$
(2)	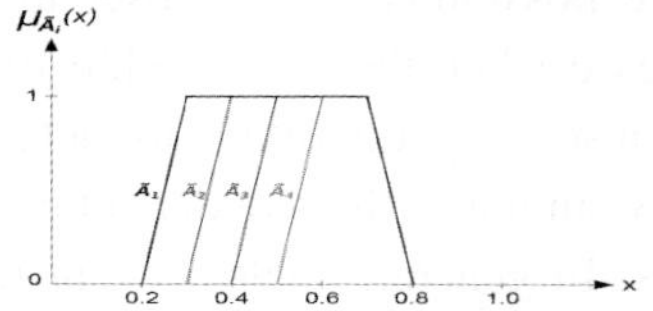	(0.2, 0.3, 0.7, 0.8)	4.0897E-05	$R(\tilde{A}_1)$
		(0.3, 0.4, 0.7, 0.8)	1.2226E-04	$R(\tilde{A}_2)$
		(0.4, 0.5, 0.7, 0.8)	2.5268E-04	$R(\tilde{A}_3)$
		(0.5, 0.6, 0.7, 0.8)	3.8533E-04	$R(\tilde{A}_4)$
(3)	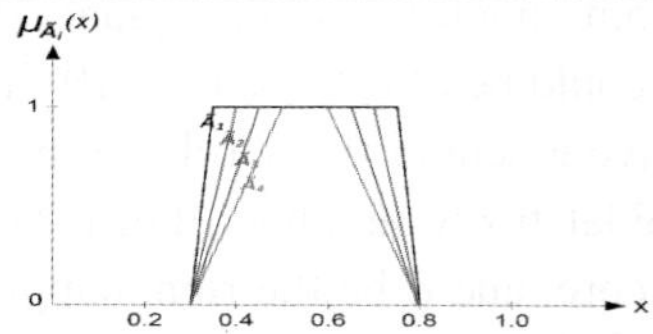	(0.3, 0.4, 0.5, 0.9)	9.5293E-05	$R(\tilde{A}_1)$
		(0.3, 0.5, 0.6, 0.9)	1.1227E-04	$R(\tilde{A}_2)$
		(0.3, 0.6, 0.7, 0.9)	1.3116E-04	$R(\tilde{A}_3)$
		(0.3, 0.7, 0.8, 0.9)	1.5207E-04	$R(\tilde{A}_4)$
(4)	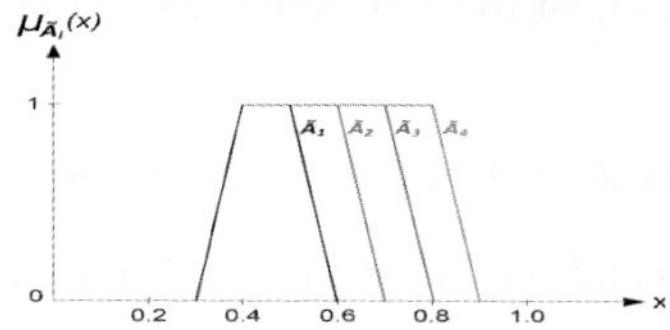	(0.3, 0.4, 0.5, 0.6)	8.1067E-05	$R(\tilde{A}_1)$
		(0.3, 0.4, 0.6, 0.7)	1.0713E-04	$R(\tilde{A}_2)$
		(0.3, 0.4, 0.7, 0.8)	1.2226E-04	$R(\tilde{A}_3)$
		(0.3, 0.4, 0.8, 0.9)	1.3207E-04	$R(\tilde{A}_4)$
(5)	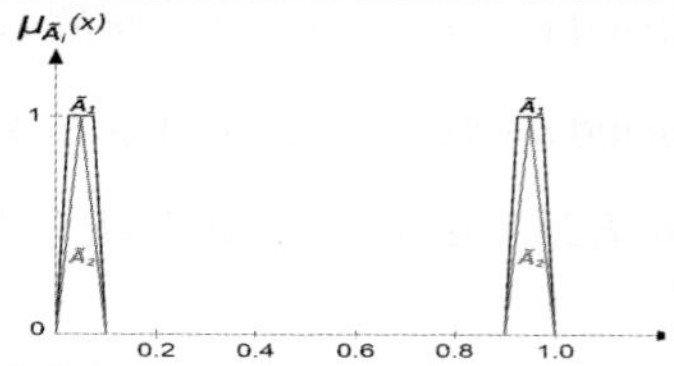	(0, 0.025, 0.075, 0.1)	1.6799E-12	$R(\tilde{A}_1)$
		(0.95, 0.97, 0.98, 1)	1.8531E-03	$R(\hat{A}_1)$
		(0, 0.05, 0.05, 0.1)	4.8098E-12	$R(\tilde{A}_2)$
		(0.95, 0.975, 0.975, 1)	1.1105E-03	$R(\hat{A}_2)$

5. Conclusion and Further Works

The fuzzy reliability approach is a good alternative for the probabilistic reliability approach to assess system reliability when quantitative historical data is inadequate or unavailable for statistical calculation. It is very important to use a suitable defuzzification technique in this approach. In this study, we develop a new area defuzzification technique and propose five fuzzy rules to be satisfied by the technique. Using five sets of fuzzy subsets, we mathematically justify the technique and results confirm that the ADT is suitable to defuzzify nuclear event failure possibilities into nuclear event reliability data. Based on the current research, we will develop the algorithm to assess nuclear event reliability data in the fuzzy reliability approach.

Acknowledgements

The work presented in this paper was partially supported by the Australian Research Council (ARC) Discovery Grant PD0880739.

References

1. H.Z. Huang, X. Tonga, and M.J. Zuo, *Posbist Fault Tree Analysis of Coherent Systems.* Reliab. Eng. Syst. Saf., 2004. **84**(1): p. 141-148.
2. H.Z. Huang, M.J. Zuo, and Z.Q. Sun, *Bayesian reliability analysis for fuzzy lifetime data.* Fuzzy Sets Syst., 2006. **157**(12): p. 1674-1686.
3. H.N. Cho, H.H. Choi, and Y.B. Kim, *A risk assessment methodology for incorporating uncertainties using fuzzy concepts.* Reliab. Eng. Syst. Saf., 2002. **78**(2): p. 173-183.
4. J.H. Purba, J. Lu, D. Ruan, and G. Zhang, *A Hybrid Approach for Fault Tree Analysis Combining Probabilistic Method with Fuzzy Numbers*, in *Artificial Intelligence and Soft Computing*, L. Rutkowski, et al., Editors. 2010, Springer: Berlin / Heidelberg. p. 194-201.
5. J.H. Purba, J. Lu, D. Ruan, and G. Zhang. *Probabilistic safety assessment in nuclear power plants by fuzzy numbers.* in *The 9th International FLINS Conference*. 2010. Chengdu: World Scientific.
6. J.H. Purba, J. Lu, D. Ruan, and G. Zhang, *Failure possibilities for nuclear safety assessment by fault tree analysis.* Int. J. Nucl. Knowl. Manag., 2011. **5**(2): p. 162-177.
7. T. Onisawa, *An Approach to Human Reliability in Man-Machine Systems Using Error Possibility.* Fuzzy Sets Syst., 1988. **27**(2): p. 87-103.
8. IAEA, *Generic component reliability data for research reactor PSA*, in *IAEA-TECDOC-930*. 1997, IAEA Austria: Vienna - Austria.
9. I.A. Papazoglou, R.A. Bari, A.J. Buslik, R.E. Hall, D. Ilberg, P.K. Samanta, T. Teichmann, R.W. Youngblood, A. EI-Bassioni, J. Fragola, E. Lofgren, and W. W. Vesely, *Probabilistic safety analysis: Procedures guide*, in *NUREG/CR-2815*. 1984, Department of Nuclear Energy, Brookhaven National Laboratory: Upton - NY.
10. T.E. Wierman, S.T. Beck, M.B. Calley, S.A. Eide, C.D. Gentillon, and W.E. Kohn, *Reliability Study: Combustion Engineering Reactor Protection System, 1984–1998*, in *NUREG/CR-5500, Vol. 10*. 2001: Idaho Falls, Idaho.
11. T.E. Wierman, S.T. Beck, M.B. Calley, S.A. Eide, C.D. Gentillon, and W.E. Kohn, *Reliability Study: Babcock & Wilcox Reactor Protection System, 1984–1998*, in *NUREG/CR-5500, Vol. 11*. 2001: Idaho Falls, Idaho.

MULTIVARIABLE ROBUST CONTROL SYSTEM: APPLICATION TO A PLASMA REACTOR

KARIMA CHAKER

Electrical Engineering Department, University of 20 Août 1955-Skikda, Route el hadeik BP26, Skikda, 21000, Algeria

ABDELKRIM MOUSSAOUI

Electrical Engineering Department, University of 08 Mai 1945-Guelma Guelma, 24000, Algeria

This paper describes a multivariable H_2-H_∞ robust control approaches, for an ill conditioned, difficult to control system, given by a plasma chemical reactor CF_4/O_2. The control of such a system resulted, in the past, in a slow, oscillatory coupled response. The goals of applying robust controls, is the achievement of a decent smooth response with minimum oscillations and overshoot. In the pursuit of this goal, we proceed to the quantification of uncertainties affecting the plasma process. Then, the robustness conditions in closed loop, to be achieved, are clearly stated. Only then, the H_2 and H_∞ algorithms are applied, after the construction of the augmented system resulting from the standard process form. To provide the advantages of the H_∞ algorithm, we applied the H_2 one and compare the results obtained between the two controls.

1. Introduction

Particular industrial applications, such as the aeronautics, chemical reactors as well as spatial structures are more than ever in the heart of the current control preoccupations. This revival is bound to the constant increase of quality requirements and performance of the controlled systems. Faced with these challenges, dynamic based control systems appears as an inescapable tool compared to graphically based control methods, not often suitable for control design of complex problems.

With the emergence of Computer-aided design (C.A.O) tools, based on the convex optimization under constraints LMI, classical frequency control methods witnessed considerable growth [2]. Nowadays, a sufficient condition to efficiently control a system is to find an algorithm that minimizes a certain control criteria satisfying the control needs [3]. One way of achieving the minimization of the mathematical control formulation, is firstly to express them on the H_2 or H_∞ norms.

The control of linear systems using H_∞ control, is a solved issue with many applications in the 1990s it is, since then, gaining ground on nonlinear system control [3], with the main target remaining, difficult control problems. Keeping in mind the main goal of these approaches that is to ensure the control of the physical process despite model/plant differences, is the core idea of "Robust Control" [4]. In this article, the process on which H_2 and H_∞ controls will be applied was studied by [1]. This multivariable process proved to be ill-conditioned and difficult to control. This reflected in a slow, oscillatory, coupled response with significant overshoot, in the time domain. Desired objectives are to improve the response time and reduce oscillations using First H_2 and then H_∞ robust controls. Section 2, presents the plasma process, leading in section 3 to the expression of the robustness condition and focuses on the system augmentation. Finally section 4 shows results and comments the obtained figures.

2. Process Description of the Plasma CF_4/O_2 Reactor

The plasma process CF_4/O_2 is a square multivariable system, with four inputs and four outputs. For more details refer to [1].The output variables to control are:

- *[F]* is the Fluor
- *[CF2]* is the Difluor of Carbon.
- *[CO2]* Dioxyde of Carbon, and
- *E/P:* E is the Electric field and P is the Pressure.

While the process inputs are given by:

- *O2* is the oxygen
- *Pr:* is the Pressure in the reactor chamber
- *Pa:* is the Feeding power
- *Db:* is the Flow rate

The process model is given by the following transfer function:

$$G(s)=\begin{bmatrix} 0.671/(1+0.105s) & 0.726/(1+0.105s) & 0.190/(1+0.105s) & 0 \\ 0 & 0.856/(1+0.053s). & 0.587/(1+0.011s) & 0 \\ 0.303/(1+0.105s) & 0.709/(1+0.105s) & 0.626/(1+0.105s) & 0 \\ 0 & -1.06/(1+0.053s) & 0.276/(1+0.053s) & 0.004/(1+0.053s) \end{bmatrix} \quad (1)$$

As stated in the previous section, the process is ill-conditioned, i.e., strongly coupled. This results in a high conditioning number of around 2300 in low frequencies. The ill-conditioning problem is emphasized by the fact that three main gains, are very close to each other's and that the minimal principal gain, is relatively far away. The Plasma gains are shown in Figure 1. This process is thus very difficult to control, indeed the temporal response obtained by the authors, using Singular Value Decomposition (SVD) method, and the Relative Gain Array (RGA) method, turn out to be slow (0.35mn), with an oscillatory coupled transitory response [1].

3. Uncertainties Quantification and Robustness Conditions

The model $G_{mod}(s)$ obtained by identification, represents only the behavior of a given functioning input/output space, where high frequency dynamics are neglected along with some system nonlinearities. The initiative which we adopt to estimate the errors between the real system and its model consists in introducing a stable transfer $\Delta(s)$ representing a multiplicative output uncertainty. Therefore, the transfer matrix involved in the case of a multiplicative uncertainty, is the additional sensibility function. We shall thus have:

$$\left\| T(s)\Delta(s)_\infty \right\|_\infty < 1 \Rightarrow \left\| \Delta(s) \right\|_\infty < 1 / \left\| T(s) \right\|_\infty . \quad (2)$$

But as:

$$\left\| \Delta(s) \right\|_\infty \leq \left\| W_t(s) \right\|_\infty . \quad (3)$$

Then

$$\left\| W_t(s) \right\|_\infty \leq 1 / \left\| T(s) \right\|_\infty . \quad (4)$$

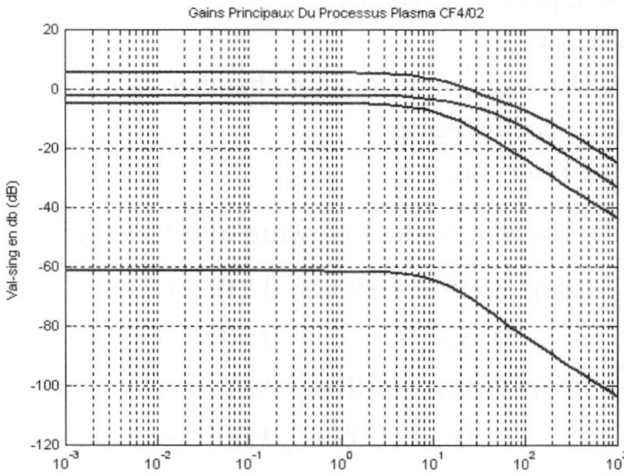

Figure 1. Plasma Principal Gains

The Inequality (4) represents therefore robustness condition on stability, which is in fact, the consequence of the application of the *small gain* theorem. The robustness condition on performance is given by:

$$\left\| W_p(s) \right\|_\infty \leq 1/\left\| S(s) \right\|_\infty . \tag{5}$$

Where $W_p(s)$ represents performance specifications and must be chosen so that the static error is null, response time in loop closed is of the same order of the open loop one and finally the response of the system is soft i.e. that there should not be important oscillations. The robustness conditions (4) and (5) must take into account the importance to find a compromise between these two objectives (Stability/performance dilemma) that is essential because any adaptation on stability engenders a performance adaptation.

3.1. *Construction of Augmented System*

Starting from the augmented system configuration given by Figure 3, the augmented system transfer matrix $P(s)$, may be deducted by Eq.6.

$$P(s) = \begin{bmatrix} W_p(s) & -W_p(s)G_{\text{mod}}(s) \\ 0 & W_t(s)G_{\text{mod}}(s) \\ I & -G_{\text{mod}}(s) \end{bmatrix} . \tag{6}$$

The transfer functions $T_{zw}(s)$ in closed loop between w and z is the cost function and is given by:

$$T_{ZW} = \begin{bmatrix} W_p S \\ W_t T \end{bmatrix} . \tag{7}$$

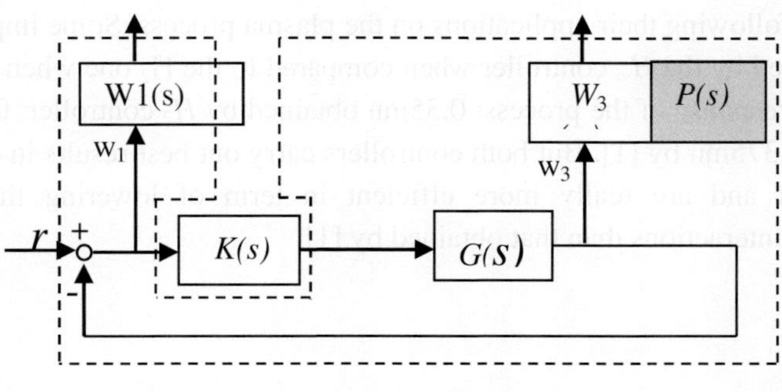

Figure 2. Two blocs criteria

The H_2 synthesis control problem consists in minimizing the transfer $T_{zw}(s)$ according to the H_2 norm, as follow:

$$\left\|T_{ZW}\right\|_2 \leq 1. \Rightarrow \left\|W_p S\right\|_2 < 1 \text{ and } \left\|W_t T\right\|_2 < 1. \tag{8}$$

Posing the H_∞ synthesis control problem consists in minimizing the transfer $T_{zw}(s)$ according to the H_∞ norm, as follow:

$$\left\|T_{ZW}\right\|_\infty \leq 1. \Rightarrow \left\|W_p S\right\|_\infty < 1 \text{ and } \left\|W_t T\right\|_\infty < 1. \tag{9}$$

In order to realize the robust controller, the H_∞ approach follows an algorithm solving the two Ricatti equations obtained from the minimization of the control problem [5].

4. Simulation Results

4.1. *Performances in Frequency Domain*

The H_2 and the H_∞ robust controllers obtained give the frequency domain results, highlighting some issues. A perfect principal gains superposition in closed loop proves a total process decoupling. We can also note a scrupulous respect of the robustness condition on stability. A respect of both robustness conditions: stability and performance, Figures 3 and 4. However, using H_∞, one can squeeze more from a particular design specifications than with H_2 and hence, cause the singular values to more precisely meet frequency-domain "loop-shaping" objectives.

4.2. *Performances in Time Domain*

The H_2 and H_∞ controllers step responses are obtained as we can see, in Figures 5 and 6, following their applications on the plasma process. Some improvements are realized by the H_∞ controller when compared to the H_2 one when calculating the time response of the process: 0.35mn obtained by H_2 controller, 0.175mn by H_∞ and 0.375mn by [1]. But both controllers carry out best results in eliminating overshoot and are really more efficient in term of lowering the transient response interactions than that obtained by [1].

5. Conclusions

It is clear from the results above that the application of H_2 and H_∞ control to the Plasma CF_4/O_2 process, succeeded in improving frequency and time domain closed loop responses with respect to previously obtained results. The two methods used in this paper, permitted to implement robust controllers achieving good performances overcoming the dilemma stability/performance. However, the H_∞ design permits to approach performance and stability objectives chosen more than the H_2 design.

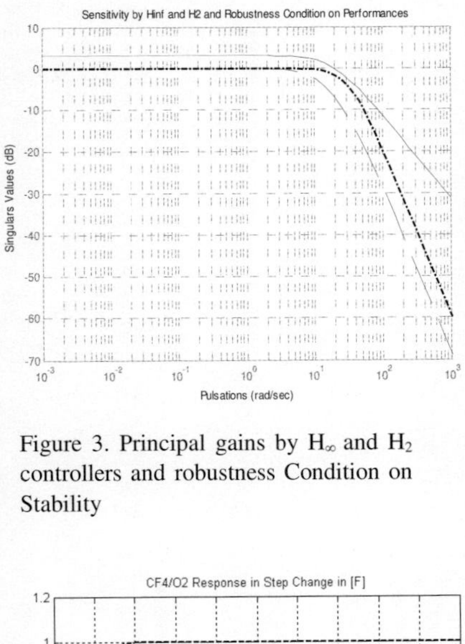

Figure 3. Principal gains by H_∞ and H_2 controllers and robustness Condition on Stability

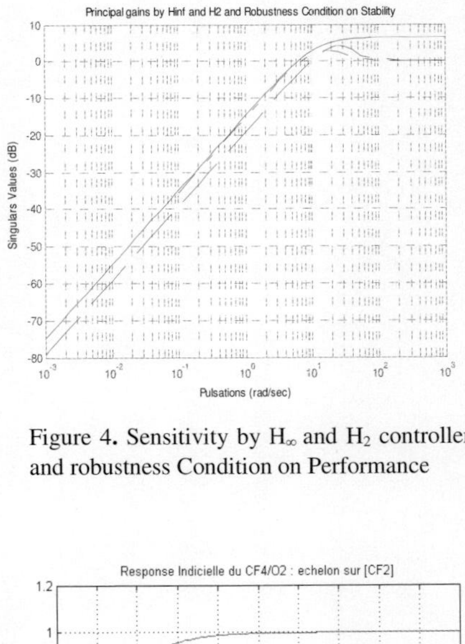

Figure 4. Sensitivity by H_∞ and H_2 controllers and robustness Condition on Performance

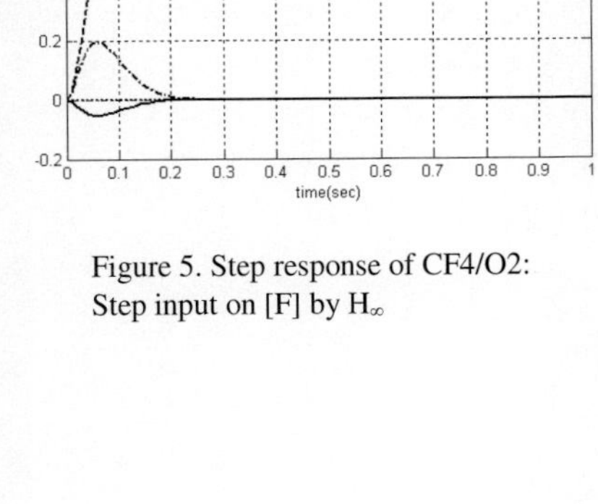

Figure 5. Step response of CF4/O2: Step input on [F] by H_∞

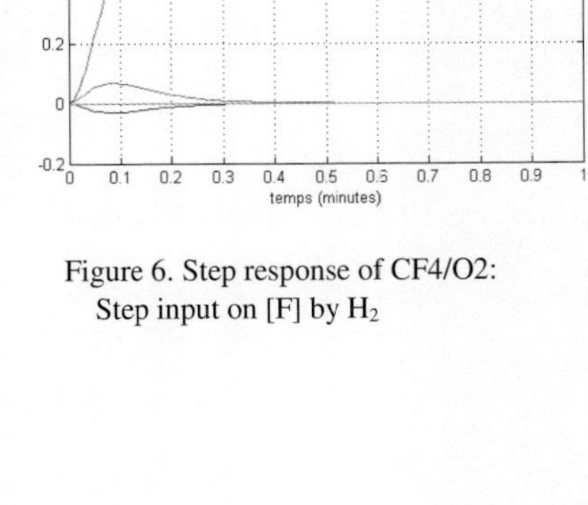

Figure 6. Step response of CF4/O2: Step input on [F] by H_2

References

1. K.J. McLaughlin, T.F. Edgar and I. Trachtenberg, *IEEE Contr. Syst.* 3 (1991).
2. S. Boyd, L. Elghaoui, E. Feron, V. Balakrishnan, **15**, SIAM *Studies in Appl. Math.*, Philadelphia, (1994).
3. S. Font, Thèse Université de Paris-Sud /Supelec, France, (1995).
4. M.G. Safonov, , *MIT, Press,* Cambridge, (1980).
5. D.J.N. Limebeer, **E.M.** Kasenally and M.G Safonov, *Report,* Imperial College, (1988).

APPLICATION OF GENETIC ALGORITHMS TO THE OPTIMIZATION OF THE ROBUSTNESS OF CONTROLLERS*

MATILDE SANTOS†, NICOLÁS ANTEQUERA

Computer Architecture and Automatic Control Department, University Complutense of Madrid, C/ Profesor García Santesmases s/n
28040-Madrid, Spain

When designing controllers for complex systems, it is not only necessary to control the system but to improve the robustness in order to get a better response. Some indexes allow to measure the performance of the controller, such as the gain and phase margins of the system response. In this paper a computational tool based on genetic algorithms is designed and implemented in order to optimize the robustness of a controller. So it is possible to analyze how the variation of the controller parameters influences the robustness of the system. The tool is applied to the optimization of a LQ controller for a MIMO autonomous vehicle.

1. Introduction

Controlling a system is not a simple task, especially for complex systems. Even more, when designing a controller the more complicated task is the tuning of the parameters. Soft computing techniques have been applied to these and other kinds of complex control problems with encouraging results [2, 14].

On the other hand, it is usually necessary to improve the robustness of a system in order to get a better response. Some indexes allow to measure this performance, such as the gain and phase margins of the system response.

In this paper, a computation tool based on genetic algorithms is implemented and applied not only to find good values of the controller parameters but to improve the robustness indexes [5, 9].

The optimization tool has been applied to a multiobjective problem, solving the optimization of two functions that represent the robustness of a controller simultaneously. As an example it has been applied to the optimization of a LQR for a multivariable system, an autonomous vehicle.

---

* This work has been partially support by the Spanish project DPI2009-14552-C02-01.

† Corresponding author: msantos@dacya.ucm.es

2. Autonomous vehicle description

The UAV (unmanned autonomous vehicle) to be controlled is a helicopter, as in previous works [3]. It is a strongly non-linear and highly unstable MIMO (Multi-Input Multi-Output) system.

The system is represented by an approximate non-linear model of the helicopter dynamics derived in [8]. The linearized model in hover and slow flight conditions can be expressed in the state space as,

$$\mathbf{x}' = \mathrm{A}\mathbf{x} + \mathrm{B}\mathbf{u}; \qquad \mathbf{y} = \mathrm{C}\mathbf{x} + \mathrm{D}\mathbf{u} \tag{1}$$

where the matrix A represents the internal dynamic of the system. The B matrix expresses the relationship between the inputs and states of the helicopter. The C matrix describes the observable states. In this case, the D matrix is zero.

The state vector, **x**, of the model of the helicopter is,

$$\mathbf{x} = [V^T\ \Theta^T\ \omega^T]^T \tag{2}$$

being $\mathbf{V} = [Vx\ Vy\ Vz]^T$ the linear velocity vector, $\Theta = [\varphi\ \theta\ \psi]^T$ (*roll*, *pitch* and *yaw* motion, respectively) the Euler angles vector, and the angular velocity is $\omega = [p\ q\ r]^T$. The inputs contain both the linear and the rotational forces that are highly coupled [6, 11], what makes difficult the design of a controller for this system [15]. The input vector **u** is,

$$\mathbf{u} = [T_M, T_T, a_{1S}, b_{1S}]^T \tag{3}$$

where T_M and T_T are the normalized main and tail rotor thrust respectively, and a_{1s} and b_{1s} are the longitudinal and lateral tilt of the main rotor. The output vector **y** consists of the three velocities and the heading, ψ. That is,

$$\mathbf{y} = [Vx\ Vy\ Vz\ \psi]^T \tag{4}$$

3. Robustness indexes: Gain margin and phase margin

Phase margin and gain margin are measures of stability in closed loop dynamic control systems. Phase margin indicates relative stability, the tendency to oscillate during its damped response to an input change such as a step function. Gain margin indicates absolute stability and the degree to which the system will oscillate without limit given any disturbance.

The gain margin (GM) can be determined from the open loop Bode plot by finding the frequency ω where the phase is -180º,

$$\mathrm{Phase}[G(j\omega)] = -180 \tag{5}$$

The system is unstable when $|GM| < 1$

The phase margin is easily determined from the open loop Bode plot. The open loop transfer function will cross the 0 dB line at some frequency, ω_c, open loop crossover frequency, at which

$$PM = \theta(\omega_c) + 180 \tag{6}$$

where $\theta(\omega_c)$ is the phase (in degrees) at ω_c.

Both of them must be positive, but a low phase or gain margins will lead to undesirable oscillations in the step response.

To obtain the robustness of MIMO systems, the following functions can be used: the sensitivity function $S = (I + L)^{-1}$, and the complementary sensitivity function $T = L(I + L)^{-1}$ or the balanced sensitivity function (S + T), where L is the open loop gain matrix and I the identity matrix.

This sensitivity may be calculated at the inputs and outputs of the actuator. At the inputs, L = HG and at the outputs, L = GH, where H is the controller transfer function matrix and G is the plant transfer function matrix. We will use the maximum of the singular values (σ) of L to obtain the gain and the phase margins (7) and (8), respectively, being the parameter *a,* defined by the expression $a = 1/\sigma$, where $a \leq 1$,

$$GM = [1/(1+a), 1/(1-a)] \tag{7}$$

$$PM = \pm\, 2sin^{-1}(a/2) \tag{8}$$

4. GA tool for the optimization of the robustness of controllers

The multiobjective optimization is carried out by applying the epsilon-constraint method [7]. Most generating methods work by transforming the multiobjective problem into a sequence of parameterized single-objective problems such that the optimum of each single-objective problem corresponds to a Pareto-optimal solution [1, 10, 12]. That is, given *n* functions to be optimized, this strategy first optimizes one of them. When the obtained result for this function is satisfactory enough, it proceeds with the following function, taking into account that the previously optimized function should not be reduced in its optimum value in more than a percentage named epsilon (ε).

In this paper we use a heuristic method, Genetic Algorithms (GA), to solve the resulting single-objective problems [4], thereby results in a metaheuristic whose aim is to find a set of approximate Pareto-optimal solutions accessible to the single-objective heuristic.

The problem we want to solve is the optimization of the robustness of a controller, that is, the gain margin and the phase margin, or according to section 3.3, the sensitivity functions that will be the fitness functions. In the specifications of the system response, we can set the values for the GM, PM or for both simultaneously.

The optimization tool has a graphic interface where the following values can be specified by the user,

- Stability criteria to be optimized (GM, PM, both)
- State space matrixes of the system (A, B, C, D)
- GM or PM Limit values, i.e, if only one of them is to be optimized, the desired value that we want to achieve and that will stop the algorithm.
- Differential (%), i.e., the epsilon constraint. Once the value of one margin has been obtained, it cannot be changed in more that this value when optimizing the other one.
- Configuration parameters of the genetic algorithm (population size, mutation probability (%), etc.)

At each iteration, the algorithm shows the best individual. Once the procedure has finished, the evolution of the individuals is displayed. The values of the GM, PM, and the eigenvalues are recorded for each iteration so at any time the user knows the values of the eigenvalues that give certain stability values.

5. Example of LQR optimization by genetic algorithms

The LQR [13] has been designed applying the expression,

$$J[u] = \int_0^\infty \left(x^T Qx + u^T Ru\right) dt \tag{9}$$

where Q is the quitar matrix. In its diagonal, the values of the velocities are weighted by 10 and the rest are equal to one. The weighting matrix R of the LQ controller has been obtained taking into account the range of the control signal, i.e, T_M, T_T, a_{1S} and b_{1S}.

The system response when using the LQ controller is shown in Figure 1.

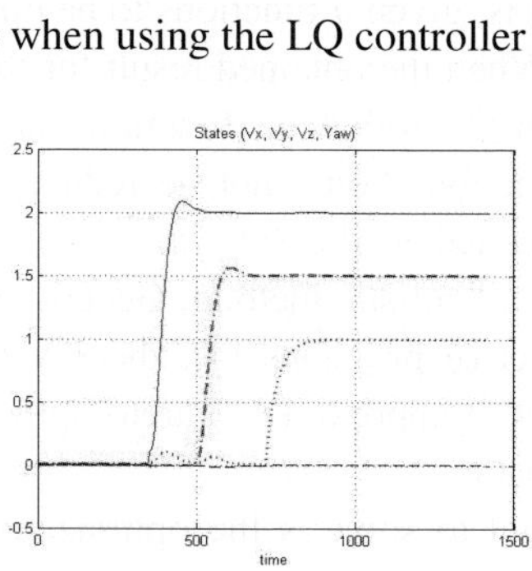

Figure 1. System response with the LQ controller (velocities and yaw for some step changes)

The states variables represented are: Vx (solid), Vy (lines), Vz (dots) and yaw angle (line-dot). It is difficult to see the yaw behaviour because of the magnitude order. The system follows the references when they are abruptly changed. The response is satisfactory but it presents overshoot.

Besides the values of the robustness indexes, for inputs and outputs, are computed for the system controlled with the LQR (Table 1). As it is possible to see, the GM is quite low, and the PM is lower than the required 30º. Although the response is not bad, it can be improved if the controller is better tuned.

Table 1. System robustness with the LQ controller

Actuators	σ	$a = 1/\sigma$	GM (dB)	PM (deg)
Input	3.8922	0.2569	[-1.9862, 2.5793]	±14.7613
Output	3.9756	0.2515	[-1.9489, 2.5166]	±14.4502

5.1. *GA optimization of the robustness of the LQ controller*

The initial population of the GA are the diagonal values of Q (8 values) and R (4 values). Initially the diagonal values of the desired matrixes Q_d and R_d are set to 1. A hard constraint has been added as matrix R must be positive.

The size of the population is 10 individuals. Mutation probability is the value set for the user (%) multiply by the number of genes of the individual. The crossover rate is 1, and the tournament strategy is applied for the selections of the parents.

The limits for the GM and PM are fixed to 4 dB and 30º, respectively. The GA will not stop until it reaches those values. The tolerance is 5% (differential), meaning that once we have obtained a good values of GM, it cannot be reduced in more that 5% when optimizing the PM.

The initial GM was 1.2414 dB, and it has reached the value of 5.95 dB. The initial value of the PM was 8.1769º and it reaches up to 38º.

The Q and R matrixes obtained by the GA are shown below.

$$Q=\begin{pmatrix} 504 & 0 & 0 & 0 & 0 & 0 & 0 & 0 \\ 0 & 474 & 0 & 0 & 0 & 0 & 0 & 0 \\ 0 & 0 & 62 & 0 & 0 & 0 & 0 & 0 \\ 0 & 0 & 0 & 256 & 0 & 0 & 0 & 0 \\ 0 & 0 & 0 & 0 & 179 & 0 & 0 & 0 \\ 0 & 0 & 0 & 0 & 0 & 323 & 0 & 0 \\ 0 & 0 & 0 & 0 & 0 & 0 & 238 & 0 \\ 0 & 0 & 0 & 0 & 0 & 0 & 0 & 275 \end{pmatrix} \quad R=\begin{pmatrix} 216.31 & 0 & 0 & 0 \\ 0 & 427.89 & 0 & 0 \\ 0 & 0 & 423.57 & 0 \\ 0 & 0 & 0 & 451.59 \end{pmatrix}$$

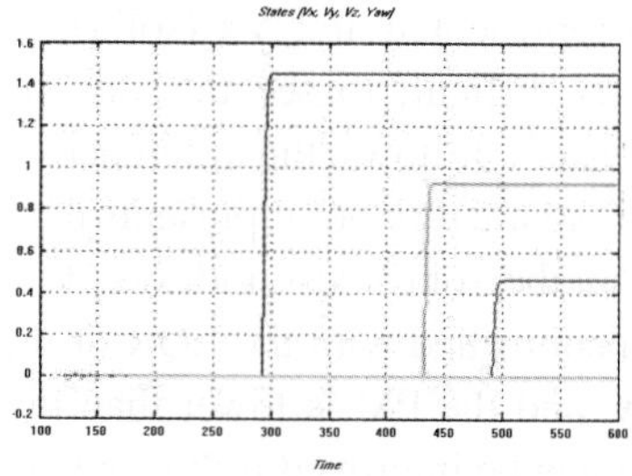

Figure 2. System response with the LQ controller optimized by GA (Vx: blue, Vy: green, Vz: red).

The system response (Figure 2) has improved as it is quicker and it does not have overshoot.

6. Conclusions

In this paper, Genetic Algorithms have been applied to the optimization of the controller parameters of a MIMO complex system, an autonomous helicopter, to get a better response of the system in terms of stability. The multiobjective optimization has been carried out on two variables simultaneously that determine the robustness of the system response, the Gain and Phase Margins.

References

1. F. Alonso, M. Santos, Inform Sciences **180**(6) 896-910 (2010).
2. F. Alonso, M. Santos, Eng. Appl. Artif. Intel. **25**, 189-199 (2012).
3. N. Antequera, M. Santos, J.M. Cruz. Proc. IEEE ETFA 719-724 (2006).
4. K. Deb, Multi-Objective Optimization using Evolutionary Algorithms, Wiley &Sons (2002).
5. S.A. Ghoreishi, M.A. Nekoui, S.O. Basiri, Optimal IJIIP **2**(1) 63-74 (2011).
6. A. González, R. Mahtani, M. Béjar, A. Ollero, World Automat. Congress **15**, 399-404 (2004).
7. Y. Haimes, L. Lasdon, D. Wismer, IEEE Trans SMC **1**, 296–297 (1971).
8. B. Horowitz, J. Liebman, C. Ma, T.J. Koo, A. Sangiovanni-Vincentelli, S. Sastry, IEEE **91**,1, 198-211 (2003).
9. M. Jamshidi, L. dos Santos Coelho, R.A. Krohling and P.J. Fleming, Robust Control Systems with Genetic Algorithms, CRC Press (2003).
10. M. Laumanns, L. Thiele, E. Zitzler, Eur. J. Oper. Res. **169** 932–942 (2006).
11. U-L. Ly, Stability and Control of Flight Vehicle, Univ. Washington (1997).
12. G. Mavrotas. Appl Math and Comput **213**, 455–465 (2009).
13. S. Mobayen, et al. Int. J. Phys. Sci. **6**(30) 6958-6966 (2011).
14. M. Santos, Rev. Iberoam. Automática & Inform Ind. **8**(4) 283-296 (2011).
15. H. Shim, T.J. Koo, F. Hoffman, S. Sastry. Proc. 37th IEEE CDC (1998).

A HYBRID ENSEMBLE APPROACH FOR PROCESS PARAMETER ESTIMATION IN OFFSHORE OIL PLATFORMS

PIERO BARALDI[A*], FRANCESCA MANGILI[A], ENRICO ZIO[B,A]

[A]*Dipartimento di Energia, Politecnico di Milano, Italy*

[B]*Chair on System Science and the Energetic challenge, European Foundation for New energy – Electricite de France, EcoleCentrale, Paris and Supelec, Paris, France*

GIULIO GOLA[C,D], BENT H. NYSTAD[C]

[C]*Institutt for energiteknikk, OECD Halden Reactor Project, Norway*

[D]*IO Center for Integrated Operations, Trondheim, Norway*

In offshore oil platforms, choke valve erosion is a major issue. The difference between the theoretical and the actual values of the valve flow coefficient is commonly used during operation as the indicator of the choke erosion. The actual value of the flow coefficient is analytically calculated as a function of measured and allocated parameters. The latter are typically obtained using a physics-based model proved to be inaccurate for some operation conditions. As a consequence, inaccuracies are introduced in the evaluation of the erosion evaluation and this undermines the possibility of using it for prognostic purposes. To overcome this hurdle, a hybrid ensemble of physics-based and Kernel Regression (KR) models is proposed. The method is verified on real measurements.

1. Introduction

In this paper, we consider the degradation of choke valves located topside at wells in offshore platforms on the Norwegian Continental Shelf [1]. The difference between the actual valve flow coefficient and its theoretical value is retained as the choke valve erosion indicator. While the theoretical value of the valve flow coefficient depends only on the choke opening, its actual value is analytically calculated as a function of the pressure drop through the choke and the mass flow rates of oil, water and gas. The pressure drop is directly measured on a daily basis, whereas the values of the mass flow rates are allocated with respect to a single well by a physics-based model based on the measured total production and on the values of physical parameters such as pressures and temperatures related to the specific well. Oil, water and gas flow rates are actually measured only during the

* Piero.baraldi@polimi.it

well tests (WTs), which are carried out throughout the valve life. In [2] it has been shown that in correspondence of some operation conditions the mass flow rates provided by the physics-based model, which is fed by noisy measurements, tend to be excessively noisy, thus undermining the possibility of using them for the evaluation of the choke valve degradation.

In this work, an ensemble of Kernel Regression (KR) models [3] which learn the relationships between the process parameters from the well test data and provide an estimate of the mass flow rates, has been devised to complement the physics-based model when the latter is not accurate. In order to exploit the KR models only when actually needed, a hybrid ensemble approach has been developed, in which the KR estimates of the mass flow rates and the corresponding physics-based allocations are aggregated by a local fusion technique [4-5].

The paper is framed as follows: Section 2 describes the traditional procedure for choke valve erosion assessment; in Section 3, the hybrid ensemble approach is presented and results discussed; finally, conclusion and potential perspectives for future work are drawn in Section 4.

2. Choke valve erosion assessment

In oil and gas industries, choke valves are used to control flow rates and protect the equipment from unusual pressure fluctuations. They are based on a throttle mechanism made of two circular disks, one fixed and one rotating, each with a pair of circular openings to create variable flow areas (see Figure 1).

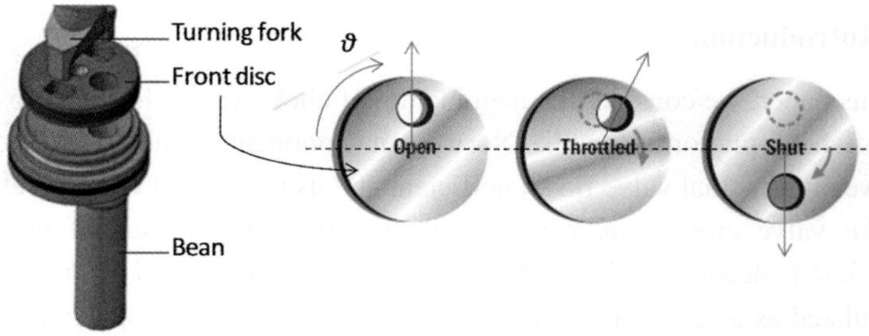

Figure 1. Typical scheme of a choke valve: by rotating the disk the flow is throttled (picture taken from www.vonkchokes.nl).

Combined mechanisms of erosion-corrosion cause choke lifetimes of less than one year [6]. To avoid failures, effective erosion management strategies must be implemented, requiring the development of reliable models to estimate the erosion and predict the lifetime of choke valves [7-8].

A common indicator of the valve flow capacity is the flow coefficient C_V. For a specific valve opening θ, erosion produces a gradual increase of the valve area available for flow transit, thus determining an increase of C_V. For this reason, knowing the value of the flow coefficient is fundamental for assessing the erosion state of the choke. During operation, C_V is not directly measured but computed as a function of the pressure drop through the choke, ΔP, the oil, water and gas flow rates, $\dot{m}_o$, $\dot{m}_w$ and $\dot{m}_g$, respectively, and the valve opening θ [9]. Thus, for a correct assessment of the choke erosion state, it is fundamental to have reliable measurements or estimates of the parameters ΔP, θ, $\dot{m}_o$, $\dot{m}_w$ and $\dot{m}_g$ used to compute the flow coefficient C_V. Nevertheless, only the pressure drop ΔP, and the valve opening θ are measured during daily standard inspections (SI), whereas measures of water, oil and gas flows rates are taken downstream of the choke with a multiphase flow separator only during WTs. On daily bases, the values of $\dot{m}_o$, $\dot{m}_w$ and $\dot{m}_g$ for a single well are allocated based on the measured total production and on the values of physical parameters, such as pressures and temperatures related to the specific well.

3. A hybrid ensemble approach to estimate the mass flow rates

By comparing, for some historical patterns, the allocated values of $\dot{m}_o$, $\dot{m}_w$ and $\dot{m}_g$ provided by the physics-based model with the corresponding WT measurements, we noticed that in some circumstances corresponding to specific operating conditions in which the physics-based model does not work properly, the allocation are noisy and unreliable [2]; for this reason, a hybrid ensemble approach is here developed. The ensemble is fed with the measures of ΔP and θ and the allocated values of $\dot{m}_o$, $\dot{m}_w$ and $\dot{m}_g$, and provides improved estimates of $\dot{m}_o$, $\dot{m}_w$ and $\dot{m}_g$ (see Figure 2). The ensemble is constituted by the physics-based model previously described and by four data-driven models. These latter are Kernel Regression (KR) models [10], which learn the relationships between the parameters ΔP, θ, $\dot{m}_o$, $\dot{m}_w$ and $\dot{m}_g$ from a training set of WT data. KR models are auto-associative, i.e., they return as outputs the estimates of the same parameters they have been fed with in input. Such estimates are obtained as a weighted average of the training patterns, with weights decreasing in function of the distance between the test and training patterns: by use of a Gaussian kernel function [10], training patterns closer to the test pattern considered, are weighed more in the averaging because they are more similar to it. Diversity between the KR models of the ensemble is obtained by varying the parameters used to compute the distance between train and test patterns: model 1 uses all five parameters, whereas models 2, 3 and 4 use ΔP, θ and alternatively $\dot{m}_o$, $\dot{m}_w$ or $\dot{m}_g$.

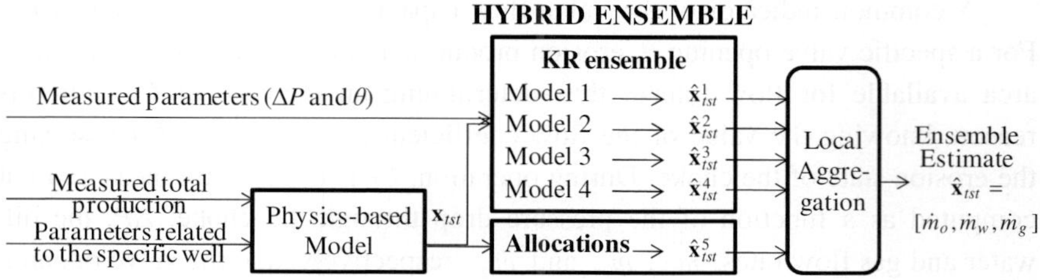

Figure 2. scheme of the hybrid ensemble approach.

In [2], it was verified that the performance of the individual models varies with the values of the input parameters: using an ensemble of models allows obtaining accurate results in all situations while increasing the robustness of the final estimates. Indeed, the ensembles of diverse models is capable of achieving performance superior to that provided by each individual model [11].

Different techniques for the aggregation of the outcomes of the individual models have been proposed in the literature, from statistics methods like the simple mean and the median [11-12], to weighed averages of the model outcomes based on the global or local performances of the individual models [4-5]. In the latter ones, the aggregation is guided by the local performance of each model, i.e., its reconstruction accuracy on training patterns similar (and for this reason also called neighbors) to that under test. These methods rely on the idea that each model can perform well in some regions of the parameter space and poorly in others. In this work, a local performance-based technique is applied for estimating the mass flow rates. This allows exploiting the four KR models only when they actually outperforms the physics-based model, thus avoiding affecting its accuracy when it actually outperforms the KR models.

For each parameter p to estimate, the local performance aggregation approach here adopted assigns to the generic model m in the ensemble a weight $w_{p,m}$ proportional to the model performance evaluated considering the inverse of its mean square estimation error over the $N = 3$ training patterns closest to the test pattern $\mathbf{x}_{tst}$. The estimation error made by model m in providing the estimate $\hat{x}_i^{p,m}$ of the p-th parameter is obtained by comparing the estimate $\hat{x}_i^{p,m}$ to the corresponding WT measurement $x_i^{p,WT}$. The resulting weights are:

$$w_{p,m} = \left[\frac{1}{N} \sum_{i=1}^{N} \left(\hat{x}_i^{p,m} - x_i^{p,WT} \right)^2 \right]^{-1} \tag{1}$$

The final estimate $\hat{x}_{tst}^{p}$ of parameter p is obtained as the average of the multiple model estimates $\hat{x}_{tst}^{p,m}$ weighted with the weights $w_{p,m}$.

3.1. *Results*

The hybrid ensemble is here applied to estimate the mass flow rates $\dot{m}_o$, $\dot{m}_w$ and $\dot{m}_g$ from patterns collected at five different wells. Notice that the model used to estimate the mass flow rates at well w is trained by considering only WT data from the same well, since it has been observed that well behaviors vary from one well to another. In practice, when estimating the test pattern $\mathbf{x}_{tst}=\mathbf{x}_k$ for a specific well w, only the WT patterns $\mathbf{x}_j$, j=1,…,k-1 previously collected during the life of the w-th well are used by the KR model to generate the estimates $\dot{m}_{o,w,g}$. Due to the fact that degraded choke valves are replaced, patterns collected for a single well may refer to different chokes.

The ensemble performance is evaluated by considering the Root Mean Square Error (RMSE) between the estimated values $\dot{m}_{o,w,g}$ and the corresponding WT measurements, normalized in the range [0,1]. Similarly, the RMSE is computed also for the physics-based model and for an ensemble made only by the KR models (data-driven ensemble). Table 1 compares the RMSE values.

Table 1. Comparison of the performance of the hybrid ensemble with that of the physics-based model and of the data-driven ensemble

	RMSE (10^{-2})					
	Well 1	Well 2	Well 3	Well 4	Well 5	average
Physics-based	7.118	2.742	1.056	3.248	5.796	4.435
Data-driven ensemble	4.438	3.892	3.395	4.208	6.277	4.441 (+0.1%)
Hybrid ensemble	3.877	2.984	1.194	3.665	5.480	3.623 (-18%)

The results show that the data-driven ensemble outperforms the physics-based model only for well 1, producing a 38% reduction of the RMSE, whereas in the other four wells the physics-based model generates more accurate estimates than the data-driven ensemble. The hybrid ensemble mediates between the physics-based and data-driven approaches: it slightly reduces the good performance of the physics-based model on wells 2, 3 and 4, but, at the same time, enhances significantly the performance of the data-driven ensemble on wells 1. This confirms that the local-performance aggregation technique is able to automatically distinguish those cases in which the physics-based model works properly from those in which it is convenient to integrate the data-driven models.

4. Conclusions

In this paper, a hybrid ensemble approach has been propounded for estimating process parameters related to the erosion state of choke valves in offshore oil platforms. In the application considered, the hybrid ensemble is constituted by one physics-based model and four data-driven models based on auto-associative kernel regression. The aggregation of the outcomes of the models in the ensemble is done by a local-performance technique.

The results obtained from the application of the approach on a number of choke valves have confirmed that in those circumstances where the physics-based model is inaccurate, an ensemble approach with data-driven models can increase the accuracy of the estimates. The hybrid ensemble correctly favors the most accurate between data-driven and physics-based models. This allows improvement of the parameter estimate.

Acknowledgments

The authors wish to thank Erling Lunde and Morten Løes at Statoil ASA for proving us with the operational data and the IO Center for Integrated Operations in the Petroleum Industry for partly funding this research project.

References

1. B.H. Nystad, G. Gola, J.E. Hulsund, D. Roverso, *Proc. PHM Society*, Oct. 11-13 (2010).
2. P. Baraldi, E. Zio, F. Mangili, G. Gola, B.H. Nystad, *IJCIS* (Submitt.2011).
3. C.G. Atkeson, A.W. Moore, S. Schaal, *Artif. Intell. Rev.* **11**, 11 (1997).
4. P. Bonissone, F. Xue, R. Subbu, *Appl. Soft Comp.* **11**(2), 1529–1539 (2011).
5. P. Baraldi, A. Cammi, F. Mangili, E. Zio, *IEEE Trans. on Nucl. Sci.* **57**(2), 793 (2010).
6. L. Ngkleberg, and T. Sontvedt,*Wear* **186-187**(2), 401 (1995).
7. P.M. Birchenough, D. Cornally, S.G.B. Dawson, P. McCarthy, S. Susden. *SPE Eur. Petro. Conf.*, Oct. 25-27 (1994).
8. P. Vaidya, Annual Conf. of the PHM Society, Oct. 10-16 (2010).
9. Metso Automation. Flow Control Manual. 4th edition (2005).
10. M.P. Wand, W.R. Schucany,*Canad. J. of Statist.* **18**, 197 (1990).
11. P. Baraldi, E. Zio, G. Gola, D. Roverso, M. Hoffmann, *Ann. of Nucl. Energy* **38** (2-3), 212 (2011).
12. R. Polikar, *IEEE Sign. Process. Mag.* **59**, 59 (2007).

HEURISTIC GP OPTIMIZATION TECHNIQUE FOR DESIGN SYNTHESIS OF MONOTONIC OBJECTIVES

SAYED M. METWALLI[†]

Professor Emeritus, Mechanical Design and Production Department,
Faculty of Engineering, Cairo University, Cairo 12316, Egypt
e-mail: metwallis2@asme.org

This work presents a developed novel optimization process that can be used for the synthesis of mechanical components and systems. One sample application to the design optimization of space frames is presented in detail. Several other applications are also indicated. The main objective in the sample application is to minimize the space frame volume and satisfy stress constraints. The optimization effectively uses the devised Heuristic Gradient Projection (HGP) technique to synthesize the optimum design of the Midi-Bus frame. Results indicate a striking improvement over available designs and remarkably faster convergence over other design and optimization techniques. The technique can moreover be effectively applied to other large 3D-space frame synthesis and optimization. The developed synthesis methodology is also applied to other designs as a tailored optimization paradigm for a better theory of design realization.

1. Introduction

Optimization process has been endeavored for the previous decades since the advent of computers. Several numerical techniques are available to perform local and global optimization in basic textbooks [26, 33]. Most of those techniques are general and can handle multi-objective functions and multitude of constraints. They may not necessarily reach the constrained optimum all the time. This has been the reason for continuous need of more optimization techniques and even more techniques for tailored optimization process.

Several techniques have been developed in the field of optimization by the author and with his collaborators [1-32, 34-37]. The design applications have always been the drive rather than the abstract mathematical procedures. The concept has also been used in other applications such as modal analysis utilization in fault diagnoses [31]. Developed techniques employ heuristic factors stemming from the basic understanding of each application, its main

[†] PhD, Life Fellow ASME, Fellow SME.

independent and dependents variable and their relational effect on the objective function or constraints. Results have always been robust, fast and effective.

In early development the objective function has been shallow and required a large integration time. The developed technique had to follow the ridge path of the function with newly developed exterior penalty function [6, 8, 9, 13, 17, 19, 28, 29, 30]. The heuristic nature in that stemmed from the objective function topology. The other development is the optimization of rotating flywheel for a uniformly stressed configuration [14, 16, 20]. The heuristic Gradient Projection approach (GP) has been initiated in a heuristic sense to change the thickness by a percentage increment as a linear gradient to satisfy the constraint. It has formally been developed latter as an efficient optimization technique for tailored constrained problems [26]. Applications of heuristic GP in 3D space frame [2, 3, 9, 34] and MEMS resonator [34] have also been presented.

2. Heuristic optimization technique

To demonstrate the development of the technique, an application in the field of 3D frames is presented herein. In that regard, the best utilization of material is achieved by loading the members to the maximum allowable constraints limits. This would satisfy these constraints and also minimizes the frame volume or weight. If some constraints are violated, we may like to move in a direction combining the negative constraint gradient (i.e. parallel to $\boldsymbol{\nabla g}$) for violated constrains, gradient tangent for active constraints, constraint gradient for inactive constraint and minimizing the objective function all at the same time. Projecting the constraint gradients onto the objective gradient vector and projecting the objective gradient vector onto all active constraint tangents can implement this. This last projected gradient vector consists of the gradient vector itself minus the component $\boldsymbol{u}_i$ of the constraint gradient (i.e. parallel to $\boldsymbol{\nabla g}_i$). Therefore, the projected gradient vector $\boldsymbol{f_p}$ is given by

$$f_p = \nabla f - G_r\, u \tag{1}$$

where $\boldsymbol{G_r} = [\, \boldsymbol{\nabla g_1} \quad \boldsymbol{\nabla g_2} \quad \dots \quad \boldsymbol{\nabla g_r}]$ and all the "r" active constraints are considered. The projected vector, however, must be orthogonal to the constraint gradient such that

$$G_r^T f_p = 0 \tag{2}$$

To move in the projected vector and minimizing the objective function, one should also have the projected vector satisfying the following relation.

$$f_p^T\, \nabla f(x) < 0 \tag{3}$$

Substituting Eq. (1) into Eq. (2), one obtains

$$u = [G_r^T G_r]^{-1} G_r^T \nabla f \tag{4}$$

and
$$f_p = \{ I - G_r [G^T{}_r G_r]^{-1} G^T{}_r\} \nabla f = \mathbb{P} \nabla f \tag{5}$$

where $\mathbb{P}$ is the projection matrix. To perform a minimization one proceeds in the $-f_p$ direction as a regular gradient projection method [26, 33].

Obviously for convex constraints, the search direction should depart into the infeasible region and one must step into the $-\nabla g_i$ to converge. If the component of $-\nabla g_i$ is included in the new combined gradient projection search direction $-\nabla g_i^T \nabla f$, the process will be automatic. Therefore, this is suitable for convex constraints. For concave constraints, one may need to step into the ∇g_i to converge. If the component of ∇g_i is also included in the new combined search direction $\nabla g_i^T \nabla f$, the process will also be automatic [26].

This process is most suitable to apply for the optimization of 3D space frames where the objective function is the total volume and the constraints are to satisfy allowable stress constraints. The objective function for 3D space frames with n members of circular cross-section is therefore

$$f(d) = \Sigma\, \pi d_i^2\, l_i / 4, \quad i=1 \rightarrow n \tag{6}$$

Without any loss of generality, if only removing trivially loaded members can topologically optimize the 3D frame, the objective function will then be

$$f(d) = \Sigma\, a_i\, d_i^2, \quad i=1 \rightarrow n \tag{7}$$

In the local coordinates, the stress in each member is evaluated numerically [3]. Without key loss of generality, the bending stress constraints may be:

$$g_i(d) \approx b_i\, d_i^{-3} - c_i \leq 0 \tag{8}$$

where the c_i represents allowable stress constraint in each member i. Through the continuous iteration, however, the coupling or stiffness feedback will be accounted for. This is accomplished by recalculating the ensued forces and moments. The resulting von Mises stress is then used instead of the initial bending consideration. If, however, the members were mainly subjected to axial loading, Eq. (8) would instead be mainly in d^{-2}. If Eq. (8) is used, conversion will be slower and no instability will occur.

The gradient vectors for the objective function and the constraints are then:

$$\nabla f_i(d) = \alpha_i\, d_i \tag{9}$$

$$\nabla g_i(d) = -\beta_i\, d_i^{-4} \tag{10}$$

The projected gradient vector can be obtained as follows.

$$\nabla g_i^T \nabla f = -\gamma_i d_i^{-3} < 0 \qquad (11)$$

From Eq. (2) of active constraints and applying homogeneous coordinates,

$$f_{pi} = -\gamma_i d_i^4 \qquad (12)$$

By the relation in (4), the direction of HGP can be such that

$$G^T \nabla f_i (d) = -\kappa_i d_i^{-3} \qquad (13)$$

To implement the procedure and considering the slackness variables, we use the recursive heuristic GP direction such that:

$$S_{pi}(d_i) \cong \sigma_{i,equiv} (d_{i,new})^{-3} - \sigma_{i,const.} (d_i)^{-3} \qquad (14)$$

which can be implemented by recursively using

$$d_{i,new}{}^3 = d_i{}^3 \frac{\sigma_{i,equiv}}{\sigma_{i,const.}} \qquad (15)$$

The scheme of this heuristic GP method is then as follows:

1- For each element, read stress value; if the equivalent stress is ≠ the allowable stress, then go to step 2; else go to step 5.

2- For each element, read its diameter.

3- Find a new diameter based on

$$d_{i,new} = d_i \sqrt[3]{\frac{\sigma_{i,equiv}}{\sigma_{i,const}}} \qquad (16)$$

where $d_{i,new}$ is the required step optimum diameter, d_i is the previous input diameter, $\sigma_{i,equiv}$ is the calculated equivalent von Mises stress, and $\sigma_{i,const.}$ is the maximum allowable stress constraint.

4- Check stress values; if, for all elements, the equivalent stress equals the maximum stress constraint, then go to step 5; else go to step 1.

5- End the iterative procedure; the obtained structure is optimum.

The relation defined in Eq. (16) satisfies the constraint; in addition, it reduces the objective function. Therefore, these steps chase the constraint boundary while reducing the objective function. This is tantamount to the process of gradient projection method but does not require following the constraint tangents. This is the reason of coining the heuristic gradient projection (HGP) name to this method.

3. Results and Discussion

The optimum design of 3D frame of a Midi-Bus has been obtained by HGP. The developed HGP optimization process results in reducing the total number of members by a percentage of 47% - form 297 to 156 member and has approximately 48% from the initial volume in only 46 FE iterations [2, 3]. In another application, the optimization of other 3D-frames is performed using Fuzzy logic in the HGP [35]. Only stress constraints were described using fuzzy logic while deflection constraints were not. The total number of function evaluation as well as objective function was further decreased (by 24% and 3.2%, respectively) relative to the HGP method.

General Heuristic Gradient Projection (GHGP) is used in a more general form to satisfy the stress constraints for macro and MEMS applications [9]. Results showed that the proposed hybrid technique (GHGP) converges to the optimum solutions faster by 8 %. The MEMS weight is also decreased by 23.7 %. For a macro level, the GHGP improved the solution time by 33.3 %.

Other applications are carried out and underway such as rolling bearing optimization or fault diagnoses that demonstrate the efficiency of the method relative to other optimization techniques [1, 31]. The convergence of the HGP is in five iterations against selected 500 generations for GA procedure [1].

4. Conclusion

The present work introduced a new novel procedure of synthesizing and optimizing different systems. The developed Heuristic Gradient Projection Method (HGP) is applied in the optimization stage by modifying components dimensions in order to satisfy the stress or other constraints to get the best synthesized design. The FEM simulation can automatically analyze these modified elements. Eliminating redundant elements also modifies the design. This subsequent stage of the optimization process has reduced iterations through the developed synthesis in order to achieve the optimum in short period of time.

It should be indicated that the proposed technique is a local optimization technique. It might partially sacrifice global optimality in favor of obtaining a quick result. The multi-modality, which may be present, can be handled by several runs from random starting points using the proposed technique, and confirming that they all lead to the same optimum or to other global optimum.

References

1. M. Abbas, H. and S. Metwalli, *DETC2011-47624* (2011).

2. M. Abd El Malek, M. Senousy, H. Hegazi, S. Metwalli, *DETC2005-85348* (2005)
3. M. Abd El Malek, *Cairo U.* (2003)
4. W. Abd El-Hady, S. Metwalli, *DETC2007-35012* (2007).
5. S. Badran, A. Nassef, S. Metwalli, *Thin-Wall Struct.,* 47, 1331 (2009).
6. S. Elmaghraby, S. Mewalli, C. Zorowski, *Math. Prog. Study,* 11, 150 (1979)
7. S. Elmoselhy, B. Azzam, S. Metwalli, H. Dadoura, *SAE J. Comm. Vehc.* 115-2, 226 (2006)
8. A. Elzoghby, S. Metwalli, G. Shawki, *Opt. Th. Appl.*, 30, 161, (1980).
9. M. Hanafy, S. Metwalli, *DETC2009-87012* (2009).
10. H. Hegazi, A. Nassef, S. Metwalli, DETC2002/DAC-34107 (2002).
11. S. Metwalli, K. Afimiwalla, W. Mayne, *73-ICT-48*, (1973).
12. S. Metwalli, R. Mayne, *77-DAC-9*, (1977).
13. S. Metwalli, *Top. Fl. Film Bear. Rot. Bear. Sys. Des. Opt.* 231, (1978).
14. S. Metwalli, E. Afify, *Cur. Adv. Mech. Des. Prod.* 41, (1981).
15. S. Metwalli, G. Shawki, M. Sharobeam, *Mech. Trans. Auto. Des.* 105, 249 (1983)
16. S. Metwalli, G. Shawki, M. Mokhtar, M. Seif, *Mech. Trans. Auto. Des.* 106, 54, (1984)
17. S. Metwalli, G. Shawki, A. Elzoghby, *Mech. Trans. Auto. Des.* 107, 477 (1985)
18. S. Metwalli, *Mech. Trans. Auto. Des.* 108, 197 (1986)
19. S. Metwalli, *86-DET-129*, (1986).
20. S. Metwalli, S. Elmi, *Cur. Adv. Mech. Des. Prod.* (1989).
21. S. Metwalli, *Mach. Des. Fund.* (1990).
22. S. Metwalli, M. Radwan, A. Elmeligy, *19th Des. Auto. Conf.* (1993)
23. S. Metwalli, M. Radwan, A. Elmeligy, *Int. Comp. Eng. Conf.*, 767, (1994)
24. S. Metwalli, E. El Danaf, *Des. Auto. Conf.*, (1996).
25. S. Metwalli, *Optimum Design* (1998).
26. S. Metwalli, M. Salama, R. Taher, *Comp. Ind. Eng*. 35, 603 (1998).
27. S. Metwalli, H. Hegazi, *DETC99/CIE-9139* (1999).
28. S. Metwalli, H. Hegazi, *DETC2001/CIE-21680* (2001).
29. S. Metwalli, *Cur. Adv. Mech. Des. Prod.* 3, (2004).
30. S. Metwalli, *16th CIRP*, 48 (2009).
31. S. Metwalli, *ICSV17*, (2010).
32. A. Nassef, H. Hegazi, S. Metwalli, *DETC2000/DAC-14295* (2000).
33. S. Rao, *Engineering Optimization*, (2009).
34. M. Senousy, H. Hegazi, S. Metwalli, *DETC2005-85353* (2005).
35. M. Senousy, H. Hegazi, S. Metwalli, *DETC2008-49563* (2008).
36. M. Shalaby, H. Hegazi, A. Nassef, S. Metwalli, *DETC2003/DAC-48770* (2003).
37. G. Shawki, S. Metwalli, M. Sharobeam, *Mech. Trans. Auto. Des.* 106, 376 (1984).

QUADCOPTER SEE AND AVOID USING A FUZZY CONTROLLER

M. A. OLIVARES-MENDEZ[1], LUIS MEJIAS[2], PASCUAL CAMPOY[1], IGNACIO MELLADO-BATALLER[2]

[1]Computer Vision Group, DISAM, Polytechnic University of Madrid
Centro de Automática y Robótica (CAR) UPM-CSIC, Madrid, Spain
[2]Australian Research Centre for Aerospace Automation (ARCAA)
Queensland University of Technology, Brisbane, Australia

Unmanned Aerial Vehicles (UAVs) are fast growing sector. Nowadays, the market offers numerous possibilities for off-the-shelf UAVs such as quadrotors. Until UAVs demonstrate advance capabilities such as autonomous collision they will be segregated and restricted to flight in controlled environments. This work presents a visual fuzzy servoing system for obstacle avoidance using UAVs. To accomplish this task we used the visual information from the front camera only. This image is processed off-board and the information is send to the Fuzzy Logic controller which then send commands to modify the orientation of the aircraft. Results from flight test are presented with a commercial off-the-shelf platform.

1. Introduction

In the last decade the number of aerial platforms in the market has been steadily increasing. Small UAVs such fixed-wing, helicopters, quadcopters, octocopters, etc are now available in many toy stores. They are cheap, easy to use and easily interfaced with most smartphones like the AR-Drone-Parrot [1]. There are also more sophisticated models such as the ones offered by Ascending Technologies [2]. The potential use in civilian tasks gives to these platforms the opportunity to become principal players for applications such as surveillance, monitoring, vigilance and inspection [3,4]. Before their adoption in civil airspace the see and avoid problem -among others- must be addressed first. Some authors have proposed approaches to this problem [5].

There exist numerous control techniques for quadcopter control but particularly to our application, Soft-Computing techniques offer some advantages such as model-free implementation (these techniques works well even when dynamic model to control is unknown) and easy controller tuning procedures. These advantages make these techniques an attractive approach for the researchers in the area of field robotics.

The paper is organised as follows. The visual algorithm that uses color detection to identify the obstacle to avoid, is presented in Section 2. The visual servoing approach using a heading controller developed with Fuzzy Logic, is described in Section 3. The behaviour of the controller in real tests with an AR.Drone-Parrot quadcopter are presented in Section 4. Finally, conclusions and future works are presented in Section 5.

2. Visual System

In order to get feedback of the environment, the front camera of the quadcopter is used. The visual information is sent to the ground station for visual processing, then the outcome of this processing is used for control purposes. The obstacle avoidance task using visual servoing is based on the idea of detecting the object to avoid, track it keeping it at the right or left side of the image until a maximum yaw angle is reached and the object is overtaken. We approach the problem of tracking by exploiting the colour characteristic of the target. The target used has a particular color (orange) that then is tracked on consecutive images. The Camshift algorithm is used to track a defined color on an image sequence. This process is deeply explained in [6].

3. Fuzzy Controller

For the control of the aircraft we developed a controller that send yaw commands to the aircraft. This section describes this controller.

The controller is based on Fuzzy Logic and was implemented using the Miguel Olivares' Fuzzy Software (MOFS). This software has been used in a wide range of control applications in aerial [7] and mobile robots [8]. Detailed information of this software can be found in [9]. The yaw controller has two inputs and one output. The first input is angle between the quadcopters, the object to avoid and the right or the left side of the image, as shown in Figure 1(a). The second input is the measure the evolution of this angle between the last two frames, as is shown in Figure 1(b). The output of the controller is the desired yaw angle that the quadcopter need to turn to keep the object at the desired position, see Figure 1(c).

4. Results

Flight test were conducted using Parrot-AR.Drone platform. Communication routines were deloped to send and receive information from the vehicle. A typical orange traffic cone was selected as the object to avoid. We used a VICON motion tracking system [10] to record accurately the trajectory of vehicle with the maximum precision. This information was used for 3D plotting, and no data was used for the control of the aircraft. As mentioned before in this paper, the only information used by the Yaw-controller is the visual information.

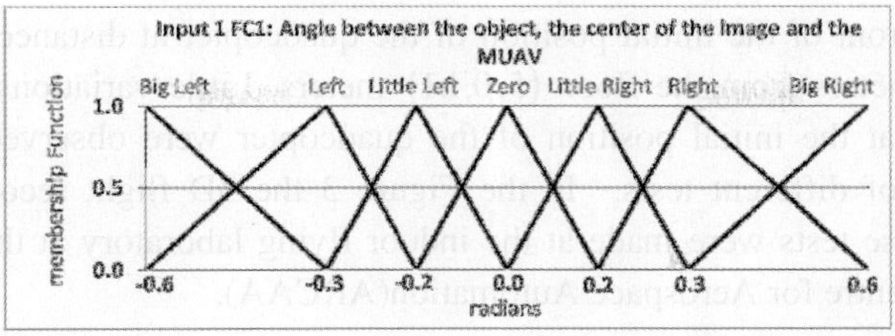

(a) First input. Estimation of the deviation of the object from the centre of the image capture from the MUAV.

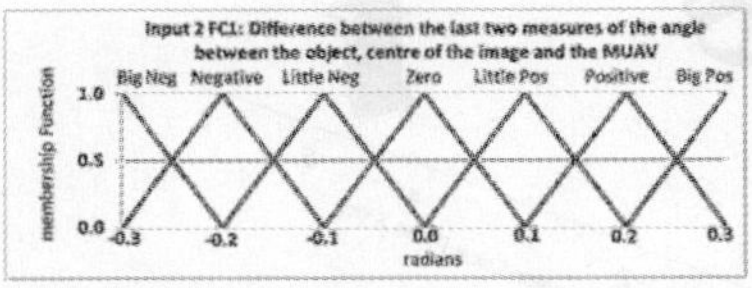

(b) Second input. Difference between the last two measures.

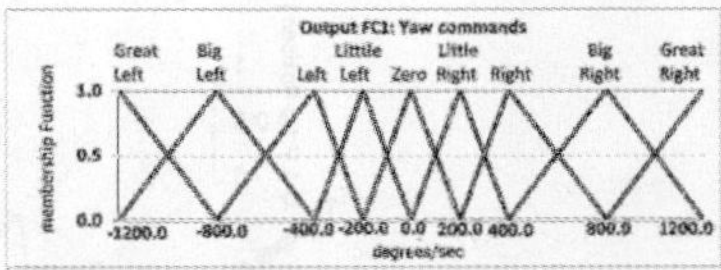

(c) Output. Velocity commands to change the heading of the MUAV

Figure 1. Definition of the Yaw controller.

4.1. *Quadcopter System*

The quadcopter system used for this work is the commercial Parrot AR.Drone. This is a four-rotors aircraft with two cameras onboard, one at the front (forward-looking) which has been used in this work, and other at the bottom (downward-looking). The aircraft is connected to a ground station by a wi-fi connection. A extended explanation of this platform is presented in [1] . Figure 2 shows the control loop diagram.

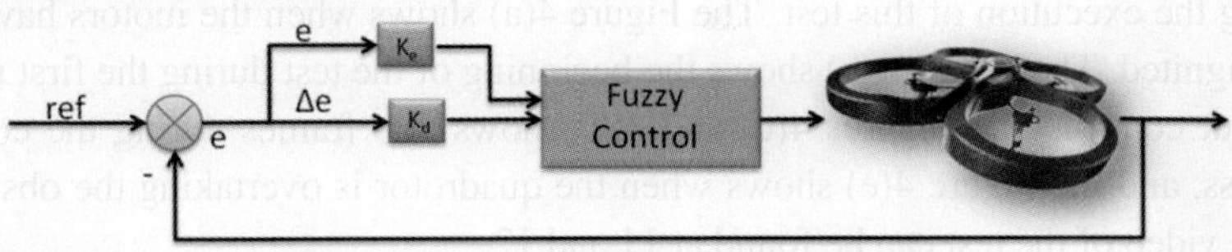

Figure 2. Schematic diagram of the control Loop.

4.2. *Flight Test*

The flights test were performed with constant forward speed (constant pitch angle). No roll commands were sent during the experiments. The altitude was set to a constant value of $0.8m$ and is was controlled by the internal altitude controller of the AR.Drone.

The position of the quadcopter is calibrated at the beginning of the test, being the initial position the point $(0,0,0)$ meters. The obstacle to avoid is located in front of the initial position of the quadcopter at distance of 6 meters and at 1.1 meters from the floor $(5,0,1.1)$ meters. Little variations of no more that 10 cm at the initial position of the quadcopter were observed during the executions of different tests. In the Figure 3 the 3D flight reconstruction is shown. These tests were made at the indoor flying laboratory at the Australian Research Centre for Aerospace Automation(ARCAA).

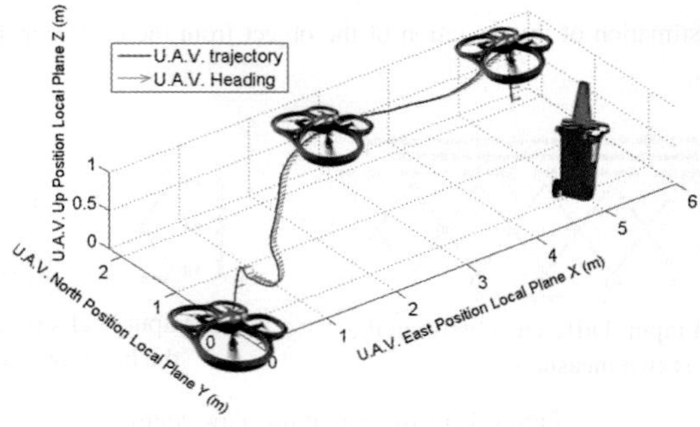

Figure 3. 3D flight reconstruction of the flight path. The obstacle to avoid is a orange traffic cone located at the position (5,0,1.1).

Once the quadrotor takes-off, it flies for 1 meter towards the obstacle in open loop. Then the control process is activated, and then during the next following 5 seconds the controller is sending commands to the aircraft. The image processing and control task are finished after the quadrotor reaches its maximum allowable yaw angle. After this point the aircraft will go forward without any yaw commands. The Figure 4 shows some images captured from the onboard camera during the execution of this test. The Figure 4(a) shows when the motors have not been ignited. The Figure 4(b) shows the beginning of the test during the first meter without control. The Figures 4(c) and 4(d) shows two frames during the control process, and the Figure 4(e) shows when the quadrotor is overtaking the obstacle. A full video of the test can be found at 11 and 12.

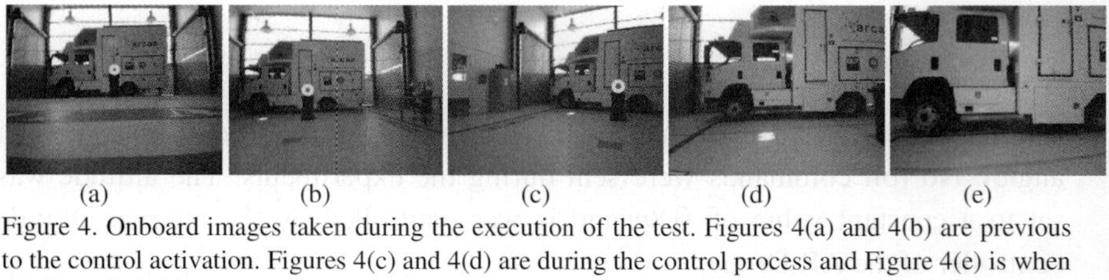

(a) (b) (c) (d) (e)

Figure 4. Onboard images taken during the execution of the test. Figures 4(a) and 4(b) are previous to the control activation. Figures 4(c) and 4(d) are during the control process and Figure 4(e) is when the obstacle has been overtaken.

The behavior of the controller is represented in the Figure 5 which shows the evolution of the error during the test. The red line step represent the moment in which the image processing start. The measure of the step is 25 degrees, but at the moment when the step is applied the aircraft was looking at the opposite side increasing the step command to 35 degrees. To evaluate the behavior of the controller we use the error estimator of the root mean-square error (RMSE). The lower value this error estimator of $RMSE = 9.57$ degrees. The quick response of the controller shown in this Figure corroborate the excellent behavior of the optimized-controller. The Video of this test can be found at [13] .

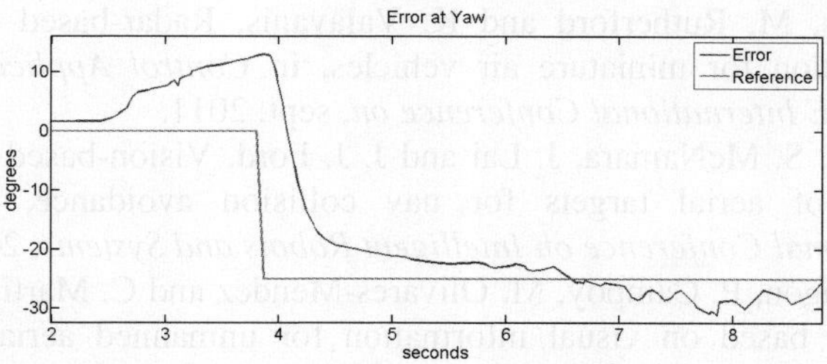

Figure 5. Evolution of the error during a real test.

5. Conclusions

This paper presented a micro-UAV visual servoing approach for see-and-avoid. A Fuzzy Logic controller has been developed to automatize the collision avoidance. This controller acts changing the heading of the aircraft, keeping the obstacle to avoid at the right side (or left) of the image until the object can be overtaken. Excellent results have been obtained in real tests using the commercial quadcopter AR.Drone-Parrot with a quick response and a low error estimation. We are in the process to extending this approach to 3D, adding altitude control. Also, we plan to study different optimization algorithms to improve the controller and reduce the RMSE.

Acknowledgment

The work reported in this paper is the consecution of several research stages at the Computer Vision Group - Universidad Polit\'{e}cnica de Madrid and at the Australian Research Centre for Aerospace Automation - Queensland University of Technology (ARCAA-QUT). The authors would like to thank to Caja Madrid for the mobility research grant of one of the authors. This work has been sponsored by the Spanish Science and Technology Ministry under the grant

CICYT DPI2010-20751-C02-01 and by the IRSES Program Marie Curie FP7 - PIRSES-GA-2009-230797 - ICPUAS (International Cooperation Program for Unmanned Aerial Systems Research and Development).

References

1. Ar.drone parrot http://ardrone.parrot.com, (2010).
2. Ascending technologies http://www.asctec.de, (2010).
3. O. Meister, N. Frietsch, C. Ascher and G. Trommer, Adaptive path planning for a vtol-uav, in Position, *Location and Navigation Symposium, 2008 IEEE/ION* , may 2008.
4. A. Moses, M. Rutherford and K. Valavanis, Radar-based detection and identification for miniature air vehicles, in *Control Applications (CCA), 2011 IEEE International Conference on*, sept. 2011.
5. L. Mejias, S. McNamara, J. Lai and J. J. Ford, Vision-based detection and tracking of aerial targets for uav collision avoidance, in *IEEE/RSJ International Conference on Intelligent Robots and Systems*, 2010.
6. I. Mondragon, P. Campoy, M. Olivares-Mendez and C. Martinez, 3d object following based on visual information for unmanned aerial vehicles, in *Robotics Symposium, 2011 IEEE IX Latin American and IEEE Colombian Conference on Automatic Control and Industry Applications (LARC)*, oct. 2011.
7. M. Olivares-Mendez, I. Mondragon, P. Campoy and C. Martinez, Fuzzy controller for uav-landing task using 3d-position visual estimation, in *Fuzzy Systems (FUZZ), 2010 IEEE International Conference on*, july 2010.
8. M. Olivares-Mendez, I. Mellado, P. Campoy, I. Mondragon and C. Martinez, A visual agv-urban car using fuzzy control, in *Automation, Robotics and Applicatios (ICARA), 2011 IEEE International Conference on*, December 2011.
9. I. Mondragón, M. Olivares-Méndez, P. Campoy, C. Martínez and L. Mejias, Autonomous Robots **29**, 17 (2010), 10.1007/s10514-010-9183-2.
10. VICON, www.vicon.com (2012).
11. Computer vision group-upm http://www.vision4uav.eu, (2012).
12. Youtube channel computer vision group-upm http://www.youtube.com/colibriprojectUAV, (2012).
13. Computer vision group-upm. see and avoid test video http://www.vision4uav.eu/ ?q=researchline/SeeAndAvoidFuzzy, (2012).

DECISION FUZZY SYSTEM ON THE STATE OF DELAYED VEHICLES*

MATILDE SANTOS

Computer Architecture and Automatic Control Department, University Complutense of Madrid, C/ Profesor García Santesmases s/n 28040-Madrid, Spain

MANUEL G. ROMANA

Civil Engineering School – Caminos, Technical University of Madrid 28040-Madrid, Spain

This paper shows how an intelligent system based on fuzzy logic has been designed, in order to determine if a vehicle is delayed while travelling on a two-lane highway. This state is a subjective consideration of the driver, and therefore, many fuzzy variables have been defined to take into account these vague perceptions. A new classification of the state of the vehicle in rural highways is also proposed. The results of some surveys made to two-lane roads users proved that the results are reasonable and useful for estimating the proportion of delayed drivers, the drivers desire to overtake, and on advising the drivers on the possibility of overtaking.

1. Introduction

Overtaking is one of the potentially most dangerous maneuvers that can be performed when driving. Its attempt is strongly related with the perception of the driver of being delayed. Overtaking involves both the desire to overtake and the availability of a long gap on the opposing traffic.

Besides, vehicle delay (through a service measure called percent time spent following) is one of the main factors used to estimate the level of service on two-lane highways, and therefore the quality of the traffic flow in these facilities. Delay is something felt (or not) by each driver. Many deterministic methods have been applied to estimating the percentage of delayed vehicles, when it is obvious that it is a probabilistic variable: all drivers with a short headway (say, below 1,5 s) are delayed, all drivers with a long headway (say, over 4 s) are not delayed. Between these headways the proportion of delayed

* This work has been partially support by the Spanish project DPI2009-14552-C02-01.

drivers grows from 0 to 1. But there are many different subjective factors that influence the delay and very different perceptions of this issue by the drivers.

In this work an intelligent adaptative system is proposed to bridge this gap using approximate reasoning to identify delayed drivers. Drivers' subjectivity has been taken into account in defining when a vehicle is delayed.

The system calculates and weighs some factors that have influence on the driver delay such as the quality of the environment, the state of the vehicle and the driver, etc., and uses these measurements to estimate the delay. Through delay, the desire to overtake can be inferred, thus making it possible to assess the possibility of making aggressive or unsafe passing maneuvers.

2. Vehicle Delay: Definition and Determination by Fuzzy Logic

In the Highway Capacity Manual (TRB, 1965), the most widely used reference for traffic engineering, the level of service was defined as a qualitative measurement that describes the vehicular traffic conditions. It takes into account some factors such as speed, travel time, the maneuverability, traffic interruptions, comfort of the passengers, active and passive safety, etc.

The last edition of the Highway Capacity Manual (TRB, 2011) establishes the way of calculating the level of service of two lanes roads used for mobility using two variables: average travel speed (ATS) and percent time spent following (PTSF). When the speed is not important, such as in touristic routes or short distances, the only issue to be considered is the vehicle delay, measured through PTSF. So, in order to know the level of service, it is necessary to calculate the percentage of vehicles that are delayed on a particular road.

Nevertheless, the estimation of whether a vehicle is delayed is only measured through a debated surrogate variable: vehicles travelling with headways under 3 s. This surrogate measure suggested by the Highway Capacity Manual was measured only in one study, and is still pending widespread validation. Different approaches have been proposed in order to estimate this critical threshold (Romana, 1995; Corupe et al. 1975; Fletcher, 1968).

But there are many subjective factors to be considered to determine if a driver is delayed that are not taken into account by these traditional methods like the critical interval one. Besides, these methods suppose that all the drivers behave the same, which is clearly not the case.

Therefore, in this paper we propose the application of fuzzy logic in order to incorporate the subjectivity of these perceptions when making the decision of overtaking. Some variables such as road conditions, weather, speed of the vehicle coming on the opposing lane, etc, are taken into account without the necessity of taking accurate measurements of those values. On the other hand, the intelligent system that has been developed considers a wide range of different factors as a whole, and not only one or two variables as other methods.

2.1. *Fuzzy System Description*

A fuzzy MISO (Multi-Input Single-Output) system has been designed and developed in order to assess if a vehicle is delayed. This general system consists of seven different MISO fuzzy subsystems that are related. The outputs of some of them are the inputs of others. In this way, the different factors are properly modeled.

In fact, 23 different **input variables** have been taken into account, such as: Day/night, weather, road, section length, driver motivation, driver rest state (hours of sleep the night before), driver tiredness, car state, tires, mileage, car age, speed of preceding vehicle, gap, available gap in opposing direction. Some of them have been proposed in the literature. For example, Romana (1995) talks about safety, desire of overtaking, distance between vehicles, environmental conditions, type of vehicle, gap between vehicles, etc. Murat (2006) models the delay but for traffic intersections; nevertheless, he takes into account the environment. Valverde *et al.* (2010) have developed an intelligent system that calculates the risk on roads based on three different fuzzy models: the environment, the vehicle and the driver.

In addition, a survey was made to drivers on two-lane road conditions. Different factors were proposed and the driver had to weight the importance of each of them in order to overtake. Some personal features of the driver were also asked. This way of knowledge acquisition has been used by Pattnaik and K.R. Kumar (1996).

The **outputs of the system** are two factors that are considered to be the most relevant: the possibility of overtaking and the driver's desire of overtaking. Both of them, together with the speed of the vehicle and the speed of the previous vehicle, are the inputs of the final crisp system that calculates if the vehicle is delayed.

The fuzzy subsystem *Overtaking_Desire* includes, among others, the fuzzy variable *safety_perception_bytheuser*. This variable is, at the same time, the output of another fuzzy system whose inputs come from three fuzzy subsystems called Environment, Vehicle, and Driver, following the system developed by Valverde *et al.,* 2010.

The fuzzy subsystem *Overtaking_Possibility* requires as inputs both crisp and fuzzy values. For example, if overtaking is allowed (crisp), or the fuzzy driver's perception of the safety of overtaking, the estimation of the gap, the distance to other vehicles, etc. Some of these fuzzy values are the outputs of the corresponding fuzzy subsystems (for instance, the fuzzy system *Gap*).

The **fuzzy rules** have between 3 and 7 antecedents, depending on the linguistic input variables of the fuzzy subsystem. The rules were generated like in a decision tree. This allowed to prune some branches, so fewer rules than all the possible ones were necessary.

Take, as an example, the following rule,

IF (safety) IS **low** AND (Previous_vehicle_speed) IS **fast**
THEN (Overtaking_desire) IS **not**

In this case, this rule includes the 144 possible combinations of low safety and high previous vehicle speed with other variables such as Vehicle_type, Road_type, Time_in_queue, etc. Those two variables are the most important ones in this rule.

On the contrary, there are rules for which many variables must be evaluated in order to get the consequent, such as in the following,

IF (safety) IS **middle** AND (Previous_vehicle_speed) IS **fast** AND (Previous_vehicle_size) IS **big** AND (Vehicle_size) IS **big** AND (Road_type) IS **main** AND (Time_in_queue) IS **small** AND (Vehicle_speed) IS **fast** THENs (Overtaking_desire) IS **not**

All the rules are evaluated in parallel. The fuzzy subsystems are Mamdani type except Overtaking_possibility and Overtaking_desire that are Sugeno-type.

The t-norm product has been chosen to implement the AND operator, as it is usual in engineering applications.

3. New Proposal of Delayed Vehicles Classification

The final output of the entire system indicates if the vehicle is delayed, free, isolated or in line. This is the only completely crisp module of the general system and it is called *Vehicle_State*. Its four inputs are the crisp values of the vehicle speed and previous vehicle speed, and the fuzzy outputs of subsystems *Overtaking_desired* and *Overtaking_possibility*.

The output is one of the four possible states of the vehicle. The new proposal of classification is based on rules like the following,

- If the speed of the previous vehicle is higher than the considered vehicle by more than 4 km/h, the state of the vehicle is ISOLATED.

- If the previous vehicle is slower than the vehicle under study, or its speed is higher but the difference is still lower than 3 km/h, then
  - If Overtaking_desire is not, the vehicle is IN LINE
  - If Overtaking_desire is yes and Overtaking_possibility is not, the vehicle is DELAYED
  - If Overtaking_desire is yes and Overtaking_possibility is yes, the vehicle is FREE

We propose a new definition of those states as follows.

- **Isolated vehicle:** Driver travels at desired speed, without influence from other vehicles in the same direction.
- **Free vehicle:** the driver chooses the speed but he may be influencing other vehicles. For example, vehicles behind can be delayed. It is also possible that he or she travels quite close to the preceding vehicle, although this one is faster. This is the case when the vehicle has just been overtaken; the distance between the vehicles is small but both are circulating at desired speeds.
- **Vehicle in line**: its speed is close to the desired speed, slightly lower. Nevertheless, the driver does not desire to overtake it or cannot do it because of the road and vehicle conditions. The driver does not think it is worthy to overtake.
- **Delayed vehicle:** The speed is clearly lower than desired because of the preceding vehicle. The driver would like to overtake but he cannot do it because it is not allowed in that stretch of the road or there are vehicles in the opposing lane.

We think this classification is closer to reality than others and therefore more useful to estimate the level of service of a road. Most of the pervious approaches considered only two possible states, free and delayed vehicles, (Buckley, 1968) or ambiguous cases (Ovuworie, Darzentas, and Mcdowell, 1980), all of them based on measurements such as the critical interval (Rozic, 1992).

4. Conclusions

A fuzzy system has been designed in order to find the status of a vehicle when circulating on rural highways, particularly if it is delayed. This system includes subjective information that must be taken into account when dealing with the perception of the driver. We cannot forget that "vehicle delayed" is a subjective state of the driver.

Besides, at an intermediate stage the system can be used to assess the desire of the driver to overtake, keeping in mind safety and other issues related.

The algorithm performance was tested asking the interviewed persons on whether they would want to overtake and if they felt delayed in specific combinations of road situations and vehicle characteristics. A total of 100 surveys were conducted, using 20 for the system adjustment and 80 for the validation. The fuzzy system was asked independently the same questions, and the answer was the same in over 90% of the instances.

Finally, a new classification of the state of a vehicle has been proposed. This proposal considers many different fuzzy variables and allows to forecast the desire and possible reaction of the drivers.

This system is currently being used to assess the level of service of two lane roads in a research project in Madrid. In the future it is hoped that a probability curve delay-headway can be developed to estimate the real number of delayed drivers in a two-lane road in Spain.

Acknowledgments

Authors would like to acknowledge the collaboration of Sergio Martín Moreno.

References

1. D.J. Buckley. *Transportation Science*, **2**(2) (1968).
2. E.G. Corupe, E.H. Geissler, G.K. Hunter et al. *Systems Design* (1975).
3. J.H. Fletcher. *Traffic Engineering and Control*, **3** (1968).
4. S.Y. Murat. *Transportation Research Part C*, **14**, 316-314 (2006).
5. G.C. Ovuworie, J.Y. Darzentas, M.R.C. Mcdowell, *Traffic Engineering and Control*, **8-9** (1980).
6. S.B. Pattnaik, K.R. Kumar, *Civil Engineering Systems* **14**, 87-110 (1996).
7. M. G Romana. PhD, Civil Engineering School, Technical University of Madrid, Spain (1995).
8. P. Rozic. Capacity of Two-Lane, Two-Way Rural Highways: The New Approach. 71st TRB Annual Meeting, Washington D.C. (1992).
9. TRB (Transportation Research Board). Highway Capacity Manual. National Research Council, Washington, D.C. (1965).
10. TRB (Transportation Research Board). Highway Capacity Manual. National Research Council, Washington, D.C. (2011).
11. L. Valverde, M. Santos, and V. López, Intelligent Decision Making Systems, Proc. 4th Int. ISKE Conference, 326-331 (2010).

VISUAL INTERPRETATION OF FABRIC TACTILE PROPERTIES USING FUZZY INCLUSION DEGREE AND ANFIS

ZHEBIN XUE[A,B], XIANYI ZENG[A,B], LUDOVIC KOEHL[A,B], YAN CHEN[C]

[a]*Université de lille 1 Sciences et Technologies, 59655, France*
[b]*GEMTEX, ENSAIT, 2 allée Louise et Victor Champier, BP30329, F-59056 Roubaix Cedex 1, France*
[c]*Department of Clothing Design and Engineering, College of Textile and Clothing Engineering, Soochow University, Suzhou, Jiangsu Province 215021, P. R China*

The current study is aimed to investigate how fabrics' tactile properties can be perceived through products' visual representations. Two major steps are taken to unveil the visual compensatory mechanism. The first step is to describe the mechanism. A novel method based on rough sets and fuzzy sets theories is proposed to extract principal visual features for each tactile property. In this part, the single-to-single and multiple-to-single correlations are studied by applying this method. The second step is to quantify the explored mechanism. A mathematical model between each tactile property and the corresponding principal visual features is established using an adaptive network-based fuzzy inference system (ANFIS). This model has been proved to be capable of predicting fabrics' tactile properties from the perceived visual features with a satisfactory accuracy.

1. Introduction

Recent years, much research work has been done to investigate how it is possible to provide a user with a completely reliable sense of fabric through a virtual experience. One remarkable step forward is the development of many haptic force feedback devices, such as the HAPTEX system [1]. These inventions are aimed to build an interaction between users and virtual fabrics using simulation techniques. But till now, there is not a commonly accepted device that is capable of comprehensively quantifying people's complicated sense of fabric hand.

In fact, "fabric hand" was defined as "the subjective assessment of a textile material obtained from the sense of touch" [2]. It is believed that the most effective way to study fabric hand should be from the perspective of human's natural perception. For a long time, fabric hand is studied on the basis of real touch experiments, during which the evaluators are introduced to directly handle the textile fabrics and give assessments according standardized techniques and criteria. But, online shopping happens in a non-haptic environment. In this case, comparatively, vision can work as a barrier-free source of perceptual information. Besides, according to the physiological psychology, over 70% of the daily information to be processed by our brain is perceived through the eyes [3]. On this premise, we are considering the possibility to study fabric tactile properties from visual information. In our previous study, we found that most of fabrics' tactile properties can be well interpreted through some specific visual representations (video or image displays). On this basis, here, we are going to unveil the mechanism lying beneath this finding, which we call the visual compensation of fabrics' tactile information. Sensory experiments were organized on a number of flared skirts being displayed in video clips. Two sets of sensory data concerning the skirts' visual features and fabric hand, respectively, were obtained from a panel of assessors. In order to discover the visual compensatory mechanism, measures should be taken to find the underlying relations between these two sensory data.

In the present study, two steps are taken to investigate the visual compensatory mechanism of fabrics' tactile properties. The first step is to find for each tactile property the visual features that have the most significant impact. From the computational point of view, this step is in fact the process of feature selection which is aimed to reduce the complexity of the system. Being different from mechanical measurements, there exists frequent and widespread sensory cooperation during human perception. So, the correlation to be concerned in this study involves two aspects, one is the single-to-single correlations between one tactile property and any visual feature; the other is the multiple-to-single correlations between one tactile property and several principal visual features, which resembles the basic structure of the visual compensatory mechanism. In this step, we proposed a new mathematical methodology which integrates the ideas of both rough sets [4] and fuzzy sets theories [5]. Based on this methodology, a series of methods are developed to study both the single-to-single and multiple-to-single sensory correlations being mentioned above. Rough sets have high capacity in data structure analysis while the fuzzy theory can effectively process the uncertainty of semantic data and model human expert reasoning. The proposed methods can well detect the nonlinear structure of the sensory data, while being safe to use a comparatively small number of samples. On this basis, the second step is to quantify the multiple-to-single correlations, which finally unveils the visual compensatory mechanism for each tactile property. A predictive model is established based on an Adaptive Network-based Fuzzy Inference System (ANFIS). This system has absorbed the advantages of both neural network and fuzzy inference system. By setting up intuitively reasonable initial membership functions, a learning process is launched to generate a set of fuzzy if-then rules to describe the input-output behavior of the data system. By employing ANFIS, the complicated relations between different sensory data can be modeled with reliability and interpretability.

2. Feature Selection Based on a Fuzzy Inclusion Measure

For simplicity, we only study the single-single relation (between one tactile and one visual feature) in this section.

The formalization of the present problem is given below. Let $U=\{e_1, e_2, \cdots, e_{18}\}$ be the set of 18 samples. The corresponding evaluation scores for any visual (e.g., pleat size) and tactile descriptor (e.g., stiffness) are denoted as $(c(e_1) \ldots c(e_{18}))^T$ and $(d(e_1) \ldots d(e_{18}))^T$, respectively. In the following discussion, the visual feature perception C is taken as condition variable and the fabric hand perception D as decision variable. From the previous two vectors of C and D, we obtain two partitions for the visual and tactile results, i.e. $U/C=\{X_0, X_1, \cdots, X_{10}\}$ and $U/D=\{Y_0, Y_1, \cdots, Y_{10}\}$. $X_i \in U/C$ $(i=0,1,\cdots,10)$ (or $Y_j \in U/D$ $(j=0,1,\cdots,10)$) is the class of samples in which the evaluation scores are all i (or all j). In practice, some X_i (or Y_j) can be empty if its index i (or j) does not exist in the evaluation scores of all the samples e_k's.

According to the rough sets theory, the inclusion degree is aimed to measure the extent to which the partition of the condition set is consistent with that of the decision set. The basic point to compute inclusion degree is to establish the rough membership function, which measures degrees of inclusion of decision classes into subsets of the universe. Let $S=(U, C \cup D)$ be a complete decision table, $X_i \in U/C$ an equivalence class and $U/D=\{[e_j]_D : u \in U\}$. For any object (sample) $e_j \in U$, the membership function of e_j in X_i is denoted by

$$\delta_{X_i}(e_j)=\frac{|X_i \cap [e_j]_D|}{|X_i|}, \quad (i=0,1,\cdots,10 \;\; ; j=1,2,\cdots,18) \tag{1}$$

where $0 \le \delta_X(e_i) \le 1$.

$[e_j]_D$ is the set of samples of which all the evaluation scores in D are equal to $d(e_j)$.

On the basis of the above membership function, a consistency measure of X_i with respect to D is defined as:

$$Cons(X_i, D)=1-\frac{4}{|U|}\sum_{j=1}^{|U|}\delta_{X_i}(e_j)(1-\delta_{X_i}(e_j)) \tag{2}$$

where $0 \le C(X_i, D) \le 1$.

The general inclusion degree of the condition attribute set C with respect to the decision attribute set D is defined as follows:

$$I(C,D)=\sum_{i=0}^{10}\frac{|X_i|}{|U|}\left(1-\frac{4}{|U|}\sum_{j=1}^{|U|}\delta_{X_i}(e_j)\left(1-\delta_{X_i}(e_j)\right)\right),\ (j=1,2,\cdots,18) \quad (3)$$

Where $\delta_{X_i}(e_i)$ is the membership function of e_j in X_i.

In practice, an inclusion degree based on the previous crisp partition of samples might lead to serious information loss. For any specific class, it does not make difference between samples close to it and those far from it. For any sample e_k not belonging to a class X_i, we have its membership value equaling zero, indicating that its adhesion to this equivalence class is regarded as null. Evidently, in the definition of the inclusion degree, it is more reasonable to consider the degrees of adhesion of the samples to the equivalence classes so that samples close to a class are more important than those far from it. According to this idea, we modify the previous inclusion degree using the concept of fuzzy partition.

According to the method introduced above, the fuzzy inclusion degree *I*(*C*, *D*) quantifies the extent to which the partition of the condition set (or, the visual result set) is consistent with that of the decision set (or, the fabric hand result set). However, the ordinal consistency between the sets is not well taken into consideration. Actually, in an information system, the element's different positions in the two sets may lead to big differences between the knowledge to be represented respectively. Therefore, it is significant to include another index to measure the ordinal difference between the data obtained from different sensory modalities. For this purpose, Kendall's rank coefficient, denoted as tau, is introduced in our study [6].

Next, an aggregation criterion was proposed to integrate the previous two indices, Fuzzy inclusion degree (*FID*) and Kendall's tau (*RC*), so as to constitute a general consistency measure (*GCM*) to investigate the overall correlation between different sensory data. This criterion should be both robust to noise and easy for knowledge interpretation. For this purpose, a fuzzy inference system is designed from rules provided by 6 experts. For example, a fuzzy rule can be described as the following expression:

***IF** FID is medium (M) and RC is small (S), **THEN** GCM is small (S).*

After applying this aggregation criterion, a general consistency measure (*GCM*) was constituted to measure the single-to-single correlation between each tactile property and any visual feature.

3. Predictive Modelling Using ANFIS

In Section 2, several principal visual features are extracted and proved to have the most significant cooperative impact on each tactile property. On this basis, here, we are going to establish a predictive model to quantify these multiple-to-single relationships. Given that we have a relatively small number of samples, and there exists much uncertainty and imprecision in sensory data, an adaptive network-based fuzzy inference system (ANFIS) is employed to solve the current modelling problem.

In the current study, the eighteen samples are characterized by eleven tactile properties, for each of which an ANFIS model is developed. To be specific, the eighteen samples are taken as the training data for this model. In addition, another three samples which were selected and prepared in the same way as the eighteen samples are included as the testing data to verify the modeling results. Let's take the tactile property 'liveliness' (D2) as an example. Its value is regarded as the output of the ANFIS model, while the selected visual features E20 (drape), E3 (fitness to body shape) and E7 (expending extent from below the hip), are taken as the inputs.

From the ANFIS model, we obtain two following rules for D2:

1) if (E20 is low)and(E3 is low)and(E7 is medium),
then D2=0.6532*E20+0.3938*E3+0.6362*E7+0.1997

2) if (E20 is medium)and(E3 is low)and(E7 is low),
then D2=0.4918*E20+0.3418*E3+0.4515*E7+0.1586

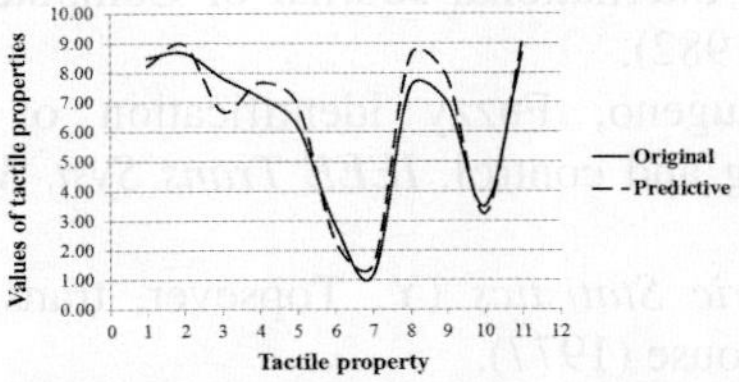

Figure 1. Comparison of real and predictive results on tactile properties for one testing sample.

Figure 1 illustrates the comparison between the original (in dash lines) and the estimated perceptions (in solid lines) on the eleven fabric tactile properties for each testing sample. We can observe that the established model has a good predictive capacity. Besides, the calculated predictive errors over most of the tactile properties for each testing sample don't exceed 1 which is regarded as a relatively satisfactory result in sensory science. Therefore, it is concluded that, as far as is concerned in this study, the developed ANFIS model is capable of well predicting fabrics' tactile properties from the perceived visual features of the samples. So far, the visual compensatory mechanism of fabrics' tactile properties is successfully unveiled.

4. Conclusion

So far, the discovery of the visual compensatory mechanism, or the visualization of fabrics' tactile properties for the specific skirts being concerned in our study has been accomplished. A total of seventeen visual features have been proved to be responsible for the interpretation of fabrics' tactile properties through video representations. Our study is a significant step forward on the way of investigating the possibility of visually manipulating the tactile information of textile products. In the future, when you enter an online apparel store, with all the desired information even including the tactile properties about the apparel items being visualized to you, your virtual purchasing experience would be greatly enriched.

References

1. N. Magnenat-Thalmann, P. Volino, U. Bonanni and I.R. Summers, From physics-based simulation to the touching of textiles: the HAPTEX project. *The International Journal of Virtual Reality*. 6(3) 35-44 (2010).
2. S.I. Ali and S. Begum, Fabric softeners and softness perception. *Ergonomics*. 37(5), 801-806 (1994).
3. Y. Wang, S. Li and J. Mao, *Computer Image Processing and Recognition Technology*. Higher Education Press, Beijing, PRC (2001).
4. Z. Pawlak, Rough sets, International Journal of Computer and Information Sciences, 11, 341-356 (1982).
5. T. Takagi and M. Sugeno, Fuzzy identification of systems and its applications to modeling and control. *IEEE Trans Syst, Man and Cybernet*, 15,116–132 (1985).
6. S. Siegel, *Nonparametric Statistics* (Y. Topsever, trans.), Ankara: Ankar University Publishing House (1977).

FUZZY LOGIC ASSISTED HELICOPTER UAV FLIGHT CONTROL

A. J. VAN DER WAL*
Netherlands Defence Academy, P.O.Box 10000,
1780 CA Den Helder, The Netherlands

The use of small-sized helicopters as UAV for professional applications is rapidly increasing. Flying a helicopter is a task that requires a great amount of experience and skill. This is due to the strong coupling that exists between the six degrees of freedom, resulting in the 12 dimensions needed to describe the dynamics of a helicopter. Therefore, to perform even simple flight movements a helicopter pilot has to carefully adjust more than one control simultaneously. The application of fuzzy controllers in a ground-based user interface enables fixed-wing pilots and even non-flying operators to fly a helicopter UAV mission in a controllable way. This paper investigates the feasibility of using fuzzy logic to assist a ground-based UAV pilot.

1. Introduction

The use of small-sized helicopters as UAV for professional applications is rapidly increasing. Examples of such applications include military reconnaissance and police surveillance, surveying, etc. The strong nonlinear coupling among the degrees of freedom and the amount of experience and skills required for safe flight control of small-size remote-controlled helicopters, have stimulated the development of a flight assistant that helps inexperienced operators in flying a successful mission. This is even the more true for small helicopters, because of their smaller inertia and the associated smaller time scales of the dynamics involved. The present research is aimed at the development of an intelligent user interface that is capable to navigate a small helicopter safely using elementary commands that are given by a non-expert user. The intelligent interface must take all the necessary actions that the experienced pilot would take to control the helicopter and ensure the implementation of the desired flight path within a safe flight envelope. Guidance systems have been developed and implemented for model helicopters by different research groups, *e.g.* Linkøping University [1], Swiss Federal Institute

* E-mail: AJ.vd.Wal@nlda.nl

of Technology in Zürich [2] and Tokyo Institute of Technology [3]. The use of soft-computing theory, such as fuzzy logic covers a broad scope, ranging from theoretical work in *e.g.* the foundations of quantum mechanics [8], to industrial applications in pattern recognition and sensor fusion (for a review see: Refs. [9],[10]), mission-critical applications [11], and nuclear reactor control [12]. M. Sugeno of the Tokyo Institute of Technology has conducted the first work in this field [3] already in 1995 and has developed an autonomous helicopter using fuzzy logic control, but it is difficult to locate any specific publication with details on the implementation of the actual fuzzy logic controllers. The mathematical model of the helicopter used in our calculations can be found elsewhere [4-7]. In this paper we will discuss the hovering control, since take-off and hovering at relatively low altitude are the very first skills that any helicopter pilot learns.

2. Helicopter dynamics

Helicopter flight is complicated due to the strong nonlinear coupling of the various degrees of freedom of the helicopter. The work of a helicopter pilot is therefore more difficult than that of a pilot of a fixed-wing airplane. The helicopter pilot simultaneously operates three different controls, which manipulate the angle of attack of the main and tail rotor blades. The prime role of the main rotor is to provide the lift force that allows the helicopter to hover and fly and is controlled by two controls, the "collective" and the "cyclic". The tail rotor produces lateral thrust in the same way as the main rotor of the helicopter does. It changes the amount of thrust that is produced by changing the angle of attack of the tail rotor blades. The tail rotor is directly connected with the main rotor and therefore also has a constant angular velocity that depends on the gear ratio of the gearbox. The tail rotor is primarily needed to counteract the top axis moment that is exerted on the helicopter body by the movement of the main rotor blades. The secondary effect of the tail rotor is to enable the helicopter to rotate about the main rotor's shaft axis (yaw). The helicopter controls can thus be divided into two groups, the controls responsible for the manipulation of the main rotor (the *collective* and the *cyclic*), and the ones that are responsible for the tail rotor (*tail pedals*).

2.1. *Hovering*

Hovering at a certain altitude with a given heading has been implemented. The main rotor thrust is strongly coupled with the angular momentum produced about its shaft axis and therefore influences the heading. It is commonly observed at helicopter take-off that the helicopter slightly rotates (yaws) about the main rotor shaft axis, before the pilot can stabilise and bring the heading back to the initial heading. Similar phenomena are observed when the pilot tries to change direction while having low forward speed. The difference in the starboard and portside contribution to the lift force due to cyclic control command not only results in a change in direction, but also to loss of height, because additional thrust is needed to compensate for the inclination and subsequent reduction of the effective rotor blade surface. When using the tail rotor to compensate the yaw torque, the result is an excess of force in the direction, for which the tail rotor is meant to compensate, that will tend to make the helicopter drift sideways. The pilot adjusts the cyclic to maintain the helicopter's position over a point on the ground and adjusts the collective to maintain a fixed altitude (especially important when close to the ground), simultaneously adjusting the foot pedals to maintain the direction. External disturbances (*e.g.* wind) further complicate the hovering manoeuver.

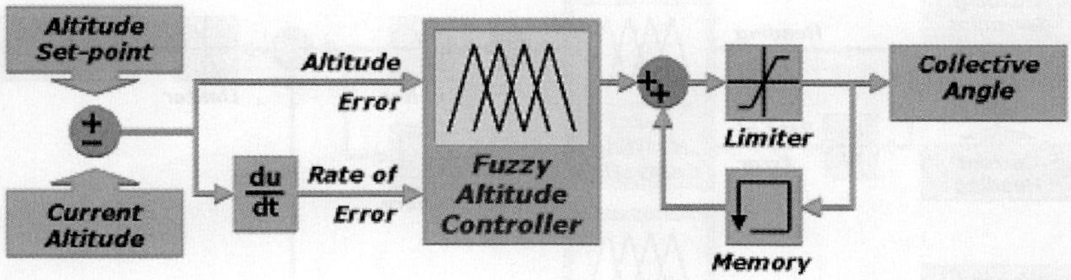

Figure 1. Architecture of the fuzzy altitude controller.

2.2. *Main rotor collective fuzzy controller: Altitude*

The main rotor control system consists of a fuzzy logic controller (Figure 1) that controls the main rotor collective command according to the required vertical displacement. The output of the fuzzy altitude controller is incremental. The limiter placed after the output ensures that the output value will not exceed the actual physical limits of the helicopter model. Integral action via the memory loop is required to distinguish between the several hovering positions. The inputs of the controller are the "altitude error", which represents the difference between the current and the required altitude of the helicopter model, and the "altitude rate of error", the rate at which this error changes with time. The controller has the structure of a fuzzy PD (proportional and differential) controller.

2.3. *Pedals controller for tail rotor: Heading*

The tail rotor control system consists of two parallel fuzzy logic controllers (Figure 2), one to control the yawing angular velocity of the helicopter and the other to control its heading (angular position). The need for the two controllers arises from the fact that we have two different control objectives, corresponding to two different control regimes. The first control objective has to do with safety and staying within the operational flight envelope. The second control objective is maintaining the desired heading. We note that high angular velocities about the main rotor axis can produce instability of the helicopter system. Once the velocity of the helicopter is controlled and does not introduce any instability into the system, it is possible to implement the positioning control for obtaining the required heading. The fuzzy heading controller has a similar structure as the fuzzy altitude controller.

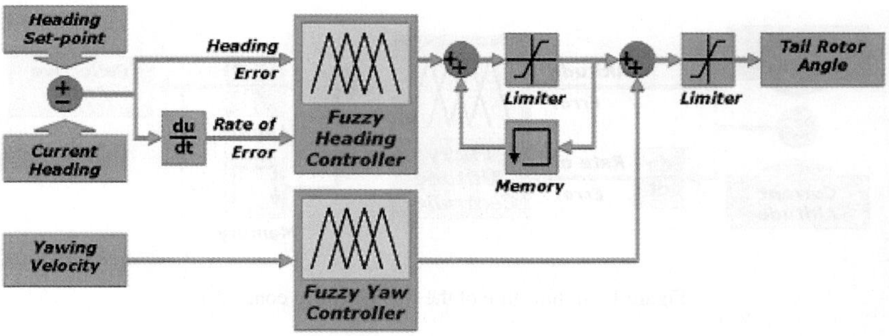

Figure 2. Architecture of the combined fuzzy heading and yaw controller.

3. Experiments

The first test was to determine the ability of the controller to make the helicopter to take off, and to change altitude and heading according the user commands. The fuzzy altitude controller should manipulate the pitch angle of the rotor blades in such way that the helicopter body will not start revolving about the main rotor axis and that the helicopter model will be able to reach a certain altitude with a desired heading as quickly as possible without interfering with the safe flight envelope. For this test the initial conditions of the model helicopter are starting from the ground (0 m) with zero heading (0°). The initial set-point for the fuzzy logic controller was to bring the helicopter to an altitude of 4 m with a heading of 10° (point A). In Figure 3 the resulting trajectories are presented. The helicopter model ascends quickly and reaches the desired altitudes without any noticeable overshoots. The heading shows some fluctuations, overshooting and undershooting due to the changes of the imposed moment from the main rotor.

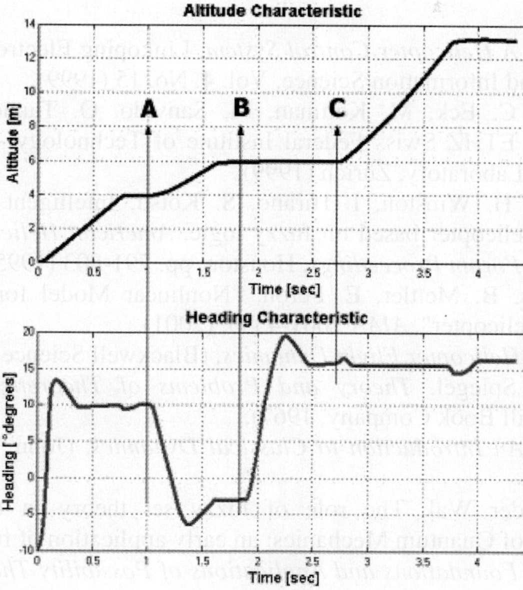

Figure 3. Performance of hovering simulation: simultaneous changes in altitude and heading. A,B, and C correspond to stable hovering with setpoints [altitude (m), heading (°)] of [4,10],[6,-3], and [6,16], respectively.

4. Conclusions

In the present work we have shown that fuzzy logic controllers are capable of controlling two of a helicopter's coupled degrees of freedom. The incorporation of both basic mechanics and the experience of a helicopter pilot into a fuzzy rule base has been experimentally demonstrated by successful take-off, hovering with defined heading and controlled soft landing. The system is open for extensions that could further enhance its performance (supervisory fuzzy controller, learning control, speed control). Future research is aimed to also control the other degrees of freedom of the helicopter in order to fly a fully autonomous mission. Finally it should be understood that this work describes laboratory-scale experiments and that in order to apply these controllers for real UAV missions ICAO certification must be obtained.

Acknowledgments

The help of Dr. Dimitrios Iakovou with the simulations is gratefully acknowledged.

References

1. E. Skarman, *A Helicopter Control System*, Linkøping Electronic Articles in Computer and Information Science, Vol. **4**, No. 15 (1999).
2. J. Chapuis, C. Eck, M. Kottman, M. Sanvido, O. Tanner, *Control of Helicopters*, ETHZ Swiss Federal Institute of Technology - Measurement and Control Laboratory, Zürich (1999).
3. M. Sugeno, H. Winston, I. Hirano, S. Kotsu, Intelligent control of an unmanned helicopter based on fuzzy logic, *American Helicopter Society / 51st Annual Forum Proceedings*, Houston, pp. 791-803 (1995).
4. V. Gravilets, B. Mettler, E. Feron, "Nonlinear Model for a Small-Size Acrobatic Helicopter", *AIAA* 2001-4333, (2001).
5. G. Padfield, *Helicopter Flight Dynamics*, (Blackwell Science Ltd, 1996).
6. Murray R. Spiegel, *Theory and Problems of Theoretical Mechanics*, (McGraw-Hill Book Company, 1967).
7. G. Sposito, *An Introduction to Classical Dynamics*, (John Wiley & Sons, 1976).
8. A. J. van der Wal, The role of fuzzy set theory in the conceptual foundations of Quantum Mechanics: an early application of fuzzy measures, (invited), in *Foundations and Applications of Possibility Theory*, Vol. **8** of Advances in Fuzzy Systems, World Scientific, Singapore, 234-245 (1995).
9. A. J. van der Wal, Applications of fuzzy logic in industry, (invited), in *Proc. FLINS '94*, Mol, 91-95 (1994).
10. A. J. van der Wal, The potential of fuzzy logic applications in industry, (invited),in *Fuzzy Logic Foundations and Industrial Applications*, Kluwer, Boston, 275-312 (1996).
11. A. J. van der Wal, The importance of soft-computing networks for mission systems: a tutorial, *AGARD, NATO RTO-MP-3*, Monterey CA (1998).
12. D. Ruan and A. J. van der Wal, *Int'l Journal of Information Science* **110**, Vol. 3-4,151-177 (1998).

MAKING PRESENTATIONS USING FUZZY LOGIC IN A TOUR-GUIDE ROBOT*

J. JAVIER RAINER†, RAMÓN GALÁN, AGUSTÍN JIMÉNEZ

Intelligent Control Group. Universidad Politécnica de Madrid
C/ José Gutiérrez de Abascal 2. 28006, Madrid, SPAIN

In the last years, numerous attempts for develop autonomous robots with intelligent and cognitive skills are been made. This paper proposes an automatic presentation generator for a robot guide who let it make better presentations, with predefined durations, and, also, it is shown how the use of fuzzy logic makes possible the evolution of the criteria for the paragraphs, from an initial knowledge to a real one that agrees with the opinions of public. A presentation is composed by a number of paragraphs, which have associated a set of criteria in the ontology, used by the generator in order to select or reject the paragraphs included in the presentation. The modelling of the quality index of the presentation is made using fuzzy logic and it represents the *beliefs* of the robot about what is good, bad, or indifferent about a presentation. The *beliefs* of the robot continue to evolving in order to coincide with the opinions of the public. Updating of the criteria, the learning, is made using genetic algorithms.

1. Introduction

The ability for learning is what makes robots more intelligent each time and capable of performing more tasks. Autonomous robots are intelligent machines capable of performing tasks by themselves, without direct human control [1]. Probably, one of the first real world applications of indoor service robots has been mobile robots serving as tour guides in museums or exhibitions. We have developed our own interactive mobile robot, called Urbano, specially designed to become a tour guide in exhibitions [2]. Such application is extremely interesting for researchers because it allows the advance in knowledge fields such as autonomous navigation in dynamic environments, human robot interaction, cognitive skills, etc.

This paper describes a learning mechanism, capable of giving a guide-robot the ability to make presentations. Each presentation is made up of several

* This work is founded by the Spanish Ministry of Science and Innovation (ROBONAUTA: DPI 2007-66846-C02-01 and ARABOT: DPI 2010-21247-C02-01) and supervised by CACSA whose kindness we gratefully acknowledge.
† javier.rainer@upm.es.

paragraphs and each paragraph has a set of criteria that determine, according to the restrictions, whether the paragraph will be included in the presentation or not. The evolution of these criteria will ensure the agreement with the opinions of the public. In order to achieve this, robot will need to make a representative number of presentations and receive some feedback from the public.

The knowledge is based on domain-specific concept words ontology. Ontologies have been known in computer science as consensual models of domains of discourse, usually implemented as formal definitions of the relevant conceptual entities [3].

2. Urbano system overview

Urbano is a B21r platform from iRobot, equipped with a four wheeled synchrodrive locomotion system. It is a SICK LMS200 laser scanner horizontally mounted on top, used for navigation and SLAM, with a mechatronic face and a robotic arm used to express emotions such as happiness, sadness, surprise or anger. It is also equipped with two sonar rings and one infrared ring, which allows detecting obstacles at different heights, used for obstacle avoidance and safety. The platform has also two on-board PCs and one touch screen.

The software is structured in several executable modules to allow for a decoupled and independent development; they are connected via TCP/IP. Most of these executable modules are conceived as servers or service providers, as face control, arm control, navigation systems voice synthesis and recognition, and web server. A client-server paradigm is used, the central module being the only client called the Urbano Kernel. It is responsible of managing the whole system [2].

URBANO has a technology based on distributed application software. The recent version is an agent based on architecture which uses a specific CORBA approach as an integration tool.

The knowledge server is a Java application developed using the libraries of Protégé-OWL API, capable of reading and editing files in ".owl" format. Thus, the knowledge is stored in ".owl" files, in the form of ontologies.

APG agent

APG agent software has been developed in order to be integrated in the architecture based on the agents, which constitute Urbano software. APG software is activated when receiving a request from a user selecting a file with the pattern to be developed. APG will request all the information from the knowledge server agent, using the necessary restrictions to generate a file with the best presentation to be used by the system in the next performance.

All presentations have been stored with their quality indices (QI), and in some cases, when the request is very similar to a presentation with high quality indices already performed, this last presentation will be used, instead of generating a new one.

3. Paragraphs

The paragraph is the minimum element of expression. It is assumed that the size of the paragraphs is such that coincidences or references between them are irrelevant, and that each paragraph expresses a meaningful content.

For each of the paragraphs there is a script specifying robot facial expressions and arm movements, tone and type of voice to use, position in the floor plan, and various details for interaction with the public.

However, in the ontology, each paragraph is "related" to a certain category, (e.g., in a paragraph describing the painting 'Las Meninas' by Velazquez: It is_a Painting, It is_a Description, It is_a Adult Level, Painted_by Diego Velazquez, etc. These relationships are used for selecting the specific paragraph).

4. Patterns and trees

The pattern defined by the user points out to the elements that should take part of the presentation. For each item, the following elements are established: its identification, its priority, its numerical order, and its reference to the theme within available knowledge. When a limitation of time and a very large pattern are present, the priority index indicates the most important content to be included. The opposite may happen when some time is left. Then, the system incorporates content not initially considered. Patterns are represented using XML language, which guarantees an easy use with different tools and programming languages [4].

APG requests from the knowledge server the paragraphs available for each topic. The knowledge server will submit one or more paragraphs, since for the same concept may have several versions. Thus, the pattern indicated in Figure 1 shows that the first item of the presentation will be "museum presentation". It is identified at the knowledge server as "Greetings". It may involve three possible paragraphs, which will be included in the tree of possibilities, as it is shown in Figure 2.

Some global restrictions, including the level of audience, will be used to prune the tree, eliminating the paragraphs that do not fit the requested level. The

APG agent generates all possible paragraph combinations and then, using a set of fuzzy rules, estimates the QI from each presentation, and it will select the presentation with the highest index.

The presentation generated in this way analyses the estimated time for its execution, and when greater than anticipated, it eliminates the paragraphs with the least priority. Otherwise, if there is enough time, it includes some socially-oriented paragraphs: jokes, sports, political or local events comments.

```
<pattern>
<id>children_visit_museum</id>
<date_creation> ... </date_creation>
<date_lastused>...</date_ lastused>
<restrictions>
<age><10</age>
<technical_level>low</technical_level>
<time>60</time>
</restrictions>
<item>
<item_id>Museum_ Presentation</item_id>
<item_order>1</item_order>
<item_priority>10</item_priority>
<item_data>Greetings</item_data>
</item>
<item>
<item_id>Tour_Guide</item_id>
<item_order>2</item_order>
<item_priority>20</item_priority>
<item_data>Tour_Guide</item_data>
</item>
<item>
<item_id>Painting_presentation</item_id>
<item_order>3</item_order>
<item_priority>40</item_priority>
<item_content>
<item_data_id>key_picture</item_data_id>


                    <subitem> title</subitem>
          <subitem>date</subitem><subitem_multiple>
          description</subitem_multiple>
            <subitem_multiple>period</subitem_multiple
>
</item_content>
</item>
</pattern>
```

Figure 1. XML Pattern Example

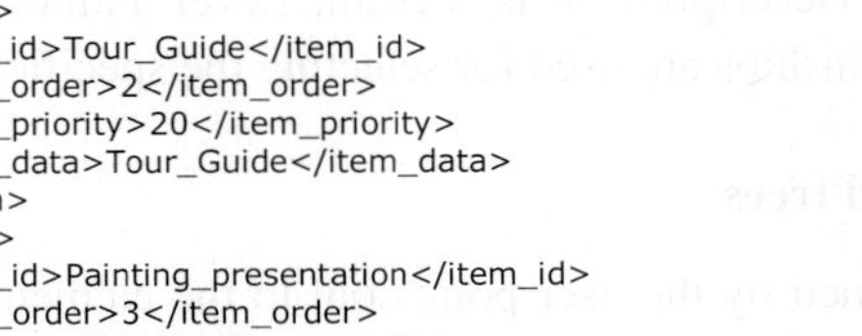

Figure 2. Tree Data Structure

5. Decision-making process for electing the paragraphs

Decision-making is a part of the paradigm proposed by Zadeh [5], currently examined in [6]. It is defined as the cognitive process of selecting a course of action out of multiple alternatives. Fuzzy set approaches to decision-making are usually most appropriate when human evaluations and the modelling of human knowledge are needed.

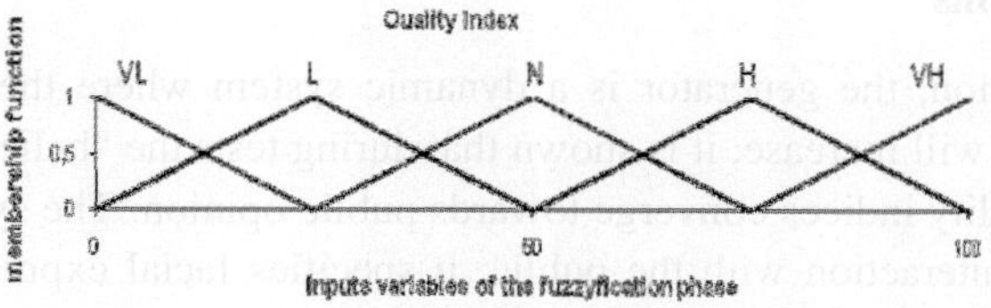

Figure 3. Inputs variables of the fuzzyfication phase

The fuzzy system will generate QI. All quality criteria information available and their influence on the QI is stored in the knowledge server ontology.

Five linguistic terms have been defined: VERY_HIGH (VH), HIGH (H), NORMAL (N), LOW (L), VERY_LOW (VL). The fuzzyfication phase uses the membership function and applies it to initially equidistant triangles, but in the learning phase their centres can vary. The exit variable quality_index is also modelled with five terms and triangular functions. The technique of centroid method is used in the defuzzyfication phase.

The number of possible presentations increases enormously when the categories expressed in the pattern are too generic, but it leads to higher quality presentations.

6. Learning process

One important characteristic of this proposal is the robot learning ability. Initially it is thought that the robot will have a small number of quality criteria available to evaluate some presentations as good, and others as bad, corresponding to the minimum level of education for a human guide. In order to learn, the robot needs some kind of feedback. That is achieved by a simple questionnaire that the public will fill out after attending a presentation.

An adjustment of linguistic terms and the membership functions will allow QI to be the closest to the average expressed by the public. For a greater accuracy, a genetic algorithm is simultaneously used over several presentations as an attempt to offset the local minimal for a presentation.

Learning is realized when new criteria of quality are incorporated in the paragraphs. The genetic algorithm readjusts the rules when it produces a disparity between public opinion and the QI, and it is used for the evolution of the rules. There are parts of the questionnaires intended to eliminate cases generated when the evaluation of the criteria runs contrary to public opinion, while still generating a correct QI.

7. Conclusions

As a conclusion, the generator is a dynamic system where the quality of the presentations will increase; it is shown that during tests the "beliefs of the robot" about the quality indices converge towards public opinion. The system also adds abilities for interaction with the public: it specifies facial expressions and arm movements of the robot, tone and type of voice to use, position in the floor plan, and several other details. It presents a novel approach in the use of computer ontology to represent the corpus with which the robot works, and the quality criteria for a presentation.

This work will help in the design of tour-guide robots capable to offer more affective learning and more dynamic tour-guide visits. At the same time, it advances in a new visitor-oriented approach.

References

1. G. Bekey, Autonomous Robots: from biological inspiration to implementation and control, MIT Press books, 2005.
2. D. Rodríguez-Losada, F. Matia, R. Galán, M. Hernando, J. M. Montero, and J. M. Lucas, Urbano, an Interactive Mobile Tour-Guide Robot. Advances in Service Robotics. Ed. H. Seok. In-Teh, 2008, pp 229-252.
3. M. Uschold and M. Grüninger, "Ontologies: Principles, Methods, and Applications," Knowledge Eng. Rev., vol. 11, 1996, pp. 93–155.
4. M. Y. Maarouf and S. M. Chung, "XML Integrated Environment for Service-Oriented Data Management," 20th IEEE International Conference on Tools with Artificial Intelligence, 2008.
5. L. A. Zadeh, "Fuzzy sets," Information and Control, 8, 338–353, 1965.
6. L. A. Zadeh, "Is there a need for fuzzy logic?" Information Sciences, 178(13), 2751–2779, 2008.

A SMALL WORLD OPTIMIZATION FOR MULTIPLE OBJECTS OPTIMIZATION OF MIXED-MODEL ASSEMBLY LINE PROBLEM WITH PARALLEL MACHINE

ŞÜKRAN ŞEKER*

Department of Industrial Engineering, Yildiz Technical University, Istanbul, Turkey

MESUT ÖZGÜRLER

Department of Industrial Engineering, Yildiz Technical University, Istanbul, Turkey

Mixed Model Assembly Line (MMAL) are widely used in manufacturing industries, becauseMMAL fulfills the diversified demand from small-high product. This paper presents an integer programming model for the mixed model assembly line problem (MMALBP). The proposed mathematical model minimizes the cycle time and select assembly sequences. The proposed model includes parallel machines in some stations. Integer programming formulation is presented for balancing and scheduling decisions simultaneously for mixed model assembly line problem which have parallel machines in stations. The proposed approach is illustrated with a computational example problem.

1. Introduction

An assembly line is a manufacturing process in which components are consecutively assembled to an unfinished product depending on a set of tasks to produce a final product. The tasks are allocated to stations according to a given precedence relationship among tasks, and they are performed in a certain time called as the task time. At each station, certain tasks are repeatedly performed in a limited time called as the cycle time. The problem of assigning tasks to stations in such a way that one or more objectives are optimized subject to some specific constraints is called as the assembly line balancing (ALB) problem [1].

Simple assembly lines are being replaced by mixed model assembly lines due to growing trend for greater product variability and shorter life cycles. Mixed model ALB includes a complication as a result of blockage and starvation caused by the arrival of different models to the line, having different assembly time requirements at each station [2].

* sseker@yildiz.edu.tr.

Formally, a mixed model assembly line balancing problem(MMALBP) can be stated as follows: given *K* models, the set of operations associated with each model, the processing time of each operation (operation time), and the set of precedence relations which specify the permissible orderings of the operations for each model, the problem is to assign the operations to an ordered sequence of stations such that precedence relations of each model are satisfied and some performance measures are optimized [3].

In a mixed-model line, an operation may be carried out on only a subset of the products assembled in the line. This implies that task times in each workstations may considerably vary. For this reason also the assembly sequence of the model units is important with respect to the efficiency of a line. Thus, the MALBP is also connected to a sequencing problem. Another additional decision by which the MALBP may be accompanied concerns the utilization of parallel stations that perform the same task set. The aim of using parallel stations is often to perform tasks with processing time larger than the cycle time [4].

If we discuss extended literature about MMAL, the balancing and the sequencing problem are heavily interdependent. While the line balance decides on the assignment of tasks to stations and thus determines the work content per station and model, the production sequence of a given model mix is arranged on this basis with regard to minimum overloads. The amount of overload by itself is a measure of efficiency for the achieved line balance. That is why some authors have proposed a simultaneous consideration of both planning problems. A simultaneous approach is, however, only viable under very special conditions as both planning problems usually have completely different time frames.

In ALB the mixed-model case is usually transformed to the single-model case by the use of a joint precedence graph. Here, the process times of tasks which vary for certain models are averaged with regard to the occurrence of respective models in the estimated model mix and are then composed to form a unique precedence graph. [5]. In this paper balancing and sequencing are obtained simultaneously with minimizing transportation times between stations. We assume that there are parallel machine in stations. So we introduced mixed integer programming model based Tadeusz Sawik's models to solve our problem.

2. Integer Programming For Loading and Routing In MMAL

This paper has presented a mathematical model for MMAL and it can be categorized as an Mixed Integer Programming model using Tadeusz Sawik's [6,7] models about mixed model assembly line. Below, the mixed integer programming model is presented for balancing and select assembly sequences in MMAL.

2.1. *Assumptions*

The assumptions of the proposed model to solve problem are as follows:

1. Each assembly task must be assigned to at least one station (alternative assignments are allowed);
2. There are parallel machine in some station
3. Total space required for the tasks assigned to each station must not exceed the station's finite work space available;
4. Each product must be routed to the stations subject to precedence relations defined by its assembly plan;
5. Revisiting of stations is not allowed;
6. Each station can perform at most one task at any given time;
7. Transfer times between stations aren't negligible.

2.2. *Notation*

i= assembly station $i \in I=\{1,...,m\}$
j= assembly task $j \in J=\{1,...,n\}$
p=parallel machine at station i {p=1,….,mi}
k= product, $k \in K=\{1,..., v\}$
s= assembly sequence, $s \in S=\{1,...,w\}$

Input parameters

a_i= working space of station *i*
t_{jk}= assembly time for task *j* of product *k*
q_{il}= transportation time from station *i*to station *l*
I_j= the set of stations capable of performing task *j*
J_k= the set of tasks required for product *k*
R_s= the set of immediate predecessor-successor pairs of tasks (j,r) for assembly sequence $s \in S$ such that task j must be performed immediately before task r.
T_s= the set of tasks in assembly sequence s.

The following decision variables are introduced to model the loading and routing problem

Lambda= the weight factor(0<lambda<1)

u_s=1, if assembly sequence $s \in S$ selected, othewise 0

x_{ij}=1, if task *j* is assigned to station $i \in I_j$ otherwise x_{ij}=0

zipj=1, if task *j* is assigned to parallel machine p in station $i \in I_j$ otherwise zipj= 0

yiljs=1, 1, if for assembly sequence s, after completion of task *j* on station *i,* product is transferred to station *l* >= *i*to perform next task; otherwise *yiljs*= 0

Pmax=maximum station workload (cycle time)
Qsum=represents the weighted sum of total assembly and transportation time.

Objective

$\lambda P_{\max} + (1-\lambda) Q_{sum}$

Subject to

$$\sum_{k\in K}\sum_{j\in J_K} t_{jk} z_{ipj} \le P_{\max} \quad (1)$$

$$\sum_{i\in I}\sum_{l\in I}\sum_{s\in S}\sum_{j\in T_s} q_{il} y_{iljs} = Q_{sum} \quad (2)$$

$$\sum_{i\in I} x_{ij} \ge 1 \; j\in J \quad (3)$$

$$\sum_{i\in I_j}\sum_{l\ge i} y_{iljs} = u_s ; s\in S, j\in T_s \quad (4)$$

$$\sum_{j\in J} x_{ij} \le m_i a_i \; i\in 1 \quad (5)$$

$$\sum_{j\in J} z_{ipj} \le a_i , i\in I \quad (6)$$

$$x_{ij} \le \sum_{p=1}^{m} z_{ipj} \le m_i x_{ij}, i\in I, j\in J \quad (7)$$

$$\sum_{i\in I} i x_{ij} \le \sum_{i\in I} i x_{ir} \, j, r\in R_s \quad (8)$$

$$y_{iljs} \le x_{ij} ; i\in I_j , l\ge i, l\in l_r , (j,r)\in R_s , s\in S \quad (9)$$

$$y_{iljs} \le x_{lr} ; i\in I_j , l\ge i, l\in l_r , (j,r)\in R_s , s\in S \quad (10)$$

The objective function minimization of the weighted maximum workload Pmax and sum of transportation time Qsum. Constraint (1), (2) shows the workload of the bottleneck station with parallel machines and the total transportation time, **re**spectively. Constraint (3) ensures that each task is assigned to at least one station. Constraint (4) ensures for each product and assembly sequence selected that all of its required tasks be allocated among the stations. Constraint (5) includes total flexibility capacity of all parallel machines at each station. Constraint (6) the number of tasks assigned to each machine doesn't exceed the number of part feeders available. Constraint (7) ensures that each task is assigned to at least one machine and not more than all mi parallel machines of such a station i where the task has been assigned. Constraint (8) maintains the precedence relations among the part types to be assembled and ensures that a product does not revisit any station in a unidirectional flow system. Constraint (9) and (10) ensure that each product successively visits such stations where the required tasks may be assembled subject to precedence relations defined by the assembly sequence selected.

3. Illustrative Example

In this section numerical example and computational results are presented to illustrate application of the mathematical programming formulation. The system is made up of 3 identical stations, in which n = 20 task types are assembled to produce k= 3 product types.

There is two parallel machines in each station. Hence, the transportation times qil from station i to station l are q12 = q21 = 2, q13 = q31 = 4, q23 = q32 = 2. The assembly times tjk :t1=4, t2=2, t3=2, t4=2, t5=4, t6=2, t7=3, t8=5, t9=2, t10=4, t11=2, t12=3, t13=2, t14=2, t15=3, t16=3, t17=2, t18=2, t19=3, t20=2.The set of stations capable of performing task *j* are seen with IJ
IJ=[{1,3},{1,3},{1,3},{2,3},{2,3},{2,3},{2,3},{2,3},{2,3},{2,3},{2,3},{2,3},{2,3},{2,3},{2,3},{2,3},{2,3},{2,3},{2,3},{2,3}];

The set of tasks required for product *k* are as the following.
JK = [{1,2,3,4,6,8,9,11,12,14,15,18,19}, {1,2,4,5,6,7,9,10,13,16,17,19,20}, {2,3,4,5,7,8,9,10,11,12,14,18,20}];

Transfer times between stations are equal 2 unit times. The available working space for each station are; a1=10, a2=10, a3=10.

This multi-objective problem is solved using ILOG OPL optimization software. Results of computational experiments are reported. This results are showed in table 1.

Table 1.The computational results

lambda	Pmax-Qsum	Qsum	Sequence
0	62	64	1,2,4,3,6,8,9,11,12,14,15,18,19
0.5	45	64	1,2,5,4,6,7,9,10,13,16,17,19,20
1	45	68	2,4,3,5,7,8,9,10,11,12,14,18,20

4. Conclusion

Flexible assembly lines involve some serially connected stations which are capable of performing a set of assembly tasks. In this study, we propose a mathematical model that can solve balancing and sequencing problems simultaneously in MMALs at the same time. Mixed integer programming formulation is solved in ILOG OPL optimization software. Results of computational experiments are reported. We hope that this research encourages further work in these directions.

References

1. U. Özcan, B. Toklu, Balancing of mixed-model two-sided assembly lines,Computers & Industrial Engineering 57, 217–227,(2009)
2. E. Çevikcan, B. Durmuşoğlu, M. Ünal, A team-oriented design methodology for mixed model assembly systems,Computers & Industrial Engineering 56, 576–599, (2009)
3. W. Xu, T. Xiao, Mixed Model Assembly Line Balancing Problem With Fuzzy Operation Times And Driting Operations, Proceedings of the 2008 Winter Simulation Conference, (2008).
4. L. Tiacci, S. Saetta, Process-oriented simulation for mixed-model assembly lines, (2007)
5. N. Boysen, M. Fliedner, A. Scholl, Assembly line balancing: Which model to use when?, Int. J. Production Economics 111, 509–528, (2009).
6. T. Sawik, Flexible Assembly Line Balancing with Alternate Assembly Plans and Duplicate Task Assignments, (2008).
7. T. Sawik, An LP-based approach for loading and routing in a Flexible assembly line, Int. J. Production Economics 64, 49-58 (2000) .

A HYBRID BEES ALGORITHM FOR SOLVING A ROBOTIC ASSEMBLY LINE BALANCING PROBLEM*

SLIM DAOUD[1,2,†], FAROUK YALAOUI[1], LIONEL AMODEO[1], HICHAM CHEHADE[1], PHILIPPE DUPERRAY[2]

[1]*Université de Technologie de Troyes, Institut Charles Delaunay (STMR UMR CNRS 6279), LOSI, 12 rue Marie Curie, 10010 Troyes, France*
[2]*ARIES Packaging, Technopole de l'Aube en Champagne, 10430 Rosières, France*

This paper presents a hybrid bees algorithm with a fuzzy logic controller (FLC-BA) to solve a pick and place robotic assembly line balancing problem. It aims to find the suitable tasks and components in order to define the gripping strategies for each robot. Bees algorithms are designed using the natural foraging behavior of honey bees. A computational study is carried out and the results are compared with the results of an exact method deduced by Cplex. The experimental results show the advantages and the efficiency of the FLC-BA within a short computation time. This is the main advantage of our study as in our industrial application we are bounded by the time.

1. Introduction

In this paper, we are interesting by developing a modified bees algorithm with fuzzy logic controller to solve a robotic assembly line balancing problem (RALB) in order to maximize the efficiency of the line. Rubinovitz and Bukchin in 1991 were the first to introduce the robotic assembly line balancing (RALB). The bees algorithm (BA) is a population-based search algorithm first developed in 2005 by Pham *et al.* [1]. It mimics the gathering food of honey bees and it performs a kind of neighborhood search coupled with random search. Dereli and Sena Das [2] have developed a hybrid bee(s) algorithm for solving container loading problems. Since the parameters settings of this algorithm is difficult, a fuzzy logic controller is therefore proposed. This latter, allows to enhance the number of bees recruited for *e* elite patches (*nep*), number of bees recruited for *m-e* best patches (*nsp*) and size of patches (*ngh*) in order to improve the search ability. We have adapted a similar FLC of Lau *et al.* [3] to tune the BA with the difference of the membership functions parameters.

---

* This work is supported by ARIES Packaging (France) and national association of technical research (ANRT) in France.

† slim.daoud@utt.fr

Our contribution is to develop FLC-BA to RALB. For our knowledge, this method has never been used in this field. This algorithm allows choosing the best location of the final product and component to assign to each robot.

The remainder of this paper is organized as follows: the second section presents the problem description. The resolution method which is a Bees algorithm with FLC is presented in the third section. Computational experiments and numerical results are presented in the fourth section before ending the paper by a conclusion and perspectives for future works.

2. Problem description

As mentioned before, we are interested in developing bees algorithm coupled with FLC to define the gripping strategies for each robot for the studied robotic assembly system. The main objective of this work is to balance all tasks for each robot and to maximize the line efficiency E by maximizing the gripping products. The robots in our robotic systems seize the components (on the left side of the robot) from the moving conveyor belt and place them in specific locations (right side of the robot). Furthermore, the assembly of the final product requires the execution of K assembled tasks. The final assembled product is composed of n layers and for each layer we have K_α components. The order in which the tasks have to be performed is ensured by precedence constraints. In the precedence graph, all the tasks are presented like nodes. Our robotic assembly line consists of I serial workstations, each one equipped with a robot.

We have developed a mixed integer linear program to model the studied problem and to evaluate bees algorithm. This model is strictly linear and consists of maximizing the line efficiency E (productive part of the line capacity) which is the number of seizing components by the robots. Thus, the Cplex solver can be used here to solve the problem

3. Resolution methods

This section describes the bee algorithm coupled with a fuzzy logic controller that we have developed to solve our problem. FLC-BA is a new approach applied to the robotic systems which realize pick and place operations.

3.1. *Bees algorithm*

The solutions encoding is presented as a matrix of KM dimensions. KM represents all position points of the final assembled product. Each position point is assigned to robot number i between 1 and I and component number j between

1 and *J*. Each solution encoding (bee) *h* is assigned a fitness related to the number of seized products.

By performing the 'waggle dance', different bees foragers can share the information about the visited patch of flowers. Bees algorithm is based on several parameters which must be initialized: number of scout bees (*n*), number of elite bees (*e*), number of patches selected out of *n* visited (*m-e*), number of bees recruited for regions visited by elite bees (*nep*), number of bees recruited for the other selected patches (*nsp*), size of patches (*ngh*) and a stopping criterion. The description of BA is as follows: it begins by initializing the *n* scout bees which are placed randomly in a search space. After that, the fitness of the patches visited is evaluated. Bees with highest fitness value are chosen as elite bees and a neighborhood search was performed around them. The remaining bees in scout population are assigned randomly. For neighborhood search, we adapt the shift neighborhood for component representation by changing one number of components of the solution. The swap neighborhood is adapted for robot number by interchanging two elements of robot number of the solution.

3.2. *Fuzzy logic controller*

In this paper, we apply a fuzzy logic controller (FLC) to better set the number of bees recruited for *e* best patches (*nep*), number of bees recruited for *m-e* patches (*nsp*) and size of patches (*ngh*). A similar FLC used by *Lau et al.* [2] is presented in this paper to guide the genetic algorithm.

Fuzzification: In this study, the input of FLC are: $f_a(t)=f(t-1)-f(t-2)$ where $f(t-1)$ and $f(t-2)$ are the average values of the objective function at iterations $t-1$ and $t-2$ respectively and $d(t-1)$ is the sum of the hamming distance between each solution encoding of the entire population at iteration $t-1$.

Decision making: The decision is obtained by combining the different linguistic terms obtained by the membership function of $f_a(t)$ and $d(t-1)$. The membership functions are presented by triangular form.

Defuzzification: It aims to calculate the variation of *nep*, *nsp* and *ngh*. In this step, we use the center of gravity defuzzification method. The reason for choosing this method is that give a good results as presented in the work of Li *et al.* [4].

3.3. *FLC-Bees algorithm*

In order to improve the performance of the bees algorithm, a fuzzy logic controller (FLC) is proposed to up-date the number of bees recruited for regions visited by elite bees (*nep*), the number of bees recruited for the other selected patches (*nsp*) and size of patches (*ngh*) each ten consecutive generations. This is due to provide sufficient time for the modification in the algorithm. The average fitness value of the population and diversity are taken as the input of the fuzzy controller. Then, the three fuzzy controllers will determine the values of $\Delta nep(t)$, $\Delta nsp(t)$ and Δngh(t) as the outputs. At generation *t*, *nep(t)*, *nsp(t)* and ngh(t) will be recalculated based on equations 1 , 2 and 3:

$$\text{nep}(t) = \text{nep}(t-1) + \Delta\text{nep}(t) \tag{1}$$

$$\text{nsp}(t) = \text{nsp}(t-1) + \Delta\text{nsp}(t) \tag{2}$$

$$\text{ngh}(t) = \text{ngh}(t-1) + \Delta\text{ngh}(t) \tag{3}$$

The value of the *d(t)* is the degree of population diversity at iteration *t*. Lau *et al.* [2] defines it as the average of the sum of the hamming distance which is the difference of all pairs of solution encoding.

$$d(t) = \frac{1}{\frac{N(N-1)}{N}} \sum_{h=1}^{N} \sum_{l=1}^{N} \sum_{r=1}^{KM} \frac{\delta(g_{h,r}, g_{l,r})}{KM} \tag{4}$$

Where *N* is the population size, the value of *KM* is the solution encoding length which is equal to the number of the points and $g_{h,r}$ is the values of the two elements of the rth point of the hth solution encoding [3]. Therefore $\delta(g_{h,r}, g_{l,r}) = 1$ if the two points of the different solution encoding are not the same, 0 otherwise.

Based on work of Pham and Kalyoncu [5], The FLC-BA algorithm is represented below:

Initialize the search with a population of n scout bees
Check the precedence constraint of assembled task[k]
If assembled task[k] = Robot i then assembled task [k+1] = {Robot i or Robot j}
$\nabla k \in pred(h)$ *and* $\nabla (i,j) \in \{1,\ldots,I\}$
If *t = MaxIter* ***then*** *Stop*
Else
Evaluate the fitness
Select good sites (m) for neighborhood search
Select elite sites (e) from selected good sites (m)
Do neighborhood search with patch size ngh from e sites by elite bees (nep) and evaluate the fitness
Do neighborhood search with patch size ngh from m-e sites by other bees (nsp) and evaluate the fitness
Select best from each site
Assign the remaining bees (n-m) to random search and evaluate the fitness

Obtain a new population of scout bees from these neighborhood search and random search
If** t mod 10 =0 **Then
The values of f(t-1)-f(t-2) and d(t-1) are considered as the input of FLC, update the nep, nsp and ngh with FLC
End If
Run until the stooping criterion is met (t=MaxIter)

4. Computational results

To evaluate the bees algorithm coupled with fuzzy logic controllers (FLC-BA), we have developed an exact method based on Cplex. The two methods are applied to different robotic systems structures. All methods are applied to different robotic assembly line structures. We realize computational experiments by studying robotic assembly systems structures which are different by the number i of robots, k the number of position points and n the number of layers of the final assembled product. The first structure (S1) represents a system with 2 robots, 4 or 8 position points and 2 or 4 layers whereas (S2), (S3) and (S4) refer respectively to systems with 4, 6 and 8 robots with always the same number of position points and layers.

Table 1. Numerical Results

	Robots	Points position	Number of layers	Number of product (E)			Execution time		
				Cplex	BA	FLC-BA	Cplex(s)	BA(s)	FLC-BA(s)
(S1)		4	2	**66**	**66**	**66**	1.83		
	2		4	**65**	**65**	**65**	1.21	0.121	0.177
		8	2	**66**	**66**	**66**	2.13		
			4	**65**	**65**	**65**	4.12		
(S2)		4	2	**80**	**80**	**80**	5.98		
	4		4	**80**	**80**	**80**	6.76	0.201	0.357
		8	2	**80**	**80**	**80**	3.54		
			4	**80**	**80**	**80**	6.81		
(S3)		4	2	**80**	**80**	**80**	12.45		
	6		4	**79**	**80**	**80**	10.86	0.389	0.466
		8	2	**80**	**80**	**80**	**11.78**		
			4	**80**	**80**	**80**	**13.7**		
(S4)		4	2	**80**	**80**	**80**	**24.82**		
	8		4	**78**	**78**	**78**	**21.28**	0.475	0.6
		8	2	**79**	77	**79**	**25.91**		
			4	**78**	76	**78**	**26.78**		

For each structure, we have tested four instances by changing at each time the number of position points and the number of layers. We adopt the proportion between the average workloads of different robots $\delta = 20\%$. The required time to assemble J components C_{max} depends on the structure, for (S1) and (S2) $C_{max} = 33.6$ (s) and for (S3) and (S4) $C_{max} = 16$ (s). The number of final assembly products to simulate the performance of the line is fixed at 80. This number is taken from an industrial case in the starting phase of a robotic line of dairy food. The industrial application of our method requires an execution time less than **600 ms**. Based on work of Pham and Kalyoncu [4], BA parameters are: n=20, m=10, e=10, nep=10, nsp=7 and ngh=0.01.

Table 1 presents the comparison between the BA, FLC-BA and Cplex. We may notice that for the 14 first instances, the BA has succeeded to reach the optimal solutions that have been obtained with the exact method. For the FLC-BA, it can reach all optimal solutions. Furthermore, the execution time shows that the BA method is faster than FLC-BA. This is due to the fuzzy logic procedure included in the FLC-BA algorithm. But, both algorithms have satisfied the industrial constraint about the execution time algorithms. This advantage is more obvious when the size of the structures is more important. In general, we may reduce the execution times by 90% if we apply the BA or FLC-BA algorithms instead of the exact method based on Cplex. We notice that the FLC-BA is better to BA in terms of rapidity and quality of obtained solutions. This fact represents a great interest taking in consideration that in our industrial application we are bounded by the execution time to manage the functioning of a real industrial robotic assembly system.

5. Conclusion

The tuning of a fuzzy logic controller for a bees algorithm is considered in this paper. The objective of our study is to maximize the efficiency of robotic assembly system for an industrial application which requires a small execution time. The numerical results show the performance of FLC-BA. It has been able to get the optimal solutions with a small execution time. The perspectives of our work are first to apply other neighborhood search for our algorithm. The application of the local search may also be one of our perspectives as to compare its results with those of the fuzzy logic controllers.

References

1. D.T. Pham, A. Ghanbarzadeh, E. Koc, S. Otri, S. Rahim and M. Zaidi, *Technical note* (2005).
2. Dereli. T, Sena Das. G, Applied Soft Computing, *11*, (2011).
3. Lau. H. C. W., Chan, T. M;, Tsui, W. T., & HO, G. T. S. Cost, *IEEE Transactions on Knowledge and Data Engineering*, (2009).
4. Li. X, Yalaoui. F, Amodeo. L, Chehade. H., *FLINS*, (2010).
5. D.T.Pham and M.Kalyoncu, *IEEE-INDIN*, (2009).

THROUGHPUT ANALYSIS OF TWO-MACHINE-ONE-BUFFER LINE MODEL: A COMPARATIVE STUDY

YASSINE OUAZENE*, ALICE YALAOUI AND HICHAM CHEHADE

Laboratoire d'Optimisation des Systèmes Industriels, UMR-STMR-CNRS 6279
Université de Technologie de Troyes, 12 rue Marie Curie,10010 TROYES Cedex, France.

The two-machine line model is often used as a building block to analyze and develop approximation methods for wider and more complex production systems. Therefore, many analytical formulations have been presented in the literature. In this paper, we are interested on the throughput calculation of a two-machine production line considering the general case where the two machines have different failure and repair rates. Therefore, we have introduced an analytical formulation based on the construction and analysis of an equivalent machine. The presented model is compared with the Alden's model based on different reliability scenarios and different system configurations. The performances of both models are reported and commented.

1. Introduction

Regardless of the production line kind, the system evaluation is a common step of an optimized design. It is the step where the performances of the combined elements of the system are examined. In fact, often the performance of a production line is expressed according to its global production rate or throughput. The two-machine models are often used as a construction base to analyze and develop approximation methods for wider and more complex production lines. Various two-machine models have been presented in literature; Li *et al.* [1] have presented a comparative study with eight different models. They have considered the case with deterministic processing times, unreliable machines and finite intermediate buffers. The authors have presented the major differences and their impacts among these models, how they differ in system performance and the common nature of all the models.

In this paper, we are interested by the particular case of asynchronous line with operation-dependent failure model. Alden [2] has proposed an exponential operation-dependent-failure model to calculate the throughput of a two-machine line model in which, each machine is characterized by a constant production

* yassine.ouazene@utt.fr

capacity and exponential reliability parameters (failure rate and repair rate). This model is also presented by Blumenfeld and Li [3] and Li *et al.* [1]. In the Alden's model, a particular case with no simultaneous downtime is considered. Indeed, while one machine is down, the other machine does not fail and its production capacity is reduced to its isolated production rate or capacity. This assumption is an approximation to account for the possibility of both machines being down. In the presented work, we are interested by the general case. We consider the classical assumptions common to the Alden's and Gershwin's models, but we consider also the case with the possibility of both machines being down.

The rest of this paper is organized as follows: Section 2 presents the problem formulation and details the two presented models. Section 3 presents the experimental results. Finally, section 4 summarizes the contribution of this paper.

2. Problem formulation

We consider two unreliable machines (work stations) separated by intermediate finite buffer. The system is subject to unavailability due to the finite buffer and/or to the failure and repair times of the machines. We consider also the classical assumptions defined below:

- The processing times (or machining times) of pieces are deterministic.
- The failure is state dependent. A machine cannot fail when it is starved or blocked.
- The first machine cannot be starved and the second one cannot be blocked.
- Both failure and repair times are exponentially distributed and statistically independent.
- The buffer does not fail and its transition time is equal to zero.

We, also, adopt the following notations:

N	the capacity of the buffer
ω_i	the processing rate of machine M_i
λ_i	the average failure rate of machine M_i
μ_i	the average repair rate of machine M_i
$\alpha = {}^{\omega_1}/_{\omega_2}$	the processing rates fraction (ratio)
$P_j; j \in \{0,1,\ldots,N\}$	the probability to have j pieces in the buffer

2.1. *Alden's model*

This model is detailed by Alden [2] and in Blumenfeld and Li [3]. This model was compared with that one of Gershwin in Li *et al.* [1]. The authors concluded that the two models present similar performances.

In addition to the classical assumptions cited below, no simultaneous downtime is considered. When one machine is down, the other machine does not fail, but its speed is reduced to its isolated production rate defined by equation (1).

$$\omega_i^* = \omega_i \times \frac{\mu_i}{\lambda_i+\mu_i} \tag{1}$$

The system throughput is calculated based on the balance equations in steady state of the system (see Alden [2], Blumenfeld and Li [3] for details).

$$\psi = \begin{cases} \omega_2 \times \big(P(U) + P(SB)\big) + \omega_2^* \times P(E) & \text{if } \omega_1 > \omega_2 \\ \omega_2 \times \big(P(U)\big) + \omega_2^* \times P(E) & \text{if } \omega_1 = \omega_2 \\ \omega_2 \times \big(P(U) + P(SS)\big) + \omega_2^* \times P(E) & \text{if } \omega_1 < \omega_2 \end{cases} \tag{2}$$

Such as:

$P(U)$: the probability that both machines are processing.

$P(SB)$: the probability that the system is blocked.

$P(SS)$: the probability that the system is starved.

$P(E)$: the probability that the second machine is processing and the first machine is down.

For the case with identical production rates, the probabilities above are calculated based on the estimation of the expected times that the system spent in each of the possible states. These expected times are expressed by the following equations:

$$P(U) = \frac{1}{\lambda_1+\lambda_2} \tag{3}$$

$$P(SB) = \frac{\lambda_2}{\lambda_1+\mu_1} \tag{4}$$

$$P(SS) = \frac{1}{\mu_2} \times P_0 \tag{5}$$

$$P(E) = \frac{N\times\lambda_2\times(\lambda_2+\mu_2)}{\omega_2\times\mu_2\times(\lambda_1+\lambda_2)} \times P_0 \tag{6}$$

$$P_0 = \frac{\omega_1\times\lambda_1\times\mu_1}{\omega_1\times\mu_1\times(\lambda_1+\lambda_2)+N\times\lambda_1\times\mu_2\times(\lambda_1+\mu_1)} \tag{7}$$

2.2. *Equivalent machine model (EMM)*

In this model, we consider the case where the two machines can be down and repaired simultaneously. Also, if the buffer is empty, both machines are up and $\omega_2 > \omega_1$, then the second machine has to slow down its production rate to ω_1. If the buffer is full, both machines are up and $\omega_1 > \omega_2$, then the first machine has to slow down its production rate to ω_2.

Considering the buffer as an isolated system between two reliable machines, Songgui and Changjun [4] have analyzed the buffer transitions using steady states of Markov processes. To do so, they have considered a birth-death process with *(N+1)* states {0, 1... N} and ω_1, ω_2 are respectively the birth and the death transition rates.

The differential equations for the probability that the system is in state *j* at time *t* are:

$$\begin{cases} P_0'(t) = -\omega_1 \times P_0(t) + \omega_2 \times P_1(t) \\ P_j'(t) = \omega_1 \times P_{j-1}(t) - (\omega_1 + \omega_2) \times P_j(t) + \omega_2 \times P_{j+1}(t) \\ P_N'(t) = \omega_1 \times P_{N-1}(t) + \omega_2 \times P_N(t) \end{cases} \tag{8}$$

For the steady state of system operation, the probability to be in the jth state of the buffer is given by equation (9). This probability is obtained by simplifying the system of equations above with considering that $\sum_{j=0}^{N} P_j = 1$.

$$P_j = \begin{cases} \frac{\alpha^j \times (1-\alpha)}{1-\alpha^{N+1}} & \text{if} \quad \alpha \neq 1 \\ \frac{1}{N+1} & \text{if} \quad \alpha = 1 \end{cases} \tag{9}$$

The construction and analysis of the equivalent machine is based on the analysis of the possible states of the buffer, especially the empty and full buffer states. The system throughput $\psi = \min\{\rho_1, \rho_2\}$ is defined as the bottleneck between the effective production rates of the two equivalent machines.

$$\rho_1 = \begin{cases} \frac{\omega_1 \times \mu_1 \times (1-\alpha^N)}{(\mu_1 \times (1-\alpha^{N+1}) + \lambda_1 \times (1-\alpha^N)} & \text{if } \alpha \neq 1 \\ \frac{\omega \times \mu_1 \times N}{(\mu_1 \times (N+1) + \lambda_1 \times N} & \text{if } \alpha = 1 \end{cases} \tag{10}$$

$$\rho_2 = \begin{cases} \frac{\omega_2 \times \mu_2 \times \alpha(1-\alpha^N)}{(\mu_2 \times (1-\alpha^{N+1}) + \lambda_2 \times \alpha(1-\alpha^N)} & \text{if } \alpha \neq 1 \\ \frac{\omega \times \mu_2 \times N}{(\mu_2 \times (N+1) + \lambda_2 \times N} & \text{if } \alpha = 1 \end{cases} \tag{11}$$

So, the system throughput can be formulated as follows:

$$\psi = \begin{cases} \omega_1 \times min\left\{\frac{\mu_1 \times (1-\alpha^N)}{\mu_1 \times (1-\alpha^{N+1}) + \lambda_1 \times (1-\alpha^N)}; \frac{\mu_2 \times (1-\alpha^N)}{\mu_2 \times (1-\alpha^{N+1}) + \lambda_2 \times \alpha(1-\alpha^N}\right\} & \text{if } \alpha \neq 1 \\ \omega \times min\left\{\frac{\mu_1 \times N}{\mu_1 \times (N+1) + \lambda_1 \times N}; \frac{\mu_2 \times N}{(\mu_2 \times (N+1) + \lambda_2 \times N}\right\} & \text{if } \alpha = 1 \end{cases} \tag{12}$$

3. Computational experiments

In this section, we compare the two models presented above among different configurations. Two representative scenarios of reliability are considered: short frequent failures and long infrequent failures. The different configurations of each scenario are described by the reliability parameters values of both machines given in Table 1. The comparison is made up by calculating the gap between the system throughputs, obtained by each model for all the considered configurations, using on equation (13).

$$gap(\%) = 100 \times \frac{\psi_{Alden} - \psi_{EMM}}{max\{\psi_{Alden}, \psi_{EMM}\}} \quad (13)$$

Without loss of generality, we consider that the production rate of the machines is equal to the unit for all the configurations.

Table 1. Reliability parameters of different configurations

Scenarios	Short frequent failures				Long infrequent failures			
-	λ_1	μ_1	λ_2	μ_2	λ_1	μ_1	λ_2	μ_2
I1: Identical machines	0.1	0.4	0.1	0.4	0.01	0.04	0.01	0.04
I2: Equivalent non identical machines 1	0.125	0.5	0.1	0.4	0.0125	0.05	0.01	0.04
I3: Equivalent non identical machines 2	0.1	0.4	0.125	0.5	0.01	0.04	0.0125	0.05
I4: First machine is a bottleneck	0.15	0.5	0.1	0.4	0.015	0.05	0.01	0.04
I5: Second machine is a bottleneck	0.1	0.4	0.15	0.5	0.01	0.04	0.015	0.05

Tables 2 and 3 present the numerical results of the comparison between the two models in short frequent failures and long infrequent failures scenarios. The obtained results illustrate some differences between the two models. In fact, in the first case and for the different configurations the *EMM* model underestimates the system throughput comparing to the Alden's model. In the second scenario, the *EMM* overestimates the system throughput. Also, we notice that the relative difference between the two models is decreasing according to the buffer size in both scenarios. This relative difference is more important in the second scenario. Also, an important remark is that, contrary to the Alden's model, the system throughput calculated by the *EMM* model is a constant function according to the ratio $/\mu$.

4. Conclusion

In this paper, we have presented a mathematical formulation to calculate the system throughput of a buffered two-machine production line in an operation-dependent-failure context. This mathematical formulation is based on the analysis of the buffer states using a birth-death process. We have also compared this model to the Alden's model among different configurations and different reliability scenarios. This comparison made possible the evaluation of the impact of the non consideration of possible failure state of both machines in the model of Alden. Future extension of this work may be to consider the presented model as a base to estimate the system throughput for wider and more complex production lines.

Table 2. Comparison of the two models with short frequent failures scenario

Buffer size	I1 gap (\%)	I2 gap (\%)	I3 gap (\%)	I4 gap (\%)	I5 gap (\%)
10	2.78	2.06	3.99	5.04	3.66
20	1.11	0.71	1.84	4.12	1.65
30	0.65	0.38	1.18	3.94	1.04
40	0.46	0.25	0.87	3.88	0.76
50	0.34	0.18	0.68	3.85	0.60
60	0.28	0.14	0.56	3.84	0.49
70	0.23	0.11	0.48	3.83	0.42
80	0.2	0.09	0.42	3.83	0.36
90	0.16	0.08	0.37	3.83	0.32

Table 3. Comparison of the two models with long infrequent failures scenario

Buffer size	I1 gap (\%)	I2 gap (\%)	I3 gap (\%)	I4 gap (\%)	I5 gap (\%)
10	-6.40	-9.05	-3.57	-8.10	-3.63
20	-7.56	-9.57	-5.11	-7.85	-5.16
30	-7.11	-8.74	-4.94	-6.58	-5.01
40	-6.5	-7.87	-4.55	-5.41	-4.63
50	-5.93	-7.11	-4.15	-4.43	-4.24
60	-5.42	-6.46	-3.80	-3.61	-3.89
70	-4.99	-5.91	-3.49	-2.94	-3.58
80	-4.61	-5.45	-3.22	-2.37	-3.31
90	-4.29	-5.05	-2.99	-1.88	-3.08

References

1. J. Li, D. Blumenfeld, N. Huang and J. Alden, *International journal of Production Research* **44**, 1375 (2006).
2. J. Alden, General Motors Research and Development Center, Report RD-9434 (2002).
3. D. Blumenfeld and J. Li, *Mathematical Problems in Engineering* **3**, 293 (2005).
4. S. Songgui and J. Changjun, Journal of *Systems Science and Systems Engineering* **6**, 53 (1997).

MIXED MODEL ASSEMBLY LINE BALANCING WITH SUB-ASSEMBLY SELECTION IN AUTOMOTIVE INDUSTRY

RIFAT GURCAN OZDEMIR[1]

Istanbul Kultur Univeristy, Department of Industrial Engineering, Istanbul, Turkey

UFUK KULA

Sakarya University, Department of Industrial Engineering, Sakarya, Turkey

In this study, mixed model assembly line balancing and sub-assembly selection problems are solved jointly. The mixed model assembly line balancing problem requires that tasks are assigned to workstations with balanced workloads. In sub-assembly selection problem, manufacturer determines which components of the products are assembled in sub-assembly area and which ones are assembled on the products directly in the line. Sub-assembly alternatives, task times, precedence relations and other system specific data are input to determine which tasks are assigned to sub-assembly area and which ones are assigned to the line for two conflicting objectives of minimizing cycle time and minimizing total cost. A recursive two stage goal programming approach is developed to solve the above mentioned problem. The developed model is then implemented in a real life automotive manufacturer's assembly line balancing problem.

1. Introduction

In mass production systems, an important problem is the assembly line problem. A large proportion of the studies related to assembly line systems deal with the determination of the set of tasks to be assigned to each work station under the cycle time and the precedence constraints, which is known as the simple assembly line balancing problem (SALBP). The assumptions of SALBP are very restrictive with respect to real-world assembly line systems (Becker and Scholl, 2003). Two versions of the well-known simple assembly line balancing problem i.e., SALBP-1 and SALBP-2 have intensively been studied in the case of deterministic demand. SALBP-1 is present when a new assembly line system has to be installed and the external demand can be well estimated since the cycle time and, hence, the production rate have to be specified as fixed parameters. SALBP-2 leads to the maximization of the production rate of an existing

[1]rg.ozdemir@iku.edu.tr

assembly line. Either SALBP-1 or SALBP-2 assumes that the demand is deterministic i.e., it is a well estimated value with no variation or possibility of shortage, hence, the shortage cost is negligible. However, the demand variation affects the decision on selecting of correct system parameters (desired cycle time or required number of stations). Therefore, in this study the external demand is assumed to be a random variable but follows a certain probability distribution. For a very comprehensive survey on SALB solution methods Becker and Scholl (2006) and Boysen et al. (2007). Amen (2006) solved assembly line balancing problem based on cost-oriented model formulations and he showed upper and lower bounds of the problem. Andres et al., (2008) developed a solution method for balancing and scheduling tasks in assembly lines with sequence-dependent setup times. Bukchin and Rabinowitch (2006) developed a solution approach for the mixed-model assembly line-balancing problem for minimizing duplication costs using branch-and-bound algorithm.

Subsequent sections of the paper are organized as follows. The goal programming formulation is presented in section 2, a numerical example for illustrating proposed model given in section 3, and finally paper is concluded in section 4.

2. The Goal Programming Formulation

The problems of assembly line balancing and sub-assembly selection are solved simultaneously using a goal programming approach. Our formulation assumes that each component has a unique set of tasks (i.e. there is no common task between any pair of components).

The following parameters and variables are defined for the formulation of the goal programming model:

m = model index, $m = 1,\ldots,H$

i = task index, $i = 1,\ldots,N$

j = station index, $j = 1,\ldots,M$

k = component index, $k = 1,\ldots,K$

t_{im} = task time of task i of model m.

o_k = total fixed and variable costs of outsourcing component k.

w = total in-house production cost associated with the time performed to accomplish a product.

f = fixed cost of establishing a station in the line.

E_k = set of tasks required to be performed to make the component k.

SIP_i = set of immediate predecessors of task i.

SIP_k = set of immediate predecessors of component k.

And the following variables are defined for every station j and for every task i:

C = the cycle time along the line.

$$X_{ij}=\begin{cases}1, & \text{if task } i \text{ is assigned to station } j.\\ 0, & \text{otherwise}\end{cases}$$

$$Y_{kj}=\begin{cases}1, & \text{if component } k \text{ is supplied to station } j.\\ 0, & \text{otherwise}\end{cases}$$

$$R_j=\begin{cases}1, & \text{if station } j \text{ is open}\\ 0, & \text{otherwise}\end{cases}$$

$$Min Z_1=\sum_{j=1}^{M}\sum_{k=1}^{K}o_k Y_{kj}+w\sum_{j=1}^{M}\sum_{i=1}^{N}t_i X_{ij}+f\sum_{j=1}^{M}R_j \tag{1a}$$

$$Min Z_2=C \tag{1b}$$

Subject to:

$$\sum_{j=1}^{M}X_{ij}=1 \qquad ,\forall i\notin\left\{\bigcup_{k=1}^{K}E_k\right\} \tag{2}$$

$$\sum_{j=1}^{M}X_{ij}=1-\sum_{j=1}^{M}Y_{kj} \quad ,\forall k \text{ and } i\in E_k \tag{3}$$

$$\sum_{j=1}^{M}Y_{kj}\le 1,\forall k \tag{4}$$

$$\sum_{j=1}^{M}j\cdot X_{hj}\le\sum_{j=1}^{M}j\cdot X_{ij} \qquad ,\text{for } h\in SIP_i \tag{5}$$

$$\sum_{j=1}^{M}j\cdot X_{ij}\le\sum_{j=1}^{M}j\cdot Y_{kj} \qquad ,\text{for } i\in SIP_k \tag{6}$$

$$\sum_{j=1}^{M}j\cdot Y_{kj}\le\sum_{j=1}^{M}j\cdot X_{ij} \qquad ,\text{for } k\in SIP_i \tag{7}$$

$$\sum_{j=1}^{M}j\cdot Y_{gj}\le\sum_{j=1}^{M}j\cdot Y_{kj} \qquad ,\text{for } g\in SIP_k \tag{8}$$

$$\sum_{m=1}^{H}\left(\sum_{i=1}^{N}t_{im}X_{ij}+\sum_{k=1}^{K}t_{km}Y_{kj}\right)\le C\cdot R_j \qquad ,\forall j \tag{9}$$

$$X_{ij},Y_{kj},R_j=0,1 \qquad ,\forall i,j,k \tag{10}$$

In equation 1a, our objective is to minimize the total cost which involves the variable and fixed outsourcing cost per period and total in-house production cost per period, and total installation cost for stations. In equation 1b, global cycle time for all models is minimized. Constraint (2) ensures that each task which is not part of any component should be assigned to a station. Constraint set (3) ensures that each task of an unassigned component should be assigned to a station. Constraint set (4) ensures that each component can be supplied to only one station. Constraint sets (5 – 8) show the precedence relations among tasks, between tasks and components and among components, respectively. Constraint et (9) ensures that station content cannot exceed the given cycle time. Finally, constraint set (10) defines the decision variables.

3. Numerical Example

In this section we will illustrate our proposed model over an example problem. The data for this numerical example is gathered form a TV manufacturer. The line in practice is configured as being three segments. The first segment involves all necessary tasks to build a TV set, the second segment is configured just for testing and the last segment is for packing final product. The sub-assembly selection problem arises in the first section, thus in this paper we has a focus on the first segment for implementation of the method. Figure 1 presents the all assembly tasks that should be performed to make a product in the first segment, and also precedence relations and times are given in the same figure. Tasks 4 and 5, tasks 6, 7 and 8, and tasks 10, 11 and 12 are performed to make sub-assemblies SA1, SA2 and SA3, respectively.

Task	Time (sec.)
1	15
2	22
3	16
4	10
5	12
6	12
7	14
8	15
9	17
10	12
11	14
12	15
13	23
14	15
15	8

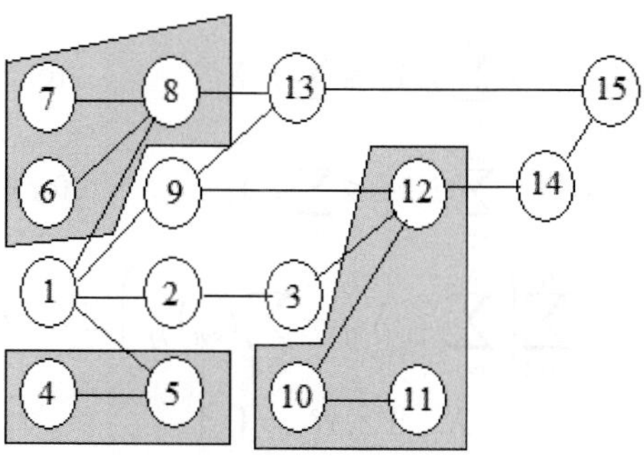

Figure 1. Times and predecessors of all assembly tasks.

The model presented in the previous section of the paper is coded in GAMS 23.1 and solved using CPLEX 12.0 for obtaining a solution to joint problems of line balancing and sub-assembly selection. The solution to the problem for given components and tasks is presented in the following figure.

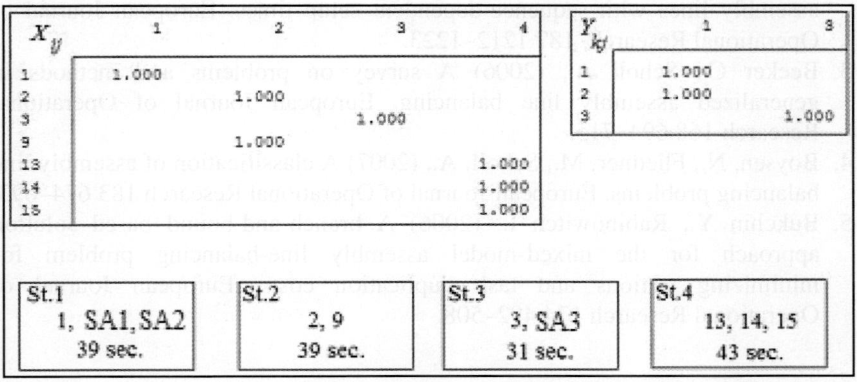

Figure 2. Solution to the first stage of the problem for given components.

As seen in figure 2, four stations are open and maximum load attained from among all stations is realized to be 43 seconds. All three components are decided to be outsourced and components 1 and 2 are supplied to the first station and component 3 is supplied to the third station of the line. Objective function value for this solution is obtained as $192,000. The execution timqe for solving the given problem optimally is only 740 seconds. The early experiments for larger problems show that using CPLEX solver for finding solution is efficient when the number of tasks is under 600.

4. Conclusion

This paper has considered a two-stage goal programming model to solve the problem of line balancing and sub-assembly selection. In the first stage of the linear model, tasks are assigned to stations for minimizing total cost under given cycle time constraint. In the second stage, cycle time is decreased iteratively to its minimum value with the same total cost found in the first stage. A numerical study has been provided for the proposed approach. The early results show that the idea of consideration of sub-assembly selection with line balancing problem is highly significant and provide a new perspective to the decision maker in assembly process.

References

1. Amen M., (2006) Cost-oriented assembly line balancing: Model formulations, solution difficulty, upper and lower bounds. European Journal of Operational Research 168 747–770.
2. Andres C., Miralles C., Pastor R., (2008) Balancing and scheduling tasks in assembly lines with sequence-dependent setup times. European Journal of Operational Research, 187 1212–1223.
3. Becker C., Scholl A., (2006) A survey on problems and methods in generalized assembly line balancing. European Journal of Operational Research 168 694–715.
4. Boysen, N., Fliedner, M., Scholl, A., (2007) A classification of assembly line balancing problems. European Journal of Operational Research 183 674–693
5. Bukchin Y., Rabinowitch I., (2006) A branch-and-bound based solution approach for the mixed-model assembly line-balancing problem for minimizing stations and task duplication costs. European Journal of Operational Research 174 492–508.

MATHEURISTICS FOR THE MULTI-ITEM LOT-SIZING WITH VEHICLE ROUTING PROBLEM

HEITOR LIBERALINO*

Departamento de Informática,
Universidade do Estado do Rio Grande do Norte,
Mossoró, RN, Brazil

CHRISTOPHE DUHAMEL

Université de Technologie de Troyes,
Troyes, Champagne-Ardenne, France

ALAIN QUILLIOT

Laboratoire d'Informatique, Modélisation et Optimisation des Systèmes,
Université Blaise Pascal
Aubière, Auvergne, France

Our problem consists in scheduling both a production distributed on many sites and the transportation of items between those sites. The goal is to optimize the global solution. The production model adopted is a multi-item, multi-level, Capacitated Lot-Sizing Problem. The transportation model is based on the Vehicle Routing Problem with discretized time. We present four heuristics based in matheuristics to solve it. Computational results are presented to evaluate the efficiency of the four heuristics.

1. Introduction

Moving production to foreign countries where the manpower is cheaper is a politic adopted by many industries. This has been partly made possible by a lot of improvements on transportations systems. Meanwhile, transportation costs may strongly impact the total cost of a product, and substantial savings can be obtained by planning both the production and the transportation.

The integrated Multi-item Lot-Sizing and Vehicle Routing Problem (M-LSVRP) described here considers the production of components in different sites and the inter-site transportation of these items. This approach can model AGV systems as well. The consumption of intermediate components is

* heitorliberalino@uern.br

necessary to produce a final item (assembly Bill of Materials). Our production model is based on the Capacitated Lot-Sizing Problem (CLSP) [9] where demands have to be satisfied at each period of time and the setup times. The transport model is based on the Vehicle Routing Problem (VRP). The time scale is discretized in periods for both production and transportation, one being a fraction of the other [8]. The complexity of this problem is NP-Hard and to solve it we developed some matheuristcs that explores the mathematical model, improving both aspects of the problem.

The article is organized as follows: Section 2 presents a bibliographic review for the problem. Section 3 presents the four matheuristics for M-LSVRP and Section 4 discusses the results obtained on a set of 200 randomly generated instances. Conclusions and perspectives are presented in Section 5.

2. Bibliographic Review

Since 70 years the integration of production and transportation was seriously taken into account [7]. Thus, one of the frequent questions about such integrated models is the time scale for production and transport. Hahm and Yano [8] integrate production and deliveries in the Economic Lot and Delivery Scheduling Problem (ELDSP). They showed that the time scale of both problems can be adjusted by a multiplication factor.

Chandra and Fisher [4] consider a scenario where a site produces a number of items and stores the finished ones to be delivered by a fleet of vehicles. The authors consider the production as a CLSP and the routing as multi-period local deliveries. A decomposition, where each problem is solved by known heuristics, allows a reduction between 3% and 20% of the operating costs.

Recent methods can be found in Bard and Nananukul [2], that uses a solve a reactive tabu search with a path relinking to solve a single-item production problem in a single site to satisfy the demands of a set of customers, and in Boudia *et al.* [3] that presents a memetic algorithm with population management (MA|PM) and compares it with previous work using GRASP and path relinking methods. It solves the problem in one stage and it is shown to produce the best results, with a higher computing time yet.

3. Heuristics

M-LSVRP is a NP-hard problem since it combines several located lot-sizing problems, which are also NP-Hard. It can be modeled using three sets of variables: a first set dedicated to the transportation problem, a second set for the production problem on each site, and a third set for the input/output constraints that define the interface of each site. Then the proposed four matheuristics are directly based on the MIP model structure. They mainly relax arc decision variables on the transportation block as well as setup decision variables on the production block.

3.1. *Relax-and-Fix Based Methods*

Relax-and-Fix (RF) heuristics are well-known in production scheduling problems. It was first proposed by [5] to solve the resource allocation problem for production planning and scheduling with period overlapping setups. Later, some more variations were proposed, as in [6].

Our heuristics first compute the minimal number of periods to satisfy the first demand. Then they iterate using parameters as the quantity of decision variables in a subset expressed as a percentage of the planning horizon, the type of variables like setup variables (y) and/or the arc decision variables (g), the execution time at each iteration, and the overlap defined as the number of periods whose variables are recalculated in the next iteration.

Our first RF heuristic follows the classic method: in the first iteration, part of decision variables are treated as integers and the remaining is relaxed. After solving the first problem, the variables once treated as integers are fixed to their values and part of the relaxed variables becomes integers to set a new problem. The method follows relaxing and fixing variables until all the relaxed variables have been fixed, thus providing the heuristic solution. If period overlap is considered, only a part of the integers variables will be fixed at each iteration, the other part will be recalculated.

The second method is called *Insert-and-fix* (IF). It considers the subsets once treated as relaxed as "inactive". At each iteration, the size of the problem increases with the insertion of a new subset. The subsets whose variables were considered are fixed (or overlapped). IF takes advantage of the available time. Special attention must be paid to the process since dropping some variables from the original problem may lead to infeasibility. Thus, upon infeasibility, new subests are inserted into the partial problem and the computation is performed again. If infeasibility happens at the end of the process, IF is said to fail solving the problem.

The last RF heuristic is called *Fractional Relax-and-fix* (FRF). It combines both previous methods, keeping their good features: a limited comptutional time (IF) and the advantage provided by the linear relaxation (RF). In the first iteration is considered a problem with integers and real subsets. Between the 2nd and the n-2th iterations FRF works with fixed, integer and real variables, and some variables from the original problem will be inactive. The n-1th iteration is similar to the previous except that all variables will be active. In the last iteration, an integer problem is solved.

Figure 1 shows the general scheme of each heuristic.

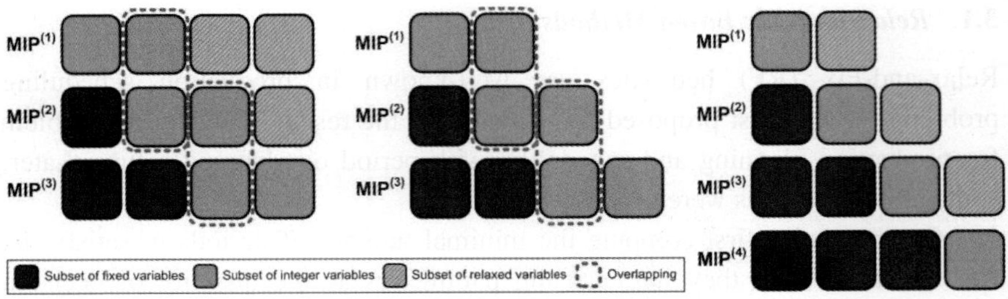

Figure 1(a). Relax-and-Fix. Figure 1(b). Insert-and-Fix. Figure 1(c). Fractional RF.

Figure 1. Evolution of the RF methods.

3.2. *Dive-and-Fix (DF)*

The *Dive-and-fix* heuristic (also known as *Diving*) is a strategy that explores the branch-and-bound tree, performing variables rounding and fixing according to some rules. This process allows simplifying the problem.

Our method uses a time limit per iteration and one of the two decision rules to round and fix variables: a tolerance that decreases if no variable has been fixed recently, and a given interval [a, b] in which the considered variables can be rounded and fixed. Values for a and b are the two bigger fractional values in an iteration. Once the production decision is a dominant factor in the whole problem, fixing variables to 0 may easily lead to infeasibility. So, the variables are fixed only to 1.

4. Computational Results

Our set of problems is composed by 200 randomly generated instances. They were built according to the ranges defined by Armentano and Toledo [1] for production costs, and adding values for capacity (100 units), consummation of components, capacity and setup times (1 to 3 units), fixed transportation cost (5 to 95), unit transportation cost (12 to 45), fixed vehicle cost (1000) and travel duration (1 to 3 periods).

The tests were made on a machine with an Intel Core 2 Quad processor and 4GB of RAM. The code was done in C++ on Windows XP. The instances are divided into 20 groups. Each one corresponds to 3 to 6 sites, with the maximum of 2 items non-final each one, and to 12, 24, 36, 48 or 60 time periods.

The CPU time per RF iteration has a significant impact on the overall performance. It means that giving more time to Cplex increases the chances to get a better solution for the current iteration. The fixing is more adequate wrt. the relaxation. On the other hand, the percent of time horizon may seem secondary: it would be natural to select a large value, but the amount of time required for Cplex would then quickly exceed its time limit. Thus the good strategy is to select a reasonable CPU time per iteration, then to identify the corresponding good settings for the percent of time horizon considered. Those parameters are set here in 450s and 25% of integer variables per iteration.

Table 1 shows the value founded by the IBM Cplex solver on the model in a limit of 18000s (3h), and the results from the four heuristics. For each group (sites × periods) of instance, the average mean gap and CPU time are reported.

Table 1. Results.

	Relax-and-Fix		Insert-and-Fix		Fractional RF		Dive-and-Fix	
Instance	gap	time	gap	time	gap	time	gap	time
3×12	2.09	2.80	4.46	1.77	2.45	2.41	0.24	2.00
3×24	0.31	0.47	1.87	0.22	1.64	0.41	0.38	0.61
3×36	0.06	0.12	0.28	0.03	0.46	0.08	0.34	0.16
3×48	0.07	0.11	0.09	0.03	0.26	0.07	0.52	0.10
3×60	0.52	0.11	0.36	0.07	0.18	0.09	2.72	0.14
4×12	7.01	1.18	1.45	0.89	0.76	1.05	0.67	0.91
4×24	0.91	0.38	3.03	0.51	3.84	0.66	0.37	0.73
4×36	0.49	0.12	0.73	0.06	1.37	0.07	4.31	0.32
4×48	1.17	0.07	1.10	0.05	1.97	0.07	4.01	0.23
4×60	1.87	0.08	0.73	0.08	1.81	0.09	-	0.29
5×12	6.91	0.83	0.40	0.72	0.61	0.53	2.59	0.30
5×24	2.42	0.09	5.55	0.04	5.08	0.05	20.22	0.16
5×36	1.56	0.09	2.43	0.05	2.91	0.07	-	0.21
5×48	1.70	0.09	2.91	0.07	7.07	0.09	-	0.07
5×60	4.16	0.08	4.96	0.06	6.92	0.07	-	0.04
6×12	6.70	0.05	4.84	0.03	7.42	0.03	8.49	0.05
6×24	7.99	0.05	9.01	0.02	10.01	0.03	-	0.07
6×36	6.49	0.06	6.23	0.04	8.99	0.06	-	0.05
6×48	7.88	0.08	6.76	0.07	13.96	0.08	-	0.02
6×60	9.35	0.09	6.15	0.06	31.75	0.07	-	0.02

Clearly RF is globally dominant in terms of solution quality. However all methods behave differently with respect to the size of the problem. Insert-and-Fix is well suitable for the larges problems. Fractional Relax-and-Fix shows the equilibrium between RF and IF for most of the problems. Dive-and-Fix is well suitable for small and medium size problems but it cannot solve large instances.

5. Conclusions and Perspectives

In this paper several MIP-based heuristics were presented to solve the integrated multi-item, multi-level, lot-sizing and vehicle routing problem (M-LSVRP). They are able to find good solutions in a reasonable amount of time.

Each heuristic shows a particular behavior depending on the size of the problem. Those behaviors provide some information to understand the details of the problem and then improve the solution methods. There are many possibilities for future works once the model is very flexible.

References

1. V. A. Armentano and F. M. B. Toledo, *A Lagrangean-based Heuristic for the Capacitated Lot-sizing Problem in Parallel Machines*, **European Journal of Operation Research**, 1070 (2006).
2. J. F. Bardand and N. Nananukul, *The Integrated Production-Inventory-Distribution-Routing Problem*, **Journal of Scheduling**, 257 (2009).
3. M. Boudia and C. Prins, *A Memetic Algorithm with Dynamic Population Management for an Integrated Production-distribution Problem*, **European Journal of Operational Research**, 703 (2009).
4. P. Chandra and M. L. Fisher, *Coordination of Production and Distribution Planning*, **European Journal of Operational Research**, 503 (1994).
5. C. Dillenberger, L. F. Escudero, A. Wollensak and W. Zhang, *On Practical Resource Allocation for Production Planning and Scheduling with Period Overlapping Setups*, **European Journal of Operational Research**, 275 (1994).
6. A. Federgruen, J. Meissner and M. Tzur, *Progressive Interval Heuristics for Multi-Item Capacitated Lot Sizing Problems*, **Operations Research**, 490 (2007).
7. F. Glover, G. Jones, D. Karney, D. Klingman and J. Mote, *An Integrated Production, Distribution, and Inventory Planning System*, **Interfaces**, 21 (1979).
8. J. Hahm and C. A. Yano, *The Economic Lot and Delivery Scheduling Problem: the Single Item Case*, **International Journal of Production Economics**, 235 (1992).
9. Y. Pochet and L. A. Wolsey, *Production Planning by Mixed Integer Programming*, **Springer**, (2006).

SCHEDULING JOB AND PREVENTIVE MAINTENANCE TO OBTAIN TRADE-OFF BETWEEN CMAX AND AVAILABILITY OF 2-LEVEL FLOW-SHOP

F. HNAIEN* AND F. YALAOUI

ICD-LOSI, University of Technology of Troyes, UMR STMR-CNRS.
12 rue Marie Curie, 10010 Troyes, France
(e-mail: {faicel.hnaien, farouk.yalaoui}@utt.fr).

We consider simultaneously the scheduling problem and the preventive maintenance tasks of a two-level permutation Flow-Shop. Our objective is to optimize simultaneously both criteria. For this we have chosen to study the minimization of *Cmax* to the scheduling problem and minimizing the unavailability of system (flow-shop) for maintenance problem. We consider the F_2 //(*Cmax* , unavailability) case. For this, we present some properties of the optimal solution based on Johnson algorithm.

1. Introduction

The majority of scheduling studies considers the deterministic environment and assumes the total availability of resources. However in reality the human and material resources may be unavailable for various reasons. The majority of the literature with maintenance scheduling adopts this approach often called "scheduling with machine availability constraints. In the case of preemptive tasks, [1] studied the case of Flow-Shop with two machines minimizing *Cmax* objective function. The two machines each undergo a single period of unavailability. The author has shown that the problem is NP-hard in the weak sense whatever the machine affected by the unavailability. To our knowledge there is only [2] which looked at the issue more than two machines and the majority of studies are restricted to the case of two machines given the difficulty of the problem. In [3], the author have studied the case of permutation Flow-Shop where periods of maintenance must be performed in a predefined interval. This study has the advantage of reducing the conflict between the two services, but they promote the maintenance, to the detriment of production. Recently [4] have studied a model of bi-objective problem of parallel machines using reliability models to take into consideration the maintenance aspect. The two criteria are the minimization of downtime of machines and minimization $Cmax$. For the standard of maintenance is the unavailability of the whole system is optimized to be a threshold of reliability for each machine. Two genetic algorithms were developed to obtain approximation of optimal Pareto.

2. Problem Formulation

This section describes separately the scheduling problem and the problem of planning preventive maintenance. We assume that tasks are available at the beginning of production and preemptions are not allowed. We focus on preventive maintenance. The maintenance problem is to determine maintenance schedules for each machine with minimal downtime. The availability $A_s(t)$ of the Flow-Shop is as follows (1):

$$A_s(t) = \prod_{i=1}^{2} A_i(t) \tag{1}$$

where: $A_i(t)$: is the availability of a machine M_i ($i = 1, 2$).

Here, we consider only those machines whose failure rates and repair rates are constant. In other words, we assume that the downtime (resp. repair time) of a machine M_i is represented by a probability distribution to exponential distribution with parameter failure rate λ_i (resp. the rate parameter repair μ_i. We also assume that the preventive maintenance actions are used to restore the machine to "as good as new". Taking into account these assumptions, from the first instant $t = 0$, the availability of a machine M_i at time t is given by the following expression [5,6]:

$$A_i(t) = \frac{\mu_i}{\lambda_i + \mu_i} + \frac{\lambda_i}{\lambda_i + \mu_i} E(t) \tag{2}$$

where: $E(t) = exp^{(-(\lambda_i + \mu_i)(t - pm))}$ pm: the completion time of the last maintenance action on M_i.

The two objectives to minimize are the following (3) and (4):

Minimize the makespan:

$$F1 = Min \quad Cmax \tag{3}$$

Minimize the system unavailability:

$$F2 = Min \max_{t \in T}(\overline{A}_s(t)) \tag{4}$$

Where :

$c_{i,j}$	The completion time of each task i $(i=1,2,...,n$ on M_j $(j=1,2)$
$Cmax = c_{n,2}$	The maximum completion time of all tasks
$t_{i,j}$	The start time of preventive maintenance tasks on M_j $(j=1,2)$
$T_j = \{t_{i,j}\}$	The set of start time of preventive maintenance tasks on M_j $(j=1,2)$
$T = \{T_j\}$	The set of all start time of preventive maintenance tasks
$\bar{A}_s(t) = 1 - A_s(t)$	Unavailability of system s

Thus, two decisions must be taken simultaneously. The first decision is to find the best assignment of 2 tasks on 2 machines to minimize $Cmax$. The other decision is to decide when to perform preventive maintenance actions to minimize downtime. The two objective functions are generally not the same optimal solution (in fact they are contradictory). Minimize in this context means identifying a set of trade-off solutions.

2.1. *First Objective*

We consider only the non-resumable case $(nr(M_j), j=1,2)$ We will give some properties of the optimal solution for the minimizing non-resumable flowshop case with 2 machine and 2 jobs: $F_2 / nr(M_1, M_2) / C_{max}$. Let the following notations:

$p_{i,j}$	The processing time of job i on machine M_j $(j=1,2)$
$J_{i,j}$	The job i on machine M_j $(j=1,2)$
PM_j	The preventive maintenance job on machine M_j $(j=1,2)$
$g_j = \dfrac{1}{\mu_j}$	The duration of preventive maintenance job on machine M_j $(j=1,2)$
$\Delta g = g_2 - g_1$	The difference between the duration of preventive maintenance jobs
C_{max}^{JO}	the completion time obtained by johnson algorithm

In this section we will consider only 2 preventive maintenance tasks (PM_j) and 2 production job $(J_{i,j})$ $(i, j = 1,2)$. Each task of maintenance start after the first production job (see Fig.1).

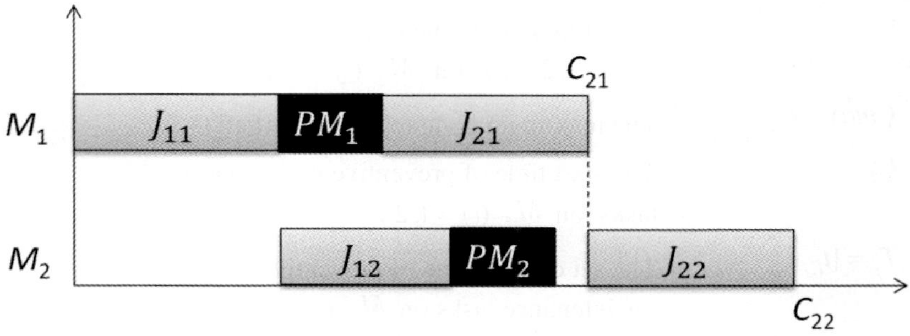

Figure 1. Example of $F_2 / nr(M_1, M_2)), 2\, jobs / (C_{max}, unvavailability)$

Proposition 1. Johnson's Algorithm is optimal for $F_2 / nr(M_1, M_2), 2\, jobs //$ C_{max} under modified processing time $p'_{i2} = p_{i2} + \Delta g,: \forall i = 1,2$ job i precedes job j if $\min(p'_{j2}, p_{i1}) < \min(p'_{i2}, p_{j1})$

Proof. If we place the production job 1 before production job 2, we will note in the following $F1$ by $F1^{1-2}$. Otherwise, If we place the production job 2 before production job i, we will note $F1$ by $F1^{2-1}$.

$$F1^{1-2} = p_{22} + p_{11} + \max(g_1 + p_{21}, g_2 + p_{12})$$
$$F1^{2-1} = p_{12} + p_{21} + \max(g_1 + p_{11}, g_2 + p_{22})$$

Let $\Delta F1 = F1^{2-1} - F1^{1-2}$

$$\begin{aligned} \Delta F1 &= p_{12} + p_{21} + \max(g_1 + p_{11}, g_2 + p_{22}) - p_{22} - p_{11} - \max(g_1 + p_{21}, g_2 + p_{12}) \\ &= \max(g_1 - p_{22}, g_2 - p_{11})) - \max(g_1 - p_{12}, g_2 - p_{21}) \\ &= \min(p_{12} - g_1, p_{21} - g_2) - \min(p_{22} - g_1, p_{11} - g_2) \\ &= \min(p_{12} - g_1, p_{21} - g_2) + g_2 - (\min(p_{22} - g_1, p_{11} - g_2) + g_2) \\ &= \min(p_{12} + \Delta g, p_{21}) - \min(p_{22} + \Delta g, p_{11}) \end{aligned}$$

Thus: $\Delta F1 < 0$ if and only if: $\min(p_{12} + \Delta g, p_{21}) - \min(p_{22} + \Delta g, p_{11}) < 0$

Finally, job i precedes job j if $\min(p_{j2} + \Delta g, p_{i1}) < \min(p_{i2} + \Delta g, p_{j1})$

Corollary 1.1. If $g_1 = g_2 = g$ and $p_{ij} \neq p_{kj}, \forall i \neq k, i, k, j = 1,2$. Johnson's Algorithm is optimal for $F_2 / nr(M_1, M_2), 2\, jobs) // C_{max}$

Proof. From proposition 1, $\Delta F_1 < 0$ if and only if:

$$\min(p_{12} + \Delta g, p_{21}) - \min(p_{22} + \Delta g, p_{11}) < 0$$

If $\Delta g = 0$ $\Delta F_1 < 0$ if and only if: $\min(p_{21}, p_{12}) < \min(p_{22}, p_{11})$ Thus, Johnson's Algorithm is optimal for $F_2 / nr(M_1, M_2), 2\, jobs) // C_{max}$ under equal preventive maintenance durations.

2.2. *Second Objective*

For the second objective $F2 = Min\ \max_{t \in T}(\overline{A}_s(t))$ For this, we will explain the calculational of the the unavailability with an example of 2 jobs and where the job production 1 is placed before job 2: $A_1^{1(1-2)}$, $A_1^{2(1-2)}$, $A_1^{3(1-2)}$, $A_2^{1(1-2)}$, $A_2^{2(1-2)}$ and $A_2^{3(1-2)}$, and so we calculate $F2^{1-2}$ (see fig.2).

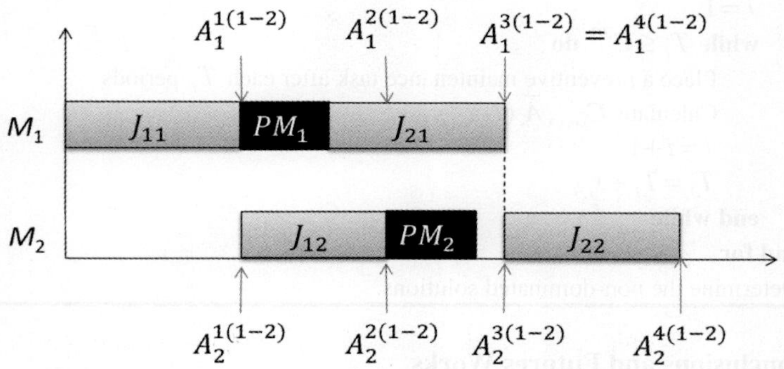

Figure 2. Example of unavailability calculation

Let: $\gamma_i = \dfrac{\mu_i}{\lambda_i + \mu_i}, \forall i = 1,2$

$$A_1^{1(1-2)} = \gamma_1 + (1-\gamma_1) \times exp^{-(\lambda_1+\mu_1)p_{11}}$$
$$A_2^{1(1-2)} = \gamma_2 + (1-\gamma_2) = 1$$
$$A_1^{2(1-2)} = \gamma_1 + (1-\gamma_1) \times exp^{-(\lambda_1+\mu_1)A1}$$
$$A_2^{2(1-2)} = \gamma_2 + (1-\gamma_2) \times exp^{-(\lambda_2+\mu_2)p_{12}}$$
$$A_1^{3(1-2)} = \gamma_1 + (1-\gamma_1) \times exp^{-(\lambda_1+\mu_1)p_{21}}$$
$$A_2^{3(1-2)} = \gamma_2 + (1-\gamma_2) \times exp^{-(\lambda_2+\mu_2)A2}$$
$$A_1^{4(1-2)} = \gamma_1 + (1-\gamma_1) \times exp^{-(\lambda_1+\mu_1)p_{21}}$$
$$A_2^{4(1-2)} = \gamma_2 + (1-\gamma_2) \times exp^{-(\lambda_2+\mu_2)p_{22}}$$

Where:

$$A1 = p_{12} - g_1 \quad if \quad p_{12} > g_1 \quad and \quad p_{11} \quad otherwise$$
$$A2 = p_{12} \quad if \quad p_{21} + g_1 < p_{12} + g_2 \quad and \quad 0 \quad otherwise$$

2.3. *Bi-Objective Optimization*

We will present now a bi-objective heuristic *BJA* based on the the previous properties (see Algorithm 1).

Algorithm 2.1. *BJA*: Bi-objective Johnson algorithm under unavailability

Scheduling the n production tasks using Johnson algorithm
$S_j = \{s_{i,j}, i = 1,...,n; j = 1,...,m\}$ The set of the job sorted on machine j in the increasing order of $p_{i.i}$
$T_i = \max(p_{i.i}), \forall j = 1,...,m$: Preventive maintenance frequency on machine j
for $j = 1,...,m$ **do**
    $i = 1$
    while $T_i \leq C_n^{JO}$ **do**
        Place a preventive maintenance task after each T_i periods
        Calculate $C_{max}, \bar{A}_s(T_i)$
        $i = i + 1$
        $T_i = T_i + s_{i.i}$
    end while
end for
Determine the non-dominated solutions

3. Conclusions and Futures Works

We are interested in the joint optimization of scheduling and maintenance for the Flow-Shop case. We proposed the F_2 //($Cmax$, unavailability) scheduling problem. The bi-objective optimization is proposed to find a compromise for the service scheduling and service maintenance. Our decision variables are the order of tasks on production machines and the date of preventive maintenance for each machine. For this problem, we have established some properties of the optimal solution for the particular case of two production jobs and two preventive maintenance tasks. Our future research will be focus to implement the BJA and bi-objective metaheuristics like NSGA-II algorithm.

References

1. C. Lee, Operations Research Letters 20, 129 (1997).
2. R. Aggoune, Minimizing the makespan for the ow shop scheduling problem with availability constraints (In Operational Research Peripatetic Post-Graduate Programme, ORP3, Paris, 2001).
3. J. Kaabi, C. Varnier and N. Zerhouni, APII-JESA (French journal) 37, 641 (2003).
4. A. Berrichi, L. Amodeo, F. Yalaoui, E. Ch^atelet and E.Mezghiche, Computers & Operations Research 37, 1584 (2010).
5. C. Ebeling, An introduction to reliability and maintainability engineering (McGraw-Hill, USA, 1997).
6. A. Villemeur, Reliability, availability, maintainability and safety assessment (Wiley, USA, 1991).

A MULTI-OBJECTIVE METHOD TO SOLVE A CONTAINER TERMINAL PROBLEM

FARAH BELMECHERI-YALAOUI[†]

Paris Descartes university
143 avenue de versaille, 75016, Paris (France)

FAROUK YALAOUI AND LIONEL AMODEO

Charles Delaunay Institute, University of Technology of Troyes
12 rue marie curie, 10010, Troyes (France)

The river and maritime transport represents an attractive alternative to land and air transport. The containerization allows the industries to save costs thanks to the standardization of dimensions. The container terminal has to manage container traffic at the crossroads of land road, railway. In this paper, we propose to optimize, simultaneously, the storage problem and the quayside transport problem. In a space storage, it exists several blocks and each one has its storage cost. The first aim is to minimize the cost storage of containers. These latter are loaded into vessels, the vehicles have to transport the containers from blocks to quays (of vessels). So, the second aim consists to minimize the distance between the space storage and the quays. The optimization methods of operations research in container terminal have become more and more important in recent years. Objective methods are necessary to support decisions. To solve this multiobjective problem, we develop a resolution method which is a metaheuritic approach called multiobjective ant colony optimization. The second resolution method is the multiobjective ant colony optimization with a local search.

1. Introduction

One of the most used transportations is the river and maritime transport. It represents an attractive alternative to land and air transport. Indeed, several industries (companies) try different opportunities for reducing costs of this transport. First solution adopted is the containerization which became widespread to river and maritime transport of goods. This solution allows the industries to save costs thanks to the standardization of dimensions, the security of goods, and the fast handling operations for loading/unloading. However, container can induce to high investment costs in storage. The seaport has to

[†] farah.yalaoui@gmail.com.

manage container traffic at the crossroads of land road, railway…etc. It has also a role for managing: a stock of containers on its site, and a platform of quaysides. These managements solve a set of decision problems, and the aims are to reduce the costs of transport. The objective methods are necessary to support decisions. Concerning the storage allocation problem, containers are stacked in several levels and the whole storage area is separated into blocks. Kim and Park [4] propose two heuristic algorithms to outbound containers arriving at a storage yard and a sub-gradient optimization technique. The storage space allocation problem is also studied by Zhang et al [6]. For the transport problem, the authors Kim and Kim [3] minimize the total travel distance of straddle carriers in the yard. They propose a beam search algorithm. Belmecheri et al. [1] have also proposed a model to optimize the empty containers in the container terminals.

2. Problem description

In this study, we assume only one planning horizon. We consider during a certain number of days, some containers of customers (industries) are received by the container terminal (maritime port), then they are placed in the storage space then they are loaded in vessels.

The objectives are:

- The minimization of the cost storage,
- The minimization of total distance traveled by vehicles between blocks storage and quays of vessels.

3. Resolution methods

To solve this multiobjective problem, we develop a resolution methods based on metaheuritic approach called Ant colony algorithm.

3.1. *MultiObjective Ant Colony Optimization-MOACO*

The MultiObjective Ant Colony Optimization (MOACO) has proved its effectiveness in several works. MOACO is even used in combinatorial problems as the works of Dugardin et al [2]. They exist several versions of ant based algorithms developed to solve combinatorial problems. Here, we utilize the Ant Colony System. The multiobjective optimization takes different criterion simultaneously. So, the number of pheromone matrices is equal to the number of objectives to be optimized. Therefore, two pheromone matrices are used in our works.

3.1.1. *Solution construction for storage cost optimization*

The first colony called ACO1 (Ant Colony Optimization 1) affects the containers in the space storage. Each ant is deposited randomly on a point. This point represents the demand of one customer which is a set of containers. This set is selected to be assigned for one block of storage. The points to be visited are chosen based on the application of the state transition rules. An ant a chooses a customer i for storing its containers co_i to block b (referred arc is (i, b)).

3.1.2. *Solution construction for distance optimization*

When the containers are assigned to the storage blocks, we apply the second colony called ACO2 (Ant Colony Optimization 2) to optimize the distance traveled by vehicles for transporting these containers to quays. Each demand of one customer is chosen randomly to affect its containers to quay q (here, the referred arc is (i, q)).

3.2. *The proposed algorithm: MOACO*

The algorithm MOACO for optimizing the cost of storage and distance simultaneously is given in MOACO algorithm (figure 1). The process is applied T and A times which are respectively the number of iterations and ants. Each ant a constructs one solution Sol^1_a for storing the containers of customers I in blocks B. Then, this solution is evaluated ($Cost^1_a$). The second task of ant a concerns the construction of one solution Sol^2_a for transporting the containers from blocks to quays Q, then the $Cost^2_a$ is computed. After that, we compute the complete solution $Cost_a$ which is a vector with coordinates $(Cost^1_a, Cost^2_a)$. When all ants A construct the solutions at iteration t, each solution can update the new Pareto front F which is the set of non dominated solutions $F = \{F_1, F_2, F_r, F_R\}$. The F_r, which is initially a non dominated solution, is composed by coordinate x (y), where $F_r.x = Cost^1_a$ $(F_r.y = Cost^2_a)$.

```
Data: B, I, Q
FOR {t ∈ T}
    FOR {a ∈ A}
        //To build a complete solution;
        Part1
        Construct storage solution with ACO1 procedure Sol^1_a;
        Evaluate storage solution Cost^1_a;
        Part2
        Construct distance solution with ACO2 procedure Sol^2_a;
        Evaluate distance solution Cost^2;
        Evaluate complete solution Cost_a;
        ENDFOR
        Construct the non dominated solutions of ants;
        Update the different fronts F_r of colony;
        Update the fronts with current and preceding iterations for the new best front;
        Update pheromone for storage solutions;
        Update pheromone for distance solutions;
        Return the best Pareto front;
ENDFOR
```

Figure 1. MOACO algorithm

3.3. *Pheromone update (storage and distance part)*

When ants have built solutions (stored and transported containers), the pheromone update process is applied to find the best Pareto front. For the storage part (distance part), the solutions having arcs *(i, b)* or *(i, q)* in the Pareto front are updated in increasing or decreasing the amount of pheromone quantity.

3.4. *MOACO-LS*

The second proposed approach is MOACO with a local search, called MOACO-LS. This one is applied on non dominated solutions in each iteration t. The moves consists to relocate a set of containers of customer i with those of customer j in checking the capacity of block and quay containing co_i and the block and quay having co_j. The aim is to verify the behavior of MOACO when the local search is added.

4. Computational results

The computational experiments are carried out in order to evaluate the performances of the developed algorithms. The problems are generated randomly for size I equal to 10, where we consider 5 blocks and 4 quays. When the 10 problems are created, we apply the MOACO and MOACO-LS. In order to compare the two algorithms, several tests are made to evaluate the good values of different parameters. Final values are determined as a compromise between the quality of the final solutions and the convergence time needed. The stopping criterion for the algorithms is a given number of generations and is fixed at 500.

Instances	NbrCo	CapB	CapQ	nF1	nF2	μ	C(F2, F1)	C(F1, F2)
Pb10-1	66	78	104	4	3	-2.23	0.5	0
Pb10-2	54	81	85	1	2	4.2	1	0
Pb10-3	56	89	86	2	3	-5.01	0	0.33
Pb10-4	48	81	99	5	5	-2.63	0.2	0.6
Pb10-5	48	84	92	3	2	2.19	0	0
Pb10-6	55	88	93	4	2	-3.03	0	0.5
Pb10-7	61	72	75	4	6	-9.12	0.75	0.17
Pb10-8	65	77	86	1	2	1.49	0	0
Pb10-9	54	79	88	2	4	3.32	0	0.5
Pb10-10	59	69	78	3	3	-6.41	0	0.67
Mean				2.90	3.20	-1.72	0.25	0.28

Table 1. Comparison between MOACO (F1) and MOACO-LS (F2) algorithms

Table 1 shows a comparison between the best fronts with non dominated solutions obtained by each algorithm. Three comparison criteria are used: the number of solutions *nF* in an optimal front *F*, the distance proposed by Riise [5] μ , the Zitlzler [7] measure *C*. *nF1* is the number of solutions in the optimal front obtained by the MOACO and *nF2* for the MOACO-LS. The distance of Riise is presented by μ, *C(F1, F2)* and *C(F2, F1)* stand for the Zitzler measure [7]. *C(F2, F1)* is a measure for front *F1* of the MOACO and *C(F1, F2)* for that of front *F2* of the MOACO-LS. To understand this table, we explain the lecture of the first line. The non dominated solutions (*nF1*) in the MOACO Pareto front are equal to 4 and 3 for MOACO-LS. The negative value of μ (-2.23) shows that MOACO-LS front is under the MOACO front. In all problems, the mean of *nF1* is 2.90 which is less than *nF2*. We can see also that, in several cases, the non dominated solutions of front *F1* are improved by *F2*.

5. Conclusion

We have studied a mutliobjective container terminal problem with an ant colony optimization. The two objectives are the minimization of cost storage and the minimization of distance transport. The multiobjective metaheuristics developed are a multiobjective ant colony optimization algorithm (MOACO) and a multiobjective ant colony optimization algorithm with local search (MOACO-LS). The tests are applied on several instances. We have compared the results obtained by the two approaches with a complete enumeration. We have noticed that the MOACO-LS is better than MOACO. Many different metaheuristic approaches could be developed to solve the problem studied such as the particle swarm intelligent, the genetic algorithm. Each approach has a specific behavior to explore the research space of solutions. In our future works, new methods will be developed to compare our results.

References

1. F. Belmecheri, T. Cagniart, L. Amodeo and F. Yalaoui, Modelling and optimization of empty container reuse: a real case study, in IEEE: proceedings of the International Conference of Computers and Industrial Engineering, (Troyes-France, 2009).
2. F. Dugardin, F. Yalaoui and L. Amodeo. New efficient methods to solve reentrant hybrid flow shop scheduling problem. European Journal of Operational Research 203, 22-31 (2010).
3. K. H. Kim and K. Y. Kim. Routing straddle carriers for the loading operation of containers using a beam search algorithm. Computers and Industrial Engineering 36, 106-136 (1999).
4. K. H. Kim and K. T. Park, A note on a dynamic space-allocation method for outbound containers. European Journal of Operational Research 148, 92-101 (2003).
5. A. Riise. Comparing genetic algorithms and tabu search for multiobjective optimization. In IFORS conference, (Edinburgh-UK, 2002).
6. 6. C. Zhang, J. Liu, Y. W. Wan, K. G. Murty and R. J. Linn. Storage space allocation in container terminal. Transportation Research-B 37, 883-903 (2003).
7. E. Zitzler and L. Thiele. Multiobjective evolutionary algorithms: A comparative case study and the strength pareto approach. IEE transaction on evolutionary computation 3, 257-271 (1999).

FUZZY CHANCE CONSTRAINED PROGRAMMING FOR WAREHOUSE LOCATION PROBLEM WITH IMPRECISE CUSTOMER DEMAND

CEREN BUKET KAYI, ERHAN BOZDAG
Industrial Engineering Department, ITU, Macka
Istanbul, 34376, Turkey

In this paper we applied fuzzy chance constrained programming with uncertain parameters to warehouse location problem. The problem studied in the paper has uncertain demand which necessitates the utilization of fuzzy numbers for representing the related demands. Fuzzy values being embedded in the objective function, in turn compels the implementation of credibility measure in the proposed model. The model is a hybrid intelligent algorithm integrating fuzzy simulation and genetic algorithm. Lastly, a numerical example is illustrated in order to demonstrate the validity of the algorithm.

1. Introduction

Given a set of m warehouses and n stores, the incapacitated warehouse location problem consists of choosing a subset of warehouses that minimizes the fixed costs of the warehouses and the transportation costs from the warehouses to the stores [1]. Also determining the location of the distribution center(s) is one of the most crucial elements in the success and efficiency of the distribution network [2]. The warehouse location problem has been solved with linear-programming, branch and bound, genetic algorithms, and local search methods in the literature. For example, Michel and Hentenryck [1] developed a simple and efficient tabu search algorithm which finds optimal warehouse location very quickly and with very high frequencies in deterministic conditions. Since deterministic conditions are very seldom in real world, methods that cope with uncertainties have been investigated in the literature. There are two types of warehouse location problems with uncertainty. The first category involves problems that are characterized by random variables and include a number of uncertain parameters such as customers' demand and operating costs. On the contrary, the second category of warehouse location problems involves uncertain parameters that are characterized by imprecise or vague information, and are dealt with fuzzy sets theory. Wem and Kang [3] presented the expected cost minimization model and chance maximization model with random fuzzy demands. Wang and Watada [4] proposed a facility location model with fuzzy random parameters and a swarm intelligence approach.

In the aforementioned first category of warehouse location problem, stochastic programming offers various means of considering the objectives and constraints with stochastic parameters. Chance constrained programming models stochastic decision systems with the assumptions that the constraints will hold at least α of the time, where α is referred as the confidence level. Fuzzy programming is mathematical programming with fuzzy parameters and is related the second category of the warehouse location problems [2]. This paper presents a fuzzy chance constrained programming model based on credibility measure for the warehouse location selection with fuzzy demands.

2. Methodology of the Proposed Model

Proposed model is a hybrid intelligent algorithm integrating fuzzy simulation and genetic algorithm. Before detailed explanation of the methodology, a brief description of the subjected facility location problem will be given.

A company with n stores (s_i) requires to decide the location of m warehouses (w_j), aiming the minimum distance from the stores, taking into consideration their fuzzy demands (ξ_i). The distance function given in the Eq. (1) is used to form the distance-demand function for i_{th} store, which is given in the Eq. (2). The overall distance-demand function is given in the Eq. (3).

$$d_{ji}(w_j, s_i) = \sqrt{(x_i - x_j)^2 + (y_i - y_j)^2} \quad i = 1, \dots n \quad j = 1, \dots m \tag{1}$$

$$f_i(w_1, \dots w_m, \xi_i) = \min(d_{1i}(w_1, s_i), \dots d_{mi}(w_m, s_i)) * \xi_i \quad \forall i \tag{2}$$

$$F(w, \xi) = \sum_{i=1}^{n} f_i(w_1, \dots w_m, \xi_i) \tag{3}$$

2.1. *Fuzzy Simulation*

Objective function of the described problem includes fuzzy parameters ξ_i with membership functions $\mu_i(u_i)$. Therefore it is needed to redesign problem with respect to credibility measure given in Eq.(6), which is the average of the possibility and necessity measures given in Eq.(4) and Eq.(5) respectively.

$$Pos\{\xi \geq r\} = \sup_{u \geq r} \mu(u) \tag{4}$$

$$Nec\{\xi \geq r\} = 1 - \sup_{u < r} \mu(u) \tag{5}$$

$$Cr\{\xi \geq r\} = \frac{1}{2}[Pos\{\xi \geq r\} + Nec\{\xi \geq r\}] \tag{6}$$

New model constructed with the credibility measure is given as following Eq.(7) where the F_0 is the distance-demand value, defined by decision maker. Model maximizes credibility measure where distance-demand value is less than F_0.

$$\max Cr\{F(w,\xi) \leq F_0\}$$

subject to:

$$x_j \geq \min_{1\leq i\leq n}(x_i), for\ 1 \leq j \leq m \tag{7}$$

$$x_j \leq \max_{1\leq i\leq n}(x_i), for\ 1 \leq j \leq m$$

The following formula is suggested as the fuzzy approximation of the credibility measure by Shao and Ji [5].

$$L = \frac{1}{2}(\max_{1\leq k\leq N}\{\mu(u_k)|F(w,u_k) \leq F_0\} + \min_{1\leq k\leq N}\{1-\mu(u_k)|F(w,u_k) > F_0\}) \tag{8}$$

Following fuzzy simulation algorithm is generated to calculate the approximate value for the credibility measure [5].

Step1. k=1, Possibility=0, Necessity=1.

Step2. Randomly generate the demand for n stores u_{ik} from fuzzy variables ξ_i. Take the infimum of their membership value as following and call *Infimum.*

$$\mu(u_k) = \mu_1(u_{1k}) \wedge \mu_2(u_{2k}) \wedge \ldots \wedge \mu_n(u_{nk}) \tag{9}$$

Step3. Calculate $F(w,u_k)$. If $F(w,u_k) \geq F_0$, then go to Step4, else go to Step5.

Step4. If Infimum>Possibility, then Possibility=Infimum. Go to Step6.

Step5. If (1-Infimum)<Necessity, then Necessity=1-Infimum. Go to Step6.

Step6. k=k+1. If k<N, go to Step2, else go to Step7.

Step7. $L = \frac{1}{2}(Possibility + Necessity)$. Return L.

2.2. *Genetic Algorithm*

In this section implementation of genetic algorithm to this problem will be described.

Initialization Process: Population size is needed to be defined as an integer value, and will be called *pop_size* in this paper. Generation of the first population is done as following; an interior point on feasible region is chosen as v_0, and for each warehouse w_j, *x-y* coordinates are obtained as $w_j=v_0+M*d$, where M is a large positive number, and d is a direction generated randomly for each warehouse location. And uniformly lessen that large positive number M, till being in feasible region [6].

Fitness Value Calculation: Fuzzy objective function values are calculated by fuzzy simulation for each chromosome, as fitness value of that chromosome.

Natural Selection: After obtaining fitness values, chromosomes are ranked with respect to their fitness values and rank-probabilities calculated by Eq.(10) where j is the rank of the chromosome x_j and a is a parameter in the genetic system. Then cumulative probabilities calculated by Eq.(11). Subsequently using roulette wheel method new population is generated as following algorithm; generating a random number r in [0,1], if $p_{i-1} < r \le p_i$, then i^{th} chromosome is selected. Repeat that procedure for *pop_size* time.

$$Rank\_Probability(x_j) = a(1-a)^{j-1}, \quad j = 1,2,\dots,pop\_size. \qquad (10)$$

$$p_0 = 0, \quad p_i = \frac{\sum_{j=1}^{i} Rank\_Probability(x_j)}{\sum_{j=1}^{pop\_size} Rank\_Probability(x_j)}, \quad i = 1,2,\dots,pop\_size. \qquad (11)$$

Crossover Operation: Crossover probability should be defined as P_c. For each chromosome a random number r is generated to decide if that chromosome be a parent or not. If $r < P_c$, then chromosome x_i is selected as a parent. Subsequently parents divided into pairs as $(\grave{x}_1, \grave{x}_2), (\grave{x}_3, \grave{x}_4), \dots$, and crossover operation is done for pairs. For the pair $(\grave{x}_1, \grave{x}_2)$, a random number c is generated in (0,1) and two children are produced as $x_{child1} = c\grave{x}_1 + (1-c)\grave{x}_2$ and $x_{child2} = (1-c)\grave{x}_1 + c\grave{x}_2$.

Mutation Operation: Mutation probability should be defined as P_m. Similar to the crossover procedure, for each chromosome a random number r is generated to decide if that chromosome mutates or not. If $r < P_m$, then chromosome x_i is selected for mutation operation. Similar to the initialization, a mutation direction is selected as d, and new chromosome is generated as $x_{newi} = x_i + Md$.

The Overall Genetic Algorithm Procedure: Algorithm used in this paper is given below.

Step1: Initialization of population by checking their feasibility. A chromosome with zero credibility is considered as the initial value of best solution. k=0.

Step2: Generation of fuzzy simulation to obtain fitness values.

Step3: Ranking of chromosomes. First ranked one should be checked. If it is better than the best solution, then best solution is replaced with it.

Step4: Natural selection of new population.

Step5: Crossover operation.

Step6: Mutation operation by checking feasibility of new chromosomes.

Step7: k=k+1. If *k<Iteration number*, go to *Step2*, else go to *Step8*.

Step8: Return the best solution.

3. Numerical Example

It is considered a facility location problem of an organization that requires to locate 2 warehouses which will serve to 7 stores. In the European side of Istanbul, location of 7 stores are taken from map, and rescaled to [1, 100] and [1, 45] in x and y coordinates respectively. Fuzzy demands are generated by decision makers as fuzzy normal numbers. All those values about stores are given at following summary table. Target objective function value, F_0 is determined as 15000, and the interior point coordinates v_0 is determined as (40, 20) by decision maker.

Table 1. Coordinates and demands for stores.

Stores	Rescaled Coordinates		Demand - Fuzzy Normal Variables	
	x	y	μ	σ
Beylikduzu	5,32	10,88	300	25
Bakırkoy	50,47	2,44	390	40
Fatih	65,78	12,17	420	30
Cevahir	75,84	25,33	360	45
Profilo	77,75	27,12	240	20
Akmerkez	82,91	29,23	285	30
Istinye Park	84,43	38,23	405	50

Genetic algorithm parameters are tuned with trial and error. Iteration numbers are selected 10000 and 500 for genetic algorithm and fuzzy simulation respectively. Number of individuals in population is 30. Fuzzy parameter, crossover probability and mutation probability are 0.2, 0.4 and 0.2 respectively.

Hybrid intelligent algorithm was run, locations for two warehouses were found as (52.09, 7.88) and (80, 30.67) and given in the Figure 1. While first warehouse supplies Beylikduzu, Bakirkoy and Fatih stores, the second one supplies the rest of stores. Maximum credibility is 0.86 and distance-demand value is 27,473.

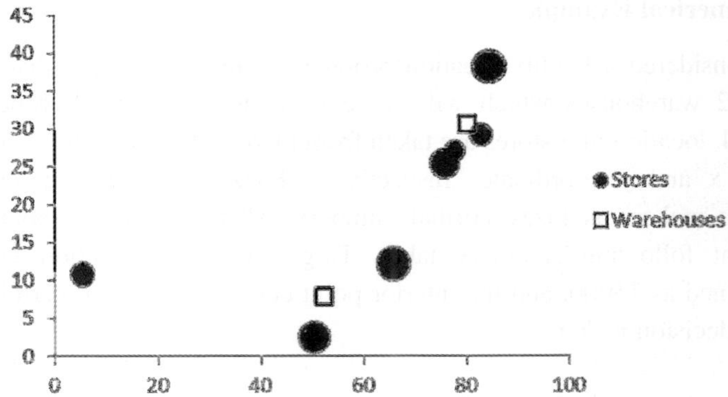

Figure 1: Locations for stores and warehouses

4. Conclusion

In this paper, a fuzzy chance constrained programming model based on hybrid intelligent algorithm integrating fuzzy simulation and genetic algorithm was constructed for selection of warehouse locations. In order to illustrate use of the model a numerical example was presented. The proposed model has been observed to be conveniently applicable for solving warehouse location problems involving uncertain demands. Cost minimization can be achieved through the model in the presence of fuzziness.

References

1. L. Michel and P.V. Hentenryck, "A simple tabu search for warehouse location", *European Journal of Operational Research*, 157, 576-591 (2004).
2. S. Liu, F.T.S. Chan and S.H. Chung, "A study of distribution center location based on the rough sets and interactive multi-objective fuzzy decision theory", *Robotics and Computer-Integrated Manufacturing*, 27, 462-433 (2011).
3. M. Wen and R. Kang, "Some optimal models for facility location-allocation problem with random fuzzy demands", *Applied Soft Computing*, 11, 1202-1207 (2011).
4. S. Wang and J. Watada, "A hybrid modified PSO approach to VaR-based facility location problems with variable capacity in fuzzy random uncertainty", *Information Sciences*, (in press).
5. Z. Shao and X. Ji, "Fuzzy multi-product constraint newsboy problem", *Applied Mathematics and Computation*, 180, 7-15 (2006).
6. B. Liu and K. Iwamura, "Chance constrained programming with fuzzy parameters", *Fuzzy Sets and Systems*, 94, 227-237 (1998).

AUTHOR INDEX

Z. M Sbartaï; 94

ACKNOWLEDGEMENTS

Special thanks go to Organizing Committee and Scientific Committee for their invaluable contributions to this conference.

Honorary Chair:
Lotfi A. Zadeh (USA)

General Chair:
Da Ruan (Belgium)

Organizing Chair:
Cengiz Kahraman (Turkey)

Organizing Co-chairs:
Tunc Bozbura (Turkey)

Local Organizing Committee:
(ordered alphabetically)

Abdullah Çağrı Tolga
Ahmet Beşkese
Ahmet Can Kutlu
Ayberk Soyer
Başar Öztayşi
Fatih Tüysüz
Hülya Behret
Ihsan Kaya
Irem Uçal Sarı
Murat Durucu
Nihan Cetin Demirel
Ozgur Kabak
Pınar Dursun
Seda Uğurlu
Selcuk Çebi
Şeyda SerdarAsan
Sezi Çevik Onar
Tolga Kaya
Tufan Demirel
Tuncay Gürbüz
Umut Asan

International Scientific Committee
(ordered alphabetically)

R. A. Aliev (Azerbaijan)
U. Asan (Turkey)
Z. Ayağ (Turkey)
P. Baraldi (Italy)
H. Behret (Turkey)
J. Benítez (Mexico)
A. Beşkese (Turkey)
Z. Bien (Korea)
S. Birgün (Turkey)
P. Bonissone (USA)
H. Bouchachia (Austria)
B. Bouchon-Meunier (France)
T. Bozbura (Turkey)
H. Bustince (Spain)
G. Buyukozkan (Turkey)
O. Castillo (Mexico)
S. Cebi (Turkey)
G. Chen (China)
F. Chiclana (UK)
C. Cornelis (Belgium)
B. De Baets (Belgium)
M. De Cock (Belgium)
G. de Cooman (Belgium)
R. De Moraes (Brazil)
N. C. Demirel (Turkey)
T. Demirel (Turkey)
A. Deshpande (India)
D. Dubois (France)
P. F. Fantoni (Norway)
J. Feng (China)
F. Gomide (Brazil)
P. Guo (Japan)
M. M. Gupta (Canada)
T. Gurbuz (Turkey)

W. Halang (Germany)
R. Hampel (Germany)
F. Herrera (Spain)
E. Herrera-Viedma (Spain)
J. W. Hines (USA)
C. Huang (China)
A. Inoue (USA)
R. John (UK)
O. Kabak (Turkey)
J. Kacprzyk (Poland)
E. Karsak (Turkey)
N. Kasabov (New Zealand)
I. Kaya (Turkey)
T. Kaya (Turkey)
E. Kerre (Belgium)
G. Klir (USA)
G. Kolaczek (Poland)
T. Li (China)
Z. Li (Germany)
J. Liu (UK)
Y. Liu (China)
L. Kóczy (Hungary)
V. Lopez (Spain)
J. Lu (Australia)
L. Magdalena (Spain)
L. Maguire (UK)
A. Maisseu (France)
L. Martinez (Spain)
P. Melin (Mexico)
J. Montero (Spain)
M. Nachtegael (Belgium)
V. Novak (Czech Republic)
T. Onisawa (Japan)
A. F. Ozok (Turkey)
B. Öztayşi (Turkey)
G. Pasi (Italy)
S. Patnaik (India)
M. Santos Penas(Spain)
C. M. N. A. Pereira (Brazil)
I. Perfilieva (Czech Republic)
H. Prade (France)
D. Radojevic (Serbia)
G. Resconi (Italy)
R. A. Ribeiro (Portugal)
D. Roverso (Norway)
L.D. Santos Machada (Brasil)
Ş. Serdar Asan (Turkey)
Y. Shen (Belgium)
A. Soyer (Turkey)
H. Tanaka (Japan)
E. Tolga (Turkey)
A. C. Tolga (Turkey)
E. Trillas (Spain)
I. Truck (France)
I. B. Türkşen (Turkey)
F. Tuysuz (Turkey)
Z. Ulukan (Turkey)
A. Ustundag (Turkey)
Ariën J. van der Wal (Holland)
P. P. Wang (USA)
D. Xu (UK)
Y. Xu (China)
R. R. Yager (USA)
F. Yalaoui (France)
J. B. Yang (UK)
H. Ying (USA)
B. Yuan (USA)
X. Zeng (France)
G. Zhang (Australia)
H. J. Zimmermann (Germany)
E. Zio (Italy)